Document Image Analysis

Lawrence O'Gorman
Rangachar Kasturi

IEEE Computer Society Press
Los Alamitos, California

Washington • Brussels • Tokyo

Library of Congress Cataloging-in-Publication Data

O'Gorman, Lawrence
Document image analysis / Lawrence O'Gorman, Rangachar Kasturi
p. cm.
Includes bibliographical references (p) .
ISBN 0-8186-6547-5
1. Image processing—Digital techniques.
I. Kasturi, Rangachar, 1949– . II. Title.
TA1637.036 1995
006.4'2—dc20 94-32859
CIP

Published by the
IEEE Computer Society Press
10662 Los Vaqueros Circle
P.O. Box 3014
Los Alamitos, CA 90720-1264

IEEE Computer Society Press Order Number 6547-01
IEEE Catalog Number EH0410-1
Library of Congress Number 94-32859
ISBN 0-8186-6547-5 (paper)

Additional copies can be ordered from

IEEE Computer Society Press
Customer Service Center
10662 Los Vaqueros Circle
P.O. Box 3014
Los Alamitos, CA 90720-1264
Tel: (714) 821-8380
Fax: (714) 821-4641
Email: cs.books@computer.org

IEEE Service Center
445 Hoes Lane
P.O. Box 1331
Piscataway, NJ 08855-1331
Tel: (908) 981-1393
Fax: (908) 981-9667

IEEE Computer Society
13, avenue de l'Aquilon
B-1200 Brussels
BELGIUM
Tel: +32-2-770-2198
Fax: +32-2-770-8505

IEEE Computer Society
Ooshima Building
2-19-1 Minami-Aoyama
Minato-ku, Tokyo 107
JAPAN
Tel: +81-3-3408-3118
Fax: +81-3-3408-3553

Technical Editor: Yutaka Kanayama
Production Editor: Lisa O'Conner
Cover artist: Alex Torres

Printed in the United States of America by KNI, Incorporated
98 97 96 95 5 4 3 2

The Institute of Electrical and Electronics Engineers, Inc.

Contents

Preface

In the late 1980s, the prevalence of fast computers, large computer memory, and inexpensive scanners fostered an increasing interest in document image analysis. With many paper documents being sent and received via fax machines and stored digitally in large document databases, the interest grew to do more with these images than simply view and print them. Just as humans extract information from these images, research was performed and commercial systems were built to read text on a page, find fields on a form, and locate lines and symbols on a diagram.

Today, the results of research work in document image analysis can be seen and felt every day. Character recognition scanners are used by the post offices to route mail automatically. Engineering diagrams are extracted from paper for computer storage and modification. Fingerprint images are routinely analyzed by computer for matching. In the future, applications such as these will be improved, and other document applications will be added. For instance, the millions of old paper volumes now in libraries will be replaced by computer files of page images that can be searched for content and accessed by many people at the same time—and will never be misshelved. Business people will carry their file cabinets in their portable computers, and paper copies of new product literature, receipts, or other random notes will be instantly filed and accessed in the computer. Signatures will be analyzed by the computer for verification and security access. Musical scores and other symbolic and diagrammatic documents will be read and their contents recognized.

This book describes some of the technical methods used in document image analysis that have grown out of the fields of digital signal processing, digital image processing, and pattern recognition. The objective is to give the reader an understanding of what approaches are used for application to documents and how these methods apply to different situations. Since the field of document image analysis is relatively new, it is also dynamic, so current methods have room for improvement, and innovations are still being made. In addition, there are rarely definitive techniques for all cases of a certain problem.

In the tutorial notes that begin each chapter, we attempt to identify major problem areas and to describe more than one method applied to each problem, along with advantages and disadvantages of each method. This overview gives an understanding of the problems and also of the nature of trade-offs that so often must be made in choosing a method. With this understanding of the problems and a knowledge of the methodology options in place, the reprints included here then describe methods in detail. The reprints have been chosen on the basis of their applicability to the problems addressed here, their general utility, and their ease of reading and use. That is to say, a particular reprint may not describe the latest or fastest method; when information on greater speed or more focused applicability is desired, the References and Bibliography contained herein can be used. We have included 34 papers, of which 19 describe mainly techniques and 15 describe mainly systems or applications (though there is overlap either way). Twenty-six of the papers were published in 1990 or later. We also include a comprehensive bibliography of post-1986 papers, annotated by subject area.

This book is aimed at those with some rudimentary knowledge of digital signal processing. Knowledge of digital image processing would make the book more straightforward. Although we attempt to describe the techniques using basic concepts, we cannot review the basics of computers or signal processing. For instance, the reader will be expected to understand that an image must be spatially sampled for computer storage and processing and that filtering is the process used to modify the contents of the image, such as for noise reduction. Having said this, the systems presented in this book are described from a high-enough level to be generally understandable and, in addition, to motivate understanding of some of the techniques.

The book is organized in the sequence in which document images are usually analyzed. After document input by digital scanning, pixel processing is performed. This level of processing includes operations that are applied to all image pixels, including noise removal, image enhancement, and segmentation of image components into text and graphics (lines and symbols). Feature-level analysis treats groups of pixels as entities and includes line and curve detection and shape description. The last two chapters separate text and graphics analysis. Text analysis includes optical character recognition and page format recognition. Graphics analysis includes recognition of components of engineering drawings, maps, and other diagrams.

We would like to express our sincere thanks to Professors George Nagy, Jonathan Hull, Sargur Srihari, Anil Jain, and Thomas Nartker for their invaluable help during the preparation of this book. We also appreciate the constructive critiques provided by the anonymous reviewers. Our thanks to Professor Rao Vemuri and Jon Butler for evaluating and accepting our proposal and manuscript of this text and to Catherine Harris and her staff at the IEEE Computer Society Press for assembling all the material in such a short time to ensure timely publication. We are indebted to the assistance provided by Juan Arias, Chan-Pyng Lai, Tarak Gandhi, Teri Rudy, and Gopinath Rajgopal in compiling the Bibliography. Our many thanks to the authors for providing reprints and for permitting us to use their work. We trust that the readers will find this volume helpful in their work and we welcome any comments, corrections, and suggestions.

Lawrence O'Gorman
Rangachar Kasturi
October 1994

Chapter 1
Introduction

1.1: Introduction

Traditionally, transmission and storage of information have been by paper documents. In the past few decades, documents increasingly originate on the computer; however, in spite of this trend, it is unclear whether the computer has decreased or increased the amount of paper documentation. Documents are still printed out for reading, dissemination, and markup. The oft-repeated cry of the early 1980s for the "paperless office" has now given way to a different objective: dealing with the flow of electronic and paper documents in an efficient and integrated way. The ultimate solution would be for computers to deal with paper documents as they deal with other forms of computer media. That is, paper would be as readable by the computer as magnetic and optical disks are now. If this were the case, then the major difference—and the major advantage—would be that, unlike current computer media, paper documents could be read by both the computer and people.

The objective of document image analysis is to recognize the text and graphics components in images and extract the intended information as a human would. Two categories of document processing can be defined (see Figure 1). *Textual processing* deals with the text components of a document image. Some tasks here are recognizing the text by optical character recognition (OCR); determining the *skew* (any tilt at which the document may have been scanned into the computer); and finding columns, paragraphs, text lines, and words. *Graphics processing* deals with the nontextual line and symbol components that make up line diagrams, delimiting straight lines between text sections, company logos, and the like. Because many of these graphics components consist of lines, such processing includes line thinning, line fitting, and corner and curve detection. (It should be noted that the use of "line" in this book can mean straight, curved, or piecewise straight and/or curved lines. When straightness or curvedness is important, it will be specified.) Besides text and graphics, pictures are a third major component of documents, but—except for recognizing their location on a page—further analysis of these images is usually the task of other image-processing and machine vision techniques, so we do not deal with picture processing in this book. After application of the above-described textual and graphics analysis techniques, the megabytes of initial data are culled to yield a much more concise semantic description of the document.

It is not difficult to find examples of the need for document analysis. Look around the workplace to see the stacks of paper. Some may be computer generated, but if so, inevitably by different computers and software such that even their electronic formats are incompatible. Some will include both formatted text and tables as well as handwritten entries. There are different sizes, from a 3.5" × 2" (8.89 centimeters × 5.08 cm) business card to a 34" × 44" (86 cm × 111 cm) engineering drawing. In many businesses today, imaging systems are being used to store images of pages to make storage and retrieval more efficient. Future document analysis systems will recognize types of documents, enable the extraction of their functional parts, and facilitate translation from one computer-generated format to another. There are many other examples of the use of and need for document systems. Glance behind the counter in a post office at the mounds of letters and packages. In some US post offices, over a million pieces of mail must be handled each day. Machines to perform sorting and address recognition have been used for several decades, but there is a need to process more mail, more quickly and more accurately. As a final example, examine the stacks of a library, where row after row of paper documents are stored. Loss of material, misfiling, limited numbers of each copy, and even degradation of materials are common problems, and the problems may be alleviated by document analysis techniques. All of the above examples serve as applications ripe for the potential solutions of document image analysis.

Although document image analysis has been in use for several decades (especially in the banking business, for computer reading of numeric check codes), the area has grown much more rapidly in the late 1980s and early 1990s, largely because of the greater speed and lower cost of hardware now available. Since fax machines have become ubiquitous, the cost of optical scanners for document input has dropped to a level

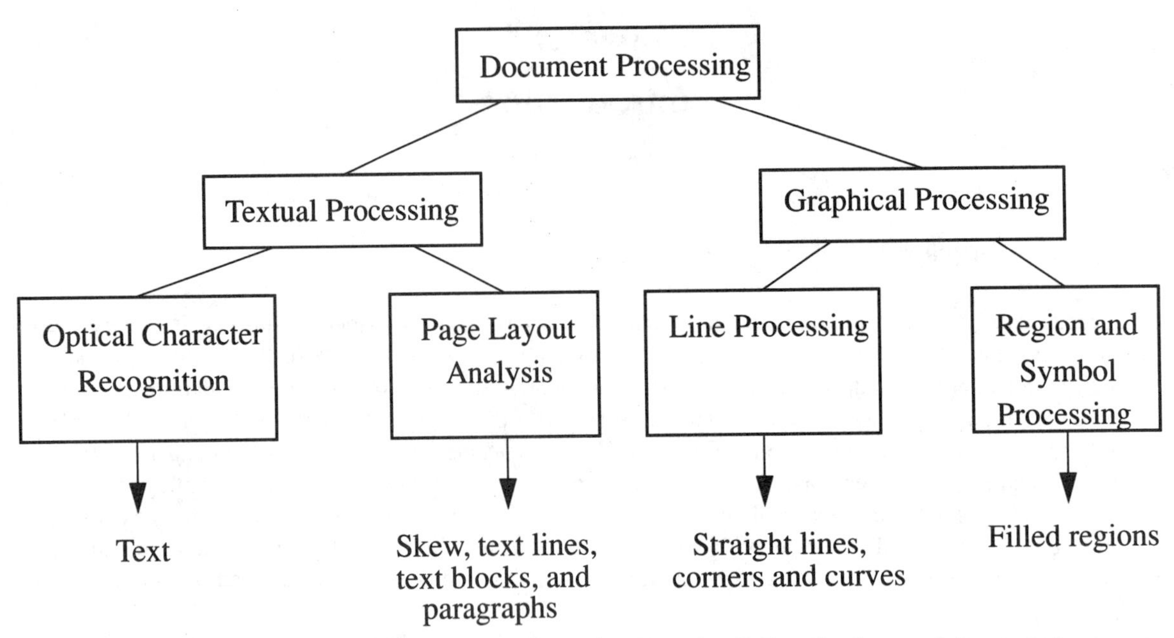

Figure 1. A hierarchy of document-processing subcategories listing the types of document components dealt with in each subcategory.

affordable to even small businesses and individuals. Although document images contain a relatively large amount of data, even personal computers now have adequate speed to process them. Also, computer main memory is now adequate for large document images, but more importantly, optical memory is now available for mass storage of large amounts of data. This improvement in hardware, as well as the increasing use of computers for storing paper documents, has led to an increasing interest in improving the technology of document processing and recognition. An essential complement to these hardware improvements is the advances being made in document analysis software and algorithms. With OCR rates now in the mid- to high 90-percent range, and other document processing methods achieving similar improvements, these advances in research have also driven document image analysis forward.

As these improvements continue, document systems will become increasingly more evident in the form of everyday document systems. For instance, OCR systems will be more widely used to store, search, and excerpt from paper-based documents. Page layout analysis techniques will recognize a particular form or page format and allow its duplication. Diagrams will be entered from pictures or by hand and then will be logically edited. Pen-based computers will translate handwritten entries into electronic documents. Archives of paper documents in libraries and engineering companies will be electronically converted for more efficient storage and instant delivery to a home or office computer. Although it will be increasingly the case that documents are produced and reside on a computer, the fact that there are very many different systems and protocols, along with the fact that paper is a very comfortable medium for us to deal with, ensures that paper documents will be with us to some degree for many decades to come. The difference will be that these documents will finally be integrated into our computerized world.

1.2: Hardware advancements and the evolution of document image analysis

The field of document image analysis can be traced back through a computer lineage that includes digital signal processing and digital image processing. Digital signal processing, whose study and use were initially fostered by the introduction of fast computers and algorithms such as the fast Fourier transform in the mid-1960s, has as its objective the interpretation of one-dimensional signals such as speech and other audio. In the early 1970s, with larger computer memories and still faster processors, image processing methods, and systems were developed for analysis of two-dimensional signals, including digitized pictures. Special fields of image processing are associated with a particular application—for example, biomedical image processing with medical images, machine vision with processing of pictures in industrial manufacturing, and computer vision with processing images of three-dimensional scenes used in robotics.

In the mid- to late 1980s, document image analysis began to grow rapidly, again predominantly because of hardware advancements enabling processing to be performed at a reasonable cost and time. Whereas a speech signal is typically processed in frames 256 samples long and a machine vision image size is 512 × 512 pixels, a document image is from 2,550 pixels × 3,300 pixels for a business letter digitized at 300 dots per inch (12 dots per millimeter) to 34,000 pixels × 44,000 pixels for a 34 "× 44" E-sized engineering diagram digitized at 1,000 dpi.

Commercial document analysis systems are now available for storing business forms, performing OCR on typewritten text, and compressing engineering drawings. Document analysis research continues to pursue more intelligent handling of documents, better compression (especially through component recognition), and faster processing.

1.3: From pixels to paragraphs and drawings

Figure 2 illustrates a common sequence of steps in document image analysis. This sequence is also the organization of chapters in this book. After data capture, the image undergoes pixel-level processing and feature-level analysis; text and graphics recognition are then treated separately.

1.3.1: Data Capture

Data capture is performed on a paper document, usually by optical scanning. The resulting data are stored in a file of picture elements, called *pixels*, that are sampled in a grid pattern throughout the document. These pixels may have values: OFF (0) or ON (1) for binary images, 0 to 255 for gray-scale images, and three channels of 0 to 255 color values for color images. At a typical sampling resolution of 300 dpi, a 8.5"×11" page would yield an image of 2,550 pixels × 3,300 pixels. It is important to understand that the image of the document contains only raw data that must be further analyzed to glean the information. For example, Figure 3 shows an image of the letter "e." This is a pixel array of OFF or ON values whose shape is known to humans as the letter "e"; however, to a computer it is just a string of bits in computer memory.

1.3.2: Pixel-level processing (Chapter 2).

This step includes binarization, noise reduction, signal enhancement, and segmentation. For gray-scale images with information that is inherently binary, such as text or graphics, binarization is usually performed first. The objective in methods for binarization is to automatically choose a threshold that separates the foreground and background information. An example that has seen much research is the extraction of handwriting from a bank check, especially a bank check containing a scenic image as a background.

Document image noise occurs from image transmission, photocopying, or degradation due to aging. *Salt-and-pepper noise* (also called *impulse noise*, *speckle noise*, or just *dirt*) is a common form of noise on a binary image, consisting of randomly distributed black specks on a white background and white specks on a black background. It is reduced by performing filtering on the image where the background is "grown" into the noise specks, thus filling these holes. Signal enhancement is similar to noise reduction but uses domain knowledge to reconstitute expected parts of the signal that have been lost. Signal enhancement is often applied to graphics components of document images to fill in gaps in lines that are known to be continuous otherwise.

Segmentation occurs on two levels. On the first level, if the document contains both text and graphics, they are separated for subsequent processing by different methods. On the second level, segmentation is performed on text by locating columns, paragraphs, words, titles, and captions; on graphics, segmentation usually includes separating symbol and line components. For example, in a page containing a flow chart with an accompanying caption, text and graphics are first separated. Then the text is separated into that of the caption and that of the chart. The graphics is separated into rectangles, circles, connecting lines, and so on.

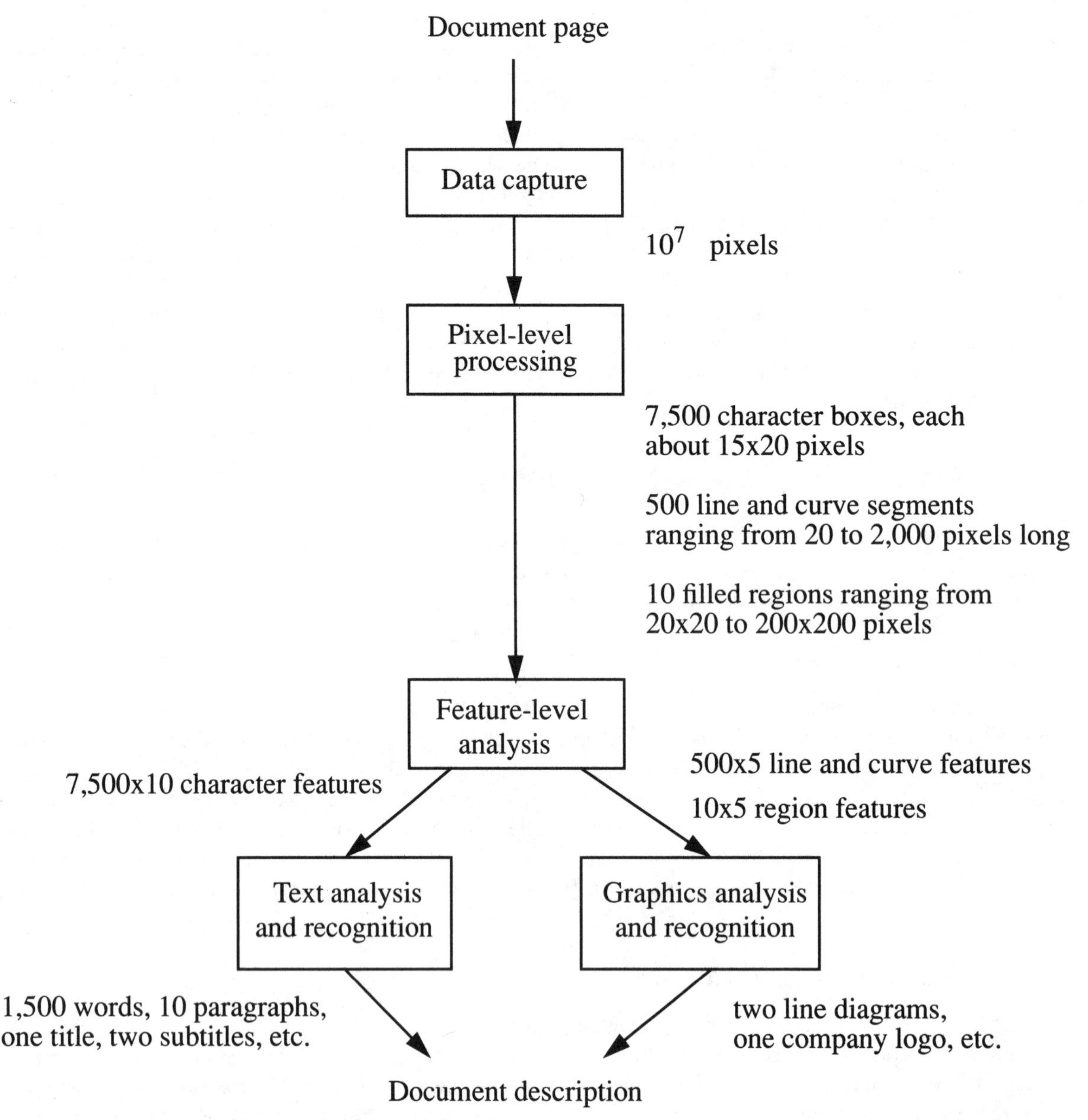

Figure 2. A typical sequence of steps for document analysis, along with examples of intermediate and final results and the data size.

1.3.3: Feature-level analysis (Chapter 3)

In a text image, the global features describe each page and consist of skew, line lengths, line spacing, and so on. There are also local features of individual characters, such as font size, number of loops in a character, number of crossings, and accompanying dots, which are used for OCR.

In a graphical image, global features describe the skew of the page, line widths, range of curvature, minimum line lengths, and so on. Local features describe each corner, curve, and straight line, as well as rectangles, circles, and other geometric shapes.

1.3.4: Analysis and recognition of text and graphics (Chapters 4 and 5)

The final step is the recognition and description step, where components are assigned a semantic label and the entire document is described as a whole. It is at this stage that domain knowledge is used most extensively.

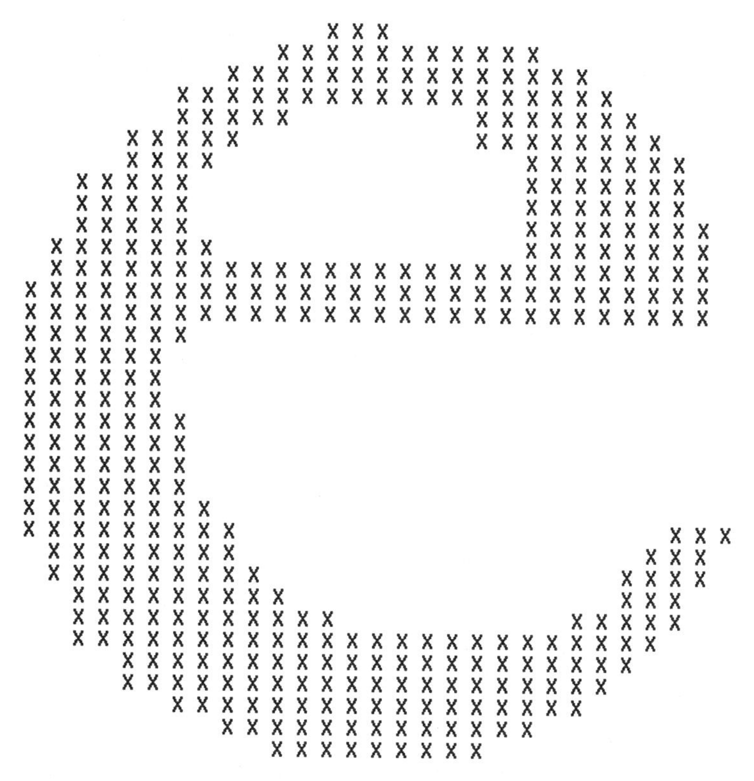

Figure 3. A binary image of the letter "e" is made up of OFF and ON pixels, where the ON pixels are shown here as "X"s.

The result is a description of a document as a human would give. For a text image, we refer not to pixel groups, or blobs of black on white, but to titles, subtitles, bodies of text, footnotes, and so on. Depending on the arrangement of these text blocks, a page of text may be a title page of a paper, a table of contents of a journal, a business form, or the face of a piece of mail. For a graphical image—an electrical circuit diagram for instance—we refer not to lines joining circles and triangles and other shapes but to connections between AND gates, transistors, and other electronic components. The components and their connections describe a particular circuit that has a purpose in the known domain. It is this semantic description that is stored most efficiently and used most effectively for common tasks such as indexing and modifying particular document components.

Chapter 2
Pixel-Level Processing

2.1: Introduction

Data capture of documents by optical scanning or by digital video yields a file of picture elements, or pixels, that is the raw input to document analysis. These pixels are samples of intensity values taken in a grid pattern over the document page, where the intensity values may be OFF (0) or ON (1) for binary images, 0 to 255 for gray-scale images, and three channels of 0 to 255 color values for color images. The first step in document analysis is to perform processing on this image to prepare it for further analysis. Such processing includes thresholding to reduce a gray-scale or color image to a binary image, reduction of noise to reduce extraneous data, and thinning and region detection to enable easier subsequent detection of pertinent features and objects of interest. This pixel-level processing (also called *preprocessing* and *low-level processing* in other literature) is the subject of this chapter.

2.2: Thresholding

(*Keywords*: thresholding, binarization, global thresholding, adaptive thresholding, intensity histogram)

In this treatment of document processing, we deal with images containing text and graphics of binary information. That is, these images contain a single foreground level that is the text and graphics of interest and a single background level with which the foreground contrasts. We will also call the foreground *objects*, *regions of interest*, or *components*. (Of course, documents may also contain true gray-scale or color information, such as in photographic figures; however, besides recognizing the presence of a gray-scale picture in a document, we leave the analysis of pictures to the more general fields of image analysis and machine vision.) Although the information is binary, the data (in the form of pixels with intensity values) are likely to not have only two levels, but rather to have a range of intensities, as a result of nonuniform printing or nonuniform reflectance from the page or as a result of intensity transitions at the region edges located between foreground and background regions. The objective in binarization is to mark pixels that belong to true foreground regions with a single intensity (ON) and background regions with a different intensity (OFF). Figure 1 illustrates the results of binarizing a document image at different threshold values. The ON values are shown in black and the OFF values are shown in white in our figures.

For documents with a good contrast of components against a uniform background, binary scanners are available that combine digitization with thresholding to yield binary data. However, for the many documents that have a wide range of background and object intensities, this fixed threshold level often does not yield images with a clear separation between the foreground components and the background. For example, when a document is printed on differently colored paper, when the foreground components are faded because of photocopying, or when different scanners have different light levels, the best threshold value will also be different. For these cases, there are two alternatives. One is to determine empirically the best binarization setting on the scanner (most binary scanners provide this adjustment), and to do this each time an image is poorly binarized. The other alternative is to start with gray-scale images (having a range of intensities, usually from 0 to 255) from the digitization stage and then use methods for automatic threshold determination to better perform binarization. While the latter alternative requires more input data and processing, its advantage is that a good threshold level can be found automatically, ensuring consistently good images and precluding the need for time-consuming manual adjustment and repeated digitization. The following discussion presumes initial digitization to gray-scale images.

If the pixel values of the components and those of the background are fairly consistent in their respective values over the entire image, then a single threshold value can be found for the image. This use of a single threshold for all image pixels is called *global thresholding*. Processing methods are described below that automatically determine the best global threshold value for different images. However, for many documents a

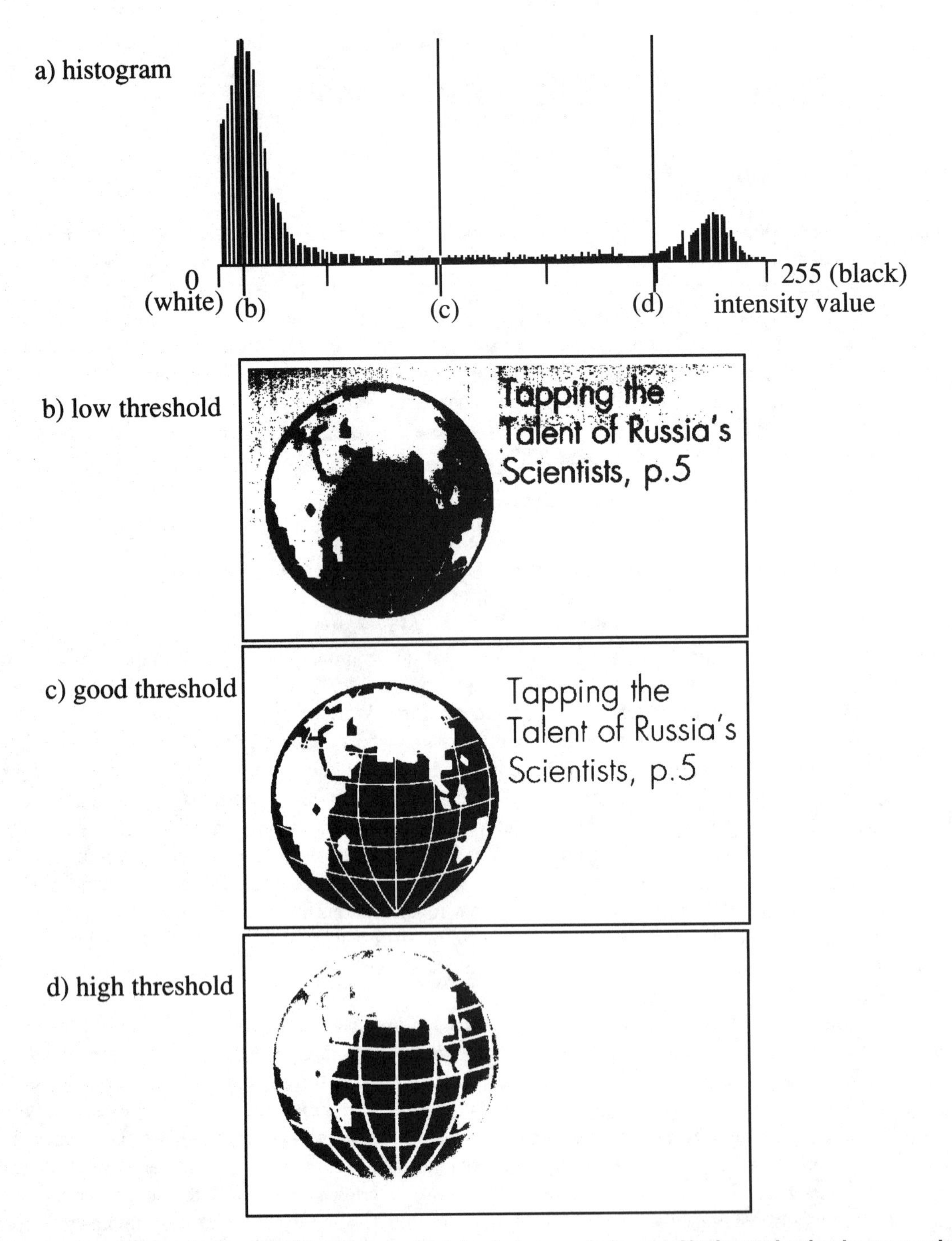

Figure 1. Image binarization. (a) Histogram of original gray-scale image. Horizontal axis shows markings for threshold values of images below. The lower peak is for the white background pixels and the upper peak is for the black foreground pixels. Image binarized with (b) is at too low a threshold value, with (c) is at a good threshold value, and with (d) is at too high a threshold value.

single global threshold value cannot be used even for a single image because of nonuniformities within foreground and background regions. For example, for a document containing white background areas as well as highlighted areas of a different background color, the best thresholds will change by area. For this type of image, different threshold values are required for different local areas. This use of different threshold values for different local areas, called *adaptive thresholding*, is described below.

2.2.1: Global thresholding

The most straightforward way to automatically select a global threshold is by use of a histogram of the pixel intensities in the image. The intensity histogram plots the number of pixels with values at each intensity level (see Figure 1 for a histogram of a document image). For an image with well-differentiated foreground and background intensities, the histogram will have two distinct peaks. The valley between these peaks can be found as the minimum between two maxima, and the intensity value there is chosen as the threshold that best separates the two peaks. Since peak-and-valley detection methods are prone to locating local maxima and minima other than those desired, one method [Rosenfeld, and De La Torre, 1983] fits a convex hull to the bimodal histogram and then detects the valley as the maximal distance between the hull and bin heights.

There are a number of drawbacks to global threshold selection based on the shape of the intensity distribution. The first is that images do not always contain well-differentiated foreground and background intensities as a result of poor contrast and noise. A second is that, especially for an image of sparse foreground components, such as for most graphics images, the peak representing the foreground intensities will be much smaller than the peak of the background intensities. This difference often makes it difficult to find the valley between the two peaks. In addition, reliable peak-and-valley detection methods are separate problems unto themselves. One way to improve this approach is to compile a histogram of pixel intensities that are weighted by the inverse of their edge-strength values [Mason et al., 1975]. Region pixel intensities with low edge values will be weighted more highly than intensities of boundary and noise pixels with higher edge values, thus sharpening the histogram peaks caused by these regions and facilitating threshold detection between them. An analogous technique is to highly weight those intensities of pixels with high edge values and then choose the threshold at the peak of this histogram, corresponding to the transition between regions [Katz, 1965, Watanabe et al., 1974; Weszka, Nagel, and Rosenfeld 1974; Weszka, Rosenfeld 1979]. This method requires peak detection of a single maximum, which is often easier than valley detection between two peaks. This approach also reduces the problem of a large discrepancy in size between foreground and background region peaks, because edge pixels are accumulated on the histogram instead of region pixels; the difference between a small- and large-size area is a linear quantity for edges versus a much larger squared quantity for regions. A third method uses a Laplacian weighting. The Laplacian is the second-derivative operator, which highly weights transitions from regions into edges (the first derivative highly weights edges). This method will highly weight the border pixels of both foreground regions and their surrounding backgrounds; thus, the histogram will have two peaks of similar area. Although these histogram-shape techniques offer the advantage that peak-and-valley detections are intuitive, still peak detection is susceptible to error due to noise and poorly separated regions. Furthermore, when the foreground or background region consists of many narrow regions—such as for text—edge and Laplacian measurement may be poor because of very abrupt transitions (narrow edges) between foreground and background.

A number of techniques determine classes by formal pattern recognition techniques that optimize some measure of separation. One approach is minimum-error thresholding [Kittler, and Illingworth, 1986; Ye, and Danielson 1988]. (See Figure 2.) Here, the foreground and background intensity distributions are modeled as normal (Gaussian or bell-shaped) probability density functions. For each intensity value (from 0 to 255, or a smaller range if the threshold is known to be limited to it), the means and variances are calculated for the foreground and background classes, and the threshold is chosen such that the misclassification error between the two classes is minimized. The latter method is classified as a parametric technique because of the assumption that the gray-scale distribution can be modeled as a probability density function. This is a popular method for many computer vision applications, but some experiments indicate that documents do not adhere well to this model; thus, results with this method are poorer than those with nonparametric approaches [Abutaleb, 1989]. One nonparametric approach is Otsu's method [Otsu, 1979; Reddi, Rudin, and Keshavan 1984]. First, calculations are made of the ratio of between-class variance to within-class variance for each potential threshold value. The classes here are the foreground and background pixels, and the purpose is to find the threshold that maximizes the variance of intensities between the two classes and minimizes them within each class. This ratio is calculated for all potential threshold levels; the level at which the ratio is maximum is the chosen threshold. A similar approach to Otsu's employs an information theory measure, entropy, which is a measure of the information in the image expressed as the average number of bits required to represent the information [Abutaleb, 1989; Johannsen and Bille, 1982; Kapur, Sahoo, and Wong, 1985; Pun, 1981]. Here, the entropy for the

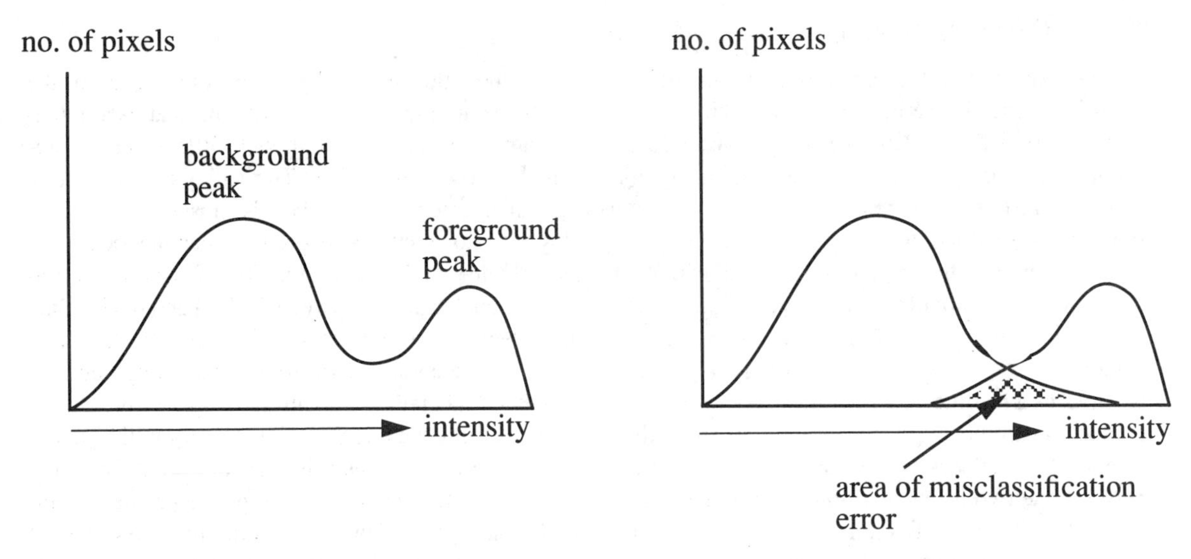

Figure 2. Illustration of misclassification error in thresholding. On the left is an intensity histogram showing foreground and background peaks. On the right, the tails of the foreground and background populations have been extended to show the intensity overlap of the two populations. It is evident that this overlap makes it impossible to correctly classify all pixels by means of a single threshold. The minimum-error method of threshold selection minimizes the total misclassification error.

two classes is calculated for each potential threshold, and the threshold where the sum of the two entropies is largest is chosen as the best threshold. Another thresholding approach is by moment preservation [Tsai, 1985]. This is less popular than the methods above; however, we have found it to be more effective in binarizing document images containing text. For this method, a threshold is chosen that best preserves moment statistics in the resulting binary image as compared with the initial gray-scale image. These moments are calculated from the intensity histogram—the first four moments are required for binarization.

2.2.2: Adaptive thresholding

A common way to perform adaptive thresholding is by analyzing gray-level intensities within local windows across the image to determine local thresholds [Casey and Wong, 1990; Kamel and Zhao, 1993; Wong, 1978]. White and Rohrer [White and Rohrer, 1983] describe an adaptive thresholding algorithm for separating characters from background. The threshold is continuously changed through the image by estimating the background level as a two-dimensional running average of local pixel values taken for all pixels in the image (see Figure 3). Mitchell and Gillies [Mitchell and Gillies, 1989] describe a similar thresholding method where background white-level normalization is done by first estimating the white level and subtracting this level from the raw image. Then, segmentation of characters is accomplished by applying a range of thresholds and selecting the resulting image with the least noise content. Noise content is measured as the sum of areas occupied by components smaller and thinner than empirically determined parameters. Looking back at the results of binarization for different thresholds in Figure 1, it can be seen that the best threshold selection yields the least visible noise.

The main problem with any adaptive binarization technique is the choice of window size. The chosen window size should be large enough to guarantee that the number of background pixels included is large enough to obtain a good estimate of average value but not so large as to average over nonuniform background intensities. However, the features in the image often vary in size such that there are problems with fixed window size. To remedy this problem, domain-dependent information can be used to check that the results of binarization give the expected features (for instance, a large blob of an ON-valued region is not expected in a page of smaller symbols). If the result is unexpected, then the window size can be modified and binarization applied again.

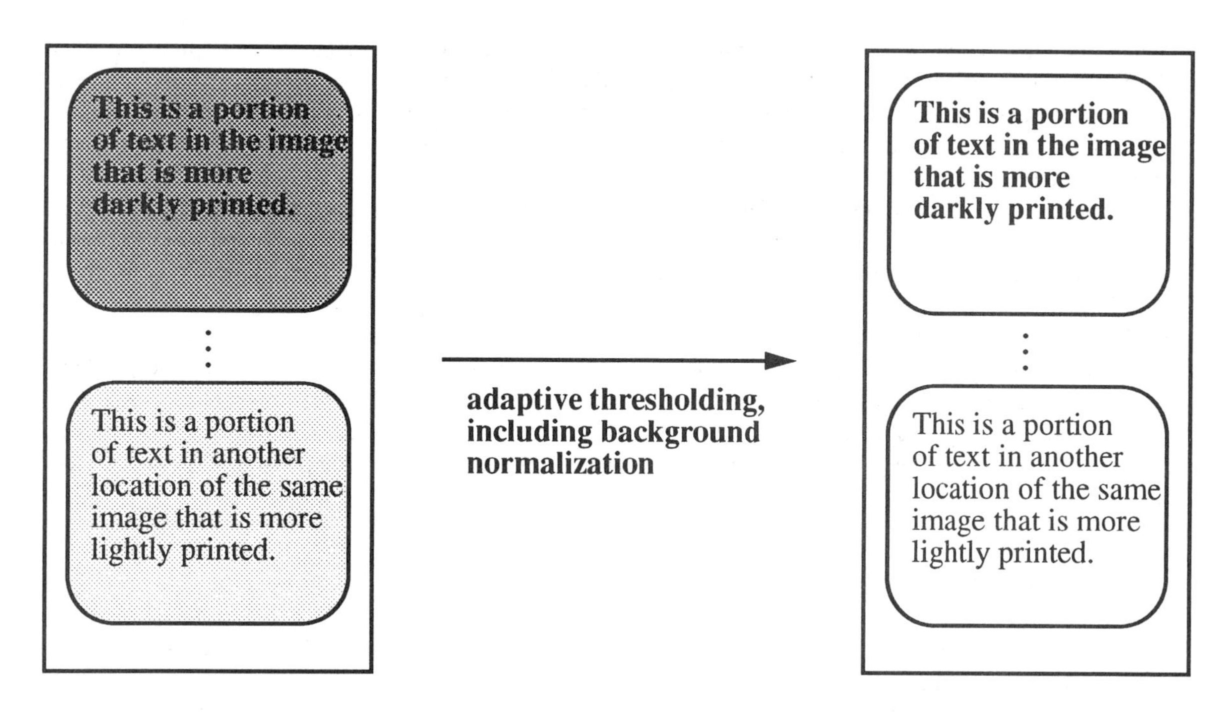

Figure 3. Diagram illustrates adaptive thresholding by background normalization. Original image on the left has portions of text with different average background values. The image on the right shows that the backgrounds have been eliminated leaving only ON on OFF.

2.2.3: Choosing a thresholding method

Whether global or adaptive thresholding methods are used for binarization, one can never expect perfect results. Depending on the quality of the original, there may be gaps in lines, ragged edges on region boundaries, and extraneous pixel regions of ON and OFF values. The assumption that processing results will not be perfect is generally true with other document-processing methods and indeed with image processing in general. The recommended procedure is to process as well as possible at each step of processing but defer decisions that do not need to be made until later steps, to avoid making irreparable errors. In later steps, there is more information as a result of processing to that point, and this information provides greater context and higher-level descriptions to aid in making correct decisions and, ultimately, recognition. Deferment, when possible, is a principle appropriate for all steps of document analysis (except, of course, the last).

A number of different thresholding methods have been presented in this section. No single method is best for all image types and applications. For simpler problems, where the image characteristics do not vary much within the image or across different images, the simpler methods will suffice. For more difficult problems of noise or varying image characteristics, more complex (and time-consuming) methods usually will be required. Commercial products vary in their thresholding capabilities. Today's scanners usually perform binarization with respect to a fixed threshold. More sophisticated document systems provide manual or automatic histogram-based techniques for global thresholding. The most common use of adaptive thresholding is in special-purpose systems used by banks to image checks. The best way to choose a method at this time is by first narrowing the choices by the method descriptions and then just experimenting with the different methods and examining their results.

Because there is no "best" thresholding method, there is still room for future research here. One problem that requires more work is to identify thresholding methods or approaches that best work on documents with particular characteristics. Many of the methods described above were not formulated for documents in particular, and their performance on them is not well known. Documents have characteristics, such as very thin lines, that will favor one method above another. Related to the question of which method is the best to use is how best to quantify the results of thresholding. For text, one way is to perform optical character recognition on the binarized results and measure the recognition rate for different thresholds. Another problem that requires further work is that of multithresholding. Sometimes documents have not two but three or more levels of intensities. For example, many journals contain highlighting boxes within the text, where the text is set against a background of a different gray level or color. While multithresholding capabilities have been claimed for some of the methods discussed above, not much dedicated work has been focused on this problem. The reader is referred to surveys providing reviews and more complete comparisons of thresholding methods [Fu and Mu, 1981; Sahoo et al., 1988; Weszka, 1978].

In our work, we have found that the moment preservation technique outperforms the other global-thresholding methods for documents, so we have included here a paper by Tsai that this method describes. The second paper by Kamel and Zhao describes adaptive-thresholding techniques for character thresholding on nonuniform backgrounds. This paper also gives background on and comparison of these adaptive methods.

2.3: Noise reduction

(*Keywords*: filtering, noise reduction, salt-and-pepper noise, filling, morphological processing, cellular processing)

After binarization, document images are usually filtered to reduce noise. Salt-and-pepper noise (also called *impulse noise*, *speckle noise*, or just *dirt*) is a prevalent artifact in poorer quality document images (such as poorly thresholded faxes or poorly photocopied pages). The noise appears as isolated pixels or pixel regions of ON noise in OFF backgrounds or OFF noise (holes) within ON regions and as rough edges on characters and graphics components (see Figure 4). The process of reducing this phenomenon is called *filling*. The most important reason to reduce noise is that extraneous features will otherwise cause subsequent errors in recognition. Also, noise reduction reduces the size of the image file, and this in turn reduces the time required for subsequent processing and storage. The objective in the design of a filter to reduce noise is that it remove as much as possible of the noise while retaining all of the signal.

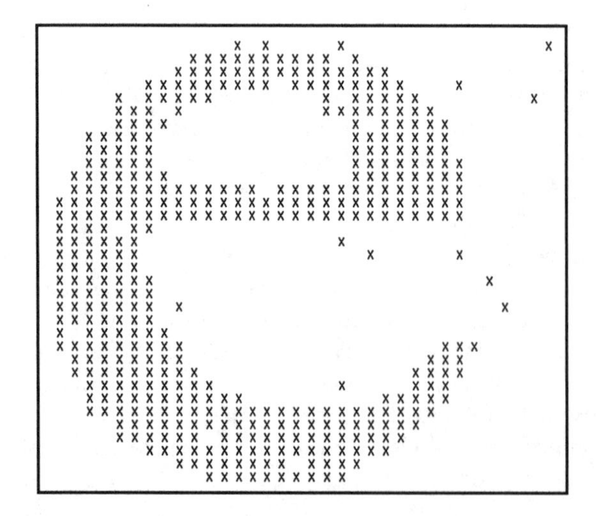

Figure 4. Illustration of the letter "e" with salt-and-pepper noise. On the left, the letter is shown with its ON and OFF pixels as "X"s and blanks. On the right, the noisy character is shown.

2.3.1: Morphological and cellular processing

Morphological [Haralick 1987, Haralick 1992, Serra 1982] and cellular processing [Preston 1979] are two families of processing methods by which noise reduction can be performed. (These methods are applicable to much more than just the noise reduction application mentioned here, but we leave further description of the methods to the references.) The basic morphological or cellular operations are erosion and dilation. *Erosion* is the reduction in size of ON regions, most simply accomplished by peeling off a single-pixel layer from the outer boundary of all ON regions in each erosion step. *Dilation* is the opposite process, where single-pixel, ON-valued layers are added to boundaries to increase their size. These operations are usually combined and applied iteratively to erode and dilate many layers. In one of these combined operations, called *opening*, one or more iterations of erosion are followed by the same number of iterations of dilation. The result of opening is that boundaries can be smoothed, narrow isthmuses broken, and small noise regions eliminated. The morphological dual of opening is *closing*, which is combining one or more iterations of dilation followed by the same number of iterations of erosion. The result of closing is that boundaries can be smoothed, narrow gaps joined, and small noise holes filled. See Figure 5 for an illustration of morphological processing.

2.3.2: Text and graphics noise filters

For documents, more specific filters can be designed to take advantage of the known characteristics of the text and graphics components. In particular, we desire to maintain sharpness in these document components, not to round corners and shorten lengths, as some noise reduction filters will do. In the simplest case, single-pixel islands, holes, and protrusions can be found by passing a 3×3 window over the image that matches these patterns [Shih 1988]; then these are filled. For noise larger than one pixel, the kFill filter can be used [O'Gorman 1992].

In lieu of including a paper on noise reduction, we describe the kFill noise reduction method in more detail. Filling operations are performed within a $k \times k$-sized window that is applied in raster-scan order, centered on each image pixel. This window comprises an inside $(k - 2) \times (k - 2)$ region, called the *core*, and the $4(k - 1)$ pixels on the window perimeter, called the *neighborhood*. The filling operation entails setting all values of the core to ON or OFF, dependent upon pixel values in the neighborhood. The decision on whether or not to fill with ON (OFF) requires that all core values must be OFF (ON) and depends on three variables, determined from the neighborhood. For a fill-value equal to ON (OFF), the n variable is the number of ON (OFF) pixels in the neighborhood, the c variable is the number of connected groups of ON pixels in the neighborhood, and the r variable is the number of corner pixels that are ON (OFF). Filling occurs when the following conditions are met:

$$(c = 1) \text{ AND } [(n > 3k - 4) \text{ OR } ((n = 3k - 4) \text{ AND } (r = 2))]$$

The conditions on n and r are set as functions of the window size k such that the text features described above are retained. The stipulation that $c = 1$ ensures that filling does not change connectivity (that is, it does not join two letters together or separate two parts of the same connected letter). Noise reduction is performed iteratively on the image. Each iteration consists of two subiterations, one performing ON performing fills and the other OFF fills. When no filling occurs on two consecutive subiterations, the process stops automatically. An example is shown in Figure 6.

The kFill filter is designed specifically for text images to reduce salt-and-pepper noise while maintaining readability. It is a conservative filter, erring on the side of maintaining text features versus reducing noise when these two conflict. To maintain text quality, the filter retains corners on text of 90° or less, reducing rounding that occurs for other low-pass spatial filters. The filter has a k parameter (the k in "kFill") that enables adjustment for different text sizes and image resolutions, thus enabling retention of small features such as periods and the stick ends of characters. Since this filter is designed for fabricated symbols, text, and graphics, it is not appropriate for binarized pictures where less regularly shaped regions and dotted shading (halftone) are prevalent. A drawback of this filter—and of processes that iterate several times over the entire image in general—is that the processing time is expensive. Whether the expenditure of applying a filter such

a)

X X X
X [X] X
X X X

structuring element
with origin at center

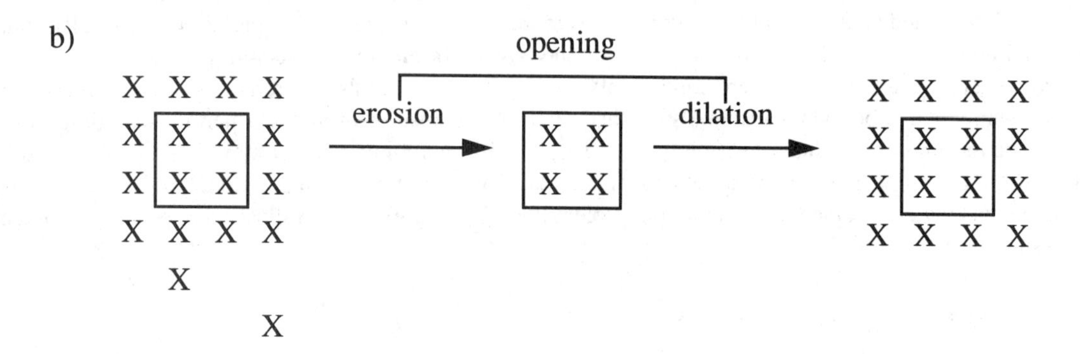

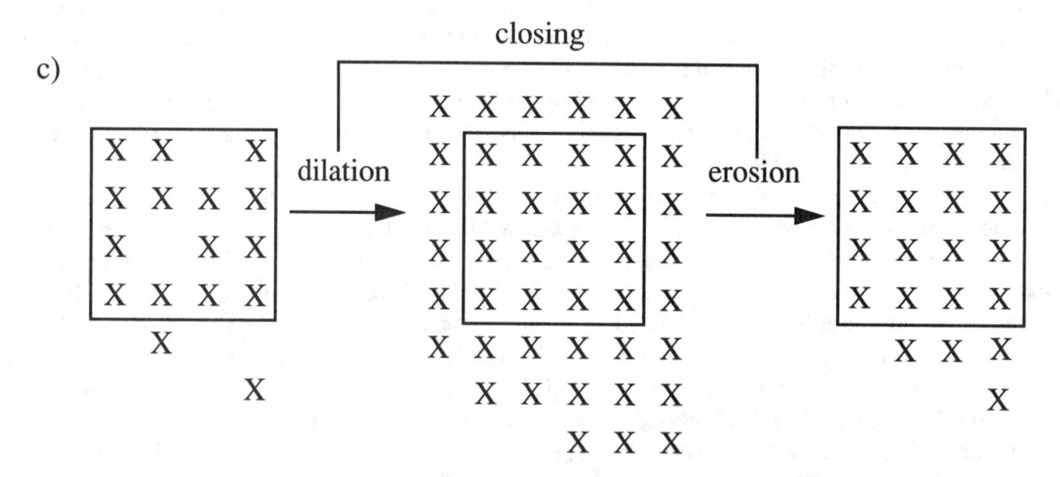

Figure 5. Morphological processing. The structuring element in (a) is centered on each pixel in the image, and pixel values are changed as follows. For erosion, an ON-valued center pixel is turned OFF if the structuring element is over one or more OFF pixels in the image. For dilation, an OFF-valued center pixel is turned ON if the structuring element is over one or more ON pixels in the image. In (b), erosion is followed by dilation; that combination is called *opening.* One can see that the isolated pixel and the spur have been removed in the final result. In (c), dilation is followed by erosion; that combination is called *closing.* One can see that the hole is filled, the concavity on the border is filled, and the isolated pixel is joined into one region in the final result.

as this one in the preprocessing step is justified depends on the input image quality and the tolerance for errors due to noise in subsequent steps.

Most document-processing systems perform rudimentary noise reduction by passing 3×3-size filter masks across the image to locate isolated ON and OFF pixels. For more extensive descriptions of these techniques in document systems, see [Modayur 1993] for the use of morphology in a music-reading system and [Story 1992] for the use of kFill in an electronic library system.

2.4: Thinning and distance transform

(*Keywords*: thinning, skeletonizing, medial axis transform, distance transform)

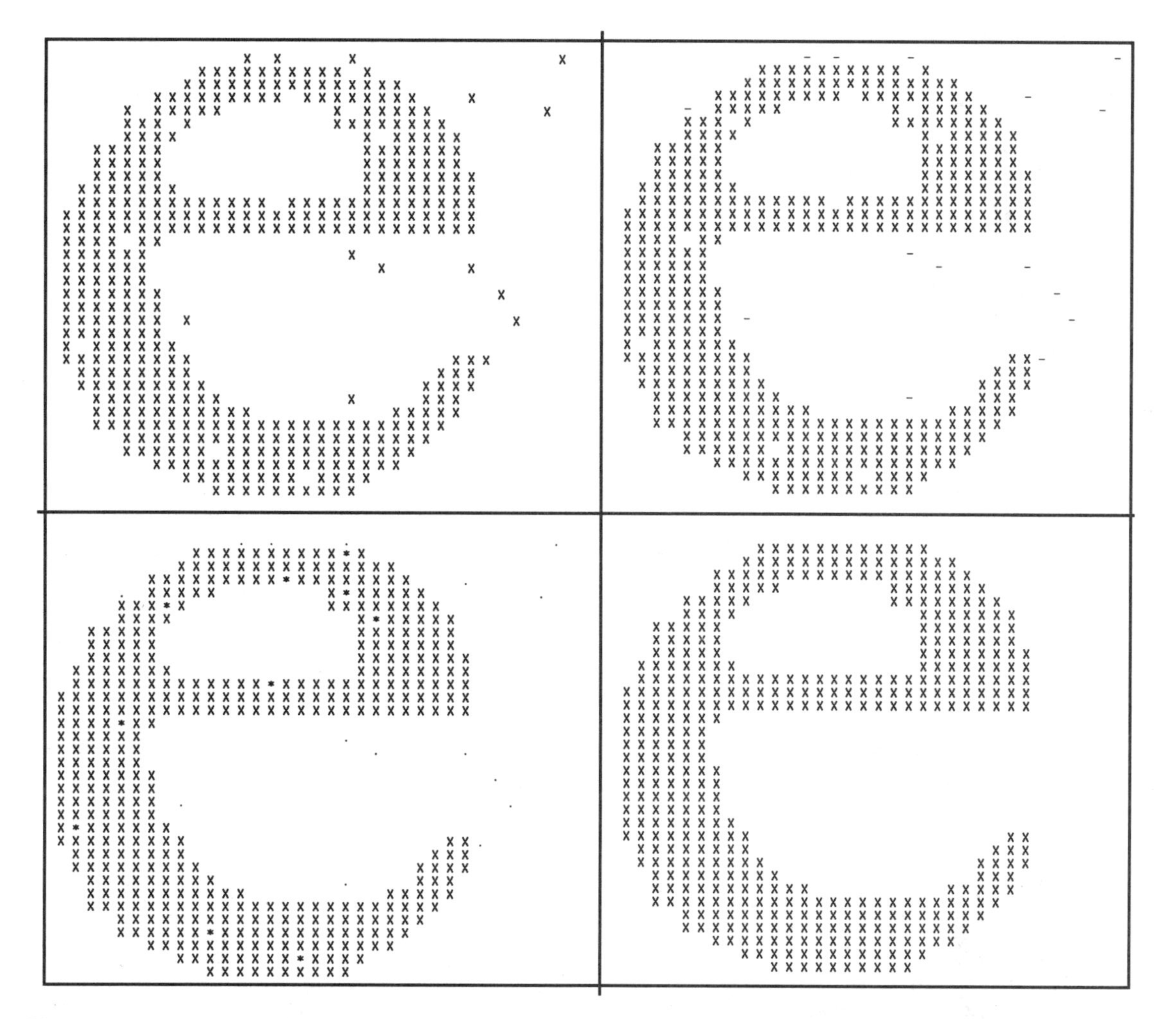

Figure 6. Results of kFill filter for salt-and-pepper noise reduction. The original noisy image is shown at the top left. The top right shows ON-pixel removal, the bottom left shows OFF-pixel filling, and the final image is shown at the bottom right.

2.4.1: Thinning

Thinning is an image-processing operation in which binary-valued image regions are reduced to lines that approximate the center lines, or *skeletons*, of the regions. The purpose of thinning is to reduce the image components to their essential information so that further analysis and recognition are facilitated. For example, the same words can be handwritten with different pens, giving different stroke thicknesses, but the literal information of the words is the same. For many recognition and analysis methods where line tracing is done, it is easier and faster to trace along 1-pixel-wide lines than along wider ones. Although the thinning operation can be applied to binary images containing regions of any shape, it is useful primarily for "elongated" shapes versus convex, or "blob-like," shapes. Thinning is commonly used in the preprocessing stage of such document analysis applications as diagram understanding and map processing. In Figure 7, some images are shown whose contents can be analyzed well due to thinning, and their thinning results are also shown here.

Note should be made that thinning is also referred to as *skeletonizing* and *core-line detection* in the literature. We will use the term *thinning* to describe the procedure, and *thinned-line*, or *skeleton*, to describe the results. A related term is the *medial axis*, the set of points of a region in which each point is equidistant from its two closest points on the boundary. The medial axis is often described as the ideal that thinning approaches. However, since the medial axis is defined only for continuous space, it can be only approximated by practical thinning techniques that operate on a sampled image in discrete space.

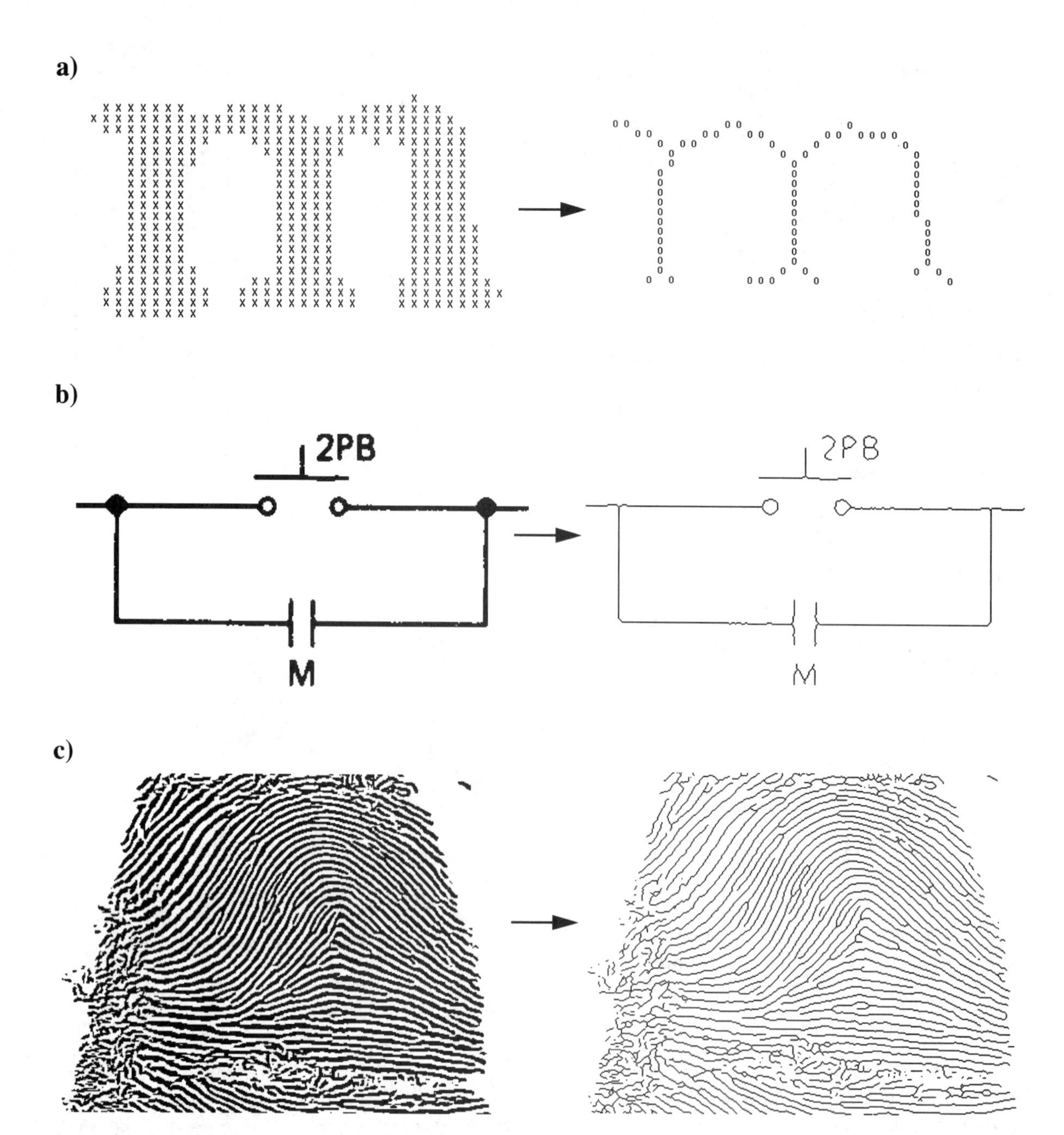

Figure 7. Original images are on the left and thinned-image results are on the right. (a) The letter "m." (b) A line diagram. (c) A fingerprint image.

The thinning requirements are formally stated as follows:

(1) Connected image regions must thin to connected line structures.
(2) The thinned result should be minimally eight-connected (explained below).
(3) Approximate end line locations should be maintained.
(4) The thinning results should approximate the medial lines.
(5) Extraneous spurs (short branches) caused by thinning should be minimized.

That the results of thinning must maintain connectivity as specified by Requirement (1) is essential. This requirement guarantees the number of thinned connected-line structures to be equal to the number of connected regions in the original image. By Requirement (2), we stipulate that the resulting lines should always contain the minimal number of pixels that maintain eight-connectedness. (A pixel is considered eight-connected to another pixel if the second pixel is one of the eight closest neighbors to it.) Requirement (3) states that the locations of endlines should be maintained. Since thinning can be achieved by iteratively removing the outer boundary pixels, it is important not to also iteratively remove the last pixels of a line. This removal would shorten the line and not preserve its location. Requirement (4) states that the resultant thin lines should best approximate the medial lines of the original regions. Unfortunately, in digital space, the true medial lines can be only approximated. For instance, for a 2-pixel-wide vertical or horizontal line, the true medial line should run at the half-pixel spacing along the middle of the original. Since it is impossible to represent this in digital image space, the result will be a single line running at one side of the original. With respect to Requirement (5), it is obvious that noise should be minimized, but it is often difficult to say what is noise and what is not. We do not want spurs to result from every small bump on the original region, but we do want to recognize when a somewhat larger bump is a feature. Although some thinning algorithms have parameters to remove spurs, we believe that thinning and noise removal should be performed separately. Since one person's undesired spur may be another's desired short line, it is best to perform thinning first and then, in a separate process, remove any spurs whose length is less than a specified minimum.

The basic iterative thinning operation is to examine each pixel in an image within the context of its neighborhood region of at least 3×3 pixels and to "peel" the region boundaries, one pixel layer at a time, until the regions have been reduced to thin lines. (See [Hilditch 1969] for basic 3×3 thinning and [O'Gorman 1990] for generalization of the method to $k \times k$-sized masks.) This process is performed iteratively; on each iteration, every image pixel is inspected, and 1-pixel-wide boundaries that are not required to maintain connectivity or endlines are erased (set to OFF). In Figure 8, one can see how, on each iteration, the outside layer of 1-valued regions is peeled off in this manner and, when no changes are made on an iteration, the image is thinned.

Unwanted in the thinned image are isolated lines and spurs off longer lines that are artifacts due to the thinning process or noise in the image. Some thinning methods [Arcelli 1985] require that the binary image is noise filtered before thinning, because noise severely degrades the effectiveness and efficiency of this processing. However, noise can never be totally removed and, as mentioned, it is often difficult to distinguish noise from the signal in the earlier stages. In [Jagadish 1989], an opposite approach is taken: to delay noise reduction until after thinning, when more information has been found. After thinning, segments of image lines between endpoints and junctions are found, and descriptive parameters (such as length, type as classified by junctions or endlines at the ends of the segment, average curvature, and absolute location) are associated with them. This descriptive and contextual information is then used to remove the line artifacts.

Instead of iterating through the image for a number of times that is proportional to the maximum line thickness, thinning methods have been developed to yield the result in a fixed number of steps [Arcelli 1985, Arcelli 1989, Sinha 87]. These methods are computationally advantageous when the image contains thick objects that would otherwise require many iterations. For these noniterative methods, skeletal points are estimated from distance measurements with respect to opposite boundary points of the regions (see the next subsection, on distance transformation). Some of these methods require joining the line segments after thinning to restore connectivity and also require a parameter estimating maximum thickness of the original image lines so that the search for pairs of opposite boundary points is spatially limited. In general, these noniterative thinning methods are less regularly repetitive, not limited to local operations, and less able to be pipelined, compared to the iterative methods; these factors make their implementation in special-purpose hardware less appropriate.

Algorithms have also been developed for extracting thin lines directly from gray-level images of line drawings by tracking along gray-level ridges—that is, without the need for binarization [Watson 1984]. These algorithms have the advantage of being able to track along ridges whose peak intensities vary throughout the image, such that binarization by global thresholding would not yield connected lines. However, a problem with tracking lines on gray-scale images arises: following false ridges (the gray-scale equivalent of spurs), which results in losing track of the main ridge or requires computationally expensive backtracking. Binarization and thinning are the methods most commonly used for document analysis applications because they are well understood and relatively simple to implement.

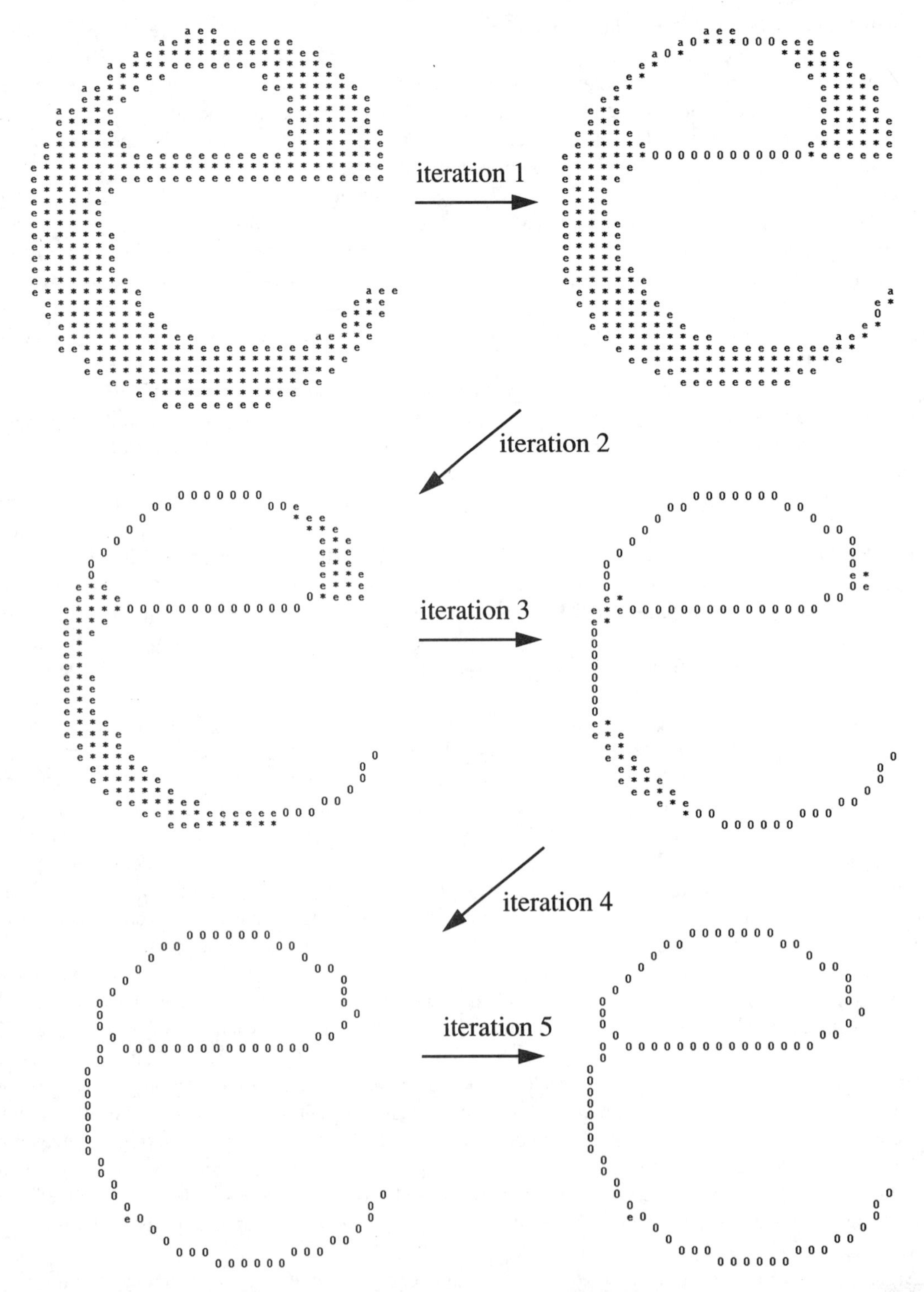

Figure 8. Sequence of five iterations of thinning on letter "e." On each iteration, a layer of the outer boundaries is peeled off. On iteration 5, the end is detected because no pixels change.

For a good background on thinning techniques, see [Pavlidis 1982]. For comparisons of the methods, see [Tamura 1978, Naccache 1984, Pal 1992] for some of the iterative techniques, [Guo 1989] for some parallel techniques, and [Lam 1992] for the most recent comprehensive survey. For a reference on thinning applied specifically to documents, see [Eckhardt 1991].

2.4.2: Distance transformation

The distance transform is a binary-image operation in which each pixel is labeled by the shortest distance from it to the boundary of the region within which it is contained. One way this technique can be used is to determine the shortest path from a given interior point to the boundary. It can also be used to obtain the thinned image by retaining only ridges of maximum local-distance measures. This thinned image, complete with distance values, has more information than simply the thinned image without distance information. It can be used as a concise and descriptive representation of the original image from which line widths can be obtained or the original image can be reconstructed. Results of the distance transform are shown in Figure 9.

There are two general approaches to obtaining the distance transformation. One is similar to the iterative thinning method described above. On each iteration, boundaries are peeled from regions. But, instead of setting each erased pixel value to OFF, it is set to the distance from the original boundary. Therefore, on the first iteration (examining 3 × 3-size masks), erased boundaries are set to 0. On the second iteration, any erased core pixels will have a distance of 1 for vertical or horizontal distance to a boundary point or $\sqrt{2}$ for diagonal distance to a boundary point. On the third and subsequent iterations, each pixel's distance value is calculated as the sum of the smallest distance value of a neighboring pixel plus its distance to that neighbor. When a core can no longer be thinned, it is labeled with its distance to the closest boundary.

The second approach requires a fixed number of passes through the image. To obtain the integer approximation of the Euclidean distance, two passes are necessary. The first pass proceeds in raster order, from the top row to the bottom row, left to right on each row. The distances are propagated in a manner similar to that above, but because the direction of the raster scan is from top left to bottom right, these first-iteration values are only intermediate values—they contain only distance information from above and to the left. The second pass proceeds in reverse raster order—from bottom right to top left—where the final distance values are obtained taking into account the distance information from below and to the right as well. For further treatments of distance transformations, see [Arcelli 1992, Borgefors 1984, Borgefors 1986].

The iterative method is a natural one to use if iterative thinning is desired. As mentioned above, thinning is only appropriate for elongated regions, and if the distance transform of an image containing thick lines or more convex regions is desired, the fixed-pass method is more appropriate. The fixed-pass method can also be used as a first step toward thinning. For all of these distance transformation methods, since distance is stored in the pixel, attention must be paid so that the distance does not exceed the pixel word size, usually a byte. This problem is further exacerbated since floating point distance is approximated by integer numbers usually by scaling up the floating point number (for example 1.414 would become 14). This word size consideration is usually not a problem for images of relatively thin, elongated regions but may be a problem for larger regions.

Thinning is available on virtually all commercial graphics analysis systems. The particular method varies—there are many, many different thinning methods—but it is usually iterative and uses 3 × 3-or 3 × 4-size masks. Most systems take advantage of a fast table look-up approach for the mask operations, implemented in software or in hardware for more specialized (faster and more expensive) machines. Because of the diverse thinning methods, we include a survey paper on thinning by Lam, Lee, and Suen.

2.5: Chain coding and vectorization

(*Keywords*: chain code, Freeman chain code, Primitives Chain Code (PCC), line and contour compression, topological feature detection, vectorization)

2.5.1: Chain coding

When objects are described by their skeletons or contours, they can be represented more efficiently than simply by ON- and OFF-valued pixels in a raster image. One common way is by chain coding, where the ON pixels are represented as sequences of connected neighbors along lines and curves. Instead of storing the absolute location of each ON pixel, the direction from its previously coded neighbor is stored. A neighbor is any of the adjacent pixels in the 3 × 3-pixel neighborhood around that center pixel (see Figure 10). There are two

```
                        1 1 1
                    1 1 1 2 1 1 1 1 1 1 1
                1 1 1 3 3 3 3 3 3 3 2 2 1 1 1
            1 1 1 3 2 1 1 1 1 1 1 1 3 3 2 2 1 1
            1 2 3 1 1               1 2 4 2 2 1 1
        1 1 1 2 1                   1 1 1 4 4 2 1 1
        1 2 3 1                         1 2 3 5 2 1 1
    1 1 1 3 1                           1 2 3 6 2 2 1
    1 2 4 2 1                           1 2 3 6 3 2 1
    1 2 5 2 1                           1 2 3 4 7 2 1 1
  1 1 2 5 2 1 1                         1 2 3 7 3 2 2 1
  1 2 2 3 5 2 1 1 1 1 1 1 1 1 1 1 1 1 1 1 4 5 3 3 3 2 1
1 1 2 3 6 2 3 3 3 3 3 3 3 3 3 3 3 3 3 3 3 2 2 2 2 2 2 1
1 2 2 3 6 2 1 1 1 1 1 1 1 1 1 1 1 1 1 1 1 1 1 1 1 1 1 1
1 2 3 6 2 1 1
1 2 3 6 2 1
1 2 3 6 2 1
1 2 3 6 2 1
1 2 3 6 2 1 1
1 2 3 6 2 2 1
1 2 3 6 3 2 1
1 2 3 6 3 2 1
1 2 2 3 6 2 1 1
1 1 2 3 6 2 2 1 1                                   1 1 1
  1 2 2 3 6 2 2 1                                 1 1 1
  1 1 2 3 6 3 2 1 1                           1 1 2 1
    1 2 3 3 5 2 2 1 1                         1 3 1
    1 2 2 2 3 6 2 2 1 1 1                   1 1 1 2 1
    1 1 1 2 3 3 5 2 2 2 1 1 1 1 1 1 1 1 1 1 1 2 3 1
        1 2 2 2 3 5 3 2 2 2 2 2 2 2 2 2 2 2 4 2 1
        1 1 1 2 2 2 5 5 5 3 3 3 3 3 3 4 4 4 1 1
            1 1 1 2 2 2 3 6 6 6 6 5 5 2 1 1 1
                1 1 1 2 2 2 2 2 2 2 1 1 1
                    1 1 1 1 1 1 1 1 1
```

```
                        - - -
                    - - - - - - - - - - -
                - - - 3 3 3 3 3 3 3 - - - - -
            - - - 3 2 - - - - - - - 3 3 - - - -
            - - 3 - -               - - 4 - - - -
        - - - 2 -                   - - - 4 4 - - -
        - - 3 -                         - - - 5 - - -
    - - - 3 -                           - - - 6 - - -
    - - 4 - -                           - - - 6 - - -
    - - 5 - -                           - - - - 7 - - -
  - - - 5 - - -                         - - - 7 - - - -
  - - - - 5 - - - - - - - - - - - - - - - 4 5 - - - - -
- - - - 6 - 3 3 3 3 3 3 3 3 3 3 3 3 3 3 3 - - - - - - -
- - - - 6 - - - - - - - - - - - - - - - - - - - - - - -
- - - 6 - - -
- - - 6 - -
- - - 6 - -
- - - 6 - -
- - - 6 - - -
- - - 6 - - -
- - - 6 - - -
- - - 6 - - -
- - - - 6 - - -
- - - - 6 - - - -                                   - - -
  - - - - 6 - - -                                 - - -
  - - - - 6 - - - -                             - - 2 -
    - - - - 5 - - - -                           - 3 -
    - - - - - 6 - - - - -                   - - - 2 -
    - - - - - - 5 - - - - - - - - - - - - - - - 3 -
        - - - - - 5 - - - - - - - - - - - - 4 2 -
        - - - - - - 5 5 5 - - - - - - 4 4 4 - -
            - - - - - - - 6 6 6 6 5 5 - - - -
                - - - - - - - - - - - - -
                    - - - - - - - - -
```

Figure 9. Distance transform. Top letter "e" shows pixel values equal to the distance to the closest border, midline pixels (shown below also) show the total width at the midline.

3	2	1
4	X	0
5	6	7

Figure 10. For a 3 × 3-pixel region with the center pixel denoted as X. The figure shows codes for chain directions from the center pixel to each of eight neighbors: 0 (east), 1 (northeast), 2 (north), 3 (northwest), and so on.

advantages of coding by direction versus absolute coordinate location. One is in storage efficiency. For commonly sized images larger than 256 × 256, the coordinates of an ON-valued pixel are usually represented as two 16-bit words; in contrast, for chain coding with eight possible directions from a pixel, each ON-valued pixel can be stored in a byte or even packed into three bits. A more important advantage in this context is that, since the chain coding contains information on connectedness within its code, this feature can facilitate further processing, such as smoothing of continuous curves, and analysis, such as feature detection of straight lines.

Further explanation is required regarding connectedness between pixels. The definition of *connected neighbors* that we will use here is called *eight-connected.* That is, a chain can connect from one pixel to any of its eight closest neighbors in directions 0 to 7 in Figure 10. Other definitions of *connectedness* are also used in the literature. The most common is *four-connected,* where a pixel can be connected to any of its four closest neighbors in the directions 0, 2, 4, or 6. An advantage of the four-connected chain is that each of its four chain directions has the same distance, 1 pixel spacing, versus eight-connected chains, where there are two distances, 1- and $\sqrt{2}$-pixel spacings. The primary advantage of the eight-connected chain is that it more closely represents diagonal lines and portions of lines, that is, it is not limited to the horizontal and vertical stair-steps to which the four-connected code is limited. Also eight-connected codings yield more concise representations.

The Freeman chain code is a widely used chain-coding method [Freeman 1974]. Coding is accomplished in the following manner. A raster search is made from the top left of the image to the bottom right, examining each pixel. When an ON-valued pixel is found, the coordinate location of this pixel is stored, that pixel is set to OFF in the image, and chain coding is begun. The direction code is stored for each connection from the current pixel to a neighboring pixel and the current pixel is set to OFF. The chain coding is done for all connected pixels until the end of a line or until a closed loop rejoins. If there is a branch in the line, one of the branching neighbors is chosen arbitrarily. The end of the chain is indicated by adding a code that is the inverse direction of the previous code (for example, 0, 4 or 1, 5), and since consecutive codes of opposite directions are otherwise impossible, this indicates the chain end. See Figure 11 for an example of a Freeman-chain-coded line structure.

Although the Freeman chain code is highly effective for compression of line images, it was designed for contours, without any provision for maintaining branching-line structures (each branch is coded as a separate chain). This is fine for compression, but for image analysis it is important to retain the complete line structure with all its branches and to know the topology at each junction. In [Harris 1982], the chain code is supplemented by coding with pixel values of 0, 1, 2, and 3 for background pixels, terminal (end-of-line) pixels, intermediate pixels, and junction pixels, respectively. This improvement allows line segments and their interconnections to be easily determined.

Another method, the Primitives Chain Code, also preserves topological features [O'Gorman 1992]. PCC contains codes for the following features: ends of lines, bifurcation and cross junctions, and breaks indicating the beginning of coding within a closed contour. With these additional features, subsequent pattern recognition steps are facilitated. Also, this code usually results in higher compression than the other chain code techniques, because it limits the number of code words to the number of eight-connected possibilities and packs them efficiently. PCC has been applied to analysis of fingerprints, maps, and engineering diagrams, all of which contain branches as important features. PCC coding is accomplished in a similar manner to that for Freeman chain coding. That is, a raster scan for ON-valued pixels is first done, and any lines are followed and

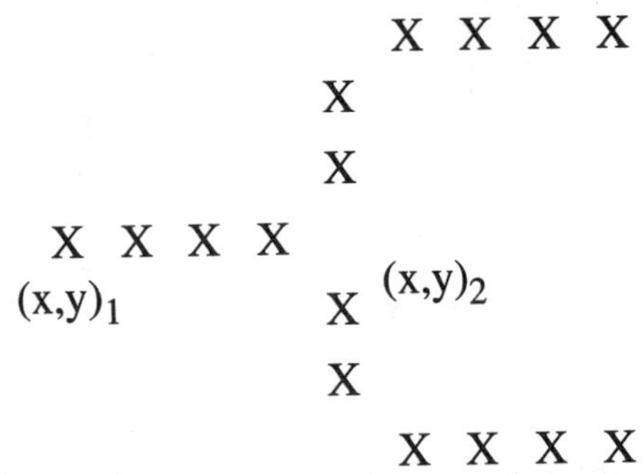

```
                    X X X X
                  X
                  X
     X X X X
   (x,y)1         X (x,y)2
                  X
                    X X X X
```

Figure 11. Pixels of a branching-line structure.The Freeman chain code for this is: {$(x, y)_1$, 0, 0, 0, 1, 2, 1, 0, 0, 0, 4, $(x, y)_2$, 6, 7, 0, 0, 0, 4}, where the top branch is coded first, then the bottom beginning with the pixel at $(x, y)_2$.

encoded. The difference is that features are also encoded, and because of the presence of codes for these features, different line topologies can be recognized from the code. For the example in Figure 11, the PCC code is $(x, y)_1$, 41, **b**, 15, 41, **e**, 61, 41, **e**. The bold characters indicate PCC features and the nonbold numbers indicate portions of up to three chain directions between the features. Here, the features are seen easily as a branch junction (b) followed by two endlines (e). The other codes are table look-up values for the connections between features: 41 is equivalent to {0,0,0}, 15 is equivalent to {1,2,1}, and 61 is equivalent to {6,7,0} in Freeman chain coding.

2.5.2: Vectorization

An alternative to thinning and chain coding is representing image lines by the straight-line segments that can be drawn within the original thicker lines. This approach is called *vectorization.* In one vectorization approach [Ramachandran 1980], horizontal runs within lines are first found, adjacent runs on subsequent scan lines are grouped together, and piecewise-straight segments are fit along these connected regions. In [Ejiri 1990], vectorization is performed at the initial digitizing level by hardware that locates and tracks long straight- line segments (see Figure 12). One advantage of vectorization is that, since long lines are searched and found, there are fewer spurs than usually result by thinning and chain coding. It should be noted that these straight-segments found by vectorization will not usually correspond to complete straight lines as intended in the original drawing. This effect is due in part to the stair-stepping that results from straight lines being digitized at slight angles. To determine complete line features, subsequent analysis is necessary. An example of post-processing of vectorized data to better match objects found against the object model is described in [Hori 1992].

Vectorization is intermediate in purpose between the thinning and chain-coding procedures and polygonalization, which will be described in the next chapter. Thinning and chain coding attempt to represent exactly the line paths, polygonalization attempts to approximate line features, and vectorization yields something between the two. If the vectorization tolerance parameter is 0, the results are closest to those of thinning and chain coding. If the tolerance is such that a piecewise-straight approximation of straight and curved lines results, the result is similar to polygonalization. In practice, the term *vectorization* is often used loosely for techniques that span this range. To avoid this confusion we use *chain coding* and *polygonalization* as terms to describe families of methods and use *vectorization* only for the so-called *hardware vectorizers* mentioned here.

A final note should be made on automatic and human-assisted vectorization. Often, the image quality is such that automatic line extraction is too error-prone to be useful. For instance, for conversion of engineering drawings to CAD format from older blueprints, the cost of a human operator correcting all the errors made by the computer may be higher than that of manually entering the entire diagram in the first place. For cases such as this, a computer can be used to facilitate manual conversion. One example of this is Fastrak [Fulford 1981], an interactive line-following digitizer in which a human operator simply "leads" the computer along the path of the line, using an input device such as a mouse. Virtually all practical commercial systems for diagram entry employ a combination of image analysis and computer-aided manual correction.

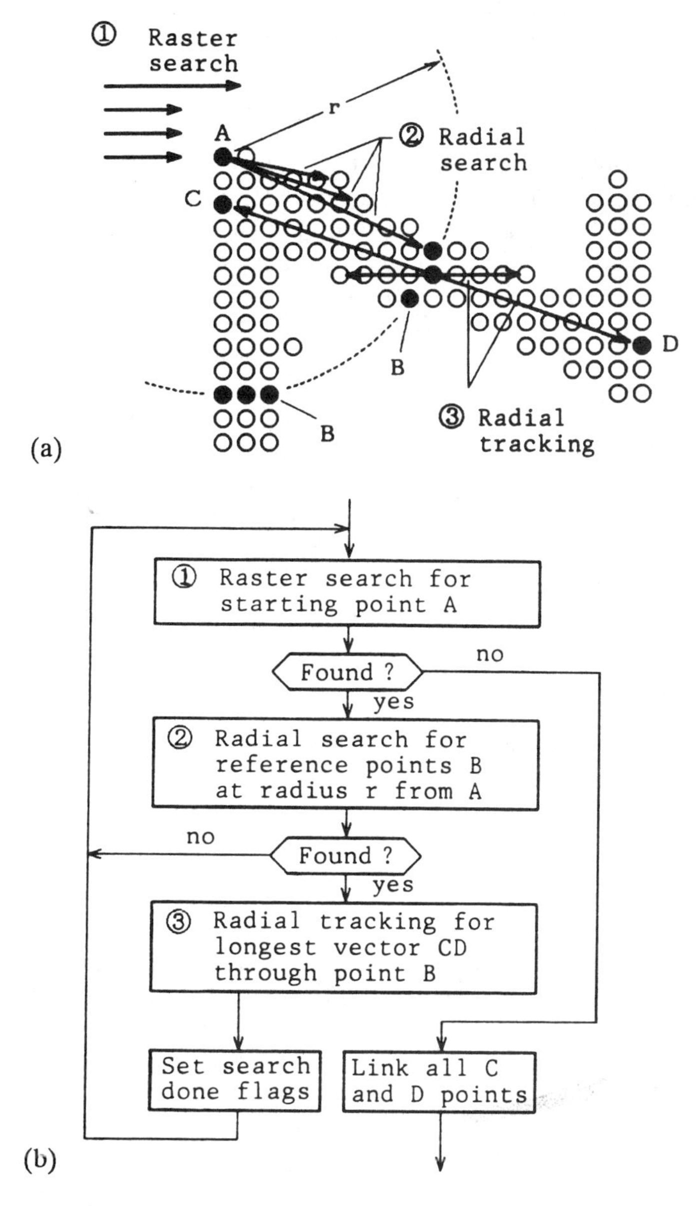

Figure 12. The diagram shows the radial-searching procedure to determine long, straight lines for vectorization [Ejiri 1990].

Chain coding or vectorization is performed in virtually every commercially available system for storing and analyzing graphics documents. The Freeman chain code is most prevalent, although systems often incorporate proprietary vectorization techniques. We include here a recent vectorization paper by Hori and Okazaki, that illustrates application to maps.

2.6: Binary region detection

(*Keywords*: region extraction, binary segmentation, region growing, blob detection, blob coloring, connected component labeling, contour detection, line adjacency graph)

Before feature-level analysis can take place, segmentation must be performed to detect individual regions (also called *objects* or *blobs*) in the image. For gray-scale and color images, this is sometimes difficult, because the objects may blend into the background or there may be overlap. However, for binary images, it is a more straightforward procedure to find each group of neighboring ON pixels. Thus, the character "a" comprises one region and the character "i" two regions. A dashed line comprises many regions. Once regions have been found, features can be determined from them, and recognition can be made of the region as one, or part of a, particular character or graphics component.

Region detection is analogous to chain coding preceded by thinning, in that the objective of both procedures is to yield a single representation of a group of neighboring ON pixels. The difference is the applicability of each method. It has been mentioned that thinning and chain coding can be applied to any shape, but they are most useful for elongated shapes. Region detection can also be applied to any shape, but it is useful especially for the rounder shapes where the results from thinning are less useful. For example, a circular disk will thin to a single point (under perfect thinning conditions), and the size information will be lost in this thinned image, whereas region detection will retain this size information. Region detection is also useful for hole detection. For example, thinning an image of Swiss cheese will result in many connected lines representing the cheese connections; however, this may not be as useful as the result of contour detection, which will have one contour for the boundary of the cheese and contours for each of the holes. As a counterexample, for the letter "H," thinning will yield the pertinent topological information that the symbol is made up of two vertical lines joined by a horizontal line, whereas region detection will yield only a single region with no topological information and will require additional feature detection for character recognition. Since many document components are made from line strokes, thinning and chain coding often are used. However, if the purpose is just to locate document components, or if the objects are truly blob-shaped, region detection is appropriate.

2.6.1: Contour detection

Contour detection can be thought of as the reciprocal operation of thinning. Whereas thinning yields the inside skeletons, contour detection yields the outside boundaries (or *contours*). Since a single contour envelopes a single region, contour detection can be used for region detection. Contours are comprised of boundary ON-valued pixels that border OFF-valued pixels. These contours can be found easily by examining each pixel within a 3×3 window; if the center pixel is ON and at least one of its neighborhood pixels is OFF, then the center pixel is a contour pixel and it is set to ON; all other pixels are set to OFF. Each resulting contour can then be chain coded to represent each region. Usually chain coding is done in one direction (counterclockwise) around ON regions (counterclockwise) and in the opposite direction (clockwise) around holes. The result of contour detection is illustrated in Figure 13. (See [Cederberg 1980, Pavlidis 1982] for contour-tracing algorithms.)

The contour image can be used in a number of ways. For example, the number of contours gives the number of regions (both ON-valued regions and OFF-valued holes within ON-valued regions). The centroid of the boundary pixels gives a measure of the region location. The length of a contour indicates the enclosed region size. The length and enclosed area can be used to give a measure of how elongated or "fat" the region is. Curvature and corner features can be determined from the contour to determine the region shape. (For other feature detection methods applicable to contours, see the next chapter).

2.6.2: Region labeling

Regions can also be found by other techniques based on interior (versus contour) pixels. One way to locate regions is to perform connected-component labeling (or *region coloring*) which results in each region of an image being assigned a different label (or color) [Ballard 1982, Danielsson 1982, Dillencourt 1992, Ronse 1984, Rosenfeld 1982]. The method involves a two-pass process where pixels are individually labeled. First, the image is scanned in raster order, and then each pixel is examined. For each ON-valued pixel, the previously labeled pixels to the left and above are examined. If none is ON, then the pixel is set to a new label value. If one of these is ON, then the current pixel is given the same label as that pixel. If more than one are ON, then the pixel is set to one of the labels, and all labels are put in an equivalence class (that is, they are stored for later merging). At the end of this pass, the number of connected components is the number of equivalence classes, plus the number of labels not in an equivalence class. The second pass consists of first merg-

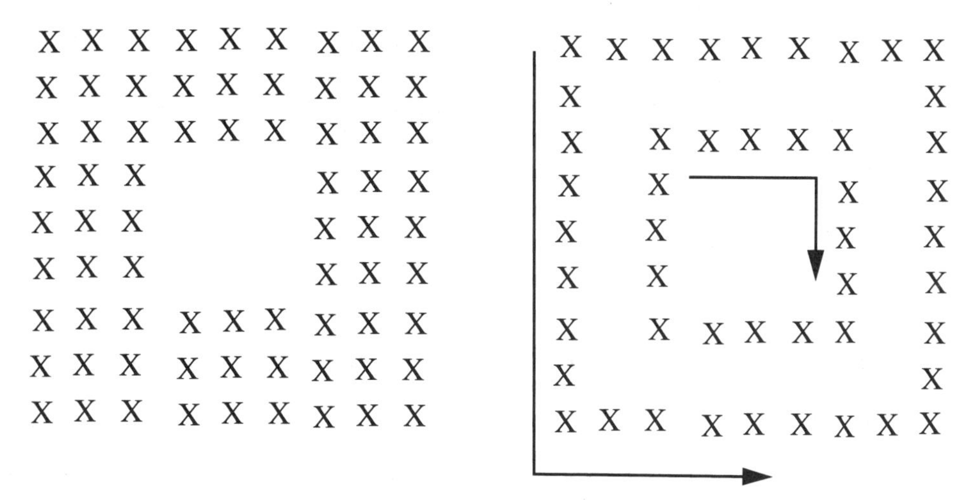

Figure 13. On the left is a region where the ON-valued pixels are represented by "X"s. On the right is the contour of this region, where only the outer and inner boundaries remain ON. The arrows around the contour show the direction of coding.

ing all labels in each equivalence class to be the same label and then reassigning all labeled pixels in the image to these final labels. In practice, the equivalence classes can be efficiently maintained as a tree data structure and merging can be performed by a set-union algorithm such as the union-find algorithm [Aho 1974].

A method that relates to connected-component labeling and to thinning is the line adjacency graph [Pavlidis 1982]. First, runs of ON pixels are found, where a *run* is a group of same-valued, adjacent pixels on the same row. Then, for each run, if there is a run on an adjacent row in which one or more pixels are neighboring, these runs are said to be *graph nodes*, and they are joined by *graph branches*. If these branches are

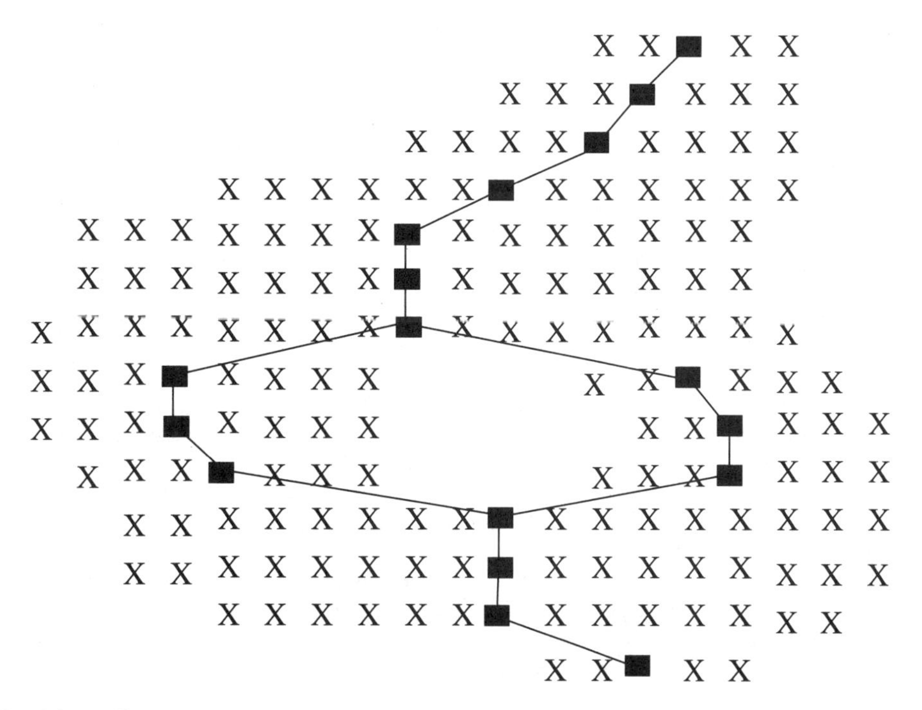

Figure 14. Line adjacency graph of region. The ON pixels are shown as "X"s. The middle of each row of ON pixels is shown as a filled box. The lines that join these midpoints of each ON row shows the line adjacency graph.

visualized as segments drawn between the middle points of each pair of runs, the branches will form a structure similar to a skeleton (Figure 14). This skeleton will retain topological information on holes and will approximate large nodules in the shape, but because the method is dependent upon the orientation, it cannot be used to obtain a good approximation of the medial lines or the endlines.

Both contour detection and coloring operations are usually available on graphics analysis systems to be applied to the nonthin components. We leave further description to the references.

References

* Papers marked with an asterisk are included as reprint papers in this book.

Abutaleb, A.S., "Automatic Thresholding of Gray-Level Pictures Using Two-Dimensional Entropy," *Computer Vision, Graphics, and Image Processing*, Vol. 47, No. 1, July 1989, pp. 22–32.

Aho, A., J.E. Hopcroft, and J.D. Ullman, *The Design and Analysis of Computer Algorithms*, Addison-Wesley, Reading, Mass., 1974.

Arcelli, C., and G.S. di Baja, "A One-Pass Two-Operation Process to Detect the Skeletal Pixels on the 4-Distance Transform," *IEEE Trans. Pattern Analysis and Machine Intelligence*, Vol. 11, No. 4, July 1989, pp. 411–414.

Arcelli, C., and G.S. di Baja, "A Width-Independent Fast Thinning Algorithm," *IEEE Trans. Pattern Analysis and Machine Intelligence*, Vol. PAMI-7, No. 4, July 1985, pp. 463–474.

Arcelli, C., and G.S. di Baja, "Ridge Points in Euclidian Distance Maps," *Pattern Recognition Letters*, Vol. 13, 1992, pp. 237–243.

Ballard, D.H., and C.M. Brown, *Computer Vision*, Prentice-Hall, Englewood Cliffs, N.J., 1982, pp. 151–152.

Borgefors, G., "Distance Transformations in Arbitrary Dimensions," *Computer Vision, Graphics, and Image Processing*, Vol. 27, No. 2, Aug. 1984, pp. 321–345.

Borgefors, G., "Distance Transformations in Digital Images," *Computer Vision, Graphics, and Image Processing*, Vol. 34, No. 3, June 1986, pp. 344–371.

Casey, R.G., and K.Y. Wong, "Document Analysis Systems and Techniques," in *Image Analysis Applications*, R. Kasturi and M. M. Trivedi, eds., Marcel Dekker, New York, N.Y., 1990, pp. 1–36.

Cederberg, R., *On the Coding, Processing, and Display of Binary Images*, doctoral dissertation, Linkoping Univ., Linkoping, Sweden, 1980.

Danielsson, P.E., "An Improved Segmentation and Coding Algorithm for Binary and Nonbinary Images," *IBM J. Research and Development*, Vol. 26, 1982, pp. 698–707.

Dillencourt, M.B., H. Samet, and M. Tamminen, "A General Approach to Connected-Component Labeling for Arbitrary Image Representations," *J. ACM*, Vol. 39, No. 2, Apr. 1992, pp. 253–280.

Eckhardt, U., and G. Maderlechner, "Thinning for Document Processing," *Proc. Int'l Conf. Document Analysis and Recognition*, 1991, pp. 490–498.

Ejiri, M., *et al.*, "Automatic Recognition of Engineering Drawings and Maps," in *Image Analysis Applications*, R. Kasturi and M.M. Trivedi, eds., Marcel Dekker, New York, 1990.

Freeman, H., "Computer Processing of Line Drawing Images," *ACM Computing Surveys*, Vol. 6, No. 1, 1974, pp. 57–98.

Fu, S. K., and J.K. Mu, "A Survey on Image Segmentation," *Pattern Recognition*, Vol. 13, No. 1, 1981, pp. 3–16.

Fulford, N. C., "The Fastrak Automatic Digitizing System," *Pattern Recognition*, Vol. 14, 1981, pp. 65–74.

Guo, Z., and R.W. Hall, "Parallel Thinning with Two-Subiteration Algorithms," *Comm. ACM*, Vol. 32, No. 3, Mar. 1989, pp. 359–373.

Haralick, R.M., and L.G. Shapiro, *Computer and Robot Vision*, Addison-Wesley, Reading, Mass., 1992.

Haralick, R.M., S.R. Sternberg, and X. Zhuang, "Image Analysis Using Mathematical Morphology," *IEEE Trans. Pattern Analysis and Machine Intelligence*, Vol. 9, No. 4, July 1987, pp. 532–550.

Harris, J.F., *et al.*, "A Modular System for Interpreting Binary Pixel Representations of Line-Structured Data," in *Pattern Recognition: Theory and Applications*, J. Kittler, K. S. Fu, L. F. Pau, eds., D. Reidel Publisher, Norwell, Mass., 1982, pp. 311–351.

Hilditch, C.J., "Linear Skeletons from Square Cupboards," *Machine Intelligence,* Vol. 4, 1969, pp. 403–420.

*Hori, O., and A. Okazaki, "High Quality Vectorization Based on a Generic Object Model," in *Structured Document Image Analysis*, H.S. Baird, H. Bunke, K. Yamamoto, eds., Springer-Verlag, New York, N.Y., 1992, pp. 325–339.

Jagadish, H.V., and L. O'Gorman, "An Object Model for Image Recognition," *Computer*, Vol. 22, No. 12, Dec. 1989, pp. 33–41.

Johannsen, G., and J. Bille, "A Threshold Selection Method Using Information Measures," *Proc. 6th Int'l Conf. Pattern Recognition*, IEEE CS Press, Los Alamitos, Calif., 1982, pp. 140–143.

*Kamel, M., and A. Zhao, "Extraction of Binary Character/Graphics Images from Gray-Scale Document Images," *CVGIP: Graphical Models and Image Processing*, Vol. 55, No. 3, 1993, pp. 203–217.

Kapur, J.N., P.K. Sahoo, and A.K.C. Wong, "A New Method for Gray-Level Picture Thresholding Using the Entropy of the Histogram," *Computer Vision, Graphics, and Image Processing*, Vol. 29, No. 3, Mar. 1985, pp. 273–285.

Katz, Y.H., "Pattern Recognition of Meteorological Satellite Cloud Photography," *Proc. 3rd Symp. Remote Sensing Environment*, IEEE CS Press, Los Alamitos, Calif., 1965, pp. 173–214.

Kittler, J., and J. Illingworth, "Minimum Error Thresholding," *Pattern Recognition*, Vol. 19, No. 1, 1986, pp. 41–47.

*Lam, L., S.-W. Lee, and C.Y. Suen, "Thinning Methodologies — A Comprehensive Survey," *IEEE Trans. Pattern Analysis and Machine Intelligence*, Vol. 14, No. 9, Sept. 1992, pp. 869–885.

Mason, D., et al., "Measurement of C-Bands in Human Chromosomes," *Computers Biology and Medicine*, Vol. 5, 1975, pp. 179–201.

Mitchell, B.T., and A. M. Gillies, "A Model-Based Computer Vision System for Recognizing Handwritten ZIP Codes," *Machine Vision and Applications*, Vol. 2, 1989, pp. 231–243.

Modayur, B.R., et al., "MUSER — A Prototype Music Score Recognition System Using Mathematical Morphology," *Machine Vision and Applications*, Vol. 6, No. 2, 1993, pp. 140–150.

Naccache, N.J., and R. Shinghal, "SPTA: A Proposed Algorithm for Thinning Binary Patterns." *IEEE Trans. Systems, Man, and Cybernetics*, Vol. SMC-14, No. 3, May/June 1984, pp. 409–418.

O'Gorman, L., "Image and Document Processing Techniques for the RightPages Electronic Library System," *Proc. Int'l Conf. Pattern Recognition,* IEEE CS Press, Los Alamitos, Calif., 1992, pp. 260–263.

O'Gorman, L., "$k \times k$ Thinning," *CVGIP*, Vol. 51, 1990, pp. 195–215.

O'Gorman, L., "Primitives Chain Code," in *Progress in Computer Vision and Image Processing*, A. Rosenfeld and L. G. Shapiro, eds., Academic Press, New York, N.Y., 1992, pp. 167–183.

Otsu, N., "A Threshold Selection Method from Gray-Level Histograms," *IEEE Trans. Systems, Man, and Cybernetics*, Vol. SMC-9, No. 1, Jan. 1979, pp. 62–66.

Pal, S., and P. Bhattacharyya, "Analysis of Template Matching Thinning Algorithms," *Pattern Recognition*, Vol. 25, No. 5, 1992, pp. 497–505.

Pavlidis, T., *Algorithms for Graphics and Image Processing*, Computer Sci. Press, New York, N.Y., 1982, pp. 116–120 (LAG) and pp. 142–148 (contour detection).

Pavlidis, T., *Algorithms for Graphics and Image Processing*, Computer Sci. Press, New York, N.Y., 1982, pp. 195–214.

Preston, K., Jr., *et al.*, "Basics of Cellular Logic with Some Applications in Medical Image Processing," *Proc. IEEE*, Vol. 67, No. 5, May 1979, pp. 826–855.

Pun, T., "Entropic Thresholding: A New Approach," *Computer Vision, Graphics, and Image Processing*, Vol. 16, No. 3, July 1981, pp. 210–239.

Ramachandran, K., "A Coding Method for Vector Representation of Engineering Drawings," *Proc. IEEE*, Vol. 68, 1980, pp. 813–817.

Reddi, S.S., S.F. Rudin, and H.R. Keshavan, "An Optimal Multiple Threshold Scheme for Image Segmentation," *IEEE Trans. Systems, Man, and Cybernetics*, Vol. SMC-14, No. 4, July-Aug. 1984, pp. 661–665.

Ronse, C., and P.A. Divijver, *Connected Components in Binary Images: The Detection Problem*, Research Studies Press Ltd., Letchworth, England, 1984.

Rosenfeld, A., and A.C. Kak, *Digital Picture Processing*, Academic Press, New York, N.Y., 1982, pp. 241–242.

Rosenfeld, A. , and P. De La Torre, "Histogram Concavity Analysis as an Aid in Threshold Selection," *IEEE Trans. Systems, Man, and Cybernetics*, Vol. SMC-13, 1983, pp. 231–235.

Sahoo, P.K., et al., "A Survey of Thresholding Techniques," *Computer Vision, Graphics, and Image Processing*, Vol. 41, No. 2, Feb. 1988, pp. 233–260.

Serra, J., *Image Analysis and Mathematical Morphology*, Academic Press, London, U.K., 1982.

Shih, C-C., and R. Kasturi, "Generation of a Line-Description File for Graphics Recognition," *Proc. SPIE Conf. Applications of Artificial Intelligence*, Vol. 937, SPIE, Bellingham, Wash., 1988, pp. 568–575.

Sinha, R.M.K., "A Width-Independent Algorithm for Character Skeleton Estimation," *Computer Vision, Graphics, and Image Processing*, Vol. 40, No. 3, Dec. 1987, pp. 388–397.

Story, G., *et al.*, "The RightPages Image-Based Electronic Library for Alerting and Browsing," *Computer*, Vol. 25, No. 9, Sept. 1992, pp. 17–26.

Tamura, H., "A Comparison of Line Thinning Algorithms from Digital Geometry Viewpoint," *Proc. Int'l Joint Conf. Pattern Recognition*, IEEE CS Press, Los Alamitos, Calif., 1978, pp. 715–719.

*Tsai, W.H., "Moment-Preserving Thresholding: A New Approach," *Computer Vision, Graphics, and Image Processing*, Vol. 29, No. 3, Mar. 1985, pp. 377–393.

Watanabe, S., et al., "An Automated Apparatus for Cancer Processing CYBEST," *Computer Vision, Graphics, and Image Processing*, Vol. 3, No. 4, Dec. 1974, pp. 350–358.

Watson, L.T., *et al.*, "Extraction of Lines and Drawings from Grey Tone Line Drawing Images," *Pattern Recognition*, Vol. 17, No. 5, 1984, pp. 493–507.

Weszka, J.S., "A Survey of Threshold Selection Techniques," *Computer Vision, Graphics, and Image Processing*, Vol. 7, No. 2, Apr. 1978, pp. 259–265.

Weszka, J.S., and A. Rosenfeld, "Histogram Modification for Threshold Selection," *IEEE Trans. Systems, Man, and Cybernetics*, Vol. SMC-9, No. 1, Jan. 1979, pp. 38–52.

Weszka, J.S., R.N. Nagel, and A. Rosenfeld, "A Threshold Selection Technique," *IEEE Trans. Computers*, Vol. C-23, 1974, pp. 1322–1326.

White, J.M., and G.D. Rohrer, "Image Thresholding for Optical Character Recognition and Other Applications Requiring Character Image Extraction," *IBM J. Research and Development*, Vol. 27, No. 4, July 1983, pp. 400–411.

Wong, K.Y., "Multi-Function Auto-Thresholding Algorithm," *IBM Tech. Disclosure Bull.*, Vol. 21, No. 7, 1978, pp. 3001–3003.

Ye, Q.-Z., and P.-E. Danielson, "On Minimum Error Thresholding and Its Implementations," *Pattern Recognition Letters*, Vol. 7, No. 4, Apr. 1988, pp. 201–206.

Extraction of Binary Character/Graphics Images from Grayscale Document Images

MOHAMED KAMEL[1] AND AIGUO ZHAO[2]

Pattern Analysis and Machine Intelligence Laboratory, Department of Systems Design Engineering, University of Waterloo, Waterloo, Ontario, Canada N2L 3G1

Received November 27, 1991; revised October 16, 1992; accepted December 14, 1992

The extraction of binary character/graphics images from grayscale document images with background pictures, shadows, highlight, smear, and smudge is a common critical image processing operation, particularly for document image analysis, optical character recognition, check image processing, image transmission, and videoconferencing. After a brief review of previous work with emphasis on five published extraction techniques, viz., a global thresholding technique, YDH technique, a nonlinear adaptive technique, an integrated function technique, and a local contrast technique, this paper presents two new extraction techniques: a logical level technique and a mask-based subtraction technique. With experiments on images of a typical check and a poor-quality text document, this paper systematically evaluates and analyses both new and published techniques with respect to six aspects, viz., speed, memory requirement, stroke width restriction, parameter number, parameter setting, and human subjective evaluation of result images. Experiments and evaluations have shown that one new technique is superior to the rest, suggesting its suitability for high-speed low-cost applications.

1. INTRODUCTION

Documents are the primary information medium in the information society today. Common symbols representing information on documents are handwritten or machine-printed characters and graphics. With the availability of computers, video terminals, optical disks, fax machines, laser printers and scanners, these symbols can be generated, transmitted, accessed, stored, displayed, printed out and reabsorbed automatically. As a substitute for manual handling, document image analysis systems can recognize the symbols on documents and organize them in either data files which can be manipulated by various word processing and graphics softwares or databases. Various such systems have recently appeared or proposed in both academic journals and commercial markets. The systems range from PC-based OCR (Optical Character Recognition) software packages [1] to special systems which include automatic character recognition and graphics vectorization systems for engineering drawings [2, 3], maps [4], and automatic mail sorting systems [5, 6].

The application of these systems is greatly limited by their requirements for good quality documents. However, clean good quality documents are seldomly encountered in real applications. Some common problems which contribute to poor-quality documents are: (1) highlighted characters in various colours, (2) smeared or smudged characters/graphics, (3) nonuniform change in colours due to long term storage, (4) poor writing or printing quality, and (5) shadows due to poor lighting conditions when the document image is captured. Although more processing can be made in later phases, these problems are mainly handled in the first phase of a document analysis system. In this phase, a grayscale image is first captured and digitized using a scanner or a video camera, then a binary image is extracted from the original grayscale image using certain extraction technique. These poor-quality documents are also often handled by videoconferencing systems which transmit document images for human recognition. It is sufficient to represent a character/graphics image in a binary image which will be more efficient to transmit and process instead of the original grayscale image. Although often in good quality, checks provide us a similar problem. In the perspective electonic check image processing systems which are being developed by some manufacturers, once a check arrives, its image is transmitted, stored and accessed electronically, and the characters on it will be automatically recognized using OCR readers. At the first phase of such systems, a binary character image is extracted from the original grayscale check image and the background picture is suppressed as much as possible. Although sophisticated extraction techniques have been published [3, 7–9], our experiments have shown that further effort to solve these problems is still necessary.

This paper is concerned with the extraction of binary

[1] To whom correspondence should be addressed: (519)888-4645 (phone) and/or mkamel@watnow.uwaterloo.ca (e-mail).

[2] Mr. Zhao is now with Atlantis Scientific Systems Group Inc., Ottawa, Canada.

"Extraction of Binary Character/Graphics Images from Grayscale Document Images" by M. Kamel and A. Zhao from *CVGIP: Graphical Models and Image Processing,* Vol. 55, No. 3, May 1993, pp. 203–217. Copyright © 1993 by Academic Press, Inc., reprinted with permission.

character/graphics images from grayscale document images. In Section 2, previous work with an emphasis on five published techniques is briefly reviewed. In Section 3, two new techniques are presented. In Section 4, speed, memory requirement, stroke width restriction, parameter number, parameter setting, and human subjective evaluation of result images are proposed and discussed as six aspects for evaluating and analyzing extraction techniques. With experiments on a typical check image (Fig. 1) and a common text document image with highlighted characters and shadows (Fig. 2), both new and published techniques are evaluated and analysed with respect to these six aspects. The last section includes a summary and conclusion drawn from this work.

2. REVIEW OF PREVIOUS WORK

One of the earliest but also most simple techniques for extracting a binary image from an original grayscale image is the global thresholding technique, which can be found in text books on image processing, see, for example, [10]. This technique does not consider the difference between character/graphics and background, and therefore usually cannot give satisfactory results. Later techniques consider the difference between character/graphics and background. They consider both characters and graphics as line drawings and either fully or partially enforce a restriction on stroke width by assuming that the stroke widths of all these line drawings lie in a predetermined range. Based on this restriction, every character/graphics pixel is extracted according to not only its own gray level, but also the gray levels of the pixels in some approximate character-size neighborhood around the pixel. These techniques are called locally adaptive techniques [7, 11]. However, these techniques do not differentiate characters from graphics and treat the separation of characters from graphics as another problem for which some techniques have been reported in [12–16]. In this paper, we also follow this approach, and therefore will not differentiate characters from graphics.

FIG. 1. A 640 × 512 × 8 original check image.

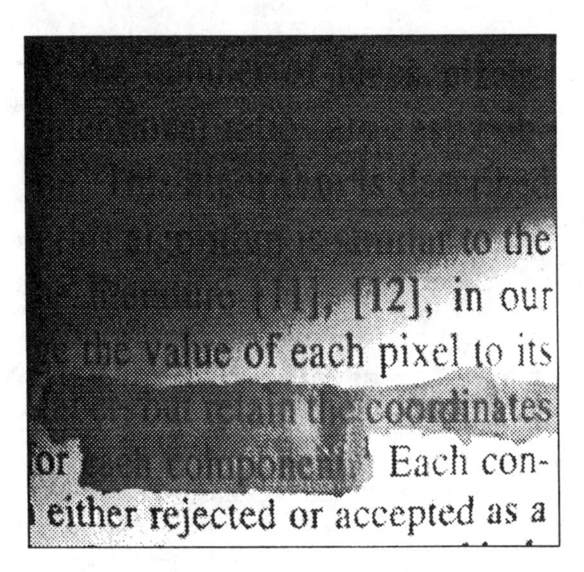

FIG. 2. A 512 × 480 × 8 original document image with highlighted characters and shadows.

A survey on initial work in this field can be found in [17]. Recent techniques originating from various application areas including check image processing [8, 18], OCR [11], blueprint image extraction [3], and videoconferencing systems [9] have been reported. An evaluation of three such techniques viz., the nonlinear function technique [8], the local contrast technique [11], and a second derivative technique which is virtually equivalent to the integrated function technique in [8], has been presented by Palumbo, Swaminathan, and Srihari [7].

Considered as part of the optical scanning process, this extraction is often implemented using minimal software and/or hardware in scanners or personal computers. Therefore, we put our research emphasis on the extraction techniques which have relative high-speed performances and minimal memory requirements, and therefore are suitable for high-speed low-cost applications. In this section, we briefly review five published techniques which will be evaluated and compared with our two new techniques later. For simplicity, unless specified, the physical meaning of every mathematical symbol introduced from now on will not be changed.

2.1. *Global Thresholding Technique*

Let $f(x, y)$ with $1 \le x \le M$, $1 \le y \le N$, be the original grayscale image with width of M, height of N and gray level range of [0, 255], $b(x, y)$ be the extracted binary character/graphics image with gray level 1 representing character/graphics and 0 representing background, and T be a predetermined constant, then the mathematical description of this technique is

$$b(x, y) = \begin{cases} 1 & \text{if } f(x, y) \leq T \\ 0 & \text{otherwise.} \end{cases}$$

2.2. *Technique by Yasuda, Dubios, and Huang (YDH Technique)*

This technique was proposed as the first phase of a check image coding scheme by Yasuda, Dubois, and Huang [18]. It uses a combination of fundamental image processing techniques which can be mathematically described as follows:

1. *Dynamic Range Expansion (normalization)*

$$f_1(x, y) = \frac{f(x, y) - min}{max - min}$$

where

$$max = \max_{1 \leq x \leq M, 1 \leq y \leq N} [f(x, y)];$$

$$min = \min_{1 \leq x \leq M, 1 \leq y \leq N} [f(x, y)].$$

2. Smoothing.

$$f_2(x, y) = \begin{cases} f_1(x, y) & \text{if } range(x, y) > T_1 \\ \sum_{(x', y') \in A(x, y)} f_1(x', y')/8 & \text{otherwise} \end{cases}$$

where

$$range(x, y) = \max_{(x', y') \in A(x, y)} f_1(x', y') - \min_{(x', y') \in A(x, y)} f_1(x', y');$$

$$A(x, y) = \{(x', y') \| x' - x| \leq 1, |y' - y| \leq 1 \text{ and } (x', y') \neq (x, y)\};$$

T_1 is a predetermined parameter.

3. *Adaptive Thresholding and Dynamic Range Expansion.* Dividing the whole image area into overlapping blocks with size of $T_2 \times T_2$, for any (x, y) in the ith block, we have

$$f_3(x, y) = \begin{cases} 255 & \text{if } maxb(i) - minb(i) < T_3 \text{ or } f_2(x, y) \geq aveb(i) \\ \dfrac{f_2(x, y) - minb(i)}{aveb(i) - minb(i)} 255 & \text{otherwise,} \end{cases}$$

where

$$maxb(i) = \max_{(x, y) \in i\text{th block}} [f_2(x, y)];$$

$$minb(i) = \min_{(x, y) \in i\text{th block}} [f_2(x, y)];$$

$$aveb(i) = \sum_{(x, y) \in i\text{th block}} [f_2(x, y)]/(T_2 \times T_2);$$

T_2, T_3 are predetermined parameters.

4. *Segmentation.*

$$b(x, y) = \begin{cases} 1 & \text{if } min3(x, y) < T_4 \text{ or } \sigma(x, y) > T_5 \\ 0 & \text{otherwise,} \end{cases}$$

where

$$min3(x, y) = \min_{-1 \leq i \leq 1, -1 \leq j \leq 1} [f_3(x - i, y - j)];$$

$$\sigma(x, y) = \left(\sum_{-1 \leq i \leq 1, -1 \leq j \leq 1} [f_3(x - i, y - j) - ave3(x, y)]^2 \right)^{1/2} \Big/ 3;$$

$$ave3(x, y) = \sum_{-1 \leq i \leq 1, -1 \leq j \leq 1} [f_3(x - i, y - j)]/9;$$

T_4, T_5 are predetermined parameters.

2.3. *Nonlinear Adaptive Technique*

This technique as proposed in [8], conceptually compares the gray level of every pixel with some average of gray levels in a neighborhood, about the pixel, whose size is approximately equal to the character-size. This average is calculated with minimal memory and high-speed using two nonlinear equations. After adjusted by a bias function, the gray level of the processed pixel is compared with the calculated average. The mathematical description of this technique is as follows:

$$f_H(x, y) = f_H(x - 1, y) + W_H[f_H(x - 1, y) - f(x, y)]$$

$$f_V(x, y) = f_V(x, y - 1) + W_V[f_V(x, y - 1) - f_H(x, y)]$$

$$b(x, y) = \begin{cases} 1 & \text{if } Z_b[f(x - L, y)] > f_V(x, y) \\ 0 & \text{otherwise,} \end{cases}$$

where

W_H, W_V are two predetermined update functions. Either W_H or W_V has a value of zero only when its argument has a value of zero, and otherwise has

	a value between zero and the value of its argument;
Z_b	is a predetermined bias function;
L	is a "look ahead" factor;
$f_H(x, y)$	is the horizontal average calculated from the gray levels of pixels at all the points (x', y) such that $1 \le x' \le x$;
$f_V(x, y)$	is the local average calculated from the gray levels of pixels at all the points (x', y') such that $1 \le x' \le x$ and $1 \le y' \le y$.

Figure 3 gives the update functions and the bias function adopted by White and Rohrer [8]. The update function provides the rate of response of the average to changes in the gray levels. Its value is zero only when its argument is zero, otherwise it has a value between zero and the value of its argument. This will guarantee that the average never exceeds the range of the input and will converge to the input for uniform gray areas. The bias function is used to offset the decision level and to eliminate the noisy background.

The major disadvantage of this technique is that it creates some black shadows in the background and extracts the left-top part edges of the large dark areas. Figures 4 and 5 illustrate this disadvantage by the black shadows in the check image and the left-top part edge of the highlight dark area in the text document image. One reason for this disadvantage is that the local average $f_V(x, y)$ is calculated based on the pixels in the left-top direction of the processed pixel only. Therefore a possible improvement can be made as follows. After extracting a binary image using the nonlinear adaptive technique, another binary

FIG. 4. Binary check character image extracted using Nonlinear Adaptive Technique.

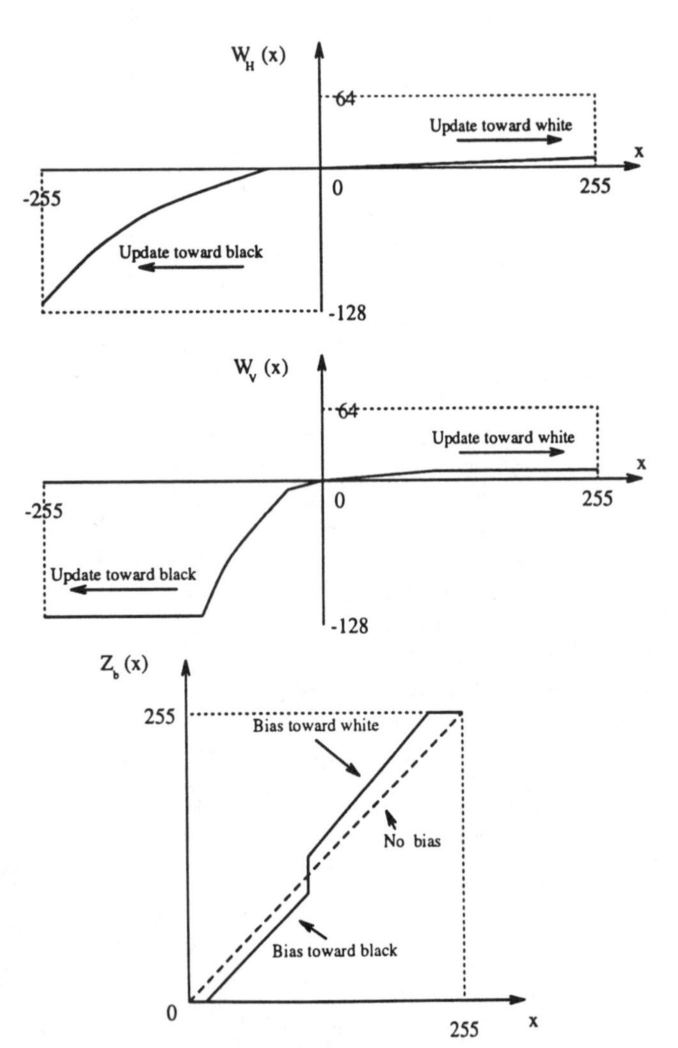

FIG. 3. The update functions $W_H(x)$, $W_V(x)$ and bias function $Z_b(x)$ used in Nonlinear Adaptive Technique.

id the number of black pixels.
limensional ratios are easily ob-
ion. This algorithm is described
h this algorithm is similar to the
he literature [11], [12], in our
ge the value of each pixel to its
label, but retain the coordinates
or each component. Each con-
either rejected or accepted as a

FIG. 5. Binary document character image extracted using Nonlinear Adaptive Technique.

image is also extracted using the same update functions and bias function but different local average which is calculated based on the pixels in the right-bottom direction of the processed pixel. Then the character/graphics image is obtained by performing an *AND* operation on the two binary images. On one hand, almost every character/graphics pixel in one of the two images is also a character/graphics pixel in the other so that it is kept in the final result image, on the other hand, most of the shadow and large dark area edge pixels in one image are not extracted again in the other image so that they are removed in the final result image. However, our experiments has shown that this technique has two disadvantages. One disadvantage is that the strokes are also significantly eroded, the other is its low speed. Although this technique can be extended by including images calculated from right-top and left-bottom directions, it will not be further researched due to its two disadvantages.

2.4. *Integrated Function Technique*

This technique as presented in [8] is virtually equivalent to the second derivative technique evaluated by Palumbo, Swaminathan, and Srihari in [7]. It first detects all the pixels lying near an edge (sharp change in gray level) and labels the pixels on the dark side with + and the pixels on the light side with −. The mathematical description of this operation is as follows:

$$S(x, y) = \begin{cases} 0 & \text{if } A(x, y) < T \\ - & \text{if } A(x, y) \geq T \text{ and } ddxy(x, y) < 0 \\ + & \text{if } A(x, y) \geq T \text{ and } ddxy(x, y) > 0 \end{cases}$$

where

$$dx(x - i, y - j) = |f(x - i, y - j) - f(x - i - 1, y - j)|;$$

$$dy(x - i, y - j) = |f(x - i, y - j) - f(x - i, y - j - 1)|;$$

T is a predetermined parameter;

$$A(x, y) = \sum_{-1 \leq i \leq 1} \sum_{-1 \leq j \leq 1} [dx(x - i, y - j) + dy(x - i, y - j)];$$

$$ddxy(x, y) = f(x + 2, y) + f(x - 2, y) + f(x, y + 2) + f(x, y - 2) - 4 \times f(x, y).$$

Considering all the pixels along with some straight line passing through the currently processed point (x, y), we get a sequences of +, − and 0. If the processed pixel is a character/graphics pixel, then this sequence should be bounded by ordered sequences as illustrated below,

$$-, +, \ldots, [S(x, y) = 0 \text{ or } +], \ldots, +, -$$

and the distance between (−+, +−) is the "stroke width" along this line. Background pixels tend not to be bounded by such ordered sequences. Therefore character/graphics pixels can be extracted out by examing the two sequences corresponding to the horizontal and vertical straight lines passing through the currently processed pixel. If these two sequences are bounded and one of two "stroke widths" is in the predetermined stroke width range, then the processed pixel is a character/graphics pixel; otherwise it is a background pixel.

2.5. *Local Contrast Technique*

In their patent for character recognition applications [11], Giuliano, Paitra, and Stringa presented this technique which is virtually a window operator. Every pixel in the output image is calculated using the following operation on the 3 × 3 × 5 pixels in a window as shown in Figure 6:

begin
  if ($f(x, y) < T_1$) then $b(x, y) = 1$;
  else
    begin
      a_1 = Average of 9 pixels in area A_1;
      $A_{2t} = \{(x, y) \mid (x, y) \text{ in } A_2 \text{ and } f(x, y) > T_2\}$;
      a_2 = Average of pixels in area A_{2t};
      if $(((T_3 \times a_2) + T_5) > (T_4 \times a_1))$ then $b(x, y) = 1$;
      else $b(x, y) = 0$;
    end;
end.

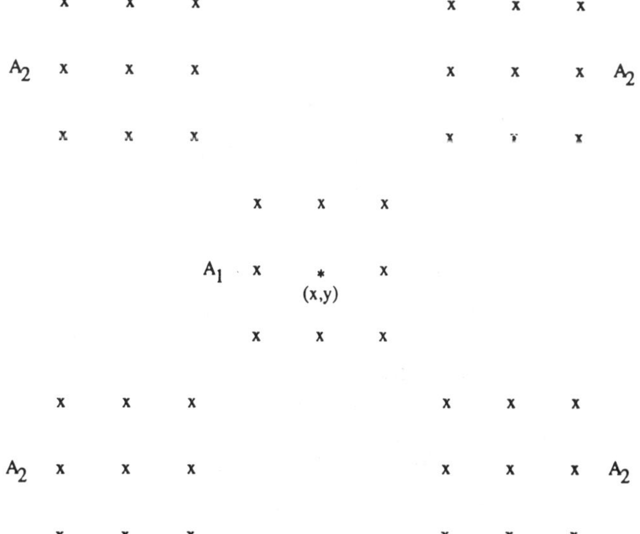

FIG. 6. Window operations in Local Contrast Technique.

where

T_1, a predetermined parameter, is used to extract all the pixels whose gray levels are less than it as character/graphics pixels (similar to the parameter in global thresholding techniques);

T_2, a predetermined parameter, is used to detect all the pixels in A_2 with gray levels not greater than it as unconsidered pixels;

T_3, T_4, T_5, three predetermined parameters, are used for comparing a_1 with a_2.

3. TWO NEW TECHNIQUES

3.1. *Logical Level Technique*

YDH technique, the nonlinear adaptive technique and the local contrast technique have a common disadvantage that they extract the edges of large dark areas. The reason for this is that they have only partially factored the stroke width restriction by a comparison of two values. One value is the gray level of the processed pixel or some local average in a small neighborhood about the processed pixel. The other value is another some local average in a bigger neighborhood (usually the approximate character-size) about the processed pixel. However, unlike the integrated function technique, they are not sensitive to noise because they use local averages.

The integrated function technique has fully factored the stroke width restriction in the sense that it can remove all the large dark areas completely. Similar to the three techniques mentioned above, this technique also uses derivatives, but further, it labels the processed pixel according to the values of its derivatives, detects all the possible character/graphics pixels using the logical bound on the ordered sequences and finally extracts all the character/graphics pixels whose vertical or horizontal "stroke widths" lie in the predetermined stroke width range. The labeling and logical detecting assure that every large dark area is a connected black blob which is removed from the image in the final extraction phase. Therefore the result images from this technique do not have the unwanted edges of large dark areas. However, this technique is very sensitive to noise because of the sensitivity of the derivatives to noise.

From the observations, we get an outline of a better extraction technique. Conceptually, the idea is to compare the gray level of the processed pixel (if the processed image is noise-free) or its smoothed gray level (if the processed image is noisy) with some local averages in the neighborhoods about a few other neighbouring pixels. Thus more than one comparison is made and the comparison results can be considered as "derivatives." Therefore the labeling, detecting and extracting using the "derivatives," the logical bound on the ordered sequences and the stroke width range can be adopted. Since local averages are not sensitive to noise, these "derivatives" should not be sensitive to noise.

Based on this idea, we develop a new technique called logical level technique. This technique predetermines the stroke width range as $[0, W]$. It processes every pixel by simultaneously comparing its gray level or its smoothed gray level with four local averages in the $(2W + 1) \times (2W + 1)$-size neighborhoods centered at the four points P_i, P'_i, P_{i+1}, P'_{i+1} shown in Fig. 7. If for certain i, the (smoothed) gray level is at least T levels below all the four local averages, then the processed pixel is extracted as character/graphics pixel. Mathematically, this technique can be described as follows:

$$b(x, y) = \begin{cases} 1 & \text{if } \bigvee_{i=0}^{3} [L(P_i) \wedge L(P'_i \wedge L(P_{i+1}) \wedge L(P'_{i+1})] \\ & \text{is true} \\ 0 & \text{otherwise} \end{cases}$$

where

W is the predetermined maximal stroke width;
P'_i $= P_{(i+4) mod 8}$, for $i = 0, \ldots, 7$;
$L(P)$ $= ave(P) - g(x, y) > T$;
T is a predetermined parameter;
$ave(P) = \sum_{-W \le i \le W} \sum_{-W \le j \le W} f(P_x - i, P_y - j)/(2 \times W + 1)^2$;
P_x, P_y are the coordinates of P;
$g(x, y) = f(x, y)$ or its smoothed value.

In order to reduce the computation, we adopt the following fast algorithms:

1. A fast algorithm to calculate local average

$$\begin{aligned} a_H(x, y) &= \sum_{-W \le i \le W} f(x - i, y) \\ &= a_H(x - 1, y) - f(x - W - 1, y) \\ &\quad + f(x + W, y) \\ a_V(x, y) &= \sum_{-W \le j \le W} a_V(x, y - j) \\ &= a_V(x, y - 1) - a_H(x, y - W - 1) \\ &\quad + a_H(x, y + W) \\ ave(x, y) &= \frac{a_V(x, y)}{(2 \times W + 1)^2}. \end{aligned}$$

2. Logical decomposition:

$$\bigvee_{i=0}^{3} [L(P_i) \wedge L(P'_i) \wedge L(P_{i+1}) \wedge L(P'_{i+1})] = \left\{ \bigvee_{i=0,2} [L(P_i) \wedge L(P'_i)] \right\} \wedge \left\{ \bigvee_{i=1,3} [L(P_i) \wedge L(P'_i)] \right\}.$$

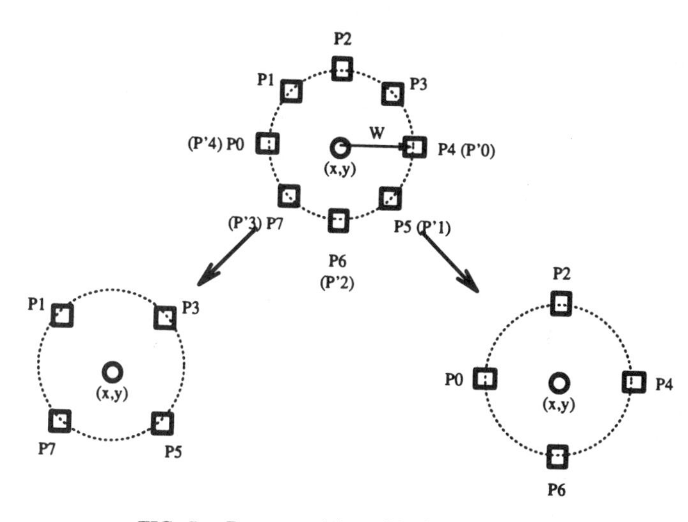

FIG. 7. Decomposition of logical detection.

The major advantages of this technique are its low noise sensitivity, full factoring of the stroke width and high speed. This technique is not sensitive to noise since it uses local averages as basic computing units. Since few of the large dark area pixels satisfy the logical restriction, large dark areas are removed with their edges together. Although usually not all the *AND* and *OR* operations are performed in both the original and fast algorithms of this technique, we can approximate the computation reduction brought by the fast algorithms. The computation reduction in calculating local averages is $[(2W + 1)^2 - 4]$ add/subtraction operations and the reduction in logical comparisons is $(3 - 2)$ *OR* operations and $(12 - 5)$ *AND* operations. Figures 8 and 9 show the result images of this technique.

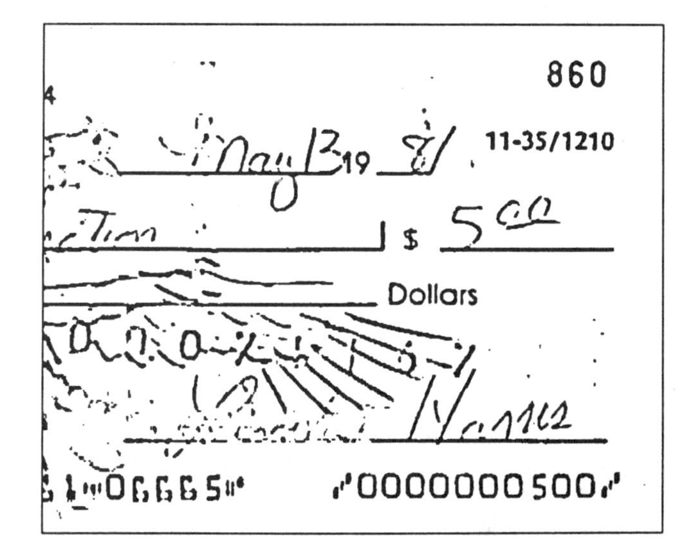

FIG. 8. Binary check character image extracted using Logical Level Technique.

ıid the number of black pixels.
dimensional ratios are easily ob-
ion. This algorithm is described
h this algorithm is similar to the
he literature [11], [12], in our
ge the value of each pixel to its
label, but retain the coordinates
or each component. Each con-
either rejected or accepted as a

FIG. 9. Binary document character image extracted using Logical Level Technique.

3.2. *Mask-Based Subtraction Technique*

This technique predetermines the stroke width range as $[W_{min}, W]$. It considers every original grayscale image as a sum of its background image and its character/graphics image, and consists of four steps. In the first step, most of the background pixels are detected using a logical filter. Considering all the pixels along some straight line passing through the processed pixel, we get a one dimensional signal sequence in which the character/graphics pixels look like some kinds of "particle noise." Therefore we get a logical filter which can be considered as an extension of the median filter for removing particle noise. For more accuracy, this filter is applied to four sequences corresponding to the four straight lines passing through the processed pixel with slopes of 0, $\pi/4$, $\pi/2$ and $3\pi/4$. The resulting image of this step is a binary image in which black pixels are possible character/graphics pixels and white pixels are background pixels. We call this binary image "mask image." In the second step, the mask image is modified by detecting more background pixels using the predetermined stroke width range $[W_{min}, W]$ (refer to Fig. 10). In the third step, for every possible characters/graphics pixel detected out in the modified mask image, its gray level of background image is estimated by a linear interpolation of the four background pixels, $f(P_L)$, $f(P_R)$, $f(P_U)$, $f(P_D)$, in the modified mask image (refer to Fig. 10). In the last step, a grayscale characters/graphics image is obtained by subtracting the estimated background image from the original image, and the binary characters/graphics image is extracted out by applying the global thresholding technique to the grayscale characters/graphics image. The mathematical description of this technique is as follows:

1. Mask image detection using logical filter

begin
  if at least W pixels of $\{f(x+i, y), -W \le i \le W\}$
    have values greater than $f(x, y)$,
      then $b_1(x, y) = 1$;
  else
    if at least W pixels of $\{f(x, y+i), -W \le i \le W\}$
      have values greater than $f(x, y)$,
        then $b_1(x, y) = 1$;
    else
      if at least W' pixels of $\{f(x+i, y+i), -W' \le i \le W'\}$
        have values greater than $f(x, y)$,
          then $b_1(x, y) = 1$;
      else
        if at least W' pixels of $\{f(x-i, y+i), -W' \le i \le W'\}$
          have values greater than $f(x, y)$,
            then $b_1(x, y) = 1$;
        else $b_1(x, y) = 0$;
end.

where

W' is the integer closest to $W/\sqrt{2}$;
$b_1(x, y)$ is 1 for a possible character/graphics pixel and 0 for a background pixel.

2. Mask image modification using stroke width range

begin
  if $b_1(x, y) = 1$;
    detect d_R, d_L, d_U, d_D;
    width $= min(d_U + d_D, d_R + d_L)$;
    if width $< W_{min}$ or width $> W$, then $b_1(x, y) = 0$;
end.

3. Background image estimation using interpolation

begin
  if $b_1(x, y) = 1$;
    detect d_R, d_L, d_U, d_D;
    if $d_R > W$ or $d_L > W$, then $f_b(x, y) = \dfrac{f(x, y-d_U)(1/d_U) + f(x, y+d_D)(1/d_D)}{(1/d_U) + (1/d_D)}$;
    else
      if $d_U > W$ or $d_D > W$, then $f_b(x, y) = \dfrac{f(x+d_{R,y})(1/d_R) + f(x-d_{L,y})(1/d_L)}{(1/d_R) + (1/d_L)}$;
      else
        $f_b(x, y) = \dfrac{f(x, y-d_U)(1/d_U) + f(x, y+d_D)(1/d_D) + f(x+d_{R,y})(1/d_R) + f(x-d_{L,y})(1/d_L)}{(1/d_U) + (1/d_D) + (1/d_R) + (1/d_L)}$;
  else
    $f_b(x, y) = f(x, y)$;
end.

4. Character/graphics image extraction

$$b(x, y) = \begin{cases} 1 & \text{if } f_b(x, y) - f(x, y) > T \\ 0 & \text{otherwise} \end{cases}$$

where T is a predetermined parameter.

The major advantage of this technique is that it has fully factored the stroke width since it examines every possible character/graphics pixel using the predetermined stroke width range. Figures 11 and 12 show the result images of this technique. This technique has some disadvantages such as low speed, and large memory requirement which will be discussed in the next section.

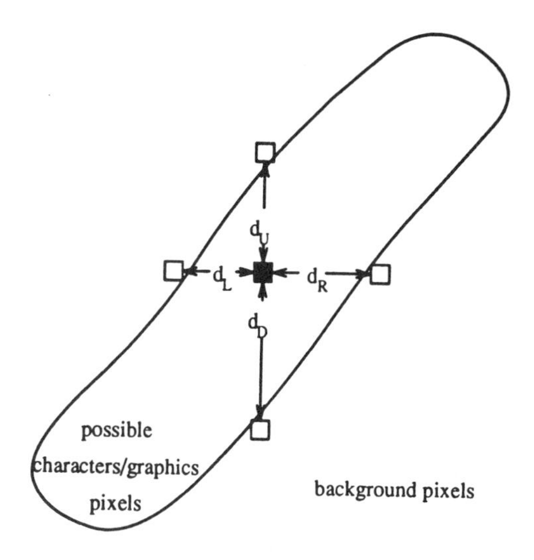

FIG. 10. Four points used in mask image modification and background image estimation.

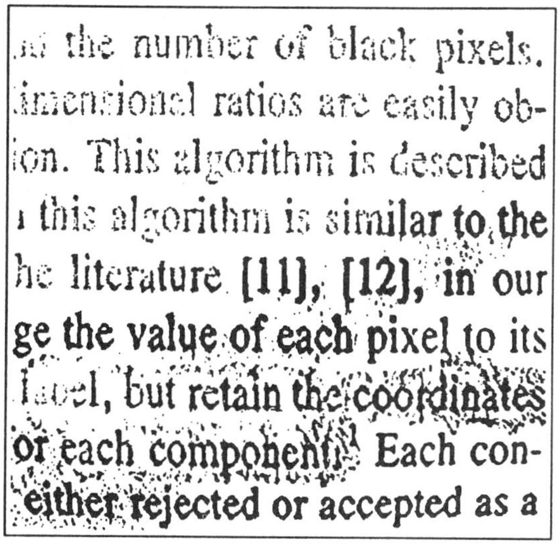

...as the number of black pixels.
...imensional ratios are easily ob-
...ion. This algorithm is described
...this algorithm is similar to the
...he literature [11], [12], in our
ge the value of each pixel to its
label, but retain the coordinates
or each component. Each con-
either rejected or accepted as a

FIG. 12. Binary document character image extracted using Mask-Based Subtraction Technique.

4. EVALUATION, ANALYSIS, AND EXPERIMENTAL RESULTS

Since there are many evaluation aspects for extraction techniques and there are conflicting requirements for these aspects, it is difficult to find a technique which is superior in every aspect. The lack of quantitative measures for some aspects such as subjective evaluation of output images and parameter setting makes the evaluation more difficult. Therefore the best evaluation is testing in practical applications.

Evaluations of three of the extraction techniques have been reported in [3] and [7]. However, these evaluations did not consider the memory requirement which is very important for low-cost applications.

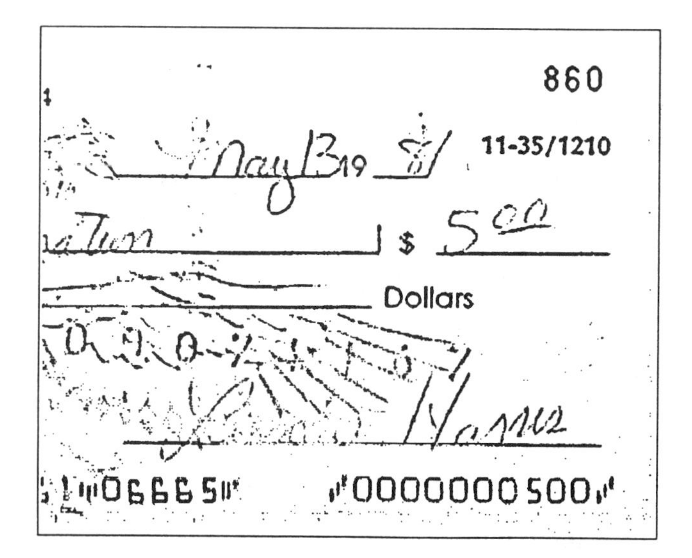

FIG. 11. Binary check character image extracted using Mask-Based Subtraction Technique.

In this section, we first discuss six evaluation aspects or measures, then systematically evaluate and analyse both new and published techniques with respect to these six aspects based on theoretical analysis and experiments. We implemented and applied the eight techniques to a number of test images. We present here two representative test images shown in Figs. 1 and 2 on which we base our evaluation. One is a typical personal check image with width of 640, height of 512, and gray level range of [0, 255]; the other is a common text document image with highlighted characters and shadows, and with width of 512, height of 480, and gray level range of [0, 255]. All the implementations and experiments were done using software written in C programming language in the UNIX operating system on micro VAX-II computer. Table 1 gives a summary of this evaluation and analysis.

4.1. *Six Evaluation Aspects*

1. *Subjective Evaluation of Result Images.* While some applications require machine recognition and the performances of the extraction technique involved can be measured in terms of recognition ratio, human subjective evaluation is still a reasonable and widely adopted evaluation aspect especially for the applications requiring human judgment. For example, in videoconferencing systems, the transmitted document images are used for manual reading. In check image processing systems, the extracted character image is also sometimes used for manual reading. Even when automated readers are expected to be used, it is still reasonable to use such measure rather than using recognition ratios as the latter may

TABLE 1
Evaluation and Analysis of Seven Character/Graphics Extraction Techniques

Technique	Subjective evaluation	Width*	CPU time	Memory requirement	Parameter number	Parameter setting
Global threshold	Worst	Not	1.1	0	1	Easy
YDH	Block lines, unwanted edges	Partly	74.8	MN	5	Easy
Nonlinear function	Pseudo-shadows unwanted edges	Partly	6.1	M	1276	Difficult
Integrated function	Noise	Fully	14.5	$(2W + 3)M$	2	Easy
Local contrast	Unwanted edges edge	Partly	103.0	$9M$	4	Difficult
Logical level	Best	Fully	7.7	$(4W - 1)M$	2	Easy
Mask-based subtraction	Better than some others	Fully	30.3	$2MN$	3	Easy

Note. Width*, how the stroke width restriction was factored. CPU time, the CPU time (represented in second) used for processing the check image. Memory, the number of registers used to store the intermediate results. M, N, the column number and line number of the processed image. W, the predetermined maximal line width.

depend on the software and/or hardware used for the recognition. In any case it is expected that good visual quality will lead to improved recognition ratios.

2. *Memory*. Since the extraction is often completed along with an optical scanning process, it is often implemented with minimal software and/or hardware within scanners or personal computers. Therefore memory is a key measure for evaluating techniques in low-cost small scale applications. This memory is used to store intermediate results and can be measured in terms of number of registers or bits. In this paper, we adopt the number of registers measure. Therefore an intermediate result data is considered to occupy the same-length register no matter what its data type is.

3. *Speed*. Speed is mostly of concern to the users, especially those who have high volume documents to be processed. Speed is mainly determined by the computation but also depends on the software/hardware implementation. Here we will use CPU time as an inverse proportion of speed.

4. *Stroke Width Restriction*. All extraction techniques except the global thresholding technique have either fully or partly factored this restriction. This restriction comes from the fact that most character/graphics encountered in real application consists of line drawings with almost uniform stroke width. However, for images which contain character/graphics of different stroke widths, factoring this restriction will require adapting the parameters to change from one region to another.

5. *Number of Parameters*. Each of the techniques considered has some parameters that need to be predetermined. For example, the global thresholding technique has one and the local contrast technique has five. Generally, this measure reflects the complexity of using the extraction technique.

6. *Parameter Setting*. Parameter setting is another measure for technique complexity. For every parameter, an initial value needs to be predetermined or set through either some estimation from the processed document or manual direct setting. For example, the only parameter of the global thresholding technique can be set either based on histogram properties [19, 20], "busyness" properties [21, 22], global edge information [23, 24] or by the user. Therefore the parameter setting for the global thresholding is easy and simple. But for the nonlinear adaptive techniques, three functions W_H, W_V, and Z_b need to be predetermined or set for every processed document by the user, and no systematic method or detailed guidance for setting these functions were provided [8]. Therefore the parameter setting for the nonlinear adaptive technique is difficult and complicated. This aspect should be given much consideration when the techniques are to be used by non-professional users.

4.2. *Analysis and Experimental Results*

1. *Global Thresholding Technique*. Global thresholding technique has the highest speed due to its most simple computation, minimal memory requirement due to no intermediate results, smallest parameter number due to only one parameter, and easiest parameter setting due to the availability of various systematic setting methods [19–24]. However, this technique can only handle "ideal" document images of which the gray level distri-

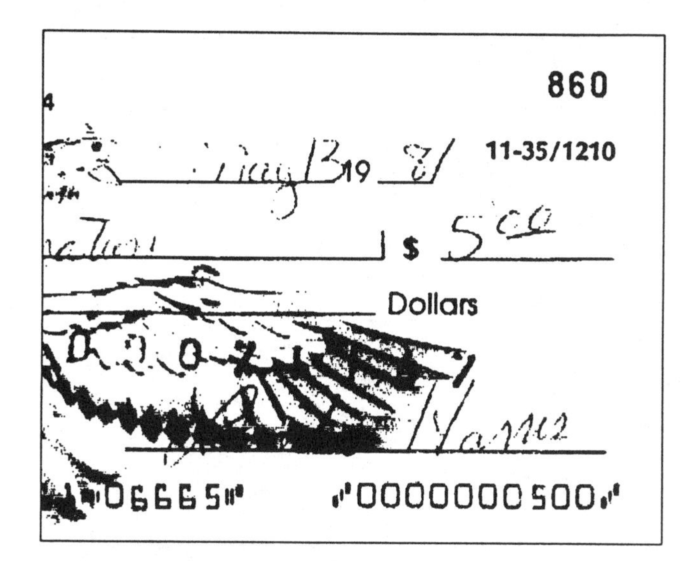

FIG. 13. Binary check character image extracted using Global Thresholding Technique.

butions of character/graphics pxiels and background pixels are well separated away from each other. It can not be applied to the poor-quality documents and checks of which the gray level distributions of character/graphics pixels are often buried within those of background pixels. Figures 13 and 14 give the results of applying this technique to the two test images. For these two result images, the parameter was manually set to the optimal value based on subjective evaluation. It is obvious that many character/graphics pixels have been lost as background pixels and many background pixels have been misextracted as character/graphics pixels. The reason for this is that this technique does not consider the difference between character/graphics pixels and background pixels and processes every pixel based on its own value without consideration of its neighbouring pixels.

2. *YDH Technique*. This technique has partially factored the stroke width restriction through dividing the processed image into blocks of size $T_2 \times T_2$. As pointed out in [18], if T_2 is too large, the background will still contain wide variations within the blocks, so that this technique does not work. On the other hand, if T_2 is too small, a big character could be mistaken to be background and removed from the processed image. However, because of the discontinuity between two adjacent blocks, this dividing may bring a pseudo line at the boundary between two blocks. Another disadvantage is that this technique also extracted the edges of a large dark area in the background such as the eagle's wing in the check image and the highlight area in the text document image. These unwanted edges and pseudo-block lines can be founded clearly in the result images of this technique shown in Figs. 15 and 16. This technique consists of a series of computation on the whole image, therefore its speed is not high and it requires approximately $M \times N$ registers to store its intermediate results. Although the parameter number of this technique is five, the parameter setting is easy since the physical meaning of these five parameters is clear, and T_2, T_4, and T_5 have been suggested by Yasuhiko, Dubois, and Huang to be 16, 128, and 20 respectively.

3. *Nonlinear Adaptive Technique*. The important advantage of this technique is its high speed. It also requires approximately M registers only. However, from its result images shown in Figs. 4 and 5, we can see that some pseudo-shadows were created in the background and that

FIG. 14. Binary document character image extracted using Global Thresholding Technique.

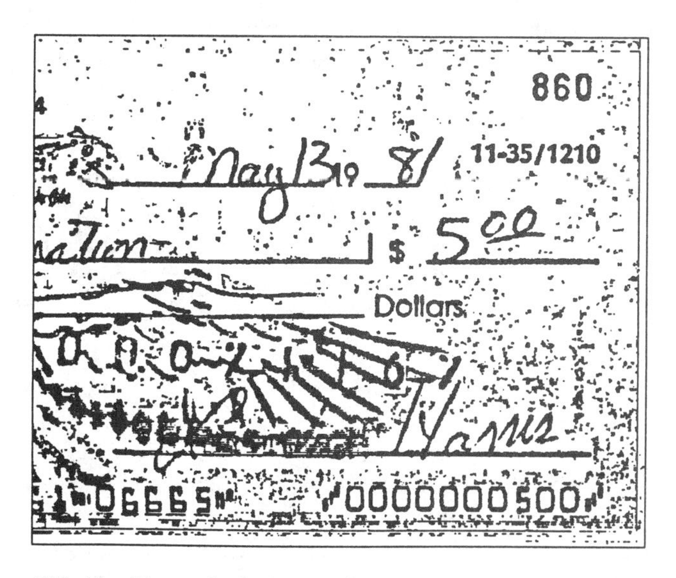

FIG. 15. Binary check character image extracted using YDH Technique.

ud the number of black pixels.
limensional ratios are easily ob-
ion. This algorithm is described
h this algorithm is similar to the
he literature [11], [12], in our
ge the value of each pixel to its
label, but retain the coordinates
or each component. Each con-
either rejected or accepted as a

FIG. 16. Binary document character image extracted using YDH Technique.

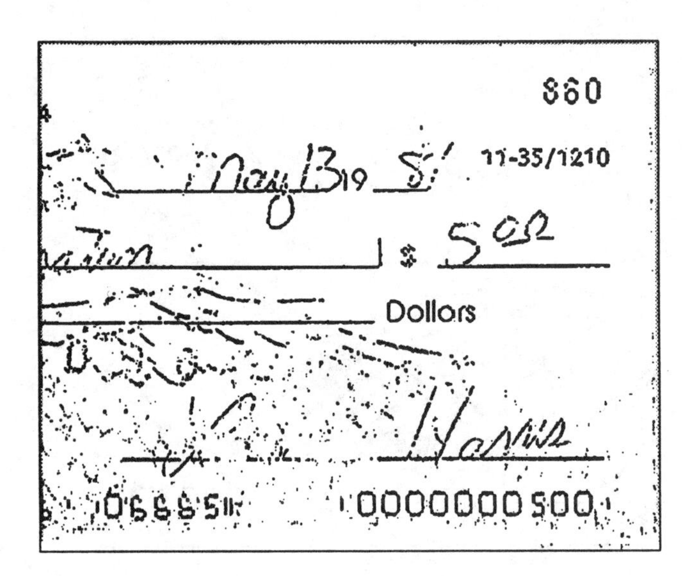

FIG. 17. Binary check character image extracted using Integrated Function Technique.

the left-top part of the dark areas in the background, such as the highlight dark area in the text document image, was also extracted. There are two reasons for this. One is that the local average $f_V(x, y)$ was calculated based on the pixels in the left-top direction of the processed pixel only. The other is that this technique has partially factored the stroke width restriction through its two update functions W_U and W_V. For different images, different values should be set to the two update functions. Although White and Rohrer [8] have given some basic suggestions for setting the update functions, the setting is still very difficult because there are no systematic methods or detailed guidances. For the two test images, although many other settings have been tried, the functions in Fig. 3 seem to give the optimal results shown in Figs. 4 and 5. Although when setting the update or bias functions, people usually choose mathematical functions rather than individual values, we can approximate the number of parameters as follows. Since the argument range of the two update functions is $[-255, 255]$ and $W_H(0)$, $W_V(0)$ have been fixed as zeros, there are 2×510 parameters $W_H(i)$, $W_V(i)$, $1 \leq |i| \leq 255$, to be predetermined. On the other hand, for the bias function, there are 256 predetermined parameters $Z_b(i)$, $0 \leq i \leq 255$. Therefore, the numbers of parameters is approximately 1276.

4. *Integrated Function Technique.* Of the five published techniques, this technique is the only one which fully factors the stroke width restriction. After detecting all the possible character/graphics pixels using first- and second-order derivatives, it extracts only these pixels which are in the bounded sequences and whose vertical or horizontal "stroke widths" lie in the predetermined stroke width range. For the result images of this technique shown in Figs. 17 and 18, the stroke width range was predetermined as $[0, W]$ where W is the maximal stroke width and had value of 4 for the check image and 6 for the text document image. We can see that the major disadvantage of this technique is the noise in its result images. This noise comes from the sensitivity of derivatives used by this technique. However, the speed of this technique is moderate (refer to Table 1), parameter number is only two (T and W), and parameter setting is easy since the physical meanings of T and W are obvious so that the user can easily set values for them according to the processed image. Since only the pixels in the current line and the $(W + 1)$ lines above and below are needed for processing a pixel, the memory requirement of this technique is $(2W + 3)M$.

ud the number of black pixels.
imensional ratios are easily ob-
on. This algorithm is described
h this algorithm is similar to the
he literature [11], [12], in our
ge the value of each pixel to its
label, but retain the coordinates
or each component. Each con-
either rejected or accepted as a

FIG. 18. Binary document charcter image extracted using Integrated Function Technique.

5. *Local Contrast Technique*. Figures 19 and 20 show the result images of this technique. Compared with the result image of other four published techniques, the images in Figs. 19 and 20 have the best performance with respect to subjective evaluation. The memory requirement is small ($9 \times M$ registers) since this technique is virtually a 9×9 window operation. However, this technique has three disadvantages. The first is its low speed (refer to Table 1). The second is its large number of parameters and difficulty of setting. Since only two out of the T_3, T_4, and T_5 are independent, the number of parameters is 4. T_1, T_2 have easily-understood physical meaning and are easy to set. For a given image, it is not clear how to set the values of T_3, T_4, and T_5. Palumbo, Swaminathan, and Srihari [7] set T_1, T_2, T_3, T_4, and T_5 as 20, 20, 0.85, 1, and 0 for their test images. But our result images using this parameter setting look very unsuccessful. After many trials, we got a setting (40, 40, 1, 1, −3) whose result images look best, shown in Figs. 19 and 20. Different parameter settings can yield quite different results for a given image. The third disadvantage is that this technique also extracts the edges of large dark areas in the background, such as the highlight dark area in the text document image.

6. *Logical Level Technique*. From the result images shown in Figs. 8 and 9, we can see that this technique has the best performance in terms of subjective evaluation. On one hand, its result images do not look as noisy as those of the integrated function technique. On the other hand, unlike the result images of YDH technique, the nonlinear adaptive technique and the local contrast technique, its result images do not contain the unwanted edges of large dark areas in the background, therefore this technique has fully factored the stroke width restriction. Compared with the integrated function technique which has also factored the stroke width restriction, this technique has the following similarities. Every logical comparison $L(P)$ is similar to the comparison $A(x, y) > T$; every $L(P_i) \wedge L(P_i')$ is similar to the bound on the ordered sequences along the line passing through P_i and P_i'. This technique implicitly assumes the predetermined stroke width range is $[0, W]$. In order to make more accurate extraction, four points P_i, P_i', P_{i+1} and P_{i+1}' are used instead of the two points P_i and P_i'. This technique has only two parameters T and W and the parameter setting is very easy because both T and W have obvious physical meaning. The memory requirement is $(4W - 1)M$ registers which can be met by usual low-cost hardware because W is usually small. In addition, this technique has very high speed which is almost equal to that of the nonlinear adaptive technique (refer to Table 1).

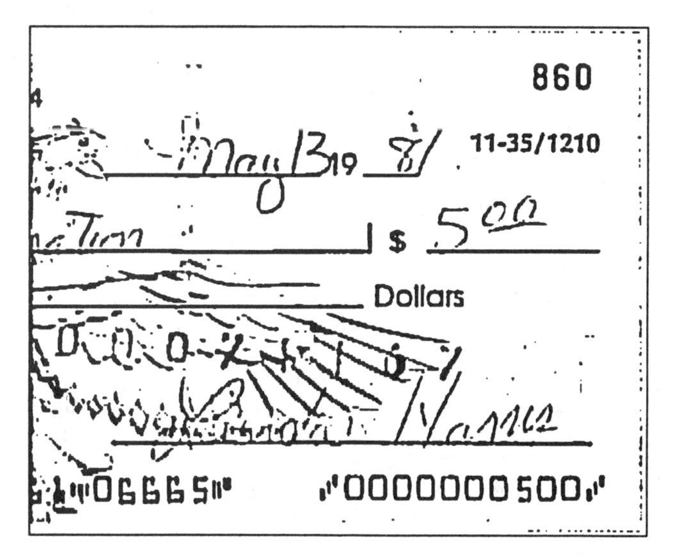

FIG. 19. Binary check character image extracted using Local Contrast Technique with ($T_1 = 50$, $T_2 = 50$, $T_3 = 1$, $T_4 = 1$, $T_5 = -15$).

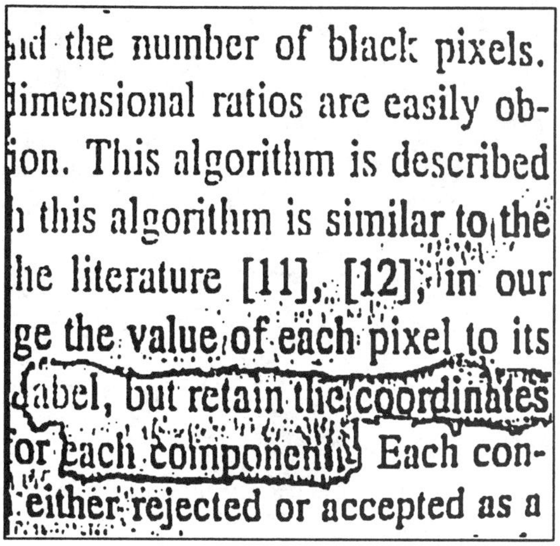

nd the number of black pixels.
imensional ratios are easily ob-
ion. This algorithm is described
n this algorithm is similar to the
he literature [11], [12], in our
ge the value of each pixel to its
label, but retain the coordinates
or each component. Each con-
either rejected or accepted as a

FIG. 20. Binary document character image extracted using Local Contrast Technique with ($T_1 = 40$, $T_2 = 40$, $T_3 = 1$, $T_4 = 1$, $T_5 = -3$).

7. *Mask-Based Subtraction Technique*. Unlike the other two new techniques which are based on the previous works, this technique comes from an independent idea. Figures 11 and 12 show the result images of this technique. Figures 21 and 22 show the modified mask images. Compared with the result images of the five published techniques, the result images of this technique look better than those of YDH technique, integrated function technique and nonlinear adaptive technique, but worse than those of local contrast technique. It has only three parameters W_{min}, W, and T whose parameter setting is easy. It has fully factored the stroke width restriction because it examines every possible character/graphics pixel using the predetermined stroke width range. However this technique seems inappropriate to handle mixed text font and size documents, which is revealed by the result images. This technique has another two disad-

FIG. 21. Modified binary check mask image obtained using Mask-Based Subtraction Technique.

vantages. One is that it requires approximately $2 \times M \times N$ registers of which $M \times N$ registers to store the binary image b_1 and other $M \times N$ registers to store the original image. The other disadvantage is its low speed (refer to Table 1). These two disadvantages restrain its use for high-speed low-cost applications.

5. SUMMARY AND CONCLUSIONS

Aiming at high-speed low-cost applications, this paper reviews five published extraction techniques and presents two new techniques, of which one is an improvement on previous work and the other is independent. Systematical evaluation and analysis of these eight techniques with experiments on images of a typical check and a poor-quality text document have shown that one of the new techniques, the logical level technique, is superior to the other seven with respect to subjective evaluation. The analysis has also shown that this technique is most suitable for high-speed low-cost applications due to its high speed, small parameter number, easy parameter setting, and minimal memory requirement.

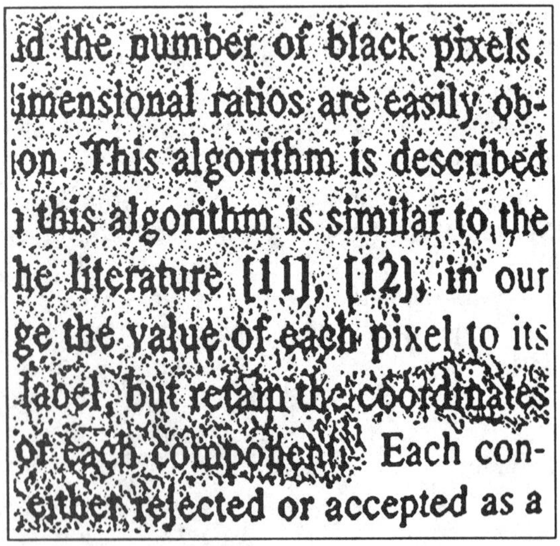

id the number of black pixels
imensional ratios are easily ob-
ion. This algorithm is described
i this algorithm is similar to the
he literature [11], [12], in our
ge the value of each pixel to its
label, but retain the coordinates
of each component. Each con-
either rejected or accepted as a

FIG. 22. Modified binary document mask image obtained using Mask-Based Subtraction Technique.

REFERENCES

1. Calera Recognition Systems, Inc., *TrueScan: Software Developer's Manual.*
2. V. Nagasamy and N. A. Langrana, Engineering drawing processing and vectorization system, *Comput. Vision Graphics Image Process.* **49,** 1990, 379–397.
3. H. H. Chang, Image thresholding for blueprint image extraction, *Conference Proceedings: Vision '87,* Detroit, Michigan, June 8–11, 1987, pp. 5-1–5-18.
4. R. Kasturi and J. Alemany, Information extraction from images of paper-based maps, *IEEE Trans. Software Eng.* **14**(5), May 1988, 671–675.
5. A. C. Downton and C. G. Leedham, Preprocessing and presorting of envelope images for automatic sorting using OCR, *Pattern Recognition* **23**(3/4), 1988, 347–362.
6. S. N. Sirhari, Feature extraction for locating address blocks on mail pieces, *From Pixels to Features*, pp. 261–273, Elsevier, Amsterdam, 1989.
7. P. W. Palumbo, P. Swaminathan, and S. N. Srihari, Document image binarization: evaluation of algorithms, in *Proceedings, SPIE Symposium on Applications of Digital Image Processing IX*, 1986, Vol. 697, pp. 278–285.
8. J. M. White and G. D. Rohrer, Image thresholding for optical character recognition and other applications requiring character image extraction, *IBM J. Res. Dev.* **27**(4), July 1983, 400–411.
9. J. L. Mitchell, W. B. Pennebaker, D. Anastassiou, and K. S. Pennington, Graphics image coding for freeze-frame videoconferencing, *IEEE Trans. Commun.* **37**(5), May 1989, 515–522.
10. R. C. Gonzalez and P. Wintz, *Digital Image Processing*, 2nd ed., Addison–Wesley, Reading, MA, 1987.
11. E. Giuliano, O. Paitra, and L. Stringa, Electronic character reading system, U. S. Patent 4,047,15 (Sept. 6, 1977).
12. L. A. Fletcher and R. Kasturi, A robust algorithm for text string separation from mixed text/graphics images, *IEEE Trans. pattern Anal. Mach. Intell.* **10**(6), Nov. 1988, 910–918.
13. D. Wang and S. N. Srihari, Classification of newspaper image blocks using texture analysis, *Comput. Vision Graphics Image Process.* **47,** 1989, 327–352.
14. D. G. Elliman and I. T. Lancaster, A review of segmentation and contextual analysis techniques for text recognition, *Pattern Recognition* **23**(3/4), 1990, 337–346.
15. W. Scherl, F. Wahl, and H. Fuchsberger, Automatic separation of text, graphics and picture segments in printed material, in *Pattern Recognition in Practice* (E. S. Gelsema and L. N. Kanal, Eds.), pp. 213–221, North-Holland, Amsterdam, 1980.
16. F. M. Wahl, K. Y. Wong, and R. G. Casey, Block segmentation and text extraction in mixed text/image documents, *Comput. Graphics Image Process.* **20,** 1982, 375–390.

17. M. Nadler, Survery: Document segmentation and coding techniques, *Comput. Vision Graphics Image Process.* **28,** 1984, 240–262.

18. Y. Yasuda, M. Dubois, and T. S. Huang, Data compression for check processing machines, *Proc. IEEE* **68**(7), July 1980, 874–885.

19. N. Otsu, A threshold selection method from gray-level histograms, *IEEE Trans. Syst. Man Cybernet.* **9,** Jan. 1979, 62–66.

20. J. S. Weszka and A. Rosenfeld, Histogram modification for threshold selection, *IEEE Trans. Syst. Man Cybernet.* **9,** Jan. 1979, 38–52.

21. F. Deravi and S. K. Pal, Gray level thresholding using second-order statistics, *Pattern Recognition Lett.* **1,** 1983, 417–422.

22. J. S. Weszka and A. Rosenfeld, Thresholding evaluation techniques, *IEEE Trans. Syst. Man Cybernet.* **8,** Aug. 1978, 622–629.

23. J. Kittler, J. Lllingworth, J. Foglein, and K. Paler, An automatic thresholding algorithm and its performance, *Proc. 7th International Conference on Pattern Recognition,* Montreal, 1984, pp. 287–289.

24. J. Kittler, J. Lllingworth, J. Foglein, and K. Paler, Threshold selection based on a simple image statistics, *Comput. Vision Graphics Image Process.* **30,** 1985, 125–147.

NOTE

Moment-Preserving Thresholding: A New Approach

WEN-HSIANG TSAI

Department of Information Science, National Chiao Tung University, Hsinchu, Taiwan 300 Republic of China

Received August 1, 1984

A new approach to automatic threshold selection using the moment-preserving principle is proposed. The threshold values are computed deterministically in such a way that the moments of an input picture is preserved in the output picture. Experimental results show that the approach can be employed to threshold a given picture into meaningful gray-level classes. The approach is described for global thresholding, but it is applicable to local thresholding as well.

I. INTRODUCTION

Image thresholding is a necessary step in many image analysis applications. For a survey of thresholding techniques, see [1 or 2]. In its simplest form, thresholding means to classify the pixels of a given image into two groups (e.g., objects and background), one including those pixels with their gray values above a certain threshold, and the other including those with gray values equal to and below the threshold. This is called *bilevel thresholding*. More generally, we can select more than one threshold, and use them to divide the whole range of gray values into several subranges. This is called *multilevel thresholding*. Most thresholding techniques [3–8] utilize shape information of the image histogram in threshold selection. In the ideal case, the histogram of an image with high-contrast objects and background will have a bimodal shape, with two peaks separated by a deep valley. The gray value at the valley can be chosen as the threshold. In real applications, such histogram bimodality is often unclear, and several methods have been proposed to overcome this problem [4–8] so that the valley seeking technique can still be applied.

Another direction of threshold selection is to evaluate the goodness of selected thresholds by a certain measure [9–13]. One way is to use entropy information to measure the homogeneity of the thresholded classes [9–11] or the independency of the classes from one another [12]. Another way is to make use of the class separability measures used in discriminant analysis [13].

In this paper, we propose another threshold selection method based on the moment-preserving principle which has also been applied to subpixel edge detection [14]. Specifically, before thresholding, we compute the gray-level moments of the input image. The thresholds are then selected in such a way that the moments of the thresholded image are kept unchanged. This approach may be regarded as a moment-preserving image transformation which recovers an ideal image from a blurred version. The approach can automatically and deterministically select multiple thresholds without iteration or search. In addition, a representative gray value can also be obtained for each thresholded class.

"Moment-Preserving Thresholding: A New Approach" by W.-H. Tsai from *Computer Vision, Graphics, and Image Processing,* Vol. 29, 1985, pp. 377–393. Copyright © 1985 by Academic Press, Inc., reprinted with permission.

In the remainder of this paper, we first describe the moment-preserving approach to bilevel thresholding. The result is next generalized to multilevel thresholding. A deterministic procedure to compute the threshold values is then described. Experimental results are also presented to support the validity of the proposed approach.

II. MOMENT-PRESERVING BILEVEL THRESHOLDING

Given an image f with n pixels whose gray value at pixel (x, y) is denoted by $f(x, y)$, we want to threshold f into two pixel classes, the below-threshold pixels and the above-threshold ones. The ith moment m_i of f is defined as

$$m_i = (1/n)\sum_x \sum_y f^i(x, y), \qquad i = 1, 2, 3, \ldots . \tag{1}$$

Moments can also be computed from the histogram of f in the following way:

$$\begin{aligned} m_i &= (1/n)\sum_j n_j (z_j)^i \\ &= \sum_j p_j (z_j)^i, \end{aligned} \tag{2}$$

where n_j is the total number of the pixels in f with gray value z_j and $p_j = n_j/n$. We also define m_0 to be 1. Image f can be considered as a blurred version of an ideal bilevel image which consists of pixels with only two gray values z_0 and z_1, where $z_0 < z_1$. The proposed moment-preserving thresholding is to select a threshold value such that if all below-threshold gray values in f are replaced by z_0 and all above-threshold gray values replaced by z_1, then the first three moments of image f are preserved in the resulting bilevel image g. Image g so obtained may be regarded as an ideal unblurred version of f.

Let p_0 and p_1 denote the fractions of the below-threshold pixels and the above-threshold pixels in f, respectively, then the first three moments of g are just

$$m_i' = \sum_{j=0}^{1} p_j (z_j)^i, \qquad i = 1, 2, 3. \tag{3}$$

And preserving the first three moments in g means the following equalities:

$$m_i' = m_i, \qquad i = 1, 2, 3. \tag{4}$$

Note that

$$p_0 + p_1 = 1. \tag{5}$$

The four equalities described by (4) and (5) above are equivalent to

$$\begin{aligned} p_0 z_0^0 + p_1 z_1^0 &= m_0, \\ p_0 z_0^1 + p_1 z_1^1 &= m_1, \\ p_0 z_0^2 + p_1 z_1^2 &= m_2, \\ p_0 z_0^3 + p_1 z_1^3 &= m_3, \end{aligned} \tag{6}$$

where m_i with $i = 1, 2, 3$ are computed by (1) or (2) and $m_0 = 1$. To find the desired threshold value t, we can first solve the four equations described by (6) above to obtain p_0 and p_1, and then choose t as the p_0-tile of the histogram of f, i.e., choose t such that

$$p_0 = (1/n) \sum_{z_j \leq t} n_j.$$

In practice, there may exist no discrete gray value which is exactly the p_0-tile of the histogram. Then, the threshold t should be chosen the gray value closest to the p_0-tile. The equations described by (6) will be solved later when we discuss multilevel thresholding. Note that z_0 and z_1 will also be obtained simultaneously as part of the solutions to (6). They can be regraded as the *representative gray values* for the below-threshold and the above-threshold pixels, respectively.

III. MOMENT-PRESERVING MULTILEVEL THRESHOLDING

In the general case, we want to threshold a given image f into more than two pixel classes. To threshold f into N pixel classes, we need $N - 1$ threshold values $t_1, t_2, \ldots, t_{N-1}$. Let z_i denote the representative gray value of the ith pixel class, and let p_i denote the fraction of the pixels in the ith class. By preserving the first $2N - 1$ moments of f and using the fact that the sum of all p_i values is equal to 1, we get the following set of $2N$ equations:

$$\begin{aligned} p_0 z_0^0 + p_1 z_1^0 + \cdots + p_N z_N^0 &= m_0, \\ p_0 z_0^1 + p_1 z_1^1 + \cdots + p_N z_N^1 &= m_1, \\ &\cdots \\ p_0 z_0^{2N-1} + p_1 z_1^{2N-1} + \cdots + p_N z_N^{2N-1} &= m_{2N-1}, \end{aligned} \tag{7}$$

which can be solved to get all p_i and z_i, $i = 0, 1, \ldots, N - 1$. The method we use to solve (7) above will be described next. Once all p_i values are obtained, the desired thresholds t_i can be found from the histogram of f by choosing t_1 as the p_0-tile, t_2 as the $(p_0 + p_1)$-tile, and more generally, t_i as the $(p_0 + p_1 + \cdots + p_{i-1})$-tile, of the histogram of f. For convenience, the equations described by (7) will be called the *moment-preserving equations*.

Based on Szego [15], Tabatabai [16] showed that the moment-preserving equations can be solved indirectly by executing the following three computation steps:

(i) solve the following linear equations to obtain a set of auxiliary values $c_0, c_1, \ldots, c_{N-1}$:

$$\begin{aligned} c_0 m_0 + c_1 m_1 + \cdots + c_{N-1} m_{N-1} &= -m_N, \\ c_0 m_1 + c_1 m_2 + \cdots + c_{N-1} m_N &= -m_{N+1}, \\ &\cdots \\ c_0 m_{N-1} + c_1 m_N + \cdots + c_{N-1} m_{2N-2} &= -m_{2N-1}; \end{aligned} \tag{8}$$

(ii) solve the following polynomial equation to obtain the representative gray values $z_0, z_1, \ldots, z_{N-1}$:

$$z^N + c_{N-1} z^{N-1} + \cdots + c_1 z + c_0 = 0; \tag{9}$$

(iii) substitute all z_i values obtained above into the first N moment-preserving equations described by (7) and solve the resulting equations to get $p_0, p_1, \ldots, p_{N-1}$.

To guarantee that the N solutions of the polynomial equation (9) are all distinct and real-valued, at least N distinct gray values must exist in the input image f [15]. For N less than 5, analytic solutions to (9) can be obtained. For N no less than 5, as is well known, no close-form solution to (7) exists. Numerical analysis procedures like the Newton's method need to be applied. The complete analytic solutions of the three computation steps above for $N = 2$, 3, and 4 are summarized for reference in the Appendix.

IV. AN ILLUSTRATIVE EXAMPLE

The effectiveness of moment-preserving thresholding is demonstrated with an example in this section. Figure 1a shows the pixel gray values of a given image, which includes two areas with roughly constant gray values (the leftmost 4 columns and the rightmost 4 columns) and a blurred boundary (the central 4 columns). From the distribution of the gray values, it is reasonable to expect the following results:

(1) bilevel thresholding—the leftmost 6 columns as a thresholded class and the rightmost 6 columns as the other;

(2) trilevel thresholding—the leftmost 4 columns as a class, the boundary portion (central 4 columns) as a second class, and the rightmost columns as the third;

(3) quaterlevel thresholding—the leftmost 4 and the rightmost 4 columns as two classes, and the central left 2 (columns 5 and 6 from the left) and the central right 2 columns as the other two classes.

a

10	8	10	9	20	21	32	30	40	41	41	40
12	10	11	10	19	20	30	28	38	40	40	39
10	9	10	8	20	21	30	29	42	40	40	39
11	10	9	11	19	21	31	30	40	42	38	40

b

12	12	12	12	12	12	38	38	38	38	38	38
12	12	12	12	12	12	38	38	38	38	38	38
12	12	12	12	12	12	38	38	38	38	38	38
12	12	12	12	12	12	38	38	38	38	38	38

c

10	10	10	10	25	25	40*	25	40	40	40	40
10	10	10	10	25	25	25	25	40	40	40	40
10	10	10	10	25	25	25	25	40	40	40	40
10	10	10	10	25	25	40*	25	40	40	40	40

d

10	10	10	10	19	19	31	31	40	40	40	40
19*	10	10	10	19	19	31	31	40	40	40	40
10	10	10	10	19	19	31	31	40	40	40	40
10	10	10	10	19	19	31	31	40	40	40	40

FIGURE 1

Using the analytic solutions provided in the Appendix, we get the computational results as follows:

(1) bilevel thresholding—
(i) representative gray values:

$$z_0 = 12; \ z_1 = 38;$$

(ii) class fractions:

$$p_0 = 0.498; \ p_1 = 0.502;$$

(iii) selected threshold:

$$t_1 = 27;$$

(2) trilevel thresholding—
(i) representative gray values:

$$z_0 = 10; \ z_1 = 25; \ z_2 = 40;$$

(ii) class fractions:

$$p_0 = 0.361; \ p_1 = 0.277; \ p_2 = 0.362;$$

(iii) selected thresholds:

$$t_1 = 18; \ t_2 = 30;$$

(3) quaterlevel thresholding—
(i) representative gray values:

$$z_0 = 10; \ z_1 = 19; \ z_2 = 31; \ z_3 = 40;$$

(ii) class fractions:

$$p_0 = 0.311; \ p_1 = 0.191; \ p_2 = 0.190; \ z_3 = 0.308;$$

(iii) selected thresholds:

$$t_1 = 11; \ t_2 = 27; \ t_3 = 37.$$

The thresholded images are shown in Figs. 1b, c, and d which are exactly the expected results mentioned previously except only three pixels (marked with * in Figs. 1c and d. This means that the moment-preserving principle is indeed feasible for meaningful image thresholding. This fact is further verified by experimental results described next.

V. EXPERIMENTAL RESULTS

The proposed approach has been tried on a lot of images. Some results are shown in Figs. 2–7. Each image tested is of the size 80 by 60. Each figure shown includes a tested image in (a), the histogram of the image in (b) with computed threshold values

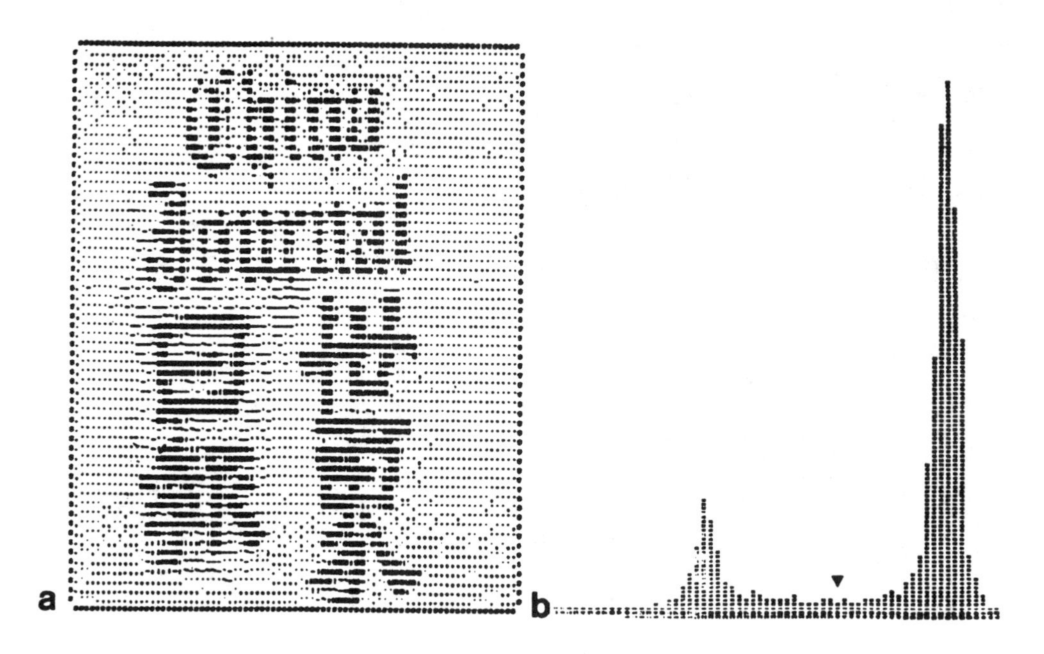

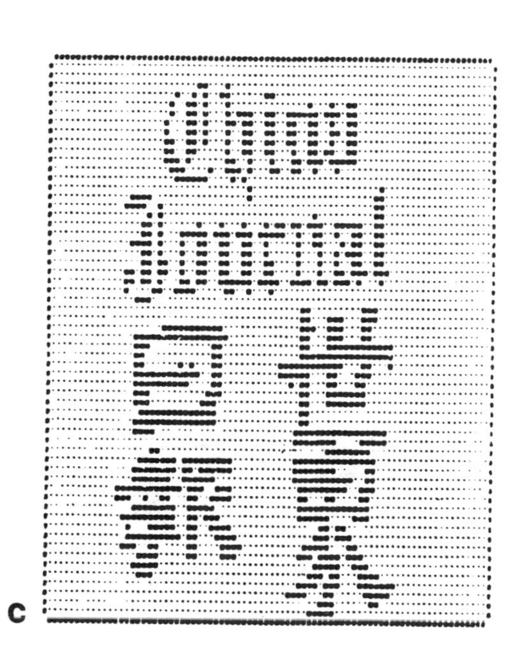

FIGURE 2

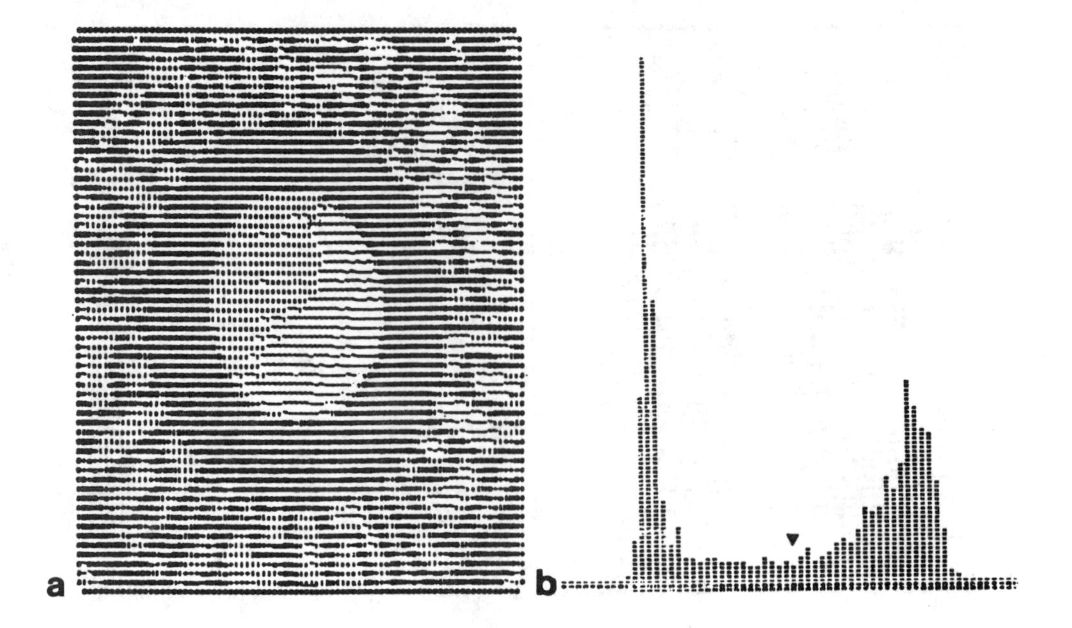

FIGURE 3

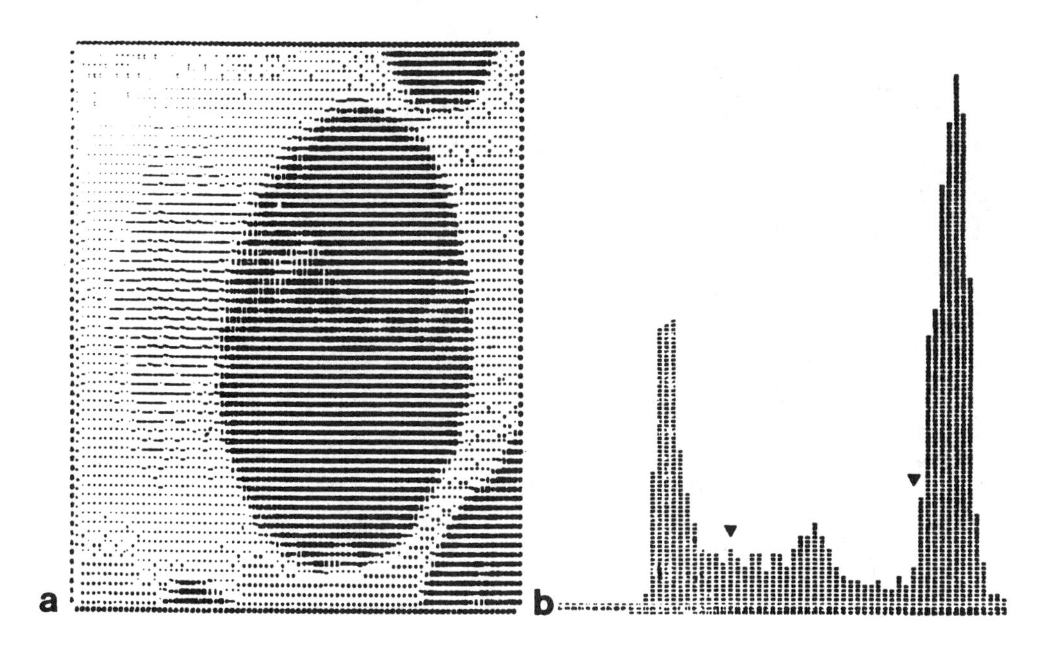

FIGURE 4

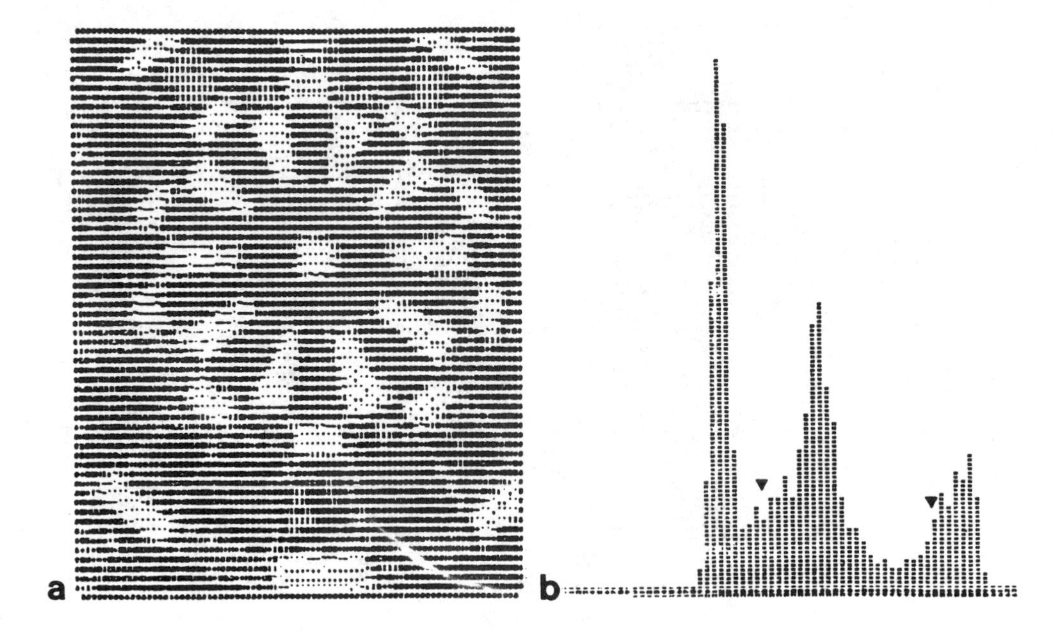

FIGURE 5

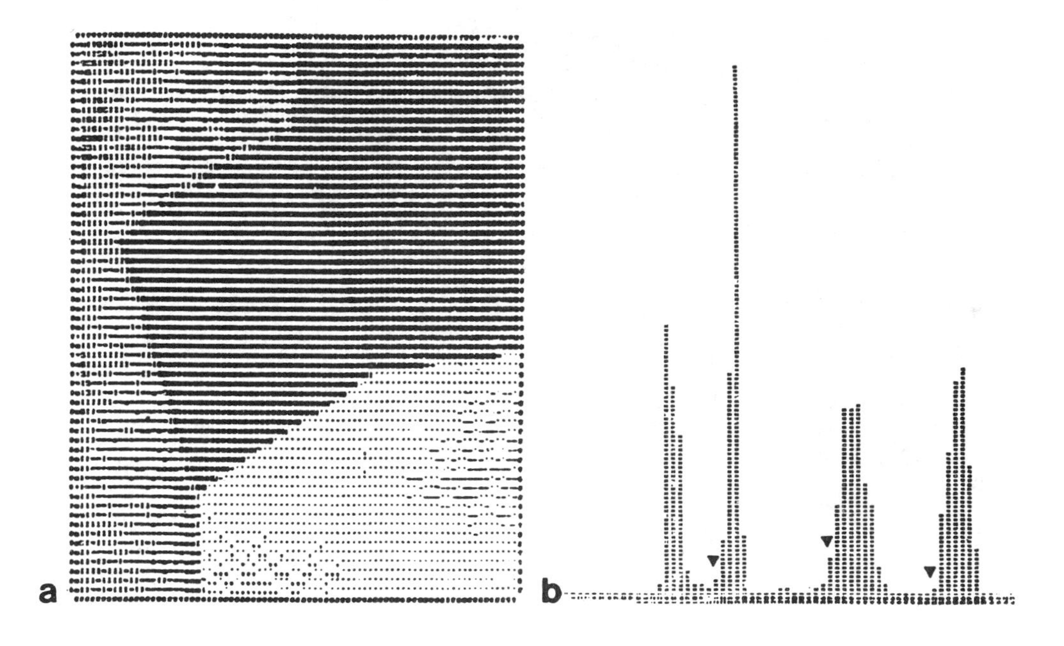

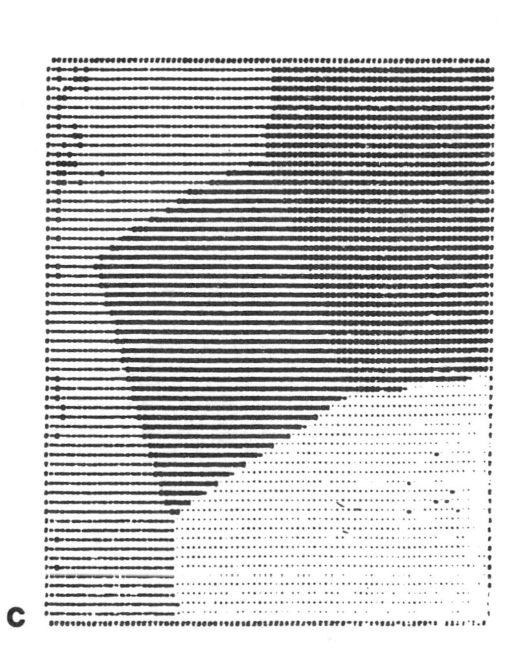

FIGURE 6

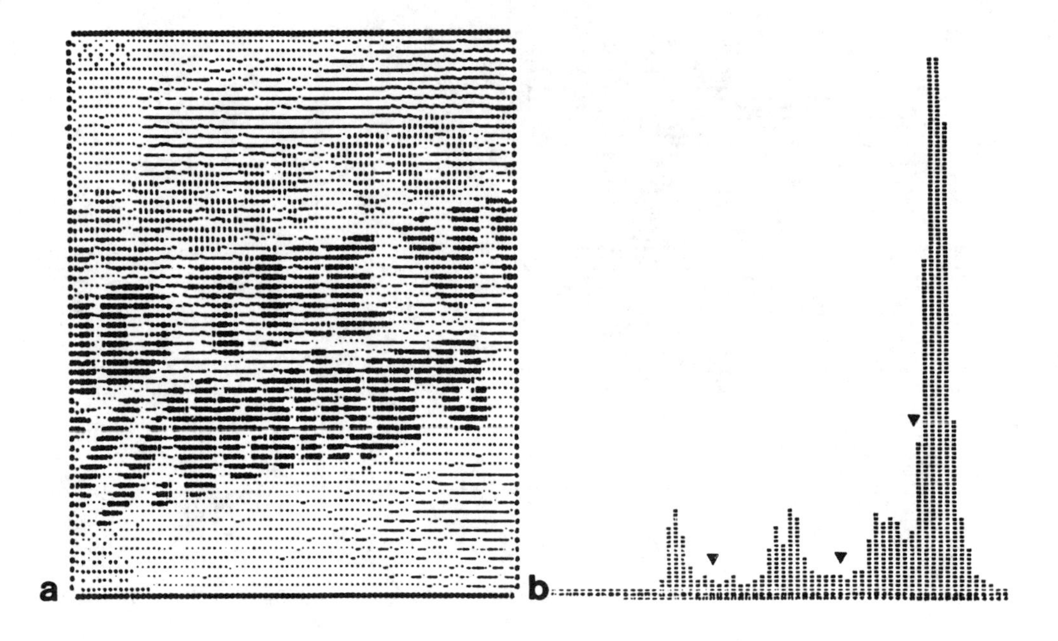

FIGURE 7

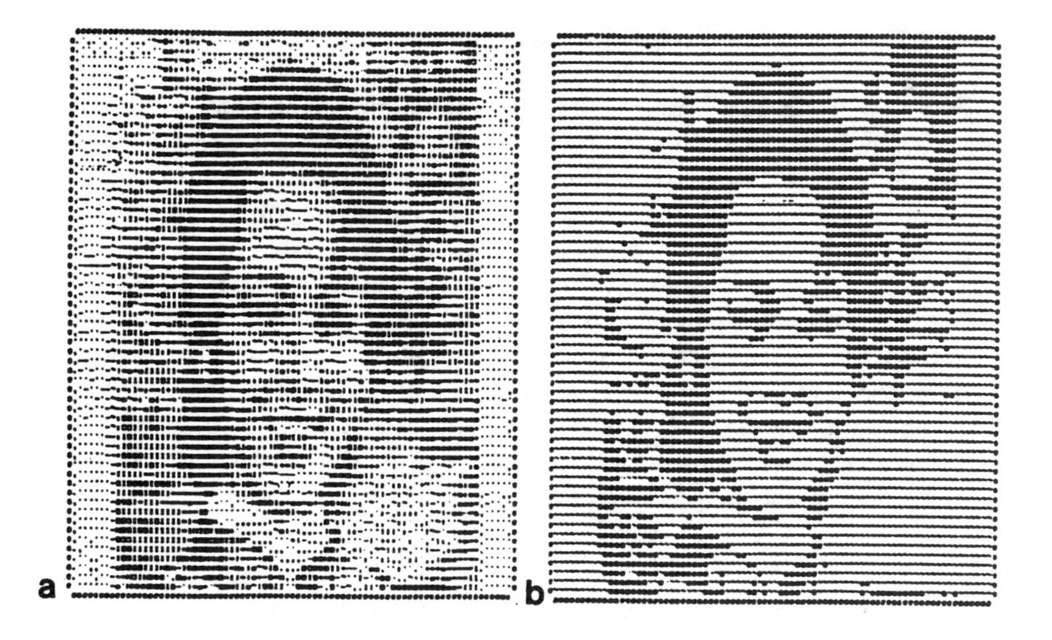

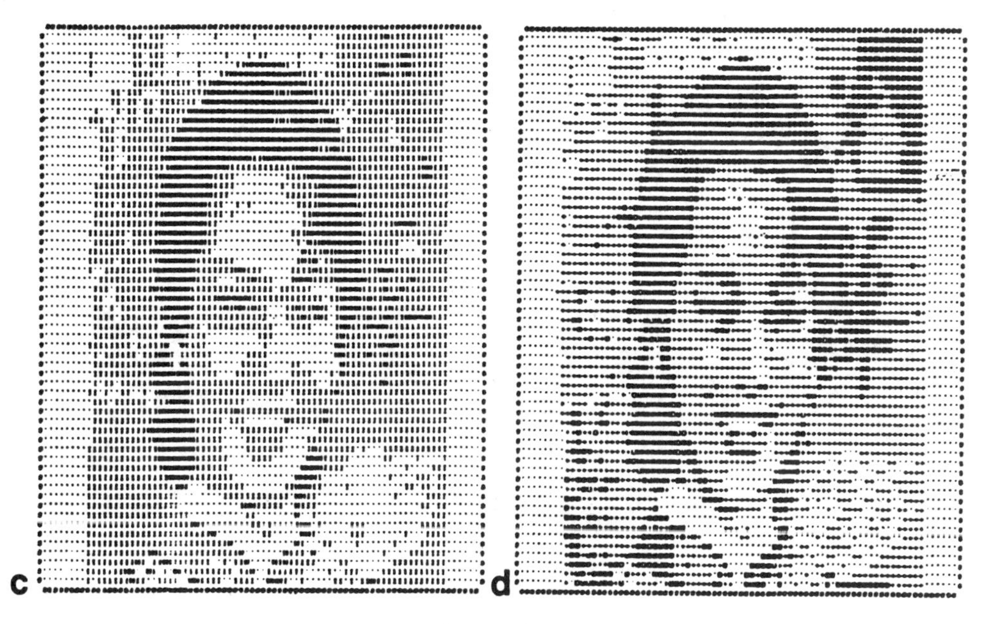

FIGURE 8

marked by "▼", and the thresholded result in (c) with pixels filled with representative gray values.

Figures 2 and 3 are the results of bilevel thresholding applied to some characters and a pattern. From Figs. 2b and 3b, we see that the computed threshold values are close to the bimodal histogram valleys. Similar results are also found in other cases described in the following. Figures 4 and 5 are the results of trilevel thresholding applied to a cell image and a three-color wheel pattern. Either of the input images shown in Figs 4a and 5a includes roughly three classes of gray values. And the thresholded images shown in Figs. 4c and 5c indicate that the pixel classes have been successfully thresholded. Similar situations can also be found in Figs. 6c and 7c which are the results of quaterlevel thresholding applied to Figs. 6a and 7a. Either image shown in Fig. 6a or 7a includes roughly four classes of pixel gray values. Finally, we include Fig. 8 to show the results of applying different-level thresholding to a single picture, the image of a girl.

APPENDIX

The solutions z_i and p_i ($i = 0, 1, \dots N-1$) to the moment-preserving equations for N-level thresholding ($N = 2, 3, 4$) are summarized here for reference convenience. Detailed derivations of the solutions are not included. Interested readers are referred to any mathematical handbook like [17]. In the following, m_0 denotes the value 1 and m_i with $i > 0$ are computed according to Eq. (1) or (2). After all p_i values are obtained, the ith threshold value t_i is selected as the $(\sum_{k=0}^{i-1} p_k)$-tile of the input picture histogram.

A.1. Bilevel Thresholding Solutions

(i)

$$c_d = \begin{vmatrix} m_0 & m_1 \\ m_1 & m_2 \end{vmatrix};$$

$$c_0 = (1/c_d)\begin{vmatrix} -m_2 & m_1 \\ -m_3 & m_2 \end{vmatrix};$$

$$c_1 = (1/c_d)\begin{vmatrix} m_0 & -m_2 \\ m_1 & -m_3 \end{vmatrix}.$$

(ii)

$$z_0 = \left(\tfrac{1}{2}\right)\left[-c_1 - \left(c_1^2 - 4c_0\right)^{1/2}\right];$$

$$z_1 = \left(\tfrac{1}{2}\right)\left[-c_1 + \left(c_1^2 - 4c_0\right)^{1/2}\right].$$

(iii)

$$p_d = \begin{vmatrix} 1 & 1 \\ z_0 & z_1 \end{vmatrix};$$

$$p_0 = (1/p_d)\begin{vmatrix} 1 & 1 \\ m_1 & z_1 \end{vmatrix};$$

$$p_1 = 1 - p_0.$$

A set of more compact solutions are also provided in [14].

A.2. Trilevel Thresholding Solutions

(i)

$$c_d = \begin{vmatrix} m_0 & m_1 & m_2 \\ m_1 & m_2 & m_3 \\ m_2 & m_3 & m_4 \end{vmatrix};$$

$$c_0 = (1/c_d)\begin{vmatrix} -m_3 & m_1 & m_2 \\ -m_4 & m_2 & m_3 \\ -m_5 & m_3 & m_4 \end{vmatrix};$$

$$c_1 = (1/c_d)\begin{vmatrix} m_0 & -m_3 & m_2 \\ m_1 & -m_4 & m_3 \\ m_2 & -m_5 & m_4 \end{vmatrix};$$

$$c_2 = (1/c_d)\begin{vmatrix} m_0 & m_1 & -m_3 \\ m_1 & m_2 & -m_4 \\ m_2 & m_3 & -m_5 \end{vmatrix}.$$

(ii)

$$z_0 = -c_2/3 - A - B;$$
$$z_1 = -c_2/3 - W_1 A - W_2 B;$$
$$z_2 = -c_2/3 - W_2 A - W_1 B,$$

where A, B, W_1, and W_2 are as follows:

$$A = \Big\{\big(c_0/2 - c_1c_2/6 + c_2^3/27\big) - \Big[\big(c_0/2 - c_1c_2/6 + c_2^3/27\big)^2 + \big(c_1/3 - c_2^2/9\big)^3\Big]^{1/2}\Big\}^{1/3};$$

$$B = -\big(c_1/3 - c_2^2/9\big)/A;$$
$$W_1 = -1/2 + i\big(\sqrt{3}/2\big);$$
$$W_2 = -1/2 - i\big(\sqrt{3}/2\big);$$
$$i = \sqrt{-1}.$$

(iii)

$$p_d = \begin{vmatrix} 1 & 1 & 1 \\ z_0 & z_1 & z_2 \\ z_0^2 & z_1^2 & z_2^2 \end{vmatrix};$$

$$p_0 = (1/p_d)\begin{vmatrix} m_0 & 1 & 1 \\ m_1 & z_1 & z_2 \\ m_2 & z_1^2 & z_2^2 \end{vmatrix};$$

$$p_1 = (1/p_d)\begin{vmatrix} 1 & m_0 & 1 \\ z_0 & m_1 & z_2 \\ z_0^2 & m_2 & z_2^2 \end{vmatrix};$$

$$p_2 = 1 - p_0 - p_1.$$

A.3. *Quaterlevel Thresholding Solutions*

(i)

$$c_d = \begin{vmatrix} m_0 & m_1 & m_2 & m_3 \\ m_1 & m_2 & m_3 & m_4 \\ m_2 & m_3 & m_4 & m_5 \\ m_3 & m_4 & m_5 & m_6 \end{vmatrix};$$

$$c_0 = (1/c_d)\begin{vmatrix} -m_4 & m_1 & m_2 & m_3 \\ -m_5 & m_2 & m_3 & m_4 \\ -m_6 & m_3 & m_4 & m_5 \\ -m_7 & m_4 & m_5 & m_6 \end{vmatrix};$$

$$c_1 = (1/c_d)\begin{vmatrix} m_0 & -m_4 & m_2 & m_3 \\ m_1 & -m_5 & m_3 & m_4 \\ m_2 & -m_6 & m_4 & m_5 \\ m_3 & -m_7 & m_5 & m_6 \end{vmatrix};$$

$$c_2 = (1/c_d)\begin{vmatrix} m_0 & m_1 & -m_4 & m_3 \\ m_1 & m_2 & -m_5 & m_4 \\ m_2 & m_3 & -m_6 & m_5 \\ m_3 & m_4 & -m_7 & m_6 \end{vmatrix};$$

$$c_3 = (1/c_d)\begin{vmatrix} m_0 & m_1 & m_2 & -m_4 \\ m_1 & m_2 & m_3 & -m_5 \\ m_2 & m_3 & m_4 & -m_6 \\ m_3 & m_4 & m_5 & -m_7 \end{vmatrix}.$$

(ii)

$$z_0 = \left(\tfrac{1}{2}\right)\left\{-(c_3/2 + A) - \left[(c_3/2 + A)^2 - 4(Y + B)\right]^{1/2}\right\};$$

$$z_1 = \left(\tfrac{1}{2}\right)\left\{-(c_3/2 + A) + \left[(c_3/2 + A)^2 - 4(Y + B)\right]^{1/2}\right\};$$

$$z_2 = \left(\tfrac{1}{2}\right)\left\{-(c_3/2 - A) - \left[(c_3/2 - A)^2 - 4(Y - B)\right]^{1/2}\right\};$$

$$z_3 = \left(\tfrac{1}{2}\right)\left\{-(c_3/2 - A) + \left[(c_3/2 - A)^2 - 4(Y - B)\right]^{1/2}\right\},$$

where A, B, and Y are as follows:

$$A = \left(\tfrac{1}{2}\right)\left(c_3^2 - 4c_2 + 8Y\right)^{1/2};$$

$$B = (c_3 Y - c_1)/A;$$

$$Y = c_2/6 - C - D;$$

$$C = \left[G + (G^2 + H^3)^{1/2}\right]^{1/3};$$

$$D = -H/C;$$

$$G = \left(\tfrac{1}{432}\right)\left(72c_0c_2 + 9c_1c_2c_3 - 27c_1^2 - 27c_0c_3^2 - 2c_2^3\right);$$

$$H = \left(\tfrac{1}{36}\right)\left(3c_1c_3 - 12c_0 - c_2^2\right).$$

(iii)

$$p_d = \begin{vmatrix} 1 & 1 & 1 & 1 \\ z_0 & z_1 & z_2 & z_3 \\ z_0^2 & z_1^2 & z_2^2 & z_3^2 \\ z_0^3 & z_1^3 & z_2^3 & z_3^3 \end{vmatrix};$$

$$p_0 = (1/p_d) \begin{vmatrix} 1 & 1 & 1 & 1 \\ m_1 & z_1 & z_2 & z_3 \\ m_2 & z_1^2 & z_2^2 & z_3^2 \\ m_3 & z_1^3 & z_2^3 & z_3^3 \end{vmatrix};$$

$$p_1 = (1/p_d) \begin{vmatrix} 1 & 1 & 1 & 1 \\ z_0 & m_1 & z_2 & z_3 \\ z_0^2 & m_2 & z_2^2 & z_3^2 \\ z_0^3 & m_3 & z_2^3 & z_3^3 \end{vmatrix};$$

$$p_2 = (1/p_d) \begin{vmatrix} 1 & 1 & 1 & 1 \\ z_0 & z_1 & m_1 & z_3 \\ z_0^2 & z_1^2 & m_2 & z_3^2 \\ z_0^3 & z_1^3 & m_3 & z_3^3 \end{vmatrix};$$

$$p_3 = 1 - p_0 - p_1 - p_2.$$

Note that in all the three types of thresholding above, we assume all z_i values, after they are obtained in step (ii), are sorted into an increasing order (with z_0 as the smallest) before they are substituted into step (iii) to compute p_i values.

REFERENCES

1. A. Rosenfeld and A. C. Kak, *Digital Picture Processing*, Vol. II, Academic Press, New York, 1982.
2. J. S. Weszka, A survey of threshold selection techniques, *Comput. Graphics Image Process.* 7, 1978, 259–265.
3. J. M. S. Prewitt and M. L. Mendelsohn, The analysis of cell images. *Ann. New York Acad. Sci.* **128** 1966, 1031–1053.
4. C. K. Chow and T. Kaneko, Automatic boundary detection of the left views from cineangiograms, *Comput. Biomed. Res.*, **5** 1972, 388–410.
5. S. Watanabe and CYBEST group, An automated apparatus for cancer preprocessing, *Comput. Graphics Image Process.* **3**, 1974, 350–358.
6. J. S. Weszka, R. N. Nagel, and A. Rosenfeld, A threshold selection technique, *IEEE Trans. Comput.* **C-23**, 1974, 1322–1326.
7. D. P. Panda and A. Rosenfeld, Image segmentation by pixel classification in (gray level, gradient) space, *IEEE Trans. Comput.* **C-27**, 1978, 875–879.
8. J. S. Weszka and A. Rosenfeld, Histogram modification for threshold selection, *IEEE Trans. Systems Man Cybernet.* **SMC-9**, 1979, 38–52.
9. G. Leboucher and G. E. Lowitz, What a histogram can really tell the classifier, *Pattern Recognition* **10**, 1978, 351–357.
10. T. Pun, A new method for gray-level picture thresholding using the entropy of the histogram, *Signal Process.* **2**, 1980, 223–237.
11. T. Pun, Entropic thresholding: A new approach, *Comput. Graphics Image Process.* **16**, 1981, 210–239.
12. G. Johannsen and J. Bille, A threshold selection method using information measures, in *Proc. 6th Int. Conf. Pattern Recognition*, Munich, Germany, 1982, pp. 140–143.

13. N. Otsu, A threshold selection method from gray-level histogram, *IEEE Trans. Systems Man Cybernet.* **SMC-9**, 1979, 62–66.
14. A. J. Tabatabai and O. R. Mitchell, Edge location to subpixel values in digital imagery, *IEEE Trans. Pattern Anal. Mach. Intell.* **PAMI-6**, 1984, 188–201.
15. G. Szego, *Orthogonal Polynomials*, Vol. 23, 4th ed., Amer. Math. Soc., Providence R. I., 1975.
16. A. Tabatabai, *Edge Location and Data Compression for Digital Imagery*, Ph.D. dissertation, School of Elect. Engrg., Purdue University, Dec. 1981.
17. C. E. Pearson (Ed.), *Handbook of Applied Mathematics*, 2nd ed., Van Nostrand–Reinhold, New York, 1983.

Thinning Methodologies—A Comprehensive Survey

Louisa Lam, Seong-Whan Lee, *Member, IEEE*, and Ching Y. Suen, *Fellow, IEEE*

***Abstract*— This article is a comprehensive survey of thinning methodologies. It discusses the wide range of thinning algorithms, including iterative deletion of pixels and nonpixel-based methods, whereas skeletonization algorithms based on medial axis and other distance transforms will be the subject matter of a subsequent study.**

This self-contained paper begins with an overview of the iterative thinning process and the pixel-deletion criteria needed to preserve the connectivity of the image pattern. Thinning algorithms are then considered in terms of these criteria as well as their modes of operation. This is followed by a discussion of nonpixel-based methods that usually produce a center line of the pattern directly in one pass without examining all the individual pixels. Algorithms are considered in greater detail and scope here than in other surveys, and the relationships among them are also explored.

***Index Terms*— Parallel thinning, sequential thinning, skeleton, skeletonization, thinning.**

I. Introduction

IN THE EARLY years of computer technology, it was realized that machine recognition of patterns was a possibility, and together with this arose the need for reducing the amount of information to be processed to the minimum necessary for the recognition of such patterns. It seems that the earliest experiments in data compression were conducted on character patterns in the 1950's. In [37], it was found that an averaging operation over a square window with a high threshold resulted in a thinning of the input image, and in [61], the "custer" operation was an early attempt to obtain a thin-line representation of certain character patterns. The thinned characters were used for recognition in [107], [32], and [2]. This practice has been widely used since then (for example, [18], [117], [67], [121], and [69]), and integrated circuits have been designed for this purpose [96].

During these years, many algorithms for data compression by thinning have been devised and applied to a great variety of patterns for different purposes. In the biomedical field, this technique was found to be useful in the early 1960's, when a "shrink" algorithm was applied to count and size the constituent parts of white blood cells in order to identify abnormal cells [58], [94]. Since that time, applications in this area have included analyses of white blood cells [36] and chromosomes [51], [52], automatic X-ray image analysis [95], and analysis of coronary arteries [81]. In other sectors, thinned images have found applications in the processing of bubble-chamber negatives [73], the visual system of an automaton [43], fingerprint classification [74], quantitative metallography [68], measurements of soil cracking patterns [83], and automatic visual analyses of industrial parts [75] and printed circuit boards [133].

This wide range of applications shows the usefulness of reducing patterns to thin-line representations, which can be attributed to the need to process a reduced amount of data as well as to the fact that shape analysis can be more easily made on line-like patterns. The thin-line representation of certain elongated patterns, for example, characters, would be closer to the human conception of these patterns; therefore, they permit a simpler structural analysis and more intuitive design of recognition algorithms. For example, a partially abstract, graph-like skeleton has been considered to be a link between the abstract description of a letter and its physical representation as a character [28]. In addition, the reduction of an image to its essentials can eliminate some contour distortions while retaining significant topological and geometric properties. In more practical terms, thin-line representations of elongated patterns would be more amenable to extraction of critical features such as end points, junction points, and connections among the components (e.g., see [56]), whereas vectorization algorithms often used in pattern recognition tasks also require one-pixel-wide lines as input. Naturally, for a thinning algorithm to be really effective, it should ideally compress data, retain significant features of the pattern, and eliminate local noise without introducing distortions of its own. To accomplish all that using the simplest and fastest algorithm is the challenge involved.

Perhaps as a result of its central role in the preprocessing of data images or perhaps because of its intrinsic appeal, the design of thinning or skeletonization algorithms has been a very active research area. About 300 articles have been published on various aspects of this subject since its inception; therefore, it is felt that a comprehensive survey of this area is due. There have been other comparisons and survey articles in the field ([120], [53], [30], [111]), and some authors have included comparisons with other results in their publications (for example, [122], [77], [26], and [24]). The aim here is to collect most (if not all) of the articles in this field published in English, to classify and discuss them according to the methodologies employed, and to include areas not usually covered in the other surveys—morphological skeletonization, 3-D skeletonization, and grey-scale thinning. All of these will

Manuscript received August 6, 1990; revised June 21, 1991. This work was supported by the Natural Sciences and Engineering Research Council of Canada, the National Network of Centres of Excellence program of Canada, and the FCAR program of the Ministry of Education of the province of Québec. Recommended for acceptance by Associate Editor C. Dyer.

L. Lam and C. Y. Suen are with the Centre for Pattern Recognition and Machine Intelligence, Concordia University, Montreal, Québec, Canada H3G 1M8.

S. -W. Lee is with the Department of Computer Science, Chungbuk National University, Cheongju, Chungbuk, Korea.

IEEE Log Number 9201772.

Reprinted from *IEEE Trans. Pattern Analysis and Machine Intelligence,* Vol. 14, No. 9, Sept. 1992, pp. 869–885.

be discussed under a consistent set of terminologies (where variations will be mentioned) in the hope that such a unified treatment would be helpful.

This article is organized as follows. Section II contains an overview of the thinning process and pixel deleting criteria. In Sections III and IV, sequential and parallel thinning algorithms, respectively, will be discussed, whereas the noniterative nonpixel-based methods will be considered in Section V. This article will conclude with some observations and remarks in Section VI. In a separate follow-up article, we will consider medial axis and other distance transforms, extensions of thinning algorithms to 3-D and grey-scale images, and we will include a brief discussion of morphological skeletonization.

II. Overview of Thinning

The term "skeleton" has been used in general to denote a representation of a pattern by a collection of thin (or nearly thin) arcs and curves. Other nomenclatures have also been used in different contexts. For example, the term "medial axis" is also used to denote the locus of centers of maximal blocks that are also equivalent to the local maxima of chessboard distance [101]. Some authors also refer to a "thinned image" as a line-drawing representation of a pattern that is usually elongated. In recent years, it appears that "thinning" and "skeletonization" have become almost synonymous in the literature, and the term "skeleton" is used to refer to the result, regardless of the shape of the original pattern or the method employed. In this paper, we will observe the general usage of the term "skeleton"; however, we will refer to methods involving determination of centers of maximal blocks as "medial axis transformations," whereas the reduction of generally elongated patterns to a line-like representation will be considered as "thinning." This seems to be a reasonable way to distinguish between different methodologies while recognizing that the results can be quite similar in appearance.

In this section, we consider some general aspects of iterative thinning algorithms or, more precisely, the algorithms that delete successive layers of pixels on the boundary of the pattern until only a skeleton remains. The deletion or retention of a (black) pixel p would depend on the configuration of pixels in a local neighborhood containing p. According to the way they examine pixels, these algorithms can be classified as sequential or parallel. In a sequential algorithm, the pixels are examined for deletion in a fixed sequence in each iteration, and the deletion of p in the nth iteration depends on all the operations that have been performed so far, i.e., on the result of the $(n-1)$th iteration as well as on the pixels already processed in the nth iteration. On the other hand, in a parallel algorithm, the deletion of pixels in the nth iteration would depend only on the result that remains after the $(n-1)$th; therefore, all pixels can be examined independently in a parallel manner in each iteration.

In the following discussion (as in the literature), the terms iteration and cycle are used interchangeably, as are the terms subiteration and subcycle. The term pass has been used to denote either iteration or subiteration, but it will be used to mean subiteration here.

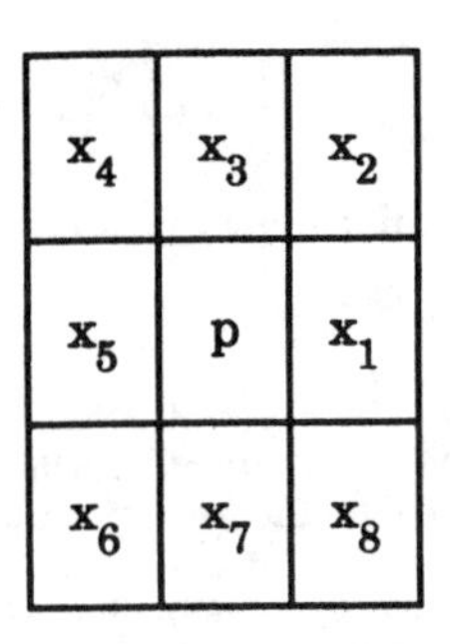

Fig. 1. Pixels of $N(P)$.

The following notations will be adopted in the discussion below. It is understood that a pixel p examined for deletion is a black pixel, and the pixels in its 3×3 window are labeled as shown in Fig. 1.

The pixels $x_1, x_2, \ldots, x_8$ are the 8-neighbors of p and are collectively denoted by $N(p)$. They are said to be 8-adjacent to p. The pixels x_1, x_3, x_5, x_7 are the 4-neighbors of p and are 4-adjacent to p.

We will use x_i to denote both the pixel and its value 0 or 1, and x_i is called white or black, accordingly. The number of black pixels in $N(p)$ is denoted by $b(p)$. A sequence of points $y_1, y_2, \ldots y_n$ is called an 8-path (4-path) if y_{i+1} is an 8- (or 4-) neighbor of $y_i, i = 1, 2, \ldots, n-1$. A subset Q of a picture P is 8- (or 4-) connected if for every pair of points x, y in Q there exists an 8- (or 4-) path from x to y consisting of points in Q. In this case, Q is said to be an 8- (or 4-) component of P. The order of connectivity of P is the number of components of its complement $\bar{P}$, and if this order is 1, we say that P is simply connected; otherwise, P is multiply connected.

A pixel p is 8- (or 4-) deletable if its removal does not change the 8- (or 4-) connectivity of P. The pixels considered for deletion are contour pixels. It has been suggested [38] that the satisfying duality of P and $\bar{P}$ having different types of connectivity would eliminate the paradoxes of P and $\bar{P}$, being both connected or both disconnected [98]. Since it is desirable for a skeleton to have unit width, the choice would be to adopt 8-connectivity for P and 4-connectivity for $\bar{P}$. This is assumed in the rest of the paper unless otherwise stated, and the reader is referred to the original publications for the (usually) analogous formulations using the other metrics. One consequence of this choice is that connectivity can be preserved by deleting only those pixels of P that are 4-adjacent to $\bar{P}$. Therefore, contour pixels are usually defined as those having at least one white 4-neighbor ([104] and [18] are exceptions). We will assume the common definition of contour pixels, unless otherwise stated, and note that contour points have also been called edge points [77] and border points [62]. Noncontour black pixels are said to be interior points. We will also consider p to be an end point and retained if $b(p) = 1$; this will be referred to as the end point (or end pixel) condition. This condition is applied differently by some authors: p may be retained when there are two or three consecutive black pixels on one side of $N(p)$ [77], the condition may be applied only after the first two iterations [57], or it may be omitted entirely in order to avoid spurious branches [18].

Most of the differences between algorithms occur in the tests implemented to ensure connectedness. This property has been defined in terms of crossing number, connectivity number, and pixel simplicity.

There are two definitions of the crossing number of a pixel. Rutovitz [103] first proposed this useful measure of connectivity as the number of transitions from a white point to a black point and vice versa when the points of $N(p)$ are traversed in (for example) counterclockwise order. Therefore, this crossing number can be defined as

$$X_R(p) = \sum_{i=1}^{8} |x_{i+1} - x_i|$$

where $x_9 = x_1$, and it is equal to twice the number of black 4-components in $N(p)$. Deletion of p would not affect 4-connectivity if $X_R(p) = 2$ since the black pixels in $N(p)$ are 4-connected in these cases. However, since disjoint 4-components can be 8-connected, skeletons obtained using this crossing number can contain 8-deletable pixels, and these skeletons are sometimes said to be imperfectly 8-connected [24] or are more than one pixel wide. In order not to belabor this point, we will assume that this will be understood in later sections and that skeletons obtained this way would usually need postprocessing to replace 4-connectedness with 8-connectedness. For this purpose, various algorithms can be applied, for example, those in [100] or [8].

Hilditch [52] defined the crossing number $X_H(p)$ as the number of times one crosses over from a white point to a black point when the points in $N(p)$ are traversed in order, cutting the corner between 8-adjacent black 4-neighbors. Therefore

$$X_H(p) = \sum_{i=1}^{4} b_i$$

where

$$b_i = \begin{cases} 1 \text{ if } x_{2i-1} = 0 \text{ and } (x_{2i} = 1 \textit{ or } x_{2i+1} = 1) \\ 0 \text{ otherwise} \end{cases}$$

and $X_H(p)$ is equal to the number of black 8-components in $N(p)$ except when p has all black 4-neighbors, in which case $X_H(p) = 0$. Obviously, for both definitions of crossing number, a pixel having all black 8-neighbors would have crossing number 0 as would an isolated pixel. If $X_H(p) = 1$, deletion of p would not change the 8-connectedness of the pattern.

Another difference between the crossing numbers $X_H(p)$ and $X_R(p)$ is that the condition $X_H(p) = 1$ would also imply that p must also be a contour point (having at least one white 4-neighbor), whereas $X_R(p) = 2$ does not ensure such a condition since this would be satisfied if p has exactly one white corner neighbor. In order to avoid deleting p in this case (and creating a hole), another condition is needed (for example, $b(p) \leq 6$) to ensure that p is a contour pixel.

An equivalent but more readily computable form of the crossing number $X_H(p)$ is the 8-connectivity number of [134] and [135] defined as

$$N_c^8(p) = \sum_{i=1}^{4} (\overline{x}_{2i-1} - \overline{x}_{2i-1}\overline{x}_{2i}\overline{x}_{2i+1})$$

where $\bar{x}$ implies "not x," whereas the 4-connectivity number

$$N_c^4(p) = \sum_{i=1}^{4} (x_{2i-1} - x_{2i-1}x_{2i}x_{2i+1})$$

represents the number of 4-connected components containing the black 4-neighbors of $N(p)$.

Furthermore, $N_c^8(p)$ is equal to the number of times the pixel p would be traversed in a contour-following algorithm for a connected component [135]; therefore, pixels that are retained (when $N_c^8(p) > 1$) would be pixels that are traversed more than once in the tracing process. At the same time, deletable pixels have often been called simple, and it is proved [99] that in a simply 8-connected pattern, a nonisolated contour pixel p is simple if and only if $N(p)$ has only one black component, which is equivalent to $X_H(p) = 1$.

An alternative approach to preserving topology during thinning is that the genus of P (and $\bar{P}$) should remain invariant [124], where the genus of P is defined as the number of connected components of P minus the number of holes of P. For any pixel p, its effect on the genus G can be determined completely from the configuration of $N(p)$. If deletion of p does not change G, p is said to be "simple." Since there are 256 configurations of $N(p)$, these combinations can be examined, and the ones where p is simple can be stored in a look-up table. It can be readily verified that the concept of "simple" based on genus is equivalent to that based on connectivity.

Pixels with connectivity number $N_c^8(p)$ greater than one also belong to the category of multiple pixels [85]. These pixels are considered to occur where a pattern "folds" onto itself, and they include end points of branches, strokes that are two pixels in width, and pixels that should be assigned to the skeleton on the basis of the connectivity criteria. Consequently, these pixels are retained in the thinning process. This notion will be considered in more detail when the algorithms that utilize them are discussed.

According to their modes of operation and the pixel testing criteria used, many thinning algorithms can be broadly classified according to the scheme shown in Table I. Under this general scheme, sequential algorithms can operate by processing only contour pixels or by raster scanning and parallel algorithms by using 4, 2, or 1 subiterations. Both classes of algorithms can ensure connectedness by finding the crossing numbers $X_R(p)$ or $X_H(p)$ by matching against thinning windows, by deleting only simple pixels, or by retaining multiple pixels. These procedures, as well as other less widely used methods, will be discussed more fully in the next two sections.

III. Sequential Thinning Algorithms

Using sequential algorithms, contour points are examined for deletion in a predetermined order, and this can be accomplished by either raster scan(s) or by contour following.

TABLE I
GENERAL CLASSIFICATION SCHEME

OPERATION		Pixel Testing Criteria			
		$X_R(p)$	$X_H(p)$	Window matching	Multiple/simple pixels
SEQUENTIAL	Contour Pixels	Arcelli [5] Wang [129]		Beun [18] Pavlidis [86] Chu [26]	Pavlidis [85] Arcelli [10]
	Raster Scanning	Arcelli [8]	Hilditch [52] Yokoi [134]		Arcelli [14]
PARALLEL	4-subcycle		Rosenfeld [100] Hilditch [53]	Stefanelli [115]	Arcelli [6]
	2-subcycle	Deutsch [35] Zhang [136] Chen [21]	Suzuki [118] Guo [49]	Stefanelli [115]	
	1-subcycle	Rutovitz [103] Holt [54]	Chen [22]	Chin [25]	

Contour following algorithms can visit every border pixel of a simply connected object [99], and of a multiply connected picture, if all the borders of the picture and holes are followed [134]. Therefore, such methods, which have been utilized previously without proof, have been shown to be valid in this sense. Border following algorithms are also given in these papers. These algorithms have an advantage over raster scans because they require the examination of only the contour pixels instead of all the pixels in P and $\bar{P}$ in every iteration. Some algorithms that use contour tracing are seen in [10], [85], [26], [126], [130], and [47], and the contours are traced using the Freeman chain codes in [64].

When a contour pixel p is examined, it is usually deleted or retained according to the configuration of $N(p)$. To prevent sequentially eliminating an entire branch in one iteration, a sequential algorithm usually marks (or flags) the pixels to be deleted, and all the marked pixels are then removed at the end of an iteration. This generally ensures that only one layer of pixels would be removed in each cycle.

To avoid repetition in the following discussion, we will assume that a pixel p considered for deletion satisfies all the following properties unless otherwise stated:

1. p is a black pixel.
2. p is not an isolated or end point, i.e., $b(p) \geq 2$.
3. p is a contour pixel, i.e., p has at least one white 4-neighbor.

A seminal algorithm in sequential thinning is that of Hilditch [52], which utilized the crossing number $X_H(p)$. The pattern is scanned from left to right and from top to bottom, and pixels are marked for deletion under four additional conditions that were also stated explicitly in [78] and [111] but are included here for completeness:

H1: At least one black neighbor of p must be unmarked.
H2: $X_H(p) = 1$ at the beginning of the iteration.
H3: If x_3 is marked, setting $x_3 = 0$ does not change $X_H(p)$.
H4: Same as H3, with x_5 replacing x_3.

Condition H1 was designed to prevent excessive erosion of small "circular" subsets, H2 to maintain connectivity, and H3-H4 to preserve two-pixel wide lines. The author also extended the method to thinning of grey-scale chromosome images.

This algorithm has also been implemented for binary images by other researchers using various techniques. Cellular logic operations based on 3×3 neighborhoods are used in [48]; since a change in the value of p may induce a change only in the pixels of $N(p)$ in the next iteration, only these pixels need to be processed. In [80], it was found that a sequential thinning method based on raster scan and 3×3 operations can be implemented using a pipeline structure to reduce the memory and processing time required. In [91], connected horizontal runs of black pixels are coded in interval tables (the use of pointers in these tables implies that only a change of pointers would be required when a shift operation is used on local neighborhoods). The image that remains after one iteration is found by inspection of the overlapping intervals in the preceding, current, and next lines of the image. Alternatively, contour pixels can be queued, with the change of a contour pixel causing all its 4-neighbors to be queued and processed in the next iteration [126]. The crossing number $X_H(p)$ together with a rough estimate $K(p)$ of the convexity at p are used as criteria for deletion of p in [97]. $K(p)$ is the maximum number of 4-connected white points in $N(p)$ and is intended as a discrete equivalent for the notion of curvature. Under this scheme, p is deleted if $K(p) < K_T$ for some threshold K_T, and conditions H2-H4 hold. Increasing the threshold K_T results in skeletons less affected by contour protrusions (at the risk of excessive erosion).

The thinning criteria of [52] are extended to $k \times k$ windows ($k \geq 3$) in [84], where the center "core" of $(k-2) \times (k-2)$ pixels can be deleted together if the boundary pixels in the window have Hilditch crossing number 1 and if they contain more than $(k-2)$ 4-connected white pixels and more than $(k-2)$ black pixels. For every black pixel, its $k \times k$ windows are examined in the order of decreasing k until $k < 3$ or the core is deleted. With larger values of k, thicker layers of pixels can be deleted in one iteration; therefore, fewer iterations would be required to obtain skeletons of thick patterns. However, this increase in speed is achieved at the expense of "coarser" results (noisy skeletons); in the thinning of elongated patterns, the use of a larger k can be detrimental to processing speed. The same thinning algorithm has also been implemented in parallel in this work by using four subiterations and examining one type of border pixels (north, south, east, or west) in each subiteration.

In another early algorithm [134], every black pixel p is examined and labeled according to the rule

$$L(p) = \begin{cases} 2 \text{ if } x_3 = 0 \text{ or } \bar{x}_3 + \bar{x}_5 + x_7 = 0, \\ 3 \text{ if } \bar{x}_3 + x_5 = 0 \text{ or } \bar{x}_3 + \bar{x}_5 + \bar{x}_7 + x_1 = 0. \end{cases}$$

Then, two raster scans in opposite directions are performed to remove pixels labeled 2 and 3, respectively, provided these pixels are not end points and have connectivity number 1 (where 4- or 8-connectivity can be used). Although the algorithm is simple and yields connected skeletons, vertical strokes that are an even number of pixels wide and open at one end can be completely eliminated.

In [18], the different criteria of [104] are used for contour or edge points. For example, p is a west edge point if there is at most one black pixel in the first column of $N(p)$ and at least 3 black pixels in the rest of $N(p)$. In this case, an extra

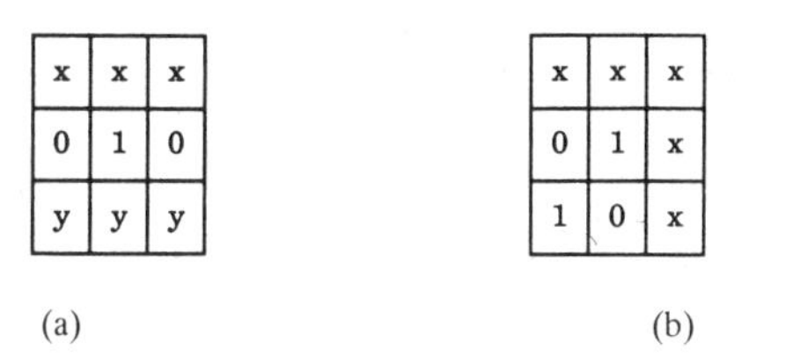

Fig. 2. Break point configurations; at least one pixel in each group marked x or y must be nonzero.

condition had to be imposed to prevent an interior pixel from being marked as an edge point. Edge points are marked in one raster scan after which the marked points are compared against six 3×3 windows (those shown in Fig. 2 and their 90° rotations) and deleted if they are not break points, i.e., their removal does not create breaks in the pattern. It was found that elimination of the end-point condition resulted in less spurious tails. The procedure is repeated until no pixels are deleted or for a preset maximum of three iterations, after which a single "clean-up" scan is applied to remove all black pixels that are not end points, break points, or interior points. The results obtained from this procedure were not very different from those of thinning carried through to the end, probably because this algorithm was designed and tested on (almost thin) character patterns.

The preceding algorithm assumes a smoothing preprocessing phase; this is extended in [26], where smoothing is performed before each iteration. In each iteration, contour pixels satisfying the usual stated conditions are marked and then examined for deletion as in [18] but with the endpoint condition. When no more pixels can be deleted, a final adjustment phase is introduced in which a skeletal pixel would be moved to one of its 4-neighbors if the latter pixel has a greater 8-distance from the background. In the process, connectivity of the skeleton is maintained while skeletal points are moved closer to the medial line of the original pattern.

The classical thinning algorithms mentioned in [87] also use the windows shown in Fig. 2 together with their 90° rotations to determine skeletal pixels. (For the sake of brevity, we use 90° rotations to also include 180° and 270° rotations when different windows are generated). Since these windows, in effect, only represent connectivity preservation criteria, their sequential application to a pattern could result in excessive erosion or serious shortening of branches. To overcome this difficulty, a mixture of parallel and sequential operations is proposed. Only one type of contour point is processed in each subiteration, skeletal pixels are marked for retention through all subsequent iterations, and a pixel q to be deleted is marked so that it is considered to be a black pixel when its neighboring pixels are checked in the current iteration. These conditions are designed to preserve connectivity and prevent disappearance of two-pixel wide lines, and they also retain end points without requiring a specific condition.

A variation of this algorithm is based on the compilation of an extensive list of thinning rules and consideration of 5×5 neighborhoods [42]. These larger neighborhoods are also used in [69], where the pattern is raster scanned, and deletion of the center pixel is based on comparison with 20 25-b templates stored on a chip.

The SPTA [77] is a sequential algorithm that uses two raster scans per cycle, where the first is left to right, and the second is top to bottom. In each scan, p is marked for retention (p is a safe point) if one of the following is true:

N1: $N(p)$ satisfies one of the windows in Fig. 2 or its rotations.

N2: $N(p)$ contains exactly two 4-adjacent black points.

These conditions are equivalent to a west contour point p being safe if the boolean expression

$$x_1(x_2 + x_3 + x_7 + x_8)(x_3 + \overline{x}_4)(x_7 + \overline{x}_6) = 0.$$

In the first scan, west contour points are marked, then east contour points that are not west safe points, and so on. Condition N2 is intended to prevent excessive erosion of diagonal strokes two pixels wide, but it can lead to noisy branches [65]. In [16], a modification was proposed to SPTA to eliminate (in most cases) the last pass when no pixels would be deleted, and the modified algorithm was implemented on a multiprocessor using the methods of a) function and b) data decomposition. In a), each processor executes one scan, and each completed row is moved to the processor performing the next scan, whereas in b) each processor scans a portion of the image with overlapping boundaries.

The Rutovitz crossing number $X_R(p)$ is used to determine pixel deletion in [5], [7], and [8]. In these algorithms, a slightly different definition of contour pixel is used; here, a contour pixel is a black pixel having at least one white 8-neighbor. This condition together with the use of $X_R(p)$ require an additional condition ($F = x_1x_3x_5x_7 = 0$) to ensure that holes would not be created when contour points are deleted. Complete conditions for deletability of p while maintaining connectivity are given in [5]; briefly, these useful conditions are as follows:

1. If $X_R(p) = 0$ or 8, p is not deletable.
2. If $X_R(p) = 2$, p is deletable iff $F = 0$ and p is not an end point.
3. If $X_R(p) = 4$, p is deletable iff $F = 0$ and one of the four corner pixels is 0 with 1's on both sides. The latter condition is equivalent to $\sum_{i=1}^{4} x_{2i-1}\overline{x}_{2i}x_{2i+1} = 1$.
4. If $X_R(p) = 6$, p is deletable iff one of its 4-neighbors is 0 and the other three are 1's belonging to distinct 4-components, or $\sum_{i=1}^{4} x_{2i-1} = 3$.

However, use of the above conditions by themselves would give spurious end points and erode corners. Therefore, a solution is proposed in [5] to test p for deletion according to the configuration of contour pixels in $N(p)$. A "second level" crossing number $CNN(p)$ represents the number of transitions from a contour pixel to a noncontour one (and vice-versa) when the points of $N(p)$ are traversed in order, and procedures are derived for the deletion of p based on $X_R(p)$, $CNN(p)$, and the number of 8-adjacent pairs of contour pixels among the 4-neighbors of p.

Further work in preserving significant contour protrusions or prominences in the thinning process are developed by Arcelli and Sanniti di Baja [7], [8], where prominences are first detected and labeled. These significant protrusions are defined as connected subsets of the contour that are beyond a threshold distance from the interior or core (the noncontour pixels of P). They are retained while the other contour pixels with $X_R(p) = 2$ are removed iteratively according to the conditions in [5] as long as they are not necessary for maintaining connectedness between the core and the prominences. When a set S_f with an empty core is obtained, a label $e(p) = 2(x_5+x_3)+x_1+x_7+1$ is assigned to each remaining pixel p in S_f. The nonmaxima pixels under this label are removed, resulting in a skeleton that is, at most, two pixels wide; then a final pass can be made to destroy 4-connectedness in favor of 8-connectedness.

The skeletons obtained this way can be unduly influenced at the end of a branch by the presence of a sharp protrusion on one side (see Fig. 21a in [3]). In [3], a compression phase was also implemented to represent each 3×3 window of P by a grey-scale value derived from the number of black pixels in the window. The reduced grey-scale image is then thresholded with the connectivity constraints of Fig. 2 to a binary image on which thinning is performed by an implementation of [8] modified to preserve T junctions with length greater than one pixel [109]. The skeleton is then expanded to the original scale.

When contour pixels p are traced sequentially and labeled, multiple pixels, where a pattern "folds" onto itself, can be easily determined [85]. Pixel p is multiple if at least one of the following conditions holds:

P1: p is traversed more than once during tracing (connectivity number > 1).

P2: p has no neighbors in the interior.

P3: p has at least one 4-neighbor that belongs to the contour but is not traced immediately before or after p.

Since P1 includes pixels with a connectivity number greater than 1, P2 includes end points, and P3 contains lines two pixels wide, the concept of multiple pixels is quite inclusive. However, if the contour is traced repeatedly and only the multiple pixels from every tracing are retained, the result may not be a connected skeleton. Therefore, in [85], multiple pixels are called skeletal, as are 8-neighbors of skeletal pixels from a previous iteration. These pixels form a skeleton that may be too thick (and thus requires editing), but the algorithm was proved to be correct—it does terminate and produces connected skeletons.

In [86] and [87], the characterization of multiple pixels is redefined in terms of local neighborhoods and can therefore be determined in sequence or parallel by comparison against a set of masks. This requires the addition of window (c) to those of Fig. 2 to produce the neighborhood patterns of Fig. 3 and their 90° rotations for multiple pixels.

The definition of multiple pixels is slightly modified in [88]. It is also proposed that a combination of sequential and parallel operations may be more efficient for images where most of the pixels do not require much processing. For such images, the pattern can be divided into fields where each is assigned to a processor. Each processor operates on the pixels of its field sequentially, and when certain steps have been completed, it waits until all other processors have completed the same steps

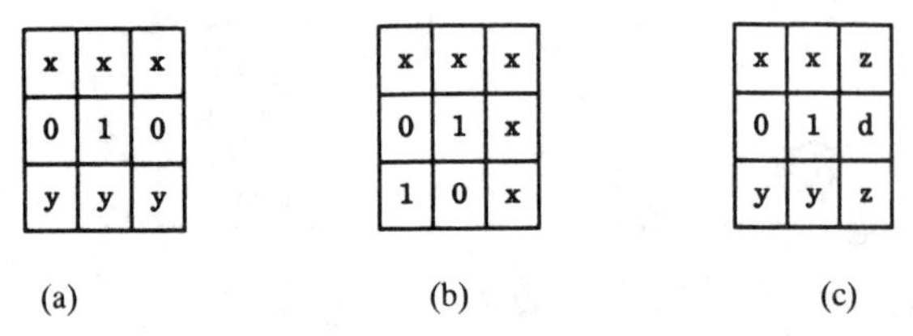

Fig. 3. Configurations of multiple pixel p. Each group of pixels marked x or y must contain a nonzero element. In (c), at least one of the pixels marked z must be nonzero; if they are both nonzero, pixels marked x or y can have any value. The pixel marked d is a contour pixel.

so that the processors can be resynchronized.

The above algorithm essentially ensures skeleton connectedness by detecting and assigning multiple pixels M to the skeleton S and then finding and assigning to S suitable (skeletal, nonmultiple) pixels to connect M to the interior of P. Since this may result in unacceptably thick skeletons when P is not initially almost thin, an alternative algorithm that deletes the nonmultiple pixels from the contour C and retains the remaining set as S is proposed [10]. At the same time, contour information is used to determine whether a contour pixel should be a) regarded as noise and not labeled multiple or b) considered to be a significant convexity and assigned to S even though it is not multiple. These are accomplished by computing the n codes [45] of contour pixels from the Freeman chain codes obtained in contour tracing. The 1-code c_i of pixel p_i is the difference (modulo 8) between the chain codes at p_i, and for $n > 1$, the n code

$$c_i^n = nc_i + \sum_{k=1}^{n-1} (n-k)(c_{i-k} + c_{i+k})$$

determines the curvature of the contour at p_i. The value of c_i is used to determine a), and if c_i^n exceeds a threshold, then p_i is assigned to S so that it can represent a significant convexity. Naturally, if the threshold value is lower, the algorithm would be more sensitive to contour protrusions. Other works to preserve such prominences have already been discussed [7], [8].

In [11]–[13], the concept of multiple pixels is developed from another point of view; it is thought of as being the opposite of curve simplicity. In the continuous plane, a (closed) curve is simple if and only if it never crosses itself; consequently, a simple curve divides the plane into two connected sets called the inside and the outside. This notion is extended to the contour C of a digital pattern P for a simply connected pattern in [11] and [12] and for a multiply connected figure in [13]. In particular, C is considered to be simple, provided that it neither touches nor overlaps itself, and this global concept was found to be equivalent to the following local conditions. If we consider $P - C$ to be the inside of C and $\bar{P}$ to be the outside of C, then C is simple, provided that every p in C satisfies the following conditions:

A1: $N(p) \cap \bar{C}$ consists of one (8-connected) component on the inside and one (4-connected) component on the outside.

A2: $N(p) \cap \bar{C}$ contains at least two pixels that are horizontally or vertically aligned: one belonging to the outside and the other to the inside.

The pixels satisfying the conditions A1 and A2 are said to be regular, and those that are not regular coincide with multiple pixels where contour arcs either coincide or are adjacent to each other. Furthermore, since A1 is computationally complicated, it is shown in [13] that equivalently to A1 and A2, a contour pixel p is multiple if it satisfies at least one of the following conditions:

A3: Neither the horizontal nor the vertical neighbors of p are such that one belongs to $P - C$ and the other to $\bar{P}$.

A4: $N(p)$ has three consecutive points such that the intermediate one is a diagonal neighbor and belongs to C, whereas the other two belong to $\bar{P}$.

Conditions A3 and A4 have the advantage of being local conditions that are easily verified once the contour pixels are located, and this is the method employed in [14]. By testing for multiple pixels on the 4-distance transform of P, each successive contour is denoted by its label on the distance transform; therefore, thinning can be accomplished in one raster scan. During this pass, however, the already-visited neighbors of a detected multiple pixel q must be examined again to verify whether any of them has been induced to become a multiple pixel by the constraint to maintain connectedness. The skeleton consists of all the multiple pixels detected, and it can be reduced to unit width.

The Freeman chain codes are also used in [47] to detect features such as 90 and 45° corners and T junctions in the contours. This information is then used to retain certain pixels in the contour-stripping process in an attempt to preserve such features. Otherwise, this method deletes the contour pixel p if $X_H(p) = 1$ and $2 \leq b(p) \leq 6$.

In [132], the four types of contour pixels (east, north, west, and south) are placed in buffers. Each type of buffer point is sequentially processed and checked against windows for connectedness and end-point preservation. If a point is removable, its 4-neighbors are examined for inclusion in the next contour and placed in the appropriate buffers. The process is repeated until the buffers are empty.

The algorithms discussed so far are based on an examination of contour pixels for deletion or retention. A different implementation method to produce a skeleton is that of contour generation or the iterative generation of a new contour inside the existing one until only a skeleton remains. This process is based on the direction of the contour pixels in [131] and [64]. When contour pixels are followed in sequence, three such consecutive pixels would form an angle θ with its vertex at the current pixel p. The interior pixel in $N(p)$ closest to the bisector of θ is considered to be a point on the next contour [131], and p is deleted. When this procedure is repeated until no interior pixels are left, a pseudo skeleton is formed by the last contour. This simple method is greatly refined and expanded in [64] by using the Freeman chain codes of the contour pixels to generate the new contours. These chain codes are used, together with breakpoint and endpoint considerations, to derive a set of rules for the generation of the new contour. When new contour pixels are determined, their chain codes are also generated. The end pixel conditions of algorithms [77], [6], and [136] can be incorporated into this algorithm [65] to simulate the behavior of the previous algorithms, at least at branch ends. This algorithm can also be implemented in a distributed environment [66] by assigning nonoverlapping subsets of the pattern to different processors for thinning and then synchronizing information about the borders at the end of each iteration. A similar procedure of successive contour generation using chain codes is implemented in [127]. In this recent work, a contour pixel is deleted if its 4-neighbors on the "inward" side (away from the background) are all black, in which case, the chain codes for the corresponding section of the new contour are derived from a predefined set of rules related to local curvature. These rules are simpler than those of [64], and it is claimed that the same results are obtained when pixels are processed in the same order.

In general, when the pixels of P are processed sequentially, there is no problem in preserving the connectivity of P and $\bar{P}$ when suitable 3×3 local operations are used. Therefore, the requirement of topological preservation is met by these algorithms. However, it would require more global information to preserve the geometric features that are significant for a shape analysis of P. For this purpose, a skeleton obtained from a raster scan of P is seldom meaningful [5], whereas an algorithm based on sequential following of the contour elements may lead to better results. This latter method can allow some correlation between the shape of the skeleton and the external contour of the pattern that may not be possible to achieve by local operations only. Of course, these more global considerations would lead to increases in computation time and more complicated procedures, and much of the complexity in sequential algorithms (for example, [8], [10] and [85]) is the result of efforts made to preserve more subtle geometric features.

IV. Parallel Thinning Algorithms

In parallel thinning, pixels are examined for deletion based on the results of only the previous iteration. For this reason, these algorithms are suitable for implementation on parallel processors where the pixels satisfying a set of conditions can be removed simultaneously. Unfortunately, fully parallel algorithms can have difficulty preserving the connectedness of an image if only 3×3 supports are considered; for example, a horizontal rectangle two pixels in width may completely vanish in such a thinning process. Therefore, the usual practice is to use 3×3 neighborhoods but to divide each iteration into subiterations or subcycles in which only a subset of contour pixels are considered for removal. At the end of each subiteration, the remaining image is updated for the next subiteration. Four subcycles have been used in which each type of contour point (north, east, south, and west) is removed in each subcycle [115], [101], [17]. These have also been combined into two subiterations [115], [136], [21], [118], [49] with one subiteration deleting the north and east contour points and the other deleting the rest, for example. Other two-subcycle algorithms have been devised to operate on alternate subfields of the pattern that is partitioned in a checkerboard manner. Recently, one-subiteration algorithms have been implemented, but these invariably have to use information from a larger context in order to preserve connectivity. These algorithms will be examined in greater detail below.

A fundamental parallel algorithm was proposed in [103], where a pixel p is deleted iff all the following are true:

R1: $b(p) \geq 2$
R2: $X_R(p) = 2$
R3: $x_1x_3x_5 = 0$ or $X_R(x_3) \neq 2$
R4: $x_7x_1x_3 = 0$ or $X_R(x_1) \neq 2$.

This algorithm is also described in [115] with the added condition that $b(p) \leq 6$ to ensure that p has a white 4-neighbor; therefore, deletion of p would not create a hole. This is a one-subiteration algorithm that uses information from a 4×4 window, and it does yield connected skeletons that are insensitive to contour noise [115] but can result in excessive erosion [116], [79], [70], [39].

Since disjoint 4-components may be 8-connected, the removal of all pixels satisfying the above conditions does not reduce diagonal lines to unit pixel width. Additional conditions were added specifically to address this problem and to allow for deletion of pixels p with $X_R(p) = 4$ when p lies on a diagonal line two pixels wide [33], and this was proved to preserve connectedness [34]. Due to the asymmetric nature of conditions R3 and R4, the skeleton would not lie centrally; therefore, 180° rotations of these rules were introduced [35] to result in the following complete set of rules for the removal of p:

D1: $X_R(p) = 0, 2, \text{or } 4$
D2: $b(p) \neq 1$
D3: $x_1x_3x_5 = 0$
D4: $x_1x_3x_7 = 0$
D5: If $X_R(p) = 4$, then in addition, a) or b) must hold:
  a) $x_1x_7 = 1, x_2 + x_6 \neq 0$, and $x_3 + x_4 + x_5 + x_8 = 0$
  b) $x_1x_3 = 1, x_4 + x_8 \neq 0$, and $x_2 + x_5 + x_6 + x_7 = 0$
D6-D8 are 180° rotations of D3-D5.

It was also suggested that two subiterations should be used, where the first one deletes pixels satisfying D1-D5, and the second deletes according to D1, D2, and D6-D8. Of course, this algorithm deletes isolated pixels. The complexity of the rules D1-D5 led to the observation [44] that if two consecutive 4-neighbors of p are 1's, then the value of the corner pixel q in between has no effect on whether p should be deleted; therefore, q can be considered to be 1. In such cases, p can be deleted if its modified crossing number is 2 provided $b(p) > 1$. It should be noted that this modification of the crossing number is the corner-cutting process in [52]; therefore, the resulting number is actually twice $X_H(p)$, and, of course, it measures 8- rather than 4-connectivity.

The crossing number $X_R(p)$ is also used as a criterion for a filling operation before thinning. In [123], a white pixel q with $X_R(q) \geq 2$ is considered to be a connecting point and is filled. Then, a 4-subiteration algorithm is implemented in which end points (having exactly one black 4-neighbor) and pixels p with $X_R(p) > 2$ are considered skeletal, whereas nonskeletal contour pixels are deleted.

The much-cited work of Zhang and Suen [136], which is also summarized in [46], is an implementation of a subset of conditions D1-D8 in two subiterations. In the first subiteration, p is deleted if it satisfies the following conditions:

Z1: $2 \leq b(p) \leq 6$
Z2: $X_R(p) = 2$
Z3: $x_1x_3x_7 = 0$
Z4: $x_1x_7x_5 = 0$.

In the second subiteration, Z3 and Z4 are replaced by their 180° rotations. Therefore, the first subcycle deletes 4-simple pixels on the south and east borders as well as north-west corner pixels, whereas the second subcycle deletes pixels with opposite orientations. This is a simple and efficient algorithm that is also immune to contour noise [21]; however, two-pixel-wide diagonal lines can be seriously eroded, and 2×2 squares would vanish completely [70], [71]. It was suggested that condition Z1 should be replaced by $3 \leq b(p) \leq 6$ in order to retain such structures. As can be expected, such a modification creates problems of its own in retaining extraneous pixels; therefore, an additional pass was proposed [128] in order to thin to unit thickness, further increasing the computation time.

Although excessive erosion of diagonal lines is not a topological problem, complete disappearance of a 2×2 square in the thinning process renders the algorithm invalid from the topological point of view. This was shown [39] to be the only possible configuration where the algorithm is not valid, and a modification to [136] was suggested so that all conceivable cases could be correctly processed. The modification depends on the number of black 4-neighbors of p when $X_R(p) = 2$. If p has none or one such neighbor, no change would be required. If p has two or three such neighbors and p satisfies the existing conditions, then p can be deleted in the first subiteration if $x_1 = 0$ or $x_7 = 0$ and in the second if $x_3 = 0$ or $x_5 = 0$.

In [129] and [130], conditions D1, D2, and D8 from [35] are implemented sequentially with minor differences. The more recent algorithm does not delete isolated pixels ($X_R(p) = 0$). It requires $2 \leq b(p) \leq 6$, and it does not preserve 4-connectivity. Therefore, a pixel p at the vertex of a right angle formed by two 4-neighbors would be deleted, thus leading to condition $x_2 + x_6 \neq 0$ being unnecessary in rule D8(a) and analogously for D8(b). These authors also implemented a two-subcycle parallel algorithm that deletes (in the first subiteration) east or south boundary points and northwest corner points with $X_R(p) = 2$ as well as contour pixels with $X_R(p) = 4$ satisfying condition D8; pixels with opposite orientations are deleted in the second subiteration. Another two-subiteration implementation of conditions D1–D8 of [35] is that of [21]; it contains the same minor differences as [129] and [130], and it claims to be an improvement over [136] in eliminating excessive erosion. In this work, the rules are implemented by table lookup; each of the 256 configurations of $N(p)$ is converted into an address in the thinning tables (one for each subiteration), where the new value of p (0 or 1) is defined.

In order to use only one subiteration per cycle, information from a 4×4 window containing p is considered [54] in order to determine its deletion. For this algorithm, p is an edge point (edge p is true) iff $N(p)$ contains between two and six 4-connected black pixels, i.e., $X_R(p) = 2$ and $2 \leq b(p) \leq 6$.

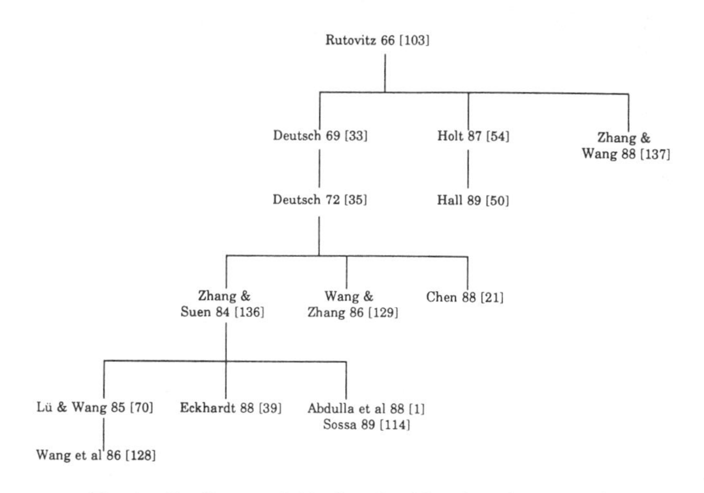

Fig. 4. Family tree of thinning algorithms based on Rutovitz.

Then, p is deleted iff condition H is true, where

$$\begin{aligned} \mathrm{H}: \ & (\text{edge } p) \wedge (\sim \text{edge } x_1 \vee \overline{x}_3 \vee \overline{x}_7) \\ & \wedge (\sim \textit{edge } x_7 \vee \overline{x}_5 \vee \overline{x}_1) \\ & \wedge (\sim \text{edge } x_1 \vee \sim \text{edge } x_8 \vee \sim \text{edge } x_7) \end{aligned}$$

With this condition, edge information on neighboring pixels is used to prevent the disappearance of vertical lines that are two pixels wide. This algorithm is implemented [55] on SIMD and MIMD machines with a modification to reduce redundancy in the edge computation of elements with common neighbors. Condition H is shown [137] to be almost equivalent to R1-R4 of [103], where the crossing number instead of edge information of neighboring pixels is considered. The only differences are that the earlier work stipulates $b(p) \geq 2$, and the conditions of [54] would add the condition "or $b(x_3) \notin [2,6]$" to R3 (and "or $b(x_1) \notin [2,6]$" to R4). In turn, [137] also proposes (without results) to modify the conditions R1-R4 by changing R1 to $2 \leq b(p) \leq 6$ and the second parts of R3 and R4 to $y_3 = 1$ and $y_1 = 1$, respectively, where y_1 is the east neighbor of x_1, and y_3 is the north neighbor of x_3. The first modification is also suggested by [50] to preserve diagonal lines in [54].

Besides all the modifications mentioned above, [1] and [114] also suggest procedures to reduce, to unit thickness, the skeletons obtained by [136]. Therefore, the algorithm originally proposed by Rutovitz and modified by Deutsch has been implemented with various levels of changes in a number of articles, and the relationships between them are shown in Fig. 4.

Despite the somewhat baffling array of modifications proposed to the algorithm originating from Rutovitz, there are many similarities among the results. Basically, the algorithms of [103], [136], [54], [39], and [137] have the common deficiency of possible excessive erosion in the thinning of diagonal lines. The problem may depend on whether the lines are at 45 or 135° to the horizontal, and lines that are an even number of pixels in width seem to be more vulnerable to erosion. The simple modification of [70] and

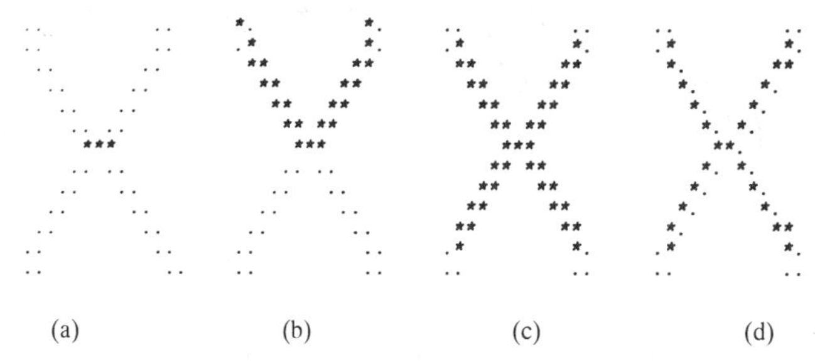

Fig. 5. Results of thinning by different algorithms based on Rutovitz: (a) See [54], [103], [136], and [138]; (b) see [39]; (c) see [50] and [70]; (d) see [128].

[50] solves this problem by retaining extraneous pixels and creating skeletons of more than unit width. The skeleton is then thinned (incompletely) by a second pass in [128]. Fig. 5 shows a typical example of thinning a problematic pattern using these algorithms.

Necessary and sufficient conditions for preserving topology while deleting border points in parallel are given in [100]. If only 3×3 local neighborhoods are considered, then an operation that deletes (for example) north border pixels would preserve topology iff the only north border points it removes are simple and have at least two 1's as 8-neighbors. It is proved [124] that this operation would not change the genus number of P. This algorithm was implemented in [101], where 4-connected skeletons are observed to contain more noisy branches.

Equivalent conditions for the simplicity of a border point are given in [6], where it is shown that a north border point with at least two 1's as 8-neighbors would be simple iff $x_5\overline{x}_7x_1 = 0$ and $\sum_{i=1}^{4} \overline{x}_{2k-1}x_{2k}\overline{x}_{2k+1} = 0$. These conditions are combined into one in [102].

Another 4-subiteration algorithm is that of [17], where skeletal pixels found in each iteration are assigned the iteration number for the subsequent calculation of object width. The values of skeletal pixels from previous iterations are left unchanged, whereas those of interior pixels are incremented by one in each iteration. In the nth iteration, pixel p is tested for deletion as a north border element if $x_3 = 0$, p has value n, and $x_7 \neq 0$. Such a border element p is considered skeletal if there exists $x_i \in N(p)$ such that $x_i > 0$ and $N(x_i) \cap N(p) = \{p, x_i\}$; otherwise, p is assigned the value 0 and deleted. This algorithm preserves topology but was found to be sensitive to boundary noise and the order of the subiterations.

Parallel algorithms using both four and two subiterations have also been implemented by means of matching against windows analogous to those of Fig. 2. Stefanelli and Rosenfeld [115] implemented two such algorithms and proved that they preserve topology. In each subcycle, the final (skeletal) pixels are stored. In order to avoid deletion of contour points that are also final points, all the contour points are first deleted, after which the final points are added. When four subcycles are used and north border points are considered for deletion, the final point conditions are those shown in Fig. 6 together with the 90° rotations of (a) and (b). In this and subsequent figures, pixels that are left blank may be 1 or 0 or are "don't care" pixels.

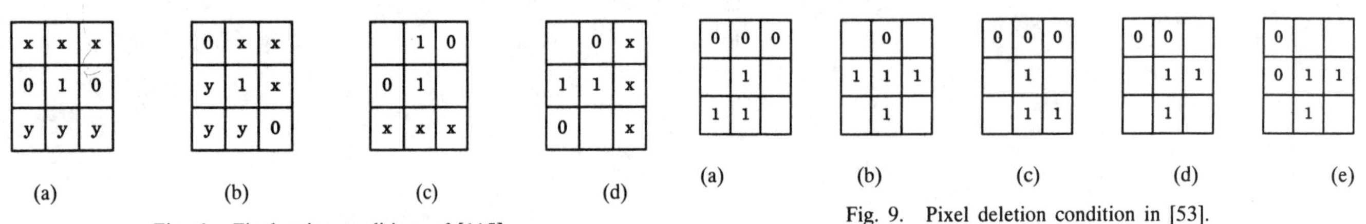

Fig. 6. Final point conditions of [115].

Fig. 9. Pixel deletion condition in [53].

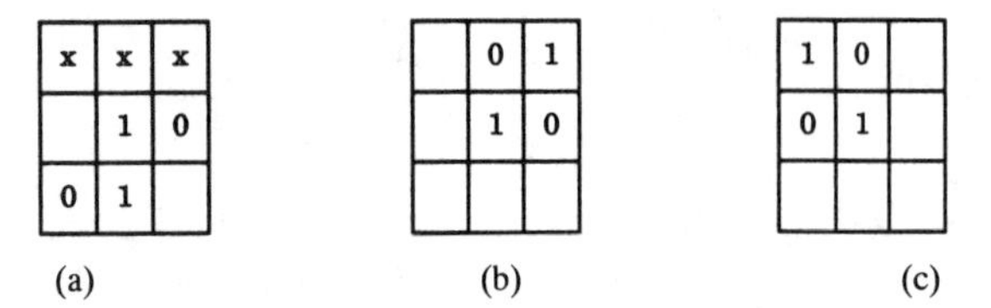

Fig. 7. Additional final point conditions [115].

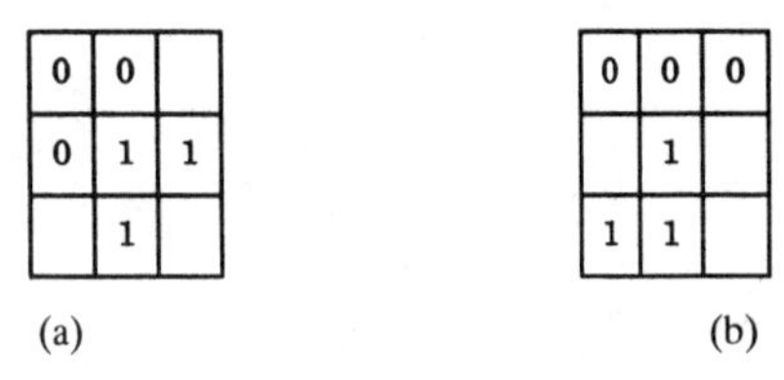

Fig. 8. Pixel deletion conditions in [4].

The difference between window (b) here and that in Fig. 2(b) implies that the present algorithm would produce imperfectly 8-connected skeletons, as is indeed the case. The two-subcycle algorithm combines deletion of south and east contour points in the first subcycle and the rest in the second. For the second subcycle, the final point conditions are the ones given above in Fig. 6 together with those shown in Fig. 7. Conditions (b) and (c) are added to ensure connectivity and preserve diagonal lines; however, thin horizontal or vertical lines are not preserved. The processing speed of this algorithm was found to be comparable with that of [103], but the use of only 3×3 windows here allows for easier implementation on a cellular network.

Another algorithm using thinning windows is [4], where the masks for pixel deletion are those shown in Fig. 8 together with their 90° rotations. Each mask is applied in parallel to P, and the masks are applied in the order (a), (b), followed by their 90° rotations, and so on. Therefore, this algorithm removes pixels from eight borders in the order nw, w, etc. The asymmetric nature of mask (b) results from the requirement that P should not vanish completely in the parallel process. This algorithm was proved to operate correctly; it was especially suitable for implementation on parallel processors that have the ability to extract "1" and "0" elements having a predetermined number of "1" or "0" elements in chosen neighborhood positions, and it was implemented on a Clip 4 parallel processor [53]. Since not all deletable pixels were removed, other masks were added to produce those of Fig. 9 and their rotations.

In addition, it was found [53] that superior results are obtained when removal is restricted to pixels that do not only satisfy the criteria so far but would have satisfied them at the start of the current iteration. This ensures that just one layer of pixels is removed all around P and results in a more predictable algorithm with fewer anomalies at the corners where two layers of pixels may be removed otherwise. The author calls this a border parallel operation, whereas the original is border sequential. Implementation of a border parallel algorithm on the Clip 4 processor does not involve extra computation since the 1's and 0's are detected in separate steps in any case; therefore, it is necessary only to test the 1's in the pattern at the beginning of the iteration, whereas the 0's are tested in the pattern obtained so far. The longer computation time required is the result of deleting fewer pixels in each cycle. In the same paper, it is also shown that the border parallel conditions of Fig. 9 are equivalent to the four-subiteration border parallel deletion of contour pixels with crossing number $X_H(p) = 1$. Fig. 10 shows the results of using this crossing number to thin a pattern by the border parallel and border sequential algorithms. Fig. 10(a) shows the results after one iteration when the contour pixels are deleted according to the sequence north, east, south, and west. Typically, the border-sequential algorithm deletes more pixels at the corners; as a result, the end pixel condition may be encountered sooner with the preservation of these pixels leading to more short (and possibly noisy) branches. For this reason, the right end of the horizontal stroke is retained by this algorithm in the final skeleton shown in Fig. 10(b), thus illustrating the tradeoff that is sometimes involved when different thinning algorithms are used.

By using look-up tables, a simulation of the border parallel conditions of Fig. 9 has been implemented [112], [113], in which mask (c) and its rotations are omitted and endpoint conditions added. To prevent excessive erosion of two-pixel-thick lines, additional information is used so that the current pixel is retained if its preceding neighbor pixel has been removed.

Two recent papers [118], [49] have implemented parallel thinning using the crossing number $X_H(p)$ to examine pixels for deletion in two subiterations. In [118], the (approximately) 4×4 window used is shown in Fig. 11. Pixels removed in the first subcycle are a) deletable north contour pixels, and b) west contour pixels that are deletable after the pixels in a) have been removed. It is this latter requirement that produces more consistent results while making it necessary to increase the size of the neighborhood and the complexity of the algorithm. In this subcycle, p is deleted iff

S1: $X_H(p) < 2$,

S2: $(b(p) > 2) \vee [(b(p) = 2) \wedge \sum_{i=1}^{8} x_k x_{k+1} = 0]$, and

S3: $(x_3 = 0) \vee [(x_5 = 0) \wedge S4 \wedge S5]$, where

S4: $x_2 \vee \bar{x}_8(y_1 \vee y_2 \vee y_3) \vee \bar{y}_2 y_3 = 1$, and

S5: $\bar{x}_4 \vee y_5 \vee \bar{x}_2 y_4 = 1$.

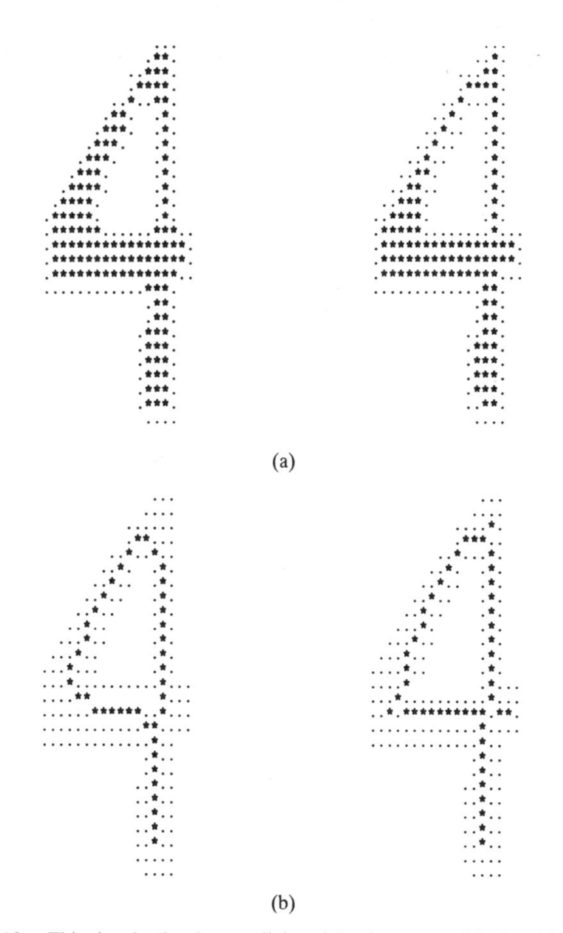

Fig. 10. Thinning by border parallel and border sequential algorithms of [53]: (a) After one cycle; (b) final skeleton. The figures to the left in both (a) and (b) show the border parallel (four cycles), and the figures to the right show the border sequential (three cycles).

	y_5	y_4	
x_4	x_3	x_2	y_3
x_5	p	x_1	y_2
x_6	x_7	x_8	y_1

Fig. 11. Neighborhood of [118].

For the second subcycle, the conditions are S1, S2, and the 180° rotations of S3-S5.

In [49], two such algorithms are implemented. For the first one, the image is divided into two distinct subfields in a checkerboard pattern, and each subiteration deletes pixels p in a subfield iff p is a contour pixel, $b(p) > 1$, and $X_H(p) = 1$. This procedure results in noise spurs and zigzagging vertical or horizontal lines. The other algorithm is a modification of [136] to thin to 8-connected skeletons and retain diagonal lines and 2×2 squares. Under this scheme, p is deleted iff

G1: $X_H(p) = 1$

G2: $2 \leq \min\,\{n_1(p), n_2(p)\} \leq 3$, where $n_1(p) = \sum_{i=1}^{4} x_{2k-1} \vee x_{2k}$ and $n_2(p) = \sum_{i=1}^{4} x_{2k} \vee x_{2k+1}$ represent the number of 4-adjacent pairs of pixels in $N(p)$ containing one or two black pixels, and

G3: $(x_2 \vee x_3 \vee \bar{x}_8) \wedge x_1 = 0$ in the first subiteration and its 180° rotation in the second.

Labeling schemes have also been suggested for use in parallel thinning. In [41], one-subcycle thinning is accomplished by recoding the pattern pixels to incorporate connectivity information from a 5×5 window into the coded pixels in $N(p)$. Two initial scans are used to recode the pixels into core, interior, rim, and skeleton points ($c, i, r,$ and s, respectively). The basic rule is to replace r pixels by b (background) pixels if a horizontal or vertical irb sequence is present, with exceptions made to preserve connectivity and end points. This is similar to the "ideal" method for thinning proposed in [40], where p is D perfect if it belongs to a horizontal or vertical configuration ipb, where i is an interior point, and b is a background pixel. (Actually, D-perfect points satisfy one of the two conditions for regular points in [11] and [12]). An I-perfect point is defined analogously in terms of a diagonal alignment with the added condition that the two pixels 4-adjacent to both p and b must both be white. This paper suggests that parallel deletion of pixels that are simple and perfect (D or I perfect) would produce well-defined pseudo skeletons that are invariant to translation and rotation. Retention of endpoints is obviously achieved because they are never perfect.

Since it is impossible to determine whether a neighbor of a border point is an interior point if only 3×3 windows are used, additional information is incorporated [125] from outside the window by defining nonend points. If SC is the set of simple contour points, and $B = P - SC$, then $p \in P$ is a nonend point iff

T1: p is adjacent to a point in B,

T2: every neighbor of p in P is in B or adjacent to a point in $B \cap N(p)$,

T3: the points in $B \cap N(p)$ are connected in $N(p)$, and

T4: $b(p) \geq 2$.

The operation that removes simple, nonend, contour points in parallel is shown to preserve topology.

In order to extend the limitations imposed by 3×3 local operations, hybrid schemes have also been proposed to combine distance transforms with thinning. The use of distance transforms provides more global information. It has also been argued that since thinning operations can preserve connectivity while distance transforms possess the reconstruction capability, combining the two operations would be logical if both properties are desired [125]. The suggested procedure is that for each iteration, the set I of interior points of the existing patterns P should be determined, and its expansion $E(I)$ is defined as $E(I) = \{q | q \in I$ or has a neighbor in $I\}$. Then, the set of local maxima (in convexity) can be obtained as $P - E(I)$ [9], and contour points can be deleted unless they are local maxima, nonsimple, or nonend points (defined above). The remaining contour points are added to the skeletal set, and the process is repeated with I as the existing pattern. It is also argued [63] that such a hybrid scheme may be most efficient since a distance transform can be used initially to remove the bulk of the exterior

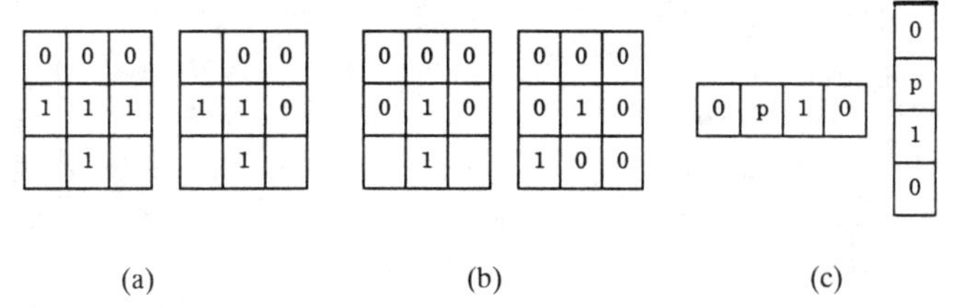

Fig. 12. Templates of [25]: (a) Thinning; (b) trimming; (c) restoring.

pixels in a fixed number of passes, and then, a peeling algorithm can be used to thin the remaining image to unit width.

The one-subiteration algorithms of [103], [54], [41], and [125] have already been discussed, where all utilize information from a larger than 3×3 neighborhood to maintain connectedness. Other algorithms of this group are [25] and [22]. In [25], eight thinning, two restoring, and eight trimming templates are applied in parallel in one pass in which pixels matching the 3×3 thinning or trimming, but not those of the 1×4 or 4×1 restoring templates, are removed. These templates are those of Fig. 12 and 90° rotations of (a) and (b).

The thinning templates delete border and corner points and the restoring templates retain 2-pixel-wide lines, whereas the trimming templates reduce noise spurs at the obvious expense of shortening branches. This algorithm is also implemented [20] using only 4×4 thinning and restoring templates (by filling in with don't care pixels). Only two iterations are used in different directions, after which postprocessing was applied to obtain a medial line that is not usually one-pixel wide.

In [22], a pseudo one-subcycle algorithm is implemented that essentially uses information from a 5×5 window. Despite a rather involved formulation employing many new terms that will not be included here, a pixel p is basically a candidate for deletion iff

C1: $X_H(p) = 1$ (equivalent to $LC(p) = T_b - 1(\neq 0)$), and
C2: (a) the black pixels in $N(p)$ are 4-connected $(E(p) = LC(p))$, or
(b) p belongs to a two-pixel-wide diagonal line.

The conditions for C2(b) had been given in [33] and modified in [21], and it is the latter scheme that is used here and stated as $S_4(p) = 2$ and $L(p) = 1$. If p satisfies the conditions for deletion, then its 4-neighbors are checked for possible deletion. If these are not deletable or connectivity would not be changed, then p is removed. The information is stored in look-up tables (as in [21]) for the checking process. In addition, a specially designed thinning condition can be added [23] to the algorithm to better preserve L-shaped patterns by removing more pixels from a concave corner than before. The condition is that p can be deleted if its four corner neighbors contain one white pixel while the other three have all black 8-neighbors. In [24], comparisons are made among [54], [25], and [22] in terms of thinning conditions and a connectivity-preserving function; The method in [22] is considered to be superior because it has more complete thinning conditions and a smaller ratio of cases guarded for connectivity preservation (hence, fewer iterations would be required).

For parallel algorithms, much of the attention in recent years has been focused on processing speed, and this has led to many comparisons either in terms of computing time or the number of iterations or subiterations used [118], [20], [49], [24]. However, some comparisons of the number of iterations may not be entirely valid [50]. A one-pass algorithm ([54], for example) is usually achieved by means of a larger support than 3×3, and this creates the need for computation of intermediate results or edge information on neighboring pixels. For this reason, a common framework is needed to define an "iteration" so that parallel algorithms can be meaningfully compared. The proposed scheme [50] assumes that in one iteration, each element of a mesh computer can compute any logical function of a 3×3 neighborhood, and the parallel speed of an algorithm is measured by the number of such iterations needed. Using this criterion of measurement, the parallel speed of [54] is not an improvement over that of two-subiteration algorithms such as [136].

In conclusion, it would appear that, as opposed to sequential algorithms, much of the complexity of parallel algorithms results from the need to preserve connectedness while using parallel operations in a small local neighborhood. This is a problem particular to the nature of parallel thinning, and it has been addressed by using subiterations (and serializing the procedure to some extent) or by enlarging the neighborhood to be examined. In either case, intermediate results are actually computed and used in some form in order to preserve more global structures.

V. Noniterative Thinning Methods

In the preceding sections, we discussed algorithms that produce a skeleton by examination and deletion of contour pixels. In this section, the algorithms to be considered are nonpixel based; they produce a certain median or center line of the pattern directly in one pass without examining all the individual pixels. Since the algorithms that accomplish this by medial axis or distance transforms are discussed in a separate paper, we are mainly concerned here with methods that determine center lines by line following or from run length encoding. It has been argued that this is the way human beings would perform thinning and that it is possible to retain global features and maintain connectedness in the process [15].

The simplest category of these algorithms determines the midpoints of black intervals in scan lines and connects them to form a skeleton. These methods have the advantage of being computationally efficient, but they also have natural disadvantages and would create noisy branches when the strokes are nearly parallel to the scan line [87]. They are valid in certain specialized applications where the scanning direction can be approximately perpendicular to that of the stroke. For example, in [75], the objects are rectangular with only small undulations along two parallel and much longer sides; therefore, a useful median line can be obtained by maintaining a constant scanning direction perpendicular to these sides. In other applications, the scanning directions are variable by rotations of 90 [74] or 45° [81]. The scanning in [60] is in both the x- and y-directions and the direction of smaller length (within certain constraints) would be selected.

Some algorithms obtain approximations of skeletons by connecting strokes having certain orientations. For example, four pairs of window operations are used in four subcycles [76] to test for and determine the presence of vertical, horizontal, right, or left diagonal limbs in the pattern. At the same time, the operators also locate turning points and end points by a set of final point conditions, and these extracted points are connected to form a line segment approximation of the skeleton. In [108] and [110], the boundary pixels are first labeled according to the above four local orientations. For each boundary pixel, a search is made for the same kind of label on the opposite side of the boundary (within a maximum stroke width) in the direction perpendicular to that given by the label. The midpoints of these pairs are then connected to form a skeleton. These methods are not appropriate for general applications since they are not robust, especially for patterns with highly variable stroke directions and thicknesses.

In [89], the vector form of a skeleton is obtained from the run length coding of P. Consecutive horizontal black intervals are grouped if they have approximately the same width and are roughly collinear. Based on a width-versus-height criterion, each such group is represented by either a horizontal vector or a vector joining the midpoints of the first and last intervals. Since this criterion alone is not sufficient, certain complex rules for a "compound vectorization" are devised to consider the relationship between groups to produce the final result. Obviously, the skeletons cannot be expected to preserve the geometric properties of the pattern, but this vector representation is convenient for use in pattern recognition. A recent thinning method [119] also produces a graph-like representation of a pattern by dividing the pattern into small units called meshes, on which partial recognition can be performed, and then merging the meshes to form partial graphs.

Another set of algorithms determine the center line of P by tracking the two contours of each curve simultaneously. In [31], the edge trackers move under the constraint of maintaining minimum distance between them. In [105], elongated ribbon-like simply connected objects without protrusions are tracked by approximate trapezoids using two pointers to follow the two contours of P. The midpoint of the base of the next trapezoid is considered to be skeletal, where the next base is either the side opposite the present base or a diagonal, depending on whether the diagonals are almost equal in length. This algorithm is extended [106] to objects containing protrusions by generating skeletons of the protrusions and backbone separately and then joining them together.

The above algorithm considers a simply connected pattern P to be a many-sided polygon, and this approach is also adopted in [72] to construct a median line. For every vertex v of P, its opposite line segment L is determined as the side of P closest to v. Then, a "projecting" line L_v is drawn from v to L, where L_v is the following:

1) The bisector of the interior angle θ at v if $\theta \leq 180°$
2) the two normal lines of the vectors forming v, otherwise.

If L_v intersects L within a threshold distance, its midpoint is added to the median line. We note that the lines of 1) are called pseudonormals in [19], where a discrete version of the symmetric axis of an object is theoretically derived using them.

Lines of elongated objects are followed by rectangular windows of variable size in [15], where the window can shrink or grow in size according to the width of the line at the location of the window. The skeleton is the unit width line connecting the centers of successive windows. In [82], a combination of thinning and stroke tracking is used on Chinese characters. Contour pixels are first sequentially examined, and p is deleted if

O1: $N(p)$ has at least one interior pixel,
O2: $N(p)$ has, at most, three contour pixels, and
O3: the contour pixels in $N(p) \cup \{p\}$ are traced consecutively.

When no more pixels can be deleted, the remaining pattern is traced using a 2×2 square, and the skeleton is the loci of these squares. The tracing is repeated if necessary.

Apart from these contour tracking methods, skeletal pixels have also been determined by a heuristic approach [59]. The deletion of a pixel p is determined by the local pattern density $d(p)$ and the density $d^*(p)$ of $d(p)$, where $d(p) = \sum_{i=1}^{8} x_i$ and $d^*(p) = \sum_{i=1}^{8} d(x_i)$. p would be retained if $d(p) \leq t_1$ and $d^*(p) \leq t_2$, where t_1 and t_2 are arithmetically consistent thresholds that can be set in a learning process.

Similar density functions based only on the 4-neighbors are considered in [138], where the values of an addition matrix are determined iteratively on black pixels p by

$$A_{(1)}(p) = \sum_{i=1}^{4} x_{2i-1}, \text{ and}$$

$$A_{(n)}(p) = \sum_{i=1}^{4} A_{(n-1)}(x_{2i-1}) \text{ if } A_{(n-1)}(p) = 4(n-1),$$

when $n > 1$.

The number of directions (out of four) in which the pixel has a maximum value (according to the final addition matrix) is its comparative degree, and the matrix of these degrees is thinned according to the sequential procedure of [137], together with the added condition of having a small degree. The addition matrix also contains information that allows for an approximate reconstruction of the pattern.

Using a different approach, skeletons can also be obtained from Fourier descriptors, at least for patterns that are not closed or overlapping and that have constant width [90]. For such patterns, the contour is traced to obtain a closed curve from which Fourier descriptors can be extracted. Fourier descriptors of the skeleton can then be determined, and the skeleton can be constructed from a finite set of harmonics. However, it would appear that highly mathematical methods can be used to obtain skeletons of only rather idealized patterns.

VI. Conclusions

In the literature, there appears to be a general agreement about the requirements that should be met by a thinning algorithm. These include preservation of topological and geometric properties, isotropy, reconstructibility, and high processing speed. Whereas the nonpixel-based thinning methods of Section V are efficient in terms of the number of operations required, it would seem impossible for them to preserve more detailed features of patterns since they are usually based on locating certain critical points and connecting them. These procedures are useful in applications where the detection of such points would suffice, one possible example being feature extraction in OCR. In general, however, the emphasis should be placed on the development of parallel algorithms for processing speed, especially when parallel image processing structures become increasingly available.

Reconstructibility, or the ability to regenerate the original pattern from the skeleton, is one objective measure of the accuracy with which a skeleton is representing the pattern. This criterion is generally satisfied by algorithms based on medial axis and other distance transforms by virtue of the fact that their skeletal pixels are also the centers of maximal blocks with known radii, but this is usually not the case with thinning algorithms. (Some exceptions are [86], [88], [138] and [77] to a certain extent, all of which use labeling schemes to retain the distances of skeletal pixels from the background).

Complete isotropy or invariance under rotation seems almost impossible to achieve in iterative algorithms. In sequential algorithms, the result depends on the order in which pixels are examined, and in parallel algorithms that remove one or two types of border points in each subiteration, the resulting skeleton depends on the order of the subiterations. Examples of nonisotropic results from these algorithms are shown in [8]. At the same time, medial axis transforms are not invariant under rotation due to a lack of algorithms based on true Euclidean distance maps so that very different results can be obtained from right-angled corners with one side parallel to an axis and at 45° to it. The use of these transforms simply transfers the problem from the thinning algorithm to the distance function, and the result can be just as idiosyncratic [53].

Maintaining connectedness and topology in thinning appears to have been resolved through various means. In sequential algorithms, it is sufficient to examine 3×3 local neighborhoods from the viewpoint of crossing number, for example. Parallel algorithms resolve the problem by dividing each cycle into subiterations or by considering a larger neighborhood in one subiteration.

Preservation of geometric properties, however, appears to be a more difficult problem. The main difficulty is that to achieve simplicity of algorithm and/or processing hardware, it is desirable to consider only small local neighborhoods, but these neighborhoods are incapable of providing global, structural information of the kind that is needed (for example) to distinguish between noise spurs and genuine end points. To prevent excessive erosion and creation of spurious end points at the same time, various attempts have been made to eliminate the end-point condition [18], make the condition more broadly applicable [70], [50], or apply the condition only at the later stages of thinning [118]. However, every modification of this type involves a tradeoff when uniformly applied, and additional information is really needed to make finer distinctions between cases. For this reason, various criteria using distance transforms have been introduced ([101], [8], and [27], for example). Some such means would be needed to propagate more global information to contour pixels, and it may allow for faster deletion of pixels in the initial stages.

Ultimately, the particular geometric properties a skeleton should preserve may be problem or application dependent. Algorithms based on medial axis transforms that possess the reconstruction capability would be well suited to applications such as data compression for storage and facsimile transmission [29]. However, the skeletons obtained by these methods are very sensitive to contour noise since branches can originate to many convexities on the boundary, depending on the distance transform used. Of course, this property is what renders them capable of exactly recreating the original pattern, but they would not be useful for pattern recognition where patterns with a wide variety of insignificant local contour differences can belong to the same class. For this latter application, it is much more important for the pattern to be represented by a collection of arcs lying along the center lines of the main curves of the pattern, whereas small perturbations of the contour should be ignored. At the same time, it is recognized that the thinning of blob-like patterns may result in skeletons that cannot preserve the original shape. In this respect, two types of skeletons are actually obtained from the different methods of thinning and medial axis transforms, and the choice should depend on the application.

In conclusion, it should be mentioned that comparison of the quality of skeletons remains a largely subjective, visual decision since the concepts of (connected) medial line, noisy branch, and excessive erosion have not been precisely defined. In addition, it is not objectively or quantitatively clear as to how a skeletal branch should accurately reflect the shapes of the two contours it represents. Comparisons are inevitably made when a new (or a modification of an existing) algorithm is proposed, but it is often based on the inconclusive evidence of one or two patterns. More general comparisons have been made from a digital geometry point of view [120], and in [92] and [93], comparisons have been made between skeletons obtained from thinning algorithms and reference skeletons prepared by human subjects. However, a more objective framework for the measurement of skeleton quality remains to be developed.

Acknowledgment

The authors are grateful to Associate Editor C. Dyer and the three reviewers; their detailed comments have been very helpful in the revision of this article.

References

[1] W. H. Abdulla, A. O. M. Saleh, and A. H. Morad, "A preprocessing algorithm for handwritten character recognition," *Pattern Recogn. Lett.*, vol. 7, no. 1, pp. 13–18, 1988.

[2] T. M. Alcorn and C. W. Hoggar, "Pre-processing of data for character recognition," *Marconi Rev.*, vol. 32, pp. 61–81, 1969.

[3] C. J Ammann and A. G. Sartori-Angus, "Fast thinning algorithm for binary images," *Image Vision Comput.*, vol. 3, no. 2, pp. 71–79, 1985.

[4] C. Arcelli, L. Cordella, and S. Levialdi, "Parallel thinning of binary pictures," *Electron. Lett.*, vol. 11, no. 7, pp. 148–149, 1975.

[5] C. Arcelli and G. Sanniti di Baja, "On the sequential approach to medial line transformation," *IEEE Trans. Syst. Man Cybern.*, vol. SMC-8, no. 2, pp. 139–144, 1978.

[6] C. Arcelli, "A condition for digital points removal," *Signal Processing*, vol. 1, no. 4, pp. 283–285, 1979.

[7] C. Arcelli and G. Sanniti di Baja, "Medial lines and figure analysis," in *Proc. 5th Int. Conf. Pattern Recogn.*, 1980, pp. 1016–1018.

[8] ——, "A thinning algorithm based on prominence detection," *Pattern Recogn.*, vol. 13, no. 3, pp. 225–235, 1981.

[9] C. Arcelli, L. P. Cordella, and S. Levialdi, "From local maxima to connected skeletons," *IEEE Trans. Patt. Anal. Machine Intell.*, vol. PAMI-3, no. 2, pp. 134–143, 1981.

[10] C. Arcelli, "Pattern thinning by contour tracing," *Comput. Graphics Image Processing*, vol. 17, pp. 130–144, 1981.

[11] C. Arcelli and G. Sanniti di Baja, "Finding multiple pixels," in *Image Analysis and Processing* (V. Cantoni, S. Levialdi, and G. Musso, Eds.). New York: Plenum, 1986, pp. 137–144.

[12] ——, "On the simplicity of digital curves and contours," in *Proc. 8th Int. Conf. Patt. Recogn.* (Paris, France), 1986, pp. 283–285.

[13] ——, "A contour characterization for multiply connected figures," *Patt. Recogn. Lett.*, vol. 6, no. 4, pp. 245–249, 1987.

[14] ——, "A one-pass two-operation process to detect the skeletal pixels on the 4-distance transform," *IEEE Trans. Patt. Anal. Machine Intell.*, vol. 11, no. 4, pp. 411–414, 1989.

[15] O. Baruch, "Line thinning by line following," *Patt. Recogn. Lett.*, vol. 8, no. 4, pp. 271–276, 1988.

[16] H. Beffert and R. Shinghal, "Skeletonizing binary patterns on the homogeneous multiprocessor," *Patt. Recogn. Artificial Intell.*, vol. 3, no. 2, pp. 207–216, 1989.

[17] A. Bel-Lan and L. Montoto, "A thinning transform for digital images," *Signal Processing*, vol. 3, no. 1, pp. 37–47, 1981.

[18] M. Beun, "A flexible method for automatic reading of handwritten numerals," *Philips Tech. Rev.*, vol. 33, no. 5, pp. 89–101; 130–137, 1973.

[19] F. L. Bookstein, "The line-skeleton," *Comput. Graphics Image Processing*, vol. 11, pp. 123–137, 1979.

[20] N. G. Bourbakis, "A parallel-symmetric thinning algorithm," *Patt. Recogn.*, vol. 22, no. 4, pp. 387–396, 1989.

[21] Y. -S. Chen and W. -H. Hsu, "A modified fast parallel algorithm for thinning digital patterns," *Pattern Recogn. Lett.*, vol. 7, no. 2, pp. 99–106, 1988.

[22] ——, "A systematic approach for designing 2-subcycle and pseudo 1-subcycle parallel thinning algorithms," *Patt. Recogn.*, vol. 22, no. 3, pp. 267–282, 1989.

[23] ——, "A 1-subcycle parallel thinning algorithm for producing perfect 8-curves and obtaining isotropic skeleton of an L-shape pattern," in *Proc. Int. Conf. Comput. Vision Patt. Recogn.* (San Diego, CA), 1989, pp. 208–215.

[24] ——, "A comparison of some one-pass parallel thinnings," *Patt. Recogn. Lett.*, vol. 11, no. 1, pp. 35–41, 1990.

[25] R. T. Chin, H. -K. Wan, D. L. Stover, and R. D. Iverson, "A one-pass thinning algorithm and its parallel implementation," *Comput. Vision Graphics Image Processing.*, vol. 40, pp. 30–40, 1987.

[26] Y. K. Chu and C. Y. Suen, "An alternative smoothing and stripping algorithm for thinning digital binary patterns," *Signal Processing*, vol. 11, no. 3, pp. 207–222, 1986.

[27] N. Chuei, T. Y. Zhang, and C. Y. Suen, "New algorithms for thinning binary images and Chinese characters," *Comput. Processing Chinese Oriental Languages*, vol. 2, no. 3, pp. 169–179, 1986.

[28] C. H. Cox, P. Coueignoux, B. Blesser, and M. Eden, "Skeletons: A link between theoretical and physical letter descriptions," *Pattern Recogn.*, vol. 15, no. 1, pp. 11–22, 1982.

[29] E. R. Davies and A. P. N. Plummer, "A new method for the compression of binary picture data," in *Proc. 5th Int. Conf. Patt. Recogn.*, 1980, pp. 1150–1152.

[30] ——, "Thinning algorithms: A critique and a new methodology," *Pattern Recogn.*, vol. 14, no. 1, pp. 53–63, 1981.

[31] J. -D. Dessimoz, "Specialized edge-trackers for contour extraction and line-thinning," *Signal Processing*, vol. 2, no. 1, pp. 71–73, 1980.

[32] E. S. Deutsch, "Preprocessing for character recognition," in *Proc. IEEE NPL Conf. Patt. Recogn.* (Teddington), 1968, pp. 179–190.

[33] ——, "Comments on a line thinning scheme," *Comput. J.*, vol. 12, 1969, p. 412.

[34] ——, "Toward isotropic image reduction," in *Proc. IFIP Congress* (Ljubljana, Yugoslavia), 1971, pp. 161–172.

[35] ——, "Thinning algorithms on rectangular, hexagonal, and triangular arrays," *Comm. ACM*, vol. 15, no. 9, pp. 827–837, 1972.

[36] A. R. Dill, M. D. Levine, and P. B. Noble, "Multiple resolution skeletons," *IEEE Trans. Patt. Anal. Machine Intell.*, vol. PAMI-9, no. 4, pp. 495–504, 1987.

[37] G. P. Dinnen, "Programming pattern recognition," in *Proc. West. Joint Comput. Conf.* (New York), 1955, pp. 94–100.

[38] R. O. Duda, P. E. Hart, and J. H. Munson, "Graphical-data-processing research study and experimental investigation," AD650926, pp. 28–30, Mar. 1967.

[39] U. Eckhardt, "A note on Rutovitz' method for parallel thinning," *Patt. Recogn. Lett.*, vol. 8, no. 1, pp. 35–38, 1988.

[40] U. Eckhardt and G. Maderlechner, "Thinning algorithms for document processing systems," in *Proc. IAPR Workshop Comput. Vision: Spec. Hardware Ind. Applications* (Tokyo, Japan), 1988, pp. 169–172.

[41] A. Favre and H. Keller, "Parallel syntactic thinning by recoding of binary pictures," *Comput. Vision Graphics Image Processing*, vol. 23, pp. 99–112, 1983.

[42] G. Feigin and N. Ben-Yosef, "Line thinning algorithm," *Proc. SPIE*, vol. 397, pp. 108–112, 1984.

[43] G. E. Forsen, "Processing visual data with an automaton eye," in *Pictorial Pattern Recognition* (G. C. Cheng, R. S. Ledley, D. K. Pollock, and A. Rosenfeld, Eds.). Washington DC: Thompson, 1968, pp. 471–502.

[44] J. G. Fraser, "Further comments on a line thinning scheme," *Comput. J.*, vol. 13, pp. 221–222, 1970.

[45] G. Gallus and P. W. Neurath, "Improved computer chromosome analysis incorporating preprocessing and boundary analysis," *Phys. Med. Biol.*, vol. 15, no. 3, pp. 435–445, 1970.

[46] R. C. Gonzalez and P. Wintz, "The skeleton of a region," in *Digital Image Processing*. Reading, MA: Addison-Wesley, 1987, pp. 398–402.

[47] V. K. Govindan and A. P. Shivaprasad, "A pattern adaptive thinning algorithm," *Pattern Recogn.*, vol. 20, no. 6, pp. 623–637, 1987.

[48] F. C. A. Groen and N. J. Foster, "A fast algorithm for cellular logic operations on sequential machines," *Pattern Recogn. Lett.*, vol. 2, no. 5, pp. 333–338, 1984.

[49] Z. Guo and R. W. Hall, "Parallel thinning with two-subiteration algorithms," *Comm. ACM*, vol. 32, no. 3, pp. 359–373, 1989.

[50] R. W. Hall, "Fast parallel thinning algorithms: Parallel speed and connectivity preservation," *Comm. ACM*, vol. 32, no. 1, pp. 124–131, 1989.

[51] C. J. Hilditch, "An application of graph theory in pattern recognition," in *Machine Intell.* (B. Meltzer and D. Michie, Eds.). New York: Amer. Elsevier, 1968, pp. 325–347, vol. 3.

[52] ——, "Linear skeletons from square cupboards," in *Machine Intell.* (B. Meltzer and D. Michie, Eds.). New York: Amer. Elsevier,1969, pp. 403–420, vol. 4.

[53] ——, "Comparison of thinning algorithms on a parallelprocessor," *Image Vision Comput.*, vol. 1, no. 3, pp. 115–132, 1983.

[54] C. M. Holt, A. Stewart, M. Clint, and R. H. Perrott, "An improved parallel thinning algorithm," *Comm. ACM*, vol. 30, no. 2, pp. 156–160, 1987.

[55] C. Holt and A. Stewart, "A parallel thinning algorithm with fine grain subtasking," *Parallel Comput.*, vol. 10, pp. 329–334, 1989.

[56] S. H. Y. Hung and T. Kasvand, "Critical points on a perfectly 8- or 6-connected thin binary line," *Pattern Recogn.*, vol. 16, no. 3, pp. 297–306, 1983.

[57] M. I. Izutsdkiver, "Algorithm for the initial processing of an ensemble of symbols in the recognition process," *Auto. Remote Contr.*, vol. 35, no. 8, pp. 1292–1298, 1974.

[58] N. Izzo and W. Coles, "Blood-cell scanner identifies rare cells," *Electron.*, vol. 35, pp. 52–55, Apr. 1962.

[59] R. N. Jones and M. C. Fairhurst, "Skeletonisation of binary patterns: A heuristic approach," *Electron. Lett.*, vol. 14, no. 9, pp. 265–266, 1978.

[60] K. Kedem and D. Keret, "A fast algorithm for skeletonizing lines by midline technique," in *Proc. Int. Comput. Sci. Conf.*, (Hong Kong), 1988, pp. 731–735.

[61] R. A. Kirsch, L. Cahn, C. Ray, and G. J. Urban, "Experiments inprocessing pictorial information with a digital computer," in *Proc. East. Joint Comput. Conf.* (New York), 1957, pp. 221–229.

[62] T. Y. Kong and A. Rosenfeld, "Digital topology: Introduction and survey," *Comput. Vision Graphics Image Processing*, vol. 48, pp. 357–393, 1989.

[63] J. T. Kuehn, J. A. Fessler, and H. J. Siegel, "Parallel image thinning and vectorization on PASM," in *Proc. Int. Conf. Comput. Vision Patt. Recogn.*, 1985, pp. 368–374.

[64] P. C. K. Kwok, "A thinning algorithm by contour generation," *Comm. ACM*, vol. 31, no. 11, pp. 1314–1324, 1988.

[65] ——, "Customising thinning algorithms," in *Proc. IEEE Int. Conf. Image Processing Applications*, 1989, pp. 633–637.

[66] ——, "Thinning in a distributed environment," in *Proc. 10th Int. Conf. Patt. Recogn.* (Atlantic City, NJ), 1990, pp. 694–699.

[67] L. Lam and C. Y. Suen, "Structural classification and relaxation matching of totally unconstrained handwritten Zip-code numbers," *Patt. Recogn.*, vol. 21, no. 1, pp. 19–31, 1988.

[68] C. Lantuejoul, "Skeletonization in quantitative metallography," in *Issues in Digital Image Processing* (R. M. Haralick and J. C. Simon, Eds.). Amsterdam: Sijthoff and Noordoff, 1980, pp. 107–135.

[69] Y. Le Cun *et al.*, "Handwritten digit recognition: Applications of neural network chips and automatic learning," *IEEE Commun. Mag.*, pp. 41–46, Nov. 1989.

[70] P. S. P. Wang, "An improved fast parallel thinning algorithm for digital patterns," in *Proc. Int. Conf. Comput. Vision Patt. Recogn.*(San Francisco), 1985, pp. 364–367.

[71] ——, "A comment on a fast parallel algorithm for thinning digital patterns," *Comm. ACM*, vol. 29, no. 3, pp. 239–242, 1986.

[72] M. P. Martinez-Perez, J. Jimenez, and J. L. Navalon, "A thinning algorithm based on contours," *Comput. Vision Graphics Image Processing*, vol. 38, pp. 186–201, 1987.

[73] B. H. McCormick, "The Illinois pattern recognition computer—Illiac III," *IEEE Trans. Electron. Comput.*, vol. EC-12, no. 6, pp. 791–813, 1963.

[74] B. Moayer and K. S. Fu, "A syntactic approach to fingerprint pattern recognition," *Pattern Recogn.*, vol. 7, pp. 1–23, 1975.

[75] J. L. Mundy and R. E. Joynson, "Automatic visual inspection using syntactic analysis," in *Proc. Int. Conf. Patt. Recogn. Image Processing*, 1977, pp. 144–147.

[76] I. S. N. Murthy and K. J. Udupa, "A search algorithm for skeletonization of thick patterns," *Comput. Graphics Image Processing*, vol. 3, pp. 247–259, 1974.

[77] N. J. Naccache and R. Shinghal, "STPA: A proposed algorithm for thinning binary patterns," *IEEE Trans. Syst. Man Cybern.*, vol. SMC-14, no. 3, pp. 409–418, 1984.

[78] ——, "An investigation into the skeletonization approach of Hilditch," *Pattern Recogn.*, vol. 17, no. 3, pp. 279–284, 1984.

[79] ——, "In response to 'A comment on an investigation into the skeletonization approach to Hilditch,'" *Patt. Recogn.*, vol. 19, no. 2, p. 111, 1986.

[80] A. Nakayama, F. Kimura, Y. Yoshida, and T. Fukumura, "An efficient thinning algorithm for large scale images based upon pipeline structure," in *Proc. 7th Int. Conf. Patt. Recogn.* (Montreal), 1984, pp. 1184–1187.

[81] T. V. Nguyen and J. Sklansky, "A fast skeleton-finder for coronary arteries," in *Proc. 8th Int. Conf. Patt. Recogn.* (Paris, France), 1986, pp. 481–483.

[82] H. Ogawa and K. Taniguchi, "Thinning and stroke segmentation for handwritten Chinese character recognition," *Patt Recogn.*, vol. 15, no. 4, pp. 299–308, 1982.

[83] J. F. O'Callaghan and J. Loveday, "Quantitative measurement of soil cracking patterns," *Patt. Recogn.*, vol. 5, pp. 83–98, 1973.

[84] L. O'Gorman, "$k \times k$ thinning," *Comput. Vision Graphics Image Processing*, vol. 51, pp. 195–215, 1990.

[85] T. Pavlidis, "A thinning algorithm for discrete binary images," *Comput. Graphics Image Processing*, vol. 13, pp. 142–157, 1980.

[86] ——, "A flexible parallel thinning algorithm," in *Proc. Int. Conf. Patt. Recog. Image Processing* (Dallas, TX), 1981, pp. 162–167.

[87] ——, *Algorithms for Graphics and Image Processing*. Rockville, MD: Comput. Sci., 1982, pp. 195–214.

[88] ——, "An asynchronous thinning algorithm," *Comput. Graphics Image Processing*, vol. 20, pp. 133–157, 1982.

[89] ——, "A vectorizer and feature extractor for document recognition," *Comput. Vision Graphics Image Processing*, vol. 35, pp. 111–127, 1986.

[90] E. Persoon and K. S. Fu, "Shape discrimination using Fourier descriptors," *IEEE Trans. Syst., Man Cybern.*, vol. SMC-7, no. 3, pp. 170–179, 1977.

[91] J. Piper, "Efficient implementation of skeletonisation using interval coding," *Patt. Recogn. Lett.*, vol. 3, no. 6, pp. 389–397, 1985.

[92] R. Plamondon and C. Y. Suen, "On the definition of reference skeletons for comparing thinning algorithms," in *Proc. Vision Interface 1988* (Edmonton, Canada), 1988, pp. 70–75.

[93] ——, "Thinning of digitized characters from subjective experiments: A proposal for a systematic evaluation protocol of algorithms," in *Computer Vision and Shape Recognition* (A. Krzyzak, T. Kasvand, and C. Y. Suen, Eds.). Singapore: World Scientific, 1989, pp. 261–272.

[94] K. Preston, "The CELLSCAN system—A leucocyte pattern analyzer," in *Proc. West. Joint Comput. Conf.* (Los Angeles, CA), 1961, pp. 173–183.

[95] K. Preston, M. J. B. Duff, S. Levialdi, P. E. Norgren, and J. -I. Toriwaki, "Basics of cellular logic with some applications in medical image processing," *Proc. IEEE*, vol. 67, no. 5, pp. 826–857, 1979.

[96] M. C. Rahier and P. G. A. Jespers, "Dedicated LSI for a microprocessor controlled hand carried OCR system," *IEEE J. Solid-State Circuits*, vol. SC-15, no. 1, pp. 14–24, 1980.

[97] S. Riazanoff, B. Cervelle, and J. Chorowicz, "Parametrisable skeletonization of binary and multi-level images," *Patt. Recogn. Lett.*, vol. 11, no. 1, pp. 25–33, 1990.

[98] A. Rosenfeld and J. L. Pfaltz, "Sequential operations in digital picture processing," *J. ACM*, vol. 13, no. 4, pp. 471–494, 1966.

[99] A. Rosenfeld, "Connectivity in digital pictures," *J. ACM*, vol. 17, no. 1, pp. 146–160, 1970.

[100] ——, "A characterization of parallel thinning algorithms," *Inform. Contr.*, vol. 29, no. 3, pp. 286–291, 1975.

[101] A. Rosenfeld and L. S. Davis, "A note on thinning," *IEEE Trans. Syst. Man Cybern.*, vol. 25, pp. 226–228, 1976.

[102] A. Rosenfeld and A. C. Kak, *Digital Picture Processing (2nd ed.)*. New York: Academic, 1982, vol. II, ch. 11.

[103] D. Rutovitz, "Pattern recognition," *J. Roy. Stat. Soc.*, vol. 129, Series A, pp. 504–530, 1966.

[104] P. Saraga and D. J. Woollons, "The design of operators for pattern processing," in *Proc. IEEE NPL Conf. Patt. Recogn.* (Teddington), 1968, pp. 106–116.

[105] B. Shapiro, J. Pisa, and J. Sklansky, "Skeletons from sequential boundary data," in *Proc. Int. Conf. Patt. Recog. Image Processing* (Chicago, IL), 1979, pp. 265–270.

[106] ——, "Skeleton generation from x, y boundary sequences," *Comput. Graphics Image Processing*, vol. 15, pp. 136–153, 1981.

[107] H. Sherman, "A quasitopological method for the recognition of line patterns," in *Proc. Int. Conf. on Inform. Processing* (Paris, France), 1959, pp. 232–238.

[108] R. M. K. Sinha, "Primitive recognition and skeletonization via labeling," in *Proc. Int. Conf. Syst. Man Cybern.* (Halifax, Canada), 1984, pp. 272–279.

[109] R. M. K. Sinha and C. J. Ammann, "Comments on fast thinning algorithm for binary images," *Image Vision Comput.*, vol. 4, no. 1, pp. 57–58, 1986.

[110] R. M. K. Sinha, "A width-independent algorithm for character skeleton estimation," *Comput. Vision Graphics Image Processing*, vol. 40, pp. 388–397, 1987.

[111] R. W. Smith, "Computer processing of line images: A survey," *Patt. Recogn.*, vol. 20, no. 1, pp. 7–15, 1987.

[112] M. Del Sordo and T. Kasvand, "A near-neighbor processor for line thinning," in *Proc. Int. Conf. Acoust. Speech Signal Processing*, 1985, pp. 1523–1525.

[113] ——, "Neighborhood look-up tables for skeletonization," in *Proc. 4th Scand. Conf. Image Anal.* (Trondheim, Norway), 1985, pp. 663–670.

[114] J. H. Sossa, "An improved parallel algorithm for thinning digital patterns," *Patt. Recogn. Lett.*, vol. 10, no. 2, pp. 77–80, 1989.

[115] R. Stefanelli and A. Rosenfeld, "Some parallel thinning algorithms for digital pictures," *J. ACM*, vol. 18, no. 2, pp. 255–264, 1971.

[116] R. Stefanelli, "A comment on an investigation into the skeletonization approach of Hilditch," *Patt. Recogn.*, vol. 19, no. 1, pp. 13–14, 1986.

[117] C. Y. Suen, M. Berthold, and S. Mori, "Automatic recognition of handprinted characters," *Proc. IEEE*, vol. 68, no. 4, pp. 469–487, 1980.

[118] S. Suzuki and K. Abe, "Binary picture thinning by an iterative parallel two-subcycle operation," *Patt. Recogn.*, vol. 10, no. 3, pp. 297–307, 1987.

[119] T. Suzuki and S. Mori, "A thinning method based on cell structure," in *Proc. Int. Workshop Frontiers Handwriting Recogn.* (Montreal, Canada), 1990, pp. 39–52.

[120] H. Tamura, "A comparison of line thinning algorithms from a digital geometry viewpoint," in *Proc. 4th Int. Conf. Patt. Recogn.* (Kyoto, Japan), 1978, pp. 715–719.

[121] C. C. Tappert, C. Y. Suen, and T. Wakahara, "The state of the art in on-line handwriting recognition," *IEEE Trans. Patt. Anal. Machine Intell.*, vol. 12, no. 8, pp. 787–808, 1990.

[122] J. -I. Toriwaki and S. Yokoi, "Distance transformation and skeletons of digitized pictures with applications," in *Progress in Pattern Recognition* (L. N. Kanal and A. Rosenfeld, Eds.). New York: North-Holland, 1981, pp. 189–264.

[123] E. E. Triendl, "Skeletonization of noisy hand-drawn symbols using parallel operations," *Patt. Recogn.*, vol. 2, pp. 215–226, 1970.

[124] Y. F. Tsao and K. S. Fu, "Parallel thinning operations for digital binary images," in *Proc. Int. Conf. Patt. Recog. Image Processing* (Dallas, TX), 1981, pp. 150–155.

[125] ——, "A general scheme for constructing skeleton models," *Inform. Sci.*, vol. 27, no. 1, pp. 53–87, 1982.

[126] L. J. Vliet and B. J. H. Verwer, "A contour processing method for fast neighborhood operations," *Patt. Recogn. Lett.*, vol. 7, no. 1, pp. 27–36, 1988.

[127] A. M. Vossepoel, J. P. Buys, and G. Koelewijn, "Skeletons from chain-coded contours," in *Proc. 10th Int. Conf. Patt. Recogn.* (Atlantic City), 1990, pp. 70–73.

[128] P. S. P. Wang, L.-W. Hui, and T. Fleming, "Further improved fast parallel thinning algorithm for digital patterns," in *Computer Vision, Image Processing and Communications—Systems and Applications* (P. S. P. Wang, Ed.). Singapore: World Scientific, 1986, pp. 37–40.

[129] P. S. P. Wang and Y. Y. Zhang, "A fast serial and parallel thinning algorithm," in *Proc. Eighth Euro. Meeting Cybern. Syst. Res.* (Vienna, Austria), 1986, pp. 909–915.

[130] , "A fast and flexible thinning algorithm," *IEEE Trans. Comput.*, vol. 38, no. 5, pp. 741–745, 1989.

[131] Y. Xia, "A new thinning algorithm for binary images," in *Proc. 8th Int. Conf. Patt. Recogn.* (Paris, France), 1986, pp. 995–997.

[132] W. Xu and C. Wang, "CGT: A fast thinning algorithm implemented on a sequential computer," *IEEE Trans. Syst. Man Cybern.*, vol. SMC-17, no. 5, pp. 847–851, 1987.

[133] Q. -Z. Ye and P. E. Danielsson, "Inspection of printed circuit boards by connectivity preserving shrinking," *IEEE Trans. Pattern Anal. Mach. Intell.*, vol. 10, no. 5, pp. 737–742, 1988.

[134] S. Yokoi, J. -I. Toriwaki, and T. Fukumura, "Topological properties in digitized binary pictures," *Syst. Comput. Contr.*, vol. 4, no. 6, pp. 32–39, 1973.

[135] ——, "An analysis of topological properties of digitized binary pictures using local features," *Comput. Graphics Image Processing*, vol. 4, pp. 63–73, 1975.

[136] T. Y. Zhang and C. Y. Suen, "A fast parallel algorithm for thinning digital patterns," *Comm. ACM*, vol. 27, no. 3, pp. 236–239, 1984.

[137] Y. Y. Zhang and P. S. P. Wang, "A modified parallel thinning algorithm," in *Proc. 9th Int. Conf. Patt. Recogn.* (Rome, Italy), 1988, pp. 1023–1025.

[138] ——, "A maximum algorithm for thinning digital patterns," in *Proc. 9th Int. Conf. Patt. Recogn.* (Rome, Italy), 1988, pp. 942–944.

Louisa Lam received the B. A. degree from Wellesley College, Wellesley, MA, where she was elected to Phi Beta Kappa. She received the Ph.D. degree in mathematics from the University of Toronto, Toronto, Canada.

She is currently teaching mathematics at Vanier College and conducting research at Concordia University, Montreal, Canada. Her research interests include character recognition and skeletonization algorithms.

Seong-Whan Lee (M'91) was born in Beolgyo, Korea, in 1962. He received the B. S. degree in computer science and statistics from Seoul National University, Seoul, Korea, in 1984 and the M. S. and Ph.D. degrees in computer science from the Korea Advanced Institute of Science and Technology in 1986 and 1989, respectively.

In 1987, he worked as a visiting researcher at the Pattern Recognition Division, Delft University of Technology, Delft, the Netherlands. He was a visiting scientist at the Centre for Pattern Recognition and Machine Intelligence, Concordia University, Montreal, Canada, during the winter of 1989 and the summer of 1990. Since 1989, he has been an Assistant Professor in the Department of Computer Science, Chungbuk National University, Chungbuk, Korea. His research interests include pattern recognition, computer graphics, and intelligent man-machine interfaces.

Dr. Lee was awarded a best paper prize from the Korea Information Science Society in 1986. He is a member of the governing board of the Special Interest Group on Artificial Intelligence of Korea. He is also a member of the Korea Information Science Society, the Pattern Recognition Society, the Association for Computing Machinery, and the IEEE Computer Society.

Ching Y. Suen (F'86) received the M.Sc. (Eng.) degree from the University of Hong Kong and the Ph.D. degree from the University of British Columbia, Vancouver, Canada.

In 1972, he joined the Department of Computer Science at Concordia University, Montreal, Canada, where he became Professor in 1979 and served as Chairman from 1980 to 1984. Presently, he is the Director of the new Center for Pattern Recognition and Machine Intelligence (CENPARMI) at Concordia University. During the past 15 years, he has been appointed to visiting positions at several institutions in different countries. He is the author/editor of several books including *Computer Vision and Shape Recognition, Frontiers in Handwriting Recognition*, and *Computational Analysis of Mandarin and Chinese.* His latest book is entitled *Operational Expert System Applications in Canada*, which is published by Pergamon Press. He is the author of many papers, and his current interests include pattern recognition and machine intelligence, expert systems, optical character recognition and document processing, and computational linguistics.

An active member of several professional societies, Dr. Suen is an Associate Editor of several journals related to his areas of interest. He is the Past President of the Canadian Image Processing and Pattern Recognition Society, Governor of the International Association for Pattern Recognition, and President of the Chinese Language Computer Society.

High Quality Vectorization Based on a Generic Object Model

Osamu Hori and Akio Okazaki

Toshiba Research and Development Center,
1, Komukai Toshiba-Cho, Saiwai-Ku, Kawasaki 210, Japan

We propose a new method for high quality vectorization based on a generic object model. In conventional methods, the quality of vectorization has been regarded as the same thing as digitizing accuracy, which is evaluated in terms of the error between the original line-drawing and the resultant vector sequence. However, this accuracy does not always correspond to perceptual quality. Based on the discussion about possible distortions in a vectorization process which strongly affects perceptual quality, we introduce a generic model of an object. The generic properties are described in the object model, for example, "The object is a polygon whose corners all have a right angle" or "The object is composed of several pairs of almost parallel lines". The method consists of three processes: pre-vectorization, object recognition and shape modification based on an object model. The approximation line-figure which is first defined in the pre-vectorization process, is modified so as to meet the properties described in the object model. A cost function is derived from the model so that shape modification can be formalized as an optimization problem. A relaxation method is employed to save on computation time. Object recognition is needed for object model selection. The method was applied to geographic maps for urban planning to extract building polygons. Subjective evaluation on the extracted building shapes showed that 93.6% of the buildings were vectorized with high quality and that the number of buildings with insufficient quality was reduced to one twelfth.

1 Introduction

Raster-to-vector conversion is a key technique for automatic graphic information entry into CAD systems, data-base systems, data communication systems, etc. The purpose of this conversion is to obtain a vector representation for the original image without loss in the necessary pictorial information. There are two types in this technique. In engineering drawing recognition systems [1, 2, 3], raster-to-vector conversion is used as a pre-processing technique for symbol recognition and network analysis. It is important to preserve a topological features and geometric properties for this purpose. In automatic digitizing systems, on the other hand, both perceptual quality and digitizing accuracy are important because the vector data are not only used for various graphic data manipulations, such as the measurement of the area or perimeter, but also displayed for human interaction. In conventional systems [4, 5], perceptual quality has been regarded as the same thing as digitizing accuracy. Many reports [8, 9, 10, 11] have discussed digitizing accuracy in terms of the error between the original line drawing and the resultant vector sequence. However, this accuracy does not always correspond to perceptual quality, owing to the noises mixed in the image scanning process. For example, parallel lines are not always vectorized as parallel lines by the conventional methods. Two sides of a rectangle may not meet with a right angle. These facts lead to the degradation of vectorization quality, because human perception is very sensitive to even a slight distortion. A typical example, in which this problem arises, is discovered in the vectorization for buildings and roads in geographical map images. Certain complementary knowledge is needed in order to obtain an ideal shape for human perception by eliminating such distortions It is easy graphic information entry systems involving symbol recognition to obtain high quality vector representation by replacing detected symbols with a pre-determined template. The template is the most simple model for representing such knowledge. However, this approach cannot be applied to geographic map constituents like buildings and roads because they vary greatly in shape. Pavlidis has proposed an automatic method for beatifying drawings and illustrations from the same point of view [12]. He showed basic ideas for beatification and several examples to which beatification operations are effective. His method beatifies drawings and illustrations based on certain rules. However, it is not a trivial problem to decide automatically which rules should be applied to each sub-pattern in an line drawing. He did not mention the selection problem for beautification rules. Furthermore, it is considered that more efficient beautification algorithm is needed in the practical situation, in which the size of a line drawing is quite large. He did not show a practical beatification algorithm. In this paper, a generic object model for high quality vectorization is introduced instead of beautification rules. The pattern recognition techniques are used to extract target line patterns which should be beautified using the model. The model is efficiently utilized for transforming a shape with distortions into an ideal shape for human perception. The basic idea for high quality vectorization is described in Section 2. In Section 3, the method is applied to urban planning maps to vectorize building polygons. Finally, the effectiveness of the method is shown through subjective evaluations on the experimental results in Section 4.

2 The Basic Idea

This section discusses several kinds of distortions occurring in raster-to-vector conversion, which affects the quality of the obtained vector sequence. Then, an approach to obtaining a high quality vectorized shape is described.

2.1 What Affects Vectorization Quality?

It is often found that vectorized shape is "dirty" or "undesirable" for human perception, even if the accuracy is sufficient. The factors which lower the shape quality can be considered to be as follows.

(i) Local shape distortion, which has occurred in the processing stage of line structure extraction from an input image, by a thinning operation (or other skeleton extraction techniques) and line tracing. This kind of distortion often appears in the neighborhood of cross points. Typical distortion examples are shown in Figure 1.

(ii) Location error in the extraction of feature points, such as bending points, inflection points or maximum curvature, and inadequate selection of knots in line approximation. In most case, this causes lack of smoothness.

(iii) Delicate mismatch among the directions or the lengths of the vectors forming the total shape. For example, most building shapes, shown on an urban planning map, have several parallel vectors. This is a geometrical property expected by humans. However, the vectorized shapes do not always rigidly maintain the above property due to delicate error in knot location. This error cannot be overcome by local processing techniques alone. Human perception senses a kind of "dirtiness" in the entire vectorized shape.

Of the three types, (i) and (ii) are local distortions, while (iii) is a global distortion.

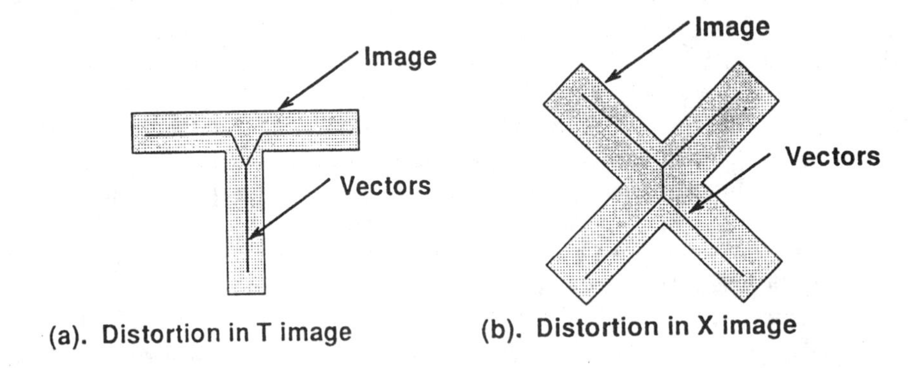

Fig. 1. Local distortion categories.

2.2 Approach

Considering the above-mentioned three distortion levels, a powerful recovering method for high quality vectorization has been designed. A functional block-diagram is shown in Figure 2. Pre-vectorization, which accomplishes line structure extraction, feature point detection and line approximation has been proposed [13] as an efficient and effective method for this processing. This method has been adopted to obtain the pre-vectorized shape without distortion caused by simple noise, such as short line segments.

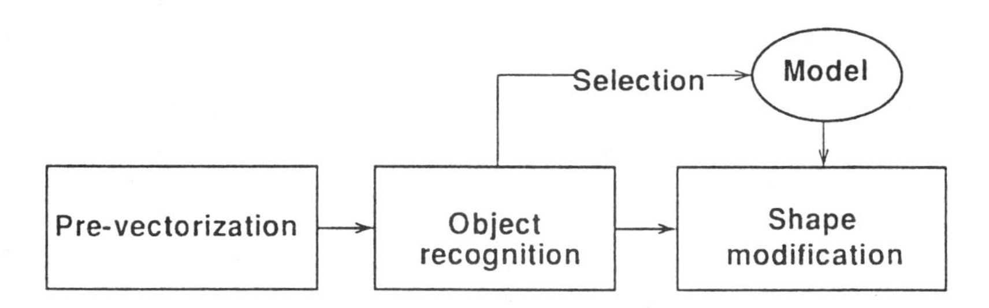

Fig. 2. Three sub-processes for high-quality vectorization.

Many techniques for recovering from local distortions (i) [8, 9, 10, 11] and (ii) [7] in the previous section have been proposed. However, they do not take distortion (iii), into consideration. It is observed that conventional methods are not applicable for line patterns such as buildings or roads in a map as shown in Figure 3(a). Mismatch in the sense of (iii) often occurs, because humans have a relatively severe shape model for these patterns in mind as shown in Figure 3(b). However, such patterns can not be strictly defined as templates because of variable shapes, although their shapes do obey certain rules. A generic model is introduced for the description of such rules. For example, generic properties such as: "The object is a polygon whose corners all have 90 degree angles," or "The object is composed of pairs of parallel lines" are described in the model. The pre-vectorized shape is improved in the form of a desirable shape which matches the generic model. Object recognition is accomplished in order to decide which object model is to be used for shape modification.

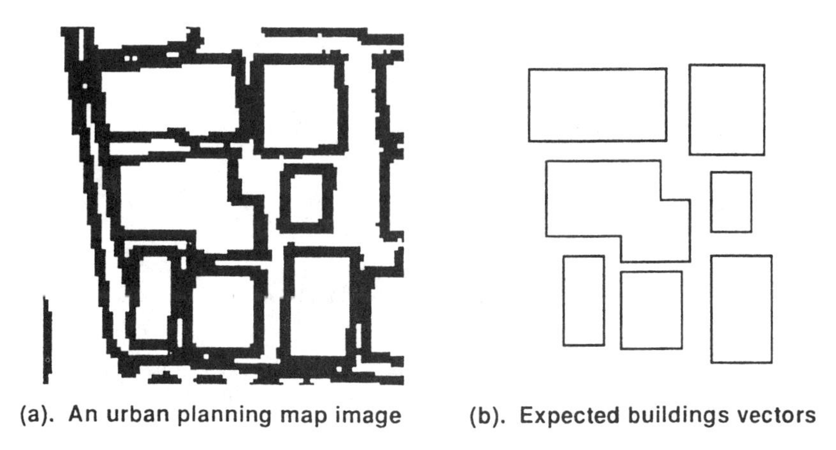

Fig. 3. Raster image and expected building vector.

Shape modification can be formalized as an optimization problem by introducing a generic object model, that is: (i) a cost function which outputs the similarity between the current vectorized shape and the generic model is defined; and (ii) each knot of the vectorized shape is shifted in a certain manner until the most optimal shape, which means minimum cost, is obtained. The optimization flow is shown in Figure 4. The initial input is a pre-vectorized shape. Knot shifting means a slight adjustment for shape quality improvement. The limit for the shifting distance is given *a priori* to preserve vectorization accuracy. The number of combinations for the knot shifting becomes larger as the shifting limit becomes larger. In that case, a strategy which reduces the number of combinations is needed to shorten processing time. An example of the strategy is shown in the next section.

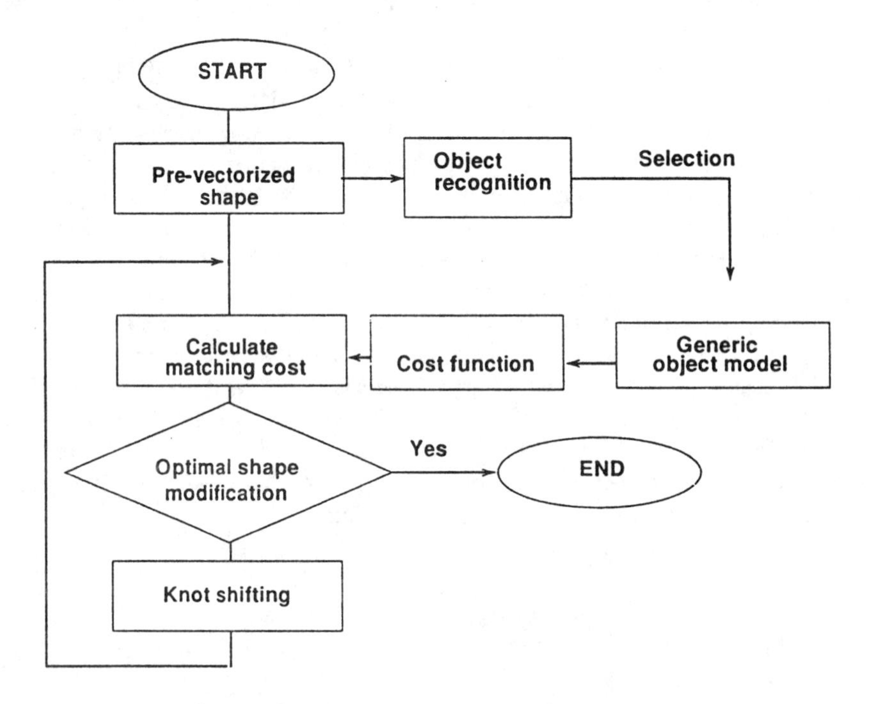

Fig. 4. Shape modification flowchart.

3 High Quality Vectorization for Buildings on Maps

In this section, the proposed method is applied to geographical maps used for urban planning. An example is shown in Figure 5. The vectorization process consists of three stages: pre-vectorization, object recognition and shape modification.

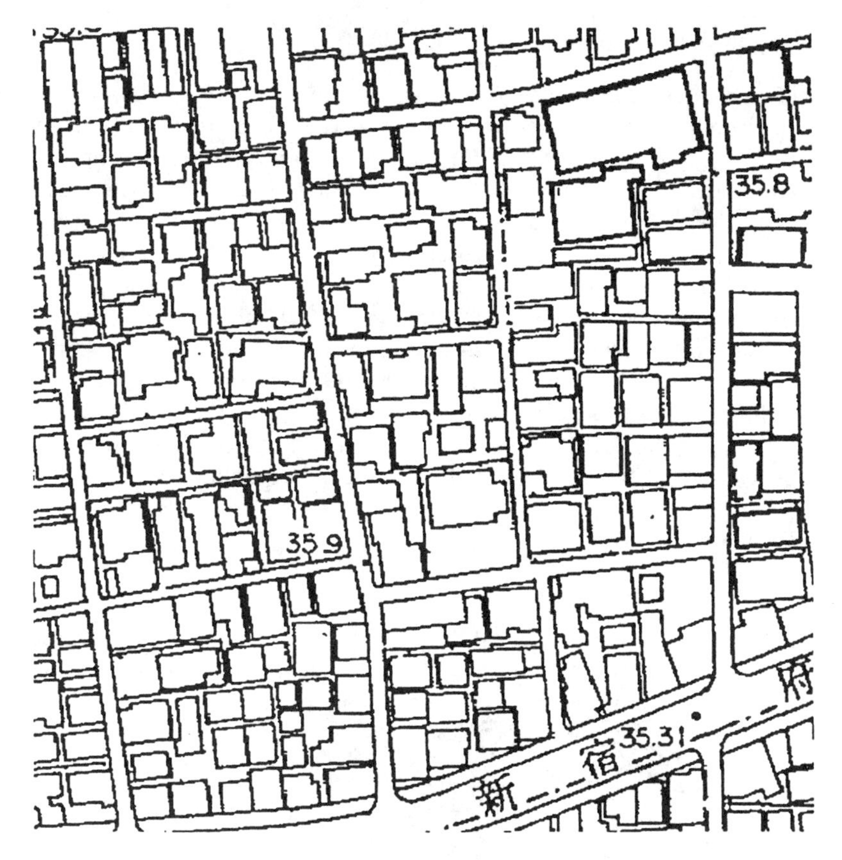

Fig. 5. Original map image for Shinjuku ward in Tokyo.

3.1 Pre-vectorization

In the pre-vectorization stage, candidate shapes for buildings are extracted and vectorized. Buildings are represented by their outline in maps. They are basically an isolated closed polygon. Candidate extraction can be accomplished by finding out closed loops. However, because building shapes touch neighboring ones and road boundaries in way places, it is not appropriate to use external contours for the closed loop extraction. Moreover, the thinning operation causes large distortions in shape when such touches occur. Therefore, internal contours are used to represent building candidates. Every internal contour is then eroded by a half of the line width, which is a predetermined value for a specific map. The resultant internal contour can be regarded as the skeleton of the original line. A vector sequence for the contour is finally generated using a line approximation algorithm.

3.2 Object Recognition

Candidate shapes for buildings include true buildings, loops surrounded by buildings, small loops in characters, and so on. In the object recognition stage, true building polygons are selected by a pattern recognition method described below. Two features, simpleness and size of a shape, are used for the recognition. The simpleness measure of a shape is defined as $area/(perimeter^2)$ [14]. The thresholds for the features are statistically determined. Figure 6 shows a two dimensional distribution of feature values for buildings and other loops. The number of samples was about 3,500, including 2,500 building polygons and 1,000 other loops. The symbol □ indicates buildings and × indicates other loops. Three lines as shown in Figure 6 which separate the distribution into two categories are found. Building polygons are effectively distinguished by the three lines. Line *a* and line *b* mean the size and simpleness threshold, respectively. Line *c* indicates that more complex buildings are allowed by the increasing size. This is because a large building is able to have a sufficiently large living space in spite of its complex shape.

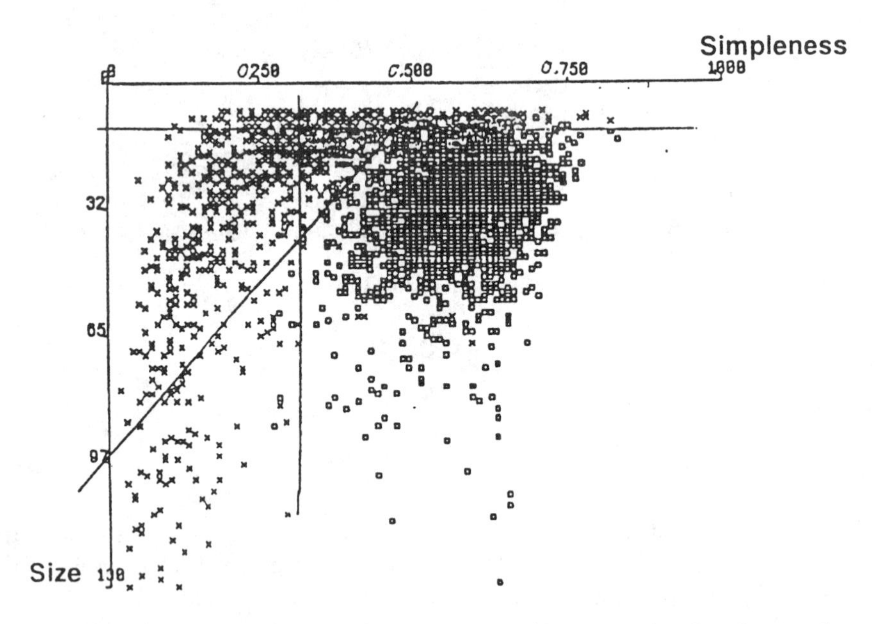

Fig. 6. Graph of relations between size and simpleness on closed polygons (□ are buildings, × are not buildings).

3.3 Shape-modification Based on a Generic Object Model

In this stage, the vector sequence obtained in the pre-vectorization stage is interactively modified so that the cost function derived from the generic model is minimized. The generic model of a building is described by the following two statements: (a) The building shape is a simple polygon. (b) The relation between the sides of a polygon is parallel or perpendicular, when and if possible. The cost function for this generic model is defined as follows. Figure 7(b) shows the histogram of the vector orientation weighted by the vector length. The horizontal axis indicates the frequency weighted by the vector length. The vertical axis indicates the vector orientation. If a building has a desirable shape, the histogram should have two sharp peaks whose distance along the vertical axis is 90 degrees. By taking the modulus of 90 for the orientation angle, Figure 7(b) is transformed into Figure 7(c) which has a single peak to simplify the evaluation. It is clear that all the vectors are either parallel or perpendicular to each other if the peak is single. Moreover, there are no redundant vectors, so the building is represented by the simplest possible polygon. Consequently, the variance for the histogram stands for the cost function.

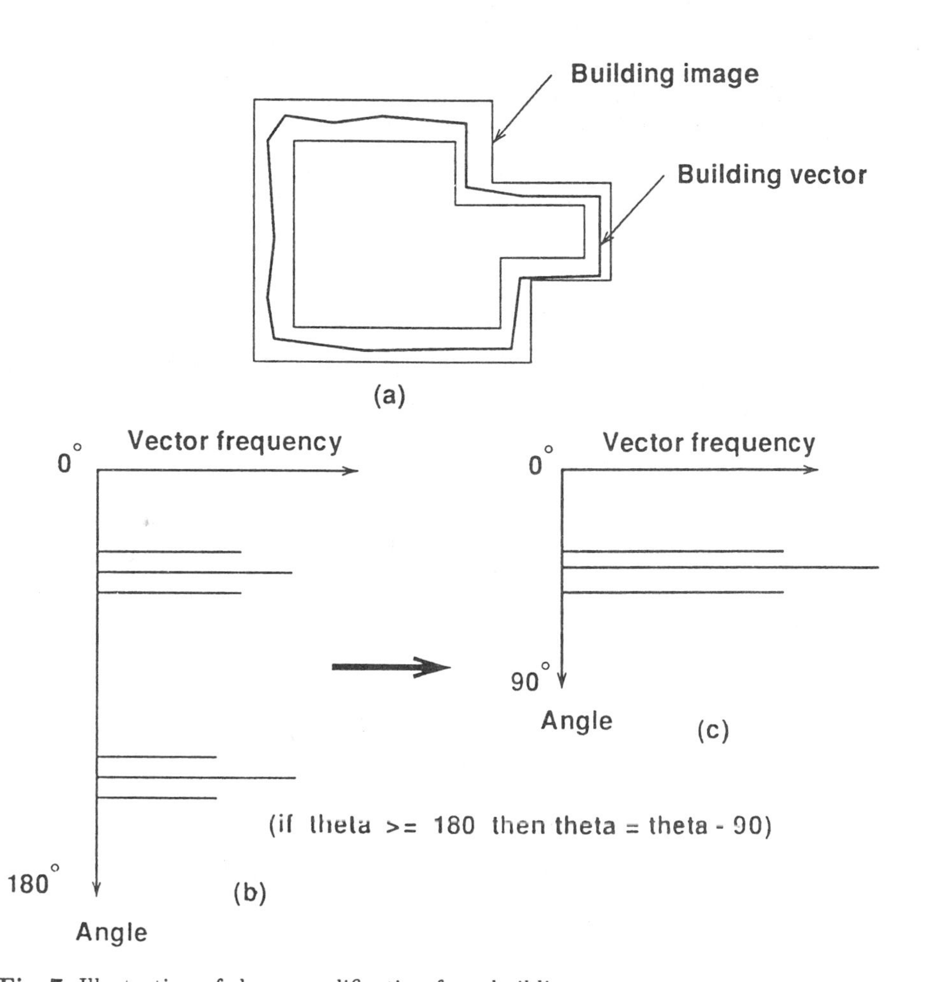

Fig. 7. Illustration of shape-modification for a building.

The next step is to select a combination of knot points for a vector sequence which minimizes the cost function. Knot points can be shifted within a certain range under the restriction that no vectors should stick out from the inside of an image of a building, as shown in Figure 8. If a building has m knot points and the shiftable range is $n \times n$ pixels, then the number of combinations would be enormous: $n^2 m$. Therefore, a relaxation method is employed to save on computation time. For a knot point, the histograms after shifting in $n \times n$ ways are evaluated and the best position for the knot point is determined. This operation is repeated for all the m knot points until a minimum variance is achieved for the histogram, as shown in the flow chart in Figure 9. The iterative k times operation reduce the calculation time to $n^2 m k$ and the histogram gradually forms a single sharp peak. The k times depends on drawings and determined empirically. Figure 10 shows an example of shape modification results. The shape of an extracted building, as shown in Figure 10(c), is modified into a good-shape, as shown in Figure 10(d). The round corner is sharpened and the relation between individual lines becomes vertical or parallel. A redundant knot point is also successfully eliminated. Histograms, as shown in Figure 10(b), indicate that the directions of lines are concentrated on a local peak after shape modification. Figure 11 shows several buildings before and after shape modification. Figure 12 shows the result of buildings extracted and modified from the image in Figure 5.

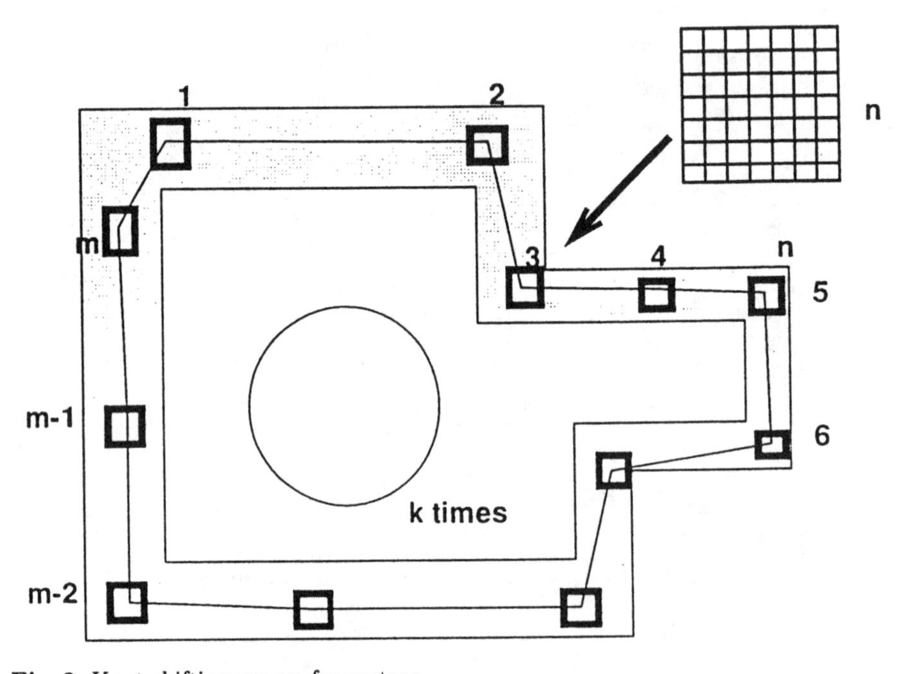

Fig. 8. Knot shifting ranges for vectors.

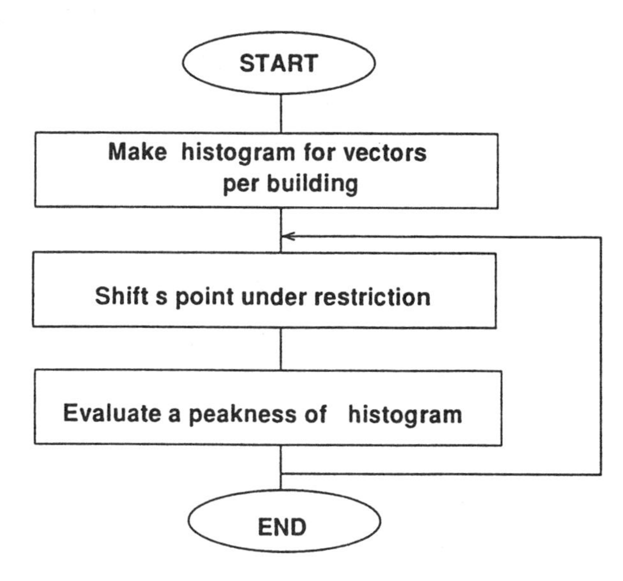

Fig. 9. Building shape modification process flowchart.

4 Experimental Results

About 65,000 buildings were extracted from twelve sheets of urban planning maps, 1000 × 600 mm in size, having a 1:2500 scale, with a resolution of 8 pixels per mm, in Shinjuku Ward, Tokyo. The recognition accuracy and the quality of building shapes were evaluated in sixteen selected local areas(64 × 64 mm size) on the above maps.

4.1 Recognition Accuracy

The number of isolated closed-polygons extracted in the first pre-vectorization process was 3,448, while the actual number of buildings was 2597. Some polygons were broken by noises and written characters on the map. Table 1 shows the number of failures in extracting actual building-polygons.

Table 1. Number of failures extracting buildings in first pre-vectorization (total: 2597).

Broken loops cause	Number	(Ratio)
Interrupted by noises	29	(1.1%)
Interrruped by characters	53	(2.0%)
Total	82	(3.2%)

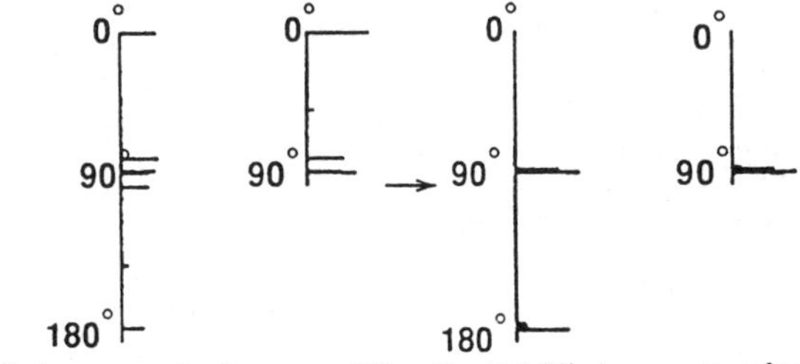

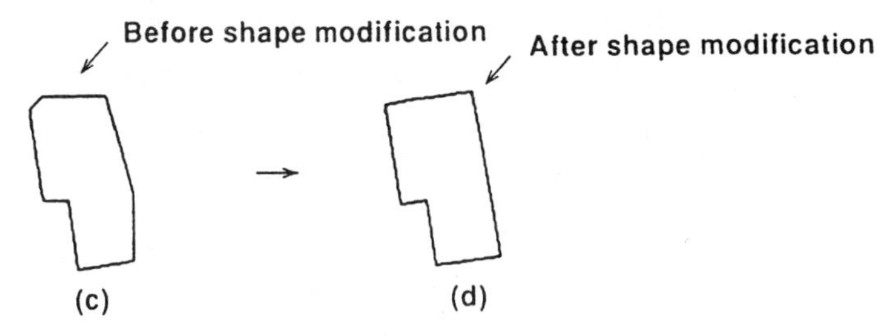

Fig. 10. Shape modification result.

After object recognition by the machine, a human checked the actual buildings. Table 2 shows the confusion matrix for object recognition. The correct building identification rate is the ratio of the number of extracted buildings to the number of actual buildings. The correct rate for other polygons is the ratio of the number of other polygons to that of incorrectly extracted other polygons. The experimental results indicate the correct ratio of buildings and other polygons were 98.2% and 93.9%, respectively. The system fails to recognize actual buildings as buildings in case that corners of buildings are not detected in pre-vectorization stage because they are not so sharp.

Table 2. Confusion matrix for building object recognition.

Human\machine	Buildings	Others	Total	Correct ratio
Buildings	2469	46	2515	98.2%
Others	58	875	933	93.9%
Total	2527	921	3448	97.0%

Pre-Vectorization should be improved in corner extraction robustness. On the other hand, most errors in elimination of other polygons are caused by misrecognition of holes in kanji characters and meaningless loops, as shown in Figure 5. In order to remove these errors, the system should deal with high level information, that is, contextual relations among elements in a map. That will be our

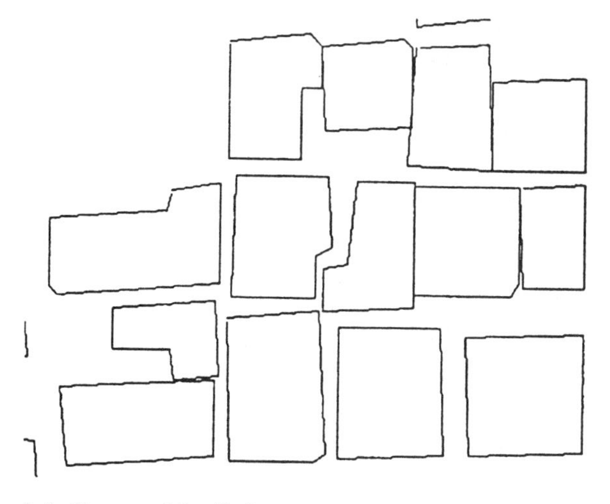

(a) Several buildings before shape modification

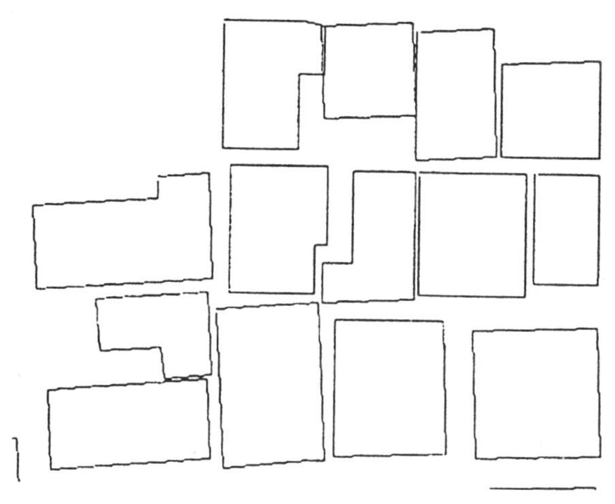

(b) Several buildings after shape modification

Fig. 11. Shape modification result.

future works. Consequently, 128 (82+46) building images failed to be extracted and 95.1% ((2597-128)/2597) of actual buildings images were extracted and recognized from these maps. 95.1% is a high rate. If all extracted buildings were acceptable, hand digitizing operation would have been reduced to one twentieth.

4.2 Shape Quality Evaluation

Shape quality evaluation was carried out by several human inspectors. A standard with four levels of quality, as shown in Table 3, was decided upon from

Fig. 12. Buildings extracted from a map of Shinjuku ward, Tokyo.

three points: the corner feature, shape quality and line location accuracy. A higher level indicated better shape quality. Building images in level 1 were not acceptable for the human perceptual criterion. Level 2 or 3 was equal to the quality for building shapes extracted with a hand digitizer. Buildings in Level 4 had perfect shapes. It is difficult to achieve Level 4 using a hand digitizer.

Tables 4 and 5 show the evaluation results for shapes before and after shape modification for the selected 885 buildings. Only 80% of the buildings were acceptable before shape modification. On the other hand, about 98% of the buildings were acceptable after shape modification while still retaining the location accuracy of the vectors. Moreover, the shapes of buildings in each level were improved in quality. The overall performance is shown in Table 6. Consequently, a high rate of 93.6% (95.1% x 98.4%) acceptable building shapes was successfully extracted by the proposed method. Hand digitizing operations were reduced to one twelfth.

Table 3. Building shape standards.

Worse ↓ Better; Not acceptable / Acceptable

Level\item	Corner feature	Shape	Location accuracy
1	With ambiguous corners	Poor	Outside line width (but not always)
2	With necessary corners	Normal	within line width
3	With sharp corners	Good	within line width
4	With minimum corners	Very good	within line width

Table 4. Evaluation of shapes of buildings before shape modification.

Level	1	2	3	4	Total
Number	178	184	196	327	885
Ratio	20.1%	20.8%	22.1%	36.9%	100%

Levels 2–4: 79.9%

Table 5. Evaluation of shapes of buildings after shape modification.

Level	1	2	3	4	Total
Number	14	51	175	642	885
Ratio	1.6%	5.8%	19.8%	72.5%	100%

Levels 2–4: 98.4%

Table 6. Ratio of available buildings.

Process	Ratio
Object recognition	95.1%
Shape modification	98.4%
Total	93.6%

5 Conclusion

A new method for high quality vectorization has been proposed, which significantly improves the shape of buildings extracted from geographic maps. A generic model for building shapes was defined and effectively translated into a cost function which drove the extraction. The method is applicable to vectorization of other objects by changing the object model and defining a corresponding cost function.

Acknowledgment

The authors wish to thank Haruo Asada for his constructive discussions and useful suggestions.

References

[1] K. Ramchandran, "Coding Method for Vector Representation of Engineering Drawings," *Proc. IEEE* 68, pp. 813–817, 1980.

[2] M. Ejiri *et al.*, "Automatic Recognition of Design Drawings and Maps," *Proc. 7th ICPR*, Montreal, pp. 1296–1305, 1984.

[3] A. Okazaki *et al.*, "An Automatic Circuit Diagram Reader with Loop-structure-based Symbol Recognition," In *IEEE Trans. on PAMI*, Vol.10 No.3, pp. 331–340, 1988.

[4] A. Maeda and J. Shibayama, "Application of Automatic Drawing Reader for the Utility Management System," *IEEE Computer Society Workshop on CAPAIDM*, pp. 139–145, 1985.

[5] S. Suzuki and T. Yamada, "MARIS, Map recognizing input system," *Proc. Int'l Workshop on Industrial Applications of Machine Vision and Machine Intelligence*, Feb. 1987, p. 214.

[6] M. Sakauchi and Y. Ohsawa, "The AI-MUDAMS, The drawing processor based on the multidimensional pattern data structure," *Proc. IEEE Computer Society Workshop on CAPAIDM*, 1985, p. 146.

[7] S. Suzuki, "Graph-based Vectorization Method for Mine Patterns," *Proc. IEEE Computer Vision and Pattern Recognition*, pp. 616–621, 1988.

[8] C. Williams, "An Efficient Algorithm for the Piecewise Linear Approximation of Planar Curves," *Computer Graphics and Image Processing* 8, pp. 286–293, 1978.

[9] C. Williams, "Bounded straight-line approximation of digitized planar curves," *Computer Graphics and Image Processing* 16, pp. 370–381, 1981.

[10] J. Sklansky and V. Gonzales, "Fast polygonal approximation of digitized curves," *Pattern Recognition* 12, pp. 327–331, 1980.

[11] Y. Kurozumi and W. Davis, "Polygonal approximation by the minimax method," *Computer Graphics and Image Processing* 19, pp. 248–264, 1981.

[12] T. Pavlidis and C. J. Van Wyk, "An Automatic Beautifier for Drawings and Illustrations," *Proc. ACM SIGGRAPH '85*, pp. 225–234, 1985.

[13] S. Shimotsuji *et al.*, "A High Speed Raster-to-vector Conversion using Special Hardware for Contour Tracing," in *Proc. IAPR Workshop on COMPUTER VISION* (Special Hardware and Industrial applications), pp. 18–23, 1988.

[14] A. Rosenfeld and A. C. Kak, *Digital Picture Processing*, 2nd Edition, pp.265–266, Academic Press, NY (1982).

Chapter 3
Feature-Level Analysis

3.1: Introduction

After pixel-level processing has prepared the document image, intermediate features are found from the image to aid in the final step of recognition. The following example demonstrates the difference between pixel data and features. A line diagram may be reduced to thinned lines and then stored as a chain code. Although this representation requires less storage size than the original image, still each code word corresponds to a pixel, and there is no interpretation of pixels other than that they are ON or OFF. At the feature level, these thinned and chain-coded data are analyzed to detect straight lines and curves. Also, this more informative representation is closer to how humans would describe the diagram: as lines and curves rather than as ON and OFF points. In addition to discussing the features describing lines in this chapter, we also cover region features such as size and shape. These features are used as intermediate descriptions that aid in the final recognition of characters for OCR and of lines and symbols for graphics analysis.

3.2: Polygonalization

(*Keywords*: polygonalization, straight-line approximation, piecewise linear approximation)

Polygonal approximation is one common approach to obtaining features from curves. The objective is to approximate a given curve with connected straight lines such that the result is close to the original but the description is more succinct. Polygonal approximation to the coastline of Britain is shown in Figure 1 [Wall 1986]. The user can direct the degree of approximation by specifying some measure of maximum error compared to the original. In general, when the specified maximum error is smaller, a greater number of lines is required for the approximation. The effectiveness of one polygonalization method compared to another can be measured by the number of lines required to produce a comparable approximation and also by the computation time required to obtain the result.

3.2.1: Iterative and sequential methods

The iterative end point-fit algorithm [Ramer 1972] is a popular polygonalization method whose error measure is the distance between the original curve and the polygonal approximation. The method begins by connecting a straight-line segment between end points of the data. The perpendicular distance from the segment to each point on the curve is measured. If any distance is greater than a chosen threshold, the segment is replaced by two segments from the original-segment end points to the curve point where the distance to the segment is greatest. The processing repeats in this way for each portion of the curve between polygonal end points, and the iterations are stopped when all segments are within the error threshold of the curve. See Figure 2 for an illustration of this method.

The iterative end point-fit algorithm is relatively time-consuming because error measures must be calculated for all curve points between new segment points on each iteration. Instead of this "overall" approach, an approach can be used in which segments can be fit starting at an end point, then fit sequentially for points along the line [Sklansky 1980, Tomek 1974, Williams 1978, Williams 1981]. That is, starting at an end point, a segment is drawn to its neighboring point, then to the next, and so on, and the error is measured for each segment. When the error is above a chosen threshold, then the segment is fit from the first point to the point previous to the pixel with the above-threshold error. This process is continued starting at the next point after the fit. The error criterion used for this approach states that the straight-line fit must be constrained to pass within a radius around each data point. See Figure 3 for an illustration of this method.

In another class of polygonalization techniques, area is used as a measure of goodness of fit [Wall 1984] (see Figure 4). Instead of measuring the distance error, a scan-along technique is used, where the area error between the original line and the approximation is accumulated. When this area measurement for a segment

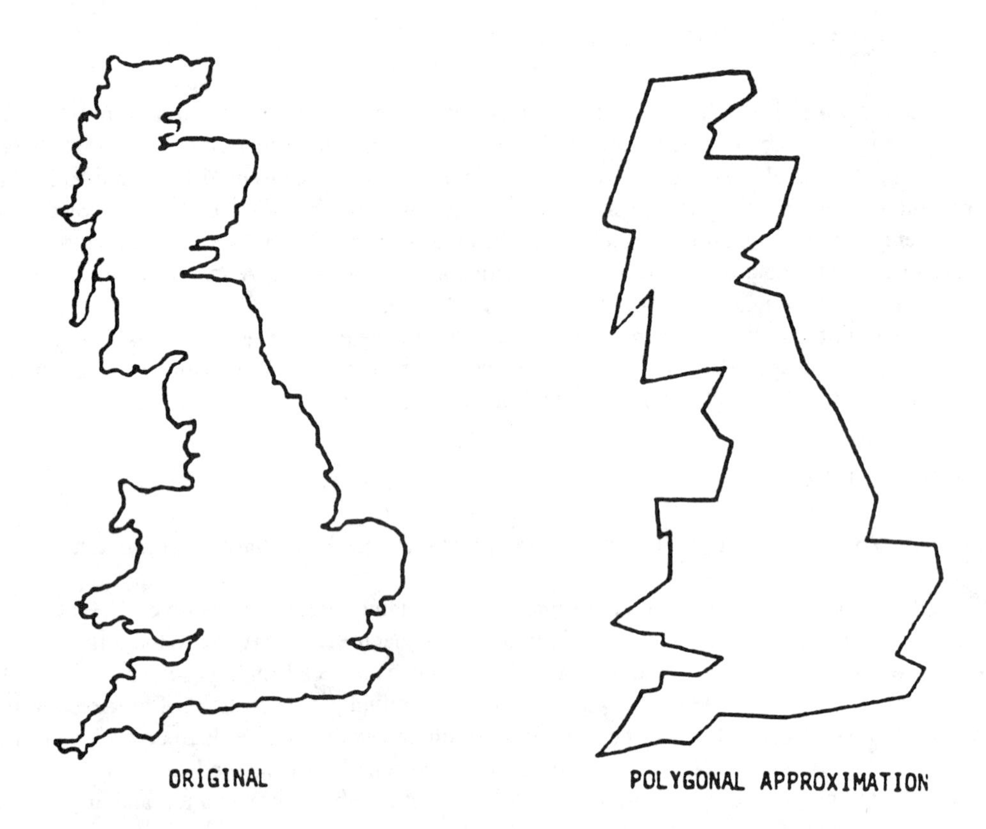

Figure 1. On the left is an outline of the coastline of Britain, and on the right is its polygonal approximation [Wall 1986].

exceeds a preset value, the line segment is drawn to the previous curve point, and a new segment is begun. A similar area measure is used in another technique [Imai 1986], where a piecewise linear approximation is made to a curve so as to cover a sequence of points by a minimum number of rectangles. The use of an area measure, versus the worst-case perpendicular-distance measure as described above, is better from the standpoint of giving a closer overall approximation to the original curve; in contrast, the use of a distance measure gives a closer approximation to sharp corners and curves, thus affording better worst-case approximation.

3.2.2: Methods for closer approximation

The methods described above are usually adequate for the purpose of polygonalization. There are more sophisticated methods that can yield better approximations, but at the expense of more computation, increased code complexity, or both. One of the drawbacks of most of the polygonal-fit techniques is that the operations are not performed symmetrically with respect to curve features such as corners and the centers of curves. The result is that the computed breakpoints between segments may be at different locations, depending on starting and ending locations the direction of operation of the line fit. Extensions can be made to some of the methods to produce fewer and more consistent breakpoint locations, but these procedures are usually iterative and can be computationally expensive. In choosing a particular method, the user must weigh the trade-off between the closeness of the fit and the amount of computation.

If a closer polygonal approximation is desired, one method is to use a minimax measurement of error [Kurozumi 1982]. Here, the line segment approximations are chosen to minimize the maximum distance

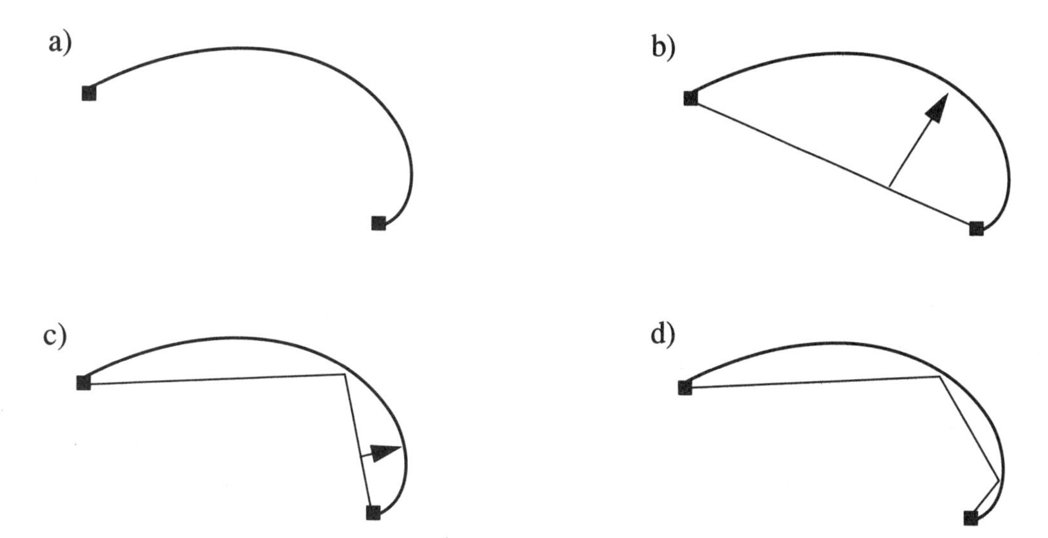

Figure 2. Illustration of the iterative endpoint-fit algorithm. (a) The original curve. (b) A polygonal approximation is made between endpoints, and the furthest perpendicular distance from this line to the curve is found (line with arrowhead). (c) Since this is greater than the threshold, two segments are fit to the curve at this point of greatest error, and the greatest distance is found between each new segment and the curve. (d) Only one error is above threshold, so that segment is split, and the result is shown where all segments are within the threshold distance of the curve.

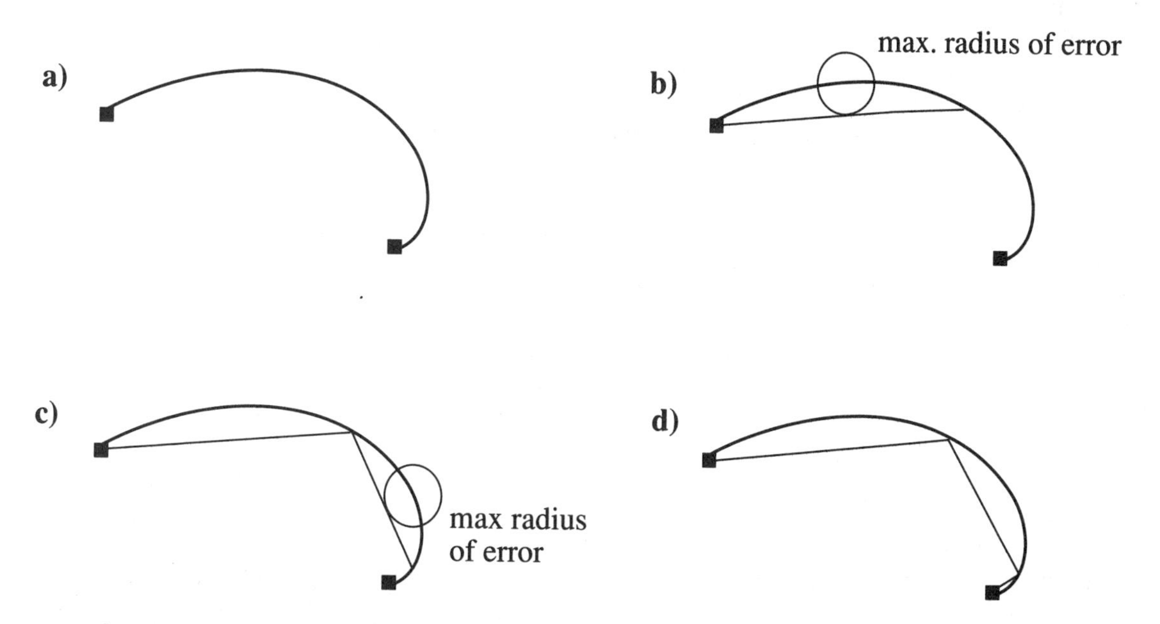

Figure 3. Illustration of sequential polygonal approximation employing distance as the measure of error. (a) The original curve. (b) A polygonal approximation is made from the first endpoint to as far a point on the curve as possible until the error (the radius of circle between the curve and the approximating segment) is above threshold. (c) A new segment is begun from the endpoint of the last, and it is drawn to as far a point on the curve as possible until the error is again above threshold. (d) A third segment goes from the previous point to the endpoint of the original curve, and this is the final approximation.

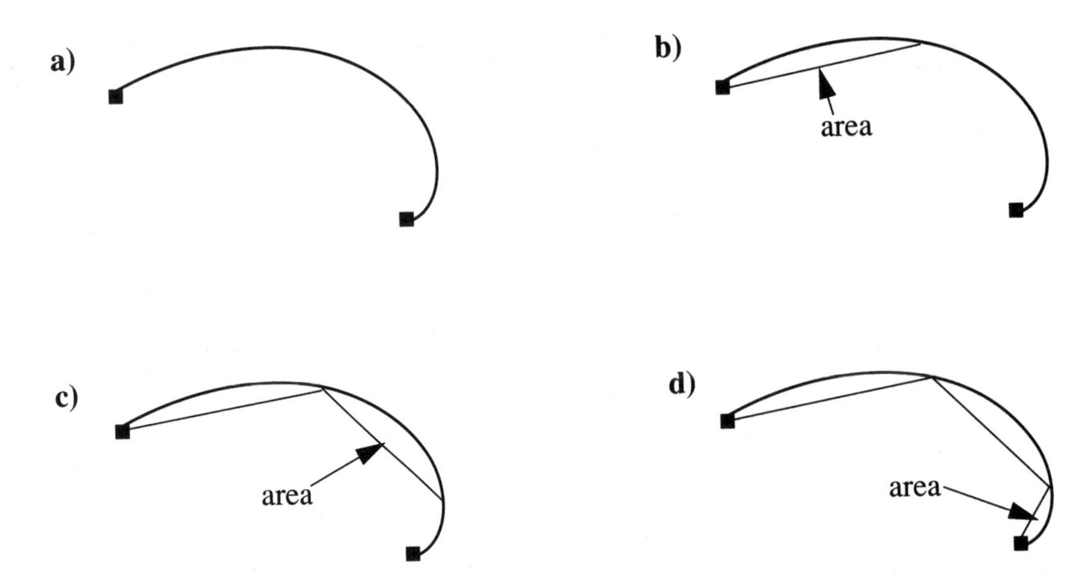

Figure 4. Illustration of sequential polygonal approximation employing area as the measure of error. (a) The original curve. (b) A polygonal approximation is made from the first endpoint to as far a point on the curve as possible until the error (the area between the curve and the approximating segment) is above threshold. (c) A new segment is begun from the endpoint of the last, and it is drawn to as far a point on the curve as possible until the error is again above threshold. (d) A third segment goes from the previous point to the endpoint of the original curve, and this is the final approximation.

between the data points and the approximating line segment. This method is especially useful for noisy curves such as curves fit to data with outlying (large-linear-error) data points. Although the minimax approach can give a better approximation, there is an added level of complexity for this method, because segment end points must be adjusted to fall within a gap tolerance of connecting segment end points.

One potential drawback of polygonal approximation is that the result is not usually unique; that is, an approximation of the same curve that begins at a different point on the curve will likely yield different results. The lack of a unique result will present a problem if the results of polygonalization are to be used for matching. One approach to reducing this problem is to find a starting point along the curve that will always yield similar results. In [Leu 1988], the polygonal approximation is started simultaneously at positions along the curve where the arcs are closer to straight lines than their neighboring arcs are. This operation is done by first calculating the maximum arc-to-chord deviation for each arc and then finding the local minimum deviation by comparing each deviation with those of the neighboring arcs. Polygonal approximation is done by fitting lines between points whose maximum arc-to-chord distance is less than a given tolerance. Because an effort is made to position the segments with respect to true features in the original curve (that is, straight lines and curves), this method straddles the boundary between polygonal approximation methods and the feature detection methods discussed in Section 3.4.

Besides polygonal-approximation methods, there are higher order curve- and spline-fitting methods that achieve closer fits; these methods are computationally more expensive than most polygonalization methods and can be more difficult to apply. We discuss these methods in detail in Section 3.3. It is important to stress that a polygonal fit just approximates the curve to some close error measure. If it is feature detection of true corners and curves that is desired, then the feature detection methods of Section 3.4 should be used rather than the methods in this section.

Commercial systems for engineering drawing and geographical-map storage will usually provide the polygonalization operation. Sometimes polygonalization is used in drawing systems, often unbeknownst to the user, to store curves more efficiently. These systems use a very small value of the maximum error of approximation so that viewers perceive the curves as smooth rather than as straight-line approximations.

One paper on polygonalization is included in this section: Wall and Danielsson's paper on fast sequential polygonalization using an area measure of error. This method is easily implemented and will provide good approximation. For smaller error (but longer computation), one of the minimax methods in the references can be used. In the paper included in the next section [Wu 1993], polygonalization methods similar to the ones described here are used in the preprocessing step.

3.3: Critical-point detection

(*Keywords*: critical points, dominant points, curvature extrema, corners, curves, curvature, difference of slopes, *k*-curvature, Gaussian smoothing)

The concept of *critical points* or *dominant points* derives from the observation that humans recognize shape in large part by curvature maxima in the shape outline [Attneuve 1954]. The objective in image analysis is to locate these critical points and represent the shape more succinctly by piecewise linear segments between critical points—that is, to perform shape recognition on the basis of the critical points. Critical-point detection is especially important for graphics analysis of man-made drawings. Since corners and curves have intended locations, it is important that any analysis locate precisely these corners and curves. It is the objective of critical-point detection to do this—as opposed to polygonal approximation, where the objective is only a visually close approximation. In this section, we will describe methods for locating critical points. In addition to corner features, we will include circular curves as a type of critical feature; the critical points for curves are the transition locations along the line from straight lines into the curves.

3.3.1: Curvature estimation approaches

One approach for critical-point detection begins with curvature estimation. A popular family of methods for curvature estimation is called the *k-curvatures* approach (also the *difference of slopes* [DOS] approach) [Freeman 1977, O'Gorman 1988, Rosenfeld 1973, Rosenfeld 1975]. For these methods, curvature is measured as the angular difference between the slopes of two line segments fit to the data around each curve point (see Figure 5). Curvature is measured for all points along a line and plotted on a curvature plot. For straight portions of the curve, the curvature will be low. For corners, there will be a peak of high curvature that is proportional to the corner angle. For curves, there will be a curvature plateau whose height is proportional to the sharpness of the curve (that is, the curvature is inversely proportional to the radius of curvature) and whose length is proportional to the length of the curve. To locate these features, the curvature plot is thresholded to find all curvature points above a chosen threshold value; these points correspond to features. Then, the corner is parameterized by its location and the curve by its radius of curvature and bounds, the beginning and end transition points from straight lines around the curve into the curve.

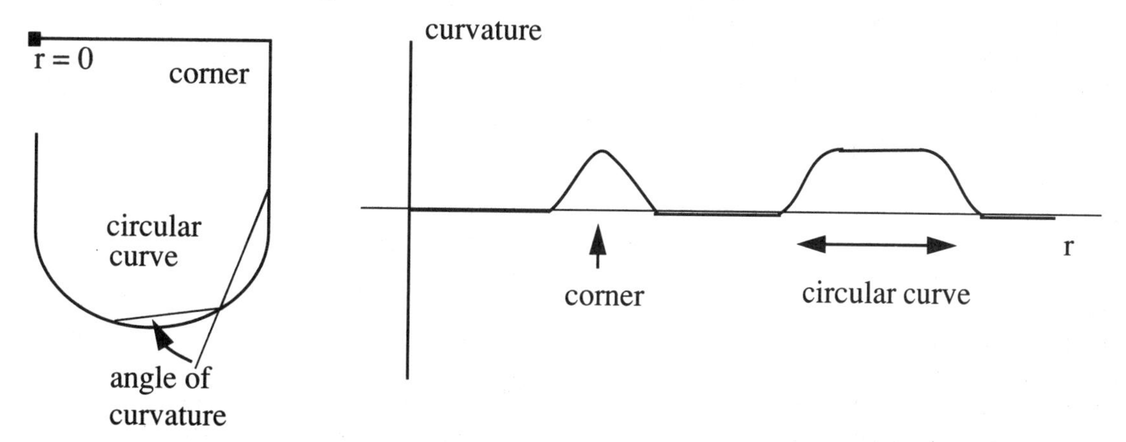

Figure 5. On the left is a curve with a corner and a circular-curve feature. On the right is the curvature plot derived from this by the DOS method. The angle of curvature is shown on the curve for one point. On the curvature plot, the first portion has 0 curvature for the horizontal straight line from r = 0. The corner is indicated by a peak on the curvature plot. The straight line from the corner to the curve also has 0 curvature. The rise and fall of the curvature plot around the plateau shows the transition point locations along the arc length into the curve. Finally, there is another 0-curvature section after the circular curve.

The user must choose one method parameter: the length of the line segments to be fit along the data to determine the curvature. There is a trade-off in the choice of this length. It should be as long as possible to smooth out effects due to noise but not so long that its length averages beyond that of a single line segment. That is, the length should be chosen as the minimum arclength between critical points.

Another approach to critical-point detection we will call *Gaussian smoothing* [Asada 1986, Pridmore 1987]. In this approach, instead of regulating the amount of data smoothing by choice of the line segment length fit to the data as above, the difference of curvatures is first measured on only a few points along the curve (typically three to five); these local curvature measures are then plotted on a curvature plot, and a Gaussian-shaped filter is applied to smooth noise and retain features. Critical points are found from this curvature plot by thresholding as before. The trade-off between noise reduction and feature resolution is made by the user's choice of the width of the Gaussian smoothing filter. Similarly to choosing the length of the fit for the *k*-curvatures method above, the Gaussian filter width is chosen as wide as possible to smooth out noise but not so wide that adjacent features merge. The user must empirically determine a width that yields a good trade-off between noise reduction and feature maintenance.

An advantage of both the DOS and Gaussian-smoothing families of methods over many other methods is that their results are symmetric with respect to curve features. That is, a feature will be located at the same point along the curve independently of the direction of curve traversal. However, a drawback of these methods is that the user must have an idea of the noise and feature characteristics to set the length of the line fit for the DOS approach or the width of the Gaussian smoothing filter for that method. This length or width is called the *region of support* because it relates to the arclength on the curve that affects the curvature measurement at any single point. The trade-off on this region of support is that the support should be as large as possible, to tend to average out noise, but should not be so large as to reduce true features. This trade-off requires the user to specify what constitutes a true feature versus noise—for example, the minimum corner angle or curve length. A problem can be encountered here if the noise or feature characteristics change for different images in an application, because this variation requires the user to adjust the method parameters for different images. Worse yet is the problem encountered if the feature characteristics have a large range in a single image—for example, if there is a very large, low-curvature circular curve and a very tight one. In this case, it is very difficult to accurately determine all critical points by the methods described.

3.3.2: Adaptive and multiscale methods

For the above methods, the choice of the method parameters determines the minimum curvature feature that can be resolved, and small-curvature features can be lost if the chosen parameter value is too large or noise points can be detected as features if the chosen parameter value is too small.

Both of these methods can be performed at multiple scales (see below), but another approach is to estimate curvature adaptively when features of different size and separation are present along the curve. The adaptive approach requires no user parameter of expected region of support; instead, the parameter is determined using local-feature data [Phillips 1987, Teh 1989]. In [Teh 1989], the region of support is adaptively determined by extending segments from a data point to points along the curve at both sides of each data point. The segments are extended from point to point, and the maximum perpendicular distance from the segment to the data is measured (similarly to the measure of error for Ramer's polygonalization method in Section 3.2.1) for each extended segment. This extension is halted when a function of the error from segment to curve and the length of the segment indicates that the region of support has reached the next feature; this final extent on the curve is the region of support. Because this region of support adapts to the data, it is as large as possible to reduce noise but not so large as to reduce true signals. Critical points are detected by thresholding high values as above. This is an advantage over the *k*-curvatures and Gaussian approaches, where the line segment length or Gaussian-smoothing width must be chosen with some knowledge of the minimum region of support of a feature. However, one drawback, in contrast to the *k*-curvatures and Gaussian-smoothing methods is that the output generated by this algorithm is sensitive to the direction of travel along the curve, resulting in critical points not necessarily being located in the center symmetrical features. Figure 6 shows an example of the results of this adaptive method and compares it to other methods, including the *k*-curvatures method.

Another way to detect different-sized features in the presence of noise is to process effectively with many regions of support—that is, at multiple scales. This approach detects features across the range of scales ex-

Figure 6. Critical-point detection for a number of different methods [Teh 1989].

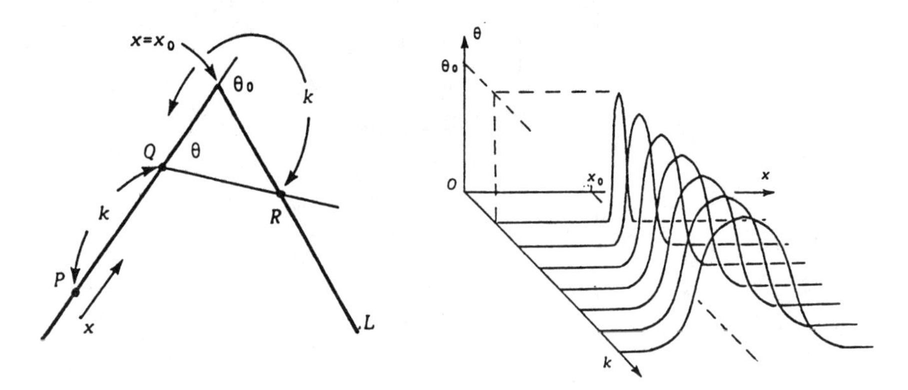

Figure 7. Multiscale *k*-curvatures method. On the left is a curve containing a corner. On the right are curvature plots for multiple scales of *k*-curvatures (that is, multiple lengths of *k*). As *k* increases, the widths of the peaks increase and the peaks decrease in height [Deguchi 1988].

amined. At a large scale, the region of support will extend a great length along the data curve, so that noise will be averaged out and large features will be detected; however, small features may be lost. At a small scale, the arclength of support will be small; thus, small features will be found but noise may be erroneously detected as features. A uniform sequence of scales is used between the largest and smallest features to determine features that will be detected best at specific scales throughout this range. One approach to multiple scales is an extension of the Gaussian-smoothing method mentioned above [Asada 1986, Mokhtarian 1986, Witkin 1983]. The method employs Gaussian smoothing of the curvature plot for a range of sizes of Gaussian-filter widths to cover the range of scales desired. Another approach is an extension of the *k*-curvatures method [Deguchi 88]. In this approach, *k*, the length of support on the arc, is the scale parameter. See the curvature plot for multiple scales of support upon the corner shown in Figure 7. The peaks on the curvature plot indicate the corner feature, and these peaks are smaller and wider as the region of support is larger.

Instead of all scales of a feature being calculated, an adaptive method can be used to find the most appropriate scale of filtering that reduces noise but keeps the desired features. In [Saint-Marc 1991], the approach taken is to first assign continuity values to points on the curve such that smooth sections have high continuity values. Then, small filter masks are applied iteratively to the data, weighted by these coefficients such that discontinuities are retained and noise within otherwise-smooth portions is reduced. In the example in Figure 8, the curve is shown with its curvature plot before and after adaptive critical-point detection, and the resulting critical points are also shown. In the first curvature plot without adaptive smoothing, there are peaks that correspond to corner and curve features, but there are also smaller peaks due to noise. The curvature plot is shown after 75 iterations of adaptive smoothing. Note that the peaks corresponding to features are retained, but those due to noise are reduced. These features can be more easily located on this plot, and their correspondence can be made to features in the original curve.

One paper, by Wu and Wang, is included on critical-point detection. The approach in this paper begins with polygonal approximation, then refines the corners to better coincide with true corners. This paper uses *k*-curvature and the Teh/Chin methods [Teh 89] in the preprocessing stage, thus providing a good description of these methods and their use.

3.4: Line and curve fitting

(*Keywords*: straight-line fit, corner detection, polynomial fit, spline fit, B-splines, Hough transform)

Lines and curves, or portions of lines and curves, can be represented by mathematical parameters of spatial approximations, or *fits*, made to them. These parameters can be used in turn to identify features in the objects for subsequent recognition or matching. This representation by parameters instead of by the original pixels is useful because the parameters can provide a useful description of an object as, for example, an object of four straight lines joined at 90° corners (a rectangle) or an object with a specific type of curve that

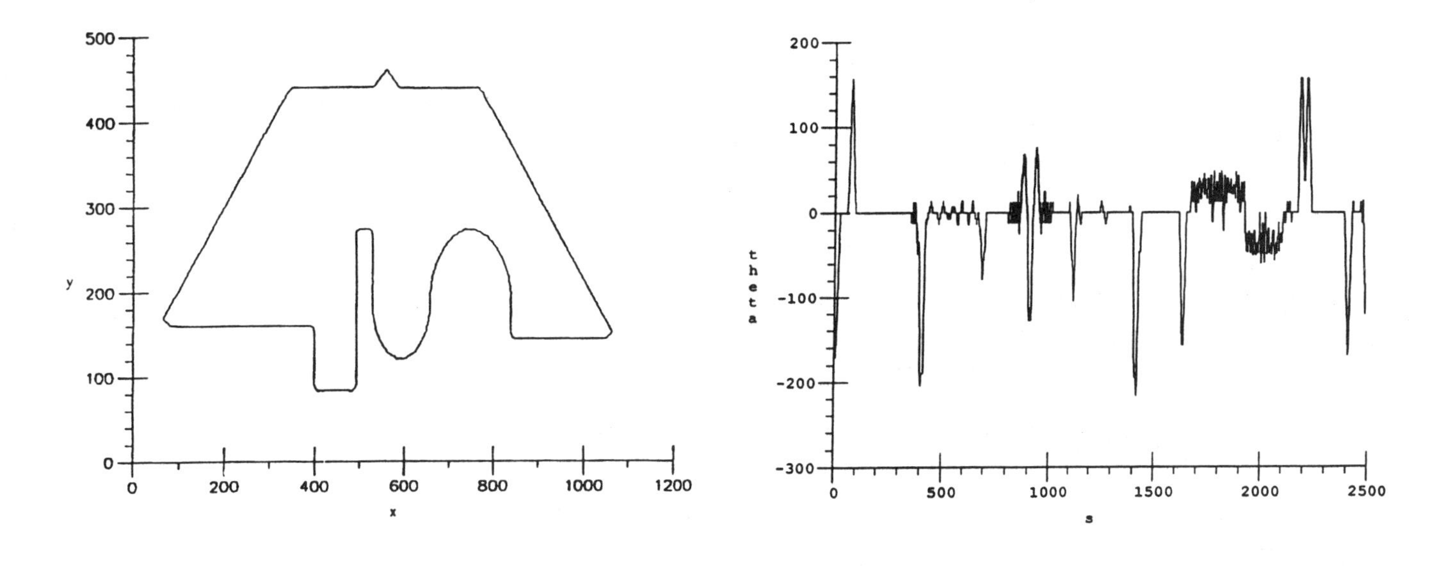

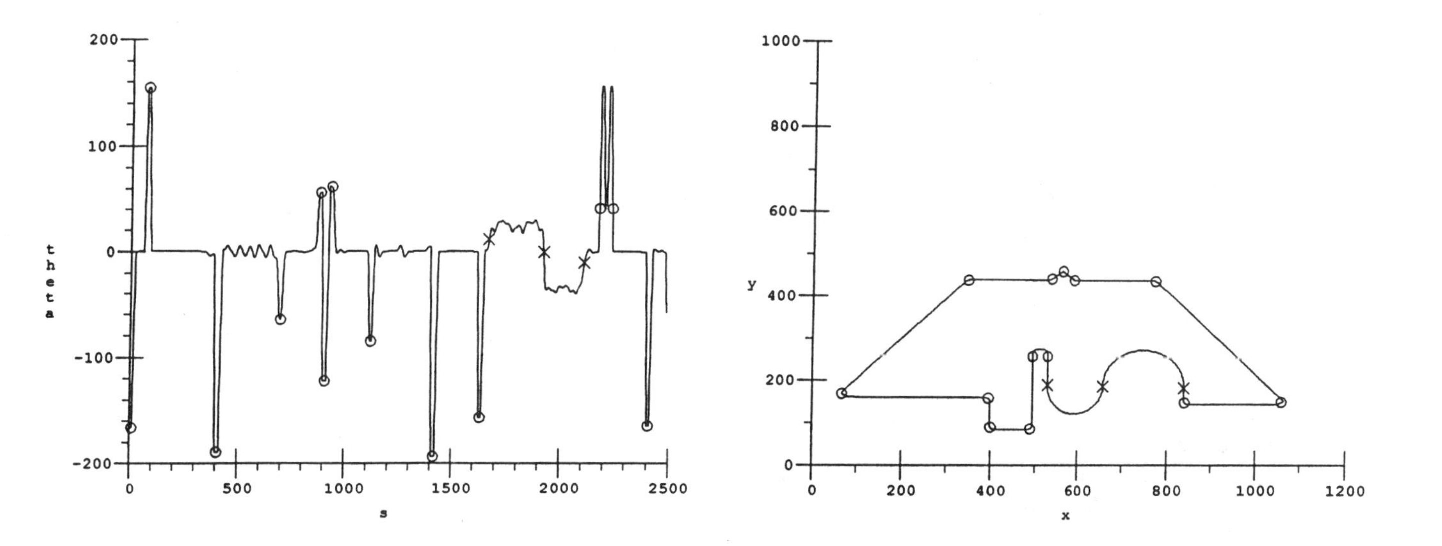

Figure 8. The data curve is shown on the top left, with a number of corner and curve features. The curvature plot shown on the top right is calculated from a small region of support; consequently, much noise is evident. The adaptively smoothed curvature plot is shown on the lower left, where discontinuities are retained and noise reduced. The final diagram on the lower right shows the locations of critical points from this curve, where the X's indicate the bounds of circular curves and the circles indicate corners. (from *Structured Document Image Analysis*, 1992, pp. 307-309. © Springer-Verlag, reprinted with permission.)

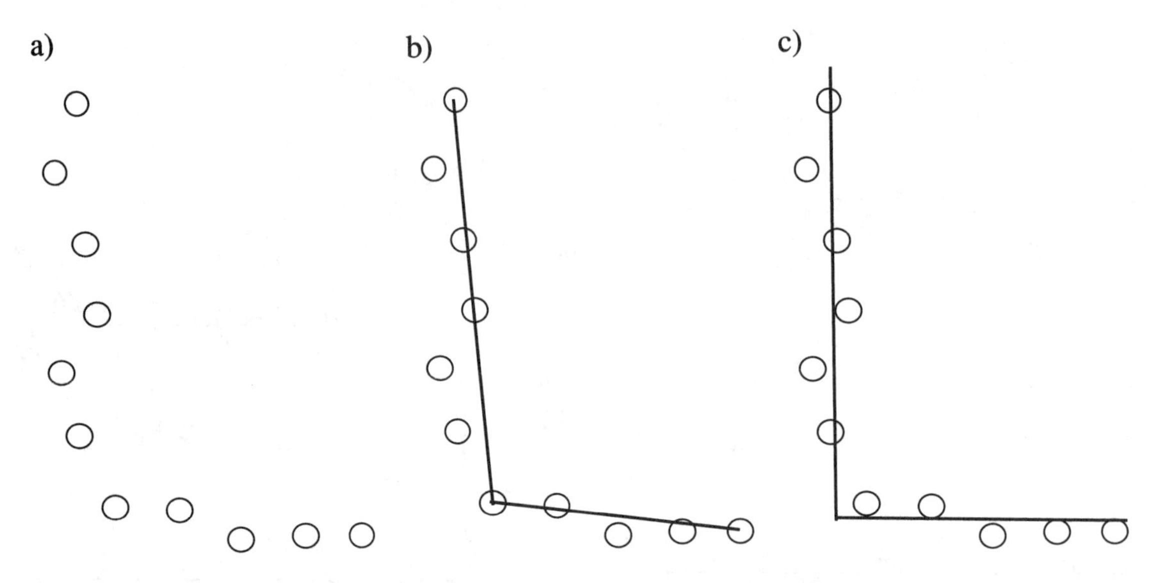

Figure 9. ON-valued pixels are shown in (a) and two corner fits are shown in (b) and (c). In (b), the fit is made exactly on data points. Consequently, the endpoints are sure to connect, but the corner angle is not as sharp as intended in the original. In (c), a least-squares fit was performed to two sets of points, and the intersection of them was taken as the corner location, yielding a sharper corner.

matches a curve of a particular object of interest. To aid the fitting process, we can use information from analyses already described to guide the fit, such as the polygonal approximations (Section 3.2) or locations of critical points (Section 3.3). It is also helpful to have some a priori idea of the type of fit and the limited shapes of objects in an application to reduce computation and ensure better fits.

3.4.1: Straight-line and circular-curve fitting

A simple way to fit a straight line to a curve is to specify the end points of the curve as the straight-line end points (see Figure 9). However, this procedure may result in some portions of the curve having a large error with respect to the fit. A popular way to achieve the lowest average error for all points on the curve is to perform a least-squares fit of a line to the points on the curve. For a line fit of equation $y = mx + b$, the objective is to determine m and b from the data points (x_i, y_i). The least-squares fit is accomplished by minimizing the sum of errors $\Sigma(y - y_i)^2$ over all data points $i + 1, \ldots n$. The solution is found by solving the simultaneous equations $\Sigma y_i = m\, x_i + bn$ and $\Sigma x_i y_i = m \Sigma x_i^2 + b \Sigma x_i$ for all data points i to obtain m and b.

There can be a problem in blindly applying the least-squares method just described to document analysis applications. The method is not independent of orientation, and as the true slope of the line approaches vertical, this method of minimizing y-error becomes inappropriate. One approach is to minimize y-error only for slopes expected to be ±45 degrees around the horizontal and to minimize x-error for slopes expected to be ±45 degrees around the vertical. A more general approach is to minimize error not with respect to x or y but with respect to the perpendicular distance to the line in any orientation. This can be done by a line-fitting method called *principal axis determination*, or *eigenvector line fitting*. The method is further detailed in [Ballard 1982, Duda 1973]; numerical-analysis texts such as [Press 1992] describe methods more generally.

A second note with respect to line-fitting methods is that they do not guarantee that end points fall on data points. For piecewise linear data, each line fit will have to be adjusted to connect the lines. This adjustment is usually done by extrapolating to the intersection of the adjacent line fit.

Especially for machine-drawn documents such as engineering drawings, circular-curve features are prevalent. These features are described by the radius of the curve, the center location of the curve, and the two transition locations where the curve either ends or smoothly makes the transition to one or two straight lines around it [O'Gorman 1988, Rosin 1989]. (See Figure 10.) One sequence of procedures in circular-curve detection begins by finding the transition points of the curve—that is, the locations along the line where a straight line makes a transition into the curve and then becomes a straight line again. Next, a decision is made on

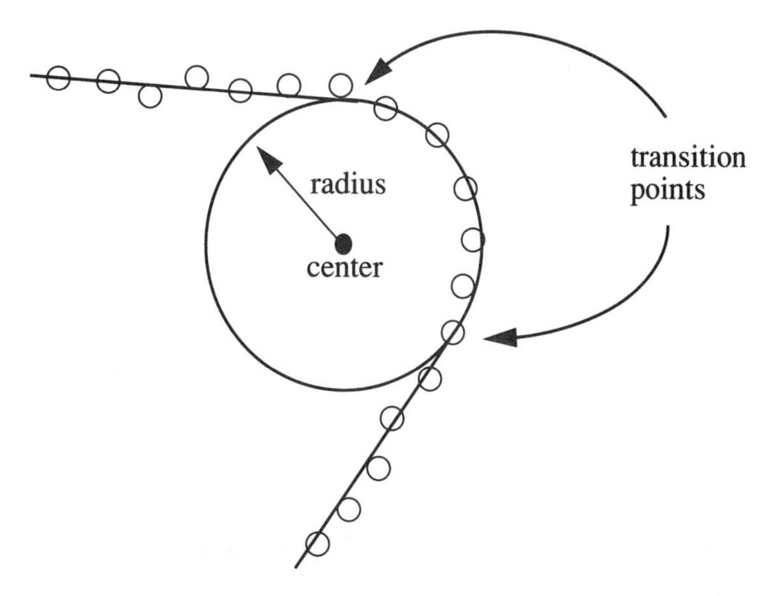

Figure 10. Circular curve fit to data points. The diagram shows parameters of this fit: two transition points from straight lines into the circular curve and the radius and center of the circle describing the curve.

whether the feature between the straight lines is actually a corner or a curve. This decision can be made on the basis of an arclength threshold. That is, if the arclength between transition points is longer than this threshold, then the feature is a curve; otherwise, it is a corner. Finally, the center of the curve can be determined with the information that—since each transition point is defined to be at a tangent to the curve—two perpendicular lines through these transition points will intersect at the curve center.

3.4.2: Other line-fitting methods, including splines and the Hough transform

If a higher order fit is desired, splines can be used to perform piecewise polynomial interpolations among data points [Foley 1982, Medioni 1987, Mokhtarian 1986, Pavlidis 1982]. *B-splines* are piecewise polynomial curves that are specified by a guiding polygon, such as is generated from polygonalization. Given two end points and a curve represented by its polygonal fit, a B-spline can be determined that performs a close but smooth fit to the polygonal approximation (see Figure 11). The B-spline has several properties that make it a popular spline choice. Besides a smooth interpolation between end points, it is also smoothly continuous (up to the second derivative) at the end points. Another property that makes the B-spline popular is that it is guaranteed to vary less than the guiding polygon of the sample points, thus preventing wildly erroneous mis-fits at singularities, which can occur with other types of fits, especially higher order fits. Although spline fits are useful to produce smoothly pleasing curves, it should be noted that first- and second-order features (straight lines and circular curves) are the most common features used for matching in document analysis, because they are the most common features found in document applications, and it is difficult to obtain a higher order fit with matching parameters for two curves, because of noise and discretization error.

A different approach for line and curve fitting uses the Hough transform [Illingworth 1988]. This approach is useful when the objective is to find lines or curves that fit groups of individual points on the image plane. This method involves a transformation from the image coordinate plane to parameter space. Consider the case where lines are the objects of interest. The equation of a line can be expressed as $r = x \cos \theta + y \sin \theta$, so there are two line parameters, the distance r and the angle θ, that define the transform space. Each (x, y) location of an ON pixel in the *image* plane is mapped to the locations in the transform plane for all possible straight lines through the point—that is, for all possible values of r and θ. When multiple points are colinear, their transformations will intersect at the same point on the transform plane. Therefore, the (r, θ) locations having the greatest accumulation of mapped points indicate lines with those parameters (see the example in Figure 12). In practice, because of discretization error and noise, points will not be exactly colinear and thus

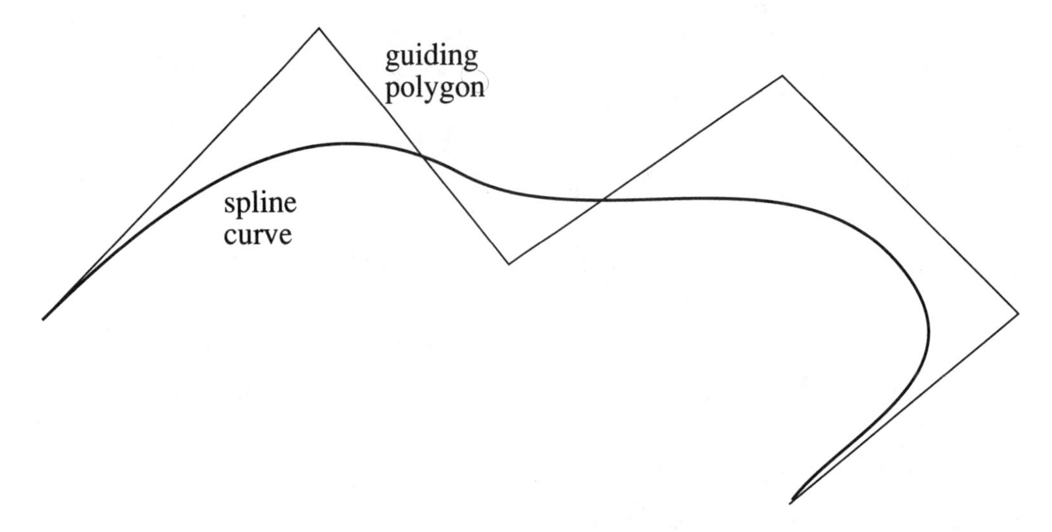

Figure 11. A spline curve is shown with its guiding polygon. The guiding polygon is from a straight-line polygonal fit to the image data. Notice that the spline lies within the corners of the polygon, thus yielding a smooth fit to the data.

will not map to exactly the same location on the transform plane. Therefore, after transformation, the detection step of the Hough method requires two-dimensional peak detection. The method can be extended to other parametric curves. For example, for a circle of equation $(x - a)^2 + (y - b)^2 = r$, the parameter space is three-dimensional, with axes a, b, and r. For the circle and higher order curves, there is an increased memory and computational expense as the dimensionality increases.

While the Hough transform is useful for fitting unconnected points, it is costly in terms of computation. In document analysis, we usually have more information than simply isolated points, and this information can be used to reduce computation. In particular, points are usually connected in continuous lines (although there may be small gaps, as in dashed lines, or gaps due to noise). For connected lines or portions of lines, compu-

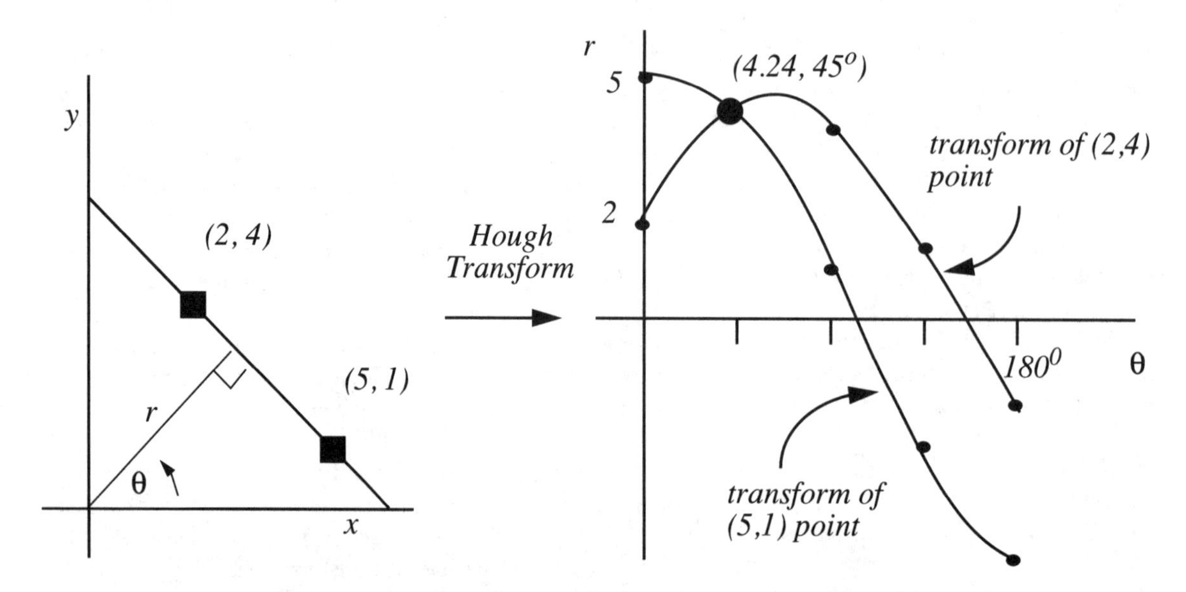

Figure 12. On the left is the image plane with two points through which a line can be drawn. On the right is the Hough parameter space for a line. Each point transforms to a sinusoid in Hough space. The point of intersection between the two sinusoids indicates the parameters of the line passing through both points. Note that in practice, the Hough plane is discrete; transform points (not curves) are plotted and accumulated, and the peak indicates the colinear point(s) in the image.

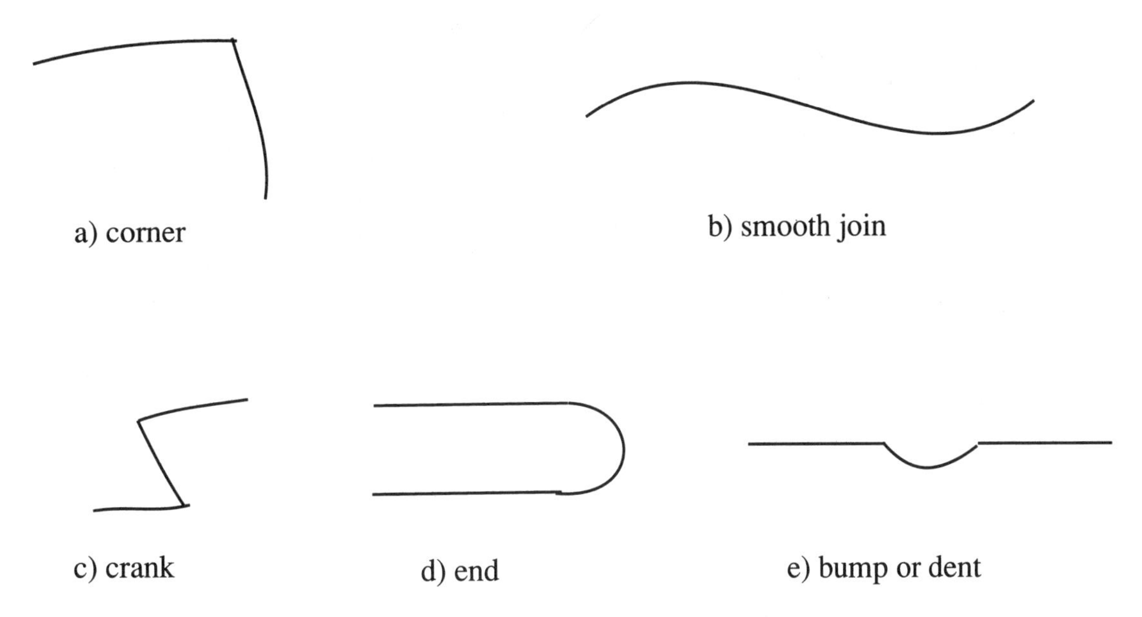

Figure 13. The curvature features proposed in [Asada 1986]: (a) corner, (b) smooth join, (c) crank, (d) end, and (e) bump or dent.

tation can be reduced greatly by mapping not to all (r, θ) possibilities but just to the one orientation (one point) indicated by the angle information of the line. Besides its computational constraints, the Hough transform is limited in that it does not yield the coordinates of the line end points and long lines are favored over shorter lines (in the peak-picking stage). For these reasons, there is often a spatial domain method that is faster and more effective in solving a problem. For example, for dashed-line detection in an engineering drawing, it is important to locate the line ends and the lengths of the dashes (when there are more than one type of dashed line). While the Hough transform could be used to perform a portion of this problem, a dashed-line detection technique designed specifically for this situation can be used to solve the whole problem [Kasturi 1990, Lai 1991]. However, the Hough transform is a general and widely useful method, and its use to determine text lines will be described in Chapter 4. Higher level descriptions can be made of line and curve objects that include compound features concatenated from straight lines and curve subparts. These descriptions are usually done with knowledge of the expected shapes or features. For example, in [Asada 1986], the features defined are corner, smooth join (two curves joined to form an "S" shape), crank ("Z" shape), end (two approximately parallel lines closed at one end by two corners or a curve), and dent (curve in an otherwise, straight line) (see Figure 13).

The line, curve, and spline methods described in this section are essential to most graphical analysis systems. As mentioned, fits for matching are usually limited to first- or second-order polynomials, whereas cubic splines are not usually used for matching but rather for obtaining pleasingly smooth curves from noisy data.

Two papers are included in this section. One paper by O'Gorman, concisely describes straight-line and circular-curve fitting to true line features. The other paper by Medioni and Yasumoto, describes corner detection and B-spline fitting.

3.5: Shape description and recognition

(*Keywords*: shape description, topology, shape metrics, convex hull, moments, projections)

When matching one object to another, the objective is to have a precise and concise description of the object so that incorrect matches can be quickly rejected and proper matches can be made without ambiguity. The nature of the description will depend on the application, in particular the shapes of the objects and the number of objects to be matched against. For instance, if the objective is to recognize text and graphics, where all the graphics consist of long lines, then discriminating descriptors would be region size or contour length. To recognize the difference between squares, triangles, and circles, the corner-and-curve detection methods of the previous section would apply. In this section, descriptions of a sampling of shape descriptors are provided, especially of those that may apply to document analysis. The purpose here is not to give a comprehensive listing of descriptors, because there can be many and they can be devised by the user for shapes of a particular application. Instead, many examples are shown to convey the approach in choosing shape descriptors. The approach is that they be appropriate to describe essential, inherent features of the shapes involved and that they act as good discriminators to differ between these shapes. Figure 14 shows a number of shapes and some appropriate shape metrics discussed in this section. For more comprehensive reviews of shape description techniques, see [Marshall 1989, Pavlidis 1978]. For an extensive collection of papers on shape analysis, see [Arcelli 1992].

One of the simplest ways to describe shapes is by shape metrics. For instance, the area measurements (number of ON-valued pixels) of connected components of a document can be used to separate large-font-size characters from smaller characters or text from larger line graphics. Instead of counting up all the pixels in a connected component, a faster method is to measure the length of the contour, especially when the shape has already been stored as the chain code of its contour. This length measurement can be used to separate small regions from large ones. Of course, contour length and region area are not directly related—they depend also on other aspects of the shape. If the fatness versus thinness of a shape is an important discriminator, then compactness—that is, the ratio of the square of the contour length over the area—can be measured. This is small for a circular contour and large for a long shape or a shape with many concavities and convexities.

Another way to describe a shape is by its moments. The first moment of a region is the average of (x, y) locations for all pixels in the region. These x and y moments are $m_x^{(1)} = (1/n)\Sigma x_i$ and $m_y^{(1)} = (1/n)\Sigma y_i$, where $i=1, \ldots, n$ is the number of ON pixels. The first moment describes the average location of an object. The second moment is the average of the squares of x and y locations, and so on. The central moments normalize the measurement with respect to the first moment to make it independent of location. The second central moments in x and y are $m_x^{(2)} = (1/n)\Sigma(x_i - m_x^{(1)})^2$ and $m_y^{(2)} = (1/n)\Sigma(y_i - m_y^{(1)})^2$. These moments can be used to indicate the nonroundness, eccentricity, or elongation of the shape; for example, if the second moments in x and y are similar in value, then the object is more round than otherwise. The third central moment gives a measure of the lack of symmetry of this eccentricity. Besides these central moments, moments of different orders can also be combined into functions that are invariant to other geometric transformations, such as rotation and scaling; these functions are called *moment invariants.* (See [Hu 1977, Jain 1989] for details on moment invariants.) Moments are less useful to describe more complicated shapes with many concavities and holes; in these cases, specific features such as the number of holes or thinned topology, are more applicable.

Shapes can be described by topological features—that is, by the number of holes and branches. For example, the letter "B" has two holes; the letter "P" has one hole and one branch. For these two examples, connected-component labeling can be used to determine the holes. However, for this example, thinning is probably more appropriate for finding both the holes (these are loops on the thinned characters) and the branches. If a shape has many convexities, its skeleton will have many branches. For example, the thinned results of the symbols "*" and "X" will have six and four branches respectively. The number of branches, the number of loops, the number of endlines, and the directions, lengths, and so on of each, are all descriptive of the original shape. Note should be made with regard to the use of thinning that, because of digitization noise and noise in the boundary of the shape, the thinned results will rarely match pixel-for-pixel against the same shape. This is why shapes are better matched on the basis of some description based on features such as number of branches and number of holes.

The contour of a shape can be used for shape description. A global description of the contour can be determined by calculating the Fourier transform of the curvature of the contour along its arc length, giving a frequency domain description of the boundary in terms of Fourier coefficients, called *Fourier descriptors.* Lower-order terms contain information on the large-scale shape—that is, on its low-frequency curvature characteristics; and higher-order terms contain information on sharper features—that is, on high-frequency curvature characteristics. Therefore, shapes can be recognized and matched on the basis of having, for example, lower high-order coefficient values for a smooth shape, such as an "O," versus larger high-order coefficient values for a shape with sharper corners such as a "W." A drawback of using Fourier descriptors is that they provide only global information; therefore, if two shapes have the same frequency information—for exam-

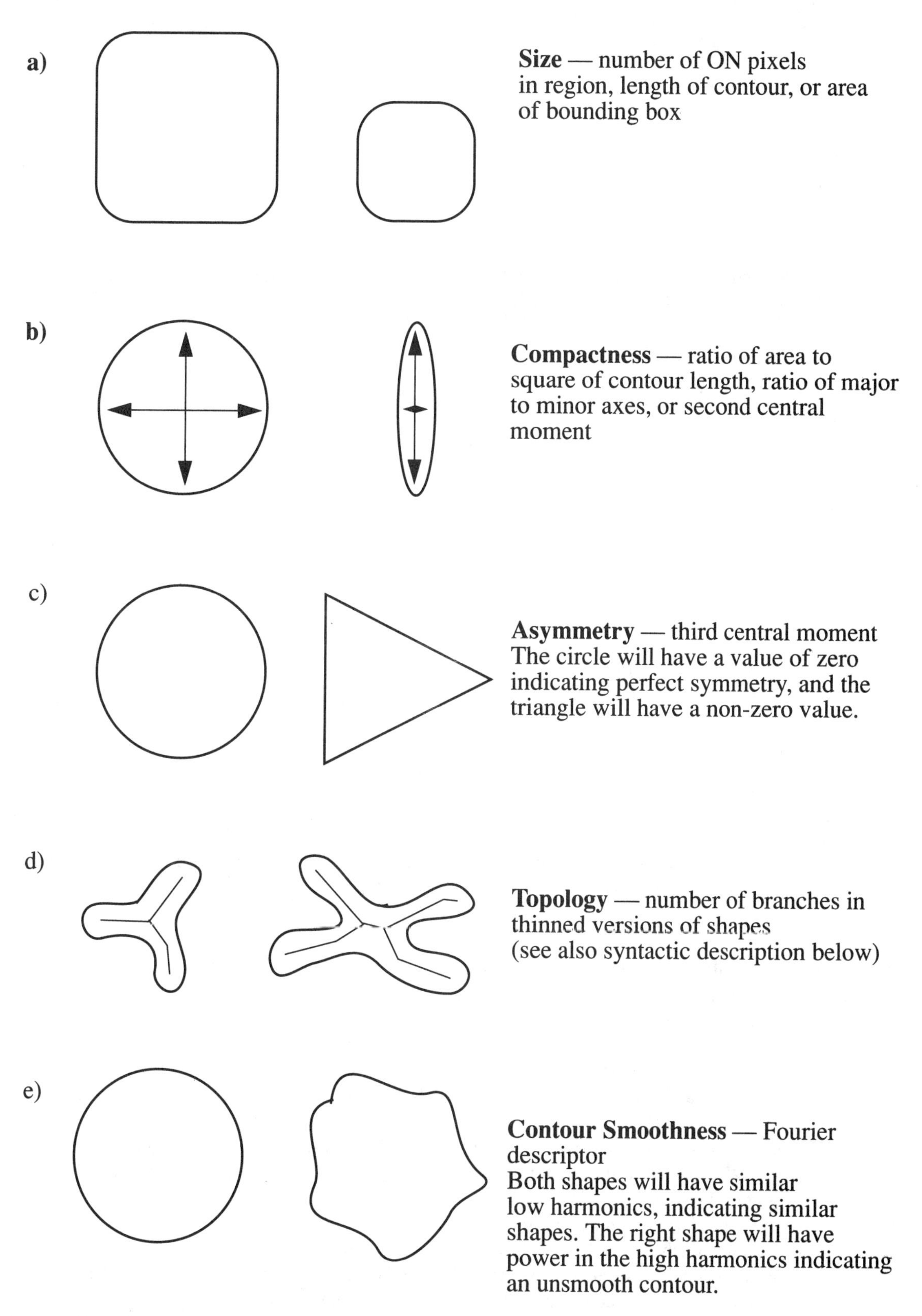

Figure 14. Shapes and suggested discriminating shape measures.

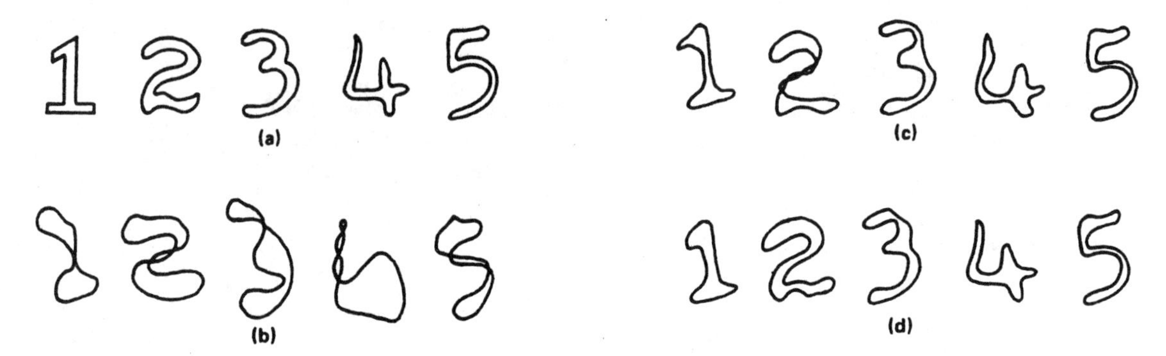

Figure 15. **Reconstructions of shapes from Fourier descriptors: (a) original, (b) using only five harmonics, (c) using 10 harmonics, and (d) using 15 harmonics. As more of the complete Fourier description is used, the shapes appear more like the originals [Brill 1968].**

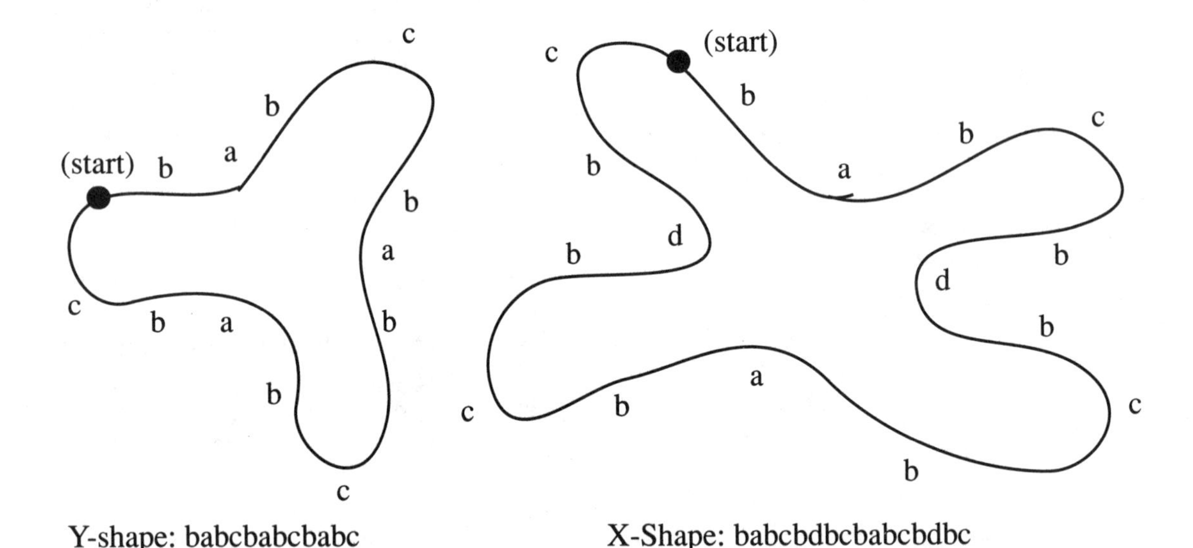

Figure 16. **Y and X shapes with coded contours. The codes describe the type of curve segments by their shapes and amounts of curvature: a-low curvature, concave curve; b-low curvature, curve or line; c-high curvature, convex curve; d-high curvature, concave curve. The two syntactic descriptions are the sequences of codes around the contour. These are different, indicating different shapes.**

ple a "P" and a "d,"—then they cannot be discriminated using Fourier descriptors alone. Figure 15 shows letters and their Fourier approximations. (For further description of Fourier descriptors, see [Jain 1989].)

Nonglobal features along contours can also be used for shape description. Critical points—or the lines, corners, and curves, between them—can be determined in the contour, as described in Section 3.2; they can then be matched against these features on other shapes. For complicated shapes, it is common to code a sequence of features found around a contour into a syntactic description that describes a "legal" sequence such as may be found for all shapes of interest. Examples of legal sequences are: (line, corner, line, corner, line, corner, line, corner) for a rectangle and (curve) for a circular object. Then, syntactic pattern recognition techniques that locate these legal sequences can be used to assist in matching. In Figure 16, two contours are shown with their syntactic descriptions.

A final example of a shape descriptor is the projection profile, or *signature*, of a shape, which is the accumulation of the number of ON pixels along all rows or columns into a one-dimensional histogram whose shape is descriptive of the original two-dimensional shape. To use the example of the letters "P" and "d" again, a horizontal projection profile (an accumulation along all rows through the letters) will indicate more ON pixels above and fewer below for "P" than "d," and vice versa. This shape measure is very dependent upon the orientation of the shape with respect to the axis of accumulation. Chapter 4 describes how the projection profile can be used to describe the "shape" (orientation and text lines) of a document page.

In commercial systems, shape descriptors are used in most aspects of document analysis. For OCR, systems use shape descriptors (in part) to distinguish characters. Lines are distinguished from characters by their shapes, with the former to be analyzed by graphics analysis tools and the latter to be analyzed by OCR. In graphics analysis, long, thin objects are distinguished from rounder objects for appropriate processing: thinning-based processing is used for the former and region-based processing is used for the latter.

We include here an excellent review paper by Pavlidis on shape analysis techniques.

References

* Papers marked with an asterisk are included as reprinted papers in this book.

Arcelli, C., L.P. Cordella, and G.S. di Baja, eds., *Visual Form: Analysis and Recognition*, Plenum Press, New York, N.Y., 1992.

Asada, H., and M. Brady, "The Curvature Primal Sketch," *IEEE Trans. Pattern Analysis and Machine Intelligence*, Vol. PAMI-8, No. 1, Jan. 1986, pp. 2–14.

Attneuve, F., "Some Informational Aspects of Visual Perception," *Psychological Rev.*, Vol. 61, 1954, pp. 183–193.

Ballard, D.H., and C.M. Brown, *Computer Vision*, Prentice-Hall, Englewood Cliffs, N.J., 1982, pp. 485–489.

Brill, E. L., "Character Recognition via Fourier Descriptors," *Proc. WESCON (Western Electric Show and Convention)*, 1968.

Deguchi, K., "Multi-Scale Curvatures for Contour Feature Extraction," *Proc. 9th Int'l Conf. Pattern Recognition*, IEEE CS Press, Los Alamitos, Calif., 1988, pp. 1113–1115.

Duda, R.O., and P.E. Hart, *Pattern Classification and Scene Analysis*, Wiley-Interscience, New York, N.Y., 1973, pp. 332–335.

Foley, J.D., and A. van Dam, *Fundamentals of Computer Graphics*, Addison-Wesley Pub., Reading, Mass., 1982, pp. 514–523.

Freeman, H., and L. Davis, "A Corner-Finding Algorithm for Chain-Coded Curves," *IEEE Trans. Computers*, Vol. C-26, NO. 3, Mar, 1977, pp. 297–303.

Hu, M.K., "Visual Pattern Recognition by Moment Invariants," in *Computer Methods in Image Analysis*, J.K. Aggarwal, R.O. Duda, and A. Rosenfeld, eds., IEEE CS Press, Los Alamitos, Calif., 1977.

Illingworth, J., and J. Kittler, "A Survey of the Hough Transform," *Computer Graphics and Image Processing*, Vol. 44, 1988, pp. 87–116.

Imai, H., and M. Iri, "Computational-Geometric Methods for Polygonal Approximations of a Curve," *Computer Graphics and Image Processing*, Vol. 36, 1986, pp. 31–41.

Jain, A.K., *Fundamentals of Digital Image Processing*, Prentice-Hall, Englewood Cliffs, N.J., 1989, pp. 377–381.

Kasturi, R., et al., "A System for Interpretation of Line Drawings," *IEEE Trans. Pattern Analysis and Machine Intelligence*, Vol. PAMI-12, No. 10, Oct. 1990, pp. 978–992.

Kurozumi, Y., and W.A. Davis, "Polygonal Approximation by Minimax Method," *Computer Graphics and Image Processing*, Vol. 19, 1982, pp. 248–264.

Lai, C.P., and R. Kasturi, "Detection of Dashed Lines in Engineering Drawings and Maps," *Proc. 1st Int'l Conf. Document Analysis and Recognition*, 1991, pp. 507–515.

Leu, J.G., and L. Chen, "Polygonal Approximation of 2-D Shapes through Boundary Merging," *Pattern Recognition Letters*, Vol. 7, No. 4, 1988, pp. 231–238.

Marshall, S., "Review of Shape Coding Techniques," *Image and Vision Computing*, Vol. 7, No. 4, Nov. 1989, pp. 281–294.

*Medioni, G., and Y. Yasumoto, "Corner Detection and Curve Representation Using Cubic *B*-Splines," *Computer Vision, Graphics, and Image Processing,* Vol. 29, 1987, pp. 267–278.

Mokhtarian, F., and A. Mackworth, "Scale-Based Description and Recognition of Planar Curves and Two-Dimensional Shapes," *IEEE Trans. Pattern Analysis and Machine Intelligence*, Vol. PAMI-8, No. 1, Jan. 1988, pp. 34–43.

*O'Gorman, L., "Curvilinear Feature Detection from Curvature Estimation," *Proc. 9th Int'l Conf. Pattern Recognition,* IEEE CS Press, Los Alamitos, Calif., 1988, pp. 1116–1119.

Pavlidis, T., *Algorithms for Graphics and Image Processing*, Computer Sc. Press, Rockville, Md., 1982.

*Pavlidis, T., "A Review of Algorithms for Shape Analysis," *Computer Graphics and Image Processing*," Vol. 7, 1978, pp. 243–258.

Phillips, T.Y., and A. Rosenfeld, "A Method of Curve Partitioning Using Arc-Chord Distance," *Pattern Recognition Letters*, Vol. 7, 1972, pp. 201–206.

Press, W.H., et al., *Numerical Recipes in C*, 2nd ed., Cambridge Univ. Press, Cambridge, U.K., 1992, pp. 671–680.

Pridmore, A. P., J. Porrill, and J. E. W. Mayhew, "Segmentation and Description of Binocularly Viewed Contours," *Image and Vision Computing*, Vol. 5, No. 2, May 1987, pp. 132–138.

Ramer, U.E., "An Iterative Procedure for the Polygonal Approximation of Plane Curves," *Computer Graphics and Image Processing*, Vol. 1, 1972, pp. 244–256.

Rosenfeld, A., and E. Johnston, "Angle Detection on Digital Curves," *IEEE Trans. Computers*, Vol. C-22, Sept. 1973, pp. 875–878.

Rosenfeld, A., and J.S. Weszka, "An Improved Method of Angle Detection on Digital Curves," *IEEE Trans. Computers*, Vol. C-24, 1975, pp. 940–941.

Rosin, P.L., and G.A.W. West, "Segmentation of Edges into Lines and Arcs," *Image and Vision Computing*, Vol. 7, No. 2, May 1989, pp. 109–114.

Saint-Marc, P., J.S. Chen, and G. Medioni, "Adaptive Smoothing: A General Tool for Early Vision," *IEEE Trans. Pattern Analysis and Machine Intelligence*, Vol. 13, No. 6, June 1991, pp. 514–529.

Sklansky, J., and V. Gonzalez, "Fast Polygonal Approximation of Digitized Curves," *Pattern Recognition*, Vol. 12, 1980, pp. 327–331.

Teh, C.H., and R.T. Chin, "On the Detection of Dominant Points on Digital Curves," *IEEE Trans. Pattern Analysis and Machine Intelligence*, Vol. 11, No. 8, Aug. 1989, pp. 859–872.

Tomek, I., "Two Algorithms for Piece-Wise Linear-Continuous Fit of Functions of One Variable," *IEEE Trans. Computers*, Vol. C-23, No. 4, 1974, pp. 445–448.

Wall, K., "Curve Fitting Based on Polygonal Approximation," *Proc. 8th Int'l Conf. Pattern Recognition*, IEEE CS Press, Los Alamitos, Calif., 1986, pp. 1273–1275.

*Wall, K., and P.E. Danielsson, "A Fast Sequential Method for Polygonal Approximation of Digitized Curves," *Computer Graphics and Image Processing*, Vol. 28, 1984, pp. 220–227.

Williams, C.M., "An Efficient Algorithm for the Piece-Wise Linear Approximation of Planar Curves," *Computer Graphics and Image Processing*, Vol. 8, No. 2, 1978, pp. 286–293.

Williams, C.M., "Bounded Straight-Line Approximation of Digitized Planar Curves and Lines," *Computer Graphics and Image Processing*, Vol. 16, 1981, pp. 370–381.

Witkin, A.P., "Scale-Space Filtering," *Proc. 8th Int'l Joint Conf. Artificial Intelligence*, Morgan Kauffman, San Mateo, Calif., 1983, pp. 1019–1022.

*Wu, W.Y., and M.-J.J Wang, "Detecting the Dominant Points by the Curvature-Based Polygonal Approximation," *CVGIP: Graphical Models and Image Processing*, Vol. 55, No. 2, Mar. 1993, pp. 79–88.

A Fast Sequential Method for Polygonal Approximation of Digitized Curves

KARIN WALL AND PER-ERIK DANIELSSON

Department of Electrical Engineering, Linköping University, S-581 83 Linköping, Sweden

Received October 27, 1983; revised February 2, 1984

A new and very fast algorithm for polygonal approximation is presented. It uses a scan-along technique where the approximation depends on the area deviation for each line segment. The algorithm outputs a new line segment when the area deviation per length unit of the current segment exceeds a prespecified value. Pictures are included, showing the application of the algorithm to contour coding of binary objects. Some useful simplifications of the algorithms are suggested.

1. INTRODUCTION

In image processing and pattern recognition digitized curves occur as contours of regions or objects, or as waveforms. Approximation of such curves results in compact descriptions which decrease memory requirements and facilitates feature extraction. Of course, the approximating segments must be chosen with care in order not to destroy the shape of the curve.

Line-fitting algorithms based on least square, Chebycheff, and nonlinear minimization procedures are ill-suited for this type of approximation since they were designed for other purposes. However, the literature contains descriptions of various algorithms for solving the problem more directly [1–11].

Williams [1, 2] and Sklansky and Gonzales [3] use cone intersection to find longest allowable approximating segments. For each of these, the maximum Euclidian distance between the segments and the points they approximate is less than a prescribed value. Circles are drawn around each point. An acceptable approximating polygon should pass through each of the circled areas. Moving from point to point, tangents are drawn to the circles to find sectors in which an acceptable segment must lie. [1 and 3] use a subset of the given points as vertices in the polygon while [2] forms segments that are tangents to some of the circles in order to produce better approximations. Cone intersection methods are relatively fast due to simplicity and sequential processing of the points. However they do not produce optimal results.

Kurozumi [4] uses a minimax method to find a polygon which satisfies the condition that the number of sides are minimum and the maximum distance between the sides and the data points is less than a prespecified tolerance. First the maximum polygon is constructed, that is, a convex polygon containing all points in the current subset. Then a side in the polygon is found and also found is the vertex giving the maximum perpendicular to it. The side must be the one for which this perpendicular is shortest. The current segment is then a line parallel to the side passing through the midpoint of the perpendicular. If the deviation is greater than the given tolerance the segment is not accepted and the previous segment is used as an approximation. Minimax methods give optimal results but are rather complex. They process the

"A Fast Sequential Method for Polygonal Approximation of Digitized Curves" by K. Wall and P.-E. Danielsson from *Computer Vision, Graphics, and Image Processing,* Vol. 28, 1984, pp. 220–227. Copyright © 1984 by Academic Press, Inc., reprinted with permission.

points only partly in sequence so several points near the current segment are needed in each step.

Like [4], Dettori [9] is based on the idea that a set of points can be approximated by a line segment if a strip exists which contains all the points and is not wider than a given value. A convex hull (similar to the convex polygon in [4]) is constructed each time a new point is added to the current set. When a strip satisfying the above conditions no longer exists, the longest segment is found, that is, a line joining the extreme points of the last accepted strip. The reasons for using this line as an approximation rather than the axis of the strip are that the polygonal line will be continuous and no new points are generated.

Pavlidis [5] finds the minimum number of segments where on each of them the error norm is less than a prespecified value. The error norm could be the maximum Euclidian distance or the integral square error. Initially the set of points is divided into arbitrary subsets. Each subset is approximated by a straight line (this method can be used even for polynomial approximations). The error norm for each is calculated. Then subsets are split and merged in order to drive the error norm under the given bound. Split and merge methods produce good results but are time-consuming since they require multiple passes through the data.

This paper introduces a new sequential method based on area deviation, suggested in [12]. Since it is sequential, the approximations are not optimal, but they are generated quickly. The segment endpoints are required to be some of the original data points which means that the approximation errors will be larger than they would be if a segment could be terminated more freely. Again, this is the price of speed. A short version of this paper was presented in [13].

2. THE NEW METHOD

The longest allowable segment is found by merging points one after another to the initial one, until a certain test criterion is no longer satisfied. The output segment will be a line from the initial point to the last point which passed the test. The criterion involves a parameter T, which can be set to different values. Greater values of T create longer segments on the penalty of a poorer approximation. The geometrical meaning of T is *the maximum allowed area deviation per length unit of the approximating segment.*

Assume that a digitized curve is given either as a sequence of coordinate pairs or as a chaincode $(x_0, y_0), a_1, a_2, \ldots, a_n$. In the latter case the starting point is (x_0, y_0) and the links a_i take values from the set $(0 \cdots 7)$ corresponding to the eight main directions in the digitized xy-plane. In either case we could formulate the following algorithm:

Initiate $i = 2$.

(a) Translate the coordinate system in order to get the origin in the starting point.

Initiate $f_i = 0$.

(b) Calculate the increments Δx_i and Δy_i and move to next point (x_i, y_i). Perform the accumulation

$$f_i = f_{i-1} + \Delta f_i, \qquad \text{where } \Delta f_i = x_i \cdot \Delta y_i - y_i \cdot \Delta x_i.$$

Calculate the length L_i of the current line segment from the starting point to (x_i, y_i)

$$L_i = \sqrt{x_i^2 + y_i^2}.$$

Apply the test

$$|f_i| \leq T \cdot L_i.$$

(c) If the test criterion is satisfied, increment i and repeat from (b). Otherwise the longest allowable segment has been found, that is, a line to (x_{i-1}, y_{i-1}). This latter point is taken as a starting point for next segment and the algorithm is repeated from (a) until all points are processed.

3. ANALYSIS

The increment Δf_i when moving from the point P_{i-1} to next point P_i is given by (Fig. 1).

$$\Delta f_i = x_i \cdot \Delta y_i - y_i \cdot \Delta x_i = \pm|\bar{L}_{i-1} \times \bar{\Delta}_i| = L_i \cdot d_i,$$

where L_i is the length of the current segment and d_i is the signed distance from the current segment to the previous point P_{i-1}. Division by 2 gives

$$\Delta f_i/2 = L_i \cdot d_i/2$$

which is the area of $OP_{i-1}P_i$.

Now, assume that the curve in Fig. 2 is to be approximated. Moving along the curve the accumulated sum f varies and after a number of steps we will reach the last point P_k which passes the test $|f| \leq T \cdot \mathrm{L}$. The output segment will then be OP_k. The sum f_k is given by

$$f_k = \sum_{i=1}^{k} \Delta f_i = \sum_{i=1}^{k} L_i d_i.$$

Division by 2 gives

$$f_k/2 = \sum_{i=1}^{k} L_i d_i/2,$$

the magnitude of which is the area deviation, that is, the difference between the areas above and below the segment OP_k, enclosed by the curve. This means that if the

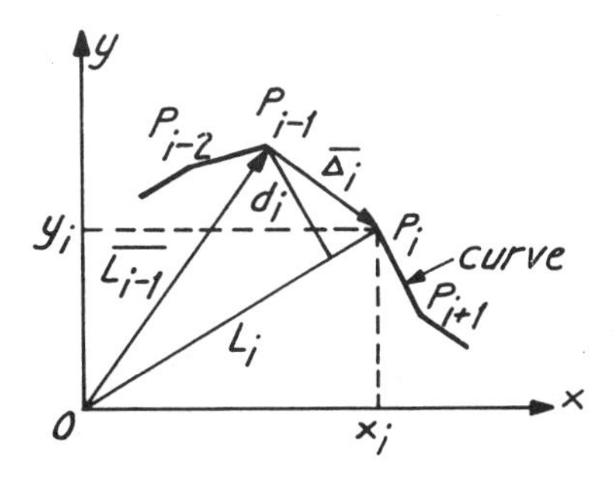

FIG. 1. The increment Δf_i is given by $\Delta f_i = L_i \cdot d_i$.

area deviation is smoothed out on OP_k, the result will be a rectangle with a maximal width of $T/2$.

Having found all segments, the area of a polygonal approximation will differ from the real object area by at most

$$|\text{object area} - \text{polygon area}| \leq T/2 \cdot \sum_{i=1}^{n} L_i,$$

where n is the number of segments.

For convex curves the *maximum Euclidian distance* $\epsilon_{\max}$ from any point to the approximating segment is less than T. This is because the worst case for such curves will be when the area between the curve and the line has the form of a triangle as shown by Fig. 3. The $\epsilon_{\max}$ can be derived as follows:

$$\epsilon_{\max} \cdot L_p = |f_p| \leq L_p \cdot T, \qquad \epsilon_{\max} \leq T.$$

For an arbitrary curve the accumulating nature of the algorithm makes $\epsilon_{\max}$ hard to derive. However, an estimation is given below. First we consider a smooth curve (Fig. 4). Assume that the curve has been approximated by the segment OR. This means that even the critical points Q and R have passed the test:

$$A + B \leq T/2 \cdot L_Q \qquad (Q \text{ passes the test}) \tag{1}$$

$$C - A \leq T/2 \cdot L_R \qquad (R \text{ passes the test}). \tag{2}$$

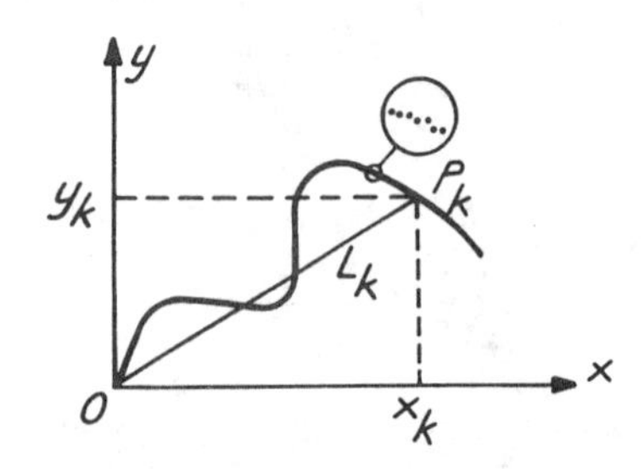

FIG. 2. P_k is the last point which passed the test.

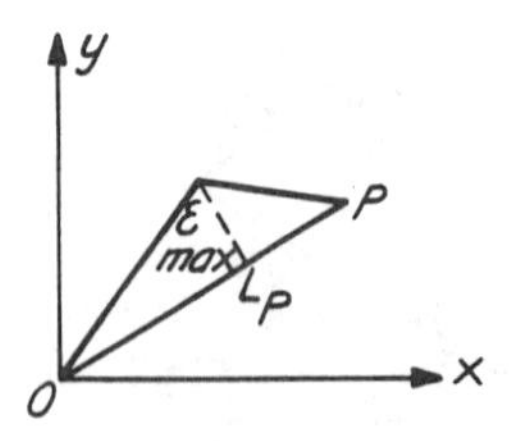

FIG. 3. The worst case for a convex curve is when the area between the curve and its approximating segment has the form of a triangle.

Adding (1) and (2) gives

$$B + C \leq T/2 \cdot (L_R + L_Q)$$
$$L_R \cdot \epsilon/2 \leq T/2 \cdot (L_r + L_Q)$$
$$\epsilon \leq T \cdot (1 + L_Q/L_R).$$

For a smooth curve, L_R is larger than L_Q. Thus ϵ_{max} is bounded by $\epsilon_{max} < 2T$.

4. PEAK TEST

For rougher curves, e.g., waveforms, the assumption $L_R \geq L_Q$ is not valid any more. Especially when the curve has narrow peaks, L_Q can be much larger than L_R which will result in a great value for ϵ_{max}, that is, narrow peaks will remain undetected by the algorithm. This is a feature which this algorithm shares with cone-intersection methods [1, 2, 3]. In most cases this will not cause any problem since narrow peaks often are considered as noise and should be ignored. However, sometimes these peaks are important features that should be retained. This indicates that if the algorithm is used for approximation of rough curves an extra test, a peak test, is needed to obtain satisfying results. We suggest:

if the length L of the segment decreases, the coordinates for the point Q for which L had a maximum are saved;

if L continues to decrease, a breakpoint will be output when the distance D between the current point and Q reaches the value of T.

The breakpoint can be either Q or the current point. Using Q as breakpoint results in a better approximation with the penalty of more calculations. With the current point as breakpoint, the peak test only demands calculation of the distance D. On the parts of the curve where L is increasing the peak test will not delay the algorithm more than the time it takes to store the coordinates and to test L. The above variation of the algorithm has an effect similar to the "altered version RMPP/LS" in [3].

5. SIMPLIFICATIONS

The test can be written $f_i^2 \leq T^2 \cdot L_i^2$ thus avoiding evaluation of the square root. Further, if L_i^2 is updated as

$$L_i^2 = L_{i-1}^2 + 2 \cdot (x_i \Delta x_i + y_i \Delta y_i) + (\Delta x_i)^2 + (\Delta y_i)^2,$$

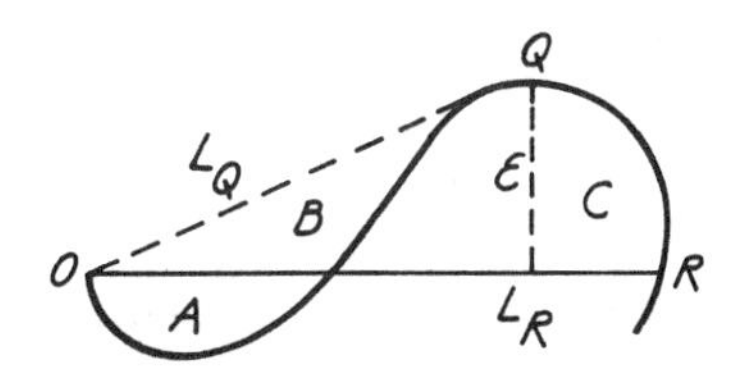

FIG. 4. $\epsilon_{max} < 2T$ for smooth curves.

only additions are needed for normal chaincode, since $(\Delta x_i)^2$ and $(\Delta y_i)^2$ are either 0 or 1.

More speed can be gained by approximating L_i with

$$L_i \approx (|x_i| + |y_i|) \qquad \text{or} \qquad L_i \approx \max(|x_i|, |y_i|).$$

Usually the first approximation creates longer, and the second one shorter, segments than those obtained from the normal version for the same parameter value. This is due to the fact that

$$|x_i| + |y_i| \geq L_i \qquad \text{and} \qquad \max(|x_i|, |y_i|) \leq L_i.$$

The main disadvantage of these approximations is that they create results which depend not only on T and the shape of the curve but also on the orientation of the line segments.

6. EXPERIMENTS

Our algorithm has been implemented on a computer. Some results are shown in Figs. 5–8. Figures 5 and 6 show a smooth and a rough curve, respectively, and their approximating polygons for different parameter values. Figure 7 shows a comparison between the behavior of the algorithm in its basic version and the simplified versions. Finally, an example of the effect of including the peak test is shown in Fig. 8.

FIG. 5. A smooth curve and its approximating polygon.

FIG. 6. The coastline of Great Britain.

7. CONCLUSION

A new method for piecewise linear approximation of digitized curves has been introduced. It can accept curves whose points are nonuniformly spaced. As it uses a scan-along technique and outputs a subset of the given points as vertices of the polygonal approximation, the amount of memory required is relatively small. It is simpler and faster than any previously published algorithm for the same purpose. Comparing it to [1] (which seems to be the fastest of the cone-intersection methods) shows that while our algorithm requires six multiplications (the simplified versions only three–four), [1] requires seven multiplications/divisions and evaluation of a square root. This holds for curves consisting of nonequidistant points. For equidis-

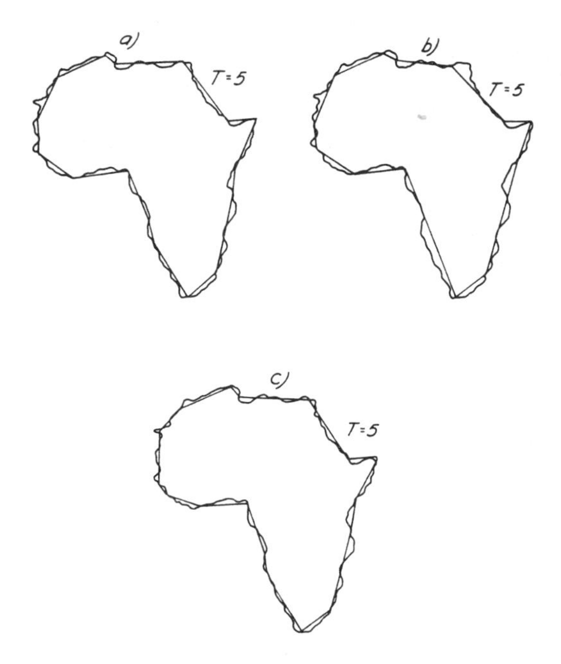

FIG. 7. The coastline of Africa (a) basic version; (b) simplified version $L \approx |x| + |y|$; (c) simplified version $L \approx \max(|x|, |y|)$.

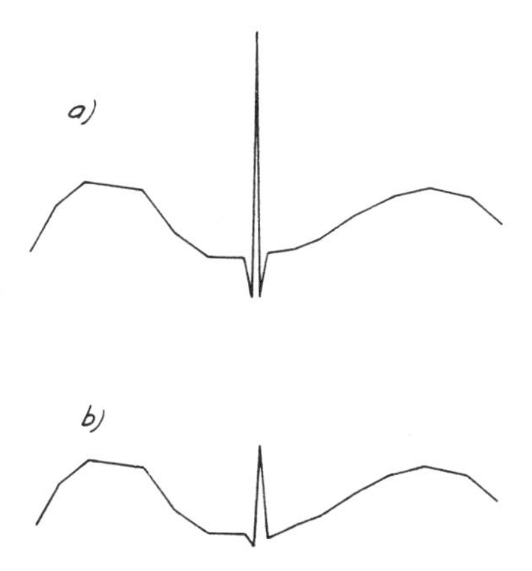

FIG. 8. The curves above are approximations of a digitized curve. The same parameter value was used in both cases but (a) was obtained from a version including the peak test, (b) from the basic version.

tant points, e.g., a chaincoded contour, the difference is greater: the number of multiplications decreases to five for [1] but to two for our algorithm (one for a simplified version).

A modified version of the algorithm could be used for another purpose, namely corner finding: Calculate the area deviation for a line segment from P_i to P_{i+j}, where j is a prespecified even integer. Divide the area deviation by the length of the current segment. The magnitude of the resulting value indicates the possibility for $P_{i+j/2}$ being a corner. By moving the segment step by step along the contour and performing the above calculations, each point will be provided with a value indicating its "strength" as a corner.

REFERENCES

1. C. Williams, An efficient algorithm for the piecewise linear approximation of planar curves, *Comput. Graphics Image Process.* **8**, 1978, 286–293.
2. C. Williams, Bounded straight-line approximation of digitized planar curves and lines, *Comput. Graphics Image Process.* **16**, 1981, 370–381.
3. J. Sklansky and V. Gonzales, Fast polygonal approximation of digitized curves, *Pattern Recognition*, **12**, 1980, 327–331.
4. Y. Kurozumi and W. Davis, Polygonal approximation by the minimax method, *Comput. Graphics Image Process.* **19**, 1981, 248–264.
5. T. Pavlidis and S. Horowitz, Segmentation of planar curves, *IEEE Trans. Comput.* **C-23**, 1974, 860–870.
6. W. Li-De, Consistent piecewise linear approximation, Proc. 6th Int. Conf. Pattern Recognition, 1982, Vol. 2, 739–741.
7. T. Pavlidis, Curve fitting as a pattern recognition problem, Proc. 6th Int. Conf. Pattern Recognition, 1982, Vol. 2, 853–857.
8. U. E. Ramer, An iterative procedure for the polygonal approximation of plane curves, *Comput. Graphics Image Process.* **1**, 1972, 244–256.
9. G. Dettori, An on-line algorithm for polygonal approximation of digitized plane curves, Proc. 6th Int. Conf. Pattern Recognition, 1982, Vol. 2, 840–842.
10. I. Tomek, Two algorithms for piecewise linear continuous approximation of functions of one variable, *IEEE Trans. Comput.* **C-23**, 1974, 445–448.
11. G. M. Phillips, Algorithms for piecewise straight line approximations, *Comput. J.* **2**, 1968, 211–212.
12. P. E. Danielsson, Polygonal approximation of digital curves, *IBM Tech. Disclosure Bull.* **24**, No. 11B, 1982, 6215–6217.
13. K. Wall and P. E. Danielsson, A new method for polygonal approximation of digitized curves, in *Proc. Third Scandinavian Conf. on Image Analysis*, pp. 60–66, Studentlitteratur, Lund, Sweden, 1983.

Detecting the Dominant Points by the Curvature-Based Polygonal Approximation

WEN-YEN WU AND MAO-JIUN J. WANG*

Department of Industrial Engineering, National Tsing Hua University, Hsinchu, Taiwan, 30043, Republic of China

Received July 7, 1992; accepted January 12, 1993

Detecting dominant points is an important step for object recognition. Corner detection and polygonal approximation are two major approaches for dominant point detection. In this paper, we propose the *curvature-based polygonal approximation method* which combines the corner detection and polygonal approximation techniques to detect the dominant points. This detection method consists of three procedures: (1) extract the break points that do not lie on a straight line, (2) detect the potential corners, and (3) perform polygonal approximation by partitioning the curves between two consecutive potential corners. Both quantitative and qualitative evaluations have been conducted. Experimental results show that the combined methods are superior to the conventional methods, and the dominant points can be properly detected by the combined methods.

1. INTRODUCTION

One important problem in computer vision is to extract some meaningful features from the images. It has been suggested that the dominant points along with the object boundary are sufficient to describe the shape of an object. The dominant points are those points which can suitably describe the curve for both visual perception and recognition. A number of methods have been developed to detect the dominant points. The two major approaches are corner detection [5, 8] and polygonal approximation [1, 3, 6, 10]. The common stages for corner detection techniques are: (1) estimate the curvature for each point on the curve and (2) locate the points which have local maximum curvatures as the corners. The detected cornes are considered as the dominant points on the curve. For polygonal approximation, sequential and iterative methods are commonly used. Sequential techniques have the drawbacks of missing some important features such as sharp corners and spikes. On the other hand, the performance of the iterative techniques is sensitive to the setting of the starting points for partitioning curves.

Ansari and Delp [2] proposed an alternative method to detect the dominant points. They first find a set of positive maximum and negative minimum curvature points on the Gaussian smoothed boundary, and then followed by a split-and-merge polygonal approximation algorithm to detect the dominant points. The authors claimed that the dominant points obtained by this method are less sensitive to the orientation of the boundary than the other methods.

In this paper, a different approach which combines corner detection and polygonal approximation is proposed. The combined method is called the *curvature-based polygonal approximation method*. A simple corner detection method is first applied to detect the potential corners. Then each curve between any two consecutive potential corners is further partitioned by polygonal approximation. The potential corners as well as the partitioned points detected are considered as the dominant points. The detailed description of the combined method is given in the next section. Section 3 presents the results of evaluation of this method. Several quantitative evaluation criteria have been used to assess the performance of the combined methods, and the results of comparison between the combined method and the conventional methods are presented. Some concluding remarks are discussed in the last section.

2. DOMINANT POINT DETECTION

In this section, we first illustrate the corner detection and polygonal approximation techniques. For corner detection, two simple curvature estimation methods are introduced. These two curvature estimation methods are the modified version of the methods proposed by Rosenfeld and Johnston [5] and Teh and Chin [8], respectively. For polygonal approximation, two partitioning point selection rules are introduced. One is based on the maximum deviation partitioning rule proposed by Ramer [3]. The other one is based on the maximum curvature partitioning rule developed in this paper.

In final, we combine the methods of corner detection and polygonal approximation to detect the dominant

* To whom correspondence should be addressed.

"Detecting the Dominant Points by the Curvature-Based Polygonal Approximation" by W.-Y. Wu and M.-J.J. Wang from *CVGIP: Graphical Models and Image Processing,* Vol. 55, No. 2, Mar. 1993, pp. 79–88.

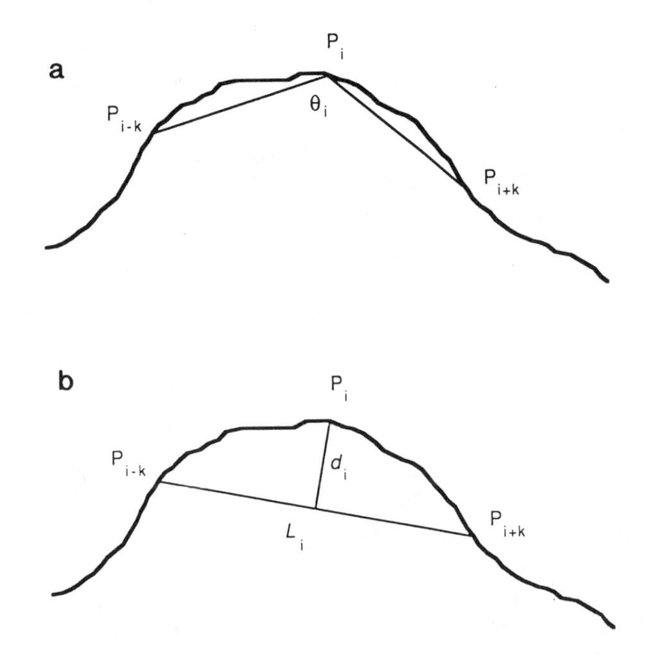

FIG. 1. Estimate curvature by using (a) cosine value cos θ_i, and (b) deviation to chord length ratio d_i/L_i.

points. The overall process of the combined method is discussed.

2.1. *Corner Detection*

The purpose of corner detection is to detect some potential feature points which can sufficiently describe the shape of the object. The common procedures for corner detection involve: (1) estimate the curvature of each point on the curve, and (2) locate the points which have local maximum curvatures as the corners. Many methods are available to estimate curvatures. In here, we introduce two methods to estimate the curvatures. The first one is proposed by Rosenfeld and Johnston [5]. They used the cosine values to estimate the curvatures. As shown in Fig. 1a, the estimated curvature c_i is set to be cos θ_i, where P_i is the ith point on the curve. Another method is a modified version of the method proposed by Teh and Chin [8]. They used the ratio of the deviation and the chord length, where the deviation is the distance between point P_i and chord $\overline{P_{i-k}P_{i+k}}$, to determine the region of the support of point P_i. The above ratio is used to estimate the curvature at point P_i, i.e., the curvature c_i is estimated by d_i/L_i (see Fig. 1b). It is the same as the Teh and Chin method except that we use a prespecified K value to speed up the computation. The above two curvature estimation methods are used in our experiments.

Once the curvatures of all the points on the curve have been estimated by the above two methods, the next step is to locate the points which have local maximum curvature. The points which the curvatures are greater than a threshold value are considered as the potential corners. They are used as the starting points for the following polygonal approximation.

2.2. *Polygonal Approximation*

Performing polygonal approximation for the object's contour can achieve both data compression and feature extraction. The approximation methods can be classified into two categories, which are sequential and iterative approaches. The sequential methods start from a point to find the longest allowable segment by merging points one after another to the starting point, until it fails to satisfy certain criteria [4, 7, 9]. The last point on the longest allowable segment is designated as the next starting point, and the procedure is repeated until the whole curve has been traced. The starting points obtained sequentially are then considered as the dominant points. This approach has the drawback that some feature points may be missed or shifted [1]. The iterative methods begin the partition of the curve by assigning two starting points. The partition will be stopped if some testing criteria are satisfied. Otherwise, the curve will be partitioned into two small curves by a point between these two starting points, and the partitioned point will be one of the two starting points for the two newly partitioned small curves. And the procedure will be continued until no further partition is necessary. A testing criterion for determining whether the curve needs further partitioning is that: if $d_i \leq T_d$ stop partitioning the curve, for all i which $s < i < e$, where s and e are the indices of the two starting points, and T_d is the deviation tolerance; otherwise, choose a point to segment the curve.

In this paper, we introduce two methods which can determine the point to segment the curve when a partition is needed. The first method is proposed by Ramer [3]. A curve P_sP_e will be segmented into two smaller curves P_sP_{max} and $P_{max}P_e$ at point P_{max}, if $d_{max} > T_d$, where P_{max} is the point which has the maximum deviation, d_{max}, from the chord $\overline{P_sP_e}$ (see Fig. 2a); otherwise, no partition is applied to the curve P_sP_e. It needs to compute the deviations d_i individually, for $s < i < e$. The process is time-consuming. Since the dominant points on a curve are usually the points with large curvature, we propose to partition the curve at the points which have large curvatures. Here, instead of partitioning the curve at the point with maximum deviation, we propose a new method which partitions the curve at the point with maximum curvature. And the needed curvatures have been computed in the corner detection stage (see Fig. 2b).

Suppose that $c_{(1)} \geq c_{(2)} \geq \ldots \geq c_{(m)}$, where m is the number of points on the curve P_sP_e. The point $P_{(j)}$ has the jth greatest curvature, $c_{(j)}$, and the deviation $d_{(j)}$ on the curve P_sP_e. The partitioning rule is that if the deviation $d_{(1)} > T_d$, then we partition the curve at the point $P_{(1)}$;

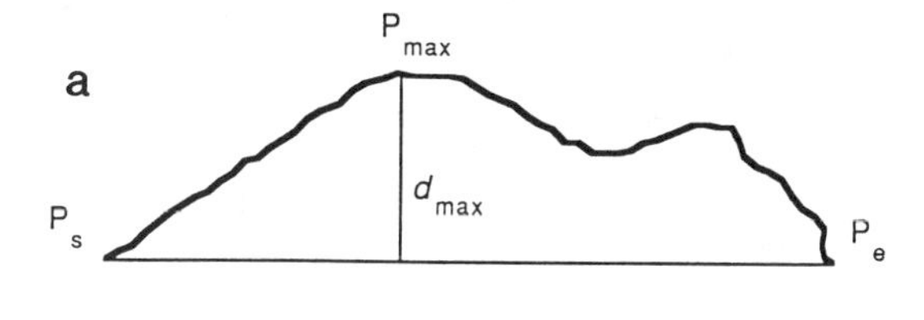

FIG. 2. Partition curve by using (a) the maximum deviation and (b) the maximum curvature.

otherwise, check the next deviation $d_{(2)}$. If the deviation $d_{(2)} > T_d$, then we partition the curve at the point $P_{(2)}$; otherwise, the next deviation $d_{(3)}$ will be checked. The checking procedure terminates when the condition $d_{(j)} > T_d$ (and $d_{(i)} \leq T_d$, for all $i = 1, 2, \ldots, j - 1$) is satisfied for certain j or when no further partition is needed. From the above rule, it is seen that the point with large curvature has high priority to partition the curve. And if all the deviations of the points are less than or equal to T_d, then the curve will not be partitioned. This termination condition is the same as that of Ramer's except that it needs not to compute the deviations individually. Only j deviations need to be computed when $d_{(i)} \leq T_d$, for all $i = 1, 2, \ldots, j - 1$, and $d_{(j)} > T_d$, where $j = 1, 2, \ldots, m$. Further, the above two polygonal approximation methods will be compared.

2.3. *Curvature-Based Polygonal Approximation*

The corner detection algorithms can detect the important feature points on the curve. But these feature points may not be able to properly represent the curves, especially for the smooth curves, e.g., a circle. Further, the performance of the iterative polygonal approximation methods is sensitive to the setting of the starting points. Different starting points may result in different approximations. The choice of the starting points is important to these methods. It is known that corners play an important role in shape representation and recognition, and they are the significant dominant points on the curve. An intuitive approach is to apply the corner points as the starting points for iterative polygonal approximation. Here, we use the results of the corner detection as the starting points for the iterative polygonal approximation. We first apply the corner detection procedure to detect potential dominant points, then we use a polygonal approximation to partition each curve. The combined methods are called the *curvature-based polygonal approximation methods*.

The combined method is very simple and effective. Because the most significant dominant points will be detected in the first step and the remaining small curves can be properly approximated by using these good starting points.

Since the points on a straight line cannot be considered as the dominant points, we must exclude these points [1]. It can be done by checking the chain codes. Suppose that the chain codes of the curve are the ordered sequences $a_1, a_2, \ldots, a_n$, then point P_i cannot be considered as a dominant point, if $a_{i-1} = a_i$. For example, point P_5 and P_6 in Fig. 3 are not considered as dominant points ($a_4 = a_5 = a_6 = 7$), and points P_4 and P_7 are considered as the possible dominant points ($a_3 = 0 \neq 7 = a_4$ and $a_6 = 7 \neq 5 = a_7$). The possible dominant points are called the *break points*, which are denoted as "○" in Fig. 3. Therefore, only the curvatures of the break points need to be estimated. And this can reduce the processing time for both curvature estimation and deviation computation.

In summary, the three stages of the curvature-based poloygonal approximation method are:

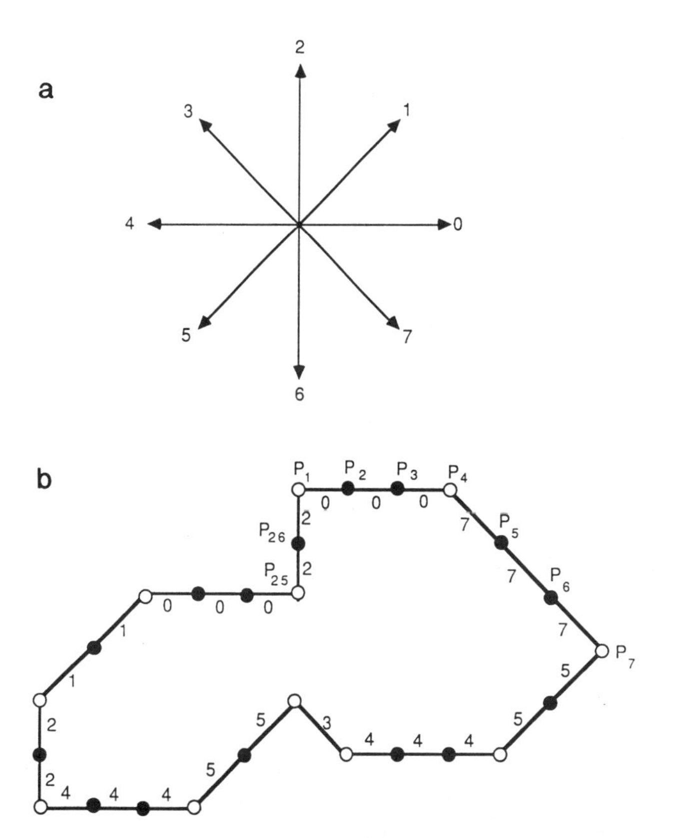

FIG. 3. Freeman's chain codes and (b) break point detection by the relationship between the points P_i and the chain codes a_i on the curve. Chain codes: 000777554443554442211100022. (○) break point; (●) linear point.

(1) perform break point detection by checking the chain codes;

(2) detect potential corners by applying corner detection method;

(3) partition the curve between each of the two consecutive corners by using the iterative polygonal approximation.

The detected corners and the partitioned points are considered as the dominant points on the curve. These dominant points will be useful for further analysis such as shape recognition and corresponding points matching in time-varying image analysis.

3. IMPLEMENTATION

In order to evaluate the combined methods, some testing images are applied. With certain performance evaluation criteria, the performance of these combined methods are discussed. In the above section, we use two approaches to estimate the curvatures in corner detection, and two different partition rules for polygonal approximation.

(1) Corner detection: (a) cosine value (*COS*), i.e., $c_i = \cos\theta_i$, and (b) deviation to chord length ratio (*DCR*), i.e., $c_i = d_i/L_i$.

(2) Polygonal approximation: (a) maximum deviation (*MAXD*), i.e., $d_{\max} > T_d$, and (b) maximum curvature (*MAXC*), i.e., $d_{(j)} > T_d$, for $c_{(1)} \geq c_{(2)} \geq \ldots \geq c_{(m)}$.

Combining the two methods involved in each of the two stages, we have four curvature-based polygonal approximation methods to detect dominant points. They are denoted as (1) *COS-MAXD*, (2) *COS-MAXC*, (3) *DCR-MAXD*, and (4) *DCR-MAXC*.

3.1. *Performance Evaluation Criteria*

From the view point of data compression, it is desirable to obtain a small number of dominant points rather than to have a large number of dominant points. However, a small number of dominant points may cause a large distortion on the curve. It seems that, there is a trade-off between the number of dominant points and the degree of distortion of the curve. In here, three quantitative criteria were used to evaluate the performance of the four methods.

(1) Compression ratio (*CR*): An intuitive index to assess the effectiveness of data compression is the number of detected dominant points. The ratio of the number of points on the curve and the number of detected dominant points is taken as the compression ratio. The compression ratio is defined as

$$CR = \frac{N}{M}, \tag{1}$$

where N is the number of points on the original curve (if the curve is closed then $N = n$, where n is the number of chain codes on the curve; otherwise, $N = n + 1$), and M is the number of dominant points on the partitioned curve. The larger the compression ratio is, the more effective in data reduction the method is.

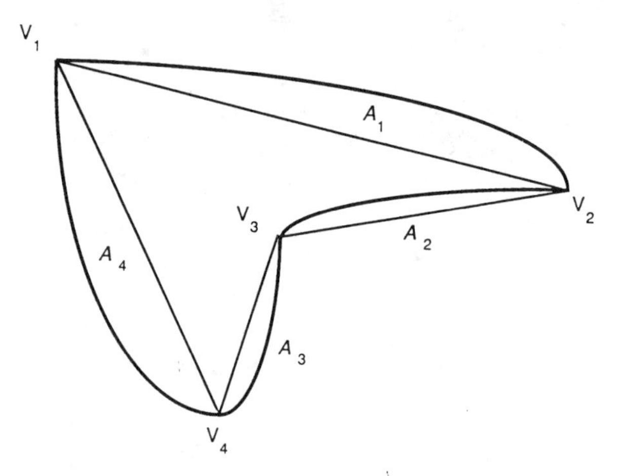

FIG. 4. Illustration of the mean area errors (V_i: dominant points; A_i: area errors).

(2) Mean area error (*MAE*): The criterion of assessing the distortions caused by dominant point detection is to evaluate the differences between the original curves and the approximated polygons. The mean area error is the average of the area differences of all the approximated segments. The mean area error is defined as

$$MAE = \frac{1}{M'} \sum_{i=1}^{M'} A_i, \tag{2}$$

where M' is the number of segments on the partitioned curve (if the curve is closed then $M' = M$; otherwise $M' = M - 1$) and A_i is the area enclosed by the original curve and the chord $\overline{V_iV_{i+1}}$ (V_i is the ith dominant point, that is, the ith vertex on the approximated polygon; see Fig. 4). The mean area error is used to evaluate the distortion caused by polygonal approximation. It can indicate whether the approximated polygon is good enough to describe the original curve or not. It is clear that the smaller the mean area error is, the better descriptive ability the method has.

(3) Normalized mean area error (*NMAE*): It is known that the approximated polygon can be the same as the original curve, if the deviation tolerance T_d is set to zero. The tolerance is controlled by T_d. It is necessary to consider the relation between the distortions and the tolerances. Here, we define the normalized mean area error as the ratio of the mean area error and the squared T_d:

$$NMAE = \frac{MAE}{T_d^2}. \tag{3}$$

FIG. 5. The testing images: gear (N = 319 and 76) and screw (N = 599).

The ratio of mean area error and T_d^2 can indicate the compensation effect of T_d on the mean area error. If two methods have the same mean area errors, then the method with a larger T_d is considered more robust. This is due to that a large T_d usually causes a large mean area error, we would prefer to choose the method which can perform well under a large T_d. Therefore, the normalized

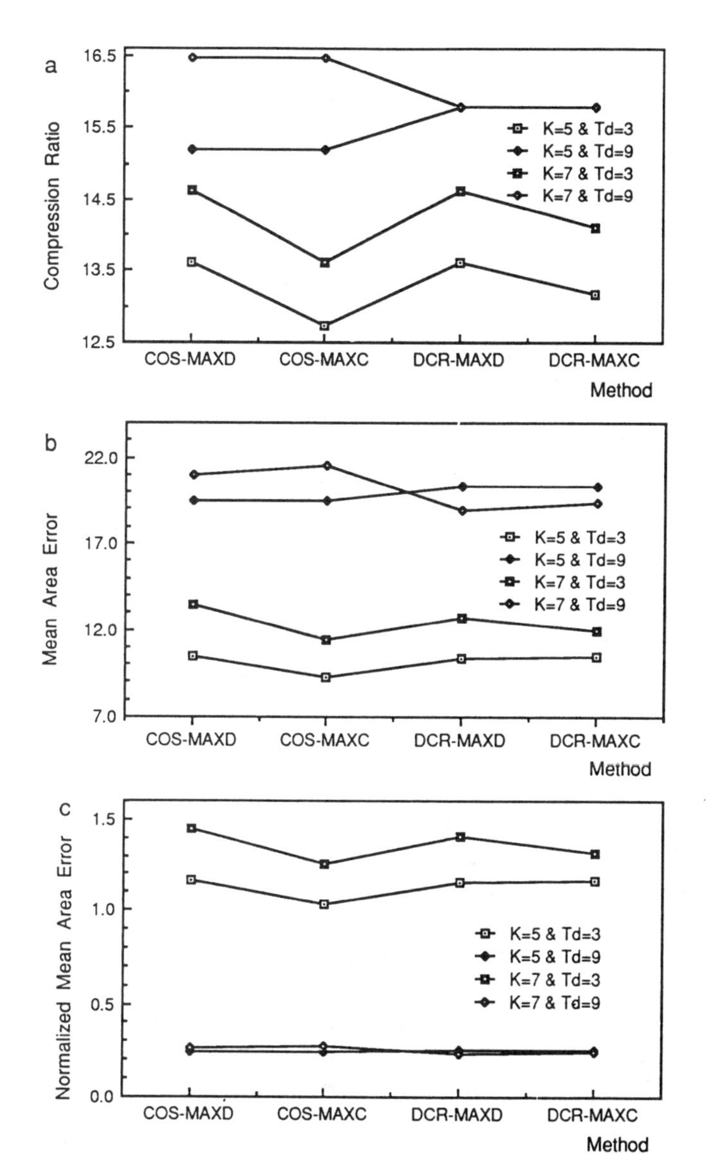

FIG. 6. Performance of the four combined methods for the gear image.

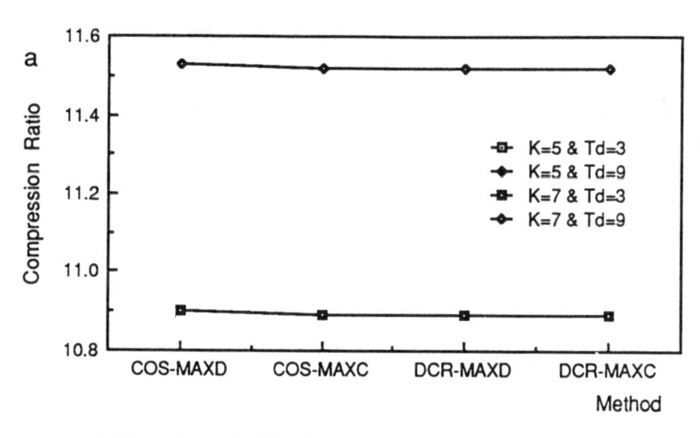

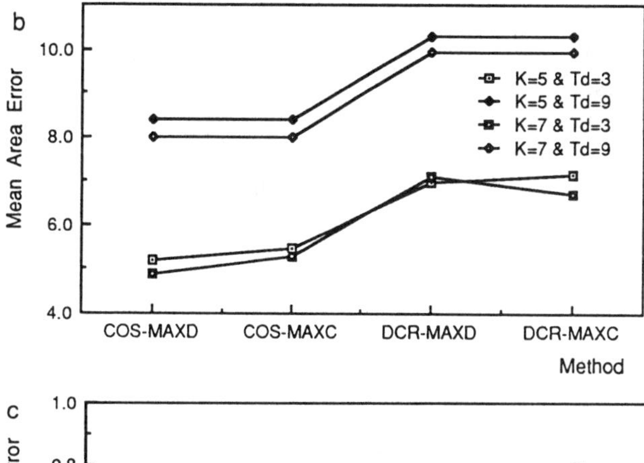

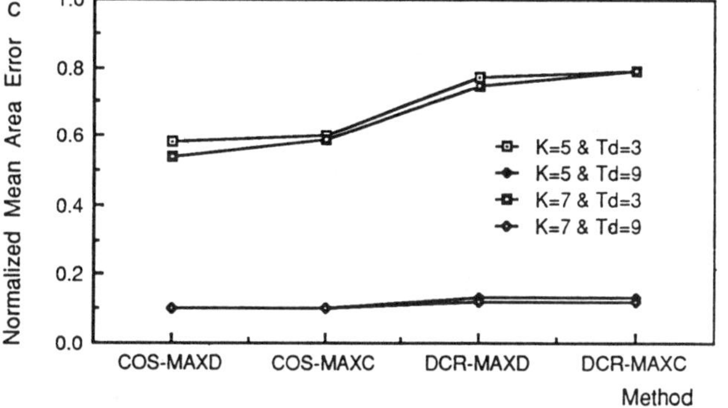

FIG. 7. Performance of the four combined methods for the screw image.

mean area error is applied to evaluate the robustness of the methods. Thus one can say that the smaller the normalized mean area error is, the more robust the method is.

3.2. *Results and Discussions*

Two testing images were used to evaluate the performance of the four curvature-based polygonal approximation methods. As shown in Fig. 5, they are the images of gear and screw. Two parameters (K and T_d) are involved in the detection process. The value of K is used to estimate the curvatures of the break points on the curve, and the tolerance T_d is for polygonal approximation. In experiment design, two levels were specified for each of the two parameters (K = 5 and 7, and T_d = 3 and 9). Figures 6 and 7 show the performance of the three quantitative performance evaluation criteria of the four methods for the gear and screw images respectively.

(1) *The Effects of Deviation Tolerances.* As expected, in Figs. 6a and 7a, a large T_d results in a large compression ratio. This is due to that a large distortion is allowed when a large T_d is specified. Further, Figs. 6b and 7b show that a large T_d results in a large mean area error. It indicates that the increase of the value T_d will increase the ability in data reduction. Meanwhile, it will also increase level of distortions. It is logical to find a trade-off between data reduction and distortion. The results in Figs. 6c and 7c, show an opposite relation between *NMAE* and T_d. As the value of T_d increases, the normalized mean area error decreases. It is somewhat interesting, because the mean area error has a positive relation with T_d. From the definition of the normalized mean area error, it is evident that the method applied with a large T_d tends to have a smaller *NMAE* than that of being applied with a small T_d, if the mean area errors are close to each other. That is, if the increasing degree of squared T_d is larger than that of *MAE*, then we will have an opposite relation between *NMAE* and T_d, while *MAE* and T_d have a positive relation. This is the cases shown in Figs. 6b and 6c (or Figs. 7b and 7c). As a result, if two methods have the same mean area error value, the method with a larger tolerance T_d tends to be more robust than the other method. Because it can detect the same dominant points under a larger deviation tolerance. Thus, we can evaluate the methods under different levels of T_d.

(2) *The Effects of the Value of K.* From Fig. 6, we can see that a large K value causes a large compression ratio, a large mean area error, and a large normalized mean area error, given the same T_d. It suggests that a large K value can have a positive effect on data reduction, but an adverse effect on shape representation. But for the screw image in Fig. 7, the setting of K value has no obvious effects on all the three performance evaluation criteria as compared to the effects caused by T_d. This is due to that the dominant points in the screw image are more obvious than those in the gear image. Hence, the results of dominant point detection of the screw image are less sensitive to the curvature values which are influenced by the value of K.

(3) *The Effects of Curvature Estimation Methods and Partitioning Point Selection Rules.* For the choice of curvature estimation methods, from Figs. 6a to 6c it is seen that the performance differences of the four methods under the three evaluation criteria are not large. It indicates that curvature estimation by using cosine value or by using deviation to chord length ratio has the similar effect for the gear image. For the screw image, as shown in Fig. 7a, the effects of the curvature estimation methods on compression ratio are also not obvious. But, in Figs. 7b and 7c, the method differences become evident. The *COS* methods have better performance than the *DCR* methods in both *MAE* and *NMAE*. Thus, the *COS-MAXD* and *COS-MAXC* methods are recommended to detect the dominant points for the screw image.

Figure 6a shows that the *COS-MAXD* (*DCR-MAXD*) method tends to cause larger compression ratio and larger mean area error than that of the *COS-MAXC* (*DCR-MAXC*) method. That is, applying the maximum deviation rule to partition a curve will result in detecting less dominant points. The maximum curvature methods have better performance in shape representation. But they have worse performance in data reduction than that of the maximum deviation methods. It suggests that if the data reduction is the most important criterion, then one should choose the maximum deviation rule in polygonal approximation. Further, from Figs. 7a to 7c, it is clear that the performance differences between maximum deviation and maximum curvature partitioning rules are not large. This is due to that most of the dominant points on the screw image are sharp, and they tend to be identified in the corner detection process. Hence, the number of the dominant points extracted in the polygonal approximation process is small. Therefore, the effects of the partitioning point selection rules is not obvious.

Overall, the most important factor that affect the quantitative performance of the methods is the setting of T_d. For the gear image, the *COS-MAXD* method has the best performance in compression ratio and the *COS-MAXC* method has less distortion. For the screw image, the *COS*-MAXD method has the best performance, because it has large compression ratio, and small mean area error as well as normalized mean area error.

In addition to quantitative evaluation, qualitative comparison is also used. Figures 8a and 8b show the approximated curves of the detected dominant points by the four combined methods with $T_d = 3$ and $T_d = 9$, respectively. As shown in Fig. 8a, all the four methods can detect the obvious feature points in the gear and screw images. In Fig. 8b, one feature point in the screw image is missed by all the four methods. This feature point is on the smooth part of the curve (compared with the other feature points), and it was not detected at the corner detection stage. Further, this point is also missed in the polygonal approximation stage due to that a large tolerance $T_d = 9$ is specified. But if the value T_d is set enough small (e.g., $T_d = 3$), the missed feature point can be successfully detected as shown in Fig. 8a.

From the above illustrations, we can see that the resolution of the approximated curves can be controlled by the setting of the deviation tolerance T_d. This can be found by comparing the approximated curves of the circles in the gear image between Fig. 8a and 8b. The circles in the Fig. 8a are better approximated than those in Fig. 8b, due to a smaller T_d value is set.

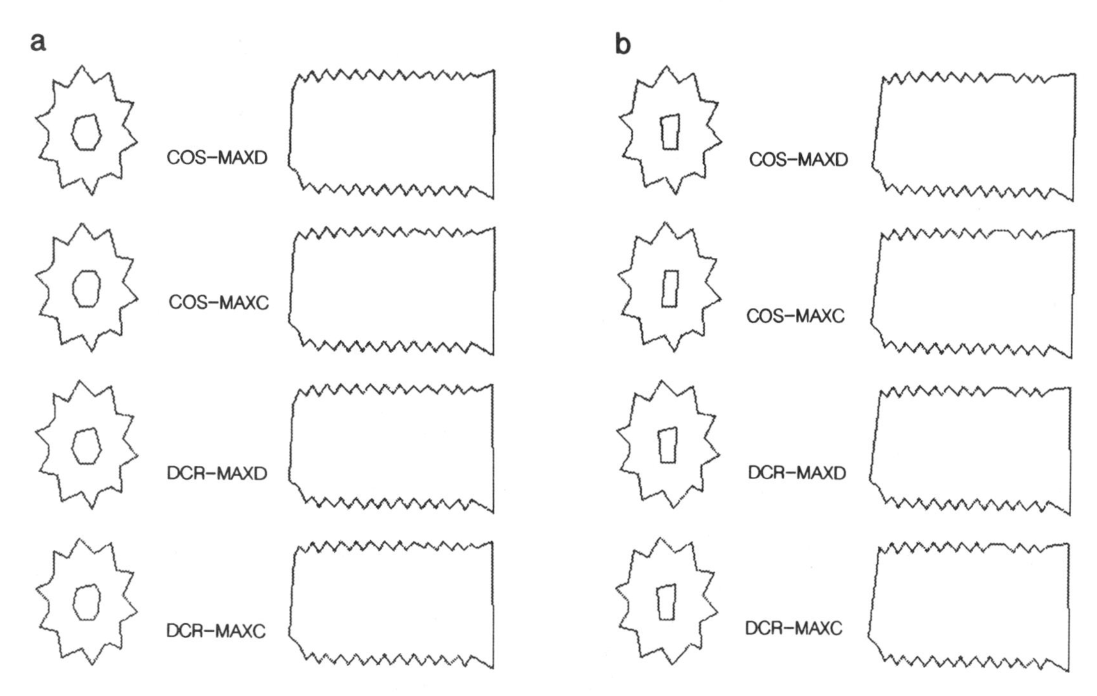

FIG. 8. The approximated polygons of the detected dominant points by the four methods with $K = 5$, and (a) $T_d = 3$ and (b) $T_d = 9$.

Two parameters, K and T_d, are involved in the curvature-based polygonal approximation methods. The results of the experiments indicate that these two parameters will influence the approximated results. In order to eliminate the effects caused by them, one can use the method proposed by Teh and Chin [8] to automatically select the *support of regions* of the points (which determines the value of K) in the corner detection process. Further, instead of checking the testing criterion $d_i \leq T_d$, for all i, one can use the *significant measures* [6], which do not need the parameter T_d for polygonal approximation. Consequently, the computation time will be increased if the above changes are made for the *CBPA* method. For further study, one may apply the above alternatives to the *CBPA* method, and compare its performance with the methods introduced in this paper.

3.3. *Comparison with Conventional Methods*

The performance of the curvature-based polygonal approximation methods has been discussed in the last section. In this section, we want to compare the combined methods with the conventional methods, to show how the combined methods can improve the effectiveness of dominant point detection. For simplicity, the *COS-MAXD* method is used to compare with the two corresponding traditional dominant point detection methods. The first one is the corner detection method which uses the cosine values as the estimated curvatures (denoted as the *COS* method), and the second one is the iterative polygonal approximation method which is based on the maximum deviation partitioning rule (denoted as the *MAXD* method). In order to assess the performance of these three methods (*COS*, *MAXD*, and *COS-MAXD*), two different types of objects (a small mechanical part and a key) were tested in the experiment. They are shown in Fig. 9. Generally speaking, a good dominant point detection method should be able to properly detect the dominant points on the boundary of an object regardless of its orientations or scalings. Here, for each testing object, by changing its orientation or size, three different versions of the object were used for evaluation. The six outer boundary images are shown in Figs. 10a–15a. Further, the images in (b), (c), and (d) of each figure present the detected dominant points (marked as dots) by the *COS*, the *MAXD*, and the *COS-MAXD* methods, respectively. Tables 1 and 2 summarize the detection results. In Tables 1 and 2, since T_d is not used in the *COS* method, the *NMAE* performance is not assessed.

As shown in Fig. 10b, the *COS* method detects four dominant points. From the view point of corner detection, the *COS* method has satisfactory performance because these four points are the main corner points. But for polygonal approximation, the four points are not suffi-

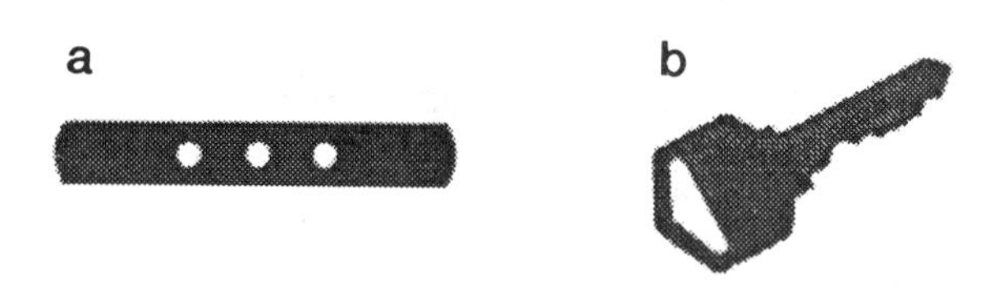

FIG. 9. Two types of testing objects: (a) mechanical part and (b) key.

TABLE 1
Detection Results for the Curves in Figs. 10a, 11a, and 12a

	Performance								
	M			*MAE*			Processing time (sec)		
Method	10a	11a	12a	10a	11a	12a	10a	11a	12a
COS	4	4	5	46.0	95.7	62.4	5.4	5.6	4.9
MAXD	6	8	7	7.9	27.9	29.0	7.6	8.2	6.9
COS-MAXD	6	6	6	8.2	42.2	31.0	2.8	3.9	3.0

cient to represent the curve in Fig. 10a due to that there are two arcs in the curve. On the other hand, both the *MAXD* and the *COS-MAXD* methods detect six dominant points (see Figs. 10c and 10d) which can sufficiently describe the curve in Fig. 10a. Figure 11a is the boundary image of the same object in Fig. 10a with a different orientation. As shown in Figs. 11b and 11d, the *COS* and the *COS-MAXD* methods have the similar results as comparing to those in Figs. 10b and 10d. However, in Fig. 11c, the *MAXD* method detects too many dominant points, this is due to that the starting points are not correctly chosen. And it also indicates that the iterative polygonal approximation method is sensitive to the orientation of the curves. Further, the curve in Fig. 12a is the reduced version of the object in Fig. 10a. From Figs. 12b–12d, we can see that both the *COS* and the *MAXD* methods detect dummy dominant points, only the *COS-MAXD* method detects the correct dominant points. Since it has the most consistent result, we can say that the combined method is less sensitive to the orientation and scaling of the object in Fig. 9a. Besides, from the results in Table 1, the *COS* method has the largest mean area error among the three methods, the other two methods have rather similar performance on mean area error. For processing time, the *COS-MAXD* method has the best performance among these three methods. This is due to that a break point detection procedure was included in the combined method. Hence, it can save the computation time both in estimating curvatures and in computing distances for the linear points.

The dominant point detection results for the key image (Fig. 13a) by the three methods are presented in Figs. 13b–13d. From these figures, again the *COS-MAXD* method gives the best performance, because all the significant dominant points have been successfully detected by the combined method. The *MAXD* method misses some dominant points and detects some dummy dominant points (as seen in the upper-left part of the curve in Fig. 13c), and the *COS* method misses a couple dominant points. In Fig 14 (which is a rotated version of Fig. 13), the results are not quite satisfactory because some dominant points are not detected. But, among the three methods, the *COS-MAXD* still has the best result. The *COS* method misses many dominant points, and the *MAXD* method misses some dominant points and detects a dummay dominant point. The reason that the *COS-MAXD* method misses several dominant points is due to that all the deviations between the undetected points and the line segment connected by two consecutively detected dominant points are less than the value of T_d in polygonal

a b c d

FIG. 10. (a) Original curve (N = 423) and the detected dominant points with K = 3 and T_d = 3 by the (b) *COS*, (c) *MAXD*, and (d) *COS-MAXD* methods.

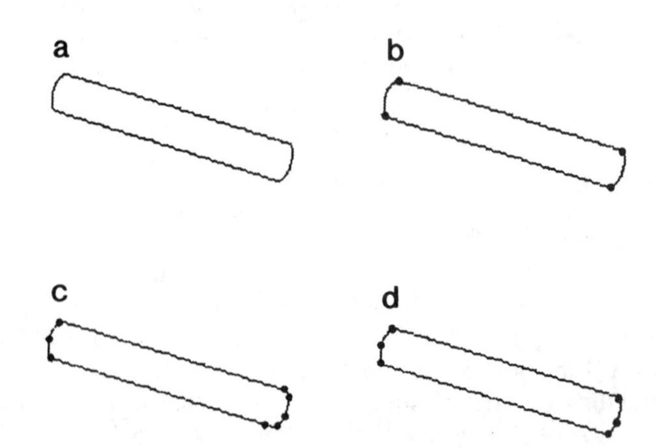

FIG. 11. (a) Original curve (N = 411) and the detected dominant points with K = 5 and T_d = 3 by the (b) *COS*, (c) *MAXD*, and (d) *COS-MAXD* methods.

TABLE 2
Detection Results for the Curves in Figs. 13a, 14a, and 15a

Method	M			Performance *MAE*			Processing time (sec)		
	13a	14a	15a	13a	14a	15a	13a	14a	15a
COS	18	12	13	22.8	104.7	166.8	5.7	5.6	6.3
MAXD	17	15	21	12.7	17.2	12.0	8.8	8.8	10.9
COS-MAXD	21	17	21	10.0	15.1	13.7	5.0	4.1	5.0

approximation. Moreover, in Figs. 15b–15d, for an enlarged image, the same phenomena are found and again the *COS-MAXD* method has the best approximation result. In Fig. 15b, some important points are undetected by the *COS* method. And the *MAXD* method detects some dummy dominant points and misses some dominant points (see Fig. 15c). Table 2 summarizes the detection results of Figs. 13–15. As it can be seen that the combined method has the best performance on both *MAE* and processing time. The reason that the *COS-MAXD* method has the smallest *MAE* is due to that it has the best approximation result. And the reason that the *COS-MAXD* method is the most efficient method is again due to the inclusion of the break point detection procedure in dominant point detection. Besides, the consistent detection results of the *COS-MAXD* method indicates that it is relatively insensitive to the orientation and scaling change of the object. And these results illustrate that the combined method can improve the robustness of the conventional methods.

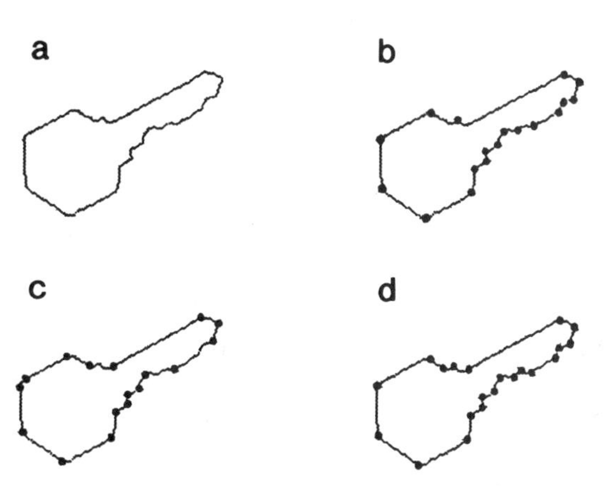

FIG. 13. (a) Original curve (N = 360) and the detected dominant points with K = 5 and T_d = 3 by the (b) *COS*, (c) *MAXD*, and (d) *COS-MAXD* methods.

4. CONCLUSIONS

This paper presents the dominant point detection by the curvature-based polygonal approximation method which is the combination of corner detection and polygonal approximation techniques. Prior to corner detection, a preprocessing is performed to eliminate the points that belong to the straight line of the curve. In short, the curvature-based polygonal approximation method involves three stages: (1) extract the break points that do not lie on a straight line, (2) detect the potential corners, and (3) perform polygonal approximation for each curve between two consecutive potential corners.

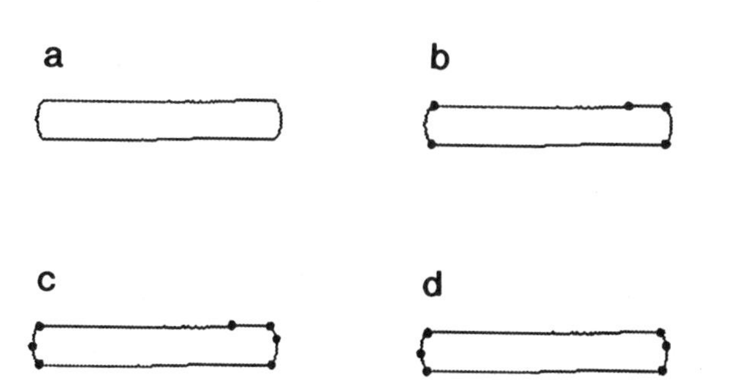

FIG. 12. (a) Original curve (N = 354) and the detected dominant points with K = 5 and T_d = 3 by the (b) *COS*, (c) *MAXD*, and (d) *COS-MAXD* methods.

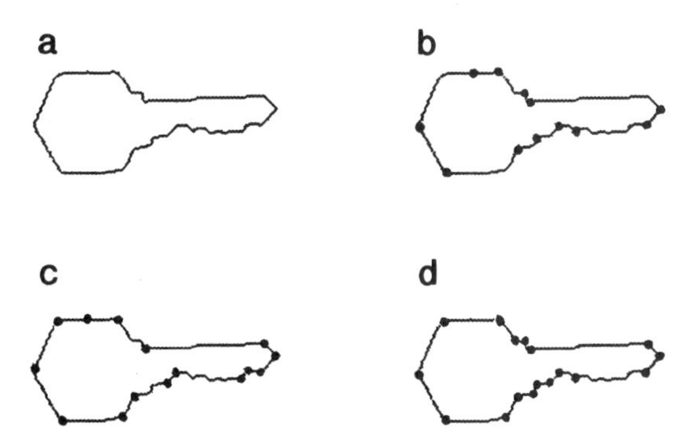

FIG. 14. (a) Original curve (N = 369) and the detected dominant points with K = 5 and T_d = 3 by the (b) *COS*, (c) *MAXD*, and (d) *COS-MAXD* methods.

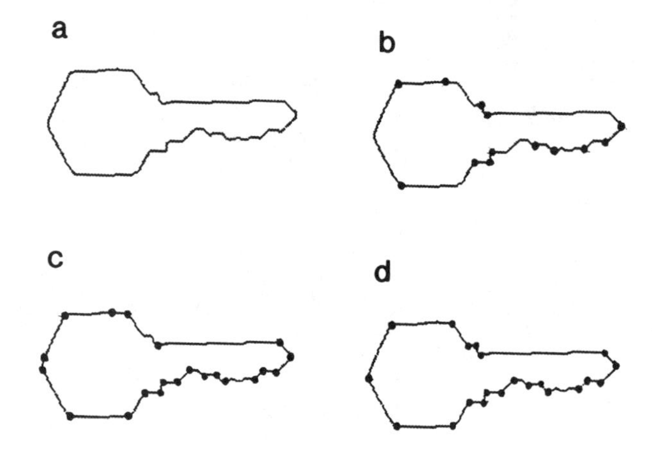

FIG. 15. (a) Original curve (N = 447) and the detected dominant points with K = 5 and T_d = 3 by the (b) *COS*, (c) *MAXD*, and (d) *COS-MAXD* methods.

From the performance evaluation of the four *CBPA* methods, it is seen that the dominant point detection can be done by combining the methods of corner detection and polygonal approximation. Further, using the maximum curvature rule for determining the partitioned points has equally good performance as that of using the maximum deviation rule. A merit of using the maximum curvature rule is that it does not need to compute all the deviations and thus saves computation time. Overall, the most significant dominant points can be detected during corner detection. And the results of corner detection can provide a set of suitable starting points for iterative polygonal approximation. Since, the *CBPA* method can preserve the significant feature points, thus it can effectively represent the curve.

Further, when comparing the *COS-MAXD* method with its corresponding corner detection and polygonal approximation methods, it is seen that the combined method has the following advantages:

(1) The combined method is more efficient in processing time than that of the conventional methods.

(2) The combined method is more robust than the conventional methods, because the dominant points detected by the combined method are very consistent for the object when changing its size and orientation.

(3) The combined method can generate more satisfactory approximation outcomes than those of the other two conventional methods.

REFERENCES

1. H. Aoyama and M. Kawagoe, A piecewise linear approximation method preserving visual feature points of original figures, *CVGIP: Graphical Model Image Process.* **53,** 1991, 435–446.
2. N. Ansari and E. J. Delp, On detecting dominant points, *Pattern Recognition* **24**(5), 1991, 441–451.
3. U. Ramer, An iterative procedure for the polygonal approximation of plane curves, *Comput. Graphics Image Process.* **1,** 1972, 244–256.
4. B. K. Ray and K. S. Ray, A new approach to polygonal approximation, *Pattern Recognition Lett.* **12**(4), 1991, 229–234.
5. A. Rosenfeld and E. Johnston, Angle detection on digital curves, *IEEE Trans. Comput.* **22,** 1973, 875–878.
6. P. L. Rosin and G. A. W. West, Segmentation of edges into lines and arcs, *Image Vision Comput.* **7**(2), 1989, 109–114.
7. J. Sklansky and V. Gonzalez, Fast polygonal approximation of digitized curves, *Pattern Recognition* **12,** 1980, 327–331.
8. C. H. Teh and R. T. Chin, On the detection of dominant points on digital curves, *IEEE Trans. Pattern Anal. Mach. Intell.* **11**(8), 1989, 859–872.
9. K. Wall and P. E. Danielsson, A fast sequential method for polygonal approximation of digitized curves, *Comput. Vision Graphics Image Process.* **28,** 1984, 220–227.
10. G. A. W. West and P. L. Rosin, Techniques for segmenting image curves into meaningful descriptions, *Pattern Recognition* **24**(7), 1991, 643–652.

Curvilinear Feature Detection from Curvature Estimation

Lawrence O'Gorman

AT&T Bell Laboratories
Murray Hill, New Jersey 07974

Abstract

One method of detecting features on digitized image lines is to first estimate the local curvature along the lines, then analyze the curvature plot for peaks corresponding to corners and curves. The DOS$^+$ method, a particular case of the difference of slopes (DOS) approach, has been shown to be effective for estimating curvature in terms of signal detectability. In this paper, relationships are established between the DOS$^+$ parameters and feature parameters to distinguish corner and curve features and accurately determine the locations of corners, transitions between straight lines and curves, and curve centers. Examples show the results of this method applied to curvilinear features and diagrams.

1. Introduction

For computer interpretation of line diagrams, and for recognition of geometrically shaped objects by their boundaries or contours, extraction of corners and curves is important. Segmentation and feature identification enable manipulation and editing of diagram components, and matching for object recognition. The subject of this paper is the effective conversion of raster images to their vector, or feature, formats.

There are a number of methods for detecting features on digitized lines. (For a more complete description of background methods than will be given here, see reference [1].) For one class of methods, detection of corner, curve, and straight line features is made from a plot of estimated local curvature along the image lines. In [1], two approaches for curvature estimation, the difference of slopes (DOS) approach [2-3] and the Gaussian smoothing method [4], are described and compared. It is shown that a particular case of the DOS approach, the DOS$^+$ method, yields the best signal detectability among these methods. This DOS$^+$ method is described here along with techniques for distinguishing curves from corner features, and calculations to determine parameters of these features.

For the DOS$^+$ method, two line segments each of arc length W and separated by arc length M, are fit to the data curve as in Figure 1. (The "+" in DOS$^+$ indicates that the gap length M is positive — as opposed to other DOS cases where the gap length may be zero or negative.) At each point along the data curve, the angle of intersection of the straight-line extensions of the W segments is measured, and the supplement of this angle is plotted in the θ-plot as a function of arc length s. Therefore, the θ-plot displays the deviation from collinearity of the W-segments and this describes the local curvature along the line.

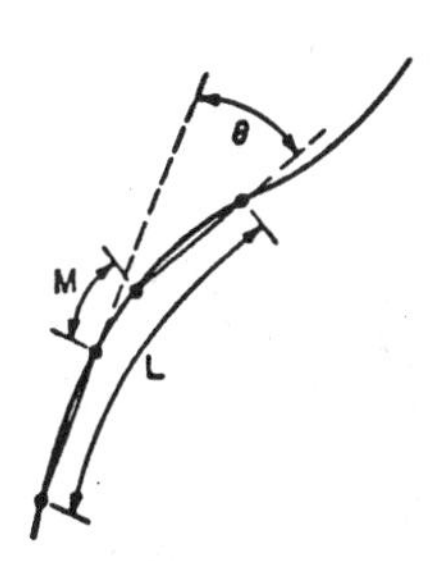

Figure 1. DOS$^+$ fit applied to curved line.

For continuous curves and noiseless conditions, the correspondence between line features and θ-plot peaks is straightforward: values of $\theta = 0$ correspond to straight line fits to the data, non-zero plateaus correspond to curves of constant curvature, and sharp peaks correspond to corners. In practice, images contain noise and sampling error, and the correspondence between line features and peaks on the θ-plot is not so straightforward. In this paper, we consider the image lines to be curvilinear — that is all features, including corners, are considered to have some degree of curvature. The relationships for determining straight line, curve, and corner features from the curvature plot as functions of the DOS$^+$ parameters are established here.

2. DOS$^+$ Parameters and θ-Plot Calculation

For the DOS$^+$ method, the two segments which are fit to the data are of length W, and the gap between segments is M. The arc length from the beginning of one segment to the end of the other is L. Therefore,

$$L = 2W+M. \tag{1}$$

(Note that in this paper we deal only with discrete space, so a line is a chain of data points, and its length is the sum of Euclidian distances from pixel to pixel.) The length L determines the feature resolution of the method. The more closely spaced are corner and curve features on image lines, the shorter L must be to reduce interference between neighboring features. On the other hand, we desire that the segment lengths, W, be as long as possible to reduce noise. The final constraint, discussed in reference [1], is that M should be positive and equal to or greater than the maximum arc length of a corner. From these constraints, we make L equal to the minimum straight line arc length separating features, $L = s_{L_{\min}}$; M is set to the maximum arc length of a corner, $M = s_{K_{\max}}$; and W is then $W = (s_{L_{\min}} - s_{K_{\max}})/2$. Implicit in these specifications is the assumption that there is a minimum straight line arc length, L, between corner and curve features, and that this is greater than the maximum arc length of a corner. As $s_{L_{\min}}$ becomes small with respect to $s_{K_{\max}}$, the ability to resolve adjacent features is reduced.

At any location $s=s_i$ along the data, the W-segments are fit from $s_i - M/2 - W$ to $s_i - M/2$, and from $s_i + M/2$ to $s_i + M/2 + W$. For fitting the W segments to the data we have found that a simple endpoint to endpoint line fit gives good results at a low computational expense. The (x,y) coordinates corresponding to these endpoints of the W-segments before and after s_i are (x_{1o_i}, y_{1o_i}), (x_{1f_i}, y_{1f_i}), (x_{2o_i}, y_{2o_i}), and (x_{2f_i}, y_{2f_i}) respectively. (The subscript i indicates the location on the curve where the curvature is being estimated. Where the meaning is clear, we will omit this subscript for readability.) If the lines are extrapolated from the W segments to the point of intersection, then the angle by which these segments deviate from collinearity is $\theta(s_i)$. The angle of orientation for each W segment is found,

$$\gamma_m = \arctan(\frac{\Delta y_m}{\Delta x_m}) + \begin{cases} 0, & \text{for } \Delta y_m \geq 0, \Delta x_m \geq 0 \\ \pi, & \text{for } \Delta y_m \geq 0, \Delta x_m < 0 \\ 0, & \text{for } \Delta y_m < 0, \Delta x_m \geq 0 \\ \pi, & \text{for } \Delta y_m < 0, \Delta x_m < 0 \end{cases} \quad (2)$$

where $\Delta y_m = y_{mf} - y_{mo}$ and $\Delta x_m = x_{mf} - x_{mo}$, for $m = 1,2$. The angular difference between W segments is,

$$\theta(s_i) = \gamma_2 - \gamma_1. \quad (3)$$

3. Feature Determination

The straight line, corner, and curve features are determined from the θ-plot. For a straight line which is sampled on the Cartesian coordinate plane, the samples may or may not be collinear. The chord property [5] states that a chain of samples is the digitization of a straight line if a line segment joining the endpoints of the chain lies everywhere within a distance of one sample from a chain sample. From this criterion, it has been shown in [1] that the maximum θ for an eight-connected straight line segment is,

$$\theta_{S_{\max}} = \arctan\frac{2}{L-M-2\sqrt{2}+2}. \quad (4)$$

Therefore we define the region on the θ-plot which is within $\pm\theta_{S_{\max}}$ of the straight-line fit, as the *zero-range*:

$$\text{zero-range: } -\theta_{S_{\max}} \leq \theta \leq \theta_{S_{\max}}. \quad (5)$$

When there is a crossing on the θ-plot into or out of the zero-range, it is called a *zero-range crossing*, s_z. That is,

$$\text{zero-range crossing: } \theta(s_z) = \pm\theta_{S_{\max}}. \quad (6)$$

For a corner or curve, the θ-plot is outside the zero-range for the length that any part of the DOS$^+$-fit encompasses a feature. This peak width between zero-range crossings s_{z_j} and $s_{z_{j+1}}$ on the θ-plot is,

$$\text{peak width: } s_{zz_j} = s_{z_{j+1}} - s_{z_j}, \quad \text{where } \{(\theta(s_i) > 0) \text{ or } (\theta(s_i) < 0), \text{ for } s_{z_j} < s_i < s_{z_{j+1}}\}. \quad (7)$$

Therefore, the length between zero-range crossings, s_{zz}, on the θ-plot is,

$$s_{zz} = \begin{cases} L + s_K - 1, & \text{for corner of length } s_K, \\ L + s_C - 1, & \text{for curve of length } s_C. \end{cases} \quad (8)$$

Ideally only the width of the peak is needed to determine the features. However in practice, due to the effects of noise and quantization, the width between zero-range crossings on the θ-plot is not a reliable measurement. Because the slope around the zero-range crossings on the θ-plot is usually small, a small deviation in θ can result in a large deviation in zero-range crossing location s_{z_j}, and thus a large change in the peak width. The peak θ value of a feature,

$$\theta_{peak_j} = \max\left\{ |\theta(s_i)|,\ s_{z_j} < s_i < s_{z_{j+1}} \right\}, \quad (9)$$

is a more reliable measurement. However it alone is not sufficient to differentiate a curve from a corner. It is best to combine these measures of peak width and peak height with some task-dependent knowledge which is usually available. For instance, often only corners of 90° and curves of much lower curvature are present. In this case, a simple threshold on the θ-plot peaks is sufficient to distinguish corners from curve features. Or, if all curves in the application are known to have long arc length, these are also easily distinguishable from corners just from equation (8).

4. Feature Description

After correspondences have been found between θ-plot peaks and data features, we wish to determine descriptive parameters of these features. These parameters are the endpoint locations of straight lines, the transition points between straight lines and curves, and the radius and center of curvature of the curves.

The equations of the two straight lines bounding a feature will first be found. Since we have assumed a straight line of length at least L bounding each feature, there will be a zero-range crossing before and after each feature peak, s_{z_j} and $s_{z_{j+1}}$ respectively. We designate the zero-range crossing which ends the preceding peak and begins the following peak as $s_{z_{j-1}}$ and $s_{z_{j+2}}$ respectively. Therefore the straight line preceding a feature peak will be found from $s_{z_{j-1}}$ and s_{z_j}, and the straight line following a feature peak will be found from $s_{z_{j+1}}$ and $s_{z_{j+2}}$. If the image coordinate corresponding to curve location s_i is denoted $(x(s_i), y(s_i))$, then the coordinates for the straight line preceding a feature are:

$$(x_{1a_i}, y_{1a_i}) = \left[x(s_{z_{j-1}} - \frac{L}{2}), y(s_{z_{j-1}} - \frac{L}{2}) \right]$$
$$(x_{1b_i}, y_{1b_i}) = \left[x(s_{z_j} + \frac{L}{2}), y(s_{z_j} + \frac{L}{2}) \right]; \quad (10)$$

and the coordinates for the straight line following a feature are:

$$(x_{2a_i}, y_{2a_i}) = \left[x(s_{z_{j+1}} - \frac{L}{2}), y(s_{z_{j+1}} - \frac{L}{2}) \right]$$
$$(x_{2b_i}, y_{2b_i}) = \left[x(s_{z_{j+2}} + \frac{L}{2}), y(s_{z_{j+2}} + \frac{L}{2}) \right]. \quad (11)$$

(Again we will omit subscript i for readability, with the understanding that curvature is estimated for each data point, s_i.) A corner is completely described by its location. This location is simply the intersection of the two straight lines bounding the corner from equations (10) and (11). The intersection gives the corner coordinates (x,y):

$$x = \frac{1}{d-e}\left[dx_{1a} - ex_{2a} + y_{2a} - y_{1a} \right] \quad (12)$$
$$y = d(x - x_{1a}) + y_{1a},$$
$$\text{where: } d = \frac{y_{1b} - y_{1a}}{x_{1b} - x_{1a}}, \quad e = \frac{y_{2b} - y_{2a}}{x_{2b} - x_{2a}}.$$

For a curve feature, we assume constant curvature and model it as a circular arc (although fits such as a spline could also be used). A curve is described by its radius of curvature, center of curvature, and locations of the two transition points between the curve and the straight lines bounding it. The coordinates of the transition points, $\{(x_{C_j}, y_{C_j}), (x_{C_{j+1}}, y_{C_{j+1}})\}$ are found from the θ-plot:

$$\text{transition 1: } s_{z_j} + \frac{L}{2} \rightarrow (x_{C_j}, y_{C_j}),$$
$$\text{transition 2: } s_{z_{j+1}} - \frac{L}{2} \rightarrow (x_{C_{j+1}}, y_{C_{j+1}}), \quad (13)$$

where the arrow indicates that the image coordinates are those which correspond to the θ-plot locations. The arc length of the curve, s_C, is the length along the data between these transition points,

$$s_C = (s_{zj+1} - s_{zj}) - L. \quad (14)$$

Once the curve has been segmented, curvature parameters must be found. A diagram of a curve is shown in Figure 2. The transitions between straight lines and curve are at $\{(x_{C_j}, y_{C_j}), (x_{C_{j+1}}, y_{C_{j+1}})\}$. If these points are not equidistant from the point of intersection of the extended lines, they must be adjusted to be so. The intersection angle of the straight line extensions is α. The osculating circle shown in dashed lines is the extension of the curve. The center point of this circle is (x_o, y_o). As in equations (10) and (11) for the corner feature, the straight lines bounding the corner are defined by $\{(x_{1a}, y_{1a}), (x_{1b}, y_{1b})\}$ and $\{(x_{2a}, y_{2a}), (x_{2b}, y_{2b})\}$. The intersection point of the lines perpendicular to the tangent lines from each transition point is the center of curvature, (x_o, y_o):

$$x_o = \frac{1}{d-e}\left[de(y_{C_{j+1}} - y_{C_j}) + dx_{C_{j+1}} - ex_{C_j}\right] \quad (15)$$

$$y_o = \frac{1}{d}(x_{C_j} - x_o) + y_{C_j}.$$

$$\text{where: } d = \frac{y_{1b} - y_{1a}}{x_{1b} - x_{1a}}, \quad e = \frac{y_{2b} - y_{2a}}{x_{2b} - x_{2a}}.$$

The radius of curvature is,

$$r_c = \sqrt{(x_o - x_{C_j})^2 + (y_o - y_{C_j})^2}. \quad (16)$$

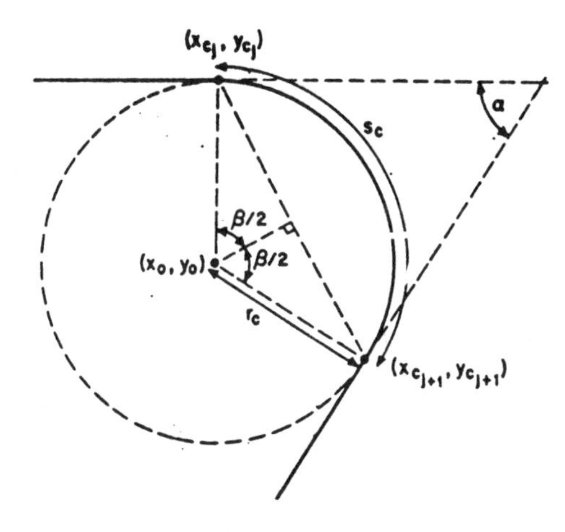

Figure 2. DOS⁺ method applied to curve feature of arc-length s_C.

Reconstruction of the data from these parameters is straightforward. Curves are connected between curve transition pairs using the equation of a circle,

$$(x - x_o)^2 + (y - y_o)^2 = r_c. \quad (17)$$

Straight lines are connected between all other curve transition points and corner locations.

5. ALGORITHM

The general algorithm is given below. In these steps, the initial and final data points are s_o and s_f respectively. It was mentioned in Section 3 that, although the given method is completely defined in theory, task-dependent knowledge can be used to enhance its effectiveness in practice. Tolerances ε_1 and ε_2 are introduced here for that purpose. If no task-dependent knowledge is assumed, these tolerances are equal to zero.

The steps of the DOS⁺ algorithm are listed below:

1. Calculate the θ-plot.
   a. For $s_i = s_o + L/2$ to $s_i = s_f - L/2$, calculate the orientation of the W segments from equation (2), and $\theta(s_i)$ from equation (3).
2. On the θ-plot, peak widths which indicate corner and curve features are found, each feature type is determined, and feature parameters are calculated.
   a. For $s_i = s_o + L/2$ to $s_i = s_f - L/2$, measure the distance between zero-range crossings, s_{zzj}, as in equation (7). If $s_{zzj} > L-1$ $(+\varepsilon_1)$, then do steps 2b and 2c; otherwise, increment $s_i = s_{i+1}$ and repeat step 2a.
   b. If $s_{zzj} \le L + s_{K_{max}} - 1$ $(+\varepsilon_2)$, then classify the feature as a corner. If $s_{zzj} > L + s_{K_{max}} - 1$ $(+\varepsilon_2)$, then classify the feature as a curve.
   c. If the feature is a corner, calculate the corner location from equation (12). If the feature is a curve, calculate the transition points from straight lines to curves from equation (13). Find the curve parameters by calculating the center of curvature from equation (15), and find the radius of curvature from equation (16).

6. Examples

Results of application of the DOS⁺ method to several curvilinear shapes are shown in Figures 3 through 5. In these figures the first diagram shows the original line structure(s). The second shows circle symbols at corner and curve transition locations and "plus" symbols at curve center locations. The third diagram shows the final straight line and curve structures with dots overlayed at the ends of features. The last diagram shows the final vectorized structures with no overlay.

Figure 3 shows a series of "corners" that go from sharp to smooth. The DOS⁺ parameters used on each of these structures are L=20 and M=5. The results show that curves can be interpreted as corners or curves depending on the degree of curvature and the DOS⁺ threshold. The DOS⁺ parameters can be adjusted for different thresholds between corners and curves as would depend on image resolution, expected amount of noise, etc.

Figure 4 shows a line structure with two curves of different curvature values. The diagrams show the transition locations between straight lines and curves and the center of curvatures as found by the DOS⁺ method.

Figure 5 is a hand-drawing outline of a girl skipping. Note that since this does not have straight lines and curves of known ranges (as an engineering drawing or industrial component outline would have), we cannot directly relate the DOS⁺ parameters to the diagram parameters. This drawing also does not obey the assumption of a straight line of length greater than L between each corner or curve feature. The results show retention of the larger features of the drawing, but details such as the pant leg are missing and the extended hand is rounded. One way to handle features of different sizes on the same diagram is to use a scale-space representation as in reference [4]. Another approach is to recalculate the curvature with smaller scale parameter values only for lengths of the curve where some measure of error from the original curve is exceeded. This latter approach is currently being investigated.

7. Summary

Curvature is estimated by the DOS⁺ method as the angular difference between two segments which are fit along line data and separated by a gap. The peaks on the curvature plot correspond to corner and curve features in the data. Relationships are established here between the DOS⁺ parameters and the feature parameters to determine corner location, transition location between straight lines and curves, and curve center and radius of curvature for circular curves.

References

1. L. O'Gorman, "An analysis of feature detectability from curvature estimation", Computer Vision and Pattern Recognition, Ann Arbor, Michigan, June, 1988.
2. A. Rosenfeld, J.S. Weszka, "An improved method of angle detection on digital curves", IEEE Trans. Comput., Vol. C-24, 1975, pp. 940-941.
3. H. Freeman, L. Davis, "A corner-finding algorithm for chain-coded curves", IEEE Trans. Comput., Vol. C-26, 1977, pp. 297-303.
4. H. Asada, M. Brady, "The curvature primal sketch", IEEE Trans. Pattern Analysis and Machine Intelligence, Vol. PAMI-8, No. 1, Jan., 1986, pp. 2-14.
5. A. Rosenfeld, "Digital Straight Line Segments", IEEE Trans. Comput., Vol. C-23, No. 12, Dec., 1974, pp. 1264-1269.

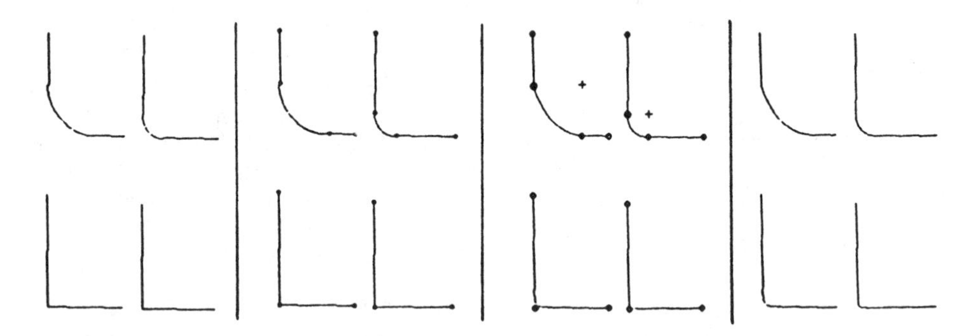

Figure 3. DOS⁺ method applied to a series increasingly smooth "corners".

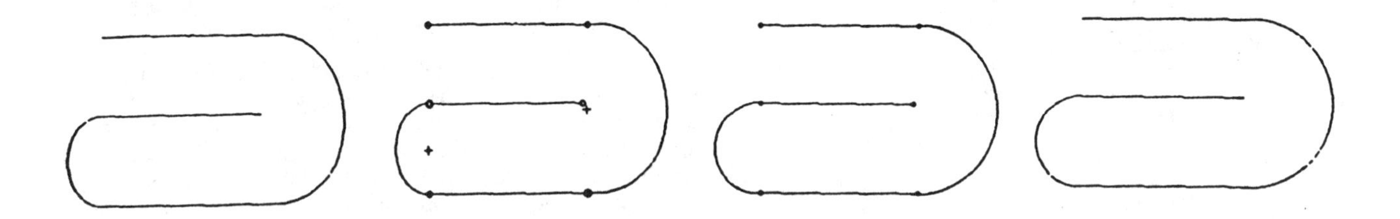

Figure 4. DOS⁺ method applied to structure with two curves of different curvatures.

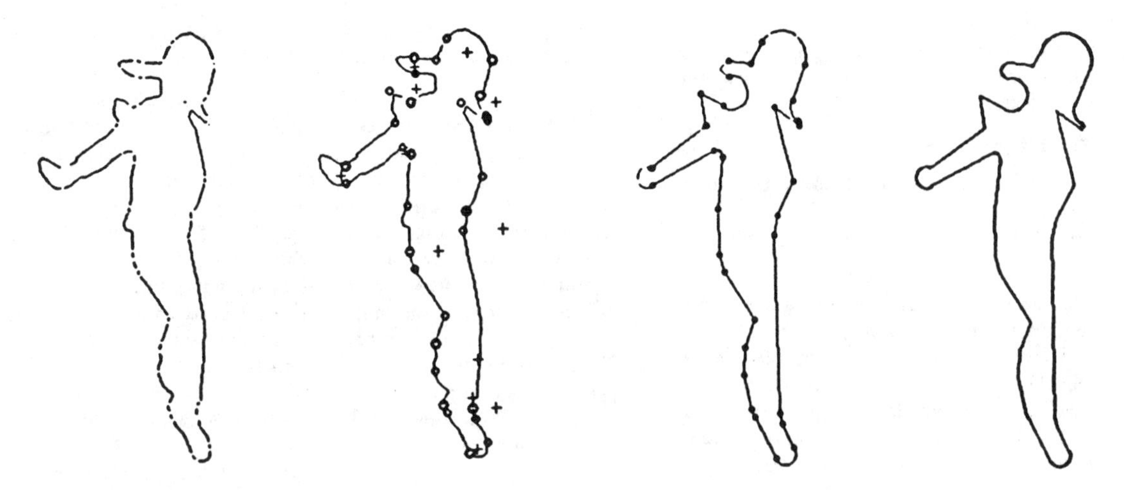

Figure 5. DOS⁺ method applied to hand-drawn outline of girl skipping.

Corner Detection and Curve Representation Using Cubic *B*-Splines*

GERARD MEDIONI AND YOSHIO YASUMOTO†

Intelligent Systems Group, PHE 234 MC 0273, University of Southern California, Los Angeles, California 90089-0273

Received December 12, 1985; accepted September 12, 1986

We propose to use *B*-Splines to represent digital curves. We have developed an efficient algorithm to locate corners and at the same time encode curve segments between them using *B*-Splines. Used in conjunction with our subpixel edge detector [1], it allows us to obtain accurate position of the corners, as needed in many registration problems such as stereo matching and motion parameter estimation. In addition to corners, we detect points of significant curvature between them. The resulting representation is a good approximation of the original, in the sense that it makes interesting points explicit, and achieves significant data compression.

1. INTRODUCTION

Corner detection and digital curve representation are important research areas in computer vision and image processing. The first step in image analysis is to encode significant changes in gray level and represent them in a convenient form. The resulting primitives are used as features which may be used at subsequent levels such as stereo matching or motion estimation.

In the method described here, we perform two tasks:

First, we *extract corners* of the digital curve formed by connected edge points using a cubic *B*-Spline, then discard most of the points not marked as corners and *encode the curve* between two corners using *B*-Splines again. This technique presents two major advantages: it only requires one iteration, and all processing is done locally, it is therefore suitable for parallel processing.

Various corner detectors have been developed and studied in [2–4]. According to a recent report [5], these methods are equivalent. Furthermore, it says that extracting edges from gray level images prior to performing corner detection is better as it eliminates "noise" corners in the background. A corner is an interest point in terms of the curvature measurement. The curvature of a continuous curve is defined as the changing ratio of the tangent along the arc, but the extension to *discrete* curves is not quite straightforward, as discussed in [6, 7]. Usually the k-curvature, or curvature obtained by using k points on each side, is computed for a discrete curve, but this still fails to detect and localize what humans perceive as corners, as small quantization errors create large variations of the computed curvature. To successfully detect corners, some smoothing must be performed as described in [8].

*This research was supported, in part, by the Defense Advanced Research Projects Agency and was monitored by the Air Force Wright-Patterson Aeronautical Laboratories under Contract F33615-84-K-1404, Darpa Order No. 3119.

†Yoshio Yasumoto is an Engineer for Matsushita Electric Industrial Co., Ltd. and was a visiting scholar at USC.

Langridge [9] fits cubic polynomials to a digital curve to detect slope discontinuities, but does not explicitly encode the resulting curve between corners.

The general idea behind the algorithm is to fit a *B*-Spline to the curve, then to let the points move around their original position, and measure their displacement. Points that move a lot and have high curvature are marked as corners and are treated as multiple control points, points between corners with high curvature are used as control points, all other points are discarded.

Many methods of curve representation have been proposed for data compression and coding. *B*-Splines are commonly used in rendering a curve in computer graphics, CAD-CAM systems, and also for curve fitting and shape description [10]. But very few are designed for use in computer vision. Recent work includes spline approximation of lines in images by Ishimura [11], parallel architecture for contour extraction by Weems [12], and pseudo coding of line figures by Nakajima [13]. Ishimura uses a modified dynamic programming technique to find the knots of a spline, Weems uses an iterative method to locate the interest points, and Nakajima uses a complicated vector finder to obtain some feature points. A major drawback of these methods is that they are very time-consuming.

The next section introduces the notation and describes the algorithm used to detect corners and represent curve portions between them; Section 3 presents results on three different types of images, and compares our method with more traditional techniques of corner detection and curve representation.

2. FORMULATION

2.1. Curvature

Evaluating the slope and curvature of a discrete curve is not simple, because the value of the slope is quantized. In order to reduce the effects of quantization on slope or curvature, we need to introduce some smoothing. We can take k points on the left and right sides of a given point, and get the k-slope and k-curvature. Since curvature is defined as the rate of change of the slope, the k-curvature can be obtained by taking the difference between the left and right k-slopes.

Although a corner is defined as a point where curvature goes through a discontinuity or an extremal value, the above smoothing procedure is not appropriate to detect corners. Asada [8] uses Gaussian filters to smooth the curvature in order to get not only corners but smooth joins. He defines corner, smooth join, crank, end, and bump–dent as significant primitives to represent interesting changes in curvature. Varying the value for σ in the Gaussian filter permits to control the amount of smoothing. The method is very powerful to describe curve but involves computing fourth derivatives.

After we obtain a connected set of points, it is useful to introduce the following notation: Let S be a curve in parametric form,

$$\begin{aligned} x &= f(t) \\ y &= g(t), \end{aligned} \tag{1}$$

where t is a parameter.

The *slope* of a curve S at a given point A $(t = t_1)$ is

$$\left[\frac{dy}{dx}\right]_{t=t_1} = \left[\frac{\dfrac{dg}{dt}}{\dfrac{df}{dt}}\right]_{t=t_1} \tag{2}$$

And the *arc length* of S between point A (t_1) and B (t_2) is

$$\int_{t_1}^{t_2} \sqrt{(df/dt)^2 + (dg/dt)^2}\, dt. \tag{3}$$

The *curvature* $C_v(t_1)$ at point A $(t = t_1)$ is the derivative of the slope with respect to the arc length

$$\begin{aligned} C_v(t_1) &= \frac{d^2y}{dx^2} \Bigg/ \left(1 + \left(\frac{dy}{dx}\right)^2\right)^{3/2} \\ &= \left(\left(\frac{df}{dt}\right)\left(\frac{d^2g}{dt^2}\right) - \left(\frac{dg}{dt}\right)\left(\frac{d^2f}{dt^2}\right)\right) \Bigg/ \left(\left(\left(\frac{df}{dt}\right)^2 + \left(\frac{dg}{dt}\right)^2\right)^{3/2}\right) \end{aligned} \tag{4}$$

We can fit a cubic polynomial to the curve S between point A and B, where the parameter t varies between 0 and 1,

$$\begin{aligned} x &= f(t) = a_1 t^3 + b_1 t^2 + c_1 t + d_1 \\ y &= g(t) = a_2 t^3 + b_2 t^2 + c_2 t + d_2. \end{aligned} \tag{5}$$

Once these coefficients are calculated, the curvature at point A $(t = 0)$ is derived from (4) as follows:

$$C_v(0) = 2\frac{c_1 b_2 - c_2 b_1}{\left(c_1^2 + c_2^2\right)^{3/2}}. \tag{6}$$

2.2. *B*-Splines

We use cubic B-Splines for both calculating curvature and representing a curve. A cubic polynomial is the lowest-order representation of curve segments which can provide continuity of position and slope at a point where two curve segments meet, and at the same time ensures that the ends of the curve segment pass through specified points.

The parametric cubic B-Spline with equally spaced breakpoints [10] is defined by

$$\begin{aligned} x(t) &= TM_s G_{sx} \\ y(t) &= TM_s G_{sy} \end{aligned} \tag{7}$$

where

$$M_s = \frac{1}{6}\begin{bmatrix} -1 & 3 & -3 & 1 \\ 3 & -6 & 3 & 0 \\ -3 & 0 & 3 & 0 \\ 1 & 4 & 1 & 0 \end{bmatrix}. \tag{8}$$

In these equations, $T = (t^3, t^2, t, 1)$, and G_{sx}, G_{sy} are the geometry matrices determined by the location of neighboring points.

To approximate the control points P_1 through P_n by a series of B-Splines, we use a different geometry matrix between each pair of adjacent points. Let us assume that the ith point P_i has coordinates (x_i, y_i). The approximation from P_i to P_{i+1} uses

$$G_{sx}^i = \begin{bmatrix} x_{i-1} \\ x_i \\ x_{i+1} \\ x_{i+2} \end{bmatrix} \qquad G_{sy}^i = \begin{bmatrix} y_{i-1} \\ y_i \\ y_{i+1} \\ y_{i+2} \end{bmatrix}. \tag{9}$$

If we compute the product TM_s, we obtain

$$TM_s = \tfrac{1}{6}\left[(-t^3 + 3t^2 - 3t), (3t^3 - 6t^2 + 4), (-3t^3 + 3t^2 + 3t + 1), t^3\right]. \tag{10}$$

Postmutiplying this by G_{sx} yields

$$\begin{aligned} x(t) = TM_sG_{sx} = x_{i-1}(-t^3 + 3t^2 - 3t + 1)/6 + x_i(3t^3 - 6t^2 + 4)/6 \\ + x_{i+1}(-3t^3 + 3t^2 + 3t + 1)/6 + x_{i+2}t^3/6 \end{aligned} \tag{11}$$

or

$$\begin{aligned} x(t) = (-x_{i-1} + 3x_i + 3x_{i+1} + x_{i+2})t^3/6 + (x_{i-1} - 2x_i + x_{i+1})t^2/2 \\ + (-x_{i-1} + x_{i+1})t/2 + (x_{i-1} + 4x_i + x_{i+1})/6. \end{aligned} \tag{12}$$

Therefore, we get four coefficients in (5)

$$\begin{aligned} a_1 &= (-x_{i-1} + 3x_i - 3x_{i+1} + x_{i+2})/6 \\ b_1 &= (x_{i-1} - 2x_i + x_{i+1})/2 \\ c_1 &= (-x_{i-1} + x_{i+1})/2 \\ d_1 &= (x_{i-1} + 4x_i + x_{i+1})/6. \end{aligned} \tag{13}$$

Using (6), we can calculate the curvature at a given point only from the position of neighboring points $i - 1$, i, and $i + 1$, since a_1 and a_2 disappear

$$C_v = 4\frac{(x_{i+1} - x_{i-1})(y_{i-1} - 2y_i + y_{i+1}) - (y_{i+1} - y_{i-1})(x_{i-1} - 2x_i + x_{i+1})}{\left((x_{i+1} - x_{i-1})^2 + (y_{i+1} - y_{i-1})^2\right)^{3/2}}. \tag{14}$$

2.3. Corner Detection

In order to detect a corner, we have to smooth the curve as mentioned in the previous section. In this paper, we use the displacement value between the original position of a point and the interpolating spline to decide whether this point is a corner or not. Since we use cubic *B*-Splines with equally spaced breakpoints, we can calculate both curvature and displacement only from the position of each point. A displacement of a given point i from the cubic B-Spline can be expressed as

$$\begin{aligned} \delta_x &= d_1 - x_i = x_{i-1}/6 - x_i/3 + x_{i+1}/6 \\ \delta_y &= d_2 - y_i = y_{i-1}/6 - y_i/3 + y_{i+1}/6. \end{aligned} \tag{15}$$

To evaluate the *cornerness* of a given point i, we re-compute the curvature by fitting a B-Spline through the displaced points. In other words, we use $P_i + \delta$ instead of P_i to obtain a new B-Spline, where $\delta = (\delta_x, \delta_y)$. To calculate the new curvature C_v', we substitute P_{i-1} through P_{i+1} in (15),

$$\begin{aligned} P_i &\to P_{i-1}/6 + 2P_i/3 + P_{i+1}/6 \\ P_{i-1} &\to P_{i-2}/6 + 2P_{i-1}/3 + P_i/6 \\ P_{i+1} &\to P_i/6 + 2P_{i+1}/3 + P_{i+2}/6. \end{aligned} \tag{16}$$

Thus we obtain

$$C_v' = 2\frac{c_1'b_2' - c_2'b_1'}{\left(c_1'^2 + c_2'^2\right)^{3/2}}, \tag{17}$$

where

$$\begin{aligned} b_1' &= \frac{x_{i-2}}{12} + \frac{x_{i-1}}{6} - \frac{x_i}{2} + \frac{x_{i+1}}{6} + \frac{x_{i+2}}{12} \\ b_2' &= \frac{y_{i-2}}{12} + \frac{y_{i-1}}{6} - \frac{y_i}{2} + \frac{y_{i+1}}{6} + \frac{y_{i+2}}{12} \\ c_1' &= \frac{(x_{i+1} - x_{i-1})}{3} + \frac{(x_{i+2} - x_{i-2})}{12} \\ c_2' &= \frac{(y_{i+1} - y_{i-1})}{3} + \frac{(y_{i+2} - y_{i-2})}{12}. \end{aligned} \tag{18}$$

Similarly, we get a second displacement $\delta' = (\delta_x', \delta_y')$

$$\begin{aligned} \delta_x' &= x_{i-2}/36 + x_{i-1}/18 - x_i/6 + x_{i+1}/18 + x_{i+2}/36 \\ \delta_y' &= y_{i-2}/36 + y_{i-1}/18 - y_i/6 + y_{i+1}/18 + y_{i+2}/36. \end{aligned} \tag{19}$$

And, the total displacement $\delta_t = \delta + \delta'$ is

$$\begin{aligned} \delta_{tx} &= x_{i-2}/36 + 2x_{i-1}/9 - x_i/2 + 2x_{i-1}/9 - x_{i-2}/36 \\ \delta_{ty} &= y_{i-2}/36 + 2y_{i-1}/9 - y_i/2 + 2y_{i+1}/9 + y_{i+2}/36. \end{aligned} \tag{20}$$

At this point we have performed all necessary calculations, using the position of the neighboring five points P_{i-2} through P_{i+2}. The condition for a given point i to be a corner is as follows:

(1) the displacement δ_t is larger than a given threshold d_c,

(2) the curvature C_v' is larger than a given threshold c_c,

(3) the curvature C_v' is a local maximum.

2.4. Representation

If a given curve is open, we keep both ends and treat them just like corners. At all corners and ends, the representation must reflect the slope discontinuity. This is achieved by using a different *B*-Spline segment between any pair of corner points, which is equivalent to treating each corner as an end point in an open curve. Therefore, to represent an open digital curve, we need $(n + 1)$ *B*-Spline segments if we have 2 ends and $n - 1$ corners; for a closed curve, we need n segments for n corners.

A list of corner points is not a sufficient representation of a digital curve (closed smooth curves are not represented!), so we need more control points to guide the spline and keep it close to the original data. This is achieved very simply at the same time as we evaluate points for cornerness: we set an additional threshold on curvature c_t and mark all points with curvature larger than c_t as significant curvature points (SCPs). These points are used as control points of the spline. It should be noted that this scheme provides a very efficient encoding of straight lines.

3. RESULTS

In this section we demonstrate the usefulness of our corner detector and *B*-Spline representation on an aerial image, a synthesized drawing, and a laboratory image. We compare our corner detector with the *k*-curvature method, an iterative cubic polynomial approximation [14], and a corner detector based on the facet model [4]. We also compare our *B*-Spline representation with the Nevatia–Babu linear approximation [15].

3.1. Edge Detection

Our corner detector operates on edges of an image. The performance of the edge detector therefore affects the corner detector directly. We have used zero crossings of the convolution with Laplacian of Gaussian masks with pixel and subpixel precision [1].

Figure 1 shows an original gray level aerial image (a part of some airport facilities) and the obtained edges with pixel and subpixel precision. The original image of Fig. 1a has 64×64 pixels and 8 bits accuracy. For the subpixel edge detection (shown in Fig. 1c), the original image is expanded to 128×128 after convolution with a 9×9 Laplacian–Gaussian mask. As clearly seen, edges with subpixel precision are better than the ones obtained with pixel precision (shown in Fig. 1b). In particular, some gaps are filled using $\frac{1}{2}$ pixel accuracy.

The total computation time for convolution, zero crossing detection, and subpixel fitting is less than 10 sec in SAIL on a PDP10.

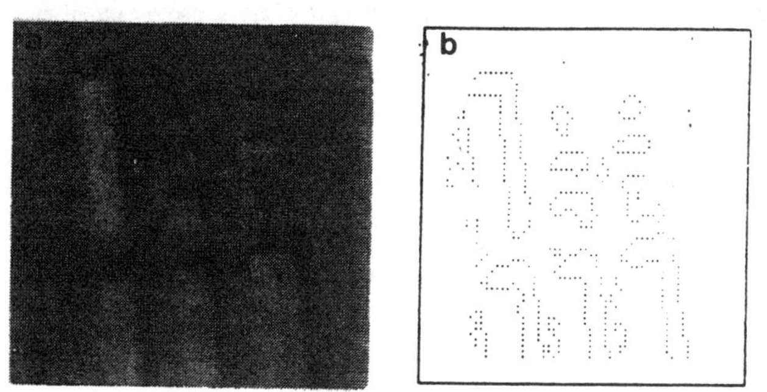
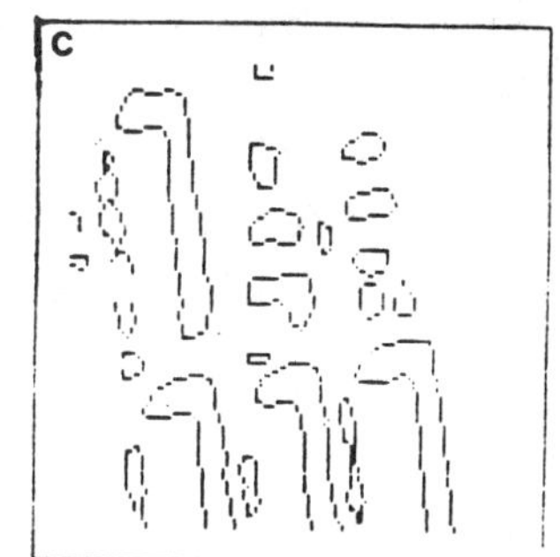

FIGURE 1

3.2. Corner Detector

We show the results of our cubic B-Spline corner detector and compare it with an iterative cubic polynomial approximation, the k-curvature method, and a corner detector that operates on edges obtained by the facet model [5].

Figure 2a–e shows the corners obtained by

(a) our B-Spline method,

(b) the iterative cubic polynomial method on edges with subpixel precision,

(c) computing the 3-curvature,

(d) the iterative cubic polynomial on edges with pixel precision,

(e) the facet model corner detector.

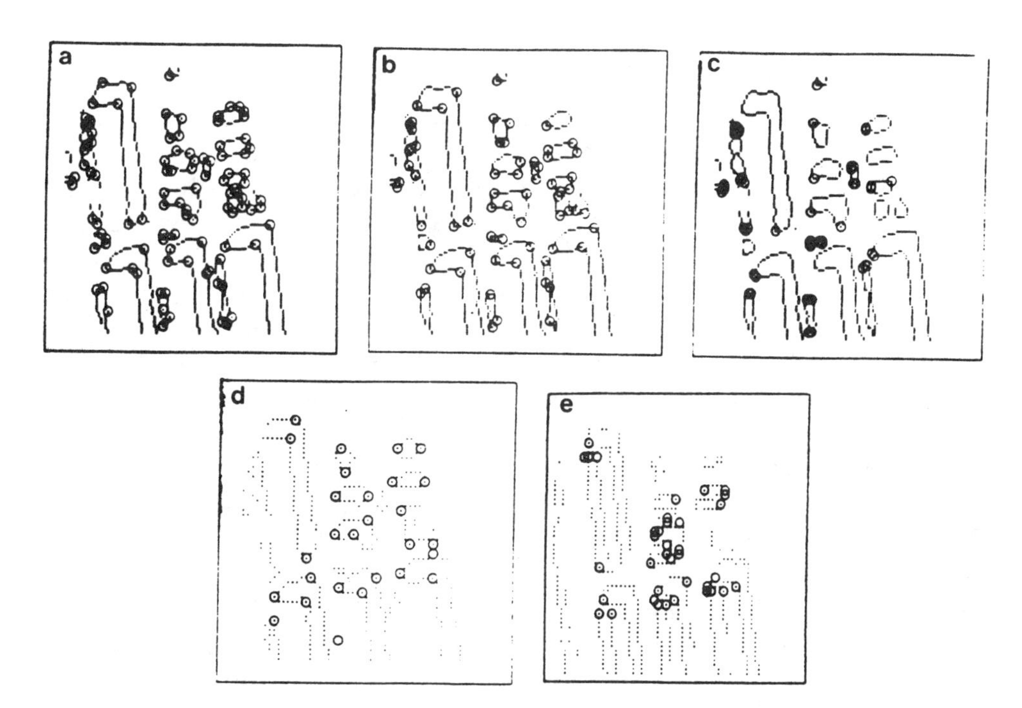

FIGURE 2

On these figures, the corners are represented by circles superimposed on the image of all edges.

Each method has a set of associated thresholds, such as value of curvature above which a point is declared to be a corner. In the k-curvature method, left and right curvature are defined, but not curvature, so we used a threshold on the change of slope instead. For each method, we chose the threshold(s) that gave the "best" subjective results.

Let us first explain briefly how to compute the k-curvature. Given any point of a digital curve, we can take k points on each side of this point if this point is at least k points apart from an end point or if the curve is closed. We can define the k-curvature at every point which satisfies this condition. Here we take the 3-curvature, so we need to consider 7 points at a time. In this example, we set the threshold of change of slope to 60°.

Figure 2c shows the result of the 3-curvature method and we can see that, even though many points are detected as corners, the ones corresponding to the shadow of large buildings are missed. For this example, the computation time was 3 seconds.

Figures 2b and d show the results of a cubic polynomial approximation. In this method, we take 5 points into account and fit a cubic polynomial to these points. This method is similar to cubic B-Spline corner detector, but is iterative. In this example, we set the thresholds as: curvature threshold $c_c = 0.5$, displacement threshold $d_c = 0.15$, and the number of iterations is 5. This method gives much better results than the 3-curvature method, as shown in Fig. 2b, where we can see almost all possible corners are detected in the four large shadow regions. The computation time for this method is 45 sec. The curvature threshold c_c relates to the number of corners we detect, although the displacement threshold does not affect it so much. Furthermore, we have to notice that the difference between Fig. 2b (subpixel) and Fig. 2d (pixel): The results are much better in the subpixel edge image because the edges are much better.

Figure 2f shows the results of the facet model corner and edge detector. Based on the facet model [16], we use a 5 × 5 window and 16 5 × 5 coefficients masks. In this example we use edges with pixel resolution and have to compare Fig 2e with Fig 2d. By using a polynomial fit, we can obtain both edges and corners at the same time, but edges are of poorer quality than those of Fig. 2d. The corner detection depends on edges so heavily that it is affected by this poor edge quality. We set the gradient threshold value to 12, and if a zero-crossing occurs in a direction of $\pm 15°$ of the gradient direction within a circle of a unit pixel length centered in a given pixel, then the pixel is declared to be an edge point. At the same time, if the curvature exceeds 0.4, we declare this edge point to be a corner. The computation time to detect edges and corners is 41 sec. As this time includes edge detection, the facet model corner detector is faster than cubic polynomial approximation.

Figure 2a shows the results obtained by our cubic B-Spline corner detector detailed in the previous section. In this example we set the curvature threshold $c_c = 0.4$, and the displacement threshold $d_c = 0.2$. As shown in Fig. 2a, the number of detected corners is larger than the one from the cubic polynomial approximation and some corners are close to each other in the small curve. But in one of the four large shadows, the six possible corners are obtained only with this method. As this method is not iterative, the computation time is only 3 sec, the same as the 3-curvature method and 15 times faster than the cubic polynomial approximation. Nonetheless, the quality of the results is quite comparable to the cubic polynomial approximation. Another advantage of the technique is that *B*-Splines can also be used for curve representation, at no extra cost, as described in the next subsection.

3.3. Representation

First, we explain how to represent the original curve by corners and *B*-Spline using one of large shadows mentioned before. Figure 3a shows the original shadow and *B*-Spline representation. In this figure, corners are marked by large circles and significant curvature points (SCPs) are denoted by solid squares. The original curve has 142 points and through cubic *B*-Spline approximation we obtain 6 corners and 45 significant curvature points (SCPs) for representation. In all examples, we set the threshold $c_t = 0.3$.

To draw a curve from only corners and SCPs, we make six segmented *B*-Splines with SCPs in this example. We store only the position of corners and SCPs, so we can achieve considerable data compression. In this example a, as the number of corners and SCPs is 51, the data compression rate is 0.36. Other examples show a screwdriver as a more complicated curve, and a synthesized circle as a simpler one. The number of points in the original curve, detected corners, SCPs, and data

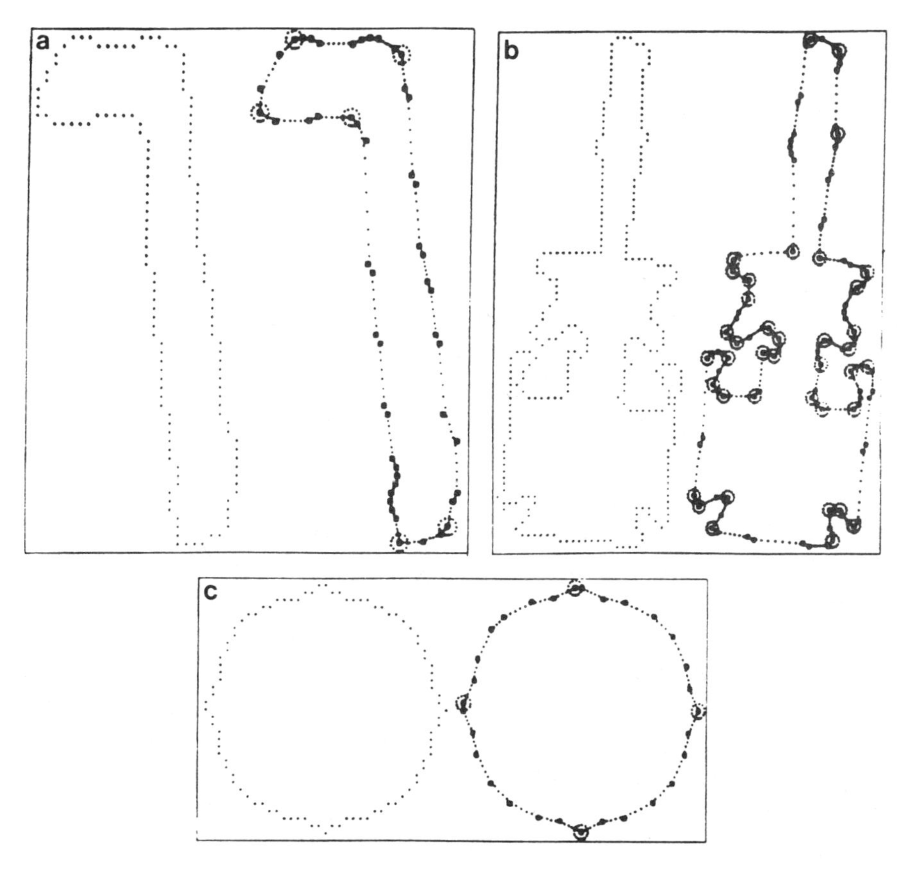

FIGURE 3

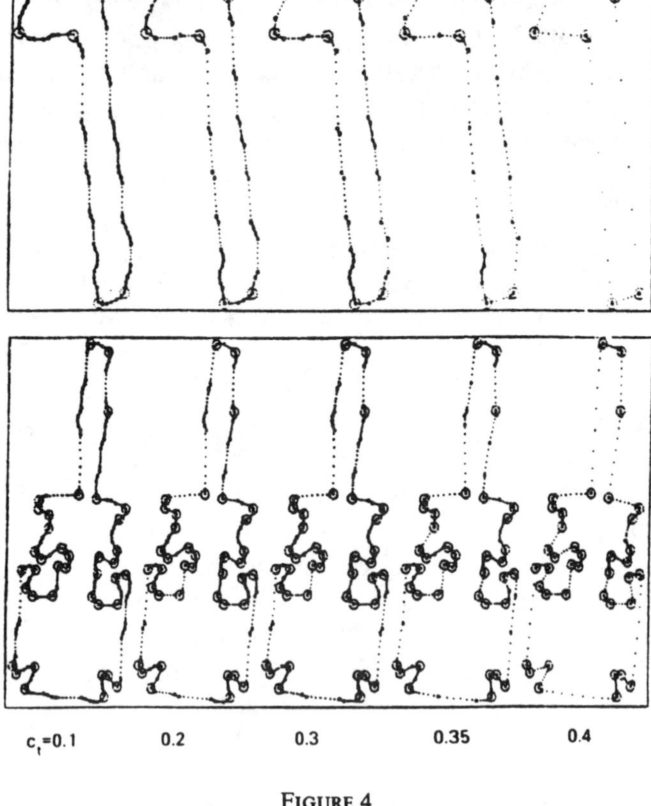

FIGURE 4

compression rate are 267, 39, 84, 0.46 for the screwdriver and 93, 4, 26, 0.32 for the circle, respectively.

Figure 4 shows the influence of the threshold c_t. The value of c_t is 0.1, 0.2, 0.3, 0.35, 0.4, from left to right, respectively. The left-end figure shows the case when a large number of SCPs remain. From left to right, the compression rate is increasing and the right end figure shows very few SCPs. Thus, this figure is very similar to the one obtained by polygonal representation. The "best" value of c_t depends on the purpose of the curve representation and on the type of curve.

We compare our *B*-Spline representation with a more traditional method. Figure 5 shows the original curve, our *B*-Spline representation, and the Nevatia–Babu linear representation (polygonal), from left to right, respectively. The line segments approximating the edge points are obtained by recursively fitting a line between two endpoints and breaking it at the point farthest away from the line, if it is above a preset threshold (two pixels here). This method achieves excellent data compression but creates vertices that may not be perceptually significant. It is also a global method (its time complexity grows with the length of the curve).

Figure 6 shows two more examples of curve encoding using *B*-Splines. These images represent a smooth circle and a combination of arcs and lines. For a circle, the *B*-Spline representation gives an even smoother shape than the original one as it reduces quantization noise. For the combination of arcs and lines, only two corners are detected but 56 SCPs draw an original shape successfully. The number of the original points, corners, SCPs, and compression rate are 92, 8, 16, 0.26 for the circle and 232, 2, 56, 0.25 in the combination, respectively.

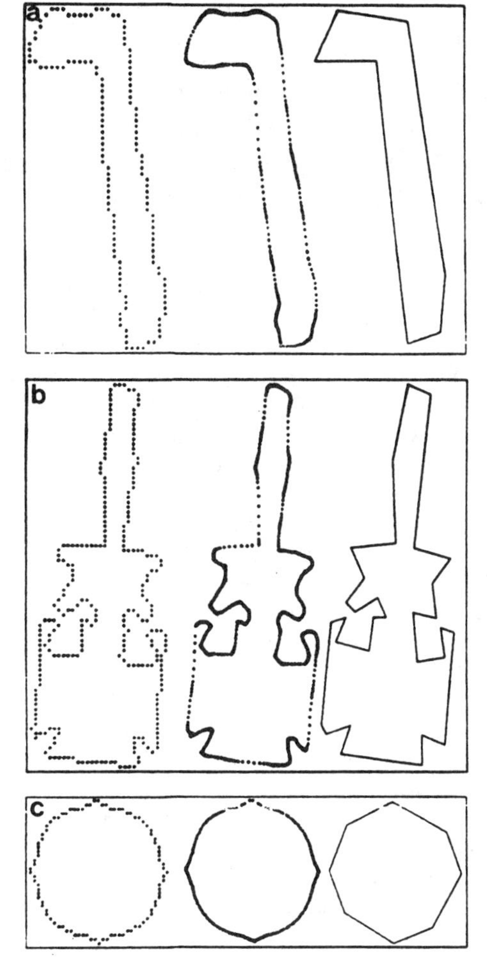

FIGURE 5

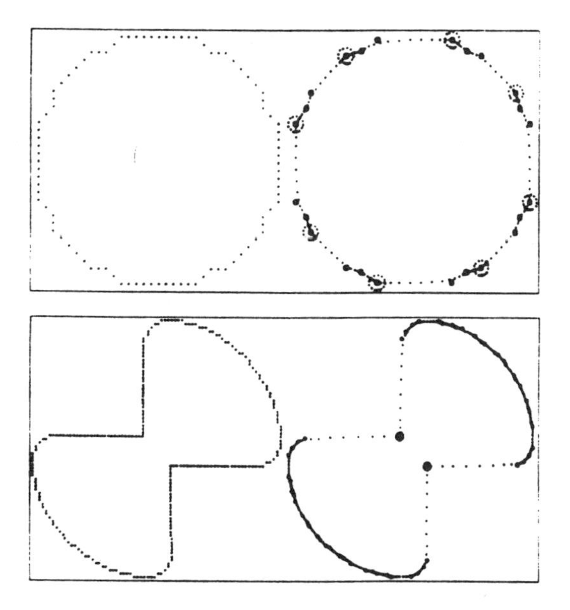

FIGURE 6

4. CONCLUSION

We have developed a powerful corner detector and curve representation technique using *B*-Splines. Our method is noniterative and generates corners and a *B*-Spline representation *at the same time*. This representation, together with the detected corners, achieves significant data compression, and can be used in further stages of processing, such as feature matching of images.

REFERENCES

1. G. Medioni and A. Huertas, *Edge Detection with Subpixel Precision*, Technical Report USCISG 106, University of Southern California, 1985.
2. L. Dreschler and H. H. Nagel, Volumetric model and 3-d trajectory of a moving car derived from monocular tv-frame sequence of a street scene, in *International Joint Conference on Artificial Intelligence, August 1981, Vancouver, Canada.*
3. L. Kitchen and A. Rosenfeld, Gray-level corner detection, *Pattern Recognition* **4** (6), April 1983.
4. Oscar A. Zuniga and Robert M. Haralick, Corner detection using the facet model, in *Proceedings of IEEE Conference on Computer Vision and Pattern Recognition at Washington, D.C.* July 1983, 30–37.
5. Mubarak A. Shah and Ramesh Jain, Detecting time-varying corners, *Comput. Vision, Graphics and Image Process.* **28**, 1984, 345–355.
6. Rosenfeld A. and A. Kak, *Digital Picture Processing*, 2nd ed., Academic Press, New York, 1982.
7. M. Nagao, Denshi tsushin gakkai, *Inst. Electronics Comput. Eng. Japan* **1** (4), 1983.
8. H. Asada and M. Brady, The curvature primal sketch, in *Proceedings of the 2nd IEEE Workshop on Computer Vision: Representation and Control*, Annapolis, MD, May 1984, 8–17.
9. D. J. Langridge, Curve encoding and the detection of discontinuities, *Comput. Vision, Graphics, Image Process.* **20**, 1982, 58–71.
10. T. Pavlidis, *Algorithms for Graphics and Image Processing*, Computer Science, Rockville, MD. 1982.
11. N. Ishimura, T. Hashimoto, S. Tsujimoto, and S. Arimoto, Spline approximation of line images by modified dynamic programming, *Trans. IECE Japan* **68** (2), February 1985, 169–176.
12. C. Weems, D. Lawton, S. Levitan, E. Riseman, A. Hanson, and M. Callahan, Iconic and symbolic processing using a content addressable array parallel processor, in *Proceedings of the IEEE Conference on Computer Vision and Pattern Recognition at San Francisco, 1985.*
13. M. Nakajima, Takeshi Agui, and Kazuki Sakamoto, Pseudocoding method for digital line figures, *Trans. Inst. Electronics Commun. Eng. Japan* April 1985, **68**-*D* 4, 623–630.
14. Y. Yasumoto, Corner detection using cubic polynomial with subpixel accuracy, *Matsushita Electric Engineering Documentation*, 1985.
15. R. Nevatia and K. R. Babu, Linear feature extraction and description, *Comput. Graphics Image Process.* **13**, 1980, 257–269.
16. R. M. Haralick, Digital step edge from zero crossing of second directional derivatives, *IEEE Trans. Pattern Anal. Mach. Intelligence* **6** (1), January 1984, 58–68.

SURVEY

A Review of Algorithms for Shape Analysis *, †

THEODOSIOS PAVLIDIS

Department of Electrical Engineering and Computer Science, Princeton University, Princeton, New Jersey 08540

Received September 20, 1976

Algorithms for shape analysis are reviewed and classified under two criteria: whether they examine the boundary only or the whole area, and whether they describe the original picture in terms of scalar measurements or through structural descriptions.

1. INTRODUCTION

The problem of shape discrimination is a central one in pattern recognition and as such has received considerable attention. The size of the relevant literature is immense since shape is discussed, at least implicitly, in most papers dealing with recognition of characters (including numerals, Chinese characters etc), waveforms, chromosomes, cells, machine parts, etc. In the sequel we will restrict ourselves to the study of the shape of *plane objects*, i.e., we will assume that we deal only with segmented pictures [82, 85] and we will not consider the question of inferring the shape of three-dimensional projections, even though the latter is an extremely interesting subject. In essence we will examine algorithms for the analysis of the shape of "silhouettes." However this is still an enormous subject.

We may attempt to introduce some structure in our review by classifying the methodologies used.

It is possible to do so according to many criteria. One is whether the algorithm used traces the boundary only while ignoring the interior of the object or whether it examines the points of the interior as well. A typical instance of the latter is the medial axis transformation (MAT) [6, 7, 59, 75]. Among the boundary tracers we may distinguish between those which proceed sequentially along the points and those which "skip across." The difference is not just aesthetic. Usually sequential boundary scanning algorithms require time linearly proportional to the number of boundary points while the other type may require time proportional to their square. A typical example of the former is the parsing of the boundary according to a simple grammar, e.g., [43], and similarly for the evaluation of a

* Research supported by NSF Grant ENG72-04133.

† A preliminary version of this paper was presented at the Engineering Foundation Conference on Algorithms for Image Processing, Franklin Pierce College, Ringe, N. H., August 8–13, 1976.

single coefficient of the Fourier transform, e.g., [99]. An example of the latter is parsing according to context sensitive grammars as well as certain algebraic methods for evaluating the MAT [56].

Not too surprisingly, techniques which skip along the boundary tend to be related to those which scan the interior of a figure. We will return to this point later but for the moment we chose the names *external* and *internal* to refer to local boundary followers and global boundary or area examiners, respectively.

Another distinction can be made on the basis of *scalar transform* and *space domain* techniques. The distinction is somewhat subtle; thus the MAT is a space domain technique because it transforms one picture into another picture rather than into an array of scalar features. As far as subsequent processing is concerned scalar transforms are most appropriate as input to classical statistical pattern recognizers while space domain techniques are most appropriate for producing input to syntactic and/or structural pattern classifiers [22–24, 27, 28, 64, 66].

Finally, we may talk about *information preserving* and *information nonpreserving* techniques depending on whether it is possible to reconstruct the picture from the shape descriptors or not. The distinction is fuzzy because some techniques allow only imperfect reconstruction. Furthermore any pattern recognition encoding must perform some information reduction. We will resolve this by examining whether there is a set of parameters under which a given technique can give a replica of the original picture which is arbitrarily close to it. Thus the ratio of the square of the perimeter to the area [5] does not allow reconstruction under any circumstances except for trivial cases where the general shape is known and it is simple (e.g., if it is a rectangle or a regular polygon, etc). On the other hand, a Fourier expansion allows reconstruction provided that sufficient coefficients have been evaluated. In essence we distinguish between schemes where the loss of information can be controlled and those where it cannot. In this paper we will emphasize primarily the information preserving techniques.

The terms *normative* and *generative* have also been suggested for these two types [98]. However, we prefer the present terminology as more descriptive and because the term *generative* should best be reserved for methods where the *physical process generating the shape* is taken into account in the design of descriptors. This approach has been emphasized by Grenander in the context of pattern synthesis [28], as well as in many theoretical biology studies, e.g., [49].

It should be pointed out that many of the publications on shape have dealt with information nonpreserving techniques. In particular, they emphasize properties such as symmetry, elongation, angularity, etc. [5, 37, 41, 42]. Such properties give very useful information about the shapes of simple objects but fail to do so for complicated ones. The description of the latter must often be given in terms of very local characteristics [84].

We start with a compact review of the literature and then we discuss the main shape criteria. The following notation will be used:

> $f(x, y)$ or $f(i, j)$ will denote the pictorial brightness function or matrix.
> $bx(t)$ and $by(t)$ (or $bx(i)$ and $by(i)$) will note the list of boundary points.
> The parameter t is often, but not always, the length along the boundary.

2. INTERNAL SCALAR TRANSFORM TECHNIQUES

One of the earliest of these techniques is the method of moments [3, 29, 30, 34], which is defined as follows:

$$m(u, v) = \int f(x, y)x^u y^v dx dy.$$

The concept originated in mechanics and it can be shown that it is information preserving [34]. It can also be shown that simple linear combinations of moments have values which are invariant under a number of similarity transformations. Such "moment invariants" have been used for pattern recognition, rather than the moments themselves. The major disadvantage of this methodology is that although the first few moments convey significant information for simple objects, they fail to do so for more complicated ones. Furthermore, the computational requirements are quite substantial.

Another possibility is to take the two-dimensional Fourier transform (FT) of the characteristic function of the object. By this we mean a function equal to zero all over the visual field except on the points of the object, where it equals one. The coefficients of the transform convey shape information but their computation is quite expensive.

Finally, we may mention the use of binary masks to extract features conveying shape information [60].

3. EXTERNAL SCALAR TRANSFORM TECHNIQUES

These involve mostly the FT of the boundary. This can be expressed in terms of tangent angle versus arc length, or as complex function $bx(t) + jby(t)$. The first approach has been used by Zahn and Roskies [99], and others (e.g., [85]); the second, by Granlund [26], Persoon and Fu [74], Richards and Hemami [79], etc. We review here briefly the Zahn and Roskies formulation. Let ϕ_k be the rotation of the tangent to a boundary at the kth point in comparison to the direction of the tangent at an initial point. Let l_k be the arc length between these two points. The function ϕ_k versus l_k can be normalized and be made to depend on a parameter t ranging from 0 to 2π. Let

$$a(t_k) = \phi_k + t_k,$$

where

$$t_k = (2\pi/L)l_k.$$

L is the total length so that

$$a(0) = a(2\pi) = 0$$

It can be verified that if the data points form a regular polygon then $a(t)$ is identically zero and the same is true for a circle. Figure 1 shows this function for an L-shaped object. The coefficients of the FT of $a(t)$, or similar functions, are used for shape description.

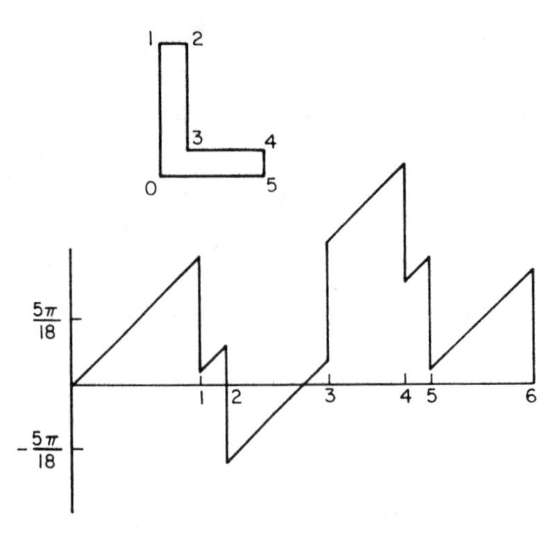

FIG. 1. Examples of the normalized angle versus boundary function $a(t)$ (bottom) for an L-shaped polygon (top) [99].

Strokes or lobes extending from the main body of an object produce peaks which, if repeated, cause a high component at a certain frequency of the FT. Thus such methods are helpful where the counting of such elements is important for shape description. The major advantage of these algorithms is that they are rather easy to program and are backed by a well-developed theory, that of Fourier transforms. Their major disadvantage is that of all transform techniques, the difficulty of describing local information [78].

4. INTERNAL SPACE DOMAIN TECHNIQUES

The medial axis transformation (MAT), proposed by H. Blum, is the earliest and most widely studied technique among these [6, 7, 56, 59, 75, 81]. In that approach a "full" figure is transformed into a line drawing in the following manner: Let S be a set in the plane and let B be its boundary. If X is a point which belongs to S then it is always possible to find its closest neighbor belonging to B. If X has more than one such neighbor then it is said to belong to the medial axis of S [6, 7]. Figure 2 shows a few simple examples. The line drawing may be labeled with the distance of each of its points from the boundary and this allows the reconstruction of the original figure as an envelope of circles centered on the medial axis, or *skeleton.*

This skeleton may be used to derive information about the shape of the original figure, but except for some rather gross properties (e.g., elongation [81]), the process of doing so is by no means straightforward. The computation of the skeleton can be quite time consuming [81] and very sensitive to noise as can be seen by comparing the first and last two drawings in Fig. 2. Both difficulties may be reduced by first performing a *polygonal approximation* of the original contour. This can certainly remove noise and also allows relatively fast computation of the skeleton [56]. However, obtaining such an approximation can be quite sufficient in itself for shape description. Thus one can decide trivially whether the figure is a triangle, a square, or a rectangle. We will return to this point later.

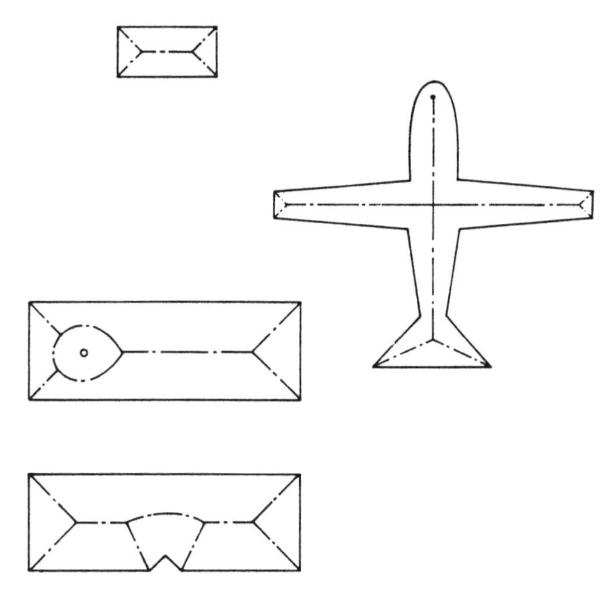

FIG. 2. Examples of the medial axis transformation (shown by -·-·- lines) [6]. An arbitrarily small hole or dent creates substantial distortions.

We may also mention that related to the MAT are numerous thinning algorithms [85].

A number of techniques related to integral geometry has been proposed by various authors [39, 40, 61, 62, 86, 92, 96]. The object is intersected by a number of chords in different directions and locations and the length of the intersections is used in various ways. The choice of the chords may be random, and length statistics can be used for shape description. However, this approach meets certain major analytical difficulties [96]. Rutowitz has used radial chords all passing through a common point to describe the shape of chromosomes [86]. Klinger *et al.* [40] have also used a similar method. Pavlidis used chords in two orthogonal directions to obtain *integral projections* and to describe the shape of typewritten characters [62].

Another set of techniques is based on decomposition. In these the original figure is expressed as the union of certain of its subsets. The shape of the latter may be simpler and therefore some of the simpler descriptors will be applicable. The earliest discussion of such techniques in the literature can be found in the work of Frishkopf and Harmon [21] and of Eden [15] where cursive script is decomposed into strokes. The extrema of a character in the X and Y directions were used as break points in the first case while Eden's paper describes a generative method.

Most of the subsequent schemes have emphasized the concepts of convexity and have assumed polygonal approximations of the original object. One of them is the decomposition into primary convex subsets (PCS) [63, 65, 66]. Figure 3 shows some examples. The method can be summarized briefly as follows: If P is a polygon then each one of its sides defines a halfplane H_i which lies in the same part of the plane as P with respect to the side. The intersection of all these halfplanes forms the kernel of P, i.e., the locus of all points from which all the vertices

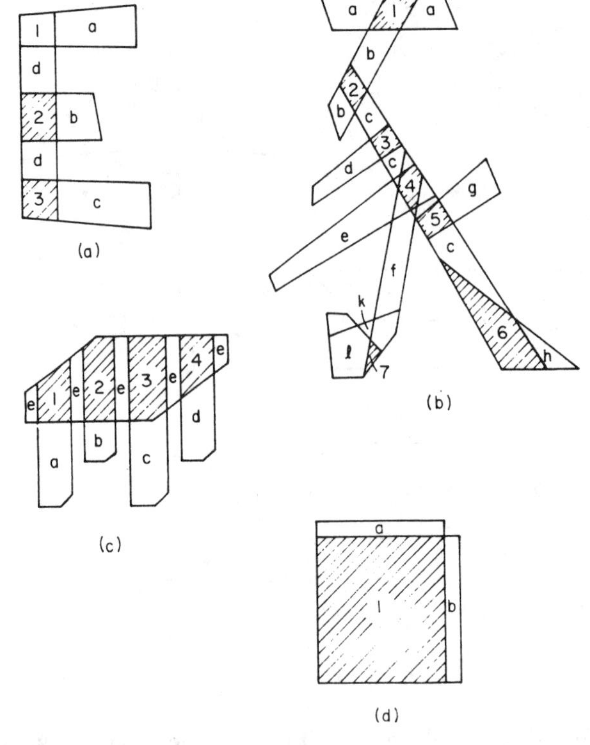

FIG. 3. Examples of decompositions into primary convex subsets [63]. These are marked by letters while the nuclei (or kernels) are marked by numbers and are shaded.

are visible [95]. By taking the intersection of successively fewer such halfplanes we can form increasing sequences of convex sets. A primary convex set is defined as the largest element in such a sequence which is also a subset of P [63, 66]. It can be shown that the union of the primary convex subsets forms a cover for P [63].

It is not too surprising that a certain relationship exists between the MAT and the PCS. Indeed, let us define as *primary branches* those branches of the MAT graph which connect nodes of degree 2 or greater at both ends. Then for many shapes there is an one-to-one correspondence between primary branches and primary convex subsets.

The expression of a polygon through its primary convex subsets often gives intuitively appealing decompositions (Fig. 3a–c). However, this is not always the case (Fig. 3d). Requiring that the primitive elements be convex is also somewhat too strict a condition since it excludes slightly curved strokes which are met often in cursive script. Its computational requirements are also substantial.

A decomposition technique which attempts to overcome these problems has been proposed by Feng and Pavlidis [17, 24]. Figure 4 gives a few examples. The dividing lines of the polygon are those connecting certain pairs of concave vertices. The primitive elements are either convex sets or *spirals,* which are defined as polygons which have all their concave vertices adjacent to each other. The method is quite fast but sometimes it looses strokes as seen from the last example in Fig. 4.

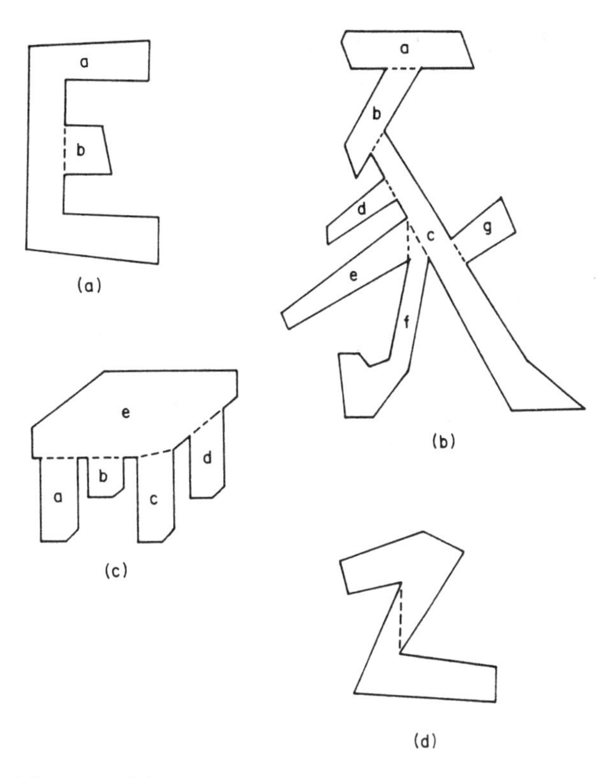

FIG. 4. Examples of decompositions at concave angles [17]. Dividing lines are shown by – – –. Each component is marked by a letter.

The output of both methods can be expressed through a juxtaposition graph [17, 65, 66]. Their results have a number of desirable features, which are also shared to some extent by the MAT:

(1) They are translation and rotation invariant and insensitive to registration. Rotation invariance can be controlled through the description of the juxtaposition relations in the final graph. The insensitivity to registration is important for many practical applications including optical page readers, mail sorters, cell counters, etc.

(2) To a large extent they are size invariant. Problems may occur only when some of the objects in a picture are so small as to be of the same order of magnitude as what is considered noise for others.

(3) They produce usually "anthropomorphic" descriptions and therefore they can be quite useful for feature extraction.

(4) They produce data structures which are particularly appropriate for syntactic or structural pattern recognition, which is natural since the methodology of decomposition is itself structural.

The main disadvantage of the methodology is that the programs implementing it tend to be quite complex. Such complexity does not necessarily imply slow processing but it may impose certain difficulties during research and development. It is our opinion that this is an unavoidable problem with any reasonable shape description scheme. After all we are trying to imitate a very complex mechanism, the human visual and perceptual processes.

5. EXTERNAL SPACE DOMAIN TECHNIQUES

The earliest among these are the syntactic descriptions of Ledley [43] and those based on the Freeman chain code [18, 19, 25]. Extensive studies of the subject have been made by Fu [22–24]. Pavlidis has suggested that before the application of such techniques, preprocessing of the data through functional approximation should be performed [64, 73]. Davis has used hierarchical curve fitting for the description of contours and in particular for detecting corners and sides [13]. Moayer and Fu [54, 55] have used tree grammars to describe local shape in the context of automatic fingerprint recognition.

The main disadvantage of boundary encodings is that points which are geometrically close together can be encoded quite far apart in the string. In the example of Fig. 5 there is no simple way to describe the neck AB or the gap CD through the boundary string. However, they have many advantages, including fast algorithms for their analysis, small storage requirements, and the availability of well-developed general methodologies such as the theory of formal languages. We shall discuss some of their major features in detail in the next two sections.

6. THE IMPORTANCE OF CURVATURE AND CORNERS

Some of the earliest theories of vision have stated that curvature maxima are important in shape perception [4, 14]. It is not too surprising that such maxima and measures of curvature in general play an important role in most shape analysis algorithms. Ledley has used boundary segments of constant curvature for primitives [43]. Strokes and lobes also are associated with regions of high curvature and therefore the latter express themselves through the coefficients of the FT. Similar conclusions are obtained on the basis of bending energy [98].

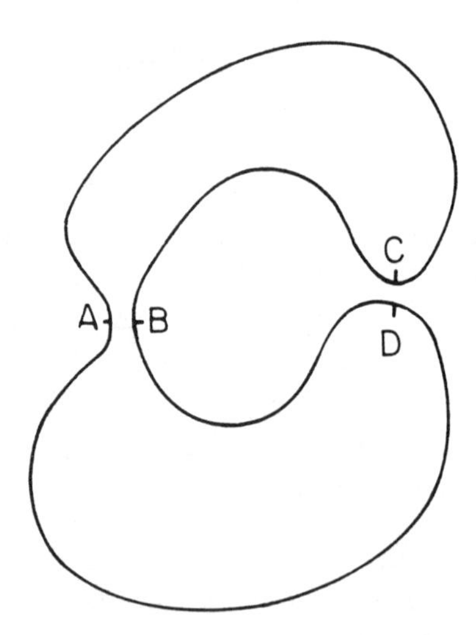

FIG. 5. Illustration of a problem with boundary scanning methods. Although A is close to B and C to D on the plane, they are far apart along the boundary.

The relevance of "intrusions" and "protrusions," both locations of high curvature (negative or positive), has also been made clear in the literature [84]. It can also be shown that each curvature maximum corresponds to a node of degree 1 of the MAT graph and therefore it plays a central role in determining its form.

It is probably true that if the outlines of objects were noise-free then a direct evaluation of curvature would go a long way toward determining their shape. However, in the presence of noise one may have to distinguish between "essential" and "inessential" curvature maxima. A number of algorithms have been suggested for finding the essential curvature maxima, e.g. [83]. Probably the most interesting of them are those based on functional approximation.

Closely related to curvature as a shape descriptor is the concept of a *corner*, which is defined as a point of infinite curvature in the case of continuous curves [20]. The situation is not quite as clear for discrete data. It is probably best to define a corner as a point where it is necessary to switch descriptions of the curve in terms of smooth curves. The latter may be defined as lines or quadratic arcs.

There is a remarkable result from the theory of splines which has a bearing on these questions.

THEOREM [53]. *Let $f(t)$ be a function to be approximated by a spline with variable knots while minimizing the integral square error. If k is the degree of the polynomials forming the spline and n the number of its knots, then the asymptotic distribution (as n tends to infinity) of the knots follows the $(k + 1)$st derivative of $f(t)$.*

For continuous piecewise linear approximations where the error is evaluated along the normal to the curve the asymptotic distribution follows the *curvature* (McClure, 1975, personal communication). At first this result may not seem very interesting since one has to go to an infinite number of segments before placing the knots at curvature maxima. However computational experience indicates that for a small number of knots their distribution follows pretty much what one would call essential curvature maxima [68]. Part of the reason for this property is that when the number of knots is increased new ones are placed between the old ones whose position remains more or less fixed [68]. If no continuity constraints are imposed then the integral square error is minimized when knots are placed either near curvature maxima (with continuity there resulting as a bonus of the optimal location [69, 71]) or near points of inflection (curvature zeros) with the approximation being symmetric around the curve.

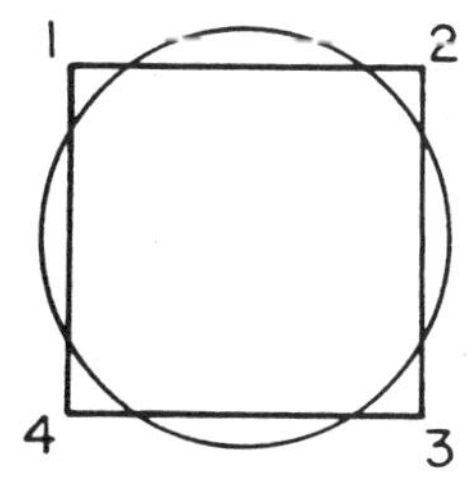

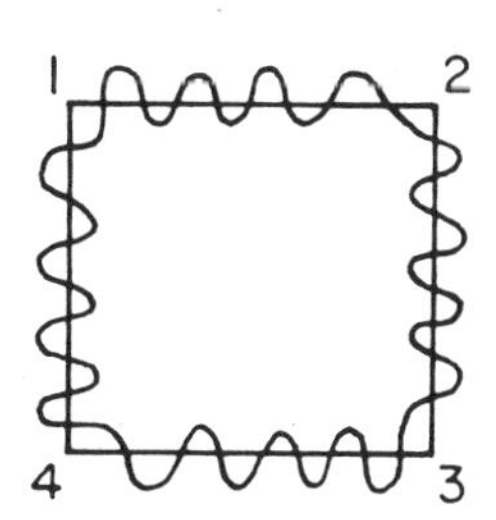

FIG. 6. Illustration of a problem with functional approximation methods. The error of approximation (by most norms) is the same for both figures even though in the right figure the approximation is more "natural" than in the other.

However, the results of such approximations leave open some questions. We illustrate this by an example using the objects of Fig. 6, a noisy square and a circle. For a given tolerance on the L_2 error norm the approximating polygon will be in both cases a square. In the first case the corners are "real," but not so in the second. One obvious way to distinguish between the two is to examine how sharply the minimum of the error norm depends on the chosen locations of corners. Such a dependence will be very steep in the case of the square but not so in the case of the circle. Mathematically this is manifested through the matrix of the second derivatives of the error norm with respect to the location of the breakpoints. If the minimum is "steep" then the matrix has significant diagonal dominance and a large determinant. Although some analytical expressions for these derivatives have been obtained [70, 71] there are problems in their evaluation on noisy data.

Some valuable information may be obtained by keeping track of the error norm during the iterations. Another alternative is to consider variations at the locations after termination. This, however, may be rather expensive computationally. The recent methods of Davis, Freeman, and Rosenfeld [13, 20], as well as some earlier ones [83], are essentially of this type, the major difference being that they start from "scratch" rather than from an initial approximation.

The previous discussion suggests that polygonal approximations should have certain merits for shape analysis. Indeed this has been recognized by many investigators independently of the arguments based on the theory of splines [2, 14, 36, 56, 47, 52, 66, 76–78]. It is rather obvious that higher-order approximations would also be desirable, but their evaluation is quite expensive computationally [12]. A possible solution is to use polygonal approximations as a basis for higher-order ones [9].

At this point we may say a few words about the relevance of convexity in shape analysis. One possible definition of a convex object is that of having no points of negative curvature along its boundary. Thus, given the importance of curvature in shape perception, it is not surprising that convex contours have been found of interest in the shape analysis literature (see for example Section 4 above). To this factor we should add the many desirable mathematical properties of convex sets [16, 95]. Generally speaking the recognition of the shape of convex objects is much easier than the general case and there exist a number of techniques applicable only to this class, while some of the general ones perform much better if their input is restricted to it [5, 44, 51, 52, 74, 75].

7. SYNTACTIC TECHNIQUES

These techniques seem quite promising because they attempt to reproduce the descriptions of human observers. We will attempt here a general formalism. Let

$$V = v_1, v_2, \ldots, v_n$$

be a representation of a boundary. v_i may be an element of a chain code [18, 19], a side of a polygonal approximation [68, 73], a quadratic arc [12, 43], etc. One can then establish a set of production rules which will produce this string from certain nonterminal symbols. In general such grammars tend to be quite complex if one wants to guarantee certain global properties such as boundary closing.

However, simpler grammars can suffice if only local properties are considered. The peak detector developed by Horowitz [33] is a simple example of this philosophy. Other examples are the waveform parser of Stockman *et al.* [94] and the detection of circuit board faults by lexical analysis of chain codes proposed by Jarvis [35]. Local grammars will result in replacing the string V by another string

$$S = s_1, s_2, \ldots, s_m,$$

where s_j is a descriptor of a high-order structure, e.g., stroke, lobe, concave arc, etc. It may contain a number of attributes such as length of a stroke, its orientation, etc. In general m will be much smaller than n. For the purpose of this paper we may limit our attention to this level since subsequent parsing of the string S may be heavily problem-related. Using the original model of Ledley we may think of the V_i's as arcs of constant curvature and of the S_i's as "arms" and "concavities." Further transformations would result in the classification of the original outline as that of a normal or abnormal chromosome, etc.

One promising avenue is the combination of approximation and syntactic techniques. We will illustrate this by discussing the case of a corner detector. Let us assume that it looks for a part of the boundary formed by two straight lines forming a significant angle with each other. If the data have any noise then the parser will be required to detect general directions along the boundary. Thus deviations from linearity would have to be evaluated as significant or not. If a functional approximation preprocessor is used this will eliminate most if not all of the noise. By choosing appropriately tight tolerances we may be sure that no "real" corners are missed. The syntactic analyzer would not worry about "overall" directions since these are either directly available or readily computable. Thus it will search only for significant local changes in the angle. Quadratic arcs would produce polygonal approximations with equal sides and angles which could be easily detected. For example, we may have the following grammar:

Nonterminals	= SIDE, CORNER, TURN,
Terminals	= side (long side of polygon resulting from approximation) break (short side of above polygon),
Semantics	= $A(a, b)$ angle of sides a, b. If a or b are breaks $A = 0$.
SIDE	$\rightarrow$ (SIDE)(side) A(SIDE, side) $< \epsilon$,
CORNER	$\rightarrow$ (SIDE)(TURN)(SIDE) A(SIDE, SIDE) $> \theta$,
TURN	$\rightarrow$ empty/break/(break)(break).

Figure 7 gives an illustrative example. In general there will be a gap between the thresholds ϵ and θ.

8. CONCLUSIONS

It seems that it is possible to develop mathematical algorithms which analyze silhouettes and produce results compatible with human intuition. The combina-

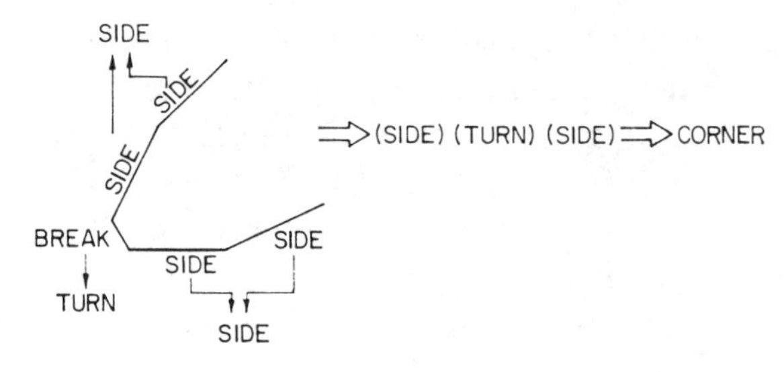

FIG. 7. Illustration of the combination of functional approximation and a syntactic technique [73].

tion of curve or line fitting techniques with syntactic methods seems particularly promising [33, 43, 73, 94].

Techniques which are based on the space domain seem very attractive and popular but they often lack the mathematical elegance of transforms. However, the theory of splines, especially those with variable knots [11, 44, 51, 53, 78], seems to offer a strong mathematical foundation for the space domain techniques [67, 69, 71].

Further progress in syntactic algorithms and especially parsing of stochastic or fuzzy grammars [22–24, 27] is also very desirable.

ACKNOWLEDGMENTS

I want to thank Professors K. S. Fu and A. Rosenfeld for many helpful comments on a first draft of this paper. A talk based on this paper was given at the Engineering Foundation Conference on Algorithms for Image Processing, Franklin Pierce College, Ringe, New Hampshire, August 8–13, 1976. The rewriting of the original draft benefited tremendously from the discussions at that meeting.

REFERENCES

1. M. Aiello, C. Lami, and U. Montanari, A system for computer measurements and karotyping of wheat metaphases, in *Proceedings of the First International Joint Conference on Pattern Recognition, Washington D.C., Oct. 1973*, pp. 205–219.
2. A. Albano, Representation of digitized contours in terms of conic arcs and straight-line segments, *Computer Graphics and Image Processing* **3**, 1974, 23–33.
3. F. L. Alt, Digital pattern recognition by moments, *J. Assoc. Comput. Mach.* **11**, 1962, 240–258.
4. F. Attneave, Some informational aspects of visual perception, *Psychol. Rev.* **61**, 1954, 183–193.
5. J. W. Bacus and E. E. Gose, Leucocyte pattern recognition, *IEEE Trans. Systems, Man and Cybernetics* **SMC-2**, 1974, 513–526.
6. H. Blum, A transformation for extracting new descriptions of shape, in *Symposium on Models for the Perception of Speech and Visual Form,* M.I.T. Press, 1964.
7. H. Blum, Biological shape and visual science, I. *J. Theor. Biol.* 1973, 205–287.
8. B. A. Blesser, T. T. Kuklinski, and R. J. Shillman, Empirical tests for feature selection based on a psychological theory of character recognition, *Pattern Recognition* **8**, 1976, 77–85.
9. L. P. Chang and T. Pavlidis, *Functional Approximation with Variable-Knot Variable-Degree Splines,* Tech. Report No. 201, Princeton University, Feb. 1976.
10. Y. P. Chien and K. S. Fu, Recognition of X-ray picture patterns, *IEEE Trans. Systems, Man and Cybernetics* **SMC-4**, 1974, 145–165.

11. A. K. Cline, Scalar and planar valued curve fitting using splines under tension, *Commun. ACM* **17**, 1974, 218–220.
12. D. B. Cooper and N. Yalabik, *On the Cost of Approximating and Recognizing a Noise Perturbed Straight Line or a Quadratic Curve Segment in the Plane,* Tech. Report, Brown University, March 1975.
13. L. S. Davis, *Understanding Shape. I. Angles and Sides,* TR-376, University of Maryland, Computer Science Center, 1975.
14. R. O. Duda and P. E. Hart, *Pattern Classification and Scene Analysis,* Chap. 9, Wiley, New York, 1973.
15. M. Eden, Handwriting and pattern recognition, *IRE Trans. Information Theory* **IT-8**, 1962, 160–166.
16. G. Ewald and G. C. Shephard, Normed vector spaces consisting of classes of convex sets, *Math. Z.* **91**, 1966, 1–19.
17. H. Y. Feng and T. Pavlidis, Decomposition of polygons into simpler components: Feature extraction for syntactic pattern recognition, *IEEE Trans. Computers* **C-24**, 1975, 636–650.
18. H. Freeman, On the encoding of arbitrary geometric configurations, *IEEE Trans. Electronic Computers* **EC-10**, 1961, 260–268.
19. H. Freeman, Boundary encoding and processing, in *Picture Processing and Psychopictorics* (B. S. Lipkin and A. Rosenfeld, Ed.), pp. 241–266, Academic Press, New York, 1970.
20. H. Freeman and L. S. Davis, *A Corner Finding Algorithm for Chain Coded Curves* TR-399, University of Maryland, 1975.
21. L. S. Frischkopf and L. D. Harmon, Machine reading of cursive script, in *Proceedings of the Symposium on Information Theory* (C. Cherry, Ed.), pp. 300–316, Butterworths, London, 1961.
22. K. S. Fu and P. H. Swain, On syntactic pattern recognition, in *Software Engineering* (J. Tou, Ed.), Vol. 2, pp. 155–182, Academic Press, New York, 1971.
23. K. S. Fu, *Syntactic Methods in Pattern Recognition,* Academic Press, New York, 1974.
24. K. S. Fu, (Ed.), *Applications of Syntactic Pattern Recognition,* Springer–Verlag, New York/ Berlin, 1976.
25. G. Gallus, Contour analysis in pattern recognition for human chromosome classification, *Appl. Biomed. Calcolo Elettr.*, 1968, pp. 95–108.
26. G. H. Granlund, Fourier preprocessing for hand print character recognition, *IEEE Trans. Computers* **C-21**, 1972, 195–201.
27. U. Grenander, "Foundations of Pattern Analysis," *Quart. Appl. Math.* **27**, 1969, 1–55.
28. U. Grenander, *Pattern Synthesis: Lectures in Pattern Theory,* Vol. 1, Springer–Verlag, 1976.
29. V. E. Giuliano, P. E. Jones, G. E. Kimball, R. F. Meyer, and B. A. Stein, Automatic pattern recognition by a gestalt method, *Inform. Contr.* **4**, 1961, 332–345.
30. M. J. Hannah, *Generalized Automated Pattern Recognition: Pattern Classification by Moment Invariants,* M. Sc. Thesis, Industrial Engineering Department, University of Missouri, Columbia, June 1971.
31. L. D. Harmon, Automatic recognition of print and script, *IEEE Proc.* **60**, 1972, 1165–1176.
32. S. L. Horowitz and T. Pavlidis, Picture processing by graph analysis, in *Proceedings of the Conference on Computer Graphics, Pattern Recognition and Data Structure, Los Angelos,* 1975, pp. 125 129.
33. S. L. Horowitz, A general peak detection algorithm with applications in the computer analysis of electrocardiograms, *Commun. ACM* **18**, 1975, 281–285.
34. M. K. Hu, Visual pattern recognition by moment invariants, *IRE Trans. Information Theory* **IT-8**, 1962, 179–187.
35. J. F. Jarvis, Regular expressions as a feature selection language for pattern recognition, 3-IJCPR.
36. T. Kaneko and P. Mancini, Straight line approximation for boundary of left ventricular chamber from a cardial cineangiogram, in *Proceedings of the 6th Annual Princeton Conference on Information Sciences and Systems, 1972,* pp. 337–341.
37. L. Kaufman *et al., Contour Descriptor Properties of Visual Shape,* Final Report on Contract AF19 (628)-5830, Sperry Rand Research Center, Sudbury, Mass., Sept. 1967.

38. R. Kirsch, Computer interpretation of english text and picture patterns, *IEEE Trans. Electronic Computers* **EC-13**, 1964, 363–376.
39. A. Klinger, Pattern width at a given angle, *Commun. ACM* **14**, 1971, 21–25.
40. A. Klinger, A. Kochman, and N. Alexandridis, Computer analysis of chromosome patterns: Feature-encoding for flexible decision making, *IEEE Trans. Computers*, **C-20**, 1971, 1014–1022.
41. P. A. Kolers, The role of shape and geometry in picture recognition, in *Picture Processing and Psychopictorics* (B. S. Lipkin and A. Rosenfeld, Eds.), pp. 181–202, Academic Press, New York, 1970.
42. D. J. Langridge, On the computation of shape, *Frontiers of Pattern Recognition* (S. Watanabe, Ed.), pp. 347–365, Academic Press, New York, 1972.
43. R. S. Ledley, High speed automatic analysis of biomedical pictures, *Science*, **146**, 1964, 216–223.
44. E. H. Lee and G. E. Forsythe, Variational study of nonlinear spline curves, *SIAM Rev.* **15**, 1973, 120–133.
45. E. T. Lee, Shape-oriented chromosome classification, *IEEE Trans. Systems, Man, and Cybernetics* **SMC-5**, 1976, 629–632.
46. E. T. Lee, Shape-oriented classification storage and retrieval of leucocytes, *Math. Biosci.*
47. P. M. Lindsay and D. A. Norman, *Human Information Processing*, Academic Press, New York, 1972.
48. B. S. Lipkin, Introduction: Psychopictorics, in *Picture Processing and Psychopictorics* (B. S. Lipkin and A. Rosenfeld, Eds.), pp. 3–36, Academic Press, New York, 1970.
49. S. Løvtrup and B. von Sydow, D'Arcy Thompson's theorems and the shape of the molluscan shell, *Bull. Math. Biophys.* **36**, 1974, 567–575.
50. K. Maruyama, *A Problem in Form Perception: Odd Shape Detection*, Tech. Report R-71-490, University of Illinois, Dec. 1971.
51. D. E. McClure, Problems and methods of nonlinear feature generation in pattern analysis, in *Proceedings of the 8th Princeton Conference on Information Sciences and Systems* (1974), pp. 244–247.
52. D. E. McClure and R. A. Vitale, Polygonal approximation of plane convex bodies, *J. Math. Anal. Appl.* **51**, 1975, 326–358.
53. D. E. McClure, Nonlinear segmented function approximation and analysis of line patterns, *Quart. Appl. Math.* **33**, 1975, 1–37.
54. B. Moayer and K. S. Fu, A syntactic approach to fingerprint pattern recognition, *Pattern Recognition* **7**, 1975, 1–23.
55. B. Moayer and K. S. Fu, A tree system approach for fingerprint pattern recognition, *IEEE Trans. Computers* **C-25**, 1976, 262–274.
56. U. Montanari, Continuous skeletons from digitized images, *J. Assoc. Comput. Mach.* **16**, 1969, 534–549.
57. U. Montanari, A note on minimal length polygonal approximation to a digitized contour, *Commun. ACM* **13**, 1970, 41–47.
58. U. Montanari, Heuristically guided search and chromosome matching, *Artif. Intelligence* **1**, 1970, 227–245.
59. J. C. Mott-Smith, Medial axis transformations, in *Picture Processing and Psychopictorics* (B. S. Lipkin and A. Rosenfeld, Eds.), pp. 267–283, Academic Press, New York, 1970.
60. G. Nagy, Feature extraction on binary patterns, *IEEE Trans. Systems Sci. Cybernet.* **SSC-5**, 1969, 273–278.
61. Y. Nakimoto, K. Nakato, Y. Uchikura, and A. Nakajima, Improvement of chinese character recognition using projection profiles, in *Proceedings of the First International Joint Conference on Pattern Recognition (1973)*, pp. 172–178.
62. T. Pavlidis, Computer recognition of figures through decomposition, *Inform. Contr.* **14**, 1968, 526–537.
63. T. Pavlidis, Analysis of set patterns, *Pattern Recognition* **1**, 1968, 165–178.
64. T. Pavlidis, Linguistic analysis of waveforms, in *Software Engineering* (J. Tou, Ed.), pp. 203–225, Academic Press, New York, 1971.
65. T. Pavlidis, Representation of figures by labelled graphs, *Pattern Recognition* **4**, 1972, 5–17.

66. T. Pavlidis, Structural pattern recognition: Primitives and juxtaposition relations, in *Frontiers of Pattern Recognition* (S. Watanabe, Ed.), pp. 421–451, Academic Press, New York, 1972.
67. T. Pavlidis, Waveform segmentation through functional approximation, *IEEE Trans. Computers* **C-22**, 1973, 689–697.
68. T. Pavlidis and S. L. Horowitz, Segmentation of plane curves, *IEEE Trans. Computers* **C-23**, 1974, 860–870.
69. T. Pavlidis, Optimal piecewise polynomial L2 approximation of functions of one variables, *IEEE Trans. Computers* **C-24**, 1975, 98–102.
70. T. Pavlidis, Fuzzy representations as means of overcoming the overcommittment of segmentation, in *Proceedings of the Conference on Computer Graphics, Pattern Recognition and Data Structure, Los Angeles, Calif., May 1975*, pp. 215–219.
71. T. Pavlidis, *Polygonal Approximations by Newton's Method*, Tech. Teport No. 194, Computer Science Laboratory, Princeton University, October, 1975.
72. T. Pavlidis and F. Ali, Computer recognition of handwritten numerals by polygonal approximations, *IEEE Trans. Systems, Man, and Cybernetics* **SMC-5**, 1975, 610–614.
73. T. Pavlidis, *Structural Pattern Recognition*, Springer–Verlag, Berlin–Heidelberg–New York, 1977.
74. E. Persoon and K. S. Fu, Shape discrimination using fourier descriptors, in *Proceedings of the Second International Joint Conference on Pattern Recognition, Copenhagen, 1974*, pp. 126–130.
75. O. Philbrick, *A Study of Shape Recognition Using the Medial Axis Transformation*, Report No. 288, Air Force Cambridge Research Laboratories, November 1966.
76. J. M. S. Prewitt and M. L. Mendelsohn, Analysis of cell images, *Ann. N. Y. Acad. Sci.* **128**, 1966, 1035–1053.
77. U. Ramer, An iterative procedure for polygonal approximation of plane curves, *Computer Graphics and Image Processing* **1**, 1972, 244–256.
78. J. R. Rice, *The Approximation of Functions*, Vol. 2, Chap. 10, Addison–Wesley, Reading, Mass., 1969.
79. C. W. Richards, Jr. and H. Hemami, Identification of three-dimensional objects using fourier descriptors of the boundary curve, *IEEE Trans. Systems, Man and Cybernetics* **SMC-4**, 1974, 371–378.
80. W. S. Rosenbaum and J. J. Hilliard, Multifont OCR postprocessing system, *IBM J. Res. Develop.* **19**, 1975, 398–421.
81. A. Rosenfeld and J. L. Pfaltz, "Sequential Operations in Digital Picture Processing," *J. Assoc. Comput. Mach.* **13**, 1966, 471–494.
82. A. Rosenfeld, *Picture Processing by Computer*, Academic Press, New York, 1969.
83. A. Rosenfeld and E. Johnston, Angle detection on digital curves, *IEEE Trans. Computers* **C-22**, 1973, 874–878.
84. A. Rosenfeld and J. S. Weszka, Picture recognition and scene analysis, *Computer* **9**, 1976, 28–38.
85. A. Rosenfeld and A. C. Kak, *Digital Picture Processing*, Academic Press, New York, 1976.
86. D. Rutovitz, Centromere finding: Some shape descriptors for small chromosome outlines, *Machine Intelligence* **5**, 1970, 435–462.
87. D. Rutovitz, An algorithm for in-line generation of a convex cover, *Computer Graphics and Image Processing* **4**, 1975, 74–78.
88. J. Sklansky, Recognition of convex blobs, *Pattern Recognition* **2**, 1970, 3–10.
89. J. Sklansky, R. L. Chazin, and B. J. Hansen, Minimum-perimeter polygons of digitized silhouettes, *IEEE Trans. Computers* **C-21**, 1972, 260–268.
90. J. Sklansky, Measuring concavity on a rectangular mosaic, *IEEE Trans. Computers* **C-21**, 1974, 1355–1364.
91. J. Sklansky, On filling cellular concavities, *Computer Graphics and Image Processing* **4**, 1975, 236–247.
92. R. J. Spinrad, Machine recognition of hand printing, *Inform. Contr.* **8**, 1965, 124–142.
93. W. Stallings, Approaches to chinese character recognition, *Pattern Recognition* **8**, 1976, 87–98.
94. G. C. Stockham, L. N. Kanal, and M. C. Kyle, Design of a waveform parsing system, in

Proceedings of the First International Joint Conference on Pattern Recognition, 1973, pp. 236–243.

95. F. A. Valentine, *Convex Sets*, p. 157, McGraw–Hill, New York, 1964.
96. E. Wong and J. A. Steppe, Invariant recognition of geometric shapes, in *Methodologies of Pattern Recognition* (S. Watanabe, Ed.), pp. 535–546, Academic Press, New York, 1969.
97. I. T. Young, The classification of white blood cells, *IEEE Trans. Biomed. Eng.* **BME-19**, 1972, 291–298.
98. I. T. Young, J. E. Walker, and J. E. Bowie, An analysis technique for biological shape, *Inform. Contr.* **25**, 1974, 357–370.
99. C. T. Zahn and R. Z. Roskies, Fourier descriptors for plane closed curves, *IEEE Trans. Computers* **C-21**, 1972, 269–281.

Chapter 4
Text Analysis and Recognition

4.1: Introduction

There are two main types of analysis that are applied to text in documents. One is optical character recognition (OCR) to derive the meaning of the characters and words from their bit-mapped images. The other is page layout analysis to discover formatting of the text and, from that, to derive meaning associated with the positional and functional blocks in which the text is located. These operations may be performed separately, or the results from one analysis may be used to aid or correct those from the other. OCR methods are usually distinguished as being applicable to either machine-printed or handwritten character recognition. Page layout analysis techniques are applied to formatted, machine-printed pages, and a type of layout analysis (forms recognition) is applied to machine-printed or handwritten text occurring within delineated blocks on a printed form. Both OCR and page layout analysis are covered in this chapter, and the last section contains descriptions of document systems incorporating both of these types of analysis.

We include in this introductory section a paper, by Jain and Bhattacharjee, on separation of text from graphics components. This topic straddles the bounds between text analysis (in this chapter) and graphics analysis (in chapter 5). In the introductory section of chapter 5, we include another paper on text/graphics separation [Fletcher 1988], which describes methods for the same objective, but the methods are described more from the point of view of graphics analysis.)

The other paper included in this section, by Phillips, Chen, and Haralick, describes a document database intended for researchers to test against a common and known set of data. The database contains images of machine-printed documents, each with a description of the contents that includes the text and its location. We include this paper both to show the extent of document types and as a readily obtainable (by CD-ROM) source of documents for testing.

4.2: Skew estimation

(*Keywords*: Hough transform, nearest-neighbor clustering, page orientation, projection profile, skew estimation, text lines)

A *text line* is a group of characters, symbols, and words that are adjacent, relatively close to each other, and through which a straight line can be drawn (usually with horizontal or vertical orientation). The dominant orientation of the text lines in a document page determines the *skew angle* of that page. A document originally has zero skew—that is, horizontally or vertically printed text lines are parallel to the respective edges of the paper. However, when a page is manually scanned or photocopied, nonzero skew may be introduced. Since analysis steps such as OCR and page layout analysis most often depend on an input page with zero skew, it is important to perform skew estimation and correction before these steps. Also, since a reader expects a page displayed on a computer screen to be upright in normal reading orientation, skew correction is normally done before scanned pages are displayed.

We describe the following three categories of skew estimation techniques based on their approaches:

- using the projection profile,
- fitting baselines by the Hough transform, and
- using nearest-neighbor clustering.

4.2.1: Projection profile methods

Projection profile methods are popular for skew detection. A *projection profile* is a histogram of the number of ON-pixel values accumulated along parallel sample lines taken through the document (see Figure 1). The profile may be at any angle, but often it is taken horizontally along rows or vertically along columns; such

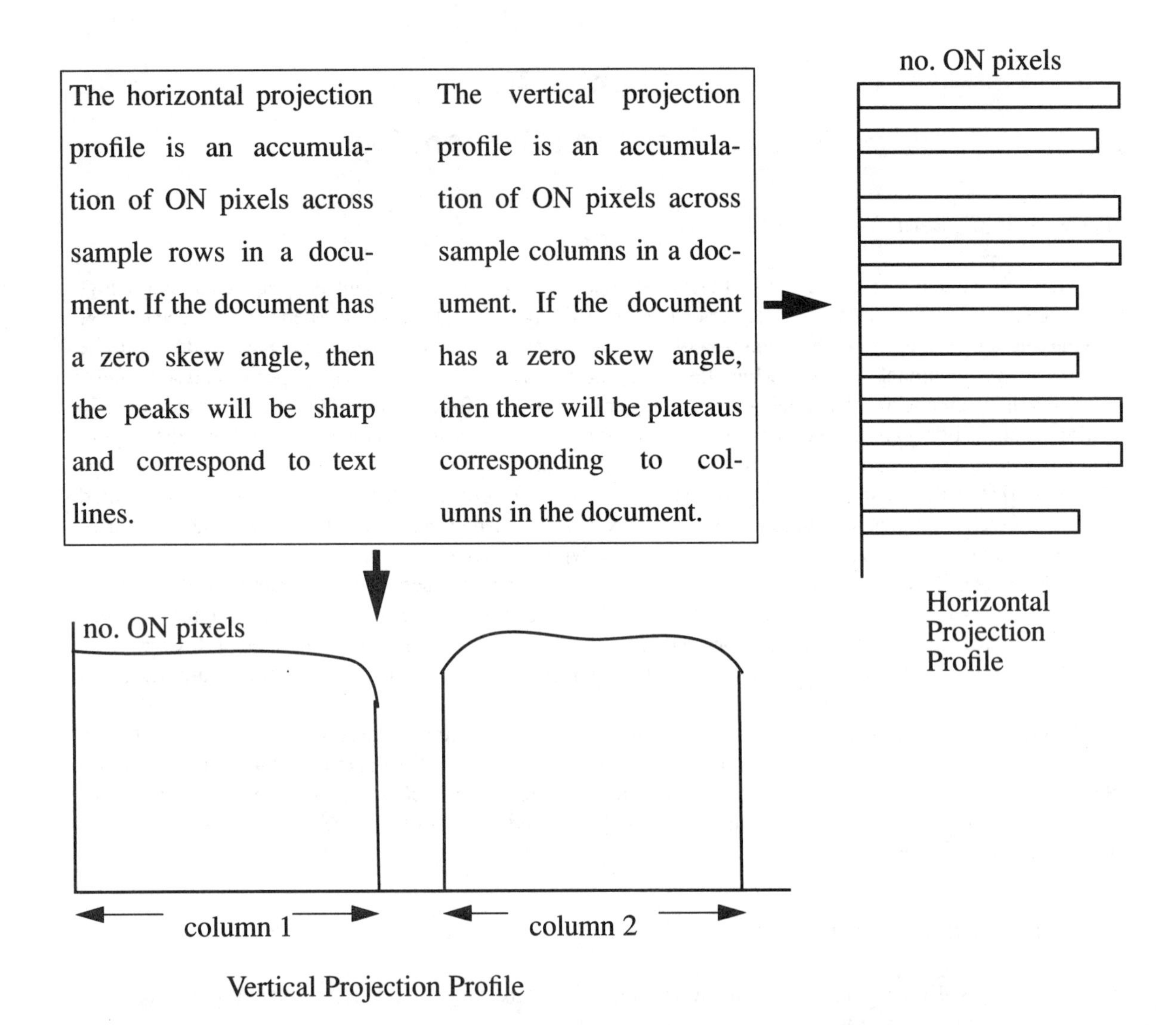

Figure 1. A portion of two columns of text are shown on the top left. The horizontal and vertical projection profiles from this text are also shown. Notice in the horizontal projection profile that each bin corresponds to a text line and in the vertical projection profile that each plateau corresponds to a column.

profiles are called *horizontal* and *vertical projection profiles*, respectively. For a document whose text lines span horizontally, the horizontal projection profile will have peaks whose widths are equal to the character height and valleys whose widths are equal to the between-line spacing. For multicolumn documents, the vertical projection profile will have a plateau for each column, with plateaus separated by valleys for the between-column spacing and the margin spacing.

The most straightforward use of the projection profile for skew detection is to compute the skew at a number of angles close to the expected orientation [Postl 1986]. For each angle, a measure is made of the variation in the bin height along the profile, and the one with the maximum variation gives the skew angle. At the correct skew angle, since scan lines are aligned to text lines, the projection profile has maximum-height peaks for text and valleys for between-line spacing. Modifications and improvements can be made to this general technique to iterate more quickly to the correct skew angle and to determine it more accurately.

One modification to the projection profile method was proposed by Baird [Baird 1987] to improve the speed and accuracy of skew detection. First, connected components are found; these are represented by points at the bottom center of their bounding boxes. A measure of the total variation (such as the peak-to-valley difference in height) is determined for various projection angles. The resulting value for each angle is a measure of the colinearity of the baseline points along each angle—the higher the variation, the closer the skew is to being zero. The peak of these measures gives the true skew angle. The accuracy of this method is typically ±0.5 degrees with respect to the correct orientation. Because detection is done using the "bottom-center" point of each component, there is an assumption that the page is roughly upright when scanned, and, partly as a result of this assumption, the method has its highest accuracy within ±10 degrees of skew.

In a faster, though less accurate, method of approximating the skew angle, shifts in projection profiles are measured. [Akiyama 1990]. For text lines that are approximately horizontal, the document is divided into equal-width vertical strips that each cover the height of the page image. Horizontal projection profiles are calculated for each vertical strip, and the skew is the angle of the shift between neighboring strips where the peaks correspond. The chosen width of the vertical strips is related to the skew; it must be small enough that sharp peaks are obtained in the profile. For larger skew, the width must be smaller, but there is a lower limit when not enough of the text line is accumulated to obtain good peaks. The authors' experiments show this technique to work well if skew is less than ±10 degrees.

Another fast method based on the measurement of shifts between strips uses vertical projection profiles determined for horizontal strips of the page [Pavlidis 1991], in contrast to the use of horizontal profiles on vertical strips by [Akiyama 1990], as described above. The locations of edges of columns for each strip are found from the locations of minima on each profile. Lines fitted along these edge locations for all intervals are perpendicular to the skew. This method is faster than most others because iterations are not required and fewer points need to be fitted; however, for the same reasons, it is not as accurate. Also, the accuracy is a function of the initially unknown skew angle and the chosen interval length. The method is said to work up to a skew angle of ±25 degrees.

4.2.2: Hough transform methods

In Section 3.3, we mention that the Hough transform was useful for straight-line detection. It can be used for a similar purpose to find skew from text line components [Hinds 1990, Nakano 1990, Srihari 1989]. As described in greater detail in Section 3.3, the Hough transform maps each point in the original (x, y) plane to all points in the (r, θ) Hough plane of possible lines through (x, y) with slope θ and distance from origin r. We mention in Section 3.3 that performing the Hough transform on individual points is expensive, but there are speedups, such as using the slope of line segments. For documents, one speedup is to compute "burst" images to reduce the number of pixel transformations to the Hough space. These horizontal and vertical bursts are runs of continuous ON pixels along rows or columns, respectively. These bursts are coded by their ON length at the location of the end of the run. Thus, the burst image has a higher value closer to the right and bottom edges of characters (for a page with small skew angle), and the total number of pixels to transform is reduced. The burst image is then transformed to Hough space. Here, each burst value is accumulated in bins at all values of (r, θ) that parameterize straight lines running through its (x, y) location on the burst image. In Hough space, the peak θ bin gives the angle at which the largest number of straight lines can be fit through the original pixels, and this is the skew angle. One limitation is that when vertical and horizontal bursts are taken, an assumption has been made that the skew is limited in range—in this case, the authors assume ±15 degrees. Also, if text is sparse, as for some forms, it may be difficult to correctly choose a peak in Hough space. Even with using the bursts to improve efficiency, the Hough transform approach typically is slower than the noniterative projection profile methods described above; however, the accuracy typically is high.

4.2.3: Nearest-neighbor methods

All the above methods have some limitation in the maximum amount of skew that they can handle. Another approach—using nearest-neighbor clustering—does not have this limitation [Hashizume 1986]. In this approach, the first step is to determine connected components. Then, the nearest neighbor of each component is found (that is, the component that is closest in Euclidean distance), and the angle between centroids of nearest-neighbor components is calculated (see Figure 2). Since intracharacter spacing is smaller within words and between characters of words on the same text lines, these nearest neighbors will be predominantly neighbors of adjacent characters on the same text lines. All the direction vectors for the nearest-neighbor con-

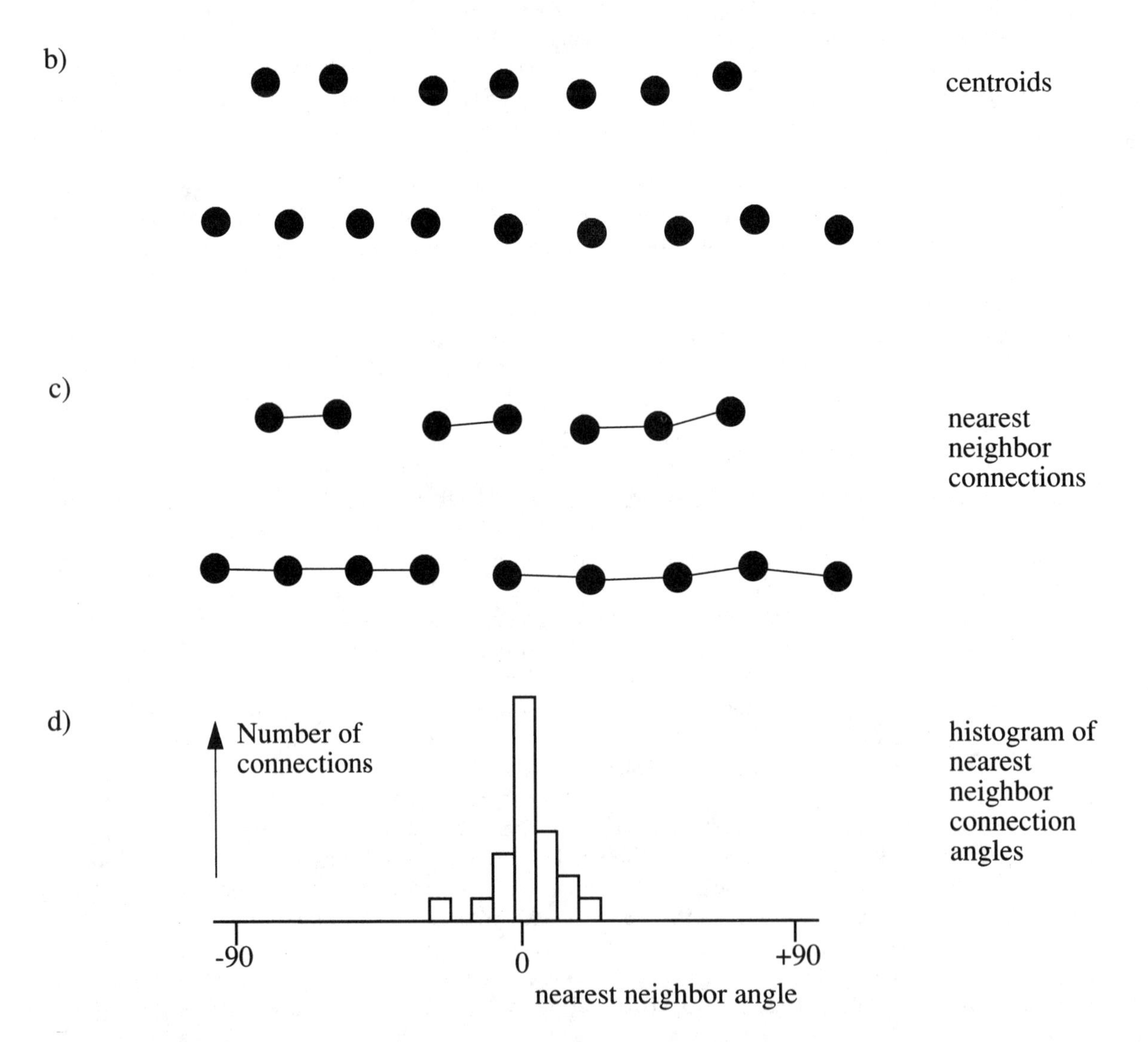

Figure 2. Diagram illustrates nearest-neighbor method. (a) Text. (b) Centroids of characters in (a). (c) Connections between closest neighbors (one neighbor each). (d) Plot of connection angles showing peak at zero, indicating zero skew angle.

nections are accumulated in a histogram, and the histogram peak indicates the dominant direction—that is, the skew angle. The cost of being able to obtain any skew angle is that this method has a relatively higher computational cost than most other methods, especially in comparison to the fast projection profile methods. The accuracy of the method depends on the number of components; however, since only one nearest-neighbor connection is made for each component, connections with noise, subparts of characters for example, the (dot on an "i"), and between-text lines can reduce the accuracy for relatively sparse pages.

An extension of this nearest-neighbor approach is to obtain not just one neighbor for each component but k neighbors, where k typically is 4 or 5 [O'Gorman 1993]. As in the one-nearest-neighbor approach in the previous paragraph, this method begins with the nearest-neighbor directions between centroids of connected components being accumulated on a histogram. However, because there are $k=5$ directions for each component and connections may be made both within and across text lines, this histogram peak is used only as an initial estimate of the skew angle. This estimate is used to eliminate connections whose directions are outside of a range of directions around the estimate; that is, they are probably interline connections. Then, a least-squares fit is made to the centroids of components that are still grouped by nearest-neighbor connections. These least-squared fits are assumed to be fits to full text lines, and this refined measurement is used as the more accurate estimate of skew angle. Because individual text lines are found here, this method encroaches on page layout analysis, in which text lines, columns, and blocks are found. Indeed, this method, called the *docstrum*, comprises both skew estimation and layout analysis. It is explained further in the next section.

The skew detection methods used in commercial systems employ primarily the projection profile. Consequently, they are limited in the amount of skew they can handle, which is usually less than 10 degrees.

Papers describing two popular and effective approaches to skew detection are included here. A projection profile approach is employed in the paper by Baird and the Hough transform is used in the paper by Hinds, Fisher, and D'Amato.

4.3: Page layout analysis

(*Keywords*: bottom-up analysis, document layout analysis, functional labeling, geometric layout analysis, structural layout analysis, syntactic labeling, text blocks, text lines, top-down analysis)

After skew detection, the image is usually rotated to zero skew angle, and then layout analysis is performed. *Structural layout analysis* (also called *physical* and *geometric layout analysis* in the literature) is performed to obtain a physical segmentation into groups of document components. Depending on the document format, segmentation can be performed to isolate words, text lines, and structural blocks (groups of text lines such as separated paragraphs or table-of-contents entries). *Functional layout analysis* (also called *syntactic* and *logical layout analysis* in the literature) uses domain-dependent information consisting of layout rules of a particular page to perform labeling of the structural blocks, giving some indication of the function of the block. (This functional labeling may also entail splitting or merging of structural blocks.) An example of the result of functional labeling for the first page of a technical paper would indicate the title, author block, abstract, keywords, paragraphs of the text body, and so on. See Figure 3 for an example of the results of structural and functional layout analysis of a document image.

Structural layout analysis can be performed in top-down or bottom-up fashion. For top-down analysis, a page is segmented from large components to smaller subcomponents; for example, the page may be split into one or more column blocks of text. Then, each column may be split into paragraph blocks, each paragraph may be split into text lines, and so on. For bottom-up analysis, connected components are merged into characters, then words, then text lines, and so on. Or top-down and bottom-up analyses may be combined. We describe some specific methods below.

4.3.1: Top-down analysis

Horizontal and vertical projection profiles can be used for layout analysis. As described in Section 4.2, text columns can be located on the vertical projection profile as plateaus separated by valleys. In a similar manner, paragraph breaks within each column can be located on the horizontal projection profile (when the paragraphs are separated by some extra blank space that is greater than the interline spacing). Within paragraphs, text lines also can be found on the horizontal projection profile. In addition, some differentiation among text blocks can be made using this approach. For instance, often the horizontal projection profile of a title will indicate larger-size text than that of the body of text. Or a footnote may have smaller text line spacing than that of the body of text. In this top-down manner, much information can be determined relating to the layout of the page.

In practice, it is not necessary to perform processing on all pixels of the original resolution image when this projection profile approach to layout analysis is used. Instead, the image may be reduced in size to im-

On the Problem of Local Minima in Backpropagation

Original Document Page

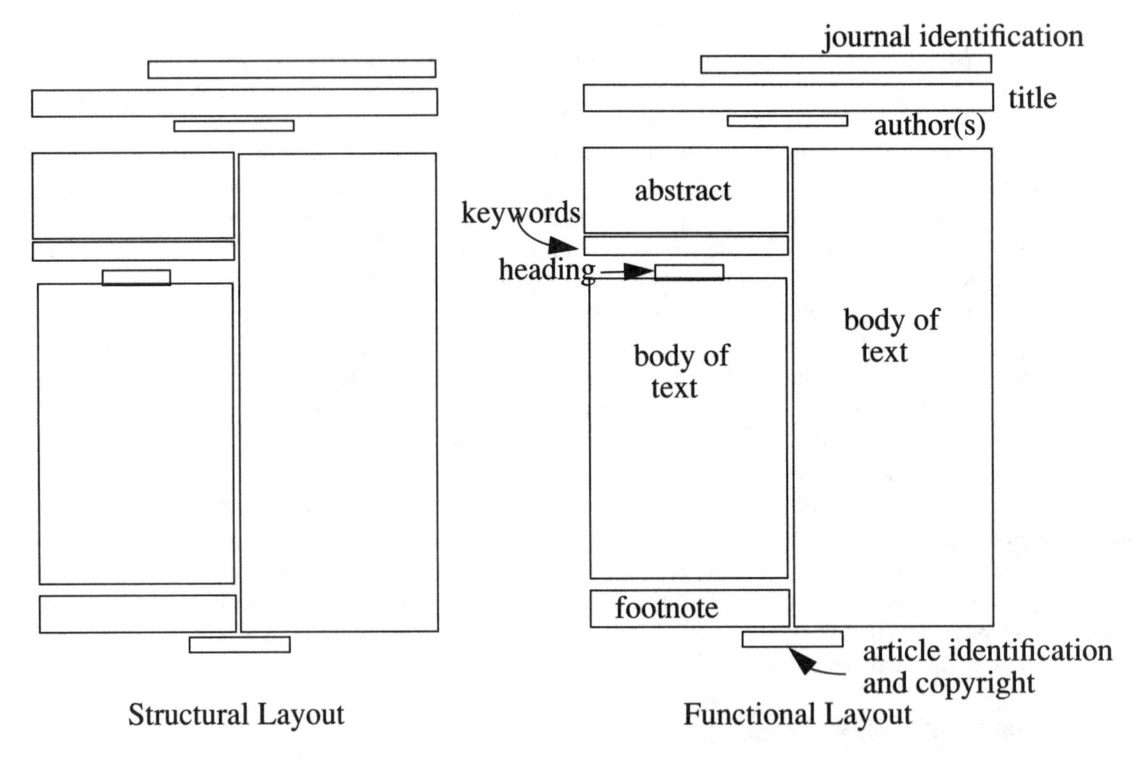

Figure 3. The original document page is shown with results from structural and functional layout analysis. The structural layout results show blocks that are segmented on the basis of spacing in the original. The labeling in the functional layout results is made with the knowledge of the formatting rules of the particular journal (In this case, IEEE Transactions of Pattern Analysis and Machine Intelligence).

prove efficiency and results. One way to do this is to smooth characters into smaller, unrecognizable (as characters) blobs. Depending on the amount of smoothing, these blobs can be of a single character, a word, or a text line. Projection profiles are constructed the same as before, but on the basis of fewer pixels. The run-length smoothing algorithm is a popular method for performing this smoothing [Wong 1982]. This method merges characters into words, words into text lines, and (sometimes) text lines into paragraphs, by "smearing" the text to join characters into blobs. This is done by inspecting white spaces between black pixels on the same lines and setting them to black if the length is less than a threshold. To merge within-word characters, the threshold is chosen to be greater than within-word character spacing and less than between-word spacing. Subsequent merging is accomplished similarly with appropriately sized thresholds. Besides reducing the number of pixels, this method can lead to better results, especially for sparse text that otherwise yields poor or noisy projection profiles. This use of horizontal and vertical projection profiles requires that the image be first skew corrected and that spacing is known and uniform within the image.

A more structured top-down method that also uses projection profiles splits the document into successively smaller rectangular blocks by alternately making horizontal and vertical "cuts" along white space, starting with a full page and continuing with each subblock [Nagy 1984, Nagy 1992b]. (See Figure 4.) The locations of these cuts are found from minima in the horizontal and vertical projection profiles of each block. Segmentation is aided by functional labeling being performed at the same time; this labeling is based on a priori knowledge of features of the expected blocks and on the characteristics of their profiles described in a document syntax. For example, a title block may have large-font titles that appear in the horizontal projection profile as wider peaks, whereas the text body will appear in the horizontal profile as narrower, periodic peaks and valleys. The results of segmentation are represented on an X-Y tree, where the top-level node is for the page, each lower node is for a block, and each level alternately represents the results of horizontal (X-cut) and vertical (Y-cut) segmentation. With the use of this technique, segmentation can be performed down to individual paragraphs, text lines, words, characters, and character fragments. An advantage here is that the combination of structural processing and functional labeling enables the process to be directed and corrected. Of course, this method requires the initial specification of block syntax, which depends upon knowledge of block features (such as font size and line spacing). It should be noted that it is critical that the image have no skew and that any salt-and-pepper noise be removed first. These two conditions are required by most segmentation methods.

Another top-down layout technique analyzes white space (that is, background area versus foreground text) to isolate blocks and then uses projection profiles to find lines [Baird 1990]. (See Figure 5.) First, all locally maximal white rectangles among the black connected components are enumerated from largest to smaller. The largest of these form a "covering" of the white background on the page, thus forming a partition of the foreground into structural blocks of text. Rectangle size may be chosen such that segmentation is performed down to a chosen level: for example, down to columns, paragraphs, or even individual characters. The page must have a Manhattan layout; that is, it must have only one skew angle and must be separable into blocks by horizontal and vertical cuts. One advantage of using background white space versus foreground text for layout analysis is the language independence that results, because white space is used as a layout delimiter in similar ways in many languages. Another advantage is that few parameters need be specified. A drawback is that the choice of what constitutes a "maximal" rectangle—for example, the longest, or the maximal area,—may be nonintuitive as well as different for differently formatted documents. A primary advantage of top-down methods is that they use global page structure to their benefit to perform layout analysis quickly. For most page formats, this is a very effective approach. However, for pages where text does not have linear bounds and where figures are intermixed both in and around text, these methods may be inappropriate. For example, many magazines crop text around an inset figure so that the text follows a curve of an object in the figure rather than a straight line. Table-of-contents pages in magazines and journals are also often formatted with inset figures, centered column entries (rather than justified), and other non-Manhattan layouts. The bottom-up techniques described below are more appropriate for these formats, with the trade-off that they are usually more expensive to compute.

4.3.2: Bottom-up analysis

Above, we explain that bottom-up layout analysis starts with small components and groups them into successively larger components until all blocks are found on the page. However, there is no single, general method that typifies all bottom-up techniques. In this subsection, we describe a number of approaches that can all be classified as bottom-up but use very different intermediate methods to achieve the same purpose. This subsection also gives an idea of complete software systems for page layout analysis.

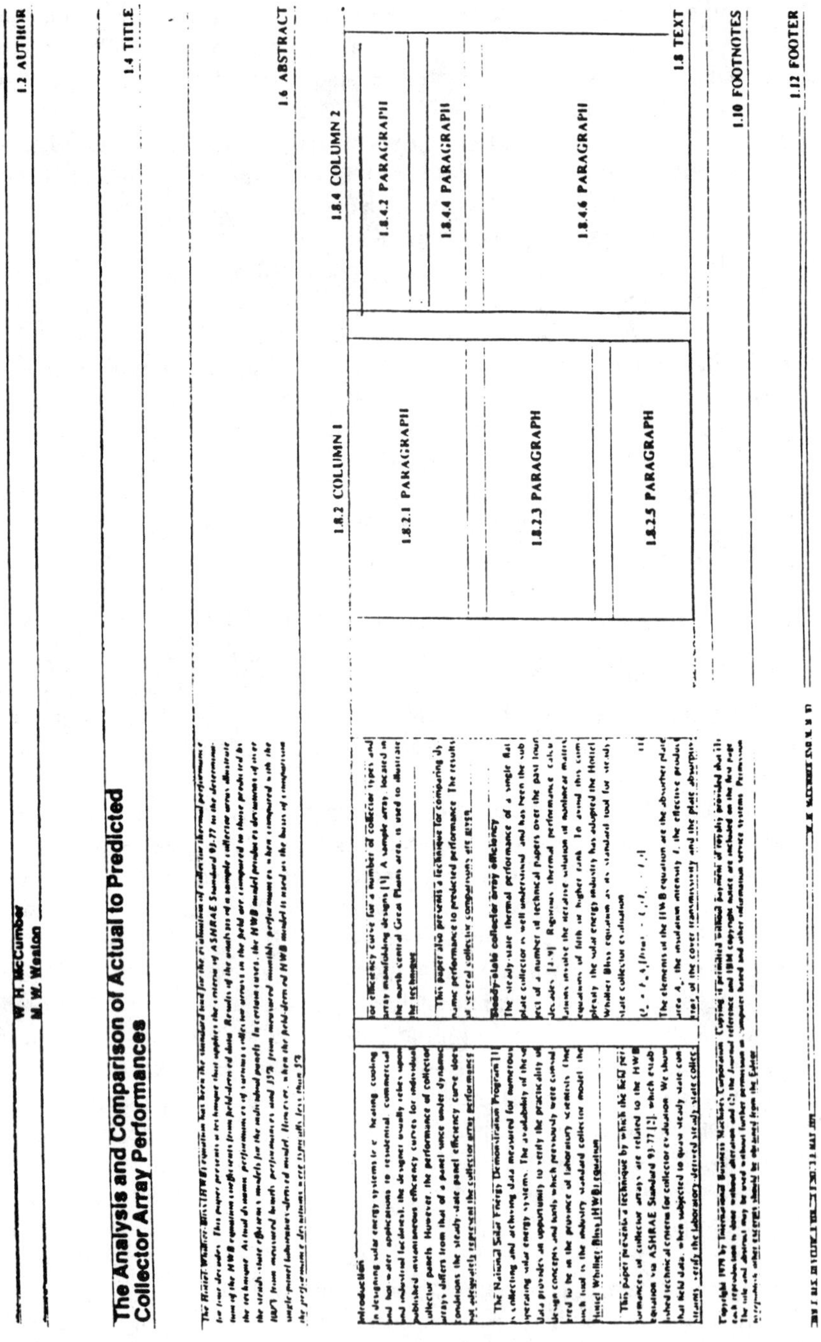

W. H. McCumber
M. W. Weston

1.2 AUTHOR

The Analysis and Comparison of Actual to Predicted Collector Array Performances

1.4 TITLE

The Hottel-Whillier-Bliss (HWB) equation has been the standard tool for the evaluation of collector thermal performance for four decades. This paper presents a technique that applies the criteria of ASHRAE Standard 93-77 to the determination of the HWB equation coefficients from field-derived data. Results of the analyses of a sample collector array illustrate the technique. Actual dynamic performances of various collector arrays in the field are compared to those predicted by the steady-state efficiency models for the individual panels. In certain cases, the HWB model produces deviations of over [illegible] from measured hourly performances and 15% from measured monthly performances when compared with the single-panel laboratory-derived model. However, when the field-derived HWB model is used as the basis of comparison the performance deviations were typically less than 5%.

1.6 ABSTRACT

1.8.2 COLUMN 1

Introduction
In designing solar energy systems (i.e., heating, cooling, and hot-water applications to residential, commercial, and industrial facilities), the designer usually relies upon published instantaneous efficiency curves for individual collector panels. However, the performance of collector arrays differs from that of a panel since under dynamic conditions the steady-state panel efficiency curve does not adequately represent the collector array performance.

1.8.2.1 PARAGRAPH

The National Solar Energy Demonstration Program [1] is collecting and archiving data measured for numerous operating solar energy systems. The availability of these data provides an opportunity to verify the practicality of design concepts and tools which previously were considered to be in the province of laboratory scientists. One such tool is the industry standard collector model, the Hottel-Whillier-Bliss (HWB) equation.

1.8.2.3 PARAGRAPH

This paper presents a technique by which the field performances of collector arrays are related to the HWB equation via ASHRAE Standard 93-77 [2], which established technical criteria for collector evaluation. We show that field data, when subjected to quasi-steady-state constraints, verify the laboratory-derived steady-state collec-

1.8.2.5 PARAGRAPH

1.8.4 COLUMN 2

tor efficiency curve for a number of collector types and array manifolding designs [3]. A sample array, located in the north central Great Plains area, is used to illustrate the technique.

1.8.4.2 PARAGRAPH

This paper also presents a technique for comparing dynamic performance to predicted performance. The results of several collector comparisons are given.

1.8.4.4 PARAGRAPH

Steady-state collector array efficiency
The steady-state thermal performance of a single flat-plate collector is well understood and has been the subject of a number of technical papers over the past four decades [4-9]. Rigorous thermal performance calculations involve the iterative solution of nonlinear matrix equations of fifth or higher rank. To avoid this complexity, the solar energy industry has adopted the Hottel-Whillier-Bliss equation as its standard tool for steady-state collector evaluation:

[illegible] (1)

The elements of the HWB equation are the absorber plate area A_p, the insolation intensity I, the effective product $\tau\alpha$ of the cover transmissivity and the plate absorptiv-

1.8.4.6 PARAGRAPH

1.8 TEXT

Copyright 1979 by International Business Machines Corporation. Copying is permitted without payment of royalty provided that (1) each reproduction is done without alteration and (2) the Journal reference and IBM copyright notice are included on the first page. The title and abstract may be used without further permission in computer-based and other information service systems. Permission to republish other excerpts should be obtained from the Editor.

1.10 FOOTNOTES

[illegible]

1.12 FOOTER

Figure 4. Page layout analysis by top-down segmentation and labeling. On the upper portion is the original page (of *IBM Journal of Research and Development*) with superimposed cuts of an X-Y tree. On the lower portion is the automatically constructed labeled tree. [Nagy, 1992d, p. 59. Copyright © Springer-Verlag, reprinted with permission.

One approach combines a number of the techniques described above [Fisher 1990]. First, the skew is found from the Hough transform, as described in Section 4.1. Then, between-line spacing is found as the peak of the one-dimensional Fourier transform of the projection profile for θ fixed at the computed skew angle. Run-length smoothing is performed, and then within-line spacing is determined by finding the peak on a histogram of these within-line lengths of white spaces (that is, intercharacter and interword spacing) and of black lengths (that is, words). Next, bottom-up merging of the text components is done by a sequence of run-length smooth-

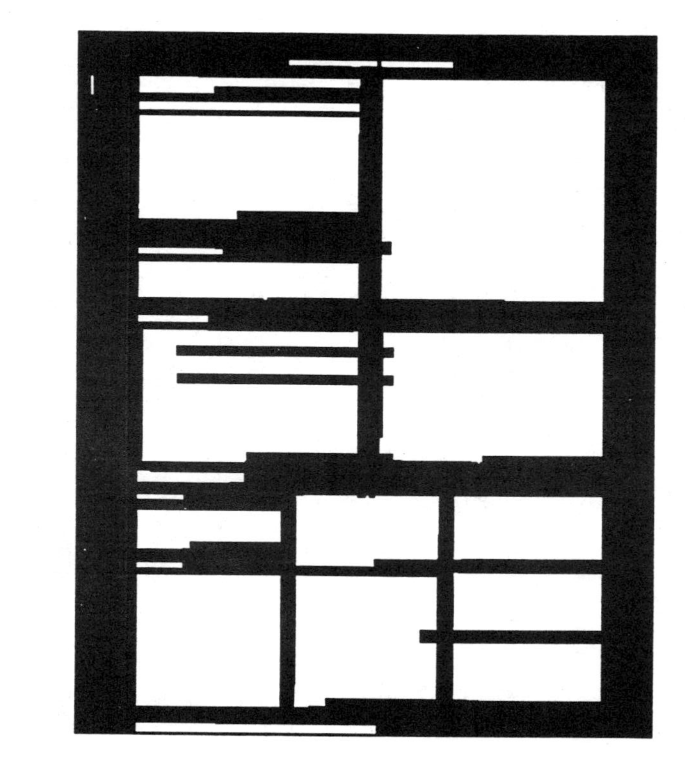

Figure 5. Page layout analysis by blank-space covers. On the left is the original page, showing bounding rectangles over components. On the right is the result of segmentation, showing black rectangles over blank regions from the original page. [Baird 1992, p. 1061]

ing operations, in the directions of the skew for words and text lines and perpendicular to the skew for paragraphs and text columns. The results are ON regions, upon which connected-component analysis is performed. Statistics are calculated on these connected-components: for example, ranges of word height, area, and length. This feature information is used to discern text blocks and to discriminate text and nontext. Esposito et al. [Esposito 1990] used a similar approach, but they first determined bounding boxes of individual characters and then operated with respect to these bounding boxes, instead of on individual pixels, to reduce the amount of computation.

The docstrum method [O'Gorman 1993] employs bottom-up *k*-nearest-neighbors clustering to group from characters into text lines and structural blocks (see Figure 6). (The following describes the complete process of layout analysis by the docstrum method, whereas we describe only the first step—skew detection—in Section 4.1.) First, for each document component, *k* nearest-neighbor connections to neighboring components are found (where *k* is usually taken to be 4 or 5). The distances and angles of these connections are compiled in histograms. Because most connections will be made between characters on the same text lines, the peak angle will indicate the skew and the peak distance will indicate the intercharacter spacing. Using these estimates, text lines are found as groups of characters and words along the page orientation. Text lines are then grouped into blocks, using the document characteristic that text lines of the same block are usually spaced more closely than text lines of different blocks.

Similar to the top-down method that uses functional labeling information in the course of performing structural layout analysis [Nagy 1992d], a combination of the two processes can be used to advantage in bottom-up analysis. In [Akiyama 1990], segmentation is performed, using field separators (lines between text regions) and then blank delimiters, as for many other methods. Then, global and local text features are determined, and blocks are found and classified using generic properties of documents. The text features used here are measures on horizontal and vertical projection profiles, horizontal and vertical crossing counts (the number of black-white and white-black transitions along a line), and bounding boxes for each character. For example, headline text is identified by the property that the larger characters have smaller crossing counts. Individual characters are identified using their bounding box size relative to that of other characters in the same region. This combination of structural layout segmentation and functional labeling has both advantages and disadvantages. Obviously, using this combination requires more from the user earlier in the processing stages to supply information on the expected features of labeled fields. Some researchers feel that proceeding as far as possible in early processing without application-specific knowledge being needed is advantageous because this practice encourages robust techniques. However, others feel that if application-specific knowledge is available and useful, then it should be used to facilitate solving the problem. In the following subsection, functional labeling is discussed independently (as a following process) of structural layout analysis.

4.3.3: Functional labeling

As described above, some of the layout analysis methods [Nagy 1992d, Akiyama 1990] perform functional labeling in the course of structural blocking. Other methods perform these steps sequentially, first obtaining structural blocks and then applying functional labels. In either case, functional labeling is performed based on document-specific rules and using many features derived from the document. These features may include the relative and absolute positions of a block on a page, the relative and absolute positions of a field within a block, general typesetting rules of spacing, and the font size and style of the text. The formatting rules used to produce the page are those that are used in the reverse process of functional labeling. One problem is the wide variation of formatting style across different documents. However, despite this problem, most humans can easily locate blocks of interest using their acquired knowledge of formatting as well as by using text recognition.

Because of the large variation of formats and because of common image-processing problems such as noise, erroneous feature detection, and imperfect text recognition results, most work in functional labeling has been restricted to particular domains. In the domain of postal recognition, much work has been done on the problem of locating addresses and address fields. [Palumbo 1992, USPS 1990]. For US mail, the objective usually is to locate the address block, then the ZIP code on the last line, and then other parts of the address block from there. This is a good example of a familiar format in which blocks at different locations and

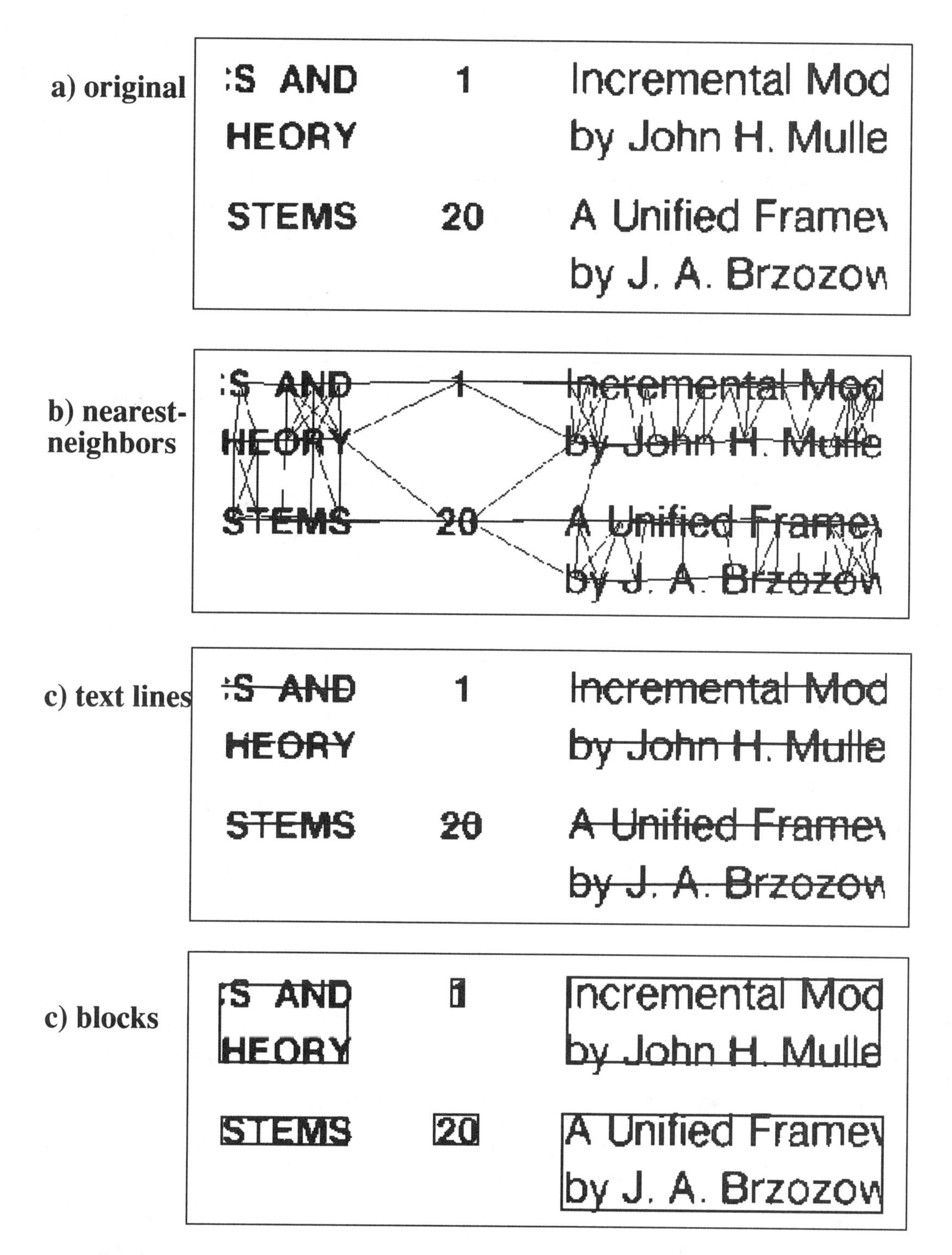

Figure 6. The docstrum method of bottom-up page layout analysis. (a) Original portion of a table-of-contents page. (b) Nearest neighbors of document components are found. (c) Least-square fits are made to nearest-neighbor groups to find text lines. (d) Structural blocks are found as groups of text lines.

fields on different lines connote different functional information. Work has been done also in the domain of forms recognition, where the problem of structural blocking is facilitated by the lines of the form delimiting blocks [Casey 1992, Taylor 1992]. Functional labeling must be performed to recognize, for example, that a block at the bottom of a column of blocks is a "total" field and that indented blocks are intermediate values. In [Dengel 1992], functional labeling is used in the domain of business letters (see Figure 7). The Office Document Architecture (ODA) standard for document formatting (discussed below) is used here to hierarchically represent the document components. In [Amano 1992], the domain of interest is Japanese journals and patents, and labeling is done in part using rules of relative horizontal and vertical placement as well as ranges on the numbers of lines per block. Even within each of these restricted domains, variations of formatting are large, thus presenting challenging problems.

Since the processes of structural and functional layout analysis are the reverse of the document-formatting process, it is logical to ask if there is some standard for producing and describing document contents that can be used to perform layout analysis and to test its results. Two of the internationally recognized standards are ODA [Horak 1985, ISO 1988] and Standard Generalized Markup Language (SGML) [Goldfarb 1990]. Although both describe documents, they are very different in their description. ODA is an object-oriented document architecture that describes both structural and functional components of a complete document. These ODA descriptions are hierarchical, so a layout may be defined as a document containing pages on which there are columns, paragraphs, and so on. The functional hierarchy describes a document, which may contain sections or chapters, and each chapter has headings, subheadings, and so on. The original pur-

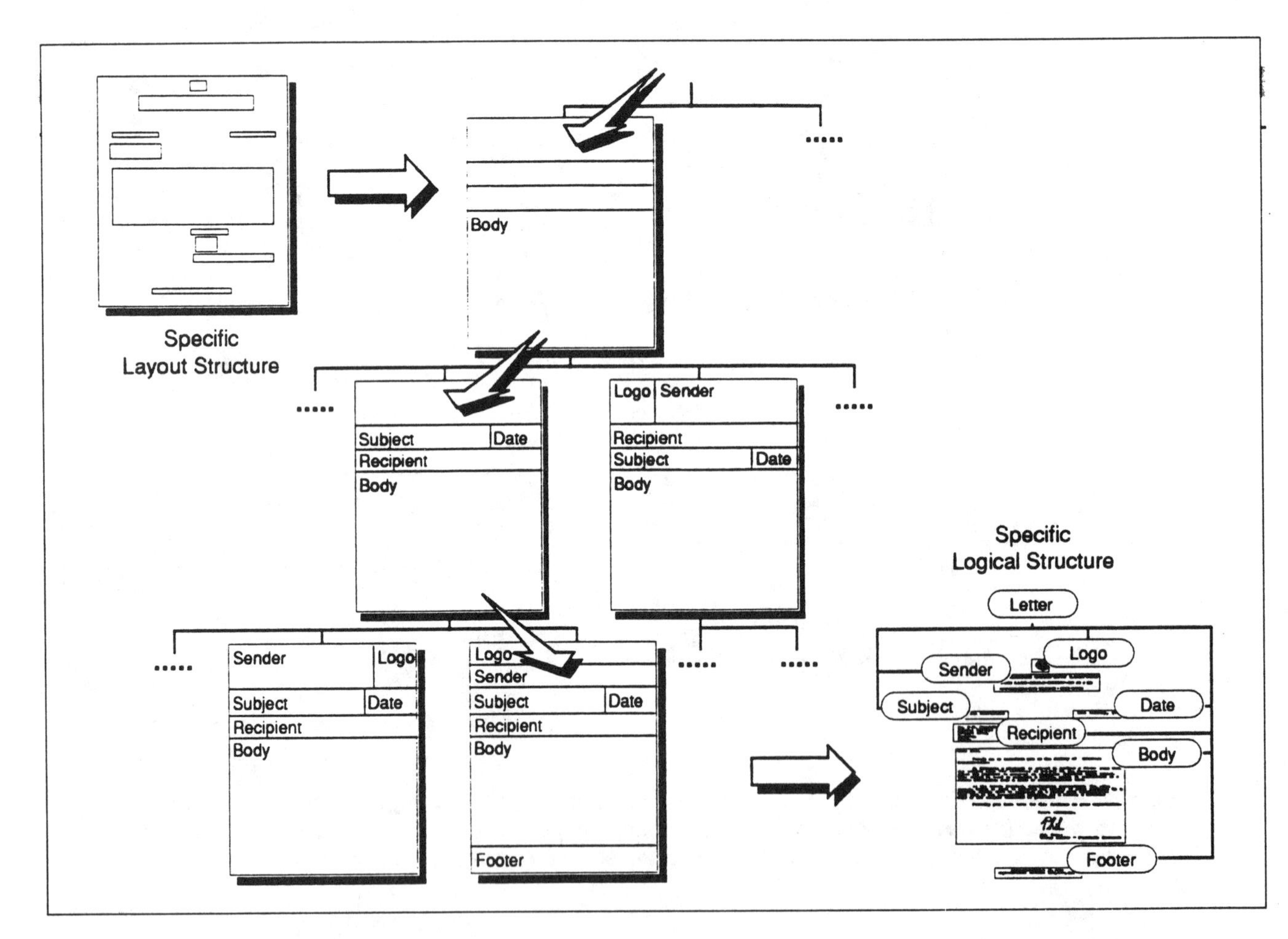

Figure 7. Results of structural layout analysis are shown on the top left, and the hierarchical process is shown whereby a business letter is functionally labeled, with final results on the lower right. [Dengel 1992, p. 65].

pose of ODA was to create a common standard for the interchange of office documents. SGML is a language for "marking up" (or annotating) text with information that describes it as functional components. It was created for publishing documents. In the publishing process, these functional components are used together with publication-specific rules to format the document in a chosen style. In contrast to ODA, SGML can be thought of more as a bottom-up description of a document, and this definition is in keeping with its purpose. Both of these standards have grown beyond their original purposes; for example, both are used for document layout analysis. However, while both are used for publishing and in document analysis, they are by no means used exclusively. Many publications use proprietary systems that are compatible with neither ODA nor SGML, and most current document recognition research today still does not produce results that conform to any standard.

The use of page layout analysis is currently very limited in commercial document systems. The only capability provided by most general systems is specifying absolute block coordinates on which to scan or perform OCR; that is, most general systems do not do automatic page layout analysis. Two exceptions to this limitation are systems for special-purpose applications: postal reading and forms reading. These systems are described more fully in Section 4.6.

We include here one paper from each subsection. The top-down paper is by Pavlidis and Zhou. They employ run-length smoothing and segmentation to examine both foreground and background regions. (Their use of background is similar to that of Baird's method [Baird 1990], described above.) The bottom-up paper is by O'Gorman, who uses the k-nearest-neighbor, docstrum approach. The final paper, by Dengel et al., describes functional labeling and classification to the ODA standard.

4.4: Machine-printed-character recognition

(*Keywords*: Character recognition, lexical analysis, omnifont recognition, optical character recognition, reading machines, text recognition, word recognition)

Recognition of machine-printed characters—or, more simply, printed characters—has traditionally been the most important goal of optical character recognition (OCR) systems. Early systems were generally limited to recognizing characters printed in one font and one size for example, typewritten text. Such systems relied upon uniform spacing of characters (that is, fixed pitch) to segment individual characters. Each segmented character was then recognized using template matching [Mori 1992] or another similar method. For such systems to perform satisfactorily, it was essential that there be no rotation, scale change, or other distortion of characters. Although such systems did serve a small niche of applications for a short period of time, it soon became evident that more sophisticated methods were needed to process documents created by laser printers and page typesetters. Thus, recent efforts have focused on designing systems for omnifont recognition.

Omnifont recognition has been an extremely challenging task to OCR designers. There are over 300 common typefaces, with commonly used character sizes ranging from 6 to 26 points and with many other variations, such as weight (for example, light, regular, bold), ligatures, and super and subscripts [Nagy 1992a]. A common paradigm for character recognition entails the following steps: Page images are segmented to locate individual characters; next, discriminating features are extracted for each character; and finally, these extracted features are matched against those of different character classes in the database. Good features are those with small intraclass variance and large interclass variance. Because of the wide variations in the characteristics of symbol sets from different font families, there is no ideal feature set that would classify all inputs with 100-percent accuracy. A large variety of feature sets have been proposed and used in character recognition systems; such systems perform reasonably well as long as the input data are within the constraints assumed during the design or training phase. Many of the techniques that we describe in earlier sections of this book have been used in various stages of such systems. OCR systems and the features used in such systems have become progressively more sophisticated as a result of extensive experimentation using several test databases that have become available in recent years [Bradford 1991, Hull 1993, Nartker 1992, Phillips 1993, Toriachi 1989, Wilson 1990, Yamamoto 1986].

We now briefly describe the steps used in character recognition systems.

4.4.1: Character segmentation

As in many other applications in computer vision, the effectiveness of early processing steps, or *low-level processing*, is critical to the final recognition results. For OCR, it is the initial segmentation of characters that can mean the difference between very good and very poor results. The objective in character segmentation is to identify each individual character within words and sentences.

A common approach to segmenting characters is to first determine a vertical projection profile for each text line. (The projection profile is described in Subsection 4.2.1 and an example of a vertical projection profile is shown in Figure 1 of that subsection.) The vertical projection profile for a text line is a histogram of the number of ON pixels accumulated vertically through words and characters along the text line. Therefore, this profile will have low peaks between words, higher peaks within words, and highest peaks for tall ascending or descending parts of letters. The ideal noiseless profile would have valleys at zero height between characters, indicating character segmentation locations. However, because of noise, characters may be joined or may be broken; also, because of italic font, there may be no vertical line clearly separating characters. Therefore, segmentation cannot be performed simply by identifying zero-value valleys between peaks along the profile.

Approaches to character segmentation usually begin with the projection profile. A threshold is set on the profile, and peaks above the threshold are said to correspond to locations of candidate segmented characters. This threshold may be adaptive; that is, its value may be raised if segmented widths are longer than expected character widths. Recognition is then performed on each candidate character, as described below. If recognition confidence is low (for either an individual character or a complete word), then resegmentation can take place to join or separate the initial candidates. This process can be performed iteratively until recognition confidence is high. (See [Tsujimoto 1992] for more detail on character segmentation.)

4.4.2: Feature discrimination

Although a bitmap of a complete character is of little use as a template for omnifont character recognition, partial templates capturing salient features in a character have been found to be quite useful. For example, templates of regions surrounding the left and right junctions between the horizontal bar and vertical strokes in characters such as H or A are useful for discriminating such characters from others that do not have such junctions. Similarly, simple features such as the number of connected components, holes, ascenders, and descenders are also useful for discriminating characters. Other simple features that have been used include the aspect ratio of the bounding rectangle, the perimeter-to-area ratio, convex hull and convex deficiencies, horizontal and vertical projection profiles, moment invariants, and so on. (See Section 3.5 for more details on these and other feature descriptors.) Since bit-map-based features are sensitive to the thickness of character strokes, features based upon the core lines obtained by thinning algorithms have also been tried extensively. Of course, such methods suffer from the artifacts introduced during the thinning step (see Section 2.4). Methods using critical points (for example, corners, high curvature points, inflection points, and junctions) along the contours or on the core lines, as well as contour-based feature representation and matching methods such as Fourier descriptors, have also been used to classify characters. (See Sections 3.3 and 3.5.) Many of these simple methods used alone are not adequate to discriminate between characters that look alike (for example o versus 0, and `l` versus 1) and those that are rotated versions of others (for example, d versus p, and 9 versus 6); they also often fail to correctly discriminate characters of different type families (for example, Times Roman font characters with serifs versus Helvetica characters without).

4.4.3: Character classification

In general, a single feature is seldom adequate for discriminating among all character classes of an alphabet. Instead, typical systems extract several features from each character image and then attempt to classify that image based on the similarity of this feature vector with that of a character class. Many well-known decision-theoretic pattern classification methods as well as syntactic and structural methods have been used [Bokser 1992, Mori 1992, Nagy 1992a]. For example, in the Bayesian pattern classification method, each character class is characterized by the conditional probability distribution given by $p(f \mid c_i)$, where f represents the feature vector and c_i represents character class i. A priori probability of various character classes are assumed to be known. Then, a measured feature vector f is assigned to class c_k, which maximizes the a posteriori probability $p(c_k|f)$. Often, the value of each feature is quantized to take on one of several discrete values to provide a degree of tolerance on measured feature values. Quite often, binary-valued features (denoting the presence or absence of corresponding features) are used. Selection of good (that is, representative) training sets is extremely important in obtaining satisfactory results. Efforts to improve the performance of such character-by-character feature extraction and classification methods have long since reached a point of diminishing return. It has now been recognized that any further improvement is obtained by exploiting contextual information [Nagy 1992c]. The classifier in such systems frequently outputs several classes for each input pattern and associates a degree of confidence with each class label. Final class assignment is made after the outputs from a string of characters have been analyzed (rather than a decision being based on a single character).

4.4.4: Use of context in character recognition

One of the most obvious methods for improving the accuracy of OCR performance is by the use of a dictionary to disambiguate the identity of one character based on the identities of neighboring characters in the same word. In such a system, the classifier outputs a rank-ordered list of several alternate labels for each symbol in a word. Those combinations that are not in the dictionary are discarded. In most cases, only one acceptable word remains after this step. In fact, it has been shown that it is possible to assign correct labels to symbols even without creating an initial list of possible labels to symbols [Nagy 1987]. In such a method, the classifier simply groups together all symbols in the document that are of similar shape (but does not attempt to recognize the characters). A small-sized dictionary is then used to substitute symbols in each group with a unique character; such substitutions are first done for short words since there are only a few possible combinations of such words in the dictionary.

Relative frequency of occurrence of different characters, character pairs, character strings (*n-grams*), and words have also been used to enhance OCR performance. Syntax checking, based on grammar, and semantic interpretation of the word with respect to its context have also been proposed as postprocessing methods to improve raw OCR [Nagy 1992a].

Sometimes, an application can have a constrained grammar and vocabulary such that the use of context is very powerful in aiding OCR. A particularly interesting example of applying OCR to a constrained-language problem was done for transcripts of chess games [Baird 1990]. The characters in this application are numbers and chess symbols describing games in the Chess Informant. A sample image of one of the games is shown in Figure 8. Each move that is recognized by OCR is checked for legality in the context of prior and later moves. By using this contextual information, an effective success rate is obtained of 99.995 percent at the character level and 98 percent at the complete-game level.

We include here five papers to represent different aspects of work on OCR of printed characters. The paper by Mori, Suen, and Yamamoto gives a comprehensive overview of and introduction to OCR technologies. The paper by Bokser presents a more detailed description of the methodologies followed in a commercial OCR system. The paper by Ho, Hull, and Srihari describes the combination of character recognition, seg-

108.* (R 76/a) **A 62**

KORTCHNOI – TRINGOV

Luzern (ol) 1982

1. d4 ♘f6 2. c4 e6 3. ♘f3 c5 4. d5 ed5 5. cd5 d6 6. ♘c3 g6 7. g3 ♗g7 8. ♗g2 0—0 9. 0—0 ♘a6!? 10. h3 [10. e4 ♗g4=] **♘c7** [RR 10... ♖e8!? 11. ♗f4 ♘c7 12. a4 ♘e4 13. ♖c1 b5! 14. ♖e1 ♖b8 15. ♘d2 g5! 16. ♘de4 gf4 17. ab5 f5 18. ♘d2 fg3 19. fg3 ♕g5 20. ♘f1 ♘b5∓ Csom – Şubă, Băile Herculane 1982; 11. ♘d2!?] **11. e4**

Figure 8. A sample text from the Chess Informant showing a game's header and opening moves [Baird 1990].

mentation, and word shape for OCR. The paper by Tsujimoto and Asada describes a document system whose major components are document analysis, understanding, and character segmentation and recognition. Finally, Baird's paper objectively details the types of noise and other defects upon which a document image is subjected, providing a model for OCR testing.

4.5: Handwritten-character recognition

(*Keywords*: Handprinted-character recognition, handwriting analysis, handwritten-character rccognition, off-linc rccognition, on-line recognition, optical character recognition, pen-based computers, text recognition, word recognition)

While many of the methods for handwritten-character recognition are similar to those for machine-printed-character recognition, there are some important differences. First, there are two distinct approaches for capturing data for handwriting recognition: on-line and off-line. In on-line recognition systems, captured data are represented as a time sequence of the position of the pen tip, thus preserving the rate of movement of the pen as well as its position. In off-line methods, data are captured by conventional scanning and digitizing methods similar to those used in printed-character recognition. A second difference is that a segmentation process to isolate individual characters in cursive text is a major component of handwriting recognition systems (whereas simple connected-component analysis methods are adequate for such separation in the case of printed-character recognition). Third, the variety of ways in which text is written by hand is essentially unlimited, since the same text written by the same writer at two different times is seldom identical in all respects. Handwriting recognition is an extremely active area of research at this time, and the reader is referred to the proceedings of the International Workshop on Frontiers in Handwriting Recognition (IWFHR), [IWFHR 1990, IWFHR 1991, IWFHR 1993].

It has been argued that additional temporal information (such as the order in which strokes are written as well as the relative velocities of writing different parts of a stroke) that is readily available in on-line systems should facilitate the design of improved recognition algorithms [Tappert 1990]. This line of research has received increased attention in recent years as a result of the availability of low-cost electronic writing tablets and pen-based computers. For large-alphabet languages, such as Chinese, a pen-based system with on-line character recognition is an attractive alternative to a keyboard system based character composition. On-line recognition systems are also useful as writer identification and signature verification systems.

As noted earlier, one of the most difficult challenges of off-line handwriting recognition is that of connected-character segmentation. One of the strategies described in [Bozinovic 1989] for solving this problem is to initially presegment the word into minimal portions along the horizontal direction. This step is followed by a sequence of operations to hypothesize various letter assignments to one or more consecutive presegments. Finally, lexical analysis (that is, matching against a dictionary) is applied in order for the complete word to be recognized. Off-line text recognition has been an important practical problem for postal applications, and research progress in this area has been described extensively in the literature.

Representative papers describing on-line and off-line work in handwriting recognition that we have included here are by Bozinovic and Srihari, and by Nouboud and Plamondon. For a discussion of the commercial state of handwriting recognition, see Section 4.7.

4.6: Document- and forms-processing systems

(*Keywords*: Document-processing, document processing systems, electronic library systems, forms-processing, forms processing systems, postal processing systems)

A document-processing system embodies many of the image- and document-processing techniques that we describe in this book. Typically, pixel-level processing is performed first, to reduce noise and to modify the pixel data into a form (such as a binary or thinned image) that best facilitates subsequent analysis. Then, pertinent features are found from this data, to describe more succinctly the pixel regions contained in the image. These features are used to recognize textual format and content. For recognition of tables and forms, graphics recognition techniques (described in the next chapter) may also be employed, to obtain information from the delimiting lines. Multiple pages introduce additional aspects of complexity with complete documents: continuing text across pages, groups of pages (for example, chapters), and so forth. Although there are document representations for handling the production of multiple-page documents (for example, ODA and SGML), this is a problem that has usually not been handled by document analysis systems.

One application of document-processing techniques is to image-based electronic libraries. Three recent systems for storage and display of technical journals include Nagy's [Nagy 1992b], the RightPages System [Story 1992], and the Document Recognition System [Amano 1992]. Some of the technical work associated with these systems includes noise reduction, structural and functional page layout analysis, OCR, OCR enhancement, and image subsampling for readable display of text. User interface design is also an important aspect of each of these systems and document systems in general.

Document processing of postal mail is an important application area that is probably the most economically important application of current document system implementation. The objectives of such systems are to locate address blocks and to read at least some portion of the addresses for automatic routing of each piece. One important aspect of this problem is that recognition and routing must be performed very quickly to compete with human operations. Some of the work in this area includes that of [Palumbo 1992] on real-time address block location, [Matan 1992] on a neural-network system for reading ZIP codes, and [Kimura 1992] on segmentation and recognition of ZIP codes.

The application of forms recognition entails document images whose formats are more restricted than those of mail or journal publications, because fields are located in specified locations that are bounded by lines or highlighting. With the knowledge of locations of blocks of interest, these blocks can be found and their contents recognized. Although this may seem simpler than nondelineated-block location, there are problems inherent in forms reading that are distinct from those of other applications. Two of these problems are locating boxes, usually by graphics recognition techniques upon the lines, and dealing with text that is misaligned with or intersecting the boxes. Also, since forms are often filled out by hand, handwriting recognition may be involved. Some work in forms recognition includes that of [Casey 1992] and [Taylor 1992]. These systems have been demonstrated on US tax forms, among others.

We include five papers describing different document systems. The papers by Casey et al. and by Taylor, Fritzson, and Pastor are on forms reading; the paper by Nagy, Seth, and Viswanathan describes an image-based electronic library; the paper by Palumbo et al. discusses a real-time postal document system, and Schürmann et al.'s paper describes a general textual-document system.

4.7: Commercial state and future trends

In the Preface, we say that the prevalence of fast computers, large computer memory, and inexpensive scanners has fostered an increasing interest in document image analysis. Hardware advances will continue to provide faster computation, greater storage, and easier computer input, with the result that tools of document analysis will improve. It is thus with some sense of dissatisfaction that we see the algorithms of the field seem to progress with less speed. However, this is due not so much to a slow rate of improvement as it is to the distance still required to reach our ultimate goal of attaining recognition results similar to a human's. This goal will not be attained soon.

Having stated that we have a long way to go to reach our ultimate goal, we can report intermediate progress that has proven to be very productive for many applications. We suggest below where progress will continue and where it is especially needed.

- Most documents are scanned today in binary form, though gray-scale scanners are becoming increasingly prevalent. In the future, more gray-scale and color scanning will be employed. The primary advantage of these is simply more-pleasing images, especially for those of gray-scale and color pictures. However, this will also enable more-sophisticated document analysis to be performed. Thresholding will be improved when more-complex algorithms can be applied digitally rather than today's simpler optical methods being used. Multithresholding can be done to separate, for instance, black text from red text from a blue highlight box from white background. Segmentation can be performed to distinguish among text, graphics, and pictures.

- Noise processing to remove salt-and-pepper noise is available on current systems. Future systems will employ more-sophisticated filters that will use some knowledge of the document components to provide recognition-based cleaning. Of course, the furthest extent of this speculation, will be to perform OCR along with font and format recognition, and then to redisplay a page exactly as before. The problem here is that the recognition must be flawless so as not to introduce more than the initial amount of noise.
- Document scanners and systems can be purchased today with software to correct up to about 10 degrees for the small amount of skew often present when documents are placed on the scanner. Future systems will be able to reorient documents from any skew angle, even from upside-down. Related to the increased ease and utility of scanning systems are the improvement and reduction of size in the hardware. Portable and hand-held scanners (usually packaged with OCR software) are currently available. In the future, scanners will be incorporated into portable computers so the portable office—computer, fax machine, copying machine, and telephone—will always be available.
- While rudimentary structural-layout analysis is already performed in some document systems (mainly to separate text columns), more-complex tasks, such as paragraph and caption segmentation, will also be performed in the future. On a further horizon is structural-layout analysis combined with functional labeling. Since the latter requires knowledge of the document, automatic training procedures will be developed to facilitate this task. Document description will be made in some standard language, such as ODA or SGML.
- For printed text, commercial OCR boxes currently yield 99.5-percent to 99.9-percent recognition accuracy of individual characters for high-quality text [Rice 1993, Nartker 1994]. This drops off to 90-percent to 96-percent recognition accuracy for full words. Although these rates may appear to be quite good at first glance, they correspond to one incorrectly recognized word for every few lines of machine-printed text. We expect to see major progress in accuracy, not by improved techniques for individual-character recognition but by the use of word- and document-based information to augment that raw recognition. Currently, dictionary lookup is used to change unrecognized words into correct ones that have similar word features. In the future, increased computer power and memory (plus fast indexing methods) will enable practical searching of large spaces so corrections can be made using syntactic and even semantic knowledge of the text. Even with these improvements, we do not expect to see flawless OCR in the near future. Near-flawless OCR will be available only for constrained situations, such as a limited character set of 0 to 9. Erroneous OCR can be tolerated in some applications, such as one to search for keywords in a paper where pertinent keywords will likely be repeated. (The University of Nevada at Las Vegas [UNLV] performs annual evaluations of OCR products [Rice 1993, Nartker 1994]. The evaluations are based on character and word recognition results for a range of document conditions, such as skew and document quality. In 1993, systems from eight vendors were tested. We have referenced the current report here, but for up-to-date evaluations of OCR systems, the latest report can be requested from UNLV.)
- Most applications for most current OCR apply to the standard character sets: in particular, numerals and the Roman, Kanji, and Arabic alphabets. However, there are other character sets and symbols that will be recognized in future systems. An example is the larger symbol set used for mathematics. A problem in mathematical equations is that character sequence is not strictly linear along a text line: There are superscripts and subscripts and bounds of integrals and summations that are above and below other symbols. One more problem with equations is simply that superscripts and subscripts are usually in a smaller font size, making the problem of recognition even more difficult.
- For handwritten text, recognition rates are lower than those for printed text. Current off-line recognition rates for untrained data are about 96.5-percent for digits (0–9), 95-percent for uppercase letters only, and 86.5-percent for lowercase letters, all for isolated characters [Wilkinson 1992]. On-line recognition rates are between recognition rates of off-line handwritten text and those of machine-printed text, but are usually highly dependent upon training and the size of the character set. Although the off-line recognition rates are unsatisfactory for many applications, there are those applications that do very well with them. For example, for the postal application, although ZIP code reading is currently only about 70-percent accurate, this still reduces by a huge amount the mail that must be sorted by hand. Besides applications like this, the most predominant applications of handwriting recognition in the short term will be highly constrained. These include trained and untrained digit recognition, recognition of presegmented block characters (from a form, for example), signature storage and recognition, and trained, single-user handwriting recognition. Many of the other comments made above for printed-text OCR apply to handwritten text also.

- Current systems use different techniques for machine-printed and handwritten-text recognition. Also different approaches are often used for different alphabets (for example, Roman, Kanji, and Arabic). Even tables and mathematical equations are most often analyzed using different modules (where they are analyzed at all in current commercial systems). A goal in character recognition is seamless and automatic multilingual, multisymbol recognition, where language and character type are recognized first, and then characters are recognized quickly and correctly.

References

*Papers marked with an asterisk are included as reprinted papers in this book.

Akiyama, T., and N. Hagita, "Automated Entry System for Printed Documents," *Pattern Recognition*, Vol. 23, No. 11, 1990, pp. 1141–1154.

Amano, T., et al., "DRS: A Workstation-Based Document Recognition System for Text Entry," *Computer*, Vol. 25, No. 7, July 1992, pp. 67–71.

Baird, H.S., "Anatomy of a Versatile Page Reader," *Proc. IEEE*, Vol. 80, No. 7, July 1992, pp. 1059–1065.

*Baird, H.S., "The Skew Angle of Printed Documents," *Proc. Conf. Photographic Scientists and Engineers*, SPIE, Bellingham, Wa., 1987, pp. 14–21.

Baird, H.S., and K. Thompson, "Reading Chess," *IEEE Trans. Pattern Analysis and Machine Intelligence*, Vol. 12, No. 6, June 1990, pp. 552–559.

Baird, H.S., S.E. Jones, and S.J. Fortune, "Image Segmentation Using Shape-Directed Covers," *Proc. 10th Int'l Conf. Pattern Recognition (ICPR)*, IEEE CS Press, Los Alamitos Calif., 1990, pp. 820–825.

*Bokser, M., "Omnidocument Technologies," *Proc. IEEE*, Vol. 80, No. 7, July 1992, pp. 1066–1078.

*Bozinovic, R.M., and S.N. Srihari, "Off-Line Cursive Script Word Recognition," *IEEE Trans. Pattern Analysis and Machine Intelligence*, Vol. 11, No. 1, Jan. 1989, pp. 68–83.

*Casey, R., et al., "Intelligent Forms Processing System," *Machine Vision and Applications*, Vol. 5, No. 3, 1992, pp. 143–155.

*Dengel, A., et al., "From Paper to Office Document Standard Representation," *Computer*, Vol. 25, No. 7, July 1992, pp. 63–67.

Esposito, R., D. Malerba, and G. Semeraro, "An Experimental Page Layout Recognition System for Office Document Automatic Classification: An Integrated Approach for Inductive Generalization," *Proc. 10th Int'l Conf. Pattern Recognition (ICPR)*, IEEE CS Press, Los Alamitos Calif., 1990, pp. 557–562.

Fisher, J.L., S.C. Hinds, and D.P. D'Amato, "A Rule-Based System for Document Image Segmentation," *Proc. 10th Int'l Conf. Pattern Recognition (ICPR)*, IEEE CS Press, Los Alamitos Calif., 1990, pp. 567–572.

*Fletcher, L.A. and R. Kasturi, "A Robust Algorithm for Text String Separation from Mixed Text/Graphics Images," *IEEE Trans. Pattern Analysis and Machine Intelligence*, Vol. 10, No. 6, Nov. 1988, pp. 910-918.

Goldfarb, C.F., *The SGML Handbook*, Clarendon Press, Oxford, United Kingdom, 1990.

Hashizume, A., P.S. Yeh, and A. Rosenfeld, "A Method of Detecting the Orientation of Aligned Components," *Pattern Recognition Letters*, Vol. 4, 1986, pp. 125–132.

*Hinds, S.C., J.L. Fisher, and D.P. D'Amato, "A Document Skew Detection Method Using Run-Length Encoding and the Hough Transform," *Proc. 10th Int'l Conf. Pattern Recognition (ICPR)*, IEEE CS Press, Los Alamitos, Calif., 1990, pp. 464–468.

Ho, T.K., J.J. Hull, and S.N. Srihari, "A Computational Model for Recognition of Multifont Word Images," *Machine Vision and Applications*, Vol. 5, No. 3, 1992, pp. 157–168.

Horak, W., "Office Document Architecture and Office Document Interchange Formats: Current Status of International Standardization," *Computer*, Vol. 18, No. 10, Oct. 1985, pp. 50–60.

Hull, J.J., "A Database for Handwritten Word Recognition Research," *IEEE Trans. Pattern Analysis and Machine Intelligence*, Vol. 16, No. 5, May 1994, pp. 550–554.

ISO, "Information Processing — Text and Office Systems — Office Document Architecture and Interchange Format," ISO 8613, Parts 1–8, Int'l. Standards Organization (ISO), Geneva, Switzerland, 1988.

[IWFHR 1990] *Proc. IAPR Int'l Workshop Frontiers Handwriting Recognition (IWFHR),* Concordia Univ., Montreal, Canada, 1990.

[IWFHR 1993] *Proc. IAPR Int'l Workshop Frontiers Handwriting Recognition (IWFHR),* SUNY-Buffalo, Buffalo, N. Y., 1993.

[IWFHR 1991] *Proc. IAPR Int'l Workshop Frontiers Handwriting Recognition (IWFHR),* World Scientific, Singapore, 1991.

*Jain, A.K. and S. Bhattacharjee, "Text Segmentation Using Gabor Filters for Automatic Document Processing," *Machine Vision and Applications*, Vol. 5, 1992, pp. 169–184.

Kimura, F., and M. Shridhar, "Segmentation-Recognition Algorithm for ZIP Code Field Recognition," *Machine Vision and Applications*, Vol. 5, No. 3, 1992, pp. 199–210.

Matan, O., et al., "Reading Handwritten Digits: A ZIP Code Recognition System," *Computer*, Vol. 25, No. 7, July 1992, pp. 59–63.

* Mori, S., C.Y. Suen, and K. Yamamoto, "Historical Review of OCR Research and Development," *Proc. IEEE*, Vol. 80, No. 7, July 1992, pp. 1029–1058.

[Nagy1992a] Nagy, G., *Optical Character Recognition and Document Image Analysis*, Rensselaer Video, Clifton Park, N.Y., 1992 (videotape).

*[Nagy1992b] Nagy, G., S. Seth, and M. Viswanathan, "A Prototype Document Image Analysis System for Technical Journals," *Computer*, Vol. 25, No. 7, July 1992, pp. 10–22.

[Nagy1992c] Nagy, G., At the Frontiers of OCR," *Proc IEEE*, Vol. 80, No. 7, July 1992, pp. 1093–1100.

[Nagy1992d] Nagy, G., "Toward a Structured Document Image Utility," in *Structured Document Image Analysis*, H.S. Baird, H. Bunke, and K. Yamamoto, eds., Springer Verlag, Berlin, 1992, pp. 54–69.

Nagy, G., and S. Seth, "Hierarchical Representation of Optically Scanned Documents," *Proc. 7th Int'l Conf. Pattern Recognition (ICPR)*, IEEE CS Press, Los Alamitos, Calif., 1984, pp. 347–349.

Nagy, G., S. Seth, and K. Einspahr, "Decoding Substitution Ciphers by Means of Word Matching with Application to OCR," *IEEE Trans. Pattern Analysis and Machine Intelligence*, Vol. PAMI-9, No. 5, 1987, pp. 710–715.

Nakano, Y. et al., "An Algorithm for the Skew Normalization of Document Images," *Proc. 10th Int'l Conf. Pattern Recognition (ICPR)*, IEEE CS Press, Los Alamitos, Calif., 1990, pp. 8–13.

Nartker, T.A., R.B. Bradford, and B.A. Cerny, "A Preliminary Report on UNLV/GT1: A Database for Ground Truth Testing in Document Analysis and Character Recognition," *Proc. Symp. Document Analysis and Information Retrieval*, Univ. of Nevada, Las Vegas, Nev., 1992, pp. 300–315.

Nartker, T.A., S.V. Rice, and J. Kanai, "OCR Accuracy: UNLV's Second Annual Test," *Inform*, Jan. 1994, pp. 40–45.

* Nouboud, F., and R. Plamondon, "On-Line Recognition of Handprinted Characters: Survey and Beta Tests," *Pattern Recognition*, Vol. 23, No. 9, 1990, pp. 1031–1044.

*O'Gorman, L. "The Document Spectrum for Structural Page Layout Analysis," *IEEE Trans. Pattern Analysis and Machine Intelligence*, Vol. 15, No. 11, 1993, pp. 1162–1173.

*Palumbo, P.W., et al., "Postal Address Block Location in Real Time," *Computer*, Vol. 25, No. 7, July, 1992, pp. 34–42.

Pavlidis, T., and J. Zhou, "Page Segmentation by White Streams," *Proc. 1st Int'l Conf. Document Analysis and Recognition (ICDAR)*, Int'l Assoc. Pattern Recognition, 1991, pp. 945–953.

*Phillips, I.T., S. Chen, and R.M. Haralick, "CD-ROM Document Database Standard," *Proc. Int'l Conf. Document Analysis and Recognition*, 1993, pp. 478–483.

Phillips, I.T., S. Chen, and R.M. Haralick, *CD-ROM English Database Design*, Dept. of Electrical Eng., Univ. of Washington, Seattle, Wash., 1993.

Postl, W., "Detection of Linear Oblique Structures and Skew Scan in Digitized Documents," *Proc. 8th Int'l Conf. Pattern Recognition (ICPR)*, IEEE CS Press, Los Alamitos, Calif., 1986, pp. 687–689.

Rice, S.V., J. Kanai, and T.A. Nartker, "An Evaluation of OCR Accuracy," in *UNLV Information Science Research Inst. 1993 Report,* University of Nevada, Las Vegas, Nev., 1993.

*Schürmann, J., et al, "Document Analysis—From Pixels to Contents," *Proc. IEEE*, Vol. 80, No. 7, July 1992, pp. 1101–1119.

Srihari, S.N., and V. Govindaraju, "Analysis of Textual Images Using the Hough Transform," *Machine Vision and Applications*, Vol. 2, 1989, pp. 141–153.

Story, G., et al., "The RightPages Image-Based Electronic Library for Alerting and Browsing," *Computer*, Vol. 25, No. 9, Sept. 1992, pp. 17–26.

Tappert, C.C., C.Y. Suen, and T. Wakahara, "The State of the Art in On-Line Handwriting Recognition," *IEEE Trans. Pattern Analysis and Machine Intelligence*, Vol. 12, No. 8, Aug. 1990, pp. 787–808.

*Taylor, S.L., R. Fritzson, and J.A. Pastor, "Extraction of Data from Preprinted Forms," *Machine Vision and Applications*, Vol. 5, No. 3, 1992, pp. 211–222.

Toraichi, K., et al., "Handprinted Chinese Character Database," in *Computer Recognition and Human Production of Handwriting*, R. Plamondon, C.Y. Suen, and M.L. Simner, eds., World Scientific, Singapore, 1989, pp. 131–148.

*Tsujimoto, S., and H. Asada, "Major Components of a Complete Text Reading System," *Proc. IEEE*, Vol. 80, No. 7, July 1992, pp. 1133–1149.

[USPS 1990] *Proc. 4th United States Postal Service Advanced Technology Conf.*, Washington, D.C., 1990.

Wilkinson et al., "The First Census Optical Character Recognition Systems Conference," Tech. Report NIST 4912, U.S. Bureau of the Census and Nat'l Inst. of Standards, Washington, D.C., 1992.

Wilson, C.L., and M.D. Garris, "Handprinted Character Database," *Standard Reference Data,* Nat'l Inst. of Standards and Technology, Gaithersburg, Md., April 1990.

Wong, K.Y., R.G. Casey, and F.M. Wahl, "Document Analysis System," *IBM J. Research and Development*, Vol. 6, No. 6, Nov. 1982, pp. 647–656.

Yamamoto, K., et al., "Recognition of Handprinted Characters in the First Level of JIS Chinese Characters," *Proc. 8th Int'l. Conf. Pattern Recognition (ICPR)*, IEEE CS Press, Los Alamitos, Calif., 1986, pp. 570–572.

Text Segmentation Using Gabor Filters for Automatic Document Processing*

Anil K. Jain and Sushil Bhattacharjee

Pattern Recognition and Image Processing Processing Laboratory, Michigan State University, E. Lansing, MI 48824-1027, USA

Abstract: There is a considerable interest in designing automatic systems that will scan a given paper document and store it on electronic media for easier storage, manipulation, and access. Most documents contain graphics and images in addition to text. Thus, the document image has to be segmented to identify the text regions, so that OCR techniques may be applied only to those regions. In this paper, we present a simple method for document image segmentation in which text regions in a given document image are automatically identified. The proposed segmentation method for document images is based on a multichannel filtering approach to texture segmentation. The text in the document is considered as a textured region. Nontext contents in the document, such as blank spaces, graphics, and pictures, are considered as regions with different textures. Thus, the problem of segmenting document images into text and nontext regions can be posed as a texture segmentation problem. Two-dimensional Gabor filters are used to extract texture features for each of these regions. These filters have been extensively used earlier for a variety of texture segmentation tasks. Here we apply the same filters to the document image segmentation problem. Our segmentation method does not assume any a priori knowledge about the content or font styles of the document, and is shown to work even for skewed images and handwritten text. Results of the proposed segmentation method are presented for several test images which demonstrate the robustness of this technique.

Key Words: document image analysis, texture, Gabor filters, bar codes, segmentation

Address offprint requests to: Anil K. Jain, Pattern Recognition and Image Processing Lab, Michigan State University, East Lansing, MI 48824-1027 USA.

* This work was supported by the National Science Foundation under NSF grant CDA-88-06599 and by a grant from E. I. Du Pont De Nemours & Company.

1 Introduction

For centuries paper has been the primary medium for journals, books, business correspondence, maps, and technical drawings. Recently, the availability and economy of computer technology has made electronic storage and digital dissemination of documents practical. Automatic optical character recognition (OCR) systems now facilitate efficient storage and retrieval of correspondence. Technical drawings need to be converted into electronic form before they can be used as input to CAD systems. Digital versions of maps are easier to update and a single representation can be viewed at different scales. Computers also provide security and facilitate authentication to a greater extent. These desirable aspects also prompt the need for converting existing paper documents to electronic form. The large volumes of paper documents available in an organization make data entry by operator keying practically impossible. Even if enough manpower were available, manual entry is prone to typographical errors. What is required is an automatic system which will scan the given document and OCR methods to convert the document image into the corresponding symbolic form. A summary of the various issues involved in the different aspects of automatic document processing can be found in Nagy (1990). Most documents contain graphics and images in addition to text. Thus, the document image has to be segmented to identify the text region so that OCR techniques may be applied only to those regions. This will reduce the load on the OCR system and also improve its accuracy.

Several methods for segmenting document images have been proposed in the literature. These methods can be broadly classified as either top-down or bottom-up approaches (Srihari 1986). Most popular top-down techniques are based on the run length smoothing (RLS) algorithm (Wong et al. 1982), also called the constrained run length (CRL) algorithm (Wahl et al. 1982), and the projection profile cuts (Wang and Srihari 1989). Bottom-up methods (Fletcher and Kasturi 1988) typically involve grouping of pixels as connected components which are merged into successively larger regions. A brief discussion of these methods follows.

The constrained run length (CRL) algorithm requires a binary image as input. In any arbitrary string of 0s and 1s, this algorithm replaces every string of contiguous 0s of length less than some predetermined value, called the *constraint*, by a string of 1s of equal length. An algorithm for text extraction by block segmentation using the CRL algorithm is presented in Wahl et al. (1982). In this work, the CRL algorithm is applied to a binary image of the document, in both horizontal and vertical directions. The logical AND of these two outputs produces a bit-mapped image which consists of blocks, where each block represents a text segment or graphics or a halftone image. A three-step linear classification scheme based on these features is used to classify the blocks into one of the following four classes: text, horizontal solid black lines, vertical solid black lines, and halftone images. This algorithm requires the user to choose a number of threshold values. Though the choice of these thresholds is not critical, there are no guidelines for selecting these values. One drawback of this method is that text embedded in graphics or images is not detected (Nadler 1984).

Projection profiles (Iwaki et al. 1987) have also been successfully used for newspaper image segmentation in the recursive x-y cuts (RXYC) algorithm (Wang and Srihari 1989). This method is based on the observation that a typical newspaper page consists of rectangular blocks. At each recursive step the projection profiles are computed along the horizontal and vertical directions for each block. The block is then subdivided by making cuts corresponding to troughs, of width greater than some reasonable threshold, in the projection profiles in the horizontal and vertical directions separately. Three features, namely, short run emphasis, long run emphasis, and extra long run emphasis, are computed for each block obtained from the RXYC algorithm. The final step is a linear classification in the three-feature space. Five classes—small characters, medium size characters, large characters, graphics, and halftone images—are defined.

The CRL algorithm performs better than the RXYC algorithm in extracting small blocks but the resulting blocks may not be rectangular in shape. The RXYC algorithm extracts large rectangular blocks, and hence is more suitable for newspaper images. Both algorithms perform poorly if the input image is skewed, i.e., when the text lines in the image are not aligned with the image coordinate system, or contain noise.

Bottom-up methods typically involve grouping of pixels as connected components which are merged into successively larger regions. One such algorithm to separate text strings from the grpahics present in a document image has been proposed by Fletcher and Kasturi (1988). An *area/ratio* filter is used to separate the large graphical components so as to reduce the number of connected components that may be candidates for classification as members of text strings. This is followed by *collinear component grouping*. Hough transform is applied to the centroids of the enclosing rectangles of the connected components to find those connected components that are collinear. Positional relationships between collinear components, an intercharacter gap threshold, and an interword gap threshold are then used to logically group the components into text strings. The algorithm performs well if the input image conforms to certain requirements about the character size, the interline spacing, the intercharacter spacing, and the digitizing resolution, as specified in Fletcher and Kasturi (1988).

In this paper we present a simple method for document image segmentation in which text regions in a given document image are identified. The method does not assume any a priori knowledge about the layout or font styles of the document. Our segmentation method is based on a multichannel filtering technique for texture segmentation. The specific filters used here are Gabor filters which have been shown to perform very well for a variety of texture classification and segmentation tasks. We find that regions of text in a document image define a unique texture which can be easily captured by a small number of Gabor filters. We use exactly the same filters which were used in earlier texture segmentation studies (Farrokhnia 1990; Jain and Farrokhnia 1991; Farrokhnia and Jain 1991). These filters are applied directly to the input scanned images. No gray-level thresholding is necessary. The texture of text regions is quite distinct from the texture of the "background" in the document images. This allows us to easily separate text from other nontext content (graphics and pictures) in the document images.

The paper is organized as follows. Section 2 briefly reviews the Gabor-filtering based texture segmentation technique proposed by Farrokhnia (1990) and Jain and Farrokhnia (1991). Section 3 describes the proposed scheme for document image segmentation. Section 4 presents the results of several experiments. Based on initial results of unsupervised segmentation, we show how filter selection and supervised classification can be used to localize 2-D bar codes in a complex document image. Finally, Section 5 contains a summary and conclusions of the study.

2 Multichannel Filtering for Texture Segmentation

Most textures can be characterized based on the local spatial frequency and orientation information present in the image. An obvious way to capture these features is to compute the power spectrum of the image. However, the Fourier transform of the textured image gives only a global estimate of the frequency content. Gabor transforms allow one to obtain local estimates of frequency content (Gabor 1946). A brief introduction to the Gabor transform and filters follows.

2.1 Gabor Filters

Physically interpreted, the Gabor transform acts like the Fourier transform but only for a small Gaussian window over the image, not the entire image. Mathematically, the one-dimensional Gabor transform can be expressed as

$$S_w(u, t) = \int_{-\infty}^{+\infty} s(x)w(x - t)e^{-j2\pi u_0 x}\, dx.$$

S_w gives the Gabor transform of $s(x)$ using the Gaussian window function $w(x)$ centered at t. Thus, a Gabor function consists of a sine wave, with particular frequency, u_0, modulated by a Gaussian function. The concept can be extended to two dimensions as a sinusoidal plane of particular frequency and orientation modulated by a two-dimensional Gaussian envelope. In spatial domain, the two-dimensional Gabor filter is given by

$$h(x, y) = exp\left\{-\frac{1}{2}\left[\frac{x^2}{\sigma_x^2} + \frac{y^2}{\sigma_y^2}\right]\right\} cos(2\pi u_0 x + \phi),$$

where σ_x and σ_y are the standard deviations of the Gaussian envelope along the x and y directions, respectively, and u_0 and ϕ are the frequency and phase of the sinusoidal plane wave along the x-direction (0° orientation). Orienting the sinusoid at an angle θ will define Gabor filters at orientation θ.

The frequency- and orientation-selective characteristics are more obvious in the modulation transfer function (MTF) of the Gabor filter, given by the Fourier transform of the above function, for $\phi = 0°$:

$$H(u, v) = A\left(exp\left\{-\frac{1}{2}\left[\frac{(u - u_0)^2}{\sigma_u^2} + \frac{v^2}{\sigma_v^2}\right]\right\} + exp\left\{-\frac{1}{2}\left[\frac{(u + u_0)^2}{\sigma_u^2} + \frac{v^2}{\sigma_v^2}\right]\right\}\right)$$

where $\sigma_u = \frac{1}{2\pi\sigma_x}$, $\sigma_v = \frac{1}{2\pi\sigma_y}$, and $A = 2\pi\sigma_x\sigma_y$.

Figure 1 shows the MTFs of Gabor filters used in this study.

These filters have been extensively used earlier for a variety of texture segmentation tasks (Clark and Bovik 1989; Farrokhnia and Jain 1991; Perry and Lowe 1989; Tan and Constantinides 1990; Turner 1986). Farrokhnia (1990) presents a survey of these techniques. Our goal here is to apply the same filters for the document segmentation problem.

2.2 Texture Segmentation

A texture segmentation algorithm using Gabor filters has been proposed by Jain and Farrokhnia (1991). The main steps of the algorithm are given below.

- Filter the input image through a bank of n even-symmetric Gabor filters, to obtain n *filtered images*. We use only the even-symmetric filters (zero phase). For a justification of the use of only even-symmetric filters, see Malik and Perona (1990).
- Compute the *feature images* consisting of the "local energy" estimates over windows of appropriate size around every pixel in each of the filtered images. The actual method for computing the feature images is described later.
- Cluster the feature vectors corresponding to each pixel using a squared-error clustering algorithm to obtain a segmentation of the original input image into K clusters or segments. The row and column pixel coordinates are used as additional features in clustering in order to get smooth segment boundaries.

For an image with N columns, where N is a power of 2, the Gabor filters with radial frequencies, $1\sqrt{2}$, $2\sqrt{2}$, $4\sqrt{2}$, . . . , $(N/4)\sqrt{2}$ cycles/image are selected. For each radial frequency, u_0, four filters, with orientations $\theta = 0°, 45°, 90°, 135°$, are selected (Jain and Farrokhnia 1991). Thus, for an image with $N = 256$, a bank of 28 filters, ($n = 28$) would be used. A filter selection scheme, involving the reconstruction of the original image using the outputs of the filters, is described in Jain and Farrokhnia (1991).

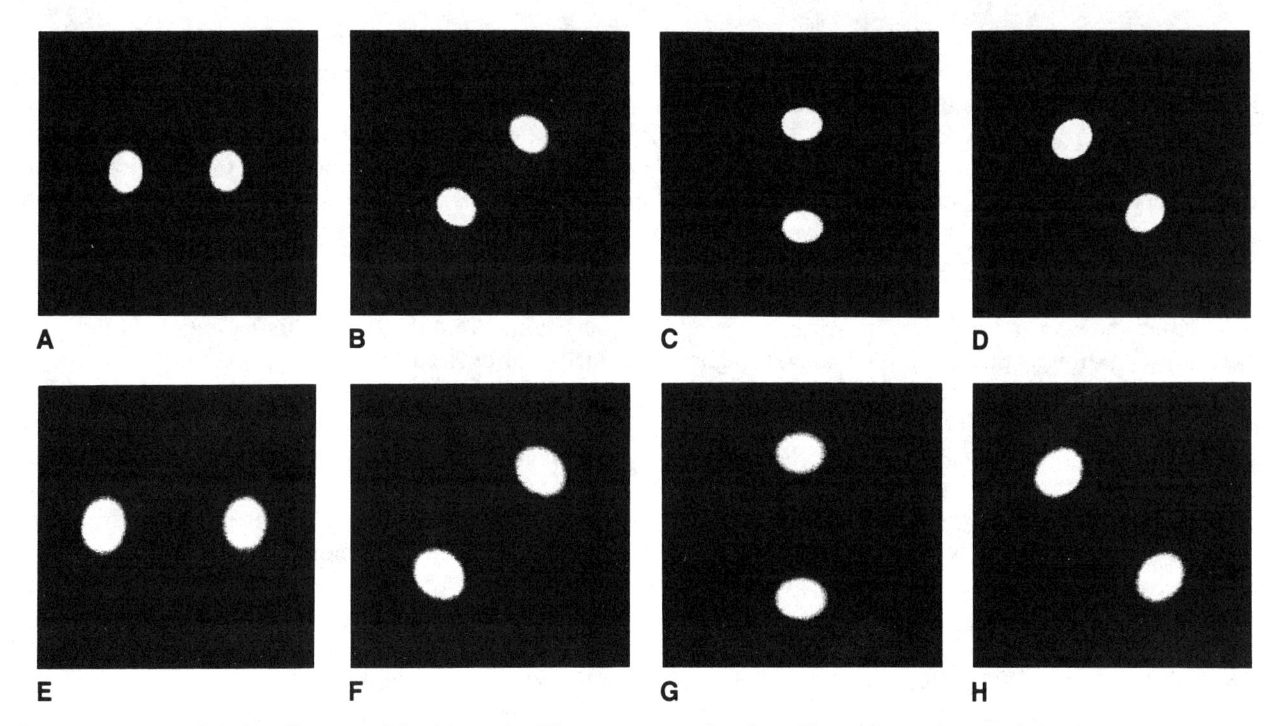

Figure 1. **MTFs of Gabor filters used in this study. The parameters for these filters, for an image of size 256 × 256 are: (A) $u_0 = 32\sqrt{2}$, $\theta = 0°$ (B) $u_0 = 32\sqrt{2}$, $\theta = 45°$ (C) $u_0 = 32\sqrt{2}$, $\theta = 90°$ (D) $u_0 = 32\sqrt{2}$, $\theta = 135°$ (E) $u_0 = 64\sqrt{2}$, $\theta = 0°$ (F) $u_0 = 64\sqrt{2}$, $\theta = 45°$ (G) $u_0 = 64\sqrt{2}$, $\theta = 90°$ (H) $u_0 = 64\sqrt{2}$, $\theta = 135°$.**

The spread of the modulating Gaussian is specified in terms of the frequency of the sinusoid. In the frequency domain the spread of the two-dimensional Gaussian is given by

$$\sigma_u = \frac{1}{2\pi\sigma_x},$$

and

$$\sigma_v = \frac{1}{2\pi\sigma_y},$$

where σ_x and σ_y are the standard deviations of the Gaussian in the spatial domain. Alternatively, the parameters σ_u and σ_v can be specified in terms of the half-peak magnitudes of the frequency bandwidth, B_r (in octaves) and the orientation bandwidth, B_θ (in degrees) (Farrokhnia 1990) as follows:

$$\sigma_u = \frac{u_0}{\sqrt{(2)}} \frac{2^{B_r} - 1}{2^{B_r} + 1},$$

and

$$\sigma_u = \frac{u_0}{\sqrt{(2)}} tan\left(\frac{B_\theta}{2}\right).$$

In our experiments, $B_r = 1$ octave and $B_\theta = 45°$. All the parameters of the Gabor filters (u_0, σ_x, σ_y, and θ) have now been specified.

The following procedure is used to compute the feature images. First, a nonlinearity is introduced into each filtered image by applying the following transformation (Jain and Farrokhnia 1991):

$$\psi(t) = \tanh(\alpha t) = \frac{1 - e^{-2\alpha t}}{1 + e^{-2\alpha t}}.$$

For $\alpha = 0.25$, this function acts almost as a thresholding transformation, similar to a sigmoid function. The output of this nonlinear stage is a set of *response images*. The texture feature, called the average absolute deviation (AAD), is computed from the response images as the mean value in small overlapping windows centered at each pixel. This is a measure of the 'texture energy' around the pixel. More formally, the feature image $e_k(x, y)$, corresponding to the kth filtered image $r_k(x, y)$, is given by:

$$e_k(x, y) = \frac{1}{M^2} \sum_{(a,b) \in W_{xy}} |\psi(r_k(a, b))|, k = 1, \ldots, n$$

where W_{xy} is a window of size $M \times M$ centered at

pixel (x, y) in the filtered image. The window size, M, is inversely related to u_0, and is the smallest odd integer larger than or equal to 5σ, where $\sigma = 0.25N/u_0$.

The values in the n feature images corresponding to a given pixel form an n-dimensional feature vector representing the pixel. To incorporate spatial context into our clustering scheme, the spatial coordinates of the pixels are used as additional features. The features are normalized to zero mean and unit standard deviation to prevent a feature with higher numerical range from dominating the other features. The squared-error clustering algorithm, CLUSTER (Jain and Dubes 1988), is used to cluster the normalized feature vectors. Due to the large number of patterns, clustering becomes computationally impractical (for a 256 × 256 image, 65,536 patterns would have to be clustered). In practice, we cluster only a subset of randomly chosen patterns (pixels) from the image, into K clusters. Specifically, for a 256 × 256 image, 4,000 patterns are selected randomly for the initial clustering. The patterns in these clusters are then used for training a minimum Euclidean distance classifier to classify the entire image into K classes.

3 Proposed Segmentation Method

The proposed segmentation method for document images is based on the multichannel filtering approach to texture segmentation described in Section 2.2. The lines of text in a document image produce a distinct texture. Nontext components in the document image, including blank spaces, are considered as regions with different textures. Thus, the problem of segmenting document images into text and nontext regions can be posed as a texture segmentation problem. Two-dimensional Gabor filters are used to extract texture features for each of these regions. The filter selection procedure proposed by Jain and Farrokhnia (1991) will select filters that emphasize all frequencies equally. Since the characters in the text regions of the document image produce rapid tonal variations, higher segmentation accuracy can be achieved, for document images, by using only the eight filters corresponding to the two highest radial frequencies in each of the four directions.

The input image shown in Figure 2 is passed through each of the eight filters to generate eight filtered images. The filter outputs are shown in Figure 3. Note that the filter responses for the text regions are different from those for the nontext regions. For example, Figure 3d shows high response (white lines) corresponding to the lines of text in the document image (Figure 2). The eight feature images obtained from the filtered images of Figure 3, respectively, are shown in Figure 4.

We seek three clusters (K = 3) in the document images; one cluster includes the patterns belonging to the text, one represents the uniform regions in the document image (blank spaces and pictures which have a slow intensity variation), and the remaining cluster corresponds to the boundaries of the uniform regions in the document. It has been experimentally ascertained that the three-cluster segmentation produces the desired result. In general, the number of clusters depends on the resolution and the image content. We can now isolate the text region in the document image by applying a mask to the original image which will pick out only those pixels that correspond to the cluster representing the text segment. For all the document image segmentation results presented below, it was visually observed that the text region corresponded to segment number two. The regions in the image identified as text usually have ragged borders. The storage and future procesisng of text is easier if the text appears in rectangular blocks. To obtain the rectangular blocks of text, we find connected components corresponding to the text regions in the segmentation result. The enclosing rectangles for these connected components are computed and the regions in the input image corresponding to the pixels included in these boxes are identified.

4 Experimental Results

This section presents the results of the proposed document segmentation method for a number of images. Two different methods have been used for digitizing the documents. In the first scheme, the document image is digitized using a CID2250 square-pixel CCD camera manufactured by CIDTEC. The spatial resolution in these images is approximately 83 pixels per inch. In the second digitization scheme we used a JX-300 flat-bed scanner, manufactured by Sharp, to obtain the image. The flat-bed scanner can be used in six different modes: 50 dots per inch (dpi), 75 dpi, 100 dpi, 150 dpi, 200 dpi, and 300 dpi. In general, the flat-bed scanner provides sharper images than the CCD camera.

4.1 Unsupervised Segmentation

We first present some results of performing unsupervised segmentation of the document images. Section 4.2 contains the results of supervised segmentation.

Figure 5 shows the segmentation result for the newspaper image shown in Figure 2, obtained using the CCD camera. The input image appears in Figure 5a. The segmentation result corresponding to the

African perfor

From the [illegible] of the [illegible] to the [illegible] singing of great [illegible] the cast of Africa [illegible] the Wharton [illegible] audience Wednesday night.

More than [illegible] dancers, singers and musicians presented a rich selection of African [illegible] and culture to a [illegible] audience.

Sporting [illegible]

[illegible] English [illegible] during the show.

[illegible]

Although the [illegible] and [illegible] African [illegible]

Tanya Garda

Figure 2. 256 × 512 newspaper image captured by a CCD camera. The resolution of the image is approximately 83 dpi.

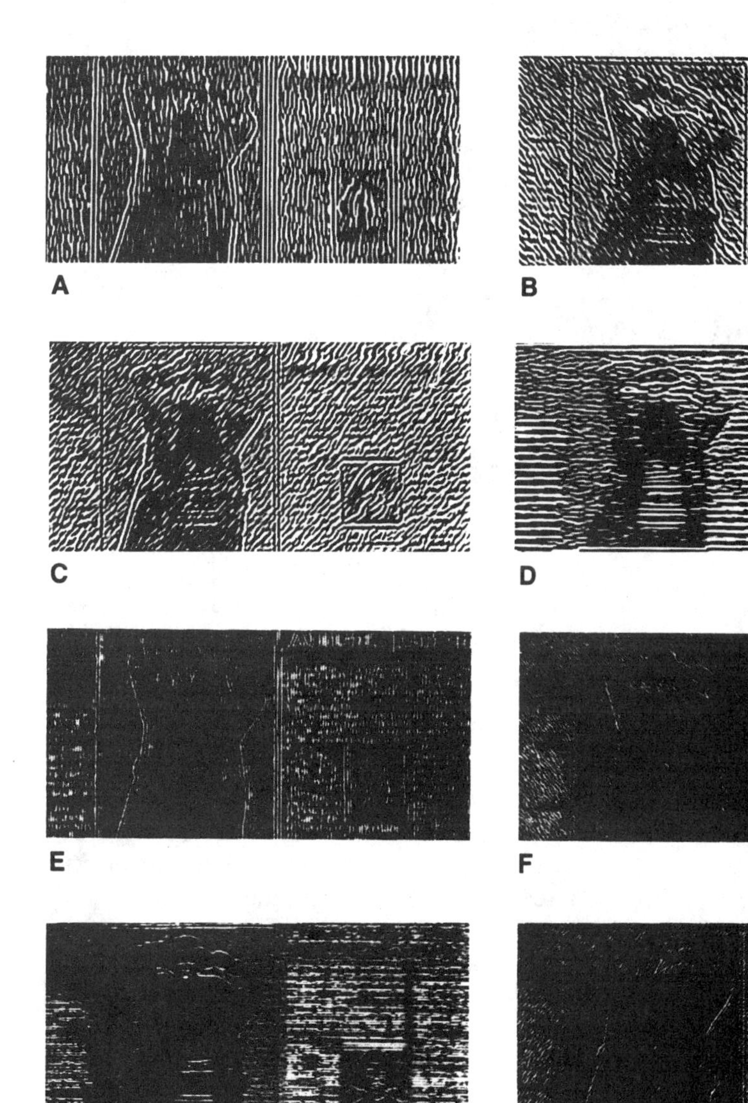

Figure 3. Eight filtered images of Figure 2. The frequency u_0 and orientation θ of the respective filters are: (A) $u_0 = 32\sqrt{2}$, $\theta = 0°$ (B) $u_0 = 32\sqrt{2}$, $\theta = 45°$ (C) $u_0 = 32\sqrt{2}$, $\theta = 90°$ (D) $u_0 = 32\sqrt{2}$, $\theta = 135°$ (E) $u_0 = 64\sqrt{2}$, $\theta = 0°$ (F) $u_0 = 64\sqrt{2}$, $\theta = 45°$ (G) $u_0 = 64\sqrt{2}$, $\theta = 90°$ (H) $u_0 = 64\sqrt{2}$, $\theta = 135°$.

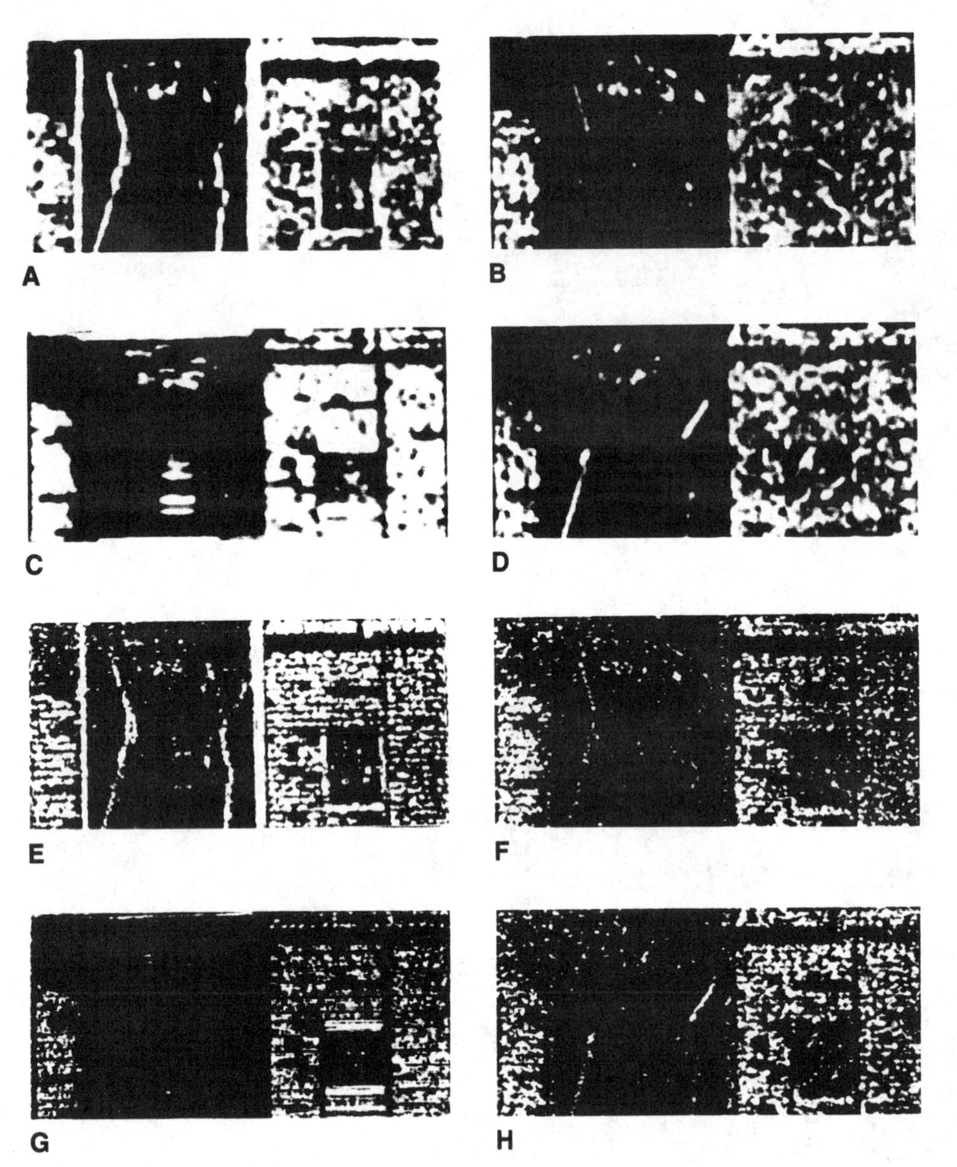

Figure 4. Eight feature images obtained from the filtered images shown in Figure 3. The images (A) . . . (H) are obtained from filtered images shown in Figure 3(A) . . . (H), respectively.

three-cluster solution is shown in Figure 5b. Finally, Figure 5c contains the extracted text region which shows the input gray levels for only those pixels which belong to the identified text-cluster. Almost all of the text region is recovered. Only the "corners" of a few characters are lost. The algorithm also detects large gaps between individual words in the text.

In Figure 6, the input image is a rotated version of the image in Figure 5. The image was rotated by 55° in this case. This value of the rotation parameter does not align the rotated image with any of the filters. The three-cluster solution and the extracted text regions are shown in Figures 6b and 6c, respectively. This demonstrates the capability of the proposed method to handle any skew that may occur due to imaging conditions or due to the presence of slanted text.

We do not present the results in Figures 5c and 6c as rectangular blocks since the text regions in these images form large connected components and the enclosing boxes for these components would include some of the nontext regions correctly rejected by the proposed algorithm. This problem occurs due to the low resolution of the input images.

Figure 7 shows the results of the segmentation algorithm on a newspaper image digitized at 150 dpi by the flat-bed scanner. Figures 7a and b, respectively, show the digitized image and the three-cluster segmentation results. It is clear from this example that at a resolution of 150 dpi, the segmentation algorithm demarcates each word, in the text, individ-

African perfor

Figure 5. Performance of the proposed segmentation method on a newspaper image. The 256 × 512 image, captured by CCD camera, has a resolution of approximately 83 dpi. (A) Input image; (B) three-cluster segmentation of (A): the black regions are the very low frequency regions, the gray color corresponds to the high frequency regions (text), and the medium frequency regions are colored white; (C) the text segments extracted from (A).

ually. The text region boundaries are more accurate here, compared to the image in Figure 5, owing to the better sharpness of the input image. Figure 7c shows the resulting rectangular blocks of text extracted using the proposed algorithm.

In Figure 8, we show the segmentation results when this algorithm is applied to the image of a printed page containing tabular data. This image was obtained using the flat-bed scanner at 150 dpi. The lines in the table form the nontextual information in the image. Unlike the previously developed block-segmentation methods mentioned in Section 1, our algorithm does not consider graphics as integral entities. Each line segment is considered separately. Thus the algorithm is able to recognize the text regions within the table. In contrast, the block segmen-

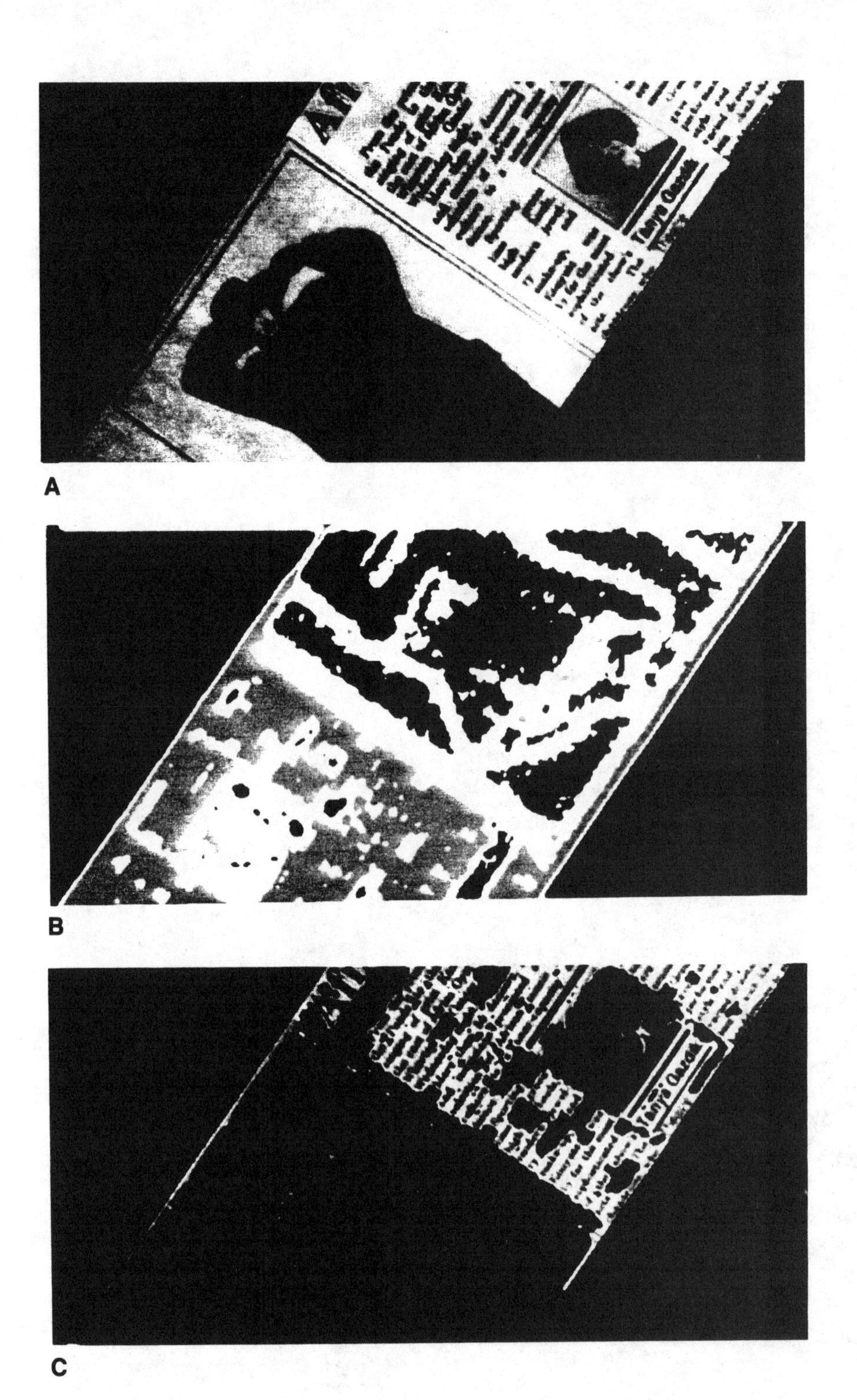

Figure 6. Performance of the proposed segmentation method on a rotated newspaper image. (A) Figure 5(A) rotated by 55°; (B) three-cluster segmentation of (A); (C) the text segments extracted from (A).

tation methods would consider the entire table as one graphical block and would fail to recognize the text contained in the block. In this example, however, the vertical line in the center of the table is incorrectly identified as text. This occurs because the high response of the two 0° orientation filters to the vertical line is comparable to the response of the filters to the text region due to the small inter-character spacings in the words. However, this long and thin vertical connected component can be easily removed by suitable postprocessing.

For all the results presented above, we performed a three-cluster segmentation. Referring to Figures 5b, 6b, 7b, and 8b, the black areas correspond to regions with very low intensity variation, usually the background; the white color represents regions with

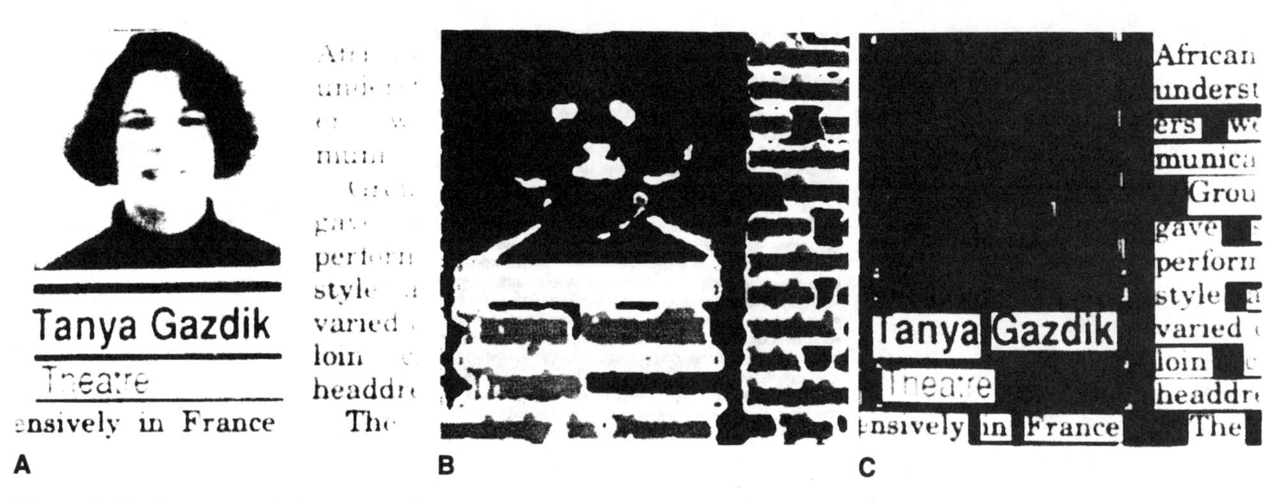

Figure 7. Performance of the proposed segmentation method on a newspaper image. The 256 × 256 newspaper image, captured by document scanner, has a resolution of 150 dpi. (A) Input image; (B) three-cluster segmentation of (A); (C) the text segments extracted from (A).

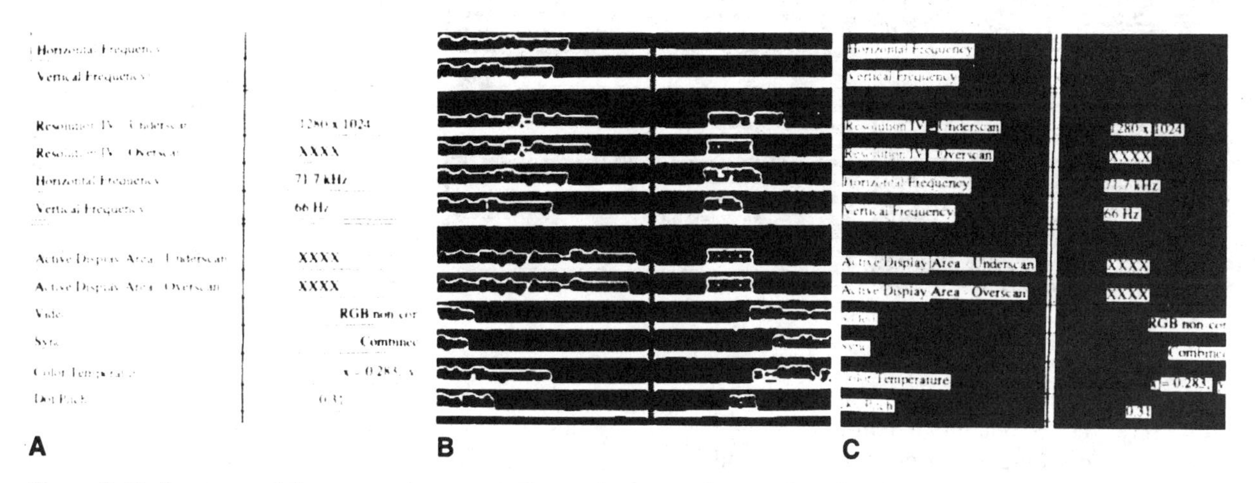

Figure 8. Performance of the proposed segmentation method on an image of a printed page containing tabular data. The 512 × 512 image, captured by document scanner, has a resolution of 150 dpi. (A) Input image; (B) three-cluster segmentation of (A); (C) the text segments extracted from (A).

medium variation; and the gray areas correspond to the high frequency regions, i.e., text. What happens when the document image contains only text? One might assume that in such cases a two-cluster solution is appropriate. Figure 9 shows the results for such an image obtained at 75 dpi using the document scanner. Figure 9b shows the two-cluster segmentation result. The three-cluster segmentation result is shown in Figure 9c. Although both segmentations identify the text regions correctly, the three-cluster segmentation often identifies individual words, and thus may be more desirable. Figure 9d shows the extracted text regions corresponding to the three-cluster segmentation.

The proposed method can easily handle handwritten text also. This is demonstrated in Figure 10. The input image shown in Figure 10 contains handwritten text. This image was digitized on our flatbed scanner at a resolution of 50 dpi. The three-cluster segmentation of this image, shown in Figure 10b, classifies most of the (nontext) lines as text. This can be attributed to the fact that the handwritten text in the input image in this case does not produce a uniform texture as the printed text does in the other examples. Therefore, the boundary between the text and nontext classes is not as sharp in the feature space. However, the four-cluster segmentation identifies most of the text correctly (Figure 10c). The clasification is even more refined in the eight-cluster segmentation, shown in Figure 10d. The extracted handwritten text shown in Figure 10e corresponds to the eight-cluster segmentation.

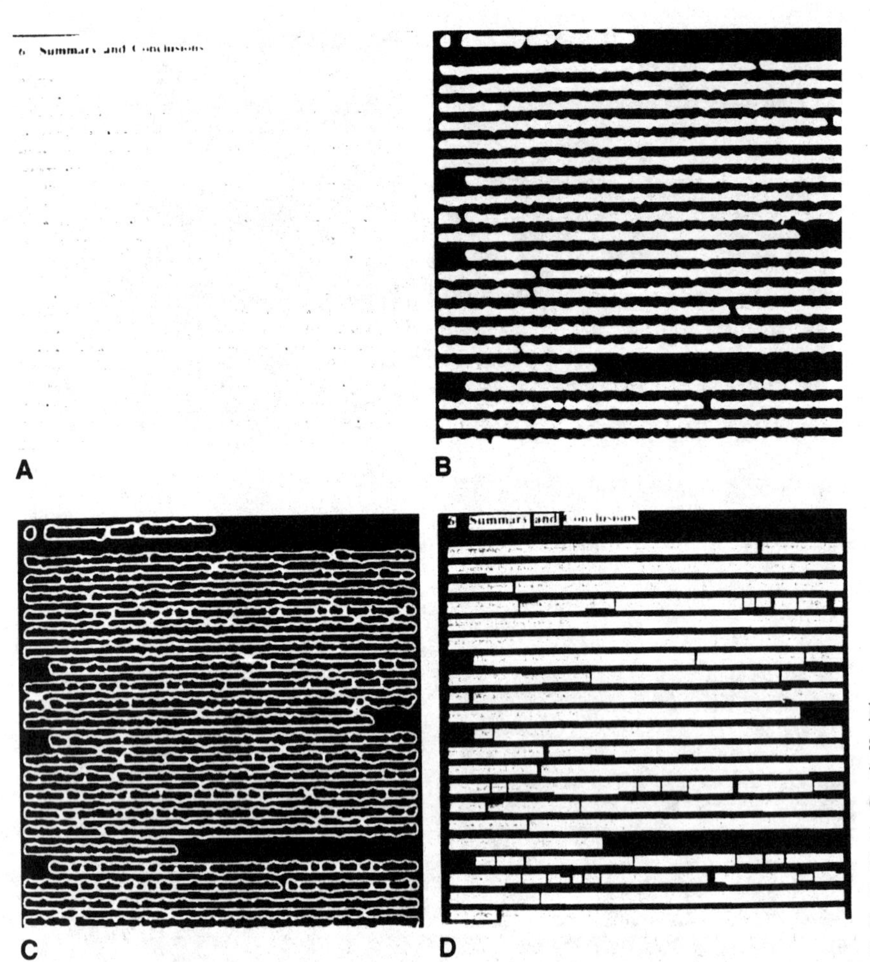

Figure 9. Performance of the proposed segmentation method on an image containing only text. The 512 × 512 image, captured by document scanner, has a resolution of 75 dpi. (A) Input image; (B) two-cluster segmentation of (A); (C) three-cluster segmentation of (A); (D) the text segments extracted from (A) using the three-cluster segmentation.

Figure 11a shows the mean feature values for the three gneeric segments, or clusters, 1, 2, and 3, corresponding to the input image in Figure 6a. Figures 11b and c show similar plots for the eight Gabor filter features corresponding to the input images in 7a and 8a, respectively. From these plots it is easy to see that the mean feature values for segment number 2 (text regions) is the highest for almost all the eight features. Thus, the mean feature values for each segment can be used as a criterion for selecting the text segment.

Chernoff faces (Chernoff 1973) provide a succinct representation of multidimensional data. Each feature vector (pattern) is represented by a cartoon face where the feature values determine the various attributes of the face. Figure 12 shows thirty Chernoff faces representing eight-dimensional feature vectors from the three clusters in the segmentation result for the image in Figure 7. These faces were generated using the S statistical software package (Becker et al. 1988). Ten sample patterns have been arbitrarily selected from each cluster. The cluster to which each pattern belongs is indicated by the number below the corresponding face. For each feature vector, features 1, ... , 8 correspond to area of the face, shape of the face, length of the nose, location of the mouth, curve of the smile, width of the mouth, location of the eyes, and separation of the eyes, respectively. It is clear from Figure 12 that the three clusters are compact and well separated.

Finally, we demonstrate the application of the proposed method for document image segmentation to a related problem—identification of bar code regions in an image. Bar codes are finding increasing use as a means of data compression (Pavlidis et al. 1990). In fact, some two-dimensional bar codes can represent several hundred characters in one square inch. The first step in an optical bar code recognition system is to capture an image containing the bar code. The captured image will usually be larger than the actual bar code region. The region in question may even be surrounded by other text or graphics. Thus, it is desirable to locate the approximate position of the bar code region, so that the bar code reader can be appropriately positioned automatically. We present results to show the effectiveness

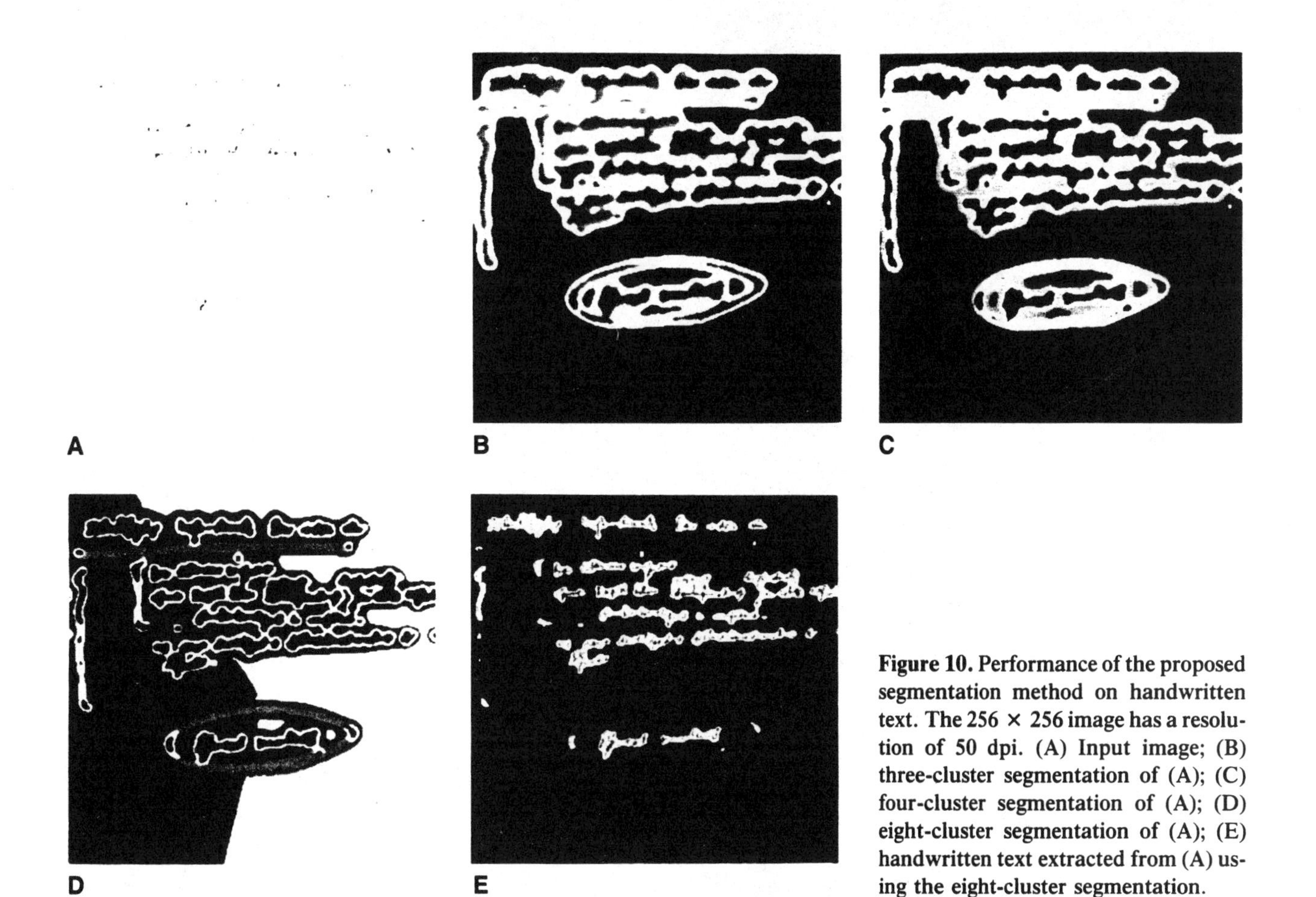

Figure 10. Performance of the proposed segmentation method on handwritten text. The 256 × 256 image has a resolution of 50 dpi. (A) Input image; (B) three-cluster segmentation of (A); (C) four-cluster segmentation of (A); (D) eight-cluster segmentation of (A); (E) handwritten text extracted from (A) using the eight-cluster segmentation.

of the proposed segmentation technique in locating the bar code region. The input image is shown in Figure 13a. This image was obtained by digitizing an article on page C1 of the *New York Times,* April 24, 1991 edition. A 512 × 512 image was acquired using the flat-bed scanner at a resolution of 75 dpi which was later reduced by subsampling to a size of 256 × 256. Thus, the effective resolution of the image in Figure 13a is 37.5 dpi. As can be seen in Figure 13b, a three-cluster segmentation fails to isolate the bar code region in the image. This is because the response of the bar code region to the high frequency filters is comparable to that of the rest of the text region. A nine-cluster segmentation, however, does isolate the bar code region from the rest of the image (Figure 13c). Some other small isolated regions are also classified into the bar code category, but these can be easily removed in a postprocessing phase using size constraints.

4.2 Supervised Classification

For many of the text-graphics segmentation problems, it is not necessary to do unsupervised segmentation. If we know the number and types of regions categories) to be identified in the image, we can design a simple supervised classification scheme. For example, in the bar code segmentation problem discussed in Section 4.1, since we are interested in localizing a specific type of region (2-D bar codes), the use of supervised classification is justified. This not only results in better classification accuracy, but it also requires less computation time. The training patterns can be randomly selected from the two classes (bar code and background). If the proposed segmentation method is to be used for bar code localization on a regular basis, the same training patterns can be used for subsequent images as long as the imaging environment remains the same.

Feature selection is an important issue in designing a classifier. It is well known that the addition of more feature does not necessarily lead to improved classification accuracy (Jain and Chandrasekaran 1982). For unsupervised classification of the bar code image, we have used eight features. The classification time can be reduced by using a subset of the eight filters. In fact, feature selection also improves the classification accuracy. Assuming that the image is appropriately aligned to the image coordinate system, four features corresponding to the filter set

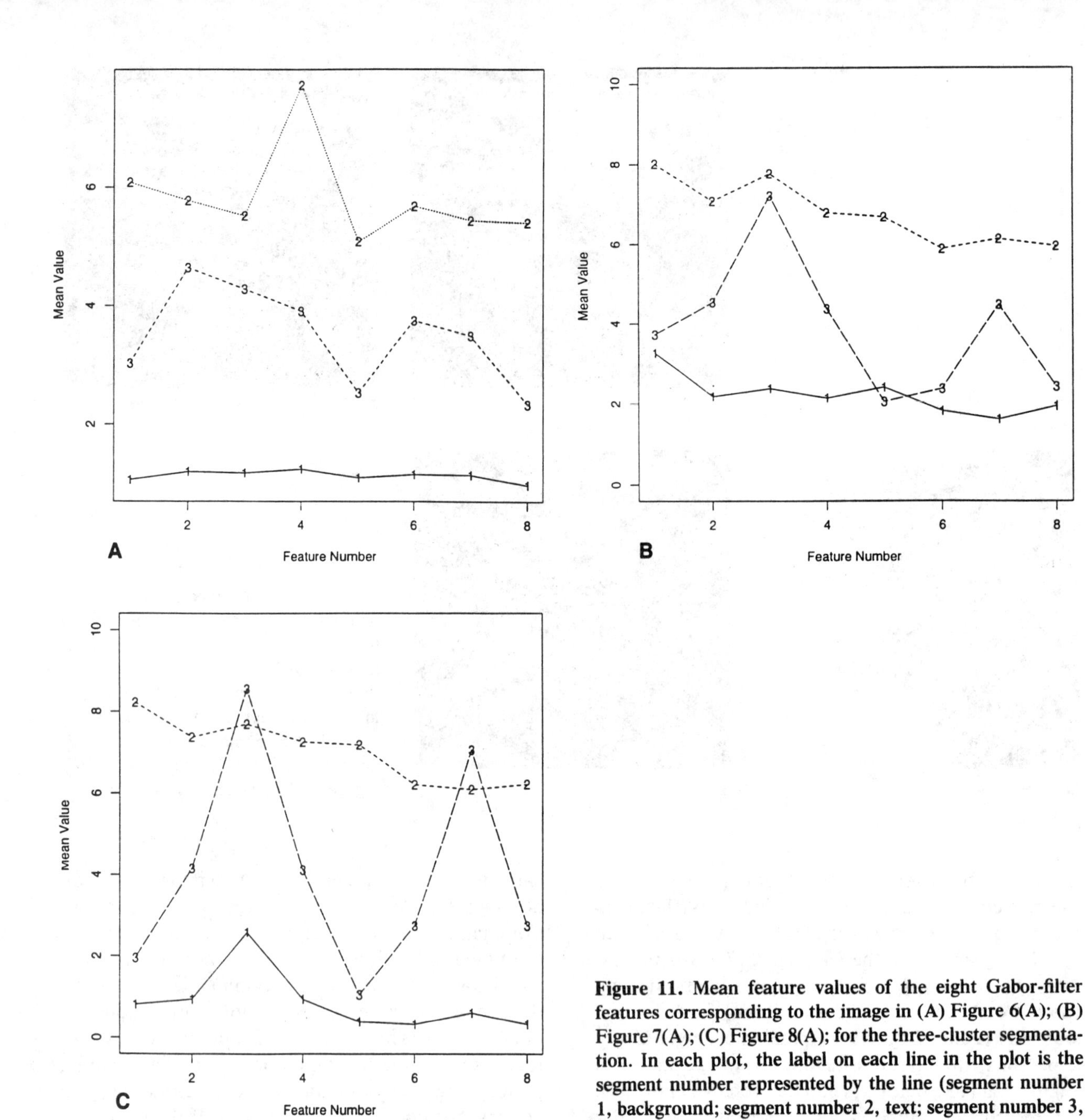

Figure 11. Mean feature values of the eight Gabor-filter features corresponding to the image in (A) Figure 6(A); (B) Figure 7(A); (C) Figure 8(A); for the three-cluster segmentation. In each plot, the label on each line in the plot is the segment number represented by the line (segment number 1, background; segment number 2, text; segment number 3, boundary region between the background and the text).

-$\{u_0 = 32\sqrt{2}, \theta = 0°; u_0 = 32\sqrt{2}, \theta = 90°; u_0 = 64\sqrt{2}, \theta = 0°; u_0 = 64\sqrt{2}, \theta = 90°\}$ were selected.

We designed a two-class classifier with 4,000 randomly chosen training samples (pixels) from bar code and background regions. The remaining pixels in the 256 × 256 image (Figure 13a) were classified using the 7-nearest neighbor (7-NN) classifier. We tried a number of classifiers and empirically determined that the 7-NN classifier gives the best results. The classification map resulting from this two-class supervised classification is shown in Figure 14a. This image clearly shows one large region associated with the bar code which is very well separated from the background. Several very small regions in the input image are also misclassified into the bar code class. These misclassified regions were removed by finding connected components in the image and merging all components smaller than a threshold (20 pixels in our experiments) with the background. This leaves us with a single connected component corresponding to the bar code region as shown in Figure 14b. We compute the bounding box corresponding to this

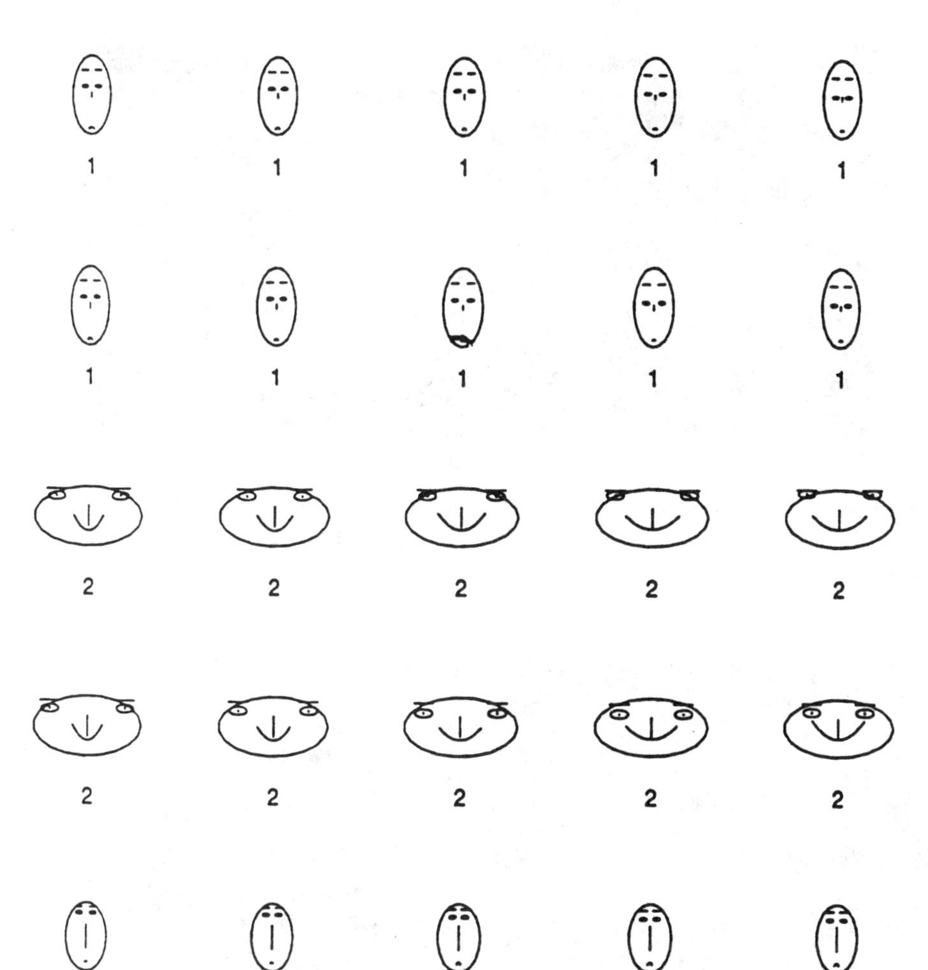

Figure 12. Chernoff faces for 30 randomly chosen eight-dimensional patterns, ten from each of the three clusters (indicated by the number below the corresponding face) in the segmentation result for the image in Figure 7(A). The features 1, . . . , 8 describe the area of the face, shape of the face, length of the nose, location of the mouth, curve of the smile, width of the mouth, location of the eys, and separation of the eyes, respectively, for each face.

component and use this box to locate the bar code in the input image. Figure 14c shows the bounding box for the component shown in Figure 14b, superimposed on the input image. The bar code has been accurately identified.

Since Gabor filters are sensitive to directionality, the supervised classification scheme described above will be able to handle only a small amount of skew (less than 10°). For rotation-invariant supervised classification isotropic filters (Coggins and Jain 1985) are more appropriate.

5 Summary and Conclusions

Automated document analysis has become an important research issue. Due to the enormous amounts of data involved, it is worthwhile to devise automatic methods for extracting the text regions in a document image so as to reduce the workload in the later stages of analysis. Most of the previous attempts reported in the literature are based on identification of connected components, which relies heavily on proper thresholding. In this paper, we present a method to identify the text regions in a document image using texture-based segmentation. Our method uses Gabor filters which have been used earlier for the general problem of texture segmentation (Farrokhnia 1990). One distinct advantage of our method is that it uses gray-level images, usually 256 gray levels, and no thresholding operation is applied to the input document images. The document image is considered as an image containing

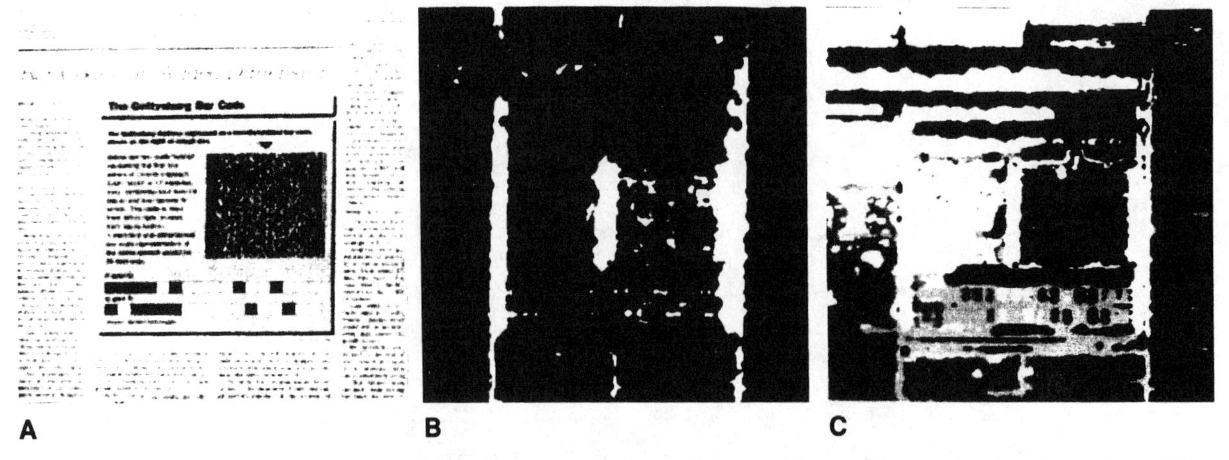

Figure 13. Performance of the proposed method in locating bar codes in an image. The results are shown for a 256 × 256 newspaper image containing a bar code. The effective resolution of the image is 37.5 dpi. (A) Input image; (B) three-cluster segmentation of (A); (C) nine-cluster segmentation of (A).

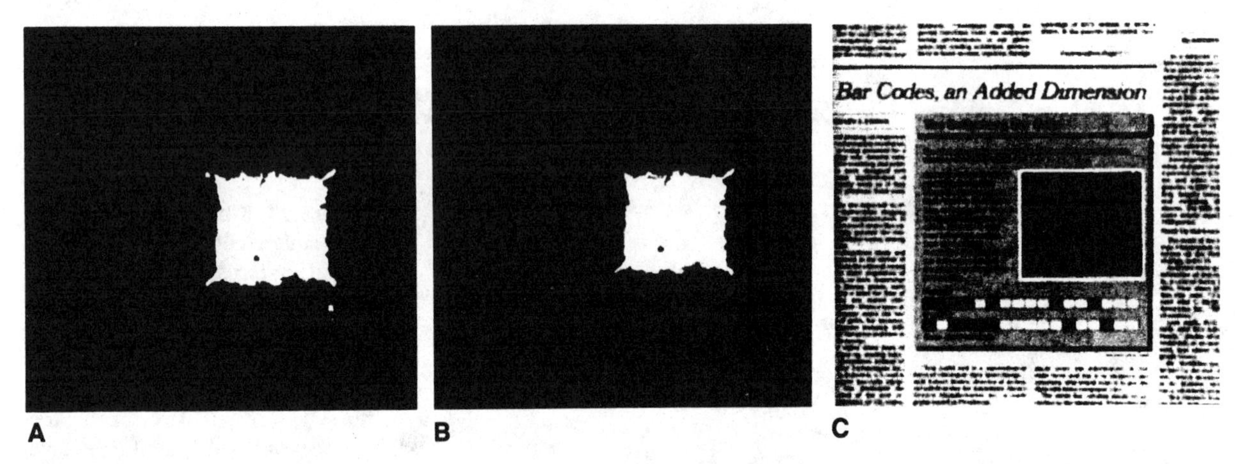

Figure 14. Supervised classification results. (A) Two-class supervised classification of Figure 13(A); (B) result after eliminating connected components smaller than 20 pixels; (C) the bounding box of the component remaining in (B), superimposed on the input image.

several textured regions, some of which correspond to the text in the document. A multichannel filtering based texture analysis approach is used to compute the characteristic features for these textures. Unsupervised classification of pixels utilizing these features produces a three-cluster segmentation of the document image. One of these segments correponds to the text regions in the document. The second segment corresponds to regions with very low frequency variations, usually the blank spaces and regions corresponding to "pictures" in the image. The third segment corresponds to regions with "medium" variations, usually the boundaries of the low frequency segments.

The performance of the proposed method on several test images is shown. Most of the previously proposed methods have strict requirements about the imaging conditions and require high resolution (~300 dpi) images. Our method is demonstrated to perform well even at lower resolutions (75 dpi). The experimental results also show that our segmentation method is robust to skew in images. The same filters can also identify handwritten text as well as two-dimensional bar code regions. However, for the handwritten-text image and the bar code image, the three-cluster segmentation does not yield correct results and we need to segment the image into a larger number of clusters. For a given segmentation problem, where we know the number of categories a priori, a supervised classification scheme is easier to implement. We have demonstrated that a 7-NN classifier can be used to localize the 2-D bar code in a newspaper image. The supervised classification scheme avoids the typical problems (how many clusters?, large computation time) associated with unsupervised segmentation. For the supervised classification, we have also implemented a two-layer feedforward artificial neural network (ANN) classifier, which matches the performance of the 7-NN classifier in classification accuracy.

Our current implementation of the proposed algorithm requires about two minutes of CPU time to process a 512×512 image on a SUN SPARCstation 2. In one particular experiment, 53.1 CPU s were required for generating the eight filtered images. The computation of feature values required 12.8 CPU s and the clustering operation required 64.6 CPU s. In practice, our proposed algorithm can be implemented as a four-stage pipeline. Such an implementation can dispense with the unnecessary overhead of reading and writing intermediate results and can reduce the total processing time significantly. The time required for generating the filtered images can be reduced drastically in a prallel implementation where all the eight Gabor filters can be applied to the image simultaneously. The filters themselves can be implemented efficiently in hardware (digital or optical). It is also possible to design special filters with frequency and orientation response which are better suited for the text segmentation task. Special hardware may be required to reduce the computational requirements of the clustering or the supervised classification stage (Ni and Jain 1985).

Acknowledgment. Professor T. Pavlidis introduced us to 2-D bar codes and encouraged us to apply our texture segmentation appraoch to the bar code localization problem.

References

Becker RA, Chambers JM, Wilks AR (1988) The New S Language. Wadsworth & Brooks/Cole, Pacific Grove, CA

Chernoff H (1973) The use of faces to represent points in k-dimensional space graphically. J. Am. Stat. Assoc. 68:361–368

Clark M, Bovik AC (1989) Experiments in segmenting texton patterns using localized spatial filters. Pattern Recognition 22(6):707–717

Coggins JM, Jain AK (1985) A spatial filtering approach to texture analysis. Pattern Recognition Letters (3):195–203

Farrokhnia F (1990) Multi-channel filtering techniques for texture segmentation and surface quality inspection. Ph.D. thesis. Dept. of Electrical Eng.. Michigan State University

Farrokhnia F, Jain AK (1991) A multi-channel filtering approach to texture segmentation. Proc. IEEE Computer Vision and Pattern Recognition Conf. Maui, June, pp 364–370

Fletcher LA, Kasturi R (1988) A robust algorithm for text string separation from mixed text/graphics images. IEEE Trans. Pattern Analysis and Machine Intelligence 10(6):910–918

Gabor D (1946) Theory of communication. J. Inst. Elect. Engr. 93:429–457

Iwaki O, Kida H, Arakawa H (1987) A segmentation method based on office document hierarchical structure. Proc. IEEE Int. Conf. Sys. Man Cybern. Alexandria, VA, October, pp 759–763

Jain AK, Chandrasekaran B (1982) Dimensionality and sample size considerations in pattern recognition practice. In: Krishnaiah PR, Kanal LN (eds), Handbook of Statistics 2, North Holland, pp 835–855

Jain AK, Dubes RC (1988) Algorithms for clustering data. Prentice-Hall, New Jersey

Jain AK, Farrokhnia F (1991) Unsupervised texture segmentation using Gabor filters. Pattern Recognition 24(12):1167–1186

Malik J, Perona P (1990) Preattentive texture discrimination with early vision mechanisms. J. Opt. Soc. Amer. A. 7(5):923–932

Nadler M (1984) A survey of document segmentation and coding techniques. Computer Vision, Graphics and Image Processing 28:240–262

Nagy G (1989) Document analysis and optical character recognition. Proc. Fifth Intl. Conf. on Image Analysis and Processing, Positano, Italy, Sept. 20–22, pp 511–529

Ni LM, Jain AK (1985) A VLSI systolic architecture for pattern clustering. IEEE Trans. Pattern Analysis and Machine Intelligence 7:80–89

Pavlidis T, Swartz J. and Wang YP (1990) Fundamentals of bar code information theory. IEEE Computer 23(4):74–86

Perry A and Lowe DG (1989) Segmentation of textured images. Proc. IEEE Computer Soc. Conf. on Computer Vision and Pattern Recognition San Diego, CA, pp 326–332

Srihari SN (1986) Document image understanding. Proc. IEEE Comput. Soc. Fall Joint Computer Conf. Dallas, Texas, Nov. 2–6

Tan TN, Constantinides AG (1990) Texture analysis based on a human visual model. Proc. IEEE Int. Conf. on Acoust., Speech, Signal Proc. Albuquerque, New Mexico, April, pp 2091–2110

Turner MR (1986) Texture Discrimination by Gabor Functions. Biol Cybern. 55:71–82

Wahl FM, Wong KY, Casey RG (1982) Block segmentation and text extraction in mixed text/image documents. Computer Graphics and Image Processing 20:375–390

Wang D, Srihari SN (1989) Classification of newspaper image blocks using texture analysis. Computer Vision, Graphics and Image Processing 47:327–352

Wong KY, Casey RG, Wahl FM (1982) Document analysis system. IBM Journal Res. Dev. 26(6):647–656

CD-ROM Document Database Standard

Ihsin T. Phillips

Department of Computer Science
Seattle University
Seattle, WA 98122

Su Chen and Robert. M. Haralick

Department of Electrical Engineering
University of Washington
Seattle, WA 98195

Abstract

The paper presents the design of a comprehensive standard document database for machine-printed documents. Our effort to produce a series of carefully ground-truthed document databases to be issued on CD-ROMs is described in detail. The databases can be utilized by the OCR and document understanding community as a common platform to develop, test and evaluate their algorithms.

1 Introduction

Systems that do OCR or any aspect of Document Image Understanding must work nearly perfectly over a broad range of document conditions and types in order to be really useful. To develop algorithms for OCR or to develop algorithms for Document Image Understanding requires that the developer have a suitable database of documents which are accurately ground-truthed so that the free parameters of the algorithms can be estimated. Customers of OCR or Document Image Understanding systems likewise must have a suitable database in order that they may accurately evaluate vendor proposed systems.

Database requirements for both the developer and the customer are nearly identical. Therefore, in order to help both developer and customer, there must be the creation of a comprehensive series of databases, each specialized to a given subset of document types, or intended as additions to already created databases.

Throughout the history of OCR research, and document layout and segmentation researchers, there has been a need for common data sets on which to develop and compare the performance of the algorithms. Some efforts have been made by researchers in the OCR community to make the data sets on which their algorithms have been tested available to the research community. Unfortunately, many researchers remain unwilling to do the same. Thus it becomes impossible for a researcher to verify published results by the process of replicating the algorithm or to compare the performance of a competing algorithm that she/he is developing. Even when data sets are available, the data sets are often tuned to the algorithm that the researcher supplying the data sets has developed. Thus the researcher obtaining the data sets has no alternative but to tune her/his algorithm to the data set and this does not promote good research. It is time for a series of comprehensive standard document databases to be constructed and made them available to researchers. Such databases would serve to provide uniform platforms on which researchers could develop and compare the recognition accuracy of their OCR and document understanding algorithms.

The cost for the creation of a document image database is relatively high due to the care that must exist in the creation process and the requirement for near perfect accuracy. Therefore, it is worthwhile to consider leveraging the cost of producing the ground truth of the document images by also including in the database some software to artificially degrade and distort document images in ways which approximate the real degradations and distortions that documents undergo as they get copied, recopied, faxed, etc. Such degradation software can be used to generate controlled degraded document pages for OCR algorithm development as well as for performance evaluations and testing of the algorithms.

Our efforts to create such a series of carefully ground-truthed databases to be issued on CD-ROMS is described in this paper. The paper presents the design of a comprehensive standard document database for machine-printed documents. An English version of the database is currently under its construction at the University of Washington and it is to be completed in July, 1993.

Reprinted from *Proc. Int'l Conf. Document Analysis and Recognition,* 1993, pp. 478–483.

2 Requirement for Machine-printed Document Database

2.1 Languages

In order to be useful for developers of OCR algorithms and document understanding systems. The document databases to be constructed should reflect the full range of machine-printed documents. The document databases should also include documents in each one of the world's major language and scripts, such as Roman, Hebrew, Arabic and Farsi, Kanji, Hangul, Devnagiri and Cyrillic. Our efforts are restricted to the Roman script and English language.

2.2 Document Types

The database series should at least include the following document types:

- Articles: journals, proceedings, books, etc.;
- Business letters and memorandums;
- Newspapers/magazines;
- Maps: street maps, terrain maps, etc.;
- Forms;
- Manuscripts;
- Engineering CAD/CAM drawings;
- Advertisements.

2.3 Document Format and Quality

For each type of document, a variety of documents of various formats and quality needed to be present in the database. These documents should be drawn according to the frequency of their usage to satisfy the requirements of performance evaluation and drawn so that sufficient samples are present of each variety to satisfy requirement of algorithm developers.

2.4 Document Page Images

The database should included both greyscale and binary document images in TIFF formats. The database should also include synthesized noise-free document images as well as various degraded document images (degraded through both the real process and simulation). All image files on the CD-ROM should be compressed to save space.

2.5 Document Page Description

Page Attributes

The page attributes should include at least the language and the script, the font information, the publication information, the page condition, the page layout, and the character orientation and reading direction of the document page.

Zone Definition and Attributes

The zone definition should specify the shape, the size and the location of each zone in a document page. The zone attributes may include zone type (i.e., text zone, figure zone, form, map, table, etc.), zone label (i.e., title, page number, paragraph, author, footnote, etc.), language, script, character orientation, reading direction, font information and text alignment format.

2.6 Ground Truth

The database should provide two types of ground truths. One is the character-based ground truth and the other is the zone-based ground truth.

The character-based ground truth will include the name, the size and the position of each individual character on the page.

The zone-based ground truth will contain the character string with the line break for each text line within a text zone.

2.7 Degradation Models

It is also necessary to develop document degradation models to provide researchers with a mechanism for introducing random perturbations on noise-free ideal images. These degradation models will be developed based on the kinds of degradation found in real-life photocopying and FAX transmission, as well as, coffee stains, ink bleeding, page aging, etc.. The document degradation software will simulate these real-life degradations. This degradation software can be used to generate controlled degraded document pages for OCR algorithm development, as well as for performance evaluations and testing of the algorithms. Such synthetically degraded images give unlimited extension to the real data sets in the database without having to provide additional ground truth for the generated documents.

2.8 Software

The database should include data compression and decompression software, OCR performance evaluation software, degradation software, as well as utility tools for the database.

3 CD-ROM English Document Database: A Case Study

In the rest of the paper, we presents the design of a machine-printed English document database. The set of document pages will be selected from various technical journals and reports.

3.1 Contents and Organization Overview

The English document database has two logical compartments: the software compartment and the document compartment. The software compartment contains the data compression and decompression software, the OCR performance evaluation software, the photocopy degradation software, and the FAX degradation software. The descriptions of this compartment is given in Section 3.2.

The document compartment contains all the document page images, the page and zone attribute record files, the zoning information files and ground truth files. The descriptions of this compartment is given in Section 3.3.

Since the document database will be packaged on a CD-ROM. The file names and directory structures will be in complete compliance with ISO 9660. The general file name conventions for the document database are given in Section 3.4.

3.2 Software Compartment Contents

Compression and Decompression Software

All the scanned and simulated documents in bitmap (TIFF) format on the CD-ROM are supplied in compressed form. The compression algorithm used is the CCITT Group IV bi-level image compression standard. The user can then use the decompression program provided in the CD-ROM to uncompress the compressed data files.

OCR Performance Evaluation Software

OCR performance evaluation software will be developed and provided. Given a list of document zone IDs and OCR outputs of the zones, the algorithm will evaluate the output of OCR algorithm against the corresponding ground truth residing in the CD-ROM. A set of contingency tables for characters and misrecognized words will be computed and output by the algorithm. The user's manual of the software package is given in [4].

Photocopy Degradation Software

Software is also provided that simulates two selected document degradation models. One is Baird's degradation model [2] and the other is currently developed by researchers at the Intelligent Systems Laboratory. Given a document file (binary image file) and degradation model parameters, the user can run the photocopy degradation program to degrade the document as desired. The requirement specification of the software is given in [3].

3.3 Document Compartment Contents

Document Page Image Files

The document compartment includes a set of document image files. Each document image corresponds to one document page. The document images can be classified into the following categories:

1. Scanned greyscale images from real documents.
2. Scanned binary images from real documents.
3. Synthesized noise-free binary image.
4. Degraded noise-free binary image.

The real document pages will also include a set of documents which are taken from the set of synthesized noise-free documents and degraded through real processes – both by successively photocopying or FAX transmission. The degraded document pages will also include a set of synthetic degraded documents (for convenience) that are degraded by the same degradation software that will be provided in the database. The source document of the degraded pages will come from a set of selected pages from category 2 and 3.

Page Attribute Files

For each document page in the database, there is a set descriptive attribute which describe the various attributes of the page. Each document page type (journal, letter/memo, etc.) has its own set of attributes. For technical journals/reports, the attributes include page condition, page bounding boxes, page contents, page layout, font information, publication information, etc. The journal page attributes definitions are given in Section 4.

The advantage of defining a set of document page attribute records per document type over a set of general page attributes for all document types is that it allows one to add another document type to the database without any change to the database design.

Zone Attribute Files

Each document page will be zoned manually according our zoning conventions [5]. Each zone in a page is associated with a set of zone attributes that describes the contents of the zone. We define a distinct zone attribute record for each document type (journal, etc.). The definitions of the zone attribute are given in Section 4.5.

Ground Truth Files

The database provides the ground truth for all text zones on all document pages. (For mathematical zones, the form of ground truth may be developed and provided. For line-art, halftone zones, etc., there will be no ground truth.) The format of the ground truth is given as character sequences: the correct character sequences (with line breaks between sequences) within the zone.

In addition, each LaTeX generated document page resides in the database, we provide a ground truth file that contains character positions of all characters on the page. The format of the ground truth is given in a character position sequence: the sequence of characters with the position (coordinates) and the size of each character in the zone. (This format only for synthesized and degraded document pages arising from the LaTeX generated documents in the database.

The ground truth files will be used by the OCR evaluation software reside in the database. The software evaluates the performance of OCR algorithms.

All special symbols will be represented in LaTeX-alike syntax. The translation table will be provided.

Bounding Box Information Files

The database also provides bounding boxes informations for the page header, page footer, live matter and each zone on a document page. The live matter of a document page is the usable area of the page between the margins [1]. Each bounding box will be represented as a rectangular region on a document page. The definitions of the bounding box information files are given in Section 4.

3.4 File Name Convention

The general file name conventions for the document database are as follows: 1) A legal file name will consist of at most 8 characters (26 capital English letters and 10 digit numbers) followed by a period and a 3-character extensions. 2) The first character of the file name must be a capital English letter.

Under our current design, the document files have additional file name constraints to make them more identifiable. For example, the first four characters of the file name represent the document page ID. The last four characters of the file name are used to identify the category of the file (scanned binary, scanned greyscale, page/zone attribute record file, page/zone bounding box record file, ground truth, etc.). The filename extension indicates the file format (.TIF for image file, .TXT for ASCII file and .TEX for LaTeX file).

4 Document Page Record Definitions

This section gives the definitions of all records that constitute record files within the document compartment of the CD-ROM.

4.1 Page Condition Record

This record includes attributes that describe the visual conditions (or qualities) of a given document page. The page condition record has the following fields:

Record Field Definitions:

- Document ID:
- Degradation type: (original)(photocopy)(fax)
- n-th copy: (noise-free)(1)(2)( )
- Visible salt/pepper noises: (yes)(no)
- Visible vertical streaks: (yes)(no)
- Visible horizontal streaks: (yes)(no)
- Extraneous symbols on the top: (yes)(no)
- Extraneous symbols on the bottom: (yes)(no)
- Extraneous symbols on the left: (yes)(no)

- Extraneous symbols on the right: (yes)(no)
- Page skewed on the left: (yes)(no)
- Page skewed on the right: (yes)(no)
- Page smeared on the left: (yes)(no)
- Page smeared on the right: (yes)(no)
- Visible page rotation: (yes)(no)
- Page rotation angle (in degree):
- Page rotation angle standard deviation:

4.2 Page Attribute Record

The following record fields define the set of descriptive attributes which describe the various attributes of a journal document page.

Record Field Definitions:

- Document ID:
- Document language: (English)

  The value for this field is English for this database. The field is provided for upward compatibility with future databases that we or others might produce in languages other than English, for ex. Kanji, Arabic etc.
- Document script: (Roman)
- Document type: (journal) (letter) (memo) (news)
- Publication Information: This attributes contains information about the name, the volume number, the issue number and the publishing date of the publication. It also has the corresponding page number of the document page from the publication.
- Multiple pages from the same article: (yes)(no)

  A flag indicating whether multiple document pages from the same article are included in the database. The document pages within the same article can be retrieved by reference to the publication name, volume and issue number of the page.
- Text zone present: (yes)(no)
- Special symbol present in text zone: (yes)(no)

  The special symbols are defined as the symbols other than the standard ASCII symbols.
- Displayed Math zone present: (yes)(no)
- Table zone present: (yes)(no)
- Half-tone zone present: (yes)(no)
- Drawing zone present: (yes)(no)
- Page header present: (yes)(no)
- Page footer present: (yes)(no)
- Max number of text columns: The number of equal-width text columns within of the live matter area of the document page.
- Page Column layout: (regular) (combined-columns)
- Character orientation: (up-right) (rotated-right) (rotated-left).

  This field gives the orientation of characters within the text line when the page is oriented to up-right position.
- Text reading direction: (left-right) (right-left) (top-down) (bottom-up)

  This field gives the text reading direction within a text line when a page is oriented to up-right position. For example, for a landscaped oriented page, the page needed to be rotated to in up-right position.
- Dominant font type: (Serif)(Sans-Serif)
- Dominant character spacing: (proportional) (fixed)
- Dominant font size (pts): (<< 9) (9-12) (13-18) (19-24) (25-36) (>> 36)
- Dominant font style: (plain) (bold) (italic) (underline) (other)

  Any combination of the font styles are allowed. The word 'dominant' is defined as the most frequently used font (type, style, size) in a given page.

4.3 Page Bounding Box Record

The page bounding box record defines the page header area, page footer area and live matter area. A page bounding box record has the following fields:

Record Field Definitions:

- Document ID:

- Header area upper-left corner coords:
- Header area lower-right corner coords:
- Live matter area upper-left corner coords:
- Live matter area lower-right corner coords:
- Footer area upper-left corner coords:
- Footer area lower-right corner coords:

4.4 Zone Bounding Box Record

The zone bounding box record defines each zone on a document page.

Record Field Definitions:

- Document ID:
- Zone ID:
- Zone upper-left corner coords:
- Zone lower-right corner coords:

4.5 Zone Attribute Record

This attribute record describes a set of attributes that are common to zones from a journal/report document page. The record has the following fields:

Record Field Definitions:

- Document ID:
- Zone ID:
- Zone content:

  The zone content can take on either of the following values: text, text with special symbols, displayed math, table, half-tone, drawing, form, ruling, bounding box, logo, map, advertisement, announcement, handwriting and others.

- Text zone label:

  The zone label can be one of the following values: text body, list item, drop cap, caption, abstract body, abstract heading, section heading, synopsis, highlight, pseudo-codes, reference heading, reference list item, footnote, author biography, page header, page footer, page number, article title, author, affiliation, diploma information, society membership information, article submission information, abstract heading, abstract body, footnote heading, keyword heading, keyword body and others.

- Text alignment within the zone:

  This attribute defines the text alignment within the zone. The types of text alignment are: left aligned, center aligned, right aligned, justified, justified hanging, left hanging.

- Font information: This attributes defines the dominant font type, character spacing, font size and font style within the zone.
- Character orientation:
- Text reading direction:
- Zone's column number:

  This attribute describes the zone's column location. A zone may be in the header area, footer area and column number 1 of 1, 1 of 2 and etc.

- Next zone ID within the same thread: (__) (nil)

  The zones of each document page can be grouped into several logical units. Within each logical unit, the reading order is sequential. We call such a logical unit as a semantic thread. This attribute is used to indicate the reading order among the zones that constitute a semantic thread. "nil" is used to indicate the end of the semantic thread.

References

[1] *Xerox Publishing Standards: A Manual of Style and Design*, A Xerox Press Book, Watson-Guptill Publications/New York.

[2] H. S. Baird "Document Image Defect Models", *Structured Document Image Analysis*, Springer Verlag, N. Y., 1992, p546-556.

[3] T. Kanungo, *Document Degradation Model Requirement Specification*, ISL Report, 1993, University of Washington.

[4] Su Chen, *OCR Performance Evaluation Software User's Manual*, ISL Report, 1993, University of Washington.

[5] I. T. Phillips and S. Chen, *English Document Database Zone Label Definitions and Examples*, ISL Report, 1993, University of Washington.

[6] I.T. Phillips, S. Chen, J. Ha and R.M. Haralick, "English Document Database Design and Implementation Methodology", *Proc. of the second Annual Symposium on Document Analysis and Information Retrieval*, April 26-28, pp. 65-104, 1993.

The Skew Angle of Printed Documents

Henry S. Baird

AT&T Bell Laboratories
Murray Hill, NJ 07974

Abstract

A new algorithm is presented for determining the skew angle of lines of text in an image of a printed page. The best estimate of skew occurs at the global maximum of an energy function on sets of projection-counts of character locations. Experiments on over 50 pages show that the method works well on a wide variety of layouts, including multiple columns, sparse tables, variable line spacings, mixed fonts, and a wide range of point sizes. Both speed and accuracy of the method are an order of magnitude better than previously reported results. The high accuracy achieved, two minutes of arc, assists rapid and reliable top-down segmentation into columns and lines of text using projection-profile splitting.

1. Introduction

Automatic segmentation of images of printed pages into columns, lines of text, and characters is a necessary early stage of document image analysis. In a recent survey, Srihari and Zack [SZ86] point out that projection profile splitting, a fast top-down segmentation technique, is vulnerable to non-zero skew angles. Bottom-up clustering techniques, by contrast, are relatively insensitive to skew angle but are slower and unreliable due to decisions made with low statistical confidence in small neighborhoods. A fast and accurate method of determining skew that is unaffected by a wide range of layout styles, such as is described here, provides a secure starting point for reliable top-down segmentation.

Let us define the *skew angle* of a printed page to be the orientation angle of its lines of text. Where several angles occur, we mean the *dominant* angle of the majority of the text lines. In spite of a great variety of layout styles in common use, a dominant skew angle is often unambiguous [NS84]. This definition includes text but deliberately excludes graphics, boxes, underlining, *etc.* We require that the user specify a maximum text size so that connected regions too large to be a character can be easily ignored; in practice, results are good as long as the number of characters overwhelms the number of fragments of graphics, noise, *etc.*

2. Required Accuracy of Skew Estimates

The bottom-up segmentation strategies described by Nagy *et al* [NSS85] and Doster [D86] require that skew angle be restricted to "a few" degrees. A reasonably careful user can place a document on the flat surface of a scanner with an error of less than 3 degrees. Thus segmentation may be at least *possible* without highly-accurate estimates of skew.

On the other hand, some problems are *more rapidly and easily solved* given an estimate accurate to a small fraction of a degree. For example, a long line (40 picas) of set-solid 6-point text, when projected horizontally, will interfere with neighboring lines at skew angles greater than about a third of a degree [Ph68]. Also, character height above baseline is useful in classification only if it can be estimated to within a fifth of the x-height. This can be guaranteed simply by selecting the median of a set of baseline heights, if skew can be controlled to within about four minutes of arc.

3. Prior Work

Trincklin [T84] gives a method based on piecewise least-squares fitting of vertical white run-lengths. The method requires a single column of text and a clean left margin. No analysis of accuracy or speed is presented.

Hashizume, Yeh, & Rosenfeld [HYR86] exploit the fact that characters are often closer to one another in the direction *along* text lines than in other directions. They therefore compute the nearest neighbor of each component, and each pair of neighbors is connected by a straight line segment. A histogram of the orientations of these line segments is computed. Such histograms often have a strongly-marked peak at the dominant skew angle. There is no discussion of runtime or analysis of error as a function of digitization error or sample size. Among the 13 examples

---

"The Skew Angle of Printed Documents" by H.S. Baird from *Proc. Conf. of the Society of Photographic Scientists and Engineers*, 1987, pp. 14–21.

reported, the average error was 1.5 degrees and the worst 4.1 degrees. The method naturally fails for wide character spacing, such as occurs in sparsely-populated tables.

Postl [Po86] has described experiments with Trincklin's method and two others of his own design. The first applies the discrete 2D Fourier transform to the image plane and examines a half plane of the power spectrum coefficients W(U,V), as follows: for each "simulated scan angle" in a given range, a scalar alignment measure is computed by integrating along a radius vector from the origin at that angle to the V-axis. This measure represents the accumulation of energy in spatial frequencies sharing that orientation angle; its global maximum reveals the dominant skew angle. Postl does not discuss accuracy or speed of the Fourier method.

Postl's second method similarly hunts for the maximum of a measure over a range of angles, but the measure can be computed directly from the image, as follows. Along every line inclined at the scan angle, compute the integral density. Then, for each pair of neighboring scan lines, compute the difference of their densities. Finally, compute the sum of squares of these differences. The skew of a 1600x1600 pixel multilevel image was determined within 0.6 degree, after 5.1 CPU seconds on a Sun III (MC 68020) programmed in optimized C. A potential advantage of both these methods is that they can operate on multilevel images, prior to binarization or segmentation into characters.

Rastogi & Srihari [RS86] describe a method using a Hough transform. For each angle in the discrete representation of Hough space, the number of large "low-high-low transitions" is counted, and the maximum count is interpreted as identifying the dominant skew. There is no discussion of runtime or accuracy. In the five examples shown, skew angle was coarsely quantized in increments of 15°.

4. An Alignment-Measure Approach

We choose to "locate" a character at the midpoint of the bottom of its bounding box. Given an orientation angle θ, project the locations of characters (abstracted as points) onto an *accumulator* line perpendicular to the projection direction. The accumulator line is partitioned into m bins, and bin size is set at 1/3 of the x-height of 6-point text. Let $c_i(\theta)$ denote the number of points projected into the ith bin at angle θ.

A real-valued *energy alignment measure* function of θ is computed as $A(\theta) = \sum_{i=1}^{m} c_i^2(\theta)$. This function displays a global maximum at the correct skew angle (*e.g.* Figures 1 & 2). Experiments suggest that *any positive superlinear* function of c_i within the summation will perform correctly. A similar observation has been made independently by Biland & Wahl [BW86] in a superficially dissimilar context, that of searching for global maxima in discrete Hough transform space.

5. Locating the Maximum

Locating the global maximum value of the alignment measure to a given accuracy by exhaustive search is straightforward and reliable, but expensive. For a document of average complexity (page [bask], 3624 characters, Figure 2), computing the measure for all 5400 angles at a resolution of 2 minutes of arc, required over 13 CPU minutes on a DEC VAX 11/750 with floating-point accelerator.

An alternative approach is an iterative sampling procedure. Unfortunately, the energy function is not always convex, unimodal, locally monotonic, or smooth enough to estimate local derivatives. Thus many classical optimization techniques ([W79] and [Br73]) cannot be applied with confidence. However, about the global maximum there always appears a roughly symmetrical peak, whose slopes are smooth with slowly-rising bases and steep tops, meeting at a sharp maximum. This peak shape is well-approximated as the intersection of two functions S^+ and S^-, of the form: $S^{\pm}(\theta) = \frac{a}{b \pm c\,\theta}$ where a, b, & $c > 0$ are real parameters (different for S^+ and S^-). I have tailored an interactive variable-step peak-finding heuristic to locate the peak. It first samples at a coarse resolution (0.7°) until the characteristic shape of a peak is detected. Then, its slopes are fitted with approximating functions of the form shown above, using an iterative non-linear separable least-squares fitting algorithm [K75] as implemented by function NSF1 in the portable mathematical subroutine library PORT 3 [F84]. The point at which the approximations analytically intersect is taken as a refined estimate of the peak maximum. A cluster of samples at finer resolution is then computed to refine its location. This is repeated at finer and finer resolutions until the target resolution of 2' of arc is attained.

6. Experimental Trials

The effectiveness of the heuristic has been tested in experimental trials on over 50 pages representing a variety of typographical and layout styles, having been selected from books, journals, theses, and typewritten pages, and including multiple columns, sparse tables, mixed fonts (both fixed and proportional pitch), various text sizes, headers, trailers, footnotes, and scanner noise at margins. All pages tested were monochrome, high-contrast, and accurately printed. Most had an unambiguous skew angle within fifteen degrees of horizontal. Page scanning, digitizing, and binarization were performed by a Ricoh IS-30 document scanner at a resolution of 300 pixels/inch giving a binary image of about 8.5 million pixels. This was segmented into black regions by connectivity analysis. Regions too large to be a character were set aside. The remaining regions often included margin noise and character fragments, without however noticeably affecting the computation of skew.

7. Accuracy

To evaluate accuracy of the computed skew, projection profile analysis was applied, first vertically to find columns, then horizontally within columns to find text lines. The characters in each line were examined at high magnification on a graphics monitor in such a way that skew errors as small as 1/10 of the x-height of the text could be easily detected. Skew was well within this bound on every page but one. In this case (page [old]) two columns had been pasted up at different skew angles, and the average of these angles was reported. Even at a resolution of 2' of arc, the maximum value is often clearly detached above its neighboring values. illustrating a marked sensitivity of the method to small angular changes. Such accuracy is about an order of magnitude better than previously published results.

8. Runtime

Most pages required samples at about 40 angles for convergence. Runtime for documents with over 1000 characters was dominated by the cost of sampling, which was in turn dominated by the cost of projections. Thus the peak-finding heuristic is about a factor of 100 times faster than exhaustive search.

The innermost projection loops, written in C, were speed-optimized using integer arithmetic, pointer arrays, and loop-unrolling. Runtime required to determine skew of a document of average complexity (page [bask], 3624 characters, Figure 2) was 8.5 CPU seconds on a DEC VAX 11/750 (with floating point accelerator but no special-purpose hardware). This does not include time to perform rough segmentation into characters, which consisted of run-length-encoding followed by connectivity analysis, requiring 95 CPU seconds on page [bask].

It is possible to compare absolute runtimes with Postl's second method, also written in optimized C (running on a Motorola 68020). Taking into account relative machine speed and number of pixels in the images, the energy alignment measure method runs about about a factor of 8 faster, and achieves 20 times the angular resolution.

9. Conclusions

I have presented a new skew-estimation method exploiting an energy function of sets of projection-counts of character locations. The approach requires that (a) text lines have been accurately printed, (b) there is a dominant text-line orientation, (c) characters have been roughly segmented, (d) loose bounds on point size are known, and (e) non-textual graphics can be ignored by some means or are overwhelmed by the number of the characters. Also, for highest accuracy, skew angle should be within about 15° of horizontal.

Otherwise, the method operates robustly on a wide variety of page layout styles. It is insensitive to many other interesting properties of pages, such as the number and location of columns, number and location of text lines, word and character spacing, font style, and (within loose bounds) point size. This shows that skew angle is, in effect, an *independent property with global support* which can, with high statistical confidence, be computed as an early stage in a top-down segmentation strategy.

The accuracy of 2 minutes of arc is about a factor of twenty better than the best prior published result. Assuming that a rough segmentation into characters has been performed, the heuristic runs about an order of magnitude faster than prior published results.

It would be pleasant to find a fast closed-form computation for skew angle. All methods known to me require a search, either slow and guaranteed or fast but heuristic.

10. Acknowledgements

Detailed and helpful comments on an earlier draft were provided by Lorinda Cherry, Jon Bentley, Peter Weinburger, and Doug McIlroy. David Gay introduced me to non-linear separable least-squares fitting. Brian Kernighan pointed out reference [Ph68]. The Ricoh scanner interface to the DEC UNIBUS was designed by Joe Condon and built by Wing Moy, and its software driver was written by Lorinda Cherry. The encouragement of Clark Maziuk and Jon Bentley has been much appreciated.

References

[BW86] Biland, H. P., & F. M. Wahl, "Cluster Estimation for Image Analysis and Object Recognition," *IBM Technical Disclosure Bulletin*, Vol 28, No. 8, (1986), pp. 3667ff.

[Br73] Brent, R. P., *Algorithms for Minimization Without Derivatives*, Englewood Cliffs, NJ: Prentice-Hall, 1973.

[D86] Doster, W., "Designing a Document Analysis System," tutorial presented at *8th Int'l Conf. on Pattern Recognition*, Paris, FRANCE, 27-31 October 1986.

[F84] Fox, P., ed., *The PORT Mathematical Subroutine Library*, Murray Hill, NJ: AT&T Bell Laboratories, 1984, Chapter "Optimization", section "NSF1".

[HYR86] Hashizume, A., P-S Yeh, & A. Rosenfeld, "A Method of Detecting the Orientation of Aligned Components," *Pattern Recognition Letters* 4 (1986), pp. 125-132.

[K75] Kaufman, L., "A Variable Projection Method for Solving Separable Nonlinear Least Squares Problems," *BIT* **15** (1975), pp. 49-57.

[NS84] Nagy, G. & S. Seth, "Hierarchical Representation of Optically Scanned Documents," *Proceedings, 7th Int'l Conf. on Pattern Recognition*, Montreal, Canada, July 30 - August 2, 1984, pp. 347-349.

[NSS85] Nagy, G., S. Seth, & S. Stoddard, "Document Analysis with an Expert System," *Proceedings, Pattern Recognition in Practice*, Amsterdam, 1985.

[Ph68] Phillips, A., *Computer Peripherals and Typesetting*, Harrow: HMSO Press, 1968.

[Po86] Postl, W., "Detection of Linear Oblique Structures and Skew Scan in Digitized Documents," *Proceedings, 8th Int'l Conf. on Pattern Recognition*, Paris, FRANCE, 27-31 October 1986, pp. 687-689.

[RS86] Rastogi, A. & S. Srihari, "Recognizing Textual Blocks Using the Hough Transform," TR 86-01, Dept. Computer Science, Univ. Buffalo (SUNY), 1986.

[SZ86] Srihari, S., & G. Zack, "Document Image Analysis," *Proceedings, 8th Int'l Conf. on Pattern Recognition*, Paris, FRANCE, 27-31 October 1986, pp. 434-436.

[T84] Trincklin, J. P., "Conception d'un systeme d'analyse de documents *etc*," Univ. de Franche-Comté, Besançon, FRANCE, 1984. Ph.D. Thesis.

[W79] Wilde, D. J., *Foundations of Optimization*, Englewood Cliffs, NJ: Prentice-Hall, 1978.

Pages (whose *images* are referred to)

[bask] *The Type Specimen Book*, New York: Van Nostrand, 1974, p. 23. 3674 characters.

[hart] Hartner & Owen, *Selected Tables in Mathematical Statistics*, vol. 1., Chicago: Markham, 1970, p. 264. 3724 characters.

[old] Glare, P. G. W., ed., *The Oxford Latin Dictionary*, Oxford: Clarendon Press, 1982, p. xxiv. 3646 characters.

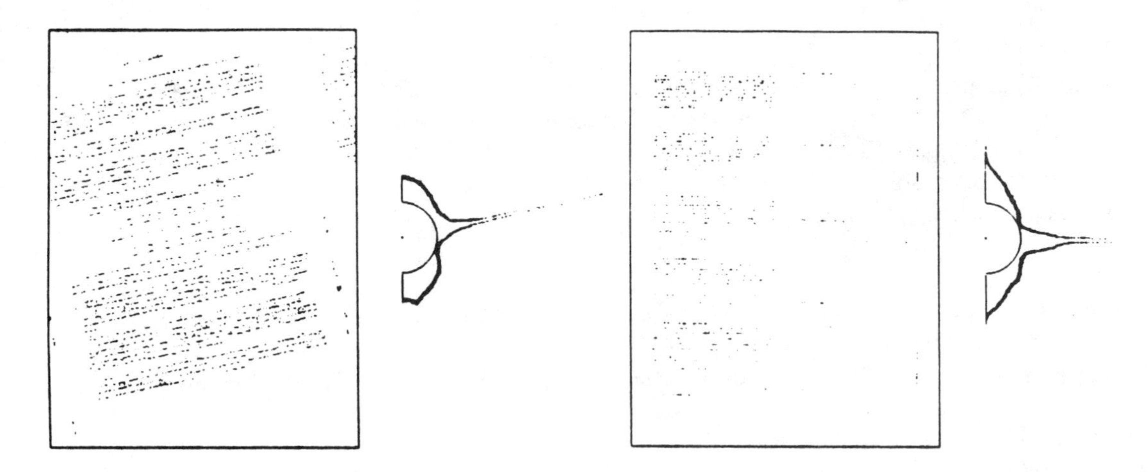

Figure 1. Log-energy alignment measure for page [hart], which was badly skewed. Notice the presence of uncorrected noise at the edges and center fold of the book. Skew angle: 12.26°.

Figure 2. Log-energy alignment measure for page [bask], illustrating the Baskerville font style over a range of text sizes from 6 to 16 point. Note the multiple columns and frequently-varying line spacing. Skew angle: -0.55 °

Explanation of the Figures. *On the left is the page area, with character locations shown as points. On the right is a graph of logarithms of alignment measure values over the full range of angles [−90°, +90°], sampled every 2' (5400 angles total). Values are plotted radially about a central point, and the global minimum value is indicated by a semi-circle. The −90° and +90° values are indicated by vertical lines. For consistency in display, the global minimum and maximum values have been forced to fixed values using an affine transform which is of course in general different for each page.*

A DOCUMENT SKEW DETECTION METHOD USING RUN-LENGTH ENCODING AND THE HOUGH TRANSFORM

Stuart C. Hinds,[1] James L. Fisher, and Donald P. D'Amato

The MITRE Corporation/Civil Systems Division
Systems Engineering and Applied Technology
7525 Colshire Drive
McLean, Virginia 22102-3481, USA

Abstract

As part of the development of a document image analysis system, we have devised a method, based on the Hough transform, for the detection of document skew and interline spacing--necessary parameters for the automatic segmentation of text from graphics. Because the Hough transform is computationally expensive, we reduce the amount of data within a document image through the computation of its horizontal and vertical black run-lengths. Histograms of these run-lengths are used to determine whether the document is in portrait or landscape orientation. A grey scale "burst image" is created from the black run lengths that are perpendicular to the text lines by placing the length of the run in the run's bottom-most pixel. This data reduction procedure decreases the processing time of the Hough transform and reduces the effects of non-textual data on the determination of skew and interline spacing.

1. Introduction

As more and more institutions and federal agencies confront an overwhelming abundance of paper and microfilm-based information, it has become increasingly important to convert such information into an efficiently stored, computer searchable form. Document image analysis is the process of identifying the components (text and non-text) of a document's structure and can be divided into three phases [1]: 1) scanning and binarization, 2) segmentation and labeling of text and figure blocks, and 3) processing of text (usually by optical character recognition) and figures. Fast top-down methods, such as the run-length smoothing algorithm (RLSA), developed by Wong et. al. [2], and projection profile cuts [3], [4], have been developed for accomplishing phase 2. Unfortunately, these methods are generally ineffective when applied to skewed documents. Although other researchers [5], [6] have successfully segmented text from figures in skewed documents, their methods involve a connected components analysis, which can be computationally intensive.

We have developed a document image analysis system based on the approach of Wong et. al. Because Wong's system is limited to aligned documents, we have added a skew detection stage to our system. While other methods for determining document skew exist (see [7] for a short review), we have based our method on the Hough transform because of its sensitivity to skew and its applicability in determining a document's interline spacing--a necessary parameter for the RLSA. Because the Hough transform is computationally expensive and slowed by noise, we have developed a procedure, based on the computation of an image's vertical black run-lengths, to reduce the amount of data and minimize the effects of noise and non-textual material within the original document image.

1 Current address: University of California at San Diego, Neuropsychology Research Lab., Childrens Hospital, 8001 Frost St., San Diego, CA 92123.

2. Review of the Hough Transform and Its Use In Document Skew Detection

The Hough transform can be used to detect lines at any orientation. It consists of mapping points in cartesian space (xy) to sinusoidal curves in $\rho\theta$ space via the transformation:

$$\rho = x\cos(\theta) + y\sin(\theta).$$

Each time a sinusoidal curve intersects another at a particular value of ρ and θ, the likelihood increases that a line corresponding to that $\rho\theta$ coordinate value is present in the original image. An accumulator array (consisting of R rows and T columns) is used to count the number of intersections at various ρ and θ values. Those cells in the accumulator array with the highest number of counts will correspond to lines in the original image. The basic computational steps in the Hough transform are as follows:

```
For (x)
   For (y)
       if (pixel is black) {
              For (θ) {
                   Calculate ρ = xcos(θ) + ysin(θ)
                   increment accumulator array at ρ,θ
              }
       }
```

It should be noted that the Hough transform is usually applied to binary images; hence the need to check if a pixel is black (i.e., part of the information within the image).

The angular resolution that the Hough transform will detect depends on how finely the columns of the accumulator array (corresponding to θ values) are spaced. The increment size between θ values and the maximum skew angle that should be detected will in turn affect the computation time of the Hough transform. In general, θ ranges either from 0 to 180 or -90 to 90 degrees in increments of one degree.

The number of rows, R, of the accumulator array (corresponding to ρ values) affects how well the Hough transform resolves lines. To detect fine lines, R should be such that each xy point along a straight column can be mapped to a unique row. Therefore, to detect every point in a rectangular image (with θ ranging from -90 to 90 degrees):

$$R = (w^2 + h^2)^{1/2}$$

where w = width and h = height.

Because text lines are actually thick lines of sparse density, the problem of determining the skew angle of text lines becomes more difficult than determining the skew angle of fine lines. Two groups of researchers have developed different methods for detecting document skew with the Hough transform.

Reprinted from *Proc. 10th Int'l Conf. Pattern Recognition,* 1990, pp. 464–468.

Fletcher and Kasturi [6] take the approach of treating the text as thick lines. Thus, the problem becomes one of reducing the resolution of ρ such that the pixels comprising the height of a character are mapped to a single row in the accumulator array. If the font size of the text in a document image is known a priori, it becomes a simple matter to calculate the appropriate reduction in resolution of ρ. However, more than one font size may be present in a document image and, most likely, none of these font sizes will be known a priori. Fletcher and Kasturi overcome this problem by first finding the connected components of an image and then separating graphics from text based on the relative frequency of occurrence of components as a function of their area. The average height of all connected components in the non-graphics set is used as an initial estimate for reducing the resolution of ρ. Fletcher and Kasturi then use an iterative procedure to improve this initial estimate. While their skew detection method is quite robust, accurate, and applicable to segmentation of text from graphics, the computation time of their procedure increases dramatically with the use of a connected components analysis on the original image.

Rastogi and Srihari [8] use a simpler approach for detecting the skew angle of a document. Their method takes advantage of the periodicity of text lines by detecting periodic variations in the counts of accumulator array cells along θ columns. This periodicity is detected by counting the number of high-low and low-high transitions along each θ column. To minimize the effects of noise and spurious data, transition height ratios must be greater than two. In addition, transition widths must exceed a value of three and transitions between counts below 20 are ignored. The θ column with the greatest number of transitions is taken as the angle of skew of the text. Although this method does not require a connected components analysis, there is no reduction in the amount of data that is input to the Hough transform. Consequently, because the Hough transform is such a computationally intensive operation, Rastogi and Srihari's method can be quite slow at determining the skew angle of a document.

3. Data Reduction by Run-length Bursts

As a means of reducing the amount of data within a binary document image (black print on a white background), we create a "burst image." The burst image is a grey scale image with each pixel's intensity representing the vertical run-length of a column of black pixels in the original binary image. More specifically, each individual vertical column (aligned with the scan axis) of L_i contiguous black pixels is replaced with L_i-1 white pixels and one non-white pixel of value L_i positioned at the end of the original black run. For illustration, Fig. 1a contains a sample image, Fig. 1b displays the burst image pixel values, and Fig. 1c shows the resulting burst image as a grey scale image. Note that the longer the column of black pixels in the original image, the darker (higher pixel value) the corresponding pixel (positioned at the bottom edge of a character or line) in the grey scale burst image.

Three important characteristics of the burst image are: 1) The number of black pixels is significantly reduced as compared to the original image. (For typical text documents digitized at 300 pixels per inch, we observe a reduction in the black pixel count by slightly more than a factor of two. This reduction factor increases with increasing non-text area populations, and is typically a factor of 11 for average newspaper pages with a mixture of pictures and text.) 2) It can be obtained readily from the compact form of an image which has been stored after application of the run-length encoding algorithm along the vertical axis. 3) Greater emphasis is given to the bottoms of text lines, thereby allowing a more accurate detection of skew and interline spacing.

In addition to decreasing the amount of data within an image and emphasizing the bottoms of text lines, run-length bursts can be used to determine the orientation mode (i.e., portrait or landscape) of a document in which text data predominates. On the average, since stroke heights have longer run-lengths than stroke widths for a portrait mode document, a histogram of its horizontal run-length burst image will contain a larger number of counts corresponding to the run-lengths of character stroke widths than would a histogram of its vertical run-length burst image. Therefore, the orientation mode of a document can be determined from a comparison of the counts of short run-lengths (which correspond to character width) in the histograms of vertical and horizontal burst images.

One advantage of using burst images to determine document orientation is that it decreases the range that θ must be searched to determine document skew. If it is determined that the document is in portrait mode, the range of θ is [-45 45] degrees. For landscape mode, the range of θ is [45 135] degrees. Therefore, for either portrait or landscape mode, the range that θ is searched is decreased by a factor of two from the standard range of [-90 90] degrees.

4. Determining Interline Spacing

As shown in [8], there is a regular rise and fall in cell counts along the accumulator array column corresponding to the correct skew angle Θ_c of the document image. Taking advantage of the correspondence between this periodicity in cell counts and the periodicity of text lines, we apply a 1D Fast Fourier Transform (FFT) to the Θ_c column to determine line spacing frequency. The resulting waveform is then high pass filtered to eliminate the D.C. component and infrequently encountered large interline values. The inverse of the frequency with the largest magnitude is then taken as the interline spacing.

5. Skew Detection Procedure

To increase the speed of our skew detection procedure, we reduce the resolution of the document image from 300 to 75 pixels per inch (ppi). Next, a vertical and a horizontal burst image is produced. The sum of the first four bins (excluding the bin corresponding to a run-length of zero) of the histograms of these images are compared to determine landscape or portrait orientation. For images digitized at 300 ppi, this assumes that stroke widths are at most 4/75 inch. The Hough transform is then applied to either the vertical burst image (in the case of a portrait mode document) or the horizontal burst image.

Within ρθ space, cells in the accumulator array are incremented by the value of the pixel currently being transformed. Unlike the standard approach of incrementing the accumulator cells by a value of one, incrementing by the pixel value de-emphasizes noise and emphasizes the contribution of the bottoms of lines of text. However, since figures and black margins usually have large run-lengths, the contribution of these large run-lengths can be greater than the contribution of the smaller but more frequent character height run-lengths. This can lead to skew detection errors and, more commonly, to errors in determining interline spacing. To eliminate the negative effects of these large run-lengths, only run-lengths between 1 and 25 pixels (1/75 to 1/3 inches) are mapped to ρθ space. In doing this, we are assuming that most of the text is equal to or less than 24 point size (1/3 inch).

To actually determine the skew angle of the input document, the accumulator array is searched for the cell with the largest value. The column that this cell belongs to is taken as the skew angle Θ_c of the document. Once Θ_c has been determined, interline spacing is found as previously described in Section 4.

To increase further the speed of the Hough transform, we assume that most document images are skewed by less than 15 degrees. Therefore, the range of θ is [-15 15] degrees for portrait mode documents and [75 105] degrees for landscape mode documents. However, to be sure of high accuracy in the segmentation and recognition process of a document analysis system, we increment θ by 0.5 degrees. The range of ρ is determined from the size of the document, as described previously.

6. Experimental Trials

Experimental trials were conducted on thirteen document images that were deemed to be representative of government forms and magazine and newspaper pages. All of the images were digitized at 300 ppi, reduced to 75 ppi, and thresholded. Two images (one newspaper and one magazine) were also scanned in landscape mode. A Sun 4-280 (a 10 MIPS reduced instruction set computer) was used for all computations.

The 13 images were grouped into five categories, namely:

- Simple text (i.e., text of a single font size only)
- Multiple font text
- Text and line drawings
- Text and grey scale images
- Forms (e.g., Internal Revenue Service forms)

An example of each document category is depicted in Fig. 2.

Our skew detection method was evaluated in terms of its processing time and its accuracy in determining skew and interline spacing. In addition, we evaluated how well our method of comparing the histograms of horizontal and vertical burst images determines the orientation (landscape or portrait) mode of a document.

Skew and Interline Spacing Accuracy

As shown in Table 1, the skew angle Θ_c was correctly determined for each document image. Interline spacing was correctly determined for all of the tested images except the line drawing image of Fig. 2c. Our method for determining interline spacing is based on finding the dominant frequency present in the Θ_c column. However, in the case of Fig. 2c, not only are there a number of interline spacings (we counted 5 different interline spacings), but none of them is truly dominant. We experimented with clipping run-lengths smaller than 25 pixels and were able to obtain some interline spacings that corresponded to an actual interline spacing present in the image. However, more experimentation and refinement of our interline spacing detection method will have to be performed to be able to detect reliably a combination of interline spacings within a document.

Orientation Accuracy

While our method for automatically determining whether a document is in landscape or portrait mode did not perform as well as hoped (as evidenced by the last column of Table 1), it was not expected to perform well with forms, where text is not always dominant. (It should be noted that the orientation of the one form with a high density of text was correctly determined.)

It is interesting that upon removal of the black margins surrounding the simple text image (of Fig. 2a) that failed the orientation test, its correct orientation was determined. Although black margins did not influence the other misclassified images (they were tested without black margins as well), the sparseness of text within the image of Fig. 2a was such that the black margins altered the expected number of counts in the histograms of the horizontal and vertical run-length burst images.

Processing Time Reduction

Absolute processing times ranged between 7 seconds for the simple text image of Fig. 2a to 102 seconds for one of the multiple font documents. However, these times were atypical; the majority of the document images took an average of 55 cpu seconds to process (including times for obtaining a horizontal and vertical burst image - about 1.5 cpu seconds to create both images). Adding this time to the average time of 12 cpu seconds for reducing a 300 ppi image to a 75 ppi image gives an average processing time for skew and interline spacing detection of 67 cpu seconds. This time compares well with the average processing time of Baird's "Alignment-Measure Approach" [7]. Baird's approach took an average of 8.5 cpu seconds to determine the skew angle of a document at an angular resolution of 1/20 degree over a 180 degree range. However, the alignment-measure approach requires a connected component analysis, which takes an average of 95 cpu seconds to segment characters into individual components. It should be noted that while our method has a faster total processing time than the alignment-measure approach (67 cpu seconds to 8.5 + 95 cpu seconds), Baird's approach has a much better angular resolution and searches over a greater range of θ.

For each category, the smallest, greatest, and average processing time reduction factors were listed in Table 2. The processing time reduction factor was determined by dividing the processing time for a Hough transform of the original 75 dpi image by the processing time for computing the vertical and horizontal burst images plus the processing time for a Hough transform of the burst image. It should be noted that the processing time for a Hough transform of either the 75 dpi or burst image includes the time to detect the document's skew angle as well as its interline spacing.

The presence of black margins within the simple text image of Fig. 2a is responsible for the large decrease in processing time in the simple text category. Discounting the simple text category, the greatest decrease in processing time was for the document images with grey scale figures. The large number of black pixels present in the figures will be greatly reduced to only a few pixels in the burst image and therefore, considerably reduce the processing time of the Hough transform. A rather high decrease in processing time was also observed for the forms, presumably because of the large number of vertical lines within these documents.

Detection of Skew and Interline Spacing in Severely Skewed Documents

Since the thirteen document images were not severely skewed when scanned and digitized, we decided to test our method on two severely skewed document images. We selected a portion of one of the grey scale images and a portion of one of the line drawing images as our test images (shown in Fig. 3). The image in Fig. 3a was skewed by 20 degrees and then rescanned. It was subsequently digitally rotated to a skew angle of 10, 0, -10, and -20 degrees. The image in Fig. 3b was digitally rotated to 45, 60, 75, 90, 105, 120, and 135 degrees.

The orientation of the image in Fig. 3a was correctly determined as portrait mode for all five rotations. For a range of [-25 25] degrees, our method correctly determined the skew and interline spacing of the image at all five rotations. For the image in Fig. 3b, the rotated versions at 0 and 45 degrees were classified as portrait mode while the other six rotated versions spanning the range of [60 135] degrees were classified as landscape mode. The range of θ was then set at [-50 50] degrees for the portrait mode images and at [40 140] degrees for the landscape mode images. The skew angle and interline spacing of each of the rotated images was correctly determined.

Although the actual skew was correctly determined for the rotated versions of Fig. 3b, a skew of 45 or -45 degrees could have easily been chosen as the correct angle of skew for the 45 degree rotated version of Fig. 3b. Our skew detection procedure can detect portrait or landscape mode orientation and the associated skew angle but it has no way of determining whether or not a document is upside down. Our next step is to supplement our procedure with a method for determining this [9].

7. Conclusions

The Hough transform can not only be used to detect document skew but interline spacing as well. However, the presence of noise can limit its effectiveness. Furthermore, the presence of figures and black margins can cause textual data to be mixed with non-textual data in the Θ_C column of the accumulator array and thus impair the detection of interline spacing. This limits the use of the Hough transform in the segmentation phase of a document analysis system. A greater limiting factor of the Hough transform, though, is its computational complexity.

By creating a burst image from the original document image, we have shown that the processing time of the Hough transform can be reduced by a factor of as much as 7.4 for documents with grey scale images. Moreover, because only small run-lengths are input to the Hough transform and because the accumulator array is incremented by the run-length associated with a pixel rather than by a factor of one, the negative effects of noise, black margins, and figures are avoided. Consequently, interline spacing can be determined more accurately.

Although the orientation mode of forms was incorrectly determined three out of four times, it should be noted that our method was not intended to work on documents that are not predominantly textual. Curiously, when the wrong run-length burst image of a form (i.e., the horizontal rather than vertical burst image of a portrait mode document) is input to the Hough transform, the resultant interline spacing is two to five times greater than normal. This result could be used as a verification of whether the correct orientation mode was determined. To determine if this verification scheme can indeed be used, further tests will have to be run on more forms. Another future task is to investigate a method for determining whether there is more than one interline spacing in a document and, if so, what their values are. This would be valuable in documents such as the one shown in Fig. 2c.

Acknowledgements

This work was made possible through MITRE-Sponsored Research funds. We would like to thank Washington Technology, the IEEE, and the Association for Computing Machinery for the use of portions of their publications.

REFERENCES

[1] S. N. Srihari, "Document image understanding," *Proc. of the IEEE Computer Society Fall Joint Computer Conference*, pp. 87-96, 1986.

[2] K. Y. Wong, R.G. Casey and F.M. Wahl, "Document analysis system," *IBM J. Res. Develop.*, vol. 6, pp. 642-656, Nov. 1982.

[3] G. Nagy, S.C. Seth, and S.D. Stoddard, "Document analysis with an expert system," *Proc. of ACM Conference on Document Processing Systems*, pp. 169-176, 1988.

[4] O. Iwaki, H. Kida, and H. Arakawa, "A segmentation method based on office document hierarchical structure," *Proc. of the 1987 IEEE International Conference on Systems, Man, and Cybernetics*, pp. 759-763, 1987.

[5] J. P. Bixler, "Tracking text in mixed-mode documents," *Proc. of ACM Conference on Document Processing Systems*, pp. 177-185, 1988.

[6] L. A. Fletcher and R. Kasturi, "A robust algorithm for text string separation from mixed text/graphics images," *IEEE Trans. on PAMI*, vol 10, pp. 910-918, Nov. 1988.

[7] H. S. Baird, "The skew angle of printed documents," *Proc. of Society of Photographic Scientists and Engineers*, vol. 40, pp. 21-24, 1987.

[8] A. Rastogi and S. N. Srihari, "Recognizing textual blocks in document images using the Hough transform," TR 86-01, Dept. of Computer Science, SUNY Buffalo, NY, Jan 1986.

[9] S. N. Srihari and V. Govindaraju, "Analysis of textual images using the Hough transform," Technical Report, Dept. of Computer Science, SUNY Buffalo, NY, April 1988.

Table 1
Skew and Interline Spacing Accuracy

Document Category	Simple Text	Multiple Fonts	Line Drawing	Grey Scale	Forms
Number of Documents	2	2	2	3	4
Number with Correctly Determined Skew	2	2	2	3	4
Number with Correctly Determined Interline Spacing	2	2	1	3	4
Number with Correctly Determined Orientation (Portrait/Landscape)	1	2	2	2	1

Table 2
Processing Time Reduction Factors Using Run-Length Bursts

Document Category	Simple Text	Multiple Fonts	Line Drawing	Grey Scale	Forms
Number of Documents	2	2	2	3	4
Lowest	1.7	2.2	2.4	5.5	3.6
Highest	15.0	2.8	7.9	10.6	4.9
Average	8.4	2.5	5.2	7.4	4.3

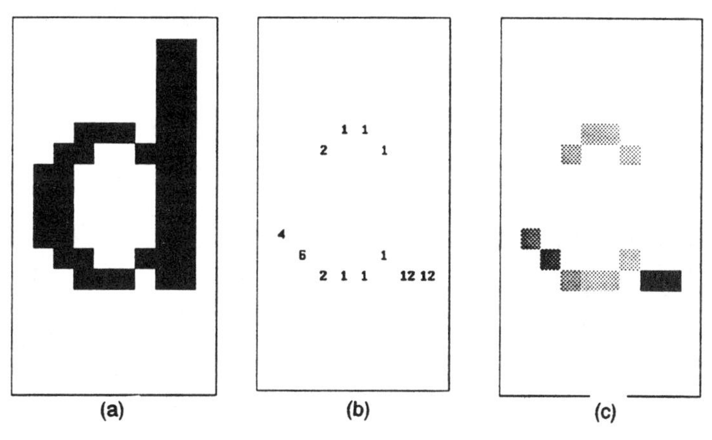

Figure 1. Creation of a burst image: (a) sample image; (b) burst image pixel values; (c) burst image as a grey scale image.

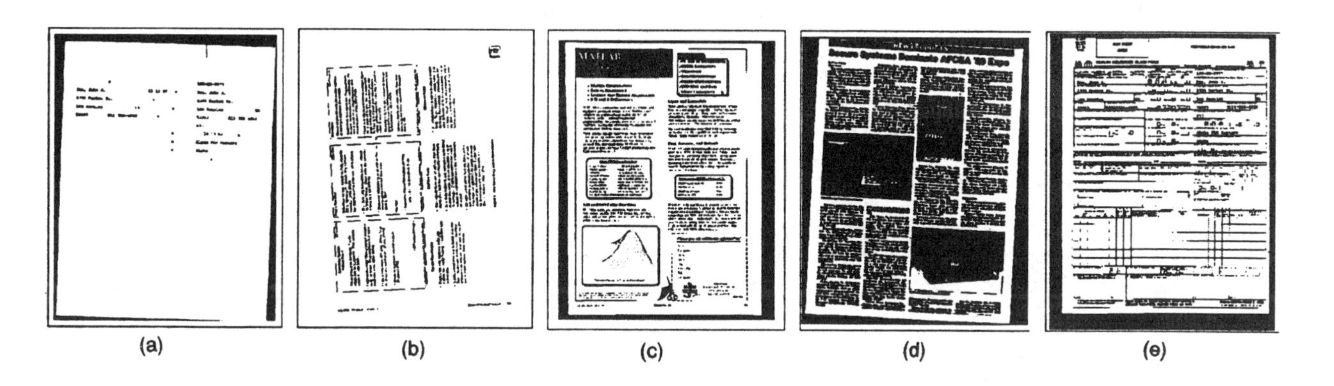

Figure 2. Examples of 5 document categories: (a) simple text; (b) multiple font text; (c) text and line drawings; (d) text and grey scale figures: (e) form

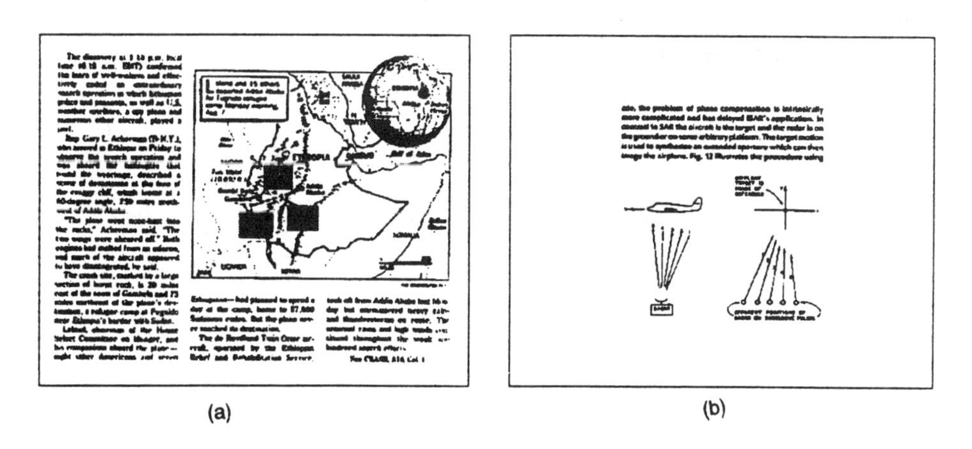

Figure 3. Document images used to test for detection of severe skew: (a) grey scale image tested at rotations of [-20,-10,0,10,20] degrees; (b) line drawing image tested at rotations of [0,45,60,75,90,105,120,135] degrees.

The Document Spectrum for Page Layout Analysis

Lawrence O'Gorman

Abstract—**Page layout analysis is a document processing technique used to determine the format of a page. This paper describes the document spectrum, or *docstrum*, which is a method for structural page layout analysis based on bottom-up, nearest-neighbor clustering of page components. The method yields an accurate measure of skew, within-line, and between-line spacings and locates text lines and text blocks. It is advantageous over many other methods in three main ways: independence from skew angle, independence from different text spacings, and the ability to process local regions of different text orientations within the same image. Results of the method shown for several different page formats and for randomly oriented subpages on the same image illustrate the versatility of the method. We also discuss the differences, advantages, and disadvantages of the docstrum with respect to other lay-out methods.**

Index Terms— **Document image processing, geometric page layout analysis, page segmentation, skew estimation, structural page layout analysis.**

I. Introduction

PREDICTIONS of the "paperless office," which were made so frequently during the early 1980's, are made far less frequently now. Perhaps it is the sense of futility engendered by the *increase* in volume of paper documents with the prevalence of computers (much of it directly due to computer output) rather than a decrease in volume. Perhaps it is the realization that paper is a legitimate form of media that is well suited for many purposes. Instead of the quest to remove all paper, a more realistic objective is to integrate it into our computer world. The ideal would be that it be as easily machine-readable as magnetic tapes and optical discs. If this were the case, paper documents would have one advantage over these other traditional computer media because they could be read by both man and machine.

There are two analyses necessary for reading documents: one is optical character recognition (OCR), and the other is page lay-out analysis. Advances have been made in OCR technology that enable text to be machine read with rates in the high 90% range. However, although text is an important part of a document, it is also essential to know where the text resides in a page. Page format establishes meaning to regions of text. For instance, when searching for a paper by author name, only the author block of the title page needs to be be examined. Names in the body of the text and in the references have different connotations. Therefore, determining the location of text by page lay-out analysis is an essential complement to OCR.

Manuscript received June 3, 1992; revised November 30, 1992. Recommended for acceptance by Associate Editor R. Kasturi.

The author is with AT&T Bell Laboratories, Murray Hill, NJ, 07974.

IEEE Log Number 9212306.

In this paper, the *docstrum* method for page lay-out analysis is described. This is advantageous over many other methods in three main ways. One is that analysis is independent of page orientation or skew. There is no need to find the skew and correct it prior to docstrum analysis; however, a precise measure of orientation is a byproduct of the analysis. The second is that the method does not require *a priori* input of character size and line spacing. Instead, these parameters are determined in the course of the analysis. The third is that this technique can be applied to an image containing subregions of different document characteristics. For instance, the docstrum can be used to segment independently oriented smaller documents (receipts, checks, index cards, business cards, etc.) in a single image.

II. Background

A. Definitions

Document lay-out analysis is performed to determine **physical structure** of a document, that is, to determine document components. These **document components** can consist of single **connected components**—regions of black (1-valued) pixels that are adjacent to form single regions—such as noise specks, dots, dashes, lines, and the parts of a character that are touching; and groups of connected components, or **blocks**, such as a complete character consisting of one or more connected components (e.g., i, é, ü), a word, text line, or group of text lines. A **text line** is a group of characters, symbols, and words that are adjacent, "relatively close" to each other and through which a straight line can be drawn (usually with horizontal or vertical orientation). The dominant orientation of the text lines determines the **skew angle**. A document originally has zero skew, but when a page is manually scanned or photocopied, nonzero skew may be introduced. Text may also have a chosen orientation that is different from the page skew (for instance, diagonally aligned text in diagrams and charts). We refer to text **orientation** rather than skew when we imply this more general case.

When document layout analysis is performed in a **top-down** manner, a page may be split into one or more column blocks of text, each column is split into paragraph blocks, and each paragraph is split into text lines, etc. Alternatively, analysis may be performed in a **bottom-up** manner, where connected components are merged into characters, then words, then text lines, etc, or both top-down and bottom-up analyses may be combined. **Document lay-out understanding** consists of **functional labeling** of blocks determined by layout analysis. These **functional components** are usually distinguished in some way by their physical features (such as by font size)

Reprinted from *IEEE Trans. Pattern Analysis and Machine Intelligence,* Vol. 15, No. 11, Nov. 1993, pp. 1162–1173.

and by the "meaning" of a block. Examples of functional components of a technical paper are title block, abstract, section titles, paragraphs, footnotes, etc.

B. Skew Estimation

Skew detection methods are described so that their differences can be contrasted to the docstrum method in Section V. We describe three categories of skew estimation techniques based on their approaches: 1) using the projection profile, 2) fitting baselines by the Hough transform, and 3) by nearest-neighbor clustering.

A commonly used method for skew detection employs the projection profile to iteratively compute skew angle [1]. The projection profile is a histogram of the number of black pixel values accumulated along parallel lines throughout the document. For a document whose text lines span horizontally, the horizontal projection profile will have peaks with widths equal to the character height, and the vertical projection profile will have wide peaks corresponding to each column of text. To determine the skew angle, the projection profile is computed at a number of angles close to the expected orientation, and for each angle, a measure is made of the total variation in the bin heights along the profile. The maximum variation corresponds to the best alignment with the text lines, and this determines the skew angle. Baird [2] proposes modifications to the projection profile method for very fast and accurate iterative convergence on the skew angle. Akiyama and Hagita [3] describe an approach that is very fast but less accurate. A document is divided into equal width vertical strips, horizontal projection profiles are calculated for each vertical strip, and skew is determined by measuring the average shift in zero crossings between strips. Pavlidis and Zhou [4] propose a method where alignment is measured of vertical projection profiles determined for horizontal strips of the page. A restriction of these methods is that they are limited to documents that have fairly small skews that are typically less that ±10°.

Srihari and Govindaraju [5] and Hinds *et al.* [6] propose similar skew detection methods based on the Hough transform. The Hough transform maps each point in the original (x, y) plane to all points in the (ρ, θ) Hough plane of possible lines through (x, y) with slope θ and distance from origin ρ. The dominant lines are found from peaks in the Hough space and thus the skew. One limitation of this approach is that if text is sparse, as for some forms, it may be difficult to correctly choose a peak in Hough space. The specified range of skew is limited to around ±15°.

A bottom-up technique for skew estimation based on nearest-neighbor clustering is described by Hashizume *et al.* [7]. In this work, the 1-nearest—neighbors of all connected components are found, the direction vectors for all nearest-neighbor pairs are accumulated in a histogram, and the histogram peak is found to obtain the dominant skew angle. This method has the advantage over the previous methods in that it is not limited to any range of skew. However, since only one nearest-neighbor connection is made for each component, connections with noise, subparts of characters (dot on"i"), and between-line connections can reduce the accuracy of this method. This is the most similar method to the docstrum method.

C. Lay-out Analysis

After skew detection, the image is usually rotated to zero skew, and lay-out analysis is performed. A major difference between the docstrum technique and the top-down techniques described here is that the former is not limited to Manhattan layouts, that is, layouts whose blocks are separable by vertical and horizontal cuts. We describe some top-down and bottom-up layout analysis approaches here so that their advantages and disadvantages can be contrasted with the docstrum in Section V.

Variants of the run-length smoothing algorithm (RLSA), which was first described by Wong *et al.* [8], are often used for lay-out analysis. The method merges characters into words, and words into text lines, and (sometimes) text lines into paragraphs by "smearing" the text to join characters into blobs. This is done by using smoothing filters whose sizes are based on intercomponent spacing. This method requires that the image is skew-corrected and that spacing is known and uniform within the image.

Nagy *et al.* [9], [10] use a top-down approach that combines structural segmentation and functional labeling. Horizontal and vertical projection profiles are used to split the document into successively smaller rectangular blocks. This segmentation is aided by performing functional labeling at the same time based on *a priori* knowledge of features of the expected blocks.

Baird [11] describes a top-down lay-out technique that first analyzes white space to isolate blocks and then uses projection profiles to find lines. As with all the top-down methods described above, the page must have a Manhattan layout.

Akiyama and Hagita [3] perform bottom-up lay-out analysis that employs both global and local text features as well as generic properties of documents in a similar fashion to Nagy's use of expected features, that is, the method yields both structurally and functionally labeled blocks; however, it is dependent on knowing features of the layout.

Fisher *et al.* [12] perform bottom-up segmentation that combines a number of the techniques described above. The skew is first found from the Hough transform, and then, between-line spacing is found as the peak of the 1-D Fourier transform of the horizontal projection profile. Within-line spacing is determined by finding the peak on a histogram of lengths of white spaces between black transitions along the skew angle. Merging of the text components is then done by a sequence of run-length smoothing operations in the directions of the skew and perpendicular to it. Esposito *et al.* [13] use a similar approach, except instead of operating with respect to pixels, they first determine character boxes and then accumulate projection profiles with respect to these.

III. The Docstrum: Formation and Analysis

A. Overview

In this paper, we describe the document spectrum, or *docstrum*. The docstrum is a representation of the document page that describes global structural features of the page and can be used for page analysis. It is based on bottom-up, k-nearest-neighbor clustering of connected components of the page and has some of the characteristics of the family of bottom-up techniques described above.

The k-nearest-neighbor pairings of connected component elements on the document page are shown in Fig. 1(b). (In addition, see [14] for a description of k-nearest-neighbors.) The k-nearest-neighbors to each component i are the k closest components $0 < j < k$, where closeness is measured by Euclidean distance in the image. Each nearest-neighbor pair $\{i, j\}$ is described by a 2-tuple $D_{ij}(d, \phi)$ of the distance d and the angle ϕ between centroids of the two components. For example, two characters in the same word form a nearest-neighbor pair whose distance is relatively small and whose angle is close to zero if its resident text line has zero skew. For each page element k, nearest neighbors are found. For $k = 5$ (which is the value we usually use), a character might make two or three pairings within a word and across word boundaries within the same line, as well as pairings with characters on upper and lower lines. These between-line pairings have larger distances than within-line pairs and angles that are approximately 90° apart. The docstrum is the plot of $D_{ij}(d, \phi)$ for all nearest-neighbor pairs on a page (see Figs. 2 and 3.) It is a polar plot with origin at the center; radial distance from this is d, and the counterclockwise angle from the horizontal is ϕ. The docstrum is so termed because of its similarity in appearance to the 2-D power spectrum and because of its analogous utility in globally describing an image—in this case a document image. Orientation and text line information can be determined directly from clusters on the docstrum plot. Text lines can be grouped to form blocks by the method to be described.

The resulting text lines and blocks from docstrum analysis are based on structural relationships of these entities. These results are used to perform functional labeling of document parts, perhaps entailing further merging of the structural blocks using application-specific information. We do not describe the functional labeling process here.

B. Preprocessing

Before the docstrum is determined, preprocessing is performed to get rid of noise and isolate document components. (We assume here that any diagrams or pictures have already been separated; therefore, this analysis is being applied to the textual parts of the page image only.) First, salt and pepper noise is reduced by applying the *kFill* filter [15] to the image to remove small regions of 0- or 1-valued pixels within text or background regions, respectively. This filter is designed for noise removal within binary text and removes spurs and smooths ragged edges on text and lines. The filter is designed specifically for binary text to remove noise while retaining text integrity, specifically to maintain corners of characters. Although the scale parameter of *kFill* can be set so that legitimate dots and periods are not removed if they are greater than a pixel in size, we usually do not require this. Instead, more noise is removed, including some dots and periods, and the loss of these two entities does not adversely affect docstrum analysis.

After *kFill* noise reduction, connected components are found. We use a technique that is different from traditional

:S AND 1 Incremental Mod
HEORY by John H. Mulle
STEMS 20 A Unified Frame\
by J. A. Brzozow

(a)

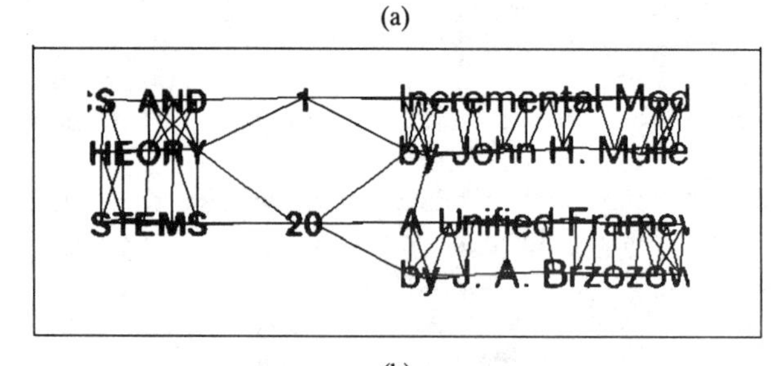

(b)

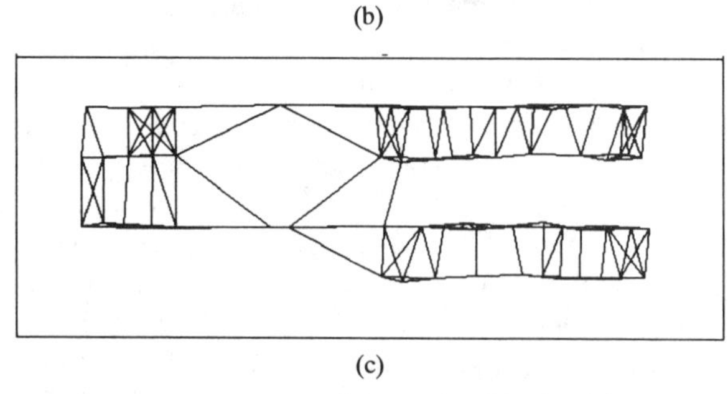

(c)

Fig. 1. Nearest-neighbor vectors for a portion of a table of contents image (the full page image is shown in Fig. 10) for k=5. The original image is in (a). In (b), overlays of the nearest-neighbor vectors are shown over the image. For each component, there are five vectors to the five components that are closest in Euclidean distance to the image plane. (Note that where less than five vectors can be seen, this is because two vectors overlay each other.) In (c), only the nearest-neighbor overlays are shown.

Fig. 2. Image of a title page (from the *IEEE Transactions on Pattern Analysis and Machine Intelligence) is used for the examples shown in Figs. 3 to 6.*

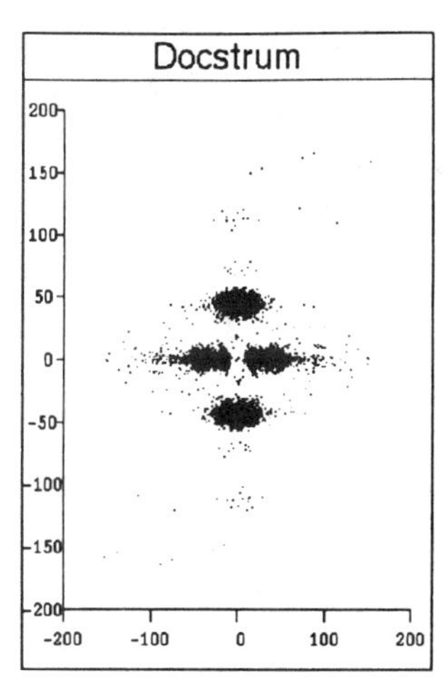

Fig. 3. Docstrum plot $D_{ij}(d, \phi)$ for the image in Fig. 2. Each point represents the distance and angle between a nearest-neighbor pair as the distance and angle from the origin at the center of the plot.

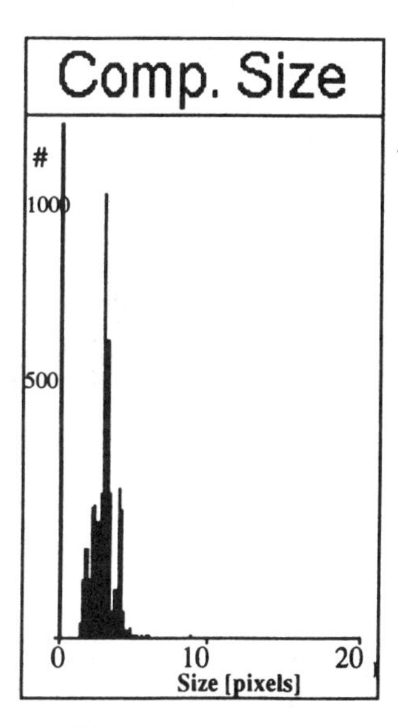

Fig. 4. Component-size histogram for the image in Fig. 2. The size is measured in pixel length as the square root of the bounding box area. The large characters of the title and small dots and noise have already been filtered out.

connected component analysis (which is described in more detail in [16] and [17]). Contours are first found within the image, and these are coded and represented by a line processing and analysis tool called thin line code (TLC). Features are calculated and stored for each TLC component, including bounding box size, contour length, moments, etc. Using these features, further noise reduction and segmentation are performed.

For pages containing text of different sizes, such as larger characters in the titles and smaller characters in the text body, the docstrum may be applied either to a subset of sizes or to each size separately. The reason for separately analyzing different ranges of component sizes before analysis is that the docstrum calculates average feature values, such as average within-line and between-line spacing, and then makes decisions based on these. The presence of a wide range of component sizes will increase the standard deviation of the averages, which may result in erroneous results. A histogram is made of component sizes, where size is that of the bounding box (Fig. 4). For a typical document, there is often a peak at small size for any remaining noise, a wider and higher peak for the distribution of characters of the predominant font size, and, often, a well-separated much smaller peak for title characters. From this data, peak detection can be done and the docstrum analysis confined to any or all of the separate component size groups. Note that this separation of different sizes need only be done for a large size range: the range of sizes between a capital letter, a small letter, and a subscript within an equation falls well within a single size range. (It should be noted that the bounding box measure is not independent of orientation; however, we have found this measure to be sufficient for our purposes because the bounding boxes of similarly shaped characters scale similarly for different orientations.)

Besides performing the separation of features in the preprocessing step, selective docstrum analysis is also performed later during docstrum analysis with respect to nearest-neighbor distances and angles. This is described in Section III-C, F, and I.

C. Nearest-Neighbor Clustering and Docstrum Plot

The first step in docstrum processing is to find the k-nearest-neighbors of each component as shown in Fig. 1. The value of k must first be chosen. Ideally, we would like to find neighbors to the right, left, above, and below each component (when those neighbors are present). This will give information on within-line and between-line spacing. We generally use a value of $k = 5$. For most components, this picks up those neighbors mentioned and an extra one for redundancy. The only disadvantage in choosing a larger value of k than the minimum is the extra computation time required. Other values of k may also be chosen for different purposes. For instance, if only text lines are ultimately desired, then between-line pairs are not needed, and $k = 2$ or 3 will be sufficient. Conversely, if text blocks are desired and between-line spacing is large compared with average within-line spacing, then larger k values are required. For example, if between-line spacing is greater than twice the average within-line spacing but less than three times, then a k value of 6 or 7 would be appropriate.

Because of the large number of components on a typical page (in the order of 1000 to 5000 for the page of a technical journal), the nearest-neighbor clustering step is the most time consuming of the nonpixel processing steps. To find the k-nearest-neighbors of each component in the most straightforward way requires $O(N^2)$ computation. We reduce this substantially by first sorting the components based on x position. Then, we find the closest k neighbors for expanding x distances until the x distance is greater than the kth closest neighbor already found.

Each nearest-neighbor pair i and j is related by distance d and the angle ϕ between components of the pair and is represented on the docstrum by $D_{ij}(d, \phi)$. We consider the vector connecting the pair of elements to be undirected; therefore, ϕ is quantized to $[0, 180°)$. For visual purposes in the docstrum plot, we replicate each point by its mirror point rotated 180°. Thus, the plot is symmetric with respect to the

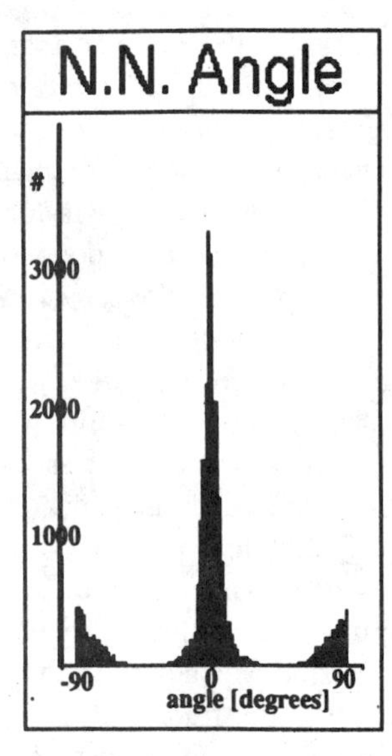

Fig. 5. Nearest-neighbor angle histogram for the image of Fig. 2. The angle range is from −90 to 90°.

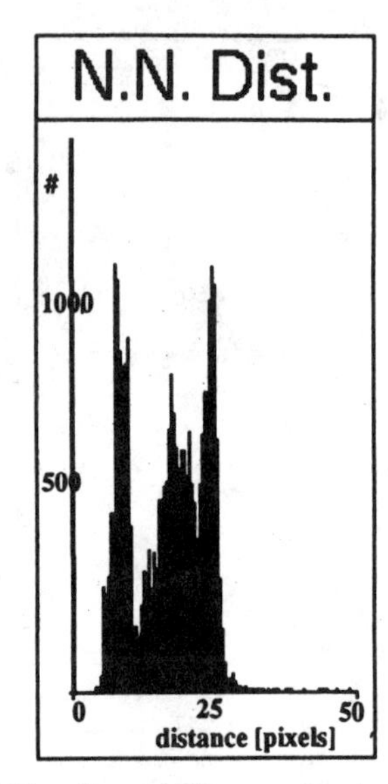

Fig. 6. Nearest-neighbor distance histogram for the image of Fig. 2. The distance is shown in pixels. Large distances are filtered out.

origin.[1] For typeset text, a number of features can be readily seen on the docstrum plot and related to the original page image. Typically, there are four to six clusters (depending on cluster separation) that form a cross pattern. For zero skew, there are two clusters each on the positive and negative x axes and one cluster each on the positive and negative y axes. Depending on the separation, the x axes clusters might appear instead as one cluster each. On the x axis, the clusters of lower distance and larger number indicates between-character spacing. The clusters of further distance and much smaller number are for between-word spacing. (We obtain the peak of one of these two clusters, usually intercharacter spacing, and refer to this distance as within-line spacing.) The y-axis clusters indicate between-line spacing. Although these features can be seen on the docstrum plot, they are not directly measured there. Instead, a particular sequence of analysis steps is performed for more reliable estimation. This is described next.

D. Spacing and Initial Orientation Estimation

Orientation, within-line spacing, and between-line spacing are estimated from the docstrum. Orientation and spacing are most easily seen when summation is carried out on the docstrum over the distance and angle variables, respectively. The histogram in Fig. 5 shows a distribution of nearest-neighbor angles for the docstrum plot of Fig. 3. The histogram in Fig. 6 shows a distribution of nearest-neighbor distances for the same docstrum plot.

Text orientation is first estimated from the nearest-neighbor angle histogram. The histogram is circularly smoothed, that is, the smoothing window is wrapped around at the ends since the angular data is circularly continuous. The peak is found from this smoothed data. To achieve a resolution of r_a°/bin, the histogram has $n_a = 180/r_a$ bins, corresponding to the angular range of [0, 180)°, and is smoothed by a rectangular smoothing window that is W_a% of the total range. For this initial estimate, we use a resolution of $r_a = 0.5$°/bin and a smoothing window of $W_a = 25\%$, giving a histogram length of $n_d = 360$ and a smoothing window length of 90. (This window shape and length are not critical because the histogram usually contains two well-separated, well-defined, and symmetric peaks. The window length should be less than 50% in order not to merge peaks and large enough to perform some smoothing in the sparsest expected histograms.) The location of the peak value is the estimate of average text line orientation or, simply, the orientation. This is an initial estimate; we find a more precise orientation measure later in the analysis.

Two spacing histograms are then compiled: one of nearest-neighbor distances for all angles within a specified angular range of the orientation estimate, which is called the within-line histogram, and the other for angles within the same angular range of the perpendicular to the orientation estimate, which is called the between-line histogram. To achieve a resolution of r_d pixels/bin, the histogram length is $n_d = (\max - \min)/r_d$, where the range of spacing values is [max, min]. The smoothing window is much more critical here than for the orientation estimate because peaks may be closely placed and are not usually symmetric. We define a tolerance that is smaller than the closest expected peak spacings t_s and make the window length a multiple of this $W_s = r_s(2t_s + 1)$. We use $r_s = 2$ pixels and $t_s = 2$ pixels. The peak found of the within-line histogram is an estimate of the character spacing; that from the between-line histogram is an estimate of the text line spacing.

E. Determination of Text Lines and Accurate Orientation Measurement

Although text line determination is synonymous with baseline determination in most of the literature, we find text

[1] It should be noted that angles are not inherently undirected. Just because component A has nearest-neighbor B does not necessarily imply the inverse; therefore, pair order is relevant, and it follows that ϕ quantization of [0, 360°) is also relevant. It has been suggested that for sparse columns of data or perhaps other formats, directedness may be important; however, we do not pursue this further in this paper.

:S AND 1 Incremental Mod
HEORY by John H. Mulle
STEMS 20 A Unified Framev
by J. A. Brzozow

(a)

:S AND 1 Incremental Mod
HEORY by John H. Mulle
STEMS 20 A Unified Framev
by J. A. Brzozow

(b)

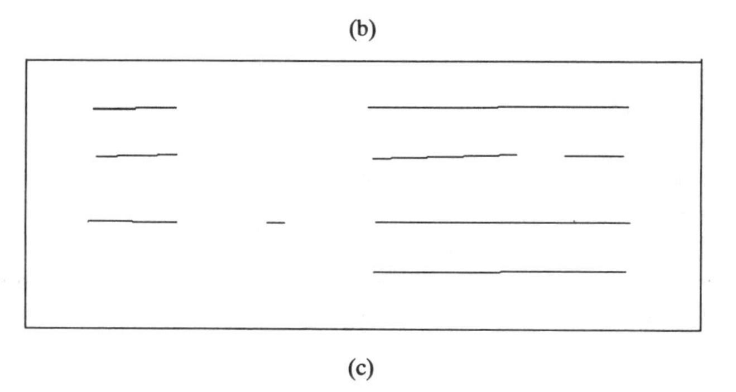

(c)

Fig. 7. Text lines for the same of a page shown in Fig. 1: (a) Original image; (b) overlays of the text lines; (c) only the overlays.

lines using connected component centroids instead. This is in keeping with our desired independence from any assumptions of page orientation, and this contrasts with the assumption in *base*line determination that the *bottom* of a character is known. If the baseline of each character could indeed be found at this point in the processing, this would yield better orientation estimates; however, the accuracy of baseline methods is reduced both by descenders and by skew. Of course, the accuracy of the docstrum orientation estimate via the centroids is also affected by descenders, as well as ascenders, but we show in Section IV that the method we describe is accurate and that its independence from orientation is advantageous.

We first perform a transitive closure on within-line nearest-neighbor pairings to obtain nearest-neighbor groups on the same text lines. A regression fit is then made to centroids of each group component to find text lines. This fits a straight line to the centroids in each group by minimizing the sum of squares of errors between the centroids and the line. Fig. 7 shows the text lines found for a portion of text.

From these text lines, we make our final estimation of orientation. This is more accurate than the initial estimate because the longer text lines substantially reduce the effects of descenders and noise.

F. Structural Block Determination

Structural blocks are groupings of one or more text lines grouped on the basis of spatial and geometric characteristics. To determine structural blocks, we extend a method first developed to find regions in contour line patterns [18], [19]. This method used three properties to determine if two lines are in the same pattern group: parallelness, perpendicular proximity, and overlap. For our application, the lines are text lines, and we extend this method by the addition of the property of parallel proximity. Therefore, if two text lines are approximately parallel, close in perpendicular distance, and they either overlap to some specified degree or are separated by only a small distance in parallel distance, then they are said to meet the criteria to belong to the same structural block.

The blocking method consists of examining pairs of text lines to determine if they should be in the same block, based on the criteria. If a pair of text lines meets the criteria, and if neither already belongs to a block, both are assigned to a new block. If one text line already belongs to a block and the other is unassigned, the unassigned text line is assigned to the block of the other. If both belong to different blocks, then these blocks are merged into one block. At the end of this process, each text line has been assigned to a block, and the blocks describe the structural layout of the page.

Although text lines, in general, are parallel to each other, our estimated text lines are only approximately parallel—to the degree of the length of the line and font size with respect to pixel resolution. For instance, a column of one- to three-digit page numbers in a table of contents page will have orientation precision that ranges from no precision for a single digit to better precision for two digits and, better yet, for three digits. Because our measured text lines are only approximately parallel, this leads to less straightforward calculation of structural blocking parameters.

Consider two text lines, which are shown as segments i and j in Fig. 8. These can be expressed as line segments with endpoints $\{(x_{Oi}, y_{Oi}), (x_{Fi}, y_{Fi})\}$ and $\{(x_{Oj}, y_{Oj}), (x_{Fj}, y_{Fj})\}$, respectively. The angular difference between the two segments is

$$\theta_{ij} = \left[\tan^{-1}\frac{\Delta y_j}{\Delta x_j} - \tan^{-1}\frac{\Delta y_i}{\Delta x_i}\right]_{180^\circ}$$

where $\Delta x = x_F - x_O$ and $\Delta y = y_F - y_O$, and $[\cdot]^\circ_{180}$ implies that this value is the smallest positive difference between angles such that $-90 \le \theta \le 90°$.

The overlap p_j is an approximation of the length that segment i would overlap onto j if both were parallel and i were translated perpendicularly onto j. The translation of point (x_{Oi}, y_{Oi}) to (x_{Aj}, y_{Aj}) onto the line collinear with segment j is given as

$$x_{Aj} = \frac{x_{Oi}\Delta x_i \Delta x_j + x_{Oj}\Delta y_i \Delta y_j + \Delta x_j \Delta y_i (y_{Oi} - y_{Oj})}{\Delta y_i \Delta y_j + \Delta x_i \Delta x_j}$$
$$y_{Aj} = \frac{\Delta y_j}{\Delta x_j}(x_{Aj} - x_{Oj}) + y_{Oj}. \tag{2}$$

Equation (2) is used when $\Delta x_j \neq 0$. If $\Delta x_j = 0$, then y_{Aj} is calculated first, and x_{Aj} is calculated from that. The translation of (x_{Fi}, y_{Fi}) to (x_{Bj}, y_{Bj}) is calculated similarly. The middle two coordinates then define the overlap segment,

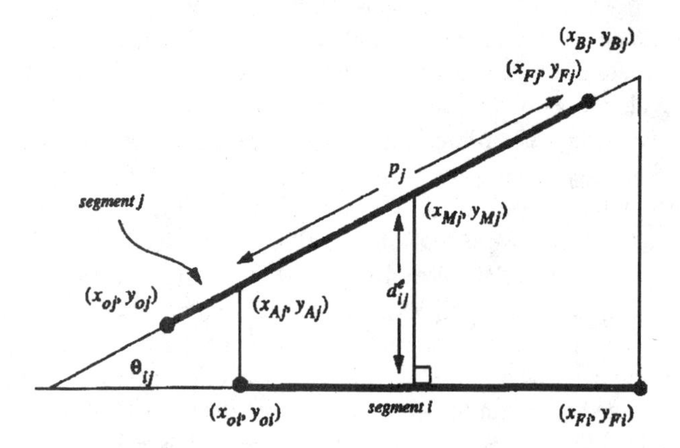

Fig. 8. Variables used in structural block determination. The two text lines, represented by segments i and j, are to be tested here to determine if they should be grouped into the same block. Their angular difference is θ_{ij}. The overlap length of segment i on segment j is p_j (and that is normalized to obtain the overlap parameter). The parallel distance between i and j is $d^a_{ij} = p_j$ in this case. The perpendicular distance betwen i and j is d^e_{ij}.

i.e., middle $\{(x, y)\} = \{(x_{Cj}, y_{Cj}), (x_{Dj}, y_{Dj})\}$. The middle coordinates are found

$$\text{middle}\{(x, y)\} = \{(x_{Cj}, y_{Cj}), (x_{Dj}, y_{Dj})\}$$
$$= \begin{cases} \{(x,y) | x \neq \min\{x\}, x \neq \max\{x\}\}, & \Delta x_j! = 0 \\ \{(x,y) | y \neq \min\{y\}, y \neq \max\{y\}\}, & \Delta y_j \neq 0 \end{cases}$$

where $\{x\} = \{x_{Oj}, x_{Fj}, x_{Aj}, x_{Bj}\}$ and $\{y\} = \{y_{Oj}, y_{Fj}, y_{Aj}, y_{Bj}\}$. From these middle points, the length of overlap is found:

$$p_j = \left[(y_{Dj} - y_{Cj})^2 + (x_{Dj} - x_{Cj})^2 \right] \frac{1}{2}.$$

These middle points are contained within both segments if they are overlapped, or they define a segment between them if they are not overlapped. We test for this and attribute positive overlap to the first case (overlapped) and negative overlap to the second case (nonoverlapped). This length is normalized by the length l_j of segment j between (x_{Oj}, y_{Oj}) and (x_{Fj}, y_{Fj}) to obtain the overlap parameter of segment i onto j

$$p_{ij} = \frac{p_j}{l_j} \text{or} - \frac{p_j}{l_j}.$$

The parallel distance d^a_{ij} between segments i and j describes the distance separating the closer opposite endpoints of each line, if they were collinear. This is the length of the overlap and is positive or negative depending on if there is or is not overlap, i.e.,

$$d_{ij}a = p_j; \text{ or } -p_j.$$

The perpendicular distance d^e_{ij} between segments i and j is the minimum, perpendicular, signed distance from segment i to the midpoint (x_{Mj}, y_{Mj}) of the overlap portion on segment j. The distance is calculated

$$d^e_{ij} = \begin{cases} \frac{(x_{Mj} - x_{Oi}) - (y_{Mj} - y_{Oi}) \Delta x_i / \Delta y_i}{(\Delta x_i^2 / \Delta y_i^2 + 1)^{\frac{1}{2}}}, & \Delta x_i, \Delta y_i \neq 0 \\ x_M j - x_O i, & \Delta x_i = 0 \\ y_M j - y_O i, & \Delta y_i = 0 \end{cases}$$

where the midpoint of the overlap portion is found $x_{Mj} = (x_{Cj} + x_{Dj})/2$, $y_M j = (y_C j + y_D j) \ / \ 2$.

G. Filtering

Filtering may be performed in docstrum analysis to eliminate components not removed during preprocessing or to iteratively apply the docstrum to separate component types of a single image. In practice, filtering is accomplished by flagging selected nearest-neighbor pairs on the basis of component size, nearest-neighbor distance, or nearest-neighbor angle whose value is out of the desired bounds so that docstrum processing is not performed on these. In this way, flags can be turned on and off for different feature values, and docstrum analysis can be performed separately on each range of values. For instance, if larger-size headline text were not eliminated during preprocessing, it could be analyzed separately on the basis that its average within-line spacing is larger than that for the body of the text. In addition, if there were a single angled line of text in a diagram, this could be analyzed separately from the body of text based on its different within-line angle.

H. Global and Local Lay-out Analysis

The analysis described thus far has pertained to global layout features, that is, the assumption has been that the analyzed image contains a single formatted page with one orientation. One of the unique features of the docstrum method is the ability to perform multiple local analyses of different layouts on a single image. There are two general cases where this is useful. One is when the image is of multiple, randomly oriented smaller documents, or "scraps," for instance, news clippings, credit card pages, business cards, store receipts, subparts of a larger document, etc. The other is when one or more regions of the same image are formatted at a different orientation. A common example of this is for mileage charts, where city names are written horizontally along the left column but rotated 90° to vertical along the top row. Diagrams also may contain angled or perpendicularly oriented text lines.

In contrast with the layout methods, where skew must be precomputed and lay-out features determined by global processing, the docstrum's bottom-up k-nearest-neighbor approach facilitates local lay-out analysis. The method is straightforward. After k nearest neighbors are found for all components, orientation is determined separately for each group of connected nearest neighbors. Components are said to be in the same group if a path can be found from one component to the other via the nearest-neighbor connections. Block determination is performed on all groups of components with the same approximate orientation. For instance, if an image consists of three scraps (two at the same orientation and one at a different orientation), then one or more groups would be found for each scrap. Blocks would be found for groups within the two scraps that are at the same orientation, and blocks would be found separately for the groups within the other scrap at a different orientation. (Continuing with this same application but following docstrum analysis, data-dependent information can be used (for the business card application, this is the business card dimensions) to group blocks within each scrap). An example of local analysis is shown in Section IV.

This method works well when groups within different scraps are well separated, that is, when the kth-nearest-neighbor does

TABLE I
RESULTS OF DOCSTRUM ANALYSIS FOR IMAGES SHOWN IN FIGS. 9 TO 13. ORIENTATIONS ARE GIVEN IN DEGREES. DISTANCES ARE GIVEN IN PIXELS (WHERE EACH PIXEL SPACING IS 1/300 IN). PROCESSING TIME IS GIVEN IN SECONDS

example	no. of components	no. of pairs	filtered pairs	prelim. orient	final orient.	within-line distance	between-line distance	no. of segments	no. of blocks	process. time
PAMI article	5021	25 105	13 147	0.6	0.1	18.2	49.2	426	18	65
JACM toc	1433	7165	4147	1.2	0.2	16.7	42.2	64	40	8
Spectrum toc	1435	7175	3780	2.4	-0.7	17.5	45.2	248	64	11
PAMI toc	1457	7285	4866	-0.6	1.5	18.1	49.4	93	49	10
cards	1505	6700	2718	-21,9	-22,16	13,25	29,67	80	6	9

not extend outside of a single scrap. When this is not the case, either the value of k should be reduced, or a filter should be applied to nearest-neighbor distance to form nearest neighbors only if the distance is less than some value that relates to the scrap size.

H. Parameters

In this section, we describe the parameter values that can be changed by the user in our implementation and their default values. The component size (area of bounding box in pixels) can have low- and high-pass thresholds. The default lowest size is 3, and the default highest size is 3 times the peak size found during analysis. The procedure can also be operated recursively for all size ranges specified. The default number of nearest neighbors is $k = 5$. If only text lines are important, that value can be reduced to 2 or 3. If between-line distances are large and blocking of text lines is desired, k should be increased such that both within-line and between-line nearest-neighbor connections are made. All processing is done with respect to the orientation angle that is estimated by analysis. There is a tolerance of ±30° around this angle, within which connections are said to be within line. This angular range can also be changed by the user. The nearest-neighbor distance can have low and high bounds as well. In the default mode, the lowest distance is zero, and the highest distance is the lesser of 3 times the within-line distance determined in the analysis or $\sqrt{2}$ times the between-line distance that is also determined at runtime. For most of our applications, the default values of these parameters are appropriate.

For the final stage of block formation—after text line determination—user parameters are more likely to be required due to the wide variety of page formats. The default maximum perpendicular distance between text lines for blocking is 1.3 times the between-line distance found by the analysis. The default maximum parallel distance between ends of text lines for blocking is 1.5 times the within-line distance, which is also found by the analysis. Both of these can be changed by the user either in relative mode (relative to spacings determined by the analysis) or in absolute mode by specifying distances in inches. An option that is available if only text lines are required or if the desired functional blocks cannot be separated structurally is to set the perpendicular distance blocking parameter value to 0 such that blocking does not occur between text lines. The overlap parameter value is set to zero (no overlap is required in the default mode), and the value of the parallelness parameter is set to the same angular range as for the orientation tolerance; the default is 30°. The final choice available to the user is to perform the analysis in global or local mode. Global mode is the default and is for the common case where the entire image has the same skew angle. Local mode is used for subimages of multiple skew angles.

IV. EXAMPLES

The docstrum has been used to segment hundreds of images (mainly of scanned journal pages for the RightPages electronic library system [20] and of business cards for a system built on the local mode docstrum to extract multiple scraps from a single image). In the RightPages system, there are issues from about 40 journals and 15 different publishers, covering over a period of about 2 yr; these represent a wide variety of page formats. The nature of images presented to the local mode docstrum also represent a wide variety of formats. These are the images on which the docstrum has been applied. In this section, we present typical examples of docstrum use and show results. In Section V, we discuss how well the docstrum has performed over these and other images analyzed and describe the conditions for which layout errors result. This information should be used to judge the appropriateness of the docstrum for formats both presented and not presented here.

For the following examples, results are tabulated in Table I. This first shows the number of components found on the page after some of the noise has been removed during preprocessing. The number of nearest-neighbor pairs is given for the chosen number of nearest-neighbors per component of $k = 5$ then given after filtering out those that were at angles outside ±30° of the estimated orientation and that were longer than three times the within-line spacing (all default values). The page orientation is given both for the preliminary estimated analysis and for the final, more accurate analysis. Within-line and between-line spacing values are shown. The number of regression segments and the final number of text blocks are given. Finally, the computation time for docstrum analysis on a Sun Microsystems SparcStation 2 is given. Note that when two numbers are given, this indicates a range of values for multiple local docstrum analyses.

The results are shown in Figs. 9 to 13. In each of these figures, subsampled versions of the original 300 dots/in images are shown at the left, and the blocks found by docstrum analysis are shown at the right. Although the text in these reduced-sized original images is not readable, the objective is to show the structural blocks on the page. For each of these examples, the smaller text of the body of the document comprises the components of interest, and the larger title text is first filtered out. The docstrum analysis could be applied in

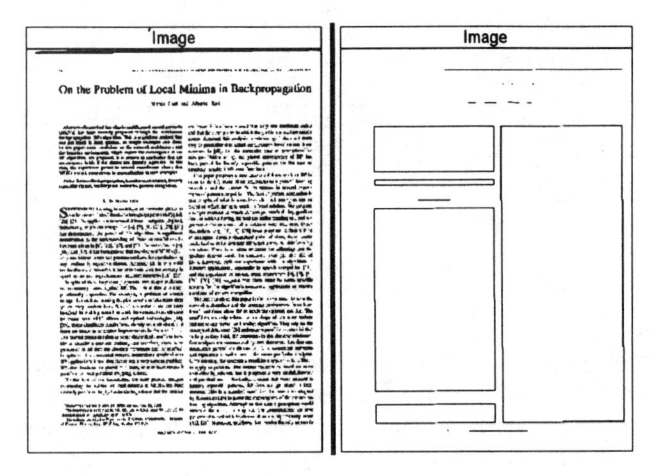
Image
Image
On the Problem of Local Minima in Backpropagation

Fig. 9. Original of a journal title page (same as in Fig. 2) is shown on the left, and its blocks found by docstrum analysis are shown on the right.

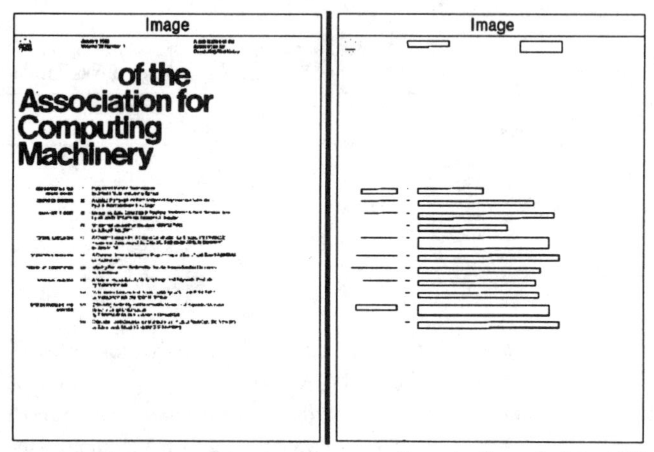
Image
Image
of the
Association for
Computing
Machinery

Fig. 10. Original of a table of contents page (from the *Journal of the ACM*) is shown on the left, and its blocks found by docstrum analysis are shown on the right. (Note that the word "Journal" was lost during binarization. It is white on a light-colored background, and the rest of the text is black.)

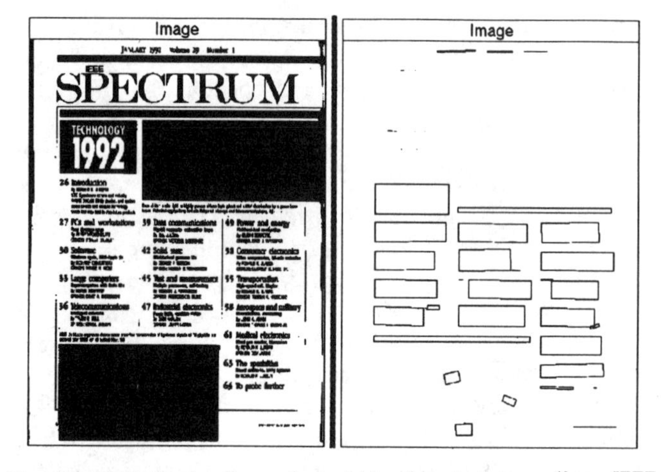
Image
Image
IEEE
SPECTRUM
TECHNOLOGY
1992

Fig. 11. Original of a three-column table of contents page (from *IEEE Spectrum*) is shown on the left, and its blocks found by docstrum analysis are shown on the right.

the same way to only the title text or to both sizes of text; however, the latter would have to be done in two iterations for the reason given in Section III-B.

The first example, which is shown in Fig. 9, is for the first page of a technical paper from the *IEEE Transactions on Pattern Analysis and Machine Intelligence*. The text of the document body constitutes the larger peak in the size histogram, and it is this size group for which analysis results are shown. In reading order, blocks are found for the following: page number, issue information, author names, abstract, keywords, section title, first column text block, footnote information, second column text block, and order number at the bottom. Although the bulk of the page has been segmented well, the author names have been segmented into four blocks. This is due to the slightly larger font size and word spacing for the authors names than the rest of the text. This would be corrected in the functional labeling stage. Note in Table I that the computation time is significantly larger than for the other examples due to the larger number of components in this article page.

Fig. 10 shows a table of contents image for the *Journal of the ACM*. Characters of larger font size in the journal title are filtered out in the preprocessing step. The three blocks at the top on the right image are for the ACM logo, the date and issue information, and the publisher information, respectively. The table of contents information is arranged with the subject heading (for one or more articles) in the left column, the page number in the middle column, and the title and author information in the blocks of the right column.

Fig. 11 illustrates a three-column table of contents for *IEEE Spectrum*. Each table of contents entry is found as a block that includes the page number, title, subtitle and author names. The two wide, short blocks are for figure captions. Note that this image has not been cleaned of noise, and neither have some of the others, and some extraneous small blocks are found. One reason for this is the presence of noise in the diagram regions, which were not presegmented. Another reason is that many of the characters are actually touching (the issue scanned here is January 1992). For the most part, blocks are correctly found here. For the few extraneous blocks, these can be recognized using journal-specific format information during functional labeling.

The table of contents page in Fig. 12 for the IEEE TRANSACTIONS ON PATTERN ANALYSIS AND MACHINE INTELLIGENCE contains article entries with equidistant spacing both within and between them. Since structural analysis alone cannot separate entries due to this equidistant spacing, blocking is performed only to join breaks within text lines and not to group between text lines. This is done by setting the perpendicular distance blocking parameter to zero. (This is the only nondefault parameter value used in these examples.) In the figure, each entry has a title, followed by dots, the author names, and page number aligned to the right margin. The separated page number text lines can be used with journal-specific information to segment each entry. In addition, note in this figure that filtering has been performed to get rid of the large font size of the journal title and the small size of the dots within each entry.

The final example in Fig. 13 illustrates local docstrum analysis. Six business cards have been scanned at various orientations on the same image. The original image is shown at the top left. The top right and bottom left images are of the nearest-neighbor connections and text lines, respectively. The

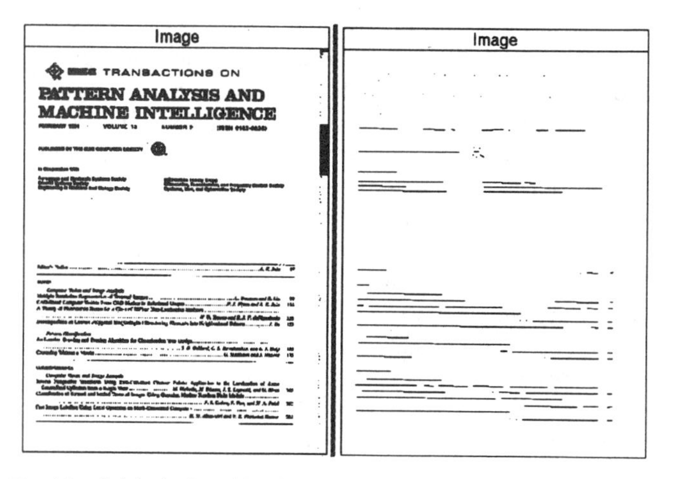

Fig. 12. Original of a table of contents page (from *IEEE Transactions on Pattern Analysis and Machine Intelligence*) is shown on the left, and its text lines found by docstrum analysis are shown on the right.

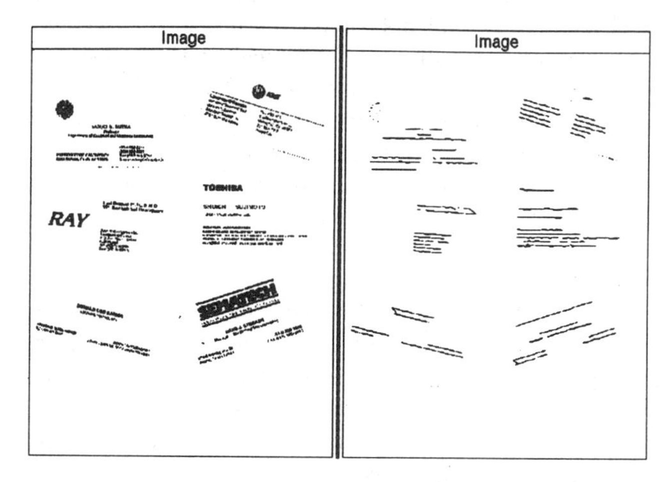

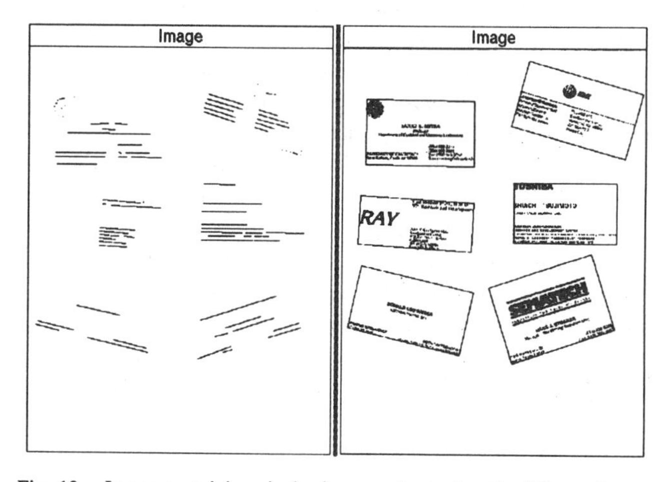

Fig. 13. Image containing six business cards, each with different formats and different orientations, is shown at the top left. The top right image shows the nearest-neighbor vectors. The bottom left image shows the text lines. The image at the lower right shows the blocks.

final image shows each block corresponding to the business cards. In Table I, the range of final orientations is given as -22 to 16° from the horizontal. After docstrum blocking, task-specific information of business card dimensions is used in this example for determining the card boundaries shown.

V. Discussion

In this section, we discuss the advantages and disadvantages of the docstrum method. We compare these with the other methods described in Section II.

One of the most important features of any algorithm is robustness with respect to input parameters. A feature of the docstrum is that spacing parameters are not required from the user. Instead, the docstrum automatically determines dominant spacing from peaks on the histogram for nearest-neighbor distances and then uses multiples of these for text line and block detection. This differs from methods that employ run-length smoothing via a fixed-sized filter window. (Note that the method described in [12] is an exception to this because the window size is a function of the character spacing that is automatically determined.) This also differs from the top-down methods where these global spacings are not automatically calculated during analysis.

Another important feature of the docstrum is independence from page orientation. One ramification of this is that skew correction need not be performed before docstrum analysis, unlike the top-down methods in Section II. Another ramification is that since most skew detection techniques are limited to a small range of skew around 0° and the docstrum is not, the docstrum is more versatile in this respect.

The method is relatively tolerant with respect to randomly distributed noise in the image. When some nearest-neighbor connections are made to noise, this usually does not affect measurement of global parameters, and if the number of nearest-neighbors is chosen large enough for some redundancy, then there is usually not a problem of gaps in text lines. An important case where noise is very destructive to the analysis is when it is not randomly distributed. For instance, we have encountered several examples where salt-and-pepper noise occurs in a line up the gutter of a book or journal being scanned. Since this noise is along a vertical line, the peak orientation can be erroneously found 90° to the correct orientation. Even though most material we scan for the RightPages project is clean, we prefer to err on the side of caution and preprocess all material as described in Section III-B.

The docstrum can be applied to text containing tables and equations as well. For tables, the larger separating lines of the table are first filtered out with the size histogram. Then, the same nearest-neighbor analysis is performed. Note that if horizontal spacing is larger than vertical spacing, the text lines will be found to be vertical columns. For equations, the major differences from regular text are the presence of subscripts and superscripts and the differently sized characters. Depending on separating distances, separate short text lines may be found for the subscript and superscript expressions (such as the bounds above and below a summation sign). Since these have the same orientation as the rest of the text, they are merged into the same blocks in the structural block determination step. As far as the differently sized characters in an equation, they are not of great enough deviation from the rest of the text size to affect the results.

One disadvantage of bottom-up versus top-down methods is that they are generally more computationally expensive. It has been shown in Table I that the method is slower with more image components. Although this is true, comparison against other methods must take into account the following. One is that the docstrum requires no previous skew correction because orientation is found as part of the analysis, unlike top-down methods. Another is that many of the skew estimation methods are iterative, requiring more iterations with larger skew or greater precision. The docstrum does not iterate to determine skew. Because the docstrum analysis measures spacing as part of the analysis, this precludes the need for some other automatic process to do so. Furthermore, all docstrum processing is done on connected components rather than pixels. Many of the other methods do the same, but for the ones that do not, pixel access is more time consuming because the ratio of pixels to components is on the order of 10^3.

The docstrum operates on ranges of global spacing values to obtain the nearest-neighbor pairs. Because larger size title text also has longer inter-character distances than smaller text on the same page, the title text and body text must be analyzed separately. This is unlike some of the top-down methods based on projection profiles where these are handled on the same step. However, performing multiple analyses of the docstrum does not require much additional time since one of the steps (for the title text) usually deals with much sparser data.

Since the docstrum measures document features via nearest-neighbor pairs, it is essential that characters be separated. For the same reason, this method is also inappropriate for joined characters of handwriting. This is true of the other bottom-up methods but different from methods using run-length smoothing.

The use of the centroid of components for estimating orientation and spacing is different from the other methods use of the "bottom" point of the bounding box. The use of the centroid enables the docstrum to be independent of orientation. Because both ascenders and descenders affect the centroid location versus just descenders for the baseline methods, initial skew estimation via docstrum nearest-neighbor connections is slightly less precise than for baseline methods for pages with small skew. However, the regression fits to text lines improve precision. Furthermore, with larger skew, the baseline methods lose precision, whereas the docstrum centroids are independent of skew. We have not undertaken a comparison of skew precision for different methods in this paper, but that would be useful future work.

Although we have found docstrum analysis up to the stage of text line detection to be very robust, block segmentation is less so. For instance, some pages are formatted with spacing between functional blocks that shrinks and expands with the amount of text on the page. When the spacing shrinks to the same size as the between-line spacing within a block, structural layout analysis alone is insufficient to perform segmentation. For these cases, we use the docstrum to the text line stage and then employ page-specific format information to perform functional segmentation. Of course, there is also a problem when an expected format changes rules. This is handled not at this stage but in the following functional labeling stage, where results inconsistent to the expected format are flagged.

As far as the feature of the docstrum of being able to perform multiple analyses on blocks of different orientations, it may be argued that most pages contain rectangular regions of singular skew. This point is quite valid; however, we have mentioned some applications where this multiorientation capability is useful. A drawback of the local versus global mode is that measurement of skews and spacings are less precise than when information from the entire page is used. This is true, but we have found that precision is adequate for the cases tested. Although this mode of analysis is not the most common, it speaks well of the versatility of the algorithm that it can handle such cases. None of the methods described in Section II have been shown to handle these. One disadvantage of this local docstrum mode is that the user must be careful that components of different orientation groups are not "too close" such that nearest neighbors are formed across these groups. This means it should be assured that the maximum distance of the kth nearest neighbor should be within the same orientation group. As mentioned in Section III-H, k may be reduced, or a filter may be put on the maximum length of the nearest-neighbors found such that connections outside of an orientation group are not made. In practice, such as for the multiple business cards application of Fig. 13, the user is told to space the cards at least a half card distance away from adjacent cards.

VI. Summary

The docstrum has been presented as a method for page layout analysis. It proceeds in a bottom-up manner by forming k nearest-neighbor pairs between image components, automatically estimating text orientation and spacing parameters, and forming text lines and blocks. Three characteristics of the method have been described: independence from skew range, independence from different text spacings, and ability to process local regions of different text orientations within the same image. It has been shown by example how the docstrum is used on different page formats. Finally, there is a discussion of the differences, advantages, and disadvantages among several families of skew detection and layout analysis methods, pointing out how the docstrum compares with each.

References

[1] W. Postl, "Detection of linear oblique structures and skew scan in digitized documents," in *Proc. 8th Int. Conf. Patt. Recogn. (ICPR)* (Paris), Oct. 1986, pp. 687–689.

[2] H. S. Baird, "The skew angle of printed documents," in *Proc. Conf. Soc. Photog. Scien. Eng.* (Rochester, NY), May 1987, pp. 14–21.

[3] T. Akiyama and N. Hagita, "Automated entry system for printed documents," *Patt. Recogn.*, vol. 23, no. 11, pp. 1141–1154, 1990.

[4] T. Pavlidis and J. Zhou, "Page segmentation by white streams," in *Proc. First Int. Conf. Document Anal. Recogn. (ICDAR)* (St. Malo, France), Sept. 1991, pp. 945–953.

[5] S. N. Srihari and V. Govindaraju "Analysis of textual images using the Hough transform," *Machine Vision Applications*, vol. 2, pp. 141–153, 1989.

[6] S. C. Hinds, J. L. Fisher, and D.P. D'Amato, "A document skew detection method using run-length encoding and the Hough transform," in *Proc. 10th Int. Conf. Patt. Recogn. (ICPR)* (Atlantic City, NJ), June 1990, pp. 464–468.

[7] A. Hashizume, P. -S. Yeh, and A. Rosenfeld, "A method of detecting the orientation of aligned components," *Patt. Recogn. Lett.*, vol. 4, pp. 125–132, 1986.

[8] K. Y. Wong, R. G. Casey, and F. M. Wahl, "Document analysis system," *IBM J. Res. Development*, vol. 6, pp. 642–656, Nov. 1982.

[9] G. Nagy and S. Seth, "Hierarchical representation of optically scanned documents," in *Proc. 7th Int. Conf. Patt. Recogn. (ICPR)* (Montreal, Canada), 1984, pp. 347–349.

[10] G. Nagy, S. Seth, and M. Viswanathan, "A prototype document image analysis system for technical journals," *IEEE Comput.*, Special issue on Document Image Analysis Systems, pp. 10–22, July 1992.

[11] H. S. Baird, S. E. Jones, and S. J. Fortune, "Image segmentation using shape-directed covers," in *Proc. 10th Int. Conf. Patt. Recogn. (ICPR)* (Atlantic City, NJ), June 1990, pp. 820–825.

[12] J. L. Fisher, S. C. Hinds, and D. P. D'Amato, "A rule-based system for document image segmentation," in *Proc. 10th Int. Conf. Patt. Recogn. (ICPR)* (Atlantic City, NJ), June 1990, pp. 567–572.

[13] R. Esposito, D. Malerba, and G. Semeraro, "An experimental page layout recognition system for office document automatic classification: An integrated approach for inductive generalization," *Proc. 10th Int. Conf. Patt. Recogn. (ICPR)* (Atlantic City, NJ), June 1990, pp. 557–562.

[14] R. O. Duda and P. E. Hart, *Pattern Classification and Scene Analysis*. New York: Wiley, 1973, pp. 103–105.

[15] L. O'Gorman, "Image and document processing techniques for the RightPages electronic library system," in *Proc. 11th Int. Conf. Patt. Recogn. (ICPR)* (The Hague, The Netherlands), Aug. 1992, pp. 260–263.

[16] L. O'Gorman, "Primitives chain code", in *Progress in Computer Vision and Image Processing* (A. Rosenfeld and L. G. Shapiro, Eds.). San Diego: Academic, 1992, pp. 167–183.

[17] H. V. Jagadish and L. O'Gorman, "An object model for image recognition," *IEEE Comput.*, vol. 22, no. 12, pp. 33–41, Dec 1989.

[18] L. O'Gorman and G. I. Weil, "An approach toward segmenting contour line regions," in *Proc. 8th Int. Conf. Patt. Recogn.* (Paris), Oct. 1986, pp. 254–258.

[19] M. Seul, L. R. Monar, L. O'Gorman, and R. Wolfe, "Morphology and local structure in labyrinthine stripe domain phases," *Sci.*, vol. 254, Dec. 13, 1991, pp. 1616–1618.

[20] G. A. Story, L. O'Gorman, D. Fox, L. Schaper, and H. V. Jagadish, "The RightPages: An electronic library for alerting and browsing," *IEEE Comput.*, pp. 17–26, 1992.

Lawrence O'Gorman received the B.A.Sc. degree form the University of Ottawa, Ontario, in 1978, the M.S. degree from the University of Washington, Seattle, in 1980, and the Ph.D. degree from Carnegie-Mellon University, Pittsburgh, PA, in 1983, all in electrical engineering.

From 1980 to 1981, he was with Computing Devices Company, Ottawa, Canada, where he worked on digital signal processing and filter design. He has been at AT&T Bell Laboratories, Murray Hill, NJ, in the Computing Systems Research Laboratory since 1984. His research interests include image processing, pattern recognition, document image analysis, and machine vision precision. Lately, he has been involved in the design of the RightPages electronic library project.

Page Segmentation and Classification

THEO PAVLIDIS AND JIANGYING ZHOU

Departments of Computer Science and Electrical Engineering, SUNY, Stony Brook, New York 11794-4400

Received October 22, 1991; accepted July 31, 1992

Page segmentation is the process by which a scanned page is divided into columns and blocks which are then classified as halftones, graphics, or text. Past techniques have used the fact that such parts form right rectangles for most printed material. This property is not true when the page is tilted, and the heuristics based on it fail in such cases unless a rather expensive tilt angle estimation is performed. We describe a class of techniques based on smeared run length codes that divide a page into gray and nearly white parts. Segmentation is then performed by finding connected components either by the gray elements or of the white, the latter forming white streams that partition a page into blocks of printed material. Such techniques appear quite robust in the presence of severe tilt (even greater than 10°) and are also quite fast (about a second a page on a SPARC station for gray element aggregation). Further classification into text or halftones is based mostly on properties of the across scanlines correlation. For text correlation of adjacent scanlines tends to be quite high, but then it drops rapidly. For halftones, the correlation of adjacent scanlines is usually well below that for text, but it does not change much with distance.

1. INTRODUCTION

When a document is to be processed by a character recognition system (OCR) it is necessary to separate text from halftones and line drawings so that time will not be wasted in attempting to interpret the graphics as text. In addition, columns of text must be identified so that the conversion is done in the right order rather than in the direction of the scan. (If a page has two columns of text and the OCR system does not know about it, it will read across the page and produce an output where the columns are interleaved.) The term *page segmentation* refers to the partitioning of a document page into columns and areas of graphics, while the term *page classification* refers to the labeling of such parts as text, halftones, line drawings, etc. The term *zoning* is often used in the OCR industry to describe both processes when the partition and classification are performed manually, and the term *auto-zoning* refers to the processes described in this paper.

[1] Some authors use the right rectangle assumption loosely, requiring only that all areas be unions of right rectangles.

In spite of considerable past research, most current commercial systems rely on manual segmentation and classification, or they require very careful document positioning for automatic processing. Both constraints are ergonomically undesirable because human intervention defies the advantages of an automated document processing system. It seems that the lack of practical systems is due to the reliance of most earlier work on the heuristic that text columns and printed images are right rectangles.[1] While this assumption is often true on the original document, it is not true for the scanned document because of tilt (skew) during scanning. A few degrees of tilt are inevitable whether documents are fed mechanically or by a human operator. This fact destroys the regular rectangle assumption for the scanned document. Significant research has been devoted in estimating the tilt angle of a document. Such methods rely on variations of the Hough transform and tend to be quite expensive computationally [1]. They are impractical for document conversion systems that consist only of software to be run on a general purpose personal computer or workstation. Classification techniques that rely on the difference in spectral characteristics between text and halftones [2] suffer from the same disadvantage. It should be pointed out that the problem of page segmentation is quite different from the problem of text/graphics separation in engineering drawings [3]. In the latter case text labels and line drawings are mixed, while in the former different types of material are segregated. In inexpensive OCR systems, text labels within images or line drawings are normally ignored. The emphasis is on obtaining quickly the main text for subsequent character recognition while other material is stored in bitmap form.

A practical page segmentation and classification system must meet two requirements: (a) it must rely on a property of page parts other than right rectangles in order to be insensitive to tilt; and (b) for the separation of text from graphics it must rely on a property that can be computed quickly and locally (i.e., without having a full block of the part to be classified).

A candidate for (a) becomes apparent if one holds a document page at some distance (or tries to read it without glasses). Areas of text and images appear as more or

"Page Segmentation and Classification" by T. Pavlidis and J. Zhou from *CVGIP: Graphical Models and Image Processing,* Vol. 54, No. 6, Nov. 1992, pp. 484–496.

less uniform gray regions. Gray areas are those that contain relatively narrow white intervals and therefore are easily identified from a run length code representation of the scanned page. The idea has been explored but only within the right rectangle assumption [4]. The term *smeared* run length code was used to describe runs where narrow white segments had been absorbed by their neighbors. If we decide to segment a page according to this model, then we have two choices. One is to perform *region growing* and identify connected components characterized by the absence of wide white intervals. The other is to perform *edge detection* and find connected runs of wide white intervals which can be used to subdivide the page. A method based on region growing is described elsewhere [5]. Here we will focus on an approach based on edge detection. We use the term *white streams* to describe the runs of wide white intervals used as separators. The idea to use white spaces as the basis for segmentation has also been used before [6], but only within the context of the right rectangle assumption.

A candidate for (b) is offered by the correlation between adjacent scanlines (or parts thereof: for example, the segments between wide white intervals). For text one expects high correlation between adjacent scanlines which drops fast as the distance between the scanlines increases. (Line drawings should have the same property but they have a much lower overall density than text so they are easily separated that way.) For halftones the correlation between adjacent scanlines is usually much lower than that for text, and furthermore, the correlation does not drop as fast. In fact, it exhibits a periodicity inherent in most of the methods used for halftoning (dither matrices in particular).

Because all the processing is local and linear (each part of a page image is processed only a fixed number of times) the methods are quite fast. A preliminary implementation in C (without any effort to optimize the code for speed) of the method described in this paper required an average of 1.5 s per page on a SPARC station. (For 10 pages the range was from 0.9 to 1.9 s.) Similar times have been obtained for the method based on region growing.

2. REVIEW OF EARLIER TECHNIQUES

Most methods described in the literature may be divided into three board categories: *top-down* (or, model-driven), *bottom-up* (or, data-driven), and *hybrid*. For a broad survey of early work, see Nadler [7].

2.1. *Top-Down Strategy*

A top-down control strategy typically starts by hypothesizing a series of interpretations at high level (e.g., that the page has a header above two columns) and attempts to verify each by a search of a tree of implied hypotheses at lower levels of detail, finally consulting evidence at the lowest level (e.g., characters or pixels). The tree search is typically depth-first and fully backtracking. The document is divided into *major* regions (e.g., text paragraphs) which are further divided into subregions. One such method is the run-length smearing technique proposed by [4]. Using recursive X–Y cuts [8] (or recursive projection profile cuts) is another way to decompose a document image into a set of blocks. Examples of such an approach can be found in Nagy and Seth [9] and Higanisho *et al.* [10].

2.2. *Bottom-Up Strategy*

Bottom-up strategies start by merging evidence at the lowest level of detail (e.g., forming words from characters) and then rise, merging words into lines, lines into columns, etc. until the entire page is completely assembled.

Examples of this approach include Toyoda *et al.* [11], Kubota *et al.* [12], Meynieux *et al.* [13], and Wang and Srihari [14]. In Toyoda's method, a set of bounding boxes of connected components of a page is first classified as *horizontal (vertical) line, text piece, abstract, title,* and *title with background texture* according to its size features. Then a merge algorithm is applied recursively to the classification to combine together the adjacent boxes within the same class until no further merge is possible. Kubota uses two segmentation techniques. The segmentation method using connected component size similar to Toyoda's is fast and simple, but it fails in separating characters which touch the lines in a line drawing. The second method, the neighborhood line density method, is quite complicated and slow, thus it is only applied to extract characters touching line drawings.

2.3. *Hybrid*

A few approaches are difficult to classify as top-down or bottom-up. Kida *et al.* [15] roughly segment an image by a sequence of horizontal and vertical projections, then use connectivity analysis to complete the segmentation; line and character spacing may be unknown but should apparently be fixed. Similarly, Akiyama and Masuda [16] and Masuda *et al.* [17] use skew-correction followed by projection, but in addition they attempt to cope with variable character- and line-spacing by applying geometric rules of character formation as well as classification.

As images are usually noisy, there are reasons for caution toward the strict application of either top-down or bottom-up strategy. In the top-down method, verification must finally depend on statistical estimation, and so they may not reliably descend to the lowest possible level of detail without triggering frequent backtracking, thus increasing the computation time. Bottom-up strategies are forced to make their earliest decisions using evidence from the smallest samples, and so they may suffer from a rapid accumulation of mistakes. H. S. Baird *et al.* [6] have suggested a different way to decompose a page. In their method, instead of developing relations among nearby foreground (black) objects, as in most prior studies, the structure of background (white space) regions is analyzed. That method is rather novel in concept in addition to its global-to-local character, that is greedy and guided by global evidence, but like other methods, it also relies on the assumption of an upright rectangle structure, thus it needs preprocessing to correct tilted components. Another approach for handling skewed images, based on the Hough transformation, is discussed in Fletcher and Kasturi [3].

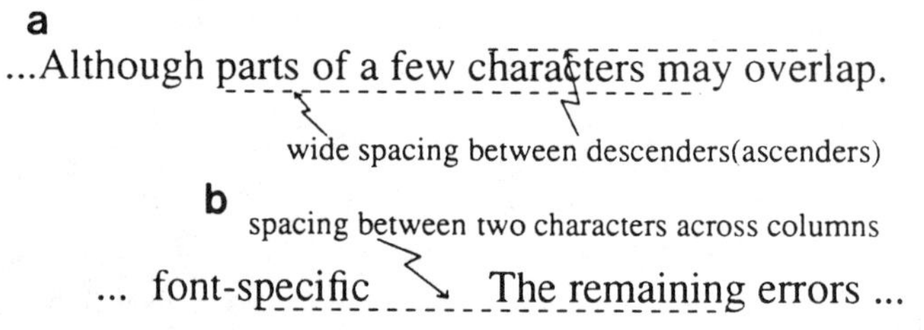

3. THE BASICS OF SEGMENTATION BY WHITE STREAMS

Our basic assumption is that columns are subregions of the input page containing ideally a unique type of data and separated by white spaces which are wide enough to be distinguished from other spacing such as the white spacing between words, etc. In addition, such white spaces form a continuous stream, while white gaps between characters or words are fragmentary, lasting no more than the height of a textline in the vertical direction. We also assume that a blank scan line or a vertical white space of size larger than a predefined value (e.g., twice as large as the nominal size of the largest text line height) separates two columns that are one above the other. In our segmentation stage, no attempt is made to find the logical layout of the page. The principal goal of our segmentation is to find the largest possible column blocks. No further restriction is imposed upon the layout of the page. Text could have various fonts and their size may vary arbitrarily within the page, although they must lie within (usually loose) known bounds. Columns could sit at different tilt angles (such as in the case when a page is distorted by a printing defect, the left column tilts toward left, and the right toward right).

While the basic idea is very simple, its implementation requires some caution because very wide white spaces may also occur, for example, between two descendents (ascendants) of letters either within a column (Fig. 1a) or across columns (Fig. 1b). In the former case they produce false starts, in the latter they distort the width of the stream. Both problems can be remedied if a *vertical projection profile* is used over a blocks of scanlines. Projection profiles have been used before for page segmentation and have been found to produce unreliable results because of skew. If the projections are computed only over a short height (typically less than 0.5″) of the text, then it will take an extremely severe tilt (over 25°) to obscure the typical column gap of 0.25″. Furthermore, we embed a horizontal smearing process [18] in forming vertical projection profiles so as to eliminate sporadic noise spots

a

...Although parts of a few characters may overlap.

wide spacing between descenders(ascenders)

b

spacing between two characters across columns

... font-specific The remaining errors ...

FIG. 1. Examples of wide white spaces that do not correspond to column separators.

which may otherwise break a long white column gap into short segments. In the current embodiment of the method a static smearing process is used. The value of the white gaps threshold should be greater than the between-character interval and smaller than the between-column interval. Gaps between words are not as critical because it is unlikely that they line up across textlines. For the resolution of our scanner (300 pdi) the column gaps are usually over 20 pixels while intercharacter gaps are 2–3 pixels. Therefore we used a threshold of 10 pixels. The smallest printed dot (1 point) is at least 4 pixels wide, hence the threshold for black dots was set to 3 pixels. The smearing process here acts as a preliminary filter. The vertical projection and the two *small region elimination processes* described later further prune away fragments caused by printing defects. The resulting vertical projection profile is very robust in revealing column gaps. The following is an outline of the process of initial segmentation.

For each block of scan lines perform the following operations.

1. Form the vertical projections and find the wide white spaces in them. We assume that these correspond to *column gaps*.

2. For each pair of column gaps identify the entity between the two gaps, the entity *column interval*. The start and end of the block of scanlines are considered to be (trivial) column gaps.

3. Except for the column intervals encountered in the first scanline block, compare the column intervals with those of the above and merge them if the conditions of Table 1 are satisfied.

At this stage the merging process is conservative and the parameters ε_1 and ε_2 equal a few pixels only (in our tests $\varepsilon_1 = 3$ pixels, $\varepsilon_2 = 3$ pixels). The ratio ε_3 is also chosen just below 1 (=0.98 in our tests).

When all the blocks have been processed, it is desirable to perform a *small region elimination* in a similar way as it is done in image segmentation. A refining process is employed to merge into adjoining major blocks the following: very narrow blocks such as those produced by

TABLE 1
Rules for Merging Column Intervals

Notation: Each column interval is a rectangle specified by the quantities X_{left}, X_{right}, Y_{top}, and Y_{bottom}. We assume that these quantities are members of a structure, so that $P \rightarrow X_{left}$ means the left endpoint of a column interval pointed by P.

Any new-found column interval pointed by P in the current block of scanlines will be merged into an existing column interval Q of the previous block(s) if the following two conditions are satisfied:

1. The two intervals are very close in the vertical direction

$$|P \rightarrow Y_{top} - Q \rightarrow Y_{bottom}| < \varepsilon_1$$

for a prespecified tolerance ε_1.

2. One of the horizontal projections of the two column intervals contains the other.

$$P \rightarrow X_{left} > Q \rightarrow X_{left} - \varepsilon_2 \quad \text{and} \quad P \rightarrow X_{right} < Q \rightarrow X_{right} + \varepsilon_2$$

or

$$P \rightarrow X_{left} < Q \rightarrow X_{left} + \varepsilon_2 \quad \text{and} \quad P \rightarrow X_{right} > Q \rightarrow X_{right} - \varepsilon_2,$$

where ε_2 is a predefined tolerance.

3. The widths of the two intervals are approximately the same,

$$\frac{\min\{P \rightarrow X_{right} - P \rightarrow X_{left}, Q \rightarrow X_{right} - Q \rightarrow X_{left}\}}{\max\{P \rightarrow X_{right} - P \rightarrow X_{left}, Q \rightarrow X_{right} - Q \rightarrow X_{left}\}} < \varepsilon_3,$$

where ε_3 is a prespecified tolerance.

isolated characters or very short text lines, and those blocks that are narrow in the vertical direction, usually containing only a fragment of a single textline. Isolated small column blocks caused by artifacts of considerable size are also removed in this process. This process involves the re-examination of the blocks that are small either in the vertical or horizontal direction, and in addition satisfy conditions 1 and 2 of Table 1, but with larger tolerance ε_1 and ε_2 (typically $\varepsilon_1 = 20$, $\varepsilon_2 = 5$).

Figure 2 shows a document page image and Fig. 3 is the result of the processing. This result may be satisfactory if there is no skew, but when the scanned page is tilted there is need for additional processing. Since we do not know in advance whether that is the case, we proceed always with skew estimation.

4. SKEW ESTIMATION AND FINAL CLUSTERING

In contrast to other approaches, the skew estimation performed at this stage is very fast because it uses the centers of column intervals and there are usually no more than a few hundred of them, as opposed to millions of pixels or thousands of scanline intervals. Any one of several methods might be used at this stage, including the Hough Transform. We used a method that has very small memory requirements and it is fast. Straight lines are fit into the column intervals in a block using a linear scan algorithm [19]. Blocks are merged if their column lines are approximately collinear and their widths are approximately the same.

The following is a formal statement of the skew estimation process. Let the name CB_class denotes a union of a group of column blocks whose centers are collinear. A CB_class is represented by quantities Y_{top}, Y_{bottom}, *width*, and a straight line $y = ax + b$, which specifies its central Y_axis. (When the central Y-axis is vertical, the equation becomes $x = c$, where c is a constant.) Let C_i be a column block formed in the segmentation process, which is represented by the quantities in Table 1 and $\Theta = \{C_i\}$ be the whole set of such column blocks found in a document page. Now, let Π be a collection of CB_classes: $\Pi = \{S_i\}$, where each CB_class entity S_i is a region comprising a portion of a text column or graphics. We are seeking a partition P which maps Θ into Π, $P: \Theta \rightarrow \Pi$.

We first define an order relation "$\leq$" in column blocks entities:

DEFINITION. The order relation "$\leq$" in column blocks is defined as follows:

$$C_1 \leq C_2 \quad \text{iff} \quad C_1 \rightarrow Y_{top} \leq C_2 \rightarrow Y_{top},$$

where C_1 and C_2 point to two column block entities.

From this definition we further define an ordered column block set:

DEFINITION. An ordered column set Ω is a set of column blocks $\{C_i\}$ which satisfy the relationship: $C_0 \leq C_1 \leq C_2 \leq \cdots$.

PARTITION ALGORITHM.

step-1.
Sort Θ into Ω, set $S_0 = \{C_0: C_0 \in \Omega\}$, and initialize $\Pi = \{S_0\}$.

step-2.
If $C_i \in \Omega$, examine C_i with $S_j \in \Pi, j = 0, 1, 2, \ldots$, C_i is merged into S_j if C_i and S_j satisfy the conditions given in Table 2 (The parameters in Table 2 are typically chosen as $\varepsilon_1 = 40$, $\varepsilon_4 = 40$ and $\varepsilon_3 = 0.85$), and the skew angle of the central Y-axis of S_j is adjusted according to the latest merge.

step-3.
If C_i can not be merged into any of S_j in the current Π, a new CB_class entity is created: $S_{n+1} = \{C_i\}$ and is inserted into Π.

step-4.
Repeat the step-2, and step-3 until all C_i in Ω have been examined.

When no further merging is possible the slope of the central Y-axis line of each element in Ω provides an estimate of the skew angle of the block. The skew angle thus

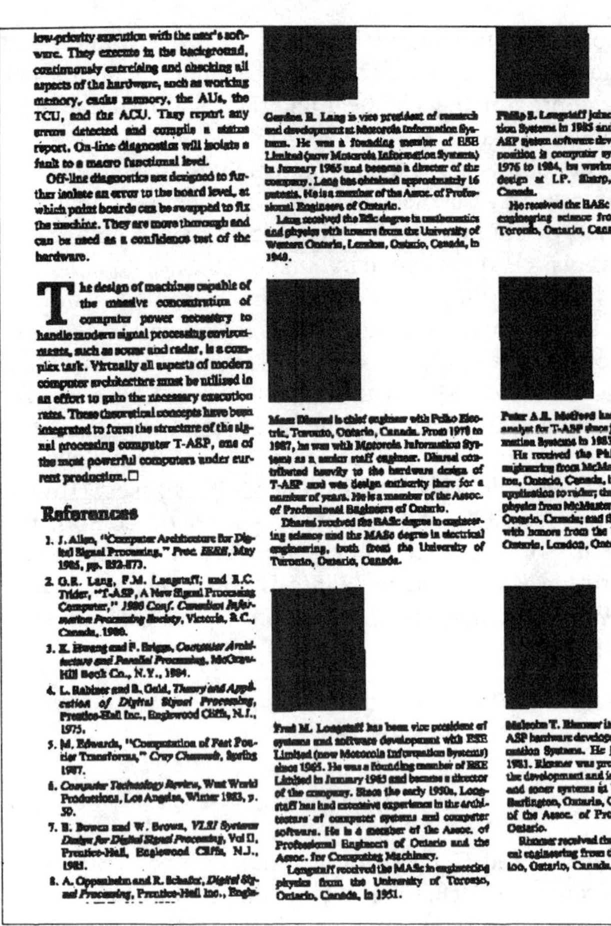

low-priority execution with the user's software. They execute in the background, continuously exercising and checking all aspects of the hardware, such as working memory, cache memory, the AUs, the TCU, and the ACU. They report any errors detected and compile a status report. On-line diagnostics will isolate a fault to a macro functional level.

Off-line diagnostics are designed to further isolate an error to the board level, at which point boards can be swapped to fix the machine. They are more thorough and can be used as a confidence test of the hardware.

The design of machines capable of the massive concentration of computer power necessary to handle modern signal processing environments, such as sonar and radar, is a complex task. Virtually all aspects of modern computer architecture must be utilised in an effort to gain the necessary execution rates. These theoretical concepts have been integrated to form the structure of the signal processing computer T-ASP, one of the most powerful computers under current production. □

References

1. J. Allen, "Computer Architecture for Digital Signal Processing," Proc. IEEE, May 1985, pp. 852-873.
2. G.R. Lang, F.M. Longstaff, and R.C. Trider, "T-ASP, A New Signal Processing Computer," 1980 Conf. Canadian Information Processing Society, Victoria, B.C., Canada, 1980.
3. K. Hwang and F. Briggs, Computer Architecture and Parallel Processing, McGraw-Hill Book Co., N.Y., 1984.
4. L. Rabiner and B. Gold, Theory and Application of Digital Signal Processing, Prentice-Hall Inc., Englewood Cliffs, N.J., 1975.
5. M. Edwards, "Computation of Fast Fourier Transforms," Cray Channels, Spring 1987.
6. Computer Technology Review, West World Productions, Los Angeles, Winter 1983, p. 59.
7. B. Bowen and W. Brown, VLSI Systems Design for Digital Signal Processing, Vol II, Prentice-Hall, Englewood Cliffs, N.J., 1983.
8. A. Oppenheim and R. Schafer, Digital Signal Processing, Prentice-Hall Inc., Engle-

Gordon R. Lang is vice president of research and development at Motorola Information Systems. He was a founding member of ESE Limited (now Motorola Information Systems) in January 1965 and became a director of the company. Lang has obtained approximately 16 patents. He is a member of the Assoc. of Professional Engineers of Ontario.

Lang received the BSc degree in mathematics and physics with honors from the University of Western Ontario, London, Ontario, Canada, in 1949.

Moez Dharssi is chief engineer with Pelko Electric, Toronto, Ontario, Canada. From 1978 to 1987, he was with Motorola Information Systems as a senior staff engineer. Dharssi contributed heavily to the hardware design of T-ASP and was design authority there for a number of years. He is a member of the Assoc. of Professional Engineers of Ontario.

Dharssi received the BASc degree in engineering science and the MASc degree in electrical engineering, both from the University of Toronto, Ontario, Canada.

Fred M. Longstaff has been vice president of systems and software development with ESE Limited (now Motorola Information Systems) since 1965. He was a founding member of ESE Limited in January 1965 and became a director of the company. Since the early 1950s, Longstaff has had extensive experience in the architecture of computer systems and computer software. He is a member of the Assoc. of Professional Engineers of Ontario and the Assoc. for Computing Machinery.

Longstaff received the MASc in engineering physics from the University of Toronto, Ontario, Canada, in 1951.

FIG. 2. An example of document page image.

FIG. 3. The result after applying the merging process to Fig. 2. The dotted lines outline the final column blocks. The labels are attached by the classification process in the second stage.

estimated does not always reflect the real skew of the block. For example, this will happen if the text in a page is not right-justified. The irregularity in a long column text usually does not affect the skew estimation, because of the averaging among column intervals. However, such irregularity may produce an erroneous skew angle when the column is short. We compute a weighted average of skew angles by using the length of a block as the weight of the corresponding angle. If the average estimated skew angle is less than a predefined value then the input page is considered to be roughly upright, and no correction is applied. Otherwise, the left and right lines of each column blocks are rotated according to the average skew angle. Figure 5 shows the result of applying skew estimation to the column blocks displayed in Fig. 4.

For the same reason that a second pass is made during the initial column block formation process, a second pass is made now to further merge blocks that have similar skew angles, touch each other in the vertical direction, have similar width, and the last interval of one overlaps significantly the first interval of the other.

TABLE 2
Rules for Merging Column Blocks

Notation: A nonlabeled column block is a column block to be merged to a CB_class. A nonlabeled column block is represented by the same quantities as column intervals in the Table I. Let P point to a CB_class, and Q point to a nonlabeled column block.

Q will be merged into P if the following three conditions are satisfied:

1. P and Q are very close in the vertical direction:

$$|P \rightarrow Y_{top} - Q \rightarrow Y_{bottom}| < \varepsilon_1 \quad \text{or} \quad |P \rightarrow Y_{bottom} - Q \rightarrow Y_{top}| < \varepsilon_1$$

2. The center point of column block Q is not far away from the central Y-axis of P:

$$\left| \frac{(Q - Y_{top} + Q \rightarrow Y_{bottom})/2) - b}{a} - \frac{Q \rightarrow X_{left} + Q \rightarrow X_{right}}{2} \right| < \varepsilon_4.$$

3. The widths of P and Q are approximately the same:

$$\frac{\min\{P \rightarrow width,\ Q \rightarrow X_{right} - Q \rightarrow X_{left}}{\max\{P \rightarrow width,\ Q \rightarrow X_{right} - Q \rightarrow X_{left}} < \varepsilon_3.$$

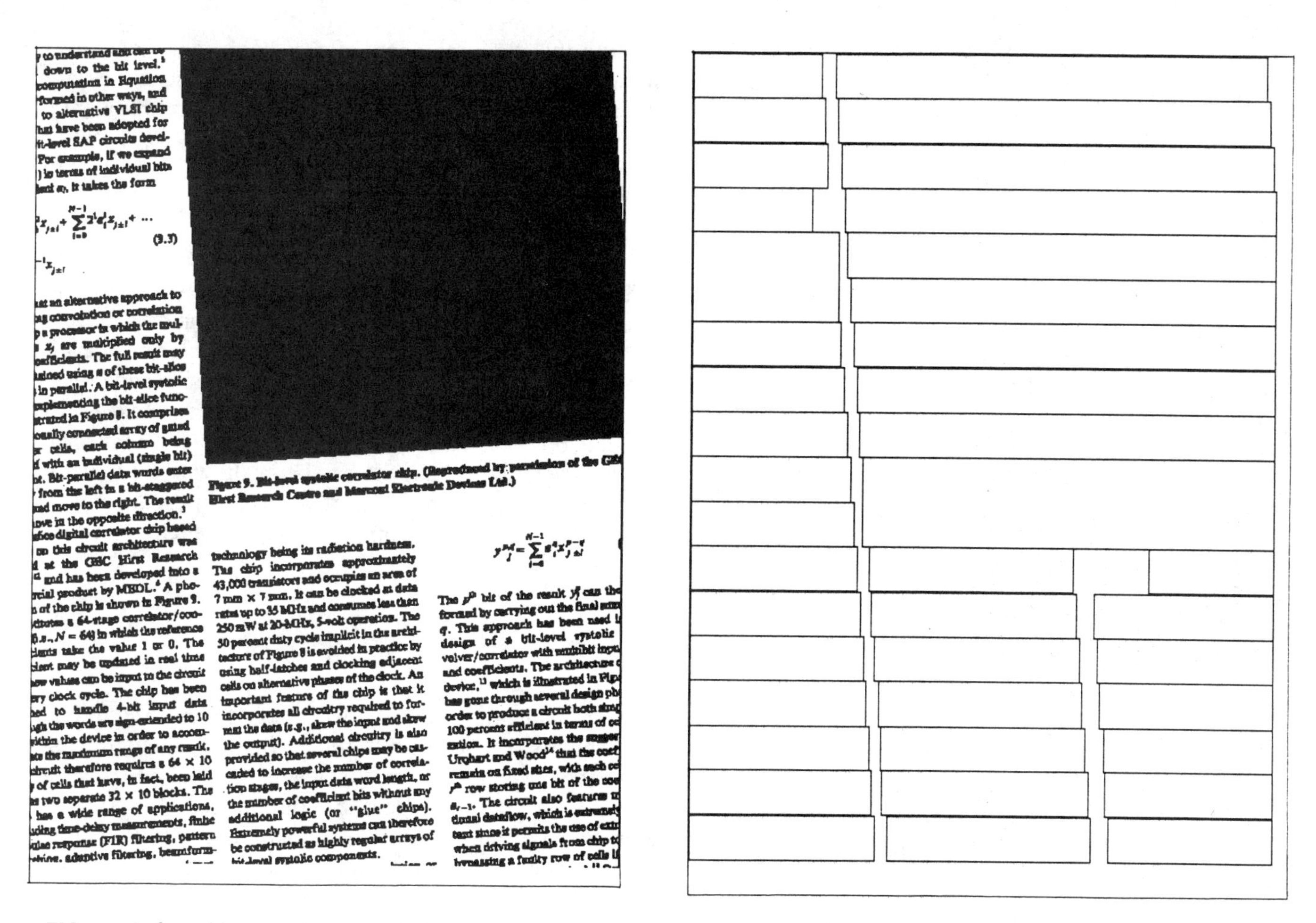

Figure 9. Bit-level systolic correlator chip. (Reproduced by permission of the GEC Hirst Research Centre and Marconi Electronic Devices Ltd.)

technology being its radiation hardness. The chip incorporates approximately 43,000 transistors and occupies an area of 7 mm × 7 mm. It can be clocked at data rates up to 35 MHz and consumes less than 250 mW at 20-MHz, 5-volt operation. The 50 percent duty cycle implicit in the architecture of Figure 8 is avoided in practice by using half-latches and clocking adjacent cells on alternative phases of the clock. An important feature of the chip is that it incorporates all circuitry required to format the data (e.g., skew the input and skew the output). Additional circuitry is also provided so that several chips may be cascaded to increase the number of correlation stages, the input data word length, or the number of coefficient bits without any additional logic (or "glue" chips). Extremely powerful systems can therefore be constructed as highly regular arrays of bit-level systolic components.

FIG. 4. *Left*: A titled page image. *Right*: Column intervals of the left document image found before applying the skew estimation process.

5. CLASSIFICATION

After a page is partitioned into coherent blocks, there is a need to classify the blocks as *text* or *illustrations*. In this paper we focus on the discrimination of halftone image regions from nonhalftone regions. Our goal is to devise a method which is reasonably fast and efficient.

Toward this end, the conventional image segmentation techniques such as Fourier analysis, although useful, are too slow to be attractive. Many authors advocated connected component analysis. A representative of this kind of methods is the analysis of the geometrical properties proposed by F. M. Wahl [20]. The method uses measurements of border-to-border distance within a connected component. These measurements provide a fairly good estimate for the mean line thickness of line shaped patterns. It has been proved to be very powerful in distinguishing line drawings from text. However, the method becomes clumsy when it comes to analyze halftone regions because the connected components of a halftone area are usually very large. This method would be rather like identifying a forest by examining each leaf of each tree. Even if it worked, the amount of computation involved in calculating those measures is prohibitively complex and time-consuming, hence preventing its use in analyzing the immense amount of data in a scanned document.

The field of statistical pattern classification offers the means for accomplishing our task. The basic idea is to extract features, i.e., measurements of quantities thought to be useful in distinguishing members of different classes. Many features serve this purpose. Examples can be found in Wahl *et al.* [4]. Clearly the ratio b/w of the total length b of black elements over the total length w of the white elements provides some useful information. Line drawings such as diagrams have a much lower ratio of b/w than text, while halftones *usually* have the highest ratio. Because of the variability of the ratio for halftones it cannot be used with confidence except to decide that something *cannot* be text or diagrams.

A very reliable discriminator for halftones versus text or diagrams is offered by the signal cross-correlation function. When the signals are binary with values of 1 and 0 equally likely, then it can be shown easily that the

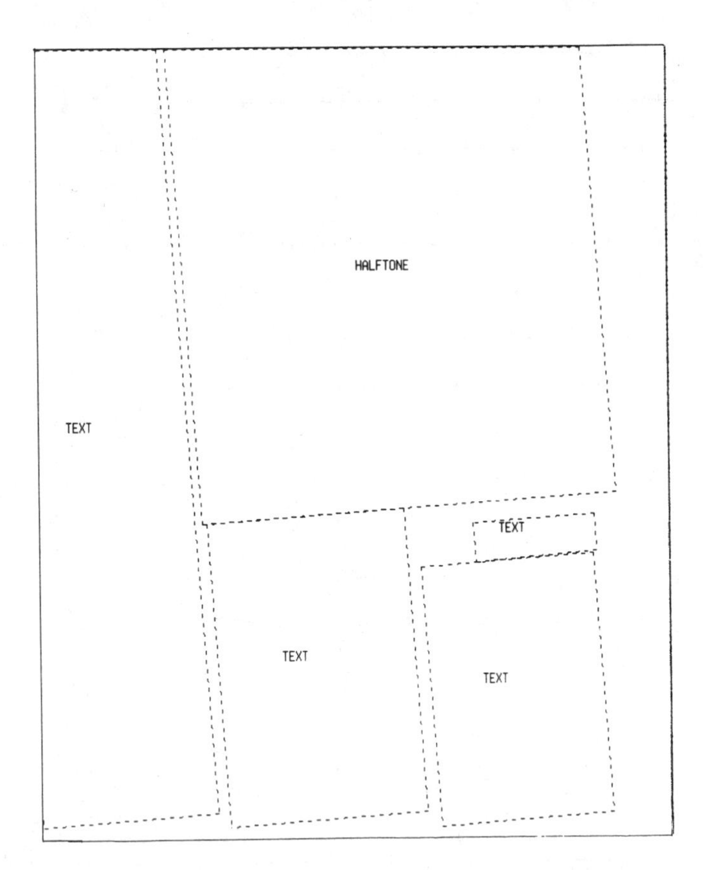

FIG. 5. Column blocks of Fig. 4 after the skew estimation and the final merging process. The labels are attached by the classification process.

terms of the cross-correlation between signals f and g are given by

$$1 - 2(f\, XOR\, g),$$

where XOR is the exclusive OR operator. We use the normalized correlation between scanlines at y and $y + r$, defined as

$$C(r, y) = \frac{1}{L} \sum_{k=0}^{L-1} [1 - 2p(y, k)\, XOR\, p(y + r, k)]$$

$$= 1 - \frac{2}{L} \sum_{k=0}^{L-1} p(y, k)\, XOR\, p(y + r, k),$$

L is the number of pixels in a scanline and $p(y, k)$ is the value of the kth (binary) pixel of scanline y. The quantity in brackets equals 1 if the two pixels have the same value and -1 otherwise. Thus if two scanlines disagree over length K (and thus, agree over $L - K$), then the sum is $L - 2K$ and $C(r, y)$ equals $1 - 2K/L$. An exactly opposite arrangement of black and white yields a sum equal to $-L$ and $C(r, y)$ equals -1, while the correlation of a scanline with itself $C(0, y)$ equals 1. $C(r, y)$ can be computed very simply from a run length code representation as well.

Note that the arguments of $C(r, y)$ are not those ordinarily used in signal processing. The two scanlines are not shifted with respect to each other, and the argument y is the position of the first scanline and r is the distance from the second scanline to the first one.

5.1. *Text and/or Line Drawing Regions*

The key observation is that for lines of text and diagrams $C(r, y)$ is a rapidly decreasing function of r, at least for small values of r, and $C(1, y)$ is quite high. For halftones $C(r, y)$ is rather flat and even exhibits periodicities. $C(1, y)$ is usually much lower than that for text. This conclusion is based on a theoretical argument confirmed by experimental results. It has been shown [21] that for text the difference in the start of black intervals among adjacent lines follows approximately a Cauchy distribution

$$P(k) = \frac{\pi}{2} \frac{1}{1 + k^2},$$

where k is the difference between starts measured in pixels. The distribution of differences in interval lengths follows the Cauchy distribution even more closely. Measured distributions suggest that differences of more than 3 pixels have a less than 10% probability to occur. Therefore adjacent scanlines will have a very high cross-correlation but as the distance increases such cross-correlation will decrease.

This observation can be illustrated by considering a line of width equal to W pixels and slope $1/m$ with respect to the horizontal. Let h be the distance between scanlines and L be the length of the bounding box containing only the segment, but no other black pixels. (L is usually much greater than the thickness of a line.) The start of intervals in adjacent scanlines will differ by mh so that

$$C_L(r, y) = \frac{L - 4rmh}{L} = 1 - 4\frac{rmh}{L}, \quad \text{if } rmh < W,$$

$$C_L(r, y) = \frac{L - 4W}{L} = 1 - 4\frac{W}{L}, \quad \text{otherwise},$$

where the subscript indicates that the cross-correlation is computed only over part of the scan line. A typical value for h is 1/300 inches and for W 1/50 inches. For a line with slope 1/2 (a 45°) L should be about 1/6 inches over the duration of the text line (about 50 scanlines) so that the above equations become

$$C_L(r, y) = 1 - \frac{r}{25}, \quad \text{if } r < 12,$$

$$C_L(r, y) = 1 - \frac{48}{100} = 0.52, \quad \text{otherwise}.$$

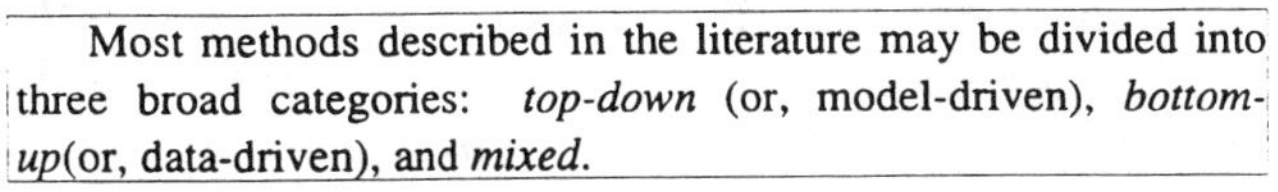

Most methods described in the literature may be divided into three broad categories: *top-down* (or, model-driven), *bottom-up*(or, data-driven), and *mixed.*

FIG. 6. A piece of document image containing three textlines.

Thus the cross-correlation drops from 0.96 for r equal 1 down to 0.52 for r equal 12. On the other hand, if the line is horizontal or vertical, $C_L(r, y)$ stays non-decreasing for a considerable interval of r and then drops sharply. Figures 6 and 7 illustrate that similar results hold over actual data.

There are a few cases when $C_L(r, y)$ takes low values between two neighboring scanlines. One such example is when one of the two scanlines is located just next to the end of a line segment, while the other intersects with the line segment. Consider the same line segment we described above. If we fix one scanline at the boundary, then

$$C_L(r, y) = \frac{L - 2W}{L} = 1 - 2\frac{W}{L}.$$

When the segment is almost horizontal (a hyphen, for example), $W \rightarrow L$, so $C_L(r, y) \rightarrow -1$, regardless of r. $C_L(r, y)$ could also have high values at some large r such as it happens for the character "o" which has a ring-like shape. In that case, a scanline which intersects with the lower part of the "o" will have a same arrangement of black and white runs with one of the scanlines which intersects with the upper part of the "o," thus produces a near-perfect correlation between those two scanlines.

The above exceptions are not essential when the correlation is calculated over a whole scanline. For example, it is unlikely that a scanline will intersect the start positions of horizontal strokes only unless there is a textline consisting entirely of hyphens and the text is perfectly

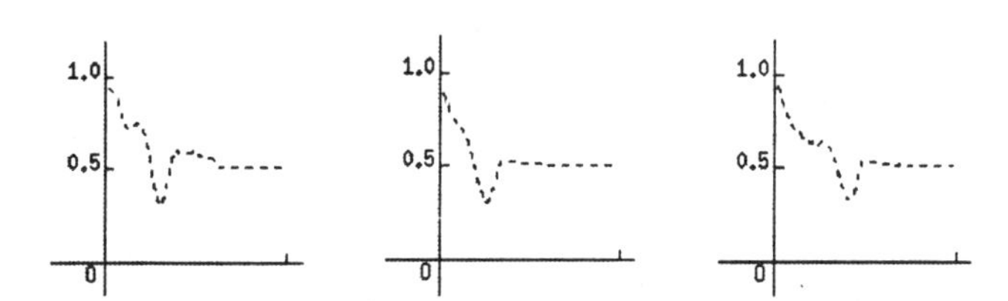

FIG. 7. $C(r, y_0)$ curves—The line correlations between scanline y_0 and scanline $y_0 + r$ computed from the image shown in Fig. 6 as a function of r. The x-abscissa is the position difference r, and the y-abscissa is the line correlation value of scanline y_0 with scanline $y_0 + r$. *Left*: y_0 is at the middle of the first textline; *Middle*: y_0 is at the middle of the second textline; *Right*: y_0 is at a position which is seven scanlines higher than the y position of the middle figure; The flat region of each curve corresponds to the correlation of a scanline in the textline region with scanlines in the blank space. The similarity of the shape of the curves reflects the relative independence of the correlation on its position.

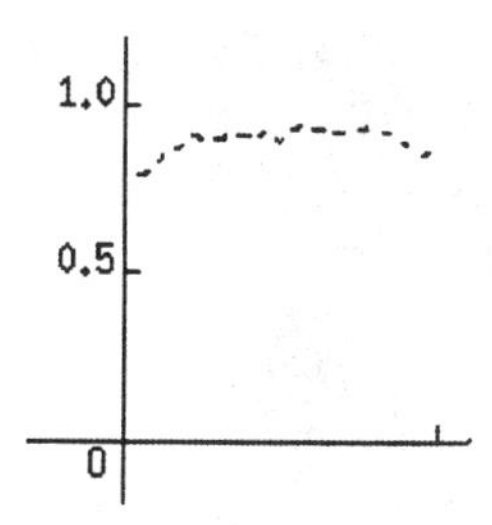

FIG. 8. $C(1, y)$—The line correlation of scanline y with scanline y + 1 as a function of y, $r = 1$, computed over the second textline of Fig. 6. The x-abscissa is the scanline location y, and the y-abscissa is the line correlation value at that scanline. The flatness of the shape of the curve once again verifies the independence of $C(r, y)$ on the position y.

aligned to the scanning direction. In general, $C(r, y)$ is independent of the position of the scanline. Figure 8 shows that indeed $C(r, y)$ varies little with y except near the top and the bottom of a text line. The independency is even more evident when one averages scanline correlations over a group of scanlines, for example, the blank spaces between textlines give very high correlation value at low r, hence smoothing out occasional variations brought by position change.

5.2. *Halftone Regions*

Halftones are produced by dot patterns that are distributed according to the average density of a region and every effort is made to avoid the appearance of regular patterns. The two most common halftoning techniques are error diffusion and ordered dithering [22].

The ordered dithering technique creates a bi-level representation of a gray level image by comparing the image intensity $P(x, y)$ to a position dependent threshold that is one of the elements of an $n \times n$ dither matrix D^n. The particular matrix element $D^n(i, j)$ to be used as a threshold depends only on the coordinates x, y:

$$i = x \text{ modulo } n, \quad j = y \text{ modulo } n.$$

Values of pixels are assigned by the rule:

$$\tilde{P}(x, y) = \begin{cases} \text{white}, & \text{if } P(x, y) > D^n(x, y), \\ \text{black}, & \text{otherwise.} \end{cases}$$

The dither threshold at each pixel in the image is obtained by repeating the dither matrix in a checkerboard fashion over the entire image. There are various techniques for selecting a dither matrix. A primary goal in the design of dither matrices is to minimize the low frequency texture.

The error diffusion technique, originally described in [23] attempts to diffuse the errors between an input grey scale image $P(x, y)$ and the displayed binary image $\tilde{P}(x, y)$

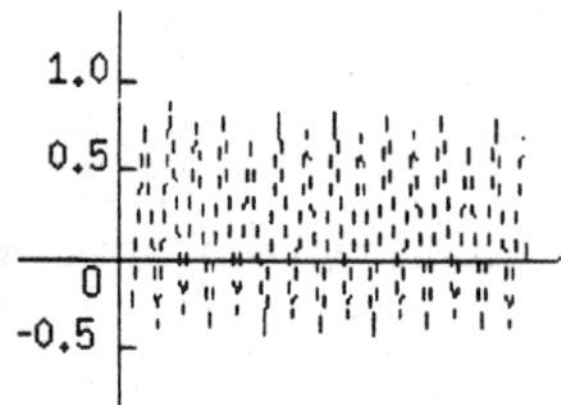

FIG. 9. *Left*: A dithered image. *Right*: The line correlation curve $C(r, y_0)$ as a function of r, y_0 is arbitrarily chosen to be at the middle of the image, $0 \le r \le 30$. As predicted, $C(r, y)$ fluctuates periodically, but has low value for small r.

amongst closely neighboring pixels. If the error $E(x, y)$ is defined by

$$E(x, y) = P(x, y) - \tilde{P}(x, y),$$

then a modified value of input intensity $P'(x, y)$ can be computed from the previous (in the sense of scanning direction) errors and current pixel intensity as

$$P'(x, y) = P(x, y) + \frac{\sum \alpha_{ij} E(x_j, y_i)}{\sum \alpha_{ij}}.$$

The coefficients $\{\alpha_{ij}\}$ define the relative contributions of the previous errors to the current intensity. The indices $\{x_j, y_i\}$ vary over a small neighborhood above and to the left of (x, y). The quality of the pictures generated by this technique depends to some extent on the characteristics of matrix $\{\alpha_{ij}\}$. An analysis conducted by R. Ulichney [22] has shown that the spectral characteristics of a well-formed dither pattern in the resulting image should exhibit flat high frequency characteristics.

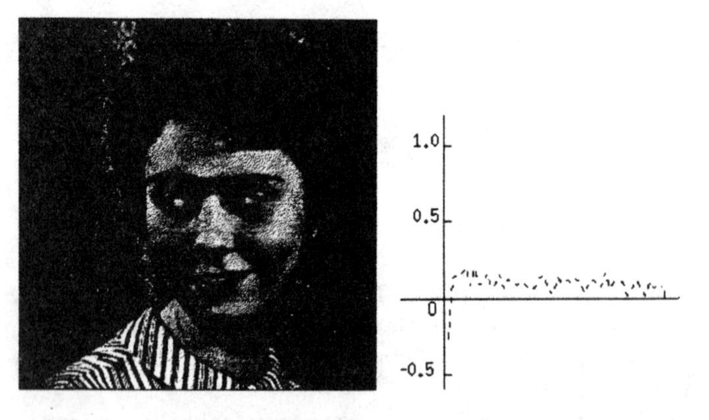

FIG. 10. *Left*: An error diffusion image. *Right*: The curve of the line correlation function, $C(r, y_0)$, $0 \le r \le 50$. The correlation value is dominantly low.

For halftones produced by dithering the line correlation will exhibit a periodic pattern with a period of the size of the dither matrix because the dithering pattern is repeated periodically. Such areas tend to have low scanline correlation between scanlines which are close to each other since the dither patterns are designed to arrange 1's and 0's runs from a scanline to next scanline in a low correlation fashion. Error diffusion results in the dots being more randomly and evenly distributed. As a result of this randomnization, adjacent scanlines have low line correlations except for the extreme cases where a region is nearly all white or nearly all black. At the same time, the line correlation between scan lines that are far from each other does not differ much from the line correlation of scanlines that are near each other. These observations are confirmed by calculations over actual data as shown in Figs. 9 and 10. In both cases, the average scanline correlation is low in comparison to the average over text area, since low correlations predominate.

The experiments suggest that the correlation function is approximately linear for small r, therefore it makes sense to fit a regression line

$$L(r) = \alpha + \gamma^* r$$

by minimizing

$$E = \sum_{i=1}^{r_0} (C(i, y) - \alpha - \gamma i)^2,$$

for some small r_0. (In our tests r_0 was chosen to be 5.) We may then discriminate text (and diagrams) from halftones by considering the slope γ of the regression line. When γ is plotted as function of y (as shown in Fig. 11) the differences are seen clearly. In the text regions $\gamma(y) \ll 0$, while in the halftone regions $\gamma(y)$ has significant larger values. An even simpler measurement of the decorrelation rate is to compute the ratio: $C(1, y)/C(s_0, y)$, which gives a rough but quick estimation of the decorrelation rate at the scanline. We found in our experiments this measurement quite satisfactory.

Depending on the setting of the scanner, the dot structure of halftone regions may not be duplicated exactly when a document page is scanned. Nevertheless the statistical character of the line correlation of halftone regions will be still exhibitively distinguishable from that of text regions. We include here experiments illustrating the influence of the states of scanners upon the line correlation of scanned halftone images (see Figs. 12 and 13).

6. PUTTING IT ALL TOGETHER

The page segmentation algorithm and the classification scheme described above have been combined to form a

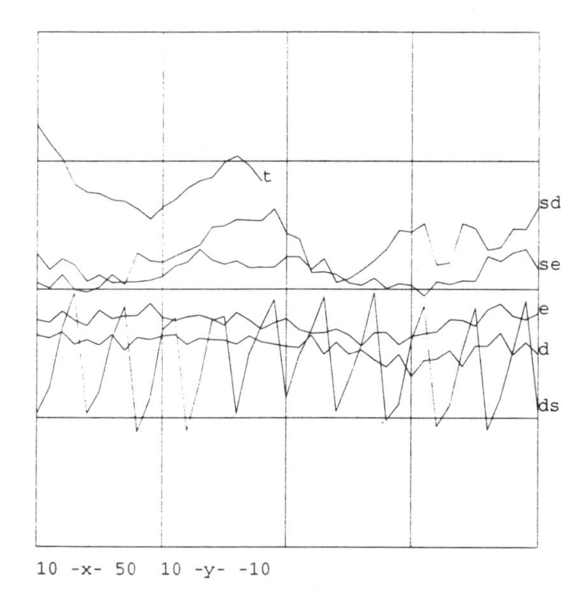

FIG. 11. The decorrelation rate $\gamma(y)$ over consecutive scanlines in text region and halftone regions, respectively. $r_0 = 5$. The horizontal axis corresponds to the scanline location y, and the vertical axis to the decorrelation rate $\gamma(y)$ at that scanline. *t*: γ curve over the second textline of Fig. 6; *d*: γ curve over the dithered image of Fig. 9; e: γ curve over the error diffusion image of Fig. 10; *sd*: γ curve over the scanned version image of Fig. 9 by a Ricoh/100 scanner; *se*: γ curve over the image of Fig. 12, which is the scanned version of Fig. 10; *ds*: γ curve over the image of Fig. 13, which is Fig. 10 reproduced by dithering the image using the built-in dithering scheme of the Ricoh/100 scanner.

single system. Once a document page is segmented into column blocks, each column block is classified as text or halftones depending on the percentage of number of lines with $C(1, y) > 0.9$, average cross-correlation between adjacent scanlines ($C(1, y)$), average correlation between scanlines with four intervening scanlines ($C(5, y)$), and the average number of black runs. In all cases y range over the set of scanlines of the column block, and all the

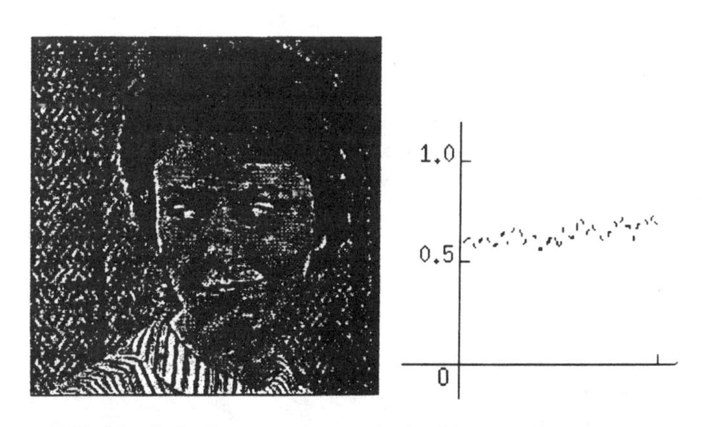

FIG. 12. *Left*: Document image obtained by scanning the image of Fig. 10 using a Ricoh 100 scanner with a resolution 300 per inch. *Right*: The correlation curve with fixed position computed from the left image. The correlation is different from that calculated from Fig. 10. The decorrelation rate $\gamma(y)$ is shown in Fig. 11.

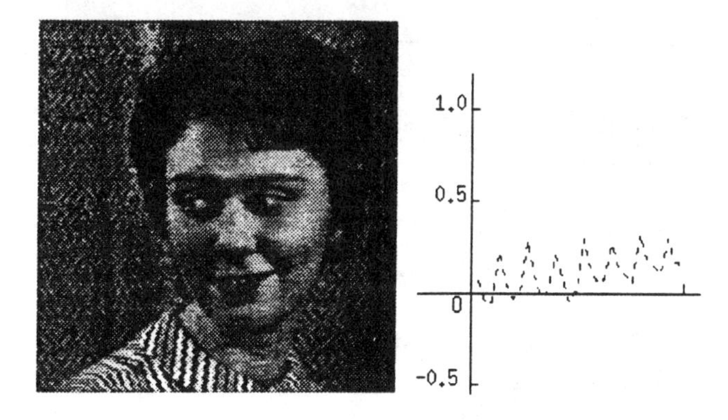

FIG. 13. *Left*: Document image obtained by scanning the image of Fig. 10 using the dithering scheme in the Ricoh 100 scanner. *Right*: The Correlation curve with fixed position. Note that the correlation increases, but it is still lower than that of a text area. The decorrelation rate $\gamma(y)$ is shown in Fig. 11.

measurements are done within the boundary of the column block. For text the first two numbers are expected to be high and the third low, while for halftones the first three numbers are all expected to be low with the second and the third being close to each other, while the fourth number is expected to be high. Because the results obtained by this initial choice were satisfactory no further attempts at parameter selection or optimization were made. A linear discrimination function, $g(\mathbf{x}) = \mathbf{w}^t\mathbf{x} + w_0$, where $\mathbf{w}$ is the weight vector, w_0 the threshold weight and $\mathbf{x}$ is the four measurements described above, is used to first classify a column block into one of the two categories: *text*, *halftone*. When a column block is classified as *text*, the ratio *b*/*w* which is the ratio of the number of black pixels to the number of white pixels in the block is used to further distinguish a diagram from a piece of text. For diagrams, *b*/*w* is expected to be low (<0.2).

The overall method was tested on a series of documents scanned by a Ricoh 100 scanner which produces a binary image of size 2512 × 3400. The scanning resolution is 300 pdi. Experiments were run over pages from magazines having multiple columns and complicated layout. Figures 14 and 15 show the results of segmentation and classification on technical documents. Figure 16 shows the result for a page from a popular magazine. The latter illustrates that our method can handle easily insets. Parts of the documents may appear over-segmented, but this presents no problem because vertically adjacent columns can be combined easily after recognition on the basis either of character size or context. The important thing is that no block contains pieces of text that should not be together.

In our experimental run of the system, the document images were scanned from regular magazines and all the parameters were predefined and fixed. The major parameters and their typical values used in the experiments are

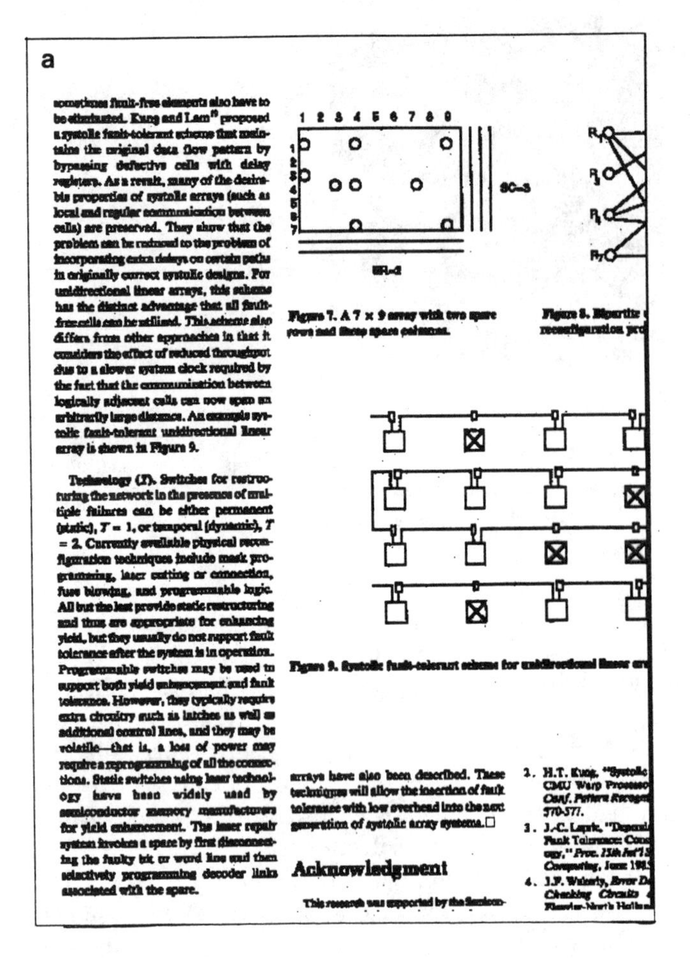

a

sometimes fault-free elements also have to be eliminated. Kung and Lam[19] proposed a systolic fault-tolerant scheme that maintains the original data flow pattern by bypassing defective cells with delay registers. As a result, many of the desirable properties of systolic arrays (such as local and regular communication between cells) are preserved. They show that the problem can be reduced to the problem of incorporating extra delays on certain paths in originally correct systolic designs. For unidirectional linear arrays, this scheme has the distinct advantage that all fault-free cells can be utilized. This scheme also differs from other approaches in that it considers the effect of reduced throughput due to a slower system clock required by the fact that the communication between logically adjacent cells can now span an arbitrarily large distance. An example systolic fault-tolerant unidirectional linear array is shown in Figure 9.

Technology (T). Switches for restructuring the network in the presence of multiple failures can be either permanent (static), T = 1, or temporal (dynamic), T = 2. Currently available physical reconfiguration techniques include mask programming, laser cutting or connection, fuse blowing, and programmable logic. All but the last provide static restructuring and thus are appropriate for enhancing yield, but they usually do not support fault tolerance after the system is in operation. Programmable switches may be used to support both yield enhancement and fault tolerance. However, they typically require extra circuitry such as latches as well as additional control lines, and they may be volatile—that is, a loss of power may require a reprogramming of all the connections. Static switches using laser technology have been widely used by semiconductor memory manufacturers for yield enhancement. The laser repair system invokes a spare by first disconnecting the faulty bit or word line and then selectively programming decoder links associated with the spare.

Figure 7. A 7 × 9 array with two spare rows and three spare columns.

Figure 9. Systolic fault-tolerant scheme for unidirectional linear ar

arrays have also been described. These techniques will allow the insertion of fault tolerance with low overhead into the next generation of systolic array systems.□

Acknowledgment

This research was supported by the Semicon-

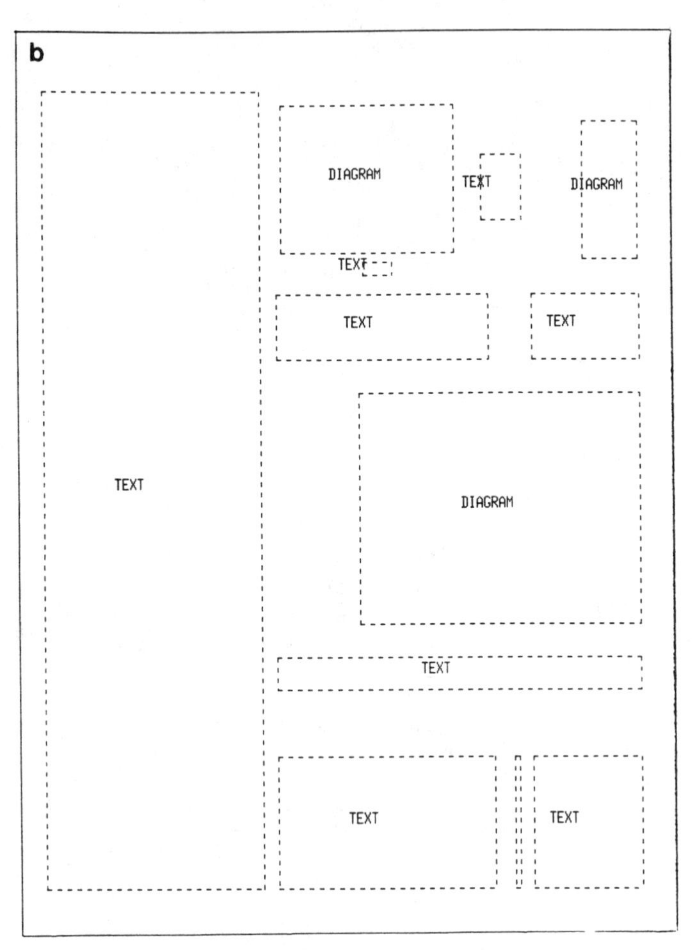

FIG. 14. (a) A document page image with graphics. (b) The segmentation and the classification result.

listed in Tables 3 and 4. In general, the parameters, such as the minimum width of column gaps, are font-size- and scanning-resolution-dependent, therefore a prescan might be needed to adjust them if different types of documents are to be scanned. Text printed in large-size font such as titles and headers often has big gaps between words, producing fragment column blocks in the segmentation (see the example of Fig. 15), so a post-segmenta-

TABLE 3
Parameters of the Segmentation Algorithm

Parameter	Range or rule for selection	Value in initial segmentation	Value in small region elimination	Value in partition algorithm
Smearing threshold (black)	Proportional to scan resolution	3 pixels (0.01″)	—	—
Smearing threshold (white)	Proportional to character space	10 pixels (0.03″)	—	—
ε_1	Proportional to the space between lines	3 pixels (0.01″)	20 pixels (0.07″)	40 pixels (0.13″)
ε_2	Proportional to the font size	3 pixels (0.01″)	5 pixels (0.02″)	— —
ε_3	0.80–1.0	0.98	—	—
ε_4	0–60 pixels (0.0″–0.2″)	— —	— —	40 pixels (0.13″)
Column gap width	Proportional to between column space	>50 pixels (0.17″)	— —	— —

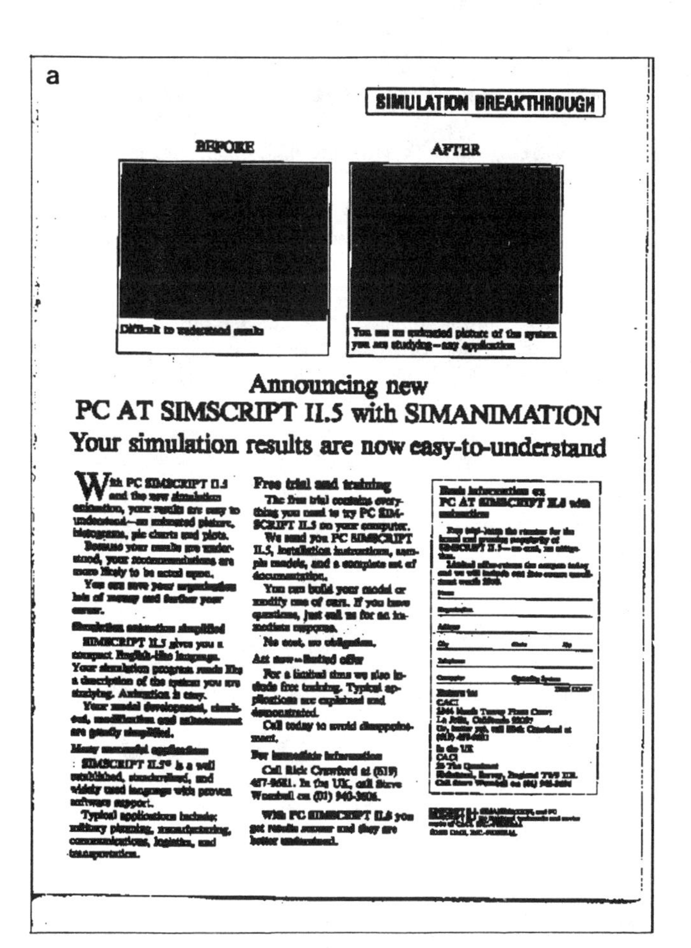

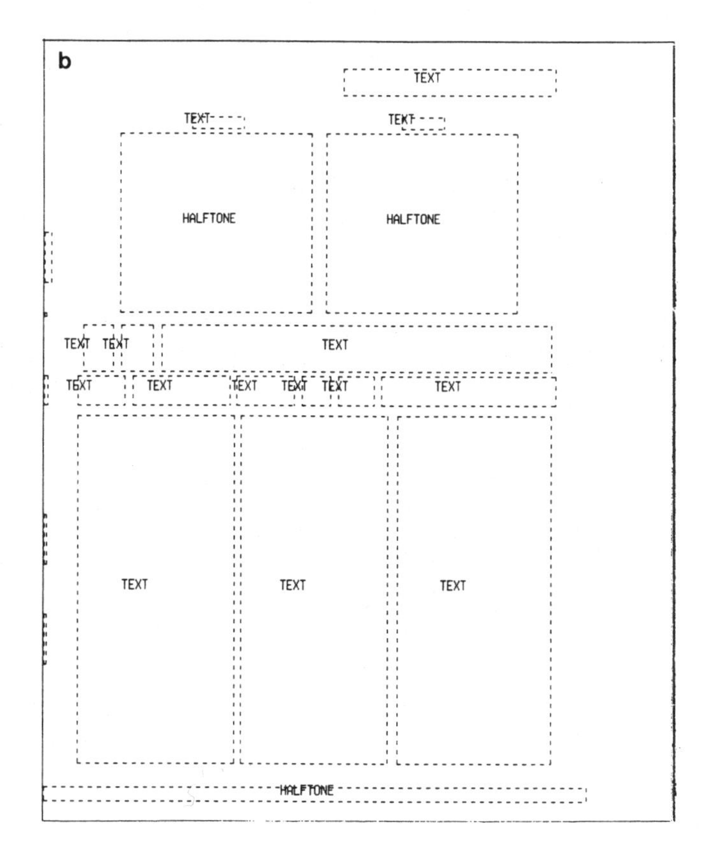

FIG. 15. (a) A document page image with complicated layout. (b) The segmentation and classification result.

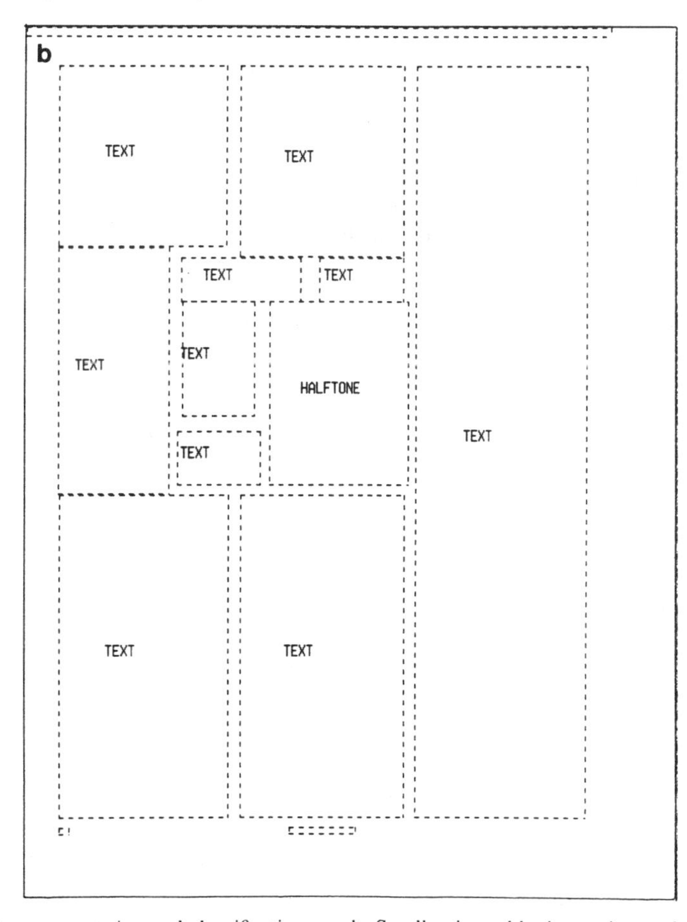

FIG. 16. (a) A document page image with halftone and insets. (b) The segmentation and classification result. Small column blocks are ignored by the classification process.

TABLE 4
Parameters of the Classification Algorithm

Parameter	Value in classification
Weight vector **w**	(5.33, −21.6, −21.50, −25.22)
Threshold weight w_0	−22.0
Ratio *b/w*	0.2

tion analysis might be necessary. During our experiments very little tuning of parameters took place.

ACKNOWLEDGMENTS

This work was supported in part by a grant from Ricoh R&D of Japan. We thank Dr. Shunji Mori, formerly of Ricoh R&D, for his encouragement of our work and for helpful discussions. The scanline correlation strategy for classification of text versus halftone was developed while T. Pavlidis was a consultant for Ricoh R&D. He acknowledges helpful interactions, in addition to Dr. Mori, with H. Nishida and T. Saitoh of Ricoh R&D in Japan and with Dr. K. Ejiri, then of the California Ricoh Research Center (now at Ricoh R&D).

REFERENCES

1. S. C. Hinds, J. L. Fisher, and D. P. D'Amato, A document skew detection method using run-length encoding and the Hough transform, in *10th International Conference on Pattern Recognition, Atlantic City, New Jersey, 16–21 June 1990,* Vol. 1, pp. 464–468, IEEE Computer Society Press.
2. W. Postl, Halftone recognition by an experimental text and facimile workstation, in *Proc. of the 6th ICPR, Munich, Germany, Oct. 1982,* Vol. 20, pp. 489–491.
3. L. A. Fletcher and R. Kasturi, A robust algorithm for text string separation from mixed text/graphics images, *IEEE Trans. Pattern Anal. Machine Intell.* **10** (6), 1988, 910–918.
4. F. M. Wahl, K. Y. Wong, and R. G. Casey, Block segmentation and text extraction in mixed text/image documents, *Comput. Vision Graphics Image Process.* **20,** 1982, 375–390.
5. T. Saitoh and T. Pavlidis, Page segmentation without rectangle assumption, *in Proc. 11th International Conference on Pattern Recognition, Le Hague, Natherland, Aug. 31–Sept. 3 1992,* vol. 2, pp. 277–280.
6. H. S. Baird, S. E. Jones, and S. J. Fortune, Image segmentation by shape-directed covers, *in "Proceedings 10th ICPR, Atlantic City, New Jersey, June, 1990,* pp. 820–825.
7. M. Nadler, A survey of document segmentation and coding techniques, *Comput. Vision Image Graphics Process.* 28, 1984, 240–262.
8. G. Nagy, S. Seth, and S. D. Stoddard, Document analysis with an expert system, *in Proceedings, Pattern Recognition in Practice, Amsterdam, June 19–21, 1985,* Vol. II.
9. G. Nagy and S. Seth, Hierarchical representation of optically scanned documents, *in Proceedings of the IEEE 7th ICPR, Montreal, Canada, 1984.*
10. J. Higanisho, H. Fujisawa, Y. Nakano, and M. Ejiri, A knowledge based segmentation method for document understanding, *Proceedings 8th ICPR, Paris, France, 1986,* pp. 745–748.
11. J. Toyoda, Y. Noguchi, and Y. Nishimura, Study of extracting Japanese newspaper article, *in 6th ICPR, October 1982,* pp. 1113–1115.
12. K. Kubota, O. Iwaki, and H. Arakawa, Document understanding system, *Proceedings 7th ICPR, Montreal, Canada, July 1984,* pp. 612–614.
13. E. Meynieux, S. Seisen, and K. Tombre, Bilevel information coding in office paper documents, *in Proceedings 8th ICPR, Paris, France, 1986,* pp. 442–445.
14. D. Wang and S. N. Srihari, Classification of newspaper image blocks using texture analysis, *Comput. Vision Image Graphics Process.* **47,** 1989, 327–352.
15. H. Kida, O. Iwaki, and K. Kawada, Document recognition system for office automation, *Proceedings 8th ICPR, Paris, France, 1986,* pp. 446–448.
16. T. Akiyama and I. Masuda, A segmentation method for document images without the knowledge of document Formats," *Trans. IECE Jpn.* **J66-D** (1), Jan. 1983, pp. 111–118. [in Japanese]
17. I. Masuda, N. Hagita, T. Akiyama, T. Takahashi, and S. Naito, Approach to smart document reader system, *in Proceedings IEEE Conf. on CVPR, San Francisco, California, June 1985,* pp. 550–557.
18. L. Abele, F. Wahl, and W. Scherl, Procedures for an automatic segmentation of text, graphic and halftone regions in documents, *in Proc. of the 2nd Scandinavian Conference on Image Analysis, Hellsinkii, 1981,* pp. 177–182.
19. T. Pavlidis, *Structural Pattern Recognition,* Springer-Verlag, Berlin/New York, 1977.
20. F. M. Wahl, *A New Distance Mapping and Its Use for Shape Measurement on Binary Image Data,"* Research report, Vol. RJ3438, IBM Research Laboratory, San Jose, CA, 1982.
21. T. S. Huang, Run-length coding and its extensions, in *Picture Bandwidth Compression* (O. J. Tretiak, Ed.), pp. 231–264, Gordon and Breach, 1972.
22. R. Ulichney, *Digital Halftoning,* MIT Press, Cambridge, MA, 1987.
23. R. Floyd and L. Steinberg, An adaptive algorithm for spatial grey scale, *SID Digest* **75,** 1975, 36–37.

From Paper to Office Document Standard Representation

Andreas Dengel, Rainer Bleisinger, Rainer Hoch, Frank Fein, and Frank Hönes
German Research Center for Artificial Intelligence (DFKI), PO Box 2080, D-6750 Kaiserslautern, Germany

The task of a paper-computer interface is to transform printed information into a symbolic representation. For integration of such interfaces into office information systems, we must consider standards for electronic document representation into which the information can be transformed. The international standard ODA[1] (Office Document Architecture) provides mechanisms to both generate and exchange electronic documents, plus achieve a fundamental common understanding of their structure.

This article presents the principles of the model-based document analysis system we call ΠODA (Paper interface to ODA) that we developed as a prototype for the analysis of single-sided business letters in German.[2] ΠODA is based on the architecture model of ODA, which has been enhanced to meet the requirements of document image analysis.

Initially, ΠODA extracts a part-of hierarchy of nested layout objects (such as text-blocks, lines, and words) based on their presentation on the page. Subsequently, in a step called *logical labeling*, the layout objects and their compositions are geometrically analyzed to identify corresponding logical objects that can be related to a human perceptible meaning (for example, sender, recipient, and date in a letter). Finally, a context-sensitive text recognition for logical objects is applied using logical vocabularies and syntactic knowledge. As a result, ΠODA produces a document representation that conforms to ODA.

Figure 1 shows all analysis steps and the ΠODA document architecture model. The individual steps are primarily sequential, but allow backtracking. Thus, each step of ΠODA can confirm or reject the results of preordered steps.

Document architecture model in ΠODA. Paper documents usually have

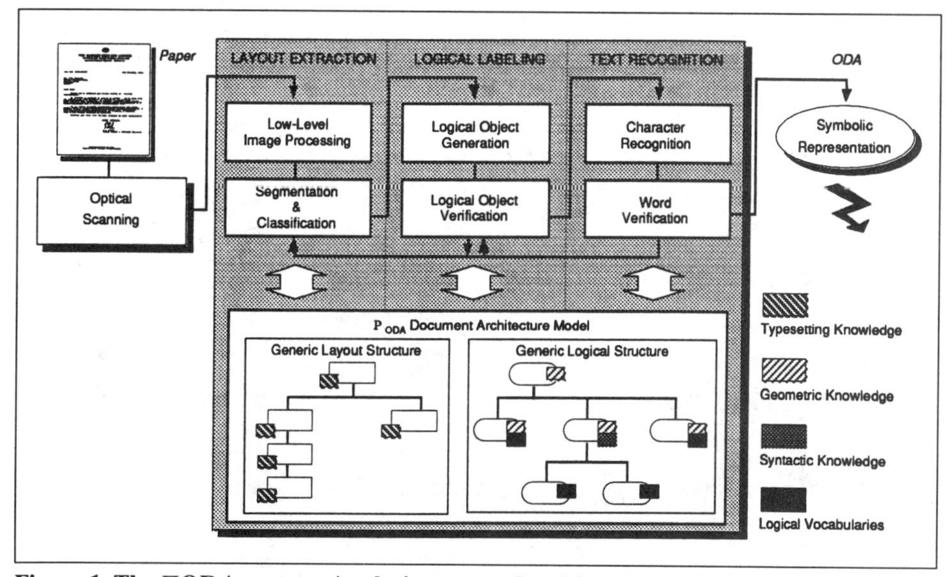

Figure 1. The ΠODA system: Analysis steps and architecture model.

Reprinted from *Computer,* Vol. 25, No. 7, July 1992, pp. 63–67.

a high degree of structure, a fact that a human reader seldom reflects on but nevertheless uses directly to filter relevant information out of printed data. Certain document classes (for example, scientific papers, letters, or forms) have characteristic structures that can be specifically described.

The ΠODA document architecture includes two generic tree structures, a layout structure and a logical structure. They provide different but complementary views of the same contents. Elements of these structures are called objects. An object that is not subdivided into smaller objects (that is, a leaf of a tree) is called a basic object; other objects are called composite objects. They are considered prototypes (classes) for specific objects (instances) specified by a set of characteristics — a pattern that is common to its members. Such a pattern includes methods for generating specific objects, methods to determine the values of attributes, and methods to control consistency among objects. This architectural basis of ΠODA is analogous to the generic structure concept of ODA.

In ΠODA, the document type "business letter" is represented by a generic logical structure and a generic layout structure that specify and aggregate relevant objects. These structures provide a means to generate an ODA-conforming representation of a given document (that is, its specific layout structure as well as its specific logical structure). Then, the specific basic objects are related to so-called content portions (that is, type text, line drawing, or photograph).

In document image analysis, the objective is to transform document structures from a nonelectronic medium, like paper, into an electronic medium. Here, questions arise regarding how to extract layout objects automatically from a given document image, how to identify corresponding logical objects, and how to recognize the contents of the document. Thus, document image analysis can be considered automatic generation of an electronic document in the sense that it electronically reproduces an existing nonelectronic document. Therefore, the generic structure concept of ODA is an ideal orientation, but one that must be enhanced according to the requirements of document image analysis. For that purpose, knowledge sources are attached to each generic class to facilitate the analysis. For example, typesetting knowledge is associated with layout classes. Geometric knowledge, logical vocabularies, as well as syntactic knowledge (grammar) are associated with logical classes. This enhanced architecture of ΠODA allows a model-based analysis of letter images.

Document image analysis. Transformation of printed business letters into an ODA-conforming format involves three ΠODA analysis steps: layout extraction, logical labeling, and text recognition.

Layout extraction. To obtain a paper-document bitmap representation, we use a 300 dots-per-inch flatbed scanner. For layout extraction, ΠODA initially performs low-level image processing, including image encoding and skew adjustment.

ΠODA uses a segmentation method called *smearing* derived from the run-length smoothing algorithm (RLSA). Smearing is an adequate technique for segmenting down to word level and is easy to implement. However, the method is sensitive to mixed font sizes.

Smearing extracts image blocks rep-

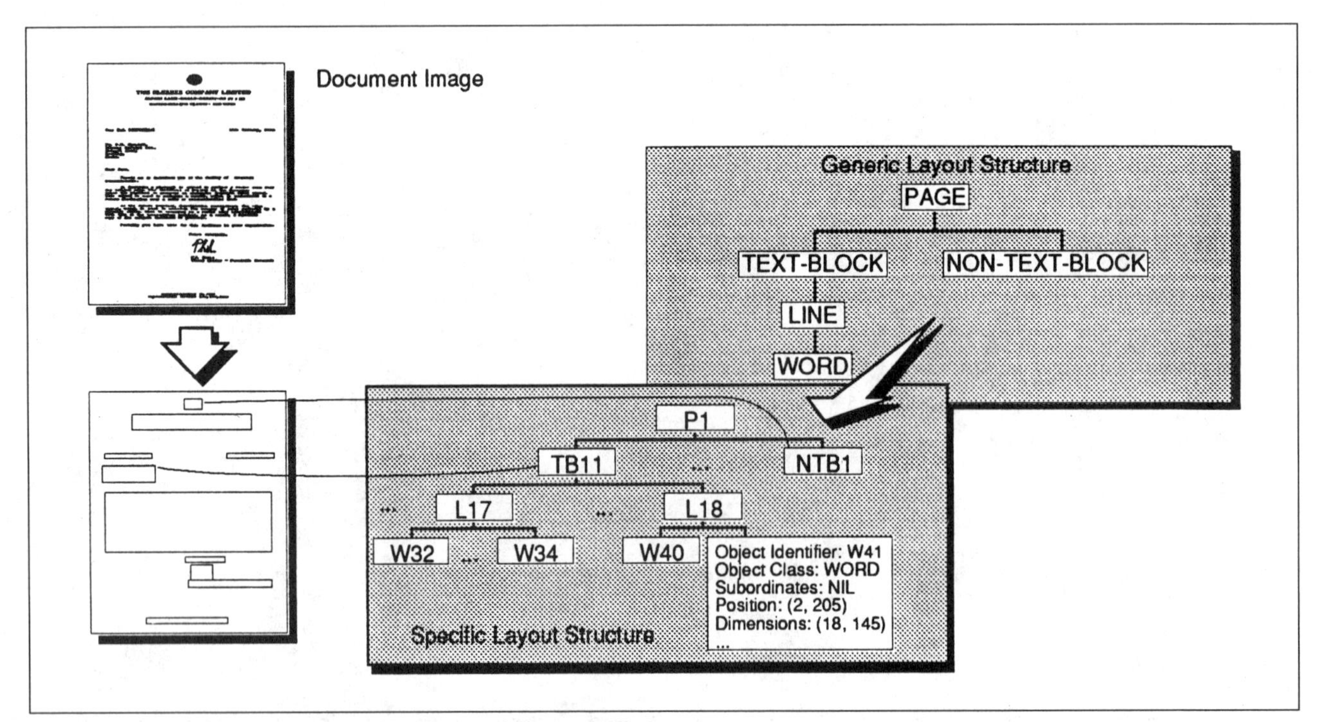

Figure 2. A document image and its corresponding specific layout structure.

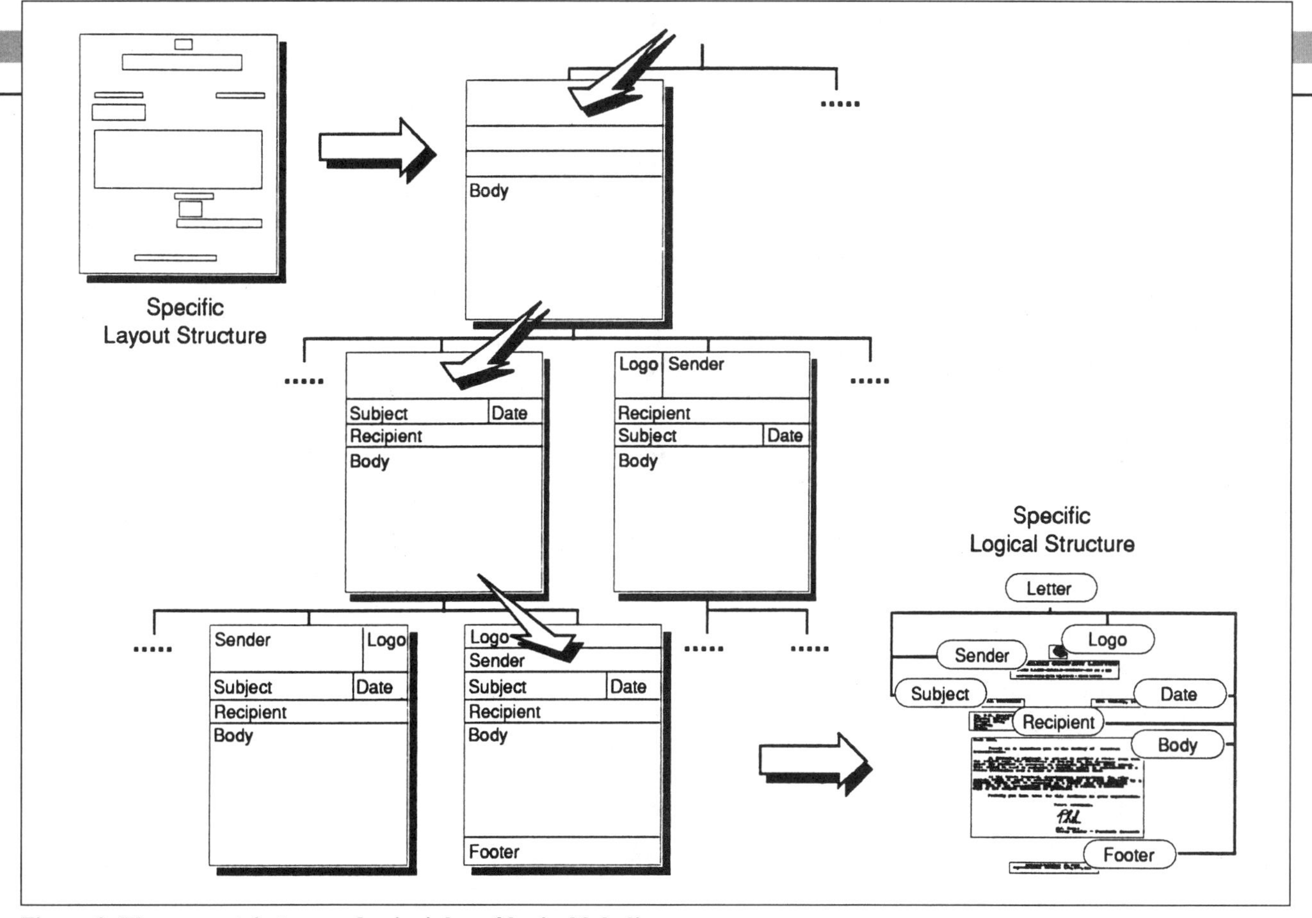

Figure 3. The geometric tree and principles of logical labeling.

resented by specific layout objects. Each object is specified by its position and dimensions, which serve as references to corresponding images. Starting with the raster image of the full page, the segmentation process recursively traverses the generic layout structure, instantiating layout objects and segmenting them into subordinates by applying the associated typesetting knowledge in the corresponding classes. Each traverse stops when a basic object is reached. Thus, the structures inside a TEXTBLOCK — LINEs and WORDs in our model — are reconstructed by means of specific layout objects (see Figure 2).

Table 1. Some geometric rules for ΠODA recipients.

Feature	Rule	Measures of Belief (MB)	Measures of Disbelief (MD)
vertical_origin	0 ≤ vertical_origin ≤ 0.25	0.89	—
	0.25 < vertical_origin ≤ 0.33	0.10	—
	0.33 < vertical_origin	—	0.99
	:		
number_of_lines	number_of_lines < 4	—	0.95
	4 ≤ number_of_lines ≤ 6	0.92	—
	number_of_lines = 7	0.02	—
	number_of_lines > 7	—	0.99

Logical labeling. Using a hypothesize and test strategy, ΠODA analyzes the specific layout objects geometrically. For that reason, arrangement alternatives as well as local geometric features of logical objects are attached to the generic logical classes as appropriate knowledge sources.

In a *geometric tree*, various arrangements of logical objects within business letters are described on different specification levels[3] (see Figure 3). The root of the tree represents the most general arrangement. Every single-sided document belongs to this class. The internal representation is organized so that logical object arrangements of parental nodes are inherited by children. For our experiments, we use a tree with about 40 terminal nodes.

In addition, an individual logical object can be described independent of the text it contains solely by its shape. Therefore, each generic logical object includes geometric knowledge capturing statistical results obtained from evaluating geometrical aspects of about 190 business letters. In particular, we examined all logical objects relative to their intrinsic geometric characteristics. The probabilities are represented in subsets of rules as measures of belief (MB), and some of them are applied as measures of disbelief (MD = 1 − MB). Table 1 reflects geometric rules for recipients in ΠODA.

In the table, MB and MD values range between 0 and 1. The values in the rules are relative to the width and height of a page. Actually, in ΠODA, 12 logical objects are each described by 11 different geometric features.

Logical labeling of a given document amounts to finding a path from the root of the geometric tree to one of its leaves, thereby matching the specific layout structure stepwise against the alternative arrangements. All nodes visited are instantiated before validating their relevance, so that their evidence can be tagged on. If the refinement in a tree node fits with the specific layout (very little cut shifting is allowed), the appropriate label is assumed to be a hypothesis. In such a case, a specific logical object is generated as an instance of the corresponding generic logical class. For verification, all layout objects associated with the logical object are compared to the geometric rules in the appropriate knowledge slot of its class. The MBs and MDs that are obtained are combined by Dempster-Shafer's rules of combination. All hypotheses are pushed into an agenda, which defines the next step in the tree. If logical labeling fails, ΠODA initiates backtracking to the layout extraction step for resegmentation. Figure 3 shows the representation of the geometric tree indicating the principles of logical labeling.

Text recognition. Here, specific basic layout objects describing word images are sent to a commercial text-recognition system. The output of that system comprises complete or fragmentary strings representing word hypotheses. Because logical labeling provides the identification of document constituents (that is, a restriction of context), the subsequent text verification can be performed in an expectation-driven manner. Consequently, word hypotheses are verified by matching corresponding strings against legal words of specific logical objects (such as a set of names as possible recipients of business letters). In particular, a letter tree (*trie*) is combined with a selective-access matrix that allows efficient matching of fragmentary input strings against legal words.[2]

For refinement of the specific logical structure and controlled selection of vocabularies, syntactic knowledge is attached to some of the generic logical classes — in particular, those representing well-structured parts of a letter (such as address and date). This knowledge determines the order in which subordinated logical objects (for example, ZIP code, city, and street) can occur and, therefore, controls the access to corresponding logical vocabularies. Vocabularies and grammars are used not only to assist the recognition process, but also to confirm or refute hypotheses of logical labeling; therefore, they can lead to a backtracking. As a result, all content portions (see the section entitled "Document architecture model in ΠODA") of type "image" related to layout objects of the word level are converted into content portions of type "text" where alternative and fragmentary results are allowed.

Figure 4 shows an example of the output produced by ΠODA. This representation of a document conforms to ODA.

Conclusions. The ΠODA system has been implemented on Sun Sparcstations. Besides transforming printed information into the international standard ODA, the design of ΠODA reveals a fundamental advantage: All stages of document image analysis, layout extraction, logical labeling, and text recogni-

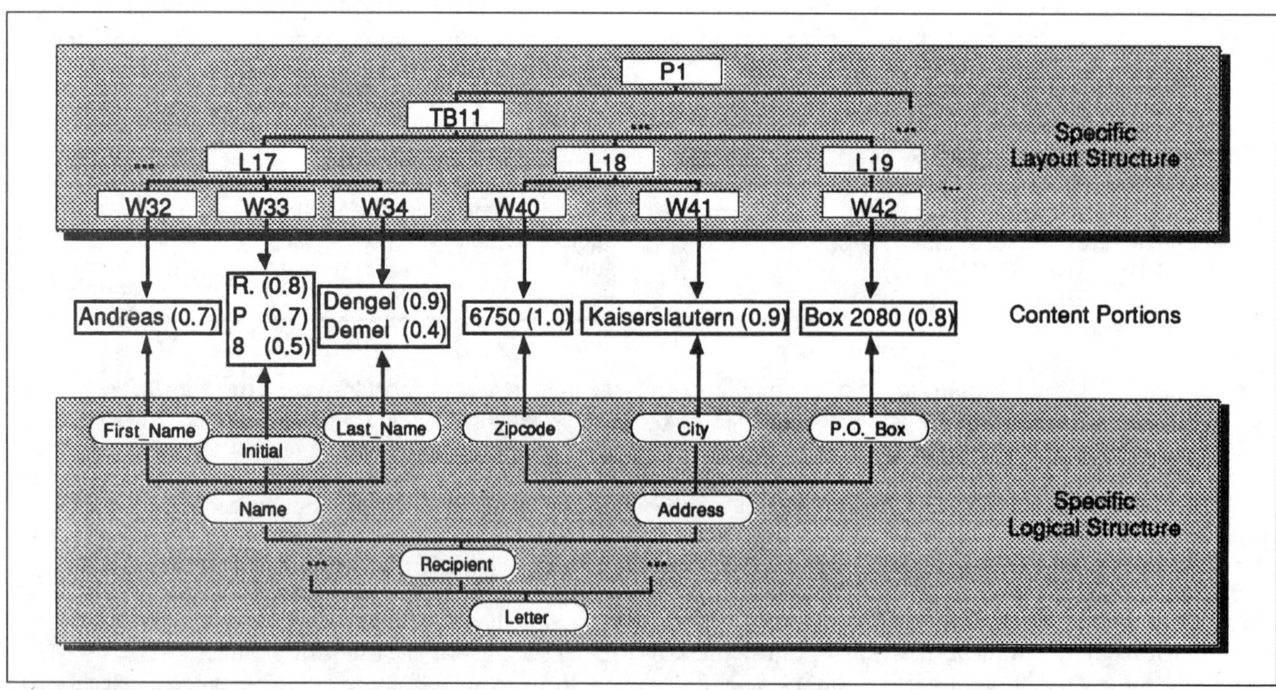

Figure 4. The representation of ΠODA results.

tion are guided by the ΠODA architecture model and additional knowledge. Obviously, such a model-based approach provides optimum flexibility because alternative analysis methods can be used for individual steps where only I/O interfaces of the distinct phases are predefined.

We are now working to eliminate existing weaknesses. One is the lack of a robust and multipurpose technique for text/graphic separation. The commercial system we use produces only one hypothesis for each character. Therefore, we are developing our own character recognition system for obtaining alternative candidates.

Our future work will also concentrate on an expectation-driven partial text analysis of logical objects, including the design of a structured lexicon. ■

Acknowledgment

This work has been supported by Grant No. FKZ-ITW-9003-0 from the German Federal Ministry for Research and Technology.

References

1. W. Horak, "Office Document Architecture and Office Document Interchange Formats: Current Status of International Standardization," *Computer*, Vol. 18, No. 10, Oct. 1985, pp. 50-60.

2. A. Dengel et al., "ΠODA: The Paper Interface to ODA," German Artificial Intelligence Research Center Research Report RR-92-02, Kaiserslautern, Germany, Feb. 1992.

3. A. Dengel and G. Barth, "Anastasil: Hybrid Knowledge-Based System for Document Image Analysis," *Proc. IJCAI 89, Int'l Joint Conf. Artificial Intelligence*, Vol. 2, Morgan Kaufmann, San Mateo, Calif., 1989, pp. 1,249-1,254.

Andreas Dengel has been with the German Research Center for Artificial Intelligence (DFKI) since 1989. He first served as assistant to the director and is currently the head of the Office Automation Research Department. His major research interests are in applications of AI and integration of methods of pattern recognition, linguistics, and psychology.

Rainer Bleisinger has been with DFKI since 1989 and serves as a research scientist. Involved in document analysis, he concentrates on modeling aspects and methods for information extraction. His major research interests are AI techniques, especially temporal reasoning, knowledge representation, and text analysis.

Rainer Hoch joined DFKI in 1991 and serves as a research scientist. His major research concerns the design of a dictionary for document analysis, document representation, hypertext, and partial text analysis strategies.

Frank Fein has been with DFKI since 1990 and serves as a research scientist. His major research interests include image processing techniques and mechanisms for analysis control.

Frank Hönes joined DFKI in 1989 and serves as a research scientist. His research interests are digital image analysis, text recognition, and AI techniques for document analysis.

Historical Review of OCR Research and Development

SHUNJI MORI, MEMBER, IEEE, CHING Y. SUEN, FELLOW, IEEE, AND KAZUHIKO YAMAMOTO, MEMBER, IEEE

Invited Paper

In this paper research and development of OCR systems are considered from a historical point of view. The paper is mainly divided into two parts: the research and development of OCR systems, and the historical development of commercial OCR's. The R&D part is further divided into two approaches: template matching and structure analysis. It has been shown that both approaches are coming closer and closer to each other and it seems they tend to merge into one big stream. On the other hand, commercial products can be classified into three generations, for each of which some representative OCR systems are chosen and described in some detail. Some comments on recent techniques applied to OCR such as expert systems, neural networks, and some open problems are also raised in this paper. Finally we present our views and hopes on future trends in this fascinating area.

Keywords— *Optical character recognition; feature extraction; template matching; structural analysis; learning; Chinese character recognition; practical OCR systems.*

I. Introduction

The history of science and technology does not flow like a straight canal, but is usually tangled like a meander. We will describe not only the main stream, but also the resulting impact, as when oxbow lakes are generated after a meander changes its direction. The history of OCR research, like that of speech recognition, is comparatively old in the field of pattern recognition. In the early days of pattern recognition research, almost everyone took the subject of OCR. One reason was that characters were very handy to deal with and were regarded as a problem which could be solved easily. However, against what was the expectation of many people, after some initial easy progress, great difficulty in solving this problem surfaced. Hence, people diversified their interests over a wide range of topics in the pattern recognition field, for example, image understanding and 3-D object recognition. Of course there were practical demands for such research. A new field always gives benefit to its pioneers, but research on these topics of pattern recognition seems to be confronted with a strong barrier. In this sense, the topic of OCR is not so exceptional, but rather is universal in that it includes essential problems of pattern recognition which are common to all other topics. In this sense, we have written its history from as general a point of view as we could. Actually the problem is most profound and indeed we realized this when writing this monograph.

On the other hand, research cannot exist without its applications in engineering. Fortunately, market demand for OCR is very strong even though word processors are prevalent. For example, a dozen leading companies in Japan sell or are preparing to sell hand-printed Kanji character readers. So far these sophisticated machines are not prevalent yet, but it is certain that if the price and performance meet the requests of the users, these machines will be widely used in offices as a very natural man–machine interface. The accumulation of OCR knowledge is reducing the gap between the users and makers, which is also helped by the rapid development of computer technology.

This historical review is roughly divided into three parts. The first part is a prelude. The second and third parts constitute the main body of the paper, i.e., research and products respectively. The research part not only has its own right to exist; it also provides a preparation for the products part, so that the reader can understand the products more easily and deeply in terms of technological development. The research part is further divided into two approaches: template matching and structure analysis. This paper shows that the two approaches are converging. That is, the template matching approach has been absorbing structure analysis techniques and now the two approaches seem to be on the verge of fusion. On the other hand, we classify commercial products into three generations, for

Manuscript received March 18, 1991. C. Y. Suen's research was supported in part by the National Networks of Centers of Excellence program of Canada and by the Natural Sciences and Engineering Research Council of Canada.

S. Mori is with the Research and Development Center, Ricoh Company, Ltd., 16-1, Shinei-cho, Kohoku-ku, Yokohama, 233 Japan.

C. Y. Suen is with the Department of Computer Science, Concordia University, 1455 de Maisonneuve Bld. W., Montreal, Quebec, H3G 1M8, Canada.

K. Yamamoto is with the Electrotechnical Laboratory, 1-1-4, Umezono, Tsukuba Science City, Ibaraki, 305 Japan.

IEEE Log Number 9203434.

each of which we choose representative OCR's and describe them in some detail. Finally we comment on expert system and neural network applications to OCR.

The description might be biased toward research and development in Japan, but it reflects the fact that research and development in OCR has been particularly active and prosperous in Japan. For many years the Electrotechnical Laboratory (ETL), in Ibaraki, has played a key role in developing OCR technology in Japan. Another reason is that many important papers have not been translated into English and so we thought that this would be a good opportunity to introduce some of them to the international community. Nevertheless, in writing this paper we have found several reference books and review papers which have been very useful, among them the books by Ullman [1], Sakai and Nagao [2], Pavlidis [3], and Mori and Sakakura book [4] and papers by Suen *et al.* [5], Schurman [6], and Couindan and Shivaprasad [7]. Other books and review papers are referred to at the appropriate places, but it is difficult to read all the numerous papers related to the topic. We thus avoided biographical description, and pursued the streams of research. Here, we do not mention the very important research fields of document analysis and cursive script recognition because other papers in this issue make reference to them. However, we might have missed some important papers or patents. Actually we found a few new papers which were very useful in preparing this paper. We would appreciate criticism on this paper from readers and we hope that it will be useful to researchers in advancing the technology of OCR.

II. Dawn of OCR

In 1929 Tausheck [8] obtained a patent on OCR in Germany and in 1933 Handel [9] did the same in the U.S. These are the first concepts of the idea of OCR as far as we know. At that time certain people dreamed of a machine which could read characters and numerals. This remained a dream until the age of computers arrived, in the 1950's. However, we think their basic idea is worth mentioning, because it is still alive. In this sense we introduce Tausheck's patent. The principle is template/mask matching. This reflects the technology at that time, which used optical and mechanical template matching. Light passed through mechanical masks is captured by a photodetector and is scanned mechanically. When an exact match occurs, light fails to reach the detector and so the machine recognizes the characters printed on paper.

Mathematically speaking, the principle is the axiom of superposition, which was first described by Euclid as the seventh axiom in the first volume of *Elements*. For humans, however, E has the same meaning as an $\mathcal{E}$ in the sense of pattern. Therefore, what is the principle of their equivalence? So far no general solution has given and yet it is the principal and central problem in pattern recognition. The seventh axiom is the first principle proposed for equivalence of shape. We will return to this problem later. We will see that the principle of superposition has been realized by employing more advanced technologies of hardware such as cathode ray tubes and analog electric circuits. Actually, this original work is the origin of the main stream of OCR technology. The "template matching method" is in a broad sense the principle of superposition.

III. Age of Cut and Try

The first commercial computer, UNIVAC I, was installed and began to work at the Bureau of Statistics in the U.S. in 1951. In terms of hardware, electronics was the basis of the age of computers. First of all, electronics made engineers regard OCR as a possible reality. However, there were strong limitations in terms of the quantity and complexity of the hardware.

A. Template-Matching Methods

The basic reduction in complexity was archieved by projecting from two-dimensional information onto one. This approach was taken by Kelner and Glauberman [10] using a magnetic shift register in 1956. An appropriately placed input character is scanned vertically from top to bottom by a slit through which the reflected light on the printed input paper is transmitted to a photodetector. It is a simple calculation using only algebraic addition to obtain a value which is proportional to the area of the black portion within the slit which segments the input character. Then the sampled values are sent to the register to convert the analog values to digital ones. Template matching is done by taking the total sum of the differences between each sampled value and the corresponding template value, each of which is normalized. The machine was not commercialized.

We note here a very important point in the matching process, concerning the general problem of registration. The template matching process can be roughly divided into two processes, i.e., superimposing an input shape on a template and measuring the degree of coincidence between the input shape and the template in the two cases mentioned above. Projection is taken horizontally or vertically, which makes the superimposing processes position invariant in one direction. This is clearly illustrated in Fig.1. When the slit is long enough to cover the input numerals, there is no change in the value of black area projected on the x axis, even if the numerals are moved vertically. However, we need to detect the starting and ending points of an input numeral to register it against the corresponding template. This is done very easily because numerals are all simply connected and there is enough space in each interval between neighboring numerals. Actually, the projection technique has been broadly used to segment the input character string and picture region of the documents, for example, in current OCR's. Such processing is called preprocessing in the terminology of OCR.

The two topics mentioned above tell us that in essence characters contain two-dimensional information. If we want to reduce the dimension to one, then we have to distort the shapes of the characters so that the machine can recognize it. Such distortion may be allowed for numerals which have a small number of characters. In this sense, MICR has a very limited application, although it is widely used by

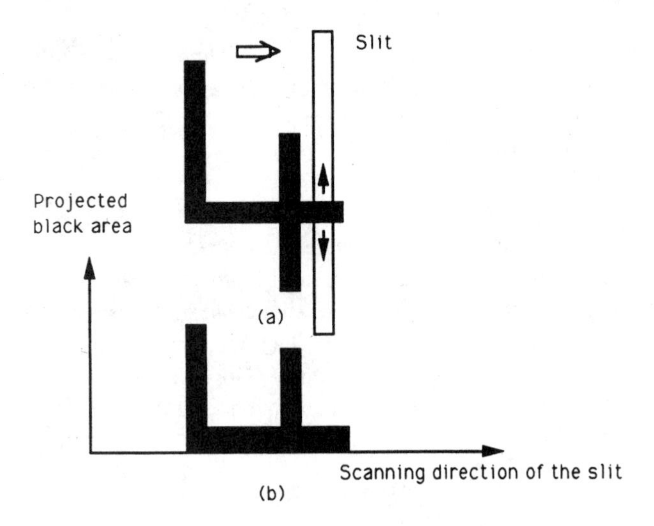

Fig. 1. Illustration of 2-D reduction to 1-D by a slit. (a) An input numeral "4" and a slit scanned from left to right. (b) Black area projected onto x axis, the scanning direction of the slit.

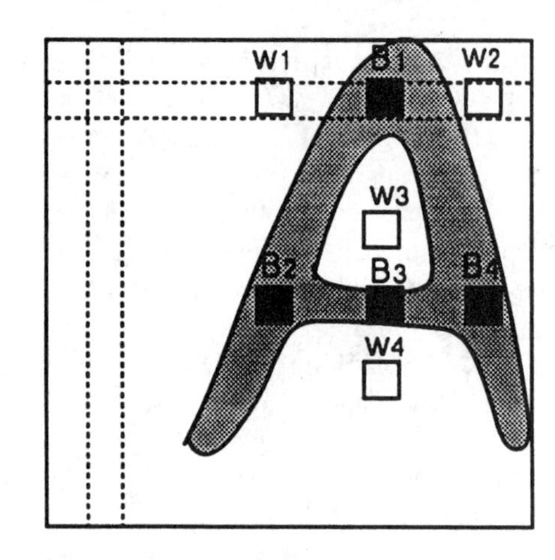

Fig. 2. Illustration of the peephole method.

banks. However, it has naturally a major problem from the man–machine interface point of view. So people proceeded to deal with two-dimensional information.

When looking at two-dimensional information, it seems natural to use optical techniques to perform template matching. Actually a very sophisticated OCR was made combining electronics and optical techniques by Hannan [11] (RCA group) in 1962. At that time RCA had the most advanced electron tube technology in the world, which was fully employed in the OCR research work. Hannan concluded his paper as follows: "In summary, the test results of this program proved that the RCA optical mask-matching technique can be used to reliably recognize all the characters of complete English and Russian fonts (91 channels are necessary)." However, no announcement was made for a commercial RCA OCR based on the techniques. The great experiment ended without a successor.

It is very natural that the advent of computers influenced the design of OCR with respect to hardware and algorithms. We introduce a logical template matching method. The simplest one is called the peephole method. First of all we assume that an input character is binarized.*Binarization* is an important preprocess in OCR technology. Ideally an input character has two levels of density, i.e., black and white, commonly represented by 1 and 0 respectively . However, real data are not always so. We will discuss this problem later. Here we note that binarization is not an easy matter to deal with.

Imagine a two-dimensional memory plane on which a binarized input character is stored and registered in accordance with some rule, with the character positioned at the top right corner, for example, as shown in Fig. 2. Then obviously for an ideal character, which has a stroke of constant size and width, black portions are always black and the same is true for the white background. Then appropriate pixels are chosen for both black and white regions so that the selected pixels can distinguish the input character from characters belonging to other classes. Looking at Fig. 2, a so-called logical matching scheme is easily constructed, which is called the peephole method.

The first OCR based on the peephole method was announced by Solatron Electronics Group Ltd. [12] and was called ERA (Electric Reading Automation) in 1957. The characters read were numerals printed by a cash register. The reading speed was 120 characters/second (chs), which was very high. This was due to the simple logic operations used. The total number of peepholes was 100, which is considerably greater than the ideal number of $\lceil log_2 10 \rceil = 4$ which would be needed to obtain stable recognition for real data.

At ETL an OCR was designed based on the same scheme by Iijima *et al.* [13] in 1958. However, the design was more systematic than ERA using three-level logic and so it was more efficient. The characters that could be recognized were 72 alphanumerals; 10×12 meshes were used. The total number of peepholes used was 44 at 10 pixels/character. Logic circuits for the ETL Mark IV computer were used. Actually the OCR was one component of a larger system which was planned at ETL at that time. It was to be an input device for a machine translator.

Autocorrelation: As we mentioned above, two-dimensional template matching has a weakness in registration. Researchers became aware of that and began to devise new methods which are shift invariant. Two methods attracted attention in particular. One is based on an autocorrelation method and the other is based on a moment method. The latter is very ambitious, aiming at both shift and rotation invariance and will be discussed later together with the Fourier series, a method which is also shift invariant.

An exact formulation of the autocorrelation method is easily derived; therefore simulation can be done before the hardware is made. This method is commonplace now. This was the first time that this approach could be taken. For example, in 1958, the IBM 7090 was announced. Rapidly this powerful computer became available at certain research centers. Horowitz and Shelton [14] of IBM did very extensive research on the autocorrelation method in 1961. They proposed a very large special-purpose shift register machine for calculating autocorrelation exactly and with high speed. In the same year, Sato *et al.* [15] of the

Japanese Government's Radio Wave Research Laboratory carried out a very systematic simulation of the autocorrelation method. The results were unfortunately disappointing. The difference between "R" and "B" is only 0.4% when normalized with a maximum output value. In addition, the differences between character pairs "K" and "R," "A" and "V," and "U" and "D" are less than 1%.

B. Structure Analysis Method

The principle underlying template matching is really only appropriate for the recognition of printed characters. However, we have another set of hand-printed/handwritten characters that need consideration. The variation of shape of handwritten characters is so large that it is difficult to create templates for them. A so-called structure analysis method has been applied to handwritten character recognition. However, in the early stages of OCR development, we note that some very primitive methods were considered in addition to template matching. The weakness of these methods was due to the constrained hardware resources of that time. We include those methods in the structural analysis method, as described in the following subsection. However, it will be shown that these methods mark a continuation of logical template matching in terms of concept. Actually, contrary to the above description, these simple methods were applied to stylized fonts, while some were applied to the recognition of constrained hand-printed characters.

In the case of the structural analysis method, there is no mathematical principle. Rather it is still an open problem and there is no sign that it will be solved in the near future. Hence, our intuition has been the most reliable weapon in attacking this problem. However, there appear to be certain informal strategies that can be used in structure analysis. First of all, we give a very general and basic idea of the conceivable strategies. Since a structure can be broken into parts, it can be described by the features of these parts and by the relationships between these parts. Then the problems are how to choose features and relationships between them so that the description gives each character clear identification. Feature extraction, therefore, has become the key in pattern recognition research .

1) Slit/Stroke Analysis: A specific description will be given along this line. It has already been mentioned that the peephole method is considered a kind of template matching method. Now we try to extend it to the structure analysis method. The peephole is not always limited to a single pixel. Instead, it can be expanded into a slit or window whereby it is not necessary to fix at a specific position on the two-dimensional plane. The logical relationship between two pixels can be extended to a general relationship between them. The above description is illustrated in Fig. 3.

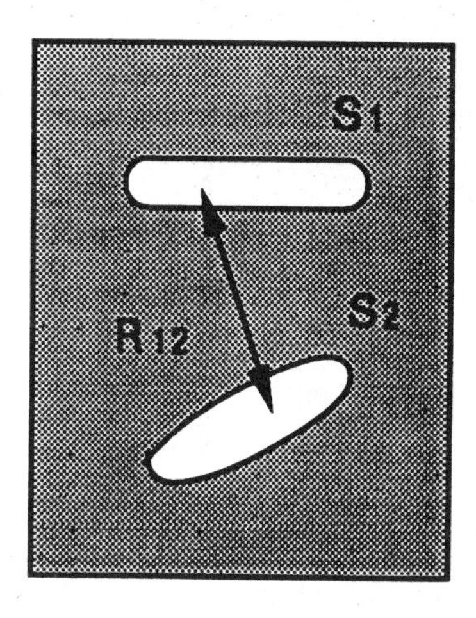

Fig. 3. Extension of the peephole method to structure analysis.

Perhaps the simplest example is the so-called cross counting technique, in which scanned lines are regarded as slits. The feature of the slit is the number of black regions in it. Rohland [16] proposed this technique in 1954, in which a vertical scan was primarily used. In 1961, Weeks [17] used this approach in a simpler fashion. In this method scanning is taken in four directions, i.e., vertical, horizontal, and two orthogonal diagonals, for each of which six equally spaced and parallel lines are used to cover a character. A crossing count, defined as the number of times each of the six rasters crossed black, is made. However, three or more crossing counts are regarded as three and so counts of 0, 1, 2, and 3 are possible. Thus, an input character is coded into a $6 \times 6 \times 4 = 144$ bit binary pattern, which is used for recognition based on a statistical decision method. The statistical decision method is a very important theoretical approach to pattern recognition in general. However, this has already been reasonablly well established and so we recommend that readers refer to the excellent book by Duda and Hart [18], to Highleyman's paper [19], and to Nagy's paper [20] concerning OCR in particular.

Here we give a rigorous description of the above scheme in a mathematical sense. The slits are usually convex regions, within each of which we can detect both topological and geometrical features. Connected components and black areas are typical features. The former is just the cross count and the latter is the length or width of the black portion. These features are given as a function of slit shape and size, and its position. In the above cases, the slit is simply given by a straight line due to a simple scanning mechanism. However, we need not be restricted to this method of scanning.

Actually Johnson [21] in 1956 and Dimond [22] in 1957 used a sonde as a slit, which is shown in Fig. 4. For numerals there seems to be two basic points around which they are written. Therefore, it is very effective to span some sondes/slits centering the two points. Thus we can count crossing times in each slit and, on the basis of this, easily distinguish the numerals. The scheme can become tougher by assigning weight to each sonde, which was done by Kamensky [23]. However, the two-point scheme is not generally applicable to many characters and so a more general scheme, free of specific points, was considered by Glucksman [24], which will be mentioned later.

So far only topological features within the slit have been described, but geometrical features have been extensively used. In particular, we note that what is referred to as run-length coding is very typical and is a special case of slit

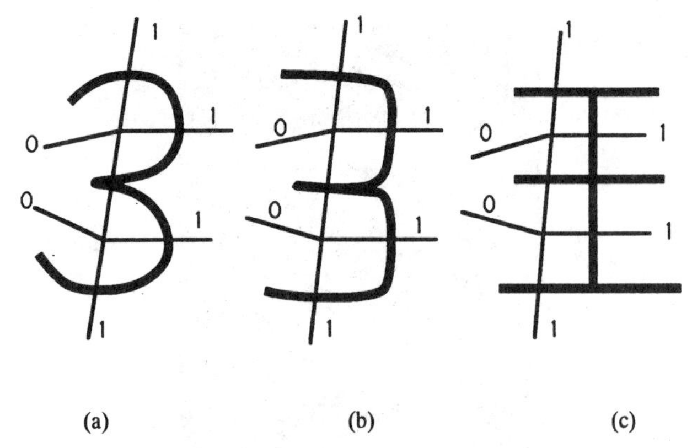

Fig. 4. Illustration of the **sonde** method; part (c) shows its weak points.

analysis. Actually run-length coding is regarded as a first step for more advanced structure analysis, which will be discussed in detail. Here, a very simple version of the geometric feature of a slit is introduced. Suppose there exists a base line such as a left/right vertical line. Then the geometrical feature is the distance from the base line to the first black pixel. If the slit is scanned from top to bottom of the character, for example, then we have a single value function, which is easy to analyze. In this case it is a convex function. Such approaches were taken by Doyle [25] in 1959. This approach can be easily expanded and will be discussed later.

Finally, as an example of the generalization of the slit/stroke analysis method, we can introduce a prototype based on an asynchronous reading method which is independent of scanning speed. This work was started by Sakai *et al.* [26] in 1963. On the other hand, the slit structure analysis approach is very effective so long as the number of categories of characters to be read is limited and the character shapes are mechanically designed as E-13B fonts in the MICR (magnetic ink character reader) used in processing checks. Some companies designed special fonts for OCR so that they could make use of the slit structure analysis, which will be described in the next section.

2) Hybrid of Template Matching and Structural Analysis: So far we have roughly classified the recognition methods into two classes, namely, template matching and structure analysis . However, there is an intermediate class between the two. In general, it is difficult to draw a clear boundary to separate the classes in the space of these recognition methods. At any rate, this trend is natural, because both methods have their own advantages. For example, template matching is very sensitive to positional change, but it can be very strong in the sense of global matching. On the other hand, structural analysis has the advantage of detecting the local stroke features of characters. From this the so-called zoned feature method was born. The general idea is that pixelwise matching is made loose, replaced by subregionwise matching, and the matching objects are local features within the subregion. Specifically, as the first step, an input character is set into a frame which is roughly partitioned into subregions, in each of which some local features are detected. Therefore the sensitivity to the position of the template matching becomes loose, because what is of concern is whether or not particular features lie within certain subregions.

For example, suppose that an input character is squeezed into a 15×15 array in which 5×5 pixels are taken as a subregion. Therefore, the array is partitioned into roughly a 3×3 array, in each of which four kinds of direction features, i.e., vertical, horizontal, and two orthogonal diagonal directions, are detected. Thus, an input character is coded into $3\times3\times4=36$ bit patterns. This is a tremendous reduction of $2^{15\times15}$ binary patterns. This example is a real one, with U.K. patent being granted to researchers at NEC [27]. A similar idea was proposed by Munson [28]. We can also find a similar idea in the study by Greanias *et al.* [29], based on which a famous OCR, the IBM 1287, was built. This machine also has the typical characteristics of the structure analysis method, i.e., a contour following method, and so it will be described in detail later.

IV. Commencement of Research

At the beginning stage it was thought that it would be easy to develop an OCR, and the introduction of a very rigorous reading machine was expected in the 1950's. Roughly speaking, the 1950's and the early half of the 1960's, the age of cut and try, were periods when researchers imagined an ideal OCR, even though they were aware of the great difficulty of the problem. Actually this is an instance of a common phenomenon which occurred in the research field of artificial intelligence in general. So some resorted to basic research, and engineers took a more systematic approach than ever. Such a trend was seen in both the template matching and the structure analysis method. Naturally it is difficult to draw a line to separate these research efforts, both chronologically and semantically. For example, basic research was done toward the end of the 1950's, as will be seen later, and research on the autocorrelation method could be regarded as basic research.

There are varieties of data for pattern recognition in general. Character data are the only specific ones in this sense. A digital picture, for example a bubble chamber photograph, represents another set of data. In such data, picture description is very important and takes precedence over identification. It was thought by some that character recognition was not very interesting, because its main purpose was class identification and so the approach was too engineering-oriented and not scientific. At that time it might have been true, but this is a misunderstanding. A description of character shape is very important when we want to proceed further. So basic research was needed and was done.

A. Template Matching Approach

1) Application of Information Theory: In logical template matching, an approach which IBM had taken for their OCR systems, the construction of the logic functions is quite tedious. Naturally a research effort on automatic generation of discriminate logic functions began at the IBM Watson Laboratory. Kamentsky and Liu [30], [31] gave such a scheme based on information theory. They gave a measure of discriminating power of a logic function, f, which was chosen randomly. Its configuration is $f(x_i, x_j, \cdots x_n)$, for which $x_i, x_j, \cdots x_n$ are randomly chosen pixels whose values are either 1 or 0. The function f has canonical form. Extensive experiments were done and this scheme seemed to be promising. Their scheme was elaborated by introducing the concept of distance. So an automatic design system for the discriminating functions was implemented on the powerful IBM computer. The IBM engineers anticipated the power of the program and applied it to their OCR design. We will continue this story later, which will provide a lesson.

2) Normalization: Iijima realized the importance of normalization in the template matching approach through his experiments and those of his group. Therefore, he thought that they should do basic research on this problem. He wrote a series of papers relating to basic research on pattern recognition, which was divided roughly into two parts. One is the theory on normalization of patterns [32] in 1962; the other is theory on feature extraction [33] in 1963. Both are very important and have exerted great influence on pattern recognition research and OCR development, particularly in Japan.

Here we introduce the theory of normalization and the related work which was done by Iijima's group. Actually the concept of blurring was first introduced into pattern recognition research by his work, whereas it was widely attributed to Marr [34] in the West. Iizima's idea was derived from his study on modeling the vision observation system. According to his model of the vision system, blurring was necessarily introduced and so normalizations of displacement and blurring were studied, in particular.

Setting reasonable conditions for the observation system, he proved that the mathematical form of the transformation must be a convolution of a signal $f(x')$ with a Gaussian kernel. That is a blurring transformation. At the first stage of their research, they tried to recover the blurring effect of an observation system, i.e., normalization of blurring [35]. However, through the experimental study of OCR, Iijima got the idea of blurring an input character rather than recovering the blurring effect. The research along this line has been carried out by Iijima and Yamazakii [36]. Since then, blurring has become a well-known preprocessing technique in Japan.

Based on this theory, Iijima advocated the analog approach to pattern recognition. Actually he derived a partial differential and integral equation for recovering displacement by introducing time [37]. This is a dynamic equation and has a form such that its behavior is controlled by the average of $f(x, t)$. The equation was actually implemented to a hardware system with only five input terminals and it was confirmed that the equation certainly works well [38]. They referred to this class of equations as a globally associated and dependent network. However, the implementation was too expensive and another approach was taken.

Mathematically speaking, the normalization can be considered a linear operator. On the other hand, data are given in discrete form. Therefore, the operator is represented necessarily by a matrix, and the desired property of the operator is given by the eigenvalue distribution on an appropriate set of eigenfunctions, which construct a subspace in a Hilbert space. Based on this idea, Mori gave a displacement matrix by d, where d takes on any real value, say 0.5 [39]. At that time, such a shifting filter had meaning, because it was very expensive to have a retina (2-D sensor) with reasonable resolution. Actually they designed a 9×9 retina. However, there was still the problem of implementing the above filter using an analog circuit because the fan-in number to the operational amplifier was too large. He therefore proposed using a much simpler approximating system as follows [40]:

$$Af(x) = f(x + d), \; A \approx (1 - d)E + dP^{-1}, \; 0 \leq d \leq 1,$$

where A is the shift filter matrix, and E and P are the identity matrix and the circulant matrix respectively. A circulant matrix can be regarded as a shift operator for a vecor under the circular boundary condition (see [41]). However, in order to make the filter we need a variable resistance element and so he suggested using an FET. The special-purpose FET was made by Mitsubishi, and Funakubo implemented the system, which was quite effective [42].

Behind these formulations, serious consideration was given to the fact that we should treat continuous functions even if they appeared in discrete form. In this sense, Mori called such functions background functions. Therefore, when we displace an input pattern by, say, 0.3 units, the background function appears real. In this sense, it is not merely an interpolation technique. Spatial network was the name given to this approach, which had been developed theoretically by Ogawa and Igarashi [43]. Now, this approach exists in practice owing to highly developed computer technology. For example, a character is represented by a continuous function and the exact formula of curvature can be easily calculated from the continuous function, which is usually a spline [44].

On the other hand, Amari introduced another view of normalization, i.e., as a process that can be done after feature extraction [45]. He investigated the condition that such a scheme holds, assuming linear feature extraction, and proved that for an affine transformation, moment features have so-called admissible properties. This meant that normalization could be conducted on the feature space. From the practical point of view, this idea provides a very important insight into normalization, in which we need not insist on using the moment feature.

3) Karhunen-Loeve Expansion: Iijima's second paper was devoted to the theory of feature extraction based on the consideration that for a given normalized pattern set denoted by $D = \{h(x, \alpha)\}$, where α is an assigned/indexing number of the individual pattern, we should choose basic functions for a coordinate system of the pattern representation space. More specifically, he constructed the following functional (mapping functions to real values):

$$J[\varphi(x)] = \frac{\int_D \omega(\alpha)(h(x,\alpha),\varphi(x))d\alpha}{\| \varphi(x) \|}$$

where $\omega(\alpha)$ is an appearance probability of α . He derived the integral equation in which the function maximizes the above functional using the variation method. However, the above formulation is the same as the Karhunen-Loeve expansion [46]. He is very good at solving integral equations and did it independently of Karhunen's [47] and Loeve's [48] work, which was done in the 1940's. K–L expansion is very attractive not only for pattern recognition but also for data compression. Therefore, many researchers used it. Among them, Watanabe was a pioneer and Fukunaga [49] was very active with him in this field. Watanabe's work was done independently of Iijima, although their work was done in almost the same period. We will show later the more advanced development of K-L expansion, which was done by both researchers from different angles. At any rate, the application of K-L expansion to OCR was explored extensively at ETL by Iijima and his group and later at the Toshiba Central Research Center. First we will introduce the work done at ETL.

Before that, the reader might note that K-L expansion is identical to principal component analysis, which was originated by Hotelling [50] in 1933. Since then it has been widely used in the field of statistical data analysis. Therefore, when a data set of $[h(x, \alpha)]$ is given by vectors, we can construct the covariance matrix and solve its eigenvectors, which form the coordinates of the given pattern space. However, we notice that the integral representation is very convenient for the theoretical work and so in this sense they are not exactly the same thing.

Now, returning to our real problem, suppose that an $N \times N$–dimensional data set is given. Say N is equal to 50; then $N \times N$ becomes the large number 2500. We must handle a 2500-dimensional vector and solve the 2500×2500 covariance matrix. This is a huge problem even if the available computer power is great. Some way must be considered to reduce its huge dimension. In this respect, we recommend Gonzalez and Wintz's book [51], which deals with data compression by K-L expansion. Here we show another approach, where a class-based correlation can be used instead of taking a pixel-based one. Note that we change the notation slightly for our convenience and that $h^r(x)$ is canonical, i.e., constructed such that the average value of $h^r(x)$, $r \in L$ in the set of L is zero. Notice that L denotes also the number of classes.

Therefore we introduce the following covariance matrix:

$$C_{nr} = \omega_n \omega_r \int_S h^n(x) h^r(x) dx.$$

Thus solving this eigenvalue problem, orthogonal eigenvectors φ_m, $m = 1, 2, \cdots L-1$, were obtained. L is usually a small number, say 36 for alphanumerals. So it is very easy to calculate eigenvectors and eigenvalues. Thus, say a 50×50–dimensional pattern space is reduced to 36, which is a very compact space. However, we need to test this theoretical result in the real field. Noguchi [52] conducted the experiment on printed Katakana data. Katakana is one of the subsets of Japanese letters and consists of 46 characters. The feature extraction rate is defined in general in terms of eigenvalues as follows:

$$T_N = \frac{\sum_{n=1}^{N} \lambda_n}{\sum_{n=1}^{L-1} \lambda_n},$$

where λ_n is the nth eigenvalue and the eigenvalues are arranged in order of magnitude. Note here that the sum up to $L-1$ is taken because one freedom is used by taking the canonical form. According to the experiment,$T_N = 0.8$ when $N = 10$ and 0.9 when $N = 20$. Therefore this scheme seemed very promising and it was expected that 31×43 dimensions of the image plane could be drastically reduced, to around 30. However, they had to realize how big a problem "noise" can create. In particular, displacement noise was the most serious, followed by the width of a stroke. On the other hand, as mentioned above, Iijima's group was constructing an OCR called the ETL pilot model [53]. They intended to make an all-analog and parallel OCR and so the analog shift filter was implemented. They conducted a systematic simulation experiment of feature extraction based on K-L expansion. The retina was a 9×9 array and 0.25 shifts in eight directions were allowed to be an acceptable displacement noise. Eventually, after several trials, they adopted only the first eigenvector and the sign of the projection value to the axis. Therefore, their discrimination scheme was the tree structure. As a result, the discrimination tree needed a total of 43 nodes. The performance of the OCR was tested using sets of 150 printed numerals and C, S, T, X, Y, and / (one font) printed on various printing conditions for each class. The recognition rate was 96%. Naturally the question arose,"Why not use similarity itself?" In that case we can use templates of more than two for each class, say three; then the necessary templates would amount to 48 (16×3). The number 48 is almost the same as 43; i.e., the complexity of the tree structure is almost the same as the similarity system in which three templates are used for every class.

Actually Iijima seriously investigated the structure of similarity in Hilbert space. First of all, he found a very important property of similarity related to blurring. Intuitively $S(f_0, f) \rightarrow 1$ seems to be reasonable when the blurring increases, but he proved that when canonical transformation is done $S(h_0, h) \rightarrow \delta < 1$. That is, even if heavy blurring is applied, essential differences between different patterns remain. Blurring is equivalent to low-pass filtering and suppressing the high-frequency part. Therefore, we can reduce sampling points by taking advantage of blurring. Furthermore, Iijima introduced the so-called multiple similarity method [54]. The method was considered to overcome the weakness of similarity against displacement noise. He investigated the trace of patterns in Hilbert space, when displacement noise is expressed by the first and second terms of its Taylor expansion expressed by $h_0(\boldsymbol{r}+\boldsymbol{d}), \boldsymbol{d} = \boldsymbol{i}d_1 + \boldsymbol{j}d_2$. The pattern $h_0(\boldsymbol{r}+\boldsymbol{d})$ moves on the same definite locus as $\boldsymbol{d}$ moves from zero. Therefore,

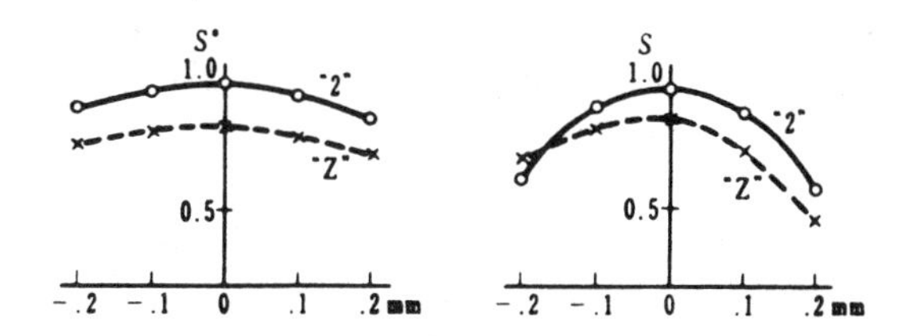

Fig. 5. Comparison of multiple similarity with ordinary similarity against displacement noise. The x axis is scaled with the displacement in millimeters. Note that in the case of ordinary similarity a reversal occurs at the displacement of 0.2 mm, and the standard width of a stroke of printed Roman is 0.35 mm.

it is definitely impossible to compensate the displacement noise by only one template. However, since the locus is known, we can avoid the loss incurred by the displacement noise by introducing more than one template and defining a new similarity in terms of the new templates. He defined the new similarity $S^*(h_0, h)$ as follows:

$$S^{*2}(h_0, h) = S^2(\varphi_0, h) + S^2(\varphi_1, h) + S^2(\varphi_2, h),$$

, where φ_0 is normalized h_0 and both φ_1 and φ_2 are constructed by the partial derivative h_0 with respect to x and y respectively. Then the following equations hold:

$$(\varphi_i, \varphi_j) = \delta_{i,j}, \qquad (i, j = 0, 1, 2).$$

The effectiveness of the multiple similarity is shown in Fig. 5. This method has become the main algorithm of Toshiba's OCR systems. Iijima's theory is not so easy to understand, but his recent book [55] is readable, although it is written in Japanese. For those who are good at linear algebra, it is a good way to illustrate the method using the concept of vector space. That is, a template vector is replaced by a subspace which is spanned by more than one orthogonal base vector. The matching is just a projection to this subspace. Actually this is almost the same as the multiple similarity method and is called the subspace method. Watanabe [56] proposed the method almost at the same time as Iijima. The subspace method has been developed by many researchers and is still an active field. We recommend Oja's fine book [57] for reference.

4) Series Expansion: In general, the K-L expansion is a special case of a series expansion. In the case of the K-L expansion, orthogonal vectors are generated from a data set. Here, we describe some series expansions obtained using built-in functions. The most typical ones are moment and Fourier expansions, while Walsh and Hadamard expansions have also been studied (see Pratt's book [58]).

a) Moment: Humans can recognize characters when viewing from any direction, with the exception of a few characters, such as "6" and "9." Recognition of shapes, independent of position, size, and orientation in the visual field, has been a goal of research. From a practical point of view in OCR in particular, orientation invariance is not as important as position and size. However, it is a very interesting research goal and work on it has been continuing since Hu's theoretical work [59]. However, we are obliged to omit the related research work, because of the limitations of space and the pessimistic prospects of using many higher order moments in practice.

When we limit the invariance to only position and size, and/or slant, the moment method is not too bad. In 1962, a first systematic experiment was conducted by Alt [60] of the National Bureau of Standards (NBS), in which one font of alphabet was used which was normalized in position, size, and slant. Therefore, moments up to the second order were used as the normalization factors, and the third to sixth moments (22 kinds) were used for classification. This was a pilot program and no recognition rate was given for the basic experiment; however he was optimistic.

In spite of the positive perspective of Alt, very litle work followed [61]. In 1987, two and a half decades after Alt's work, a very systematic and reliable experiment on the moment method was reported by Cash and Hatamian [62] at Bell Laboratories. They used central moments only up to the third order and so ten moments (ten-dimensional vector) were used for classification. The data set tested consisted of six different machine-printed fonts (62-class alphanumerals). They compared the performances of three typical similarity measures: Euclidian distance, cross correlation, and Mahalanobis distance. The results were as follows: All three similarities achieved a recognition rate over 95% for all six fonts, and the weighted and normalized cross correlation measure produced the best recognition rates: 99% for four of the six fonts for the high-quality data sets. However, the print quality seems to have been good and no rejection class was taken. Therefore, an exact evaluation from a practical point of view is difficult.

b) Fourier series: Fourier series expansion is the most popular and so naturally had been applied to character recognition systems. In the early 1960's, such research had been started by Cogriff [63]. In general, concerning the studies that followed, refer to Zahn and Roskie's paper [64]. The representation of Fourier coefficients of a boundary can be divided in two ways. The first is based on the cumulative angular function which is expanded to a Fourier series. The set $\{A_k, a_k, |k - 1, 2, \cdots\}$ is called the Fourier descriptor (FD) for a curve, where A_k and a_k are the amplitude and phase of the kth-harmonic term. The other representation was proposed by Granlund [65] and developed by Persoon and Fu [66]. A point moving along the boundary generates the complex function $u(s) = x(s)+jy(s)$, which is periodic with period S. The FD(CF) now becomes

$$a_n = S^{-1} \int_S u(t)exp(-j2\pi S^{-1}xnt)dt.$$

A fully theoretical study was conducted in particular by Zahn and Roskies. In particular, FD's have invariant properties with respect to position and scale, but we have to note that they depend on the starting point of a boundary tracing.

Now let us examine FD's properties for real characters. In the case of the cumulative angular function (CAF) expansion, a reconstruction experiment was done using Munson's data set, consisting of a 24×24 binary matrix.

It was shown that the reconstruction of a hand-printed numeral "4" is almost sufficient using up to ten harmonics (ten amplitudes and ten phases). However, the reconstructed image is not closed, which is a general feature of CAF. It was demonstrated that CF is considerably better than CAF [4]. That is, in the case of CF, reconstructed figures are closed and five harmonics are almost sufficient to differentiate between "2" and "Z." On the othr hand, at least ten harmonics are necessary in the case of CAF. The experiment used a 64×64 binary matrix.

Persoon and Fu did a recognition experiment using Munson's numeric data set. The error rate was 10.6% with no rejection class. Conspicuous misclassification occurred between "8" and "1," which was due to the fact that only the outer boundary was traced. Taking this fact into account, we may say that it was a good result, but the size of the data was so small that we could not evaluate it definitely. However, it took 14 seconds per character using CDC 6000, the most powerful computer in terms of scientific calculation at that time.

In spite of this demonstrated potential, it was more than a decade before the first systematic and reliable experiments were conducted, by Lai and Suen [67]. In their experiment, FD's were used as global features and local boundary features, such as concave and convex types, were also used as local features. The classification scheme used was a binary decision tree. The data used were Suen's data set, consisting of 100 000 hand-printed alphanumeric characters in a 64×32 binary matrix. A separate experiment was also performed in which only FD's were used, up to five harmonics, i.e., ten features. The results were recognition rates of 81.74%. However, when the combined features were used, a 98.05% recognition rate and an error rate of 1.95% were achieved. From this experiment we can conclude that FD's are very useful for global feature extraction and powerful for distortion. However, they are not so good for local feature extraction, which coincides with the reconstruction experiments shown before. Their experiment was done with careful preprocessing, starting point normalization, and extraction of inner structure (holes).

On the other hand, Fourier expansion can be applied to the 2-D plane directly. Hsu *et al.*[68] did it using a circular harmonic expansion and formalized a rotation-invariant amplitude, which was proved by a pilot experiment. It is interesting to compare this method with moment invariant method, because the circular harmonic expansion is also a powerful tool for theoretical work.

5) Feature Matching: Mathematically speaking, matching in linear space is equivalent to a correlation/inner product between two vectors. So far we have taken the implicit meaning of template matching. Now we try to expand the scheme a little. That is, each element of a vector cannot be constrained to be a scalar value, but it can be a vector. In other words, each element can have a structure, i.e., a local feature such as direction or gradient. However, we note that the structure element can be represented by a vector. Therefore, the expansion means that a higher-dimensional vector is constructed and a usual inner product is done with the high-dimensional vectors. In this sense, there is no change in mathematical form, but semantics of the element change.

Actually such expansion was done very clearly at an early stage back in 1965 by Spinrad [69]. As described later, he extracted primitive stroke features, and then tried to match simultaneously the arrangement of the strokes and their attributes also. The direction of a stroke was quantized into eight and the position of the center of a stroke was also quantized directionwise into 16 viewing the center of gravity of the whole set of strokes, such as N (north) and NE (north-east). Thus, an 8×16 matrix was constructed and at each entry of the matrix a 3×3 submatrix was set, in which the row is the quantized distance from the center of gravity and the column is the quantized stroke length. Thus the total number of the elements was $8 \times 16 \times 3 \times 3 = 1152$. Therefore, 1152-dimensional vectors were constructed, and the correlation between the unknown input vectors and each template vector, corresponding to each class, was measured. The above operation is a somewhat mechanical expansion of the correlation, but we consider "correlation" deeply apart from its mathematical form.

Correlation on a 2-D plane might convey the idea that face to face matching takes place. However, this is an illusion. This is nothing but point to point matching and a summing up of these matchings. On the other hand, the 2-D plane has its natural topology; i.e., each point on it has its neighborhood. A blurring preprocess can be interpreted as a technique to incorporate the neighboring system with correlation. The reduction of a 2-D grid plane is also one such technique. The preprocessings mentioned above are of the lowest level and are also basic. A higher-level one is a gradient in which, for example, a 3×3 neighbor system is included into one element of a vector [70]. Now we introduce work based on this principle which was done by Yasuda [71]. He attempted a one-mask recognition system for hand-printed characters where each mask for a class was constructed as follows:

1) Direction at each pixel of each sample image is extracted and quantized into four directions.
2) Size normalization is done.
3) Blurring is done.
4) An average is taken for all the samples processed.

Figure 6 gives an example of the mask for "2," in which the top row shows four mask components corresponding to four quantized directions. From the second to the bottom the blurred and size-normalized masks are shown. The blurring factor increases from the second to the bottom masks. The data set consists of unconstrained hand-printed characters written by 25 subjects. The number of classes is 41, i.e., numerals, alphabet, and some symbols. The sample size is 8200 (41×200). In the first experiment, only the numeral data set was used. No rejection class was allowed. As a result 97.1% correct recognition rate was achieved using only a single mask for each class. The secondary experiment carried out on the total data set showed a drop in the error rate of 1.4% and variations in classes are

Fig. 6. Four directional feature masks of "2" with their blurred ones being normalized [71].

conspicuous. "N," in particular, gave a 22% error rate. The experiment demonstrated the power of directional matching with blurring.

Feature matching is still sensitive to stroke positions. In this sense, considerable efforts have been made for normalization. However, so-called linear normalization is not enough and so-called nonlinear normalization has been used [72]–[74]. The basic idea is to measure busyness of lines/strokes and relocate the strokes so that the busyness become uniform based on the measurement. Tskumo *et al.* gave an interesting comparison study of normalization methods.

6) Nonlinear Template Matching: In K-L expansion, we constructed a subspace as a template instead of a one vector template. In feature matching, we constructed a very blurred template on a feature plane, which means that it creates many templates belonging to one class implicitly. Now we introduce another method to generate a set of templates explicitly and to match against the set, which was done by Sakoe [75] in 1974. First of all, he assumed that a character consists of a sum/concatenation of vectors. We can easily formalize such a shape as

$$C = \{(c_1, c_2, ..c_k, ..c_K).(i_K, j_K)\},$$

where c_k denotes the kth line segment/vector and is further expressed as $c_k = (d_k, l_k)$. Here d_k and l_k are the direction and length of the vector respectively and (i_K, j_K) denote the coordinates of the terminal point. The above expansion is very flexible when each c_k and (i_K, j_K) are changed. Therefore, we need to impose some constraints on it to represent a set of templates. The constraints are described as ranges of direction, length, and terminal points of a vector. Besides the above constraints, a weight, w_k to each kth vector is set, which is used to connect separated line segments by setting it to zero. Here, a set of the templates is denoted as B.

Now we need to consider how to define "matching" against the template set. First an input image on the $I \times J$ matrix is defined simply as follows:

$$A = \{a(i,j)\}, \qquad i = 1, 2, \cdots, I \quad j = 1, 2, \cdots, J,$$

where $a(i,j)$ is multilevel in general. Next, such practical representation has also to be made for the conceptual set of templates, too. Thus, the similarity between A and B (set), is defined as follows: $S(A, B) = max\{(A, B') \mid B' \in B\}$, where $(A, B\prime)$ denotes inner product of vectors A and $B\prime$. Here we note that a maximum template is chosen from the template set B scanning B in the range specified. By searching for the best match, we can get the best approximate pattern $B\prime$ from the set B.

By the way, Sakoe also estimated the size of the set of templates of "0," for example, as $2^{4 \times 5 \times 5} = 1640000$ where K was set to 8, the variance of length is four pixels, and the variance of position of the terminal point is 5×5. This number is enormous, which shows its degree of flexibility in some sense, but we face too big a problem on how to search the set to find the best match. Fortunately, there is an efficient technique for solving the problem, which is called dynamic programming (DP) and was invented by Bellman [76], [77]. Thanks to an additiveness of the matching estimate as shown, we can apply DP to the problem. Sakoe named the method rubber string matching and solved the problem using DP and conducted an experiment on hand-printed numerals. We will show only the results. He made standard masks represented by B, manually, with a small pilot experiment to accommodate B's. Thus 12 masks were made. On 2000 samples, the correct recognition rate was 99.5% when no rejection was allowed.

The so-called DP matching has been broadly used in the pattern recognition field. The first application to image analysis was made by Kovalevsky [78] in 1967, in which segmentation of a character string on noisy background was tried. DP matching was successfully applied to speech recognition. Sakoe was an expert in this field and built a speech recognition system at NEC based on his DP matching method [79]. With regard to the application to image analysis, we recommend Kovalevsky's book. Recently, Yamada [80] has been very active in DP matching applications to both character recognition and map recognition.

In DP matching, order is also very important because of the nature of its multistage processing. In his case, the contour of an image was approximated by a polygon, in which we can establish a very natural order of line segments along the contour. The details and results will be presented later. At any rate, an interesting aspect of DP matching lies in the fact that both template and input images are described structurally, and matching is done componentwise numerically. In this sense, the DP matching implies the two natures of template matching and structural analysis.

7) Graphical Matching: Graphical matching is a very useful technique which can be applied to all areas in the field of pattern recognition and even other fields. Stroke segments and their relationships are represented by a graph in strict mathematical sense. Therefore, graph isomorphism and subgraph isomorphism provide a basic matching theory [81]; actually they are used when the number of nodes is small. Otherwise it still has the problem of complexity. Strict and general formalization from a more practical point of view was given by Ambler *et al.* [82]. There are some algorithms which can find cliques mechanically, but this is NP-complete in general. Therefore, a practical method was considered, namely the relaxation method invented by Rosenfeld *et al.* [83]. This was first applied to shape matching by Davis [84]. The relaxation method originates from Waltz's work [85], i.e., a filtering algorithm to prune incompatible candidate labeling in labeling problems in so-called block-world understanding. Here we gave only a historical introduction, due to limited space and because graph matching is not specific to character recognition. We recommend Ballard and Brown's fine book [86], along with [3]. We note that graphical matching is an independent matching algorithm as well as DP matching. However, for our convenience, the two matching techniques were included in the template matching approach.

B. Structure Analysis Approach

We have quickly reviewed the slit/stroke analysis methods, which constitute a set of the simplest methods of structure analysis. A common feature of these methods is that they look at a character only partially. Because of the limited use of hardware at that time, researchers tried to construct their algorithms for reading machines as simply as possible. Therefore, only partial features of a characters were detected and simple relationships were used. This trend was exactly the same as that of template matching. However, as said before, character information is two-dimensional in essence. Therefore, we need to look at the full structure of a character. Actually researchers became aware of that fact from bitter experience.

There are several viewpoints to systematically see the complete structure of a character. These are classified as follows:

- thinning line analysis,
- bulk decomposition,
- consecutive slits analysis/stream following,
- contour following analysis,
- background analysis.

Among these views, thinning line analysis has been most intensively investigated. Certainly this is a very important approach and based on it many OCR systems have been made. The analysis is higher than the other analyses in terms of abstraction, except for bulk decomposition. Bulk decomposition can be regarded as being at the same level as thinning line analysis, except bulk decomposition is applicable to the general shape. In the context of OCR, bulk decomposition was considered to have overcome some of the defects of thinning line techniques. We will describe these topics when we discuss the year 1980.

1) Thinning Line Analysis: Humans look at a character as a linelike object. This view is most intuitive, at least for adults. Here, "linelike" needs to be defined strictly. A well-known definition of a line is Euclid's, which has no width. Naturally we cannot accept such a definition in digital picture representation. So a one bit width is necessary and the connectivity of a line needs to be defined. A regular grid plane is not homogeneous. Such basic definitions of the geometric concept of a digital plane were first given by Rosenfeld and Pfaltz [87] in 1966 in their epoch-making paper.

We will not go into this problem, although this is very important and gives a base for the thinning process. Instead, we will show that when looking at a character with abstracting eyes, it is not too hard to analyze the structure of a character. In 1960 Sherman [88] regarded characters as consisting of abstracted lines and constructed a graph. In his graph he ignored a node which has two outgoing lines (degree of node is 2). Therefore, feature nodes are endpoint, branching point, crossing point, and so on. The adjacency matrix represents these relationships, but so loose that many different character classes belong to one class. For example, the numerals

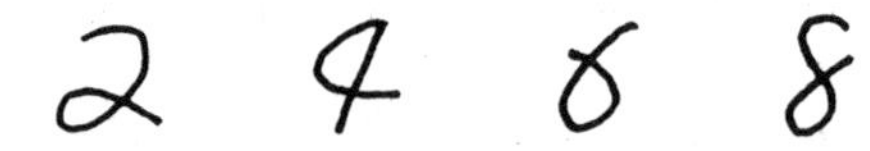

are all placed in the same class. However, this topological view is very important, because it can absorb terrible variation of character shape.

Along this line, Beun [89] of Philips did an experiment on unconstrained numerals. He used other features, such as node position relationship, to resolve the degenerated class. The experimental results were a 91.7% correct recognition rate and a 2.6% substitution rate. He stated in his paper "According to personal experience, we knew that how easy to deceive ourself and others in such experiment." This statement means two things. First, the difficulty of doing unconstrained character recognition was shown. At that time, researchers were not aware of this fact, but a few engineers were. The very ambitious of Sherman indicates that. People gradually become conservative as they gained experience. Now we have roughly three classes of hand-printed qualities, i.e., unconstrained, loosely constrained, and constrained classes. For the constrained numerals, several kinds of rules were considered so that they could be easily recognized by a machine, as mentioned in connection with second-generation OCR's. The most reasonable target is the loosely constrained class. This is because people dislike being constrained and prefer having their freedom. Too much freedom, however, causes confusion among people, and Suen [90] revealed this point by conducting psychological experiments. The other thing is that we need some common data in order to compare researchers' experimental results. Therefore several data sets have been made so far. See the paper by Suen *et al.* [5].

Now we shall describe a very important preprocessing, called thinning. While it may be regarded as preprocessing, it should be treated as more than preprocessing. At any rate, it is needed to obtain abstracted lines. Actually Beun did it [89]. An observed line usually has a width greater than that of a pixel. So the line is eroded from both sides keeping some constraints so that the line is not broken and shortened.

The idea of thinning was suggested by Kirsch [91] in 1957 and programs to describe bubble chamber pictures were written by McCormick [92] and Narashimhan [93]. For medical image applications, Rutovitz [94] tried it, as did Deutch [95] for character recognition. However, the first systematic and rigorous algorithm was given by Hilditch [96] in 1969. Since then, more than 30 variations of thinning algorithms have been proposed, some of which were compared by Tamura [97]. However, it is well known that we cannot obtain perfectly abstract lines for real lines, which include acute corners and/or intersections. This is a basic problem which cannot be avoided because of the local operations used. Therefore, for its actual application some postprocessing is done. Since it is an iterative process, it is time consuming. However, thinning is a basic preprocessing in OCR technology as well as in recognizing drawings and this is widely used in many OCR systems, as we will see. Research on this continues.

Now we move to the analysis of a set of lines after the thinning process. As mentioned above, the analysis is divided into two parts. Suppose the line set is a graph; one method is to detect nodes/singular points, and the other is to describe edges/arcs. For both purposes, a 3×3 window/mask has been extensively used. Although it is comparatively easy to detect simple and local segments, it is not so easy to obtain more global segmentation of lines. We will discuss this point further below.

Concerning the description of arcs, there is a simple technique, called chain encoding or Freeman code, developed by Freeman [98]. It is also based on a 3×3 window. We can set a 3×3 window at any point on a line, as shown in Fig. 7(a). We assume that the binary image is scanned from the left and the bottom. The scan meets a black pixel eventually. Then the line coding starts, following the line, and looking at the neighbor from the center, we can decide the next moving pixel uniquely as shown. Eight such local directions are encoded, as shown in Fig. 7(b). For example, the line is coded as 221100. This encoding scheme is naturally applicable to contour following and is extensively used. In some sense, this is the counterpart of run-length coding, which is used to code a total plane image. Both are the first stage in describing the shape in general, from which we need to describe the shape more globally. For the above example, the reader can guess easily that there exists a corner looking at the chain code.

2) Bulk Decomposition: The work of Grimsdale *et al.* [99] is surprisingly basic and valuable considering the time when the work was done. This is the first paper which dealt

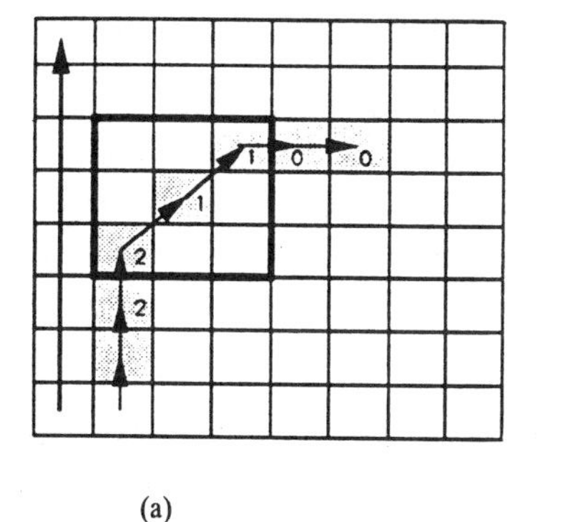

(a)

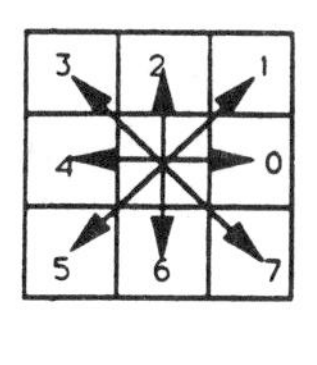

(b)

Fig. 7. Illustration of chain encoding: (a) a simple example; (b) the coding rule.

with bulk decomposition systematically. The concept of bulk decomposition is intuitively clear; i.e., a letter "L" can be regarded as consisting of vertical and horizontal lines, for example. The decomposition can be regarded as a counterpart of principal component analysis in template matching. However, it is very difficult to deal with the problem from the theoretical point of view. So they virtually began the decomposition from run-length coding. Consecutive black runs are analyzed and connected if certain conditions are satisfied. Otherwise, a new part/segment of a character is generated.

Their method is considerably practical in terms of algorithm. The first is that they took noises, breaks/gaps, and isolated small blobs into account. The second is that some proof for the rotation problem is provided. After finding each segment in a character, the final description of the character is done. This consists of two parts: first attribute description of each segment, i.e., length, slope, and curvature; second, describe the relationship between individual segments, table of joints. In this stage, merging is done so that overlapped segments are combined. Thus, it can be seen that the basic processes of structural analysis such as feature growing, called macroprocess and merging, are clearly mentioned. However, naturally it took 60 seconds for 4K instructions in their computer at that time. So their work ended up as basic research.

The work of Grimsdale *et al.* was too sophisticated at that time. So a more engineering oriented method was considered by Spinrad [69]. His method is very simple and can be regarded as an application of the slit method. It consisted of eight directed slits on a frame within which a character image was set. For example, a vertical slit is moved from the left to the right on the frame, then for a letter "D," at some movement, the slit intersects the vertical stroke of the image of "D." Thus, we can detect its vertical stroke. On the other hand, concerning the classification, his method was based on feature matching in general, which has already been described.

After Spinrad, Pavlidis [100] extended his approach from the theoretical point of view. He proposed the kth integral projection. All the black runs for any scan can be labeled by number, for example, from the left: 1, 2, 3, and so on. The kth integral projection is just the sum of the kth labeled black runs. The projection method is very resistant against noise, i.e., ruggedness of boundaries of character images. In the case of Roman alphanumerals, it is not so effective, but for Kanji the expansion of projection/profile is essentially important and broadly used in Kanji OCR. Pavlidis continued his research on bulk decomposition rigorously. As mentioned before, the abstraction level of bulk decomposition is higher than boundary/contour decomposition. Therefore, if we once succeeded in that, we could obtain a very simple and powerful description of shape in general. Pavlidis has stated the problem in his representative survey paper [101]: Thus regardless of formalism one is faced with the task of segmenting S ("irregularly" shaped subset of the Euclidian plane) into some kind of "regularly" shaped subsets.

This is a fundamental problem in pattern recognition. Regularity is very hard to prove, even if some theory were possible. However, we need some guidance which is got from a general strategy/criterion. We pick up some parts from the paper: They must conform with our intuitive notions of "simpler" components of a "complex" picture and have a well-defined mathematical characterization. This is a very important remark which should be applied to any type of shape decomposition. Along this line, he proposed convex decomposition, but it was somewhat rigid. Several researchers challenged this decomposition problem. The work continues; however, it is still in the realm of basic research. Concerning "regularity," see also the paper by Simon in this issue.

3) Stream Following Analysis: The idea of stream following seems to have been first described by Rabinow [102] in 1962. He used the term "watch bird" to describe a curve tracing circuit. However, in the same year exactly the same idea was reported in the *RCA Review* by Sublette and Tults [103], where a more concrete form and labeling technique was fully used to follow streams. The strict description of stream following was given by Perotto [104] in 1963, who was not aware of the forerunners. In any event, Rabinow and Perotto intended to make a reading system for handwritten characters, and Sublette *et al.*constructed a software system for reading multifont characters. A common point is that their target character set is rich in variation, which reflects the nature of stream following. Another interesting aspect is that RCA took a two-way strategy to OCR systems: One is an analog/hardware oriented approach, mentioned before; the other is a digital/software approach. At that time, the latter was inferior to the former, due to the slow speed of 1 chs using the large computers of that time. It was hard to test their system.

Before illustrating stream following, it may be helpful to the readers to note the order of description on a two-dimensional plane. Although we can take any coordinates on the plane, humans seem to take few coordinates unconsciously. A coordinate system is important for character recognition, in particular. In other words, humans assume natural orders on the plane, such as from top to bottom and right to left. Such an order is very important for describing shape.

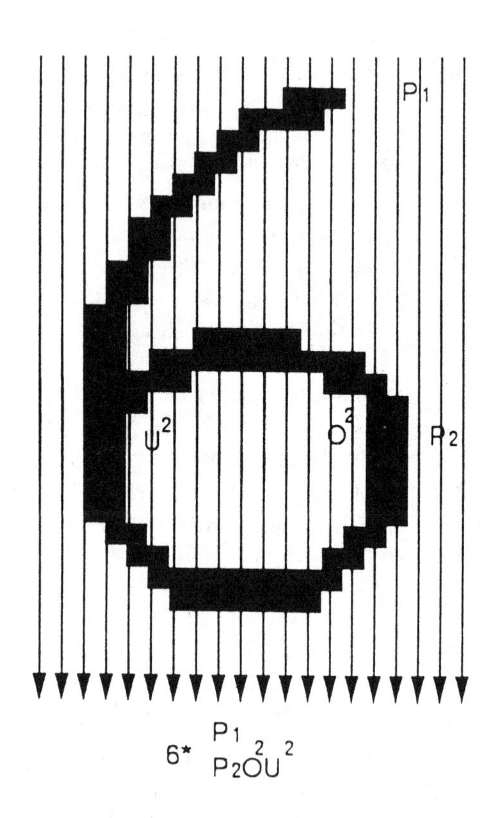

Fig. 8. Illustration of the stream following method. The image of "6" is scannned from right to left vertically.

Fig. 8 shows one example given by Perotto, which uses the order mentioned above. The order of raster scanning is from the top to bottom and from the right to left. Suppose that a slit moves from the right to left. First, there is no black component (b component) appearing in the slit, which is labeled P, as an abbreviation for principle. The coordinates of the position are stored (center point). The b component is followed. Again at once, two b components appear in the slit, both connecting to the b component detected previously. Then the state is labeled D, denoting branching of order 2. The two b components are followed. Immediately a new black component appears in the slit, which is labeled P, and its coordinates are stored. Thus, the three b conponents are followed. At the position close to the left, the state D merges to one b component, whose state is labeled U, denoting union of order 2. Finally, we find no more b components in the slit, which brings to an end the following process of the b component. According to the coordinates of the two P's, the first and the second P are labeled P_2 and P_1 respectively. Thus, "6" is labeled as shown at the bottom of Fig. 8. The order of the labels is important and reflects the order specified at the beginning.

As can be seen, from the description this method is very simple and is very strong against variations of shape. Perotto called the description a morphotopological description. However, we need to note because of the simplicity, considerably different shapes are identified as the same. Therefore, we need another raster scanning horizontally, in order to distinguish among "U," " ∩ ", and " – ." For "+" and "T," diagonal scanning is necessary. Nadler [105] proposed such scanning, but it was found to be unnecessary.

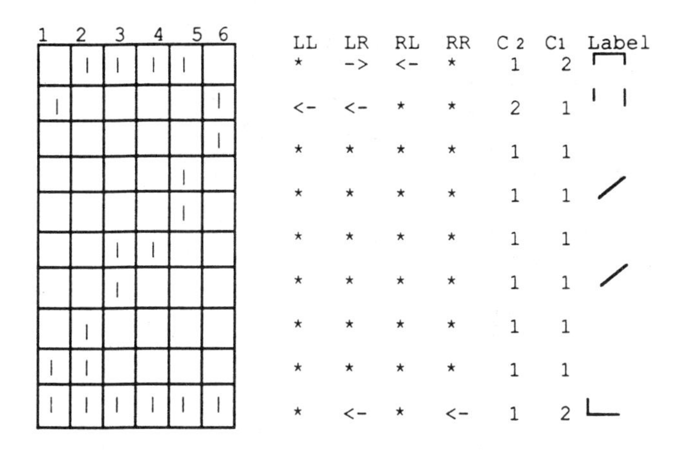

Fig. 9. Schematic illustration of the principle of feature extraction used in the NEC system.

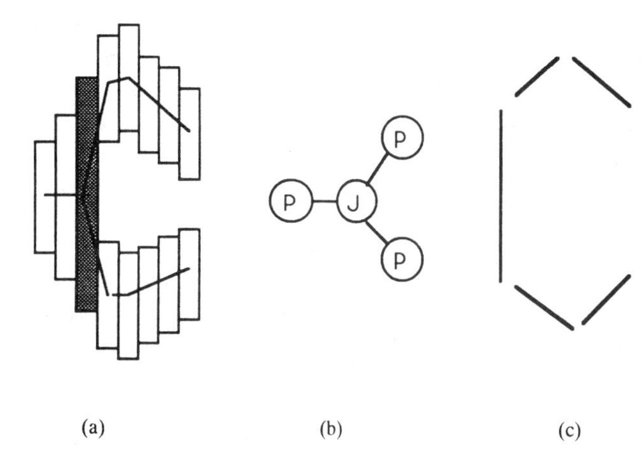

(a) (b) (c)

Fig. 10. Illustration of vectorization method based on LAG.

The description of stream following is based only on the topology of each slit. However, scanning gives run length, and so we can use profiles to extract the geometric attribute of b components. In fact, Uyehara [106] proposed a stream following method in the same year as Perotto with one modification. The term *stream following* itself was coined by him. However, there is no noise removing process and it only works well for ideal patterns. Subletts *et al.* worked on the stream following method.

Nadler [107], in 1974, gave a simple algorithm of stream following analysis, in which he used a small window of 2×1. The final description is graphlike and intuitive, being slightly redundant. The algorithm is given by the transition table/diagram of an automaton whose number of state is eight, while the original paper described a seven-state automaton. On the other hand, in 1969 Hoshino and Miyamoto [108] presented a paper on feature extraction, which was based on stream following analysis with a systematic geometric feature extraction. A schematic illustrative figure is shown in Fig. 9. The decision logic is a simple sequential symbol matching keeping the order mentioned above [109]. This is the main algorithm of the first NEC postal code number reader, which will be described in detail later. From the general point of view in image processing some algorithms had been developed based on the same idea and use run length encoding [110]–[112]. Among them the paper by Agrawala *et al.* is pedagogical and fit for computer processing, i.e., a sequential approach.

4) Vectorization: A new method of thinning using vectorization should be introduced [113] . However, we can regard it as a kind of bulk segmentation and it includes stream following. Therefore, we place its description here. First of all we note that the abstraction level of vectorization is higher than thinning and bulk segmentation . Connected lines are separated, a segmented bulk is represented by a line, and the relationships between the vectorized lines are described as a graph.

The vectorization is based on what is referred to as a LAG (line adjacency graph) [114], which is easily obtained from run length encoding, as shown in Fig. 10(a). It can be seen that the LAG exists implicitly in stream following and also in the work of Grimsdale *et al.* The LAG is transformed to a compressed LAG, called c-LAG (shown in Fig.10(b)), in which nodes of degree (1,1), where one segment is connected above and below, are mapped into single nodes. Junction nodes are connected by more than one path. Each path is examined by each segment width and the collinearity of the center of each segment. The path is then divided further into groups of segments, each of which has the same width and collinearity approximately. Merging is done for two collinear groups even if they are separated by another group between them. For example, imagine a vertical line of "E." The last example gives a feature of intersection implicitly. This method is based on run length encoding and so is very fast compared with the usual pixel-based thinning method. Based on the feature extraction and representation, Kahan *et al.* [115] had developed a recognition system for printed characters of any font and size.

5) Contour Following Analysis: We mentioned before that raster scan sets a natural order in a 2-D plane. Contour following also gives a very suitable natural order invariant to rotation. In other words, feature events along a contour of a closed shape is a circulant description. However, in a real application, for example, a starting point is fixed from the leftmost point. The contour following direction is also fixed clockwise looking at the black side. We note here that the term of contour following is used in a broad sense, i.e., including both boundary tracing and line (thin) following.

Historically speaking, the first contour following mechanism was given by using an analog device, called a flying spot scanner. This was developed by Greanias *et al.* [29] and was fully used in the IBM 1287 OCR system. A beam motion is attenuated within a black region so that the beam draws a cycloid traversing the boundary. At that time, it was very fast and flexible. Furthermore, it could absorb the coarseness of the boundary. On the other hand, its computer counterpart was very slow and suffered from boundary noise. We have already mentioned the computer-based mechanism of contour following using chain encoding. Other methods had been considered, but it became a standard method, although readers should be careful of some pathological cases. Readers can refer to Pavlidis's book [114].

The IBM 1287 was a very fine OCR system, but it was eventually replaced by an all-computer system except for input and output systems and optomechanical systems, along with electronics-oriented general trend. Coarseness on the grid plane is the first problem of contour following. Some smoothing techniques have been developed. A 3×3 window averaging process is one of them. A direct averaging technique on a contouring curve was done by Freeman [116] and Gallus *et al.* [117]. The latter is an extension of the former, both of which used chain encoding and defined local curvature based on chain encoding. They are theoretically very interesting, but are omitted here for reasons of space.

The second problem is feature selection along a contour. The third problem is how to segment a contour. The third is the most difficult and fundamental problem, as mentioned before. However, we can avoid the segmentation problem by introducing normalization of character image size and zoning the frame within which the image is squeezed or expanded. This is an ad hoc method. We introduced the zoned feature method before. That is based on a contour being segmented by forcing it into subregions of, say, 3×4 pixcels. The direction of contour following is taken as a feature. Thus, all the features describe direction and position, i.e., what direction is the contour and which partition of the 3×4 subregions is the contour in. Such features are sequentially checked from the starting position, along the contour. Concrete description is given later. However, we note that such normalization does not always work well when considering large shape variation/distortion of characters.

Therefore, the segmentation problem has been seriously considered. The simple idea is to use curvature maxima points, and this was accelerated by the famous psychological evidence by Attneave [118]. Although much work has been done, the first reliable technique was given by Rosenfeld and Johnston [119]. Analytically a curvature is given by a function of local derivatives, but actually it has the global nature suggested by Freeman. This was extensively examined by Freeman and Davis [120]. The corner feature plays a crucial role. For example, sometimes differentiating between "O" and "D" and between "5" and "S." So considerable effort has been spent in this area.

However, the methods developed do not always give stable and consistent results. Therefore an alternative approach was taken, one based on a polygonal approximation [121], which is global and very strong against noise. Theoretically speaking, the two approaches mentioned above are closely related [122]. So considerable work has also been done on polygonal approximation methods. Here we give a few representative works (Ramer [123], Pavlidis and Horowitz [124]). In particular, the work of Pavlidis *et al.* is very general. It is not constrained to make a connection. In this sense, it gives a global approximation which meets human intuition. Other polygonal approximation algorithms can be found in [3] and [4].

In fact, Pavlidis and Ali experimented with unconstrained handwritten numerals of the first file of IEEE data base

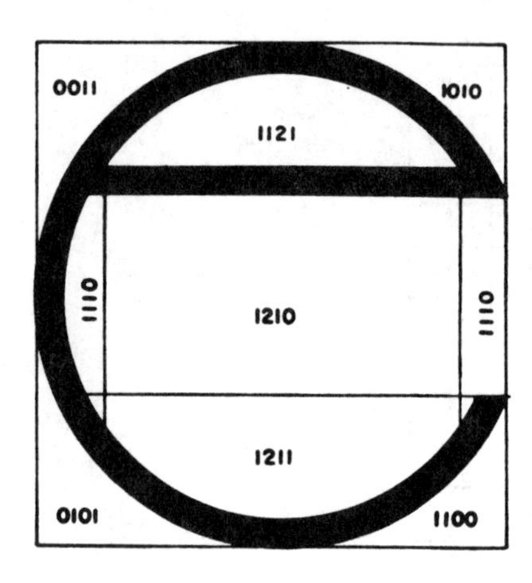

Fig. 11. Ternary coded regions. The four corners, however, are merged by taking binary code [24].

1.2.2, prepared by Munson. Input images were reasonably approximated by the polygon, i.e., not too fine and not too coarse. Therefore it is easy to detect global convexes and concavities, which are very important features. Naturally, holes are also so and are easily obtained in the course of contour following. They achieved reasonable recognition results. However, it is somewhat hard to estimate the result because Munson's data are so sloppy and small.

6) Background Analysis: A basic idea of background analysis was given by Steinbuch [125] in 1957. He assumed a potential field which is caused by electric charge whose distribution is just the same as the density distribution of character image. A simplified scheme was proposed by Kazmierczak [126] in which the potential gradient was coded.

As mentioned before, independent of them and without any analogy to physics, Glucksman took an approach of background analysis. His method is considered an extension of the sonding one, where each pixel of the background, four directed sondes, are spanned along the x and y axes. More specifically, each background pixel takes a four-digit code. This may be a binary, ternary, or higher-order code. An example is shown in Fig. 11, in which the background region is segmented into mainly seven subregions. The four corner regions have a more complex configuration, but this is meaningless and so binary code is used. As a result these subregions are effectively merged. However, no explicit region description was attempted. Classification was done using the feature vector, each element of which is a number of the ternary coded pixels. The histogram of the background features, ternary codes, was used. Theoretically, the feature vector has 81 dimensions, but 30 were sufficient in practice. An experiment on alphabetic characters of nine fonts and 52 classes of uppercase and lowercase letters was done. The total number of samples was 26 643, a large number at that time. The data sets were provided by Casey of the IBM Watson Research Laboratories. The correct recognition, rejection, and misclassified rates were 96.8%, 0.3%, and 2.9% respectively.

We continue the history, turning to algorithms. Munson of SRI [127], [28] proposed an interesting method, also

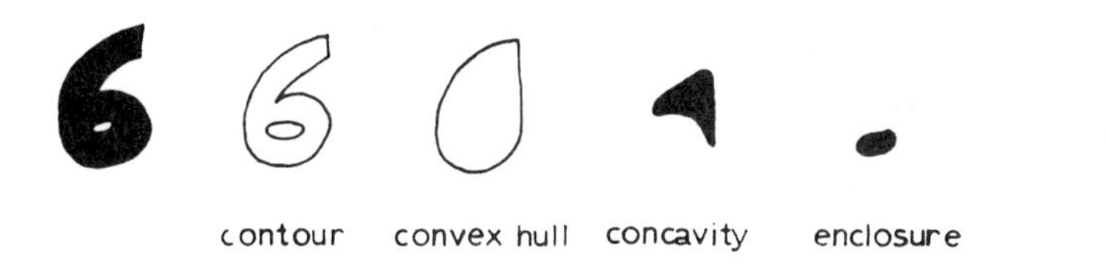

Fig. 12. Schematic illustration of Munson's idea [128].

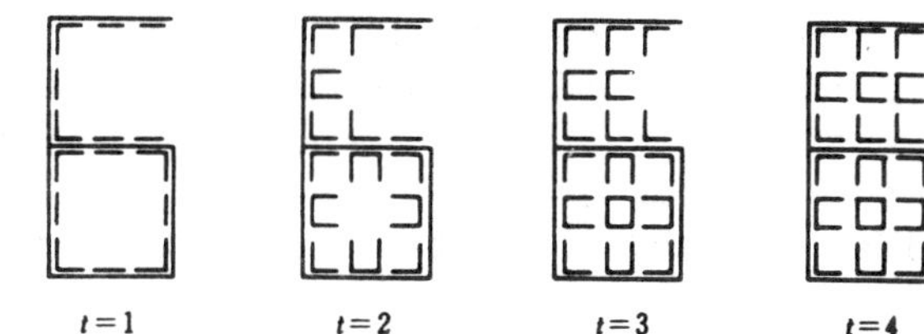

Fig. 13. Schematic issustration of the macroprocessing in the field effect method.

based on background analysis using contour following. A character image's boundary is traced and some points are marked as extreme points. Using these extreme points, a convex hull is constructed and connected regions adjacent to both the boundary and convex hull are detected as concavity regions, as shown in Fig. 12 [128]. Enclosures are detected when tracing the boundary. Concerning the detection of the extreme points, however, only a simple procedure was given. That procedure involves tracing a boundary, clockwise with a right hand system. Points turned to the right were searched and consecutive extreme points were connected with a straight line so that they lie inside the image.

Such a procedure is local and generates many extreme points; therefore some filtering technique must be provided to find global extreme points. No further structural analysis was done and both concavity and enclosure were used as a part of the components of the feature vector. This was based on which classification was done, as in Glucksman's case. The other components were profile and "spur," a kind of stroke. An experiment on a hand-printed 46-character FORTRAN alphabet was conducted, which gave a 97% correct reading rate for test data and a 3% of error rate. The data size was not specified. It is interesting to compare Glucksman's approach with Munson's. The former used only the background feature, but the latter as a part of the method. Both experiments obtained high recognition rates. We note that Glucksman's ternary coding reflects the relationship among concavities and holes implicitly, but in Munson's case, background features were used in isolation.

In the U.S. further developments were not made. On the other hand, a series of R&D efforts for OCR had been done nationwide in Japan as part of a national project, called PIPS (Pattern Information Processing System). The center of the research was ETL and it was conducted by Mori's group. The first task performed was to create a hand-printed character data base. Mori felt strongly about the need to have such a data set. A committee for it was chaired by Prof. Hiramatsu of Denki University and the principal related companies participated in the committee. Up to ten sets of data sets from alphanumerals to Kanji were assembled with the cooperation of the companies in which Yamada's and Saito's technical contributions were great. The management work was done by Mori and Yamamoto. The data sets are rich in their varieties and their volumes, and were made public internationally, which contributed to the basic R&D of OCR in Japan.

At the beginning of the OCR project at ETL, Mori established three guidelines to perform the research. These guidelines were to take structural , two-dimensional, and analog approaches. As a structural approach, they took background analysis. Three versions were developed. The first one, called system I, was considered by Mori and Oka [129], [130], in which a potential field was constructed merely to reduce the dimension low from a 60×60 to a 10×10 binary image. On the reduced plane each pixel has analog value virtually and at each 3×3 window local field was determined and labeled. However, system I was too primitive. The next version, called system II, was proposed by Mori [131], in which all 3×3 local potential fields were labeled using three digits: 0, 1, and 2. These digits just correspond to potentially different viewing from the center value. These three-level labels are transformed to symbols when their interactions take place in order to use rules which are represented by symbols in the two-dimensional configuration. Thus the iteration proceeds on the symbol field until no further change is possible. A schematic diagram of the macroprocessing is shown in Fig.13 as a typical example. Its merits are a very simple structure of masks and astonishingly strength against background and boundary noise.

However, it was not still analog in its complete sense and the research was further pushed, mainly by T. Mori, S. Mori, and Yamamoto [132], [133]. The key lies in the three-level labels, which are intermediate between analog and symbolic representations. Therefore, we can pass the symbols and use the three-level labels directly by means of transferring the digits themselves. Now the movement to system III was easy. Only the construction of the exact formalization remained, which was done by T. Mori. We skip the formulation, which is described in the cited references. A readable one appears in the book [4], which is written in Japanese.

The important point is that many rules are replaced by only two dynamic equations of density propagation. This is the strong point of the analog representation, which will be discussed later. However, the representation allowed 60 s/ch to be processed using a large scale computer of that time, for example, the TOSBAC 5600/GE 6000. Therefore, two kinds of small-scale experiments were conducted. One was for the ISO-B font data, which were very thin or of very dark print quality and consisted of 46 classes. The amount of data tested was 1300 characters and masks were made for 1, 2, 6, 8, D, Q, V, and W classes considering typical pairs which were hard to differentiate, such as (1, I), (2, Z), and (6, G). The correct recognition rate was 100%. The other was for constrained hand-printed numerals, where

2100 samples were written by 21 subjects. The correct recognition, rejection, and error rates were 98%, 0.2%, and 1.8% respectively.

On the other hand, as they proceeded to the more distorted data sets of unconstrained hand-printed alphanumeric characters, they encountered the same problem as Glucksman's case. Some subregions occurred within the same concavity region. The next step was to merge apparently separated field subregions which occur in a somewhat complex situation. The merging was easily done because the apparently separated field subregions are connected through a white line. The final step was to describe the relationship between field regions which are represented by representative fields. For example,"A" has two field regions, a hole and a concavity opened to the bottom. The relationship between the hole and the concavity regions was obtained by expanding propagation of the field regions and so if two field regions are adjacent, then the two collide with each other. The length of the collision line is an analog representation of the relationship between the field regions. An example can be seen in the book [4].

On the other hand, they found that the propagation equation includes two modes: One gives nonuniform fields as shown above; the other uniform fields, each of which is represented by any point in the field. This is similar to the special case of Glucksman's method, which is a binary coding system. At that time, it was very hard to pursue the analog system even on a computer, because it consumed too much time in its computer simulation. Therefore, they tried to make a simple version of the field effect method. One way is to use Glucksman's binary coding system. However, it generates counterintuitive regions, as mentioned before. Therefore, they needed to be merged and altered in order to meet with our intuition. Such processing was done by Zhan and Mori [134] based on a linguistic method proposed by Narashimhan [135], [136]. In 1977, Komori *et al.* [137] of NTT gave a very simple equation for these production rules. Furthermore they constructed global fields, called concentrated codes, including black strokes. Some examples are shown in Fig.14, where PS and PH stand for slant and horizontal strokes respectively. A recognition experiment was done in which a histogram of the global fields/codes was used for the classification as well as Glucksman's case. The data set was 10 000 hand-printed numerals written by 350 untrained writers. The correct recognition, rejection, and error rates were 99%, 0.6%, and 0.4% respectively.

Returning to the OCR project at ETL, the development of practical algorithms was needed, which was conducted in two ways. One was one microcomputer/programmable OCR; the other was to develop an algorithm assuming some hardware implementation to obtain a reasonable speed. Yamamoto and Mori [138] took this approach using Munson's idea. As mentioned before, in Munson's case, some difficulty was anticipated in finding global extreme points. The problem in practice was how to construct a convex hull . The definition of convex is well known: Let U be a subset of a linear space L. We say that U is convex if $\{\boldsymbol{x}, \boldsymbol{y}\} \in \boldsymbol{U}$ implies that $\boldsymbol{z} = [\lambda \boldsymbol{x} + (\lambda - l)\boldsymbol{y}] \in \boldsymbol{U}$ for all $\lambda \in [0, 1]$.

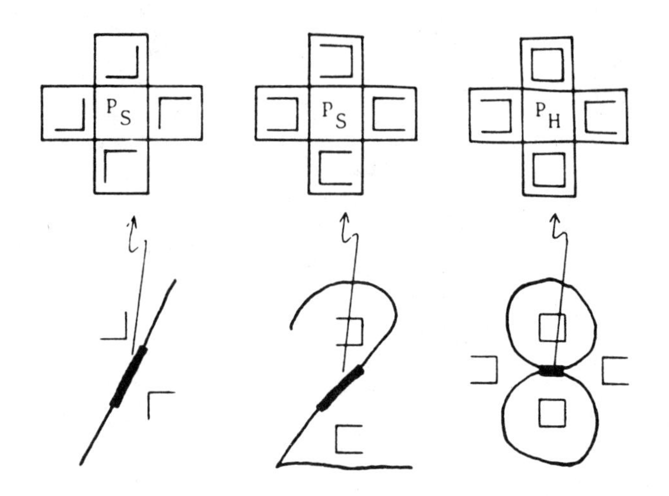

Fig. 14. Examples of the concentrated codes [137].

Table 1 Results of the Experiments on the Outermost Point Method

Set	Test Set			
Data	K	N	A	K+N+A
Mask	K	N	A	K+N+A
Correct	99.7	99.6	99.9	99.0
Reject	0.26	0.4	0.1	0.9
Error	0.04	0.0	0	0.1

The letters K, N, and A stand for katakana, numeral, and Roman alphabet, respectively [138].

This is not a constructive definition and it is of no use to construct a convex hull. Fortunately there is a constructive definition. Imagine a boundary point of the subset U of 2-D Euclidean space and a tangential line passing through it. Then further suppose a half plane whose edge is just the tangential line and includes larger U than that of the other opposite half plane. We can imagine such a half plane at every boundary point of U and select only such half planes that include only U. Then we can construct a convex hull of U by finding an envelope of the edges of the half planes. Their idea of the outermost point method is based on that constructive definition of a convex hull.

Once a convex hull is constructed, it is easy to detect concavities. However, the attributes of a concavity are important and they take three kinds of attributes, i.e., closeness, open direction, and area. The closeness is a measure of concavity, such as shallow or deep concavity. Very systematic experiments had been done using ETL data sets, ETL-3 and ETL-5. Four sets were arranged, i.e., numerals (ND), alphabet (AD), Katakana (KD), and mixed data sets of ND, AD, and KD. First of all semiautomatic masks were generated, which we will discuss later. ND, AD, and KD consist of 2000, 5200, and 9200 respectively, among which 1000, 3900, and 6900 were used as learning data respectively. The rest were used as test data sets. The results are shown in Table 1, which shows the high performance of the method. The reading speed was 0.7 s/ch using the TOSBAC 5600.

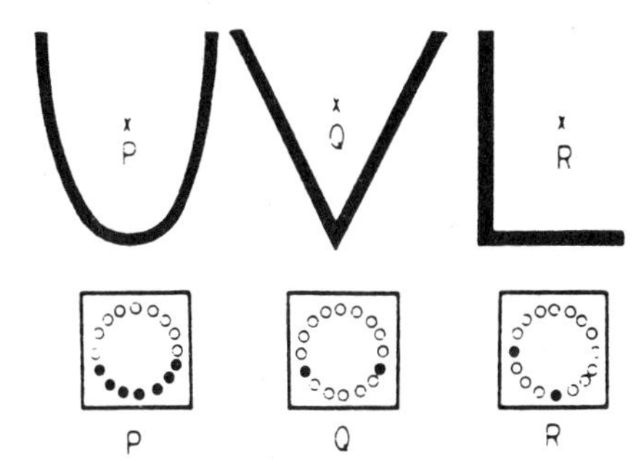

Fig. 15. Schematic illustration of the cell features.

In another direction, Oka continued basic research based purely on the field effect method. One drawback of the method was the difficulty in detecting geometric features. Actually the analog type of FEM had been combined with the Hough transformation to construct a recognition system [139]. The usual Hough transformation on the (ρ, θ) plane was not used; however, they applied edge to detect a line direction. The work [140] was done independently of O'Gorman and Clowes's [141]. At any rate, Oka [142] solved this problem using edge propagation. The general framework is based on a cellular automaton and each cell has a certain number of intracells according to the resolution of edge direction. Some illustrative and schematic examples are shown in Fig.15. The cells P, Q, and R are located at the points P, Q, and R. The system was suitable for LSI and implemented, as will be described later.

7) Syntactic/Linguistic Approach The syntactic or linguistic approach to pattern recognition has attracted many researchers in this field. In particular, Fu advocated the application of this approach to pattern recognition and developed it very actively with his colleagues [143], [144]. The approach has been included partially and/or implicitly as we have already described and will be mentioned further in real OCR systems. The basic idea has been explained before. Here we introduce it from a practical point of view. By scanning the labeled plane in a certain order, as already explained, we can construct a string of labels. The problem is how to construct an automaton which accepts the string as belonging to a class. A language whose grammar accepts the string must be constructed. A syntactic approach makes use of formal language theory, but a real image is not so simply recognized, as might be expected. Some typical problems were described by Ali and Pavlidis [145] as follows:

1) The direct syntactic analysis is faced with the need to handle the effects of noise, which causes rather complicated strings.
2) The parsing of the whole boundary requires the use of context-sensitive grammars for a description of the complete object.

In order to overcome these problems they used polygonal approximation and syntactic technique in a hierarchical manner. The former has already been mentioned. Later we will introduce another approach to problem 1. For the latter, they assumed that any closed boundary can be expressed by concatenations of semiglobal features/labels: Arc, Trus, Line, and Break. Here "Trus" means a sharp protrusion or intrusion and "Break" indicates the rest of regular shape; a Break is like a short segment having no regular shape. Thus, circular representation of a boundary is definitely segmented in this frame so that the context-sensitive problem is avoided. The production rule is very simple and intuitive, and the number of rules is 9. For example,

$$\text{TRUS} \rightarrow \text{STROKE} \mid \text{CORNER}, \text{CORNER} \rightarrow \text{LINE1+LINE2}$$

$$[\text{the-angle-of(LINE1, LINE2)} > \theta].$$

The experiment was conducted using Munson's data set, which was also used in their first version, mentioned before. The performance was reported as follows: error rate and rejection rate were 1.46% and 1.04% on the design set of 480 characters; error rate and rejection rate were 3.57% and 3.45% respectively on a testing set of 840 characters. Comparing the first version, they achieved considerably better results; the error rate of the first version was 14.7% on a testing set of almost the same size. However, their success lies mainly in the appropriately rough representation of the polygonal approximation and so it seems to be not easy to extend the method beyond numerals to a larger character set such as alphanumerals and Katakana. Concerning problem 1 above, Abe formalized a so-called penalty automaton [146]. To avoid the noise problem, counting and thresholding mechanisms were introduced into the automaton formulation.

Besides the above-mentioned problems, further difficulties exist. Given that an image is two-dimensional, a simple concatenation is not enough to represent a 2-D image; therefore, various natural high-dimensional generalizations of strings have been considered, such as Webs, plex structure, and trees. As far as we know, no concrete application of these has been reported in any OCR system. For more information we recommend the very readable book written by Gonzalez and Thomason [147].

8) Algebraic Approach: Operators for 2-D concatenation have been used by Shaw [148] in picture description language (PDL). In his scheme, a primitive is abstracted very generally by a line segment which has two points: head and tail. Four kinds of operators were defined on the primitives. We can construct letters of any shape using these operators and vectorlike primitives. In other words, we first need to determine the primitives to be used so that they are appropriate for the actual problem of representing characters. Mori [149], [150] investigated the problem from an algebraic point of view. His basic consideration was that a character set gives humans some feeling of roundness or linelike feature. This roundness or linelike feature must be reflected by the primitives that are used to generate the character set. Another guideline was that the primitives must be simple, intuitive, and strict in the mathematical sense, as described by Pavlidis (previously mentioned). He therefore introduced four kinds of monotone functions, so-called L-type primitives assuming contour following. A set of the primitives was treated as a set of integers (0,1,2,3) and a concatenation of the primitives can be represented by a "+" operator as mod 4. He derived mathematical shape descriptions based on the primitives. The point is that topological/geometrical properties are represented very simply and exactly by an algebraic form.

On the other hand, Nishida and Mori [151] further developed this line, in which the scheme is applicable to both boundary and thinned line. The primitives are four monotone functions with head and tail, including both horizontal and vertical line segments. The remaining two are "/" and "\" if schematically represented.

On these primitives, four kinds of binary operations are denoted as

$$a \xrightarrow{j} b, \qquad j = 0, 1, 2, 3,$$

where j is a characteristic number of the operator, and a and b are primitives. The operators have very intuitive meaning as well as mathematical rigor. In choosing appropriate monotone functions a and b, we can construct concavities (convexes), for example, which are open to the right, bottom, left, and top, corresponding to 0, 1, 2, 3, respectively, with a very flexible manner due to only the restriction of the monotone functions. A line segment is represented by such links of primitives. If the links are smoothly connected by a right-hand system, then they are represented by one sequence of the primitives, called a primitive sequence. That is, a line segment is compactly and hierarchically represented. In Mori's work, only quasi-topological features were considered, but both geometrical and quasi-topological features were systematically integrated, introducing singular point decomposition. Many mathematical formulas were derived, some of which are extensions of classical geometry, for example, the polygon.

Based on the theory and a new thinning algorithm [152], which is stable for simple figures of structure, a recognition experiment was conducted on a data set of 13 400 characters written by 220 subjects. The data set can be considered loosely constrained data with a variety of writing styles. Training and test sets were disjoint, selected in the ratio 1:1 in a randomized manner from the data set. Recognition, rejection, and error rates were 98.7%, 1.0%, and 0.3% respectively for the test data set. The total number of masks used was 46, in which that of "8" was 10, which was conspicuously large. This reflects the complexity in the shape of the numeral "8." They did another experiment removing poor-quality samples from the data set. This resulted in a data set containing 11 500 samples. The two methods were compared, the new method, the algebraic and quasi-topological method (AQT), based on the thinning method, and the contour following analysis. For the former, the recognition and error rates were 99.1% and 0.1% respectively using 37 masks. For the latter, the recognition and error rates were 97.6% and 0.4% respectively using 58 masks. Based on these experiments, they claimed superiority of the AQT method over the contour analysis method. We note that its flexibility is high, comparing the number of masks of the former with that of the latter.

V. Commercial OCR Products

A. Current Status

During the past ten years, a number of optical character readers have disappeared from the market place. They have been replaced by new and more powerful ones which appeared in many different forms, among them small hand-held wands with the shape of a handgun, hand-held scanners like a sweeper, page readers, flat-bed scanners, and document readers integrated with mechanical transport devices [153]–[157]. The prices of these machines range from $500 to over a million dollars. The majority of them are PC-based and cost between $2000 and $10 000 (software packages excluded). Because of space limitation, this section mainly deals with the machine capabilities and gives details of the specifications of different OCR products and lists of manufacturers, especially those in America and Europe that can be found in recent surveys compiled by the authors mentioned above.

Hand-held scanners and wands are often used to read UPC codes, magnetic data, and materials (such as tickets, price tags, checks) where severe constraints have been imposed on the printing of the codes and characters, e.g. bar codes, magnetic inked characters, and OCRA characters. They are mostly limited to the reading of numbers and just a few additional letters or symbols. Typical OCR machines can read a few 10 and 12 pitch typewritten fonts, including OCRA, OCRB, and others in single and double spacing. Depending on the complexity of the layout of the text and number of fonts treated, their reading speeds vary from five to ten pages per minute. With a scanning resolution varying from 150 to 400 DPI (dots per inch), normally these optical readers can tolerate a skew of $2 \sim 3$ degrees, or a quarter of an inch per line. A number of them can output files in the formats of common word processors in addition to the ASCII format. Most are able to reject unrecognizable characters or documents and allow manual key reentry of such characters or documents afterwards. Since the recognition rate depends heavily on the quality of the print and the preparation of the document, the resolution of the scanner, the fonts used, the complexity of the layout and format, and factors other than the recognition algorithm, many products do not state their recognition rate. Those which state such a rate, generally say that it is over 99.9% when operated under optimal conditions. Sometimes such a specification can be very misleading.

More sophisticated optical readers can function both as image scanners and as OCR machines. Such combined functions allow them to capture both pictorial and text information for storage, transport, and retrieval purposes. Software management systems also allow the user to rapidly build a full text and graphics data base from the scanned input. These readers can process documents in multifonts or omnifonts which can be typewritten, typeset, and printed by dot-matrix, line, and laser printers. They can recognize characters with font sizes from eight to 24 points at least, spaced between four and 12 lines per inch, in different formats including intermixed text and graphics. Some of them can read a large variety of fonts in regular pitch and proportional spacing, including those used in most common office correspondence and popular publisher fonts, for example OCRA, OCRB, Gothic, Courier, Elite, Pica, Orator, Narrator, Avant Garde, Helvetica, Times, Times Roman, Palatino, Bookman, Monaco, Geneva, and New Century Schoolbook. With the introduction of narrow range scanners, measuring 3 to 6 inches wide, columnar scanning is now available. With this, the optical reader can recognize multiple columns or sections of a page or mailing lists. Some are equipped with software for spell checking, and for flagging suspicious characters or words. Some can also read special symbols plus a variety of languages, including many European languages which contain vowels with different types of accents, and Oriental languages which contain thousands of characters written in different fonts.

It is interesting to note that several companies have produced low-cost software which enables users with scanners to apply it to read their documents. Common packages cost in the neighborhood of only $500. This has proved to be very convenient for a number of owners of optical scanners. Some of the optical readers are equipped with an automatic document feeder so that they can read many documents automatically. Apart from working with Personal or Apple/Macintosh computers, some optical readers can also be integrated to more advanced computers, such as PS/2, Sun workstations, and a variety of other host systems. Some have resolutions up to 600 DPI, and can read documents in different gray levels and colors. A number of them support SCSI, RS-232, and Ethernet for communication with other pieces of equipment.

One of the challenging areas of optical character recognition is the recognition of handwritten materials. However, in spite of many years of intensive research in this area [158], optical readers with such capability are rare. Even if they exist, they typically can read only hand-printed digits which must be legible, well-written, and consistently placed. Very few products can read alphanumeric handprints though some Japanese machines can read standardized alphanumerics and Katakanas hand-printed in preprinted boxes. In this respect, standardization of hand-printed characters is very important. On this topic see the paper by Suen *et al.* [159]. Recognition of characters written without any constraint is still quite remote. For the time being, recognition of handwritten characters, signatures, and cursive script seems to belong only to on-line products where typically writing tablets are used to extract real-time information and features to aid recognition. See, for example, the paper by Wakahara *et al.* in this issue. Although no complete OCR product exists, people are using OCR machines knowing that the OCR industry has become profitable recently. Now let us trace the history of OCR systems in terms of commercial products.

B. The Generations of OCR

Several commercial OCR's appeared in the beginning of the 1960's; these were outputs of the age of cut and try. The first generation can be characterized by the constrained letter shapes which the OCR's read. These simple methods were effective. The most typical one was the NCR 420 [160], for which a special font for numerals and five symbols, called NOF or bicodes, had been designed. The next typical one is the Farrington 3010 [161] of Farrington Electronics Inc. The story was the same as the NCR's OCR; i.e., they also used a special font, called Selfchek 12F,7B. IBM was also very active in developing OCR, recognizing that OCR was a very important input device for the computer. The first commercialized OCR was the IBM 1418 [162], which was designed to read a special IBM font, 407. An input character was fed to a 17 column by 10 row shift register, which made the machine resistant to vertical position variation. The letter shape looked more natural than the two fonts mentioned above. The recognition method was logical template matching; however, positional relationship was fully employed. The logic was quite complex in order to cope with document variations. The design stance of using computer technology, logic in particular, was inherited by subsequent IBM OCR systems, i.e., IBM 1428, IBM 1285, IBM 1287, and even the special-purpose OCR, the IBM 1975. In Japan, by the end of the 1960's, some mainframe companies announced their OCR's of the first generation. They included the Facom 6300A from Fujitsu and the H-852 from Hitachi, both of which used the stroke analysis method. On the other hand, the N240D-1 of NEC was somewhat similar to the IBM 1417 and also read the 407 font.

The second generation of OCR systems can be characterized by hand-printed character recognition capabilities. At the early stage they were restricted to numerals only. Such machines appeared in the middle of the 1960's and early 1970's. We introduce a few typical systems here. The first and a famous OCR system was the IBM 1287, which was exhibited at the World Fair in New York in 1965. The system was a hybrid one in terms of hardware configuration. It combined analog and digital technology. We have already mentioned it as the analog device of contour following.

The first automatic letter-sorting machine for postal code numbers was developed by Toshiba [163]. The three big companies in Japan participated in the development of the machine and Toshiba was the first winner, with the work done by Genchi, K. Mori, Watanabe, and Katsuragi. Their method is a typical structual analysis approach and is very famous. We recomend their paper, which is readily available. Two years later, NEC developed an automatic OCR sorter for postal code systems and participated in this market as well as Toshiba. The algorithm developed by NEC was based on stream following analysis, which was described above. Here we add some explanation in relation to Toshiba's system. In contrast to Toshiba's approach, an image was considerably reduced to a 6×10 matrix and size normalization was performed. Using consecutive run length encoding information, as the first step, each pixel was labeled row by row from bottom to top. A semilocal feature plane was constructed, with a crossing count pair of consecutive rows. These features were macroprocessed along each row and the resultant semifeatures were represented by 16 kinds of labels. In this sense, the scheme of the macroprocessing is similar to that of Toshiba's, but simpler. They continued the development toward better performance and the series of the machines reached NAS50 in 1975, which read numerals and alphabetical characters printed in type (multifont) as well as unconstrained numerals. For the printed characters they basically used template matching. For the numerals, they improved the algorithm taking a higher resolution of a 16×20 matrix.

The performance was reported as follows: At the early stage, they could not obtain satisfactory results. The correct sorting rate and the error sorting rate were about 60% and $5 \sim 6\%$ in 1968. However, they had improved machines to the level of $92 \sim 93\%$ correct sorting rate and $1 \sim 2\%$ error sorting rate by 1974. Here we would like to point out that the above reveals a very important element in developing OCR systems. Both a sympathetic manager and the user are key factors in developing this kind of technology. Both Toshiba and NEC people have been fortunate in this respect, aside from their intelligence and great efforts.

The second generation of OCR can also be characterized by the recognition capability of a set of regular machine-printed characters as well as hand-printed characters. In this respect, a very powerful OCR system, called RETINA [164], was developed by Recognition Equipment Inc. This system read both hand-printed and machine-printed characters. When hand-printed characters were considered, the character set was constrained to numerals, and the following letters and symbols: C, S, T, X, Z, +, and -. On the other hand, the system that could read printed characters could recognize 40 letters: a complete uppercase alphabet, numerals, and four special symbols. The user specified the font to be read by the machine. A key feature of RETINA, in terms of OCR systems, lies in its parallelism, which is associated with the name RETINA. The processing speed was very high at 2400 chs. Concerning the recognition of hand-printed characters, strokes are matched, and line crossing and corner types are also taken into account. Positional relations of detected features are examined to identify an input character. RETINA was used in practical applications in spite of its great expence.

Early in the 1970's, one big barrier to the use of OCR began to break. The first OCR which aimed at high performance and low cost was made by Hitachi [165], [166], called H 8959. As stated before, a flying spot scanner was very expensive at that time. The flying spot scanner was replaced by a laser scanner, and a special-purpose OCR processor was used instead of wired logic circuits, resulting in considerable cost reduction. This OCR was primarily for hand-printed numerals and some symbols, which was the same set for RETINA. The algorithm used belonged to thinning line analysis and actually the machine at first employed a regular thinning algorithm. The thinned line was represented by chain encoding using 3×3 masks. Singular points, such as end points and branching points, were also identified by counting matched mask(s). If the number of matched masks was one, two, three, or four, then the end point, normal line point, branching point (T type), and intersection point (X type) were identified respectively. Thus the thinned character line is decomposed into branching line segments whose terminal points are singular except for a simple loop in which the starting point was used. The coordinates of those singular points of the branching line segments were registered. Then each branch segment was further divided into four kinds of monotone line segments, called four-quadrant coding: upward right and vertically up; upward left and horizontally left; down right and vertically down; and down left and horizontally right [167]. This was a partial application of four kinds of monotone functions (not strict) as primitives, mentioned before (strict).

For a loosely constrained data set written with pencils by 60 Hitachi employees, with a sample size of 1200, the correct recognition rate and the error rate were 95.7% and 1.3% respectively. For a data set written by the same persons who had received training, a 100% recognition rate was obtained. These experiments revealed how much the recognition rates depend on the quality of the writer's group and that, in the case of a cooperative group, the recognition performance can be high without imposing severe constraints.

Hitachi improved the H-8959 by applying DP matching to branch matching. The result was the H-8957 [168], in which the branch was represented by chain encoding. The matching is nonlinear (DP matching) and so many-to-one and one-to-many are allowed. The algorithm used for DP matching is a typical and well-known method in speech recognition. The machine was primarily for FORTRAN reading, and so 47 characters of hand-printed alphanumerals and symbols were read, and context checking was taken. The recognition performance was reported: 0.06% error rate and 0.20% rejection rate were obtained with a sampling of 26 400 characters produced by six trained writers. This was so good that we can suppose that the quality of the data was exceptionally good, being close to constrained hand-printed characters.

Third-Generation OCR Systems

By the end of the 1960's, the next targets for document readers were for poor-print-quality characters, and hand-printed characters for a large category character set, such as Chinese characters. By document, we mean that it includes words such as names, addresses, and commands. These targets have been achieved partially and such commercial OCR systems appeared roughly during the decade from 1975 to 1985. Low cost and high performance are always common objectives for these systems. The dramatic advances of LSI technology were of great help to the engineers who were engaged in the development of OCR systems. Although this was common to all electric systems, in general LSI was especially important for pattern recognition systems. In the early days, the barrier had to do with the large number of components. This is being solved in general by high-speed CPU's, large available memories, and ROM's in particular, rather than by making highly parallel machines.

1) Print-Quality Proofreader To overcome the print quality problem, much work was done. We will pick two cases as representative examples. One is the ASPET/71 (TOSHIBA OCR-V100) and the other is the IBM 1975 [169]–[171] . They are in strong contrast to each other. First of all, the former is an analog type and the latter is a digital type. The name ASPET/71, derived from the analog spatial parallel system made by ETL and Toshiba, was given by Katsuragi, an excellent researcher at Toshiba and contributed significantly to the construction of the system, but died very soon after the project ended at very young age. The 71 denotes 1971, the year when the system was built completely. The algorithm is based on correlation, but a special one, called the multiple similarity method, which was mentioned in detail before. The ASPET/71 was built from analog circuits, but the actual commercial product, the OCR-V100, used digital technology fully. On the other hand, the algorithm of the IBM 1975 is based on logical matching and also was mentioned in detail before.

A problem faced by both parties of engineers was noise arising from the poor print quality of the characters. However, methods for handling the problem were very different. It was solved by theory in the former, but by human intuition in the latter. In correlation matching it is easy to make masks/dictionary by taking the average of the data belonging to the same class, which is the strong point of template matching in general. But in the case of logical matching, the average has no meaning, because it is done in linear space. So considerable effort was devoted to automatic mask making at IBM, as mentioned before. However, there was a big difference between laboratory data and field data. In fact the IBM 1975 was developed specifically to meet the needs of the U.S. Social Security Administration, which involved reports printed by more than 256 typewriter fonts with an uncontrolled and wide range of print quality. As a result, the engineers who developed the IBM 1975 gave up using the automatic mask making theory after many unsuccessful trials and relied on intuition. They wrote: "the designer imagined as generalizations of the few specific examples which he had on hand." In some sense, we may say that their approach was structural analyzed logical matching. They actually listed the features which they noticed as line segments (short, long, horizontal, vertical, slanted, etc.), line endings, various corner curvatures, gap in lines, relative positions of line segments, and so on.

ASPET/71 was planned as one product of the Ultrahigh Speed Computer System Development Project, promoted by the Japanese Government, and was achieved by cooperation between ETL and Toshiba. The great success of the project led Toshiba to develop a commercial OCR system based on the multiple similarity method, which has already been described.

Concerning the IBM 1975, besides the recognition system designed by human intuition, considerable effort had been paid to preprocessing. Binarization was crucial, particularly in logical matching, so an adaptive video thresholding system was used as a linear function of V. V was defined as the average of all video samples greater than some low threshold value within a predetermined area. Width normalization was also done, where the stroke width measurement was fed back for threshold adjustment. Postprocessing was also implemented to create a file name. (The data consisted of names and associated digits: social security number, FICA-taxable wages.)

In the recognition system, features were first measured, each of which was detected by a single-output Boolean function of about 30 inputs. A total of 96 feature detection circuits were used and so a 96-position binary feature vector was constructed for each input character. The decision process was done by comparing the feature vector of the input character with a set of reference vectors/masks and determining which reference most closely matches the input feature vector. The reference structure was ternary, including the case "don't care." Each reference was stored in ROM, where 192 (2×96) bits were used to store the main portion of the reference and 48 additional bits were used to store the reference control information. Approximately 2000 references were stored and 150 to 200 of them were processed for each input character. Remember that 256 fonts were used! The control information used a very typical technique of computer applications. A two-level decision process was specified in which a few character classes are first chosen by means of a selected group of reference vectors, and then a large number of reference vectors are used to decide among these classes.

2) hand-printed Document Reader: In September 1975, the Labor Market Center of the Labor Ministry of Japan formally decided to use Katakana OCR in its total employee insurance system in 1980. This contract attracted the competition of almost all the big electronics and electrical engineering companies. The R&D for Katakana OCR became very active and a related story can be found in the paper by Suen *et al.* [5]. That paper was written in 1979, and so a final contractor had not yet been chosen. The story will be continued here. The winner was not just any company, it was NTT. At that time, NTT was a public corporation and a telephone and telegram monopoly. The system planned by the Labor Market Center was very much involved with communication systems. NTT therefore developed OCR systems very rapidly and succeeded in developing a high-performance Katakana OCR system in time. As a result, the NTT OCR was adopted and its manufacturing was committed to NEC and OKI.

The recognition system of the NTT Katakana OCR [172] was based on background analysis, as described before. The basic idea was that relations between semiglobal features should be represented in the same form as the semiglobal features. In other words, the global features should be the natural results of macroprocessing of features. Returning to Glucksman's method, he used the ternary coding, but the system mixed semiglobal features and global features. NTT therefore first took the binary coding. According to the value of the center pixel, i.e., white or black, the codes were classified into 16 code sets or four code sets, respectively, called white point first-order features or black point first-order features, respectively. The former can be inferred from the Fig. 14, where concavities and loops are schematically represented. The latter codes were symbolized as P_V, P_H, P_S, and P_I, where V, H, S, and I denote vertical, horizontal, slant, and internal. Some representations of the global features, called secondary features, are shown also in Fig. 14 . We note here that the total coding is reduced from 16+4 to 16 in order to facilitate and economize the hardware implementation. The final stage is to construct masks using the secondary features. They made them manually, but the work was easy because the features were globally macroprocessed. However, their scheme was not enough to separate some closed categories, where the difference lies in only the geometrical feature. Contour following was used to detect an exact geometrical feature using black first-order features, which are used as auxiliary features. Concerning hardware implementation, see [173].

A large-scale simulation experiment was conducted in which a data set was written by 2000 students and clerks who were instructed to write characters within boxes keeping specified writing rules of the Japanese handprint standard. The total number of characters collected was 188 000 ((4000/category) and the number of categories was 47 (Katakana set except one letter). Half of the data set was used as learning data and the rest were used in the test. The recognition results were reported with a correct recognition rate and an error rate of 97.1% and 1.0% respectively. The number of masks used was 234 (5/category). Currently the machine, called the DT-OCR100C, reads 65 kinds of letters, numerals, Katakana, and symbols. The total number of document types handled is greater than 100.

3) Software Package as an OCR: Thanks to highly developed computer technology, very recently the recognition part of OCR has become fully implemented in software packages, which work on personal computers. Present PC's based on the 386, for example, are comparable to the large-scale computers of the early days. Of course, there are some limitations in such OCR software, in particular in its reading speed and the kinds of character sets that can be read. However, regarding printed character sets used in the West, some commercial products have appeared which have impressive performance. Here, we introduce some of them, but unfortunately no technical information is available and so the following description is based on catalog data.

The Caere Corporation sells OmniPage and Typist. Typist includes a hand-held scanner which is used with typical PC's. The PC computers use the 286 or 386 microprocessors with AT or PS/2 bus, with 640K base, and have 2 MB of expanded memory and hard disk. OmniPage works in almost the same system environment. Both OmniPage and Typist can read 11 European character sets, all nonstylized fonts including italics and boldface from 6 to 72 point sizes, and dot-matrix alphanumerics. OmniPage can read characters with an average speed of 40 cps and also underlined characters. It provides a proofing tool which can display a small bit map image of the word or character in question, adjacent to the questionable occurrence when a user jumps to the uncertain character. Some page segmentation capability is provided such as automatically differentiating columns, graphics, and text. The price of Typist is only $595 for a PC with AT bus and PS/2 bus as of 1991. This is certainly a surprising price considering its high performance.

One competitor to Omnipage is the Discover series of Kurzweil (Xerox Image System, Inc.). It reads printed characters in almost the same system environment, with 10 $\sim$ 40 cps. They claim almost the same performance/specifications as Omnipage. However, character size is limited to 8 $\sim$ 24 points and it imposes a specification on character pitch as a minimum of one dot intercharacter spacing after scanning. Regarding page segmentation, its performance is not clear. Another competiter is WordScan, produced by the Calera Corporation. Some technical infomation relating to this OCR system is contained elsewhere in this issue (refer to Bokser's paper). There are other OCR software products for reading printed characters, but for European languages, the above three seem to be major players in this market so far. Regarding other products and detailed information for OCR users we recommend [174].

4) Kanji Recognition Machine: The dominant characteristic of Kanji is its large number of categories (2000 $\sim$ 4000). Kanji was transferred from China to Japan as a subset of Chinese characters (6000 $\sim$ 50000)). Because of space limitations, two kinds of specific products will be mentioned. For a general and technical description, the reader can refer to [176].

Commercial products: Toshiba is a pioneer in Kanji OCR, which announced a Kanji OCR, the OCR-V595, in 1983, based on multiple similarity method, which was described in detail before from a historical point of view. Now Toshiba's ExpressReader 70J is popular as a reader for Japanese printed characters. It has omnifont reading and page segmentation capability. For the latter, see the paper by Tsujimoto *et al.* in this issue. Concerning Kanji it reads 4000 kinds of characters, which is virtually enough to deal with almost all kinds of characters. The character size is 6 $\sim$ 40 point and reading speed is 70 $\sim$ 100 chs, which is the highest among Kanji OCR commercial products. Its correct recognition rate is 99.5% according to the catalog.

Next we introduce two commercial products of Kanji OCR whose origin is based on the feature matching and field effect method, which was explained before. However, they are very different in terms of implementation and way of utilization. The first, the XP-70S, made by the Fuji Electric Company, is based on Yasuda's algorithm. The algorithm is based on feature matching of two kinds of feature fields: stroke directional field and edge propagation field [175]. For generation of these feature fields, Fuji developed special LSI [177], which consists of CMOS at a clock speed of 5 MHz. This gives a speed of about 115 chs. This was developed for document processing which has been printed or typed. The kinds of character sets read are rich: Roman alphabet (52 letters), Hirakana (82), Katakana (83), numerals (10), Greek letters (19), Kanji (2965), and special symbols (152). The machine is a multifont reader. Reading speed is 30 chs. The user can have his own special letters by invoking its learning mode. The machine can display a maximum of ten candidate characters, and these can be corrected manually. The price is 4.5 million yen at the time of writing. Postprocessing is also available.

The second product is the CLL-2000, made by Sanyo Electric Co. Ltd. [178]. This is a desktop hand-printed character reader which is used together with a word processor. It is small in size, and low in price (2 million yen at the time of writing). The architecture of the machine is very interesting. It consists of 21 16-bit microprocessors (8086) and so it is a multi-CPU system, which made the low cost possible. The algorithm implemented is Oka's and was explained before. It has a cellular architecture; i.e., each pixel has a cell which has eight intracells. A character is written inside a box 1 cm $\times$ 1 cm, which is sampled to a 96×96 matrix. After size normalization, it is reduced to a 60×60 matrix, on which local edge detection is done and the matrix is further reduced to a 30×30 matrix. Therefore, the feature field needs $(30\times30\times8)$ = 7200 bytes. The problem is the large number of categories, about 2400, and so the dictionary needs $7K\times2400$ = 16.8 Mbytes, which is too big in aiming at low cost. So drastic dimension reduction was done by reducing the 30×30 matrix to 7×7. Thus the size of the dictionary was reduced to about 1 Mbytes. We note that in spite of such drastic reduction of the dimension, the recognition performance does not drop too much. The CLL-2000 reads hand-printed Kanji, Roman alphabet, Hiragana, Katakana, numerals, and symbols: a total of 2377 letters. The reading speed is 2 chs. The correct recognition rate is 93%, but the machine can list ten candidates with a correct recognition rate of 99.6%. However, the data set was collected by the company. When a manuscript is read, the processing speed is twice as fast as humans, including the time to make corrections.

VI. LEARNING

Learning is a basic function of pattern recognition in animals in general and in human beings in particular. Therefore, considerable research effort has been spent and neural networks have recently attracted a lot of attention. Everyone is aware of the situation, and is interested in the near-term contributions of neural networks. However, such predictions need serious consideration and lengthy study, which is beyond the scope of this paper. In this respect we recommend the tutorial paper by Amari [179], who is known as the inventor of back-propagation and a distinguished theoretician. We can in a restricted way learn from the perspective of automatic design. In this respect, we discuss an expert system application used to design OCR. Concerning neural network applications to OCR, we will make some comments at the end of this section from the engineering point of view.

A. Automatic Design

The review paper on character recognition written by Govindan *et al.* in 1990 says that "No attempts are known to the authors in the topic of automated designs dealing with the design of recognizers suitable for structurally different character sets." Certainly research on automatic design based on a structural analysis approach is rare, in spite of its crucial importance. Naturally, however, total automatic design, which includes preprocessing, feature extraction, structure description, and mask making, is not known to the authors, and we think such total automatic design is impossible. This would insinuate an automatic creator. We are not God! The so-called connectionist seems to aim at such a learning machine. Anyway we treat this topic in a restricted way and so we can introduce very successful research which was done mainly by Yamamoto [180], [181]. We restrict automatic design to automatic mask making, which is a very tedious and laborious process, especially when dealing with unconstrained hand-printed characters. In this sense, we also need to have an expert system.

At ETL Yamamoto *et al.* became aware that topological features such as continuity, hole, and quasi-topological features are very effective in classifying characters. According to these features, a character set can be divided into a set of subclasses, each of which is denoted as ω_{jk}, where j and k mean the jth kind of letter and kth subclass in the j class respectively. Automatic mask making is divided into the following two steps:

1) taking correspondence between the input line segments and the mask's line segments;
2) taking statistics on each feature axis in the correspondence.

Step 1 is crucial. Here we assume contour following and so as an example we start it from the top left. Then because of shape variations, we cannot always obtain a stable order of line segments . However, fortunately, after the classification mentioned above is done, within each subclass the order of the line segments is very stable. Therefore, we can do step 1 easily. Since step 2 is well known, we do not mention it here. Experiments on Katakana character sets have revealed its usefulness; the average number of learned masks is 3.7 per class and naturally a 100% correct recognition rate was obtained for a 10 000 training data set. A key to automatic mask making is the good qualitative description of the shape being used. In this respect, singular points and a quasi-topological description [151] are good candidates for this purpose and in fact it is shown that automatic mask making is done effectively and strictly in both practical and mathematical senses [182]. On the other hand, another approach was taken by Baird [183], who used a Bayesian classifier by mapping structural features to numerical vectors. Usual statistical learning is done in the feature space.

B. Expert System

For sloppy data, we have to consider the construction of some kind of expert system. It is said that one human expert is assigned per famous novelist, whose writing material cannot be understood by ordinary people, without elaboration by the expert. Thus, for totally unconstrained characters some expert systems have been developed [184], [185]. Among them, Suen *et al.* recently introduced a multiple-expert system [186]. Very encouraging results have been reported. The key lies in an intelligent way of combining the expertise of these experts to reinforce their strengths and to suppress their weaknesses. More details can be found in the paper by Suen *et al.* in this issue. See also Srihari's paper in this issue.

C. Neural Network

Many different methods have been explored during the past four decades by a large number of scientists to recognize characters. A variety of approaches have been proposed and tested by many researchers in the field, as stated. Recently, the use of neural networks to recognize characters and different types of patterns has resurfaced. In Krzyzak *et al.* [187] and Le Cun *et al.* [188], back-propagation networks composed of several layers of interconnected elements are used. Each element resembles a local linear classifier which computes a weighted sum of its input and transforms it into an output by a nonlinear function. The weights at each connection are modified until a desired output is obtained. In a sense, it acts like a black box which makes use of the statistical properties of the input patterns and modifies its decision functions to produce a desired output. Owing to the lack of shape features, its output is sometimes unpredictable; hence apart from modifying the back-propagation model, some geometrical features were added to the network to enhance the recognition rate. Thus, from our experiences and informal information, when we compare a neural network OCR with a conventional OCR at the top level in this field, we cannot find any evidence that a neural network is superior to a conventional one. Some, however, advocate its superiority. One point dampening high expectations for the neural net is its poor capability for generality.

However, neural networks have been studied by researchers for several decades and some new features have been found and applied [189]. This is because each method has its own advantages and disadvantages, and it is better for us to have more information at our disposal than miss some new advancement. Recent research results call for the use of multiple features and intelligent ways of combining these various methods [190], [191]. The use of complementary algorithms to reinforce each other has also been suggested by Nadal *et al.* [192].

Here we quote Amari: "It is desirable to apply a neural network to broad area where it is effective..... However, backpropagation is so simple in terms of theory that it has its own limitation which can not be deep. It is important that we should not impose too much expectation on backpropagation. I am sure a higher level theory will appear sooner or later, because of the research being done by many researchers.... I can not now see any perspective on the new neural net theory" [179].

How does the human brain recognize different characters? Which parts of the characters is it looking for? What kinds of features do we get when we see a character? Is there some way we can extract character recognition knowledge from humans, who have become such superb character recognizers? These are some of the questions which remain unanswered. Obviously a lot more research should be done in this field before we can make computers read documents reliably and intelligently.

VII. Conclusions

We have described the stream of research and development of OCR systems, which has two large tributaries, template matching and structure analysis. We showed that both tributaries have grown to a concrete OCR technology and are merging to become a wide stream which constitutes an essential part of pattern recognition technology. On the other hand, we pursued the development of commercial OCR's and classified them into three generations. We also described some representative OCR's in some detail to show the historical development of OCR from both academic and industrial points of view.

We described many methods, some of which are somewhat related to each other and some of which are more or less independent. The important point is that we should make these methods more precise in the sense of an exact science, not a mere accumulation of empirical knowledge. The above statement is very much related to a shape model in which we need to establish an exact and flexible mathematical model of shape including a noise model which can be independent of the shape, but is intrinsically related to a given shape. The above two statements might have the relation of "chicken and egg," but it is necessary to attack both of these problems at the same time in order to establish a shape recognition technology on scientific ground. Based on these efforts, we will be able to make a machine which will approach human performance on shape recognition.

It is clear and real that, in order to make such a machine, some combination of the methods mentioned so far should be brought together. Multiple approaches and complementary algorithms will be integrated. Performance of each method and models of characters will be known by the reading machine. On the other hand, a new method is still naturally expected and because of the rapid development of modern computer technology, we can expand our creative space to a more sophisticated method, such as a morphological approach, which incorporates human expertise and the direct use of multilevel images.

In practice, it is very important that a machine gain the confidence of its users. If a serious user writes a character well, then a machine has to read it with 100% accuracy with low-cost machine/software. This means that we should specify the performance of the OCR. So far, the specification has been very loose. It sometimes does not give any confidence to a user. However, this is not an easy matter and in fact, it is very much related to the problems mentioned above. In practice, it is also related to standardization of character shapes in hand-printing. People are not so sensitive to certain confusing shapes such as "O" and "D," and so engineers should point out such issues so that users can understand the real situation. Naturally, such a standard will be less restricted, but more precise in the sense that it points out "key features" to be appreciated naturally by humans. In this connection it is very useful to have the cooperation of elementary school teachers to help children to form good habits in writing characters properly and legibly, and pay special attention to their distinctive features.

The R&D of OCR is also moving toward "word recognition," using contextual knowledge such as addresses and names. In fact such development, of postal address and name reading machines, is already such a trend. This necessarily leads the R&D of OCR to document analysis, in which characters constitute one component. Thus the R&D of OCR will expand its applications to a total document reader, posing the greatest challenge to researchers in this field.

The history tells us that OCR technology has been built by many researchers over a long period of time, consisting implicitly of something like a worldwide human research network. In such an invisible forum, people have made efforts, with "competition and cooperation," to advance the research effort. In this sense, international conferences and workshops are being organized to stimulate the growth in the area. For example the International Workshop on Frontiers in Handwriting Recognition and the International Conference on Document Analysis and Recognition will play a key role in the scholarly and practical arena.

ACKNOWLEDGMENT

The authors would like to express their thank to Prof. T. Pavlidis for encouraging them to write this extensive survey paper, to the members of the Japanese Document Standardization Committee for their important contributions, and to Dr. P. Hart and J. Cullen of the Ricoh Research Center in the U.S. for their valuable comments. They would also like to thank their colleagues, staff, and students who have contributed significantly to these research efforts during the past two decades.

REFERENCES

[1] J. R. Ullman, *Pattern Recognition Techniques*. London: Butterworths, 1973.
[2] T. Sakai and M. Nagao, *Characters and Figures Recognition Machine*. Tokyo: Kyoritsu, 1967.
[3] T. Pavlidis, *Structural Pattern Recognition*. New York: Springer, 1977.
[4] S. Mori and T. Sakakura, *Fundamentals of Image Recognition vol. I, vol. II*, vols. I and II. Tokyo: Ohm, 1986 and 1990.
[5] C. Y. Suen, M. Berthod, and S. Mori, "Automatic recognition of hand-printed characters—The state of the art," *Proc. IEEE*, vol. 68, pp.469–487, Apr. 1980.
[6] J. Schurmann, " Reading machines," in *Proc.6th IJCPR*, 1982, pp.1031–1044.
[7] V. K. Govindan and A. P. Shivaprasad, "Character recognition - A review," *Pattern Recognition*, vol. 23, no.7, pp. 671–683, 1990.
[8] G. Tauschek, "Reading machine," U.S. Patent 2 026 329, Dec. 1935.
[9] P. W. Handel, "Statistical machine," U.S. Patent 1 915 993, June 1933.
[10] M. H. Glauberman, "Character Recognition for business machines," *Electronics*, pp. 132–136, Feb. 1956.
[11] W. J. Hannan, "R. C. A. multifont reading machine," in *Optical Character Recognition*. G. L. Ficher *et al.*, Eds. McGregor & Wemer, 1962, pp.3–14.
[12] ERA, "An electronic reading automaton," *Electronic Eng.*, pp. 189–190, Apr. 1957.
[13] T. Iijima, Y. Okumura, and K. Kuwabara, "New process of character recognition using sieving method," *Information and Control Research*, vol. 1, no. 1, pp. 30–35, 1963.
[14] L. P. Horwitz and G. L. Shelton, "Pattern recognition using autocorrelation," *Proc. IRE*, vol. 49, no. 1, pp. 175–185, 1961.
[15] M. Sato, K. Yoneyama, and Y. Ogata, "Recognition of printed characters," *T. R. Radio Wave Research Laboratory*, vol. 7, no. 33, pp. 489–492, Nov. 1961.
[16] W. S. Rohland, "Character sensing system," U.S. Patent 2 877 951, Mar. 1959.
[17] R. W. Weeks, "Rotating raster character recognition system," *AIEE Trans.*, vol. 80, pt. I, *Communications and Electronics*, pp. 353–359, Sept. 1961.
[18] R. O. Duda and P. E. Hart, *Pattern Recognition and Scene Analysis*. New York: Wiley, 1973.
[19] W. H. Highleyman, "Linear decision functions with applications to pattern recognition," *Proc. IRE*, vol. 50, pp. 1501–1514, June 1962.
[20] G. Nagy, "Optical character recognition-Theory and practice," in *Handbook of Statistics*, P. R. Krishnaiah and L. N. Kanal, Eds., vol.2. Amsterdam: North-Holland, 1982, pp. 621–649.
[21] R. B. Johnson, "Indicia controlled record perforating machine," U.S. Patent 2 741 312, Apr. 1956.
[22] T. L. Dimond, "Devices for reading handwritten characters," in *Proc. Eastern Joint Computer Conf.*, 1968, pp.207–213.
[23] L. A. Kamentsky, "The simulation of three machines which read rows of handwritten Arabic numerals," *IRE Trans. Electron. Comput.*, vol. EC-10, pp. 489–501, Sept. 1961.
[24] B. A. Glucksman, "Classification of mixed-font alphabetics by characteristic loci," in *Dig. 1st Ann. IEEE Comput. Conf.*, Sept. 1967, pp.138–141.
[25] W. Doyle, "Recognition of sloppy hand-printed characters," in *Proc. Western Joint Comput. Conf.*, 1960, pp.133–142.
[26] T. Sakai, M. Nagao, and Y. Shinmi, "A character recognition system," in *Proc. Annual Conf. IECE Japan*, 1963, p. ,450.
[27] NEC (Nippon Electric Company), "Improvements in or relating to character recognition apparatus," U.K. Patent 1 124 130, Aug. 1968.
[28] J. H. Munson, "Experiments in the recognition of hand-printed text: Part I, Character recognition," in *Proc. 1968 Fall Joint Comput. Conf., AIIPS Conf.*, vol. 33, 1968, pp. 1125–1138.
[29] E. C. Greanias, P. E. Meagher, R. J. Norman, and P. Essinger, "The recognition of handwritten numerals by contour analysis," *IBM J. Res. Develop.*, vol. 7, pp. 2–13, 1963.
[30] L. A. Kamentsky and C. N. Liu, "Computer-automated design of multifont print recognition logic," *IBM J. Res. Develop.*, vol. 7 , no. 2, pp. 2–13, 1963.
[31] L. A. Kamentsky and C. N. Liu, "A theoretical and experimental study of a model for pattern recognition," in *Computer and Inf. Science*, J. T. Tou and R. H. Wilcox, Eds. New York: Spartan Books, 1964.
[32] T. Iijima, "Basic theory on normalization of pattern," *Bulletin of Electrotechnical Laboratory*, vol. 26, no. 5, pp. 368–388, 1962.
[33] T. Iijima, "Basic theory of feature extraction for visual pattern," *T. R. on Automata and Automatic Control, IECE Japan*, July 1963.
[34] D. Marr and E. Hildreth, "Theory of edge detection," *Proc. Roy. Soc. Lond. B*, vol. 207, pp. 187–217, 1980.
[35] M. Awaya and N. Funakubo, "Blur recovering circuit in observed images," in *Proc. Annual Conf. IECE Japan*, Apr. 1970, p. 2816.
[36] I. Yamazaki and T. Iijima, "Sampling of a character image," *J. of Trans. of Electron. Commun. Eng. Japan*, vol. 51-C, no. 9, pp. 428–429, Sept. 1968.

[37] T. Iijima, S. Mori, I. Yamazaki, and M. Yoshimura, "Automatic normalization circuit—A proposal of globally reciprocally dependent network," *T. R. on Automata and Automatic Control, IECE Japan*, Oct. 1963.

[38] I. Yamazaki, "Research on observation and normalization in character recognition," *Researches of the Electrotechnical Laboratory*, no. 726, Apr. 1972.

[39] S. Mori, "Theory of linear spatial network," *T. R. on Automata and Automatic Control, IECE Japan*, Oct. 1963.

[40] S. Mori, Internal Report, Electrotechnical Laboratory, June 1966.

[41] H. C. Andres and B. R. Hunt, *Digital Image Restoration*. Englewood Cliffs, NJ: Prentice-Hall, 1977, appendix B.

[42] N. Funakubo, "Studies on spatial network in printed character recognition system," *Researches of Electrotechnical Laboratory*, no. 778, Dec. 1977.

[43] A. Igarashi, E. Ogawa, and T. Iijima, "Topological analytic foundation of linear spatial network," *Trans. of IECE Japan*, vol. 53-C, no. 6, pp. 393–401, June 1970.

[44] Y. Yasumoto and C. Medioni, "Corner detection and curve representation using cubic B-splines," *Trans. of IECE Japan*, vol. J70-D, no. 12, pp. 2517–2574, Dec. 1957.

[45] S. Amari, "Invariant structures of signal and feature space in pattern recognition problems," *RAAG MEMORS* vol.4, pp.19–32, 1968.

[46] S. Watanabe, "Karhunen-Loeve expansion and factor analysis-theoretical remarks and applications," in *Trans. 4th Prague Conf. Info. Theory, Statis. Decision Funct., Random Process*, 1965, p.635.

[47] H. Karhunen, "On linear methods in probability," 1947 (English translation by I. Selin), The Rand Corporation, Dec. T-131, Aug. 11, 1960.

[48] M. Loeve, *Fonction Aleatories de Seconde Ordre*. Paris: Herman, 1948.

[49] K. Fukunaga, *Introduction to Statistical Pattern Recognition* New York: Academic Press, 1972.

[50] H. Hoteling, "Analysis of a complex of statistical variables into principal components," *J. Edu. Psychol.*, vol. 24, pp. 417–441 and 498–520, 1933.

[51] R. C. Gonzalez and P. Wintz, *Digital Image Processing*, 2nd ed. Reading, MA: Addison-Wesley, 1987.

[52] Y. Noguchi and T. Iijima, "Pattern classfication system using equi-variance characteristic papameters," *Trans. Info . Pross. Soc. Japan*, vol. 11, pp. 107–116, 1971.

[53] N. Funakubo, A. Iwamatsu, T. Suzuki, S. Mori, and T. Iijima, "ETL OCR pilot model," *Bulletin of the Electrotechnical Laboratory*, vol. 34, no. 1, Jan. 1970.

[54] T. Iijima, "Character recognition theory based on multiple similarity method," in *Proc. Annual Conf. IECE Japan*, Apr. 1970, p. 2710.

[55] T. Iijima, *Theory of Pattern Recognition*. Morishita Publishing, 1989.

[56] S. Watanabe and N. Pakvasa, "Subspace method of pattern recognition," in *Proc.1st Int. J. Conf. Pattern Recognition*, Nov.1973, pp. 25–32.

[57] E. Oja, *Subspace Methods of Pattern Recognition*. Research Studies Press, 1983.

[58] W. K. Pratt, *Digital Image Processing*. New York: Wiley-Interscience, 1978.

[59] M. K. Hu, "Visual pattern recognition by moment invariants," *IRE Trans. Inform. Theory*, vol. IT-8, pp.179–187, Feb. 1962.

[60] F. L. Alt, "Digital pattern recognition by moments," in *Optical Character Recognition*, G. L. Fischer *et al.*, Eds. Washington, DC: McGreger & Werner, 1962, pp. 159–179.

[61] T. E. Southard, "Method of recognizing single strings of non-touching rotated characters," Master's thesis, Ohio State University, Report AFOSR-TR-75–0177, Aug. 1974.

[62] G. L. Cash and M. Hatamian, "Optical character recognition by the method of moments," *CVGIP*, vol. 39, pp. 291–310, 1987.

[63] R. L. Cosgriff, "Identification of shape," Rep. 820–11, ASTIA AD 254 792, Ohio State Univ. Res. Foundation, Columbus, Dec. 1960.

[64] C. T. Zahn and R. Z. Roskies, "Fourier descriptors for plane closed curves," *IEEE Trans. Comput.*, vol. C-21, pp. 269–281, Mar. 1972.

[65] G. H. Granlund, "Fourier preprocessing for hand printed character recognition," *IEEE Trans. Comput.*, vol. C-21, pp. 195–201, Feb. 1972.

[66] E. Persoon and K. S. Fu, "Shape discrimination using Fourier descriptors," *IEEE Trans. Syst., Man, Cybern.*, vol. SMC-7, pp. 170–179, Mar. 1977.

[67] M. T. Y. Lai and C. Y. Suen, "Automatic recognition of characters by Fourier descriptors and boundary encoding," *Pattern Recognition*, vol. 14, nos. 1–6, pp. 383–393,1981.

[68] Y. Hsu and H. H. Arsebault and G. Apr., "Rotation-invariant digital pattern recognition using circular harmonic expansion," *Appl. Opt.*, vol. 21, no. 22, pp. 4012–4015, Nov. 1982.

[69] R. J. Spinrad, "Machine recognition of hand printing," *J. Info. and Cont.*, vol. 8, pp. 124–142, 1965.

[70] S. Mori, *Fundamentals of Characters and Figures Recognition Techniques*. Tokyo: Ohm, 1983.

[71] M. Yasuda, "Research on character recognition systems based on correlation," Doctoral thesis, Tokyo University, 1981.

[72] Y. Yamashita, K. Higuchi, Y. Yamada, and Y. Haga, "Classification of hand-printed Kanji characters by structural segment matching method," *Tech. Rep. IECE Japan*, PRL82–12, no. 12, p. 25, 1982.

[73] H. Yamada, T. Saito, and K. Yamamoto, "Line density equalization—A nonlinear normalization for correlation method," *Trans. IECE Japan*, vol. J67D, no. 11, pp. 1379–1388, 1984.

[74] J. Tsukumo and H. Tanaka, "Classification of hand-printed Chinese characters using nonlinear normalization and correlation methods," in *Proc. 9th IJCPR*, Nov. 14–17, 1988, pp.168–171.

[75] H. Sakoe, "hand-printed character recognition based on rubber string matching," *T. R. IECI Japan*, PRL74–20, pp. 1–10, 1974.

[76] R. Bellman, *Dynamic Programming*. Princeton University Press, 1957.

[77] R. Bellman and S. Dreyfus, *Applied Dynamic Programming*. Princeton University Press, 1962.

[78] V. A. Kovalevsky, "An optimal algorithm for the recognition of some sequences," *Cybernetics*, vol. 3, no. 4, 1967.

[79] H. Sakoe, "Dynamic programming algorithm optimization for spoken word recognition," *IEEE Trans. Acoust., Speech, Signal Process.*, vol. ASSP-26, pp. 43–49, Feb. 1978.

[80] H. Yamada, "Contour DP matching method and its applications to hand-printed Chinese character recognition," in *Proc.7th IJCPR*, 1984, pp. 389–392.

[81] J. R. Ullman, "An algorithm for subgraph isomorphism," *J. Ass. Comp. Mach.*, vol. 23, no. 1, pp. 31–42, Jan. 1976.

[82] A. P. Ambler, H. G. Barrow, C. M. Brown, P. M. Burstall, and R. J. Popplestone, "A versatile computer-controlled assembly system," *Artificial Intelligence*, vol. 6, pp. 129–156, 1975.

[83] A. Rosenfeld, R. Hummel, and S. Zucker, "Scene labeling by relaxation oprations," *IEEE Trans. Syst., Man., Cybern.*, vol. SMC-6, pp. 420–433, 1976.

[84] L. Davis, "Shape matching using relaxation techniques," *IEEE Trans. Pattern Anal. Mach. Intell.*, vol. PAMI-1, pp. 60–72, Jan. 1979.

[85] D. I. Waltz, "Generating semantic descriptions from drawings of scenes with shadows," Ph.D. dissertation, AI Lab, MIT, 1972.

[86] H. Ballard and C. M. Brown, *Computer Vision*. Englewood Cliffs, NJ: Prentice-Hall, 1982.

[87] A. Rosenfeld and J. L. Pfalz, "Sequential operations in digital picture processing," *J. Ass. Comp. Mach.*, vol. 13, pp. 471–494, 1966.

[88] H. Sherman, "A quasitopological method for the recognition of line patterns," in *Info. Process., Proc. UNESCO Conf.* (Paris), 1959.

[89] M. Beun, "A flexible method for automatic reading of hand-written numerals," *Philips Tech. Rev.*, vol. 33, no. 4, Part I, pp. 89–101; Part II, pp. 130–137, 1973.

[90] C. Y. Suen, R. Shinghal, and C. C. Kwan, "Dispersion factor: A quantitative measurement of the quality of hand-printed characters," in *Proc. Int. Conf. Cybern. and Society*, Sept. 1977, pp.683–685.

[91] R. A. Kirsch, L. Cahn, C. Ray, and G. L. Urban, "Experiments in processing pictorial information with a digital computer," in *Proc. Eastern Computer Conf.*, 1957, pp. 221–229.

[92] R. H. McComick, "The Illinois pattern recognition computer—Illiac III," *IRE Trans. Electron. Comput.*, vol. EC-12, no.5, 1963.

[93] R. Narasimhan, "Labeling schemata and syntactic descriptions of pictures," *Information and Control*, vol. 7, pp. 151–166, 1964.

[94] D. Rutovitz, "Data structures for operations on digital images," in *Digital Processing*, C. G. Chang *et al.*, Eds. Washington, DC: Thompson, 1968, pp. 105–133.

[95] E. S. Deutsch, "Computer simulation of a character recognition machine," *Post Office Elec. Eng. J.*, vol. 60, pp. 39–44 and pp. 104–109, 1967.
[96] C. J. Hilditch, "Linear skeleton from square cupboards," in *Machine Intelligence IV*, B. Meltzer and D. Michie, Eds. Edinburgh, University Press, 1969, pp. 403–420.
[97] H. Tamura, "A comparison of line thinning algorithms from digital computer view point," in *Proc. 4th Int. Joint Conf. Patt. Recog.*, 1978, pp. 715–719.
[98] H. Freeman, "Boundary encoding and processing," in *Picture Processing and Psycholopictorics*, B. S. Lipkin and A. Rosenfeld, Eds. New York: Academic Press, 1970, pp. 241–266.
[99] R. G. Grimsdale, F. H. Sumner, C. J. Tunis, and T. Kirburn, "A system for the automatic recognition of patterns," *Proc. Inst. Elec. Eng.*, vol. 106B, pp. 210–221, Dec. 1958.
[100] T. Pavlidis, "Computer recognition of figures through decomposition," *J. Info. Control.* vol. 12, pp. 526–537, 1968.
[101] T. Pavlidis, "Structural pattern recognition:Primitives and juxtaposition relations," in *Frontiers of Pattern Recognition*, S. Watanabe, Ed. New York: Academic Press, 1972, pp. 421–451.
[102] J. Rabinow, "Developments in character recognition machines at Rabinow Engineering Company," in*Optical Character Recognition*, Eds. G. L. Fischer *et al.*, Eds. McGreder & Werner, 1962, pp. 27–51.
[103] L. H. Sublettle and J. Tults, "Character recognition by digital feature detection," *RCA Review*, pp. 60–79, Mar. 1962.
[104] P. G. Perotto, "A new method for automatic character recognition," *IEEE Trans. Electron. Comput.*, vol. EC-12, pp. 521–526, Oct. 1963.
[105] M. Nadler, "Structual codes for omnifont and handwritten characters," in *Proc. 3rd IJCPR*, 1976, pp.135–139.
[106] Uyehara, "A sream-following technique for use in character recognition," in *1968 IEEE Int. Conv. Record*, part 4, Mar. 1963, pp. 64–74.
[107] M. Nadler, "Sequentially-local picture operaters," in *Proc. 2nd IJCPR*, 1974, pp. 131–135.
[108] S. Hoshino and M. Miyamoto, "Feature extraction on hand-printed numerals," in *Proc. Ann. Conf. IECE Japan*, 1969, p. 92.
[109] K. Kiji *et al.*, "Recognition of the postal code number," NEC Tech. Rep., pp. 30–38, 1969.
[110] C. B. Shelman, "The application of list processing techniques to picture processing," *Pattern Recognition*, vol. 4, pp. 201–210, 1972.
[111] R. D. Merril, "Representation of contours and regions for efficient computer search," *Commun. Ass. Comput. Mach.*, vol. 16, pp. 69–82, 1973.
[112] A. K. Agrawala and A. V. Kulkarni, "A sequential approach to the extraction of shape features," in *Proc. CGIP*, 1977, pp. 538–557.
[113] T. Pavlidis, "A vectorizer and feature extractor for document recognition," *CVGIP*, vol. 35, pp. 111–127, 1986.
[114] T. Pavlidis, *Algorithms for Graphics and Image Processing*. Rockville, MD: Computer Science Press, 1982.
[115] S. Kahan, T. Pavlidis, and H. S. Baird, "On the recognition of printed characters of any font and size," *IEEE Trans. Pattern Anal. Mach. Intell.*, vol. PAMI-9. pp. 274–288, Mar. 1987.
[116] H. Freeman, "On the digital computer classification of geometric line patterns," in*Proc. Natl. Elect. Conf.*, vol. 18, 1962, pp. 312–324.
[117] G. Gallus and P. W. Nourath, "Improved computer chromosome analysis incorporating preprocessing and boundary analysis," *Phys. Med. Biol.*, vol. 15, pp. 435–445, 1970.
[118] F. Attneave, "Some informational aspects of visual perception," *Psychol. Rev.*, vol. 61, pp. 183–193, 1954.
[119] A. Rosenfeld and E. Johnston, "Angle detection on digital curves," *IEEE Trans. Syst., Man, Cybern.*, vol. SMC-5, pp. 610–614, Nov. 1975.
[120] H. Freeman and L. S. Davis, "A corner finding algorithm for chain coded curves," *IEEE Trans. Comput.*, vol. C-26, pp. 297–303, 1977.
[121] T. Pavlidis and F. Ali, "Computer recognition of hand written numerals by polygonal approximations," *IEEE Trans. Syst., Man, Cybern.*, vol. SMC-5, pp. 610–614, Nov. 1975.
[122] C. de Boor, "Good approximation by spline with variable knots," *ISNM*, vol.1, pp. 224–256, 1972.
[123] U. E. Ramer, "An iterative procedure for the polygonal approximation of plane curve," *CGIP*, vol. 1, pp. 244–256, 1972.
[124] T. Pavlidis and S. L. Horowitz, "Segmentation of plane curves," *IEEE Trans. Comput.*, vol. C-23, pp. 860–870, Aug. 1974.
[125] K. Steinbuch, "Automatische Zeichnerkenung," *SEL Nachrichten*, Heft 3, p. 127, 1958.
[126] H. Kazmierczak, "The potential field as an aid to character recognition," in *Proc. Int. Conf. Inf. Processing*, June 1959, p. 244.
[127] J. H. Munson, "The recognition of hand-printed text," in *Proc. IEEE Pattern Recognition Workshop* (Puerto Rico), Oct. 1966, p. 115.
[128] M. D. Levine, "Feature extraction: A survey," *Proc. IEEE*, vol. 57, pp. 1391–1419, Aug. 1969.
[129] S. Mori. R. Oka, A. Iwamatsu, and T. Saito, "Reading system," Japan Patent 822486, Dec.1970.
[130] R. Oka and S. Mori, "Character recognition system-I," *T. R. IECE Japan*, PRL-71–1, Jan.1971.
[131] T. Mori, "Fundamental research with respect to feature extraction for hand-printed character recognition," *Res. Electrotech. Lab. Tokyo*, no. 762, June 1976.
[132] T. Mori, S. Mori, and K. Yamamoto, "Feature extraction method based on field effect method," *Trans. IECE Japan*, vol. 57-D, no. 5, pp. 308–315, 1974.
[133] S. Mori *et al.*, "Recognition of hand-printed characters," in *Proc. 2nd IJCPR*, Aug. 1974, pp. 233–237.
[134] Q. Zhan, "A linguistic approach to pattern recognition," Master's thesis, Waseda University, 1972.
[135] R. Narashimhan, "Syntax-directed interpretation of classes of pictures," *Commun. Ass. Comput. Mach.*, vol. 9, pp. 166–173, 1966.
[136] R. Narashimhan, "On the description, generation, and recognition classes of pictures," in *Automatic Interpretation and Classification of Images*, A. Grasselli, Ed. New York: Academic Press, 1969.
[137] K. Komori, T. Kawatani, K. Ishii and Y. Iida, "A feature concentration method for character recognition," in *Proc. IFIP Congress 77* (Toronto), Aug. 1977, pp. 39–34.
[138] K. Yamamoto and S. Mori, "Recognition of hand-printed characters by outer most point method," in *Proc. 4th IJCPR*, 1978, pp.794–796.
[139] T. Mori *et al.*, "Hand-printed alphanumerals and special symbols recognition system," *Trans. IECE Japan*, vol. 58-D, no. 8, pp. 442–449, 1975.
[140] T. Mori, S. Mori, and Y. Monden, "Pattern recognition based on coordinats and edge direction distribution," *T. R. IECE Japan*, PRL-73–11, Nov. 1973.
[141] F. O'Gorman and M. B. Clowes, "Finding picture edge through co-linearity of feature points," in *Proc. 3rd IJC on AI* (Kyoto), Aug. 1973, p. 543.
[142] R. Oka, "Cellular feature extraction from patterns," *Trans. IECE Japan*, vol. J65-D, no. 10, pp. 1219–1226, Oct. 1982.
[143] K. S. Fu, *Syntactic Method in Pattern Recognition*. New York: Academic Press, 1974.
[144] K. S. Fu, *Syntactic Method in Pattern Recognition Applications*. New York: Springer-Verlag, 1977.
[145] F. Ali and T. Pavlidis, "Syntactic recognition of handwritten numerals," *IEEE Trans. Syst., Man, Cybern.*, vol. SMC-7, pp. 537–541, July 1977.
[146] R. Abe, "Synthesis of an automaton which recognizes pattern destorted as strings of symbols," *T. R. IECE Japan*, PRL74–5, pp. 43–54, 1974.
[147] R. C. Gonzalez and M. G. Thomason, *Syntactic Pattern Recognition*. Reading, MA: Addison-Wesley, 1978.
[148] A. C. Shaw, "The formal picture description scheme as a basis for picture processing system," *Infom. and Contr.*, vol. 14, pp. 9–52, 1969.
[149] S. Mori, "A nonmetric model of hand-printed characters," *Res. Electrotech.Lab. Tokyo*, vol. 798, Aug. 1979.
[150] S. Mori, "An algebraic structure representation of shape," *Trans. IECE Japan*, vol. J64-D, no. 8, pp. 705–712, Aug. 1981.
[151] H. Nishida and S. Mori, "Structural analysis and description of curves by qusi-topological features and singular points," in *Pre-Proc. IAPR Workshop on SSPR*, 1990, pp. 310–334.
[152] T. Suzuki and S. Mori, "A description method of line drawings by cross section sequence graph and its application to thinning," *T. R. IECE Japan*, PRU90–22, June 1990.
[153] RTUA, *Buyer's Guide*, Recognition Technologies Users Association, U.S.A., vol. 13, no. 1, pp. 38–48, 1990.
[154] AIM, *The Source Book*, Automatic Identification Manufactures, Inc., U.S.A., p. 64, 1990.

[155] S. Impedovo, L. Ottaviano, and S. Occhingro, "Optical character recognition—A survey," *Int. J. Pattern Recognition and Artificial Intelligence*, 1991.

[156] C. Y. Suen, "Character recognition by computer and applications," in *Handbook of Pattern Recognition and Image Processing*, T. Y. Young and K. F. Fu, Eds. Boca Raton, FL: Academic Press, 1986, pp. 569–586.

[157] C. Y. Suen, "OCR products and manufacturers," Center for Pattern Recognition and Machine Intelligence, Concordia University, Montreal, 1991, in preparation.

[158] C. Y. Suen, Ed., *Frontiers in Handwriting Recognition*. Concordia University, Montreal: CENPARRMI, Apr. 1990.

[159] C. Y. Suen and S. Mori, "Standardization and automatic recognition of hand-printed characters," in *Computer Analysis and Perception*, vol. 1, *Visual Signals*, C. Y. Suen and De Mori, Eds. Boca Raton, FL: CRC Press, 1982, pp.41–53.

[160] R. K. Gerlach, "Wide-tolerance optical character recognition for existing printing mechanisms," in *Optical Character Recognition*, G. L. Fischer *et al.*, Eds. McGregor & Wemer, 1962, pp. 93–114.

[161] C. C. Heasly, Jr, and G. L. Fischer, Jr, "Some elements of optical scanning," in *Optical Character Recognition*, G. L. Fischer *et al.*, Eds. McGregor & Wemer, 1962, pp. 15–26.

[162] E. C. Greanias, "Some important factors in the practical utilization of optical character readers," in *Optical Character Recognition*, G. L. Fischer *et al.*, Eds. McGregor & Wemer, 1962, pp. 129–146.

[163] H. Genchi, K. Mori, S. Watanabe, and S. Katsuragi, "Recognition of hand-printed numerical characters for automatic letter sorting," *Proc. IEEE*, vol. 56, pp. 1292–1301, Aug. 1968.

[164] I. Sheinberg, "The INPUT2 document reader," *J. Pattern Recog. Soc.*, vol. 2, pp. 167–173, Sept. 1970.

[165] S. Yamamoto, M. Yasuda, Y. Miyamoto, and M. Tsutsumi, "Design of handwritten numeral recognition," *Trans. IECE Japan*, vol. 53-C, no. 10, pp. 891–898, 1970.

[166] T. Sano and T. Hananoi, "Type H-8959 optical character reader," *Hitachi Review*, vol. 54, no. 12, pp. 1077–1082,1972.

[167] M. Oka, S. Yamamoto, and S. Kadota, "Pattern feature detection system," U.S. Patent 3 863 218, Jan. 28,1975.

[168] T. Fujimoto *et al.*, "Recognition of hand-printed characters by nonlinear elastic matching," in *Proc.3rd IJCPR*, Nov. 1976, pp. 113–117.

[169] R. B. Hennis, "The IBM 1925 optical page reader, PartI: System design," *IBM J. Res. Develop.*, vol. 12, no. 5, pp. 346–353, 1968.

[170] M. R. Bartz, "The IBM 1925 optical page reader, Part II: Video thresholder," *IBM J. Res. Develop.*, vol. 12, no. 5, pp. 354–363, 1968.

[171] D. R. Andrews, A. J. Atrubin, and K. C. Hu, "The IBM 1925 optical page reader, Part III: Recognition logic development," *IBM J. Res. Develop.*, vol. 12, no. 5, pp. 364–371, 1968.

[172] K. Komori, T. Kawatani, K. Ishii, and Y. Iida, "hand-printed Katakana characters recognition by feature concentration method," *Trans. IECE Japan*, vol. J63-D, no. 11, pp. 962–968, Dec. 1980.

[173] S. Mori and C. Y. Suen, "Automatic recognition of symbols and architecture of the recognition unit," in *Computer Analysis and Perception*, vol. 1, *Visual Signals*, C. Y. Suen and R. De Mori, Eds. Boca Raton, FL: CRC Press, 1982, pp.17–40.

[174] S. Diehl and H. Eglowstein, "Tame the paper tiger," *Byte*, pp. 220–241, Apr. 1991.

[175] M. Yasuda, K. Yamamoto, H. Yamada, and T. Saito, "An improved correlation method-hand-printed Chinese character recognition in a reciprocal feature field," *Trans. IECE Japan*, vol. J68-D, pp. 353–360, Mar. 1985.

[176] K. Yamamoto, H. Yamada, and T. Saito, "Current state of recognition method for Japanese characters and database for research of hand-printed character recognition," in *Proc. Int. Workshop Frontiers in Handwriting Recognition* (Chateau de Bonas, France), Sept. 1991, pp.81–99.

[177] M. Kishi, T. Kiuchi, Y. Maruko, A. Yoshida, and Y. Hongo, "A feature extraction processor," in *Proc. Anal. Conf. IECE Japan*, 1988, D-208.

[178] H. Matsumura, K. Aoki, T. Iwahara, H. Oohuna, and K. Kogure, "Desk-top optical handwritten character reader CLL-2000," *Sanyo Tech. Rev.*, vol. 18. no. 1, pp. 3–12, Feb. 1986.

[179] S. Amari, "Mathematical foundation of neurocomputing," *Proc. IEEE*, vol. 78, pp. 1443–1463, Sept. 1990.

[180] K. Yamamoto, T. Mori, S. Mori, and J. Shimizu, "Machine recognition of Katakana and numerals," *Trans. IECE Japan*, vol. 59-D, no. 6, pp. 414–421, 1976.

[181] K. Yamamoto, "Recognition of hand-printed Hiragana characters by concave and convex features and automatically merging the dictionary," *Trans. IECE Japan*, vol. J65-D, no. 6, pp. 774–781, June 1982.

[182] H. Nishida and S. Mori, "An approach to automatic construction of structural models for character recognition," in *Proc. First ICDAR* (Saint-Malo, France), 1991, pp. 231–241.

[183] H. S. Baird, "Feature extraction for hybrid structural/statistical pattern classification," *CVGIP*, vol. 42, pp. 318–333, 1988.

[184] R. M. Brown, T. M. Fay, and C. L. Walker, "Hand-printed symbol recognition system," *Pattern Recognition*, vol. 21, no. 2, pp. 91–118, 1988.

[185] J. J. Hull *et al.*, "A blackboard-based approach to handwritten Zip Code recognition," in *Proc. U.S. Postal Service Adv. Techol. Conf.*, 1988, pp. 1018–1032.

[186] C. Y. Suen, C. Nadal, T. A. Mai, R. Legault, and L. Lam, "Recognition of totally unconstrained handwritten numerals based on the concept of multiple experts," in *Proc. Frontiers in Handwritting Recognition* (CENPARMI Concordia University), 1990, pp. 131–140.

[187] A. Krzyzak, W. Dai, and C. Y. Suen, "Unconstrained handwritten classification using modified backpropagation model," in *Proc. Frontiers in Handwritting Recognition* (CENPARMI Concordia University), 1990, pp. 155–164.

[188] Y. Le Cun *et al.*, "Constrained neural network for unconstrained handwritten digit recognition," in *Proc. Frontiers in Handwritting Recognition* (CENPARMI Concordia University), 1990, pp. 145–151.

[189] K. Fukushima and N. Wake, "Handwritten alphanumeric character recognition by neocognitron," *IEEE Trans. Neural Networks*, vol. 2, pp. 355–365, May 1991.

[190] J. J. Hull, A. Commike, and T. K. Ho, "Multiple algorithms for handwritten character recognition," in *Proc. Frontiers in Handwritting Recognition* (CENPARMI Concordia University), 1990, pp. 117–124.

[191] T. K. Ho, J. J. Hull, and S. N. Srihari, "Combination of structural classifier," in *Pre-Proc. IAPR Workshop on SSPR* , June 1990, pp.123–136.

[192] C. Nadal, R. Legault, and C. Y. Suen, "Complementary algorithms for the recognition of totally unconstrained handwritten numerals," in *Proc. 10th IJCPR*, June 1990, pp. 443–449.

[193] S. Amari, *Nikkei AI Fall Special Issue on Neural Network Applications*, p. 33, 1990.

Shunji Mori (Member, IEEE) was born in Hokkaido, Japan, in 1934. He received the B.E degree in 1956 and Dr.Engineering degree in 1978 from Hokkaido University

From 1957 to 1983 he was with the Electrotechnical Laboratory of the Ministry of International Trade and Industry of Japan. There he served as head of the Pattern Processing section from 1969 to 1981 and as supervisor of the Pattern Information Processing System (PIPS) from 1971 to 1973. In 1975 he was a visiting researcher at the Computer Vision Laboratory, University of Maryland, for half a year and from 1981 to 1982 he was a visiting professor in the Department of Computer Science, Concordia University, Montreal. From 1983 to 1987 he worked for Nippon Schlumberger as head of the Modeling Department of System Engineering. Currently he is head of the Artificial Intelligence Research Center of the Research and Development Central Research Center of Ricoh Company, Ltd. His research interests include pattern recognition, image processing, and knowledge engineering.

Dr. Mori is a member of AAAI, the Institute of Electronics, Information, and Communication Engineers of Japan, the Information Processing Society of Japan, and the Chinese Language Computer Society.

Ching Y. Suen (Fellow, IEEE) received the M.Sc. (Eng.) degree from the University of Hong Kong and the Ph.D. degree from the University of British Columbia, Canada.

In 1972, he joined the Department of Computer Science of Concordia University, Montreal, Canada, where he became Professor in 1979 and served as Chairman from 1980 to 1984. Presently he is the Director of CEN-PARMI, the new Centre for Pattern Recognition and Machine Intelligence of Concordia. During the past 15 years, he has been appointed to visiting positions at several institutions in different countries.

Dr. Suen is the author or editor of several books, with titles ranging from *Computer Vision and Shape Recognition* and *Frontiers in Handwriting Recognition* to *Computational Analysis of Mandarin and Chinese*. His latest book, *Operational Expert System Applications in Canada*, is published by Pergamon. Dr. Suen is the author of many papers and his current interests include expert systems, pattern recognition and machine intelligence, optical character recognition and document processing, and computational linguistics.

An active member of several professional societies, Dr. Suen is an Associate Editor of several journals related to his areas of interest. He is a former president of the Canadian Image Processing and Pattern Recognition Society, governor of the International Association for Pattern Recognition, and president of the Chinese Language Computer Society. Prof. Suen received the 1992 ITAC/NSERC award for his contributions to pattern recognition, expert systems, and computational linguistics.

Kazuhiko Yamamoto (Member, IEEE) received the B.E. degree in 1969, the M.E. in 1971, and the Dr.Engineering in 1983, all from Tokyo Denki University, Tokyo, Japan.

Since 1971, he has been with the Electrotechnical Laboratory of MITI studying hand-printed character recognition. He participated in the development of OCR in the Pattern Information Processing System Project from 1971 to 1979. From 1979 to 1980 he was a visiting researcher in the Computer Vision Laboratory, University of Maryland, where he worked on computer vision. He has been head of the Image Processing Section at ETL since 1986. His research interests include pattern recognition and artificial intelligence.

Dr. Yamamoto is a member of the Institute of Electronics, Information, and Communication Engineering of Japan and of the Information Processing Society of Japan.

Omnidocument Technologies

MINDY BOKSER, MEMBER, IEEE

Invited Paper

With recent technical advances, OCR is now a viable technology for a wide range of applications. Calera's OCR engine is omnifont and reasonably robust on individual degraded characters. The weakest link is its handling of characters which are difficult to segment, such as characters which are joined to adjacent characters. The engine is divided into four phases: segmentation, image recognition, ambiguity resolution, and document analysis. The features are zonal and reduce the image to a blurred, gray-level representation. The classifier is data-driven, trained off-line, and model-free. We found that handcrafted features and decision trees tend to be brittle in the presence of noise.

To satisfy the needs of full-text applications, the system captures the structure of the document so that, when viewed in a word processor or spreadsheet program, the formatting of the OCR'd document reflects the formatting of the original document. To satisfy the needs of the forms market, a proofing and correction tool displays "pop-up" images of uncertain characters.

Keywords— *Text recognition, OCR, omnifont, multifont, polyfont, feature extraction, classification.*

I. Introduction

Commercial machine-print OCR engines are still a long way from reading as well as a human. However, recent technical advances have brought OCR significantly closer to this ideal and made it a viable alternative to manual key entry for a wide range of applications. Prior to 1986, the total number of OCR systems that had been sold worldwide was on the order of a few thousand. Now that many are sold each week, either as individual software packages, board-level products, or dedicated scanner/OCR systems.

The earliest commercial page reader was developed in 1959 by the Intelligent Machine Research Corporation and could read one font in one point size. With time, multifont machines were developed that could read up to about ten fonts. The limit on the number of fonts was intrinsic to the pattern recognition algorithm, template matching [1], which compares the incoming image with a library of bit map images. The accuracy was quite good, even on degraded images, as long as the fonts in the library had been selected with some care. Two marketing paths were taken with multifont machines. One catered to the office environment by providing a font library with some of the most common typewriter fonts. The other catered to institutions such as the Post Office, the Defense Department, and credit card companies, where exceptional accuracy was required. To achieve the accuracy, customized fonts were designed to minimize the possibility of confusion between similar looking characters, such as "1" (one) and "l" (el), or "5" and "s." (The numbers on an American Express credit card are printed in a customized font called Farrington.) In 1966, an American standardized font called OCR-A and a European version called OCR-B were developed. Some attempts were made to merge the two into a single standardized font, but this never materialized and manufacturers began to develop equipment that could read both.

In 1978, Kurzweil Computer Products introduced a system which could be trained by the user to read any given font. It took several hours of training per font, but if the document was sufficiently long, training was faster than rekeying. Once the system was trained on a font, that knowledge was stored on disk so that retraining would not be required the next time. Up to nine fonts could be on line simultaneously. If the page contained pictures or multiple columns, the user was required to specify their locations.

In 1986, Calera Recognition Systems (then called Palantir) introduced an omnifont system that could read complex pages containing any mixture of nondecorative fonts without training or manual intervention. The pattern recognition algorithm was sufficiently general to handle fonts it had not seen before. User training was replaced with training in the lab and manual parsing was replaced with software that automatically located columns and pictures.

The remainder of this paper discusses Calera's OCR engine. Section II is a breakdown of the design goals entailed by "omnidocument" recognition. Section III gives a brief overview of the engine's architecture. Segmentation, image recognition, and ambiguity resolution, which are three of the four primary components of the system, are discussed in Sections IV, V, and VI respectively. The fourth primary component, document analysis, is described in Section VIII and motivated by a discussion in Section

Manuscript received February 27, 1991; revised February 4, 1992.

The author is with Calera Recognition Systems, Inc., Sunnyvale, CA 94086.

IEEE Log Number 9202522.

Reprinted from *Proc. IEEE,* Vol. 80, No. 7, July 1992, pp. 1066–1078.

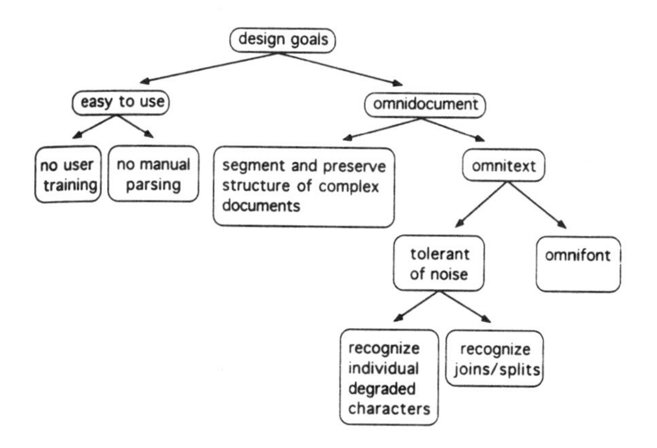

Fig. 1. Goals for the initial design of Calera's OCR engine.

VII of the markets for OCR and their requirements. The strengths and weaknesses of the current technology are described in Section IX. Finally, Section X outlines a few avenues for improvement. For reports on other successful OCR systems, see [2]–[4].

II. Design Goals

The design of Calera's OCR system was motivated by the goal of building an "omnidocument" engine which could handle documents ranging from office memos and magazine articles to spreadsheets and parts lists (Fig. 1). A further goal was that it be easy to use; that is, neither user training nor manual intervention should be required.

Omnidocument recognition required, first of all, that the engine be "omnitext," i.e. able to read the text on any page. It also required that the engine be able to segment a page with complex layout, and analyze and preserve its structure.

Reading the text on any page translated into two subgoals. First, the engine needed to be "omnifont," that is, able to recognize any of the thousands of typefaces in common use. Second, it needed to be able to recognize degraded character images.

This second subgoal was (and still is) particularly challenging. Some of the most promising approaches to omnifont recognition break down on degraded character images. These were rejected because degradation is endemic to scanned images of real-world documents. Even when the original copy is clean, the scanned image frequently is not. A few of the variables that account for variations in image quality are paper quality (e.g. grain, glossiness, color), printing technology (e.g. cloth-ribbon typewriter, laserprinter), reproduction technology (e.g. photocopy, fax), ink quality and color, scanner resolution, and scanner threshold.

There are two distinct problems raised by degradation.

First, individual characters do not always look the way they "should" (Fig. 2). An "e" may have its hole filled in, have a break in its crossbar, or be so badly blurred that it is ambiguous to a human in isolation.

Second, individual character images are frequently difficult to isolate on the scanned page. Adjacent characters may be joined to each other (Fig. 3) or overlapping (Fig. 4); a character may be split into multiple pieces (Fig. 5); or an underline may pass through the bottom of a word (Fig. 6). Joins are particularly common if the font is tightly kerned and serifed. For example, on a clean, randomly selected *Time* magazine page scanned at optimal threshhold on a good-quality scanner (HP ScanJet), 10% of the characters were joined. If the page is a dark photocopy or scanned at a low threshhold, most characters can be joined. If the page is a light photocopy or scanned at a high threshhold, most characters can be split.

Fig. 2. Examples of individual degraded characters which do not look the way they "should." For example, the serifs on the "n" are joined, the hole in the "A" is filled in, and the "d" is broken into two pieces.

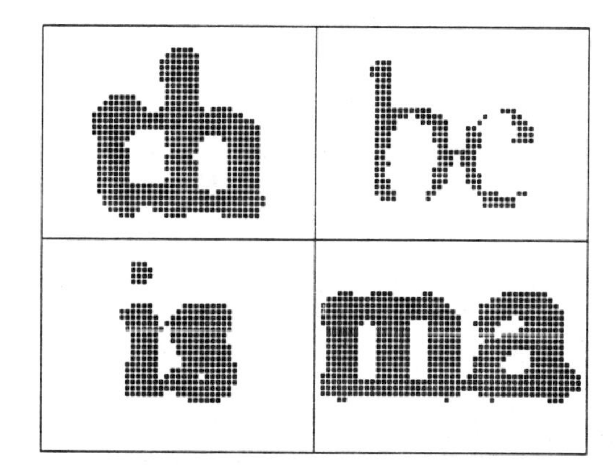

Fig. 3. Examples of joined characters. Joins are particularly common if the scanner threshhold is set too low, if the page is a dark photocopy, or if the font is serifed and tightly kerned.

Fig. 4. Examples of overlapping characters. Overlaps are particularly common in italic text.

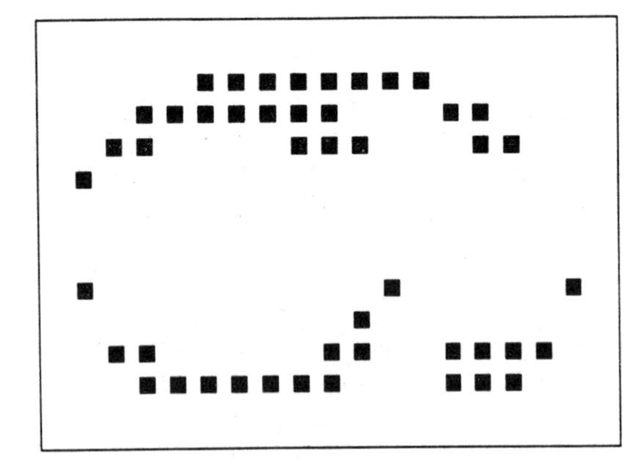

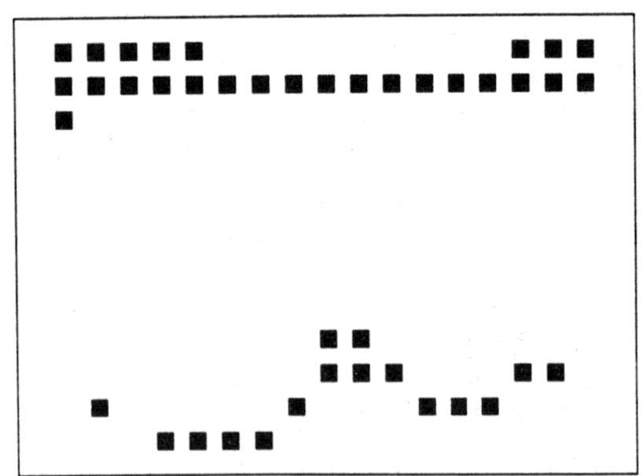

Fig. 5. Examples of split characters. Splits are particularly common if the scanner threshhold is set too high or if the page is a light photocopy.

Fig. 6. Example of a touching underline.

III. Architecture Overview

Calera's OCR engine is divided into four primary phases (Fig. 7): segmentation, image recognition, ambiguity resolution, and document analysis.

Segmentation locates individual character images on the page. Image recognition examines each character image in isolation and classifies it without (for the most part) taking contextual clues into account. When image recognition is

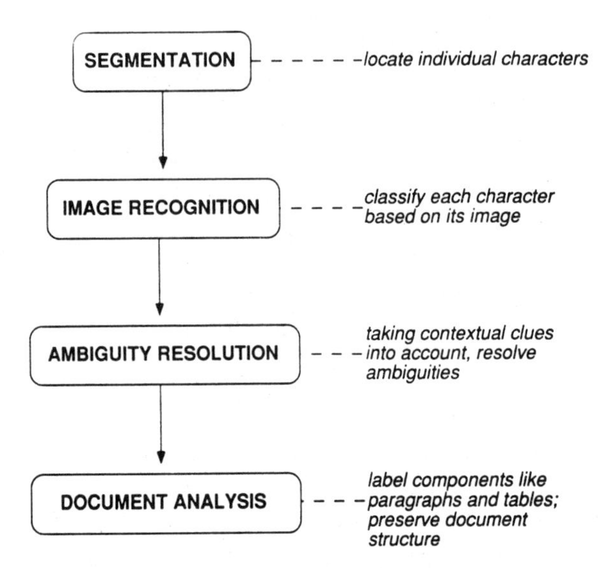

Fig. 7. Overview of the architecture of Calera's engine.

uncertain, it passes a list of character choices to ambiguity resolution, which uses contextual information to make the final classification decision. Document analysis analyzes the structure of the document and labels components such as paragraphs, tables, and lists.

The first three phases are discussed in the next three sections. The fourth, document analysis, is described in Section VII.

IV. Segmentation

Page segmentation involves determining the skew of the image, separating picture from text, and partitioning the text into columns, lines, words, and characters. Variants of the Hough transform are commonly used for detecting skew [5]. Algorithms for segmentation include projection profile cuts [6], run-length smearing [7], connected component analysis [8], and segmentation by white streams [9]. Two surveys of segmentation algorithms are given in [10] and [11]

Calera's segmentation phase locates the individual character images on the scanned document and outputs a stream of segments with ideally one complete character image per segment. Unfortunately, segmentation does not always live up to this ideal, in large part because image recognition, which comes later in the process, is sometimes required to make the best segmentation decision. To demonstrate the limitations with segmenting before recognizing, if Fig. 2 is turned upside down so that the characters are no longer readily identifiable, it is not obvious where the "th" join should be cleaved to separate it into two complete character images.

Segmentation has four failure modes. First, an output segment may contain multiple characters or only a piece of a character. Second, a segment may contain both text and noise. To the eye of the segmentation module, a floating blob may be the dot on an 'i," an accent on an "e," a stray pen mark, or scanner noise. The third and fourth failure modes are duals of each other. If a nontext region is mistaken for text, segments containing picture or noise will be passed to the recognition module. If a text region is mistaken for nontext, it will not be recognized.

The first three types of failure are not necessarily fatal. The recovery mechanism for improperly segmented text is resegmentation during the image recognition phase, which attempts to cleave segments that contain joins, glue segments that contain splits, and cleanse segments that contain noise. The recovery mechanism for outputting noise or picture segments is a clean-up module which drops sufficiently suspicious text. Coming after image recognition, it takes advantage of clues not available in the segmentation phase, such as the confidence value assigned by the recognition phase, and the position and image attributes of the segment relative to other recognized text on the page. For example, an isolated segment at the top of a page may be deemed a page number if it is recognized as a perfectly good "5" but noise if it is recognized as a low-confidence "~."

However, there is no recovery mechanism for the last failure mode. Text which is mistaken for nontext will be dropped from the OCR'd document. Therefore, the segmentation module errs on the side of outputting nontext segments in order to minimize the risk of unsegmented text.

V. Image Recognition

The mandate for the image recognition phase is to recognize the pixel image in each segment.

If the segment contains a properly segmented character image, then, ideally, the output is the set of character labels and relative confidence values that a human would assign if asked to identify the image in isolation. If, for example, the image is clearly recognizable to a human as a "b," the ideal output of the recognition phase is "b," together with a confidence value indicating certainty. If a human would deem the image probably an "b," but also possibly a "h" (Fig. 8), the ideal output is the set of choices {"b," "h"} together with confidence values indicating that, based solely on the image and not taking contextual clues into account, "b" is more likely than "h".

If the input segment contains a "thr" join, then ideally the segment is cleaved into a "t," an "h," and an "r" and each is then recognized correctly. Similarly, if a segment contains half of a broken "o," then ideally it is glued together with its other half and the newly glued segment is correctly recognized.

As with segmentation, image recognition does not always live up to its ideal. An "8" which a human would have no trouble recognizing in isolation might be deemed an ambiguous "8"/"B" or, worse, recognized as an unambiguous "B." A split "d" might be recognized as "cl." An "rn" join might be recognized as "m." The recognition module is tuned to err on the side of outputting too many choices rather than too few so that contextual clues can be brought to bear during the ambiguity resolution phase to make the best decision.

Image recognition is divided into three phases: feature extraction, classification, and resegmentation.

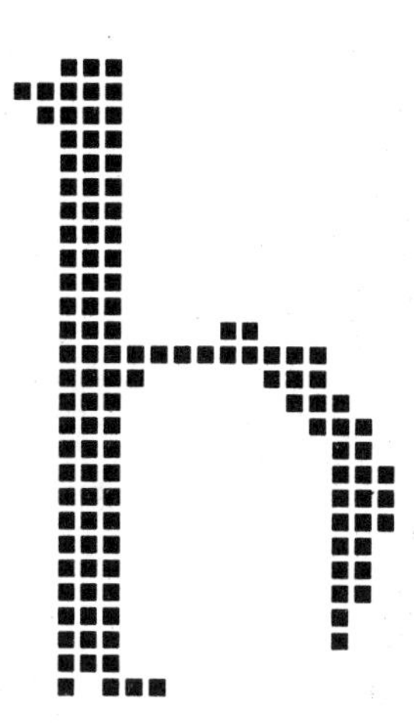

Fig. 8. An ambiguous character. It is probably a "b" but also possibly an "h." The ideal output of the image recognition phase would be {"b," "h"} together with confidence values indicating that, based solely on the image and not taking contextual clues into account, "b" is more likely than "h."

Feature extraction describes the image in each segment as a vector of fixed length.

Classification, which is trained off line [12], outputs the character label it believes is represented by the feature vector. If it is unsure, it outputs a set of choices and associated confidences. If it has reason to believe the segment might contain a join or split, it passes that opinion on to the resegmenter.

Resegmentation cleaves suspected joins, glues suspected splits, and recirculates the new segments back to feature extraction. There are a number of different techniques used for pulling apart joined characters. Some two-character joins, such as the "be" in Fig. 2, can be adequately separated by finding the column roughly in the middle with the smallest "caliper distance," which is the distance between the column's topmost and bottommost black pixels. If the image contains overlapping but nontouching characters, it can be separated into its connected components by tracing around the boundary.

Feature extraction is discussed in subsection A. Classification and off-line training are discussed in subsections B and C respectively, and motivated in subsection D.

A. Feature Extraction

Ideally, the features extracted from an image capture the "essential" characteristics of the character by filtering out all attributes which make, for example, an "e" in one font different from an "e" in another, while preserving the properties that make an "e" different from a "c." The classifier could then store a single prototype per character. Unfortunately, we know of no ideal set of features. The feature set that Calera eventually settled on was motivated by problems encountered with its initial attempts.

Under the assumption that the width of a stroke was not needed to recognize a character, we first tried extracting features from a skeletonized image [13]. However, some characters, such as the "8" and the "s" in Fig. 9, are recognizable because of variations in stroke width and can

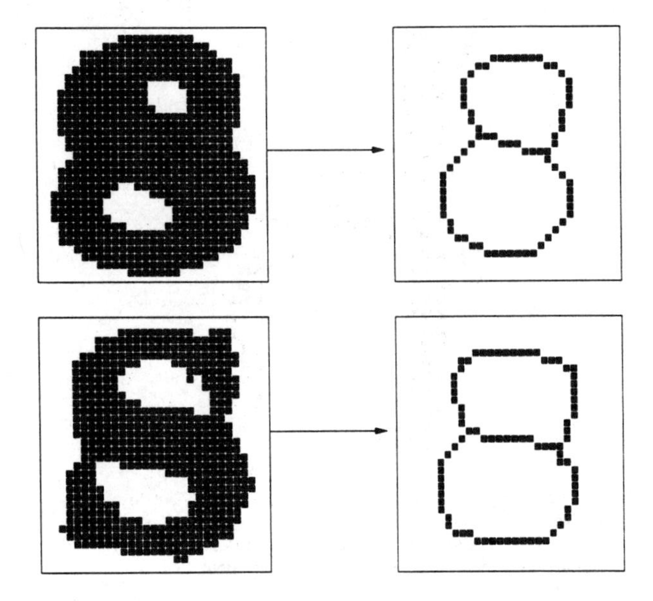

Fig. 9. Illustration of a limitation of skeletonization. The "8" and "S" are recognizable because of variations in stroke width. When the images are skeletonized, they can no longer be distinguished.

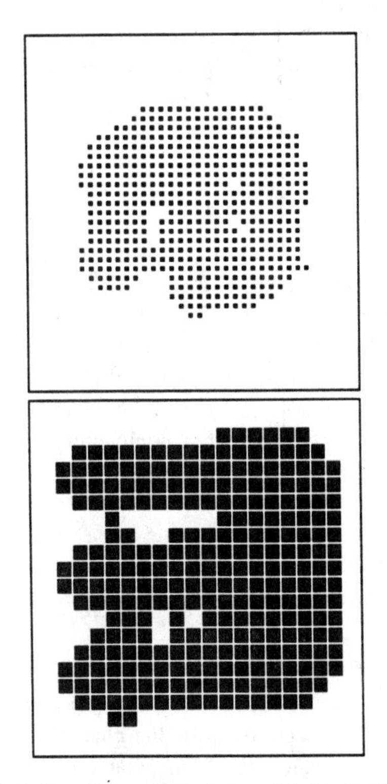

Fig. 10. Illustration of a second limitation of skeletonization.The "m" and "e" are recognizable, but have no obvious strokes, and are apt to be distorted beyond recognition when skeletonized.

no longer be distinguished after the images are skeletonized. We considered using skeletonization at the top level of a multistage classification scheme, to narrow down the choices, but decided this would not be robust. Noise can blur out the strokes in a character or break them into pieces. Images which are recognizable but have no obvious strokes, such as the "m" and "e" in Fig. 10, were apt to be distorted beyond recognition by skeletonization.

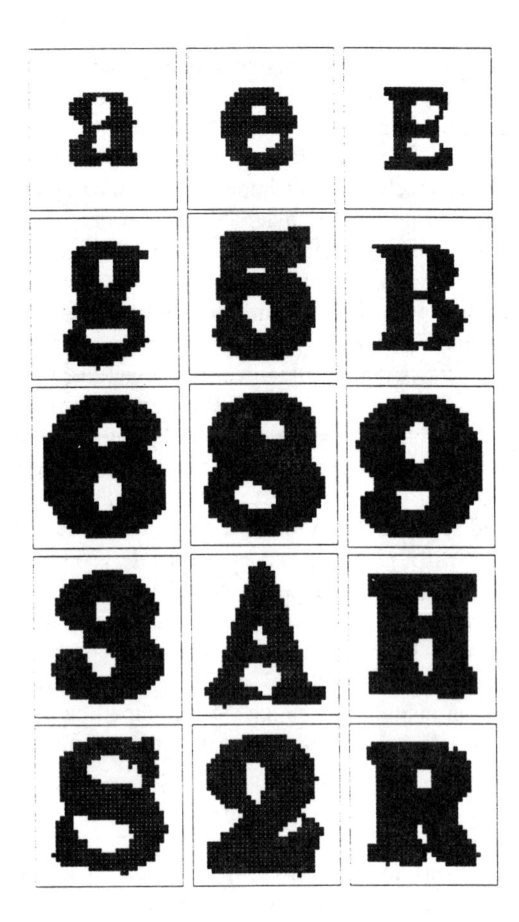

Fig. 11. Illustration of limitations found with topological and geometric features. All the characters fit the rough description "lake (hole) on top of lake." Each is recognizable because of the accumulation of subtle evidence distributed throughout the image.

Topological and geometric features were pursued as a way to aggressively filter out font-specific attributes, such as the exact position or angle of a crossbar [2],[3]. However, similar limitations were encountered when the image was not pristine. For example, the "a," "e," "E," "g," "5," "B," "6," "8," "9," "3," "A," "H," "s," "2," and "R" in Fig. 11 all fit the rough description "lake (hole) on top of lake"." This approach lost its appeal as we found ourselves continually needing to add more features to recognize degraded characters and to discriminate classes which differed in subtle ways.

The lesson learned from this experience was that handcrafted features tend to be brittle. Because of random noise, character images from real-world documents frequently do not conform to expectations, and when they do not they are apt to be misrecognized.[1]

The feature vector used at Calera is of fixed length and consists primarily of a blurred gray-scale reduction of the image. It is obtained by dividing the input image into a fixed number of zones, and assigning to each a value related to the relative density of black pixels in that zone. Several re-

[1] "Now, my suspicion is that the universe is not only queerer than we suppose, but queerer than we can suppose. . . . I suspect there are more things in heaven and earth than are dreamed of, in any philosophy." — J. B. S. Haldane

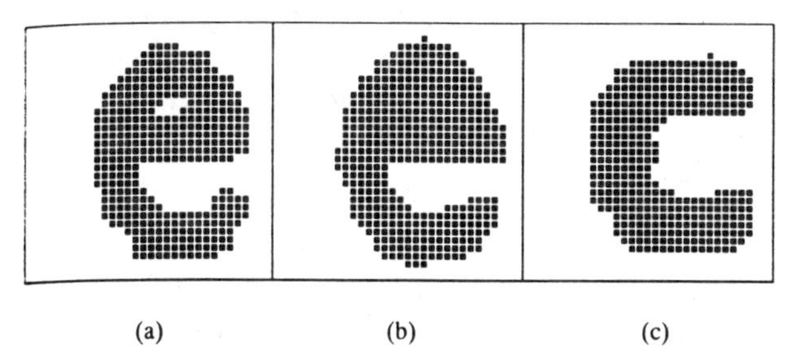

(a) (b) (c)

Fig. 12. Illustration of two properties of Calera's feature extraction. The two "e"s are topologically different but visually similar, and map to nearby vectors in the feature space. The "c" and the second "e," with its hole filled in, are topologically equivalent but easy to distinguish visually, and their distinguishing characteristics are preserved in the feature transformation.

cent papers on neural net handprint classifiers describe similar feature sets. At Bell Labs, the image is normalized to a 16 x 16 gray-scale vector before being passed to the neural net for classification [14]. In [15], the neural net input consists of zonal features combined with geometric features.

Calera's features have two properties that make them less brittle on real-world images than the geometric and topological features initially tried.

First, the transformation from image to feature vector is relatively continuous. Visually similar images, such as the two "e"s in Fig. 12, map to nearby points in the feature space even though they are topologically different. In general, the features are tolerant of noise which causes local distortion, such as a filled-in hole or a break in a crossbar.

Second, the transformation is relatively nondegenerate in the sense that visually dissimilar images do not collapse to the same vector. For example, the filled-in "e" in Fig. 12(b) and the "c" in Fig. 12(c) are topologically equivalent but easy to distinguish visually, and their distinguishing characteristics are preserved in the feature transformation.

B. Classification

The classification algorithm is based on the assumption that the feature transformation is continuous and nondegenerate. Assuming it is continuous, the set of feature vectors corresponding to all readily identifiable "e" images forms a set of regions in the feature space, which will be referred to as the true "e" territory. Assuming it is nondegenerate, the true "e" territory does not overlap the true "c" territory.

The classifier is trained in the lab to approximate the true "e" territory with "certainty regions." (The training algorithm is discussed in the next subsection.) If a feature vector created from a properly segmented "e" is passed to the classifier at run time, there are three possible outcomes:

1) If the "e" feature vector falls inside an "e" certainty region, the classifier will output the single choice "e" together with a confidence value indicating certainty. This is the best scenario.
2) If the "e" feature vector falls inside a "c" certainty region, it will be misclassified as a "c." This is the worst scenario.
3) If the "e" feature vector does not fall inside any certainty region, the classifier outputs a list of choices and associated confidences. As long as the input vector is sufficiently close to an "e" certainty region, "e" will be included on this list and the confidence assigned to "e" will be related to how close the input vector came to being inside an "e" certainty region.

As was discussed earlier, the image from which the input feature vector was generated may not contain a single, properly segmented character. The classifier has a number of mechanisms for spotting such inputs and, when spotted, relays its suspicion to the resegmentation module. Sometimes the width of the image, relative to its surrounding text, is an adequate clue. Often it is not. For example, an "li" join may be no wider than an "h." Detecting improperly segmented characters frequently requires a more subtle analysis. One technique used at Calera involves training the classifier to know about the regions in feature space that do and do not contain joins and checking, at run time, to see whether the input vector falls into suspected join territory.

Characters such as "c" and "C," which are generally indistinguishable in isolation, are collapsed to the same class before the certainty regions are generated. For characters such as "I" and "l," which are sometimes distinguishable and sometimes not, certainty regions labeled with more than one character are generated for the ambiguous regions of feature space.

The certainty regions are organized hierarchically so that only a small fraction of the regions need to be examined to classify a character. A second technique for quickly homing in on the appropriate region takes advantage of previously recognized characters on the page. Unless the document is a ransom note, two different "e"s on a document are likely to be from the same font and therefore close to each other in feature space.

C. Training the Classifier

The classifier learns to discriminate classes by generalizing from a training set. The training process is automated, so that retraining on new fonts or characters is straightforward.

Currently, the training set contains on the order of a million labeled character feature vectors, as well as vectors generated from improperly segmented text. The character images that are used to generate the training set come from a wide variety of fonts and vary in image quality, and are intended to reflect the variability the system will be expected to handle. Though the training set is large, it contains only an infinitesimal fraction of the total number of possible character feature vectors.

Certainty regions are generated by a training algorithm which attempts to cover all "e" vectors in the training set with a minimum number of "e" certainty regions in such a way that no non-"e" ("alien") training set vector is contained in an "e" certainty region. The certainty regions are hyperelliptical and vary in size.

Figure 13 outlines the training procedure. The status of all training set "e" vectors is initialized to "uncovered." Certainty regions are selected sequentially until all "e" vectors are covered. To generate the next certainty region, the algorithm finds an ellipse which contains no alien

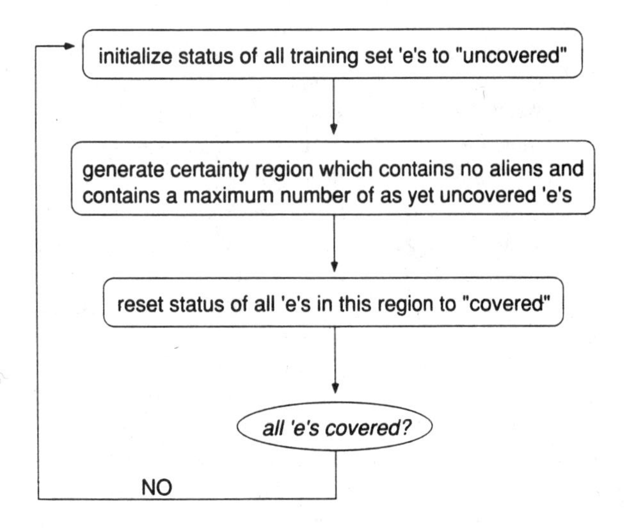

Fig. 13. Outline of off-line training: generating certainty regions for the letter "e."

vectors and contains a maximum number of as yet uncovered "e" vectors.

Ideally, the "e" certainty regions would be large enough to contain all possible "e" feature vectors and small enough to exclude all possible alien feature vectors. However, because the training set does not contain all possible character feature vectors, the "e" certainty regions are not guaranteed to cover all possible "e"s, even though they cover all training set "e"s. Similarly, the "e" certainty regions are not guaranteed to exclude all possible aliens even though they exclude all training set aliens.

Therefore the training algorithm must tread the line between greed and caution. Greed argues for making certainty regions as large as possible, subject only to the constraint that they contain no aliens from the training set. The more territory that is staked out for "e," the better the classifier will generalize to new "e"s not seen in the training set. But caution argues for smaller "e" certainty regions to minimize the risk of hard failure, i.e., the risk that a non-"e" vector will fall into an "e" certainty region at run time.

To navigate between excessive greed and excessive caution, parameters are tuned to ensure that each certainty region is surrounded by an adequately large buffer zone containing no training set aliens and that each certainty region contains no overly vast areas unpopulated by any training set vectors.

D. Discussion of Classification

Rule-based classifiers are able to discriminate characters which are strikingly different, such as "x" and "o," and have the virtue of speed. However, when the domain of character images is unconstrained, they tend not to be robust. A decision-tree classifier [16], for example, that makes intractable decisions based on selected fields of the feature vector is apt to be led down the wrong path if the image is noisy.

In order to discriminate characters such as those in Fig. 11, which are recognizable because of the accumulation of subtle evidence distributed throughout the image, we decided to use a metric-space approach which weighs the information from all the fields before making a decision. Two vectors could be considered similar, overall, even if some field deviated significantly (for example, the two "e"s in Fig. 12); two vectors could be considered dissimilar. overall, even if no single field deviated significantly (for example, the "8" and "S" in Fig. 9).

Hoping that the distribution of classes in the feature space conformed to statistical models, we first tried a parametric classifier. It worked poorly. This experience led us to drop any assumptions about the underlying distribution and look for a model-free classifier that could adapt to arbitrary distributions.

We next tried a nearest-neighbor classifier [17], which compares the input feature vector with a library of reference vectors and finds the closest. The library in effect partitions the feature space into hyperpolyhedral decision regions. Each pair of library vectors corresponds to a hyperplane decision boundary consisting of the set of points equidistant to those two vectors. The library was generated by starting with vectors from just one font and sequentially adding a training set vector to the classifier's library only if it would otherwise be misclassified.

However, we ran into the three classic problems: accuracy, speed, and memory, indicating that the feature space was too complex to be adequately described by a nearest-neighbor partitioning. Rather than leveling off to a reasonable number, the size of the library had to be continually increased as the classifier was required to recognize more and more fonts. If the classifier requires a library of 200 000 vectors to achieve acceptable accuracy on the training set, then 200 000 distances must be computed at run time to classify each input vector. Moreover, the classifier was not robust because it required an excessive amount of memory.

To understand this last claim, it is important to remember that the bottom line for a classifier is not how well it performs on the training set, but how well it performs in the field. Odds are miniscule that a vector input to the classifier at run time will exactly match a training set vector, and so accuracy on the training set does not necessarily predict accuracy on an independent test set.

The simpler the classifier, relative to the size of the training set, the greater the likelihood that its performance on the training set will generalize to new vectors not in the training set[2] [18], [19]. This is essentially Occam's razor.[3] A simple explanation of the data is more robust than a complex one. Given comparable performance on the training set, a classifier which constructs 1000 decision regions probably approximates the underlying distribution of classes more accurately than a classifier which constructs 100 000 regions.

[2] The Vapnik-Chervonenkis theorem, cited in the references, is a formal statement of this.

[3] William of Occam (1285–1349) formulated the principle of parsimony: "What can be done with fewer assumptions is done in vain with more." Also translated as "Entities should not be multiplied unnecessarily."

To illustrate this point by analogy, a polynomial of degree 2 which fits 100 points randomly selected from an unknown distribution reasonably well probably fits the underlying distribution about as well. Yet, the polynomial of degree 99 which fits the 100 points exactly probably deviates wildly from the underlying distribution. The polynomial of degree 2 is generalizing from the data; the polynomial of degree 99 is memorizing it.

The current scheme was designed to overcome the problems encountered with the nearest-neighbor approach. For comparable performance on the training set, it requires between one and two orders of magnitude less memory and is three orders of magnitude faster. Because it requires less memory, i.e. because fewer parameters are needed to outline the decision regions, it is more robust.

Trainable, model-free, classifiers such as neural nets have become the subject of intense research over the last few years. Back-propagation nets partition the feature space into hyperpolyhedral regions with fuzzy boundaries. Rumelhart [20] and Wasserman [21] are good introductions to the field. *K-d* trees are used by Omohundro in [22] to partition the feature space into hyperrectangular regions.

VI. Ambiguity Resolution

When the image recognition phase is uncertain about a character and outputs a list of choices, ambiguity resolution uses language-specific knowledge and document properties to pick the best choice. This section illustrates the kind of decision making used for resolving ambiguities.

Example 1: The recognition engine is uncertain whether the character is "c" or "C." A subline expert looks at the surrounding text, sees that the character image extends from the ascender line to the base line, and chooses "C."

Example 2: The recognition engine is uncertain whether the character is "m" or "rn." A dictionary expert looks at the word containing the ambiguous image, sees that "hamstring" is in the dictionary but "harnstring" is not, and chooses "m" [23], [24].

Example 3: The recognition engine is uncertain whether the image is a "-" or noise. A document structure expert sees that it is neatly lined up in a column of numbers containing other dashes, and decides it is a dash.

Example 4: The recognition engine is uncertain whether the character is "8" or "S." An *n*-gram expert looks at the surrounding text, decides that "$58.00" has a reasonable probability of occurrence but "$5S.00" does not, and chooses "8." The decision is based on digram and trigram probabilities [25]–[27] generated from a large corpus of "typical" English[4] text, rather than heuristics. The probability of a character trigram is estimated by combining various group-trigram, group-unigram, and character-unigram probabilities which are stored in tables generated from the English corpus. For example, the probability of "$5S" is estimated by first looking up the relative frequency of strings of the form

< currency-symbol>< digit>< capital-letter>

and then multiplying this value by

prob ("$"/currency-symbol) * prob ("5"/digit)
* prob ("S"/ capital-letter),

which is an estimate of the fraction of

< currency-symbol>< digit>< capital-letter>

trigrams whose currency symbol is "$," whose digit is "5," and whose capital letter is "S."

Example 5: The recognition engine is uncertain whether the character is "a" or "o." An *n*-gram expert looks at the surrounding text and decides that "Calera" is a bit more likely than "Colera" because "Cal" is a bit more likely than "Col." Since the *n*-gram expert is not overwhelmingly confident of its choice, it may not be given the last word when the opinions of various experts are weighed against each other. For example, if the recognition engine believes "o" is twice as likely as "a" and the *n*-gram expert believes "a" is 5% more likely than "o," and no other expert has an opinion, then "o" will be chosen.

To give one more illustration of the method used to approximate trigram probabilities, the trigram "Cal" is estimated by first looking up, in a letter trigram table which is insensitive to case, the relative frequency in the English corpus of strings of the form < c>< a>< l>, and then multiplying this by

prob (< uppercase-letter>< lowercase-letter>< lowercase-letter> / < letter>< letter>< letter>),

which is the fraction of three-letter trigrams in the corpus whose first letter is capitalized and whose next two letters are not, and which is assumed to be a reasonable estimate of the fraction of all < c>< a>< l> strings whose first letter is capitalized and whose next two letters are not.

We chose to estimate character trigram probabilities by combining several group *n*-gram probabilities, rather than by direct lookup in a single character trigram table, for two reasons.

First, significantly less memory is needed. Second, we believe it gives better estimates for low probability strings. For example, the string "$58" may have never occurred in the English corpus even though there were many occurrences of strings of the form $< digit>< digit>.

For a review article on contextual analysis techniques, see [28].

VII. Market Requirements

The most important measure of the utility of an OCR system is its character accuracy. However, even perfect character accuracy is not generally sufficient to satisfy the needs of the primary markets for OCR, which are "full-text" applications, forms processing, and automatic indexing for image data bases (Fig. 14).

In full-text applications, such as desktop publishing, the full structure of the document needs to be preserved so that when the OCR'd page is converted into a word processor or spreadsheet format, it has the same formatting as the original page. Without the correct formatting codes, OCR may be useless on a complex document, even if character

[4] Of course, for the French product, the probabilities are generated from a French corpus of text, and similarly for other languages.

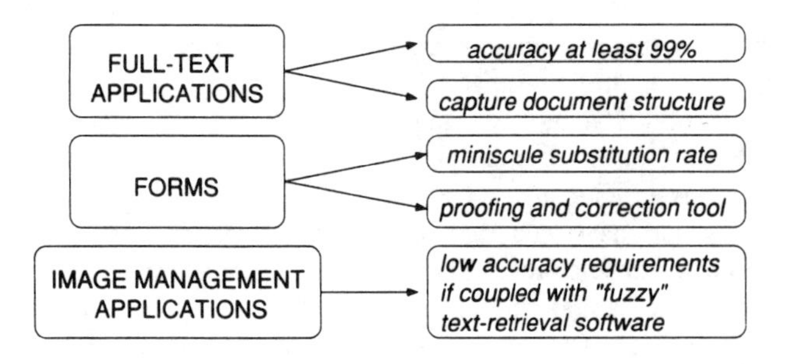

Fig. 14. Accuracy and feature requirements of the three major markets for OCR. High throughput, which is a function of recognition speed and time needed for manual correction, is also desirable.

accuracy is 100%. For example, if the columns in a table of numbers all run together, it may be faster to retype the entire table from scratch than to reformat it by hand. These applications typically require an accuracy of at least 99%.

In forms-processing applications, selected zones on the document are recognized rather than the full page. Capturing the document structure is unnecessary, but a proofing and correction tool that displays portions of the image corresponding to uncertain characters becomes essential. Because of the high volume of most such applications, physically having the paper moved to the proofing operator may not be feasible, and whereas a spell checker can usually find most OCR errors on a full-text document, it is of no help on a form that contains, for example, dollar amounts.

The critical factor in determining whether OCR is cost-effective in most forms applications is not its accuracy rate, but its substitution rate, which is the percent of errors that the software failed to flag as uncertain. In financial applications, the substitution rate must be miniscule. Virtually all errors need to be flagged so that they can be fixed by a proofing operator. An accuracy of 99.9% may be intolerable if the one error per thousand characters goes undetected. Yet in some applications, an accuracy as low as 80% is acceptable so long as the substitution rate is less than one per 50 000 characters. If the particular application has a mechanism for detecting errors, such as check sums, then a low substitution rate becomes a less critical requirement.

In image management applications, OCR is used as a substitute for manual keyword indexing. For example, all document images containing the target word "marmalade" can be retrieved by searching for the word in the OCR'd ASCII file associated with each scanned document. Capturing the document structure is unnecessary since the ASCII file is used only as an index to the scanned image.

With the availability of "fuzzy" text-retrieval software which is tolerant of recognition errors, proofing and correction are unnecessary and accuracy can be as low as 80%. For example, the image of a document containing the word "marmalade" will be retrieved even if the word was misrecognized as "rnarmalade," so long as the text-retrieval software considers "rnarmalade" a fuzzy match with "marmalade."

In all three market segments, high throughput is desirable. Throughput is the total time from scanning to final correction. If misrecognized characters are to be corrected by an operator, then speed, measured as raw characters per second, is important only insofar as it is one of the variables affecting throughput. The more critical variable is the time required for manual correction, which is affected by the error rate, the substitution rate, the percent of false positives (correctly recognized characters which are flagged), and the design of the proofing/correction tool. When confronted with accuracy/speed trade-offs, we believe it is generally best to opt for accuracy. Because of the time required to correct OCR errors, doubling the accuracy increases throughput much more significantly than doubling the speed.

VIII. Product Features

In order to meet the market requirements discussed in the previous section, additional capabilities are built into the OCR system.

Document Analysis

Document analysis is the one primary component of the system that was not discussed earlier. Its purpose is to capture the structure of a document and incorporate that structure into the OCR output so that, when viewed as an ASCII file or in a target word processor or spreadsheet, its formatting reflects the formatting of the original document. This involves

1) labeling the components of the document, such as paragraphs, tables, lists, headers, and footers;
2) establishing the reading order of the components;
3) analyzing the geometry of the page, for example, determining whether a title is centered or establishing the width of a column;
4) partitioning the text into regions that share the same font attributes, such as regions of italic, bold, and underlined text and regions that share the same point size.

The appropriate formatting codes (e.g. centering codes, hard and soft newlines) are then passed to the application software.

For references on document analysis, see [6], [10], and [29]–[35].

Autoclipping

If the user selects this option, then photographs, line art, signatures, logos, and other picture regions are automatically clipped from the document image and stored in a file so that they can be archived or pasted back into the OCR'd document.

Character Flagging and Proofing Tools (Fig. 15)

If the OCR software is not confident that it has correctly recognized a character, that character is flagged, and all flagged characters are highlighted in the proofing and correction tool. When the operator moves to the next flagged character, the portion of the image containing that character and a bit of its surrounding context pops up on the

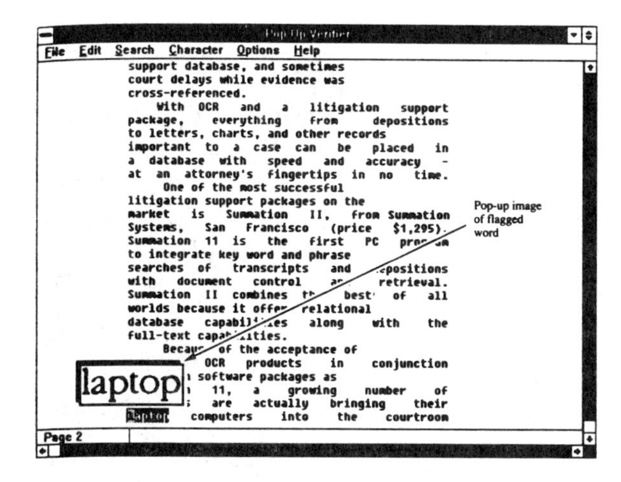

Fig. 15. Screen shot from the "pop-up verifier," which is a proofing and correction tool that displays the portion of the image corresponding to uncertain characters or to words not found in either the system dictionary or a user-supplied dictionary.

screen, so that the operator can verify its correctness and fix it if necessary. There is also the option of displaying pop-ups for all words not found in either the system dictionary or a user-supplied dictionary.

Preview Window / Zones / Templates / Numeric Mode (Fig. 16)

When a page is scanned, the user can view the image in a preview window before it is recognized, at which time any of a handful of options can be selected. The image can be rescanned at a different threshhold if it looks too dark or too light. If only selected portions of the page need to be recognized, zones can be drawn around those regions. If a zone contains only numeric data, the "numeric-mode" option can be selected, which will give significantly better recognition if the image is heavily degraded. In a forms application, a template can be constructed by zoning a sample page. The template can then be applied to any other page having the same format.

IX. Implementation

Calera's OCR engine has been implemented on a number of different platforms. Because they all share the same engine, accuracy is the same across all platforms.

WordScan is software-only and runs on the PC and Mac. Its speed depends on the speed of the host machine. On a 16 MHz 386SX, speed is about 50 characters per second (cps).[5] FaxGrabber is a newly released application built around the WordScan engine that alerts the user when an incoming fax has arrived, pastes the OCR'd text into a target word processor, and pastes the image into a target graphics program.

TrueScan is a board-level product which fits inside a PC, contains customized chips and a 68020, and runs at about 100 cps. The CDP 9000, which contains a scanner and four 68020 boards with customized chips, runs at about 250 cps. By the time this paper is in print, we expect to have released the M-Series (Fig. 17), which contains between one and four RISC-based recognition boards and runs at up to 600 cps per board.

[5] These are peak speeds, measured on a very clean set of documents. Speed on an average document is typically one-third to one-half peak speed.

An interesting pricing trend is that the cost of omnifont OCR has been dropping by a factor of about 10 every two years for the last six years. In 1986, it cost about $40 000 for a speed performance of 100 cps (CDP 3000). In 1988, the same performance cost about $4000 (TrueScan). By 1990, WordScan had a street price of about $400.

X. Current Performance

Despite the title of this paper, no OCR product today is truly omnidocument in the sense of reading all documents satisfactorily. The best products do a good job on clean documents, but they all degrade in performance—some more gracefully than others—as document quality (or scanner quality) degrades.

Figure 18 illustrates the relative strengths and weaknesses of Calera's engine. The scanned image came from a ninth-generation photocopy which contained three paragraphs printed in the same font. The paragraphs were formatted with different amounts of space between the characters. The top paragraph was formatted with condensed spacing, the middle with normal spacing, and the bottom with expanded spacing. After photocopying and scanning, most of the characters in the top two paragraphs are joined, and recognition accuracy is poor. The characters in the bottom paragraph are quite degraded, but none are joined, and accuracy is reasonably good (99.3%).

If a page has unacceptable accuracy, it is generally because the character images were difficult to segment. (On a fixed-pitch page, segmentation is not a problem because the pitch can be used to determine where one character stops and the next one begins.) Accuracy on new fonts and on degraded images of individual characters is relatively robust. We look forward to the day when recognition of joined and split characters is no longer the weakest link, and recognition of individual characters becomes the primary source of error.

It is difficult to quote meaningful and quantitative performance figures in the absence of a standardized test set. The error rate can be lower than 0.1% on clean documents, and arbitrarily high on sufficiently degraded documents. The results of a study that compared performance of various commercial OCR products are reported in [36]. Their test set was composed of 51 randomly selected pages containing a wide variety of fonts and point sizes, and exhibiting imperfections typical of real-world documents such as skew and stray marks, as well as joined, blurry, and split character images. Calera's engine, which performed the best, had an error ate of 1.4%.

The three OCR markets discussed earlier are served to varying degrees by existing products.

When combined with image retrieval software based on fuzzy string matching, current OCR is more than adequate to meet the needs of image management applications. This

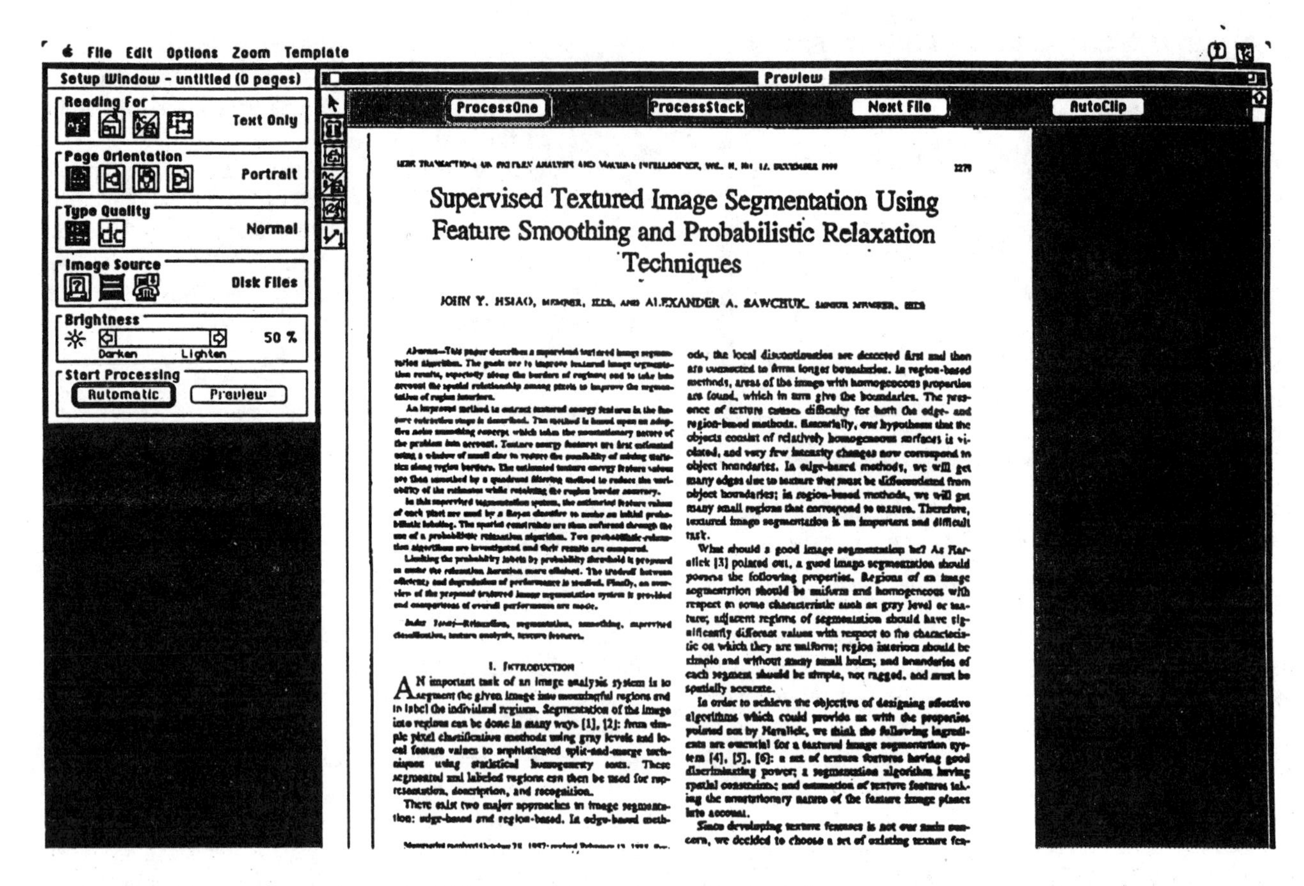

Fig. 16. Screen shot from the "preview window". The user can view the image of the page before it is recognized, at which time (a) the scanner threshhold can be adjusted, (b) picture, alphanumeric, and numeric regions can be zoned out, and (c) templates can be created for forms applications.

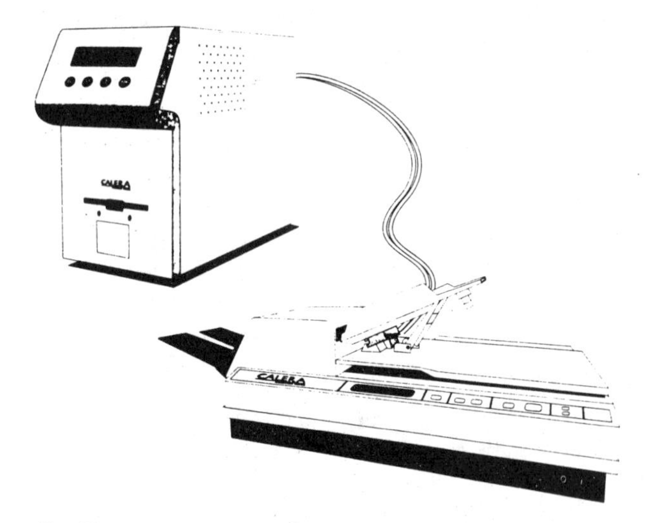

Fig. 17. Picture of the soon-to-be-released M-series, which contains between one and four RISC-based recognition boards and runs at 600 characters per second per board.

market is expected to grow as such retrieval software becomes inexpensive and available on low-end platforms.

By far the largest market for OCR today is in full-text applications. Improved recognition of degraded images and further improvements in document analysis will expand the domain of documents on which OCR is preferable to rekeying.

Significant improvements are needed to adequately meet the requirements of many forms applications. Forms pose a particularly difficult problem for OCR . They tend to be "well handled" and dirty, with poor quality print that is apt to intersect ruled lines from the form. To achieve one unflagged error in 50 000 characters, the software needs either dramatically better accuracy or better methods for assessing its own performance and knowing, without benefit of a dictionary, when it has incorrectly recognized a character.

The current approach is extendable to handprinted characters (we built a quick prototype and got quite good recognition) but is not extendable to cursive writing because, at its core, the engine is designed to recognize individual characters.

Future Work

The primary challenge for Calera is better recognition of the joined and split characters on degraded documents. Achieving this requires a control strategy that more tightly integrates segmentation and contextual analysis into the recognition process.

A number of recent papers propose such strategies. In [37], Ho *et al.* generate a consensus ranking from three classifiers, two of which incorporate contextual knowledge directly into the recognition process, and bypass the problem of segmenting joined characters by recognizing

SCANNED IMAGE	RECOGNIZED TEXT

The Palantir reader consists of a feeder,
a scanner, some switches and lights, and
our own custom circuit which does all the
work-it addresses a half-million bytes.

lb c Ewan tir ru&r conslits of a fee&r,
a scanna.s=e switches and light& and
our own custom circuit which does all thc
work-k aftems a balf-mUUon by@

image quality	recognition rate
% joins: 74%	
% overlaps: 9%	69%
% isolated: 17%	

It is near guaranteed to be able
to read any document fed in at
all, except that it fails and can't
make heads or tails of our CFO's
handwritten scrawl.

h b neu pamteed to be able
to mad any document fed in st
aU, emxpt that h fails and cawt
make beab or tails of our CFIWS
bandwritten scrawl.

image quality	recognition rate
% joins: 59%	
% overlaps: 23%	76%
% isolated: 18%	

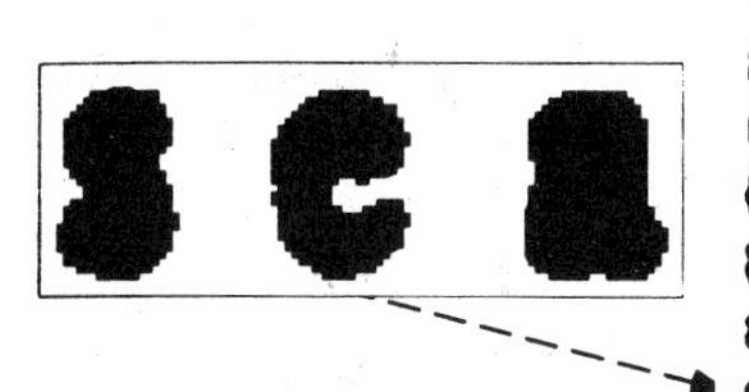

But our customer's
wants made us train
all the fonts that have
ever been used
(within limits) and
the Navy bought three
and installed one at
sea on the bridge of
the U.S.S. Nimitz.

But our customer's
wants made us train
all the fonts that have
ever been used
(witbin limits) and
the Navy bought three
and installed one at
sea on the bridge of
the U.S.S. Nimitz.

image quality	recognition rate
% joins: 0%	
% overlaps: 0%	99.3%
% isolated: 0%	

Fig. 18. Strengths and weaknesses of the recognition engine. The scanned image came from a ninth-generation photocopy scanned at 300 dpi. The top paragraph was formatted with condensed spacing, the middle with normal spacing, and the bottom with expanded spacing. After photocopying and scanning, most of the characters in the top two paragraphs are joined, and recognition accuracy is poor. The characters in the bottom paragraph are quite degraded, but none are joined, and accuracy is good. Recognition of degraded individual characters is reasonably robust. If recognition is unacceptable, it is generally because the characters were difficult to segment.

whole words. The first classifier follows the "segment then recognize then contextually postprocess" model, and is appropriate for characters that are easy to segment and recognize in isolation. The second assembles features from the individual characters in a word and matches them against a feature-tagged lexicon, and is appropriate when the word is easy to segment, but its characters are difficult to recognize in isolation. The third matches features extracted from an entire unsegmented word against a lexicon, and is appropriate when the word is difficult to segment . In [38], Hull *et al.* augment this approach with an algorithm for recognizing joined words by generating a tree of alternative segmentations.

One drawback of these approaches is that they require all words to come from a fixed lexicon. Fenrich [39] and Lecolinet and Moreau [40] present algorithms which integrate segmentation with recognition without imposing this requirement.

Higher-level contextual analysis, such as syntactic analysis [41], [42], can also play a role in improving recognition. In [43], Baird and Thompson present an interesting experiment in the use of semantic analysis to recognize degraded text from a chess encyclopedia.

We believe recognition of individual characters can be improved by combining multiple independent feature sets/classifiers, so that the weakness of one is compensated for by the strength of another. This approach has been suggested by Ho *et al.* [44] and Suen *et al.* [45]. It is also suggested by the work of Bradford and Nartker [36], who report reducing Calera's error rate with a voting scheme that combines Calera's output with the output from various other commercial OCR products.

ACKNOWLEDGMENT

The author wishes to thank J. Novicki, D. Ross, and R. Poppen for valuable feedback on an earlier draft.

References

[1] R. O. Duda and P. E. Hart, *Pattern Classification and Scene Analysis.* New York: Wiley, 1973, pp. 276–282.

[2] S. Kahan, T. Pavlidis, and H. Baird, "On the recognition of printed characters of any font or size," *IEEE Trans. Pattern Anal. Mach. Intell.*, vol. 9, pp. 274–288, Mar. 1987.

[3] H. Baird, "Feature identification for hybrid structural/statistical pattern classification," *Computer Vision, Graphics, and Image Processing*, vol. 42, pp. 318–333, 1988.

[4] H. Baird, "A 100-font classifier," in *Proc. First Int. Conf. Document Analysis and Recognition*, vol. 1, Oct. 1991, pp. 332–340.

[5] S. Hinds, J. Fisher and D. D'Amato, "A document skew detection method using run-length encoding and the Hough transform," in *Proc. 10th Int. Conf. Pattern Recognition*, 1990, pp. 464–468.

[6] G. Nagy, S. Seth, and S. Stoddard, "Document analysis with an expert system," in *Pattern Recognition in Practice II*, E. Gelsema and L. Kanal, Eds. New York: North-Holland, 1986, pp. 149–159.

[7] F. Wahl, K. Wong, and R. Casey, "Block segmentation and text extraction in mixed text/image documents," *Computer Vision, Graphics and Image Processing*, vol. 20, pp. 375–390, 1982

[8] L. Fletcher and R. Kasturi, "A robust algorithm for text string separation from mixed text/graphics images," *IEEE Trans. Pattern Anal. Mach. Intell.*, vol. 10, pp. 910–918, Nov. 1988.

[9] T. Pavlidis and J. Zhou, "Page segmentation by white streams," in *Proc. First Int. Conf. Document Analysis and Recognition*, vol. 2, Oct. 1991, pp. 945–953.

[10] Y. Tang, C. Suen, C. Yan, and M. Cheriet, "Document analysis and understanding: A brief survey," in *Proc. First Int. Conf. Document Analysis and Recognition*, vol. 1, Oct. 1991, pp. 17–31.

[11] M. Nadler, "A survey of document segmentation and coding techniques," *Computer Vision, Graphics, and Image Processing*, vol. 28, pp. 240–262, 1984.

[12] S. M. Weiss and C. A. Kulikowski, *Computer Systems That Learn.* San Mateo, CA: Morgan Kaufman.

[13] S. Lee, L. Lam, and C. Suen, "Performance evaluation of skeletonization algorithms for document image processing," in *Proc. First Int. Conf. Document Analysis and Recognition*, vol. 1, Oct. 1991, pp. 260–271.

[14] Y. LeCun, "Backpropagation applied to handwritten zip code recognition," *Neural Computation*, vol. 1 no. 4, pp. 541–551, winter 1989.

[15] H. Takahashi, "A neural net OCR using geometrical and zonal-pattern features," in *Proc. First Int. Conf. Document Analysis and Recognition*, vol. 2, Oct. 1991, pp. 821–828.

[16] M. Bernard and E. Moret, "Decision trees and diagrams," *Computing Surveys*, vol. 14, pp. 593–623, Dec. 1982.

[17] P. Devijver and J. Kittler, *Pattern Recognition: A Statistical Approach.* London: Prentice Hall, pp. 69–127.

[18] Vapnik and Chervonenkis, "On the uniform convergence of relative frequencies of events to their probabilities," *Theory Prob. Appl.*, vol. 16, pp 264–280.

[19] Y. Abu-Mostafa, "The Vapnik-Chervonenkis dimension: Information versus complexity in learning," *Neural Computation*, vol. 1, no. 3, pp. 312–317, fall 1989.

[20] D. Rumelhart and J. McClelland, *Parallel Distributed Processing.* Cambridge, MA: MIT Press, 1986.

[21] P. Wasserman, *Neural Computing: Theory and Practice.* New York: Van Nostrand Reinhold, 1989.

[22] S. Omohundro, "Efficient algorithms with neural network behavior," *Complex Systems*, vol. 1, p. 273, 1987.

[23] T. N. Turba, "Checking for spelling and typographical errors in computer-based text," SIGPLAN-SIGOA Newsletter, pp. 51–60, 1981.

[24] R. Sinha and B. Prasada, "Visual text recognition through contextual processing," *Pattern Recognition*, vol. 20, no. 5, pp. 463–479, 1988.

[25] R. Shinghal and G. T. Toussaint, "Experiments in text recognition with the modified Verterbi algorithm," *IEEE Trans. Pattern Anal. Mach. Intell.*, pp. 480–493, May 1974.

[26] E. Riseman and A. Hanson, "A contextual postprocessing system for error correction using binary *n*-grams," *IEEE Trans. Comput.*, pp. 480–493, May 1974.

[27] A. Goshtasby, "Contextual word recognition using probabilistic relaxation labeling," *Pattern Recognition*, vol. 21, no. 5, pp. 455–462, 1988.

[28] D. Elliman, "A review of segmentation and contextual analysis techniques for text recognition," *Pattern Recognition*, vol. 23, nos. 3/4, pp. 337–346, 1990.

[29] G. Nagy, J. Kanai, M. Krishnamoorthy, M. Thomas, and M. Viswanathan, "Two complementary techniques for digitized document analysis," in *Proc. ACM Conf. Document Processing Systems*, 1988, pp. 169–176.

[30] K. Wong, R. Casey, and F. Wahl, "Document analysis system," *IBM J. Res. Develop.*, vol. 26, no. 6, pp. 647–656, 1980.

[31] G. Nagy, "Toward a structured-document-image utility," in *Proc. Workshop on Syntactic and Structural Pattern Recognition*, June 1990, pp. 293–309.

[32] A. Dengel, "Document image analysis—Expectation-driven text recognition," in *Proc. Workshop on Syntactic and Structural Pattern Recognition*, June 1990, pp. 78–87.

[33] A. Dengel and G. Barth, "High level document analysis guided by geometric aspects," *Int. J. Pattern Recognition and Artificial Intelligence*, vol. 2, no. 4, pp. 641–655, 1988.

[34] J. Fisher, "Logical structure descriptions of segmented document images," in *Proc. First Int. Conf. Document Analysis and Recognition*, vol. 1, Oct. 1991, pp. 302–310.

[35] H. Fujisawa and Y. Nakano, "A top-down approach for the analysis of document images," in *Proc. Workshop on Syntactic and Structural Pattern Recognition*, June 1990, pp. 113–122.

[36] R. Bradford and T. Nartker, "Error correlation in contemporary OCR systems," in *Proc. First Int. Conf. Document Analysis and Recognition*, vol. 2, Oct. 1991, pp. 516–523.

[37] T. Ho, J. Hull, and S. Srihari, "Word recognition with multi-level contextual knowledge," in *Proc. First Int. Conf. Document Analysis and Recognition*, vol. 1, Oct. 1991, pp. 905–915.

[38] J. Hull, T. Ho, J. Favata, V. Govindaraju, and S. Srihari, "Combination of segmentation-based and wholistic handwritten word recognition algorithms," in *Proc. Int. Workshop on Frontiers in Handwriting Recognition*, Sept. 1991, pp. 229–240.

[39] R. Fenrich, "Segmentation of automatically located handwritten words," in *Proc. Int. Workshop on Frontiers in Handwriting Recognition*, Sept. 1991, pp. 33–44.

[40] E. Lecolinet and J. Moreau, "a new system for automatic segmentation and recognition of unconstrained handwritten zip codes," presented at the 6th Scandinavian Conference on Image Analysis, June 1989.

[41] C. Crowner and J. Hull, "A hierarchical pattern matching parser and its application to word shape recognition," in *Proc. First Int. Conf. Document Analysis and Recognition*, vol. 1, Oct. 1991, pp. 323–331.

[42] L. Evett, C. Wells, F. Keenan, T. Rose, and R. Whitrow, "Using linguistic information to aid handwriting recognition," in *Proc. Int. Workshop on Frontiers in Handwriting Recognition*, Sept. 1991, pp. 303–311.

[43] H. Baird and K. Thompson, "Reading chess," *IEEE Trans. Pattern Anal. Mach. Intell.*, vol. 12, pp. 552–559, June 1990.

[44] T. Ho, J. Hull, and S. Srihari, "Combination of structural classifiers," in *Proc. Workshop on Syntactic and Structural Pattern Recognition*, June 1990, pp. 123–136.

[45] C. Suen, C. Nadal, T. Mai, R. Legault, and L. Lam, "Recognition of totally unconstrained handwritten numerals based on the concept of multiple experts," in *Proc. Int. Workshop on Frontiers in Handwriting Recognition*, Apr. 1990, pp. 131–140.

Mindy Bokser (Member, IEEE) received the B.A. degree in mathematics and philosophy from Barnard College of Columbia University and the M.S. degree in mathematics from Stanford University.

In 1983 she joined Calera Recognition Systems (then called Palantir), where she has worked on developing character recognition algorithms. She is currently Director of Research and Development at Calera.

Ms. Bokser is a member of the Association for Computing Machinery.

Incorporation of a Markov Model of Language Syntax in a Text Recognition Algorithm

Jonathan J. Hull
Center of Excellence for Document Analysis and Recognition
Department of Computer Science
State University of New York at Buffalo
Buffalo, New York 14260-0001
hull@cs.buffalo.edu

Abstract

A Markov model for language syntax and its use in a text recognition algorithm is proposed. Syntactic constraints are described by the transition probabilities between classes. The confusion between the feature string for a word and the syntactic classes is also described probabilistically. A modification of the Viterbi algorithm is also proposed that finds a fixed number of sequences of syntactic classes for a given sentence that have the highest probabilities of occurrence, given the feature strings for the words. An experimental application of this approach is demonstrated with a word hypothesization algorithm that produces a number of guesses about the identity of each word in a running text. It is shown that the Viterbi algorithm can significantly reduce the number of words that can possibly match an image.

1. Introduction

Text recognition algorithms often process only images of isolated characters. This is sometimes followed by a post-processing step that uses information from a dictionary of allowable words to correct recognition errors. Additional knowledge above the level of individual words can also be used to correct for recognition errors. An example includes the use of semantics of a constrained domain (chess games) to improve character recognition [1]. the transitions between pairs of words have also been used to improve a word recognition technique [7].

Language level syntax has also been employed to improve word recognition by reducing the number of alternatives for the identity of a word [8]. This was done by using binary constraints between a group of words with the same shape and the syntactic classes that could follow them. The constraints were compiled from a training text and applied to restrict the decisions for the syntactic class of a word to be consistent with the shape of the previous word. Even this limited binary information was shown to be effective at reducing the average number of words that could match any image.

"Incorporation of a Markov Model of Language Syntax in a Text Recognition Algorithm" by J.J. Hull from *Proc. Symp. Document Analysis and Information Retrieval*, 1992, pp. 174-185, reprinted with permission.

This paper proposes to model English grammar as a Markov process where the probability that any grammatical class will be observed is dependent on the class of the previous word.[1] This model is applied to text recognition by first using a word recognition algorithm to supply a number of alternatives for the identity of each word. The syntactic categories of the alternatives for the words in a sentence are then input to a modified Viterbi algorithm that determines sequences of syntactic classes that include each word. An alternative for a word decision is output only if its syntactic class included in at least one of these sequences. The Markov model improves word recognition performance if the number of alternatives for a word are reduced without removing the correct choice.

The rest of this paper briefly introduces an algorithm for word recognition. This is followed by a description of how a Markov model of language syntax is incorporated in this model. The modified Viterbi algorithm proposed in this paper is then described. The performance of this technique in reducing the number of alternatives for words in a sample of text is then discussed.

2. Text Recognition Algorithm

The recognition algorithm that incorporates the Markov model of syntax contains the three steps shown in Figure 1. The input is a sequence of word images w_i, $i = 1,2, \cdots$. The hypothesis generation stage computes a group of possible identifications for w_i (called N_i or its *neighborhood*) by matching a feature representation of the image to the entries in a dictionary. The hypothesis testing phase uses the contents of a neighborhood to determine a specific set of feature tests that could be executed to recognize w_i. The output of hypothesis testing is either a unique recognition of w_i or a set of hypotheses that contains the word in the image. The global contextual analysis phase uses information about other words that have been recognized, such as their syntactic classification, to constrain the words that can be in N_i. The output is a neighborhood N_i^* of reduced size.

3. Syntax Model

The syntax of a sentence is summarized as the sequence of syntactic classifications for its words. Since it is known that the appearance of any tag probabilistically constrains the tags that can follow it, a Markov model is a natural representation for syntax [10]. An example of such a probabilistic constraint is given by the probabilities that certain syntactic classes follow an article in a large sample of text. The word following an article is a singular or mass noun in 51 percent of all cases and is an adjective 20 percent of the time. The other 29 percent of occurrences are scattered over 82

[1] Probabilistic information has been successfully used to automatically tag large corpora of text [2].

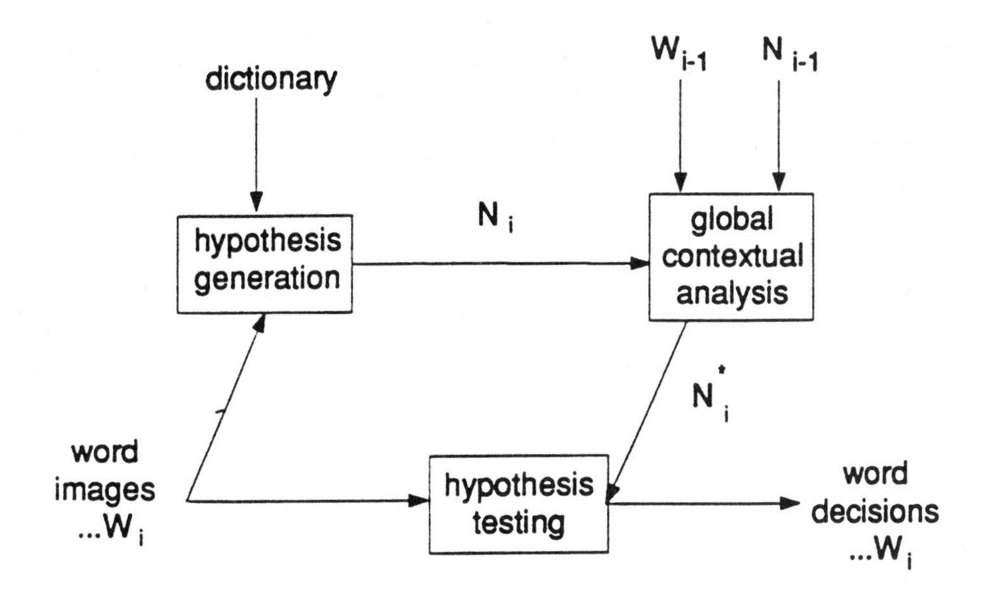

Figure 1. The text recognition algorithm.

other syntactic classes with many never following an article.

The Viterbi algorithm is a method for determining the sequence of classes with the maximum a-posteriori probability that match a given set of recognition decisions for the words in a sentence [4]. The adaptation of the Viterbi algorithm to this problem is very similar to its use in postprocessing character decisions [5]. The Viterbi algorithm has also been successfully used for speech recognition [3].

Under the assumptions that sentence delimiters (periods) are perfectly recognized and the occurrence of a word recognition error in any position is independent of the occurrence of an error in any other position, the Viterbi algorithm determines the string of syntactic classes $\bar{Z} = z_0, z_1, \ldots, z_{n+1}$ that maximize Bayes' formula:

$$P(\bar{Z} \mid \bar{X}) = \frac{P(\bar{X} \mid \bar{Z})\, P(\bar{Z})}{P(\bar{X})} \qquad 1$$

where $\bar{X} = x_0, x_1, \ldots, x_{n+1}$ is the sequence of feature vectors for the words in the sentence and $x_0 = x_{n+1} = z_0 = z_{n+1} = period$.

The independence and Markov assumptions, as well as the fact that the maximization of equation 1 is independent of $\bar{X}$ reduces the maximization of equation 1 to the maximization of:

$$\prod_{i=1}^{n+1} P(X_i \mid Z_i)\, P(Z_i \mid Z_{i-1}) \qquad 2$$

over all possible $\bar{Z}$, where $P(X_i \mid Z_i)$ is the conditional probability of the feature vector X_i taking on its value given that the corresponding syntactic class is Z_i, sometimes called a "confusion probability", and $P(Z_i \mid Z_{i-1})$ is the first order transition probability of the occurrence of syntactic class Z_i given that syntactic class Z_{i-1} has been obsereved. To avoid the combinatorial explosion inherent in calculating equation 2 for all possible strings $\bar{Z}$, the Viterbi algorithm uses a dynamic programming formulation to transform the problem into one of graph searching with a computational requirement on the order of M^2 operations, where M is the number of alternatives for any word image. The basic Viterbi algorithm is summarized:

```
Procedure Viterbi(X̄,Z̄);

classes     := NULL;
prev_cost := 1;
for Word := 1 to Numwords+1 do { examine each word in sentence }
  begin
    for j := 1 to Num_alts_Word do { each alternative for current word }
      begin
        for k := 1 to Num_alts_Word-1 do { how can it be reached from prev. word }
S1:          temp[k] := prev_cost[k] * P(X_Word | Z_j) * P(Z_j | Z_k)
        partial_cost[j] := max(temp,u);
        temp_classes[j] := concat(classes[u], Z_j);
      end
    prev_cost := partial_cost;
    classes := temp_classes;
  end
write(classes[1]);
```

where, prev_cost and classes are vectors of real values and strings, respectively, that have the same number of elements as the maximum number of recognition decisions for any word. Prev_cost holds the accumulated costs of reaching the alternatives for the previous word and word holds the corresponding strings of syntactic tags. Max(temp,u) returns the maximum of the elements in temp with the side-effect that u is set to the index of the maximum element; concat(classes[u], Z_j) returns the result of concatenating Z_j on the end of classes[u]. The algorithm operates one word at a time from left to right in a sentence. The words are numbered 0,1,...,Numwords+1, where Numwords is the number of words in the input sentence. The string of syntactic classes on the best-cost path is output from classes[1] after the last iteration.

A useful modification of the Viterbi algorithm is to allow it to find a fixed number of syntactic tag sequences, each of which have the next best cost among all possible alternatives. This modification is simply implemented by maintaining the desired number of alternative sequences and their costs at each point in the evaluation. The final result includes those sequences and their costs.

4. Experimental Investigation

Experimental tests were conducted to determine the ability of the Viterbi implementation of syntactic constraints to reduce the number of word candidates that match any image. Given a sentence from a test sample of running text, a set of candidates for each word were produced by a model of the hypothesis generation portion of the word recognition algorithm. These candidates were looked up in a dictionary to retrieve their syntactic classes as well as their confusion probabilities. The Viterbi algorithm was then run on these data, using transition probabilities from another large text. Word candidates were then removed if their syntactic classes did not appear in any of the results produced by the Viterbi.

Performance was measured by determining the average number of words present in the neighborhoods before and after the application of syntax. The number of errors were also determined. An error occurred when the correct choice for a word was not present after the application of syntax.

An example of applying the Viterbi for syntactic constraints is shown in Figure 2. The original input sentence is shown along the the top of the figure (HE WAS AT WORK.). The complete neighborhoods for each word are shown below the input words. Each word is shown along with its syntactic class and the confusion probability for that neighborhood given the syntactic class. The different syntactic classes are explained in Appendix 1. The first and third words in the sentence have only one word in their neighborhoods. The second neighborhood contains eight different words, two of which have different syntactic classes. The fourth neighborhood contains six different words, one of which has three different syntactic tags.

The transition probabilities are shown along the arcs. It is seen that some transitions, such as PPS-NNS never occurred in the training text and hence have a probability of zero. Other transitions are much more likely, such as PPS-VBD which has a probability of 0.3621.

In this case, the top choice of the Viterbi algorithm was PPS-BEDZ-IN-NP and the second choice was PPS-BEDZ-IN-NN. In the top choice, three of the four classes encompassed the correct word. The correct answer for the fourth word (NN) was only contained in the second choice of the Viterbi.

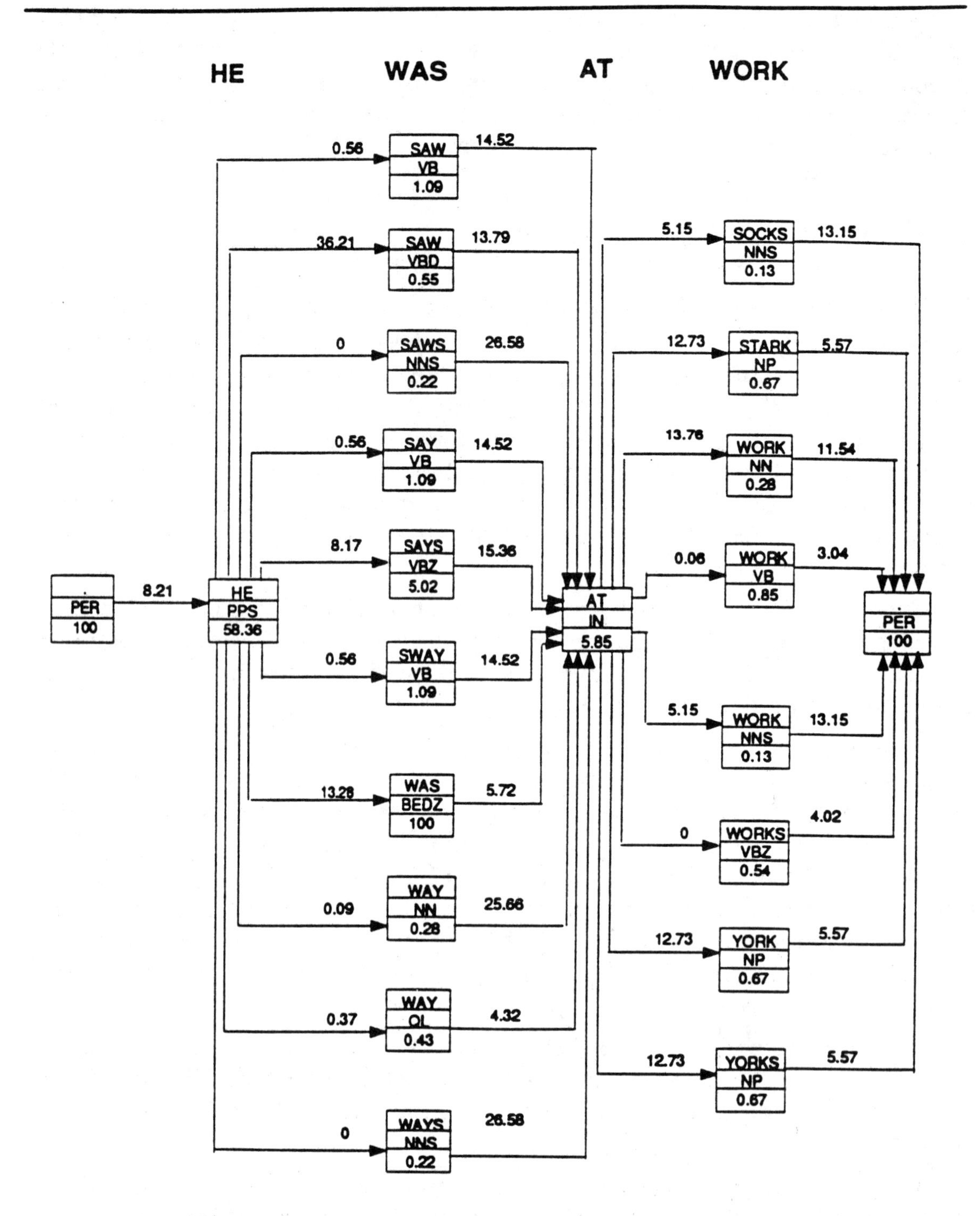

Figure 2. Example of applying the Viterbi algorithm .

4.1. Text database

A soft copy (ASCII) text sample known as the Brown Corpus was used for these experiments [9]. This text was chosen because it is large (over 1,000,000 words of running text) and every word is tagged with its syntactic class. The corpus is divided into 15 subject categories or genres. There are 500 individual samples of running text in the corpus and each one contains approximately 2000 words. The number of samples in each genre differs depending on the amount published in that area at the time the corpus was compiled. The genres and the number of samples in each one are listed in Table 1.

The syntactic tags used to describe words in the corpus are organized in six major categories:

(1) *parts of speech:* nouns, verbs, adjectives, and so on.

(2) *function words:* determiners, prepositions, conjunctions, and so on.

(3) *important individual words: not*, existential *there*, infinitival *to*, forms of *do*, *be*, and *have*.

genre	category	no. samples
A.	Press Reportage	44
B	Press Editorial	27
C	Press Reviews	17
D	Religion	17
E	Skills	36
F	Popular Lore	48
G	Belles Lettres	75
H	Miscellaneous	29
J	Learned	80
K	General Fiction	29
L	Mystery and Detective Fiction	24
M	Science Fiction	6
N	Adventure and Western Fiction	29
P	Romance and Love Story	29
R	Humor	9

Table 1. The genres and the number of samples in each genre of the Brown Corpus.

(4) *punctuation marks of syntactic importance:* ".", "(", ")", "--", ",", ":".

(5) *inflectional morphemes:* noun plurals, and possessives, verb past, present and past participle, and so on.

(6) *foreign or cited words.*

A tag sometimes has a 'U' affixed to it to indicate the word is a negation. Altogether, 84 different tags were used in the experiments described in this paper. Those tags are listed in Appendix 1.

4.2. Hypothesis Generation Algorithm

The operation of the hypothesis generation algorithm was simulated by calculating the feature description for a word from pre-defined features for the letters in the word. All the words in a dictionary with the same feature description were used as the neighborhood for an input word.

The feature description is specialized for lower case characters because the experimentation is restricted to text written in lower case. The feature description includes vertical bars of different heights, dots, and empty spaces. These features were chosen because they can be reliably computed from images of text even if the characters touch one another [6]. The complete listing of this feature set is given below:

1. A significant area at the beginning or end of a word that does not contain a vertical bar (e.g., the space to the right of the vertical bar in a "c" or an "e");
2. A short vertical bar (e.g., the leg of an "r");
3. A long high vertical bar that extends above the main bar of the word (e.g., the ascender portion of a "b");
4. A long low vertical bar that extends below the main body of the word (e.g., the descender in a "p");
5. Dots over short vertical bars (occurs in an "i");
6. Dots over long vertical bars (occurs in a "j").

When the feature description is applied to a word it yields a symbolic representation that would correspond to the sequence of occurrence of the features in an image of the word. Thus, both "me" and "may" have a symbolic representation 22221 and the neighborhood of "me" is {"me", "may"}.

4.3. Experimental Application

The Viterbi algorithm for syntactic analysis was applied to genre A (newspaper reportage) of the Brown Corpus. The first sample in the genre (A01) was used as test data and the remainder as training data (A02-A44). The transition and confusion probabilities were calculated from the training data. Because the words in the test data were assumed to be present in the image, the dictionaries were constructed from all of genre A.

Performance was measured by calculating the average neighborhood size per text word before and after the application of syntax. This statistic is defined as:

$$ANS_t = \frac{1}{N_w} \sum_{i=1}^{N_w} ns_i$$

where N_w is the number of words in the test sample and ns_i is the number of words in the neighborhood for the ith word in the text. The error rate is the percentage of words with neighborhoods that do not contain the correct choice after the application of syntax.

In the example presented in Figure 2, N_w is 4, $ns_1 = 1$, $ns_2 = 8$, $ns_3 = 1$, and $ns_4 = 6$. $ANS_t = \frac{1}{4}$ (1+8+1+6) = 4.0. After the application of syntax, ANS_t was reduced to 1.75. This is an overall reduction of about 44 percent in ANS_t with a zero percent error rate.

The results of applying syntactic constraints with the Viterbi algorithm to the 88 sentences in A01 are shown in Table 2. The reduction in ANS_t and error rates that were incurred with between one and five alternatives from the Viterbi are shown. The average neighborhood size per text word is reduced by about 50% in all cases. Surprisingly, the error rate was very low, less than about 2 percent in every instance.

no. alternatives	ANS_t before	ANS_t after	% reduction	error rate
1	2.344	1.155	51%	2.20%
2	2.344	1.186	49%	1.48%
3	2.344	1.210	48%	1.17%
4	2.344	1.231	47%	0.87%
5	2.344	1.254	46%	0.76%

Table 2. Performance of syntactic constraints.

5. Discussion and Conclusions

An approach for incorporating syntactic constraints in a word recognition algorithm was presented. The recognition algorithm produced a series of choices for the words in a sentence. Syntactic constraints were used to remove some of these choices and thereby improve recognition performance.

Syntax was modelled as a first-order Markov source and the Viterbi algorithm was used to find the sequence of syntactic classes that best fit the choices provided by the recognition algorithm. A modification of the Viterbi algorithm provided several next-best alternatives for the syntactic class sequence as well as their costs.

An experimental investigation of this approach was performed in which a simulation of the word recognition algorithm was run on a test sample of text. It was shown that the average number of choices that could match a word can be significantly reduced by the proposed algorithm for applying syntactic constraints with a very low cost in error rate.

Future work on this approach will include the use of images to generate neighborhoods. More extensive experimentation with other portions of the corpus will also be performed. This will include consideration of second order transition probabilities as well as an interface to other syntactic information. The effect of loosening the restriction that the true word must appear in the lexicon will also be considered.

References

1. H. S. Baird and K. Thompson, "Reading Chess," *IEEE Transactions on Pattern Analysis and Machine Intelligence 12* (1990), 552-559.

2. A. D. Beale, "Lexicon and grammar in probabilistic tagging of written English," *Proceedings of the 26th Annual Meeting of the Association for Computational Linguistics*, Buffalo, New York, June 7-10, 1988, 211-216.

3. V. Cherkassky, M. Rao, H. Weschler, L. R. Bahl, F. Jelinek and R. L. Mercer, "A maximum likelihood approach to continuous speech recognition," *IEEE Transactions on Pattern Analysis and Machine Intelligence PAMI-5*, 2 (March, 1983), 179-190.

4. G. D. Forney, "The Viterbi algorithm," *Proceedings of the IEEE 61*, 3 (March, 1973), 268-278.

5. J. J. Hull, S. N. Srihari and R. Choudhari, "An integrated algorithm for text recognition: comparison with a cascaded algorithm," *IEEE Transactions on Pattern Analysis and Machine Intelligence PAMI-5*, 4 (July, 1983), 384-395.

6. J. J. Hull, "Hypothesis generation in a computational model for visual word recognition," *IEEE Expert 1*, 3 (Fall, 1986), 63-70.

7. J. J. Hull, "Inter-word constraints in visual word recognition," *Proceedings of the Conference of the Canadian Society for Computational Studies of Intelligence*, Montreal, Canada, May 21-23, 1986, 134-138.

8. J. J. Hull, "Feature selection and language syntax in text recognition," in *From Pixels to Features*, J. C. Simon (editor), North Holland, 1989, 249-260.

9. H. Kucera and W. N. Francis, *Computational analysis of present-day American English*, Brown University Press, Providence, Rhode Island, 1967.

10. R. Kuhn, "Speech recognition and the frequency of recently used words: A modified Markov model for natural language," *Proceedings of the 12th International Conference on Computational Linguistics*, Budapest, Hungary, August 22-27, 1988, 348-350.

Major Components of a Complete Text Reading System

SHUICHI TSUJIMOTO AND HARUO ASADA

Invited Paper

This paper describes the document image processes used in a newly developed text reading system. The system consists of three major components: document analysis, document understanding, and character segmentation/recognition.

The document analysis component extracts lines of text from a page for recognition. This procedure finds document constituents such as photographs, graphics, and text lines. To this end, the geometric structure is obtained as a hierarchy of items on the page for modeling the relationships between characters, lines, columns, and the page.

The document understanding component extracts logical relationships between the document constituents. The geometric structure of a document, obtained in the document analysis phase, can be represented by a tree. On the other hand, the logical structure is represented by another tree. A small number of generic rules are introduced to transform the geometric structure into the logical structure.

The character segmentation/recognition component extracts characters from a text line and recognizes them. Characters which touch each other may have several candidates for their break positions, and any segmented area might possibly fit several alternative characters. Therefore, an efficient resolution of ambiguities at each stage is a crucial issue in practical text reading. The authors' approach to this is based on the heuristics of character composition as well as on recognition results for omni-fonts.

Experiments on more than a hundred documents have proven that the proposed approaches to document analysis and document understanding are robust even for multicolumned and multiarticle documents containing graphics and photographs. Experiments have also shown that the proposed character segmentation/recognition method is robust enough to cope with omni-font characters which frequently touch each other.

Keywords— *Text reader, document image processing, document understanding, character segmentation, character recognition, geometric structure, logical structure, tree transformation, multiarticle documents, multicolumned documents, omni-font characters, touching characters, OCR, high speed.*

Manuscript received March 25, 1991.

The authors are with the Research and Development Center, Toshiba Corporation, 1, Komukai Toshiba-Cho, Saiwai-Ku, Kawasaki 210, Japan.

IEEE Log Number 9202593.

I. Introduction

The range and volume of publications such as newspapers, magazines, and various types of manuals are continuously increasing. At the same time, various computer-aided text processing techniques, such as desktop publishing, text data base management, and machine translation, have become possible. Automation is widely demanded in the keyboard input area, which was conventionally manual, where large amounts of documentation must be converted into a computer-readable form for data entry. A text reader meets this need. A text reader automatically analyzes each page of a document and recognizes characters on the page for input to a computer.

It is important for such a text reader system to have the ability to deal with various kinds of document layouts and omni-font characters. Three major components are essential if this capability is to be realized: document analysis, document understanding [1], and character segmentation/recognition [2]. Document analysis is a component which decomposes a document image into several consistent items which represent coherent components of the document, such as text lines, photographs, and graphics, without any knowledge of the specific format. Document understanding is a component which extracts the logical relationships between the items just described. Character segmentation/recognition is the component which then extracts characters from a text line and recognizes them. The document analysis and document understanding components need to be robust enough to cope with multicolumned and multiarticle documents including graphics and photographs. The character segmentation/recognition component needs to have the ability to read omni-font characters which might touch each other.

Up to now, several techniques for document analysis and document understanding [3]–[14] have been proposed. Wahl *et al.* [3] presented a smearing algorithm which is widely used for document analysis. Toyoda *et al.* [4] extracted Japanese newspaper articles using domain-specific knowledge. Okamoto *et al.* [5] analyzed papers containing mathematical expressions. Baird *et al.* [6] coped with the

Reprinted from *Proc. IEEE*, Vol. 80, No. 7, July 1992, pp. 1133–1149.

extraction of chess games from the Chess Informant. Esposito *et al.* [7] classified document types (patents, scientific papers, etc.). Higashino *et al.* [8] proposed a flexible format understanding method using FDL, i.e., a format definition language. Tsuji *et al.* [9] represented a document structure with a tree. Nakano *et al.* [10] built a business form understanding system incorporating character recognition. Inagaki *et al.* [11] built a special-purpose machine for Japanese document understanding. Masuda *et al.* [12] presented a prototype of a Japanese text reader. Baird [13] proposed a versatile text reader whose layout analysis component was designed to be language-independent. Srihari *et al.* [14] surveyed the current status of document analysis. At the same time, many studies have been proposed for character segmentation/recognition, especially for the segmentation of characters which touch each other. Kahan *et al.* [15] discussed the segmentation of touching characters. Casey *et al.* [16] built a recursive segmentation and classification method for touching characters. Kooi *et al.* [17] analyzed the contour of an image of touching characters.

This paper presents new approaches to document analysis, document understanding, and character segmentation/recognition, which are essential components for a complete text reader system.

One of the essential issues for a text reader system is the ability to handle multiarticle documents containing figures and photographs. Generally, a document has a visual hierarchical structure in its layout. This is called geometric structure here, the hierarchy of which can be represented by a tree. Document analysis extracts this structure as a model for relationships between characters, lines, columns, and a page.

A description for configuration of articles and their components, called logical structure here, orders the reading sequence. The authors have defined document understanding as a transformation from a geometric structure to a logical structure. A small number of rules for this transformation have been proposed, based on the general assumption that a layout is designed according to human reading manner.

For practical systems, a text reader system must read a document accurately at high speed. To meet this requirement, character segmentation for touching characters is one of the critical problems. Touching characters have several candidates for their break positions, which are then confirmed by recursive segmentation and recognition[16], and finally by the linguistic context. Since any area segmented by the break positions might fit several alternative characters, this approach requires a very time consuming process to select real break positions. Therefore, a pruning, i.e., a mechanism for reducing processing time, is important in practical text reading systems. The authors' approach is based on the heuristics of character composition, e.g., an "m" is like a combination of an "r" and an "n," as well as on recognition results for omni-fonts.

Section II describes the hierarchical structure of a document in a logical and geometric way. Section III gives the design for robust document analysis for multicolumned documents. In Section IV, an algorithm for document understanding, namely the transformation rules, is introduced. Section V presents a character segmentation/recognition method for touching omni-font characters. Experimental results on a variety of documents are shown in Section VI.

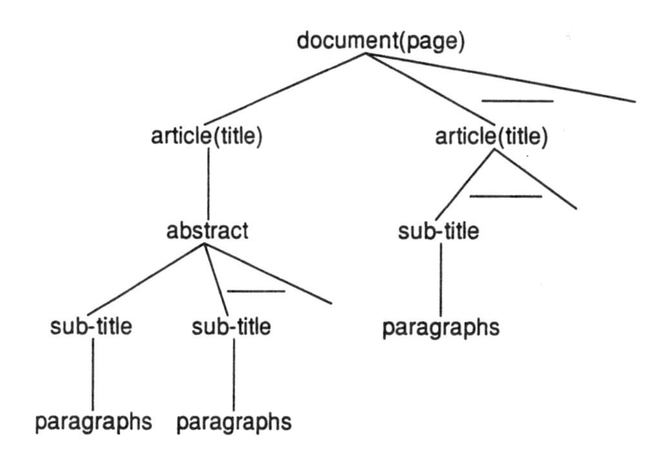

Fig. 1. Logical structure tree. Logical relationships between the items in a document can be represented by a tree.

II. Hierarchical Structure of a Document

This section describes the hierarchical structure of a document in a logical and geometric way.

A. Logical Structure

A document is normally composed of several articles, each of which consists of a title, an abstract, subtitles, and paragraphs. They are connected to each other logically in a hierarchical structure. For example, the title dominates the abstract, chapters, and sections, while subtitles dominate paragraphs. Thus, a document has a logical hierarchy. This logical structure is represented by a tree in this paper, as shown in Fig. 1.

B. Geometric Structure

A document image is composed of several blocks, each of which represents a coherent component of the document. One coherent component corresponds to a set of text lines with the same typeface and a consistent line spacing. The geometric structure, which means the geometric relationships between blocks, is also described by a tree here. Figure 2(b) shows the geometric structure tree generated from the document shown in Fig. 2(a).

In this figure, the root node represents a document page and has the value of *NULL*. Each node in the tree, except the root node, represents a set of adjacent blocks located in the same column, and has a list of blocks as a value. The list is ordered from upper to lower blocks. For example, blocks **1H**, **2B**, and **3H** in Fig. 2(a) are combined and represented as a whole by a node. The parent–daughter relationship is introduced when there is more than one block directly below another block. For example, block **4H** is the parent of blocks **1H** and **5H**. On the other hand, block **1H** is not

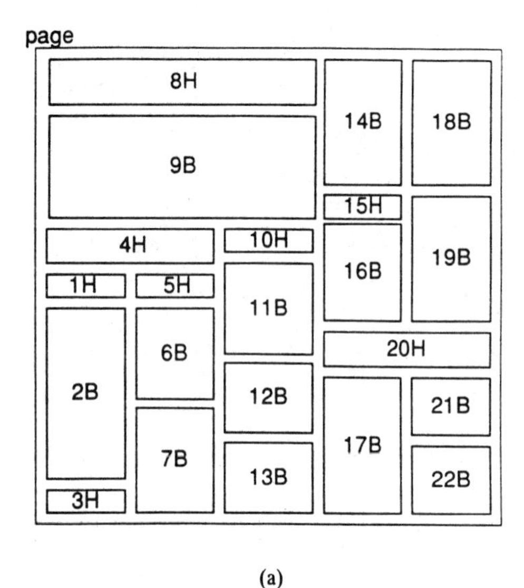

(a)

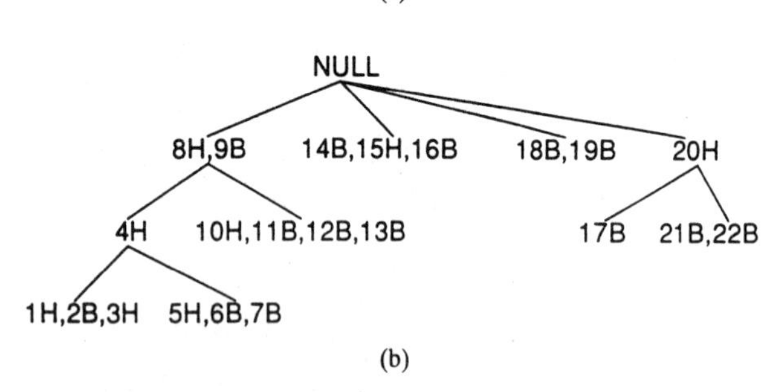

(b)

Fig. 2. Geometric structure tree. (a) Document divided into blocks. Each block represents a coherent component of a document. In this illustration, **H** indicates a *head* block, while **B** indicates a *body* block. (b) Geometric structure corresponding to (a). Geometric relationships between blocks can be represented by a tree.

the parent of block **2B** because blocks **1H** and **2B** are in the same column. Those blocks not dominated by others are directly connected to the root node as daughters. Block **20H** in Fig. 2(a) is an example of such a block because it is not dominated by either **16B** or **19B** according to the above discussion. Daughters of the root node are ordered in the sequence from left to right and top to bottom according to their location in the document. This sequence is very important in the transformation process described in Section IV.

Each block is classified as one of two categories: *head* and *body*, in order to distinguish titles from texts. This classification is carried out by examining the physical properties of a block. *Head* is the name for blocks in which there are only a few text lines; text is biased to the left or is centered. Larger type fonts are used in *head* blocks in many cases. This kind of block corresponds to titles or subtitles when a geometric structure is transformed into a logical structure. Headers, footers, page numbers, and captions also belong in this category. *Body* corresponds to blocks consisting of text lines only. It normally has a considerable number of text lines and smaller type fonts. An indentation is often found in the first text line of a *body* block. Abstracts as well as paragraphs belong to this category. In Fig. 2, **H** and **B** indicate *head* and *body* blocks, respectively.

III. Document Analysis

This section describes a document analysis which breaks down a document image into several blocks and constructs a geometric structure tree whose nodes represent a set of blocks.

The authors' approach to document analysis is described as bottom-up. This approach first extracts words from a document image, which are then merged into text lines. Text lines are then combined into blocks which usually correspond to paragraphs.

A geometric structure is generated according to the parent-daughter relationships between blocks. These relationships are established by examining the column a block belongs to, and its vertical position.

A. Run-Length Image Representation

A run-length image representation is used in the proposed system. The run-length representation is more efficient than a bit-map representation, especially in image processing by software. Accessing pixels in a bit-map image is usually very time consuming because a general-purpose computer is not designed to deal with bit-map images stored in a byte representation. Therefore, all the image processes in our system are designed to directly access run-length coded images instead of accessing bit-map images. Using the run-length representation for document image processing yields high-speed processing.

B. Text Line Extraction

The text line extraction procedure is described in this subsection. Here, the authors define a segment as a rectangular area which circumscribes a text line or a part of it.

The text line extraction process is divided into four subprocesses. The first is to extract adjacent connected components as a segment. The second classifies the segments into text lines, figures, graphics, and so on. The third is a merging process for adjacent segments which are classified into text lines. The last is another merging process for the segments in the same column defined by the column boundaries. Words are usually extracted by both the first and second subprocesses, and text lines are obtained in the third subprocess. The fourth process is added to cope with cases where a long blank between words prevents the words from being merged into a text line.

These subprocesses are detailed as follows.

1) Segment Extraction: The first subprocess in the extraction of a text line, extracts adjacent connected components as a segment by connecting two black runs whose spacing (white run) is shorter than a certain threshold (smearing algorithm, see [3]). The threshold is made very small (say 1 mm) so that words located in different columns are not merged with each other. Figure 3(a) shows a sample document image which is an example of a multicolumned, multiarticle document. The result of segment extraction for this sample image is shown in Fig. 3(b).

2) Classification of Segments: Segments are classified into text lines, figures (photographs), tables, frames, horizontal

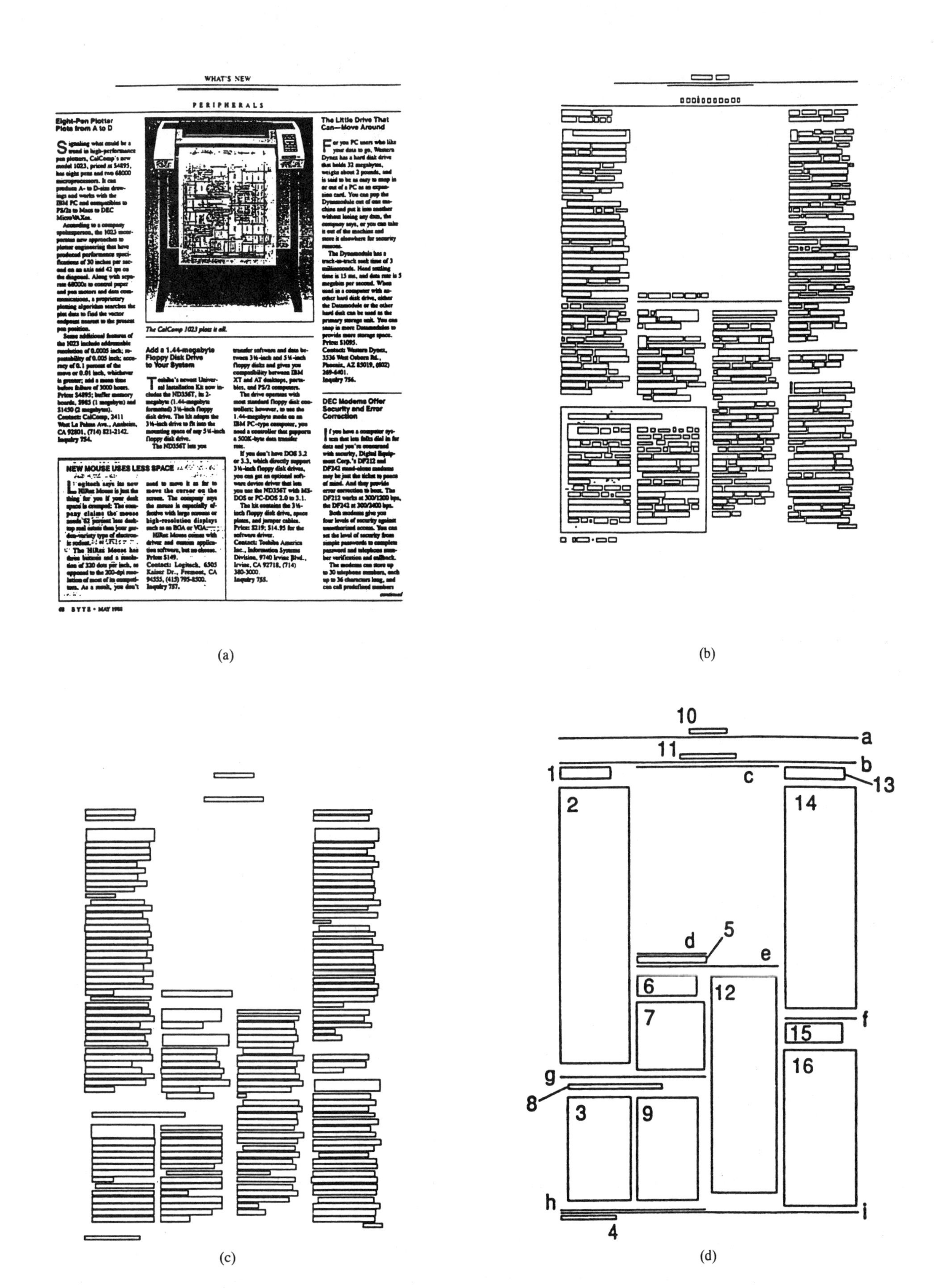

WHAT'S NEW

PERIPHERALS

Eight-Pen Plotter Plots from A to D

Signaling what could be a trend in high-performance pen plotters, CalComp's new model 1023, priced at $4895, has eight pens and two 68000 microprocessors. It can produce A- to D-size drawings and works with the IBM PC and compatibles to PS/2s to Macs to DEC MicroVAXes.

According to a company spokesperson, the 1023 incorporates new approaches to plotter engineering that have produced performance specifications of 30 inches per second on an axis and 42 ips on the diagonal. Along with separate 68000s to control paper and pen motors and data communications, a proprietary plotting algorithm searches the plot data to find the vector endpoint nearest to the present pen position.

Some additional features of the 1023 include addressable resolution of 0.0005 inch; repeatability of 0.005 inch; accuracy of 0.1 percent of the move or 0.01 inch, whichever is greater; and a mean time before failure of 3000 hours.
Price: $4895; buffer memory boards, $985 (1 megabyte) and $1450 (2 megabytes).
Contact: CalComp, 2411 West La Palma Ave., Anaheim, CA 92801, (714) 821-2142.
Inquiry 754.

The CalComp 1023 plots it all.

Add a 1.44-megabyte Floppy Disk Drive to Your System

Toshiba's newest Universal Installation Kit now includes the ND356T, its 2-megabyte (1.44-megabyte formatted) 3½-inch floppy disk drive. The kit adapts the 3½-inch drive to fit into the mounting space of any 5¼-inch floppy disk drive.

The ND356T lets you transfer software and data between 3½-inch and 5¼-inch floppy disks and gives you compatibility between IBM XT and AT desktops, portables, and PS/2 computers.

The drive operates with most standard floppy disk controllers; however, to use the 1.44-megabyte mode on an IBM PC-type computer, you need a controller that supports a 500K-byte data transfer rate.

If you don't have DOS 3.2 or 3.3, which directly support 3½-inch floppy disk drives, you can get an optional software device driver that lets you use the ND356T with MS-DOS or PC-DOS 2.0 to 3.1.

The kit contains the 3½-inch floppy disk drive, space plates, and jumper cables.
Price: $219; $14.95 for the software driver.
Contact: Toshiba America Inc., Information Systems Division, 9740 Irvine Blvd., Irvine, CA 92718, (714) 380-3000.
Inquiry 755.

The Little Drive That Can—Move Around

For you PC users who like your data to go, Western Dynex has a hard disk drive that holds 32 megabytes, weighs about 2 pounds, and is said to be as easy to snap in or out of a PC as an expansion card. You can pop the Dynamodule out of one machine and put it into another without losing any data, the company says, or you can take it out of the machine and store it elsewhere for security reasons.

The Dynamodule has a track-to-track seek time of 3 milliseconds. Head settling time is 15 ms, and data rate is 5 megabits per second. When used in a computer with another hard disk drive, either the Dynamodule or the other hard disk can be used as the primary storage unit. You can snap in more Dynamodules to provide more storage space.
Price: $1095.
Contact: Western Dynex, 3536 West Osborn Rd., Phoenix, AZ 85019, (602) 269-6401.
Inquiry 756.

DEC Modems Offer Security and Error Correction

If you have a computer system that lets folks dial in for data and you're concerned with security, Digital Equipment Corp.'s DF212 and DF242 stand-alone modems may be just the ticket to peace of mind. And they provide error correction to boot. The DF212 works at 300/1200 bps, the DF242 at 300/2400 bps.

Both modems give you four levels of security against unauthorized access. You can set the level of security from simple passwords to complete password and telephone number verification and callback.

The modems can store up to 30 telephone numbers, each up to 36 characters long, and can call predefined numbers

continued

NEW MOUSE USES LESS SPACE

Logitech says its new HiRez Mouse is just the thing for you if your desk space is cramped: The company claims the mouse needs 62 percent less desktop real estate than your garden-variety type of electronic rodent.

The HiRez Mouse has three buttons and a resolution of 320 dots per inch, as opposed to the 200-dpi resolution of most of its competitors. As a result, you don't need to move it as far to move the cursor on the screen. The company says the mouse is especially effective with large screens or high-resolution displays such as an EGA or VGA.

HiRez Mouse comes with driver and custom application software, but no cheese.
Price: $149.
Contact: Logitech, 6505 Kaiser Dr., Fremont, CA 94555, (415) 795-8500.
Inquiry 757.

48 BYTE • MAY 1988

Fig. 3. Block extraction. A domument image is analyzed using a bottom-up approach. (a) Original document image. A document image is input by a scanner at a resolution of 300 dpi. (b) Segment extraction results. Segments represent the rectangular area circumscribing a text line or a word. (c) Text line extraction result. Neighboring segments are merged into a text line after column definition. (d) Block extraction result. Vertically adjacent text lines are combined into a block.

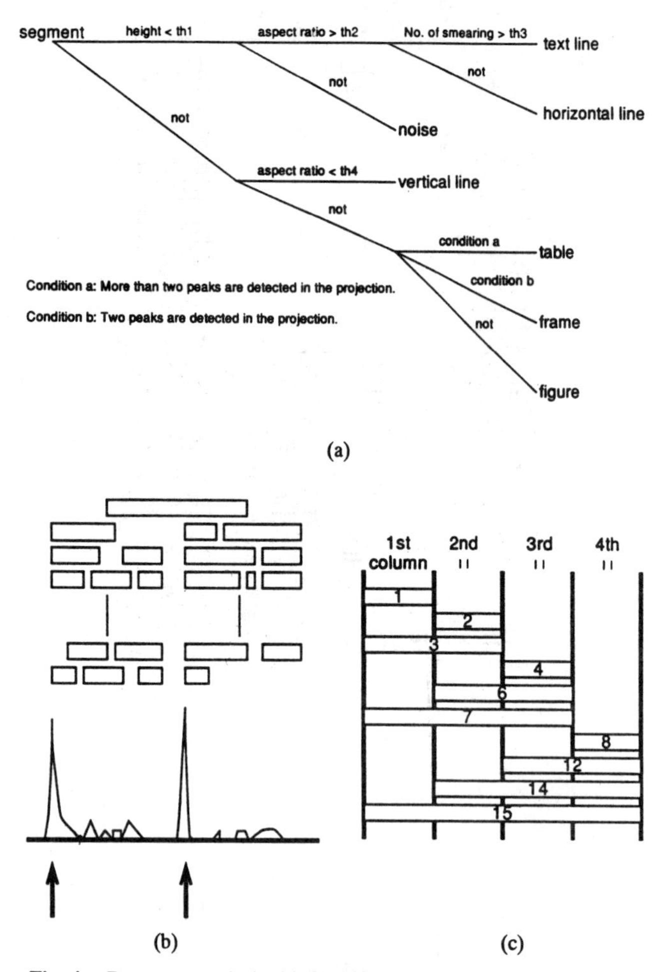

Fig. 4. Document analysis. (a) Classification of a segment. Each segment is temporarily classified by means of the physical properties of the segment. Final classification will be done later. (b) How to determine the column position. The left edges of segments contribute to the location of columns. (c) *Group number*, depending on column number. *Group number* will be introduced to describe how each segment is related to the column positions.

lines, vertical lines, and noise according to the physical properties of the segments. These physical properties of a segment are the size (width and height), aspect ratio (width/height), horizontal and vertical projection, and the number of smeared white runs [3]. Of these properties, the last one is a metric which represents the complexity of the segment.

Figure 4(a) roughly sketches the method of classifying segments, where **th1**, **th2**, **th3**, and **th4** are the given thresholds. Thresholds **th1**, **th2**, and **th4** are fixed, while threshold **th3** is defined by the segment size.

The number of smeared white runs is a metric of the horizontal white–black transition. If this metric is smaller than **th3** for a long horizontal segment, then the segment is classified as a *skewed* horizontal line; otherwise it is classified as a text line.

Frames are classified further into *text frames* and *figure frames* by examining their contents. If there are lots of text lines inside a frame, this frame is defined as a *text frame* representing a box. When figures are found inside the frame, it is a *figure frame.*

This classification is temporary, and a final classification is fixed after the merging processes. For example, noise segments are temporarily considered as text lines because they might be isolated periods, commas, or dots on the letters "i" and "j."

These classified segments play differing roles in the document analysis and understanding process. A horizontal line is used as a field separator in document understanding, and a vertical line is useful for defining column settings. A text line located below a figure (photograph), a figure frame, or a table is possibly a caption. A text frame emphasizes the independence of the text lines located within it.

3) Merging Process: The blank length between words is usually proportional to the height of the words on a document. The third subprocess makes use of this idea, merging neighboring segments which are defined as text lines if the blank space between them is smaller than a threshold which is made proportional to the words' height.

4) Text Line Extraction: In the last subprocess, columns are extracted by detecting their left edges as follows. The left coordinates of columns are defined by local maxima in a vertical projection profile, which is made by vertically adding up the left edges of those segments classified as text lines (see Fig. 4(b)). Naturally, the right edges of segments may also contribute to defining column positions, if text lines are adjusted properly at the right.

The skew of the page is corrected before the above vertical projection is obtained for the entire page.

After defining the columns, neighboring segments in the same column are merged into one text line. The result of text line extraction from Fig. 3(a) is shown in Fig. 3(c).

C. Block Extraction

In order to describe how each segment is related to the column positions, a *group number* is introduced for each segment, as shown in Fig. 4(c).

The *group number* (*g.n.*) is given by the following equation:

$$g.n. = \sum_{i=1}^{N} cp[i]^{*}2^{(i-1)}, \qquad (1)$$

where

$cp[i] = 1$ (if the segment is in the ith column),
$cp[i] = 0$ (otherwise),
N is the number of columns.

Vertically adjacent text lines with the same group number are combined into a block. Here, vertical adjacency is determined by a threshold defined by the line interval.

A *Head* or *body* label is attached to each block according to the conditions described in subsection II-B. Numerals in Fig. 3(d) indicate the blocks generated from Fig. 3(a).

D. Generating a Geometric Structure Tree

Lastly, a geometric structure tree is generated, as described in subsection II-B. Each node of the tree represents a set of adjacent blocks with the same *group number*. Parent-daughter relationships between nodes are established by examining *group numbers* and their vertical locations.

Nodes not dominated by others are directly connected to the root node. These nodes are ordered by examining their *group numbers* and vertical locations so that a block to the left and on the top precedes the others.

Generation of a geometric structure for virtual field separators, as introduced in subsection IV-C, will be described later.

IV. Document Understanding

This section defines document understanding as a transformation from a geometric structure to a logical structure. A small number of the transformation rules are introduced here. Virtual field separator techniques are also proposed for universal transformations.

A. What Is Document Understanding?

It is true that the reading order, i.e., the sequence in which a document is read, is not completely fixed until the document's contents have been examined. However, it is also generally true that the actual reading order intended by authors often influences the layout, because the author wants the reading order to be defined easily before reading starts. For example, one can easily pick out the titles in a document since they are usually emphasized in one of several ways.

The reading order is inferred by means of the titles and other region separators. This observation leads to the following approach for document understanding. The reading order is derived from the logical structure. The authors define the transformation from a geometric structure to a logical structure as document understanding. On the other hand, the reverse transformation is not unique because a logical structure could correspond to a variety of geometric structures.

B. Basic Algorithm for Tree Transformation

This subsection presents an algorithm for the geometric to logical structure transformation. The algorithm is composed of four transformation rules that define the conditions under which an element in a node list is moved. These rules are illustrated in parts (a) through (d) of Fig. 5, where **H** indicates a *head* block, **B** indicates a *body* block, and **S** indicates that a block can be either *body* or *head*. In the tree, each node is sequentially numbered in the depth-first order. This is called depth-first indexing.

Four transformation rules are described below. Through this transformation processes, a node which becomes *NULL* is deleted.

Rule (a):

If

a node (say **A**) is a terminal node, and
the first element of node **A** is a *body*, and
the preceding node (say **B**) in the *depth-first indexing* is a terminal node,

then

remove the first element from node **A**, and
append it to the last element of node **B**.

Figure 5(a) illustrates the transformation process of this rule. The rule is based on the observation that a title has a single set of paragraphs as a daughter in the logical structure. Therefore, if the parent of a terminal node containing *bodies* has several daughters, then only one of them can be the true daughter of the parent. It is reasonable that the eldest daughter represents the text dominated by the parent and that the others should be merged to her.

Rule (b):

If

a node (say **A**) is a terminal node that is not connected to the root node, and
the preceding node (say **B**) in the *depth-first indexing* is a terminal node, and
the first element of node **A** is not *NULL*, and
last element of node **B** is a *head*,

then

remove the first element from node **A**, and
append it to the last element of node **B**.

Figure 5(b) illustrates this rule. Here, element **a1** is removed and appended to node **B**. This transformation is the same as that for rule (a). The difference is that the first element of node **A** does not need to be a *body* if the last element of node **B** is a *head*.

Rule (c):

If

a node (say **A**) contains a *head* block, and
it is not the first element of the node,

then

generate a younger sister node (say **D**), and
remove the *head-body* sequence that begins with that *head* block and ends with the last element of node **A**, with daughters of node **A**, if any, and
attach them to the younger sister node **D**.

Figure 5(c) illustrates the conversion process of this rule. When a node includes more than one *head-body* sequence, a new sister node is generated for each *head-body* sequence by applying this rule recursively. This rule is mainly for extracting chapters and sections headed by a subtitle.

Rule (d):

If

there is a *head* block sequence in a node, and
it is the first part of the node,

then

generate a daughter node, and
move the *body* sequence that follows the *head* sequence to the daughter node.

Figure 5(d) shows a case in which node **A** has a single *head* sequence and a single *body* sequence. In this case, the *body* sequence is separated from node **A** and moved to a new node **C**, which is a newly generated daughter of node **A**. By this rule, each node comes to have either *head* or *body* sequence. This rule is applied after rules (a), (b), and (c) have been completed.

The next step is an interpretation process, where a label is attached to each node. Here, the labels include *title*,

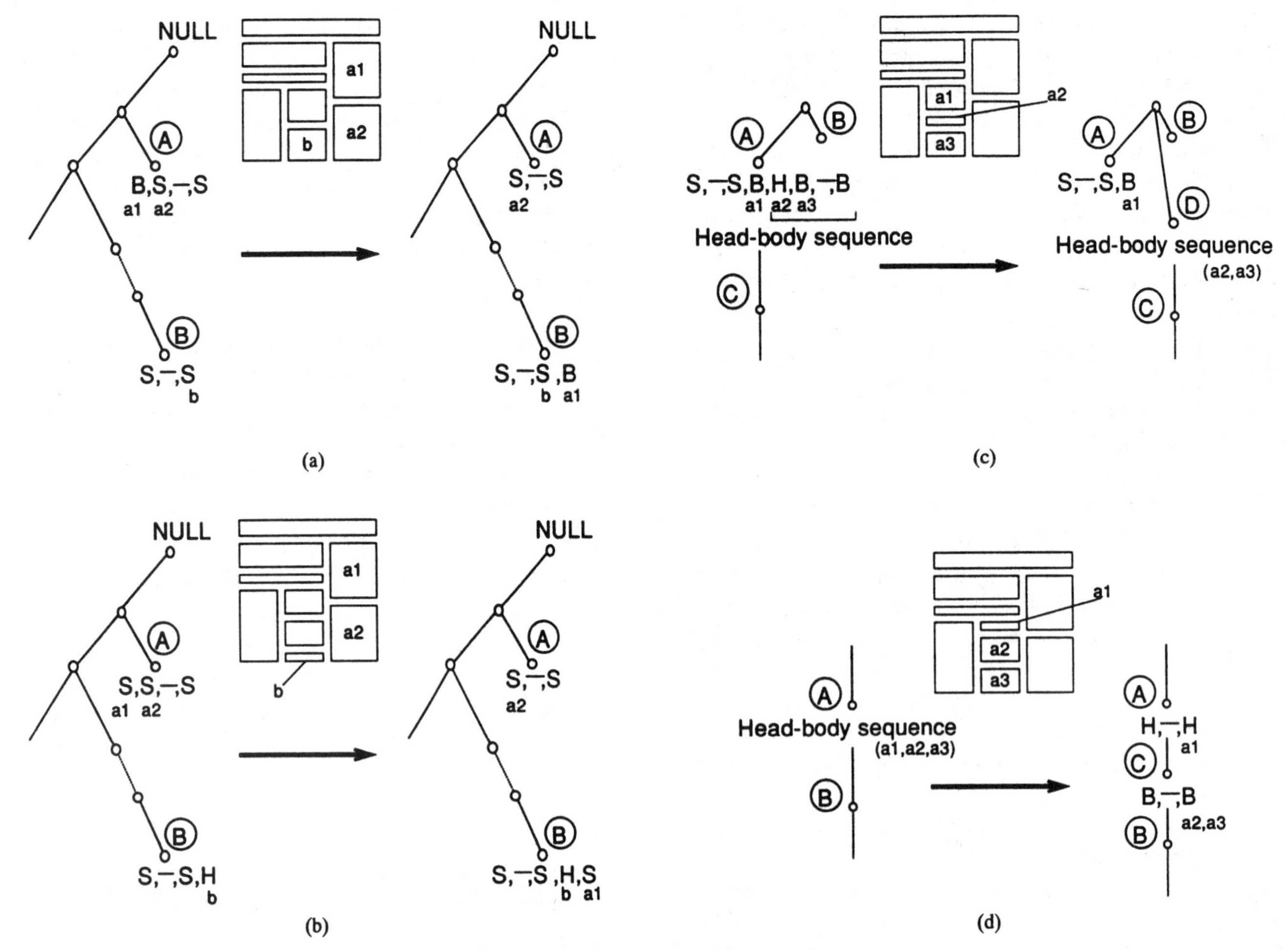

Fig. 5. Transforming a geometric structure into a logical structure using four rules. Transformation rules define the conditions for the movement of an element in a node list of a tree. In this illustration, **H** indicates a *head* block, **B** indicates a *body* block, and **S** indicates either. (a) Rule (a): This rule is based on the observation that each title has a single set of paragraphs. (b) Rule (b)—similar to rule (a). (c) Rule (c)—extracts chapters of sections for a subtitle. (d) Rule (d)—attaches a unique class (*head/body*) to each node.

abstract, subtitle, paragraph, header, footer, page number, and *caption.* If a daughter of the root node has children and she is a *head* sequence, she represents a *title.* If she has no children and she is a *head* sequence, one of the labels *header, footer, page number,* or *caption* is attached to her according to her location on the page. For example, a block which is centered and located at the bottom of a page is a *page number*. Any *head* blocks other than daughters of the root node are *subtitles. Body* blocks in terminal nodes are normally *paragraphs.* A *body* block which is the eldest and whose next sister is a *subtitle* represents an *abstract.* A *body* block with daughters also represents an *abstract.*

Figure 6 shows an example of the transformation process which generates a logical structure from the geometric structure of the document shown in Fig. 2.

C. Virtual Field Separators for a Universal Transformation

Field separators and rectangular text frames are good identification tokens for understanding document images. A field separator signals a break in the text lines, and explicitly distinguishes text lines located below it from those above it. A frame signals the independence of text lines within it.

The authors employ a virtual field separator technique which avoids the additional rules of special transformations for the field separators and frames to effectively introduce the information carried by them. This also helps the realization of a universal transformation.

In the virtual field separator technique, a *head* block with a *NULL* list is substituted for a field separator (see Fig. 7(a)). In Fig. 7, the numerals indicate that the block is a real one, while the letters indicate that the block is a virtual one; namely, it is a field separator. A field separator with a *head* label behaves as if it were a title block and it emphasizes a break in the text lines.

The upper and lower lines of a text frame are treated as virtual field separators. Nodes for these lines are connected to the root node so that text lines in the frame are independent of the others, as shown in Fig. 7(b). It should be noted that a node for the lower line does not have any blocks other than itself. A block located below the lower line and a block located above the upper line are combined in the same node. For example, **5B** and **2B** in Fig. 7(b) are combined in the same node.

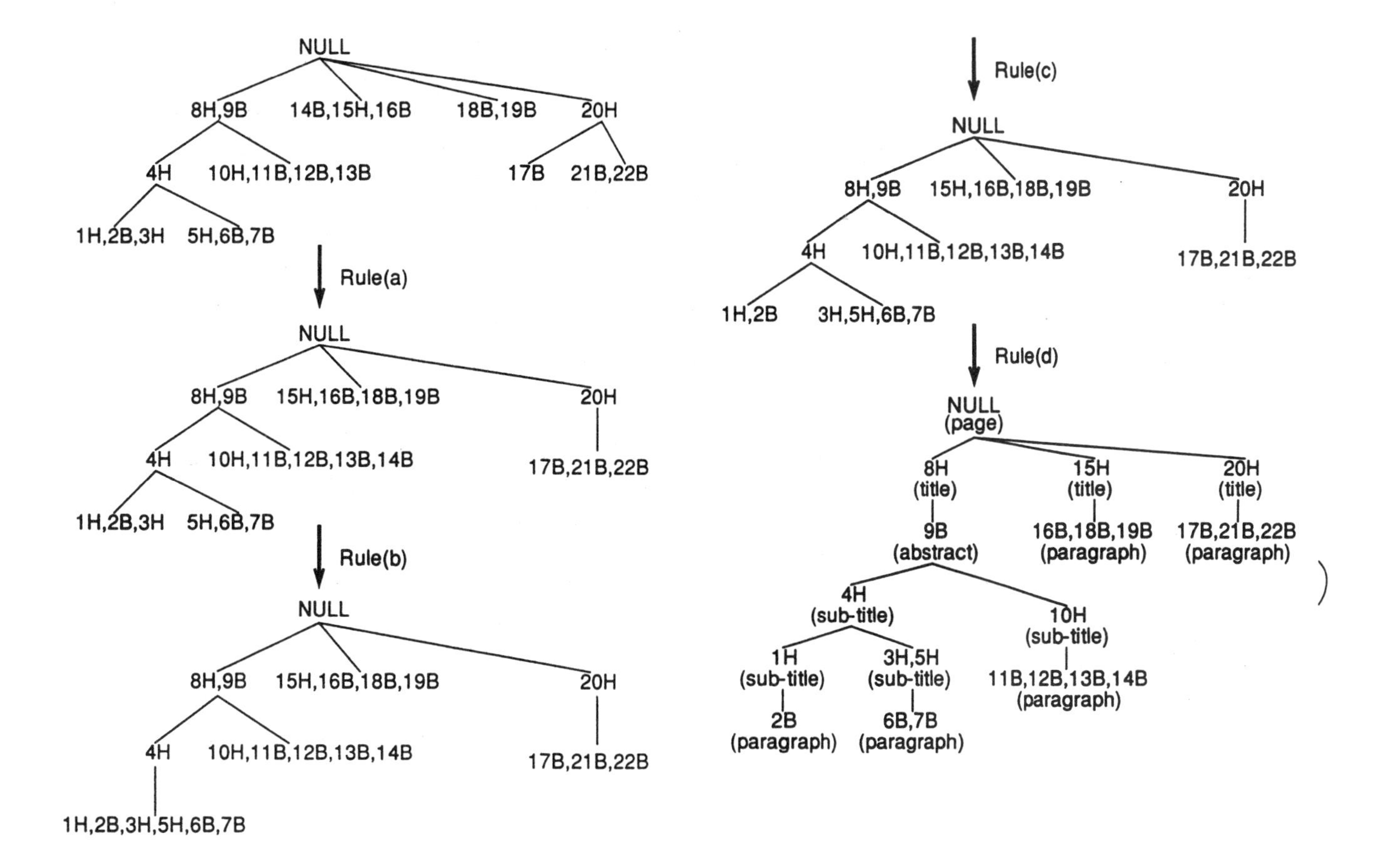

Fig. 6. Example of transformation process. The geometric structure tree of the document shown in Fig. 2 is transformed into a logical structure tree using four rules.

The virtual field separator technique is useful in treating photographs and figures, as well as their captions. In this case, a virtual frame circumscribing the figure or photograph and its caption is generated.

The virtual field separator is also defined in this virtual frame in order to distinguish text in the figure from the caption. Figure 7(c) shows the use of virtual field separators in this case. In this figure, blocks **a** and **c** are the upper and lower lines of the generated virtual frame, respectively. Block **b** is a virtual field separator for the caption.

The virtual field separator for footers is generated above them (see Fig. 7(d)), while the virtual field separator for headers is generated below them.

Virtual field separators are also used for the understanding of what we call a manual format, where title blocks are located to the left of text blocks, as shown in Fig. 7(e). In this format, each title and text pair is treated as being relevant. A virtual field separator is generated above this pair if two horizontally adjacent blocks are located in the same vertical position. Using such separators, the titles and corresponding texts are connected to each other in the logical structure tree.

In the last stage, the redundant field separators, if any, are deleted. A redundant field separator often appears when a real one happens to exist where a virtual one is generated.

This introduction of the virtual field separators does not require an increase in the number of transformation rules.

D. Geometric Structure of Virtual Field Separators

This subsection describes how a geometric structure of virtual field separators is generated. First, blocks representing headers, footers, and captions are detected. A *head* block which is located at the bottom of a page is regarded as a footer, and a *head* block at the top of a page is assumed to be a header. The existence of long horizontal lines will help the footer and header to be detected. A *head* block located below a figure is treated as a caption. Virtual field separators are generated in the way described in subsection IV-C.

At this stage, it is determined whether the page is laid out in the *manual format.* The condition for *manual format* is that a page consist of two columns and that each *head* block in the left-hand column correspond to a *body* block in the right-hand column in terms of vertical position. Virtual field separators are also generated for this *manual format.*

Virtual blocks are substituted for real and virtual field separators. Letters in Fig. 3(d) indicate several virtual blocks. Block **f** is a real field separator obtained from the original image, blocks **g** and **h** are virtual ones derived from a text frame, and blocks **c**, **d**, and **e** are also virtual ones generated for a figure and its caption. Blocks **a** and **b** are virtual ones prepared for headers, and block **i** is for a footer. Real field separators, which are located below

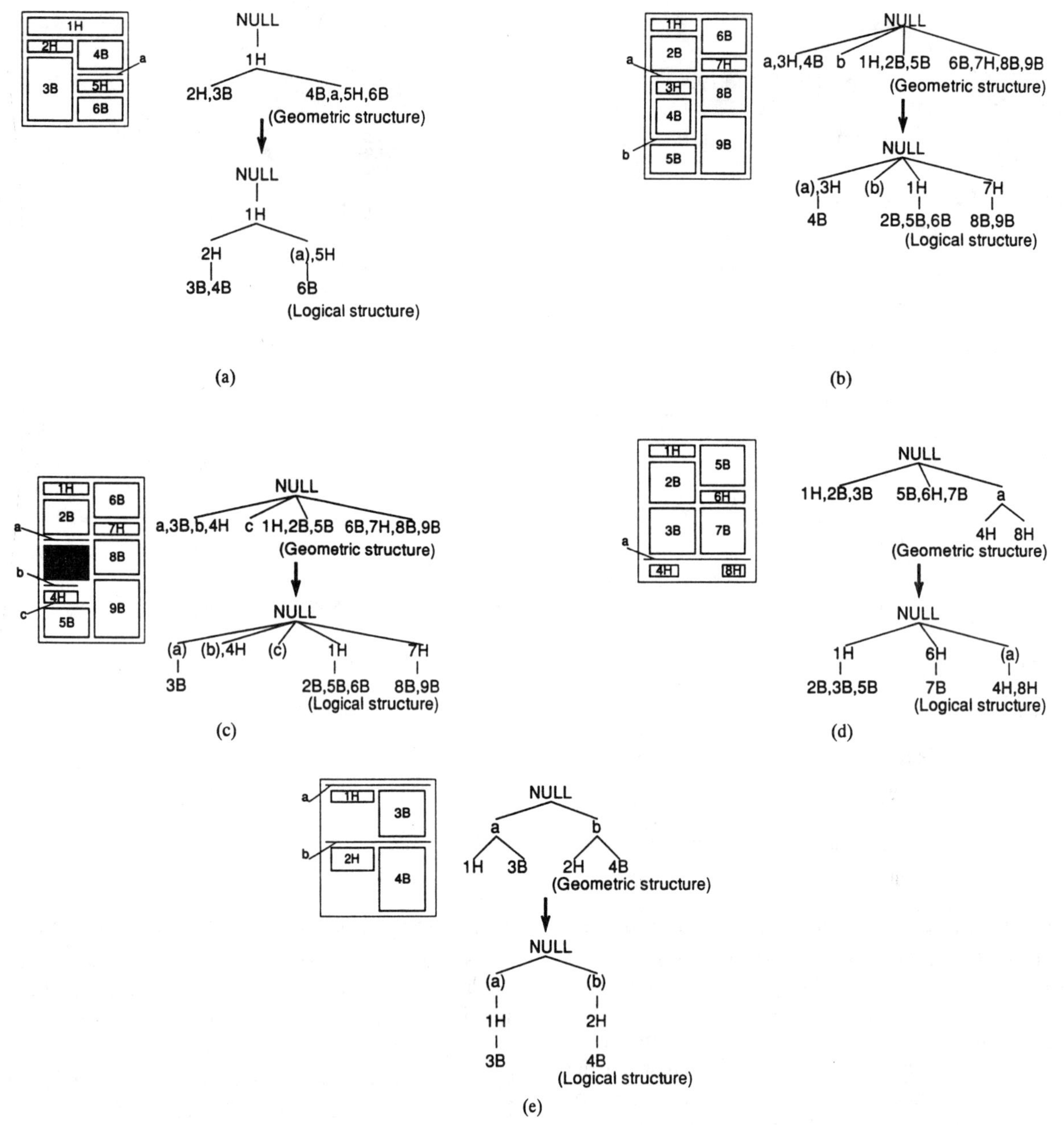

Fig. 7. Virtual field separator techniques for a universal transformation. The virtual field separator technique avoids the need for additional rules of special transformations for field separators and figures. This helps the realization of a universal transformation. Numerals in this illustration indicate real blocks, while letters indicate virtual blocks. (a) Field separator treated as a virtual block. A field separator signals a break in text lines. (b) Frame treated as two virtual field separators. A frame signals the independence of the text lines within it. (c) Virtual field separators generated for a figure and the caption. Text lines below a figure may be its caption. (d) Virtual field separator for footer. (e) Virtual field separators for *manual format*. Title blocks are located to the left of text blocks in a document laid out in a *manual format*.

blocks **10**, **11**, and **5** in the original image, are deleted because virtual ones have been generated where the real ones already exist.

Nodes for frames are connected to the root node so that they do not disturb the other node sequences. In fact, nodes for text frames are connected as eldest daughters of the root node, while nodes for figure frames are connected as youngest daughters.

Figure 8(a) shows the geometric structure tree constructed from Fig. 3(d).

E. Example of Tree Transformation

Figure 8(b) is the logical structure obtained from the geometric structure shown in Fig. 8(a) through the transformation process. Five articles are found in Fig. 8(b). The first article is in a text frame; **8** indicates the *title*, and **3** and **9** indicate *paragraphs*. Other articles are **1** (*title*), **2** (*paragraph*) and **6** (*title*), **7**, **12** (*paragraphs*) and **13** (*title*), **14** (*paragraph*) and **15** (*title*), **16** (*paragraph*). Two *headers* (**10**, **11**), a *footer* (**4**), and a *caption* (**5**) are also found.

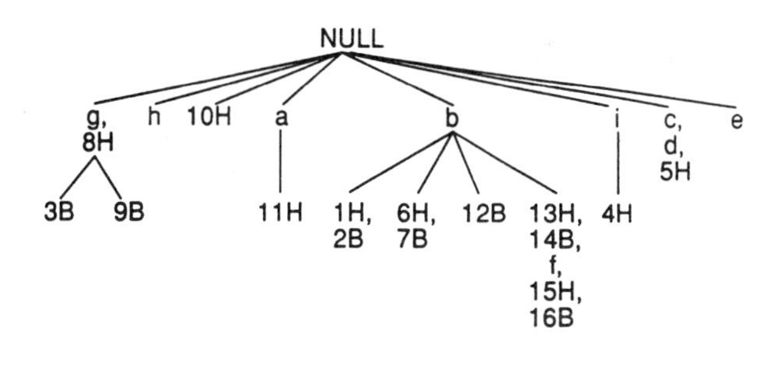

(a)

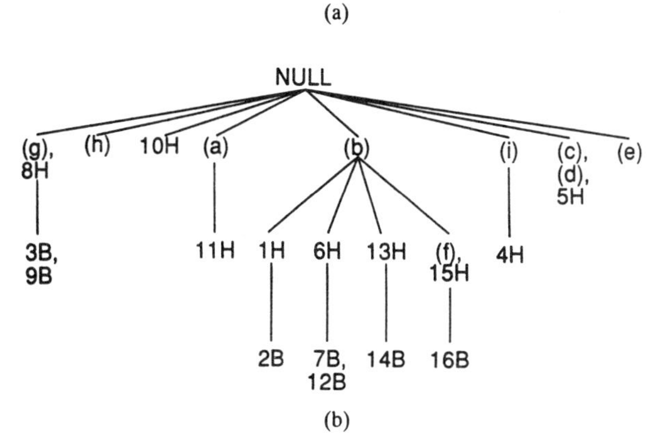

(b)

Fig. 8. Geometric and logical structure trees for the document in Fig. 3(a). Numerals in this illustration indicate real blocks, while letters indicate virtual blocks. (a) Geometric structure. (b) Logical structure. Logical structure tree is obtained from a geometric structure tree through the transformation process.

V. Character Segmentation/Recognition

Text lines are obtained in the document analysis and document understanding phase. The character segmentation phase extracts characters from these text lines. This procedure consists in extracting and recognizing characters.

The existence of touching characters makes it more difficult to design an effective character segmentation procedure because separating touching characters involves a number of ambiguities. Here, reduction of the number of such ambiguities is a crucial problem. This section presents a character segmentation/recognition method which is designed to be robust against touching characters.

A. Hierarchical Structure of a Text Line

A text line has a hierarchical structure; the text line consists of words, which in turn consist of characters.

Figure 9 shows an overview of the character segmentation hierarchy in the proposed system. First, a connected area in a text line image is defined as a component. Next, components above and below one another are combined. For example, in Fig. 9, the "i" is formed by combining two components. Components which are too small to be characters are regarded as noise, and are removed. Words are detected by examining the spaces between components.

B. Discriminating Touching Characters from a Single Character

Each component might be a single character or a pair of touching characters. Previous work has employed the concept of an aspect ratio to distinguish a component representing touching characters from a component representing a single character. However, the aspect ratio alone is not sufficient to separate proportional touching characters. For example, in Fig. 9, the single character "A" has a greater aspect ratio than "lt," which comprises two characters.

In the authors' approach, a component representing touching characters is found from the results of character recognition. First, a component is recognized as a single character, and then any component whose similarity obtained in character recognition is less than a fixed level (described in subsection V-F-4) in detail) is assumed to be a touching pair of characters. Character segmentation for touching characters will now be described.

C. The Authors' Approach to Touching Character Segmentation

One approach often adopted in the pattern recognition field is to divide the process into a number of phases and execute them sequentially by proposing several candidates. Ambiguities remain unresolved, however, in a phase where only one candidate is unsuccessfully sought as a solution.

There are generally two different approaches to solving this problem. One is an approach in which the ambiguities in each phase remain until the final phase. The other is an approach in which ambiguities in each phase are positively resolved in that phase. The former is used for problems where the final phase needs more emphasis than the other phases. The latter is suitable in cases where the phases are individually and sequentially managed. The authors' approach for touching character segmentation belongs to the former class rather than the latter.

D. Ambiguities in Touching Character Segmentation

The procedure for touching character segmentation mainly consists of three phases. First, candidates for break positions of touching characters are nominated by analyzing the touching character image. Second, the candidates are reduced by adopting recursive segmentation and recognition [16]. Last, a linguistic context confirms the break positions.

This procedure leads to several ambiguities. For example, each component image has several candidates nominated as break positions, and each segmented area may fit several alternative characters. (In Fig. 9, the "l" segment of the word "filter" fits an "l," a "t," an "I," an "i," and so on.). Another ambiguity is that an individual component may fit several possible touching characters. (In Fig. 9, the "lt" portion of the word "filter" fits the combination of an "l" and a "t," a single character "k.")

E. Resolving Ambiguities

Resolving ambiguities at each phase is very important in the segmentation of touching characters for two reasons. One is efficiency, meaning that a system which suppresses the number of possibilities to be checked can then segment touching characters at high speed. The other is power, meaning that a system which reduces ambiguities suffers fewer character recognition errors. For example, a Roman

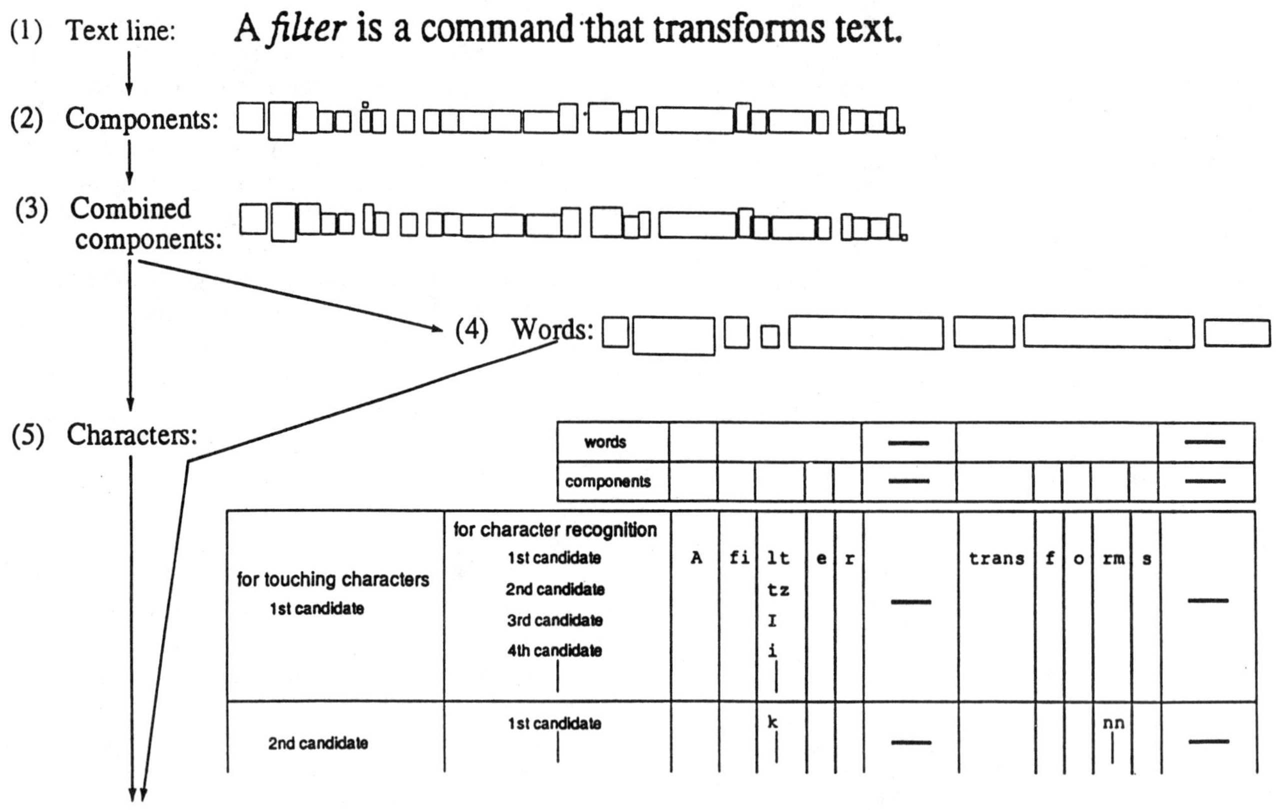

Fig. 9. Character segmentation overview. The character segmentation procedure extracts characters from a text line. Candidates for character segmentation and character recognition are deferred for a linguistic context procedure to confirm the final recognition result.

numeral "III" might be mistakenly changed into the English word "ill" if the candidate "i" remains a candidate for the first letter and the whole word is accepted by linguistic confirmation. As another example, numerals "50" and "200" might be mistakenly changed into the English words "SO" and "ZOO," respectively.

The authors employ heuristics of character composition as well as recognition results for omni-fonts to resolve ambiguities. The heuristics of character composition (e.g., a "W" looks like a combination of two "V"'s) avoids the need for extra recursive segmentation and recognition. Use of recognition results for omni-fonts reduces the number of candidates used in the linguistic confirmation process.

F. Procedure for Touching Character Segmentation

The authors propose a *break cost* as a new metric to nominate break positions in touching character images in the proposed procedure for touching character segmentation:

1) The *break cost* nominates candidates for break positions of touching characters.
2) It is examined whether each candidate is selected as a real break position or not; this is done by searching for an optimal path in a *search tree* where the recursive-segmentation-and-recognition approach [16] is adopted. Heuristics of character composition avoid the necessity for implementing a complete search.
3) Knowledge of the linguistic context is used to confirm the candidates. The number of linguistic context confirmations is reduced by employing recognition results for omni-fonts.

1) New Metric for Segmenting Touching Characters: Previous work has employed a vertical projection profile of the touching character image to find positions for separating touching characters. This profile is a function which maps the number of black pixels in each vertical column to the column's horizontal position. The authors introduce a new metric, the *break cost*, to evaluate the degree of contact.

The *break cost* is defined between each pair of neighboring columns. It is calculated by accumulating the number of black pixels vertically in the image obtained after an AND operation between neighboring columns. Figure 10 shows both the traditional vertical projection and the new projection for an input image. In this illustration, the new metric shows the ability to detect a prominent break between the left-hand and right-hand areas of the input image, while the vertical projection fails to do so.

This *break cost* consumes little processing time when the input image is represented by vertical run length.

If a typeface is *italic*, its *break cost* is calculated after the image is straightened, as shown in Fig. 11.

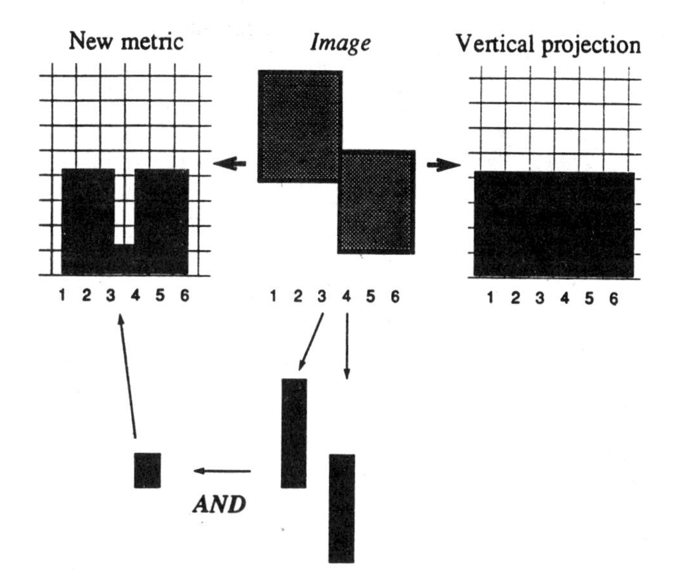

Fig. 10. *Break cost* for segmenting touching characters. *Break cost* evaluates the degree of contact. In this illustration, the new metric succeeds in exhibiting a prominent break between the left-hand and right-hand areas of the image, while traditional vertical projection fails to do so.

Fig. 11. Italic characters. (a) Italic character image. (b) Straightened italic characters. Italics are straightened before the touching character segmentation procedure is applied. The image will be returned to oblique when the character recognition procedure is applied.

2) Candidates for Break Positions: The authors assume that some variations in the *break cost* function can be found when two characters touch each other. The break position candidates are obtained by finding local minima in a smoothed *break cost* function. Several break position candidates occur as smooth local minima, while candidates are also found at both the start and end positions of touching character image. Figure 12(a) shows an example of touching characters. The *break cost* function for Fig. 12(a) is shown in Fig. 12(b), and the smoothed *break cost* function is shown in Fig. 12(c). The arrows indicate the break position candidates.

There are two kinds of candidates: one is more likely to be a real break position, and is called a preferred candidate, while the other is less likely. If the *break cost* distribution around a candidate shows a dominant sharp peak, then preference is given to this specific candidate. Preference is also given to candidates at the start and end positions of touching character image. In Fig. 12(c), thick arrows show the *preferred* candidates.

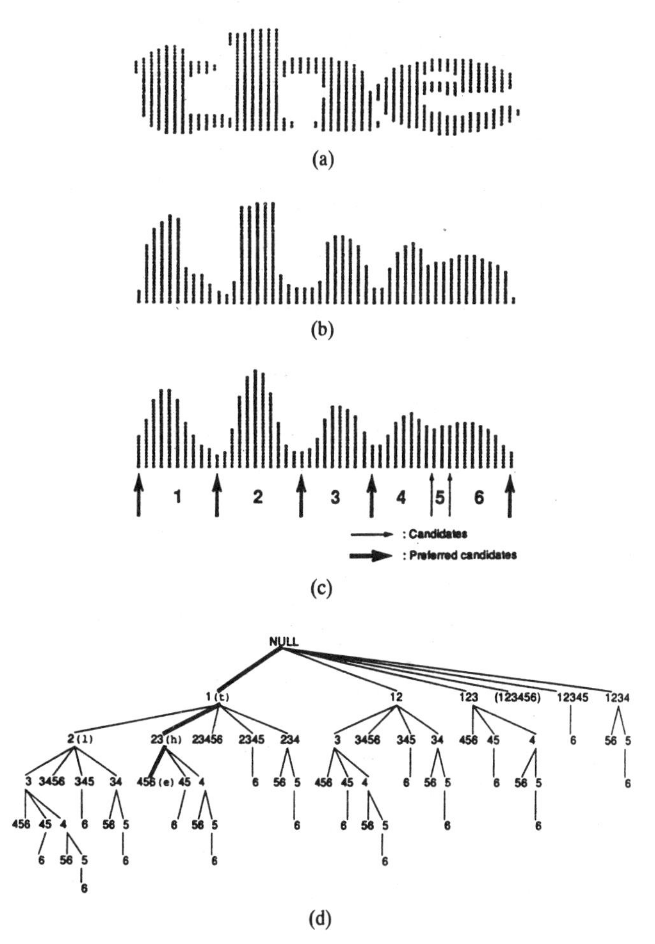

Fig. 12. Segmentation of touching characters. (a) Touching character image. (b) *Break cost.* Some changes are assumed in the *break cost* function when characters touch each other. (c) Smoothed *break cost.* The arrows indicate the candidates for break positions. The numbers indicate the areas segmented by the candidates. (d) Search tree. An optimal path representing a character segmentation result is searched for in depth-first order in the search tree through the nodes.

3) Search Tree: A search tree is introduced to represent a search order for the break position candidates. Figure 12(d) shows the search tree obtained from Fig. 12(c). In Fig. 12(c), the numbers indicate the areas of a touching character image segmented by the break position candidates.

This search tree is specified as follows:

1) Each node, except for the root node which has a *NULL* value, represents a combination of segmented areas which may correspond to a single character. For example, in Fig. 12(c), the subset {**2**, **3**} corresponds to a node which represents an "h."
2) Each node has a list of segmented areas, the first element of which is geometrically connected to the last element of the preceding node in a depth-first order. In other words, the daughters of a node represent succeeding characters. For example, the subset {**1**} has five daughters: {**2**}, {**2**, **3**}, {**2**, **3**, **4**, **5**, **6**}, {**2**, **3**, **4**, **5**}, and {**2**, **3**, **4**}.
3) The daughters of each node are sequentially ordered in the following ways:

   a) Among segmented areas ending at *preferred*

candidates, a smaller segmented area precedes the others. For example, the subset {**1**} precedes the subset {**1**, **2**} among the daughters of the root node.

b) Among adjacent segmented areas ending at *nonpreferred* candidates, a larger segmented area precedes the others. For example, the subset {**1**, **2**, **3**, **4**, **5**} precedes the subset {**1**, **2**, **3**, **4**,} among the daughters of the root node.

c) A segmented area ending at a *nonpreferred* candidate succeeds the smallest segmented area among the larger segmented areas ending at *preferred* candidates. For example, the subset {**1**, **2**, **3**, **4**, **5**} succeeds the subset {**1**, **2**, **3**, **4**, **5**, **6**} among the daughters of the root node.

This sequence is determined so that *preferred* candidates are sought in advance of *nonpreferred* candidates.

A path representing a character segmentation result is searched for in depth-first order in a search tree through the nodes. If a node is accepted as a segmented area through character recognition results, then its daughters are examined; otherwise, its sisters are checked. The criterion for character recognition acceptance is described in subsection V-F-4. The search terminates if the path arrives at a leaf node.

An example for this search is explained using Fig. 12. The subset {**1**} is recognized first, and character recognition accepts it as a "t." Consequently, its daughters are examined as follows. Subset {**2**} is accepted as an "l," but its daughter subsets {**3**}, {**3**, **4**, **5**, **6**}, {**3**, **4**, **5**}, and {**3**, **4**} are rejected. So, in this case, the path for subset {**2**} is ignored. As the next step, the sister of subset {**2**} is examined. Finally, subsets {**1**} for a "t," {**2**, **3**} for an "h," and {**4**, **5**, **6**} for an "e" are segmented from the touching characters.

4) Employing Recognition Results for Omni-fonts: A character pattern is recognized by the *multiple similarity method* [18], [19], which is designed to be insensitive to the varieties of omni-fonts. The *multiple similarity method* is a sophisticated pattern matching method which can be summarized as follows:

Given an input pattern $\boldsymbol{x} \in R^n$, the similarity value s_i for a certain category C_i is defined by

$$s_i = \sum_{j=1}^{p} (\lambda_{ij}/\lambda_{i1})(\boldsymbol{x}, \varphi_{ij})^2 / \|\boldsymbol{x}\|^2 \|\varphi_{ij}\|^2, \qquad (2)$$

where λ_{ij} and φ_{ij} are the jth eigenvalue and the jth eigenvector, respectively, obtained from a correlation matrix which is calculated from the training data belonging to C_i.

The *multiple similarity method* outputs several recognition candidates. The condition used to determine whether to reject characters is as follows:

$$\begin{array}{ll} \text{if}\,(s1 < \boldsymbol{th_1}) & \textit{rejected} \\ \text{else if}\,(s1 - s2 > \boldsymbol{th_2}) & \textit{accepted} \\ \text{else} & \textit{conflict} \end{array}$$

where similarities for the first and second candidates are denoted as $s1$ and $s2$, respectively, and both $\boldsymbol{th_1}$ and $\boldsymbol{th_2}$ are thresholds determined by statistical analysis for each character category.

Conflict suggests that both the first and second candidates are ambiguous. For example, an "l," an "I," a "1," and an "i" belong to this *conflict* category. However, despite the above rule, if the first candidate is an "l" but the second candidate is an "m," then this recognition result should be rejected.

For *conflict* characters, the images are further examined. For example, in the case of an "l," an "I," a "1," or an "i," the existence of a dot should be examined by measures such as the number of combined components described in subsection V-A. For the pair "c" and "e," the presence of a hole in the character should be examined.

Both *accepted* and *conflict* characters are regarded as being accepted. The difference between *accepted* and *conflict* is that, in employing a linguistic context, an *accepted* character fits the first candidate in recognition, while a *conflict* character fits several alternative candidates. This classification (*accepted/conflict/rejected*) enables the system to save time. Also, this classification can prevent the recognition results from being mistakenly changed by implementing a linguistic context, as described in subsection V-E. This approach makes the character segmentation method very powerful.

5) Heuristics of Character Composition: Figure 13 shows another example of touching characters where the character segmentation procedure fails. Two subsets, {**1**, **2**} for an "n" and {**3**, **4**, **5**, **6**} for an "n," are selected. If all possible combinations of subsets were to be examined, subsets {**1**} for an "r" and {**2**, **3**, **4**, **5**, **6**} for an "m" would also be selected.

Since examination of all the subset combinations would require exhaustive processing time, the authors employ heuristics of character composition to solve this problem.

As an example of such heuristics,

"m"	⟨−⟩	"r," "n"
"q"	⟨−⟩	"c," "j"
"k"	⟨−⟩	"l," "c"
"B"	⟨−⟩	"l," "3"
"H"	⟨−⟩	"I," "-," "I".
"mm"	⟨−⟩	"n," "u." "n"
"ck"	⟨−⟩	"d," "c"
etc.		

With these kinds of heuristics, another combination of subsets can be generated. In Fig. 13, the combination "nn" builds a hypothesis of another possible combination, representing an "r" and an "m." The authors have prepared more than 30 examples of such combinations. This idea makes the character segmentation method efficient.

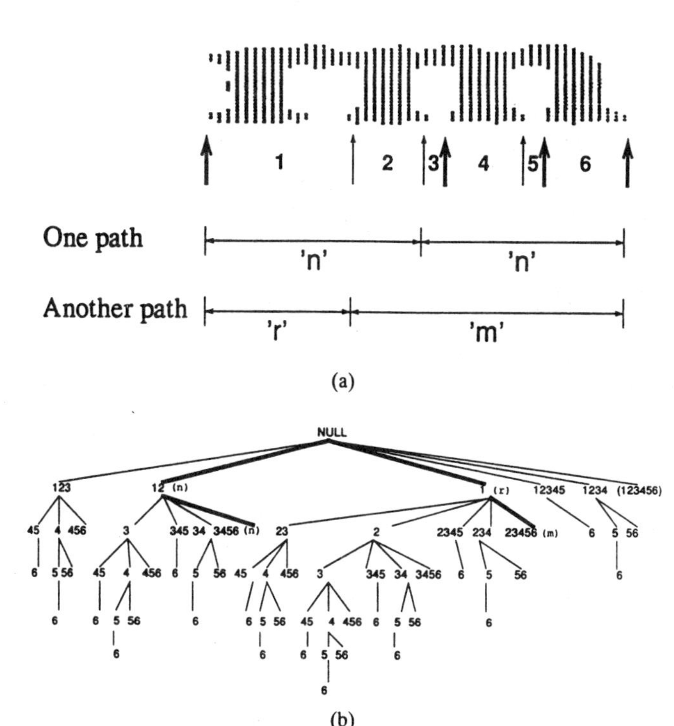

Fig. 13. Touching "r" and "m." Touching "n" and "n" may be identified from the touching characters where touching "r" and "m" should be identified. (a) Candidates for touching characters. (b) Search tree. Depth-first search stops once one path representing character segmentation result is found. Heuristics of character compositon generate another possible path. In this example, the combination "nn" builds a hypothesis of another combination, "rm."

6) Linguistic Context: All candidate combinations of touching characters are deferred for a linguistic context procedure to confirm the character segmentation result. In Fig. 9, the word "transforms" confirms that it is a combination of an "r" and "m" rather than an "n" and "n."

VI. Experimental Results

The methods proposed in this paper were implemented on a recognition board consisting of a RISC processor and memory. A scanner with a resolution of 300 dpi was directly connected to this recognition board. All procedures were realized in software alone [20].

Experiments on document analysis and document understanding were carried out on 106 documents taken from magazines, journals, newspapers, books, manuals, letters, and scientific papers. Some of the test documents with their various layouts are displayed in Fig. 14. Experiments on character segmentation were carried out on 32 of the 106 documents. These documents contained various fonts, and had many touching characters.

In experiments on document analysis and document understanding, there were 12 documents whose layouts were not correctly interpreted. This was attributed to three reasons. One was that the geometric structure was not correctly constructed because of errors in *segment* and/or *block* extraction in the document analysis process. Seven documents out of the 106 tested were not correctly interpreted for this reason. Another was because the proposed transformation

Table 1 Experimental Results of Character Segmentation

Document No.	No. Characters	No. Touching Characters	No. Errors**	No. Errors in Character Segmentation
1	6742	4756	19	8
2	6334	4773	56	43
3	4846	3452	17	7
4	3972	2828	18	4
5	3000	2181	12	3
	100%	72%*	0.5%***	
		100%		0.4%****

*Seventy-two percent of characters touched each other in the test documents.

**Recognition errors were due to both character recognition errors and character segmentation errors.

***Recogniton accuracy was 99.5%.

****Character segmentation accuracy was 99.6%.

rules did not cover all actual layouts. Four documents fell into this category. A document whose title or abstract was located in the middle of the text blocks, as in Fig. 15(a), came into this category. The last reason was that documents did not have geometrically and logically defined hierarchical structures. Figure 15(b) is a kind of table and does not have a hierarchical structure in the geometric and logical sense. These kinds of documents, however, are usually in the minority. One of the test documents was in this category.

Table 1 shows the experimental results of character segmentation for five documents where 72% of characters touched each other on average. The average recognition error rate after a linguistic context confirmation was 0.5% for these five documents. Recognition errors were due to errors in both character recognition and character segmentation.

Character segmentation failed to separate individual characters from touching characters with an error rate of 0.4%. There were three reasons for this. One was that break position candidates were not correctly nominated because of complicated contact. Figure 16(a) shows an example of tangled touching, where a two-dimensional distribution analysis of the image is required. Errors in the second document in Table 1 were caused mainly by this problem. Another reason was that insufficient examples of heuristics of character composition had been prepared. Figure 16(b) shows an example where one combination of a "T," a "1," a "J," and an "R" was accepted before the real combination of a "T," a "U," and an "R" was examined. In this example, good heuristics of character composition, e.g., a "U" could be a "1" and a "J," would generate the real combination. The last reason was that a character was rejected as a consequence of errors in character recognition though the character was

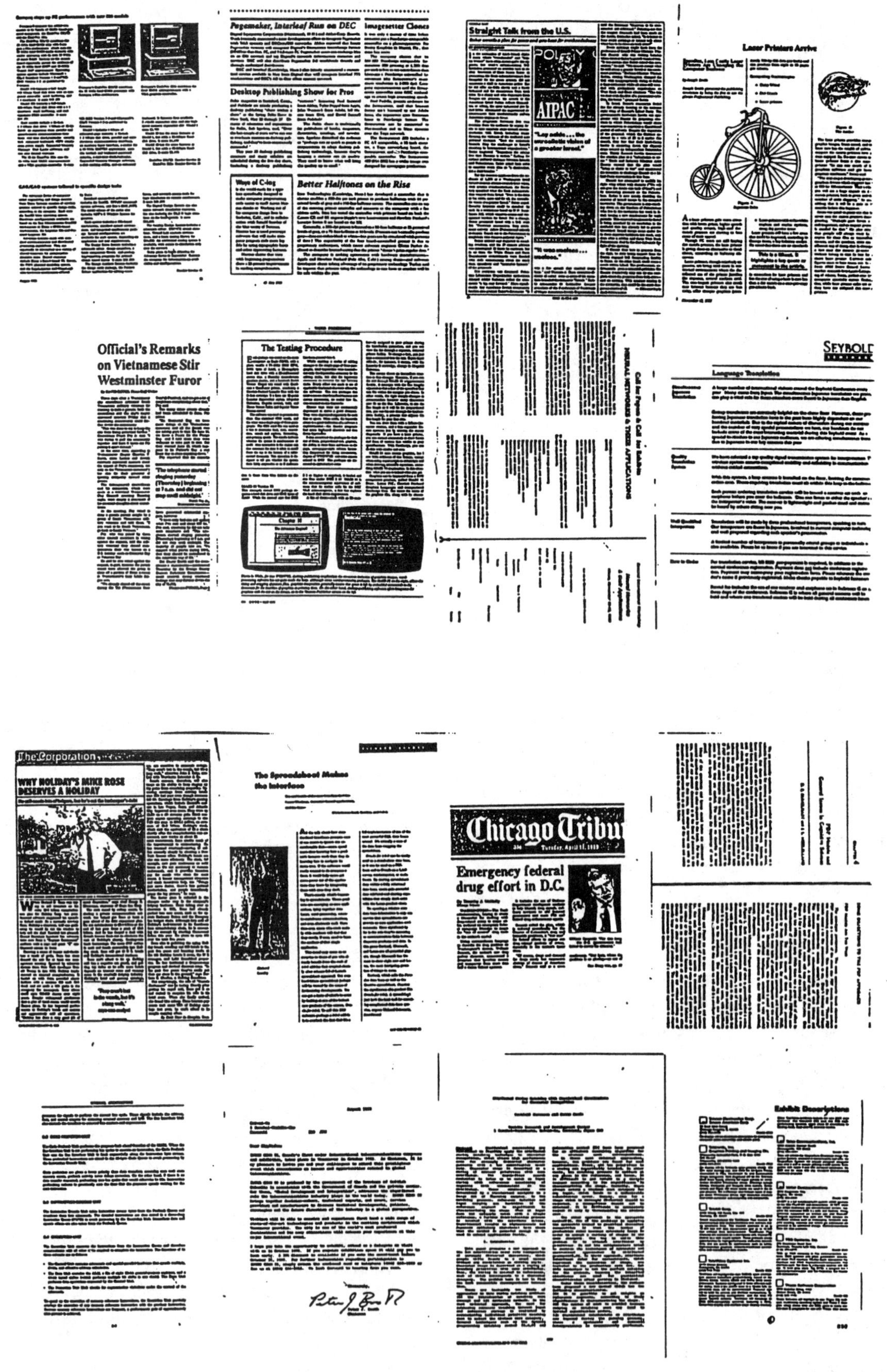

Pagemaker, Interleaf Run on DEC

Imagesetter Clones

Desktop Publishing Show for Pros

Ways of C-ing

Better Halftones on the Rise

Straight Talk from the U.S.

Laser Printers Arrive

Official's Remarks on Vietnamese Stir Westminster Furor

The Testing Procedure

Call for Papers & Call for Exhibits

NEURAL NETWORKS & THEIR APPLICATIONS

SEYBOLD

Language Translation

The Corporation

WHY HOLIDAY'S MIKE ROSE DESERVES A HOLIDAY

The Spreadsheet Makes the Interface

Chicago Tribu

Emergency federal drug effort in D.C.

Exhibit Descriptions

Fig. 14. Test samples. Experiments were carried out on documents with various layouts.

correctly segmented. Figure 16(c) shows an example where the character identifying an "a" was not accepted.

The new metric, the *break cost* for touching character segmentation, was also evaluated and compared with the traditional vertical projection. In the first document in Table 1, for example, the number of character segmentation errors

What is good dynamic range? The all-time champion is the human eye—which comes in at about 100 million to one. So what can we do to make cameras begin to approximate that performance of the human eye?

(a)

WINSRYG BUSINESS SYSTEMS

CONDEX/Fall'87 SPECIALS
Las Vegas Hilton Booth # 7066
Closeouts - Liquidations - Wholesale

P/N	Product	List Price	COMDEX Price
	RAM Products		
	640 Plus (XT) Lotus/Intel/Microsoft Expanded Memory Spec to acess memory above 640Kb + Bank Switch/RAMDisk/Diag. S/W		
#36429	w/ 0Kb	$295.	$50.0
#12103	w/ 2Mb	$995.	$239.0
23010	Dynamic Memory (XT) Long Board w/256Kb + RAMDisk, Spooler & Diag. S/W address selectable 64K to 1Mb	$489.	$35.0
23065	Wave XT Short Board w/256Kb + RAMDisk PrintSpooler & Diagnostic S/W	$499.	$40.0
#11175	Wave-Plus Short Board w/ 64Kb (same S/W as Wave XT & can hold 384Kb) address starts at 64, 128, 256 & 320Kb	$399.	$35.0
	Graphics Products		
#12450	eSCAN Image Digitizer & OCR + Cables Adaptor Board & S/W	$2,495.	$395.00
20070	Graphics Tender CGA Card w/Par. Port, PC Paintbrush S/W, RGB/Composit/TV, 320 x 200, 4 color, 640 x 200 mono + Light Pen Interface	$299.	$55.0
IBM	Monochrome Monitor Adaptor w/ Parallel Port	$250.	$45.0
ml-II	Honeywell MicroLYNX TrackBall w/Op. S/W	$159.	$35.0
20037	Graphics Master 640 x 400, 16 color; 720 x 480, 4 color; 720 x 700 mono; 128Kb, RGB-Composite-IBM mono, Light Pen Interface + PC Paintbrush Software & Graphics Master Customizing S/W	$695.	$195.00
	Communications Products		
#11001	PowerLink 3278/79 Emulator Board + Help Command/Operating/4-Session S/W with Keyboard Template	$1,099.	$125.00
#12425	5251/11 Emulator Board w/Cable & S/W	$795.	$265.00
20005	Speech Master Board uses DT1050/DIGITALKER™ (143 words) & Votrax™ Speech Synthesizer, RCA Jack for ext. speaker	$495.	$75.0
30005	Speech Master Software Support Package	$95.	$25.0
20034	Voice Recognition Board w/Shure Microphone & Operating Software	$1,000.	$175.00
22034	Voice Recognition Memory Upgrade	$300.	$50.0

page 1

(b)

Fig. 15. Examples of incorrectly interpreted documents. Twelve documents were found which were not correctly interpreted out of 106 in total. (a) Document with an abstract located in the middle of the text. The proposed rules for document understanding did not cover this layout. (b) Document not having a hierarchical structure in a geometric sense.

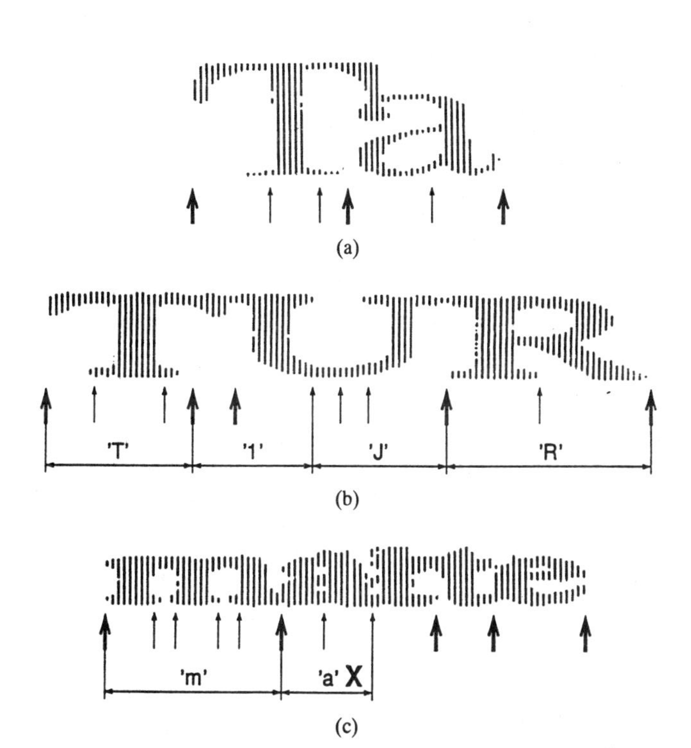

Fig. 16. Problems in segmenting touching characters. (a) Tangled touching. Candidates for break positions were not correctly nominated because of the complicated contact. (b) Insufficient heuristics of character composition. Additional heuristics of character composition (that is, a "U" can be an "1" and a "J") would be required. (c) Rejected character. Correctly segmented character "a" was rejected in character recognition.

would be 23 if traditional vertical projection were employed to evaluate the degree of contact, while the number of errors was only 8 when the new metric was employed.

Lastly, overall recognition accuracy and processing speed are described. Recognition accuracy was on the average 99.7% for photocopies of original good-quality documents after linguistic context adaptation. The authors employed a lexicon with more than 30 000 entries for the linguistic context.

The processing speed was 60 to 120 characters per second depending on the quality of the input documents. Actual processing time for the sample documents shown in Fig. 3(a) was 70 seconds, including 20 seconds for scanning, 4 seconds for run-length coding, 5 seconds for generating the geometric structure tree, i.e., document analysis, 1 second for the structure transformation, i.e., document understanding, and 40 seconds for character segmentation/recognition on a total of about 4400 characters.

VII. Conclusion

This paper describes three major components of a complete text reading system. New approaches to document analysis, document understanding, and character segmentation/recognition have been introduced. For document analysis, this paper presents a method capable of robust behavior even for multicolumned documents with graphics and photographs. This approach gives the items on the page a hierarchy which models the relationships between characters, lines, columns, and the page. For document understanding, this paper reports an attempt to build a method of understanding document layouts without the assistance of the character recognition results, i.e., the meaning of the contents. It is shown that documents have an obvious hierarchical structure in their geometry which can be represented by a tree. A small number of rules are introduced to transform the geometric structure into a logical structure which represents the semantics carried by the documents. The virtual field separator technique is employed to utilize the information carried by special constituents of documents such as field separators and frames, keeping the number of transformation rules small. For character segmentation/recognition, an efficient and powerful character segmentation method for touching characters has been presented. This approach employs heuristics of character composition as well as recognition results for omni-fonts to resolve ambiguities in segmenting touching characters. A new metric to evaluate the degree of touching is also introduced to identify each character in a group of touching characters.

Experimental results on a variety of documents have shown that the proposed methods are applicable to most of the document types commonly encountered in daily use, although there is still room for further refinement in the transformation rules for document understanding and resolution of ambiguities in segmenting touching characters.

REFERENCES

[1] S. Tsujimoto and H. Asada, "Understanding multi-articled documents," in *Proc. 10th Int. Conf. Pattern Recognition* (Atlantic City, NJ), 1990, pp. 551–556.

[2] S. Tsujimoto and H. Asada, "Resolving ambiguity in segmenting touching characters," in *Proc. 1st Int. Conf. Document Anal. and Recognition* (Saint Malo, France), 1991, pp. 701–709.

[3] F. M. Wahl, K. Y. Wong, and R. G. Casey, "Block segmentation and text extraction in mixed text/image documents," *Comput. Graphics, and Image Processing*, vol. 20, pp. 375–390, 1982.

[4] J. Toyoda, Y. Noguchi, and Y. Nishimura, "Study of extracting japanese newspaper article," in *Proc. 6th Int. Conf. Pattern Recognition* (Munich, Germany), 1982, pp. 1113–1115.

[5] M. Okamoto and A. Miyazawa,"An experimental implementation of document recognition system for papers containing mathematical expressions," in *Pre-Proc. 1990 Syntactic & Structural Pattern Recognition* (Murray Hill, NJ), 1990, pp. 335–351.

[6] H. S. Baird and K. Thompson, "Reading chess," *IEEE Trans. Pattern Anal. Mach. Intell.*, vol. 12, no. 6, pp. 552–559, 1990.

[7] F. Esposito, D. Malerba,and G. Semeraro, "An experimental page layout recognition system for office document automatic classification: an integrated approach for inductive generation," in *Proc. 10th Int. Conf. Pattern Recognition* (Atlantic City, NJ), 1990, pp. 557–562.

[8] J. Higashino, H. Fujisawa, Y. Nakano, and M. Ejiri, "A knowledge-based segmentation method for document understanding," in *Proc. 8th Int. Conf. Pattern Recognition* (Paris, France), 1986, pp. 745–748.

[9] Y. Tsuji *et al.*, "Document recognition system with layout structure generator," in *Proc. IAPR Workshop on Machine Vision Applications* (Tokyo, Japan), 1990, pp. 479–482.

[10] Y. Nakano, H. Fujisawa, and O. Kunisaki, "A document understanding system incorporating character recognition," in *Proc. 8th Int. Conf. Pattern Recognition* (Paris, France), 1986, pp. 801–803.

[11] K. Inagaki, T. Kato, T. Hiroshima, and T. Sakai, "MACSYM: A hierarchical parallel image processing system for event-driven pattern understanding of documents," *Pattern Recognition*, vol. 17, no. 1, pp. 85–108, 1984.

[12] I. Masuda, N. Hagita, and T. Akiyama, "Approach to smart document reader system," in *Proc. 1985 Comput. Vision Pattern Recognition* (San Francisco, CA), 1985, pp. 550–557.

[13] H. S. Baird, "Anatomy of a page reader," in *Proc. IAPR Workshop on Machine Vision Applications* (Tokyo, Japan), 1990, pp. 483–486.

[14] S. N. Srihari and G. W. Zack, "Document image analysis," in *Proc. 8th Int. Conf. Pattern Recognition* (Paris, France), 1986, pp. 434–436.

[15] S. Kahan, T. Pavlidis, and H. S. Baird, "On the recognition of printed characters of any font and size," *IEEE Trans. Pattern Anal. Mach. Intell.*, vol. 9, no. 2, pp. 274–288, 1987.

[16] R. G. Casey and G. Nagy, "Recursive segmentation and classification of composite character patterns," in *Proc. 6th Int. Conf. Pattern Recognition* (Munich, Germany), 1982, pp. 1023–1026.

[17] R. Kooi and W. C. Lin, "An on-line minicomputer-based system for reading printed text aloud," *IEEE Trans. Syst., Man, Cybern.*, vol. 8, no. 1, pp. 57–62, 1978.

[18] T. Iijima, H. Genchi, and K. Mori, "A theory of character recognition by pattern matching method," in *Proc. 1st Int. Joint Conf. Pattern Recognition* (Washington, DC), 1973, pp. 50–57.

[19] E. Oja, *Subspace Methods of Pattern Recognition.* Research Studies Press, 1983.

[20] S. Tsujimoto and H. Asada, "Document image processing for accurate and high speed text reading," in *Pre-Proc. 1990 Syntactic & Structural Pattern Recognition* (Murray Hill, NJ), 1990, p. 501.

Shuichi Tsujimoto was born in Osaka, Japan, on February 3, 1961. He received the B.E. and M.Sc. degrees from Osaka University, Osaka, Japan, in 1984 and 1986, respectively, for his work on data compression by means of the spline approximation.

He joined the Toshiba Corporation, Japan, in 1986 and since then has been at the Information Systems Laboratory, Research and Development Center, Kawasaki, Japan. His current research interests include document image processing, layout understanding, text processing, character segmentation, and character recognition for a state-of-the-art text reader.

Mr. Tsujimoto is a member of the Institute of Electronics, Information and Communication Engineers of Japan.

Haruo Asada was born in Osaka, Japan, in 1948. He received a B.E. degree in 1970 and an M.E. degree in 1972, both in mathematical engineering and instrumentation physics from the University of Tokyo, Tokyo, Japan.

Since 1972 he has been with the Toshiba Corporation, Japan, where he has worked on character recognition, speech recognition, pattern recognition theory, shape analysis, robot vision, and image processing hardware. During the academic year 1983–1984 he was a visiting scientist at the MIT Artificial Intelligence Laboratory, Cambridge, MA, where he worked with the robotics group on 2-D shape description. He is now a senior research scientist at the Information Systems Laboratory, Toshiba Research and Development Center. His current research interests focus on document image processing, pattern recognition theory, and shape recognition.

Document Image Defect Models

Henry S. Baird

AT&T Bell Laboratories, Computing Science Research Center,
600 Mountain Avenue, Murray Hill, NJ 07974, USA

A lack of explicit quantitative models of imaging defects due to printing, optics, and digitization has retarded progress in some areas of document image analysis, including syntactic and structural approaches. Establishing the essential properties of such models, such as completeness (expressive power) and calibration (closeness of fit to actual image populations) remain open research problems. Work-in-progress towards a parameterized model of local imaging defects is described, together with a variety of motivating theoretical arguments and empirical evidence. A pseudo-random image generator implementing the model has been built. Applications of the generator are described, including a polyfont classifier for ASCII and a single-font classifier for a large alphabet (Tibetan U-Chen), both of which which were constructed with a minimum of manual effort. Image defect models and their associated generators permit a new kind of image database which is explicitly parameterized and indefinitely extensible, alleviating some drawbacks of existing databases.

1 Introduction

Technical challenges in document image analysis arise, generally speaking, from three sources:

symbols: the set of idealized shapes that can occur, often in a hierarchy where simple symbols are assembled into more complex ones, at several levels of organization;

deformations: a range of shape variations that symbols are permitted to undergo, including geometric transformations (translation, rotation, scaling, stretching, etc.) and more complex or time-dependent distortions (*e.g.* due to the biomechanics of handwriting); and

imaging defects: imperfections in the image due to printing, optics, scanning, spatial quantization, binarization, etc.

This taxonomy, while useful, is somewhat arbitrary and may vary with the context. For example, in some applications a greatly deformed symbol is best regarded as a new symbol altogether: for example, the letter M, deformed by an extraordinarily large rotation, finally approaches W. Less obviously, a geometric transformation may be thought of as a deformation when it is large, but as an imaging defect when it is small: for example, the deformation that "grows" the roman R into its boldface variation **R** can be closely approximated by a small defect in binarization (due to blurring and thresholding). Nevertheless, in most applications there is a clear distinction between symbol formation (without respect to imaging) and image formation (without respect to symbols) — and it is this distinction that is the focus of this study.

Symbols and their deformations are the subjects of virtually all recent published research in document image analysis. Papers routinely exhibit examples of symbols, and often describe their expected deformations, but only rarely discuss the range of image defects that is tolerable. In effect, the great majority of papers assume that image defects can be neglected, or at least need not be analyzed quantitatively. Such an assumption may be justified in some applications: for example, it may permit an uncluttered exposition of the algorithm being studied. However, it is often obvious that there are classes of defects which can occur and will cause failure: and this raises the question of exactly when the method is likely to succeed.

Furthermore, when experimental trials are attempted, image defects are often found to play a critical role. Among symbolic-modeling techniques (including many syntactic and structural (SSPR) methods), image defects can trigger a rapid proliferation of slightly varying models, even when hybrid methods with statistical or Markov characteristics are used. This proliferation of models can have unpleasant effects on the space and time complexity of algorithms for recognition. I believe it is fair to say that if it were not for this problem, accurate and even fast recognition of isolated machine-printed multi-font characters would have been achieved decades ago.

Problems due to image defects often arise quite early, when extracting the most elementary shapes in the symbol hierarchy. It is not surprising that purely syntactic chain-code models (of perfectly digitized lines, squares, etc) are rarely applicable to real images — but on the other hand it is striking that reported vectorization methods are rarely based on an explicit model of imaging defects.

The inference of SSPR models, whether carried out manually or automatically, is complicated by image defects. In exact-match SSPR methods, it is often assumed that an exhaustive set of symbolic models can be specified. If defects were negligible, it might be possible to enumerate these by hand; but often in practice new cases arise throughout a long series of experimental trials.

These observations suggest that imaging defects threaten the effectiveness of many techniques for document image analysis, and that they have not yet been studied as systematically as they deserve. I suggest that an appropriate research program should include at least the following topics:

- *Parameterization:* a defect model should ideally be expressed as a function of a small number of parameters, presumably for the most part real numbers.

I will propose a preliminary model of this kind, suitable for use in studies of text, line-drawings, music, and other classes of documents.

- *Completeness,* or expressive power: this measures the probability that, given any particular defective image, there exists some choice of model parameters (and some randomization) that will duplicate the image. In this study, I will address this point only qualitatively, by exhibiting a characteristic range of effects.

- *Calibration:* the degree to which a distribution defined on the model parameters fits a given population of defective images. I have been able to measure the distribution of a few parameters; for some of the rest, I offer theoretical considerations to motivate a plausible choice.

- *Simulation.* A defect model, together with a distribution on its parameter space, can be used to generate representative sets of image samples pseudo-randomly. In some cases it may be possible to infer models from these sets automatically. This approach may be particularly well suited to SSPR methods that use hybrids of symbolic and statistical matching. I will discuss several exercises of this kind later in the paper.

- *Enumeration.* One interesting potential use of these models is to support the automatic computation of *all possible* symbolic prototypes implied by a given feature-extraction system. I am not aware of any successful effort of this kind: this may be a challenging and rewarding arena for future research. Of course, characterizing the effects of imaging defects using purely analytical means may be a difficult or impossible task. By contrast, deformations are often handled analytically by a variety of efficient and robust normalization algorithms, applied either to isolated symbols or partial structural matches.

2 A Model of Imaging Defects

In this section I describe work in progress towards a parameterized model of local imaging defects. I hope it will serve to stimulate research into more complete and fully calibrated models. It is based on approximations to the physics of the printing and imaging process ([Sch86, Eks84]).

The idealized input symbol can be thought of as a bilevel image at effectively infinite resolution. Illumination intensity is in the range [0.0,1.0], with the convention that white=0 and black=1. The model was developed for use on images of hand-written or machine-printed characters. Some parameters of the model are in typographical units: a *point* equals 1/72 inch; an *em* is equal to the nominal size of the text (usually the closest permissible vertical spacing, in points). It should be clear, however, that the model is adaptable to a wide variety of symbol types.

Resolution: The degraded image will be spatially quantized. In practice, this is the result of both the absolute size of the symbol and the scanning digitizing resolution, which I in fact specify separately as *size* (in points), and *resolution* (in pixels/inch). For experiments in character recognition I use a distribution of sizes that is uniform on [5,14] point, at a resolution of 300 pixels/inch (roughly equivalent to [10,30] pixels/x-height). Here are seven examples from 5 to 11 point, scaled to the same absolute size for clarity.

R R R R R R R

Blur: The point-spread (or, impulse response) function of the combined printing and imaging process is modeled as a circularly symmetric Gaussian filter with a standard error of *blur* in units of output pixel size. Note that *blur* < 0.7 implies effectively zero cross-talk between non-8-connected pixels: thus 0.7 may be close to the optimal hardware design for bilevel scanners, according to a theoretical study of the trade-off between spatial and intensity quantization [LPW87]. It is often difficult to find a manufacturer who will specify the point-spread function of his document scanner. For experiments in character recognition I assume *blur* is distributed normally with mean $m = 0.7$ and standard error $e = 0.3$. Here are seven images, illustrating the effects of blur parameter values $\{m - 3e, m - 2e, m - e, m, m + em + 2e, m + 3e\}$ (this set of values is also used in other illustrations below, unless noted).

R R R R R R R

Threshold: Binarization is modeled as a test on each pixel: if its intensity $\geq$ *threshold*, the pixel is black. A threshold of 0.25 guarantees that a stroke that is one output-pixel wide will not be broken under the mean blur of 0.7. For experiments on high-resolution prototypes I let threshold vary, from image to image, normally with mean 0.25 and standard error 0.04. On coarsely quantized input, other choices are appropriate (for an example, see the Applications section).

R R R R R R R

Sensitivity: Each pixel's photo-receptor sensitivity is randomized in two stages: for each char, *sensitivity* is selected, in units of intensity $\in [0,1]$; then, for each pixel, a sensitivity adjustment is chosen randomly, distributed normally with mean 0 and standard error —*sensitivity*—, and added to each pixel's intensity. For experiments in character recognition I let *sensitivity* vary from image to image, normally with mean 0.125 and standard error 0.04.

R R R R R R R

Jitter: I assume that the arrangement of the output pixel photo-receptors is an only nominally square grid: their actual location is allowed to vary slightly, in two stages: for each symbol, *jitter* is specified, in units of output pixel size; then, for each pixel in the symbol, a vector offset (x,y) is chosen (each component independently) from the normal distribution having mean 0 and standard error —*jitter*—. I have not yet been able to measure this directly, but it is clearly unlikely that two pixel centers will touch. Therefore, for experiments in character recognition I let *jitter* vary, from image to image, normally with mean 0.2 and standard error 0.1. The effects are often subtle.

R R R R R R R

Skew: The symbol may rotate (about a given fiducial point) by a *skew* angle (in degrees). In experiments on over 1000 pages of books, magazines, and letters, placed on a flatbed document scanner by hand with ordinary care, I have observed a distribution of angles that is approximately normal with a mean 0 and a standard error 0.7. The actual distribution is somewhat long-tailed: absolute skew angles greater than 2 degrees occur more often than in a true Gaussian. For experiments in character recognition I use a distribution with twice this standard error.

R R R R R R R

Width: Width variations are modeled by a multiplicative parameter *x-scale*, that stretches the image horizontally about the symbol's center. Measurements on low-quality images, such as produced by FAX machines, suggest that this is approximately normal with mean 1.0 and standard error 0.05. However, for experiments in character recognition I use a distribution uniform in the interval [0.85,1.15], for a reason peculiar to the application: I wish to model a range of font deformations (the "condensed" and "expanded" font varieties). Here are samples spaced uniformly across the range.

R R R R R R R

Height: Height variations are modeled by a multiplicative parameter *y-scale*, that stretches the image vertically about the baseline. Measurements on low-quality images, such as produced by FAX machines, suggest that this is approximately normal with mean 1.0 and standard error 0.05. However, for experiments in character recognition I use a normal distribution with mean 0 and standard error 0.02.

R R R R R R R

Baseline: In machine-printed text, the height of a symbol above the conventional baseline (or, in some writing systems, below a top-line) is often significant: this *baseline* parameter is in units of ems. In measurements on over 120,000 characters from letterpress books (printed somewhat more erratically than is usual), a long-tailed normal distribution was again observed, with a mean 0 and a standard error 0.03. For experiments in character recognition I use a distribution with twice this standard error.

R R R R R R R

Kerning: The parameter *kern* varies the horizontal placement of the image with respect to the output pixel grid. Of course, in most writing systems the horizontal position is irrelevant for segmented symbol recognition. The motivation to avoid systematic digitizing artifacts. This is easily accomplished by letting it vary uniformly in [-0.5,0.5], in units of output pixel size; here are examples spaced uniformly across the range. The effects are often subtle.

R R R R R R R

3 A Pseudo-Random Generator

I have built an image generator that simulates this defect model. Given a bilevel image and a set of model parameters, it computes a degraded bilevel image. It can also generate any specified number of images, chosen pseudo-randomly from the defect distribution model; in this case, the parameters are assumed to be independent random variables. For input, I have used high-resolution original typographical descriptions (ideal prototypes at effectively infinite resolution) as well as samples selected from images of printed text. The results are most interesting when the input is an image of a large, cleanly printed original example: the model may require some adjustment on poor-quality input.

For speed, the skew, baseline, height, and width transformations are performed on a boundary representation. Then, it is converted to a high-resolution bitmap, and the point-spread function is applied at each computed pixel center:

this is the most expensive step. Each resulting pixel intensity value is modified by the sensitivity adjustment, and binarized by comparison with the threshold.

The image defect generator is written in the C programming language. The pseudo-random number generator uses an additive-feedback algorithm [Zei69, Mar84] in a portable, machine-independent implementation. Normal distributions are truncated to 0 farther than three standard errors from the mean.

To illustrate the range of defects that can be simulated, we show a set of images at 5 point, in Figure 1. Each line holds images whose Mahalanobis distances (Euclidean distance from the mean, scaled component-wise by standard error) lie in the range shown. Baseline, jitter, and kerning are not included in the Mahalanobis distance.

[0.0,0.5]

[0.5,1.0]

[1.0,1.5]

[1.5,2.0]

[2.0,2.5]

Fig. 1. Pseudo-randomly generated samples in various ranges of mahalanobis-distance from the mean of the defect distribution.

I invite the reader to judge from the examples in Figure 1 the degree to which the model is complete. It is inevitable that an image defect model will approximate the physics of the many stages of printing and imaging. However, it is essential that it be capable of expressing a wide range of commonly occurring effects. I hope these examples will stimulate discussion of categories of defects that the model cannot yet express.

4 Applications

I now discuss two applications of the generator.

The first is a "uniformly fair" multi-font classifier for machine-printed ASCII. For each of 39 fonts, for each of 10 sizes (from 5 to 14 point), and for each of the 94 symbols in the printable-ASCII set (where available in the font), I generated 25 images, for a total of 804,500. Using half of this database (the odd point sizes), I inferred a classifier using the method of [Bai88a]. This was tested on the other half of the database. Half of the errors are due to confusions that are arguably inevitable in a multi-font environment:

Ignoring these, the success rates were 98.21% top choice and 99.45% within the top 5 choices. If the 6-point test samples are also ignored. the success rates are 99.19% top choice and 99.87% in top 5.

When performance measures such as these are quoted in technical papers, it can be difficult for the reader to judge whether they represent a significant advance over competing methods. In the present case, however. a reader is capable of replicating the test since I have listed the alphabet of symbols, an image defect model has been specified, and the fonts are commercially available (the author will supply their names on request). Only the seeds of the pseudo-random number generator remain unspecified, but the large scale of the test (over 400,000 images) will permit statistically significant comparisons no matter what seed was used.

On another point, high top-5 accuracy scores promise excellent accuracy when contextual constraints can be exploited. In particular. they allow the use of fast data-driven contextual analysis algorithms, which require *uniformly shallow* accuracy: that is, the correct interpretation for all classes must be found with high probability in a short list of alternatives supplied by the classifier. This is a more stringent standard than average top-choice or even average top-k correct. An important factor in achieving uniformly shallow accuracy is the use of training sets that are uniform in a strong sense: they should contain an equal number of samples of all symbols, over all fonts, and distorted by the same distribution of image defects. Such a *uniformly fair* sample set can only be provided by an image defect generator: attempts to collect such a set from actually occurring image populations are futile.

The second example illustrates the use of the generator to read exotic large-alphabet languages with a minimum of manual effort. A 441-page machine-printed Tibetan-to-Tibetan dictionary [R82] has been translated into ASCII with an estimated 95% accuracy, after two weeks of work by one person. The work was performed by Mr. Reid Fossey, a student of the Tibetan language, while visiting Bell Laboratories. Although trained in Tibetan, he had no prior exposure to the OCR system — and the system had received no prior training on Tibetan. I provided the tools used by Mr. Fossey. Only one change in these was required, to accommodate Tibetan's U-chen (top-line) typographic convention.

Mr. Fossey first acquired bilevel images of all the pages in the book, using a flatbed document scanner at 400 pixel/inch resolution. Next, the images were analyzed fully automatically into columns of lines of symbols. using the method of [Bai88b]. Then Mr. Fossey selected sample images of each symbol in each font style. The number of these was held to a minimum, since at most one image of each distinct symbol was required, even if it occurred at more than one size. Dr. Kurt Keutzer, a researcher in computer science at Bell Labs and also a student of the Tibetan language, performed an additional pass of training on the text to correct for oversights in selection and labeling. In Figure 2 we show twelve "original" images of the 'sku' character, selected by Mr. Fossey.

For each of these symbols. 75 distorted images (over a range of three sizes)

Fig. 2. Twelve original Tibetan character images (symbol "sku"), selected from the digitized pages.

were generated to make up the training set. Figure 3 show twelve pseudo-randomly generated training instances, using the first of the images above as the prototype:

Fig. 3. Twelve pseudo-randomly generated images, using as the prototype the first shown in Figure 2.

To compensate for the coarsely quantized input, the *threshold* parameter is Gaussian with mean 0.5 and standard error 0.4. The randomized samples show a somewhat wider range of defects than the originals.

The next few steps were completely automatic: these included the inference of a classifier, the classification of the entire document, and the translation of the symbols into user-specified ASCII (the Wylie transliteration was used). The classifier technology [Bai88a] uses a hybrid of structural shape analysis and statistical decision theory, and requires a large (> 50) and representative training set for good results.

The last step was the manual proofreading of the computer-recognized text. The set of distinct symbols encountered in the dictionary numbered 438, in two slightly different font styles. Altogether, 61,124 symbols were translated, not counting punctuation. Fossey and Keutzer, after proofreading a fraction of the output, report an accuracy of approximately 95% (not counting spacing, punctuation, or special characters used for transliterating Sanskrit).

5 Discussion

There is clearly work to be done to establish the completeness and calibration of this experimental image defect model. Still, it has already proven useful, and may be helpful as a starting point for discussion.

The fact that images generated by the model possess explicit parameterizations opens up interesting new ways to study pattern recognition methods. In principle, it should be possible systematically to explore the limitations of any given pattern recognition algorithm by regression analysis on the parameters of images on which the algorithm fails. In some cases this experience may suggest algorithm improvements; in other cases, it may at least permit *compensatory training*, in which specially designed training sets, with a greater-than-usual occurrence of troublesome defects, are used to reduce the error rate.

Standardized image databases, available to all researchers, have played an important role in driving pattern recognition technology. The existence of image defect generators makes possible an interesting new class of *implicit databases* which are indefinitely extensible. Instead of sharing a finite set of images, researchers would share a model (and its generator), so that while they are experimenting on the same distribution of images, they are not limited by an arbitrarily fixed sample size. This can be a significant advantage. For example, present neural-net learning algorithms require a large number of presentations of training images for convergence, often many times the number of distinct images available, and therefore there is often concern that the networks may be undertrained. More generally, while experienced practitioners are careful to use distinct training and test sets when measuring the performance of an algorithm, they are often forced by lack of data into a subtler but similar methodological trap: while refining the algorithm manually, they reuse the same (training and testing) data repeatedly. Ideally, they should be able to throw away each set and start from scratch: an image defect generator offers a way to do this.

The obstacles to achieving a consensus on the details of image defect models (which may be considerable, and both technical and political) should not discourage the attempt, for several reasons. The present state of affairs, in which image defects are usually ignored, is unrealistic and unwise. Explicit, quantitative models provide a foundation for scientific understanding and may be the only possible basis for reliable engineering. Investigating these models is an essential step towards the goal of mapping the range of applicability of pattern recognition methods as applied to document images.

Acknowledgements

The *jitter* parameter was suggested by George Nagy. I am grateful for stimulating discussions with him, Theo Pavlidis, and Sargur Srihari.

References

[Bai88a] H. S. Baird, "Feature Identification for Hybrid Structural/Statistical Pattern Classification," *Computer Vision, Graphics, & Image Processing* 42, pp. 318–333, 1988.

[Bai88b] H. S. Baird, "Global-to-Local Layout Analysis," *Proc. IAPR Workshop on Syntactic and Structural Pattern Recognition*, Pont-á-Mousson, France, 12–14 September, 1988.

[Eks84] M. P. Ekstrom, *Digital Image Processing Techniques*, Academic Press (Orlando, 1984).

[FL90] R. Fossey and P. Lofting, "The Typestyle Jockey: Putting the Horse Out Front in Devanagari and Tibetan," *Nordic Institute of Asian Studies Report*, Copenhagen, pp. 5–30, 1990.

[LPW87] D. Lee, T. Pavlidis, and G. W. Wasilkowski, "A Note on the Trade-off between Sampling and Quantization in Signal Processing," *J. of Complexity*, Vol. 3, pp. 359–371, 1987.

[Mar84] G. Marsaglia, "A Current View of Random Number Generators," Keynote address, *Computer Science and Statistics 16th Symp. on the Interface*, Atlanta, March 1984.

[Sch86] W. F. Schreiber, *Fundamentals of Electronic Imaging Systems*, Springer Series Information Science 15, Springer–Verlag (Berlin, 1986).

[R82] Tshe bDang rNam rGyal, ed., *Dag Yig Ma Nor Lam bZang* (The Excellent Path to Wealth Dictionary). Lhasa: Mi Rigs dPe sKrun Khang (Humanity Publishing House), 1982.

[Zei69] N. Zeirler, "Primitive Trinomials Whose Degree is a Mersenne Exponent," *Inf. Control*, 15, 1969.

Off-Line Cursive Script Word Recognition

RADMILO M. BOZINOVIC, MEMBER, IEEE, AND SARGUR N. SRIHARI, SENIOR MEMBER, IEEE

***Abstract*—Cursive script word recognition is the problem of transforming a word from the iconic form of cursive writing to its symbolic form. This paper describes several component processes of a recognition system for isolated *off-line* cursive script words; off-line means a lack of stroke order information that is often assumed in commercial systems. The approach is to transform a word image through a hierarchy of representation levels: points, contours, features, letters, and words. A unique feature representation is generated bottom-up from the image using statistical dependences between letters and features. Ratings for partially formed words are computed using a stack algorithm and a lexicon represented as a trie. Several new techniques for low- and intermediate-level processing for cursive script are described, including heuristics for reference line finding, letter segmentation based on detecting local minima along the lower contour and areas with low vertical profiles, simultaneous encoding of contours and their topological relationships, extracting features (e.g., middle loop, upper zone stroke), and finding shape-oriented events. Experiments demonstrating the performance of the system are also described.**

***Index Terms*—Cursive script recognition, hierarchical processing, multilevel perception, off-line script recognition, word recognition.**

I. Introduction

CURSIVE script recognition is the problem of transforming text from the two-dimensional spatial form of cursive (continuous or running) writing into symbolic representation. A distinction is made between on-line and off-line script for the purpose of machine recognition. The order of strokes made by the writer is available in the on-line case, whereas only the completed writing is available in the off-line case. Order information can be obtained by writing on an electronic bit pad, which causes the two-dimensional coordinates of successive points to be stored in order. The on-line case deals with a one-dimensional representation of the input, whereas the off-line case involves analysis of a two-dimensional image. This paper describes a framework and several processes for recognizing single off-line cursive script words. The input word is assumed to be from a given lexicon; members of the lexicon could presumably be determined by various contextual constraints.

Research on cursive script recognition has a long history. While the distinction between cursive writing and print is blurred (indeed, a continuum of writing styles between them can be observed), there has been extensive research on reading clearly segmented print; for a bibliography, see [28]. The most notable early work on cursive script is that of Mermelstein and Eden [24], in which on-line velocity measurements of pen strokes were utilized and letters were composed syntactically from a standard set of strokes taken from a formal handwriting model [14]. The first off-line solutions introduced wider and better use of higher-level knowledge (e.g., real/binary letter n-grams) and the notion of tentative letter segmentation [15], [26]. Solutions which used no segmentation, but recognized words as entities were also developed [16]. Methods for preprocessing operations like slant removal were developed [11]. More recently, there has been significant growth of interest in on-line script recognition. Burr [9], [10] and Tappert [30], [31] utilize dynamic programming and metrics to find incrementally the best elastic match for the observed signal against a set of stored and labeled patterns. Berthod and Ahyan [3] describe an on-line system which allows for user training, but uses this knowledge of letter formation from strokes in a syntactic-style approach. Brown [7] described an on-line system with certain off-line features, no letter-level recognition, extensive preprocessing, and adaptive capabilities. Badie and Shimura [1] sketched a syntactic system for off-line recognition. Hayes [20] developed an off-line system that used a word representation consisting of five well-defined levels and syntactic pattern matching and probabilistic hierarchical relaxation on the top two levels. The present paper gives details of several operations of an off-line cursive script recognition system also described by us in [29].

A. Organization of Paper

The organization of a cursive script word recognition system and an example of its operation are given in Section II. Sections III–VIII give details of various aspects of system design. Section III describes the preprocessing operations of smoothing, slant estimation and removal, reference line finding, and segmentation. Section IV discusses generation of the chain encoding and topological contour tracing. Section V describes the process of feature extraction or event generation. Section VI discusses the statistical and syntactic framework for combining events into letters and computing the quality of letter hypotheses. Section VII describes the strategy that exercises hierarchical control. Learning capabilities for both the su-

Manuscript received August 27, 1986; revised October 27, 1987. Recommended for acceptance by C. Y. Suen. This work was supported in part by the office of Advanced Technology of the United States Postal Service under Contract 104230-84D-096Z and in part by the National Science Foundation under Grant IRI-86-13361.

R. M. Bozinovic was with the Department of Computer Science, State University of New York at Buffalo, Buffalo, NY 14260. He is with the Department of Computer Systems, Mihajlo Pupin Institute, Belgrade, Yugoslavia.

S. N. Srihari is with the Department of Computer Science, State University of New York at Buffalo, Buffalo, NY 14260.

IEEE Log Number 8823858.

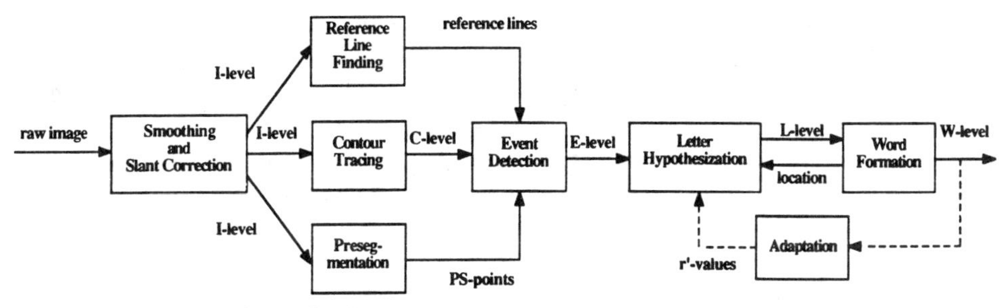

Fig. 1. Cursive script word recognition system organization.

pervised (training) and unsupervised (adaptation) cases are discussed in Section VIII. Experimental results are discussed in Section IX, and concluding remarks are made in Section X.

II. System Overview

The problem is approached as one involving a series of transformations between representation levels. The system processes the input data hierarchically (in a strict bottom-up fashion) until a certain level, and the processing is heterarchical (interaction between levels) after that (Fig. 1).

The raw image, after initial preprocessing of smoothing and slant removal, is brought to the *Image-level* (*I*-level). The slant removal and smoothing modules eliminate any significant overall slant of the strokes and remove minor contour discontinuities and roughness, bringing input to the *I*-level. An input word, *might*, is depicted in its *I*-level form in Fig. 2(a).

The *I*-level word image is input to three different operations: reference line finding, presegmentation, and contour tracing. Reference line finding and presegmentation are parts of preliminary processing, but do not produce full representations on any level; rather, they extract specific information from the *I*-level to be used later. The reference line finding operation determines three main vertical zones of script: the middle one, where the "bodies" of all letters reside, and upper and lower ones, corresponding to ascenders and descenders, which are also indicated in Fig. 2(a).

The presegmentation operation breaks up the word image horizontally into minimal portions, which are potential letters (but may turn out to be only parts thereof), each called a presegment (PS). The PS's are separated by vertical lines, each of which corresponds to a presegmentation point (PSP). This operation of presegmentation results in the first letter, m, spanning three presegments (1, 2, 3), h spanning two presegments (7, 8), and all the others spanning one each [Fig. 2(b)]. The fourth PS, between PSP 3 and PSP 4, corresponds to no letter (a ligature) and will be interpreted as such by the system, indicated with a quote (') symbol.

The contour tracing operation computes the *contour-level* (*C*-level) representation, which is a description of the image in terms of an extended chain encoding, which represents all the contours of the image and its topology. Contours are encoded by specifying a starting point, fol-

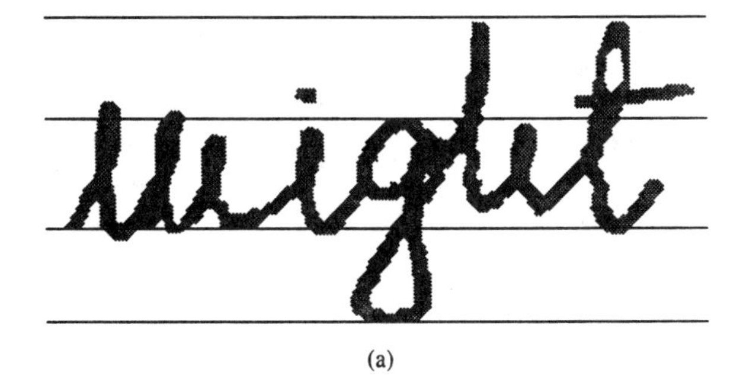

(a)

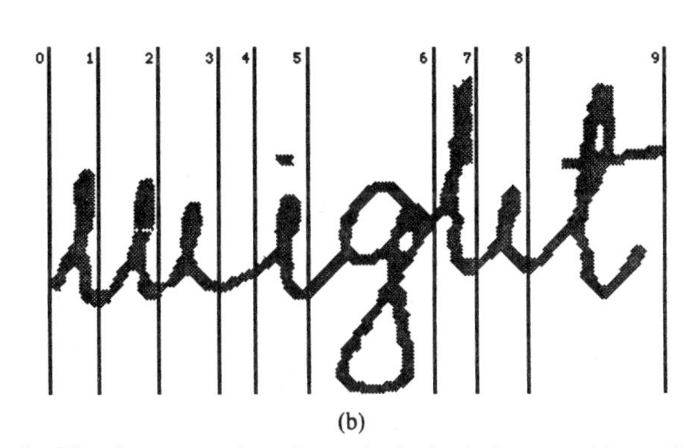

(b)

Fig. 2. *I*-level representation of a word, obtained after smoothing and slant correction. (a) With reference lines separating three zones. (b) With vertical lines corresponding to PSP's.

lowed by its chain, or sequence of *chain codes* [17]. The *C*-level representation of Fig. 2(a) represents the fact that there are six contours present: the main outside contour, followed by four inside contours ("holes") within it (three in g and one in t), and another outside contour (the dot over i).

The event detection operation takes as input the *C*-level representation, reference lines, and PSP and from these computes another exhaustive and unique description, the *event-level* (*E*-level), which is a description in terms of certain features, or *events*, and their locations. Such a description of *might* in Fig. 2 would state that between PSP 0–1 there exists an i-shape (peak), between PSP 0–2 there exists a lower curve, between PSP 1–2 there exists an i-shape, between PSP 4–5 there exists a dot, between PSP 5–6 there exists an upper curve and left curve, and so on. In PS 4, no events are registered, leading it to be interpreted as a blank or null letter. The actual *E*-level representation is a binary-valued matrix consisting of events and their locations. The columns of the matrix represent

16 events, and rows represent their location in terms of PS's, either single ones or sequences of 2.

Once the events are detected, the next step is letter hypothesization. The *letter-level* (*L*-level) consists of a series of alternative letter strings, or prefixes, that account for parts of the original *I*-level. A search procedure examines the *E*-level matrix top to bottom (i.e., left to right spatially) and also proceeds in parallel through a lexicon represented as a trie. Given a PSP, this generates an *L*-level description, one letter long, of the portion of the word from that point to at most three points to its right. Thus, starting from the left end of the word, or PSP 0, it generates a list of letter hypotheses. Each entry in the list contains a letter hypothesis, its rating on a (−1, 1) scale, and the number of PS's it covers. The entries are ordered by decreasing values of rating. For the example of Fig. 2, this list is of the form

((m 0.84 3)(n 0.73 2)(u 0.73 2)(w 0.60 3)(w 0.51 2)

(c 0.30 1)(j 0.30 1)(s 0.25 1)(s 0.25 1)(e 0.21 1)

(g 0.16 1)(v 0.10 2) · · · (f −0.74 1)).

This list is sent for word formation, or lexicon lookup, where inadmissible hypotheses are pruned, and the one remaining with the highest rating is expanded. Expansion of the best current hypothesis is done by making its rightmost PSP the starting point for new letter hypothesizing. In our example, the current best hypothesis is m, which covers three PS's. The next letter hypothesization starts from PSP 3 with the corresponding *L*-level description:

((′ 0.60 1)(u −0.46 2)(y −0.66 2)(v −0.73 2)

(n −0.74 2)(h −0.82 2)(k −0.83 2)

(x −0.83 2)(w − 0.85 2) · · · (m − 0.93 3)).

All these letter hypotheses are attached to the current top solution, *m*, from the right and sent for lexicon lookup and rating recomputation, giving a list of prefix hypotheses. The top hypothesis is replaced by its high-rated descendants in the list, and the whole list is reordered according to descending score values. The representation here has the form (prefix, rating, PS length, letter PS length(s)). The hypotheses in iteration 2 for our example are

((n 0.73 2 2)(u 0.73 2 2)(m′ 0.72 4(3 1))(w 0.60 3 3)

(w 0.51 2 2)(c 0.30 1 1)(j 0.30 1 1)

· · · (s 0.25 1 1)(e 0.21 1 1)(mu 0.19 5(3 2))

(′h − 0.31 1 1) · · ·)

where an entry such as (m′ 0.72 4(3 1)) means that the string m′ has a rating of 0.72 and accounts for the first four presegments as follows: the first three correspond to m and the next one corresponds to a ligature. In this case, m′ is the correct interpretation, but has fallen from first place and will have to wait to regain first place until it gets expanded further. This happens as the hypotheses above it get expanded, and as the evidence supporting their extensions diminishes, they drop in ranking or disappear. By iteration 4, the correct candidate resumes top position and retains it until iteration 6, as m′ig as seen below:

((m′ig 0.77 6(3 1 1 1))(un 0.71 4(2 2))

(nu 0.71 4(2 2))(w 0.60 3 3)(us 0.49 3(2 1))

(ne 0.47 3(2 1)) · · · (mu 0.19 5(3 2))).

The letter h does not get a high rating, and m′igh drops again in iteration 7. Finally, by iteration 20, all the other alternatives have diminished in rating themselves, the correct interpretation is topmost once again, and in the next step, it yields a hypothesis that is a legal lexicon word (i.e., not only a prefix) and also accounts for the whole word (nine PS's). When this happens, that word is returned as the *word-level* (*W*-level) output. The final rating for *might* is of the form

(m′ight 0.48 9(3 1 1 2 1)),

which says that the word *might* in the lexicon accounts for a segmentation of the input image into nine parts, with the first three accounting for m, the next accounting for a ligature that is not a letter, the next accounting for the letter i, the next accounting for the letter g, the next two corresponding to the letter h, and the last one corresponding to a t.

III. Preliminary Processing

In this section, we will describe the preliminary processing operations of smoothing, slant correction, reference line finding, and presegmentation. The operations depend on noise encountered and the writing constraints. Smoothing and slant correction are applied initially to the raw input to secure standardized input. Reference line finding and presegmentation are additional operations which extract important auxiliary information.

A. Smoothing

The first step is to transform input data to a form which is manageable by further processing modules. Its nature is a function of input data quality and assumptions about input characteristics. The natural input medium for off-line systems is a digitizing camera that scans prerecorded script. A digitizing tablet has different input noise than an optical digitizer.

A typical binary-valued input word obtained from a tablet is shown in Fig. 3(a). Two of the requirements imposed by certain system modules, which are generally not satisfied by these data, are continuity of the main body of script and relative contour smoothness. They can be satisfied by applying smoothing by regularization, where each pixel remains (becomes) 1 if and only if the number of 1 pixels in a window centered on it exceeds a given

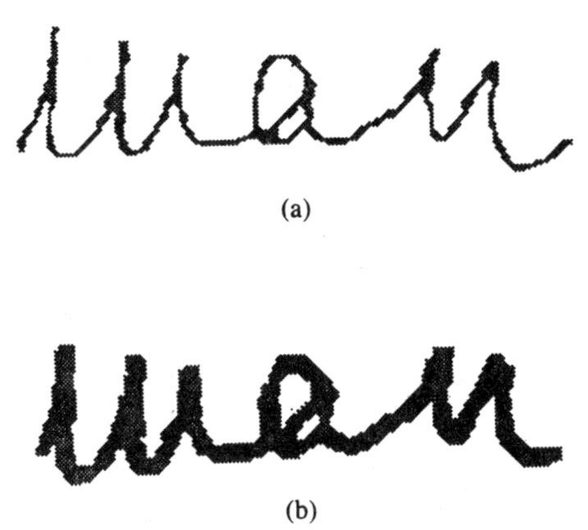

Fig. 3. (a) Sampling of a word on a bit pad 2 in long. (b) Results of smoothing (a) with a 3 × 3 window and threshold of 2.

threshold [Fig. 3(b)]. At this stage, most of the contour roughness due to sampling characteristics is removed.

B. Slant Estimation and Correction

One of the most obvious measurable factors of different handwriting styles is the angle between most longer strokes in a word and the vertical direction (y-axis), referred to as the word slant. Very poor handwriting often does not exhibit consistency in its stroke orientation and in that sense is not subject to valid slant measurement. However, a sizable portion of handwriting does.

There are two basic ways to approach this variability: to look for features that are invariant relative to slant [20], [26] or to normalize all data to a standardized form with no slant [9], [11], [14]. Normalization has certain advantages; e.g., presegmentation can be based on vertical single-run zones, a PSP can be specified by a single number, the x-coordinate, and peaks and curve orientations can be brought to a standardized form.

Our efficient slant correction algorithm is illustrated in Fig. 4 (a detailed description is in [33]). For a given word [Fig. 4(a)] first remove all horizontal lines which contain at least one run of length greater than the parameter *maxrun* [Fig. 4(b)], and additionally remove all horizontal strips of height less than the parameter *stripheight* [Fig. 4(c)]. Portions of each strip separable by vertical lines are isolated in windows [Fig. 4(d)]. For each window, the centers of gravity for its upper and lower halves are computed and connected. In case of an empty half (one with no 1 pixels) the whole window is discarded; in Fig. 4(d), only windows with angle values are valid. The slope of the connecting line defines the slope of the window, and the average for all windows defines the slope β of the word. The slant-corrected image is obtained by applying the following transformation to all 1 pixels with coordinate points x, y in the original image:

$$x' = x - y \times \tan(\beta - \text{def}), \qquad y' = y$$

where def is a parameter specifying the default (normal) slant. The results of this operation are shown in Fig. 4(e).

Slant correction needs to precede other preprocessing; i.e., it is applied before smoothing. This is because

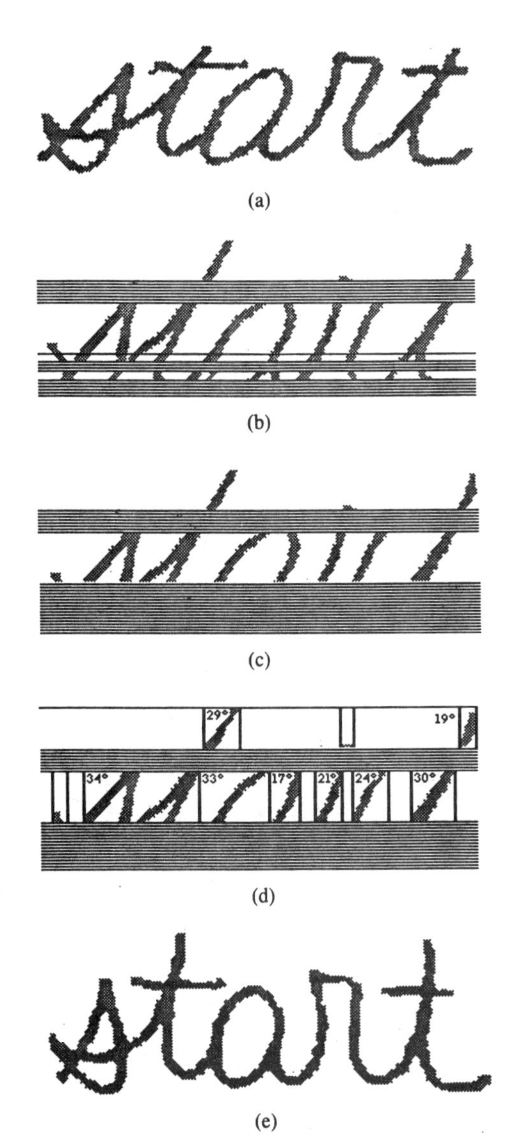

Fig. 4. Slant correction. (a) Original slanted word image. (b) Horizontal lines containing long runs are removed from consideration. (c) Remaining horizontal strips of small width are also removed. (d) Results of isolating windows from strips with lines connecting centers of gravity and slant angles shown for each window. Those with no values are disregarded. (e) Final results of slant elimination.

smoothing tends to change the image topology more than desired and the correction operation usually creates rough contours to the discrete data. Slant correction as described here is most effective for script with a fairly consistent slant within each word. Its limitation is that it globally operates on the whole word as an entity.

C. Reference Line Finding

In determining features for letter/word identification, we first determine the relative vertical extent of script portions, i.e., parts of script that are above or below the main body of the cursive line (Fig. 5). The task of detecting this main body is known as zone or baseline determination. There is psychophysical evidence of early use of such information by humans in lower-case print reading—ascenders and descenders are among the most prominent features used for defining shape for letters. When they are concatenated to determine the shape of an input word, this

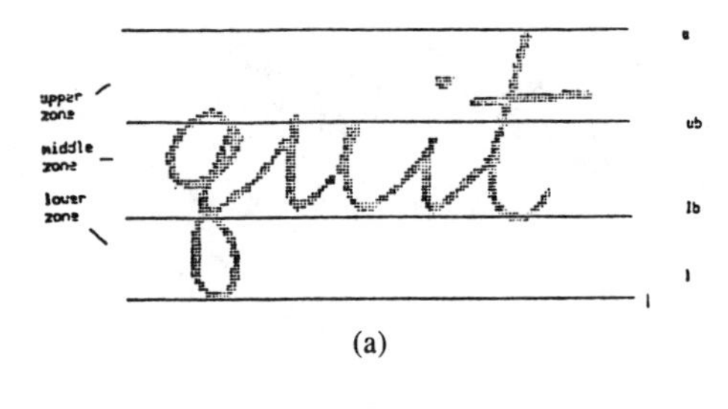

(a)

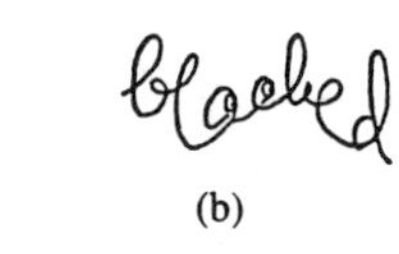

(b)

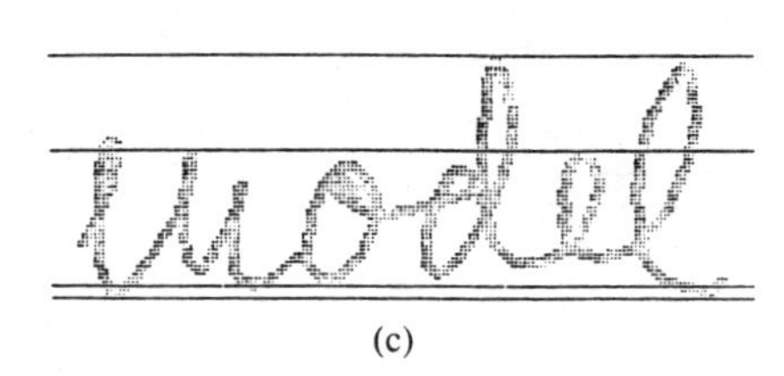

(c)

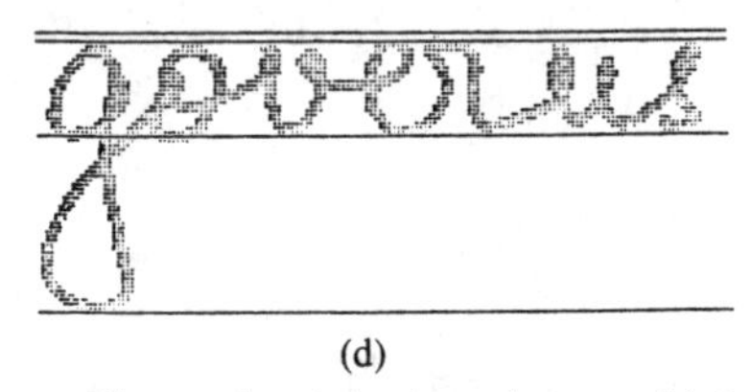

(d)

Fig. 5. (a) Reference lines and main horizontal zones. (b) Case of difficult global baseline finding: *hill-and-dale* writing. (c) Word with upper and middle zones. (d) Word with lower and middle zones.

information has been shown to be very successful in narrowing down the choice of words [4], [19], [32]. Reference line finding should ideally be performed for each letter separately, especially in cases of variable letter size and *hill-and-dale* handwriting [Fig. 5(b)]; however, such an approach requires prior segmentation.

Keeping reference line finding as a preprocessing operation, we look for four lines: the lower line, lower baseline, upper baseline, and upper line, which define three zones [Fig. 5(a)]. Some lines may overlap due to the fact that some of the outer zones may be nonexistent, e.g., with no descenders [Fig. 5(c)] or ascenders [Fig. 5(d)]. The method is based on an analysis of the horizontal density histogram. A word and its histogram of horizontal densities, called the *horizontal projection profile*, is shown in Fig. 6. A simple procedure now is to look for peaks of the first derivative of the density function and claim the lower and upper baselines at the maximum and minimum, respectively. However, such a procedure tends to be deceived by t-crosses; e.g., in the case of Fig. 6, the upper line, lower baseline, and lower line are correctly generated, whereas the upper baseline is placed just above the t-cross. Our method uses thresholds to determine the shoulders of the histogram. The k highest density values are discarded, max is the highest remaining value, the first y for which $d(y) > t \times \max$ (where d is the density and t is the threshold) is deemed the lower baseline, and the first next y such that $d(y) \leq t \times \max$ is the upper baseline. Additional heuristics alleviate a problem of accidental excursions over or under t. Results obtained by the application of this procedure are shown in Fig. 7.

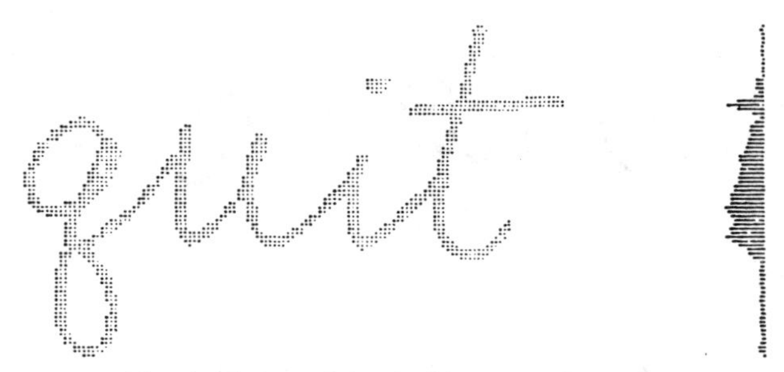

Fig. 6. Horizontal density histogram of a word.

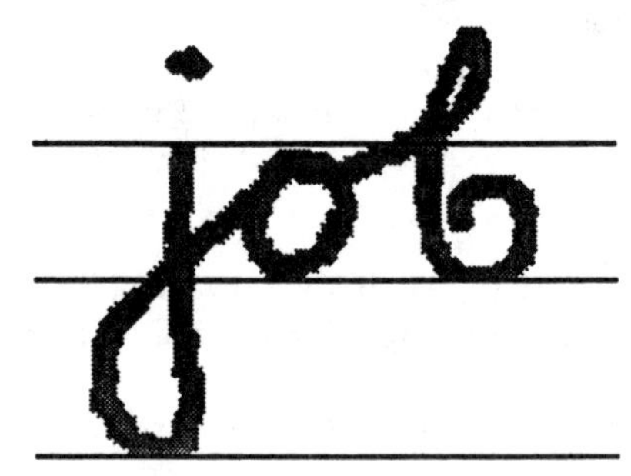

Fig. 7. Final reference line finding scheme used with correct results.

D. Presegmentation

There are two different word segmentation philosophies. One is to segment the word into what appear to be letters and apply letter recognition techniques to each segment independently. The problem here is that failure to segment the word properly is a fatal error since either the word length is erroneously estimated or suggested segments do not correspond to actual letters. One way of remedying this is by postprocessing the output with contextual string-correction algorithms [5], but these usually rely on information which can be used to avoid commitment to improper segmentation in the first place. The other approach avoids doing any kind of formal segmentation, looking at words as entities and using statistical methods to classify word samples [7], [16]. Avoiding strict segmentation is partly justified by psychophysical evidence as well, according to which such a process does not usually take place in humans. Instead, a more flexible procedure that relies on global word contour (*blob*) shape and relative positioning of identifiable features within a word is used [4], [19], [32]. The compromise solution adopted here is to do loose segmentation, i.e., to identify all *potential* points (as defined by some criterion) as PSP's [15], [26].

1) Segmentation Process: The task is to partition horizontally the word into PS's by vertical lines such that the

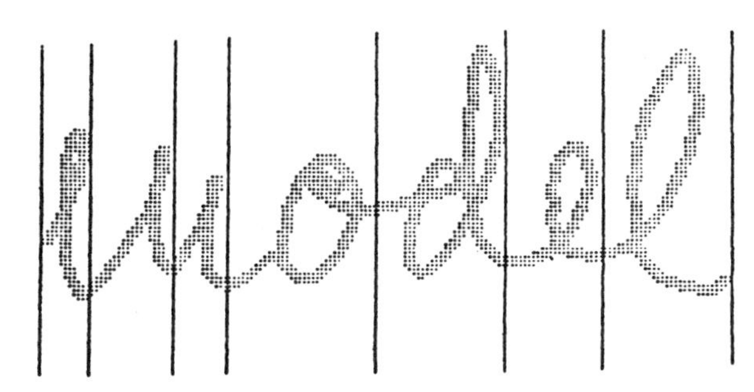

Fig. 8. The ligature between letters o and d does not possess a local minimum of its own, but the surrounding letters do, and a PSP is generated.

number of PS's is a small as possible with respect to all possible letter combinations; e.g., the partial inscription corresponding to the first letter in the word *model* in Fig. 8 could represent an m, ui, or in sequence, and we would like to have each of these potential letter borders marked as a PSP.

```
generate-PSP-left
    singlerun turned off;
    set cy to y0;
    repeat
    if (((runs(y) > 1) ∨ (density(y) > md)) & singlerun) then
    exit loop;
    else if ((runs(y) = 1) & (density(y) ≤ md) & (not singlerun)) then
    set singlerun;
    decrease cy by 1;
    until (cy = 0 ∨ cy < last PSP);
    if singlerun then
    set PSP to middle value y of singlerun stretch;
```

Our definition of PS assumes nonslanted (i.e., vertical) script, as well as connectivity of the lower contour of the word; i.e., the main continuous body of script must contain the leftmost and rightmost points in the word.

The algorithm searches for local minima along the lower contour of the inscription. These points usually correspond to PSP's when they appear on ligatures in the middle zone, e.g., PSP 3, 5, and 8 in Fig. 2. Ligatures coming off letters with lower loops (like g, j, and f) do not possess this property, but they do correspond to local minima in the lower zone of the letter just preceding them, e.g., PSP 6 in Fig. 2. On the other hand, such minima do not mark segmentation points generally in two other cases. One is when they are found in the middle of such letters as m, u, and the like, but they still correspond to the intuitive notion of a PSP. The other one is in the middle of letters a, o, b, d, and r, in which case they might be redundant, but not always, since the ligature following or preceding them might not possess a minimum of its own. Some of these points are illustrated in Fig. 8, where the PSP between o and d is generated by either of the local minima in o or d. The algorithm in formal terms follows.

```
presegment
    trace lower contour left-to-right
      and produce a chain code sequence (chain);
    start with first chain element, cc;
    repeat
      if (cc is at local min) then
        generate-PSP-left;
        generate-PSP-right;
        if (no PSP generated in this iteration) then
          emergency-generate-PSP-right;
      move to next chain element cc;
    until chain exhausted;
    compact-PSP;
```

where each PSP is defined by its y-coordinate. Each chain code cc corresponds to a contour pixel from which it emanates. The procedure generate-PSP-left is described next:

where y_0 represents the current value of y, *runs* is the number of streaks of one or more pixels of the same value, *density* is the vertical projection profile value, *md* stands for maximum density and gives the largest vertical profile of a stroke in a single run zone. Procedure *generate-PSP-right* is analogous to generation to the left with the following modifications:

- the loop termination condition is $(y - y_0 < ms)$ and
- cy is increased instead of decreased,

where *ms* is the maximum shift parameter. Further, *emergency-generate-PSP-right* is an ad hoc procedure for cases when missed segments are suspected, and is defined as

```
emergency-generate-PSP-right
    set cy to y0;
    for (cy from y0 to y0 + ms) do
      find cy with lowest density(cy);
```

Finally, *compact-PSP* is given as

```
compact-PSP
    set cps to first(leftmost) PSP;
    for current cps do
      for set {cps, all cps's such that cps' − cps < marg}
        find element msp with minimum density(msp);
```

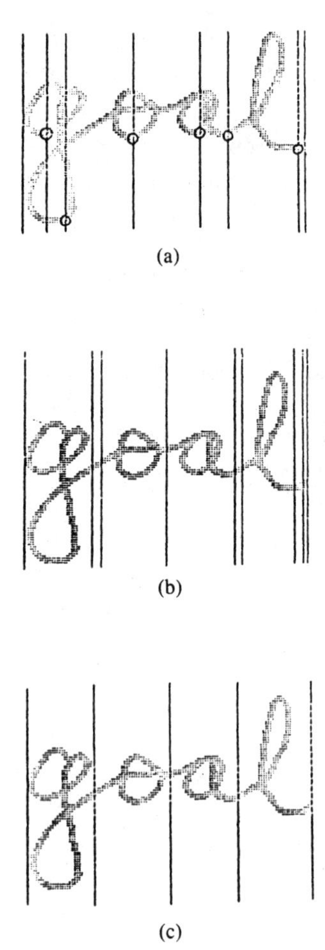

(a)

(b)

(c)

Fig. 9. (a) Local minima along a word's lower contour. (b) Presegmentation, first step (before compaction). (c) Presegmentation, second step (after compaction).

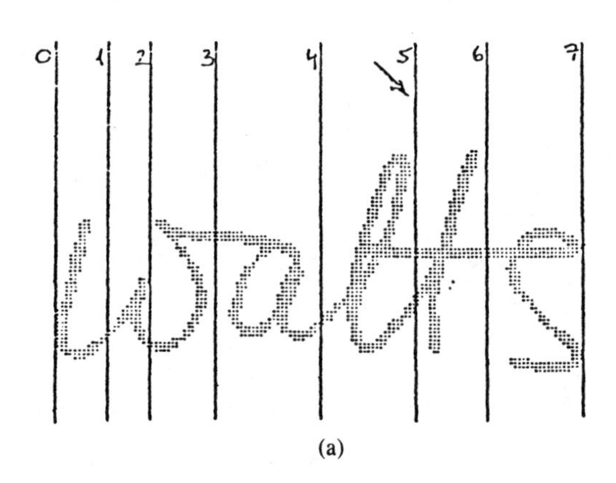

(a)

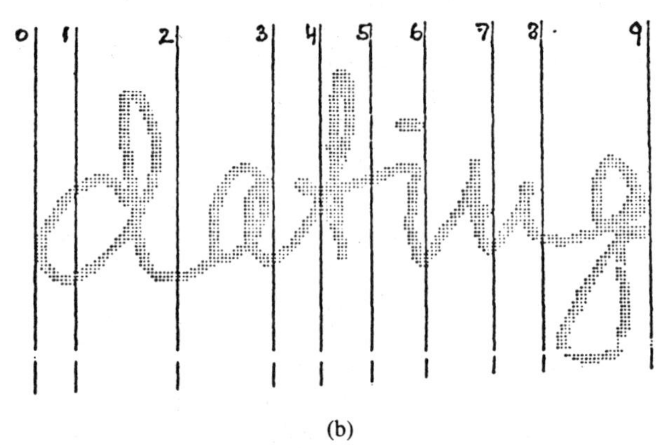

(b)

Fig. 10. (a) Justified "emergency" segmentation: PSP 5 (arrow) is not in a single-run zone. (b) Unsuccessful emergency segmentation: PSP 1 is not an intuitive PSP.

if (first iteration) **then**
  output *cps*;
set *cps* to first *cps'* such that $cps' - cps \geq marg$;
if ($cps \neq$ last *PSP*) & (not first iteration) **then**
  output *msp*;
output last *PSP*;

where *marg* is the compaction margin.

Illustrating the above procedures, Fig. 9(a) shows the local minima in the lower contour of a word image. Fig. 9(b) shows the PS points found before compaction, and Fig. 9(c) shows the results of compaction. Therefore, every minimum can naturally give rise to an intuitive PSP, either directly or less than a letter length to its left or right. We define zones in which PS points are eligible to be placed as those in which flag *singlerun* in procedures *generate-PSP-right* and *generate-PSP-left* is on. Each such zone is characterized by a continuous sequence of single vertical profiles, each with a single run and density less than *md*. From every local minimum, the algorithm looks left and right, within certain limits, to locate such zones. If found, it places a PSP in the middle of it, thus creating overall anywhere from zero to two new ones. The limits mentioned are, to the left, the immediately preceding (rightmost so far) PSP (thus ensuring monotonicity of PS points relative to the local minima that generated them), and to the right, a heuristic boundary *ms*, an upper bound of the PS length in pixels. In case of failure to generate anything to either the left or right, the algorithm assumes that a PS point has been missed and calls the "emergency" generation to the right. With the assumption of nonslanted (vertical) script, such cases are rare, but they do come up occasionally, usually with long t-slashes and t-slashes connecting through letters or with broad loops in the outer zones *overshadowing* neighboring letters. Finally, the compaction procedure is based simply on proximity.

Results of applying the presegmentation algorithm are shown in Figs. 10 and 11. The parameter values, determined empirically, were $md = 0$, $ms = 20$, and $marg = 8$. Fig. 10 shows cases where the segmentation was justified [Fig. 10(a)] and where it was not [Fig. 10(b)]. This

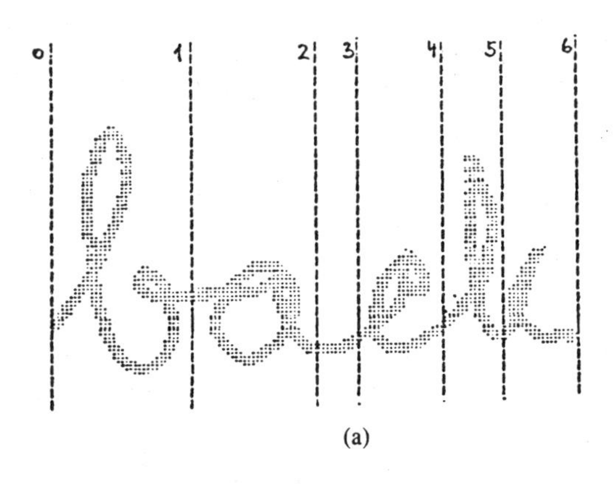

(a)

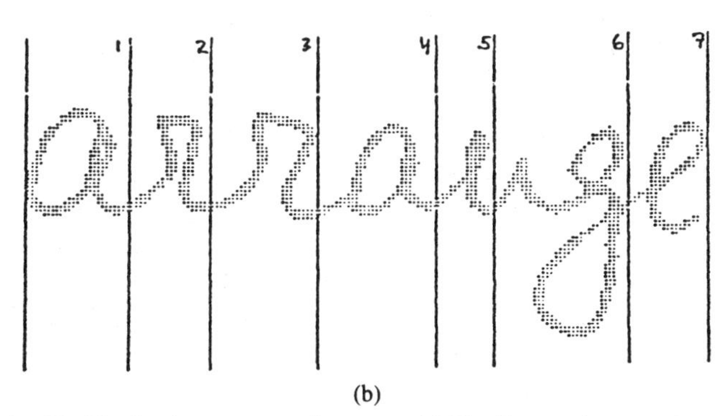

(b)

Fig. 11. Nonfatal presegmentation error. (a) The ligature between letters b and a is assigned two PS points, but the PS between them will be interpreted as a null letter and recognition will be successful. (b) Although the true segment between n and g will be missed, PS 5 is still interpreted as n, PS 6 is interpreted as g, and recognition will be successful.

segmentation method is prone to two kinds of departure from the intuitive notion of PSP's, albeit rarely. One, the nonfatal one, is redundant PS points within letters, like in Fig. 10(a), or when long ligatures are not compacted and get assigned two points [Fig. 11(a)]. Such cases, if they appear with any consistency, can be learned (see Section VIII). The fatal case involves segments missed altogether, due to either a lack of surrounding local minima or too-crowded lettering, when actual PSP's are not in a single-run zone and *ms* is too large to stop generation of the next PSP at that step. This results in fewer segments than letters or in poorly presegmented words. In certain such cases, if the number of letters does not exceed the number of PSP's, recovery by the main module is possible [Fig. 11(b)].

IV. Contour Tracing

Preliminary processing yields reference lines, PSP's, and a smoothed and slant-corrected image at the *I*-level. The next task is to transform it from this form to the *C*-level, where each word is represented as a single string of chain codes [17] and additional symbols that define contour types and beginnings. The contour types distinguished are the main outside contour of the word, inner contours within it (holes), and additional outside contours (blobs). Contour beginnings specify pixels from which each first chain code starts.

The procedure for generating the *C*-level is called *topological contour tracing* (TCT). Its purpose is to encode in a single linear string the shape and topological properties of *all* contours of an arbitrary planar binary image. The topology is described by a tree description, as shown in Fig. 12; each node corresponds to a connected component of the image or, equivalently, to a contour (border) surrounding it [8], [25]. The main idea is as follows. For each contour, upon tracing it and generating its chain to traverse boundary points, and starting from each one, scan the image to the right until either an already-traced contour is found (after which a new search is started from the next boundary pixel) or a new contour is found. In the latter case, the whole process is recursively applied to it, and some string transformations are performed, before going on to the next search. Therefore, every invocation of the main recursive routine that performs the operation described is always associated with a particular contour that it is either tracing or traversing. The following general assumptions are made.

1) A rectangular grid is used; zero pixels are considered white, and nonzero are black; connected (continuous) regions of black and white are called blobs and holes, respectively.

2) The image has a continuous white boundary.

3) *Eight-connectivity* is in effect for black pixels; only black contours are traced, and they consist of all pixels that have at least one white four-neighbor.

4) The image is initially binary (pixel values zero and one); upon being traced, each pixel has its value increased by 1, so at the end, boundary points will have value two, double boundary points will have value three, etc.

5) Tracing is done so that the black portion is always on the left.

6) The chain code used satisfies $-3 \leq s \leq 4$ and is illustrated in Fig. 13(a).

The notation used in the encoding method follows. Q is the entire current string $q_1, q_2, \cdots, q_n$ of *extended chain codes*. $Q(k)$ is the kth element of Q. Extended chain codes, sequences of which we call extended chain encodings, comprise three kinds of symbols:

- integers n, one per contour, representing the topological level of depth; the top level blobs, i.e., those from which the boundary can be reached through an all-white path, are considered to be at level three, and so on;
- coordinate pairs (x, y), also one pair per contour, which we consider to be suitably encoded to be distinct from the other symbols and representing starting points for their contours' respective chain encodings (chains).
- chain code values s, also distinctly encoded, a sequence of which makes up a contour's chain.

Each contour C is represented in Q as a sequence n, *startpt*(C), *chain*(C), where the meaning of the elements is obvious.

As an example, the image in Fig. 13(b) is represented by the string shown in Fig. 13(c). The conventional adjacency tree [8], according to the contour numbers shown, can be represented as (0 (1 (8) (2) (3 (4))) (7) (5 (6)) (9)),

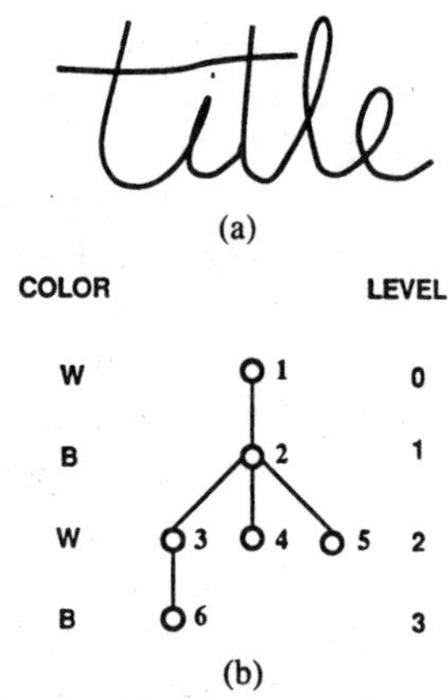

Fig. 12. (a) A word image with complex topology: here, the dot on i is a second-level outside contour. (b) The topological tree of (a): each node represents a connected component; e.g., node 1 is the white background and node 6 is the dot on i.

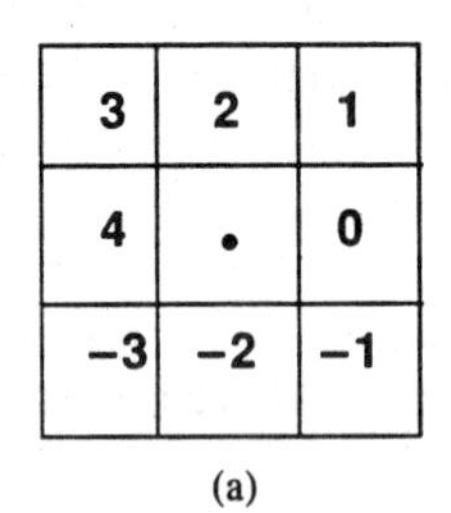

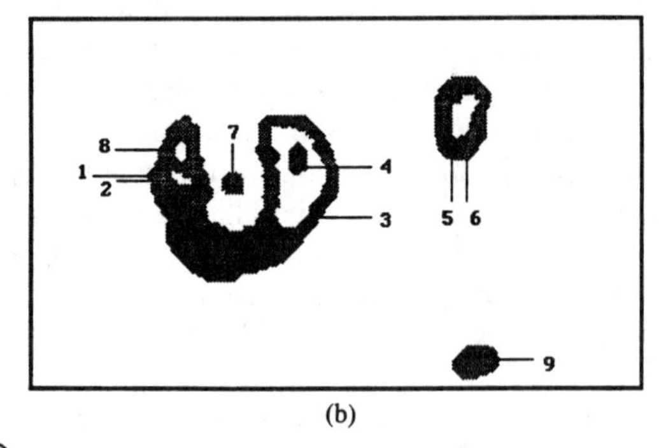

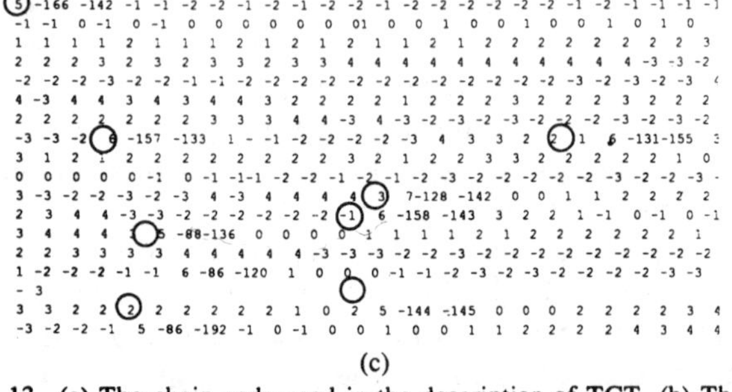

Fig. 13. (a) The chain code used in the description of TCT. (b) The operation of TCT: contour numbers specify the order in which they are detected. (c) The extended chain encoding of (a): the circled numbers are the encoded level values n that start a contour's chain, followed by encoded contour beginnings (x, y) and chain codes. The order in which contours appear in the string is 1 8 3 4 2 5 6 7 9.

where the atom (or number) in the beginning of each list identifies the contour and 0 represents the picture boundary. Thus the outermost parentheses correspond to the background, the four entries in it at the next level correspond to the four outer contours, etc., and single-element lists represent leaf nodes. The notation used in Fig. 13(c) is a different one, in which depth level is encoded by adding 4, so top-level blobs start with a 5, etc. For any contour of a certain depth, its topological ancestors are, for each lower depth value, the first contour of that depth found to the left in the string. Adjacent contours in the string with identical depth values are siblings.

V. Event Construction

A. The Event Set

The event description of a word image is in terms of certain geometric features, or *events*. Each occurrence of an event belongs to a portion of the word spanning one or two PS's (the latter is possible only for certain events). The choice of events, representing topological, spatial (zonal), curvature and cusp-shape properties, is in reasonable agreement with results from research on relevant features in human recognition of lower-case print, e.g., Bouma's classification of mutually confusable letters [4], [29]. The set of 16 events used is as follows (an asterisk indicates that the event can cover two PS's): 1) upper stroke (us); 2) long upper stroke (lus); 3) lower stroke (ls); 4) upper curve (uc, *); 5) left curve (lc); 6) lower curve (loc, *); 7) right curve (rc, *); 8) i-shape (peak, cusp) (is, *); 9) n-shape (ns); 10) long lower contour (llc); 11) large middle loop (lml, *); 12) small middle loop (sml, *); 13) upper loop (ul, *); 14) lower loop (ll, *); 15) slash (sl, *); and 16) dot (do, *).

A description of each of these events follows. An upper stroke is triggered by a substantial portion of the main contour in the upper zone of a segment, but not containing an upper loop and of length less than a certain fraction of the overall outside main contour length for that segment. In other words, the upper stroke event is mutually exclusive of both the upper loop and long upper stroke events. Analogously, a lower stroke is triggered by a substantial portion of the main contour in the lower zone of a segment not containing a lower loop. Similarly, the lower stroke event is mutually exclusive of the lower loop event. The system distinguishes simple outer strokes from outer loops because it helps to discriminate between letters that have upper/lower extensions.

The four curve events are found when there is at least one occurrence of a smoothly curved (convex or concave) contour in the middle zone such that a perpendicular drawn to its average tangent extends in the named direction. The most frequent curve to span two PS's is the lower one. The two peak events, the i-shape and n-shape (single and double peak), are based on the notion of a peak. A peak is a steep outward motion of the contour in the middle zone followed by a similar inward swing anywhere within a two-PS stretch, along either the upper or lower contour. By *outward* motion, we mean upward for the upper contour and downward for the lower one and assume an analogous definition for *inward*. Notice that in a peak an outward swing does not, according to this informal definition, have to be immediately followed by an inward one; rather, we allow for a certain flat stretch between them. Fig. 14(a) and (b) shows examples of i-shapes spanning one and two PS's and an n-shape, respectively.

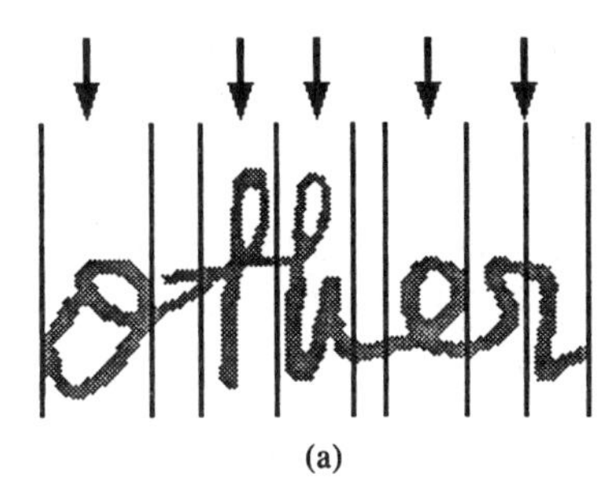

(a)

(b)

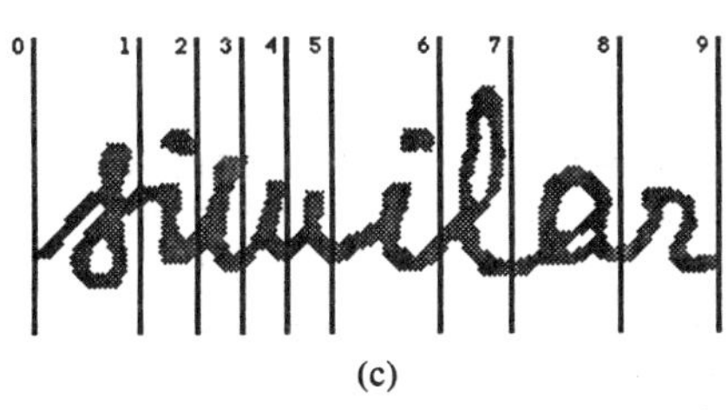

(c)

Fig. 14. (a) Some i-shape events within a word: arrows indicate presegments with single-PS i-shapes; the rightmost arrow indicates a double-PS i-shape. (b) An n-shape event in PS 6. (c) A long lower contour event in PS 1.

The long lower contour event is triggered only for segments with no outer stroke/loop events when, within a PS, the upper contour length is less than a certain fraction of the lower contour length. This event usually occurs in letter o, in s when the retrograde stroke does not cross the initial upward stroke, and for certain styles of letters r, m, and n. Fig. 14(c) illustrates this event.

Middle, upper, and lower loop events are based on loops found in the three corresponding zones. The only difference between large and small middle loops is their length. In case of multiple loops in the middle zone, only the largest one is considered; therefore, small middle loop events are not detected in the presence of large ones. Loop events are found in a multitude of letters (a, d, e, g, etc.). Slashes and dots are blobs disjoint from the main connected portion of the word, which have the requirement to appear in the upper zone. The only difference between the two is contour length. Obviously, letters i and j are prime examples of letters with a dot (or perhaps slash), but some versions of letter t might contain crosses disconnected from the rest of the letter, which is why we distinguish the two events.

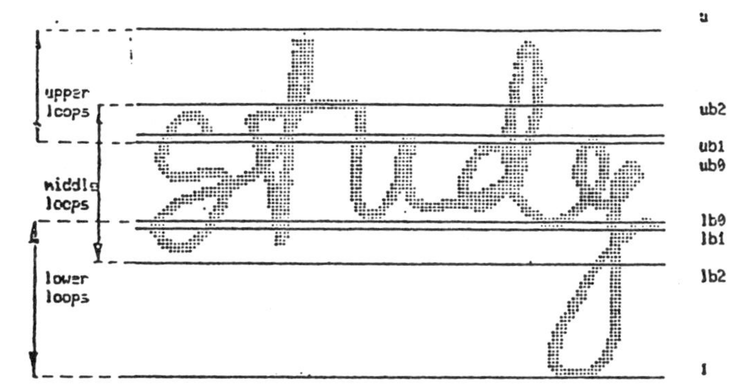

Fig. 15. Main and auxiliary baselines.

B. Event Detection

All the events are detected and assigned to various PS's in a single scan of the chain encoding Q described in Section IV. We define the lower contour of a word image as the part of the main outside contour from its starting point (lowest of the leftmost pixels) to the lowest rightmost pixel, and the remainder of the contour, which is traced right to left back to the starting point, as the upper contour.

In addition, given the upper and lower baselines, ub_1 and lb_1, we define the following auxiliary lines: $ub_0 = ub_1 - b_1$, $lb_0 = lb_0 + b_2$, $ub_2 = ub_1 + b_3$, and $lb_2 = lb_1 - b_4$ where $b_1 - b_4$ are empirically determined parameters. Fig. 15 shows the positions of the main and auxiliary baselines.

A set of points from a contour C are said to be entirely within the upper zone if, for every pixel, p: $y(p) \geq ub_0$. It is entirely in the middle zone if, for all p, $lb_2 \leq y(p) \leq ub_2$, and it is entirely in the lower zone if, for all p, $y(p) \leq lb_0$.

By event location, we mean the single-PS or two-PS stretch where the event has occurred. These two cases will be called *single-PS* and *double-PS* events, respectively. For now, we also assume only single upper and lower contours within a PS.

Event generation is carried out through the event detection modules, which operate autonomously and communicate through a global database, and by a set of compatibility rules, which are applied after preliminary event generation is accomplished. There are six modules for detecting related groups of events: outer strokes, curves, peaks, contour length comparison, loops, and blobs, for the details of which the reader is referred to [6].

VI. Letter Hypothesization

Up to and including the process of event construction, successively more refined *unique* descriptions of the *whole* word image have been generated. At the L-level, there are multiple representations competing with each other, which are not all generated at once. Competing letter hypotheses can account for exactly the same portion of the image, in which case there is only a classification difference between them, or they can only partially overlap and thus also display segmentation differences. After a series of strict hierarchical transformations, letter hypothesization is the first global module that interacts with higher ones in a heterarchical manner and, as such, when called upon might not produce a set of full word representations on the L-level; rather, it generates partial representations of the size and at the positions that are required by higher-level modules.

The letter hypothesization process consists of two steps: *letter selection* and *ratings computation* (or letter hypothesization proper). Letter selection is an operation in which certain letter hypotheses are inexpensively generated based on partial E-level data. In rating computation, all of the E-level evidence is taken into account while generating an overall likelihood score for each hypothesis. This section describes how to test hypothesized letters, or find the *a posteriori* probability of individual letter occurrences based on the observed set of events.

The objective is to compute *a posteriori* probabilities for the selected letters. For a letter l, therefore, the idea is to find $P(l \mid \bar{e})$ where $\bar{e}$ is the vector of all events which have occurred. By Bayes's decision theory we have

$$P(l \mid \bar{e}) = \frac{P(\bar{e}, l)}{P(\bar{e})} = \frac{P(\bar{e} \mid l) \times P(l)}{P(\bar{e})}.$$

The *a posteriori* probability sought depends on the *a priori* probability of l, $P(l)$. It would be desirable to eliminate this dependence since at this stage we are interested only in what an image looks like and should assume all letters equally probable; decisions concerning lexical plausibility (e.g., letter frequency) are to be made by higher-level modules. Therefore, the value $r(l, \bar{e})$, which is the rating of the letter L we are looking for, satisfies:

$$r(l, \bar{e}) = \frac{P(l \mid \bar{e})}{P(l)} = \frac{P(\bar{e} \mid l)}{P(\bar{e})}.$$

Assuming mutual independence of events from $\bar{e}$, the logarithmic rating of letter l is

$$r'(l, \bar{e}) = \ln r(l, \bar{e}) = \ln \left(\prod_i \frac{P(e_i \mid l)}{P(e_i)} \right)$$
$$= \sum_i \ln \frac{P(e_i \mid l)}{P(e_i)} = \sum_i r'(l, e_i).$$

Values $r'(l, e_i) = \ln r(l, e_i)$ measure dependence between e_i and l. For values of $r'(l, e_i)$ that are greater than zero they are positively correlated, for those less than zero they are negatively correlated, and when equal to zero the occurrence of event e_i has no bearing on the prediction of letter l. We assume that values $r'(l, e_i)$ can be easily determined and are known in advance.

The abstract set of events $\bar{e}$ is defined as follows. Assume that the number of presegments s that a letter can cover is from one to three only, which is an empirically determined fact. Then, presuming the starting PSP is fixed and thus not explicitly stated, the set of events correlated with l is

$$\bar{e}^s = \bigcup_{c \in C_s} \bar{e}^c.$$

In other words, the $\bar{e}^s$ consists of groups of events coming from different members of the set of contexts C_s. A *context* is any fixed one- or two-PS span within the hypothesized segment. Obviously, $|C_1| = 1$, $|C_2| = 3$, and $|C_3| = 5$. Notice that events that do not strictly belong to PS's inside the hypothesized segment are not included in the set $\bar{e}^s$ since presumably their influence on l is minimal. Thus, we have

$$r(l, \bar{e}^s) = \frac{P(l \mid \bar{e}^s)}{P(l)} = \frac{P(\bar{e} \mid l)}{P(\bar{e})} = \prod_{c \in C_s} \frac{P(\bar{e}^c \mid l)}{P(\bar{e}^c)}$$
$$= \prod_{c \in C_s} \prod i \frac{P(e_i^c \mid l)}{P(e^{ci})}$$

where i runs through all events within a given context c. Thus,

$$r'(l, \bar{e}^s) = \sum_{\substack{i \\ c \in C_s}} \ln r(l, e_i^c) = \sum_{\substack{i \\ c \in C_s}} r'(l, e_i^c). \quad (1)$$

According to (1), $r'(l, \bar{e})$ is computed for a given l by simply adding up all $r'(l, e_i^c)$ values corresponding to relevant events that have occurred within the contexts of l's segment. The summands in the right-hand side of (1) are referred to as r'-values, and we will drop $\bar{e}^s$ when referring to a letter's overall logarithmic rating: $r'(l)$.

We have thus far dealt only with computations for letter hypotheses that have already somehow been chosen to be rated. The purpose of the process of letter hypothesizing proper is to decide by some heuristic which letters have at least-minimal supporting evidence based on the E-level description and to send them for rating computation. For a given PSP, this is done by finding, for each segment length s ($s = 1, 3$), all letters for which there is a context c and a trigger event e_t^c within it such that $r'(l, e_t^c) > 0$. In case no events are detected for a given PS, the only hypothesis generated is a blank, with an empirically determined r'-value of 1.35.

VII. Word Hypothesization

The process of letter hypothesization generates multiple alternatives, which account for only a portion of the word at a certain location. The next step in word recognition involves finding answers to the following. 1) How do partial L-level representations get combined in an effort to account for the whole word image? 2) How do certain combinations get preferred over others in a quest for a unique solution? 3) When and where do we ask how many more letter hypotheses are to be generated? Part of the first question can be answered by specifying rules that govern these combinations, one form of which is a lexicon. The second question can be answered by stating a way of direct comparison between these combinations, and preferably in a way that does not depend on their length or position. The last problem can be solved by specifying an appropriate control mechanism. The following three sections look into the details of the three issues: lexicon representation, search strategy, and rating computation.

A. Representation of Lexicon

A well-known data structure for representing a lexicon is the trie which is defined as follows. A trie T is defined as a tree, each node of which is an ordered pair (l, e) where l is from alphabet A and $e \in \{\text{true}, \text{false}\}$, i.e., e is a Boolean end-of-word flag. The root of the trie is the only exception in that its ordered pair is (ϵ, t) where ϵ is the empty string. A certain string of letters corresponds to a node N; the letters of this string are those associated with nodes encountered in traversing the trie from the root to node N. A trie T_D is said to represent a certain lexicon D, if every lexicon word corresponds to exactly one node of T_D with $e = true$ and vice versa. We do not distinguish (unless otherwise specified) nodes from their corresponding strings (prefixes) since there is a one-to-one correspondence between the two.

With the above definition, it follows that word X is in dictionary D, i.e., $X \in D$, if and only if the expression

$$(X \in D') \,\&\, \big(e(X) = T\big)$$

is true where D' is the set of strings corresponding to the set of nodes in trie T_D and $e(X)$ is the Boolean flag value of word X.

Advantages of this representation of the lexicon are 1) storage space savings, particularly when there are numerous identical prefixes; 2) convenient knowledge representation for storing related information in its nodes; and 3) the structure naturally fits the search algorithm and hypothesis expanding rules.

B. The Search Algorithm

Given a word description in terms of events, the goal is to search the trie and come up with a full word as an answer. The search strategy employed here is modified from the so-called stack decoding algorithm [2], [21]. We have previously used the algorithm with a *contextual postprocessor* function, operating on input from an initial processor that has already made rough unique decisions about the elementary unit (e.g., letter) identities based on observed data [5]. We assumed that initial cursive script recognition was performed by a character recognition unit that partitions the word into segments and then classifies each as a character. The unit's output was assumed to transform each input letter as a memoryless channel according to a probabilistic finite-state machine. Correction was based on computing a distance between strings (observed and allowed) using the probabilistic model.

In the present implementation of the stack algorithm, word hypothesization takes place left to right. A partial word hypothesis H is defined as an ordered pair (s, p) where s and p are strings of letters and digits, respectively, and $|s| = |p|$. Therefore, s is a (possibly nonproper) legal prefix which is uniquely defined by a trie node, and if function pl gives the length of its string (letter) argument in terms of PS's covered, $p_i = pl(s_i)$, with $\Sigma_i\, p_i = pl(s)$. By hypothesis we mean such a prefix hypothesis, unless otherwise specified (e.g., letter hypothesis).

The main features of the stack algorithm are

1) heuristic search, where the hypotheses with the highest ratings are extended left to right and best-first, and

2) length-independent rating computation, performed as a function of the parent hypothesis' rating.

In practice we use a *depth-of-stack* parameter d: only the top d hypotheses are retained at each step. If no decision is reached after *maxsteps* iterations, the procedure stops and the word is considered unrecognizable (rejected). The procedure can produce alternative solutions, rather than stopping at the first one.

C. Rating Computation

The computation of ratings for prefix hypotheses should be based on the following two requirements:

1) length uniformity of values, allowing for comparison on an equal basis of hypotheses of different length, and

2) length hindependence of the computation, which is to be performed as a function of only the parent hypothesis' rating.

The way to satisfy both of these requirements is to extend the uniform scoring philosophy to the word level. So, we let $\bar{l} = l_1 \cdots l_n$ be a prefix hypothesis and also let $\bar{E}$ be the full set of occurred events for a word, i.e., its whole E-level representation. Then,

$$r'(\bar{l}, \bar{E}) = \ln \frac{P(\bar{E}|\bar{l})}{P(\bar{E})}. \tag{2}$$

$\bar{E}$ can also be viewed as the sequence $\bar{e}_1 \cdots \bar{e}_n$ of all event vectors belonging fully to each letter segment (which, we recall, can cover one to three PS's) since our earlier assumption was that double-PS events covering segment boundaries do not influence any letters and thus can be canceled out from both elements of the fraction in (2). Additionally, we assume elements of $\bar{E}$ to be mutually independent since they belong to disjoint regions of the word, and any statistical dependences stemming from lexical frequencies are invisible at this level. For similar reasons, we assume that individual event vectors $\bar{e}_i$, out of the whole string $\bar{l}$, depend only on the letter they constitute, l_i. Therefore,

$$r'(\bar{l}, \bar{E}) = \ln \frac{\prod_i P(\bar{e}_i|\bar{l})}{\prod_i P(\bar{e}_i)} = \ln \prod_i \frac{P(\bar{e}_i|l_i)}{P(\bar{e}_i)}$$

$$= \sum_i \ln \frac{P(\bar{e}_i|l_i)}{P(\bar{e}_i)} = \sum_i r'(l_i)$$

where $i = 1, \cdots, n$.

Obviously, both requirements are thus satisfied, and a new prefix rating is obtained by a single addition of the ratings for the letter and parent hypotheses.

VIII. Parameter Learning

Several parameters of the system are learned in an initial *training* phase. These parameters can gradually be altered with changes of writers and styles in a process of *adaptation*.

Hypothesizing letters depends on *r′-values*, which are the variable program parameters to be determined during training and adjusted during adaptation. The r'-values are the values $r(l, e_i^c)$ where e_i^c spans all events in various contexts. They essentially define the (probabilistic) rules of letter formation from a set of predefined and invariant primitives, namely, the event set. It is changes in these rules and, accordingly, r'-values that we presume to reflect most of the changes in writing style on the level of letter appearance. We now examine issues of their measurement and alteration.

A. Training

The ratings of letters (r-values) are learned with the user providing feedback to the system. For a given letter l and event e such that

$$e \in \bigcup_{\substack{i \\ c \in C}} e_i^c$$

where

$$C = \bigcup_{s \in S} C_s$$

is the set of all contexts ($S = \{1, 2, 3\}$), the estimation of these values is based on the formula

$$r(l, e) = \frac{p(e|l)}{p(e)} = \frac{n(l, e)/n(l)}{n(e)/n} = \frac{n(l, e) \times n}{n(e) \times n(l)}. \tag{3}$$

The various n-values in (3) are computed, after a series of word observations, as follows. For any event e_i^c such that c is a context of PS length 1 (total of six such contexts) from context subset C_s, the values n, $n(e)$, $n(l)$, and $n(l, e)$ measure the total number of PS's, occurrences of e, occurrences of l with PS length s, and joint occurrences of l and e, respectively. For every event e_i^c such that c is one of the remaining two-PS long contexts (total of three) and belongs to context subset $C_s (s = 2, 3)$, the value n, $n(e)$, $n(l)$, and $n(l, e)$ represent the respective total numbers of boundaries between PS's (number of PS's minus the number of words), occurrences of e, occurrences of l with PS length s, and joint occurrences of l and e. Consequently, it suffices to make a number of observations of various letters whose identities are known in advance, perform E-level generation (and presegmentation in the process), and keep track of all the counters on the RHS of (3). After that, values $r(l, e)$ and $r'(l, e)$ are computed, and the system is ready to operate. During training, word images and their identities are input, which leads to automatic updating of the appropriate counters, subsequent to E-level generation.

When both the numerator and denominator in (3) have a value of zero, the value of $r(l, e)$ is set to 1, and thus the corresponding value of $r'(l, e)$ is zero. In cases where only the numerator is zero, and $r'(l, e)$ should be $-\infty$, a heuristic value of -2.50 is used instead in order to reduce its negative influence. In general, the idea is to undermatch slightly the lowest computed r'-value.

This training scheme, while requiring the training set to contain all letters that are to be recognized, and in sufficient quantities to make statistical judgments about them, the selection of words themselves can be arbitrary, and isolated letter prototypes are not necessary. Segmentation of a word into letters through presegmentation and feedback allows information from each letter to be gathered independently from the others.

B. Adaptation

In the case of adaptation, there is no outside feedback, i.e., letter identities have to be decided by the system. Obviously, such information is not present at the time of letter hypothesization, but it is available at the end of the recognition procedure in the form of the claimed word identity, the only remaining problem being the reconstruction of sets of relevant events for each occurred letter. Its p string is always retained and is thus also available for the final word answer. This way, the increments for all the *n-counters* in (3) can be precisely identified, and their values can be updated. An identical procedure is used in the training part, except that the word identity comes from a different source. In cases when a word is deemed unrecognizable, no counter updating takes place. Actual recomputation of the r-values can take place at any regular interval or be triggered by some heuristic. When a new writer is installed, the counter values are reset at the original "unadapted" level, and subsequent adaptation for that writer takes place from there.

IX. Performance

A. Experiment Description

A database of off-line cursive script word images was generated by asking subjects to write words from a lexicon, with pencil on paper, so that each would fill a portion of a window on a digitizing tablet. The input words were then retraced, one at a time, on a tablet, which was interfaced to a graphics controller and host computer. In the process of transfer, the order of inscription was ignored, thus yielding a binary image for off-line recognition. The subjects were asked to write lower-case script with a fully connected main body for each word and to maintain reasonable legibility.

It is a well-known tenet of pattern recognition practice that training and testing sets should not be the same to avoid overly optimistic results. Also, for problems with large feature space dimensionality, the sizes of the training and testing sets should be roughly the same. Both these guidelines were followed in the experimentation.

In experiments where training was required, supervised learning was performed by giving as input both word images and their symbolic identities. These consisted of the string to be recognized (possibly containing blanks) and its associated array of PS lengths (i.e., the s and p arrays of the correct full hypothesis). While in this mode, the system would only perform presegmentation and E-level generation and, based on the correct solution, would uniquely update the n-counters for computation of $r(l, e)$ from (3), which were all initially set to zero. After the whole training set was run this way, the r'-values were computed, and the testing part was ready to be carried out.

The system has about 30 parameters. In particular, d is the maximum stack length, i.e., the cutoff point beyond which hypotheses are discarded. Its value has to be such that the stack is deep enough so that correct hypotheses (prefixes) are not cut off, yet too much extra depth is not desirable for time and space efficiency. A value of $d = 23$ was used in our experimentation. Larger dictionaries probably require a greater value of stack size, to be determined empirically. The parameter *maxsteps* is the maximum number of iterations allowed, after which, if no decision is reached, the word is rejected. The value of 75 was empirically determined to conform with this lexicon. Finally, *minrating* is the minimum rating that an accepted solution can have; if the solution terminating the search has a lower value, then it is rejected.

Two lexicons were used in the experiments. The main, smaller-sized lexicon used in most experiments was based on the Brown corpus [22] lexicon A05 and consisted of 710 words. The large lexicon, compiled from parts of section C of the Brown corpus, consisted of 7800 words.

B. Experimental Results

In the first experiment, only one writer was used for both training and testing. The training set consisted of 66 words where the writing naturally conformed with the constraints of being horizontal and not slanted. The test set consisted of another 64 words. Fig. 16(a) shows some of the words from the testing set that were correctly recognized. Of these, 77 percent were correctly recognized (first choice), 9 percent were incorrectly recognized, and 14 percent were rejected.

Among the cases where the correct answer was not found, in 60 percent of them a rejection took place and no answer was given. This is important since no *erroneous* learning takes place in these cases. In 8 percent of the cases, the maximum number of iterations was exceeded, and in the other 6 percent, the suggested answer had too low a rating. The ratings of correctly recognized words ranged from 0.27 (all but one from 0.37) to 0.80 and were 0.55 on the average. The mean number of iterations required for correct recognition was 18, and the average word length of the correctly found words was 5.5. Thus, the average number of iterations per letter was 3.4.

When the search procedure was extended to obtain the second answer, thus considering the first two returned words for the correct answer in 81 percent of the cases the correct answer was in the top two, while in 5 percent there was an error and in 14 percent there was a rejection. The most common reason for failure is the structure of equal events, i.e., only the occurrence of events is rewarded (or penalized), but not their absence. Therefore, if, for example, letters g and a always have a similar appearance in the middle zone, the occurrence of those events will equally favor both letters in the absence of lower-zone (loop) events, although in the case the letter is more likely to be an a. In fact, on all three second-choice recognitions that was the primary reason for the initial failure (*arose*/*a–g*, *into*/*i–j*, *person*/*s–t*).

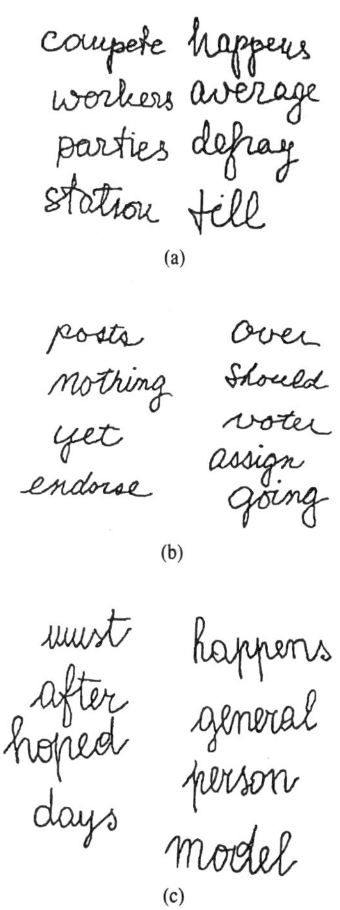

Fig. 16. Some correctly recognized words. (a) From experiment 1. (b) From experiment 4. (c) From experiment 5.

The second experiment was designed to exercise learning capabilities. The latter was programmed to carry out *latent* learning (counter updating) during the first experiment and to invoke the acquired knowledge (recompute r'-values) upon its completion. In this case, this was done by simply adding up the two sets of counters and computing the associated values of the function r'. After that, the same set of 64 words was run over again. Notice that in this case it is justified to run the *retraining set* as a test set, since it is combined with the original training set and in that sense not the only source of knowledge, and the feedback comes from inside. Thus, no improvement is guaranteed in advance (e.g., *wrong* learning, in the case of misrecognized words, could solidify the misrecognition of this input in the second pass, unless heavily counterbalanced by other evidence). In this case, 78 percent were correctly recognized (an improvement), 8 percent were incorrectly recognized, and 14 percent were rejected.

In the third experiment, conditions from experiment 1 were repeated, with the exception of using the large lexicon. The results showed some deterioration compared to the case with the small lexicon, but still in almost two thirds of the cases (64 percent) the correct word was either the best answer or at least among the top four word hypotheses.

In the fourth experiment, the input used came from a second writer. No slant restrictions were placed on the writing in this case since the slant removal module was used. The lexicon was the same as in the first two experiments. A total of 53 word samples were taken, eight of which were used for *retraining* and the remaining 45 used as the testing set. Fig. 16(b) gives a sample of successful words used for testing. The retraining was performed in the same way as the original training, except that the counters obtained from this pass were used to update (in an unweighted manner) those from experiment 1, thus just slightly modifying the original r'-values. This resulted in the correct answer being included in the top two in 64 percent of the cases, with 11 percent incorrect and 24 percent rejected.

The fifth experiment was performed under the same conditions as the fourth, but using input from a third, female, writer. Differences in the characteristics of male and female handwriting have not been documented in a computational sense so far. This time 67 word samples were taken, and 18 of them were used for retraining in the same sense as in experiment 4, except that the values to be modified were the ones existing after experiment 4. Therefore, the r'-values used in experiment 5 were based on training from the first writer, modified by writers 2 and 3. Fig. 16(c) gives some words that were correctly recognized in experiment 5. The results were as follow: 65 percent correct, 18 percent incorrect, and 16 percent rejected.

In the sixth experiment, the conditions from the first one were repeated, except for the *r'-values*, which were taken from the last one. The first writer's testing set was run on a training set based on his own data and modified by that of the other two writers. This resulted in 72 percent correct recognition (first choice). A deterioration in performance was present, but it was only a slight one. Experiment 7 tested the data from writer 3 with the system trained solely on her writing. Additional samples were obtained, and similarly to experiment 1, a set 71 words was used for training, with another 68 words for testing. Using the small lexicon, this experiment yielded a 71 percent correct recognition rate (top answer), with 24 percent incorrect, and 6 percent rejected. The training set of writer 3 was apparently more statistically consistent than that of the first writer [compare Fig. 16(a) and (c)], which resulted in faster recognition of the correct words (only 11 iterations per word), but also in a lower percentage of rejections and more errors. Again, as before, most of the errors were due to occasionally inadequate statistical representations of certain letters when certain event/letter combinations not found in the training set were encountered, which in turn is a function of a limited training set. This effect is more pronounced for statistically stable script (i.e., one where letters, particularly the more frequent ones, most of the time have more or less the same description, in terms of events in our case) since deviations from standard forms tend to be more fatal.

The results of all these experiments are summarized in Table I.

TABLE I

Experiment Number	Description	Performance (correct recognition rate)
1	First writer, small lexicon	77% first choice 81% top two choices
2	First writer, large lexicon, after adaptation	78% first choice
3	First writer, large lexicon	48% first choice 64% top two choices
4	Second writer, small lexicon, mixed training	54% first choice 61% top two choices
5	Third writer (female), small lexicon, mixed training	63% first choice 65% top two choices
6	First writer, small lexicon, mixed training	72% first choice 73% top two choices

X. Discussion

We have described several component processes and a system organization for off-line cursive script word recognition. The main contributions of this work are the following.

1) A scheme of knowledge interaction, with the appropriate control and data structures and clearly identified representation levels, was introduced.

2) Several improvements to preliminary operations in processing off-line cursive script were made.

3) A new technique for curve smoothing, detection, and orientation measurement in a single pass through its chain coded contour representation was introduced.

4) A general procedure for contour tracing of images with arbitrary topology was introduced.

5) A method for hypothesizing and rating certain objects from their (possibly unrelated) constituent features was developed. It is uniform over all features, objects, their locations, and their sizes, is computationally very simple, and was supplemented with a mechanism for dynamic modification of parameters on which the rating computation is based.

The experiments have demonstrated the system's ability to recognize a majority of cursive script words for two writers, on whose handwriting it was exclusively trained. Performance was only slightly degraded when the lexicon size was significantly increased. The script of two other writers exhibited good recognition rates when run on a set of parameters based on training with the first writer's data and slightly modified by their own samples. The recognition rates dropped only slightly when the first writer's testing data were rerun on their original training set with these two modifications. A degree of stability has been shown for parameters learned across different writing styles.

In a more general cursive script recognition environment for reading phrases and sentences, the size of the lexicon can be expected to be considerably less since the lexicon will be dynamically determined by context. In such an environment, the writing style will presumably remain fixed. Given these considerations, we have shown the correct recognition rate for isolated single words to be promising.

The approach here lies in between two different methods of using a lexicon in recognizing a word from its iconic shape. The first extremity is a method that splits the word into segments and then classifies segments into letters using a lexicon to constrain alternatives while proceeding in a left-to-right scan [27]. Such a method makes an irrevocable segmentation decision and is more appropriate to print than to cursive writing. The second extremity if a method that completely bypasses the segmentation problem [21]. In such a method, global word shape features such as ascenders, descenders, and holes are used to hypothesize words. The method described in this paper does flexible segmentation. It can be used as a method for detailed analysis once a small set of words have been hypothesized either by global context or by word-shape analysis.

Acknowledgment

The authors wish to thank the referees, in particular referee A, for making several suggestions to improve the exposition.

References

[1] K. Badie and M. Shimura, "Machine recognition of roman cursive script," in *Proc. 6th Int. Conf. Pattern Recognition*, Munich, West Germany, Oct. 1982, pp. 28-30.

[2] L. Bahl and F. Jelinek, "Decoding for channels with insertions, deletions and substitutions with applications to speech recognition," *IEEE Trans. Inform. Theory*, vol. IT-21, Sept. 1975.

[3] M. Berthod and S. Ahyan, "On line cursive script recognition: A structural approach with learning," in *Proc. 5th Int. Conf. Pattern Recognition*, Miami Beach, FL, Dec. 1980, pp. 723-725.

[4] H. Bouma, "Visual recognition of isolated lower case letter," *Vision Res.*, vol. 11, pp. 459-474, 1971.

[5] R. Bozinovic and S. N. Srihari, "A string correction algorithm for cursive script recognition," *IEEE Trans. Pattern Anal. Mach. Intell.*, vol. PAMI-4, pp. 655-663, Nov. 1982.

[6] —, "ROCS: A system for reading off-line cursive script," Dep. Comput. Sci., State Univ. New York at Buffalo, Buffalo, NY, TR 85-13, Sept. 1985.

[7] M. Brown, "Cursive script recognition," Ph.D. dissertation, Univ. Michigan, Ann Arbor, MI, 1981.

[8] O. P. Buneman, "A grammar for the topological analysis of plane pictures," *Mach. Intell.*, vol. 4, pp. 383-393, 1969.

[9] D. J. Burr, "A normalizing transform for cursive script recognition," in *Proc. 6th Int. Conf. Pattern Recognition*, Munich, West Germany, Oct. 1982, pp. 1027-1030.

[10] —, "Designing a handwriting reader," *IEEE Trans. Pattern Anal. Mach. Intell.*, vol. PAMI-5, pp. 554-559, Sept. 1983.

[11] A. Dutta, "An experimental procedure for handwritten character recognition," *IEEE Trans. Comput.*, vol. C-23, pp. 536-545, May 1974.

[12] L. D. Earnest, "Machine recognition of cursive script writing," in *Information Processing*, C. M. Popplewell, Ed. Amsterdam, the Netherlands: North Holland, 1962.

[13] M. J. Eccles, P. S. McQueens, and D. Rosen, "Analysis of the digitized boundaries of planar objects," *Pattern Recognition*, vol. 9, pp. 31-41, 1977.

[14] M. Eden, "On the formalization of handwriting," in *Proc. Symp. Appl. Math.*, vol. 12, 1961, pp. 83-88.

[15] R. Ehrich and K. Koehler, "Experiments in the contextual recognition of cursive script," *IEEE Trans. Comput.*, vol. C-24, pp. 182-194, Feb. 1975.

[16] R. Farag, "Word-level recognition of cursive script," *IEEE Trans. Comput.*, vol. C-28, pp. 172-175, Feb. 1979.

[17] H. Freeman, "On digital computer classification of geometric line patterns," in *Proc. Nat. Elec. Conf.*, vol. 18, 1962, 312-324.

[18] L. S. Frishkopf and L. D. Harmon, "Machine reading of cursive script," in *Information Theory*, C. Cherry, Ed. London, UK: Butterworth, 1961.

[19] R. N. Haber and L. R. Haber, "Visual components of the reading process," *Visible Lang.*, vol. XV, no. 2, pp. 147-181, 1981.

[20] K. Hayes, "Reading handwritten words using hierarchical relaxation," Univ. Maryland, College Park, MD, TR-783, 1979.

[21] J. J. Hull and S. N. Srihari, "A computational approach to word shape recognition: Hypothesis generation and testing," in *Proc. IEEE-CS Conf. Comput. Vision Pattern Recognition*, June 1986, pp. 156-161.

[22] F. Jelinek, "A fast sequential decoding algorithm using a stack," *IBM J. Res. Develop.*, vol. 13, pp. 675-685, Nov. 1969.

[23] H. Kucera and W. N. Francis, *Computational Analysis of Present Day American English*. Providence, RI: Brown Univ. Press, 1967.

[24] P. Mermelstein and M. Eden, "Experiments on computer recognition of connected handwritten words," *Inform. Contr.*, vol. 7, pp. 255-270, 1964.

[25] M. Minsky and S. Papert, *Perceptrons*. Cambridge, MA: M.I.T. Press, 1969.

[26] K. M. Sayre, "Machine recognition of handwritten words," *Pattern Recognition J.*, vol. 5, no. 3, pp. 213-228, Sept. 1973.

[27] S. N. Srihari, J. J. Hull, and R. Choudhari, "Integrating diverse knowledge sources in text recognition," *ACM Trans. Office Inform. Syst.*, vol. 1 no. 1, pp. 58-57, 1983.

[28] S. N. Srihari, *Computer Text Recognition and Error Correction*. Silver Spring, MD: IEEE Computer Society Press, 1984.

[29] S. N. Srihari and R. Bozinovic, "A multi-level perception approach to reading cursive script," *Artificial Intell.*, vol. 33, pp. 217-255, 1987.

[30] C. C. Tappert, "Cursive script recognition by elastic matching," *IBM J. Res. Develop.*, vol. 26, no. 6, pp. 765-771, Nov. 1982.

[31] —, "Adaptive on-line handwriting recognition," in *Proc. 7th Int. Conf. Pattern Recognition*, Montreal, P.Q., Canada, July-Aug. 1984, pp. 1004-1007.

[32] I. Taylor and M. M. Taylor, *The Psychology of Reading*. New York: Academic, 1983, pp. 183-193.

[33] G. Zachopoulos, "Slant estimation and correction for off-line cursive script," M.S. thesis, State Univ. New York at Buffalo, Buffalo, NY, 1984.

Radmilo M. Bozinovic (S'84–M'85) received the Ph.D. degree in computer science from the State University of New York at Buffalo in 1985.

He is now with the Department of Computer Systems, Mihajlo Pupin Institute, Belgrade, Yugoslavia. His interests are in pattern recognition and artificial intelligence.

Sargur N. Srihari (S'74–M'75-SM'84) received the Ph.D. degree in computer and information science from The Ohio State University, Columbus, in 1976.

He is currently a Professor in the Department of Computer Science, State University of New York at Buffalo. His research interests are in pattern recognition and document image understanding (ranging from character/work recognition to integrating visual and linguistic information in document images).

Dr. Srihari is an Associate Editor of the *Pattern Recognition* journal.

ON-LINE RECOGNITION OF HANDPRINTED CHARACTERS: SURVEY AND BETA TESTS

FATHALLAH NOUBOUD and RÉJEAN PLAMONDON*

Laboratoire Scribens, Department of Electrical Engineering, Ecole Polytechnique de Montréal, P.O. Box 6079, Station A, Montreal QC, Canada H3C 3A7

(*Received* 3 *July* 1989; *in revised form* 26 *October* 1989; *received for publication* 7 *December* 1989)

Abstract—In the first part of this paper, we present a survey of the state of the art in on-line handprinted character recognition technology. Data preprocessing and classification, and character recognition results are examined. A number of character recognition systems are compared. The second part of this paper describes beta tests carried out on a commercial system. The results are analysed with particular attention to the effects of handwriting variability and constraints imposed, and to the human factors involved.

Handprinted characters   On-line recognition   Alpha and beta tests

INTRODUCTION

Man–machine communication has to date been based on the use of the keyboard. This interface is not well-suited to users who have not mastered the keyboard and if, in addition, the number of characters is very large (as in the Chinese and Japanese alphabets, for example), this mode of communication becomes cumbersome and inefficient.

For this reason, automatic handwriting recognition systems have been developed. A number of surveys have been made of the state of the art in this field.[1-3] Automatic handwriting recognition systems are classified into two categories according to the mode of data acquisition used. On-line systems require the use of a digitizing tablet, while off-line systems use optical digitizing devices.[4,5] The mode of data acquisition has a significant effect on the other modules of the system, such as description and comparison, etc.

In the first part of this paper, we review the latest on-line recognition methods and examine the various modules which make up a recognition system (Fig. 1). A description of the preprocessing techniques (filtering, smoothing and normalizing) applied to the digitized data is followed by a discussion of the various approaches available for describing a character and the appropriate comparison techniques (dynamic programming, Euclidian distance . . .). Finally, the results obtained by different systems are provided, in addition to comparative tables.

Among the problems inherent in handprinted character recognition, and which generate recognition errors, are the variations which may occur in the writing of characters and the resemblance which exists between some of the characters. To minimize these errors, a data base containing a large number of specimens provided by many writers must be studied. The system developed by Ward *et al.*[6-8] is one of those rare alphanumeric character recognition systems where experimental analysis is supported by just such a large data base. We have, therefore, chosen to carry out Beta tests on this system to evaluate its performance and also to analyse the problems associated with handwritten character recognition.

The second part of this paper describes these tests and documents the recognition results, in addition to providing an analysis of the human factors involved.

1. THE STATE OF THE ART

1.1. *Variations in handwriting*

There may be significant differences in the way a character is written depending on the style of the writer. These variations may be in the shape of the character or in the number or the order of its

* To whom correspondence should be addressed.

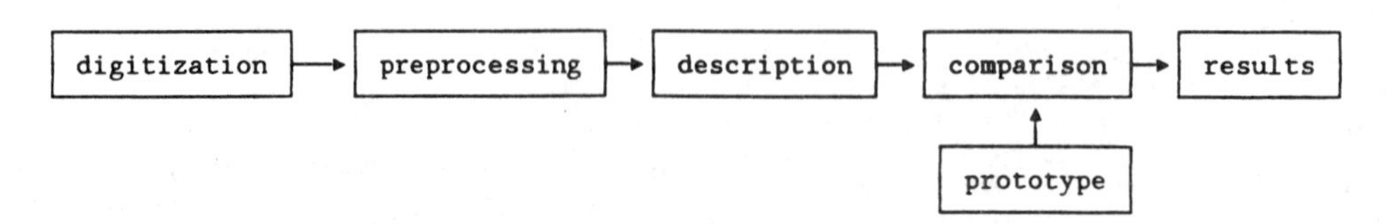

Fig. 1. Diagrammatic representation of a recognition system.

"On-Line Recognition of Handprinted Characters: Survey and Beta Tests" by F. Nouboud and R. Plamondon from *Pattern Recognition,* Vol. 23, No. 9, 1990, pp. 1031–1044. Elsevier Science, Ltd., Pergamon Imprint, Oxford, England.

North American European

Fig. 2. Writing specimens.

components. A component is a stroke between two consecutive pen lifts.[9]

Ward and Kuklinski[8] have reported that the shape of the characters may vary with the origin of the writer, for example, Europe and North America (Fig. 2) or with a tendency to write cursive handprinted characters (Fig. 3).

Handprinted characters Cursive handprinted characters

Fig. 3. Writing specimens.

The number of components may vary a great deal for one character (Fig. 4). The order of the components in a character may vary as well. Writers have a tendency to draw the vertical components before the horizontal ones, and the left components before the right.[8]

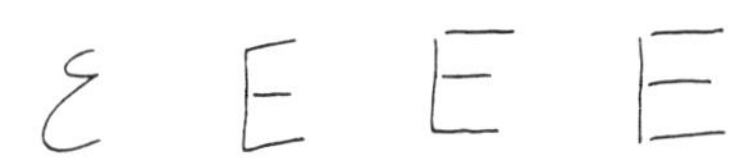

Fig. 4. Example of the character "E".

1.2. *Digitizing characters*

Digitizing tablets may play the role of a mouse (to indicate and select a point on the screen, in a menu, for example). However, they are also useful for digitizing writing, signatures and drawings.

The advent of digitizing tablets has meant a considerable increase in the number of on-line writing recognition tasks performed.[10] The technologies used are varied, but they belong to two main families. Most handwriting recognition systems use electromagnetic/electrostatic tablets which send the coordinates of the pen tip to the host computer at regular intervals. Others[11] use pressure-sensitive tablets. The advantage of these devices is that an ordinary pen may be used. However, a movable surface above the tablet is essential to accommodate hand pressure and to avoid altering acquisition data.

The resolution of the digitizing tablets is usually over 200 points per inch and the transmission speed is more than 100 points per second. Kim and Tappert[12] have studied the evolution of the recognition rate and calculation time as functions of these parameters.

There are other data acquisition devices in addition to digitizing tablets. Special pens are used[13,14] to extract the dynamic parameters of the writing and to characterize the handwritten character.

1.3. *Preprocessing*

Raw digitized data must be subjected to a number of preliminary processing steps to make it usable in the descriptive phase of character analysis. The main objectives of this processing are a reduction in the amount of information to be retained, the elimination of certain imperfections and the normalization of the data.

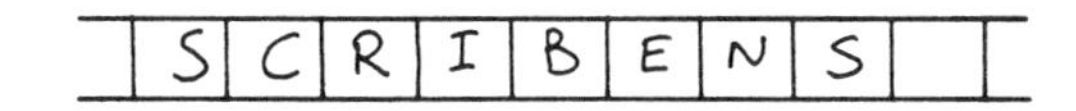

Fig. 5. Segmentation into boxes.

To separate the digitized characters, an external segmentation procedure must be carried out. The most widely used method for doing this is to use a box (Fig. 5) to contain each character.[8,15–18] Segmentation may be spatial[19] or temporal[20,21] (for example, a pause of 300 ms may mean that the character is complete and that subsequent data correspond to the next character). Burr[22] imposes the constraint that the characters (lower case letters) consist of only one stroke. The technique of hierarchic segmentation[15,23] involves surrounding the character with a rectangle.

The object of the smoothing operation is to eliminate imperfections due to the hardware and the tablet, and to trembles in the writing, hesitations, etc. This is done by replacing a point with the average over its neighbours.[11,16,21,24–27]

The data are then filtered to reduce the amount of information to be retained and to eliminate wild points.[11,16–18,26,28] Testing on the points of pronounced curvatures make it possible to keep characteristic points. In this way, for example, a "U" may be distinguished from a "V".[19–21] Tang *et al.*[29] carry out a thresholding on the extrema in x and y in order to reduce the amount of information to be retained. Tappert[19] reduces the data representing a dot at only one pair of coordinates.

Some hooks may appear at the beginning or end of a component as a result of a writer's tendency to connect the components. Their presence may be detected by sudden changes in direction and by their short length. Eliminating these strokes makes it possible to reduce the amount of information to be retained and avoid recognition errors.[15,19,21]

There are other preprocessing tasks which are designed to correct breaks in the strokes.[17,24]

The data representing the character may be normalized in terms of size,[11,15,16,22,26] orientation[22] and position.[11] Doster and Oed[23] normalize the direction codes and then reconstruct the character before proceeding to the comparison phase.

The parameters involved in the preprocessing phase (smoothing, filtering and normalizing) may be optimized after several simulation experiments[11] have been carried out.

1.4. *Recognition*

Recognition is based on the principle that variations between different characters are more important than variations between specimens of the same character. With certain pairs of characters, however, there may be confusion, for example, G and 6, I and 1, U and V. This must be taken into account in the description of the characters and in the comparison phase in order to avoid recognition errors.

1.4.1. *Description.* A character is classified by means of a description which is made up of parameters or codes extracted from the various parts of the character. These parts or segments are obtained when the internal segmentation process is completed.

There are, however, some cases[30–33] where the character is coded globally, without segmentation. The box containing the character is divided into rectangular zones, and coding consists of recording the sequence of zones touched by the pen. To enter the character "A", for example, the pen must touch the zone sequence: 6, 4, 1, 6, 4, 3, 4 (Fig. 6).

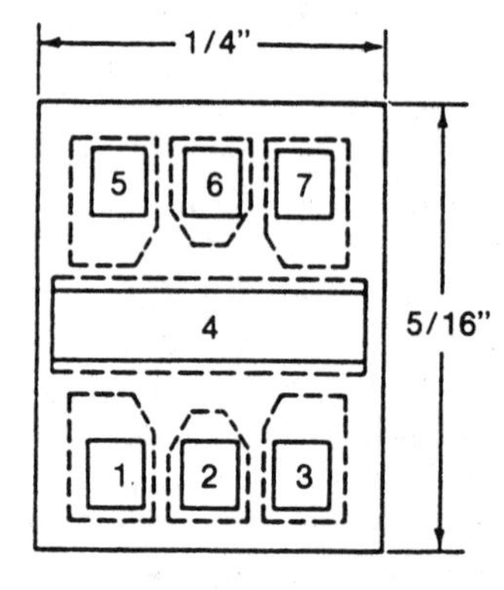

Fig. 6. Digitizing zones (according to reference 31). Reprinted with permission (© 1986 Academic Press Inc.).

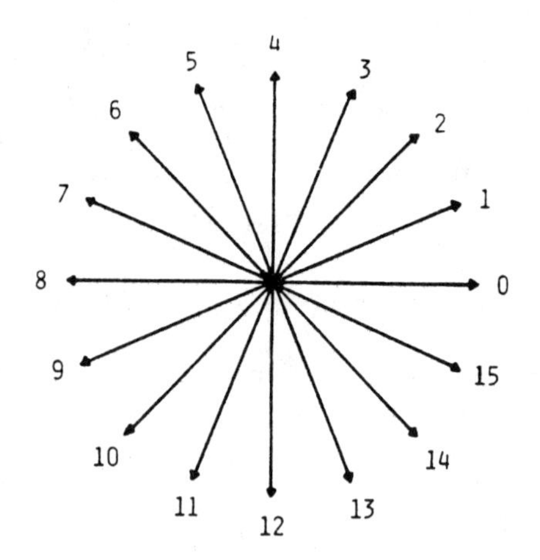

Fig. 7. Direction coding.[18] Reprinted with permission (© 1982 IEEE).

There are three ways to segment a character. The first is to segment it into elements: the vectors joining successive points.[25,28,34] The elements are generally described by their length or height, and their angles with respect to the horizontal. The angles are generally quantified using quadrants (Fig. 7).

The second way to segment a character is to identify the curvature maxima or the local extrema in x and y.[20,21,29] The segments thus obtained may be described by their position and slopes.[21] Ito and Chui[20] classify segments in straight-lines or curvilinear strokes prior to coding them by direction.

The third way to segment a character is to separate it into components where pen lifts occur.[8,11,16–19,23,24,26,29,35,36] The components may be described by a chain coding the extrema in x and y[8,29] (Fig. 8), or by the direction codes of the elements which make up the component[17–19,23,29] (Fig. 9) and its position in the character.[15,26] Hidai *et al.*[35] use the position and the direction of the component globally.

There are other approaches to describing characters. Fourier coefficients may be used for characters consisting mainly of curves[37] or concatenated straight strokes.[11] In the case of Chinese characters, which are formed primarily by straight strokes, approximation by a small number of points has proved effective.[16,38,39] Parameters specific to a given type of character, for example, Arabic characters, may be defined.[24]

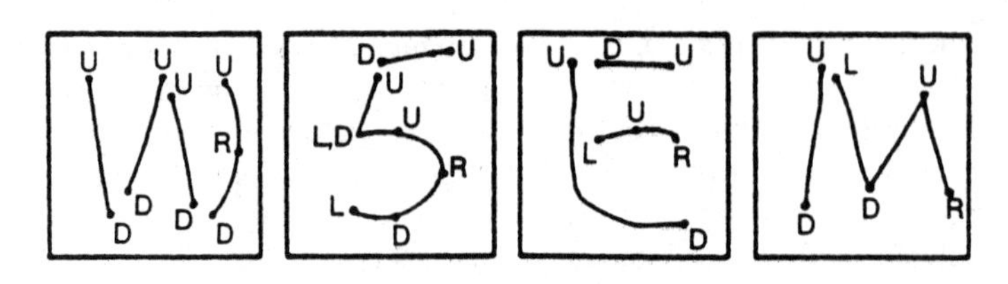

Fig. 8. Coding of extrema (according to reference 8). Reprinted with permission (© 1988 IEEE).

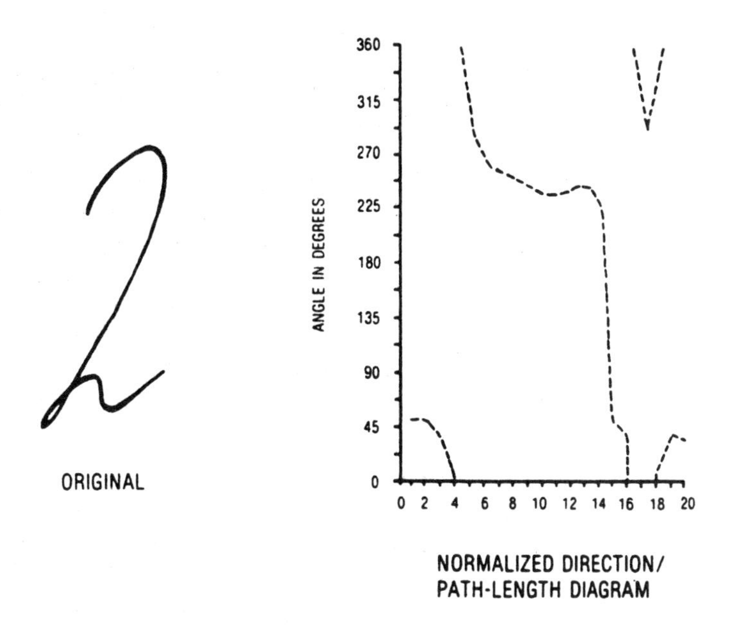

Fig. 9. Coding of directions (according to reference 23). Reprinted with permission (© 1984 IEEE).

1.4.2. *Comparison*. Comparison may be carried out in two phases. The most distant prototypes of the character are discarded in the first phase with dynamic programming[28] or linear matching.[27] The second phase compares the candidates which have been selected to complete the classification of the character. This procedure makes it possible to reduce calculation time.

Comparison techniques are closely associated with the nature of the description retained. Most of these are based on dynamic programming.[15,18,19,22,26,28,29,34,40] Thus, the vector of the variable representing the character is compared to a group of reference vectors and is assigned to the class corresponding to the minimal distance. This represents considerable calculation time, possibly as much as several seconds for each character.[13] A solution to this problem is the use of a processor dedicated to dynamic programming[28] for determining the best candidates. Another solution, proposed by Ikeda *et al.*,[26] is to restrict the use of dynamic programming to certain characters, thereby reducing the size of the dictionary and making use of information other than the shape of the components for classification purposes (number and order of components, etc.). To improve recognition results, Yoshida and Sakoe[18] artificially prepare sections common to some different characters to limit the number of variations (Fig. 10), and use local information to distinguish characters that resemble each other. In this system, by connecting components, the constraint on their number is avoided (Fig. 11).

Euclidian distance is used for the comparison in some systems.[23] Yet another comparison technique, the Bayesian Rule has been found suitable when

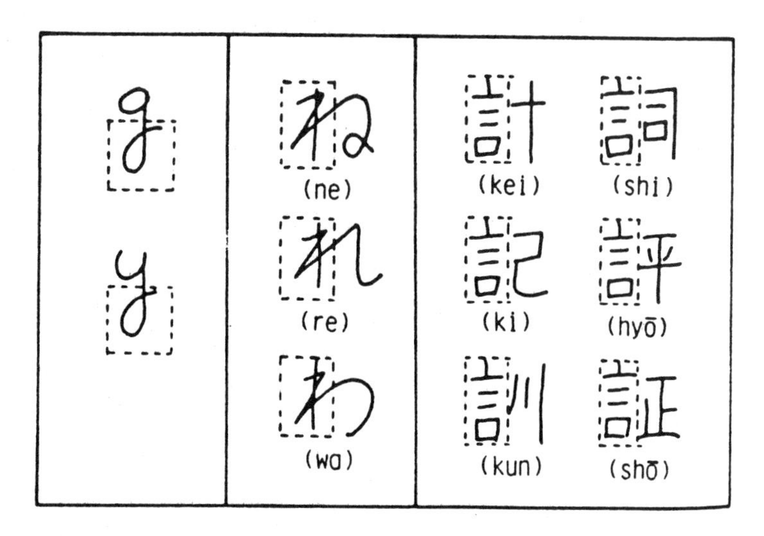

Fig. 10. Sections common to different characters.[18] Reprinted with permission (© 1982 IEEE).

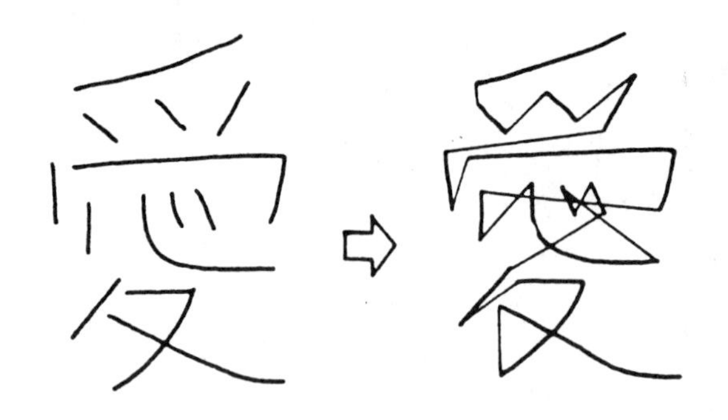

Fig. 11. Connecting the components (according to reference 18). Reprinted with permission (© 1982 IEEE).

using Fourier coefficients to describe characters.[11] In another study Odaka *et al.*[16] carry out a statistical classification using the point coordinates representing the character strokes.

Character classification may be accomplished by means of component combinations.[8] Each component of the character is identified by its coding chain. Classification consists in looking for the combination of components in the dictionary corresponding to the character. Using the allowed combinations of elements (Fig. 12a), Huh and Beus[25] have constructed the allowed components (Fig. 12b). The character is thus identified by combinations of these components.

Syntactic-statistic classification is another method of classifying characters.[17,20,21,29] The syntactic phase of the classification is formed using a grammar,[20] a decision-making tree[17] or finite-state non-deterministic machines.[29] The statistic phase uses the stroke description parameters to complete character classification.

1.5. *Results*

Experiments to test character recognition systems are based on a sizeable group of characters and writers. The writers are asked to write characters within a framework of constraints as to the number and order of the components, and the size of the characters, etc. These constraints are very stringent in some systems.[25,29] Also, calculation time is itself a constraint since even in a so-called "real-time" system, the recognition of a character requires about one second of processing time at least.* In short, the way data are acquired is quite unnatural. From today's perspective, the use of these systems requires several improvements. Recent studies have been focussed, therefore, on reducing these writing constraints, especially with regard to the order of the components,[11,17,35] their number[18] and the adaptation of the system to the writer[19,21] by means of updating.

The specifications of the various systems are summarized in the following Table 1 with their recognition results. It should be noted, however, that it is very difficult to compare the results obtained by different systems. This is because the equipment, procedures and other factors may vary considerably from one system to another.

The systems in this table process different types of characters, including alphanumeric, Chinese, Korean and Arabic characters. Recognition methods vary greatly, but dynamic programming is common to 40% of these systems. This is an indication, once again, of how powerful this comparison tool is. In fact, dynamic programming is found in a number of domains, including shorthand symbol recognition,[41] word recognition[42] and handwritten signature verification,[43–45] among others.

In this table, the recognition rates are nearly all above 92%. However, such tests are not always based on a large enough group of characters and writers. In order to be able to make generalizations about a system's performance, the data base must contain thousands of characters supplied by dozens of participants.

The only alphanumeric character recognition system which fits these criteria is the one developed by Ward *et al.*[6–8] We shall present, therefore, an experimental study (beta tests) that we have carried out on the system marketed by PENCEPT. The second part of this paper gives a description of the PENPAD 310 system and of our experiments to test its recognition algorithms.

* Average writing speed is 1.5–2.5 characters per second for the Roman alphabet, and 0.2–2.5 for Chinese characters.[3]

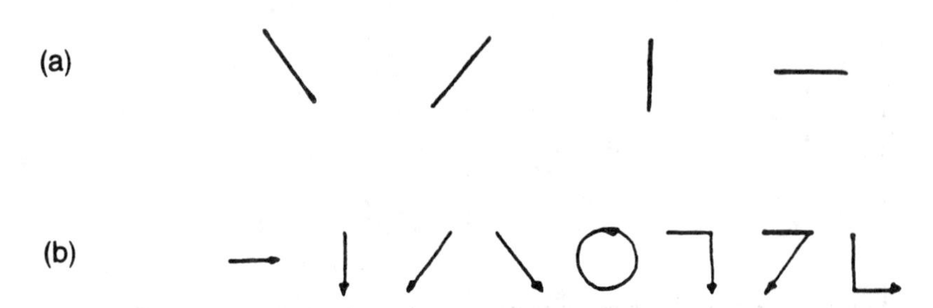

Fig. 12. (a) allowed elements (b) allowed components.[25] Reprinted with permission (© 1982 Pergamon Press plc).

Table 1. Comparison of handwritten character recognition systems

Authors	Types of characters	Feature description	Classification	Constraints	Number of writers	Number of characters	Recognition results (%)	Calculation time
Amin *et al.*[24]	Letters and Arabic numerials	Special parameters	Structural	—Comp. order —Comp. number	3	219	95.4	Real time
Arakawa[11]	Kanji, Hiragana Katakana and alphanumeric	Fourier coefficients	Bayesian Rule	—Comp. order —Comp. number	35	35 specimens per char.	99.2 to 99.7	
Baron Plamondon[33]	Alphanumeric and others + * (), etc.	Zone codes	Identification of code chains	—Boxes —Zone sequence	1	(a) 2880 (b) 960	(a) 92 (b) 100	Real time
Burr[22]	Lower case letters	—Directions —Length of elements	Elastic matching	One component per letter				Deferred time
Doster Oed[23]	All symbols	—Directions —Position of elements	Euclidian distance	—Comp. number —Comp. order				
Hidai *et al.*[35]	Kanji, Hirangana Katakana and alphanumeric	—Directions —Component position		—Comp. number	50	2500	99.5	Real time
Huh Beus[25]	Korean	Combination of elements	Combination of components	—Comp. number —Comp. order —Slow writing —Signal end of character	3	300	92	
Ikeda *et al.*[26]	Kanji, Hiragana Katakana and alphanumeric	—Directions —Position of. segments	Elastic matching	—Comp. number —Comp. order		17463	92 to 98	Real time
Ito, Chui[20]	Alphabet	—Directions of segments	Syntactic-statistic	—Comp. number	10	2340	98.3	Real time
Loy, Landau[21]	Alphanumeric	—Directions —Position of segments	Syntactic-statistics	—Comp. number —Comp. order		1000	98.8	Real time
Lu Bordersen[28]	All symbols	—Directions —Curvatures —Velocity	Elastic matching	—Comp. number —Comp. order	4	1 specimen per symbol	96	Real time

Mandler *et al.*[36]	All symbols	—Directions of elements —Position of strokes —height/width of character	Elastic matching	—Boxes —Comp. order —Comp. number	12	9360	96.2	
Odaka *et al.*[16]	Kanji, Hiragana Katakana and alphanumeric	Fixed number of points per stroke	Statistic (from point coordinates)	—Boxes —Comp. number	40	9135	99.5	Real time
Tang *et al.*[29]	Numerals	Chain code	Syntactic-statistic	—Comp. order —Comp. number —No significant trembles	30	300	89	
Tappert[19]	Upper case, lower case and numbers	—Directions —Heights —*x y* offs. from cent. of gravity	Elastic matching		6	1950	94.1	
Tappert[27]	Alphanumeric and others + * (), etc.	—Directions —*x y* offs. from cent. of gravity	Elastic matching		9	4284	97.3	Real time
Telmosse Plasmondon[14]	Numerals	Accelerometer signals	Mahanalobis' distance	—Boxes —Comp. order —Comp. number	30	15000	71.9–94.5	Real time
Ward *et al.*[6,8]	95 characters ASCII	Chain code	Combination of components	—Boxes —Comp. order —Comp. number	500	85000	93.4	Real time
Ye *et al.*[17]	Chinese	—Directions —lengths of elements	Syntactic-statistic	—Boxes —Comp. number		500	92 to 99	
Yoshida Sakoe[18]	Alphabet, Kana and others: + * (), etc.	—Directions —Lengths of elements	Elastic matching	—Boxes —Comp. order	3		99.5	Real time

2. PENPAD 310 TESTING

The two main objectives of the work described here were to test the algorithms of the tablet statistically and measure their performance, and to document and analyse the comments and observation of the participants in the experiments.

In this part of the paper, we first give a brief description of the recognition system, followed by a detailed description of the experiments, the results obtained and their analysis.

2.1. *Description of the tablet*

The PENPAD 310 system has a handwritten character or symbol recognition processor. Data acquisition is electromagnetic. When the pen is in contact with the writing surface, the coordinates of the point are sent to the recognition module at regular intervals. This module is linked to the host computer by a Serial RS-232 interface.

The characteristics of the tablet are as follows:

—active surface area: 11" × 11" (27.94 cm by 27.94 cm)
—thickness: 0.5" (1.27 cm)
—possible resolution: 1000 points per inch (394 points/cm)
—data transmission rate: 100 points per second at 9600 baud.

The PENCEPT processor is capable of recognizing the 59 characters shown in Fig. 13.

The work surface may be divided into zones. Each of these zones may be programmed in a different mode: character recognition, mouse emulation, graphics digitizing, etc. It is possible to further divide a zone into boxes. When the pen touches a box, the tablet executes the related command or sends it to the host computer.

The pen has a three-position switch which may be programmed to send commands to the tablet or to the computer.

A driver program interprets the data transmitted by the recognition module. After its installation with a list of commands programming the tablet, it remains resident in the memory with DOS.

The host computer is a COMPAQ 386 microcomputer (16 MHz). Further information about the tablet may be found in the PENCEPT documentation.[46]

2.2. *Description of the experiments*

The tests that we have carried out are of the beta type. Pressman[47] defines alpha and beta tests.

Alpha tests are carried out in the system designer's own environment and are controlled by him, whereas beta tests are carried out by the user-client in a real system application environment not controlled by the designer. The goal is to complete alpha testing of the system. Beta testing is important in that it may reveal the existence of certain problems that did not become evident during the alpha testing stage. The designer may then study these problems and improve his product.

Our experiments involved a population of 28 writers which included students, researchers and Ecole Polytechnique personnel. Both sexes were represented (14.3% female and 85.7% male). Most of the participants were French-Canadian (78.5%), and the balance were of other nationalities (14.3% European and 7.2% North-African) and their mother tongue was not French (7.2% Arabic and 7.2% Italian).

The experiments took place over a period of 3 weeks. Each of the 28 participants spent approximately one hour in the acquisition session and were paid for their time.

During a session, the writer was seated and could position the tablet to suit himself. In order to familiarize the writer with the equipment, he was asked to fill in an identification sheet using the PENPAD 310 pen. Thus, at the same time, we obtained pertinent information such as the writer's name, age, sex, profession, nationality, mother tongue, etc.[48,49] The writer then wrote nine series of characters. A series contained 59 characters that were recognizable by the PENPAD 310. The order of the characters was random and different in every series. The participant wrote down the character appearing on the screen. If the recognition result did not match the character

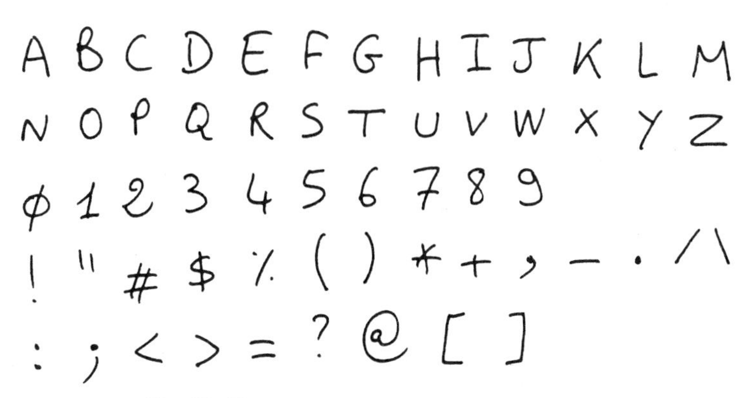

Fig. 13. Characters recognized by PENPAD 310.

on the screen, the programme asked for a validation. Thus, if the writer had written the wrong character, had not respected the imposed constraints (see Type B and Type C below), or had written a character the shape of which did not match the character on the screen (for example, an uncrossed 0, an I uncrossed top and bottom, etc.), data acquisition was restarted. At the end of the session, the comments and observations of the participant were documented.

The tablet may be configured to recognize characters either in boxes or without boxes. In addition, some characters (\/ ( ), etc.) are not easily recognized unless a few constraints are respected. An experimental study cannot be exhaustive unless these facts are taken into account. We have, therefore, defined 3 types of acquisition:

Type A: no box and without constraints.

Type B: no box and with constraints.

Type C: with boxes and constraints.

Every writer wrote three series of characters in each of these three types of acquisition.

2.3. *Results and analysis*

2.3.1. *Error rates.* The data base contains 14,868 characters:

28 writers × 9 series × 59 characters = 14,868.

There are two types of recognition errors. The first is the rejection of a character when the software cannot identify it. The second type is incorrect recognition, where the character is assigned to the wrong class. Table 2 shows the results obtained.

This table shows the definite improvement in the Type B and C results over the Type A results. This improvement is due to the important role of the constraints imposed in Types B and C. The use of boxes in Type C does not lead to any significant improvement, the small reduction in error rates being probably due rather to the increased skill of the writer after having written six data acquisition series. Types B and C may, therefore, be regrouped as follows:

Error rate without constraints	21.25%
Error rate with constraints	10.76%

The software has a tendency to assign the character to the "nearest" class, and seldom rejects it. This leads to higher incorrect recognition rates than rejection rates (to a factor of 6).

Table 2. Recognition results

	Reject rate	Incorrect recognition rate	Global error rate
Type A	2.98%	18.26%	21.25%
Type B	1.57%	9.92%	11.50%
Type C	2.46%	7.56%	10.02%
Global	2.34%	11.92%	14.26%

We did not notice any evidence of fatigue among the participants. The error rates did not increase during the final acquisition series.

The error rate associated with Type C acquisition is 10.2%. This is higher than the 6.6% rate obtained by Ward *et al.*[6-8] A higher rate was, however, predictable since it is expected that performance will be inferior in beta tests than in alpha tests. This is because alpha tests are carried out by the designer whose understanding of the algorithm specifications means that the testing conditions are optimal. Most of the characters with high error rates are the punctuation characters (, ! () ; / \). We have, therefore, calculated the global error rate only for the alphanumeric characters:

Error rate for alphanumerics:	8.53%

Table 3 shows detailed recognition results for each character: (in one column) the number of times the character is assigned to each class and the number of rejects.

The alphanumeric characters with a high error rate are:

1	recognized as %
4	recognized as Y
9	recognized as ,
G	recognized as 6
J and Y	recognized as "
P	recognized as T

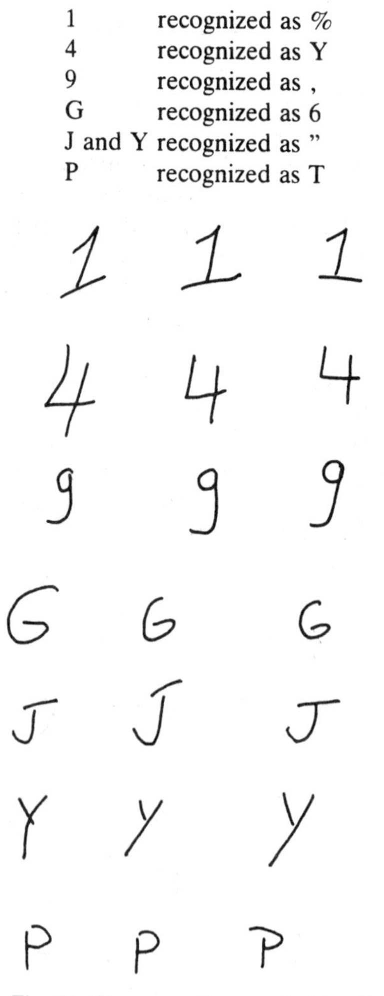

Fig. 14. Characters incorrectly identified.

Table 3. Recognition results

	A	B	C	D	E	F	G	H	I	J	K	L	M	N	O	P	Q	R	S	T	U	V	W	X	Y	Z
A	247																									
B		240						1					1					7								
C			244									2														
D				225									1			1										
E	1				247		4		2											1					1	5
F					2	243							1													
G			2				157		1																	1
H				1				238									1									
I									237																	1
J				1						182				3						1				1		
K				1				1			226		1					3			1				1	
L												237														2
M											1		233													
N											1		1	244						1						
O			1	2			3			2					248	1			2		1					
P				1							1					195										
Q							6										235									
R				1														238								
S																			238							
T		1							6	1			6	3		39		2		228			1	2	1	
U								2				1									239	5				
V				1															1		10	241				
W											1	1											250			
X																								232		
Y								1									4						1	2	198	
Z									1																	221
0	1							1		1	0								1							
1									2																	
2										1												1				15
3																									1	
4								3			1		1											2	4	
5		1								1									2					2	1	
6			2				71					2														
7				1																1					2	
8							2			1							2									
9																	4									
!																1										
"				10	2	3	3	1		45	5	1	2			10	2	1	1	12	1				39	
#							1																		1	
$		1																								
%																										
(																										
)																										
*	2	1				1			1		11			1		1	1	1						1		
+								1												2						
,												1													2	
-																										
.			1	1																						
/																				1						
:										2						1				1						
;		1																								
<																										
=										4										1						
>								1																		
?																										
@													1													
[																										
\																										
]																										
Rejet	1	7	2	7	1	5	5	2	2	12	4	7	4	1	4	3	3	9	7	3		5		10	3	7

	0	1	2	3	4	5	6	7	8	9	!	"	#	$	%	(	)	*	+	,	-	.	/	:	;	<	=	>	?	@	[	\	]
A	1	2							1									2															
B													1								1												
C							1						1			57										2				17	2		
D																																	
E		3				1								1	2										2					1	1		
F						1		3						1																			
G	1				11					1			1							1													
H																												1	1				
I															1			5						1									
J		1							1	3										11				1	3				1				2
K	1											1																					
L																										13					9		
M																																	
N														4	1						1												
O	10		1				3		1	2						5				1		1	1								2	2	
P																																	
Q	5									4														1	1				1				
R			2																														
S				1		1			1					1												1							
T		1			5					4	2			1		1		1	11						1						9		
U					1							1																					
V												14																					2
W	1																																
X																						1											
Y	1	2			23					3									4					1	3			3					
Z		9	4					1																				1					
0	219				1									2																			
1	1	198		1	2			6			1					27	71	1		75			34				1		1		5	41	3
2		3	242																														
3				233																													
4					188					4		6							1							1							
5				5		229			1																								1
6					2		243													1					1								
7			2					224											1									8				1	3
8									230	5					1						1			2	1					6			
9					1			1		193																		1	2	2			
!		2									174																		2				
"	1	8			6	5		8	2	2	64	223			27			1		4					1	2	3		11	3	8		1
#													243		1																		
$														235																			
%		19			3	1									202														3				
(																161										2					2	1	
)																	167			7								1					
*													1		1			238												1			
+												1							220														
,		1		5				1		18					1					27										15			
-																					215	1	6									6	
.			1				3	1			2									96	33	246	1		2					10			
/								1									1			7			209										
:						6			11		7													240	115		29						
;	1			1						4														1	113						1		1
<																										225							
=					1							1													1		217						
>																				1								227					
?																													230				
@																							1							177			
[																															207		
\																																188	
]																																	224
Rejet	11	5		6	8	8	2	6	4	9	2	4	5	7	15	1	13	3	15	22		3		5	8	6	2	10		20	6	13	15

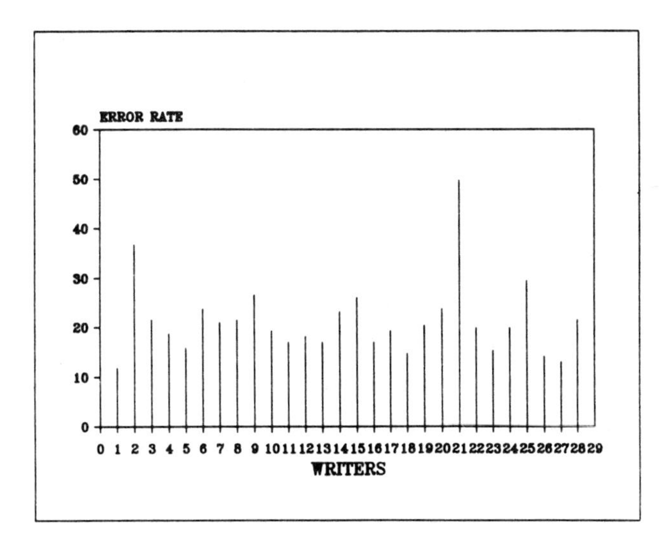

Fig. 15. Individual error rates (Type A).

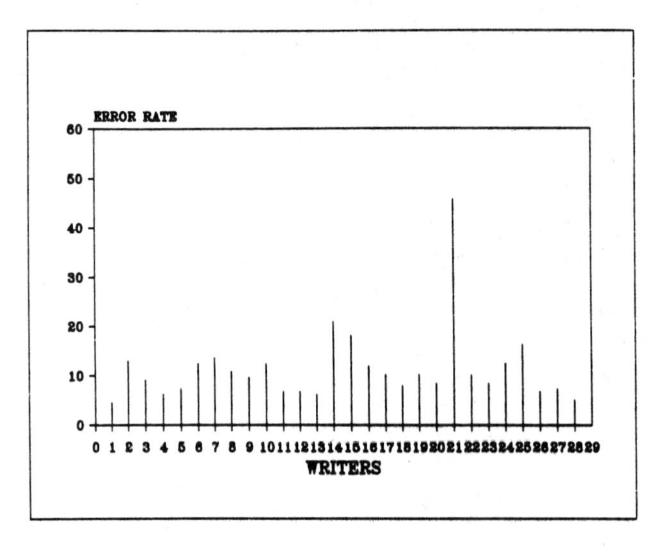

Fig. 16. Individual error rates (Type B).

Figure 14 shows samples of the characters which were incorrectly identified (Type C acquisition).

While it is easy to see that the specimen of character G really does look like character 6, there should have been no confusion about P and 4, for example. These recognition errors could perhaps have been avoided if the system had taken account of the characteristics which were evident in the images of the characters.

The participants were comfortable using the system to differing degrees and the causes of the errors varied from one participant to another. Figures 15, 16 and 17 show the error rates in relation to individual writers for each type of acquisition.

In Fig. 17, it should be noted that in the case of writers who adapted well to using the system, the error rate might be very low, for example, 2.82% for participants 24 and 28. However, for those who were not at ease with the system, the error rates were quite high, for example, 23.16% for participant 14.

The average calculation time for recognizing a character was 1.88 seconds/character using boxes (Type C) and 2.36 seconds/characters without boxes.

2.3.2. *The human factor.* Every participant was asked to fill in an evaluation form at the end of the testing. These forms contained six questions and a few lines for general comments. These were the results:

Question 1: Did you find the equipment limiting?
Answer: Very: 3.57% Somewhat: 39.28% Not very: 57.14%

Question 2: Did you find the procedure limiting?
Answer: Very: 7.14% Somewhat: 32.14% Not very: 60.71%

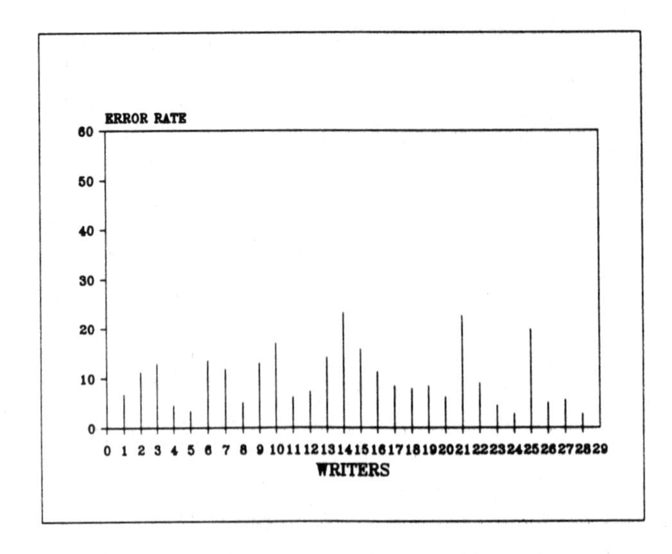

Fig. 17. Individual error rates (Type C).

Question 3: Did you find Type A acquisition natural?
Answer: Yes: 46.42% Somewhat: 35.71% Not very: 17.85%

Question 4: Did you find Type B acquisition natural?
Answer: Yes: 3.57% Somewhat: 39.28% Not very: 57.14%

Question 5: Did you find Type C acquisition natural?
Answer: Yes: 46.42% Somewhat: 39.28% Not very: 14.28%

Question 6: Would you prefer to use a keyboard?
Answer: Yes: 89.29% No: 10.71%

The pen used was thicker than an ordinary pen and was attached by a cord to the tablet. In addition, the writer had to wait until the character had been recognized before writing the next one. Although this might have been an inconvenience, most of participants did not find either the equipment or the software a hindrance (Questions 1 and 2).

In the Type A testing, the characters were written naturally. However, only 46% of the participants found this type of acquisition natural (Question 3). This may be explained by the constraint imposed on the size of the characters. In fact, if the tablet is configured in the no-box recognition mode, the characters must be at least 6 cm high to be processed properly (Fig. 18).

In Type B acquisition, constraints were introduced on some of the characters so that they would be recognized by the software. For example, the parenthesis had to be written from low to high so that it would not be confused with the C. The participants found that this type of constraint, added to the necessity of writing very large characters, made the procedure seem quite unnatural (Question 4).

The tablet for Type C was configured in such a way that the characters could be written normal size. The writers found this type of acquisition to be natural in spite of the constraints imposed on some of the characters (Question 5).

The great majority of the participants prefer the keyboard because of the slow speed of recognition and the constraints (Question 6).

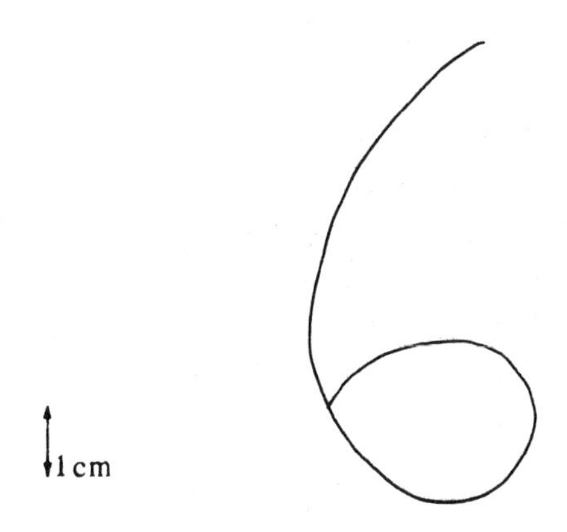

Fig. 18. Character entered without a box.

Several points were brought up in the writers' comments. The most frequent comment was on the slow recognition speed. Eleven participants hoped that the procedure would be improved in this respect. The second most frequent comment was on the number of errors. Eight participants thought that there were too many recognition errors. Seven of the participants mentioned the problem of constraints. They pointed out that it was not at all natural, for example, to draw parentheses from bottom to top. The efficiency of the equipment was mentioned by six of the participants who found that the pen was too thick and that the cord attaching it to the tablet had a tendency to become tangled which bothered them a great deal during the testing process. Two participants found the procedure tiring (compared to using a keyboard) and one could see no point to the system at all.

The comments were not all negative, however. Six participants felt that the software performed well (few errors) and one found that the procedure was not too limiting.

In designing a handwritten character recognition system, such observations must be taken into account in order to make the procedure acceptable to the user. The primary objective must be to lower the percentage of those who would prefer to use a keyboard (nearly 90% in this study).

It does seem to be very difficult, however, to find a compromise solution between writing freedom on the one hand and accurate recognition and processing speed on the other, especially when the equipment's capability is limited (a 16 MHz microcomputer). Even the PENCEPT system, which is considered to be one of the best systems available commercially at the present time, is a compromise solution in these respects.

The results obtained during this experimental study confirm the observations made in our laboratory by Jean Berthiaume.[50] His work was concentrated on perfecting a handwritten interface using software programming support designed by Beauregard and Plamondon.[51] He confirmed the existence of problems associated with writing constraints, slow calculation speed and recognition errors. The work carried out in our study enabled us to quantify his observations and to delineate these problems with greater precision.

3. CONCLUSION

In this paper, we have described the various online handprinted character recognition systems currently in existence. We have summarized the preprocessing, internal and external segmentation techniques, and the various approaches to character description and the related comparison methods. We have also carried out a series of tests on the PENPAD 310 system.

We have presented and analysed our results. The

comments of the participants and their observations on this application have also been analysed in depth.

Some research studies have focussed on the reduction of writing constraints in existing recognition systems in order to make the acquisition procedure more natural and more efficient. For practical purposes, however, these systems are only applicable to languages with very large alphabets (Chinese, Japanese, etc.) or for use by those with little keyboard skill and only wishing to enter characters occasionally.

Acknowledgements—This research has been made possible through grants from the CRSNG of Canada, the FCAR Foundation of Quebec and with the support of the Ecole Polytechnique de Montréal. Fathallah Nouboud wishes to thank the Ecole Polytechnique for its post-doctoral fellowship, and Pierre Yergeau (research assistant, Scribens Laboratory) for his valuable assistance.

REFERENCES

1. S. Mori, K. Yamamoto and M. Yasuda, Research on machine recognition of handprinted characters, *IEEE Trans.* **PAMI-6**, 386–405 (1984).
2. C. Y. Suen, M. Berthod and S. Mori, Automatic recognition of handprinted characters: The state of the art, *Proc. IEEE* **68**, 469–487
3. C. C. Tappert, C. Y. Suen and T. Wakahara, On-line handwriting recognition—A survey, *9th Int. Conf. Pattern Recognition* **2**, 1123–1132 (1988).
4. D. Gokana and D. Juvin, Classification structurelle hiérarchique en reconnaissance de caractères manuscrits, AFCET Conference, Identification et Reconnaissance des Formes, Tome 2, pp. 835–851 (1986).
5. O. Horn, J. Ciccotelli, P. Castel and R. Husson, Une méthode de reconnaissance automatique de chiffres manuscrits, AFCET Conference, Identification et Reconnaissance des Formes, Tome 2, pp. 853–864 (1986).
6. J. R. Ward and B. Blesser, Interactive recognition of handprinted characters for computer input, *IEEE Computer Graphics and Application* **5**, 24–37 (1985).
7. J. R. Ward, Issues in the validity of testing protocols and criteria for on-line recognition of handwritten text, PENCEPT report (1987).
8. J. R. Ward and T. Kuklinski, A model for variability effects in handprinting with implications for the design of handwriting character recognition systems, *IEEE Trans. Syst. Man Cyber.* **18**, 438–451 (1988).
9. R. Plamondon, A handwriting model based on differential geometry, *Computer and Human Applications of Handwriting*, R. Plamondon, C. Y. Suen, M. Simner, eds. World Scientific Publishing, Singapore (1989).
10. G. Lorette and R. Plamondon, Electronic handwriting: Trade-offs and new perspectives for education, 4th IGS Conference, Trondheim, Norway (1989).
11. H. Arakawa, On-line recognition of handwritten characters—Alphanumerics, Hiragana, Katakana, Kanji, *Pattern Recognition* **16**, 9–21 (1983).
12. J. Kim and C. C. Tappert, Handwriting recognition accuracy versus tablet resolution and sampling rate, IEEE 7th Int. Conf. Pattern Recognition, Montreal, pp. 917–918 (1984)
13. H. D. Crane and R. E. Savoie, An on-line data entry system for handprinted characters, *Computer* **10**, 43–50 (1977).
14. P. Telmosse and R. Plamondon, Caractérisation et classification de nombres manuscrits par méthode statistique, *Proc. RAI/IPAR* 86, *Identification and Pattern Recognition*, Toulouse, pp. 865–877 (1986).
15. E. Mandler, Advanced preprocessing technique for on-line script recognition of non-connected symbols, 3rd Int. Symp. Handwriting and Computer Applications, pp. 64–66 (1987).
16. K. Odaka, T. Wakahara and I. Masuda, Stroke-order-independent on-line character recognition algorithm and its application, *Rev. Elec. Com. Lab.* **34**, 79–85 (1986).
17. P. J. Ye, H. Hugli and F. Pellandini, Techniques for on-line Chinese character recognition with reduced writing constraints, 7th Int. Conf. Pattern Recognition, pp. 1043–1045 (1984).
18. K. Yoshida and H. Sakoe, On-line handwritten character recognition for a personal computer system, *IEEE Trans. Consumer Electronics* **28**, 202–209 (1982).
19. C. C. Tappert, Adaptive on-line handwriting recognition, 7th ICPR, pp. 1004–1007 (1984).
20. M. R. Ito and T. L. Chui, On-line computer recognition of proposed standard ANSI (USASI) handprinted characters, *Pattern Recognition* **10**, 341–349 (1978).
21. W. W. Loy and I. D. Landau, An on-line procedure for recognition of handwritten alpha-numeric characters, 5th ICPR, pp. 712–714 (1980).
22. D. J. Burr, Designing a handwriting reader, *IEEE Trans. Pattern Anal. Mach. Intell*, **PAMI-5**, 554–559 (1983).
23. W. Doster and R. Oed, Word processing with one-line script recognition, *IEEE Micro* **4**, 36–43 (1984).
24. A. Amin, A. Kaced, J. P. Haton and R. Mohr, Handwritten Arabic character recognition by the IRAC system, 7th ICPR, pp. 729–731 (1980).
25. Y. H. Huh and H. L. Beus, On-line recognition of handprinted Korean characters, *Pattern Recognition* **15**, 445–453 (1982).
26. K. Ikeda, T. Yamamura, Y. Mitamura, S. Fujiwara, Y. Tominaga and T. Kiyono, On-line recognition of handwritten characters utilizing positional and stroke vector sequences, *Pattern Recognition* **13**, 191–206 (1981).
27. C. C. Tappert, Speed, accuracy, flexibility trade-offs in on-line character recognition, IBM Research Report RC13228 (1987).
28. P. Y. Lu, W. Brodersen, Real-time on-line symbol recognition using a DTW processor, 7th ICPR, Vol. 2, pp. 1281–1283 (1984).
29. G. Y. Tang, P. S. Tzeng and C. C. Hsu, A microcomputer system to recognize handwritten numerals using a syntactic-statistic approach, 7th ICPR, Vol. 2, pp. 1061–1064 (1984).
30. R. M. Brown, On-line computer recognition of handprinted characters, *IEEE Trans. Elec. Comput.* **13**, 750–752 (1964).
31. R. Plamondon and R. Baron, A dedicated microcomputer for handwritten interaction with a software tool: System prototyping, *J. Microcomput. Appl.* **9**, 51–60 (1986).
32. R. Plamondon and R. Baron, On-line recognition of handprint schematic pseudocode for automatic FORTRAN code generator, 8th ICPR, Paris, pp. 741–745 (1986).
33. R. Baron and R. Plamondon, Logiciel de développement de programmes par interaction manuscrite, Technical report, Scribens Lab., Ecole Polytechnique de Montréal.
34. D. J. Burr, Elastic matching of line drawings, *IEEE Trans. Pattern Anal. Mach. Intell.* **PAMI-3**, 708–713 (1981).
35. Y. Hidai, K. Ooi and Y. Nakamura, Stroke re-ordering algorithm for on-line handwritten character recognition, 8th ICPR, Paris, pp. 934–936 (1986).
36. E. Mandler, R. Oed and W. Doster, Experiments in on-

line script recognition, 4th Scandinavian Conference on Image Analysis, pp. 75–86 (1985).
37. S. Impedovo, B. Marangelli and A. M. Fanelli, A Fourier descriptor set for recognizing nonstylized numerals, *IEEE Trans. Syst. Man. Cybern.* **SMC-8**, 640–645 (1987).
38. K. Odaka, H. Arakawa and I. Masuda, On-line recognition of handwritten characters by approximating each stroke with several points, *Trans. IECE Japan* **E-63**, 168–169 (1980).
39. K. Odaka, H. Arakawa and I. Masuda, On-line recognition of handwritten characters by approximating each stroke with several points, *IEEE Trans. Syst. Man. Cybern* **SMC-12**, 898–903 (1982).
40. R. Oed and W. Doster, On-line script recognition—A userfriendly man machine interface, COMPINT Conf. Computer-Aided Technologies, IEEE Cat., pp. 610–614 (1985).
41. C. G. Leedham and A. C. Downton, Automatic recognition and transcription of PITMAN's handwritten shorthand—An approach to shortforms, *Pattern Recognition* **20**, 341–348 (1987).
42. H. Sakoe and S. Chiba, Dynamic programming algorithm optimization for spoken word recognition, *IEEE Trans. Acoustics Speech and Signal Processing* **ASSP-26**, 43–49 (1978).
43. F. Nouboud, F. Cuozzo, R. Collot and M. Achemlal, Authentification de signatures manuscrites par programmation dynamique, PIXIM Conference, Paris, pp. 345–360 (1988).
44. F. Nouboud, Contribution à l'étude et à la mise au point d'un système d'authentification de signatures manuscrites, Thesis of Doctorat de l'Université de Caen, France (1988).
45. R. Plamondon and G. Lorette, Automatic signature verification and writer identification—The state of the art, *Pattern Recognition* **22**, 107–131 (1989).
46. Pencept Inc, Software toolkit for the PENPAD 310 and the PENPAD 320, PENCEPT Documentation (1986).
47. R. S. Pressman, *Software Engineering—A Practionner's Approach*. McGraw-Hill, New York (1987).
48. F. Nouboud and R. Plamondon, Expériences en reconnaissance en-ligne de lettres moulées, Scribens Technical Report, Ecole Polytechnique de Montréal (1989).
49. F. Nouboud and R. Plamondon, Reconnaissance en-ligne de caractères moulés: techniques et résultats comparatifs, Canadian Conference on Electrical and Computer Engineering, Montreal, pp. 761–764 (1989).
50. J. Berthiaume, Interface manuscrite pour éditeur de programmation schématique, Projet de fin d'études report, Ecole Polytechnique de Montréal (1988).
51. D. Beauregard and R. Plamondon, Schematic coding on an IBM-PC, *J. Microcomput. Appl.* **10**, 91–100 (1987).

About the Author—RÉJEAN PLAMONDON received a B.Sc. degree in physics, and a M.Sc.A. and a Ph.D. degrees in electrical engineering from Université Laval, Québec, QC, Canada in 1973, 1975, and 1978 respectively.

In 1978, he joined the Ecole Polytechnique, Université de Montréal, Montréal, QC, Canada, where he is currently an Associate Professor. In 1985–1986, he was involved in several research projects while a guest of the Computer Science Department, Concordia University, Montréal, Canada, the Motor Behavior Laboratory, University of Madison, Wisconsin, USA, the Department of Experimental Psychology, University of Nijmegen, The Netherlands and the Laboratoire de Génie Electrique de Créteil, Université de Paris Val-de Marne, France.

His research interests are the computer applications of handwriting: biomechanical models, neural and motor aspects, character recognition, signature verification, signal analysis and processing, computer-aided design via handwriting, forensic sciences, software engineering and artificial intelligence. He is the founder and director of Laboratoire Scribens at the Ecole Polytechnique de Montréal, a research group devoted exclusively to the study of these topics.

An active member of several professional societies and a senior member of IEEE ('85), Dr Plamondon is also a member of the board of the International Graphonomics Society and president of the IAPR Technical Committee on text processing applications. He is the author of numerous publications and technical reports. He is an avid swimmer and enjoys writing poetry, children's book and novels.

About the Author—FATHALLAH NOUBOUD received a Ph.D. degree in pattern recognition from Université de Caen, France in 1988. His Doctorate project consisted in the design of a handwritten signature verification system. At present, he is cooperating, as guest researcher, with Scribens Laboratory (Ecole Polytechnique de Montréal). His research interests are the computer applications of handwriting as character recognition and signature verification.

Intelligent Forms Processing System

Richard Casey,[1] David Ferguson,[2] K. Mohiuddin,[1] and Eugene Walach[3]

[1]K52/803, IBM Almaden Research Center, San Jose, CA, USA, [2]IBM Professional Services, 520 Capitol Mall, Sacramento, CA, USA, and [3]IBM Israel Scientific Center, Haifa, Israel

Abstract: This paper describes an intelligent forms processing system (IFPS) which provides capabilities for automatically indexing form documents for storage/retrieval to/from a document library and for capturing information from scanned form images using intelligent character recognition (ICR). The system also provides capabilities for efficiently storing form images. IFPS consists of five major processing components: (1) An interactive document analysis stage that analyzes a blank form in order to define a model of each type of form to be accepted by the system; the parameters of each model are stored in a form library. (2) A form recognition module that collects features of an input form in order to match it against one represented in the form library; the primary features used in this step are the pattern of lines defining data areas on the form. (3) A data extraction component that registers the selected model to the input form, locates data added to the form in fields of interest, and removes the data image to a separate image area. A simple mask defining the center of the data region suffices to initiate the extraction process; search routines are invoked to track data that extends beyond the masks. Other special processing is called on to detect lines that intersect the data image and to delete the lines with minimum distortion to the rest of the image. (4) An ICR unit that converts the extracted image data to symbol code for input to data base or other conventional processing systems. Three types of ICR logic have been implemented in order to accommodate monospace typing, proportionally spaced machine text, and handprinted alphanumerics. (5) A forms dropout module that removes the fixed part of a form and retains only the data filled in for storage. The stored data can be later combined with the fixed form to reconstruct the original form. This provides for extremely efficient storage of form images, thus making possible the storage of very large number of forms in the system. IFPS is implemented as part of a larger image management system called Image and Records Management system (IRM). It is being applied in forms data management in several state government applications.

Address reprint requests to: Richard Casey, Integrated Data Management, K52/803, Almaden Research Center, 650 Harry Road, San Jose, CA 95120-6099, USA

Key Words: Document analysis, optical character recognition, image management systems, image compression, automatic forms processing

1 Introduction

Forms processing is an essential operation in many businesses and government organizations. However, there are a number of limitations with existing manual forms processing operations. These include the cost of capture of information from printed forms, delay in accessing stored forms, vulnerability of stored forms to loss, damage, etc.

The cost of capture of information from printed forms can be very high. For example, it has been estimated that for an operator to enter the information from a form into a computer costs about $2.50 per form. A medium-sized enterprise may receive as many as 15,000 forms per day. Thus, the cost of capturing information from forms runs to several million dollars per year. For large corporations and government organizations, the cost can be orders of magnitude greater.

The form documents may be required at short notice and by multiple users belonging to an organization. A paper-based forms processing system can not support concurrent access of forms by its users. This leads to increased delay in information flow and reduced productivity. A solution to these problems is to provide an *intelligent forms processing system* that seeks to automate the manual operations in existing paper-based systems. This paper describes a combined image management and document analysis system for the automatic processing of printed forms. To be truly useful to an enterprise, an intelligent forms processing system should meet certain important requirements. The system should provide significant reduction in the cost of capture of information from printed forms, cost-effective storage

"Intelligent Forms Processing System" by R. Casey et al. from *Machine Vision and Applications*, Vol. 5, 1992, pp. 143–155.

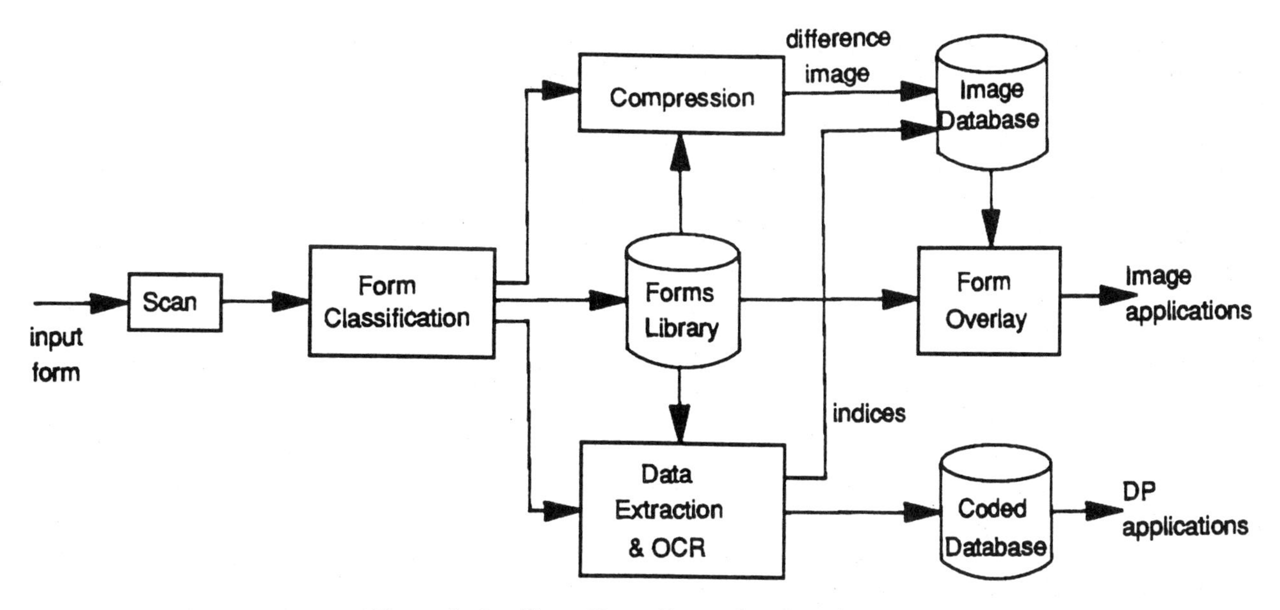

Figure 1. Intelligent Forms Processing Organization.

of form images, efficient retrieval and display of documents, and satisfactory system performance.

In addition to the cost of capture, which has already been discussed, the cost of storing the form images is an important factor in such a system. Even if the information on documents is encoded by key operations or Optical Character Recognition (OCR), the form images need to be stored for legal and operational reasons. An 8.5″ × 11″ page when scanned at 300 samples per inch amounts to about 1Mbyte of data. One can compress the image using conventional techniques and reduce the storage requirement to about 40 to 80 Kbytes, depending on the contents of the image. However, considering the volume of forms, this may still be too large to be cost-effective. We have developed efficient techniques that reduce the storage requirements by a further factor of 10, resulting in less than 10 Kbytes of storage per form.

In order to overcome the limitations inherent in a manual forms processing system, we have developed Intelligent Forms Processing System (IFPS) as an experimental prototype. This paper describes various components of IFPS. Such a system should also provide capabilities for scanning documents, storing the images in an image database, retrieving, displaying, and printing documents, etc. However, we shall limit our discussion in this paper to those aspects that are immediately relevant to intelligent forms processing. In the section to follow, the overall operation of the system will be discussed. Then the details of each of the component functions will be presented. Finally, the status of the system and an application will be described.

2 Related Work

Automated forms processing has been applied previously to "turn-around" documents, specially designed forms so constructed as to simplify automated processing. A billing statement is often preprinted with the account identification of the person billed, for example, in a font style easily read by OCR equipment. Only recently has sophisticated image analysis been applied to problems of locating areas of interest in conventional documents and seeking to extract information. Research in this area has concentrated on applications such as line drawing conversion to CAD format (Kasuri and Shih 1989), recognition of text document format and content (Nagy and Viswanathan 1991), and location and reading of addresses on mail (Mitchell and Gillies 1989). The present paper extends such research to the arena of conventional preprinted forms, which have not been constructed to facilitate reading by machine.

3 System Organization

The overall organization of intelligent forms processing is shown in Figure 1. IFPS contains two parallel paths, one for image applications such as retrieval, display, and printing of a form document, the other for data processing applications that deal with information contained on a form.

The form is initially entered into the system by an optical scanner that converts it to a bilevel digital image. The image is then subjected to an image analysis operation that determines the layout of the form.

The layout parameters are compared against a stored library of such information in order to recognize the class of the form. The system now knows which of the many types of documents that it is capable of processing is represented by the input form. The Form Library that governs the document classification process is central to a number of processing stages. It contains, for each form, the parameters needed to classify the form, to locate and extract data from it, and to perform image compression/decompression.

Once the form class is determined, the document image can be compressed for storage. This is a two-stage process. First the preprinted background form is removed by a sophisticated subtraction process. Then the remaining image, now containing mainly the data added to the form, is compressed further by conventional image encoding techniques. It can be reconstituted when desired, by reversing this process, overlaying a master form background stored in the Forms Library.

The input form is also fed to a data extraction stage, which locates and tracks any data added to specified fields of the document. Data images found are recognized as symbol codes by OCR. The coded data can be stored in a data base for use by data processing applications. Certain of the coded fields may also be used as index keys by the image storage subsystem. The overall system thus supplies access to data on the forms, as well as efficient storage and retrieval of the entire input document.

4 Forms Classification

One of the first steps in intelligent forms processing is classifying the input form into one of the known form types used by an enterprise. A simple approach might be to recognize the form number, which is usually present on most forms, using OCR. Unfortunately, the form number is not always at the same location on a form. For example, on some forms the form number might be along the bottom left edge, whereas on certain other forms it might be along the left margin. Often, the form number is printed in very small fonts, making accurate recognition difficult. On multipage forms the form number is usually present only on the first page. For these reasons, there is a need for a more general procedure for recognizing forms. One solution to this problem is to bar code the form number on forms, and to read in the number using a bar code reader. Unfortunately, the volume of forms accumulated over the years is so large that it is prohibitively expensive to bar code all the forms collected in an enterprise. Our solution to this problem is based on identifying certain features that define the form. The features that we have found to be effective are the location and length of the horizontal and vertical lines on a form.

The form classification consists of two phases—form definition and form recognition. During the form definition phase, a model for each of the unique forms used in an enterprise is generated. During the recognition phase, which is the operational phase of form classification, unknown input forms are analyzed and classified based on the models generated during form definition.

4.1 Form Definition

In the form definition phase, a model for each of the unique forms used in an enterprise is generated. The model consists of information about the location and length of significant lines, which are longer than a certain limit, the position and size of fields from which data need to be extracted, and the information required for computing the offset and skew of matching forms during the recognition phase. An interactive facility is provided for a user to scan in forms and define fields of interest on the form. The fields are identified by the opposing corners of a rectangular area. This area is used as a mask for field data extraction. The models generated during the definition phase are stored in a form library.

4.2 Form Recognition

Form recognition is the operational phase of form classification. During this phase, an input form image is analyzed and information about "significant lines" on the form are determined. The form library generated during the definition phase is searched for the closest match to the input feature set. If the match happens to be within a certain threshold, the input form is recognized to have the same form number as the matching form in the library. The threshold is chosen proportional to the number of features of the form.

This method can be generalized as a graph matching problem. However, one particular optimization of this method is to use only significant horizontal lines as features. Standard image compression algorithms, such as CCITT G4, encode horizontal runs of black and white pixels to achieve compression. When decompressing an image for analysis, the result can be specified to be returned as horizontal runs rather than black/white pixels. From these runs, it is possible to rapidly identify the horizontal lines using a process called connected component labelling.

The form recognition step should be robust enough to handle inaccuracies that are introduced

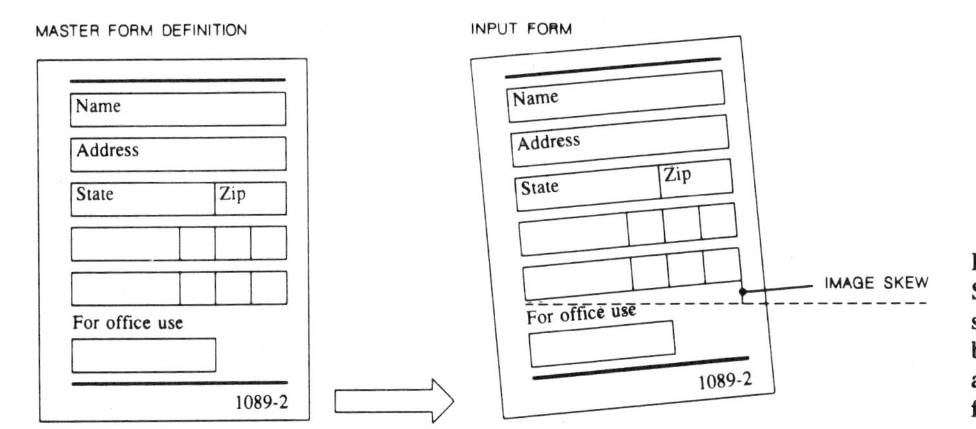

Figure 2. Example of Form Skew. Skew in IFPS is measured by the angular difference between lines in the model form and corresponding lines in the form to be read.

during scanning. The inaccuracies are form offset, skew, and scanning noise. Form offset can be along both vertical and horizontal directions, and the skew can be clockwise or counterclockwise. Offset and skew cause the field positions to be different from the model generated during the definition phase (Figure 2). These parameters are also essential for accurate registration of an input form over a matching template in the data extraction and forms drop-out stages. The offset information is computed using the shift along the horizontal and vertical directions of matching lines. The skew is computed using the slope of a long line in the form. Scanning noise causes lines to appear broken. This can lead to a line being identified as multiple lines unless appropriate corrections are applied.

5 Data Extraction

Conventionally, systems that have performed recognition of character data printed in a form's fields have imposed several significant constraints on the data to be read:

1. Data must reside entirely within the field's defined region.
2. Data must not cross or come in contact with the background form information.
3. Data must not be contaminated by extraneous marks or lines.

Typed information may be placed on the forms using a variety of typewriters or printers. It is thus difficult to be sure of the location of data on a document. Consequently, the first two limitations above significantly curtail the usefulness of conventional OCR systems, constraining their use to very specific, tightly controlled environments where forms are carefully prepared. One solution to overcome the first two constraints has been to produce forms in an ink that is invisible when the document is scanned. This removes any possibility that background form information will corrupt the data and allows for the fields to be large enough to capture even data that is not placed perfectly in the center of a field. This solution, however, has the disadvantage that new forms must be created that are tailored for the forms processing system. Unfortunately, legal requirements, cost, and the complexity involved in converting the layout of a form can make such a change either difficult or legally impossible.

The solution implemented for IFPS (Casey and Ferguson 1990) is not bound by the limitations inherent in traditional forms processing systems, and requires no modification to existing forms (although it does not preclude the changing of documents if desired). This solution comprises three steps. The first step is to register the document using skew and offset information provided by forms classification. The second step is to extract the data from the requested fields. The final step is to remove any extraneous markings that are intruding on the data.

5.1 Registration

Form skew, horizontal offset, and vertical offset affect the ability to accurately capture the data in the desired fields. Each of these variables is estimated in order to adjust the defined field coordinates so that the data may be more accurately located.

The skew of a form (Figure 2) is the rotational difference between the model form that defines a form type and the incoming form. Each field is defined to the system as a horizontal rectangle to be searched for data. A high degree of skew in the form increases the chances that background form information may reside within the rectangle. To reduce the likelihood of this occurrence, each field is adjusted to compensate for the skew. Compensation

Figure 3. **Compensating for skew by a smaller search rectangle. Size of the rectangle is reduced on a skewed form, in order to assure that no background printing is overlaid.**

for skew can be achieved by reducing the height and width of the search rectangle as the skew increases (Figure 3). This avoids the capturing of unwanted background information since the rectangle examined by the system now lies within the whitespace of the field. The reduction also increases the likelihood that some data will lie outside the rectangle. However, as discussed in the next subsection, such data fields can still be extracted correctly. Another compensation method that was considered was to rotate the field by the degree of the form skew. In this case the search rectangle does not have to be adjusted. The process, however, is significantly more compute-intensive than the solution we adopted.

The horizontal and vertical offsets denote the translation required to map the incoming form to the master form. For instance, if a line on the form appeared in the 100th row for the master form and in the 96th row for the incoming form, then the vertical offset would be 4, since it would require adjusting the incoming form down 4 rows to compensate for the difference. Such differences are generally caused by the placement of the form in the scanner or by the differences between scanners in initiating the scanning of a form. The offsets are calculated by comparing the incoming form's line descriptions to the master form's pattern (Figure 4). After the offsets have been determined, they are used to translate the fields' absolute coordinates (defined using a Forms Editor as discussed in Casey and Ferguson 1990) into the incoming form's coordinates.

5.2 Capturing the Data

Once the coordinates of a search rectangle have been mapped onto the form being processed, the data within each rectangle can be extracted. During the process of extracting the data, two types of data may be encountered: perfect data or ambiguous data (Figure 5).

Perfect data is data that resides entirely within the search rectangle. Perfect data is determined to exist when a check along the perimeter of the rectangle reveals that all pels are 0, or OFF. Perfect data can be extracted immediately. Ambiguous data exists wherever there is an ON bit along the perimeter of the search rectangle. Ambiguous data thus lies partially outside of the rectangle. Such data must be tracked beyond the rectangle and extracted using special processing as described below.

Ambiguous data that extends sufficiently far from its field may intersect a background form line or form text. Therefore, as it is tracked, a check is made for form lines. If a form line is detected, it is deleted by a line removal process (Figure 6) that eliminates the line while retaining the pixels at the intersection of data and line (see Casey and Ferguson 1990). The process is performed by deleting the line, then copying the pixels that lie just above and below the line through the region where the line is being erased.

Although resolving conflicts between the background text and the data is an important process, algorithms to solve this type of ambiguity are considerably more complex and have not yet been implemented. Simply measuring the average height of extracted data and clipping tracked data beyond this height offers an attractive method of dealing with data that does not merge far into background text.

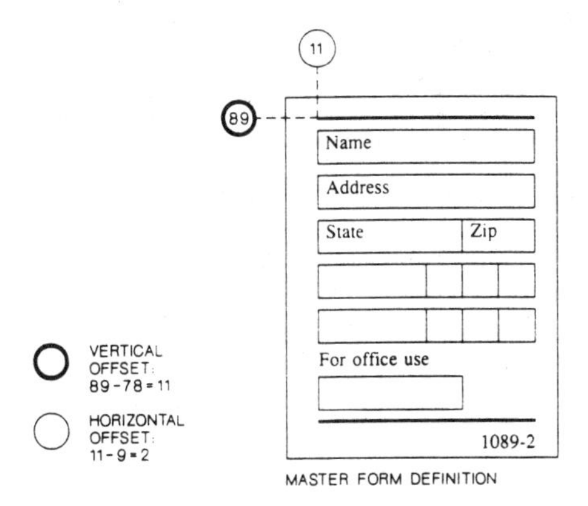

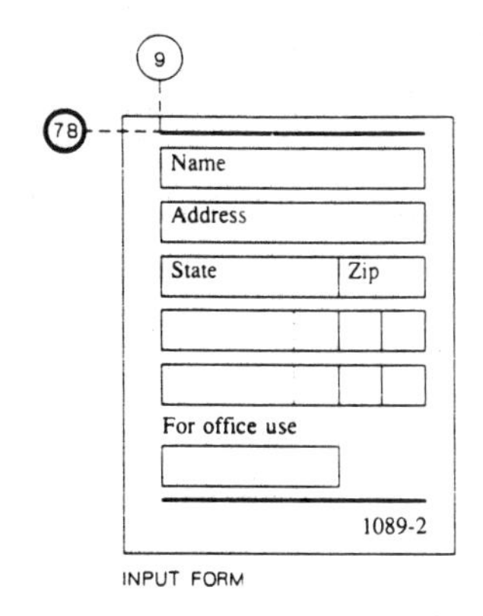

Figure 4. Form Offset. The positional difference shown is a virtual one, since form skew is first removed bya coordinate transformation, without actually rotating either image.

5.3 Removal of Nondata Images

In addition to the problems above, there is also the possibility that an extraneous marking or line that is not a part of the data intrudes upon the field. This problem can be detected when the height of ambiguous data is determined to be larger than the maximum height expected for a data character. When extraneous markings are smaller than the size of a data character, they are not currently removed. However, such small lines have been found to corrupt only a small portion of a field or to have negligible effect on the recognition of data in the field. When an extraneous line (which may be curved) is detected, a continuity-following algorithm similar to that of Clement (1981) can be applied to track the line through intersections with the data (Figure 7). The tracked line is deleted except at intersections with the data image.

As each field is processed and the data contained within the field is captured, the extracted field is placed into a new area of memory. This area can be thought of as a new or extracted image which contains only the data portion of the form image. The extracted image is then passed to Optical Character Recognition for the automatic encoding of the data.

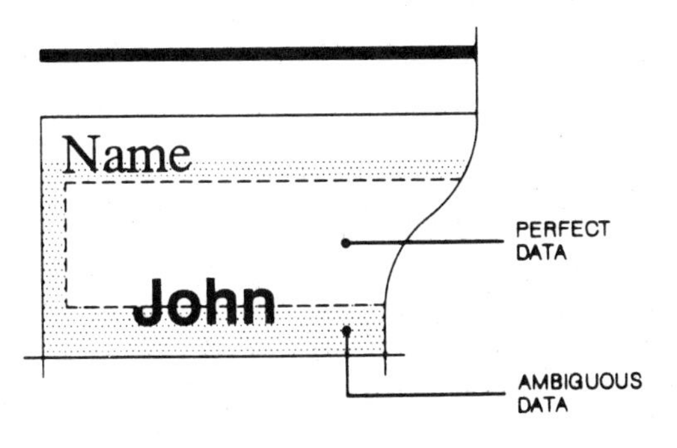

Figure 5. Examples for perfect and ambiguous data.

6 Optical Character Recognition

The objective of the data extraction process is to provide a clean image of the data to be recognized using OCR. As described above, this is done field-by-field, with the characters in each field extracted as a single image block. Before a field image can be recognized it must be segmented into individual character images so that each character can be recognized by a classifier.

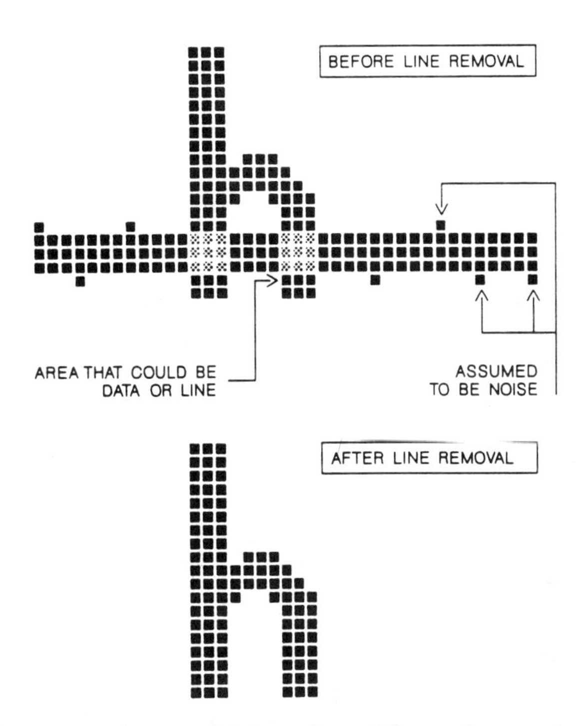

Figure 6. Line removal from data. When a line passing through character images is detected, a noise band above and below the line is first defined. Pixels adjacent to this band are assumed to belong to the character pattern, and are replicated if matching pixels exist on the other side of the band and within the band.

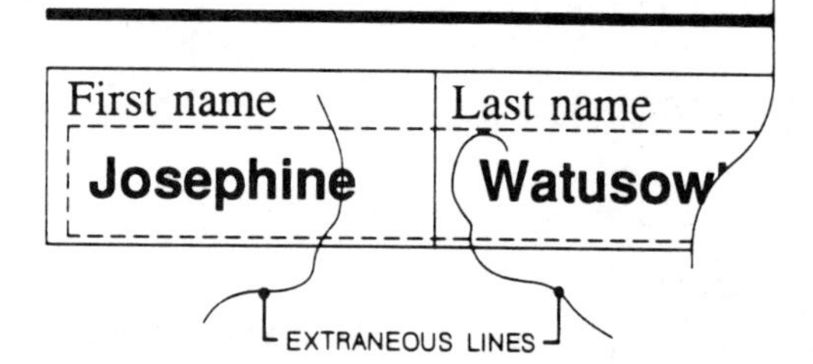

Figure 7. Removal of extraneous lines from data. Because of the possible curvature and even branching of extraneous lines, a removal process is more complex and unreliable than for straight lines previously illustrated.

6.1 Segmentation

In the process of analyzing the overall image into character patterns, the segmentation routine performs the following functions:

1. The pitch, i.e., the distance from character to character, is estimated.
2. Touching characters are separated, and broken characters are merged.
3. Skew of the typing within each field is measured.
4. The overall image is partitioned into print lines.
5. A baseline is computed for each line of print.
6. The character patterns are ordered in reading sequence.
7. The position of each character with respect to the baseline is calculated.
8. Spaces between words are detected.

The input to these operations is the set of connected components that are contained in the fields that were copied to the extracted image. A connected component is a subimage satisfying two conditions: (1) from any black pixel of the subimage there is a path consisting solely of black pixels that connects it to any other black pixel of the subimage, and (2) there is no such connection from a pixel of the subimage to any black pixel outside the subimage. Typically a connected component is represented by its "bounding rectangle," the box with vertical and horizontal edges determined by the top, bottom, left and right edges of the connected component.

Generally, the connected components correspond to individual characters. Thus, the ensemble of connected components is analyzed to determine pitch, locate baselines, and measure typing skew. Following this, touching characters are separated by searching for weak connections between left and right sections of a connected component in the neighborhood of a pitch boundary between characters. Joining of multicomponent characters, sequencing of the patterns for recognition, etc. are also done by analyzing the connected components.

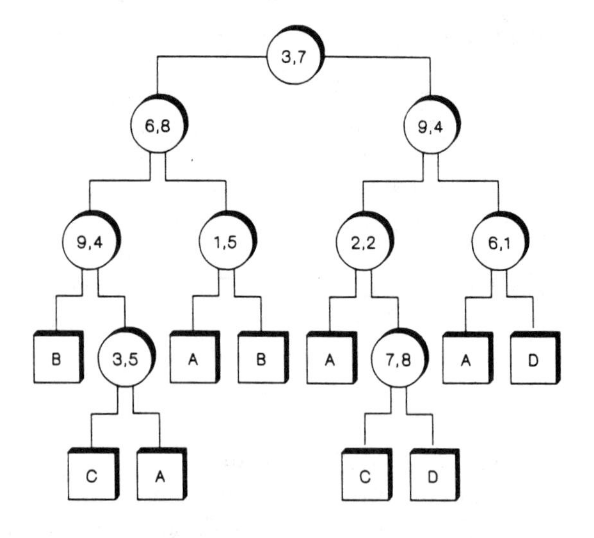

Figure 8. Decision trees for OCR. Each node of the tree identifies a pixel position in a character image. Starting from the top, the specified pixel is examined, and the left branch followed if it is white, or the right branch if it is black. This process is repeated, generating a path through the tree, until a leaf is reached, specifying the identity to be assigned to the input. The tree is designed so that the path followed consists of those pixels which give the most new information about the character, given what has been previously learned along the path.

6.2 Decision Tree Classification

An OCR classifier then examines each character pattern and returns an identification symbol, or ID (for example, an ASCII or EBCDIC code). One of the authors has shown that accurate recognition of a given font style can be obtained using decision trees that test a series of prespecified pixels on each character (Casey and Nagy 1982). These decision trees (Figure 8) are designed automatically using statistics on the probability of black and white for each pixel, where the statistics are gathered from scanned sample characters.

While there are currently many different types of OCR classifiers, the use of decision trees has several advantages that make it an attractive method for recognizing characters that are typically found on forms. These traits are:

- Ability to tolerate breaks or gaps in the character. This feature is particularly critical in environments where aging or deteriorating documents are processed. By weighing multiple classifications obtained using different sets of character pixels, a character can be recognized even though incompletely formed.

- Ability to tolerate varying degrees of darkness. If a system is to be successful in processing documents that are generated by typewriters, it must be able to tolerate darkness variations introduced by the variety and quality of ribbons used. Since the decision tree classifier does not look at the overall thickness or contour of a character, such variations are more easily tolerated.

When documents arrive from many different sources, as is typically the case with typed data on forms, a library of tree logics is needed, one for each font that may be encountered. The proper font logic for a given document is determined by trial and error on the first few lines of the document. Fonts having size characteristics matching those of the printing are tried in recognition, and each classifier provides its own estimate of the accuracy of its recognition. These estimates are evaluated to select the best classifier for reading the remainder of the image.

In a survey conducted on samples taken from the State of California's archive of over 40 million Vital Records documents, it was found that six fonts comprised more than 98% of the machine printing. In another informal study of forms used by businesses in a variety of fields, it was found that fewer than 10 fonts comprised more than 90% of machine printing.

The throughput rate of an OCR classifier is another factor that is critical to its success. In one benchmark performed on an IBM PS/2 Model 70-486 running OS/2, the decision tree classifier was able to process 60 characters per second. With the decision tree method, however, performance depends not only upon implementation, but also upon quality of the printing. Our experience shows that the rate of correct recognition remains high, over 99.5%, but the time spent in classifying degraded third generation copies is more than double that required for first copies.

In summary, IFPS combines a flexible data extraction process with decision tree OCR to achieve a versatile and comprehensive method for automatically capturing data from forms. The system can:

- Successfully process data that does not lie entirely within a field.
- Resolve ambiguity when data crosses or touches background form lines.
- Erase extraneous lines that intrude on the data.
- Recognize characters from a variety of sources with varying degrees of quality.
- Capture data with acceptable system performance.

7 Image Compression by Means of Form Drop Out

One of the fundamental problems inherent to large scale document image systems is that of compact image representation. As pointed out in an earlier section, image compression possible with conventional techniques is not adequate for this purpose. Therefore a new storage reduction technique is sought, namely that of form drop out. The basic idea behind this technique is to remove the common data in the form which is shared by all the forms of the same type, and to store only the variable data which is specific to the given form. Since the variable data or the filled-in data is much smaller in volume than the whole form, we should expect to gain an order of magnitude in compression ratio. Such an approach is utilized frequently for the effective management of electronic forms. The issue is how this simple idea can be extended to the vast body of paper forms, which must be treated as binary images.

One possible solution would be to print empty forms utilizing special color ink, which would be transparent to the conventional scanners. By this approach the form itself would be "invisible" to the scanner and only filled-in information would be entered into the computer. Such a solution would be quite effective and considerable effort has been devoted in order to develop a practical implementation of this approach (Nielsen et al. 1973; Wood and Arps 1973). Unfortunately, providing special ink for new forms excludes an effective solution for the existing archives. In addition, the necessity for the form alternation might be cumbersome and costly to the users. Also, utilization of color forms means that no form can be copied on a standard black and white copier. Accordingly, in a variety of applications it is advantageous to adopt an alternative approach, which would be valid without any alternation of the existing forms and procedures. Moreover, color forms tend to leave some degree of residue noise so that even in some cases of operating color drop out systems alternative solutions are sought.

A possible alternative would be to analyze each form and extract only certain fields where filled-in information is supposed to appear. Then, instead of keeping the entire page image, we can limit ourselves to storing only few areas of interest. This approach is quite effective in terms of providing high image compression. However, it implies that all data outside predefined fields will be lost. Hence, this approach is frequently unacceptable. Consider, for instance, how the meaning of an insurance form can be reversed if one were to disregard all the remarks and alterations which appear on page margins and between the lines.

In this section, we shall describe yet another method aimed at solving the shortcomings of both of the aforementioned techniques. The basic idea is quite simple: First, an empty form (let us call it CP for Constant Part) is scanned and stored in the system. A library of possible empty forms is created in such a manner. Then, for each filled form to be compressed, the appropriate CP will be recognized and automatically "subtracted" leaving the filled-in data (VP or Variable Part) only. Since VP contains relatively sparse data, it can be compressed and stored very efficiently. Obviously, for reconstruction purposes, VP will be retrieved decompressed and combined with the single existing copy of CP. From the user's point of view, the mode of operation will remain unaltered. At the same time, storage cost and, where necessary, data transmission time will be slashed significantly.

In the remainder of this section we shall describe in greater detail how this goal can be achieved.

7.1 Data Acquisition

The first problem that must be addressed is that of data acquisition. The scanning parameter (brightness threshold level) is separately optimized for the empty form, and for the handwritten part in a filled-in form. Since, usually, CP and VP are created in a different manner and under different conditions, they actually differ in brightness. Accordingly, proper choice of scanning threshold may substantially improve overall text legibility.

7.2 Registration

The next problem which must be addressed is that of image registration, i.e., the process of aligning an incoming form with the previously defined empty form template. In general, the geometric relationship between each form image and its template can be quite complex. Fortunately, in the overwhelming majority of practical cases excellent alignment can be achieved even if we limit ourselves to the following two processes:

7.2.1 Coarse registration. In this stage we compute and perform a linear transformation addressing global translation, rotation, and scaling. At this phase local nonlinearities are disregarded, on the other hand high robustness is achieved by allowing large global distortions. The transformation to be used can be computed by choosing a number of small characteristic patterns. By identifying these patterns on the scanned form, and relating them to their corresponding location in the template image, it is straightforward to compute a general affine transformation matrix. It is possible to decide that characteristic patterns will consist of horizontal lines, which already have been identified during the form recognition process described in an earlier section. Accordingly, the coarse registration parameters are be estimated as a by-product of form recognition process.

7.2.2 Fine registration. After the coarse registration, we are left with local nonlinear distortions. Here, we don't assume any limitation on the distortion type. However, the degree (size) is assumed to be relatively small. Therefore, these distortions can be compensated for by partitioning the image into a number of relatively small overlapping "tiles." For each such tile we compute the corresponding lateral transformation. Since each tile is small, and since it can "move" independently of its neighbors, this method can model even complex geometrical transformations quite well. Finally all the shifted tiles are combined together into a single image which closely matches the form template (Walach et al. 1989a).

7.3 Form Drop Out

The aim of this step is to remove the template (CP) out of the scanned form. Naturally, it is imperative to take into account that even for the same data sheet, two separate scans will provide significantly different data arrays. This implies that even after registration, pixel by pixel subtraction will be useless.

Our solution to this problem can be seen as a decision process which classifies each black pixel in the form as being either part of the template data or the filled-in data. The decision is based on the analysis of the pel's immediate neighborhood. The template form (CP) is removed so that scanning noise is cancelled, but at the same time filled-in information remains intact (Walach et al. 1989b).

Consider for instance the image depicted in Figure 9. (reduced image of a form of the type used in the recent Swiss Population Census). This image, using conventional compression techniques, can be compressed to 27 KB. In Figure 10 we present the empty template of the same form. After registration and form drop out, we obtain the image of Figure 11, which can be compressed to only 4.5 KB.

7.4 High Quality Binary Image Compression

The result of the subtraction phase is an image containing only the filled-in information of the original

Eidgenössische Volkszählung 1990

Personenfragebogen 180199838

Gemeinde: Baden

Name/Vorname(n)

Zählkreisnummer: 0008 Example Name

Haushaltungsnummer: 123

Bitte in Blockschrift ausfüllen: ABCDEFGHIKL

Wo Antworten vorgedruckt sind, kreuzen Sie bitte das zutreffende Feld an: X

I. Bevölkerung und Beschäftigung: Fragen an alle Personen

- Geburtsdatum und Wohngemeinde zur Zeit der Geburt — Tag Monat Jahr: 28 02 1943; Wohngemeinde zur Zeit der Geburt: RUDOLFSTETTEN; Kanton bzw. ausländischer Staat: AARGAU
- Geschlecht — 1 männlich X; 2 weiblich 2
- Zivilstand — 1 ledig X; 2 verheiratet 2; 3 verwitwet 3; 4 geschieden 4
- Heimat/Nationalität — 1 Schweizer/in X; 2 Ausländer/in 2

Fragen an Ausländer:

- Welcher ist Ihr Heimatstaat?
- Art des Ausländerausweises, Aufenthaltsstatus:
  1 Niederlassungsbewilligung (Ausweis C) 1
  2 Jahresaufenthaltsbewilligung mit Ausweis B 2
  3 Saisonbewilligung (Ausweis A) 3
  4 Asylbewerber/in 4
  5 Bewilligung des Eidgenössischen Departementes für auswärtige Angelegenheiten 5
  6 Kurzaufenthaltsbewilligung (max. 1½ Jahre mit Ausweis B oder L) oder anderer Status 6
- Weiterer Wohnort — Wird von Ihnen noch eine weitere Wohnung (Unterkunft/Zimmer) in der Schweiz bewohnt? (Ferienwohnungen bitte nicht angeben.) nein 1 ja X
- Wenn ja, geben Sie die Adresse der weiteren Wohnung/Unterkunft an: Strasse/Nummer: TOBELACHER 1; Ortschaft/Gemeinde: FISIBACH
- Für Erwerbstätige, Schüler/innen und Studenten/-innen: Gehen Sie vorwiegend von der genannten Adresse zur Arbeit bzw. zur Schule? ja 1 nein 2 Kanton: AG

A 38950 Bitte leer lassen a b c

Figure 9. Original of a Swiss population census form (reduced size). It requires 27 KB when compressed using conventional techniques.

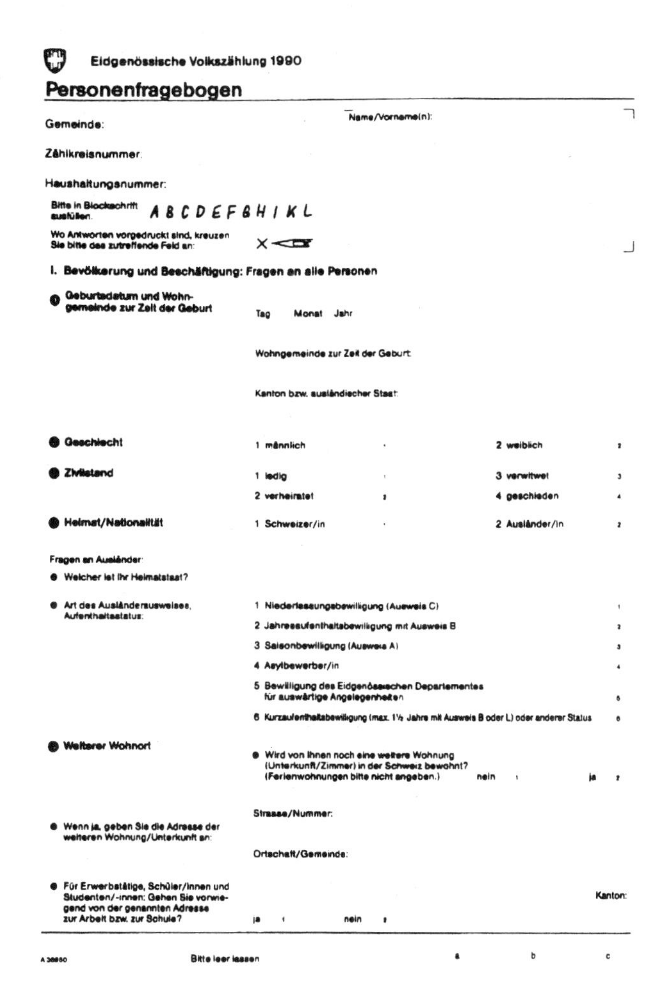

Eidgenössische Volkszählung 1990

Personenfragebogen

Gemeinde:

Name/Vorname(n):

Zählkreisnummer:

Haushaltungsnummer:

Bitte in Blockschrift ausfüllen: ABCDEFGHIKL

Wo Antworten vorgedruckt sind, kreuzen Sie bitte das zutreffende Feld an: X

I. Bevölkerung und Beschäftigung: Fragen an alle Personen

- Geburtsdatum und Wohngemeinde zur Zeit der Geburt — Tag Monat Jahr; Wohngemeinde zur Zeit der Geburt; Kanton bzw. ausländischer Staat:
- Geschlecht — 1 männlich; 2 weiblich 2
- Zivilstand — 1 ledig; 2 verheiratet 2; 3 verwitwet 3; 4 geschieden 4
- Heimat/Nationalität — 1 Schweizer/in; 2 Ausländer/in 2

Fragen an Ausländer:

- Welcher ist Ihr Heimatstaat?
- Art des Ausländerausweises, Aufenthaltsstatus:
  1 Niederlassungsbewilligung (Ausweis C) 1
  2 Jahresaufenthaltsbewilligung mit Ausweis B 2
  3 Saisonbewilligung (Ausweis A) 3
  4 Asylbewerber/in 4
  5 Bewilligung des Eidgenössischen Departementes für auswärtige Angelegenheiten 5
  6 Kurzaufenthaltsbewilligung (max. 1½ Jahre mit Ausweis B oder L) oder anderer Status 6
- Weiterer Wohnort — Wird von Ihnen noch eine weitere Wohnung (Unterkunft/Zimmer) in der Schweiz bewohnt? (Ferienwohnungen bitte nicht angeben.) nein 1 ja 2
- Wenn ja, geben Sie die Adresse der weiteren Wohnung/Unterkunft an: Strasse/Nummer:; Ortschaft/Gemeinde:
- Für Erwerbstätige, Schüler/innen und Studenten/-innen: Gehen Sie vorwiegend von der genannten Adresse zur Arbeit bzw. zur Schule? ja 1 nein 2 Kanton:

A 38950 Bitte leer lassen a b c

Figure 10. Empty template of the census form.

form. For many applications this result when compressed will be enough in terms of compression ratio. However, in the majority of practical applications, further compression may be achieved in order to attain even higher compression ratios. For these applications, it suffices to have legible text which is subjectively indistinguishable from the original without preserving pel by pel fidelity.

A trivial approach to this would be to reduce the resolution of the scanning. Indeed, practical scanner and facsimile devices do allow this option. However, while this solution might be valid for some images, such as scans of large hand-written text, for other images it will cause unacceptable distortion. Moreover, even for medium-size text, simple reduction in resolution will cause image quality deterioration, which in turn might be quite annoying to the user.

Accordingly, we have developed a method for obtaining a high compression ratio by means of a lossy, yet high-quality, technique. The basic idea is to evaluate the nature of the binary image, and limit the lossy compression to large low frequency areas, where decrease in resolution would be unnoticeable. Small high frequency areas will be compressed in a lossless manner. Moreover, even in areas were lossy compression is feasible, it wouldn't be applied indiscriminately. Instead, we construct the compression algorithm in such a manner that continuity of black and white lines and line smoothness are preserved. As a result, excellent overall image quality is maintained. Typically, this approach doubles the compression ratio compared to a scheme without the lossy compression (Walach et al. 1989b).

In Figure 12 we present the reconstruction of the image of Figure 11 after the application of the high-quality binary image compression algorithm. Figure 12 is hardly distinguishable from Figure 11, even though it requires only half the storage (2.2 KB instead of 4.5 KB). The user may view the result superimposed on the image of an empty form template as depicted in Figure 13. There is no noticeable difference between Figure 13 and the original, Figure 9,

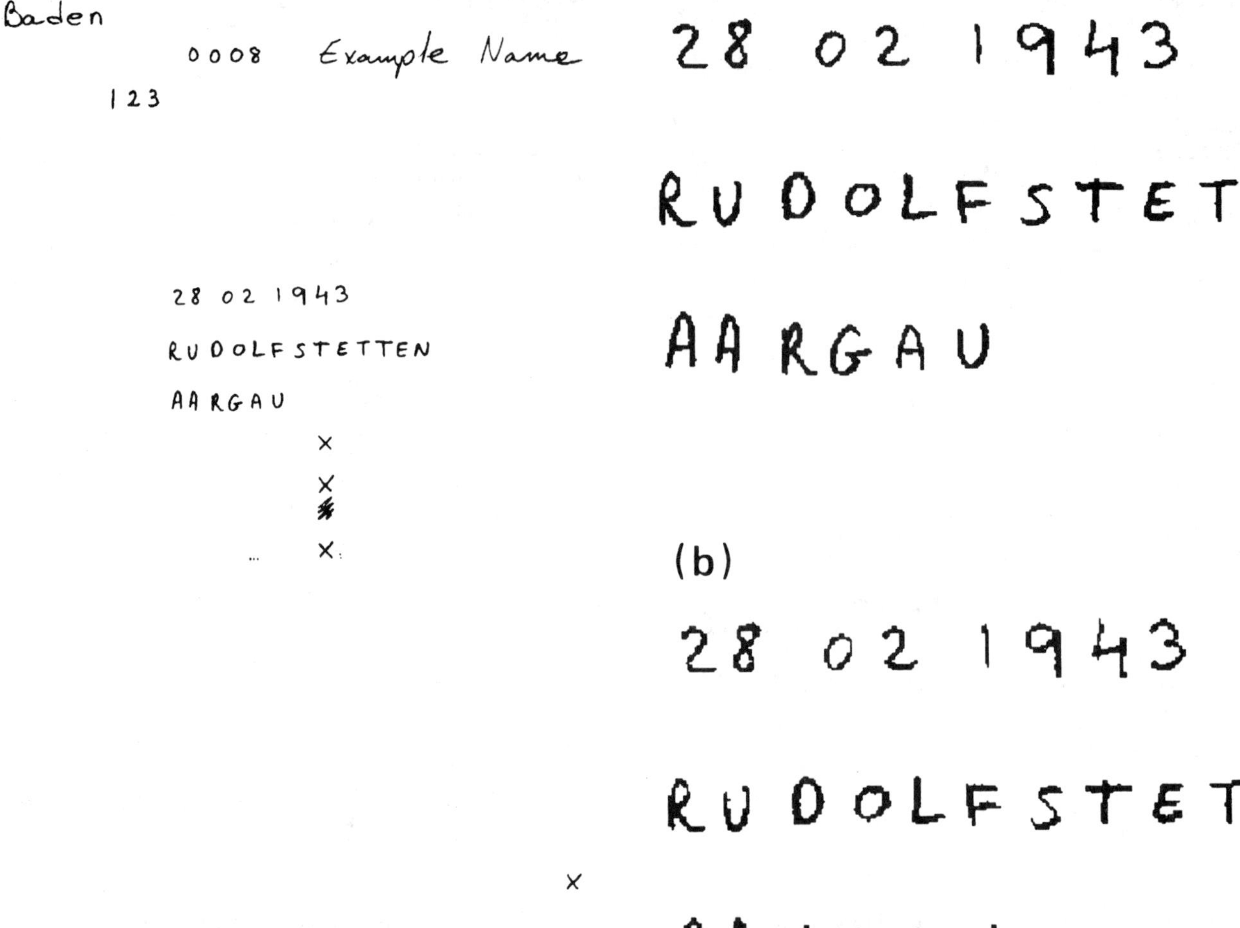

Figure 11. **Filled-in information obtained from the image of census form. It requires 4.5 KB.**

Figure 12. **Comparison of the image of Figure 11a versus a high-quality compression (b) that requires only 2.2 KB storage space.**

although space requirements have been reduced by a factor of 12.

8 A Sample Application

In this section we describe a sample application. Let us suppose that the Intelligent Forms Processing System is applied to tax form processing. During an initial setup phase, all the different tax forms are defined to the system using the form definition facility. At this time, the fields from which information will be extracted are also defined. The blank form images are stored as templates to be used in form drop-out.

As tax forms arrive at the processing center, they are collected at a central staging area. Stacks of forms are placed into a scanner which automatically feeds and scans each form. Once scanned, each form is placed into a temporary holding area until it is confirmed that the form's digitized image is acceptable. In fact, successful recognition of the form type may be used as a criterion of acceptance. Once a form's image is accepted, the form can be disposed of.

After a form is scanned into the system, it is processed by the forms processing tools. If form recognition cannot recognize the form, a warning is issued to an input administrator who manually identifies the form or rescans it. After the form is recognized, data is extracted from predefined fields on the form and recognized using OCR. This information can be used to index the form image for storage as well as

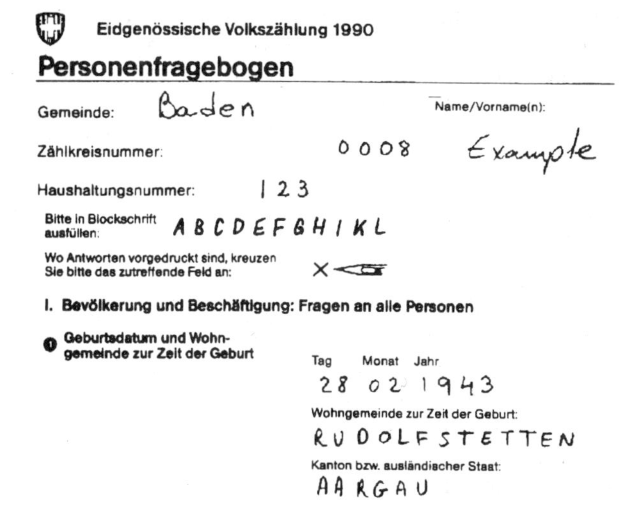

Eidgenössische Volkszählung 1990

Personenfragebogen

Gemeinde: Baden

Name/Vorname(n):

Zählkreisnummer: 0008 Example

Haushaltungsnummer: 123

Bitte in Blockschrift ausfüllen: ABCDEFGHIKL

Wo Antworten vorgedruckt sind, kreuzen Sie bitte das zutreffende Feld an: X

I. Bevölkerung und Beschäftigung: Fragen an alle Personen

1 Geburtsdatum und Wohngemeinde zur Zeit der Geburt

Tag Monat Jahr

28 02 1943

Wohngemeinde zur Zeit der Geburt:

RUDOLFSTETTEN

Kanton bzw. ausländischer Staat:

AARGAU

Figure 13. Reconstruction of the Swiss census form by overlaying the blank template (Figure 10) with the compressed data image (Figure 12).

for entry into a textual database. Typically, on a tax form, the social security number is the field that is most likely to be used for indexing the image. The form image may now be stored in an image library after compression using the form drop-out technique.

The tax processing administrator may now access the forms submitted by a particular social security number by retrieving the images indexed under that number as well as other documents related to the particular tax return. The administrator may then use other system tools such as spreadsheets, etc., to compute the tax owed by this taxpayer. At the same time, other departments such as an auditing department may have access to the forms submitted by the particular taxpayer. Thus, concurrent access to documents are enabled improving productivity and turnaround time.

9 Status

All the components described in this paper were implemented initially as part of a research prototype. These features are now available as part of a document imaging system product, called Image and Records Management, announced by IBM recently (International Business Machines 1991). The forms processing routines run under OS/2. On a 25MHZ 80486-based PS/2, the performance of the components, on the average, are listed below:

- Form Recognition: 1.7 s (including image decompression)
- Data Extract/OCR: 0.5 s per field
- Form Drop Out and Compression: 10 s
- Form Reconstruction and Display: 1.5 s

The actual times will depend on the contents of the form. But on the average, a form containing about 6 fields, can be processed through the intelligent forms processing system in about 15 s. This is significantly faster and more cost-effective than human operators. Although the form drop out and compression phase takes 10 s, the savings in storage make it well worth the additional processing time. On the contrary, form reconstruction and display take significantly less time, on the order of 1.5 s per form. It is this fast display of images that enables users to rapidly browse through documents in a document image system.

10 Future Extensions

The system described here offers a basic facility for extraction, recognition, and compression of information on printed forms. Input documents are currently limited to those which use lines and boxes to define data fields, since these characteristics are detected during processing and serve as a form "signature". However, an extension of these techniques would permit the system to operate with text-only forms, such as occur frequently in contracts and other legal transactions. In this case the system would analyze the baselines of the preprinted text to define a frame for orientation and registration. In addition, it would derive characteristics such as dimension and spacing, and perhaps perform character recognition as well, to assist in distinguishing preprinted text from data added by the user of the form.

Work is also in process to extend the recognition capability of the Intelligent Forms Processing System from typing fonts and handprinted numerics to unlimited reading of machine and handprinted data. In addition, preliminary research has been conducted in the interpretation of handwriting in cursive script. The current state of technology in script recognition is such that the process gives good accuracy only in applications where the data to be read is known to belong to a (preferably short) list of words, e.g., city names or written amounts. Fortunately, this condition is often realized in form processing, and other information read from the form may provide redundancy that can serve to verify the overall recognition process. The ultimate aim is to provide a very flexible document interpretation capability as an adjunct to the fast-growing market for document image libraries.

References

Casey RG, Ferguson DR (1990) Intelligent Forms Processing. IBM Systems Journal 29(3):435–450

Casey RG, Jih CR (1983) A Processor-based OCR System. IBM Journal of Research and Development 27(4): 386–399

Clement TP (1981) The extraction of line-structured data from engineering drawings. Pattern Recognition 14(1–6):43–52

International Business Machines (1991) Image and Records Management (IRM) System—General Information Guide, Version 1 Release 2

Kasturi R, Shih C (1989) Extraction of graphic primitives from images of paper based line drawings. Machine Vision and Applications 2:103–113

Mitchell B, Gillies A (1989) A model-based computer vision system for recognizing handwritten zip codes. Machine Vision and Applications 2:231–243

Nagy G, Viswanathan M (1991) Dual representation of segmented technical documents, First International Conference on Document Analysis and Recognition, St. Malo, France, September, pp 141–151

Nielsen DE, Arps RB, Morrin TH (1973) Evaluation of Scanner Spectral Response for Insurance Industry Documents. 6/A44 NCI Program, Working Paper 2

Walach E, Chevion D, Karnin E (1989a) Efficient Compression and Decompression of Forms by Means of the Very Large Symbol Matching. Docket. No. SZ-9-88-001, filed August 4 1989, Israel Appln. 91220, European Appln. 89810765.1

Walach E, Chevion D, Karnin E (1989b) Method for High Quality Compression of Binary Text Images. Docket SZ-9-88-002, filed August 4 1989, Israel Appln. 91221, European Appln. 89810766.9

Wood FB, Arps RB (1973) Evaluation of Scanner Spectral Response for Insurance Industry Documents. 16/A44 NCI Program, Working Paper 4

A Prototype Document Image Analysis System for Technical Journals

George Nagy, Rensselaer Polytechnic Institute

Sharad Seth, University of Nebraska at Lincoln

Mahesh Viswanathan, IBM

Intelligent document segmentation can bring electronic browsing within the reach of most users. The authors show how this is achieved through document processing, analysis, and parsing the graphic sentence.

Let's quickly calculate the requirements of electronic data storage and access for a standard library of technical journals. A medium-sized research library subscribes to about 2,000 periodicals, each averaging about 500 pages per volume, for a total of one million pages per year. Although this article was output to film at 1,270 dpi (dots per inch) by an imagesetter, reproduction on a 300-dpi laser printer or display would be marginally acceptable to most readers (at least for the text and some of the art). At 300 dpi, each page contains about six million pixels (picture elements). At a conservative compression ratio of 10:1 (using existing facsimile methods), this yields 80 gigabytes per year for the entire collection of periodicals. While this volume is well beyond the storage capabilities of individual workstations, it is acceptable for a library file server. (Of course, unformatted text requires only about 6 kilobytes per page even without compression, but it is not an acceptable vehicle for technical material.)

A 10-page article can be transmitted over a high-speed network, and printed or displayed in *image form* in far less time than it takes to walk to a nearby library. Furthermore, while *Computer* may be available in most research libraries, you may have to wait several days for an interlibrary loan through facsimile or courier. There is, therefore, ample motivation to develop systems for the electronic distribution of digitized technical material.

However, even if the material is available in digital image form, not everyone has convenient access to a high-speed line, a laser printer, or a 2,100 × 3,000-dot display. We show how *intelligent document segmentation* can bring electronic browsing within the reach of readers equipped with only a modem and a personal computer.

Document analysis constitutes a domain of about the right degree of difficulty for current research on knowledge-based image-processing algorithms. This is important because document analysis itself is a transient application: There is no question that eventually information producers and consumers must be digitally linked. But there are also practical advantages to recognizing the location and extent of significant blocks of information on the page. This is also true for segmenting and

Reprinted from *Computer,* Vol. 25, No. 7, July 1992, pp. 10–22.

Glossary

AND-OR graph (or tree). Representation of a solution strategy in which a path from the start node to the solution node requires traversing any branch at an OR node and every branch at an AND node. In a related Min-Max search used in two-person games, a path from the start node to the solution node takes the lowest cost branch at a Min node and the highest cost branch at a Max node.

Bitmap. Digital representation of an image in which points are mapped to an array of binary pixels.

Branch-and-bound. A search technique that avoids paths certain to lead to higher cost solutions than the best solution obtained so far.

CCITT (Comite Consultatif International de Telegraphie et Telephonie). International organization which promulgates standards for facsimile coding and transmission. CCITT Group 3 compression methods are used over relatively high-error-rate channels such as the telephone network. Group 4 standards are designed for low-error-rate channels such as public data networks. The compression ratio of a code is the number of bits in the bitmapped representation of the image divided by the number of bits in the coded representation.

Compiler tools. Programs that generate lexical and syntactic analysis programs for particular applications, including assembly, interpretation, compilation, and translation of computer programs. In the Unix system, Lex is a popular tool for lexical analysis and YACC (yet another compiler compiler) is used for parsing context-free languages.

Document interchange formats. Standards for coding the organization of the layout and contents of a document to facilitate transferring documents from one computer system to another. Popular examples include the Document Content Architecture (DCA), Document Interchange Format (DIF), Document Style Semantics Specification Language (DSSSL), Office Document Architecture (ODA), and Standard Generalized Mark-up Language (SGML).

Drop cap. An oversized uppercase character spanning several printed lines, often used to begin an article or a section.

Hypertext. Data structure for text or other media that includes linkages (pointers) between conceptually related items. These linkages allow considerable latitude in selecting strategies for traversing many levels of information. The linkages are either preset or created by data users.

Kerning. Separate characters placed closer together than usual by removing some of the space between them.

Layout. Physical (for example, geometric and typographic) organization of textual and graphic matter in a document, in contrast to content.

Leading. The separation between consecutive lines of text, originally achieved in printing by inserting strips of lead between rows of type.

Lexical analysis. Each token (word) belongs to a lexical category that determines what tokens may precede or follow it. In English, examples of lexical categories are verb, noun, and pronoun. Lexical analysis determines the category of a token.

Ligature. Two or more symbols, such as æ, printed as a single pattern.

OCR (optical character recognition). The technology of converting printed or written material into computer-readable (ASCII) code. The word *optical* serves to distinguish it from magnetic ink character recognition (MICR). OCR systems usually include an optical scanner, a preprocessor to locate the characters in the appropriate order, a pattern classifier, and a contextual postprocessor.

Page icon. A reduced-scale representation of the subdivision of a printed page into nested rectangles that correspond to significant layout units. It is a graphic display of the X-Y tree of the decomposition.

Point. Printer's unit of measurement, equal to about 1/72 of an inch (US, European, and Belgian points are slightly different). One pica equals 12 points. The notation 10/12 indicates 10-point type with 2-point leading (white space).

PostScript. A programming language introduced by Adobe Systems that allows device-independent specification of typeset text and illustrations for computer printers, displays, and digital phototypesetters. Such languages are generally known as page-description languages.

Syntax. Synonymous with grammar: the formal rules that determine the permissible configuration of lexical tokens such as words. Parsing (a form of syntactic analysis) determines whether a string of words forms a legal sentence. The language accepted by a grammar is the (often infinite) set of all legal sentences. A grammar is usually expressed as a set of rewrite rules, or productions, that allow replacing the string of symbols found on the left-hand side of the production by the string found on the right-hand side to generate legal sentences. Symbols that don't occur by themselves on the left-hand side and therefore cannot be replaced are called terminal symbols; all others are called nonterminal symbols. Formally, a grammar is a quintuple consisting of a starting symbol, a delimiter, a set of terminal symbols, a set of nonterminal symbols, and a set of productions.

TIFF (tagged image-file format). A family of popular formats for color, gray-level, and black-and-white images in either compressed or uncompressed form.

VGA (video graphic adapter). A format convention for 480 × 640-pixel color screen displays introduced by IBM in 1987.

X-Y tree. A spatial data structure in which each node corresponds to a rectangular block. The children of each node represent the locations of the subdivisions of the parent block in a particular (horizontal or vertical) direction. The same organization is often represented in VLSI design by means of a polar graph. Other popular hierarchical data structures for isothetic (Manhattan) spatial subdivisions include the K-D tree and the Quad tree.

Advantages of intelligent page segmentation

Smaller blocks can be reproduced on PC displays without horizontal scrolling or loss of legibility.

Selected blocks can be transmitted more quickly than an entire page can for browsing over networks.

Layout analysis helps OCR preserve the reading order.

Key-entry of selected blocks that are intractable by OCR is less costly than rekeying entire pages.

Differentiating text from graphics and halftones leads to more efficient image compression.

Straight borders between high-contrast and halftone regions are preserved for digital reprographics.

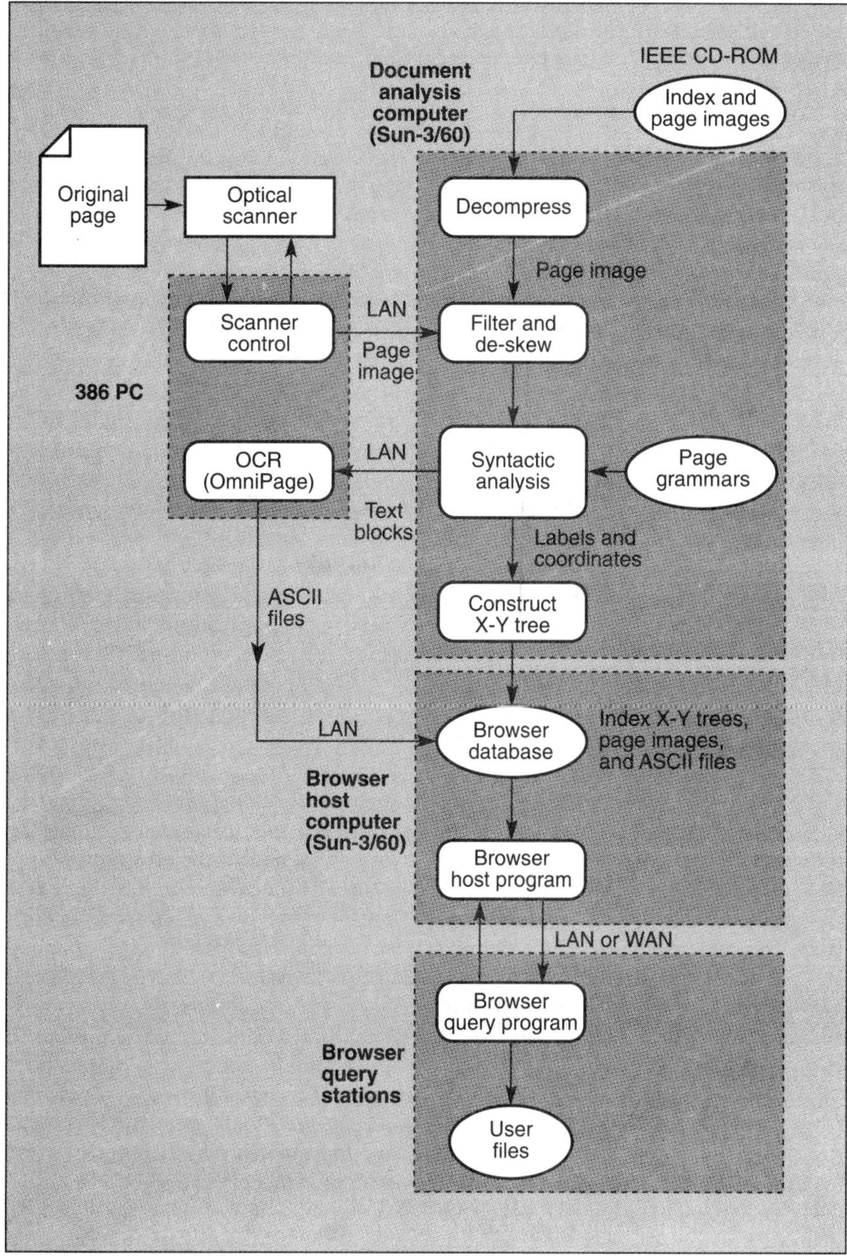

Figure 1. Gobbledoc system overview.

labeling the page image according to its logical structure, without resorting to optical character recognition (see the "Advantages of intelligent page segmentation" sidebar).

Research on OCR began at the turn of the century. Inventions for replacing telegraph operators and assisting blind readers were first demonstrated in 1914, and high-speed OCR systems have been commercially available since 1955 for specially printed forms and fixed-pitch typescript. The field of digitized page analysis started in the sixties, but only recently have the relevant technologies matured sufficiently to allow the conversion of complex typeset documents such as technical articles. Major catalysts include accurate, fast, inexpensive page scanners; high-speed computer networks; sufficient storage and processing power for digitized page images; and compression standards for digital facsimile.

Document image analysis requires the assembly of a number of software tools.[1] Gobbledoc, the system that we have developed at Rensselaer Polytechnic Institute over the last five years, integrates several existing components with a new method of image decomposition. Figure 1 shows the flow of information from the original document to the user's computer screen. A page is either scanned locally or obtained from a CD-ROM. It is segmented and labeled by using a syntactic approach, and stored in an X-Y tree. Images of text blocks are transferred to a personal computer and recognized by using OCR. The resulting ASCII output (which includes some special characters marked with escape quotes) is linked to the corresponding node of the X-Y tree. All data is then stored on the browser host computer. The browser is activated from a remote client workstation, and the user-requested information is displayed on the client terminal.

In Gobbledoc, image processing, document analysis, and OCR operations take place in batch mode when the documents are acquired. After explaining the document image acquisition process, we describe the knowledge base that must be entered into the system to process a family of page images. We show how our X-Y tree data structure converts the two-dimensional page-segmentation problem into a series of one-dimensional string-parsing problems that can be tackled by using conventional compiler tools. Next, syntactic analysis is used to divide each page into labeled rectangular blocks. Blocks labeled *text* are converted by OCR to obtain a secondary (ASCII) document representation. But such symbolic files are better suited for computerized search than for human access to the document content. Because too many visual layout clues are lost in the OCR process (including some special characters), our system preserves the original block images for human browsing. Finally, we consider storage, networking, and display issues specific to document images, and describe our prototype browser.

Partial layout specifications for *Computer* in 1991

The title lines are set in Melior 36/38 point boldface, centered.

There are one to four lines in the title: the second part of a long title is sometimes set in 24-point type.

The byline is set in Melior 12/14-point boldface, centered; affiliations may either follow the authors' names set in the same typeface or are set on the same line.

The title line precedes the byline, and the two are separated by at least 38-point leading.

Every paragraph is indented except the first and the beginning of the conclusion section, which begin with a 40/40-point drop cap.

The body type (main text) is TimesTen Roman 9/11.

The page numbers and month-year (in body type) are set flush with the margin and alternate between left and right.

Footnotes (rare) are set TimesTen Roman 7/8, separated by a hairline rule with a blank line before and after the rule.

Preprocessing

We convert each page to digital form by scanning it at 300 dpi. This rate is sufficient to preserve all significant white spaces. Since the entire X-Y tree approach is extremely sensitive to skew, we carefully align each page on the scanner bed. We recognize that this could be impossible in a production environment, where the systems could incorporate one of the excellent skew-correction methods (performed by software or hardware) already demonstrated by others.[2]

Sample pages. We also take sample pages from the CD-ROM database prepared for the IEE-IEEE by University Microfilms. The IEEE annually distributes about 130,000 pages in 300-dpi image form on CD. Other professional societies are not far behind. The IEE-IEEE database is stored in CCITT Group 4 compressed form. This is the two-dimensional Modified Modified Read (MMR) code developed as a standard for facsimile terminals operating over very-low-error-rate public data networks.[3] Currently, we decompress the page using software, but standard chips for this purpose are available. Since we have no control over the degree of skew of the CD pages, we de-skew them.

Noise. Our documents contain specks of noise caused by flawed fibers in the stock, imperfect printing, photocopying, and digitization. (High-quality journals are relatively noise-free, but we used photocopies on uncoated paper.) Such noise does not bother human readers, but it complicates automated analysis. To cope with the noise, one may construct page grammars sufficiently robust to ignore speckle. This is feasible but tedious. Instead, we filter out all specks smaller than a given size in a preliminary pass. After the analysis is completed, all specks are restored. The size threshold therefore can be quite generous. Losing a few periods or dots on the *i*'s and *j*'s does not affect the layout analysis, and they are restituted before any document component is submitted to OCR or displayed for human inspection.

Layout encoding

The first step in analyzing a specific family of documents is encoding the information that distinguishes this family from others. The required knowledge base is quite specific. It may be stated explicitly in the form of if-then rules, a form-definition language, a geometry tree of box locations, an expert system, or a formal grammar — or implicitly embedded in the processing programs. Some examples of constraints for the title page of feature articles in *Computer* as of July 1991 are shown in the accompanying sidebar. The rules for the pages from *IEEE Transactions on Pattern Analysis and Machine Intelligence (IEEE-PAMI)*, used in our experiments and shown later in Figure 6, are similar.

Publication-specific systems have been successfully demonstrated on newspaper pages, business letters, articles, and tables of contents in technical journals, patent applications, resumes, typed forms with a prespecified layout, sheet music, maps, engineering drawings, circuit diagrams, and even the periodical *Chess Informant*. A complex knowledge-based system that demonstrates the usefulness of full interaction between different components has been developed for postal address location and interpretation. Recent achievements in these areas are described in the proceedings of conferences on document analysis.[4,5]

Syntactic analysis. Just as every programming language has its fixed set of rules, each publication has predetermined layout conventions. These conventions dictate the size, position, spacing, and ordering of the blocks that correspond to logical entities on a page, as shown in the sidebar for *Computer*. Accordingly, documents that satisfy the postulated layout conventions can always be parsed into different components without OCR in the same way that syntactically correct program code can be parsed into meaningful constructs without semantic understanding.

Syntactic analysis accomplishes both segmentation and component identification simultaneously. A publication-specific *document grammar* is a formal description of all legal page formats that articles in a given publication can assume. In principle, a parse of the page will reveal whether the page is acceptable under the rules of the grammar and, if so, what labels must be assigned

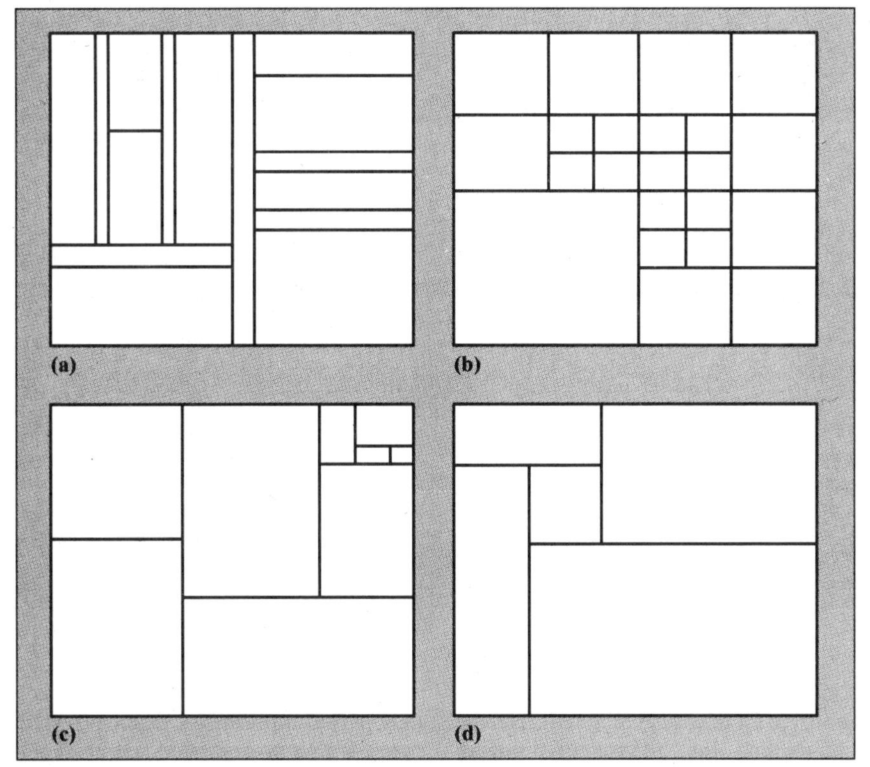

Figure 2. Hierarchical subdivision of rectangles into rectangles: X-Y tree for page segmentation (a); Quad tree for comparing or combining several images (b); K-D tree for fast search (c); example of tiling with rectangular blocks that cannot be obtained by successive horizontal and vertical subdivisions only (d).

to the various components to construe a valid graphic "sentence."

Although we use compiler tools developed primarily for formal languages, the syntactic analysis of document images exhibits many of the difficulties of parsing natural language. Layout conventions may be insufficient to identify *every* document component. For instance, text lines with equations buried in them may radically alter the expected line spacing. We must therefore ensure that minor deviations have only local effects. Furthermore, the grammar for a modern programming language is established from the start, while document grammars must be inferred indirectly, as later discussed. (A never-ending task: Journals frequently put on new faces.)

Block grammars. The document grammar for a specific journal consists of a set of block grammars. Each block grammar subdivides a block horizontally or vertically into a set of subblocks. The net result of applying the entire document grammar is therefore a subdivision of the page into *nested rectangular blocks*. Such a subdivision can be represented efficiently in a data structure called the X-Y tree[6] (Figure 2). The block grammars themselves are also organized in the form of a tree: The block grammar to be used to subdivide each block is determined recursively by the results of the parse at the level above.

Syntactic attributes

A (horizontal/vertical) *block profile* is a binary string that contains a zero for each horizontal or vertical scanline that contains only white pixels; otherwise it is a one.

A *black atom* is a maximal all-one substring. It is the smallest indivisible partition of the current block profile. A *white atom* is an all-zero substring.

A *black molecule* is a sequence of black and white atoms followed by a black atom. A *white molecule* is a white atom that separates two black molecules.

An *entity* is a molecule that has been assigned a *class label* (title, authors, figure caption). It may depend on an ordering relationship.

This approach effectively transforms the difficult two-dimensional segmentation into a set of manageable one-dimensional segmentation problems.

The syntactic formalism is theoretically well understood, and sophisticated software is available for lexical analysis and parsing of strings of symbols. Each block grammar is therefore implemented as a conventional string grammar that operates on a binary string called a *block profile*. The block profile is the thresholded vertical or horizontal projection of the black areas within the block. Zeros in the block profile correspond to white spaces that extend all the way across the block and are therefore good candidates for the locations of subdivisions.

Representing the structure of an entire page in terms of block grammars simplifies matters considerably. But each block grammar itself is a complex structure. It must accommodate many alternative configurations. For instance, to divide the title block from the byline block, the block grammar must provide for a varying number of title lines and bylines, and for changes in spacing caused by the ascenders and descenders of the letters. To simplify the design process, each block grammar is constructed in several stages, in terms of syntactic attributes extracted from profile features.[7]

Syntactic attributes. The first stage of a block grammar operates on the ones and zeros of the block profile. Strings of ones or zeros are called *atoms*. Atoms are divided into classes according to their length. A string of alternating black and white atoms is a *molecule*. The class of a molecule depends on the number and kind of atoms it contains. Finally, molecules are transformed into *entities* depending on the order of their appearance. The words *atom*, *molecule*, and *entity* were chosen because they are not specific to a particular publication or subdivision. (See the sidebar on syntactic attributes.)

The syntactic attributes that determine the parse are the size and number of atoms within an entity, and the number and order of permissible occurrences of entities on a page. Table 1 shows the expected variation in the horizontal profile of a page fragment that includes the title and byline. The assignment of symbols into larger units is accomplished by rewriting rules or *productions*. These

Table 1. Examples of syntactic profile features for *Computer* digitized at 300 dpi.

	Title	Title-byline-space	Byline
Black atom length	100-160	—	35-52
White atom length	4-20	10-40	2-14
No. atoms	1-4	—	1-6
No. entities	1	1	1
Precedence		Title before byline	

rules take into account the expected variability. The entries in the first row correspond to the maximum and minimum height (in pixels) of the title and byline lines in a feature article scanned at 300 dpi. The second row shows the expected profile height of the leading. There must be one to four title lines and one to six author lines, with the byline below the title. Each page has only a single title block and a single byline block.

In the greatly simplified example of Figure 3, a typical horizontal block profile is shown at the top. The atom lengths have been shortened to show the entire horizontal block profile.

In stage 1, runs of ones or zeros are condensed into atoms according to their length. For example, black runs of 2 to 3 are called a. In stage 2, atoms are grouped into molecules. (A run of zero to three black-white sequences of two atoms of types b,v or b,w, followed by a black atom of type b, is labelled A.) In stage 3, molecules are interpreted as *document entities.*

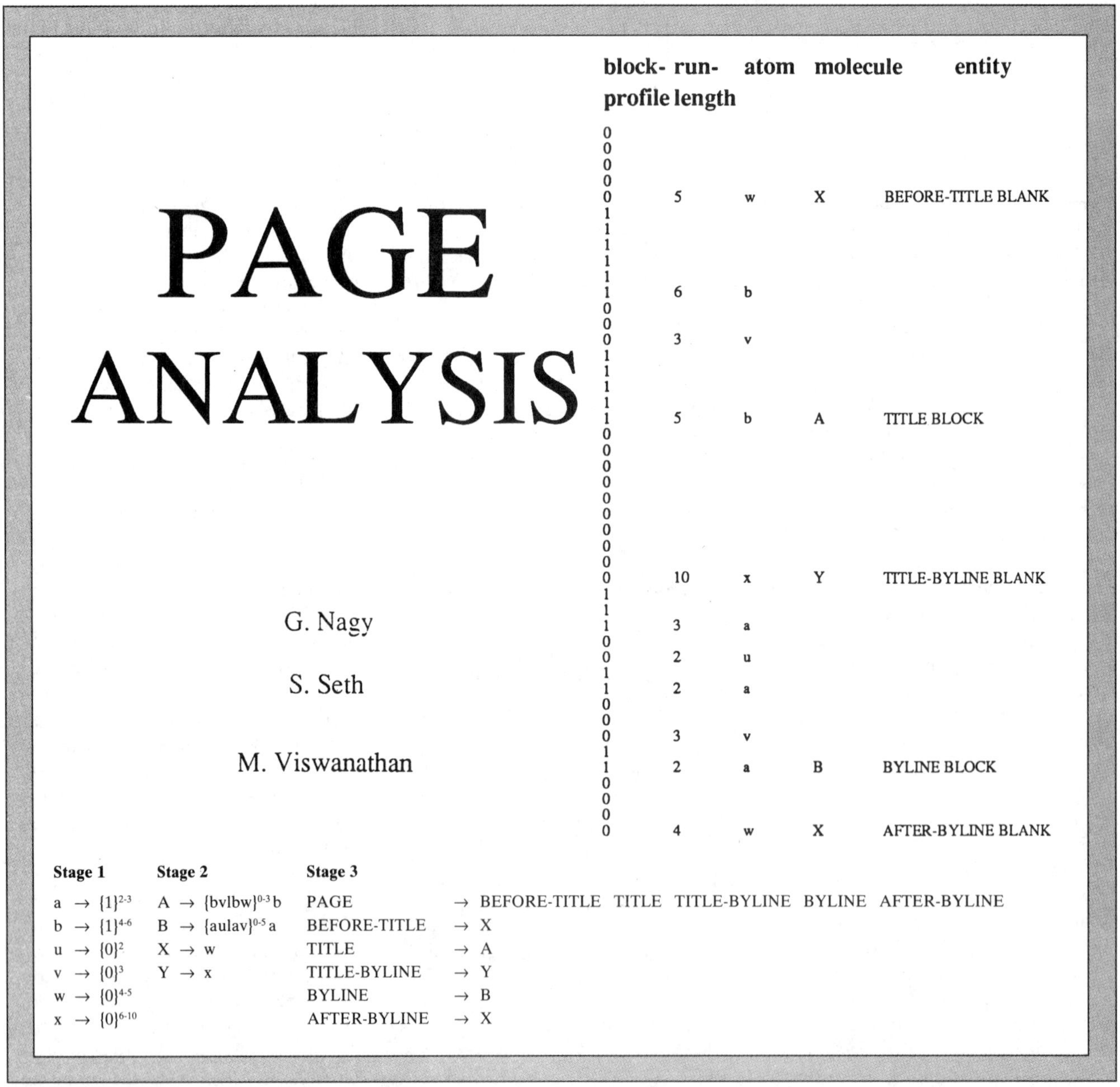

PAGE ANALYSIS

G. Nagy

S. Seth

M. Viswanathan

block-profile	run-length	atom	molecule	entity
0 0 0 0 0	5	w	X	BEFORE-TITLE BLANK
1 1 1 1 1 1	6	b		
0 0 0	3	v		
1 1 1 1 1	5	b	A	TITLE BLOCK
0 0 0 0 0 0 0 0 0 0	10	x	Y	TITLE-BYLINE BLANK
1 1 1	3	a		
0 0	2	u		
1 1	2	a		
0 0 0	3	v		
1 1	2	a	B	BYLINE BLOCK
0 0 0 0	4	w	X	AFTER-BYLINE BLANK

Stage 1	Stage 2	Stage 3	
$a \rightarrow \{1\}^{2\text{-}3}$	$A \rightarrow \{bv|bw\}^{0\text{-}3} b$	PAGE	→ BEFORE-TITLE TITLE TITLE-BYLINE BYLINE AFTER-BYLINE
$b \rightarrow \{1\}^{4\text{-}6}$	$B \rightarrow \{au|av\}^{0\text{-}5} a$	BEFORE-TITLE	→ X
$u \rightarrow \{0\}^{2}$	$X \rightarrow w$	TITLE	→ A
$v \rightarrow \{0\}^{3}$	$Y \rightarrow x$	TITLE-BYLINE	→ Y
$w \rightarrow \{0\}^{4\text{-}5}$		BYLINE	→ B
$x \rightarrow \{0\}^{6\text{-}10}$		AFTER-BYLINE	→ X

Figure 3. Illustration of a simplified block grammar for the first vertical subdivision of a title page similar to that of *Computer*.

The block grammar at the bottom specifies that the title block (molecule A) may have one to four lines of print, separated by blank lines. Each line of print (atom b) may be four to six pixels high, and each blank line may be three (v), four, or five (w) pixels high. (The height of the blank lines is divided into two ranges because of the overlap in the range of the heights of the blank lines in the title and author blocks.) Stage 3 of this block grammar ensures that the title block precedes the byline block.

Note that molecule X is sometimes interpreted as a before-title blank and sometimes as an after-byline blank, depending on its location with respect to other molecules. Stage 4 (not illustrated) merges consecutive entities with the same label separated by a wide blank, such as text paragraphs.

Block grammars based on such attributes typically have hundreds of productions for a single publication such as *IEEE-PAMI*. Defining each block grammar directly on the binary profile strings, while theoretically equivalent, would be too cumbersome in practice.

Grammar specification. Constructing the block grammars by hand is still very time-consuming and is the greatest weakness of our current approach. Available sources of information are computer-aided measurements on samples of scanned copy, publishers' style manuals, and macro definitions customized for a given publication in page formatters. We have also developed a tabular representation of layout parameters that is automatically translated into block grammars.

Appealing alternatives to our tabular representation of document layout include the Office Document Architecture (ODA), Standard Generalized Mark-up Language (SGML), and Document Style Semantics Specification Language (DSSSL) document interchange standards.[8] However, it is not easy to extract from these specifications the criteria for alternating horizontal and vertical subdivisions required by our decomposition.

Low-level block grammars tend to be much more generic (exhibit less publication-dependent variability) than high-level grammars. One can therefore often reuse existing block grammars for a new family of publications. There are, for instance, only a few paragraph formats in common use. Ideally, the initial parameters would be adjusted (learned)

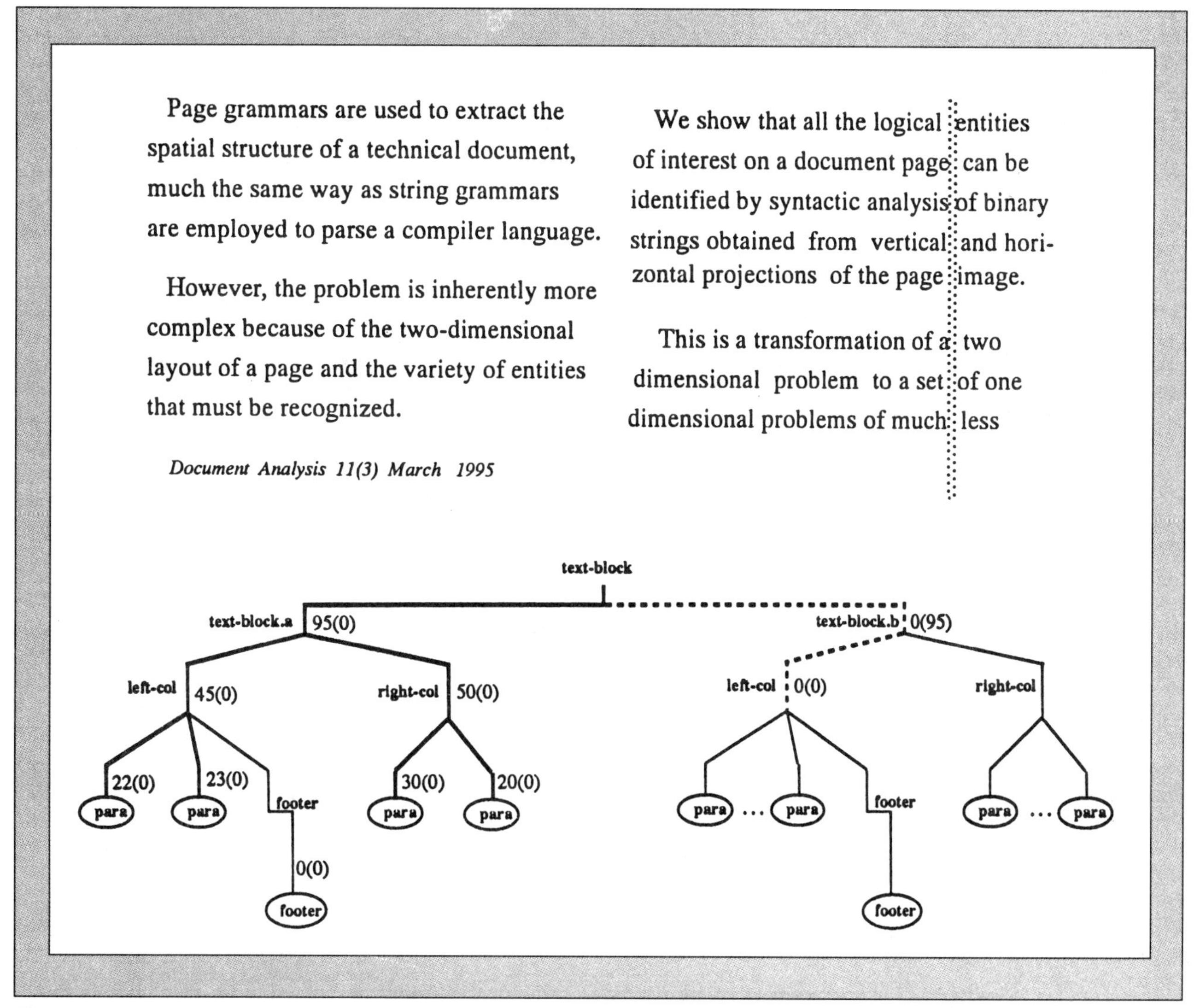

Figure 4. Simplified example of branch-and-bound analysis: a two-column text block with a *river* (accidental vertical alignment of blank spaces) (a), and a block grammar *text-block.a* dividing the block at the central gutter (b).

during actual operations; this is a subject of continuing research.

Page decomposition. We now show how the block-grammar concept can be extended to segment and label a whole page. A block grammar is intended to interpret each subblock of a given block as a particular (labeled) entity. At the subblock level, block grammars assign labels to abstract, title-block, byline-block, reference-entry, and figure-caption entities. Each of these lower level entities may have alternative grammars corresponding to different formats. During the parsing of a block, a unique label is associated with each subblock. Only blocks that are assigned labels for which no block grammars are provided are assumed to be correctly segmented and labeled. Every other *nonterminal* symbol assigned by a block grammar is equated with the start symbol of a block grammar at the level below. Unlike the four stages of each block grammar, the set of block grammars cannot be combined into a single master grammar, even in principle, because at each level a new set of block profiles must be extracted, and the location of each block boundary depends on the parse at the level above.

Several dozen different classes of logical components are isolated and identified by parsing the recursively generated block profiles. The recursive algorithm that subdivides the page verifies the correctness of label assignments in a depth-first traversal. It returns with success only if parsing is successful for the given block and all of its nested subblocks. The algorithm may try each of several alternative block grammars in turn. For example, if it is not known what part of an article a given page represents, it may be parsed with two block grammars: one designed for a title page and one for a nontitle page. The parse may therefore be considered an AND-OR tree: At each level, *every* subblock must be identified by means of one of a set of alternative grammars.

Another reason for alternative grammars is that a block profile may have two legal parses. This happens when the label of a block cannot be determined from its own profile, but only at a level below. In that case, the block must be parsed under several hypotheses, using alternative grammars.

Error control. Some portion of the page may be in an unexpected format for which no alternative grammar has been provided. Since this will yield an unrecognizable subblock at some point, the entire parse will fail. The AND-OR approach is therefore modified to avoid catastrophic failure and determine efficiently the *best possible labeling*. "Best" is defined as the *maximum cumulative area* of the labeled blocks. A *branch-and-bound* strategy avoids parsing a subblock if the maximum labeled area of the page cannot be increased over the current lower bound even with complete segmentation and labeling of the subblock (Figure 4). Subblock grammar left-col correctly divides the left column into two paragraphs and a footer, but the footer grammar fails because it expects even smaller type. Grammar right-col correctly processes the right column, so 95 percent of the area (all but the footer block) is now labeled.

In an attempt to improve on this, the second alternative top-level grammar, text-block.b (designed for narrower gutters), is applied to the entire block and divides it at the river. Now, however, left-col fails, because the lines of text do not line up across the central gutter, so the block cannot be segmented into horizontal strips at the prescribed line-spacing. The search is abandoned without attempting the rightmost branches of the tree, because even if the region to the right of the river were correctly labeled, the overall result would be inferior to the earlier parse. The numbers on the branches of the search tree show the percentage of area identified and, in parentheses, the applicable lower bound.

Typographic conventions. Our method uses a publication-specific knowledge base in the form of a document grammar. We have also studied segmentation techniques based on typesetting conventions[9] rather than on publication-specific knowledge. Previous methods generally imbed the typographic constraints in the processing routines. It is therefore difficult to take an existing program and form a clear conception of its capabilities without extensive experimentation. Further progress may depend on the development of more consistent and comprehensive knowledge bases through explicit codification of such information. Some examples of generic typesetting knowledge for technical journals set in derivatives of the Latin alphabet are shown in the sidebar called "Generic typesetting rules." The commercial software that we use for OCR incorporates such typesetting rules to locate and isolate characters in a block of text.

Generic typesetting rules

Printed lines are roughly horizontal.

The baselines of characters are aligned.

Each line of text is set in a single point size.

Ascenders, descenders, and capitals have consistent heights. In roman fonts, serifs are aligned.

Typefaces (including variants such as italic or bold) do not change within a word.

Within a line of text, word and character spaces are uniform, and word spaces are larger than character spaces.

Lines of text in a paragraph are spaced uniformly.

Each paragraph is left-justified or right-justified (or both), with special provisions for the first and last line of a paragraph.

Paragraphs are separated either by wider spaces than lines within a paragraph or by indentation.

Illustrations are confined to rectangular frames.

In multicolumn formats, the columns are spaced uniformly.

OCR

Rather than develop our own OCR, we considered commercially available products.[10-12] High-end systems consist of a combination of hardware and software. Some require only an additional processor board. Low-end systems are software only and typically run on PCs with extended internal storage. Some OCR systems can be trained for specific typefaces.

We selected the OmniPage system from Caere Corp. because of its superior performance (without user training) on typefaces and sizes used in technical journals. Nevertheless, the ASCII version produced by the OCR system is not as accurate as a keyed-in version. Unusual typefaces may baffle the system, but most of the classification errors are due to missegmented characters (particularly in small point sizes, and kerned, bold, underlined, and italicized typefaces). Formulas and equations are sometimes mangled. Lines with drop caps (such as the first letter of this article), large mathematical symbols, or superscripts may be missed altogether. Some mistakes seem irrational and may be repeated many times.

In other applications, where accuracy is essential, the text would be corrected using a spelling checker and human postediting. However, even without expensive postprocessing, the ASCII version obtained by OCR is useful not only for automated search but also because it can be converted into a standard document-interchange format[8] or copied directly into a user's word processor or desktop publishing system. (We leave aside the difficult questions of pricing, copyright law, and plagiarism.)

Text-image compression. We are also attempting to use the output of the OCR program to provide a highly compressed version of the text image. Only the first graphic instance of each character pattern is stored. When a subsequent pattern is identified by the OCR with the same alphabetic label, the bitmaps of the original pattern and the new pattern are compared, and if they match, only a pointer to the original bitmap is stored. The degree of match required governs the fidelity of reproduction. Earlier symbol-matching schemes carried out bitmap comparisons with all previous prototype patterns. We use, instead, an efficient OCR system for pattern identification. On dense all-text printed pages scanned at 300 dpi, preliminary results indicate compression ratios twice as high as CCITT Group 4.

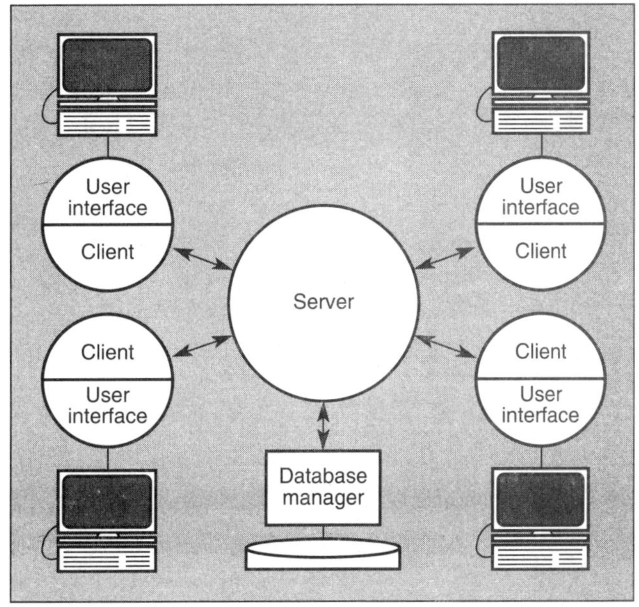

Figure 5. The remote browsing system consists of a central server connected to several query stations by either a local or a wide area network.

Document browser

Remote image browsing systems often use some form of progressive encoding and transmission. Initially, only a low-resolution image is displayed. Greater detail then emerges as additional data is transmitted. Such a scheme is almost useless for printed documents, since a low-resolution version is nearly illegible. Instead, we display parts of the page on demand, using the previously obtained subdivision into meaningful blocks.

The browser system consists of two programs on separate computers that communicate. One is a host program that handles requests from remote users and manages the database of digitized documents, labeled X-Y trees, and ASCII renditions of the textual leaf nodes of the X-Y trees. The other is a (query-station) client program that generates the user requests. The portions of the labeled image requested by the user display at the client station. For portability, we have also implemented an X-Windows version of the browser (see Figure 5).

The host program consists of the server and a database manager. The server handles the communication between the host and the query station. The query station consists of a client and a user interface. The interface transfers a user's request to the client. The client establishes a connection to the host and submits the request. The database manager parses the request received by the server at a specified port and sends the necessary data to the query station. The data received from the server is then transmitted to the user.

The user selects the desired page from a pop-up menu. A geometric representation of the page (a scaled-down labeled X-Y tree with only leaf nodes) displays. The desired block is then retrieved by "mousing" the block in the geometric representation. Using the ASCII button, the user can display either an ASCII version of the text block or a 300-dpi image.

Line wrapping. Most technical printed material is laid out in columns narrow enough for display at 300 dpi on a 480 × 640-pixel VGA screen. However, some type — often the title and abstract — spreads across the full page and cannot be displayed in image form without unwieldy horizontal scrolling. When a text block is wider than a preset width, we continue the segmentation process to the line level. We then locate interword blanks which, in each line of text, are wider than the widest intercharacter space. Each text line is divided at one or more word boundaries and consecutive segments are displayed under one another. This is a form of line-wrapping for text in *image*, rather than symbolic, form.

Linkages. The dual text-image representation provides opportunities to enhance document access. For instance, the OCR system can identify all instances of the word "Figure" in the text and in figure captions. Since we know the location of each figure and figure caption, we can derive the figure number of each illustration. Then, when the user encounters a mention of a figure in the text, depressing the mouse button brings the figure to the screen in a separate window. Alternatively, all figures mentioned in any text paragraph on the screen can be simultaneously displayed with reduced resolution.

Similar concepts apply to cited references. Again, the ASCII version is necessary to set the linkages automatically, but the references can be invoked from either image or ASCII text display. While we have experimented only with linkages within the same document, the concepts can be extended to multiple documents. Image processing plays a role here only in the identification of the coordinates of a recognized keyword in the image and in segmenting the page into blocks. Further enhancement of the information falls in the realm of hypertext.

Implementation. The decompression, de-skewing, noise removal, and syntactic analysis programs run on a Sun-3/60. The uncompressed bitmap of the entire page is stored only once. The profile extraction routines access the required blocks through the *x* and *y* coordinates stored after each parse in the X-Y tree. The parsers at each stage are C-language programs generated by the Unix compiler utilities Lex and YACC. A Unix shell program controls the recursion between levels. Generation and storage of the ASCII files is an off-line process. The MicroTek scanning program EyeStar, the OmniPage OCR system, and the CD-ROM access programs run on an Intel 80386 PC with 4 megabytes of memory.

The host and client programs of the remote browser also run on Sun workstations. All browser interactions are on-line processes. The prototype system has been tested with the host located at the Rensselaer Polytechnic Institute and the query station at the University of Nebraska, and vice versa. A single host can serve several query stations simultaneously. However, our browser does not have the elaborate information retrieval and bibliographic navigation facilities necessary for an operational system. At present, our browser contains only a few dozen pages, which can be selected from a simple menu. Furthermore, its response time is far too slow for anyone browsing in earnest. It does, however, demonstrate several functions related to document image analysis that would be desirable for full access to a technical library through a workstation.

916 IEEE TRANSACTIONS ON PATTERN ANALYSIS AND MACHINE INTELLIGENCE, VOL. 11, NO. 9, SEPTEMBER 1989

Calibrating a Cartesian Robot with Eye-on-Hand Configuration Independent of Eye-to-Hand Relationship

REIMAR K. LENZ AND ROGER Y. TSAI, MEMBER, IEEE

***Abstract*—This paper describes a new approach for geometric calibration of Cartesian robots. This is part of a trio for real-time 3-D robotics eye, eye-to-hand, and hand calibration, which use a common setup and calibration object, common coordinate systems, matrices, vectors, symbols, and operations throughout the trio, and is especially suited to machine vision community. It is easier and faster than any of the existing techniques, and is ten times more accurate in rotation than any existing technique using standard resolution cameras, and equal to the state-of-the-art vision based technique in terms of linear accuracy. The robot makes a series of automatically planned movement with a camera rigidly mounted at the gripper. At the end of each move, it takes a total of 90 ms to grab an image, extract image feature coordinates, and perform camera extrinsic calibration. After the robot finishes all the movements, it takes only a few milliseconds to do the calibration. The key of this technique is that only one single rotary joint is moving for each movement while the robot motion can still be planned such that the calibration object remains within the field of view. This allows the calibration parameters to be fully decoupled, and converts a multidimensional problem into a series of one-dimensional problem. Another key is that eye-to-hand transformation is not needed at all during the computation (a rough estimation is needed however to keep the calibration object within the field of view). Results of real experiments are reported. The approach used in viewpoint planning can be of use to the general problem of automatic determination of camera placement.**

***Index Terms*—Camera calibration, geometric vision, robot calibration, 3-D location determination, 3-D machine vision, 3-D robotics vision.**

I. INTRODUCTION

IN order for a robot to use a video camera to estimate the 3-D position and orientation of a part or object relative to its own base within the work volume, it is necessary to know the relative position and orientation between the hand and the robot base, between the camera and the hand, and between the object and the camera. These three tasks require the calibration of robot, robot eye-to-hand, and camera (see Fig. 1). These three tasks normally require large scale nonlinear optimization, special setup, and expert skills. We have developed a trio for dealing with these three tasks. The trio is:

Manuscript received December 1, 1987; revised August 15, 1988. Recommended for acceptance by O. D. Faugeras.
R. K. Lenz is with the Lehrstuhl für Nachrichtentechnik, Technische Universität München, D-8000 München 2, West Germany.
R. Y. Tsai is with the IBM Thomas J. Watson Research Center, Yorktown Heights, NY 10598.
IEEE Log Number 8928490.

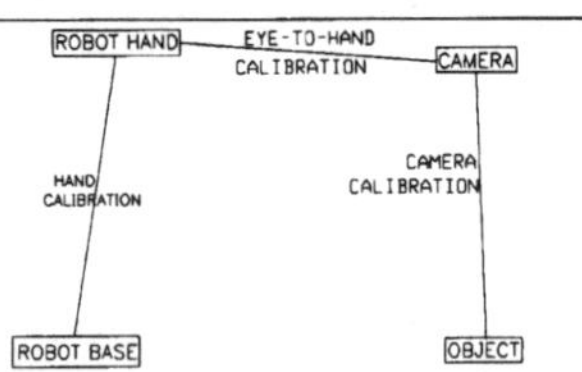

Fig. 1. For obtaining the 3-D position and orientation of the object relative to the robot world base, it is necessary to do three calibrations, namely, robot hand, eye-to-hand and eye (camera) calibration.

1) camera calibration (see [18], [19], [13]).
2) Robot eye-to-hand calibration (see [20], [21]).
3) Cartesian robot hand calibration (this paper).

Note that only the last one is restricted to Cartesian robot, although further research could be done to relieve this latter restriction (some discussion will be given in the Conclusion section). Since the camera calibration is used throughout the trio, it is summarized in Appendix A. It is a two-stage approach, and each stage requires only linear equation solving with a maximum of five unknowns. Radial lens distortion can be corrected. The whole calibration takes about 90 ms, including 25 ms for calibration and 65 ms for image grabbing and feature extraction. The robot eye-to-hand calibration method undergoes the same motion as the robot hand calibration described in this paper, and the same image feature extraction and camera extrinsic calibration are done. The actual computation time besides robot movement and camera calibration is a few milliseconds using a typical minicomputer.

This paper describes the third of the trio. It presents a new approach for robot kinematic calibration, which is easier, faster, and more accurate than comparable current techniques. The purpose of robot kinematic calibration is to determine the elements governing the transformation from the robot joint coordinates to the manipulator end effector Cartesian position and orientation, or the homogeneous transformation from a fixed robot base coordinate frame to its current end effector coordinate frame. This work actually came out as a by-product of an earlier at-

0162-8828/89/0900-0916$01.00 © 1989 IEEE

Figure 6. *IEEE-PAMI* title page extracted from the IEEE CD-ROM database. A de-skewed version is used for OCR, and the image is also filtered for syntactic analysis.

Example. A page from the *IEEE Transactions on Pattern Analysis and Machine Intelligence*, extracted from the CD-ROM and printed on an AppleWriter from its PostScript representation, is reproduced in Figure 6. The image is de-skewed and filtered before generating a labeled X-Y tree for the page. The X-Y tree contains the coordinates of all segments in the page along with the assigned labels. The 12 subimages in the example that correspond to text are converted into a TIFF (tagged image-file format) and transferred to the PC using a network file-transfer protocol. In the example, these include page number, header, title, byline, abstract, keywords, section titles, two paragraphs of text, figure caption, footnote, and footer. OmniPage is run separately on each subimage and the corresponding ASCII output files are returned to the Sun system.

When the browser is activated, it gives the user successive choices of publication name, volume, issue, and page number. The browser control buttons are shown at the top left corner of Figure 7 on the next page. The final selection prompts the display of a labeled geometric representation of the X-Y tree of the selected page, as shown in the bottom left corner of the screen. Now, the user can choose between ASCII and image renditions of the processed blocks. Multiple requests are entertained, as is evident from Figure 7, which shows the

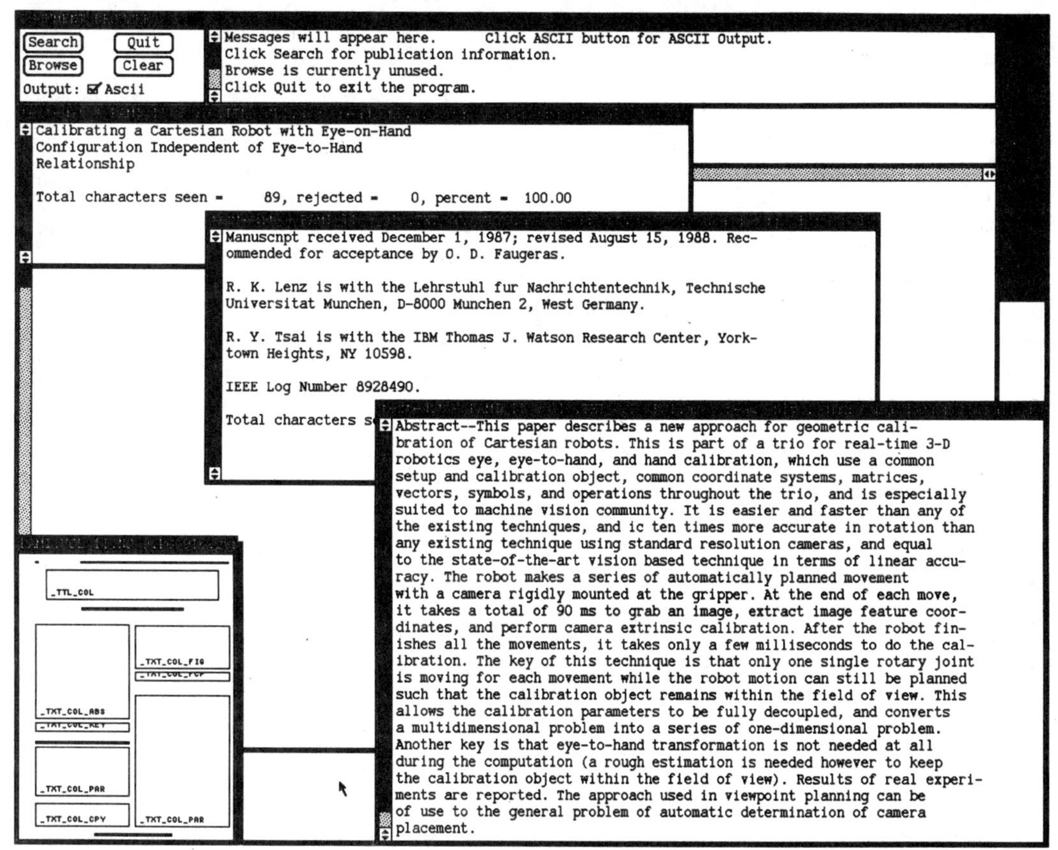

Figure 7. ASCII display of three *IEEE-PAMI* text blocks processed by OmniPage.

ASCII renditions of the title, footnote, and abstract blocks. Note that the word "Manuscript" in the footnote block is incorrectly recognized, as is the word "is" in the seventh line of the abstract. OmniPage records the number of characters and the number of rejects but cannot, of course, calculate the number of substitution errors.

Figure 8 shows the 300-dpi version of the simple line drawing of the sample page. Since the page image is currently available on the IEEE CD-ROM only in binary form, photographs would also be displayed with extreme contrast. The image can be either scrolled or zoomed if its size exceeds the screen size. Figure 9 shows both the image and ASCII displays of a text block — the figure caption.

Experimental results. We have successfully processed 21 photocopied pages from the *IBM Journal of Research and Development* and 20 pages from *IEEE-PAMI*. Since these pages were used for development and improvement of the page grammars, we then randomly selected 12 pages from each journal for an independent test. Nine of the *PAMI* pages were segmented and labeled perfectly. Three had minor mistakes. Of the 12 *IBM Journal* pages, seven were segmented perfectly, three had minor mistakes, and two missed about one quarter of the page. All mistakes could be corrected by simple modifications of the block grammars, but it is clear that several design-and-test cycles would be required for acceptable performance. An interactive step, similar

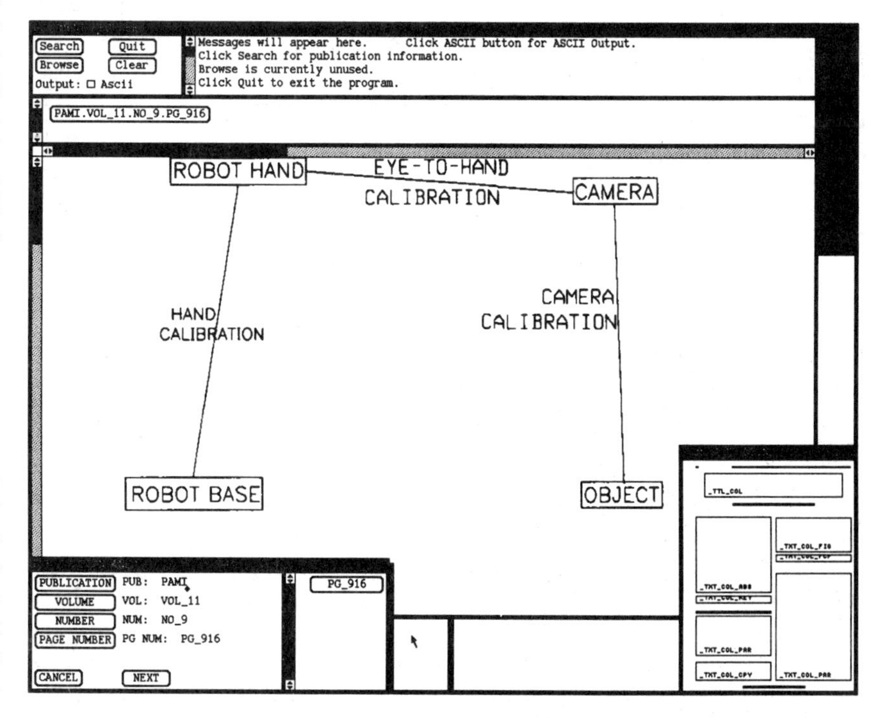

Figure 8. Image display of a figure block. This block was selected by clicking on the topmost block in the right-hand column of the geometric representation (bottom right). If the figure is small enough, it is shown at full 300-dpi resolution on the screen; otherwise it is reduced or scrolled.

to postediting in OCR, could also be invoked when the algorithm fails. Since failures can be flagged by the system itself, the overall throughput would not be greatly affected.

The bulk of the two to three minutes required to process each new page is consumed in recursive profile extraction. The algorithms were coded with little regard to efficiency; for instance, we used shell scripts whenever possible. We have recently implemented a profile extraction algorithm on a 32 × 32-processor DAP (distributed array of processors) computer. Initial comparisons show that the time required to extract the horizontal and vertical profiles of a 2,000 × 3,000-pixel image is reduced to one tenth.

Although an ASCII representation of technical documents is adequate for many purposes, a faithful rendition of the original layout is highly desirable for human access. Not only is this rendition necessary for graphics, equations, and tables whose computer representation is not yet standardized, but also preservation of the original layout and typography enhances legibility compared with OCR output, which preserves only some of this information.

We have demonstrated a prototype version of a system, based on syntactic document analysis and OCR, that can provide useful remote access to stored technical documents. The two aspects that differentiate our system from others are (1) X-Y tree data structures that are particularly suitable for printed matter, and (2) syntactic analysis of image blocks at increasing levels of refinement.

We are now interfacing our prototype browser with the Rensselaer Polytechnic Institute library information system, Infotrax. The system already includes all IEEE indexing data, including abstracts, since 1988. Infotrax yields, for each article of potential interest, the publication title, volume, issue, and page number. Our sample documents are already indexed with the corresponding information.

In addition to its use for information retrieval, we intend to adapt our system to provide input for automated or interactive indexing. Once an article is processed, only ASCII versions of such relevant fields as title, author, and abstract would be forwarded to the indexing station. The body of the text would remain available for subsequent full-text searches, as opposed to searches on selected index terms only.

Longer-term research objectives include developing improved methods for the acquisition of publication-specific knowledge bases, possibly including some form of learning. We are, however, also investigating to what extent documents can be analyzed by using generic typesetting knowledge only. ■

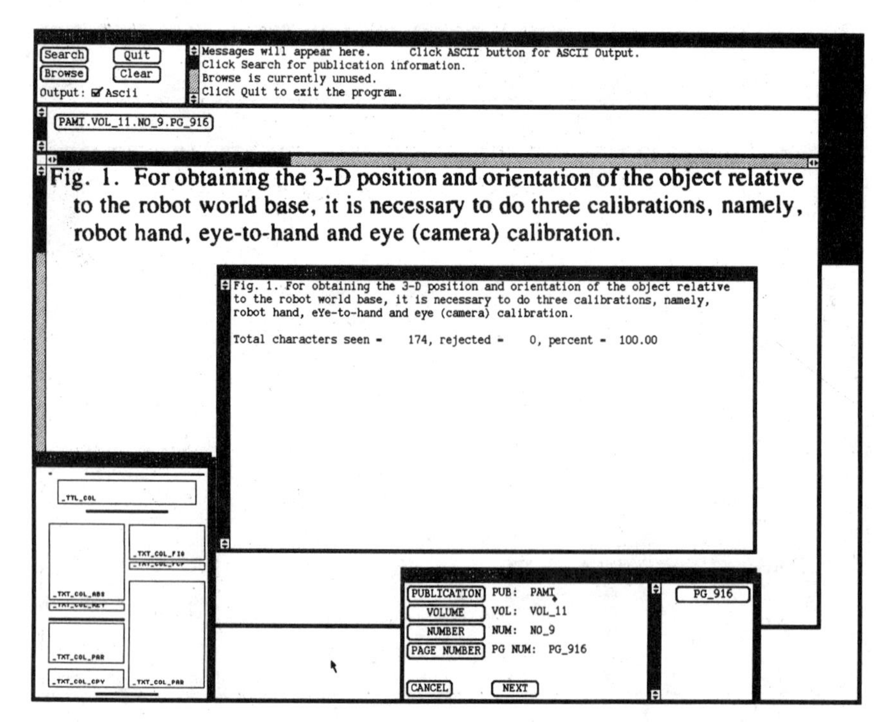

Figure 9. Image and ASCII versions of the figure-caption block. The lowercase "y" in "eye-to-hand" in the last line was incorrectly identified as uppercase.

Acknowledgments

Former and current students who have contributed extensively to the project are J. Kanai, E. McCaughrin, R. Sathyamurthy, N. Shirali, and J. Waclawik. Our collaborators at Rensselaer include M. Krishnamoorthy and T. Spencer, and Associate Director of Rensselaer Libraries P. Molholt. We are pleased to thank H.S. Baird (AT&T Bell Laboratories) for document-compression software and E. Annese (Olivetti Corp., Italy) for de-skewing software. The work was supported in part by US West Advanced Technologies and the US Department of Education College Library Technology and Cooperation Grants Program.

References

1. R.G. Casey and K.Y. Wong, "Document-Analysis Systems and Techniques," *Image Analysis Applications*, R. Kasturi and M. M. Trivedi, eds., Marcel Dekker, New York, 1990, pp. 1-36.

2. H.S. Baird, "The Skew Angle of Printed Documents," *Proc. SPSE 40th Conf. Symp. Hybrid Imaging Systems*, SPSE, Springfield, Va., 1987, pp. 14-21.

3. K.R. McConnell, D. Bodson, and R. Schaphorst, *Fax: Digital Facsimile Technology and Applications*, Artech House, Norwood, Mass., 1989, p. 96.

4. *Structured Document Image Analysis*, H.S. Baird, H. Bunke, and K. Yamamoto, eds., Springer-Verlag, 1992, in press.

5. *ICDAR, Proc. First Int'l Conf. Document Analysis and Recognition*, AFCET, Paris, 1991.

6. G. Nagy and S. Seth, "Hierarchical Representation of Optically Scanned Documents," *Proc. Seventh Int'l Conf. Pattern Recognition*, Order No. M545 (microfiche), IEEE CS Press, Los Alamitos, Calif., 1984, pp. 347-349.

7. M. Viswanathan, "Analysis of Scanned Documents — A Syntactic Approach,"

in *Structured Document Image Analysis*, H. Baird, H. Bunke, and K. Yamamoto, eds., Springer-Verlag, 1992, in press.

8. J.-M. De La Beaujardiere, "Well-Established Document Interchange Formats," in *Document Manipulation and Typography*, J.C. van Vliet, ed., Cambridge University Press, 1988, pp. 83-94.

9. R. McLean, *The Thames and Hudson Manual of Typography*, Thames and Hudson, London, 1980.

10. R. Bradford and T. Nartker, "Error Correlation in Contemporary OCR Systems," *Proc. First Int'l Conf. Document Analysis and Recognition*, AFCET, Paris, 1991, pp. 516-523.

11. V. Garza et al., "OCR Product Comparison," *Infoworld*, Oct. 22, 1990, pp. 73-90.

12. L. Grunin, "OCR Software Moves into the Mainstream," *PC Magazine*, Oct. 30, 1990, pp. 299-350.

George Nagy has been professor of computer engineering at Rensselaer Polytechnic Institute since 1985. Before that he was professor of computer science at the University of Nebraska at Lincoln, where he worked on remote sensing applications, geographic information systems, computational geometry, and human-computer interfaces. He also conducted research on various aspects of pattern recognition and OCR at the IBM T.J. Watson Research Center for 10 years. In addition to digitized document analysis and character recognition, his interests include solid modeling, finite-precision spatial computation, and computer vision.

Nagy received the BEng degree in engineering physics and the MEng degree in electrical engineering from McGill University and the PhD degree in electrical engineering (on neural networks) from Cornell University in 1962. He is a senior member of the IEEE, a member of the Computer Society, and a member of ACM.

Sharad Seth is a professor in the Computer Science and Engineering Department at the University of Nebraska at Lincoln. Besides document analysis, his research has been in testing and testable design of microelectronic circuits.

Seth received the BEng degree in 1964 from the University of Jabalpur, India, the MTech degree in 1966 from the Indian Institute of Technology, Kanpur, and the PhD degree in electrical engineering in 1970 from the University of Illinois at Urbana-Champaign. He serves on the editorial boards of *IEEE Transactions on Computer-Aided Design of Integrated Circuits and Systems*, and the *Journal of Electronic Testing: Theory and Applications* (*JETTA*). Seth is a senior member of the IEEE, a member of the Computer Society, and a member of ACM.

Mahesh Viswanathan is a development staff member with IBM Storage Systems Products Division, San Jose, California. His research interests include printing and document analysis.

Viswanathan received the BSc degree in physics in 1980 from Loyola College, Madras, India, and the BE degree in electrical engineering in 1984 from the Indian Institute of Science, Bangalore. He received the MS degree in electrical engineering from San Diego State University in 1986 and the PhD degree in computer and systems engineering in 1990 from Rensselaer Polytechnic Institute.

Readers can contact George Nagy at Electrical, Computer, and Systems Engineering, Rensselaer Polytechnic Institute, Troy, NY 12180; (518) 276-6078; fax (518) 276-6261; e-mail gnagy@mts.rpi.edu.

Postal Address Block Location in Real Time

Paul W. Palumbo, Sargur N. Srihari, Jung Soh, Ramalingam Sridhar, and Victor Demjanenko
State University of New York at Buffalo

A postal automation system locates destination address blocks on letter mail pieces with a high success rate. Pipelining and multiprocessor techniques achieve real-time processing speed.

In 1990, some 162 billion pieces of mail were received and delivered by the US Postal Service, and the volume is expected to reach 200 billion annual pieces by 1995. Manually sorting mail using ZIP (zone improvement plan) codes requires time-consuming labor; to reduce costs, the postal service is now relying more on automated sorting. (See the "Postal automation" sidebar for details.)

As part of the automated sorting process, the address recognition unit (ARU) locates and reads the destination address. The address block location (ABL) subsystem of the ARU uses the digital image of the entire piece of mail to determine the location of blocks of text that may contain the destination address. The ARU then presents the digital subimage corresponding to the destination address block (DAB) to the address interpretation subsystem, which segments the block into lines and words, locates the candidate ZIP code and other "words," and "reads" each word to determine the destination code (for example, ZIP + 4 code). This article focuses on the development of an ABL subsystem with improved performance for the next-generation US Postal Service ARU.

An ABL subsystem has been under development at the USPS Center of Excellence for Document Analysis and Recognition (CEDAR) since 1985. This subsystem (hereafter referred to as a system) uses an image of the entire mail piece to determine the coordinates of the minimum bounding rectangle enclosing the destination address block. For each candidate DAB, the ABL system determines the line segmentation, global orientation (0, 90, 180, or 270 degrees), block skew, an indication of whether the address appears to be handwritten or machine printed, and a value indicating the degree of confidence that the block actually contains the destination address.

Initially, the CEDAR ABL system was based on a blackboard data structure invoking many specialized image processing and block analysis tools using a rule-based system.[1] The major emphasis of this early (non-real-time) system was on ABL performance rather than processing throughput. All image processing operations were written in C, and the rule-based system was written in Lisp. The processing time required for the system was seven to eight minutes for letters and longer for magazines and parcels.

Reprinted from *Computer*, Vol. 25, No. 7, July 1992, pp. 34–42.

A real-time ABL system requires a processing throughput of between 1/18th and 1/9th of a second, depending on mail piece length and transport belt speed. The average throughput should be 1/13th of a second. Because many of the original algorithms were difficult to implement under real-time constraints, we redesigned them to work within the constraints or removed them from the system. For example, the blackboard system had nine specialized feature-extraction tools that would require significant real-time hardware. The system we describe in this article, however, uses only two feature-extraction tools that are relatively easy to implement in real time. In addition, we replaced the rule-based system with a more simplistic evidence combination approach.

Glossary

ABL — Address block location
ADTH — Adaptive thresholding tool
BAT — Block analysis tool
BLCS — Block-splitting tool
CEDAR — USPS Center of Excellence for Document Analysis and Recognition
COMB — Combination tool
COVF — Consistency verification tool
DAB — Destination address block
EVHP — Evidence combination tool
HEUR — Spatial heuristics tool
HSEG — Handwriting segmentation tool
HSEG1 — Column blanker and connected component locator
HWMP — Handwriting/machine-print discrimination tool
LAYO — Block layout tool
LOCA — Block location tool
LSEG — Text line segmentation tool
MSEG — Machine-printed segmentation tool
MSEG1 — Connected component locator
OCR — Optical character recognition
POST — Postprocessing tool
PSEG — Postcard segmentation tool
SIZE — Block size tool
USPS — United States Postal Service
ZIP — Zone improvement plan
ZIPM — ZIP code merging tool

Postal automation

To sort letter mail automatically, the US Postal Service has been using optical character recognition (OCR) machines since the 1970s. Postal OCR systems can handle up to 45,000 pieces per hour and automatically sort roughly 60 percent of the mail (mostly machine printed). Mail pieces are encoded with a five- or nine-digit ZIP code in a bar-code format in the lower right corner of the mail piece. Future processing then requires simply reading the bar code rather than reinterpreting the mail piece.

Figure A shows the major components of a postal OCR system designed by the US Postal Service Center of Excellence for Document Analysis and Recognition. Postal workers place mail pieces in a feeder mechanism that presents one mail piece at a time to the main transport belt. The gap between each mail piece is determined by the feeder to be approximately 3.5 inches. The transport belt moves the mail pieces at 120 inches per second past an image scanner, through a three-second delay, past the bar-code sprayer, and finally into an appropriate sorting bin. The scanner captures a 300-pixel-per-inch gray-scale image using a time-delay integration charge-coupled device array. The three-second delay corresponds to the processing time allocated for locating and reading the destination address so the correct code can be sprayed as a bar code.

During this delay an address recognition unit locates and interprets the destination address block. The unit includes subsystems for address block location and address interpretation. The unit uses postal directories to extend the printed information to sort the mail piece to nine digits, even when the address doesn't explicitly contain a "ZIP+4" code.

The goal of the USPS-supported research program is to improve OCR performance by developing new address-recognition and image-lift units to be used with the existing transport mechanism. The 1995 goal is *delivery point encoding* of 90 percent of all mail with a machine-printed destination address and 50 percent of all mail pieces with a handwritten destination address. Delivery point encoding is a much finer level of sort than the current OCR systems can perform (typically an 11-digit code rather than a five- or nine-digit ZIP code), and it requires large national address databases. To achieve these demanding goals, the performance of both the address block location (ABL) and address interpretation subsystems needs to be significantly improved.

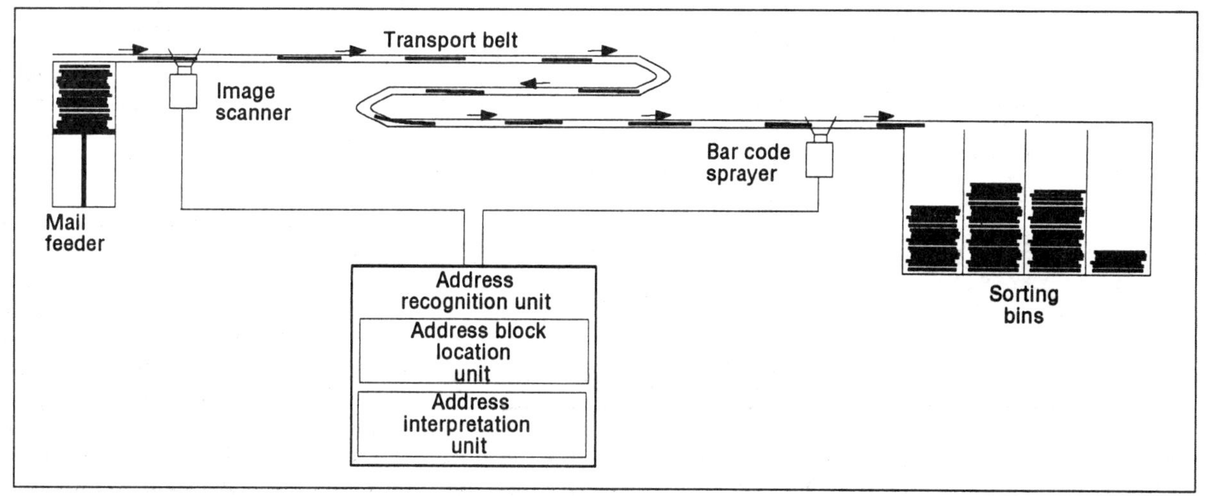

Figure A. Postal optical character recognition unit design.

System description

The current system (shown in Figure 1 and described in Table 1) is based on bottom-up or data-driven processing. It extracts primitive information from the image and groups the information into possible DABs. This framework is very conducive to real-time operation. The data presented to the real-time ABL system is an image quantized at the relatively low resolution of 100 points per inch. We selected this low resolution to reduce image processing time, but even at this resolution, a standard-size letter (4 1/2 by 9 1/2 inches) produces approximately 400,000 points of data (pixels).

This large amount of data is analyzed using fast, specialized image processing hardware. The hardware extracts interesting features from the image for further analysis with software running on general-purpose processors. The processors perform the actual grouping (segmentation), classify the print method of the text within the segmented block, and select the destination address block from the blocks segmented by the system. Pipelined image processing hardware along with multiprocessing of symbolic data provides the speed, minimum latency, and flexibility required for the system.

Since image processing hardware typically is faster than general-purpose processors, a pool of processors sustains the real-time throughput. At any given time, one processor services data transferred to it from image processing hardware and other processors are at various stages of locating a candidate DAB on different mail pieces. This multiprocessor approach permits easy addition and subtraction of processors within the pool and relatively easy software development. To control this hybrid system, a system controller determines the next processor to be used from the pool and communicates the results to the address interpretation subsystem.

Instead of analyzing all the segmented blocks at the same time, the system independently ranks the candidate blocks from each of four primary segmentation tools. The four ranked lists are subsequently combined into one list, and the top-ranked block becomes the system choice for the DAB. This approach yields better performance than analyzing all the blocks together, especially when the system attempts to find blocks that satisfy spatial relations (for example, the spatial relation between the return address and the destination address).

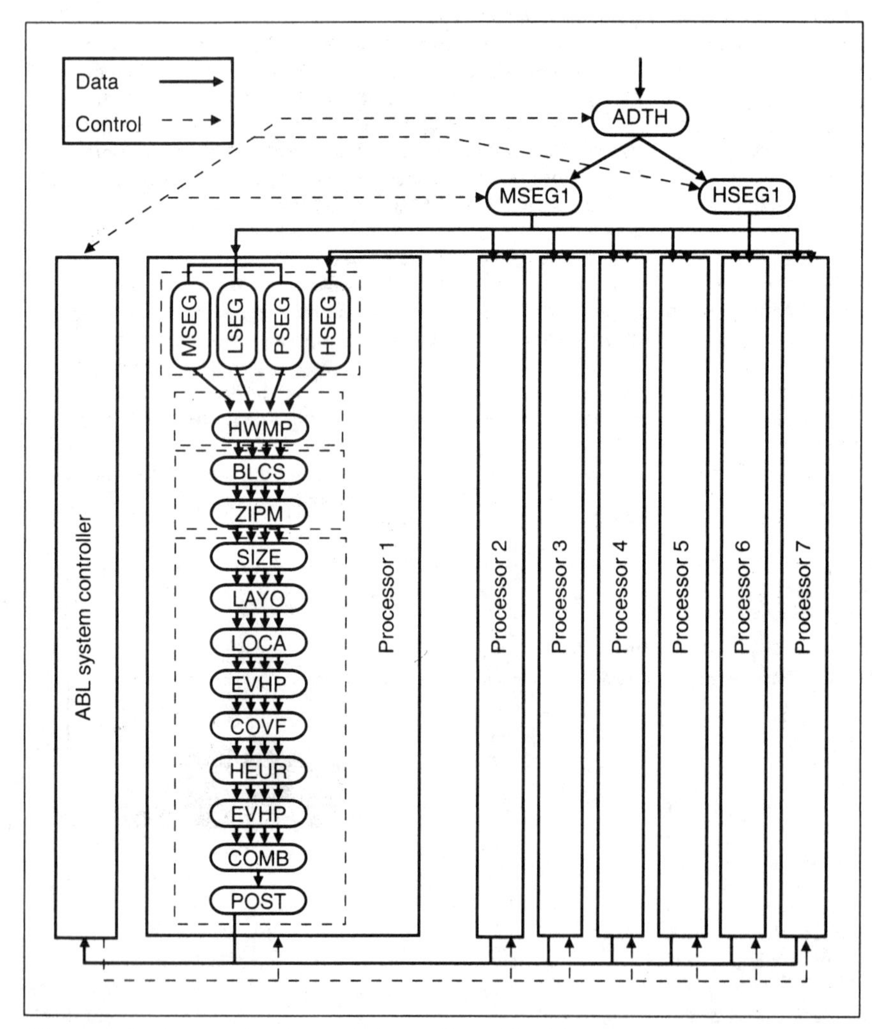

Figure 1. System design for real-time address block location.

Image processing. The image processing operations required by the CEDAR real-time ABL system are thresholding and component location. Specialized image processing hardware[2] minimizes processing time and latency (see the "MaxBus processing" sidebar) and includes off-the-shelf and custom-designed modules.

Adaptive thresholding. The adaptive thresholding tool[3] (ADTH in Figure 1) converts a gray-level image into a two-valued (binary) image. A wide variety of image quality is encountered in this application — for example, image quality is affected by colored and textured envelopes, poor illumination, and digitizing noise. We determined empirically that a locally adaptive thresholding method would perform best. The algorithm convolves the input image with an 8×8 contrast mask and thresholds the convolution output using a lookup table to produce a binary image. This tool is presently implemented with two commercially available boards manufactured by Datacube.

Connected-component location. After the image has been converted to binary form, information is extracted using connected-component analysis. A connected component is a set of all foreground pixels immediately adjacent to each other. Typically, in a machine-printed DAB under ideal digitizing conditions, each alphanumeric character is a separate connected component. The extent of each connected component is then used by the symbolic processing portion of the system to locate candidate DABs. The problem of locating connected components has been widely

Table 1. Description of tools in the real-time ABL system.

Category	Tool	Description
Image processing	ADTH	Adaptive thresholding to convert a gray-level image into a binary image using local contrast.
	MSEG1	Locates connected components from a binary image corresponding to machine-generated characters and some handwritten characters.
	HSEG1	Locates connected components from a binary image corresponding to handwritten characters and some machine-generated characters.
Primary segmentation	MSEG	Bottom-up segmenter that groups components from a binary image into primarily machine-generated words, lines, and blocks.
	LSEG	Bottom-up segmenter that groups components from a binary image first into lines, then into blocks.
	PSEG	Bottom-up postcard segmentation tool.
	HSEG	Bottom-up segmenter that groups components from a binary filtered image into primarily hand-generated words, lines, and blocks.
Print method	HWMP	Classifies a segmented block as being printed either by hand or by machine.
Segmentation refinement	ZIPM	Merges a small block to the lower right of a destination address candidate when it appears to be a ZIP code.
	BLCS	Splits a large machine-generated or handwritten text block into several smaller text blocks.
Block analysis	SIZE	Uses block features (aspect ratio, length, height, number of text lines, and number of components) to classify how likely a block is to be a destination address.
	LAYO	Uses characteristic layout of text lines in destination address to estimate the likelihood of a block being the destination address.
	LOCA	Uses the location of a block to determine the likelihood of this block being the destination address, return address, or postage.
	EVHP	Pools together the evidence generated by various tools and generates labeling hypotheses.
	COVF	Verifies the consistency of labeling hypotheses among neighboring blocks by using spatial relations.
	HEUR	Uses spatial heuristics or rules of thumb to choose the destination address from a list of candidates.
	COMB	Selects the best candidate block from among the highly ranked blocks.
	POST	Adjusts the segmentation of the top-ranked candidate destination address block, if needed.

MaxBus processing

The MaxBus, developed by Datacube, is a point-to-point bus that transfers an image between image processing boards. The bus permits modular design of the image processing stages required for each specific application. Each pixel is processed in scan-line order, synchronized to a 104-nanosecond pixel clock. Horizontal and vertical synchronization signals indicate the start of each scan line and image frame, respectively. At every pixel clock, the bus clocks an image pixel in and out of the image processing hardware in a one-dimensional raster data stream. The bus implements two-dimensional image operations, such as image convolution, by buffering several scan lines of image data to create a two-dimensional window for processing.

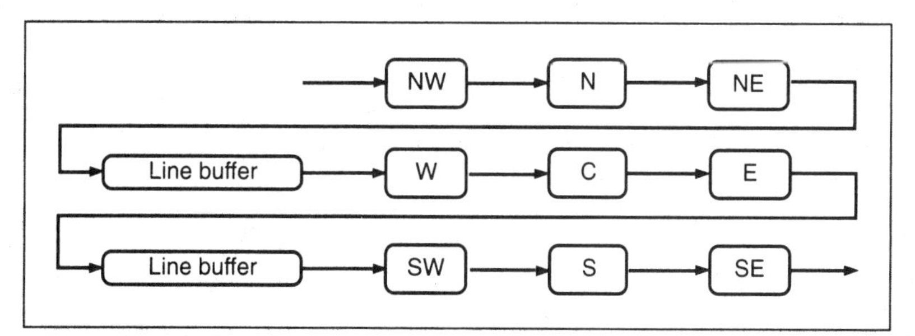

Figure B. Two-dimensional processing using MaxBus.

Figure B shows how a 3 x 3-pixel window is produced from a one-dimensional stream of pixels. Pixels are clocked into the window, starting with the northwest element, and move to the north and northeast elements before being buffered in a line buffer for nearly a complete scan line. A pixel traverses through the other elements of the window before being discarded when it leaves the southeast element. This serpentine path generates a two-dimensional operation using a one-dimensional data stream.

explored.[4,5] We needed an algorithm that used only one pass at real-time rates to conform to the MaxBus specification. The CEDAR custom algorithm[6] is a fast, minimal latency, one-pass design that accepts MaxBus input. The hardware for the connected-component locator comprises three custom boards and a custom VLSI content-addressable memory chip.[7]

Two separate system modules locate connected components: MSEG1 and HSEG1. Although they work best on extracting features from machine-printed and handwritten addresses, respectively, both modules process all images, irrespective of their actual print method. The primary difference between these two tools is that MSEG1 locates connected components in the binary image while HSEG1 locates connected components in the binary image *after* vertically "whiting-out" every 32nd column. Thereby, HSEG1 attempts to break up cursive script characters to reduce the size of the components but without using any underlying image information. More complex filtering algorithms have been tested, but this simple method produces acceptable results and is easy to implement in hardware. The connected-component data produced by these two tools is sent to the general-purpose processors using a direct memory access interface over the VMEbus.

Symbolic processing. After the connected-component location tools convert the image into symbolic data, all subsequent processing is performed on general-purpose processors. We programmed these processors with software written in C using Wind River Systems' VxWorks real-time operating system.[8] The symbolic processing tools are divided into four main functions: component grouping, print method classification, segmentation refinement, and analysis of the segmented blocks.

Component grouping. The component grouping tools cluster connected components into text lines and text blocks. The four bottom-up primary segmentation approaches use different algorithms and produce different segmentations of the same mail piece. The segmentation tools are designed for machine-printed addresses (MSEG), formation of better address text lines (LSEG), postcards (PSEG), and handwritten addresses (HSEG). This four-segmentation scheme makes the system more robust and easier to tune. The size of components and distance between them are used to group the components into blocks and lines. All components linked together in this manner are grouped into the same line or block candidate.

Print method classification. To help the downstream OCR system recognize characters, a handwriting/machine-print discrimination tool (HWMP) determines the print method of each block segmented by the ABL system. The algorithm used in the real-time ABL system relies on the frequency of different component heights in the block. It assumes a block with many different heights is handwritten and a block with uniform component heights is machine printed.

Segmentation refinement. The blocks from the four primary segmentation tools undergo a split-and-merge segmentation refinement to further improve their quality.

Sometimes, it is difficult for the primary segmentation tools to avoid producing larger blocks than desired for the DAB because the address is close to extraneous matter (such as a date, an attention line, advertising text, or other markings or graphics on the mail piece). The block-splitting tool (BLCS) splits these large machine-printed or handwritten blocks, which are usually too tall compared with their length to be the DAB. In general, the tool determines where to split a block on the basis of lines with uniform line height, distance between lines, and line-end alignment.

Machine-printed address blocks often have extra space between the city/state names and the ZIP code or between the city name and the state/ZIP code. During the connected-component grouping, this excessive space results in text being excluded from the main body of the address. The ZIP code merging tool (ZIPM) recovers from this inevitable exclusion of the ZIP code by using relaxed block size tests to identify the syntax of most possible DABs. Similarly, it chooses possible state and ZIP code blocks according to block size and number of components inside the block. For each pair of DAB and ZIP code candidates, the tool checks a set of conditions, including relative locations and the ratio of candidate block heights. If the conditions are satisfied, the tool unifies the two blocks and creates evidence for the new block being the DAB. We designed the tool primarily to merge machine-printed ZIP codes but improved it to merge handwritten ZIP codes as well.

Block analysis. A block analysis tool (BAT) is one of a collection of relatively simple tools that analyze the candidate blocks produced by the four primary segmentation sources (MSEG, LSEG, PSEG, and HSEG) and the two secondary segmentation sources (BLCS and ZIPM) to produce a minimum bounding rectangle enclosing the potential address, an associated confidence value, and other properties if the address is found. If the system cannot make a good guess at the address, it produces a reject code. Also, each segmented block is interpreted at four orientations (0, 90, 180, and 270 degrees) to determine the global orientation of the mail piece. A BAT's objectives vary:

(1) extraction of individual block features (size, layout, and location),
(2) application of heuristics based on spatial relationships between blocks (consistency verification and spatial heuristics), and
(3) integration of block features into block labeling hypotheses (evidence combination).

The block size tool (SIZE) uses statistics on DABs derived from a mail statistics database: the mean and standard deviation of the length, height, aspect ratio, number of lines, and number of characters. The tool uses a size score for each segmented block. The closer a block's features are to its statistical mean, the higher the block's score and the more evidence assigned for the block being the DAB.

Besides the size features, DAB text lines have some typical layout characteristics. The block layout tool (LAYO) uses three such characteristics: left-end alignment, line-height variation, and block density. Address lines often have their left ends aligned. Many machine-printed addresses are exactly aligned, but handwritten addresses are often more or less aligned. The LAYO measures the line-height variation by computing the ratio of the standard deviation to the mean of line heights. This ratio is usually relatively small for address blocks. The block density is the average ratio of line length to block length. The higher this measure, the denser the block and the more likely it is to contain an address. The LAYO evaluates these measures for candidate blocks and generates supporting evidence if left justification, low line-height variation, or high block density is found.

The system uses the location of a candidate block on the mail piece to estimate the likelihood of the block being the DAB, return address block, and postage. In letter mail, DABs are likely to be in the bottom two thirds of the mail piece, return address blocks in the upper left corner, and postage in the upper right corner. The block location tool (LOCA) uses the position of a block and a mail statistics database to compute a location value that compares the block location with "typical" blocks. It uses the location value to associate evidence with a block in support of its being the DAB as well as in support of its being the return address block and postage.

The evidence combination tool (EVHP) integrates pieces of evidence generated for a block into a single block-labeling hypothesis. An evidence frame consists of an attribute (for example, location, layout, or ZIP merging) and four confidence values corresponding to the likelihoods of the block being the DAB, return address block, postage, and extraneous matter. The tool combines these multiple pieces of evidence into a single labeling hypothesis that also has these four confidence values. By sorting the hypothesis confidence values for the DAB, the tool develops a ranked list of candidate DABs. The EVHP is invoked twice, once after the extraction of individual block features and again after the operations of the tools that use spatial relations between blocks. The method to combine pieces of evidence is based on the Dempster-Shafer theory,[9] which uses a commutative combination rule, making the order of combined evidence frames irrelevant.

Some spatial relationships generally hold between blocks on a mail piece. For example, the DAB is below and to the right of the return address block. For each pair of a DAB candidate b_d and a return address block candidate b_r, the consistency verification tool (COVF) checks whether the labeling hypotheses for b_d and b_r are consistent with the spatial relation between them. If the actual spatial relation agrees with a typical DAB-return address block spatial relation, then the DAB confidence of b_d and the return address block confidence of b_r are increased. If the relation disagrees (for example, the DAB candidate b_d is above and to the left of the return address block candidate b_r), then the corresponding DAB and return address block confidence values are decreased. Thus, the consistency verification tool strengthens or weakens confidence values in hypotheses based on the interblock relations.

The spatial heuristics tool (HEUR) examines the spatial relation of the top-ranked block and other blocks with high confidence values for their being the DAB. The goal is to detect an incorrect top choice and select an alternate block. This is achieved by applying a rule of thumb to a pair consisting of top-ranked b_1 and a highly ranked block b_h. For instance, if b_1 is above b_h, then b_1 might be the return address block, while b_h might be the DAB. The tool generates corresponding evidence pieces to reflect this situation. If b_1 is to the right of b_h and has bigger characters inside, then b_1 may be an advertising text block and b_h the DAB. Thus, this tool applies simple rules obtained from experience with mail piece images to cases where multiple blocks have high confidence values for being the DAB.

Each of the four segmentation streams produces an independently derived ranked list of candidate blocks. The combination tool (COMB) merges these lists into one. At present, it selects the top candidate for the destination address from each list, but a more elaborate scheme could use the top, say, three choices from each list. The tool ranks the properties of these top choices and determines the choice most likely to be the destination address. If the top choices from different streams produce nearly the same segmented region, they are treated as one candidate.

The postprocessing tool (POST) attempts to correct for minor segmentation problems in the top-ranked candidate DAB. It removes extra material printed to the left of the address, such as the words "To" and "Pay to the order of." This tool also uses top-down knowledge to correct any possible minor segmentation problems.

System controller. The CEDAR real-time ABL system is a mixture of both specialized pipelined hardware and multiprocessors. A system controller implemented on a dedicated processor controls image processing boards and routes information in and out of general-purpose processors. System controller software configures the image processors by setting the direct memory access address where the connected components will be written. A scheduling algorithm determines routing, selecting the next processor to use and time-out conditions to maintain a constant flow of images through the system. For example, if all the processors are in use when the next image is about to come through, the scheduling algorithm must free a processor to accept the symbolic information from the incoming mail piece or discard the mail piece.

The scheduling algorithm interrupts the job that has been processing for the longest time. However, if this job has not been processing very long (even though it is the oldest one in the system), the incoming mail piece is discarded or "sunk." Sinking prevents processing from being continually interrupted without any significant image-receiving processing time. Mail pieces that are sunk or interrupted by the scheduling algorithm produce no DAB candidates and require manual sorting. The scheduling algorithm minimizes the number of time-outs and provides an effective system throughput rate.

Implementation

The system described here has been integrated and runs at rates of up to 10.9 mail pieces per second. All the image processing hardware, including the custom boards, has been designed, implemented, and tested. The symbolic software runs on four 20-megahertz Sparcengines using VxWorks and controlled by another Sparcengine. In this section, we describe the timing analysis and location performance of the system. All performance evaluations used 2,000 letter mail images (63 percent machine-printed destination addresses and 37 percent handwritten destination addresses) that could not be sorted to nine-digit accuracy by current postal OCR systems.

Timing analysis. The image processing hardware time, based on the Datacube MaxBus standard, is directly proportional to the image size processed (see the MaxBus sidebar). Each pixel

Table 2. Image and symbolic processing times (in seconds).

	Moments		Extremes		Percentile		
	Mean	Standard Deviation	Minimum	Maximum	90	95	99
Image	0.041	0.009	0.022	0.059	0.047	0.047	0.059
Symbolic	0.169	0.084	0.046	1.106	0.246	0.277	0.470
Total	0.210	0.085	0.086	1.133	0.288	0.320	0.504

Table 3. Time-outs at various image-throughput rates (pieces per second).

	Throughput Rates			
	8.2	9.8	10.5	10.9
Time-outs	5	7	15	22

can be processed within 104 nanoseconds, with additional pixel delays for horizontal and vertical image blanking. In the worst case (the maximum letter size), a 6 1/8 × 11 1/2-inch area corresponds to 704,950 pixels when digitized at 100 points per inch and requires only 0.076 seconds including overhead. The latency time through the image processing tools is roughly 10 scan lines (1,228 microseconds). Table 2 shows the image processing time for the 2,000 test images. In general, these times are relatively short. The symbolic processing time of the implementation is also relatively small (see Table 2). We are constantly improving the symbolic processing algorithms to decrease the processing time even further.

We simulated the system design using the SES/Workbench design specification, modeling, and simulation package. The simulation provided information about design effectiveness which helped guide the implementation. Using this simulation, we could relatively easily determine the effect on the system of different scheduling algorithms and timing requirements.

Table 3 shows the number of image time-outs at various image-throughput rates (ranging from 8.2 to 10.9 pieces per second). Images are presented to the ABL system using a real-time disk subsystem, which limits throughput rates to 10.9 pieces per second. The subsystem will eventually be replaced by an image lift under development. Faster processors would permit more complex algorithms and a decrease in the number of system processors.

Location performance. The segmentation results with the test images show that the line-based segmenter (LSEG) is the best of the four primary segmentation tools. As Table 4 shows, it correctly segments (groups) 82.7 percent of the images. The machine-printed address segmenter (MSEG) has a relatively low location performance for handwritten DABs but performs well on machine-printed DABs. Three of the primary segmentation tools (MSEG, LSEG, and HSEG) are used on all images, but the PSEG, BLCS, and ZIPM tools are used only on blocks that meet these tools' preconditions. The overall segmentation performance — that is, at least one of these six segmentation tools correctly locates the DAB — is 96.7 percent. For images in which at least one tool located the DAB, print-method discrimination was correct 96.8 percent of the time.

Once the blocks have been correctly segmented, the BATs must select one as the system's choice for the DAB. As Table 4 shows, the system correctly segments and locates the DAB for 89.0 percent of these images (90.0 percent for machine-printed DABs, 87.2 percent for handwritten DABs). The strict guidelines for this grading were developed by the USPS. In most failures, the system locates most of the destination address but not all of it (for example, a ZIP code is missing), or the system includes extra material such as part of a cancellation or meter mark.

Figure 2 shows how the real-time ABL produces a top-ranked DAB. Figure 2a shows the connected components located by the MSEG1 tool. (HSEG1 output is not shown.) Figures 2b through 2e show the primary segmentation results from MSEG, HSEG, LSEG, and PSEG, respectively. The darker shaded blocks in Figures 2b through 2e represent the top-ranked block for the BATs using that segmentation stream. Figure 2f shows the system's overall top-ranked block (with line segmentation), derived from the top-ranked blocks from the individual segmentation streams. In this example, two segmentation tools (HSEG and PSEG) could correctly segment and rank the destination address.

The CEDAR real-time address block location system determines candidates for the location of the destination address from a scanned mail piece image. With 20-MHz Sparc processors, the average time per mail piece for the combined hardware and software system components is 0.210 seconds. The system locates 89.0 percent of the addresses as the top choice. Recent developments in the system include the use of a top-down segmenta-

Table 4. System performance (success and failure shown in percent).

Grade	Segmentation							Overall Performance
	MSEG	LSEG	HSEG	PSEG	BLCS	ZIPM	Combined	
Success	62.5	82.7	76.7	55.2	49.4	91.6	96.7	89.0
Failure	37.5	17.3	23.3	44.8	50.6	8.4	3.3	11.0

tion tool, address syntax analysis using only connected component data, and improvements to the segmentation refinement routines. This has increased top choice performance to 91.4 percent. Additional improvements are planned, including the use of character recognition and the use of glassine window-detection information; together, these improvements should enhance ABL performance to meet USPS goals. ■

Acknowledgments

We thank Rajesh Dixit, Chandima Edirisinghe, David Fishback, Parag Gokhale, Jeon-Man Park, Yong-Chul Shin, and Madhusudhan Srinivasan of CEDAR as well

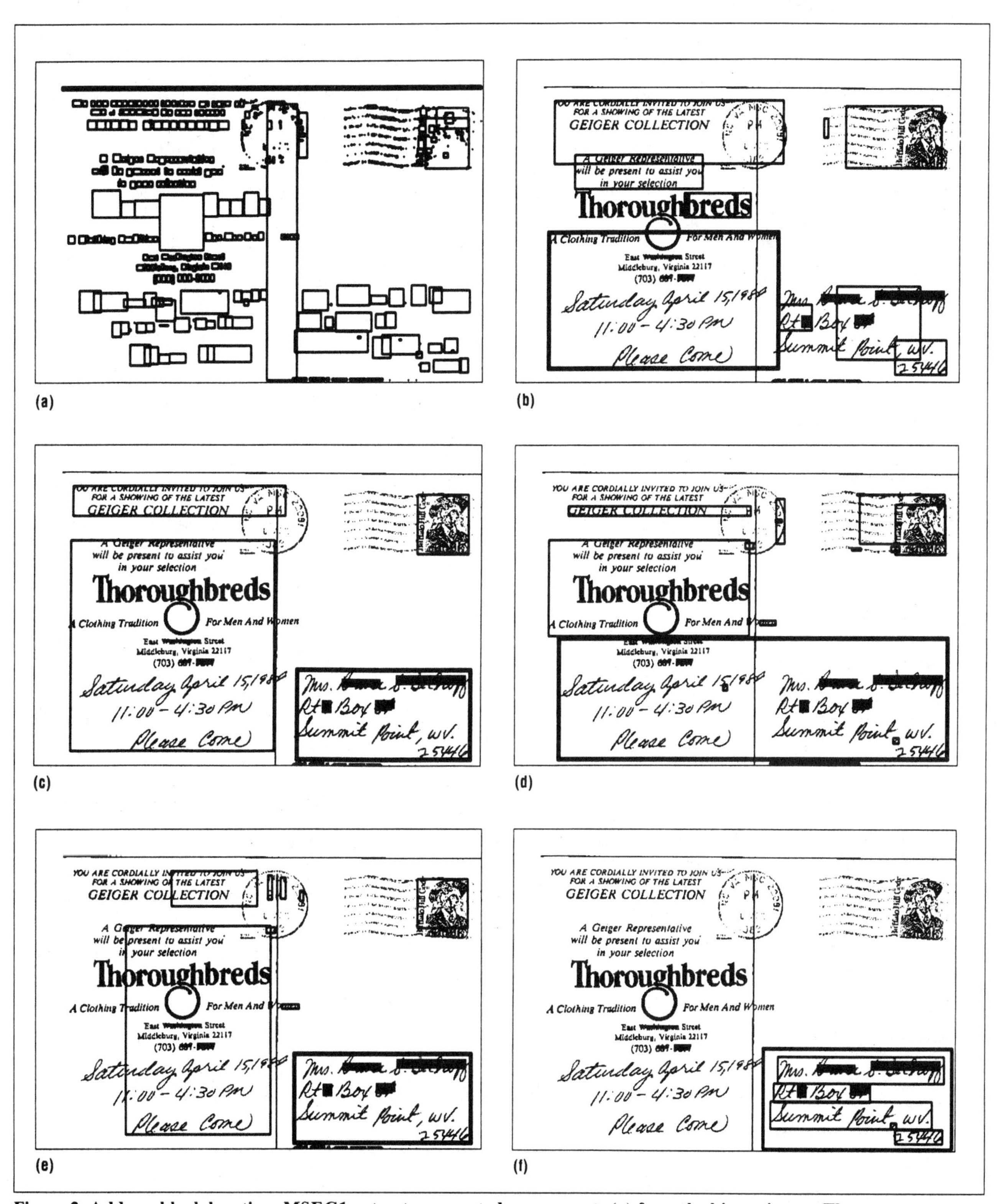

Figure 2. Address block location. MSEG1 extracts connected components (a) from the binary image. The components are grouped into possible blocks by the four primary segmentation tools MSEG (b), HSEG (c), LSEG (d), and PSEG (e). Then, the system produces the top-ranked destination address (f).

as Charlie Stenard of AT&T Holmdel for their assistance on this project. Steve Chahal and Carl O'Connor from the USPS Office of Advanced Technology also provided valuable suggestions for the development of CEDAR's ABL system.

This work was supported by the USPS Technology Resource Department under Task Order 104230-88D-2267.

References

1. C.H. Wang and S.N. Srihari, "A Framework for Object Recognition in a Visually Complex Environment and its Application to Location Address Blocks on Mail Pieces," *Int'l J. Computer Vision*, Vol. 2, No. 2, Sept. 1988, pp. 125-151.

2. J. Walker, "Pipeline Processing for Machine Vision," *Electronic Image East*, BIS CAP Int'l, Boston, 1989, pp. 579-584.

3. P.W. Palumbo, P. Swaminathan, and S.N. Srihari, "Document Image Binarization: Comparison of Techniques," *Proc. SPIE Symp. Digital Image Processing*, SPIE, Bellingham, Wash., Vol. 697, 1986, pp. 278-285.

4. R. Lumia, L. Shapiro, and O. Zuniga, "A New Connected-Components Algorithm for Virtual Memory Computers," *Computer Vision, Graphics, and Image Processing*, Vol. 22, No. 2, May 1983, pp. 287-300.

5. C. Ronse and P.A. Devijver, *Connected Components in Binary Images: The Detection Problem*, John Wiley, Letchworth, UK, 1984.

6. V. Demjanenko et al., "Real-Time Connected Component Analysis for Address Block Location," *Proc. USPS Advanced Technology Conf.*, USPS, Washington, D.C., 1990, pp. 1,059-1,071.

7. Y.C. Shin et al., "A Special-Purpose Content Addressable Memory Chip for Real-Time Image Processing," *IEEE J. Solid State Circuits*, Vol. 27, No. 5, May 1992, pp. 737-744.

8. J. Fiddler, E. Stromberg, and D.N. Wilner, "Software Considerations for Real-Time RISC," *Digest of Papers Compcon Spring 90, 35th IEEE Computer Soc. Int'l Conf.*, IEEE CS Press, Los Alamitos, Calif., Order No. 2028, 1990, pp. 274-277.

9. J.A. Barnett, "Computational Methods for a Mathematical Theory of Evidence," *Seventh Int'l Joint Conf. Artificial Intelligence*, Morgan Kaufmann, San Mateo, Calif., 1981, pp. 868-875.

Paul W. Palumbo is the project manager for the CEDAR Real-Time Address Block Location Project at the State University of New York at Buffalo. He has worked on postal research at SUNY Buffalo since 1984. His research interests are image processing, segmentation, and parallel processing.

Palumbo received his BA and MS degrees at SUNY Buffalo in 1983 and 1985, respectively. He is a member of the IEEE, the IEEE Computer Society, and the ACM.

Sargur N. Srihari is director of the USPS Center of Excellence for Document Analysis and Recognition and holds the Pattern Recognition Professorship in the Computer Science Department at SUNY Buffalo.

Srihari received his BSc in physics and mathematics from Bangalore University in 1967, his BEng in electrical communication engineering from the Indian Institute of Science, Bangalore, in 1970, and his MS and PhD in computer and information science from Ohio State University in 1971 and 1976, respectively. He is a senior member of the IEEE and a member of the IEEE Computer Society, the American Association for Artificial Intelligence, the Pattern Recognition Society, and the ACM. He received a New York State/United University Professions Excellence Award in 1991.

Jung Soh is a research assistant in the Department of Computer Science at SUNY Buffalo. His research interests are in real-time control structures for knowledge-based vision systems.

Soh received a BS in electrical engineering from the University of Wisconsin-Madison and an MS in computer science from SUNY Buffalo.

Ramalingam Sridhar is an assistant professor of electrical and computer engineering at SUNY Buffalo. His research interests include language-directed computer architecture, asynchronous processor design, real-time computer architecture, special-purpose processor architectures, and VLSI design. He is involved in the development of direct-execution processors and real-time postal address recognition systems.

Sridhar received his BE in electrical engineering from the University of Madras, India, in 1980 and his MS and PhD in electrical and computer engineering from Washington State University in 1983 and 1987, respectively. He is a member of the IEEE Computer Society, and the ACM.

Victor Demjanenko is an assistant professor of electrical and computer engineering at SUNY Buffalo. He is also president of Tree Technologies Corp., Buffalo, which specializes in product development and serves as consultant to other organizations. His research interests include real-time image processing, custom computer architectures, network-operating systems, and microprocessor-based control and diagnostic systems. He is involved in the development of real-time postal address reading machines and in the integration of vibrational signal acquisition and diagnostic signal processing.

Demjanenko received his BS, MS, and PhD in electrical engineering from SUNY Buffalo in 1980, 1981, and 1983, respectively. He is a member of the IEEE Computer Society.

Readers can contact Palumbo at the Center of Excellence for Document Analysis and Recognition, 226 Bell Hall, State University of New York, Buffalo, NY 14260. His e-mail address is palumbo@cs.buffalo.edu.

Extraction of Data from Preprinted Forms

Suzanne Liebowitz Taylor, Richard Fritzson, and Jon A. Pastor
Paramax Systems Corporation, Valley Forge Labs Research and Development, 70 East Swedesford Road, Paoli, Pennsylvania

Abstract: The widespread use of printed forms for data acquisition makes the ability to automatically read and analyze their contents desirable. The components of a forms analysis system include conversion from paper to an image through scanning, image enhancement, document identification, data extraction, and data interpretation. This paper describes techniques for manipulating electronic images of forms in preparation for data interpretation. A combination feature extraction/model-based approach is used for forms identification, registration, and field extraction. Forms identification is implemented with a neural network. The system is demonstrated on United States Internal Revenue Service forms.

Key Words: document processing, neural networks, image registration, region extraction, forms processing

1 Introduction

Improvements in the availability and affordability of scanning and mass storage devices have resulted in a strong movement away from the paper storage of documents. Information originally received in paper form must be converted to electronic form. In applications where some analysis of the document is required prior to storage, adding automated evaluation of the document's content from its electronic image reduces the need for human intervention in executing simple and repetitive tasks. The challenge of automating such processes arises from the variety and complexity of input documents.

Printed forms make up a significant percentage of submitted paperwork which must be rapidly processed and stored. Although printed forms have a "simple" layout in that they contain information that is common to every particular type of a form, processing heterogeneous batches of preprinted forms is still a challenging process which requires identifying the particular type of form, registering the input image with the model of the form, extracting information from the form, and interpreting that information. Interpretation can be achieved by converting the appropriate portions of the image to text using printed or handwritten character recognition (Suen 1980; Nagy 1982; Mantas 1986; Davis and Lyall 1988).

This paper describes a system for manipulating preprinted forms, specifically when a unique machine-readable identifier, such as a bar code, is not present. We specifically address the processing of preprinted forms which use printed horizontal and vertical guide lines as field delimiters. One familiar example is the set of United States Internal Revenue Service (IRS) forms. In this case, the algorithms to analyze the documents are less complex than those for more free-form documents (Wong et al. 1982; Akiyama and Hagita 1990; Elliman and Lancaster 1990). However, the large number of different forms and variations due to the printing and scanning mechanism prevent the use of a simple matching scheme. The processes of a forms analysis system, which are illustrated in Figure 1 include:

- *Scanning:* conversion from paper to an electronic image
- *Enhancement:* processing the image to improve subsequent operations, for example, noise removal, thresholding, skew correction
- *Form Identification:* determining the particular form type
- *Region Extraction:* locating and identifying important fields of information on the form
- *Data Interpretation:* converting extracted regions from the image representation to a text representa-

Address reprint requests to: Dr. Suzanne Taylor, Paramax Systems Corporation, Valley Forge Labs Research and Development, 70 East Swedesford Road, Paoli, PA 19301-0517, USA.

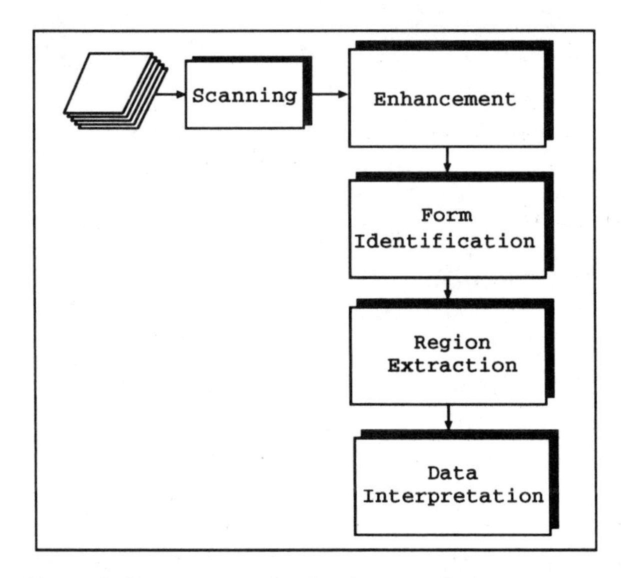

Figure 1. Image processing for forms analysis.

tion, such as ASCII (American Standard Code for Information Interchange)

Currently, our system prepares the form for interpretation, although optical character recognition (OCR) of printed labels was employed for automatic indexing. Indexing determines a unique identifier for the document instance (for example, the IRS form belongs to *J. Taxpayer*).

Our primary test case was a set of IRS tax forms. Superficial analysis of the IRS forms recognition problem might lead to the conclusion that it is appropriately addressed by matching strategies: there is a fixed set of distinct forms that one can distinguish by comparing the input image with models of each form. Implementing such a strategy, however, presents a number of practical problems.

First, one can make few correct assumptions about the layout of any particular IRS form, since forms may be obtained from the IRS, photocopied, or printed by a tax preparation package. Even forms obtained from the IRS are not necessarily identical; there are several "official" printers used by the IRS, and each of these appears to use different artwork. Therefore, different images of the same form can vary not only in size, scale, and position on the page, but also in content and structure. Any recognition strategy must therefore be based on features of the form that are relatively insensitive to these variations.

Second, some IRS forms are similar in structure, and thus require fine comparisons, necessitating high resolution in the scanned images of the forms. High resolution gray-scale images are required by subsequent processing of the form (e.g., character extraction and recognition), once it has been identified. In order to preserve sufficient detail in handwritten characters and small distinguishing features, it is typically necessary to scan forms at 300 ppi (pixels-per-inch). However, for forms identification, we would like to eliminate elements, such as characters that would cause difficulties distinguishing between forms of the same type. For our application, we found it sufficient to use 75 ppi images for the feature extraction necessary for forms recognition and registration. Using our gray-level scanner with 64 gray levels, this results in images that are on the order of 600 by 800 (eight-bit) pixels. Since there are over 160 different IRS forms and schedules, it is clearly impractical to attempt to match on full-size images. Therefore, we developed an approach which combines the data reduction of feature extraction with the precision of model-based approaches.

Form identification is implemented with a neural net as described in Section 2.1. The keystone of forms analysis is the computation of a set of geometric features which is used for both the input representation to the neural network and for image registration with the form model. In Section 2.2 we detail the form model and the feature set. An interactive window-based program developed to assist the user in creating the forms model is described in Section 2.3. Document registration and field extraction are discussed in Section 2.4. Document indexing in the form of label extraction for IRS forms is presented in Section 2.5. Image skew correction, an important image preprocessing step, is detailed in Section 2.6. Results are presented in Section 3.

2 System Overview

The shaded area of Figure 2 highlights the functional modules of forms processing that we detail in the following subsections. After the document is scanned and converted to an electronic image, it is transformed to a feature vector representation based on line-crossing features. This representation is used as input to a neural network for identification as well as for image registration to its model after the specific form type is determined. Once the input is registered, the important fields are checked for information content, and if they are not empty, their contents are extracted. In the specific case of IRS forms, document indexing is implemented by locating the preprinted label on the form, correcting the label image for skew and reading the label with optical character recognition (OCR) software. If no label is present, then the handwritten or typed information is extracted during the field location and extraction

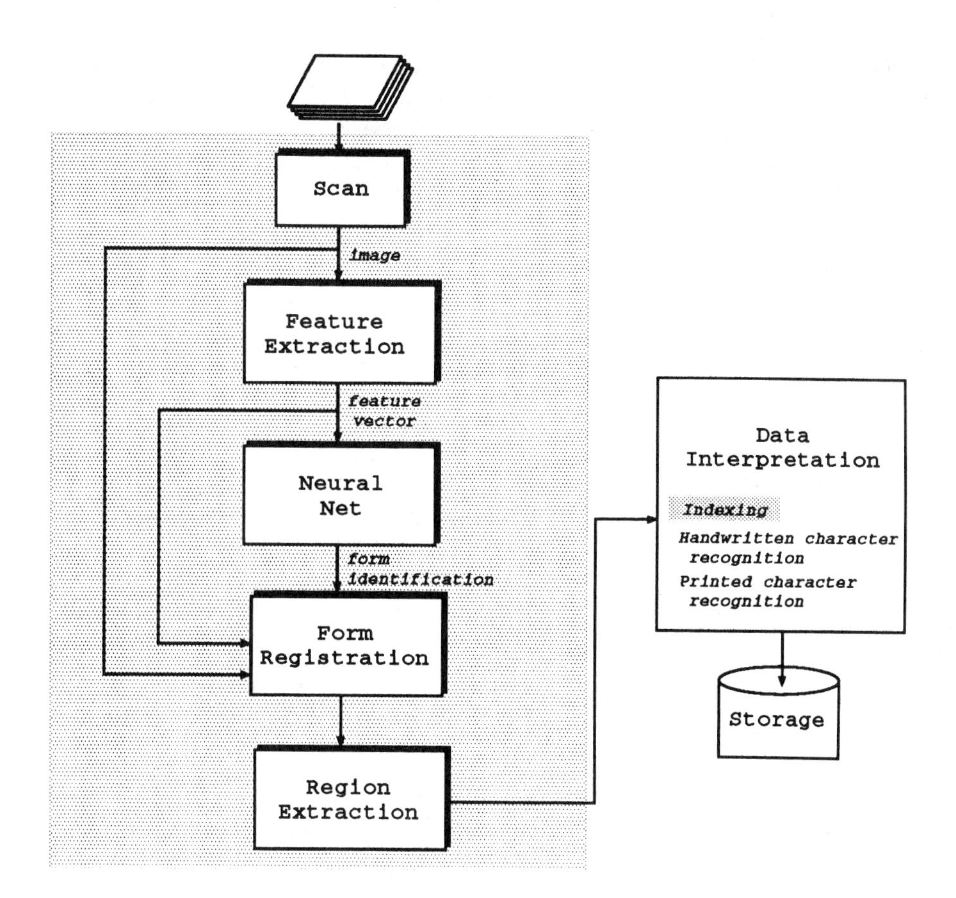

Figure 2. Feature extraction/model-based approach to forms analysis. Shaded areas are detailed in this paper.

stage (Section 2.4). These regions can be used as input to printed or handwritten character recognition modules.

2.1 Form Identification

The large size of the scanned forms suggests that a recognition strategy must include some means of image size reduction by one of: semantic (extraction of a region of the image that provides adequate discrimination among forms), pixel-level (e.g., image compression), or geometric (analysis at a higher level of abstraction than raw pixels, but below the level of semantic features) methods.

As in most neural network applications, proper input representation is the key to successful implementation. We evaluated all three reduction strategies, and experimented with several variants of two of them (Pastor and Taylor 1991) before concluding that a geometric feature representation was the better approach.

Semantic techniques proved impractical as a primary technique, because there is no single distinguishing feature that is present on all IRS forms. Many, but not all, forms have identifying information in the upper-left or lower-right corner, but even among forms that do include identifying information, it is not uniform-sized text, and does not appear in the same position from form to form, thus making it difficult to locate, segment, and recognize. While local high-level semantic features are impractical as a primary means of form recognition, they can be useful given a tentative identification of the form, since features may be sought in known locations.

Pixel-level reduction was employed by block-averaging the image. We attempted to reduce the dimension of the input and keep enough detail to discern different forms; however, we were unable to find an adequate tradeoff between a sufficient size reduction to make the network tractable and the retention of enough information to make the distinctions between forms.

We examined several types of geometric features before choosing one based on the distribution of line crossings. In a similar spirit, Casey and Ferguson (1990) use horizontal and vertical lines on the forms themselves for both determining the form's type and registration by matching to a set for form models. However, it was difficult to find a suitable description of line patterns for input to a neural network; we wanted to avoid extensive matching in determining the form type. The line-crossing features are also used for image registration and field description. Once the features have been identified, the location of each occurrence of each of the nine *visible* types of line intersections (Section 2.2, Figure 3) is recorded. We partition the form image into nine equal rectangular bins (three horizontal by three vertical), and count the number of occurrences of each type of *visible* line intersection in each bin, resulting in an 81-element feature vector. The choice of three vertical and three horizontal bins was based on the results of hierarchical cluster analysis on many different partitionings of the image. Partitioning the image captures the spatial distribution of the features, without placing undue emphasis on precise location, thus providing some shift-, scale-, and rotation-invariance (although skew correction of the image, as described in Section 2.6, is performed before calculation of the features). As an added benefit, computing this feature vector has virtually no incremental cost, since the features that it uses are derived via simple arithmetic operations from features that are already being computed for use by another component of our system (Section 2.3).

We employed a simple feed-forward network trained using back-propagation (Rumelhart et al. 1986). After experimenting with one and two hidden-layer models, we found that the two hidden-layer models produced better results. Floating-point state values were used and the output coding was "one-of-N." This decision scheme chooses the class which corresponds to the highest value of the N output nodes. The values of the output nodes are retained to permit checking of the second- and lower-ranked form types if form registration fails on the first type chosen.

2.2 Form Model

We developed a robust form model which captures enough information for input image registration and region extraction without excess clutter. The model has two components: a list of corner-point features which are used for data registration and a list of rectangular data fields used for data extraction. The list of corner points provides markers for the input image registration. Each corner feature is formed in the image by the intersection of two straight line segments which define the rectangular fields on the form.

Each rectangular data field is described by its location and type of contents. The location is specified by the coordinates of the upper-left and lower-right corner points relative to the origin at the upper-left-hand corner of the image. The contents are categorized as either a simple binary field (i.e., a "check if yes" field), or a field containing numeric, alphabetic-only, or mixed alphanumeric data.

The particular corner features used for document registration describe the line-crossing patterns of the form. Each possible line crossing on the form is labeled as one of the first nine types illustrated in Figure 3. The tenth "invisible" type describes imaginary corners which are needed in order to complete the description of a field, but are not visible on the form itself (e.g., point A in Figure 4). Such points are either formed from a single line on one side boundary of a field without a perpendicular intersector (as shown in point A in Figure 4) or when a field does not contain any visible line boundaries. Although the registration algorithm is based on matching corner points, it does not depend on matching *all* corner points. Once we have successfully registered the document, we extract all regions using our standard model, whether they are defined by visible or invisible corner points.

Table 1 contains an excerpt from our IRS 1040 field description file. The first field described is the IRS label field which is located at pixel (88,69) and contains both numeric and alphabetic content. The locations are relative to the upper-left corner in a 150 ppi image of the form. Lines beginning with **POINT** describe corner points in the image and include an index number for each of these points, the pixel location of the corner point, and the type of corner point (Figure 3). Many of these **POINT** entries, for example, lines one and two in Table 1, describe the upper-left and lower-right corner points of the "box" bounding fields on the form. Lines beginning with **FIELD** contain the field name; the field type, *alpha* for fields containing only letters, *numeric* for fields only containing numbers, *alphanumeric* for fields containing both letters and numbers, and *mark* for fields which permit only a check mark; and indices to the two corner points. Extra corner points which are not used for the field descriptions are listed at the end of the file and are used for form registration.

Each visible corner point is located by correlating a 75 ppi version (reduced by a factor of two from our original scan of 150 ppi) of the form image with four 9×9 pixel templates. These four templates, shown in Figure 5, respond to the presence of the corner types: *cornerLB, cornerLT, cornerRB,* and *cornerRT*. Notice that the actual corner point is not located at the center of the template, but rather at a variable offset from the center. For example, using the *cornerRT* template, the actual corner intersection is at a point 3 pixels below and 3 pixels to the right of the center point. The strong response is always noted at the actual corner pixel location, not at the location of the pixel corresponding to the center of the template. These 9×9 templates detect corners formed from lines of at least 7 pixels in length. Since reduction of the image to 75 ppi re-

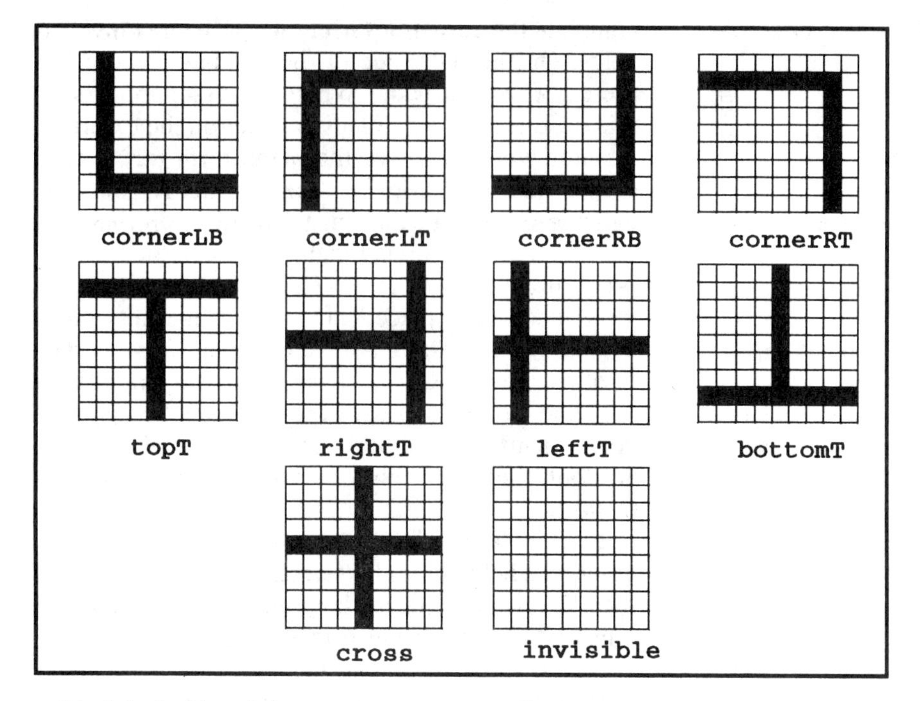

Field defined by two corner points.

A.

10.

B.

A.: Cross-junction
B.: Corner-junction
Label: Q10
Data: alpha

Figure 3. Ten line corner junctions used as features for document identification and registration. Example of a field definition using these corner points.

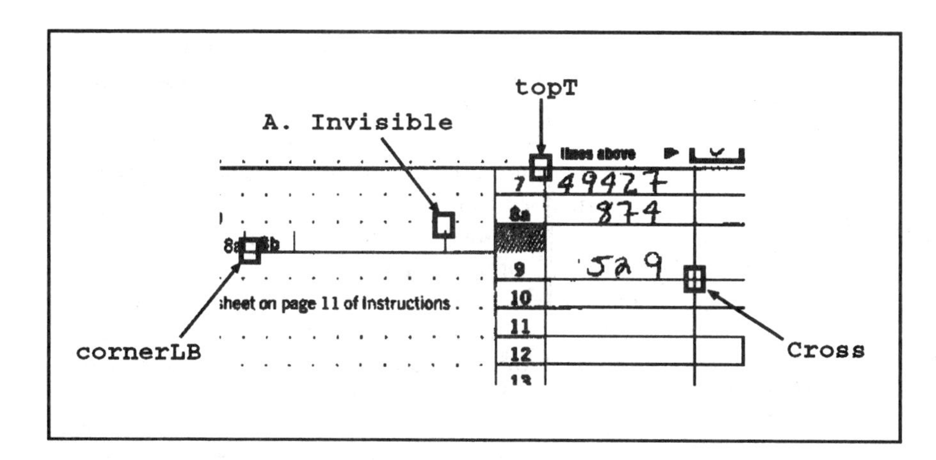

Figure 4. Segment of an IRS 1040 form with labeled corner types.

sulted in one pixel width lines, the templates of Figure 5 were designed to detect 1 pixel width corners. If this method is to be implemented on a different preprinted form which uses horizontal and vertical guide lines, it may be necessary to modify the templates to search for corners with greater than one pixel thickness.

The other five visible types are found by combining multiple high correlation responses for a corner pixel in the image from two or more of the corner templates according to the following rules:

1. if ($cornerLB \wedge cornerRB$) then *bottomT*
2. if ($cornerLT \wedge cornerRT$) then *topT*
3. if ($cornerRB \wedge cornerRT$) then *rightT*
4. if ($cornerLB \wedge cornerLT$) then *leftT*
5. if ($cornerLB \wedge cornerRT$) $\vee$ ($cornerRB \wedge cornerLT$) then *cross*

From the correlation output peaks and the application of the above rules, a list of corner points, their pixel locations, and corner type is completed. This

Table 1. Field description of forms

POINT	1	88	69	topT
POINT	2	443	170	invisible
FIELD	Label	alphanumeric	1	2
POINT	3	113	196	topT
POINT	4	136	208	rightT
FIELD	Q1	mark	3	4
POINT	5	113	208	leftT
POINT	6	136	221	rightT
FIELD	Q2	mark	5	6
POINT	7	113	221	leftT
POINT	8	136	234	rightT
FIELD	Q3	mark	7	8
POINT	9	113	234	leftT
POINT	10	136	259	rightT
FIELD	Q4	numeric	9	10
	⋮		⋮	
POINT	119	491	56	topT
POINT	120	355	56	bottomT
POINT	121	491	69	bottomT
POINT	122	446	94	rightT
POINT	123	88	94	leftT

list of points is then used to register the document image to its model.

2.3 Form Model Interactive Program

Manual specification of all the corner types and field descriptions for the form model is a tedious process. To aid the user, we developed a window-based tool **Mform** written with Motif[1] and the X Window System.[2] **Mform** has both interactive and noninteractive functions.

First, the user runs the noninteractive portion with several images of a particular type of form as input. Each of these images is correlated with the four corner templates. The coordinates of strong output peaks are marked along with the appropriate corner type. The rules of Section 2.2 are applied to corners with multiple peaks in each image; results from each of the correlation outputs from each input image are merged to form the final model with its corners marked and labeled. This final list of corner features is available as input for the interactive portion of **Mform.**

The results from the noninteractive portion of **Mform** are superimposed on a form image for the user to examine. The interactive interface is used to complete the corner description and define the form fields. On the right side of the window (Figure 6), there is a "mouse pad." When the cursor appears in this region, a cursor will also appear on the external monitor which displays the blank form image and corner markings. Mouse clicks in this region will indicate corners in the image to be modified. The specific modification is indicated by clicking on appropriate buttons to remove corner markers which were incorrectly placed, add corner markers, and change the descriptions of corner markers (as defined by the field **Corner Type** at the top left side of the window). The corner type is selected by a pull-down menu. The addition and removal of corners are controlled by the buttons **Add Corner** and **Delete Corner.**

In a similar fashion, the two corner points which describe a field are set by selecting the corner points with the mouse and pushing either the button **Set Top Left Corner** or **Set Bottom Right Corner.** The field label is typed in, and the field type is selected from a pull-down menu as either **mark, text, numeric,** or **mixed,** where **text** indicates an alphabetic field, and **mixed** indicates an alphanumeric field. After modification of the corner and field list are complete, the corner point list and the region specifications are saved for image registration and field extraction.

2.4 Document Registration and Field Extraction

The registration procedure is a key component of the field extraction process. It is responsible for aligning the visible points on a newly scanned form with the internal database of points which describe a particular form type.

The registration problem is due to the inability to predict the precise location of the form on the scanner bed when the image is formed. Although two scanned points will always be the same distance from each other on different scans (barring any extreme scaling in the image), they may on each scan have different locations with respect to the upper-left-hand corner origin of the page. The registration procedure works with two lists of points, one which describes all the visible points on the "standard" form and one which describes the observed points on the current form instance. Each point is described by its X and Y location on the page relative to the top-left corner and its "type" (as defined in Figure 3).

The result of the registration process is a two dimensional displacement (i.e., a pair of X and Y values) which, when applied to the list of newly observed points, will produce a new list which is precisely aligned with the standard list. The basic

[1] Trademark of the Open Software Foundation, Inc.

[2] Trademark of the Massachusetts Institute of Technology.

0	2	−2	0	0	0	0	0	0
0	2	−2	0	0	0	0	0	0
0	2	−2	0	0	0	0	0	0
0	2	−2	0	0	0	0	0	0
0	2	−2	0	0	0	0	0	0
0	2	−2	0	0	0	0	0	0
0	4	0	−2	−2	−2	−2	−2	−2
0	2	4	2	2	2	2	2	2
0	0	0	0	0	0	0	0	0

cornerRT

0	0	0	0	0	0	−2	2	0
0	0	0	0	0	0	−2	2	0
0	0	0	0	0	0	−2	2	0
0	0	0	0	0	0	−2	2	0
0	0	0	0	0	0	−2	2	0
0	0	0	0	0	0	−2	2	0
−2	−2	−2	−2	−2	−2	0	4	0
2	2	2	2	2	2	4	2	0
0	0	0	0	0	0	0	0	0

cornerRB

0	0	0	0	0	0	0	0	0
0	2	4	2	2	2	2	2	2
0	4	0	−2	−2	−2	−2	−2	−2
0	2	−2	0	0	0	0	0	0
0	2	−2	0	0	0	0	0	0
0	2	−2	0	0	0	0	0	0
0	2	−2	0	0	0	0	0	0
0	2	−2	0	0	0	0	0	0
0	2	−2	0	0	0	0	0	0

cornerLT

0	0	0	0	0	0	0	0	0
2	2	2	2	2	2	4	2	0
−2	−2	−2	−2	−2	−2	0	4	0
0	0	0	0	0	0	−2	2	0
0	0	0	0	0	0	−2	2	0
0	0	0	0	0	0	−2	2	0
0	0	0	0	0	0	−2	2	0
0	0	0	0	0	0	−2	2	0
0	0	0	0	0	0	−2	2	0

cornerLB

Figure 5. Four corner templates used for detecting corner types.

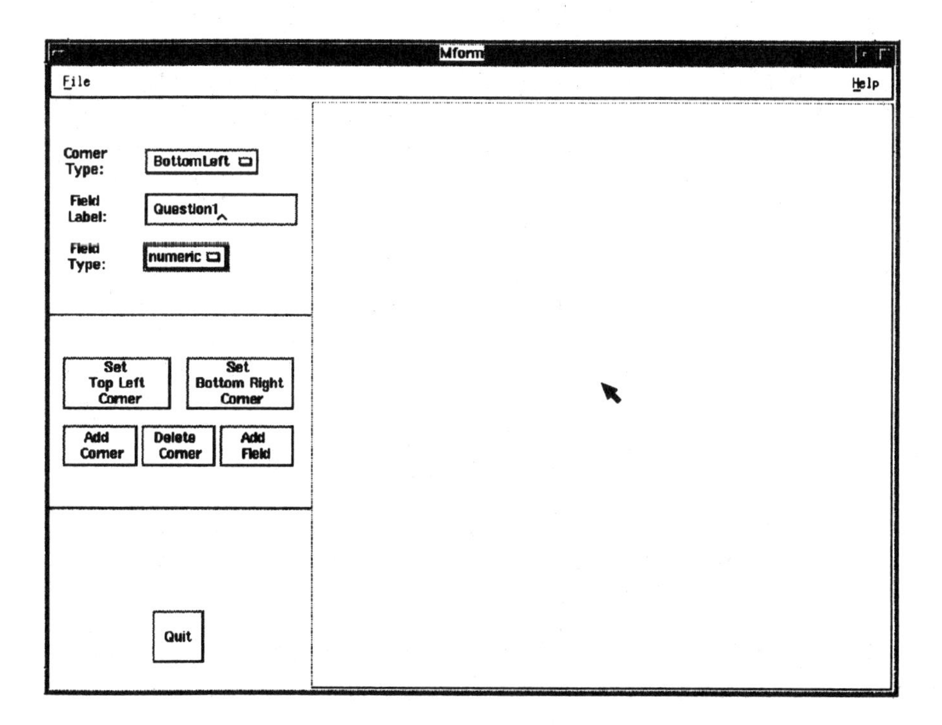

Figure 6. User interface for form model creation.

procedure is to try all reasonable displacements of the new form and select the one which produces the best results. All "reasonable" displacements of the top form consists of the set of approximately 675 pairs of points (for an IRS form) from (origin − 25, origin − 25) through (origin + 25, origin + 25). For each of these, the program attempts to find a mapping of each of the newly observed points to a corresponding one in the standard set. A point is mapped to a corresponding one if it falls within three pixels horizontally or vertically. Point scores are summed so each successful point match raises the score for that displacement.

The program does not simply select the single displacement with the highest score. Our experience has shown that it is usually possible to select an inaccurate displacement using that technique, since forms with large sets of clearly marked fields (e.g., tables created by vertical and horizontal lines) can achieve high alignment scores if they are registered "off by one column width or row height." To counter this effect, we first find a good average direction for the displacement and then select a best displacement in that direction. Specifically, this means the program first finds the direction of movement most likely to produce a good match by selecting the quadrant with the highest cumulative score. The four quadrants consist of the four sets of displacements whose values are (+,+), (+,−), (−,+) and (−,−). After selecting a quadrant, the program selects the best alignment within that quadrant.

After the proper alignment of the image with the model is complete, the coordinates of each field of interest in the model are displaced by the appropriate amount. In this way, we are able to extract all fields whether or not their boundary points were used for registration. Once the fields have been located in the image, each is checked for contents by comparing the number of text pixels (pixels whose gray level is below a threshold) within the interior of the field region to a threshold. On a 150 ppi image, the thresholds we used were: 10 pixels for a small region less than 150 pixels in area, 40 pixels for a small region which is less than 500 pixels but greater than 150 pixels in area, and 60 pixels for a region greater than 500 pixels in area. If the field is not empty, the boundary lines are removed and the data are extracted. We search the image space slightly above and below the field for chaaracters which extend outside the boundaries. This commonly occurs in alphabetic text fields containing both upper- and lowercase letters.

If this technique is to be implemented with character recognition, it would be necessary to scan the images at a higher resolution than 150 ppi, shrink the images to 75 ppi for recognition and registration, and then perform region extraction on the 300 ppi version. Because 300 ppi images are cumbersome to work with, we tested the extraction technique using 150 ppi images.

2.5 IRS Label Extraction

For the special case of IRS forms, we developed a method for detecting the presence and location of the preprinted label. The label is supposed to be placed within a specified region; however, it is not uncommon for the label to extend over the boundaries and to be affixed skewed. Therefore, in addition to locating the preprinted label, we correct for skew before input to the OCRA (OCRA is the particular character type) optical character recognition software. The label is detected by searching, through template matching, for the asterisk pattern that is common to all preprinted IRS labels (Figure 7). We search for single asterisks, and if enough are detectd, a least-squares fit determines the connecting line. If this line is not horizontal, the label is skewed, and we rotate the region accordingly. The character recognition is implemented with Unisys proprietary OCRA character recognition software. If not enough asterisks are detected, or if the ones detected do not form a straight line, then we know that no preprinted label exists. In this case, contents of handwritten label information can be extracted during the region extraction using the technique presented in Section 2.4. Each line of the label can be described as a separate field.

2.6 Paper Skew Correction

The aforementioned techniques in Section 2.1–2.4 for form recognition and form registration are sensitive to document skew. Before any corner features are calculated, we detect any rotational deviation of the form image and, if necessary, rotate the image before subsequent processing. First, the image is thresholded so that the paper is "light" and the scanner background and ink are "dark".

A bounding box is found for the image (see Figure 8). An edge finding routine selects points along the edge of the bounding box and searches inward until encountering the "lighter" paper background region. This generates a set of points along the edge of the page which are fit to a line (using a least-squares fit) and describe the edge of the paper.

The first edge to be detected is the top edge. It is found twice. Once, approximately, using a set of points clustered around the middle of the top side of the bounding box. Then a better sample set, one that is more "spread out" along the line, is generated and the procedure is repeated. The bottom and two

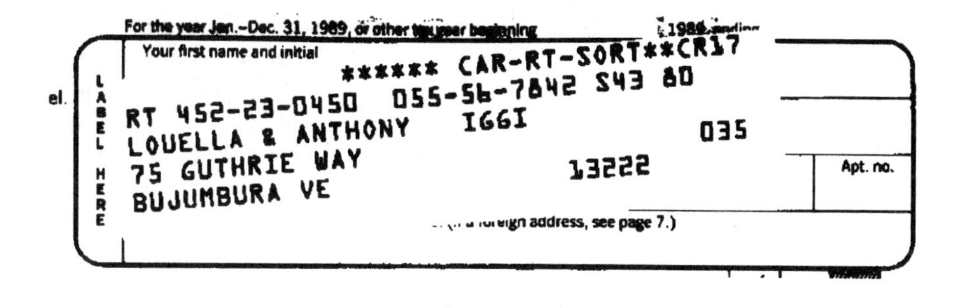

a.

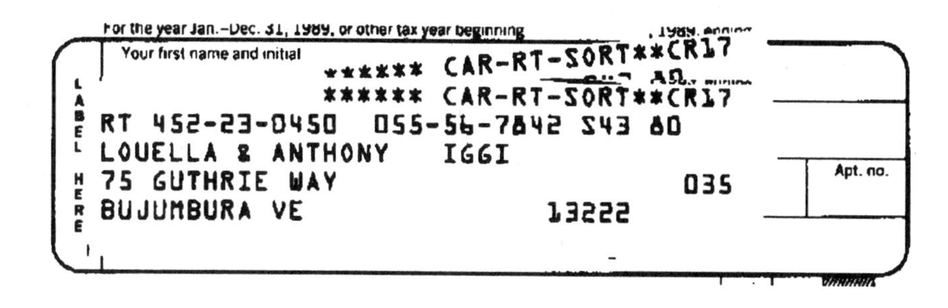

b.

Figure 7. IRS preprinted label before skew correction (a) and after skew correction (b).

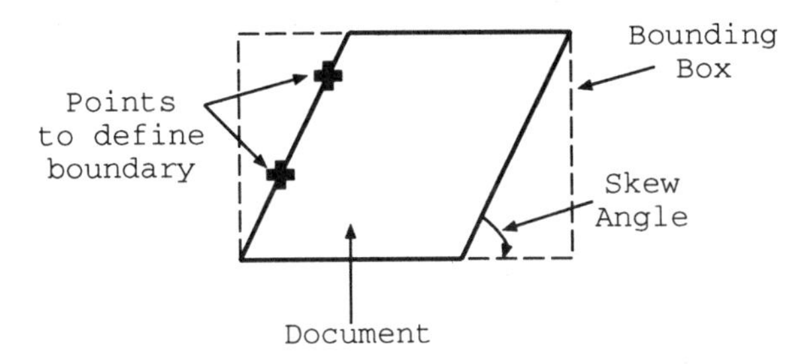

Figure 8. Detection and correction of image skew from enclosing rectangle.

side edges are found in a similar fashion. The angle of skew is easily calculated from the angle the "bottom" boundary of the page makes with a horizontal line; the image is rotated counterclockwise by this amount.

Using the detected page boundaries, we can approximate any scaling that has occurred and correct by expanding or reducing the image to fit into the expected paper size.

3 Results

Since there is no public database of filled-in IRS forms, we generated our own training and test data. Each scanned image occupies an average of about 1.4 megabytes of storage; therefore, generation and maintenance of a database have presented a significant challenge. The training and test data images were scanned from a combination of sources, including both original IRS forms and photocopies, and both blank and hand-filled-in forms. Scanning was deliberately haphazard, with shift and rotation introduced casually. Photocopies made at scales ranging from 0.9 to 1.1 were also included in our data set. The training and test sets for our experiments were mutually exclusive and selected randomly from the complete data set.

Images are captured off-line on a 64 gray-level Fujitsu M3191 Image Scanner and stored on optical disk. Image processing is performed using a set of routines written in C, some of which rely on an Androx ICS-400XM9™ signal-processing board in a Sun Microsystems SPARC™ workstation. The neural network is implemented with a dedicated accelerator board (the HNC AnzaPlus™ DP) in a networked Sun 3/160 workstation. In production, the scanner output would be fed directly to the preprocessor whose output would then be fed to a co-processor hosting the neural network.

3.1 Forms Identification

Our 232-image neural net database contains 11 samples of 21 different forms. Nine samples of each form (189 images) were used for training and two samples of each form were used for testing. Training on the 189-image training data base ceased when the change in the mean-square-error from one presentation of training data to the next was less than 10^{-4}. Initially, we compared networks with one and two hidden-layers, and with various numbers of nodes in each layer. We had the most success using nets with two hidden layers containing 40 nodes in each layer and thus adopted this architecture.

With this architecture, we achieved 98% (41 out of 42) accuracy on the test set. This figure was measured by using the largest response output node. Some "correct" identifications were made with the activation level of the best output node below 0.5, or with the ratio of the highest activation level to the second-highest only slightly better than 1 : 1. Weak identifications can be resolved in the form registration process by finding the best match to the form models in question. Saving the confidences of the classification allows us to try again with a model of the second or third choice form type if a match is not made in the initial registration stage. At any point, the system can reject the form and designate it for manual processing.

The forms that are misidentified, or identified weakly, are typically either very similar in structure to those for which they are mistaken, or have a structure that is difficult for the line-crossing algorithm to interpret. An example of the former is illustrated in Figure 9: forms 1120 and 1120-A are visually quite similar, and their line-crossing profiles will be almost identical. An example of the latter is form 1040EZ, which uses shaded boxes rather than lines to define input areas and poses a difficult challenge to the line-crossing algorithm (as well as few points for image registration).

3.2 Data Extraction

Data extraction was tested on the first and second pages of IRS form 1040, the first page of IRS form 1040A and a U.S. Department of Defense (DoD) Security Clearance form. The IRS data base consisted of 100 images of 1040 forms and 35 images of 1040A forms manually completed by our colleagues. In order to test the technique on a non-IRS form, we developed a model for a DoD Security Form which was tested on four filled-in versions. All models were created with the assistance of **Mform.**

We defined 59 different fields for the IRS 1040A form, 69 different fields for the front page IRS 1040 form, 63 fields for the back page of the IRS 1040 form, and 136 fields for the DoD Security Clearance form. It takes approximately 4 min to scan and transfer the data for an 8.5 × 11 form at 150 ppi and six grey levels on our scanner. The total time to process the form (after form type identification) is approximately 10 CPU s on a SPARCstation II.

This includes 3.5 s to load the image and correct for any skew. The most intensive operations are the correlations to locate the corner features. If the form is an IRS form with a preprinted label, an additional CPU s is needed to find the label and rotate it. We tested label location successfully on labels rotated up to 40°. The security form, which has more fields, takes slightly over 10 CPU s to process. A breakdown and summary of the average CPU times (based on eight arbitrary runs of 1040 IRS forms) to process the forms is given in Table 2. Figure 10 shows a completed IRS form with the extracted regions.

With this technique we were able to achieve document registration for every image in the data base. The total number of regions examined was on the order of 15,000 regions. To estimate error rates in region extraction, we randomly chose 23 forms which had a total of 1587 fields, 381 of which were filled-in. Of these 381 fields only 2 regions were not extracted. We also had 5 empty regions which were extracted as full. These errors occurred in regions above and below filled-in alphanumeric handwritten text fields where ascenders and descenders from the filled-in region spilled into fields above and below. These portions of the characters may be cut off during region extraction of a filled-in field or cause data to be extracted in regions which should be empty. We eliminated some of this problem by searching for characters continuing into neighboring regions above and below (as discussed in Section 2.4).

4 Conclusions

We have demonstrated a prototype system for forms identification, region extraction and registration of preprinted forms which use a line-crossing signature to define fields. These steps are an important requirement for the interpretation of the data in the forms through handwritten or machine-printed character recognition.

The line-crossing signature is used to give a new feature representation for both identification of the particular form type and registration of the input image once the form type is known. Identification of the IRS forms is accomplished with a neural network. The neural network is well suited for this problem, since we can use the line-crossing features as an appropriate reduced representation of the data, and we have to distinguish among many different types of forms. With the 1-of-N output coding scheme of the neural net, we may rank the identification results of the forms. If registration of the anticipated form model fails, we can retry with the next choice of forms. Success of the image registration using the line-crossing features guarantees the location of all data fields on the form image, even those which cannot be defined by *visible* markings on the page.

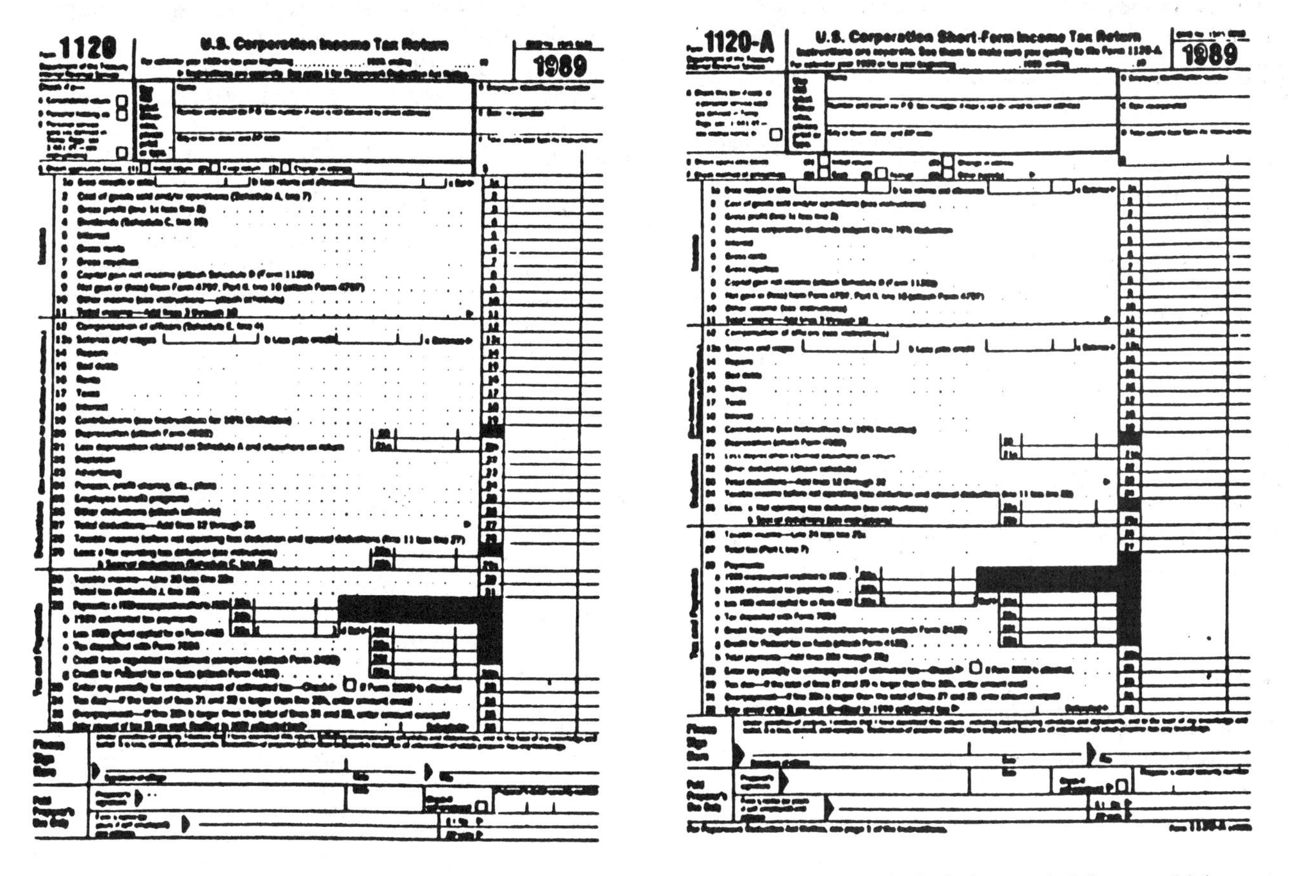

1120 U.S. Corporation Income Tax Return 1989

1120-A U.S. Corporation Short-Form Income Tax Return 1989

Figure 9. IRS Forms 1120 (left) and 1120-A (right); the quality of these images is typical of photocopied forms, which must be recognized with the same accuracy as official IRS printed versions of the same form.

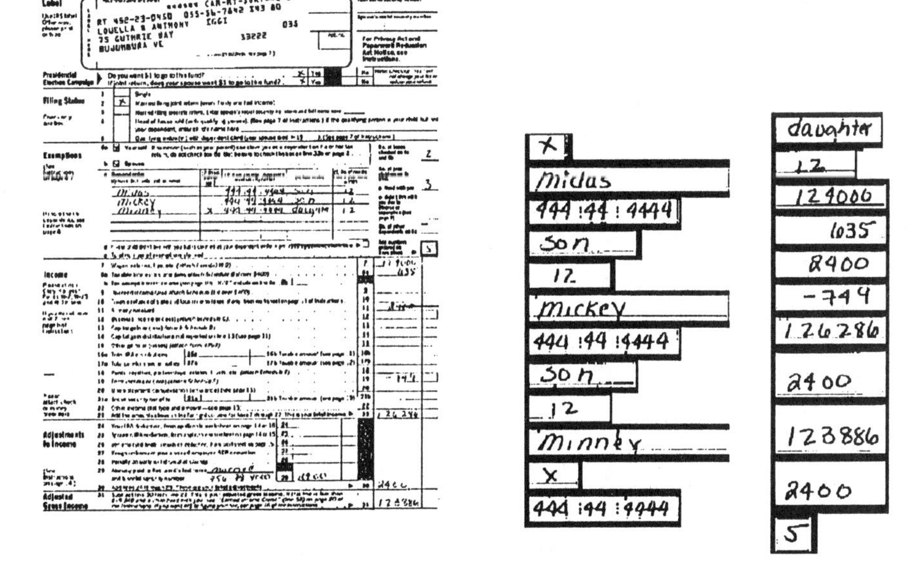

1040 U.S. Individual Income Tax Return 1989

RT 452-23-0450 035-36-7442 193 80
LOUELLA A ANTHONY EGGI
25 GUTHRIE WAY
BUJUMBURA VE 53222

Figure 10. Sample IRS 1040 tax form with extracted fields.

Table 2. Average time to process form images in CPU s. IRS forms are based an average of eight runs of 1040 forms

	IRS 69 fields	DoD 136 fields
Image loading and setup	3.5	3.5
Finding label	1.1	—
Calculating corner features	4.1	4.3
Image registration and field extraction	1.2	2.4
Total	9.9	10.2

Acknowledgments. This research was funded by Paramax Systems Corporation Independent Research and Development. The authors would like to thank Dan Harrington of Unisys Corporation for providing the OCRA software.

References

Akiyama T, Hagita A (1990) Automated entry system for printed documents. Pattern Recognition 23(11):1141–1154

Casey RG, Ferguson DR (1990) Intelligent forms processing. IBM Systems Journal 29(3):435–450

Davis RH, Lyall J (1988) Recognition of handwritten characters—a review. Image and Vision Computing 4(4):208–218

Elliman DG, Lancaster IT (1990) A review of segmentation and contextual analysis techniques for text recognition. Pattern Recognition 23(3/4):337–346

Mantas J (1986) An overview of character recognition methodologies. Pattern Recognition. 19(6):425–430

Nagy G (1982) Optical Character Recognition—Theory and Practice. In: Krishnaiah PR, Kanal LN (eds) Handbook of statistics, vol. 2. North-Holland, Amsterdam, pp 621–649

Pastor JA, Taylor SL (1991) Recognizing structure forms using neural networks. Proceedings International Joint Conference on Neural Networks

Rumelhart DE, Hinton GE, Williams RL (1986) Learning internal representations by error propagation. In: Rumelhart DE, McClelland JL (eds) Parallel distributed processing: Explorations in the microstructure of cognition. MIT Press, Cambridge, MA

Suen CY (1980) Automatic recognition of handprinted characters—the state of the art. Proceedings of IEEE 68(4):469–487

Wong KY, Casey RG, Wahl FM (1982) Document analysis system. IBM Journal Research and Development 26(6):647–656

Document Analysis — From Pixels to Contents

JÜRGEN SCHÜRMANN, NORBERT BARTNECK, THOMAS BAYER, JÜRGEN FRANKE, EBERHARD MANDLER, AND MATTHIAS OBERLÄNDER

Invited Paper

The paper presents the conceptual framework for solving the task of document analysis, which, in essence, consists in the conversion of the document's pixel representation into an equivalent knowledge network representation holding the document's content and layout. The overall system is structured into several levels of abstraction. Starting on the pixel level, the formation of elementary geometric objects is described on which layout analysis as well as the definition of character objects is based. Character recognition accomplishes the mapping from geometric object to character meaning in ASCII representation. On the subsequent level of abstraction words are formed and verified by contextual processing. Modeled knowledge about complete documents and about how their constituents are related to the application form the highest level of abstraction. The various problems arising at each stage are discussed. The dependencies between the different levels are exemplified and technical solutions put forward.

Keywords— *Pattern recognition, document analysis, character recognition, layout analysis, contextual processing, document modeling, document understanding.*

I. Introduction

Document analysis aims at the transformation of any information presented on paper and addressed to human comprehension into an equivalent symbolic representation accessible to any kind of computer information processing. This is a rather ambitious goal and it remains quite open as to which degree of perfection compared with human literacy will ever be reached. Actually, only a certain number of rather specialized tasks belonging to the above-mentioned category of problems are seriously approached and can be considered at least partially solved.

However, with the advent of inexpensive high-resolution scanning devices and the rapidly increasing availability of computing power at the office workstation, fundamental prerequisites for document analysis are fulfilled. We are convinced that in the following years document analysis will play an increasing role in the area of office automation. We have gained fast access to the document's image, resulting in millions of pixels in simple black and white fashion, in gray value, and even in color.

Within the stream of raw input data, meaningful objects and their relations must be detected. This cannot be done without having set up a world of model objects and corresponding relations explicating which conceptual entities may be encountered during analysis and how they fit together. Thus, a network of notions and conceptions is established, reflecting a mixture of general knowledge about documents, script, and typography and second knowledge which maps the relevant aspects of the application. A third and no less important ingredient to the knowledge base on which the analyzing processes are operating is the collection of procedures for image processing, iconic-to-symbolic transformation, and classification and search, which are common to different applications.

It should be made clear from the beginning, however, that document analysis is only in simple cases suited for pure straightforward sequential operation. The document analysis task must be structured into several levels of interpretation and requires a combination of bottom-up and top-down approaches. At the lower levels ambiguities are quite frequent which can be resolved only at higher levels. At every level, the generation and handling of alternatives have to be provided with the goal of minimizing their number without missing the correct one. This makes the blackboard architecture of knowledge-based systems best suited for document analysis purposes. Because of the large number of objects to be handled, the blackboard must be organized as the common data base from which the analyzing procedures can select intermediate results according to appropriate queries and where they lay down their results.

For the time being, document analysis has captured an acknowledged position only in certain market segments. From our point of view the most dominant is postal automation, followed by form reading for banking and administration purposes and page reading for general text input. The conceptual work reported in this paper has had a relevant share in bringing about the commercial success of a number of document analysis products.

Manuscript received February 1, 1991; revised November 4, 1991.

The authors are with the Daimler–Benz Institute for Information Technology, Wilhelm-Runge-Str. 11, D-7900 Ulm, Germany.

IEEE Log Number 9202964.

Reprinted from *Proc. IEEE,* Vol. 80, No. 7, July 1992, pp. 1101–1119.

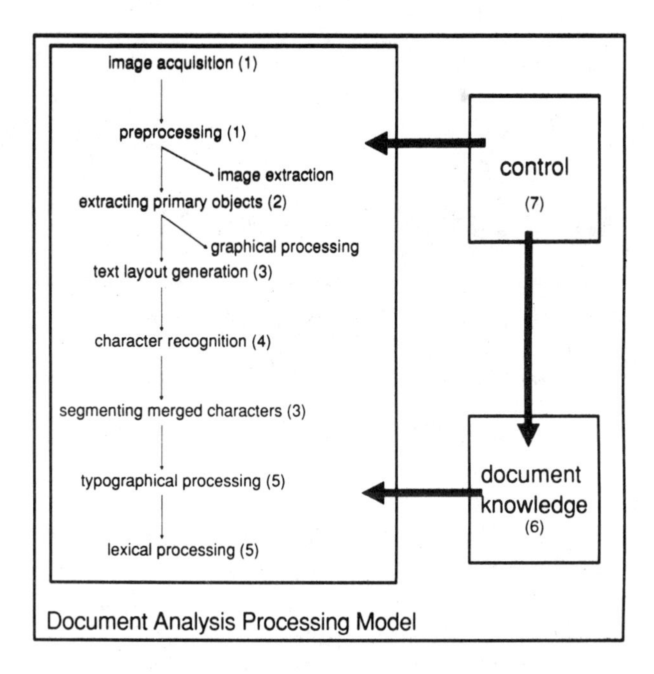

Fig. 1. The processing model of document analysis: A typical processing sequence is generated by the module control exploiting the document knowledge and specific algorithms. The numbers in parentheses denote the section in which the task is discussed.

The address reader processes the limited region of the address on letters at a very high speed and obtains the zip code along with the city name. The syntax of the address is rather restricted and the content — zip code and city name — is verified by large dictionaries. The address reader has to cope with medium to poor print quality including handwritten script. Similar conditions hold for the form reader. Again, it has to process medium to poor print quality at high speed, and the structure of the form is known in advance. However, the structure of the form and the relations between portions of the data are more complex than that of an address. In contrast to both, the document reader processes a full page comprising thousands of characters in high or medium print quality. Moreover, it may include graphics as well as pictures. The structure of the document is complex and not restricted by well-defined universal syntax.

What follows sketches an approach to the solution of document analysis tasks having evolved within an industrial research group during consistent work in this area over a long period of time. We go through the different steps of the document analysis process and outline the principal ideas as if it were a single chain of pipelined operations. In Fig. 1 a typical processing sequence is illustrated; each step is annotated by the section number in which it is discussed.

II. Image Acquisition

The first step in document analysis is to acquire a digitized raster image of the document using a suitable scanning system. Resolution in gray value and local resolution are parameters, which must be carefully chosen since they influence the analysis strongly.

The required local resolution depends on the character size expected. In standard applications, such as the analysis of office documents or address reading in letter sorting systems, a local resolution of 200 to 250 dpi has proved adequate for character recognition. For the analysis of documents with smaller fonts and the analysis of technical documents with very thin strokes (< 0.08 in.) it is necessary to scan the document with a resolution up to 400 or 500 dpi.

Normally, the interesting parts of a document (especially when analyzing text documents) are of binary nature: dark characters on a light background. But in "real-life" documents the "light" background is not really white, e.g. because of the use of colored paper or because of copying a document with a low-quality copy device. Also, "dark" text parts are not really black because of bad print or copying quality, or the use of colored ink, etc..

So, a scanning device may be necessary which acquires the image in a gray-value form digitizing the brightness of a single picture element (pixel) to $N \gg 2$ (in typical applications 256) different shades of gray. Nevertheless, the text parts are mostly darker than the background they are written on, and the document image ultimately has to be binarized.

Binarization is accomplished by comparing the gray values with a given threshold. But a decision of where to place the threshold between bright background and dark text parts cannot be made on a general and global basis, because the gray values representing background and those representing text may change from document to document and, in many cases, even between different document parts.

Approaches for binarization, applied in current reading systems, consider a certain local neighborhood (LN) around every pixel to determine the local threshold adaptively. In this LN the distribution of gray values is determined (expressed e.g. by its mean value, its maximum value, its ranking order) and as a function of this distribution it is decided whether the pixel has to be set to black or to white. The size of the considered LN depends on the expected size of typical text objects and should be at least twice the largest possible stroke width of text objects. For smaller sizes of the LN it may happen that pixels belonging to a character stroke are considered as background, because the gray-value distribution shows a homogenous form inside the LN — a typical hint at background.

Applications exist where the binarization not only is a matter of gray values and their distributions, but also where configurations of the different gray values (texture) play a role in deciding what the interesting text parts are and what the background is.

A typical example is found in letters where the address is written on patterned paper (Fig. 2). Here, it is not possible to separate the text parts from the patterned background by simply considering the distribution of gray values in a LN. In such a case it becomes necessary to analyze the texture and to derive features which describe the different pixel configurations of text parts and the background texture in order to be able to distinguish between them and to binarize the document in an adequate way. Otherwise, the

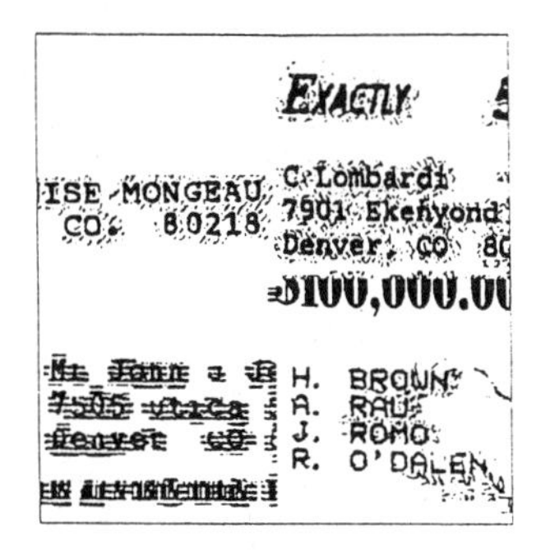

Fig. 2. Text part of a document disturbed by patterned paper and the result of texture analysis.

background patterns also would be binarized as text parts — often connected to the characters — resulting in a severe disturbance of the subsequent analysis. Binarizing such patterned documents is still a point of research [1].

Since background patterns on letters or commercial documents are often printed in colors, it may be suitable to use scanning devices which can acquire multispectral signals (red, green, blue) in order to obtain distinguishing information between text and patterned background. Again, the task of preprocessing is to transform the multispectral information of every pixel into the binary attribute — background or text. The use of color scanners in the described form is also a point of research and not yet part of current systems.

Binarization in one of the described forms is an indispensable step of preprocessing for any document analysis system. Further preprocessing steps that are specific for address reading are steps for removing the noise caused by dithered photos or badly binarized background [1]. A typical operation working on the binary raster image and doing that in a very efficient manner is the morphologic operation of opening [2], where all connected components below the size of a given structural element are removed and the larger objects remain substantially unchanged.

III. Primary Objects

For a long time pixels were considered the basic level in image processing and analysis. All low-level image operations were primarily based on pixel manipulation. Regarding an image as a two-dimensional array of either black or white dots seems a quite natural approach. It also allows for easy formulation of all sorts of algorithms which again produce raster images. This is commonly called iconic image processing. In contrast to this, the forming of higher-level entities from whole groups of pixels is called symbolic processing. Often, such symbolic entities already have an intuitive "meaning." For instance, the raster image might have been partitioned into regions representing text paragraphs. But sometimes such symbolic entities plainly are "regions of interest," no matter what they comprise or contain in particular.

Unfortunately, the number of pixels is very large. It grows quadratically with respect to image resolution. On the one hand, high resolution is most desirable in order

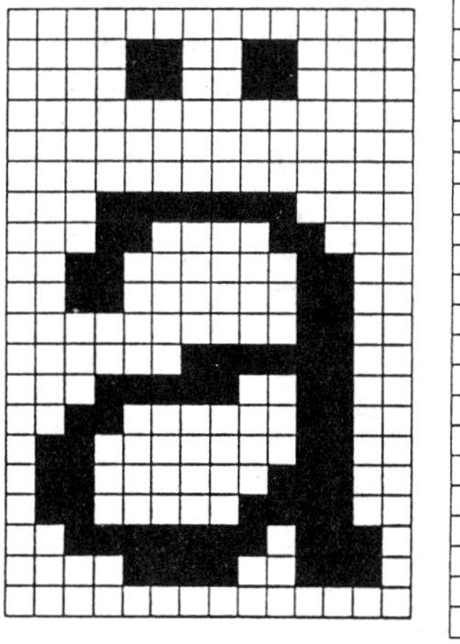

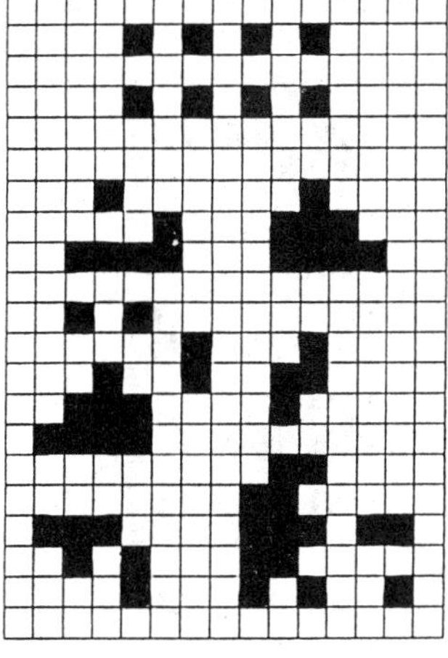

Fig. 3. Original image and its corresponding corner image.

to capture all details of a document, but on the other hand it puts a heavy burden on the processing algorithms. Many higher-level processing steps therefore assume so-called primitive objects. They do not handle individual pixels. Possible candidates of image primitives are edges, skeleton lines, and regions of homogenous texture. Many of these object species are based on heuristics, more or less vaguely defined. But for low-level or mid-level processing it is preferable to have an image representation free from any semantic connotations, but offering the advantage of reduced computing complexity.

Some coding schemes (such as one-dimensional run length coding) help saving memory since most images carry lots of redundant data. Extended areas of equal color are an example of such redundancy. A closer inspection reveals that each binary image, B, has a corresponding corner image, C. It is obtained by mapping certain 2×2 pixel combinations, called the corners, onto black pixels in the corner image and all other combinations onto white pixels. Fig. 3 gives an example. Fig. 4 shows all possible situations. There exists only one more image with an identical corner image, which is just B's inverted image. Given the corner image and the color of just a single pixel, the original image can be unambiguously reconstructed.

Usually, the number of corners is much smaller than the total number of pixels. This is particularly true for images containing text or graphics. Dithered photographic constituents, however, can prove awkward. Preprocessing steps that eliminate such "noise" are then very helpful (cf. subsection III-A).

A corner image is expected to be sparse with very few black pixels. Hence it should rather be encoded as a list of coordinate value pairs. Since a corner set fully represents the original image, it is possible to implement such standard operations as XOR, CUT, ZOOM, and SKEW on the corner set representation. As a consequence, the raster image could be discarded provided the necessary operations have been reimplemented on corner sets. We can take advantage of the smaller cardinality of corner sets compared with the total number of pixels. This makes the algorithms run faster.

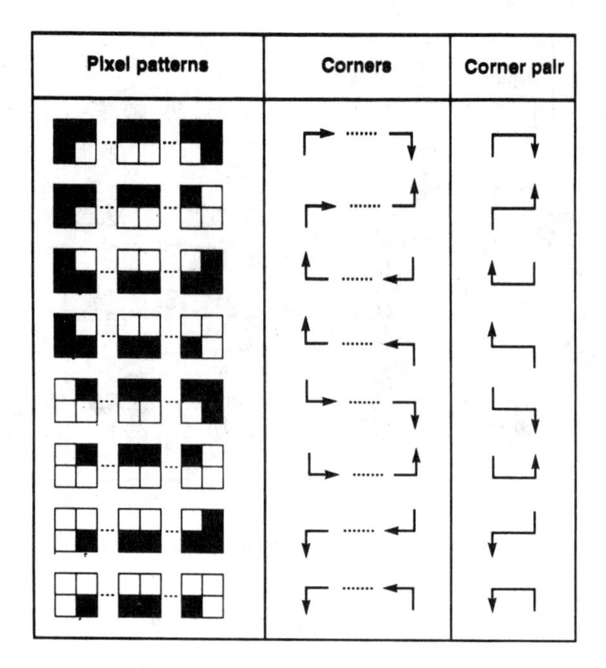

Fig. 4. Types of corners and corner pairs.

We had first defined corners as pixel configurations. Another interpretation is this: A corner is the transition of a vertical edge into a horizontal edge (or vice versa). Now, every horizontal edge must return to vertical direction before the image margin is reached. This is the underlying reason why corners can always be grouped into pairs, as was illustrated in Fig. 4. Interestingly, there are only eight corner pair types to be distinguished.

Any closed sequence of edge pieces is called a contour. A contour completely describes the shape of a region of connected pixels with uniform color, a connected component. Corner pairs can be considered as the building blocks of contours. They can be extracted from the raster image very fast and efficiently by a single-pass algorithm [3,4]. Single-pass or single-sweep means that each scan line of the image and also each corner pair in that line is accessed only once.

Since contours are always closed, the notion of inside and outside arises. Some contours may be inside, others outside of a certain contour. The nesting relation of the contours represents the image topology. The digit "8," for example, consists of one black region with two white regions within. As a consequence each region is fully described by exactly one contour plus the information which contours are "inside" or "outside." These references to inner contours can be generated during single-pass processing, too.

Geometric features such as the bounding rectangle, area, and center of gravity can be easily computed from the contour code [5]. The particular application determines which features are actually needed. The region's first- and second-order moments provide a good basis. Many other features can be derived from these values.

The results of connected component analysis, the primary objects, can be stored in a data base. The complete description of a primary object consists of three parts:

- the contour code
- the hierarchy (i.e., references to inner and outer objects)
- the geometric features.

This data base offers a flexible and versatile basis for statistical evaluation, object selection, grouping, and finally layout analysis, performed by subsequent image processing and interpretation algorithms.

Finally it should be mentioned that the connected component method can be extended from binary to multivalued images. See [4] for details.

IV. Layout Analysis

The layout structure of a document considers the geometric properties of the document's primitive objects and the relations between them. Three different layout classes are generally distinguished: text, graphics, and photographic images. The layout analysis has the task of relating the primitive objects of the document to one of these three classes and to establish further structures within one class.

It is important to note that analyzing the layout structure is one of the first processing steps that interprets primitive objects, e.g. as being textual constituents or part of a line drawing—in contrast to the generation of connected components being a mere transformation into another image representation. If this interpreting processing step fails, further steps will carry the burden of its errors and it will become difficult to correct these errors in subsequent processing steps. Consider an incorrect clustering of character hypotheses to words: a dictionary access for verifying classification results must necessarily fail, since "words" broken into pieces or put together are considered. On the other hand, if it works successfully, one is likely to obtain correct final results. Therefore, the hypotheses have to be verified or rejected during subsequent analysis. Although this section covers the analysis of the layout structure independently of higher recognition levels, it is obvious that the low-level analysis must be interwoven with higher levels. Subsection VIII-B gives a more detailed insight into this concept of processing.

This section mainly focuses on analyzing the structure of text, e.g. in order to find lines and character objects. It is assumed that the text is either typeset or typewritten. Structuring graphics are briefly discussed, together with the extraction of photographic images.

A. Structuring in Graphics, Image and Text

The distinction between graphics, (photographic) image and text parts — as obvious and natural it seems to be — can be quite difficult in many applications. The task is simple when there are closed, nonoverlapping — in the optimum case rectangularly shaped — regions of different content which can first be separated and then be classified as text, graphic, or image regions. However, the question is how to handle the case when text and graphics interflow, when constituents of a line drawing can be legitimately called text as e.g. the lettering of a coordinate axis or when text is printed or written on a pictorial background, which may happen on financial check forms.

The task of recognizing the graphic and photographic document parts can be separated in two different cases:

1) to find out closed (mostly rectangular) areas consisting mainly of photos and to extract the whole segment;
2) to extract photographic and graphic parts settled around or even connected to the text information of interest (e.g. the address on a letter).

For the first task a suitable approach [6] is to make use of the regular horizontal structure of text areas and the inhomogenous structure of photographic and graphic parts. Applying several steps of smearing (closing the white gaps) in the horizontal and vertical directions, the text parts (lines) become connected regions of nearly rectangular shape. The graphic and photographic parts, however, become large inhomogeneous connected regions with arbitrarily structured shapes. Geometric features measuring the rectangularity of these connected components can be determined and can be used to separate text areas from nontext areas.

For the second task — eliminating photographic and graphic parts that are close or connected to the text information — this method is not suitable because the smearing operation could connect the text parts and the photographic parts and would pose the danger of the text parts being eliminated together with the photographic parts. Methods applicable for this task [1] are either working on the raster image, filtering out small connected components of dithered photos by local neighborhood operations, or using the description of the connected components, separating the possible text objects from photographic and graphic objects according to their geometric features.

Graphic parts disturbingly connected to the text information often appear in applications of address reading or form reading, where e.g. the address or the check amount is written on a given base line. A possible approach separating the characters from the base line is to examine the contours of the connected components, to approximate them by straight lines, to eliminate those lines belonging possibly to base lines (because of the unique direction), and to restore the remaining lines into connected components possibly referring to characters.

Beside the interpretation of the textual content of documents, a second important application of document analysis is the analysis of technical drawings containing text and graphic information of equal importance. The primary task in those applications is to distinguish between text and graphic parts in order to apply different analysis steps. The most common way to make a first separation is to consider the size of the bounding rectangles of the connected components. It works successfully in many cases since text on technical drawings mostly consists of printed or hand-printed characters. However, in cases where characters are connected to line structures or where small graphic parts have nearly the size of characters, this technique fails. The decision has thus to be revised in a subsequent analysis step, e.g. if the separation between linelike and regionlike graphic parts yields no acceptable result or if a graphic object — considered as a character in the first step — is classified with very low certainty. For the analysis of the graphic parts of a technical drawing, the next processing step would be vectorization, transforming the given representation into a description of suitable geometric primitives as straight lines and circle arcs. These primitives are the basic elements of a vector-oriented description of technical drawings. They build the basis for further analysis steps as a model-based recognition of higher-level graphic symbols. A more detailed discussion of the many research activities in this field [7], [8], however, is beyond the scope of this paper.

The set of object primitives not being hypothesized as part of a photographic image or graphic is hypothesized to represent textual constituents. Before the structure of the text is obtained, a test is carried out to find out whether the document is skewed. The method is based on the technique mentioned in [9], which is adopted on a specialized Hough transform reducing the parameter space to one dimension. The algorithm calculates histograms — based on connected components rather than on pixels — in steps of e.g. 0.5°. The energy is calculated for each histogram and the histogram with the maximum energy value specifies the skew angle. After rotating the image by the skew angle the text can be structured into its lines and words.

The many variations of line finding algorithms can be roughly distinguished by two categories: global and local approaches. The principle of local methods is to start with some seed object and then jump from one object to the next. For this technique, a distance measure is needed plus some threshold values in order to define in geometric terms what the next right neighbor should be.

In contrast to the local approach, global methods are based on features common to a group or cluster of objects. For instance, the row coordinate of the object's center is a feature that is very similar for all objects within the same line.

The drawback of the local methods is that a specific metric has to be chosen first and must then be completely relied on. This works well in some, but not in all cases. The mere Euclidean distance measure may fail, because the vertical line offset is often smaller than the horizontal character spacing. Consequently, a kind of weighting must be introduced. Also, punctuation or diacritic marks within the text may easily mislead the local algorithm. Many further heuristics are necessary to get the method working in a reliable way.

A very well known global method for line finding is the Hough transform [10]. The principal idea is that each individual pixel may be part of many potential (geometric) lines, which have different slopes, but have this particular point in common. Since on a digital computer a discrete resolution is always used, only a limited number of different slope values are possible. This leads to a finite two-

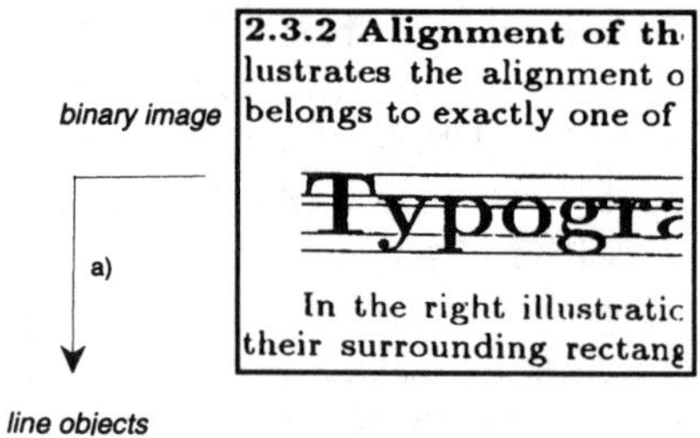

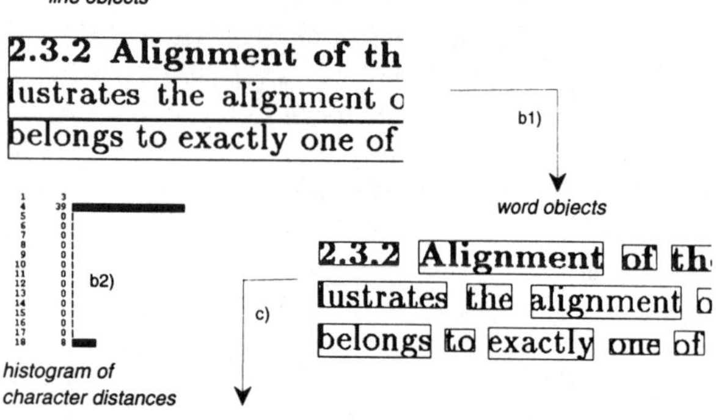

Fig. 5. A portion of a binary image including a graphic object, lines, words, and characters generated. The histogram shows the distribution of the character spacing in pixels after filtering: two distinct peaks represent the intracharacter spacing and the intercharacter spacing.

dimensional parameter space with a form similar to a raster image. Peaks in this parameter space correspond to lines in the original image. But the constituents of these lines need not be all connected. Dashed lines can be recognized in the same manner as solid lines. It is also possible to regard the individual, well-separated character objects as the basic constituents of a (text) line.

A special case of the Hough transform is the projection of the whole image onto the vertical axis. The parameter space reduces to one dimension. This is applicable if we assume the document to be well aligned after skew correction. As long as the ratio between line spacing and line skew is such that the peaks on the projection axis do not overlap, this method is feasible.

A third approach to line finding might be expected from morphology, e.g. by dilatation and thinning methods. But the objections are the same as with the local methods. The meaning of gaps, for instance, are context specific. How far should a region grow?

Having once generated a line, the words within a line are determined by inspecting the intercharacter spacing. A histogram over this spacing contains two significant peaks, one for the interword spacing and one for the intraword spacing.

Fig. 5 shows the result of a line finding algorithm for a small document portion along with an example of a histogram for detecting words in a line. The algorithm employs the one-dimensional Hough transform, thus calculating the vertical profile of the row coordinates of all connected components. This histogram is filtered in order to extract well-defined discrete peaks. After having sorted the objects within one peak from left to right the histogram of the distances between two consecutive objects can be computed, defining finally the words in a line.

B. Building Character Objects

The operations described in this subsection are restricted to typewritten character objects. Owing to the nature of handwritten characters, the application of more heuristic rules is necessary when structuring handwritten text into character objects.

The bases for building character objects of typewritten text are the connected components extracted from the binary image (cf. Section III). The first task is to identify the geometric region where a character is supposed to be present. If the lines and the words are generated beforehand, the desired region is limited to the position of these objects. Otherwise, it becomes rather crucial to hypothesize character objects in such an unrestricted environment.

Knowing what region to investigate, the assumption holds in many cases that one character corresponds exactly to one black connected component, e.g. the (undistorted) characters "a," "b," "c," "d," etc. — a very helpful assumption which facilitates life very much. Then, the task of building character objects is completely accomplished.

In the following, a method for generating character objects — a primarily typographical approach — is presented which relies on more or less good print quality of the document and is based on the set of connected components. Not only is the pure character object built; also the line structure is generated and typographical information is attached to each character.

When starting the algorithm, almost no reliable knowledge is available. Even the height and the width of the letters are unknown as well as special measures such as base-line skip. So, this approach must be able to perform several steps simultaneously:

1) determining the set of connected components of exactly one line of text;
2) assigning a typographical label to each component;
3) building letter images from connected components.

All letters found in European alphabets may be associated with one of eight typographical classes according to the position within the five writing lines (see Fig. 6): lowercase letters (e.g. m, n, o; but not: l, f, g), uppercase letters (e.g. A, B, P, l, f), lowercase with descenders (e.g. p, g, q), full range symbols (e.g. (,), {), symbols below the base line (e.g. _), symbols at the base line (e.g. ., ,), symbols around the middle line (e.g. -, =) and above the top line of lowercase letters (e.g. ", ¯). According to the height of the members, the classes may be grouped into two superclasses: symbol (the first four typographical classes) and satellite (the remaining four classes). For regular documents this class discrimination is sufficient. In special cases, for instance mathematics, the set of classes must be extended.

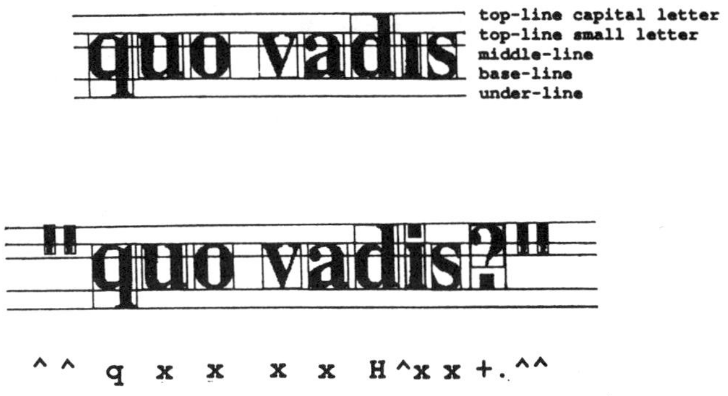

"quo vadis?"

^ ^ q x x x x H^x x +.^^

Fig. 6. The definition of the five writing lines and the geometric relationships

The algorithm generates line by line beginning from the top of the document. Using the height histogram of the connected components, a classification into the two superclasses can be done. According to the current position and the height of the symbol-class members, an analyzing window for the document is defined. This window comprises a set of connected components containing at least all letters of one line, but must not contain components of the next but one line. Thus, the region may comprise the connected components of up to two complete lines.

Setting aside the members of the satellite class, all others are ordered according to the coordinates of the left side of the bounding rectangle — determined during the connected component analysis. Each connected component in this sequence is hypothesized as a member of one of the four symbol classes. For each such hypothesis, the remaining three writing lines are calculated from the two given writing lines (top and bottom of the bounding rectangle). Applying a measurement, d, to the five writing lines of two consecutive character positions, a measure of fit is determined. Unfortunately, it is often impossible to come to the right decision based on the local distance measure. Therefore, only the best combination over the whole line will give reliable results. This can be achieved by the Viterbi algorithm without evaluating each possible combination [11].

The Viterbi algorithm uses the so-called trellis defining all possible transitions from one state into the consecutive one. For each connected component a trellis column exists containing four states. All transitions from one column to the next are allowed, i.e., 4×4 fittings for each pair of consecutive letters! The final score of a path through the trellis is given by the combination of the local distance measurements, d. The Viterbi algorithm accomplishes the task of finding the path with minimal costs — the path with best fitting symbol classes.

Keeping in mind that the analyzing window may contain symbols of two consecutive lines, the Viterbi algorithm must be able to skip those components completely which do not belong to the current line. This leads from a regular trellis to an irregular one. The number of consecutive skips, however, must be limited in order to limit complexity. In practice it is sufficient to skip up to three components in sequence. When investigating the intermediate results in the columnwise processing, another problem becomes evident: Over longer periods, several equally good alternatives may lead to a final path which is the best one by pure coincidence. Therefore, several paths are traced simultaneously (n-best Viterbi [12]), which can be done in an efficient manner within one pass.

The objects of a solution path constitute a line. Objects belonging to the line below should have been omitted by the algorithm. The components of the solution path fit into the writing line model. After constructing a line consisting of the objects of the path, the remaining objects, i.e., the objects in the satellite class, are inserted. Simultaneously, a typographical label is attached, which is derived from the relative positions within their writing lines. Horizontally overlapping parts are combined to letter objects if the typographical class membership is appropriate. The typographical class for the resulting character object is given by the union of the typographical classes of the connected components involved.

The character images, together with their typographical class labels, are now ready to be fed into the character classification module.

In the following, further approaches for building character objects are discussed. In particular, problems are focused on if the assumption that one character corresponds exactly to one connected component does not hold. Two kinds of problems can be distinguished in this scope, both falling into two different cases:

1) Several connected components constitute one character:
   - a character defined to consist of more than one component,
   - broken characters.
2) One connected component represents several characters:
   - ligatures,
   - merged characters.

Problem of the first class include characters broken up in more than one connected component. For example, a character typed in a dot matrix font style is represented by a set of discrete dots. Dependent on the print quality and the scanner resolution, the character is composed by a set of connected components—for matrix font style in the worst case by each single dot. Additionally, some characters belong in this class of problems which are—by definition—made up of several constituents; examples of these characters are the lower case letter "i," composed of the base stroke and the "i" dot, and certain characters of different languages bearing diacritic marks, e.g. the French accents. In contrast to the characters accidentally broken in pieces, the number of the segments is exactly defined and each of the segments of these characters can be treated like a complete character.

In the second set of problems, there are two subsets. First, there are ligatures which are intentionally typed as an entity. They should be treated as special character objects and no attempt should be made to separate them into their constituents. Hence, ligatures should be included in the set of classes the classifier is adapted to. Also, merged characters have this property of being represented by a single connected component but containing more than one character. In contrast to the ligatures, merged characters occur accidentally; they may emerge from too low a resolution rate, from scanning errors, or from distortions in the document.

Considering the first problem of the first set, two different approaches are possible. First, the connected components building a character are combined into one binary image by reconstructing the primary objects and merging the images. This image is then passed to the character classifier for recognition. Consequently, these merged patterns must be included in the learning set of the classifier to allow a correct treatment. The second approach treats the segments independently; they are passed to the classifier, recognized, and combined to one character according to rules which take into account the geometric relations of the segments and their recognition results.

The advantage of the first approach is that it works even in the case where the segments are merged beforehand rather than being separated. This character is then considered as a single character and is passed as an entity to the classifier. The rule-based approach would fail here. However, the rule-based approach allows an introduction of new (e.g. diacritic) characters without introducing completely new character classes into the classifier system, which would require an adaptation phase. If the classifier is able to recognize the segments, a character composed of some segments can be defined by the combination rules.

When a character is broken into several pieces—the second problem of the first set above—the pieces cannot be treated as a complete character, in contrast to the latter examples. Thus, it is not possible to follow the rule-based approach just explained. Moreover, a simple merging of small pieces as described in the first method is also not feasible, since it is not clear beforehand which pieces belong to one character and which pieces are part of a different character.

Thus, a more advanced technique is necessary to solve this kind of problem: grouping pieces of characters to an entity representing a character. Surprisingly, the algorithm presented in the following is also able to segment merged characters into its proper constituents, although this task seems to be quite the opposite of grouping pieces to an entity. However, the reason lies only in the point of view taken: When regarding the image containing the merged characters as pieces of pixel columns, the task of segmenting characters then turns to be a task of grouping pixel pieces to a character.

The algorithm is discussed in [13] in detail for the application of segmenting merged characters and is briefly explained here. It performs a "hypothesize and test" cycle. In each cycle entities are hypothesized by grouping different pieces to a single object; these hypotheses are tested by several evaluation algorithms, called experts.

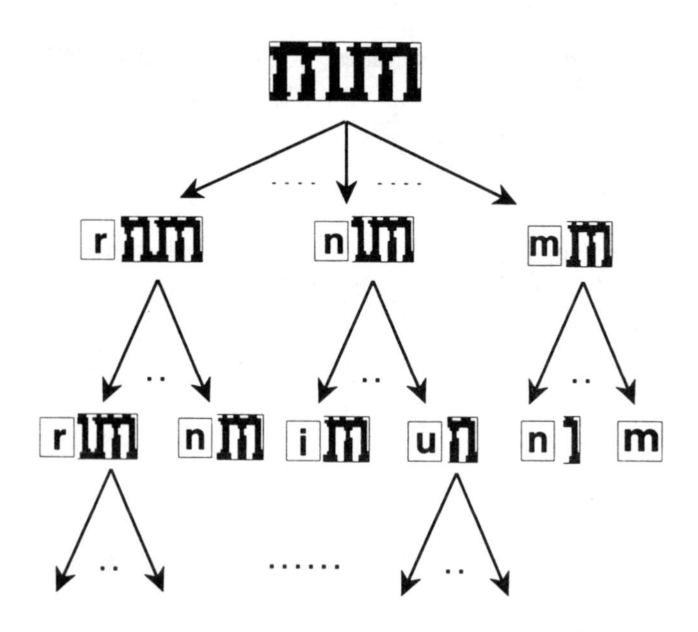

Fig. 7. Search space comprising character hypotheses.

The underlying system architecture is a blackboard; it contains a set of experts that evaluate each hypothesis as to whether it represents a valid character. These experts comprise the character classifier and a set of algorithms for context processing (cf. Section VI). The blackboard structure allows a processing that involves backtracking rather than a straightforward processing: previously taken decisions can be canceled and alternative grouping results can be pursued. Thus, the grouping process can be regarded as a search process looking for the correct solution in an efficient manner by applying as many knowledge experts as necessary. This search is controlled by a best-first graph search method known as the A(*) algorithm (see [14] and [15]). The search space is illustrated in Fig. 7 for the application of segmenting merged characters.

When segmenting merged characters, two cases have to be distinguished. If the text is printed with proportional spacing, segmentation is difficult, since it is not known beforehand how many characters are enclosed and where the positions for cutting are located. In contrast to this, minor problems arise when the text is typed with fixed character spacing: the width can be deduced from the characters correctly segmented and the pattern can be cut at these equidistant positions. The grouping algorithm is the appropriate tool only for proportional fonts. The algorithm has been tested on several thousand character patterns in different fonts and sizes containing from two to seven merged characters. The search space remains during segmentation in a reasonable size, around ten nodes for two merged characters and around 70 nodes for patterns consisting of five to seven characters.

V. Single Character Recognition

In document analysis, a main task is the reconstruction — classification — of the textual information. After some theoretical reflections on the classification task, the two main parts of the classification — the normalization and the proper classification — are described.

Classification is an act of abstraction. It establishes a mapping from a normally high-dimensional feature space into the discrete space of class labels. This mapping — in the case of character recognition — is the inverse of printing or writing. The mapping is many-to-one and should be capable of producing the correct label irrespective of what the current character looks like. The dominant problem is to cope with the variability in character image appearance. The regions in the feature space in which the feature vectors of one character class appear are complex shaped regions which must be mapped into one and the same point in the space of character labels.

Variations in appearance may be caused by a number of different sources. Some may be modeled by regular transformations. It is a generally accepted fact that character class membership is independent of character image size — at least within certain limits. There are some other influences which leave the character meaning unchanged — such as translation, rotation, slant, and stroke width. Even if there are exceptions requiring additional measures, it sounds reasonable to look for features which are size-invariant, slant-invariant, etc., and to base the classification procedures on them. The simplest way of providing features with such invariance properties is to introduce normalization procedures between the step of gaining the first set of raw features and the recognition step.

This approach divides the total classification process into preprocessing and recognition, and implements the mapping from the raw image representation to character label in two sequential mapping steps. The positioning of the border between preprocessing and recognition is an engineering design decision. Also, extreme cases are conceivable in which each one almost totally replaces the other.

In a reasonable design preprocessing applies normalizing transformations and reduces certain well-defined variations as far as possible. The inevitably remaining part of variability is left to learning from examples to be provided by statistical adaptation of the subsequent recognition system.

A. Normalization of Character Objects

The wide variety of different character shapes which is not produced by statistical processes (scanning, writing) but deliberately created by the writer or the font designer, or even systematically produced, can be reduced by the inverse transformation. The goal of these inverse transformations is the reduction of the intraclass variance. Some of these systematic transformations are (see also Fig. 8):

1) rotation (scanning with a sloping position, writing in graphics),
2) slant (italic and slanted script, handwriting),
3) stroke width (boldface, lightface, and broad script, handwriting)
4) size (headline, indices, handwriting).

To reduce the influence of these sources of variability, the underlying transformations must be detected and the parameters of the inverse transformations must be determined. Therefore, the images of the characters must be analyzed as well as the relations of the images to each other.

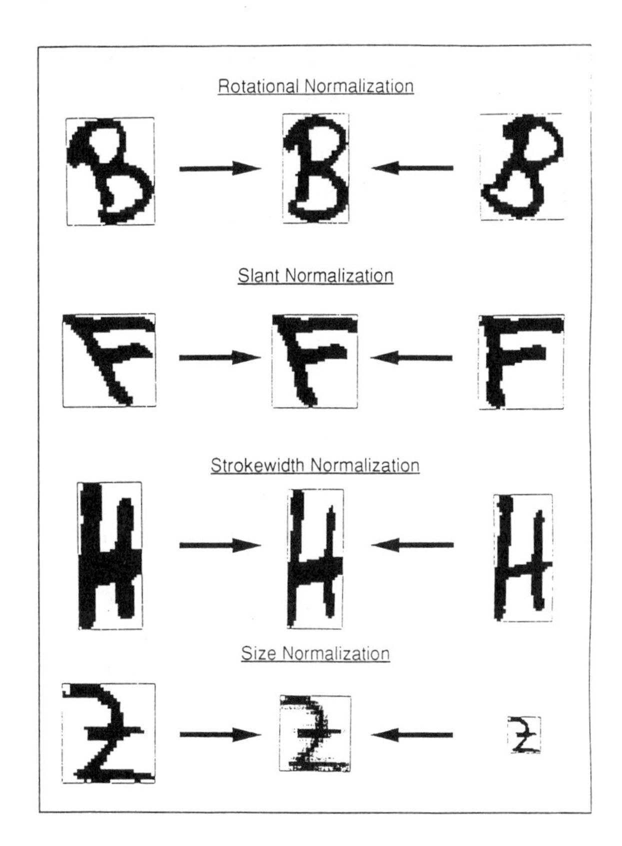

Fig. 8. Overview of normalization procedures.

Rotation: If the base line of the written text differs by a certain angle from the horizontal scanning axis, then a rotation by this angle normalizes the document. Rotation at this stage can be omitted if a global skew angle has been detected for the whole document and hence, has been eliminated before (cf. subsection IV-A).

Slant Normalization: Especially in handwritten documents, the characters are often written with slant. This does not correspond to a rotation of the character, but is more similar to a shear transformation. For each character the parameter of the shear normalization must be determined. Unfortunately, the value expected by humans requires a knowledge of the character class of the underlying character. But for the classification task it is not necessary to determine exactly what the human viewer would call the slant angle. If the parameter of the shear transformation is determined in such a way that it reduces the intraclass variability, then this problem is sufficiently solved for the classification task.

Therefore, determining the regression line of the character image — distribution of black pixel — is a mathematically well defined operation which can be performed in every case. In some cases, this value even corresponds to the value humans expect, for instance for the character "H." Normalizing all characters in this way, the variability of the output images decreases within each character class.

Stroke Width Normalization: Determining the real stroke width of a character is not a trivial task since the stroke width can differ within a single character. If the stroke width of the character should be normalized very precisely, the partial strokes having the same stroke width would have to be detected in order to calculate the proper current stroke widths. But this would be a very time consuming task.

Rather, the real stroke width is approximated in a more global sense. It is assumed that each character is written with a fixed stroke width and that each character shape can be modeled as a line (rectangle) with length l and width w. Then the black pixel areal, A, equals $l \times w$ and the circumference, C, of the shape is equal to $2(l + w)$. From the preprocessing (cf. Section III — connected components) the values of A and C are already known for all connected components. Solving these equations, the stroke width, w, of the character model can readily be determined as well as the virtual stroke length. The proper normalization is then performed either by erosion or dilatation procedures on the binary image. Although the assumption of constant stroke width — even for one character — is often not fulfilled, characters from the same class are nevertheless normalized to very similar shapes and this is the aim of the procedure. Thus, after applying the procedure given above, the mental concept stroke width varies between the different character classes but not for different instances of the same class.

Size Normalization: The class membership of a character normally does not depend on its size. Even the aspect ratio may vary. Therefore, it is reasonable to standardize both. To condense the given binary shape information, the $m \times n$ binary image is transformed into a 16 × 16 grey-value image. Each grey value corresponds to the average of the black and white pixels in the corresponding area of the binary image.

All of the normalization procedures introduced above can be carried out on different representation forms of connected components, for example on the borderline chain code, the corner list, a row coding, or in the pixel raster itself. Each of these representations may be more or less well suited for executing the normalization procedures.

The normalization procedures simplify the classification and enhance the performance. However, some drawbacks of the normalization must be taken into account. For example, some character images are indeed size and aspect ratio dependent, at least with reference to their neighbors ("C" and "c"). Therefore, it is important for a document analysis system to store the original values of the images (height, width, slant, stroke width, angle) or the parameters for the different normalization procedures for later use by postprocessing algorithms.

B. Classifier Design

The isolated and properly preprocessed raster image of the single character is sent to the classifier in order to receive the, hopefully, correct class label or at least an ordered set of class labels accompanied with confidence values.

The raster image is viewed as a vector $\boldsymbol{v}$ of measurement data which constitutes the input variable to the classifier system. The classifier output variable is called $\boldsymbol{d}$.

In mathematical terms, the operation of the classifier is to implement a mapping from measurement space $\{\boldsymbol{v}\}$ to decision space $\{\boldsymbol{d}\}$. Whereas the measurement space is given by the data type of the input variable $\boldsymbol{v}$, the decision space must be suitably determined in order to get the capability of representing class labels and confidence values. From these considerations, the decision space gets the form of a K-dimensional Euclidean space, with K being the number of classes to be discriminated. The K orthogonal basis vectors spanning the decision space are taken as representatives for the K classes. The K components of the discriminant vector $\boldsymbol{d}$ have the property of being class-membership indicators approaching zero if the character image clearly does not belong to that class, and one if it clearly belongs to it.

This conception is general enough to embrace any of the known recognition schemes, even those based on topological features and deriving the decision from going through a set of rules. Following that approach, the features are determined based on a set of prescriptions suitably chosen from the given input variable $\boldsymbol{v}$ and therefore constitute a heuristically defined mapping from measurement space $\{\boldsymbol{v}\}$ into feature space $\{\boldsymbol{f}\}$. This first mapping is followed by a second one from feature space $\{\boldsymbol{f}\}$ into decision space $\{\boldsymbol{d}\}$, which again is accomplished by the set of rules and the corresponding inference mechanism. The functional type of the overall mapping, however, in such a case is far from being a simple linear transformation from $\{\boldsymbol{v}\}$ to $\{\boldsymbol{d}\}$ and is distinctly beyond a closed formulation in mathematical terms.

The conception of constructing a mapping from measurement space $\{\boldsymbol{v}\}$ into decision space $\{\boldsymbol{d}\}$ is nevertheless a valuable guideline for classifier design. The idea is to determine a suitable type of mapping function, $d(\boldsymbol{v})$, with sufficient flexibility provided by a large enough number of coefficients contained in that function. Additionally, a rule must be constructed in order to determine how these parameters have to be adjusted so as to produce a classifier system capable of solving the current recognition problem as described by the set of learning samples.

This conception turns the task of classifier design into a problem of mathematical optimization. The optimization is carried out on the basis of the given learning set, which represents a list of pairs $[\boldsymbol{v}, \boldsymbol{y}]$, with $\boldsymbol{v}$ the given measurement vector. Vector $\boldsymbol{v}$ shows what the character to be recognized looks like and $\boldsymbol{y}$ denotes the desired value, which ideally should be the outcome of the classifier function $d(\boldsymbol{v})$.

Common mean-square optimization leads to the minimization of

$$\boldsymbol{E}\left\{|\boldsymbol{y} - d(\boldsymbol{v})|^2\right\}$$

with respect to the set of classifier coefficients defining the current form of $d(\boldsymbol{v})$. $\boldsymbol{E}\{..\}$ stands for the mathematical expectation which for practical purposes is replaced by arithmetic mean.

There are different types of conceivable discriminant functions $d(\boldsymbol{v})$, resulting in different "families" of classifier systems with different properties.

The first and most general way of following these lines of argumentation is to insert an arbitrary analytical function $d(\boldsymbol{v})$ in the above given optimization criterion, and to solve the corresponding problem of calculus. Under the conditions given here, the resulting $d(\boldsymbol{v})$ is the vector of *a posteriori* probabilities

$$d_{\text{opt}}^T = (p(\text{class}-1|\boldsymbol{v}),\ p(\text{class}-2|\boldsymbol{v}), \cdots p(\text{class}-K|\boldsymbol{v})),$$

which, according to the considerations of statistical decision theory, contains all of the necessary information for deciding on the class membership and the number of alternatives to be forwarded to the higher levels of processing. The components of d in this case are the best class-membership indicators one can imagine. By setting those components with negligibly small values down to zero, the set of alternatives together with their confidence values is easily obtained.

By using functions $d(\boldsymbol{v})$ which are defined by sets of coeffients the above given optimization problem is changed into an ordinary minimum problem with respect to the set of coefficients. Depending on the family of functions used for the mapping, the minimum problem can be solved analytically or at least by steepest descent gradient search procedures. Such constrained mapping functions $d(\boldsymbol{v})$ can no longer generate true *a posteriori* probabilities but rather approximations thereof. They can be constructed in such a way that the solution with increasing numerical effort converges to the optimum one. This property is called that of being a universal approximator, which is true for both families of functions discussed in what follows. In order to exhibit sufficient discriminative power, the mapping function $d(\boldsymbol{v})$ in all but the simplest applications must be nonlinear.

The family of functions presently attracting most of the attention in the pattern recognition community is that of the feedforward neural networks. Solving the optimization problem by steepest descent gradient techniques leads to the well-known back-propagation rule. Except for linear networks (no hidden layer) a direct solution is not known.

We prefer another type of discriminant function — polynomials. This leads to the concept of the polynomial classifier, which clearly separates the linear and nonlinear constituents of the function $d(\boldsymbol{v})$. Polynomials can be easily made nonlinear but with the remarkable property that even a nonlinear polynomial remains linear in its coefficients. Because of this property, the optimization problem for the polynomial classifier can be directly solved, leading to a set of linear equations for determining the polynomial coefficient matrix. This, however, does not exclude gradient descent type solutions, which may be appropriate in certain cases.

The direct solution is unambiguous and has some additional advantages since it renders an ordering scheme for

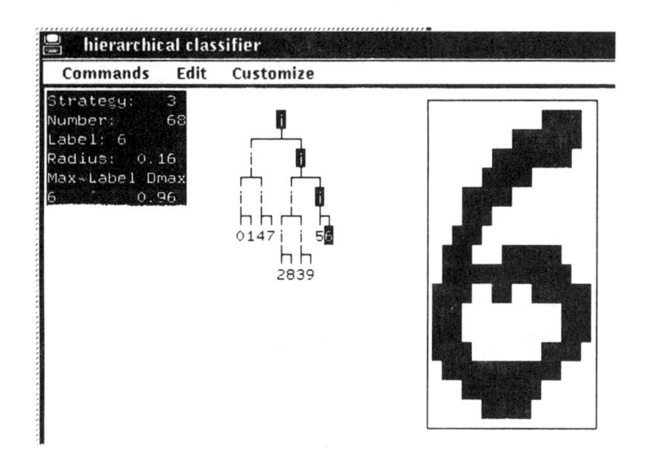

Fig. 9. Hierarchical classifier for the numerals 0,...,9. The raster image of a "6" is classified by activating those classifier nodes—indicated by inverse exclamation marks which are emphasized by a black cursor. The estimated *a posteriori* probability of being a "6" is 0.96%.

all of the polynomial terms — the features of this type of classifier — according to their contribution to solving the recognition task at hand. At the same time those polynomial terms are discarded which turn out to be linearly dependent upon others and are therefore absolutely redundant. The details of this technique are described in [16].

The polynomial classifier approach as just sketched leads to a single-stage procedure. The measurement vector $\boldsymbol{v}$ is in one step mapped into the discriminant vector $\boldsymbol{d}$ with components representing estimations for *a posteriori* probabilities or confidence values for all of the classes to be discriminated. This operation is mainly a matrix vector multiplication and can be executed in parallel up to a very high degree.

There are, however, possibilities to partition the set of class labels into subsets and construct a hierarchy of polynomial classifiers (see Fig. 9). This structure operates very similarly to a decision tree but with the difference that no hard decisions are forwarded along the tree branches but rather conditional probabilities to be multiplied until, at the tree leaves, estimations for the *a posteriori* probabilities are obtained [17]. This kind of classifier architecture enables the computing effort to be directed to those branches of the classifier tree which will most probably generate the relevant results, thus omitting redundant operations. A further advantage is that most of the individual classifiers operate on smaller sets of class labels, leading to increased overall recognition accuracy.

By combining several hierarchical classifiers with different tree structures, a further generalization of the classifier architecture can be obtained. One remarkable design is that of the classifier net constructed from as many individual classifiers as there are pairs of class labels. Each of them is trained only from examples of the two attached classes. The estimations of the individual classifiers are combined in order to again obtain estimations for the *a posteriori* probabilities as with all of the other architectural concepts.

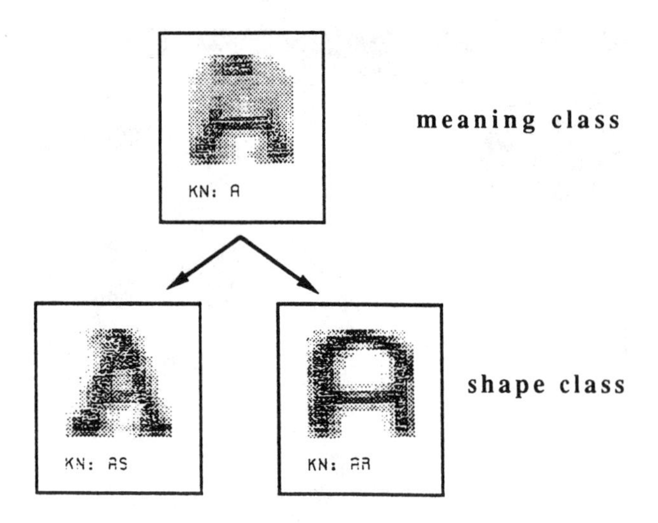

Fig. 10. The division of the class "A" into two shape classes illustrated with the corresponding mean images of the characters.

C. Practical Remarks

It should be mentioned that the term class label refers equally well to the ordinary conception of the class label — in the sense of ASCII-code — as for suitably determined shape classes introduced in order to represent different appearances of the same character class (see Fig. 10). Introducing such subclasses improves existing classifiers because some character classes have totally different shape representations. The region of one class in the feature space is thus divided into separate territories. Therefore, the character features for one class are not unimodally distributed. Since the approximation functions and thus the separation surface cannot become arbitrarily complex, it is convenient to divide those classes into different shape classes.

On the other hand, there are character classes such as the capital "O" and digit "0." It is not reasonable and is sometimes not possible to distinguish such character classes with almost identical shapes like these. This discrimination can only be done — with sufficient performance — by using contextual knowledge, either in the processing stage of character building (cf. subsection IV-B) or in a postprocessing stage (see subsection VI–A). Consequently, these classes are combined to a superclass for single-character classification.

VI. Contextual Processing

Having applied the algorithms described in the previous sections, a large set of objetcs has been generated: connected components, graphical primitives, image objects, character objetcs, etc. It must be pointed out that each object—besides the connected components—represents only a hypothesis of a specific concept. None of the decisions made can be considered to be definitely true.

Most of the hypotheses are generated by inspecting local properties. For example, the character classifier calculates features from the bit map of the underlying character object without considering the properties of the objects around itself. When calculating the character label it is not verified if the label fits into a word context; nor is it clear if the object classified is a character at all.

Contextual processing attempts to overcome the shortcomings of decisions made on the basis of local properties. The local scope is now extended to the investigation of certain relations and constraints between objects. Continuing the example of character labels, the label of a character object having the best certainty score can be withdrawn in favor of a label that fits into the word context. Although the introduction of contextual processing will definitely improve the recognition rate, it is no remedy for poor algorithms: if the set of hypotheses generated to one object does not contain the true solution, contextual processing cannot repair this defect. And furthermore, contextual processing does not change the fact that its resulting decisions are again only hypotheses, although the decision may be considered more trustworthy than the local one previously made.

In the domain of document analysis contextual processing is mainly concerned with geometric information, linguistic information, and dictionaries (for an overview, see [18]). The first subsection covers the geometric aspect: the theme of typography and the Viterbi algorithm is taken up again (cf. subsection IV-B). The following subsection deals with linguistic information, with dictionaries and techniques for using them for improving the recognition rate on word level. An extended use of contextual information is sketched in Sections VII and VIII, where document representation and control aspects are discussed.

A. Typography

Typography and the Viterbi algorithm are discussed in this subsection from a point of view different from that in subsection IV-B, since the Viterbi algorithm here employs the geometric properties as well as the recognition results rather than merely the geometric aspects.

Application of typographical knowledge is necessary when using a classification system utilizing the benefits of size normalization. However, it entails a drawback: capital and lowercase letters with very similar shapes cannot be distinguished. Although this depends on the font, the following characters commonly have similar shapes in their lowercase and uppercase appearances: k, o, p, s, u, v, w, x, y, z. When applying the classifier to patterns of these classes the set of labels returned by the classifier comprises the labels of the lowercase as well as the uppercase characters along with certainty scores hardly varying. For example, if the capital "S" is classifed, the set (s, S, 5) is likely to represent the classifier result.

This ambiguity can be resolved through typographical knowledge, either in an early stage of processing (see subsection IV-B) or in a later stage. In the early stage the typographical class already calculated selects appropriate alternatives. On the other hand, when applying a different approach for generating the layout structure — as described in subsection IV–A — the information about typographical classes of characters is not provided. Then, the following contextual processing determines the correct labels in the later stage.

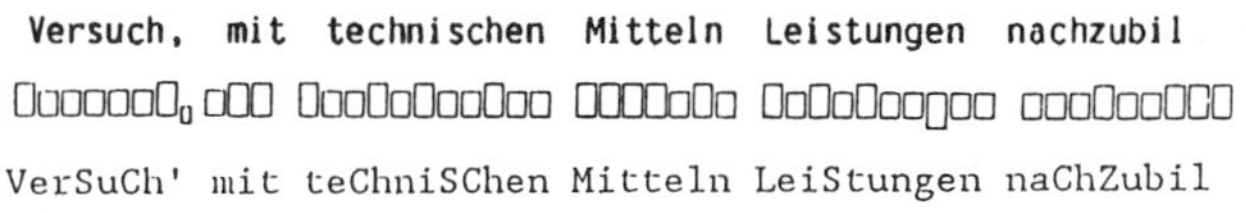

Versuch, mit technischen Mitteln Leistungen nachzubil

VerSuCh' mit teChniSChen Mitteln LeiStungen naChZubil

Fig. 11. A text line: (a) original text line; (b) bounding rectangles of the characters; (c) character labels before typographical processing.

Figure 11 illustrates the problem to be solved then. It shows a line of text in a document having been classified. The second line displays the bounding rectangles of the characters and the third line contains the first alternative of the recognition results of each character.

The recognition results for some of the characters listed above are ambiguous because of the effects of size normalization. The task is now to sort the list of alternatives in such a way as to obtain the correct label as first alternative, and discard the other ones. For example, the recognition result of the seventh character has been the list (S, s, 5) and will be reduced to (s), since only the lowercase character fits into the geometric context of the line.

The Viterbi algorithm is used again along with specific typographical rules and the geometric data of characters. In subsection IV–B the principles of the Viterbi algorithm were introduced along with eight different typographical classes. These classes are derived only from their height and their position on the sheet. At this stage of analysis — after classification — the recognition results are available; thus, a more precise partitioning in 12 typographical classes is possible.

In order to solve the problem, two supplementary data sources must be provided: the first holds the typographical class label to each character label, e.g. label "s" corresponds to typographical class 1. The second source keeps the definition of the five writing lines to each typographical class (cf. subsection IV-B). This information about the lines of each typographical class has been derived in advance and is passed to the algorithm each time it is applied.

The application of the Viterbi algorithm is very similar to the way described in subsection IV-B: the algorithm processes a linelike structure (e.g. a word, a number of words, a full line), with each column representing one character position, and the identical distance function, d, is used to obtain costs. However, in this application certainty scores in the range of [0...1] are used, these being obtained by calculating $\exp(-d)$ rather than costs expressed by d. Consequently, these certainties are multiplied along a path and not added.

The structure of the trellis is slightly different from that described earlier. Skips are not necessary in the graph since the line structure is already generated and is supposed to be correct. Furthermore, in contrast to the graph of subsection IV-B the number of nodes is dependent on the recognition results of a character and therefore, is not defined to be the maximum number of classes. Rather, a column of the trellis comprises the typographical classes which are defined by the labels of the classifier.

When applying this algorithm on the data shown in Fig. 11, the list of alternatives for each character is reduced to labels belonging to the proper typographical class, and the misclassifications are withdrawn.

B. Dictionaries

A first step in language specific content processing is the verification of word hypotheses using a dictionary. Its purpose is a further reduction of ambiguity and an increase of reliability of the final output. However, practice is not as simple. The following list briefly highlights what should be kept in mind:

- No dictionary of a living language can ever be complete. Therefore, a miss does not imply that the hypothesis is wrong. The reverse is not true either. A hit does not guarantee that the word has been interpreted correctly.
- Layout word items are not the same as logical word items. (Think of hyphenated words due to line breaks!)
- Lookup procedures should be fault tolerant with respect to both
    1) typing errors (which usually are caused by humans)
    2) recognition errors (which are caused by the computer).
- In contrast to a word processor's spell-checker, a word verifier of a recognition system should be capable of dealing with multiple character alternatives.
- Character alternatives and segmentation error correction models can dramatically increase the search space. Simple generate-and-test procedures will not work in those cases. It would be cheaper to match every entry of the dictionary against the classifier output. Of course, that would not be a good strategy for very large dictionaries.
- Error correcting features may also degrade the overall results! Also, increasing the size of the dictionary (in an attempt to make it complete!) may worsen the effect.
- Most of efficient standard access methods (e.g. binary search, trie approach) impose an ordering on the data. However, lexical order is rarely the most suitable one. For instance, a primary ordering according to word length leads to much better performance.
- Representation of dictionary entries need not be confined to character symbols only. They may be supplemented by word features (e.g. word shape [19]).
- Choice of optimization criterion makes a difference. Improving word recognition rate does not necessarily imply improvement of character recognition rate.
- In the case of a typing error, it is not clear what the computer is supposed to do. Should it merely reproduce the characters actually printed on the paper or should it guess the intended word? What should we expect from a good secretary?

Starting from the idea of binary search, we have developed a family of algorithms especially suited for handling alternatives and typing errors such as letter omission, insertion, and reversal. The task of a lookup procedure is to find all words that can be constructed from the classifier alternatives and also appear in the dictionary. Of course, it would be wasting time if each word hypothesis were searched individually. The improvement lies in sharing search paths as far as possible. The algorithm was therefore called simultaneous binary search (SBS).

Here is an example with a nonobvious solution:

aiincat0m
prlmeeion
o j clea

The columns represent the alternatives for each character position. The theoretical number of letter combinations is $3 \times 2 \times 3 \times 2**2 \times 3**4=5832$. The lookup procedure was applied to a lexicon with 25 000 words and examined only 66 until the only possible solution was found: PRINCETON. Also "wild card" characters (rejects) can be used. The input ?OS?O?, for example, produces the four city names COSMOS, MOSCOW, ROSCOE and BOSTON, using 348 accesses.

The following shows a hard-to-see spelling error: MASSACHUSSETS. A total of 577 dictionary entries had been retrieved until the correct output, MASSACHUSETTS, was found. One "S" had been eliminated and one additional "T" inserted. However, the maximum numbers of applicable insertions and deletions were both restricted to 1. If a maximum of 2 had been allowed, the effort would have increased to 3781.

It should be emphasized that the SBS algorithm does not use any heuristics, although these could be integrated without difficulty. The ranking of the character alternatives imposed by the classifier would be a good cue for cutting unlikely candidates. In general, heuristics reduce search effort, but possibly at the expense of losing legal solutions. Hence, as far as an algorithmic method keeps feasible, it should be preferred, because it guarantees an always complete solution set.

VII. Document Modeling

In the previous sections concepts and tools were presented which are essential when analyzing documents. The layout structure of the document has been generated, the single characters have been classified, and postprocessing techniques have improved the recognition result.

At this stage the system only accomplishes the task of a typist: The symbolic description obtained by the analysis holds the same information as an ASCII file that would have been keyed in manually. In many cases — this page may be an example — this result is exactly what was desired: a letter by letter transformation into an ASCII string.

On the other hand, sometimes one would like to obtain more detailed information about the contents of a document. Consider the title page of an IEEE scientific publication, e.g. the Transactions on Software Engineering (see Fig. 12). The set of contributions published in this journal will be read automatically in such a way that the author, the title, and the page are known for each contribution. Having extracted this information, the references may be inserted into a data base keeping all information of IEEE publications. Not only the sequence of characters is desired, but also a certain degree of the semantics of the text in order to extend the data base automatically.

The two examples show that the results of an analysis may differ significantly. The symbolic description level achieved resembles the degree of understanding acquired by a document analysis system: the pure character string on the (low) character level versus rather abstract logical objects. However, the representation of documents outlined in the following and the processing model described in the next section do not claim to fulfill the highest level of understanding, and thus the full comprehension of arbitrary text portions. Rather, portions of those documents can somehow be understood in which the layout structure is directly connected to the logical structure. This close neighborhood between content and layout enables the analysis system to deduce the logical content from the layout content without deeper investigation of logical structures.

When aiming at understanding documents on such an abstract level, certain logical concepts—certain labels—must be attached to objects of the document rather than a simple character content. Referring to the IEEE example, the main conceptual entities specify the blocks that denote the reference of the publication. Further concepts are the three constituents of this concept, the title of the contribution, the authors, and the page where the contribution can be found. The analysis system that knows these concepts generates the symbolic description step by step, guided by the description of the document. The result of the analysis is denoted in Fig. 12: Understanding of this document means interpretation of the constituents of the document within these concepts — the IEEE document contains a specific number of contributions each consisting of a list of authors, a title, and a page number.

These logical objects must be introduced into the analysis system as conceptual entities to define the problem domain in which a document can be understood. In the following a representation language, which makes the conceptual entities available to the analysis system, is briefly presented. A more detailed discussion can be found in [20].

When modeling conceptual entities of documents, it is helpful to have a look at the representation scheme ODA as a standard proposed by different organizations (see [21]). This language separates the objects into a set of layout objects and in a set of logical ones which are linked by certain relations. The layout objects like "block" or "page" are described by their geometric properties, whereas logical objects refer to the content of text; examples are "headline," "paragraph," "captions," and "references." However, ODA provides a data structure that aims at a general representation of documents and at interchange of documents rather than for purposes of document analysis. For example, there is no possibility in ODA to represent different hypotheses to one object in an adequate way.

Hence, a specific language, named FRESCO (frame representation language for structured documents), has

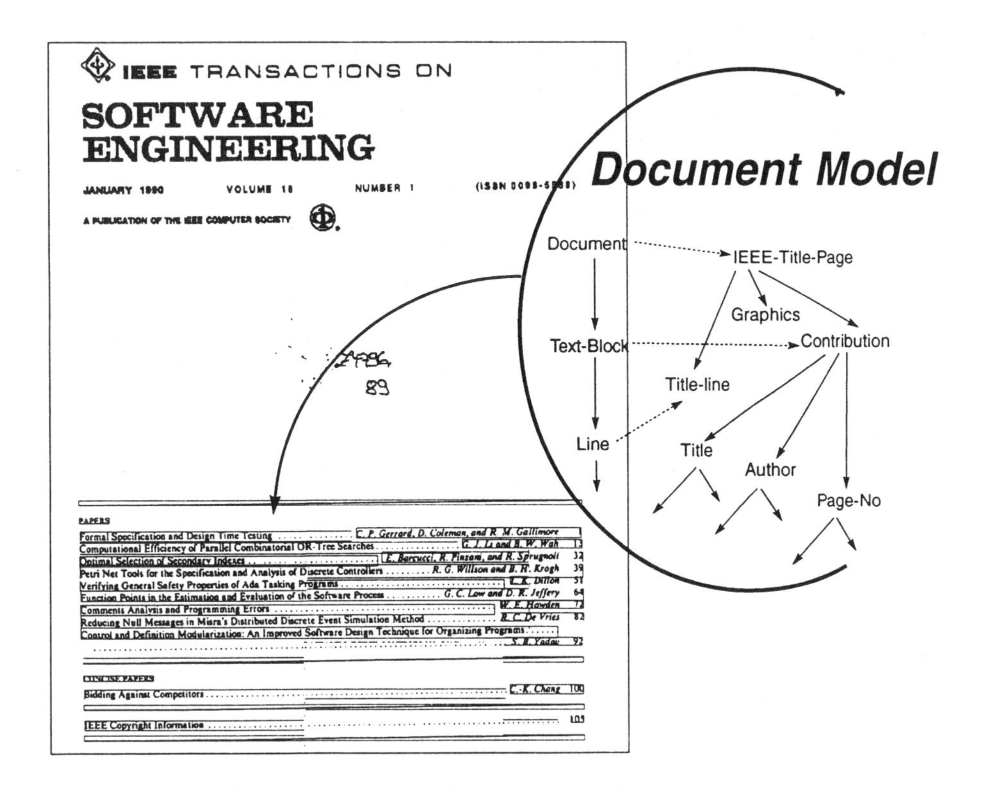

Fig. 12. Title page of the IEEE TRANSACTIONS ON SOFTWARE ENGINEERING. The concepts outline the model of this document. The dashed lines denote the subclass relationship, the straight lines the part relationship. For reasons of simplicity, no attributes and constraints are illustrated.

been developed for representing concepts of structured documents. Conceptual entities are represented in an object-oriented style; each concept defines a class which is described by three properties (which are represented as classes as well):

1) a set of attributes
2) a set of relationships between concepts
3) a set of constraints.

Examples of concepts are "character" and "word" (layout objects) and "title" and "address" (logical objects). Attributes describe the local properties of a concept, such as the extent of the bounding rectangle of a character or the font style. Two relationships, "has-part" and "is-subclass," link the concepts and generate two orthogonal hierarchies useful for subsequent processing. Constraints between attributes and components express certain relationships which must be satisfied, e.g., that objects are left-justified.

FRESCO is specifically designed for modeling entities of structured documents. Geometric properties, for example "left-justified," "same-vertical-line-spacing," or "centered," and content properties are predefined and a constituent of the language. Along with the attribute, relationship, and constraint classes, these objects represent the language primitives of FRESCO.

As is usual in object-oriented systems, FRESCO provides an inheritance mechanism along the relation subclass. This means that the properties enumerated above and the values of the properties of a concept are passed to a concept that is a subclass of it. For example, the concept "sender" inherits all properties of the concept "address." Thus, this mechanism facilitates the definition of new document classes, since the descriptional information must be specified only once in the more general classes. FRESCO models the conceptual entities of structured documents from the very low level of connected components (layout object) to high level logical objects, for example the document class "IEEE journal."

When modeling different document classes, a net of concepts is generated. The concepts are linked by two kind of arcs, representing the part- and subclass relation. FRESCO provides at the moment an extensive generic layout model. Within this model the layout object classes "document," "text-block," "line," "word," "character," "segment," and "connected component" are defined and are linked by the "has–part" relationship. For each class a set of constraint rules specify the construction of itself. For example, the constraints for the class "line" define how words must be ordered to build a line: equal-word-spacing, same-horizontal-baseline, no-vertical-overlapping, left-to-right-sequence.

The more specific logic documents and their constituents are added to these layout objects via "is-subclass" links. Exploiting the inheritance mechanism, each logic class, say an city line of an address block, captures all layout attributes, constraints, and parts of its generic layout class "line." Only the specific logic properties must be added, for example, that the city string must match the zip code. Currently, FRESCO includes several specific logic document classes, such as an address block, recipient, date, the

IEEE title page, and a general business letter. The lowest class level of logical objects is the word level, sometimes the character level. The word level is in general the lowest level a meaning can be assigned to, for example the concept "author name" or "city name." However, for the concept "date" the lowest logic level is the class "character" since a character "1" in the string "1/15/91" denotes the month Januray and it represents no distinct layout word.

VIII. Control Aspects

As mentioned in Section IV, the analysis system should involve high-level recognition tools as well as low-level ones. However, the dependencies between modules may suggest a hierarchically organized system in the sequence as they are discussed in this contribution. Applying the algorithms sequentially one after the other is likely to succeed only in very restricted environments, and with a lot of inherent assumptions: text is typed very cleanly, no noise, no distortions, horizontal lines, etc. In many cases, the algorithms have to be selected carefully according to the data, and the data obtained by the algorithms cannot be interpreted uniquely. Rather, the results are hypothesized and have to be verified or falsified in subsequent processing steps. For example, the objects generated in the layout step are only hypothesized as lines, words, and characters; they are verified when the characters are recognized and the words can be found in a dictionary. Therefore, a control strategy should be based on a data management that allows different hypotheses and backtracking to be pursued.

The following two subsections deal with controlling the analysis process and the prerequisites allowing such a control scheme. First, a flexible data model is presented, providing a management of data in different analysis states. In the second subsection, a specific control strategy is outlined configuring specific tools for the analysis and involving document models.

A. An Experimental Environment

In pattern recognition a twofold problem has to be faced: complex algorithms and complex data. Since many algorithms are rather expensive, there is strong inclination toward optimizing programs and data structures in a physical rather than a logical sense. This might still be indispensable for commercialized products, but for experimental or research vehicles the emphasis must be put on logical conciseness and simplicity. Much expertise from low-level processing up to high-level interpretation has to be integrated into one workable system. This can only be accomplished if an overall data model exists. This does not simply mean standardized data structures; rather a general conception (or paradigm) has to be developed that leads to well structured and manageable systems without knowing every detail. Object-oriented programming (OOP) surely has become a buzzword, but there is more behind it. In fact, some ideas are not far away from well-known concepts in pattern recognition.

The mysterious notion "object" corresponds to the no less mysterious notion "pattern." They both lack a definition that everyone could agree on. Yet, there are some things that can be said about objects and patterns.

Objects usually have attributes; patterns have features. Feature extraction algorithms determine which values they have. It is an essential concept that patterns belong to classes, as objects do in OOP. The problem of classification is to find out the class of a pattern. The decision itself is based on feature values. To find out the meaning of a pattern, we have to determine the class it belongs to. In OOP this would be called its "is-a" relationship. Analogous to multiple inheritance in OOP, a pattern can also belong to more than one class. ("Character '1' is-a digit and is-a machine-printed character.")

Objects may be related to others, thereby forming complex patterns. The task of pattern analysis is to detect and describe these structures. Document analysis has to solve both problems, classification and analysis.

The following outlines the most important concepts incorporated into an experimental environment at the Daimler-Benz Research Center, in Ulm, Germany. There, it serves for all kinds of document analysis research activities. The basic platform is the persistent object storage layer.

Patterns or objects are represented by automatically generated identifiers, also called surrogates. Uniqueness of identifiers is guaranteed. Attributes (or features) are represented by conventional data structures, e.g. bit maps, numbers (integer or floating point), arrays, and records. Relations between patterns are represented using binary relations (called links) with a specified link type. Independently of their type, all links are gathered into one ternary relation (a table with three columns). Each entry comprises two object identifiers and one link-type identifier. From this table all relationships an object is involved in can be determined at a glance.

For each attribute type (e.g. image, bounding box), a pair of access functions is supplied. They are of the form PutImage(id,img) and GetImage(id,img), for example. Thus, the implementation of the internal storage is completely hidden from the analysis algorithms. The object identifier is all that is needed to access data.

In order to support retrieval of links, two elementary functions are supplied, since each link can be used in forward and backward directions as well. Object x might be the successor line of object y. Given y, x can be obtained, but also vice versa.

Adding and deleting a single link are the elementary structure manipulating operations. Sequences of these operations that will preserve certain kinds of structural integrity are encapsulated within higher-level procedures. Examples of basic structures are "sets" or "ordered sets." Recursive application of these constructs, for example "sets of sets" or "sequences of sequences," leads to hierarchical or "tree" structures. Also, all kind of nonhierarchical structures can be be modeled. New object structures can be introduced by adding a module with appropriate procedures (e.g. AddLeftSon, GetRightSon) for binary tree handling.

The environment can be always enhanced by adding further attribute types. Since this need not affect existing attribute types, this feature is called orthogonal extensibility.

Finally, the object storage system is capable of modeling its own history. If a certain instance of the data base has been named, it can be reestablished at all times. The set of instances, internally handled like ordinary objects, form an acyclic directed graph. Each instance corresponds to its own context or view of the data base. Since switching the context costs nearly nothing, an important prerequisite for implementing tentative control strategies has been met. In addition, it provides the possibility of inspecting intermediate states of the analysis afterwards. Typically, such major steps as photo extraction, layout analysis, and character classification give hints when to create a new instance.

B. Control Strategy for Document Understanding

The control strategy presented here is based on the data model discussed in the previous section. Additionally, it involves a document model expressed within FRESCO and a set of algorithms necessary for analyzing documents.

Traditionally, processing models guided by a syntactic description of objects are divided into a preprocessing part and an interpretation part. Symbols are extracted by preprocessing and passed as primitives to an interpreter matching these primitives with conceptual entities and constraints of the model in a parsing process. However, if the preprocessing step extracts symbols which do not match with the model, the interpretation will finally fail. One reason is that backtracking — a common technique during interpretation — does not pass the control back to the preprocessing phase in order to change the set of symbols and restart with the new set.

In the processing model proposed here an attempt is made to bridge this gap between structural approaches and preprocessing by using a blackboard architecture. All algorithms—called knowledge sources (KS's), from image algorithms up to algorithms using linguistic information—are collected in one set and are configured by the control algorithm according to the current state of the anlysis and the descriptions of the KS. Since different KS's may be applicable in one state — e.g. different methods for generating the layout structure — various analysis states arise that span a search space. Within this space backtracking is allowed. Hence, the distinction between preprocessing and interpretation is completely abandoned.

The algorithms comprise all procedures needed in the domain of document image analysis. A KS contains a set of descriptional information about its properties and the procedure that accomplishes the required task. The knowledge about the way of solving a problem is contained in the code of a procedure rather than represented in a declarative manner, as in FRESCO. The set of properties is as follows:

1) input-classes: contains a list of concepts and the corresponding attributes the procedure needs as input data. For example, the value for the KS connectivity-analysis is the pair (binary-image bit map).
2) attribute: this slot contains the name(s) of the attribute(s) which the procedure calculates. For example, the action extract-image yields the attribute bitmap.
3) output-classes: contains concept(s) to which the attribute(s) are attached. For example, the KS extract-image contains the concept binary-image.

Thus, a KS calculates attributes of a concept using attributes of objects calculated before. The remaining properties refer to the control parameters of the algorithm, and to the cost associated with the execution and the relevance of the KS.

Although a blackboard architecture is used, this system is slightly different from common blackboard systems. Not only are rules for selecting and applying a KS specified (cf. the description of the KS); a complete document model is involved.

On the one hand, this model helps, in a very precise and clear manner, to select the appropriate KS, since it makes the system focus on certain data objects being expected in a certain analysis state. Consider the document example of the IEEE paper in Fig. 12. When expecting the page number, why should a general-purpose classifier be used rather than the KS which is responsible for digits? On the other hand, the document model allows an interpretation of the data in terms of its conceptual entities. Without them, the result would comprise only character meanings; with them, an understanding to a certain degree of abstraction is possible.

Currently, the control strategy is guided by the document model in a top-down manner. A document concept is specified and is expanded according to its parts as long there is no KS which calculates the data required for this concept. When a KS is applied, it returns a set of values interpreted as values of attributes in FRESCO; to these attributes and to concepts related to these attributes, instances are generated, added to the blackboard, and tested as to whether they fulfill the various tests and constraints. Different hypotheses arise — apart from the case where more than one KS can be applied in a state — if the results obtained by the algorithms can be interpreted in different ways depending on the concepts in the document model. Thus, instances corresponding to the concepts of the document model are calculated step by step by different KS's until an instance to the goal concept is generated.

Since each state in the search space generated during analysis is accompanied by a certainty score calculated by functions evaluating the matching of the data with the model, the search space is controlled by the best—first graph search algorithm A(*) (see [14] and 15]).

IX. Conclusion

What we have presented in the foregoing sections has been a rather rapid excursion through the realm of machine document reading and understanding — where so many problems are encountered which are easily solved by humans but are so burdensome to computers. We have tried to outline the most fundamental concepts and algorithms which are — at least to our experience — building blocks of any technical reading system.

The tasks to be solved here exhibit so much resemblance to human skills that the engineers and computer scientists trying to master them may be inclined to follow their own conceptions as to how they themselves manage to solve them and how their own brains works. Whereas ideas of this kind may be a valuable source of inspiration and a good point to start with, they usually tend to oversimplify what really happens in natural brains. We feel it necessary to put those approaches onto a sound theoretical basis, to relate the cognitive tasks to the worlds of mathematics and engineering sciences, to replace heuristics by systematics, and to rely on mathematical optimization rather than intuition and ingenuity.

Although at first glance, document analysis seems to be predominantly an application of character recognition, it should have been shown that even if the recognition of properly isolated single characters obviously is a core operation, it serves only a subordinate purpose within a network of numerous other cognitive functions — background/foreground discrimination, finding layout primitives, isolating character images, modeling the application's conceptual hierarchies, and matching image constituents with conceptual entities. Character recognition is merely one of those — and certainly one of the best understood — subtasks of the overall system. It is called by the model-driven analysis procedure whenever an image constituent is encountered which seems to be a character image. Owing to its well-defined nature, namely mapping character images to character meaning by suitably designed mapping functions, it can be accomplished by extraordinary systematic and regular approaches, and thus needs only the smallest part of the overall design effort.

The document analysis system, as a whole, is also an example of how statistical and model-based approaches of pattern recognition fit together in order to establish a nontrivial pattern recognition system. Once elementary objects have been identified and are described by feature vectors, they are recognized as entities by statistical procedures. Thus they are given intelligibility and sense in the form of labels, further processing being made in knowledge bases, frame structures, inference mechanisms, and all the instruments of model-based reasoning. There are many cases where the labeling results are ambiguous. Thus, we come out with sets of alternatives accompanied by confidence values instead of definite decisions, and are consequently led into the field of uncertain reasoning in order to find the ranked list of most probable interpretations.

The building blocks of document analysis systems as presented here can be arranged to serve a variety of applications. Some of these have already come into existence and been used in document analysis products which have proved the efficiency of those concepts. Others remain to be invented.

One point seems to be important in this context. Even if document analysis systems belong to the category of intelligent systems capable of deriving interpretations from raw observational data, they are in many cases not intelligent enough to completely automize jobs which up to now have been done by humans. It is necessary to design an efficient symbiosis between man and machine intelligence. The document analysis system can take over the routine part of the job, but provisions must be made for situations in which human decisions are necessary in cases where the competence of the machine is overloaded. It is not that the human simply makes up for machine imperfections. Especially in document generation, processing, and manipulation, there are many cases where the human is doing the creative part of the job and is therefore indispensable. Human work can, however, be substantially supported by document analysis functions which are activated in the same way that the author activates the functions of text processing systems through his own intentions.

References

[1] N. Bartneck, "Methods for photo noise extraction in postal applications," in *Proc. Advanced Technology Conference*, USPS, 1990, pp. 297–310.

[2] J. Serra, *Image Analysis and Mathematical Morphology*. London: Academic Press, 1982.

[3] E. Mandler and M. F. Oberländer, "Ein single—pass Algorithmus für die schnelle Konturkodierung von Binärbildern," in *Proc. 12th DAGM—Symposium Mustererkennung*, Aalen, 1990, pp. 248–255.

[4] E. Mandler and M. F. Oberländer, "One-pass encoding of connected components in multi-valued images," in *Proc. 10th Int. Conf. Pattern Recognition (ICPR)* (Atlantic City, NJ), 1990, pp. 64–69.

[5] N. Bartneck, "Ein Verfahren zur Umwandlung der ikonischen Bildinformation digitalisierter Bilder in Datenstrukturen zur Bildauswertung," dissertation, TU Braunschweig, 1987.

[6] S. Srihari and D. Wang, "Classification of newspaper image blocks using texture analysis," *Computer Vision, Graphics and Image Processing* vol. 42, pp. 327–352, 1989.

[7] R. Kasturi *et al.*, "A system for interpretation of line drawings," *IEEE Trans. Pattern Anal. Mach. Intell.*, vol. 12, pp. 978–990, Oct. 1990.

[8] K. Tombre and P. Vaxiviere, "Interpretation of mechanical engineering drawings for paper-CAD conversions," in *Proc. IAPR Workshop Machine Vision Applications*, Nov. 1990, pp. 203–206.

[9] H.S. Baird, "The skew angle of printed documents," in *Proc. SPSE 40th Conf. Symp. Hybrid Imaging Systems* (Rochester, NY), May 1987, pp. 21–24.

[10] A. Rastogi and S.N. Srihari, "Recognizing textual blocks in docment images using the Hough transform," Report TR 86-01, Department of Computer Science, SUNY at Buffalo, 1986.

[11] G.D. Forney, "The Viterbi algorithm," *Proc. IEEE*, vol. 61, pp. 268–278, Mar. 1973.

[12] T. Bayer and M. Oberländer, "Ein erweiterter Viterbi Algorithmus zur Berechnung der n-besten Wege in zyklenfreien Modellgraphen," in *Mustererkennung*, G. Hartmann, Ed. Berlin: Springer Verlag, 1986, pp. 56–60.

[13] T. Bayer, "Segmentation of merged character patterns with AI-techniques," in *Proc. 5th Scandinavian Conf. Image Analysis* (Stockholm), 1987.

[14] J.N. Nilsson, *Principles of Artificial Intelligence*. Berlin: Springer Verlag, 1982.

[15] J. Pearl, *Heuristics: Intelligent Search Strategies For Computer Problem Solving*. Reading, MA: Addison-Wesley, 1984.

[16] J. Schürmann, *Polynomklassifikatoren für die Zeichenerkennung.* Munich, Germany: Oldenbourg Verlag, 1977.
[17] J. Schürmann and W. Doster, "A decision theoretic approach to hierarchical classifier design," *Pattern Recognition*, vol. 17, no. 3, pp. 359–369, 1984.
[18] S. Srihari, Ed., *Computer Text Recognition and Error Correction.* Silver Spring, MD: IEEE Computer Science Press, 1985.
[19] J.J. Hull, "Word shape analysis in a knowledge based system for reading text," in *Proc. 2nd IEEE Conf. Artificial Intell. Appl.* (Miami, FL), 1985.
[20] T. Bayer, "Representation of structured documents in a frame system," in *Pre-Proc. SSPR Workshop*, 1990, pp. 47–56.
[21] W. Horak, "Office document architecture and office document interchange format" (Current Status of International Standardization), *IEEE Computer*, pp. 51–60, Oct. 1985

Jürgen Schürmann was born in Frankfurt/Oder, Germany, in 1934. He received the Dipl.-Ing. and Dr.-Ing. degrees from the Technical University of Berlin in 1960 and 1968, respectively.

In 1963 he joined Telefunken Research, now part of Daimler-Benz Research, Ulm, Germany. His pioneering work in pattern recognition laid the foundations for the pattern recognition activities of AEG-Electrocom, which have led to a well-recognized market position, primarily in the address reading field. He has authored and coauthored a number of scientific publications including two textbooks on statistical pattern recognition. Currently he heads the Pattern Recognition Group within Daimler-Benz Research working in the fields of document and image analysis, speech recognition and understanding, and signal processing. His research interests include statistical pattern recognition, with special emphasis on the relations to artificial neural networks and uncertainty concepts, image analysis, speech recognition, and radar signal processing.

Since 1974 Dr. Schürmann has been teaching pattern recognition at the Technical University of Darmstadt and became an associate professor there in 1981.

Norbert Bartneck was born in Aschaffenburg, Germany, in 1956. He received the Dipl.-Ing. degree in electrical engineering from the Technische Hochschule Darmstadt in 1981 and the Dr.-Ing. degree from the Technical University Braunschweig in 1987.

In 1981 he joined the AEG research center in Ulm, which is now part of Daimler-Benz Research. He is working in the area of document analysis. His main activities have been in the early processing stage of image analysis, e.g. binarization, noise reduction, and connectivity analysis. His current research interests encompass knowledge-based algorithms for specific postal analysis tasks (e.g. address block location) and analysis of graphic documents.

Thomas Bayer was born near Nürnberg, Germany, in 1961. He received the Dipl.Inform. degree in computer science from the University of Erlangen in 1986.

In 1986 he joined the Pattern Recognition Department of the Research Institute of AEG in Ulm, which is now the Daimler-Benz Research Center for Information Technology. He has been working in the domain of document analysis, where he first was concerned with contextual analysis and character segmentation. His current interests are expert systems, configuration systems, fuzzy techniques, and object-oriented document modeling.

Jürgen Franke was born near Hannover, Germany, in 1953. He received the Dipl.-Math degree from the University of Hannover, Germany.

In 1981 he joined the Pattern Recognition Department at the Research Center of AEG-Telefunken in Ulm, which is now part of Daimler-Benz Research. He has been working on character recognition, especially for address- and form-reader applications. His research focuses on classification algorithms (polynomial classifier, hierarchical classifier, nets of classifiers, generation of confidence values for postprocessing).

Eberhard Mandler was born near Frankfurt, Germany, in 1955. He received the Dipl.-Ing. degree in electrical engineering from the Technische Hochschule Darmstadt, Germany, in 1981.

In 1981 he joined the AEG research center in Ulm, Germany, which is now part of Daimler-Benz Research. He was first engaged in speech enhancement and speech recognition. Since 1984 he has been working in the area of script recognition. His current research interests are script recognition, belief functions, and approximate reasoning.

Matthias Oberländer was born in Schweinfurt, Germany, in 1959. He received the Dipl. Inform. degree in computer science from the University of Erlangen, Germany, in 1985.

In 1985 he joined the document analysis group of the AEG-Telefunken Research Center, Ulm, which is now part of Daimler-Benz Research. He is responsible for design of the experimental environment and algorithm integration. His current focus is on data bases, object-oriented analysis, and design and programming. He has also worked on statistical feature extraction, layout analysis, fast contour coding algorithms, and contextual postprocessing. He is also interested in the relationship between pattern recognition, object orientation, and philosophy.

Chapter 5
Graphics Analysis and Recognition

5.1: Introduction

In this chapter, we deal with methods for analysis and recognition of graphics components in paper-based documents. Graphics recognition and interpretation are important topics in document image analysis since graphics elements pervade textual material, with diagrams illustrating concepts in the text, company logos heading business letters, and lines separating fields in tables and sections of text. The graphics components that we deal with are the binary-valued entities that occur along with text and pictures in documents. We also consider special-application domains in which graphical components dominate the document; these include symbols in the form of lines and regions on engineering diagrams, maps, business charts, fingerprints, musical scores, and so on. The objective is to obtain information that semantically describe the contents within images of document pages.

We have stated that a high-level description of the document is the objective. Why is a high-level description needed and how is it used? One application is in the conversion of paper diagrams to a computer-readable form. Although simply digitizing and storing the document as a file of pixels make it accessible by computer, much more utility is usually desired. For example, for modification of the elements of the diagrams, a higher-level description is required. For translation to different computer-aided design formats and for indexing of components, the diagram must be semantically described. Even if storage is the only requirement, a semantic representation is usually much more compact than storage of the document as a file of pixels. For example, an English letter is stored as its 8-bit ASCII representation in lieu of its larger-size image. Likewise, for a compression of two to three orders of magnitude, a graphics symbol such as a company logo or electrical "AND" gate can have a similarly compact "codeword" that essentially indexes the larger-size image.

Document image analysis can be important when the original document is produced by computer as well. Anyone who has dealt with transport and conversion of computer files knows that compatibility can rarely be taken for granted. Because of the many different languages, proprietary systems, and changing versions of CAD and text-formatting packages that are used, incompatibility is especially prevalent. Because the formatted document (that viewed by humans) is semantically the same, independent of the language of production, this form is a "protocol-less protocol." If a document system can translate between different machine-drawn formats, the next objective is to translate from hand-drawn graphics. This process is analogous to handwriting recognition and text recognition in OCR. When machines can analyze complex hand-drawn diagrams accurately and quickly, the graphics recognition problem will be solved, but there is still much opportunity for research before this goal will be reached.

A common sequence of steps taken in document image analysis for graphics interpretation is similar to that for text. First, preprocessing, segmentation, and feature extraction methods such as those described in earlier chapters are applied. An initial segmentation step that is generally applied to a mixed text/graphics image is that of text and graphics separation. An algorithm specifically designed for separating text components in graphics regions irrespective of their orientation is described in [Fletcher 1988]. This is a Hough transform-based technique that uses the heuristic that text components are colinear. Once text has been segmented, typical features extracted from a graphics image include straight lines, curves, and filled regions. After feature extraction, pattern recognition techniques are applied: both structural pattern recognition methods to determine the similarity of an extracted feature to a known feature using geometric and statistical means and syntactic pattern recognition techniques to accomplish this same task using rules (a grammar) on context and sequence of features. After this mid-level processing, these features are assembled into entities with some meaning—or semantics—that is dependent upon the domain of the particular application. Techniques used for this include pattern matching, hypothesis and verification, and knowledge-based methods. The semantic interpretation of a graphics element may be different depending on domain; for example, a line may be a road on a map on an electrical connection of a circuit diagram. Methods at this so-called high level of processing are sometimes described as artificial intelligence techniques.

Most commercial OCR systems will recognize long border and table lines as being different from characters, and so will not attempt to recognize them as characters. Graphics analysis systems for engineering drawings must discriminate between text and graphics (mainly lines). This is usually accomplished very well; however, some confusion arises when characters that adjoin lines are interpreted as graphics; or when small, isolated graphics symbols are interpreted as characters. One paper is included here, by Fletcher, on text and graphics separation. (Also see paper by Jain included in Chapter 4.)

5.2: Extraction of lines and regions

(*Keywords*: line graphics, region graphics, segmentation)

The amount and type of processing applied to graphics data in a raster image are usually application-dependent. If the graphics image is a part of a predominantly textual document and the objective is simply data archiving for later reconstruction, then use of any one of the well-known image compression techniques [Jain 1989] is adequate. On the other hand, if information is to be extracted from data for such applications as indexing image components from a pictorial database [Grosky 1989], modifying graphics in a CAD system [Karima 1985], or determining locations in a geographical information system, then extensive processing to extract objects and their spatial relationships is necessary. Such a level of description is attained through a series of intermediate steps. After the graphics data have been segmented, the next step is to perform processing to locate line segments and regions. Extracting their location and attributes is an important step for graphics interpretation. In this section, we describe methods for extraction of lines and regions.

We first define how lines and regions differ in this context. A group of ON-valued pixels can be a line or a region. Consider a drawing of a barbell, with two filled disks joined by a line of some thickness. The thick line between the disks can be represented by its core line (from thinning). The two disks are regions that can be represented by their circular boundaries (from contour detection). The representation of this drawing by boundaries for regions and core lines for the line has advantages in later steps of graphics analysis. Although lines can have thicknesses and regions can sometimes be described by the core lines that form their "skeleton," the distinction between the two is usually obvious based on meaning within the domain.

To represent an image by thin lines and region boundaries, the lines and regions must first be segmented. One approach to segmentation is by erosion and dilation operations [Bow 1990, Harada 1985, Modayur 1993, Nagasamy 1990]. (See also Section 2.3, where these morphological operations are described in the context of noise reduction.) Erosion and dilation usually proceed iteratively, deleting or adding a one-pixel thickness from the boundary on each iteration until the specified thickness is reached. For graphics segmentation, erosion precedes dilation, and the thickness removed is specified to be greater than the maximum line thickness in the image so that lines disappear. Thus, the result of erosion is an image containing only eroded regions. Then dilation is performed only around these remaining regions. If the image result of these erosion and dilation steps is combined by a logical AND operation with the original image, only the regions remain. (Note that since dilation is not an exact inverse of erosion, the image should be dilated by a thickness of one or two pixels more than that by which it is eroded, to completely recover the regions.) The difference image obtained by subtracting this region image from the original image contains the remaining thin-line components. After segmentation, the boundaries are easily found in the region image using contour-following methods. Lines are thinned to obtain their core lines.

There are other methods for segmentation; some simultaneously extract boundaries and core lines. In [Shih 1989] several picture decomposition algorithms for locating solid symbols are evaluated. In one method, Wakayama's Maximal Square Moving (MSM) algorithm [Wakayama 1982], squares are fit to the ON-valued pixel regions. The union of these squares (possibly overlapping with one another) exactly covers these regions. Thus, knowing the center and size of these maximal squares, it is possible to exactly reconstruct the original image. Also the centers of these squares may be connected to form an approximation to the core line; however, since these squares are generally of different sizes, their centers are generally not connected. A system of pointers is necessary to keep track of squares that are adjacent to one another in order to generate the core lines and boundaries. This process, along with the procedure for keeping track of the squares and their attributes as they are being grown, is quite unwieldy. A simpler method, in which core lines can be found simultaneously with boundaries, is the $k \times k$ thinning method [O'Gorman 1990].

We include no papers in this section. Instead, we leave descriptions of graphics extraction techniques to the systems papers in the following section, and we refer to papers on text and graphics separation included in the Sections of 4.1 and 5.1.

5.3: Graphics recognition and interpretation

(*Keywords*: electronic-circuit-diagram conversion, engineering-drawing analysis, fingerprint analysis, graphics recognition, line-drawing analysis, line-drawing interpretation, map conversion, map reading, music reading, raster-to-vector conversion, vectorization)

Recognition of graphical shapes and their spatial relationships is an important final task in most document analysis systems. The recognition process consists essentially of two main steps: processing the input image to obtain representational primitives (as described in the previous sections) and matching these primitives against similar primitives derived from known models. Techniques used for the latter step of matching are strongly application-dependent. To recognize isolated symbols of fixed size and orientation, simple template matching applied directly to a bit-mapped image may be adequate. However, for many applications, this simple technique is inappropriate, and features as described above must be extracted. In certain applications, it may be adequate to approximate closed contours by polygons. In other applications, more-sophisticated algorithms that hypothesize possible matches, compute scene/model transformations, and verify the hypotheses are used. In more-complex images—such as maps and engineering drawings—context-dependent, knowledge-based graphics recognition and interpretation techniques have been used. Different algorithms exhibit varying degrees of flexibility, accuracy, and robustness.

In this section, various graphics recognition and interpretation techniques are described, including hierarchical decomposition and matching, interpretation based on structural analysis of graphics, and recognition using contextual information and domain-dependent knowledge. The techniques described here make use of the line and feature data obtained using techniques described in the previous sections. Recognition algorithms that operate directly on bit-mapped data, as well as those that are based on well-known techniques such as signature analysis and Fourier descriptors, are not described here (see Section 3.5). Many of the methods described here were developed originally for use in a specific application domain, although they can be adapted for use in other domains; thus, we organize this section by application domain.

5.3.1: Recognition of graphical shapes in line art

Many business documents routinely include line art in the form of, for example, organizational charts, bar graphs, block diagrams, flow charts, and logos. Techniques such as polygonal approximation are useful for recognizing simple, isolated shapes found in such documents. For complex shapes or for different shapes that are interconnected, structural analysis methods are useful.

Structural analysis methods, for the most part, use a bottom up-approach in which first the input pixel image of the document is processed to obtain vectorized line and feature data. Then, some form of structural analysis is performed on this vectorized data to extract sensible objects. What is a *sensible object* depends on the domain of the application. It could be an electrical-circuit symbol, a standard geometric figure like a rectangle or a pentagon, a component in an engineering drawing, and so on. Then the extracted object may be compared with a database of generic objects for recognition. *Structural analysis* refers to the actual search that is performed on the input data for extracting meaningful objects. This analysis stage can be very time-consuming because of the large number of line segments and other feature primitives present in the vectorized data and the enormous number of ways in which they could be combined to potentially form sensible objects. A simple "brute-force" but highly inefficient method would be to systematically check all possible combinations to see if any of them form an object of interest in our domain. Clearly, what is needed at this stage is an intelligent method that would make use of heuristics or context-sensitive information to speed up the analysis. The various structural-analysis-based graphics recognition systems described in the literature differ primarily in the way in which they perform this analysis of vectorized line and feature data.

For recognition of line art such as flow-charts, tables, and block diagrams, structural analysis of lines and their interconnections is required to generate meaningful and succinct descriptions. For example, the flow chart shown in Figure 1 contains only 57 line segments (excluding hatching lines and filling patterns); however, these line segments form 74 different closed loops [Kasturi 1990]. A meaningful interpretation would be to identify the seven simple shapes and then to describe other lines as interconnecting segments. Similarly, an unconstrained algorithm could interpret a table as a collection of many rectangles, whereas it would be more useful to describe it as a rectangle with horizontal and vertical bars. In case of electrical schematics, it would be necessary to separate lines belonging to symbols from connecting lines.

An algorithm for generating loops of minimum redundancy is described in [Kasturi 1990]. In this algorithm, first all terminal line segments are separated, since they cannot be a part of any closed loop. This procedure removes the two short-line segments at the top and bottom of Figure 1. Then, all self loops are separated. Note that self loops are formed by line segments that are not part of any other loop. For example, in Figure 1, the loops representing small-size fillings and the top rectangle are self loops; additional processing is required to locate other loops in the graphics network, as described below. This process is repeated, if necessary, to delete the short segment at the bottom of the top rectangle in Figure 1. Then, a heuristic search algorithm is applied to the remaining network to identify simple closed loops that are made up of no more than a predetermined number of line segments. This search algorithm is described with the help of Figure 2. In this figure, the current point C has been reached by following line segments AB and BC. The objective is to assign priorities to line segments at C for continuing the search for a closed loop. Line segment *j* is assigned the highest priority, since it is colinear with BC. The next priority is assigned to *k* (angles a_1 and a_2 are equal), since it has the potential to form a regular polygon. The next priority goes to the segment *l* that is parallel to AB (potential parallelogram or trapezoid). Final priority goes to segment *m,* since it forms the sharpest convex corner at C. If none of these segments continues the path to form a closed loop, segment *n* is chosen during the final search for a closed loop. This loop-finding algorithm employs a depth-first search strategy. It correctly identifies the seven simple shapes present in the figure. In particular, the algorithm traces the outline of the hatched rectangle and thus separates hatching patterns from enclosing lines. The extracted loops are then compared with a library of known shapes, shown in Table 1, for recognition and description. Those shapes not recognized are analyzed to verify if they can be described as partially overlapped known shapes. All other segments are described as interconnecting lines or hatching patterns. The system outputs all recognized shapes, their attributes, and their spatial relationships. The output generated by the system corresponding to the test image of Figure 3 is shown in Tables 2 and 3. This algorithm has been extended to describe graphics made up of circu-

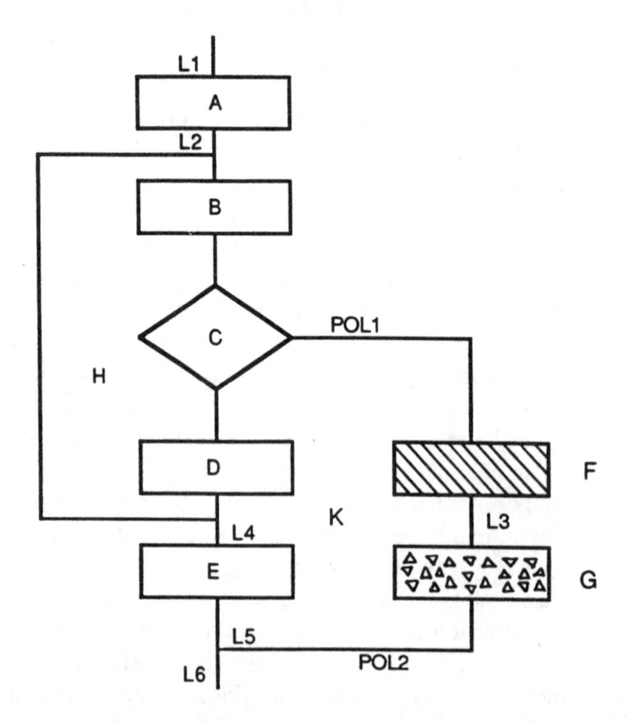

Figure 1. A test image containing a flow chart, illustrating graphics segmentation and recognition.

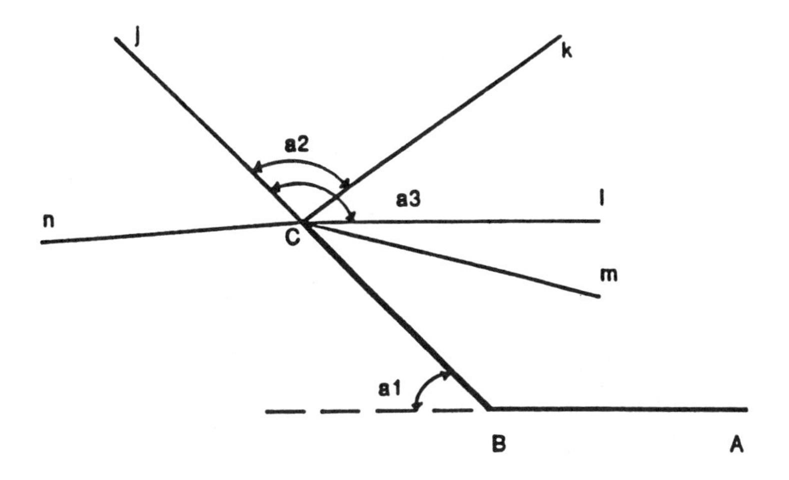

Figure 2. The assignment of priorities for determining closed loops in the diagram shown in Figure 1 (see the text for detailed information).

Table 1. The library of known shapes for loop detection and recognition.

Known Shapes	Attributes
Triangle	P1, P2, P3
Rectangle	P, W, H, ϕ
Rhombus	P, L, Θ, ϕ
Parallelogram	P, L1, L2, Θ, ϕ
Trapezoid	P, L1, L2, H, Θ, ϕ
Regular Pentagon	P, L, ϕ
Regular Hexagon	P, L, ϕ
Quasi-Hexagon	P, L1, L2, Θ, ϕ

ϕ: Orientation of longer side (major axis in case of rhombus and quasi-hexagon) passing through P with respect to x axis

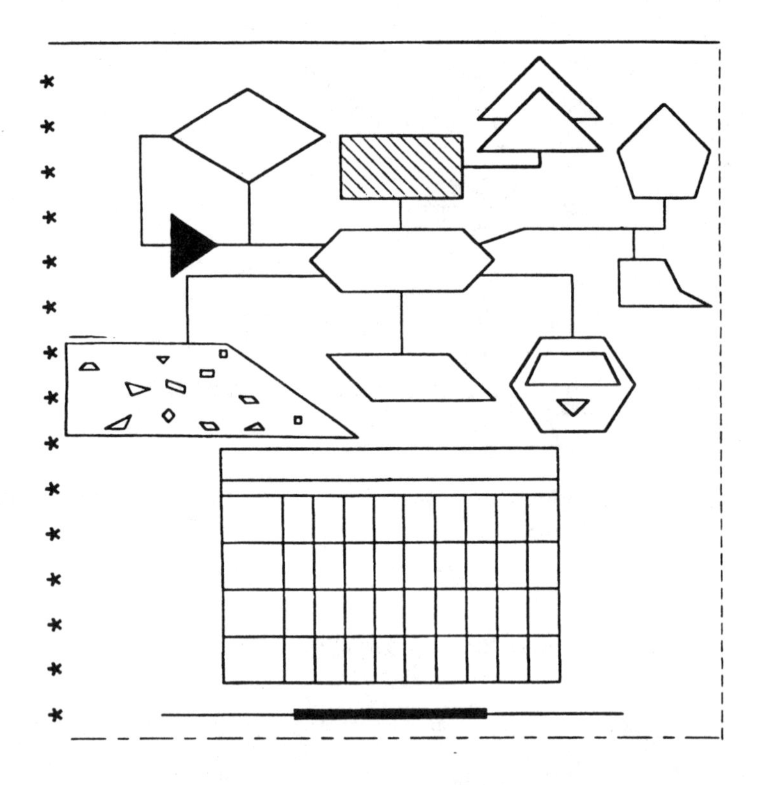

Figure 3. A test image containing different shapes, with the results of graphics recognition shown in Tables 2 and 3 [Kasturi 1990].

lar arc segments [Bow 1990]. Some of these techniques have also been used in a map-based geographic information system [Kasturi 1989]. For a description of a method for interpretation of tables, see [Chandran 1993].

Occasionally, it is necessary to recognize graphical shapes in which a portion of the shape is missing or hidden behind other shapes. Such a situation calls for methods that do not require complete object boundaries. Also, when the number of possible different shapes to be recognized is large, it may be efficient to represent complex parts as a combination of already-known simple shapes along with their spatial relationships. An object recognition system that creates a library of parts by hierarchical decomposition, described in [Ettinger 1988], is suitable for use in such situations. In this system, the library organization and indexing are designed to avoid linear search of all the model objects. The system has hierarchical organization for both structure (whole object to component subparts) and scale (gross to fine features). Object representation is based on the curvature primal sketch [Asada 1986] (see Section 3.4). Features used are corner, end, crank, smooth-join, inflection, and bump, which are derived from discontinuities in contour orientation and curvature. Subparts consist of subsets of these features, which partition the object into components. The model libraries are built automatically, using the hierarchical nature of the model representations. The recognition engine is structured as an interpretation tree [Grimson 1984]. A constrained search scheme is used for matching scene features to model features. Many configurations are pruned in the search space early in the search process, using simple geometric constraints such as orientation difference, distance, and direction between pairs of features.

5.3.2: Conversion of engineering drawings and maps

[Antoine 1992] describes REDRAW (REading DRAWings), a system for interpretation of different classes of technical documents. It uses a priori knowledge to achieve interpretation at a semantic level. Here, the aim

Table 2. Recognized objects in the diagram shown in Figure 3.

	Object	Attributes
1	Regular Hexagon	P: (1258, 1081), L = 133, ϕ = 1.51
2	Parallelogram	P: (727, 1268), L1 = 292, L2 = 146, Θ = 46.1, ϕ = 0
3	Trapezoid	P: (73, 1081), L1 = 718, L2 = 390, H = 217, Θ = 90.68, ϕ = -0.16
4	Rhombus	P: (339, 1791), L = 220 Θ = 30.1, ϕ = -0.45
5	Trapezoid	P: (1220, 1194), L1 = 221, L2 = 150, H = 68.6, Θ = 65, ϕ = 0.77
6	Triangle	P1: (1295, 1153), P2: (1363, 1151), P3: (1325, 1122), Isoceles
7	Triangle	P1: (1396, 1826), P2: (1106, 1831), P3: (1256, 1971), Isoceles
8	Rectangle	P: (457, 1044), W = 835, H = 564, ϕ = -0.4, Table
9	Quasi-Hexagon	P: (692,1497), L1 = 440, L2 = 303, Θ = 89.1, ϕ = 3.0
10	Parallelogram	P: (765, 1790), L1 = 297.1, L2 = 148, Θ = 89.0, ϕ = -1.35, Single hatch: a1 = 135, d1 = 30
11	Traingle	P1: (1256, 1897), P2: (1399, 1755), P3: (1108, 1753), Isoceles
12	Polygon, irregular	Number of segments: 6, Center: (1556, 1706), Coordinates of vertices....

is to build a general model-driven recognition system that can be completely parameterized. The model contains specific knowledge for each document class. The interpretation system is driven by this model, using a general tool box for low-level operations. Two applications, city map and mechanical-drawing interpretation, are described in the paper. The a priori knowledge about the domain induces a particular interpretation process for each document class. Among the low-level operations, the paper describes a method for extraction of parallel lines (hatched areas). A single-scan algorithm in horizontal and vertical directions determines an order relation for each extremity point for each segment. From these relations, classes of parallel lines in a same area are deduced by assuming that parallel lines from a hatched area must overlap each other in the two main directions. A prototype system for extracting higher-level structures for knowledge-based analysis of mechanical drawings is described in [Vaxivière 1992].

[Joseph 1992] describes a schema-driven system, called Anon, for the interpretation of engineering drawings. This paper's approach is based on the combination of schemas describing prototypical drawing constructs with a library of low-level image analysis routines and a set of explicit control rules. The system operates directly on the image, without prior thresholding or vectorization, and combines the extraction of primitives (low-level) with their interpretation (high-level). Anon integrates bottom-up and top-down strategies into a single framework. The system has been successfully applied to piece-part drawings.

Table 3. Recognized lines and their attributes in the diagram shown in Figure 3.

	Objects	Spatial Relationships		
11	Triangle	Overlaps Object 7		
1	Regular Hexagon	Encloses Objects 5 and 6		
10	Parallelogram	Single Hatch		
3	Trapezoid	Small Shape Fillings		
Lines and Their Interconnections				
Line	Head	Tail	From	To
Single Segment Lines				
L1	(1404,1748)	(1399,1755)	--	11
L2	(910,1416)	(911,1271)	9	2
⋮	⋮	⋮	⋮	⋮
Lines With Multiple Segments				
PL1	S1: (330,1531) S2: (260,1537) S3: (265,1791) S4: (332,1790)	(260,1537) (265,1791) (332,1790) (339,1791)	--	4
PL2	S1: (1064,1715) S2: (1252,1719)	(1252,1719) (1252,1750)	10	11
⋮	⋮	⋮	⋮	⋮

One of the major tasks in automated conversion of mechanical-engineering drawings is separating dimensioning lines and their associated text from object lines. Algorithms for performing this segmentation are described in [Lai 1993]. The complete system includes algorithms for text/graphics separation; recognition of arrowheads, tails, and witness lines; association of feature control frames and dimensioning text with the corresponding dimensioning lines; and detection of dashed lines, sectioning lines, and other object lines. Similar methods have been applied to interpret telephone system manhole drawings [Arias 1993].

A method for interpreting the three-demensional shape of an object corresponding to multiple orthographic views in an engineering drawing is described in [Lysak 1990]. This technique is based on a bottom-up approach in which candidate vertices and edges are used to generate a set of possible faces, which are in turn assembled into enclosures representing the final object [Wesley 1981]. A minimum of two views is required, and a maximum of six orthogonal views in a standard layout can be accommodated. All possible interpretations consistent with the input views are found, and inconsistent input views are recognized. The method can handle input views with small drafting errors and inaccuracies.

[Ejiri 1990] applies structural-analysis methods for recognition of engineering drawings and maps. To process large-scale-integration cell diagrams, solid lines and two types of broken lines in any one of six colors are recognized using the color digitizer. A loop-finding routine facilitates detection of overlapped lines denoted by a special mark. Structural-analysis methods have also been used to recognize characters and symbols in logic circuit diagrams, chemical-plant diagrams, and mechanical-part drawings.

A system for automatic acquisition of land register maps is described in [Boatto 1992]. This system uses the semantics of land register maps extensively to obtain correct interpretation.

5.3.3: Conversion of electronic-circuit diagrams

[Fahn 1988] describes a topology-based component extraction system to recognize symbols in electronic-circuit diagrams. The objective is to extract circuit symbols, characters, and connecting lines. Picture segments are detected using segment-tracking algorithms. These segments are then approximated using a piecewise linear-approximation algorithm. A topological search is done to form clusters of symbols or characters. A decision tree is used to assign picture segments to a class of primitives. Segments are clustered into component symbols using a context-based depth-first search method. The system—designed to recognize circuit symbols in four orientations and connection lines that are horizontal, vertical, or diagonal—has also been used to generate fair copies of hand-drawn circuit diagrams. In [Okazaki 1988], a loop structure analysis system is described for recognition of circuit symbols.

A model-based learning system for recognizing hand-drawn electrical-circuit symbols in the absence of any information concerning the pose (that is, the translation, rotation, and scale) is proposed by [Lee 1992]. A hybrid representation, called an *attributed graph* (AG), incorporates structural and statistical characteristics of image patterns. It is used for matching an input symbol with respect to model symbols. Model AGs are created interactively using a learning algorithm. The poses of input object AGs are estimated based on a minimum-square error transform, and they are classified based on minimum distance. An average error rate of 7.3 percent is reported. Another approach to recognition is using relaxation. [Bunke 1981] uses this method to analyze electrical schematics. First, a vector with probabilities of possible interpretations is assigned to each vertex in the schematic. A *vertex* is a location where either several line segments touch each other or a line segment ends. The probabilities of the interpretations at various vertices are successively changed by the relaxation process in order to achieve unique and consistent labeling. This recognition engine depends on effective propagation along the matching vertices, good initial assignments, and convexity of the configuration space and the domain. The convexity requirement removes the danger of identifying local maxima/minima as global, but it is difficult to satisfy.

5.3.4: Other applications

Document image analysis techniques have been used to solve problems in many other interesting and innovative applications, such as classification of fingerprints, musical-score reading, and extraction of data from microfilm images of punched cards. These applications are briefly described in this subsection.

- *Fingerprint classification:* The processing of fingerprint images can be classified as a graphics analysis application, although it differs from most other document applications in that fingerprints contain natural (versus machine-made) features. There are two levels of features in a fingerprint. On the "macroscopic" level, a fingerprint can be classified by the overall pattern of ridges (the lines making up the pattern). There are three main classifications of patterns at this level: a whorl (where the ridges form a bull's-eye pattern); a loop (where the ridge lines start at a lower corner, flow up, and around, back to the same corner in the opposite direction from where they started); and an arch (where the ridges flow from one lower corner, up, and then down to the opposite corner). There are subclassifications of these patterns, but these are few: usually about 10 in all. Also at the macroscopic level, one can identify singular points: the core, or the center of the fingerprint pattern (for example, the center of the bull's-eye), and the delta (usually at the lower side of a fingerprint where three macroscopic patterns abut). On the "microscopic" level, a fingerprint can be classified by the individual line features, called *minutiae*. These are ridge endings and bifurcations (branches in the ridges). A general sequence of fingerprint processing for matching is described as follows. First, the gray-scale fingerprint image undergoes pixel-level processing to reduce noise, enhance the ridges, and binarize [O'Gorman 1989]. Upon this binary image, the macroscopic pattern and singular points can be detected, usually by forming a direction map image and matching this with a pattern [Mehtre 1989, Mehtre 1993, Srinivasan 1992]. (For example, a whorl pattern will have all perpendicular directions of the ridges pointing approximately in the direction of the core.) Following this step, the fingerprint is thinned, then chain coded to locate line endings and bifurcations. Then, it is described by its macroscopic pattern and a feature

Figure 4. The fingerprint image on the left is processed and thinned to yield the result on the right. The macroscopic pattern of this print is an arch. The core is at the apex of the ridge lines, to the right of center. The delta is to the lower left of the core, where a triangular meeting of patterns occurs. One can also locate many bifurcation and line-ending minutiae.

vector of minutiae locations relative to the core and delta and/or relative to each other. Finally, this information is used for matching. Figure 4 shows a fingerprint example (the same as that shown in Section 2.4); here, some of the features are pointed out.

- *Musical Score Readings:* A top-down approach is used for recognizing printed piano music in [Kato 1992]. Although not all symbols in music are handled by this system, it is aimed at recognizing complex music notations where symbols are drawn with high density. The system uses musical knowledge and constraints to overcome the difficulties involved in processing such complex notations, both at the pattern recognition and the semantic-analysis stages. [Carter 1992] describes the initial processing steps for segmenting musical symbols from word-underlay and decorative graphics in early 17th-century madrigals notated in white mensural notation. Application of graph grammars for extracting information content in complex diagrams using music notation as an example is described in [Fahmy 1993]. The efficacy of morphological processing techniques for shape recognition to recognize music symbols is demonstrated in [Modayur 1993].
- *Extraction of data from microfilm images of punched cards:* The United States National Archives has a large collection of microfilm reels containing information about World War II enlisted personnel in the form of images of punched cards. Unfortunately, original punched cards or other forms of computer-readable data are not available—only their images are. It is impractical to rekey all this information, although the information stranded in these reels is not usable unless it is converted to a computer-readable form. A document image analysis system that includes semantic-analysis steps to convert these microfilm images is described in [Kumar 1992].

We include here five papers describing applications mentioned in this section. The papers by Vaxivière and Tombre and by Joseph and Pridmore describe mechanical-drawing recognition. Lai and Kasturi's paper is on recognition of dimension graphics in engineering drawings. The paper by Boatto et al. describes map recognition and Carter's paper discusses musical-score recognition.

5.4: Commercial state and future trends

In the mid-1980s, there was much hope for commercial products to automatically convert the vast archives of engineering drawings from paper to electronic form. Where these products simply binarized, compressed, and stored the data, they have been successful. However, where they promised component recognition, they have been less so. From this experience, the realization has come that the promise of completely automatic products was premature. Today's products for graphics recognition depend on human interaction to make decisions where the machine cannot, and the well-designed system is one that makes use of human aid most easily and quickly.

Although completely automatic graphics recognition will not be realized in the near future, there have been commercial successes in this field. We list some of these here, as well as suggestions on where progress will continue and where it is especially needed.

- For the archives of paper-based engineering drawings, products that simply digitize and compress—that is, perform no recognition—are a popular choice. Slightly more ambitious are products that employ some simple analytical methods, such as thinning and contour following, to store drawings in a more-compact form. It is now realized that the most ambitious products—ones that perform some recognition—require human interaction, as mentioned above. The up-front costs involved in interactive digitization are deemed to be worthwhile for applications where a drawing will subsequently undergo much updating or when it must be merged into more-recent drawings done in electronic (CAD) format. Future work in this area will include faster hardware for digitizing and vectorizing, better noise reduction techniques for separating components, fast matching and indexing techniques for recognizing components, and better human interaction tools for checking and correcting computer results.
- Computer graphics packages for drawing on personal computers and pen-based computers often contain some recognition techniques. These translate rough, hand-drawn geometric shapes into straight lines, perfect circles, and spline-based smooth curves. In the future, these packages will recognize more-complex shapes.
- Graphics libraries currently contain symbols for particular applications (for example, electronic components for electrical drawings and mechanical components for mechanical drawings). As recognition becomes more reliable, more general-purpose machines will be built that are similar to multifont OCR machines. These machines will recognize the character set(s) from symbols in the image and perform component recognition accordingly. Also, graphics recognition will be merged with OCR for increased recognition of graphics components such as company logos, lines in tables and forms, and diagrams.
- Drawings that are more complex than planar diagrams include shading and renditions of three-dimensional shapes. Computer vision images also possess these characteristics—that is, two-dimensional images of three-dimensional objects—and this similarity between planar drawings and more-complex diagrams suggests that as graphics recognition deals with these types of drawings, the distinction between the two areas will become blurred.

References

* Papers marked with an asterisk are included as reprinted papers in this book.

Antoine, D., S. Collin, and K. Tombre, "Analysis of Technical Documents: The REDRAW System," in *Structured Document Image Analysis*, H.S. Baird, H. Bunke, and K. Yamamoto eds., Springer-Verlag, New York, N.Y., 1992, pp. 385–402.

Arias, J.F., et al., "Interpretation of Telephone System Manhole Drawings," *Proc. 2nd Int'l Conf. Document Analysis and Recognition*, IEEE CS Press, Los Alamitos, Calif., 1993, pp. 365–368.

*Boatto, L., et al., "An Interpretation System for Land Register Maps," *Computer*, Vol. 25, No. 7, July 1992, pp. 25–33.

Bow, S., and R. Kasturi, "A Graphics Recognition System for Interpretation of Line Drawings," in *Image Analysis Applications*, R. Kasturi and M.M. Trivedi, eds., Marcel Dekker, New York, N.Y., 1990.

Bunke, H., and G. Allerman, "Probabilistic Relaxation for the Interpretation of Electrical Schematics," *Proc. Conf. Pattern Recognition and Image Processing (PRIP),* IEEE CS Press, Los Alamitos, Calif., 1981, pp. 438–440.

*Carter, N.P., "Segmentation and Preliminary Recognition of Madrigals Notated in White Mensural Notation," *Machine Vision and Applications*, Vol. 5, No. 3, 1992, pp. 223–229.

Chandran, S., and R. Kasturi, "Structural Reconfiguration of Tabulated Data" *Proc. 2nd Int'l Conf. Document Analysis and Recognition*, IEEE CS Press, Los Alamitos, Calif., 1993.

Ejiri, M., *et al.*, "Automation Recognition of Engineering Drawings and Maps," in *Image Analysis Applications*, R. Kasturi and M.M. Trivedi, eds., Marcel Dekker, New York, N.Y., 1990, pp. 73–126

Ettinger, G.J., "Large Hierarchical Object Recognition Using Libraries of Parameterized Model Sub-Parts," *Proc. Computer Vision and Pattern Recognition*, IEEE CS Press, Los Alamitos, Calif., 1988, pp. 32–41.

Fahmy, H., and D. Blostein, "A Graph Grammar Programming Style for Recognition of Music Notation," *Machine Vision and Applications*, Vol. 6, No. 2, 1993, pp. 83–99.
Fahn, C.S., J.F. Wang, and J.Y. Lee, "A Topology-Based Component Extractor for Understanding Electronic Circuit Diagrams," *Computer Vision, Graphics, and Image Processing*, Vol. 44, 1988, pp. 119–138.
*Fletcher, L.A., and R. Kasturi, "A Robust Algorithm for Text String Separation from Mixed Text/Graphics Images," *IEEE Trans. Pattern Analysis and Machine Intelligence*, Vol. 10, No. 6, Nov. 1988, pp. 910–918.
Grimson, W.E.L., and T. Lozano-Perez, "Model-Based Recognition and Localization from Sparse Range or Tactile Data," *Int'l J. Robotics Research*, Vol. 3, No. 3, 1984, pp. 3–35.
Grosky, W.I., and R. Mehrotra, *Computer*, Vol. 22, No. 12, 1989 (special issue on image database management).
Harada, H., Y. Itoh, and M. Ishii, "Recognition of Free-Hand Drawings in Chemical Plant Engineering," *Proc. IEEE Workshop Computer Architecture for Pattern Analysis and Image Database Management*, IEEE CS Press, Los Alamitos, Calif., 1985, pp. 146–153.
Jain, A.K., *Fundamentals of Digital Image Processing*, Prentice-Hall, Englewood Cliffs, N.J., 1989.
*Joseph, S.H., and T.P. Pridmore, "Knowledge-Directed Interpretation of Mechanical Engineering Drawings," *IEEE Trans. Pattern Analysis and Machine Intelligence*, Vol. 14, No. 9, Sept 1992, pp. 928–940.
Karima, M., K.S. Sadhal, and T.O. McNeil, "From Paper Drawings to Computer Aided Design," *IEEE Computer Graphics and Applications*, Vol. 5, No. 2, Feb. 1985, pp. 24–39.
Kasturi, R., et al., "Map Data Processing in Geographical Information Systems," *Computer*, Vol. 22, No. 12, Dec 1989, pp. 10–21.
Kasturi, R., et al., "A System for Interpretation of Line Drawings," *IEEE Trans. Pattern Analysis and Machine Intelligence*, Vol. 12, No. 10, Oct. 1990, pp. 978–992.
Kato, O., and S. Inokuchi, "A Recognition System for Printed Piano Music," in *Structured Document Image Analysis*, H.S. Baird, H. Bunke, and K. Yamamoto, eds., Springer-Verlag, New York, N.Y., 1992, pp. 435–455.
Kumar, S.U., and R. Kasturi, "Text Data Extraction from Microfilm Images of Punched Cards," *Proc. 11th IAPR Int'l Conf. Pattern Recognition (ICPR)*, IEEE CS Press, Los Alamitos, Calif., 1992, pp. 230–233.
*Lai, C.P., and R. Kasturi, "Detection of Dimension Sets in Engineering Drawings," *IEEE Trans. Pattern Analysis and Machine Intelligence*, Vol. 16, No. 8, 1994, pp. 848–854.
Lai, C.P., and R. Kasturi, "Detection of Dimension Sets in Engineering Drawings," *Proc. 2nd Int'l Conf. Document Analysis and Recognition*, IEEE CS Press, Los Alamitos, Calif., 1993, pp. 606–613.
Lee, S.W., "Recognizing Hand-Drawn Electrical Circuit Symbols," in *Structured Document Image Analysis*, H.S. Baird, H. Bunke, and K. Yamamoto, eds., Springer-Verlag, New York, N.Y., 1992, pp. 340–358.
Lysak, D., and R. Kasturi, "Interpretation of Line Drawings with Multiple Views," *Proc. 10th Int'l Conf. Pattern Recognition (ICPR)*, IEEE CS Press, Los Alamitos, Calif., 1990, pp. 220–222.
Mehtre, B.M., "Fingerprint Image Analysis for Automatic Identification," *Machine Vision and Applications*, Vol. 6, No. 2, 1993, pp. 124–139.
Mehtre, B.M., and B. Chatterjee, "Segmentation of Fingerprint Images—A Composite Method," *Pattern Recognition*, Vol. 22, No. 4, 1989, pp. 381–385.
Modayur, B.R., et al., "MUSER: A Prototype Musical Score Recognition System Using Mathematical Morphology," *Machine Vision and Applications*, Vol. 6, No. 2, 1993.
Nagasamy, V., and N.A. Langrana, "Engineering Drawing Processing and Vectorization System," *Computer Vision, Graphics, and Image Processing*, Vol. 49, 1990, pp. 379–397.
O'Gorman, L., "*kxk* Thinning," *Computer Vision, Graphics, and Image Processing*, Vol. 51, 1990, pp. 195–215.
O'Gorman, L., and J.V. Nickerson, "An Approach to Fingerprint Filter Design," *Pattern Recognition*, Vol. 22, No. 1, 1989, pp. 29–38.
Okazaki, A., et al., "An Automatic Circuit Diagram Reader with Loop-Structure-Based Symbol Recognition," *IEEE Trans. Pattern Analysis and Machine Intelligence*, Vol. 10, No. 3, May 1988, pp. 331–341.
Shih, C-C., and R. Kasturi, "Extraction of Graphic Primitives from Images of Paper-Based Drawings," *Machine Vision and Applications*, Vol. 2, 1989, pp. 103–113.
Srinivasan, V.S., and N.N. Murthy, "Detection of Singular Points in Fingerprint Images," *Pattern Recognition*, Vol. 25, No. 2, 1992, pp. 139–153.
*Vaxivière, P., and K. Tombre, "Celesstin: CAD Conversion of Mechanical Drawings," *Computer*, Vol. 25, No. 7, July 1992, pp. 46–54.
Wakayama, T., "A Core Line Tracking Algorithm Based on Maximal Square Moving," *IEEE Trans. Pattern Analysis and Machine Intelligence*, Vol. PAMI 4, No. 1, Jan. 1982, pp. 68–74.
Wesley, M.A., and S. Markowsky, "Fleshing Out Projections," *IBM J. Research and Development*, Vol. 25, No. 6, Nov. 1981, pp. 934–954.

A Robust Algorithm for Text String Separation from Mixed Text/Graphics Images

LLOYD ALAN FLETCHER AND RANGACHAR KASTURI, MEMBER, IEEE

Abstract—**An automated system for document analysis is extremely desirable. A digitized image consisting of a mixture of text and graphics should be segmented in order to represent more efficiently both the areas of text and graphics. This paper describes the development and implementation of a new algorithm for automated text string separation which is relatively independent of changes in text font style and size, and of string orientation. The algorithm does not explicitly recognize individual characters. The principal components of the algorithm are the generation of connected components and the application of the Hough transform in order to group together components into logical character strings which may then be separated from the graphics. The algorithm outputs two images, one containing text strings, and the other graphics. These images may then be processed by suitable character recognition and graphics recognition systems. The performance of the algorithm, both in terms of its effectiveness and computational efficiency, was evaluated using several test images. The results of the evaluations are described. The superior performance of this algorithm compared to other techniques is clear from the evaluations.**

Index Terms—**Connected component analysis, document analysis, image processing, image segmentation, image understanding, text and graphics separation, text recognition.**

INTRODUCTION

THE move from paper-based documentation towards computerized storage and retrieval systems has been prompted by the many advantages to be gained from the "electronic document" environment. Document update and revision is efficiently achieved in the computerized form. For efficient processing and storage of documents, however, it is necessary to generate a description of graphical elements in the document rather than a bit-map in order to decrease the storage and processing time. Thus, increasing emphasis is being placed on the need for the realization of computer-based systems which are capable of providing automated analysis and interpretation of paper-based documents [1]–[3]. Much of the attention paid to automated document analysis systems in the literature has been in relation to engineering drawings and diagrams [4], [5]. Such systems provide a means for originating technical information (text and graphics) in a digital form suitable for interactive graphics editing, reproduction, and distribution.

Manuscript received August 8, 1986; revised June 12, 1987. Recommended for acceptance by J. Kittler. This work was supported in part by the National Science Foundation under Grant ECS-8307445.

L. A. Fletcher was with the Department of Electrical Engineering, the Pennsylvania State University, University Park, PA 16802. He is now with Bell Communications Research, 331 Newman Springs Road, Red Bank, NJ 07701.

R. Kasturi is with the Department of Electrical Engineering, the Pennsylvania State University, University Park, PA 16802.

IEEE Log Number 8823861.

The initial processing stage in an automated document analysis system requires conversion of paper-based graphics/text to a digital bit-map representation. Wherever the primary goal of the automated document analysis system is interpretation of graphic data, text strings present within the digitized document must first be separated from the graphics in order that subsequent processing stages may operate exclusively on the graphic information. The extracted text may be stored separately for input to a character recognition system for later retrieval or revision. Since document types vary widely in style and content of both graphic and text data, an algorithm to perform text string removal must be able to accommodate documents containing text of various font styles and sizes. Further, the documents may contain text strings which are intermingled with graphics, and text characters which are similar in size or shape to graphics. In general, text strings may be of any orientation in the image; not simply horizontal or vertical but possibly diagonally aligned.

Several algorithms for text string separation have been reported in the literature [6]–[9]. However, many of these algorithms are very restrictive in the type of documents they can process and are therefore not useful in a general automated document analysis system. For example, the combined symbol matching algorithm [7] is sensitive to changes in text font style and size: the algorithm must be run once for each font type using different parameters, words of less than three characters which are embedded in longer strings are not removed, and the string removal is only along a specified orientation. The Block Segmentation technique [8] broadly classifies regions into text or graphics; i.e., characters which lie within predominantly graphics regions are classified as graphics. The Bley algorithm [9] is also sensitive to variations in text font style and size; the algorithm breaks connected components into subcomponents, which makes it difficult to process the components for graphics recognition. Thus there is no single algorithm which is robust enough to segment images containing mixed graphics and text, with multiple font styles and sizes and strings of arbitrary orientation. This paper describes a new robust algorithm for text string separation from mixed text/graphics images.

A ROBUST ALGORITHM FOR TEXT STRING SEPARATION

A robust algorithm for text separation has been designed to separate text strings from graphics, regardless of string orientation and font size or style. The algorithm uses simple heuristics based on the characteristics of text strings. The algorithm performs no character recognition and can only separate characters from text on the basis of size and orientation. Thus, for proper operation, the document should meet the following requirements:

1) The range in text font sizes within an image should be such that the average height of the characters belonging to the largest font is not greater than five times the average height of characters belonging to the smallest character font.

2) The interline spacing between parallel lines of characters should not be less than one quarter of the average height of characters in the line having the larger font size.

3) The resolution of the digitizer used should be such that each text character forms an isolated component not connected to graphics or other characters.

4) The largest character in any phrase should not have an area greater than five times the average area of all characters in the phrase.

5) The intercharacter gap between characters in the same word should not be greater than the average local character height. Also, the interword gap between words of the same phrase should not be greater than 2.5 times the average local character height.

If these constraints are satisfied for any particular image, the performance of the algorithm will be highly reliable. However, acceptable segmentation is obtained even if some of these constraints are not strictly satisfied. For example, the algorithm will correctly segment text strings even when some of the characters are eight connected to each other, as long as there are at least three components in each string.

The algorithm described here is designed to accomplish the following without explicitly recognizing individual characters:

- Locate potential text strings and separate them from the large graphics.
- Examine the strings and separate all phrases that contain at least three characters.
- Generate two images, one each for the text strings and graphics.

The algorithm consists of the following steps:

- Connected component generation.
- Area/ratio filter.
- Collinear component grouping.
- Logical grouping of strings into words and phrases.
- Text string separation.

These processing steps are described in detail in the following sections. A test image of 2048 × 2048 pixels, shown in Fig. 1, is used to demonstrate the performance of this algorithm.

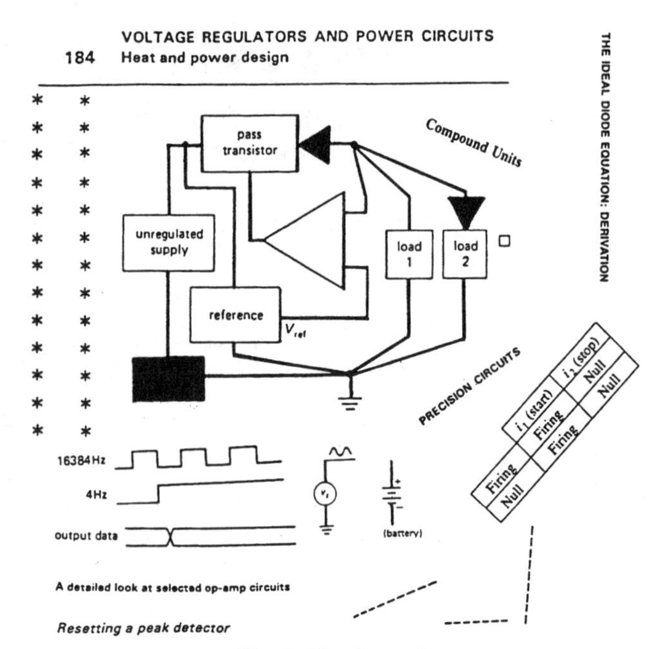

Fig. 1. Test image 1.

Connected Component Generation

Connected component generation involves grouping together black pixels which are eight connected to one another (assuming a black image on white background). In this technique, eight connected pixels belonging to individual characters or graphics are enclosed in circumscribing rectangles. Each rectangle thus identifies a single connected component. The output of the connected component generation algorithm is an information array which specifies: the maximum and minimum coordinates of the circumscribing rectangles of connected components, the coordinates of the top and bottom seeds of each connected component, and the number of black pixels. Black pixel density and dimensional ratios are easily obtained from this information. This algorithm is described in detail in [10]. Although this algorithm is similar to the algorithms described in the literature [11], [12], in our algorithm we do not change the value of each pixel to its corresponding component label, but retain the coordinates of top and bottom seeds for each component.[1] Each connected component is then either rejected or accepted as a member of a text string based on its attributes (size, black pixel density, ratio of dimensions, area, position within the image etc.).

The enclosing rectangles corresponding to the connected components of Fig. 1 are shown in Fig. 2. Note that the largest graphic is enclosed by a single rectangle. Also, the long line near the top of the image is enclosed by a rectangle. This rectangle has a height which is significantly greater than the thickness of the line since the line is oriented at a slight angle to the image axis.

Area/Ratio Filter

By initial examination of connected component attributes, the working set of connected components can be reduced to one which contains a higher percentage of

[1]An image typically contains more than 256 components. Labeling each pixel with its corresponding component number would require at least 2 bytes per pixel. This would double the memory requirement to 8 MB for a 2048 × 2048 pixel image and significantly deteriorates system performance (through increased page faults). However, this necessitates multiple passes in the text string removal stage to remove all pixels belonging to a particular connected component.

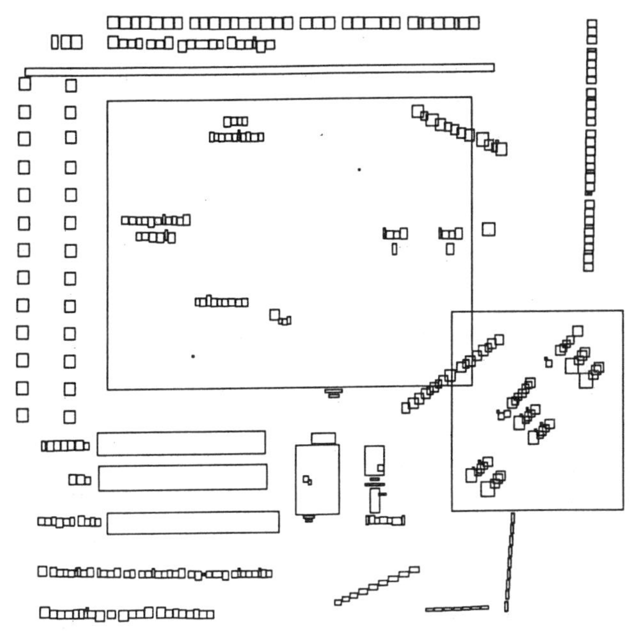

Fig. 2. Rectangles enclosing the connected components of test image 1.

characters. In general, a mixed text/graphics image will produce connected components of widely varying areas. The larger connected components represent the larger graphic components of the image. It is desirable to locate and discard large graphics in order to restrict processing to components which are candidates for members of text strings, thereby improving both accuracy and processing speed.

By obtaining a histogram of the relative frequency of occurrence of components as a function of their area, it is possible to set an area threshold which broadly separates the larger graphics from the text components. By correct threshold selection, the largest of the graphics can be discarded, leaving only the smaller graphics and text components as members of the working set of connected components. The area threshold selection procedure must ensure that the threshold lies outside that part of the histogram which broadly represents the set of text characters. The threshold itself is determined using the area histogram (as opposed to presetting it to a value) although *a priori* knowledge about the types of documents being processed (such as the amount of text data compared to graphics), if available, is helpful in determining this threshold. For documents that contain a substantial amount of text and some large graphics, the most populated area will, in general, represent mostly text components or, at least, small graphics. By ensuring that the threshold is set above the most populated area, A_{mp}, the possibility of discarding members of the text character set is avoided. This alone is not adequate to process images that contain text strings of different sizes (it is likely that the most populated area corresponds to the smallest sized characters). Thus, a second parameter, the average area A_{avg}, is computed. The area threshold is then set at five times the larger of the two parameters A_{mp} and A_{avg}. The histogram is searched to locate components that are larger than this threshold. Since text characters of the same size may have different areas, the histogram is "blurred" such that neighboring areas are grouped together in order to determine the most populated area.

A similar type of filtering approach is taken with the connected component dimensional ratio attribute. Very long lines are unlikely to be text characters, and should therefore be discarded. Isolated straight lines within the image may be discarded on the basis of dimensional ratio. If a connected component has a ratio of less than 1:20 or greater than 20:1, it is discarded.

Collinear Component Grouping

Text strings are composed of characters which are oriented along the same straight line. In order to logically connect or group together characters into strings, it is necessary to determine which characters in the image actually lie along any given straight line. By grouping components into strings associated with a particular line, the components can be ordered according to their distance along the line. Further grouping may then take place by examining the distance between characters. By comparing the intercharacter distance with the interword gap and intercharacter gap thresholds, the string can be segmented into logical character groups, that is, into logical words and phrases. Only if components belong to a logical character group can they be regarded as members of a valid text character string. Such components are identified and are separated from the image. These steps are explained in detail in this and the following sections.

The Hough transform is a line to point transformation [13] which, when applied to the centroids of connected components in an image, can be used to detect sets of connected components that lie along a given straight line. A line in the Cartesian space (x, y) given by the equation

$$\rho = x \cos \theta + y \sin \theta \tag{1}$$

is represented by a point (ρ, θ) in the Hough domain. Similarly, every point (x, y) in the Cartesian space maps into a curve in the Hough domain. Thus to locate all the connected components that are collinear, the Hough transform is applied to the centroids of the rectangles enclosing each connected component. All the curves corresponding to the collinear components intersect at the same point (ρ, θ), where ρ and θ specify the parameters of the line.

In the discrete case, the Hough domain is a two-dimensional array representing discrete values of ρ and θ. The resolution along the θ direction is set to one degree. The resolution along the ρ direction is treated as a variable R. Optimal selection of R is critical to correct grouping of connected components. This is because in any text string, the centroids of all characters belonging to a phrase are not necessarily collinear. For any phrase, character heights vary (upper, lower case) and character positions vary in relation to each other (ascenders, descenders). In general, the centroids of characters belonging to the same phrase will lie in the range $\pm\delta_c$ from the axis of the line connecting the phrase members. However, δ_c is a function of the text string height. Thus, it is important to select a Hough domain resolution, before applying the transform, which will cause the majority of component centroids belonging to a phrase to be grouped into the same cell. The optimal resolution is dependent on the heights and relative positions of characters within a string. Too large a value for R may cause several parallel strings to be grouped into a single cell (over-grouping). If R is too small then connected components belonging to the same string may be grouped into different cells (undergrouping). The value of R should depend on the average local character height, H_a, for a line l. However, since H_a cannot be determined until characters have actually been detected as belonging to a collinear string, and since R must be known before such grouping, an initial estimate for R is made based on the average height, H_{ws}, of all connected components in the current working set. The resolution is then set to: $R = 0.2 \times H_{ws}$.

The resolution R allows for some perturbations in the collinearity of components belonging to text strings. However, this will not be adequate to group all ascenders and descenders in a character string. Further, the value of R, which is based on average character height, will not be optimal for grouping text strings of different heights. Thus, clustering in the Hough domain may become necessary for each primary cell. Here, "primary cell" refers to the cell containing the majority of the connected components in a string. In order to group together *all* components belonging to a string, neighboring cells must be clustered into the primary string. The required degree of clustering, δ_{clus}, will depend on the local character height. However, when strings are initially extracted, 11 cells centered around the primary cell are grouped for examination. Once the string has been extracted, the local clustering factor can be determined.

The average local character height for a string H_a is used to determine the degree of clustering around the primary cell. This will ensure that all ascenders and descenders in a string are included in the text string grouping. The clustering factor is then set to H_a/R.

Since the vast majority of documents contain text strings which are parallel to the document axes, it is advantageous to remove these strings first. To allow for orientation errors during digitization, a tolerance of five degrees is allowed in processing horizontal and vertical lines. Strings at other orientations are processed after considering the horizontal and vertical strings. Furthermore, more populated cells in the Hough domain are processed before processing less populated cells. This order of string processing gives greatest weight to longer strings. If shorter phrases or words were treated before longer ones, a degradation in performance would be observed.

In summary, extraction of strings from the Hough domain is performed as follows:

1) Calculate the average height of components H_{ws} in the working set.

2) Set the Hough domain resolution R to $0.2 \times H_{ws}$. Set a counter to zero.

3) Apply the Hough transform to all components in the working set for θ in the range: $0° \leq \theta \leq 5°$, $85° \leq \theta \leq 95°$, $175 \leq \theta \leq 180°$. (In our implementation ρ is allowed to have negative values and $\theta_{\max} = 180°$.) Set the running threshold $RT_c = 20$.

4) For each cell having a count greater than RT_c, perform steps 5–10.

5) Form a cluster of 11 ρ cells (constant θ), including the primary cell, centered around the primary cell.

6) Compute the average height of components in the cluster H_a.

7) Compute the new clustering factor $f_{clus} = H_a/R$.

8) Re-cluster $\pm f_{clus}$ cells centered around the primary cell, including the primary cell.

9) Perform string segmentation (explained in detail in a later section).

10) Update the Hough transform by deleting the contributions from all components discarded in step 9.

11) Decrement RT_c by one. If RT_c is greater than 2, go to step 4.

12) If the counter is equal to 1 then stop. Otherwise compute the Hough transform for the remaining components for θ in the range: $0° \leq \theta < 180°$. Reset RT_c to 20. Increment the counter by one. Go to step 4.

Step 10 above is performed in order to "refresh" the Hough domain. If the Hough array is not continually updated, discarded characters contribute to cell counts despite having been removed. This can lead to a severe degradation in processing speed since information about discarded components is unnecessarily extracted from the Hough domain. Thus, the refreshing of the Hough domain is designed to improve performance.

Logical Grouping of Strings into Words and Phrases

Fig. 3 shows the connected components associated with the line l_{pr}. Note that the centroids of these components do not necessarily lie on the line l_{pr}. It is necessary to order these components according to their distance along the line [from a reference point, say $(0, y_0)$]. That is, for each component the distance along the line l_{pr} is calculated.

In order to locate words and phrases, the positional relationships between connected components are examined. It is the intercharacter gap and interword gap thresholds, T_c and T_w, which are used to determine component grouping. These parameters are functions of the character height. In computing this, an average of the height of several local characters should be used rather than simply using the height of the component currently under examination. For example, the component under examination may be a period at the end of a string, which would have very small height. For this reason the four neighboring (two on each side) characters are used, along with the current component, to determine the thresholds T_c and T_w. These operations are explained in the following paragraphs.

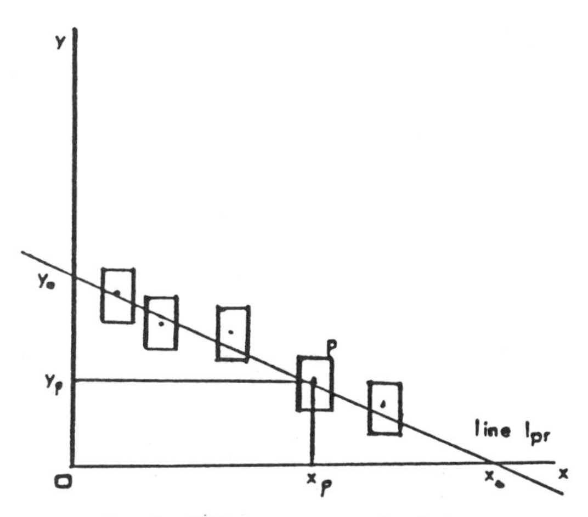

Fig. 3. Collinear component ordering.

Associated points of the line l_{pr}, once ordered, become part of a data structure containing all information about components belonging to all collinear strings. An entry is made in the string information array for each associated point, pointing to the connected component to which each point belongs. The information contained in the connected component array is then used to compute the edge-to-edge distance, D_e, for every pair of adjacent components. The magnitude of D_e belonging to the associated point i, is the distance between the two closest edges of the connected components associated with collinear component elements i and $i + 1$.

By placing a window over each connected component associated with the string information array, and considering four nearest neighbors, the average height of connected components within the window H_c is determined. In computing this average the vertical dimension of the enclosing rectangles is used for lines oriented between 0 and 45 degrees and the horizontal dimension is used for strings oriented at angles greater than 45 degrees. T_c is the intercharacter gap threshold, set at $T_c = H_c$. The interword gap threshold T_w is set at $2.5 \times H_a$. Two adjacent connected components separated by an edge-to-edge distance D_e of less than or equal to T_c belong to the same word group. Word groups separated by no more than T_w belong to the same phrase group. Connected components which belong to neither of these types are labeled as isolated characters. Note, however, that single characters which are embedded in a multiword phrase become part of the phrase group. Isolated words of less than N ($N = 3$ in our experiment) characters are not removed from the image. Further, when several isolated characters satisfy the interword gap to form a phrase containing more than three characters, the condition is checked and the characters are not removed.

The string segmentation algorithm is summarized as follows:

1) Define the data structure *String* of which each element is an associated point belonging to the same line. Define the data structure *Groups*, a vector of 100 elements. Each element of *Groups* contains four fields. The *Type* field may have the label "i," "w," or "p" (initialized to "i"), specifying the group type. The fields *Head* and *Tail* serve as pointers to the first and last connected components within the group. The last field *Num* contains the number of connected components belonging to the group. The data structure *Phrases* is identical to *Groups* except that it does not possess the *Type* field.

2) For each connected component associated with the currently active string, determine T_c and T_w based on the neighborhood average character height H_c. If $String_i(D_e) \leq T_c$ then elements i and $i + 1$ belong to the same word group. If the current group is a new one, set $Groups_{gn}(head) = i$ and $Groups_{gn}(tail) = i + 1$, where gn is the number of groups currently detected in the string. Otherwise set $Groups_{gn}(tail) = i + 1$. Set $Groups_{gn}(num) = Groups_{gn}(num) + 1$. If the group type is not "p," set $Groups_{gn}(type) =$ "w." Continue the T_c threshold test until the current group is terminated. If the T_c test fails, terminate the group entry by advancing the value of gn by one.

3) Compare $String_i(D_e)$ with T_w. If $D_e < T_w$ then the previous group and the current connected component belong to the same phrase. Make entries in *Phrases* to point to the member groups. Set $Groups_{gn}(type) =$ "p" and $Groups_{gn-1}(type) =$ "p." Begin a new group entry for the associated connected component i. Repeat step 2. If the T_w threshold is not satisfied then the adjacent connected components currently under consideration do not belong to the same phrase: terminate any active entry in *Phrases*. Begin a new group by repeating step 2.

4) Once all entries in *String* have been classified, search for two or more consecutive "i" groups present within a phrase. There should never be more than two consecutive isolated characters embedded within the same phrase. Remove any such sequences by splitting or truncating groups as necessary.

The results of the string segmentation are two data structures which provide information about words belonging to a particular phrase and information about the connected components which make up any words within a phrase.

The problem of "destructive line overlap" occurs if shorter strings, which overlap longer strings, are processed before the longer strings. For example, consider the strings shown in Fig. 4. All the components in Fig. 4(a) belong to at least one phrase. Thus, all of these components must be removed. However, if short strings are removed first, we get Fig. 4(b). Finally, after considering the longer strings, we get Fig. 4(c). Thus, it is important to consider longer strings before shorter strings for grouping and removal. However, these strings belong to different collinear component clusters in the Hough domain. Therefore, it is necessary to examine all clusters to separate longer strings before removing shorter strings.

The string processing procedure thus gives weight to a phrase group depending on the number of connected components belonging to the particular group. In examining cells for the extraction of collinear component sets, the

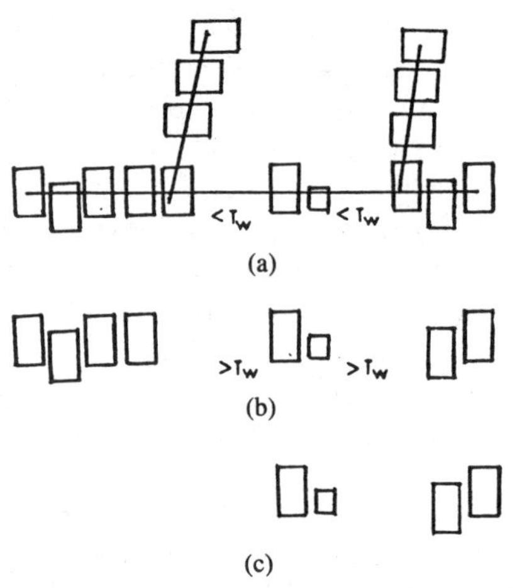

Fig. 4. Effect of processing shorter strings before longer strings: (a) characters oriented along three lines; (b) result of processing shorter strings first; (c) final result: some of the characters are not removed.

running threshold RT_c is used. The running threshold is also used as the threshold for the phrase population. Character groups are only processed if the group population exceeds both RT_c and N. This avoids the confusion which can result from destructive line overlap. Further refinements are also applied to discard disproportionately larger connected components, possibly due to graphics, from text strings.

Words or phrases which consist of several repeated characters, such as dotted or dashed lines, should not be removed from the image. Such components are in fact part of the graphics rather than actual text strings. Groups of such repeated characters possess a high degree of consistency in both ratio and black pixel density. Hence, repeated strings can be detected and discarded by calculating the variance in ratio and density for each phrase of a collinear component set. If both the variance in ratio and density fall below a given threshold for three or more consecutive characters, then these repeated characters of the phrase group are discarded from the working set and the current collinear component set, and are not deleted from the image.

Text String Separation

Once all strings have been extracted from the Hough transform, all connected components corresponding to these strings are deleted from the image array. This involves replacing black pixels, belonging to marked connected components, with white pixels. It is important that only those black pixels which originally formed a particular connected component should be deleted, not simply all black pixels which lie within the area of the circumscribing rectangle. It is for this reason that the seeds of the connected components are retained during connected component generation. Knowing the location of $Seed_t$ and $Seed_b$ of the connected component allows all black pixels which are eight connected to them to be detected and marked for deletion. Using a line by line ''scan'' method similar to that used in the connected component formation process described earlier, runs of eight connected black pixels are merged together. If a black pixel belongs to a group to which one of the seeds also belongs then the pixel is marked for deletion. The scanning procedure is repeated four times, twice for each seed. The procedure is begun with the top seed, moving south; then starting from the bottom seed and scanning north; then east from the top seed again; finally the component is scanned west from the bottom seed. This multidirectional scan ensures that all possible eight connected groups are detected. All black pixels marked during this procedure are then replaced with white pixels.

The algorithm is implemented in Pascal. The program was run on a DEC VAX 11/785 running the VMS operating system. The complete program includes over 1000 lines of code. The following section presents the results obtained by application of the algorithm to several test images.

Experimental Results

In order to test and evaluate the performance of the algorithm, several test images were required. Several documents were digitized to 2048 × 2048 pixels using a Microtek Image Scanner (interfaced to an IBM PC-AT) with a resolution of 300 pixels per inch. The scanned images in compressed form were transfered to the VAX using KERMIT communications software. The results of the algorithm application to each image are illustrated and discussed in [10]. In this paper we limit the discussion to two test images.

The image shown in Fig. 1 has several characteristics found in different types of mixed text/graphics images: various text font styles and sizes; text strings enclosed by graphics; text strings with various orientations; graphics of various size and shape. In general, most document types do not possess such a wide variation in text font styles, sizes, and orientations within the same document. However, if the algorithm could be made robust to such variations in image characteristics, then it would also be robust when applied to more uniform images.

The result of the removal of all phrases of three or more characters is shown in Fig. 5. Note that several of the characters in the diagonally oriented table, such as ''g'' and ''p'' are not removed. These letters are in fact eight connected to the table itself, as can be seen from the connected components in Fig. 4. Thus, these characters are part of the graphics and therefore cannot be isolated for identification as characters. Note that the series of asterisks have not been removed since they constitute repeated ''isolated character'' phrases. Although the inter-component gap satisfies the interword gap threshold, it does not satisfy the inter-character gap threshold. As a result, the string is discarded since it is a phrase made up of single character words. Of the three dashed lines in the bottom right of the image, two strings have been partially removed. The ratio consistency of these strings was not sufficient to prevent their removal as text. The consistency in both ratio and density of the other dashed line

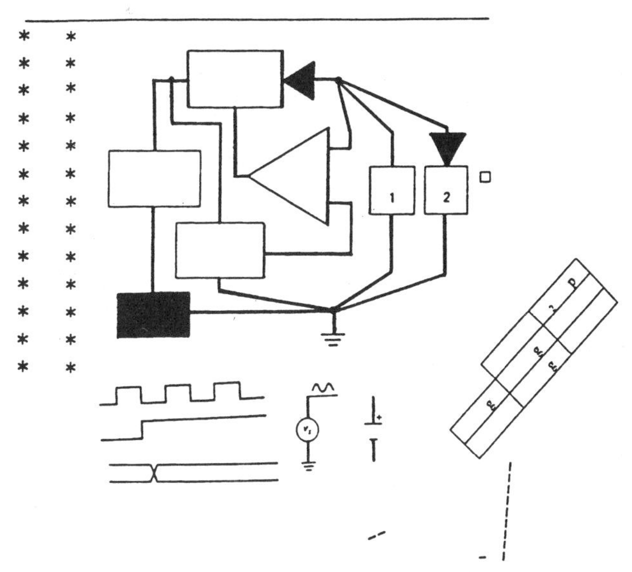

Fig. 5. Output after text string separation.

was high enough for it to be retained as a graphic. Note that three of the small graphics in the neighborhood of the battery figure have been deleted. These were grouped together as a three component phrase since the inter-word gap threshold was satisfied.

The image of Fig. 6 has a large number of text strings. In this respect this image complements the first test image. Due to ink spread in the original document several characters are connected to each other. For example, the characters s and e in most of the occurrence of the word ''Base'' are connected to each other as can be easily verified in the connected components diagram shown in Fig. 7. In several instances the word contains only two components. The output image after text extraction is shown in Fig. 8. All character strings, including those that are eight connected to each other, have been removed with one notable exception: the letters ''S'' and ''k'' in the work ''Stack'' in second column. In this word the letters a and c are connected to each other making it a string of four characters. The two connected components in the middle of the string are removed as parts of vertical strings (longer than 4 characters) before the algorithm looks for strings which contain four characters. Thus the letters S and k are left behind. The reasons for other components which are not removed are as follows: two of the 0's in the string ''000'' are connected to each other making it a two character string: the intercharacter gap between the string ''SI'' at the top of the second column and its previous character is large compared to the height of the previous character (the + sign is made up of two disjoint components of relatively small heights; see Fig. 7). Similar comments hold good for the strings 10 and 11 that are next to ''='' signs. Other strings which have not been removed contain less than three characters.

	no displacement	signed byte displacement	unsigned word displacement	no memory reference
r/m =	mode = 00	mode = 01	mode = 10	mode = 11
000	Base Relative Indexed BX + SI	Base Relative Indexed BX + SI + DISP	Base Relative Direct Indexed BX + SI + DISP	Register Identification in R/M field for Register-to-Register Instructions
001	Base Relative Indexed BX + DI	Base Relative Direct Indexed BX + DI + DISP	Base Relative Direct Indexed BX + DI + DISP	
010	Base Relative Indexed Stack BP + SI	Base Relative Direct Indexed Stack BP + SI + DISP	Base Relative Direct Indexed Stack BP + SI + DISP	
011	Base Relative Indexed Stack BP + DI	Base Relative Direct Indexed Stack BP + DI + DISP	Base Relative Direct Indexed Stack BP + DI + DISP	
100	Implied SI	Direct Indexed SI + DISP	Direct Indexed SI + DISP	
101	Implied DI	Direct Indexed DI - DISP	Direct Indexed DI + DISP	
110	Direct Direct Address	Base Relative Direct Stack BP + DISP	Base Relative Direct Stack BP + DISP	
111	Base Relative BX	Base Relative Direct BX + DISP	Base Relative Direct BX + DISP	

Figure 2.9 Format of the I8086/8088 two-operand instruction such as ADD with variations. Memory-to-memory instructions are not provided. Note that the instruction can occupy two, three, or four bytes. (Reprinted by permission of Intel Corporation. Copyright 1980.)

Fig. 6. Test image 2.

Fig. 7. Connected components of test image 2.

Fig. 8. Output after text string separation.

Performance Evaluation

The results of the application of the text string separation algorithm, given in the previous section, demonstrates the robustness of the algorithm to images of various types. In this section the algorithm is evaluated in terms of processing speed.

The time required to generate connected components depends not only on the number of components within the image, but also on the complexity of the information represented by the black pixels. In general, however, processing time increases in direct proportion to the number of connected components in the image. The time required for the application of the Hough transform to the active set of components also increases with the number of connected components. However, the time for a single Hough transform consumes only about five percent of the total processing time for an entire image.

The algorithm is implemented in several distinct steps. In Table I, a breakdown of the processing times required for several operations applied to Test Image 1 is given. Times for operations on both the whole image (reading, writing, connected component generation and deletion) and operations on character strings (segmentation, area thresholding) are given. Steps 1–7 and step 12 in the table are operations performed on the whole image array. Steps 8–11 are performed on a single collinear component set. The CPU time for each of these operations is given for a collinear component set of 15.

Several string operations typically require processing times of only a few tenths of a second. These processing times vary with the number of connected components in a collinear component set. Although individually the string processing steps require relatively small processing time, when several hundred collinear component sets are processed (as in a typical image) the cumulative processing time becomes significant and is directly related to the number of components in an image and the complexity of the image. The processing time increases as the cell population threshold decreases, since more cells become available for extraction. Many more collinear component sets of lower population are processed than sets of higher population. As the population threshold RT_c decreases, the processing time for the working set rapidly increases.

TABLE I
BREAKDOWN OF CPU TIMES FOR PROCESSING TEST IMAGE 1 AND FOR PROCESSING STEPS FOR A COLLINEAR COMPONENT SET OF POPULATION 15

Processing step	CPU time (secs.) for each operation	CPU time (secs.) for whole image
1. Read 2K image	-	48.5
2. Write 2K image to disk	-	50.8
3. Generation of CC's (395)	-	47.7
4. Initialize *Info* array	-	0.14
5. Filter out CC's based on area and ratio	-	0.47
6. Hough transform (393 CC's, 0°, 90°)	-	9.3
7. Hough transform (270 CC's, all angles)	-	18.9
8. Clustering and *String* extraction	0.54	243.3
9. *String* element initialization	0.04	94.8
10. String grouping	0.42	148.5
11. String refinement	0.11	94.3
12. Connected component deletion	-	8.50
Total		764.31

Algorithm Improvements

The algorithm is highly CPU intensive. In order to be useful in automated document processing, the processing time should be minimized. A hardware implementation of the algorithm would achieve a reduction in the required CPU time. However, certain modifications might be made to the algorithm itself to improve performance. For example, collinear component set extraction from the Hough domain involves, effectively, reapplication of the Hough transform to all active components for a given cell with values of ρ and θ. The required processing time for these steps could be reduced drastically if each Hough domain cell contained information about the components which belong to it. This would require cells to maintain a list of possibly up to one hundred component entries. Due to the limited memory on our computer, it was not possible to implement this feature.[2]

It is possible that the algorithm may be made even more robust to rapid changes in image characteristics over small areas of the image. Several of the processing steps can be made more adaptive by using several iterations in order to gather more accurate statistics about strings within the image (e.g., calculation of the thresholds T_w and T_c based on local component height). Such modifications will result in a degradation in processing speed. There is a definite tradeoff between the robustness of the algorithm and the processing speed. Making the algorithm more adaptable will result in longer processing times.

Some problems are encountered when dealing with dotted or dashed lines. The consistency in density and ratio in such graphics strings is not always as high as might be expected. Also, the intercharacter gap is not always consistent, resulting in the classification of some dashed lines as character strings. Some further processing may be required in order to fully accommodate the detection of repeated graphic strings.

[2]Our system has 8 MB of memory. Executing a program containing more than one 2048 × 2048 byte variable array resulted in significant deterioration in performance of the system due to increased page faults. With a system having more memory, larger data structures could be used to greatly enhance performance.

SUMMARY AND CONCLUSIONS

A new algorithm for text string separation from mixed text/graphics images has been presented. The algorithm is robust to changes in text font style and size within an image. The algorithm also accommodates the separation of text strings of any orientation. The algorithm adapts to changes in text characteristics within the image in order to achieve optimal performance. The algorithm is highly reliable provided that documents conform to several constraints on their characteristics. The algorithm presented has several advantages over previously applied techniques. With improvements made to the algorithm in terms of processing speed and efficiency of data representation, the application of the algorithm will become highly appropriate for use in document analysis systems.

REFERENCES

[1] K. Y. Wong, R. G. Casey, and F. M. Wahl, "Document analysis system," *IBM J. Res. Develop.*, vol. 6, pp. 642–656, Nov. 1982.

[2] R. N. Slater, "Automating data base capture for CAD/CAM," *Comput. Graphics World*, vol. 48, pp. 45–53, Oct. 1984.

[3] M. Karima, K. S. Sadhal, and T. O. McNeil, "From paper drawings to computer aided design," *IEEE Comput. Graphics Applications*, vol. 5, pp. 24–39, Feb. 1985.

[4] C. L. Huang and J. T. Tou, "Knowledge based functional symbol understanding in electronic circuit diagram interpretation," *Aplications of Artificial Intell. III, Proc. SPIE*, vol. 635, pp. 288–299, 1986.

[5] H. Bunke, "Automatic interpretation of lines and text in circuit diagrams," in *Pattern Recognition Theory and Applications*, J. Kittler, K. S. Fu, and L. F. Pau Eds. Boston, MA: D. Reidel, 1982, pp. 297–310.
[6] L. T. Watson, K. Arvind, A. W. Ehrich, and R. M. Haralick, "Extraction of lines and regions from grey tone line drawing images," *Pattern Recognition*, vol. 17, pp. 493–506, 1984.
[7] W. H. Chen, W. K. Pratt, E. R. Hamilton, R. H. Wallis, and P. J. Capitant, "Combined symbol matching facsimile data compression system," *Proc. IEEE*, vol. 68, pp. 786–796, 1980.
[8] F. M. Wahl, M. K. Y. Wong, and R. G. Casey, "Block segmentation and text extraction in mixed text/image documents," *Comput. Vision, Graphics, Image Processing*, vol. 20, pp. 375–390, 1982.
[9] H. Bley, "Segmentation and preprocessing of electrical schematics using picture graphs," *Comput. Vision, Graphics, Image Processing*, vol. 28, pp. 271–288, 1984.
[10] L. A. Fletcher, "Text string separation from mixed text/graphics images," M.S. thesis, Dep. Elec. Eng., Pennsylvania State Univ., Aug. 1986.
[11] A. Rosenfeld and A. C. Kak, *Digital Picture processing*, vol. 2, 2nd ed. New York: Academic, 1982.
[12] J. P. Foith, C. Eisenbarth, E. Enderle, H. Geisselmann, H. Ringshauser, and G. Zimmermann, "Real-time processing of binary images for industrial applications," in *Digital Image Processing Systems*, L. Bolc and Z. Kulpa, Eds. Berlin: Springer-Verlag, 1981.
[13] W. K. Pratt, *Digital Image Processing*. New York: Wiley, 1978, pp. 523–525.

Lloyd Alan Fletcher was born in Birmingham, England, in 1962. He received the B.S. degree in physics with microelectronics and computing from the University of Leicester, England, in 1984. From August 1984 until August 1986 he attended the Pennsylvania State University, where he was a teaching and research assistant, receiving the M.S. degree in electrical engineering in 1986. At Penn State his research interests included computer vision and digital image analysis.

Since September 1986 he has been a Member of Technical Staff with Bell Communications Research, Red Bank, NJ. He is currently working in the Switching Analysis and Reliability Technology Center, where he is involved with the development of criteria to assure the reliability and quality of digital switching systems and telecommunications equipment deployed by the regional Bell telephone companies.

Rangachar Kasturi (M'82) was born in Bangalore, India, in 1949. He received the B.E. degree in electrical engineering from Bangalore University in 1968 and the M.S.E.E. and Ph.D. degrees from Texas Tech University, Lubbock, in 1980 and 1982, respectively.

Dr. Kasturi is an Associate Professor of Electrical Engineering at the Pennsylvania State University, where he was an Assistant Professor from 1982 to 1986. From 1978 to 1982 he was a Research Assistant at Texas Tech University and was engaged in research in multiplex holography and digital image processing. From 1976 to 1978 he was the Engineering Officer at the Visvesvaraya Industrial and Technological Museum, Bangalore, and from 1969 to 1976 he was with the Bharat Electronics Ltd., Bangalore, as a Research and Development Engineer. His current research interests are in the applications of image analysis and artificial intelligence techniques for map data processing, graphics recognition, and document analysis. He has directed several projects sponsored by NSF, Digital Equipment Corporation, AT&T, and the Applied Research Laboratory. He has published a number of papers in journals and conference proceedings. He is the author of a book chapter and is coeditor of a book on image analysis applications to be published by Marcel Dekker.

Dr. Kasturi is a member of OSA, SPIE, Eta Kappa Nu, and Sigma Xi.

Celesstin: CAD Conversion of Mechanical Drawings

Pascal Vaxivière
Ecole Supérieure des Sciences et Techniques de l'Ingénieur de Nancy and Centre de Recherche en Informatique de Nancy

Karl Tombre
Institut National de Recherche en Informatique et Automatique and Centre de Recherche en Informatique de Nancy

A prototype CAD conversion system extracts higher level structures for knowledge-based analysis. It recognizes such entities as screws, ball bearings, and shafts.

Mechanical engineering companies that implement computer-aided design and manufacturing systems must convert their archives of paper drawings to a format suitable for CAD. Digitization of a drawing creates an image of several million black and white pixels that represent the original drawing with varying accuracy, depending (among other things) on the resolution of the scanning device and the quality of the original. However, this raster image information is not directly suitable for CAD systems, which operate with basic structures such as lines and curves. Therefore, many commercial systems[1] have been developed to convert raster images to vector representations suitable for vector editing. These systems may also provide additional modules for text/graphics separation, optical character recognition for the text layer, or semi-automated symbol recognition using simple template-matching techniques. But no commercial system actually provides higher level modules that "understand" a drawing's technical elements. Developing such a system is difficult because of the number of specific CAD/CAM applications (architecture, mechanical engineering, city maps, electrical schemes, and so on) and their diversity of contextual knowledge.

Technical drawings are based on schemes specific to the main activity of the company that uses them. Often, the company's CAD system library stores specifications for screws, bearings, gearboxes, and so on for quick retrieval and insertion into new drawings. The Celesstin system presented in this article is a working prototype that converts a mechanical engineering drawing into a format suitable for CAD (see the sidebar titled "Celesstin prototype"). In developing this system, we stressed recognition of technical entities. Mere vectorizations, such as those offered by existing commercial systems, are too low level to be really useful in a CAD context. Our system is based on the assumption that even when the

Reprinted from *Computer*, Vol. 25, No. 7, July 1992, pp. 46–54.

vectorized drawing is distorted, it can be correctly interpreted by using higher level expertise — that is, knowledge about the representation rules used in technical drawings and about the manufacturing technology associated with the represented objects.

Celesstin prototype

Celesstin is a prototype system for converting mechanical engineering drawings into the format of the Dassault Systems' Catia CAD system. Celesstin has been under development for several years at the engineering school of Ecole Supérieure des Sciences et Techniques de l'Ingénieur in Nancy, France. At the end of each academic year in June, a new integrated version of the system is released with the most up-to-date versions of the algorithms designed in the project, completely integrated with a user-friendly interface and the appropriate modules for connecting the program to the host Catia CAD system. Celesstin runs on an IBM 9370-60 machine with a 5080 graphics workstation.

The aim of the Celesstin project is to prove the feasibility of knowledge-based interpretation of drawings. As a prototype, Celesstin has several limitations:

- The input paper drawings must be relatively clean, without lots of "noise."
- Celesstin only processes representations of assemblies such as gearboxes made of rotating axes and of motion transmission and guiding devices.
- Vectorization is limited to straight lines. Celesstin does not recognize circular arcs.
- The system's knowledge rules are strongly based on the drawing's compliance with the French Afnor standard, especially the stipulation that thin and thick lines can be discriminated.

We have tested this prototype only on a limited set of drawings. Hence, it would be pretentious to claim recognition or rejection rates or to propose the use of this system in production. Our goal with Celesstin is to prove that a working production-quality system can be built, not to build one.

Guided tour of Celesstin

Celesstin integrates several modules in a blackboard-based interpretation system, as Figure 1 shows. Once a drawing has been digitized, a first processing step separates the text and the dimensioning lines from the pure graphics part. Celesstin vectorizes the graphics part and assembles the resulting lines into blocks, the basic elements for the technical entities that it creates. The result is transferred to the CAD system. Celesstin tries to match the extracted entities with the corresponding models from the CAD library. It puts the remaining blocks and lines into different layers of the CAD description.

Scanning and preprocessing. Technical drawings are input using a 600-dpi scanner. Then the user can apply image preprocessing tools, the most important of which clean up the bitmap. The user can choose from several sets of cleaning masks — morphological filter operators[2] that eliminate isolated black pixels or small black or white blobs.

Vectorization. The vectorization module converts the image bitmap into a set of lines. For Celesstin we adapted the method proposed by Lin et al.,[3] which consists of splitting the image into meshes and looking at the intersection between the lines and the mesh borders. The result of vectorization is a description of the drawing as chains of vectors having given thicknesses and intersecting at junction points.

Postprocessing. Celesstin analyzes the data structure resulting from vectorization to extract two line-thickness classes: thin lines and thick lines. Celesstin does this automatically by histogram analysis of line thicknesses over the whole drawing. Some line types are very important in technical drawings because they symbolize higher level information such as symmetries, hidden contours, or the presence of matter. Therefore, a procedure searches the vectorization result for specific elements such as dashed and dot-dashed lines and hatching areas (regularly spaced, parallel thin lines across a closed polygon drawn in thick lines).

An interactive tool lets the user correct misinterpreted lines or change line attributes. Figure 2 shows an original raster image and a screen dump of the workstation during the vector correction phase. The vectorization results (Figure 2b) are displayed with different color codes for the different line types: thick lines in white, thin lines in pink, contours of black blobs in red, and dashed and dot-dashed lines recognized as single elements. The user can correct interactively the attributes of the segments (selected line in blue).

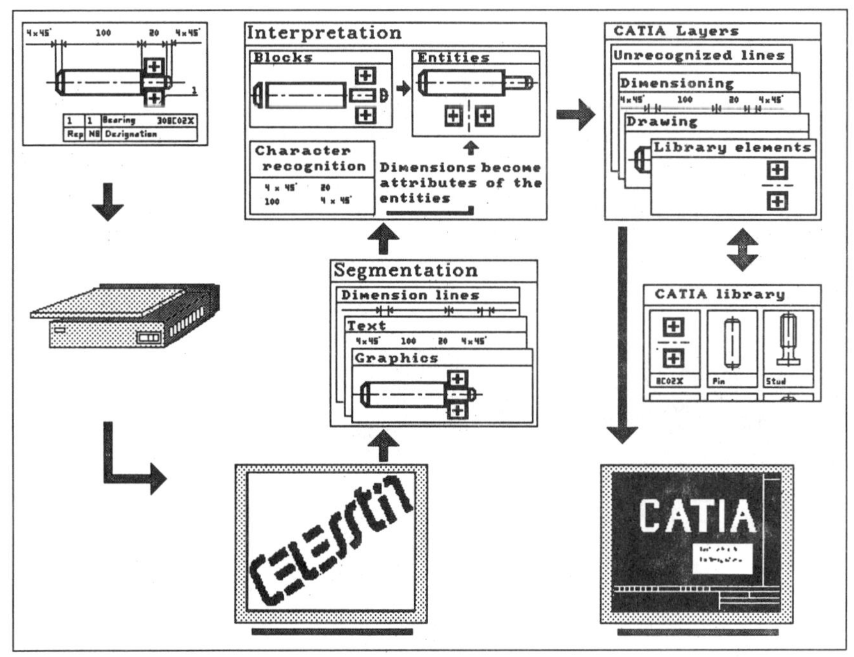

Figure 1. Overview of the Celesstin system.

Vectorization

Vectorization is a very important stage in a technical document interpretation system. It consists of converting a raster image to an image described as chains of vectors and junctions between these chains.

Many vectorization methods have been proposed, but the only approach implemented in commercial systems and thus developed as production-quality code uses skeletonizing, followed by a polygonal approximation of the skeleton.[1,2] A major drawback of this approach is that skeletonizing is based on a view of the image that is too local. This may result in distorted vectors and erroneous choices when the system computes the position of the junction points or looks for circular arcs. Instead of this method, we adapted to mechanical engineering the diagram processing algorithm proposed by Lin et al.[3]

The method is based on a priori knowledge of the maximal thickness of a line and the minimum distance between two lines. Such knowledge is easily available in mechanical engineering, where the following rules apply in France (French standard NF E 04-103):

- The thickness of a thin line must be one fourth that of a thick line.
- The distance between two thick lines must be at least 0.8 mm.
- The distance between two hatching lines varies between 1.5 and 2 mm.

The method consists of splitting the binary image into $n \times n$ meshes, where n is larger than the maximal line thickness and smaller than the minimum distance between two lines. Thus, at most one line in a given direction should cross each mesh, and any line will not overlap more than two meshes. Figure A shows the principle: The vectorization algorithm looks only at the borders of the meshes and approximates the lines by straight segments crossing each mesh.

Lin et al. compiled a library of 51 characteristic meshes that apply in most cases. All the meshes in a given image cannot be coded by an element of this library, so we have an extra "?" code, which corresponds to approximately 5 percent of the meshes in a given image vectorization. We added to the original method a recursive splitting of the meshes into smaller ones, until all the meshes can be given a known code. The mesh is split orthogonally to the first border that contains at least a sequence black-white-black. Celesstin does the splitting in the white interval between the two black intervals. If possible, it splits so that no ambiguity remains on the opposite side of the mesh. Otherwise, it splits arbitrarily. Celesstin applies the splitting operation recursively to the new meshes created, until no unknown sequence remains. For more efficient line following, it keeps only links between split meshes that are overlapped by a common line (see Figure B).

From this point, we further improved the method by going progressively from a "geographic" splitting of the image as a set of regular meshes to a "geometric" splitting where each mesh contains characteristic geometric features. To locate more precisely the angles between two lines, which are very important in technical drawings, we want an angle to be in the middle of a mesh. Such precise localization prevents oblique lines from becoming "stairway-like" during vectorization.

This requirement led us to the concept of *dynamic* meshes, which arrange themselves in the most convenient way for line following. First, by applying offsets to the mesh structure, Celesstin recenters — on a single mesh — lines that overlap two neighboring meshes. In addition, to reduce the total number of meshes used to describe a drawing, Celesstin lets dynamic meshes "grow" by merging two adjacent meshes having the same code into a single mesh. Figure C shows how this reduces the number of characteristic meshes from 51 to 16, corresponding to the following families of cases: junction/intersection, angle, straight line, and black blob.

Finally, from the mesh data structure, Celesstin creates a line structure by following lines from one mesh to the other. During line following, if Celesstin finds two neighboring meshes covered by the same black area, it considers the area as a black blob and codes it by its contour. The result of this vectorization process is a data structure describing the image as segments of different types (thick line, thin line, and contour of black blob) and junctions between these segments.

Figure A. Principle of the vectorization method.

References

1. R.W. Smith, "Computer Processing of Line Images: A Survey," *Pattern Recognition*, Vol. 20, No. 1, Jan. 1987, pp. 7-15.
2. V. Nagasamy and N.A. Langrana, "Engineering Drawing Processing and Vectorization System," *Computer Vision, Graphics and Image Processing*, Vol. 49, No. 3, Mar. 1990, pp. 379-397.
3. X. Lin et al., "Efficient Diagram Understanding with Characteristic Pattern Detection," *Computer Vision, Graphics and Image Processing*, Vol. 30, No. 1, Apr. 1985, pp. 84-106.

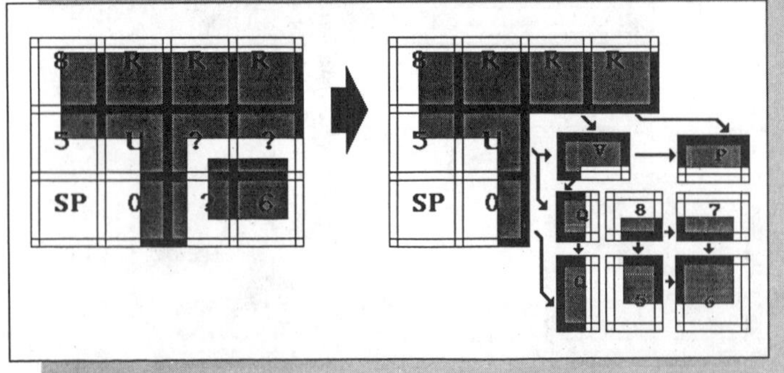

Figure B. Mesh splitting and links between split meshes.

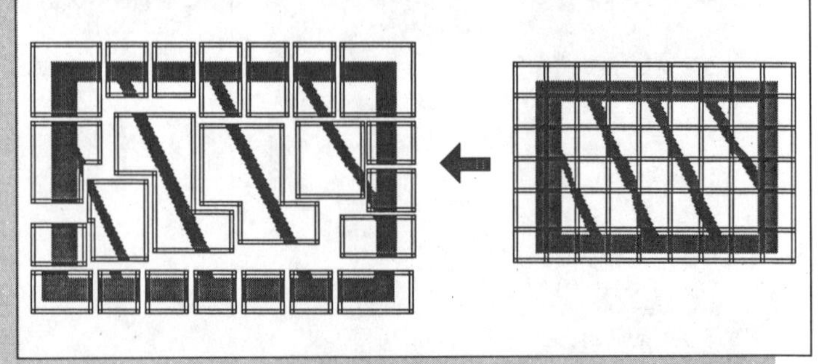

Figure C. Dynamic meshes.

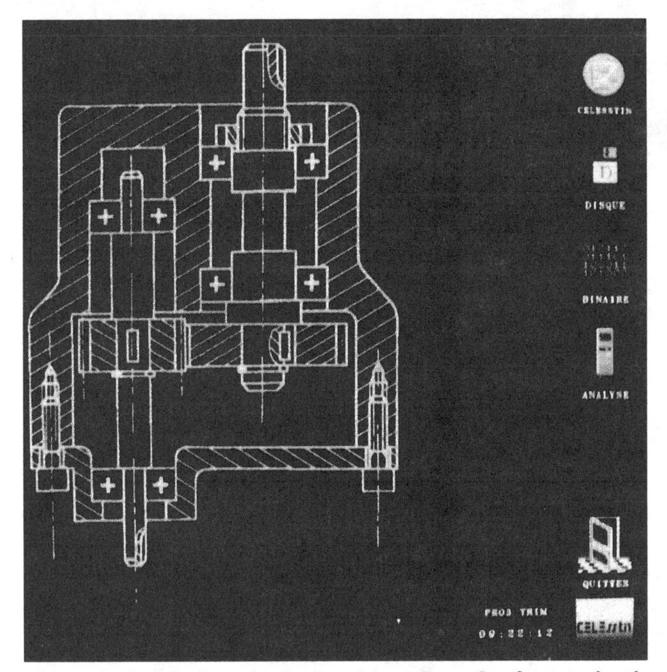

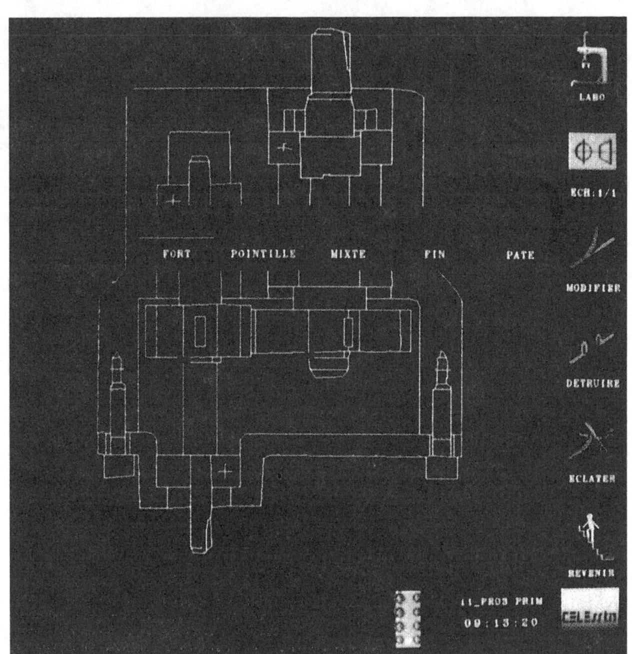

Figure 2. Original raster image (a) and result of vectorization with interactive editing facility (b).

Blocks. Because the structural level of single line vectors is too low to be really useful in the interpretation phase, we decided to create a new description of the vectorized drawing. The basic element of this description is a minimum closed polygon made of thick lines, called a *block*. All the thin lines attached to a block are its *attributes*.

Interpretation. The interpretation module works on the drawing's block description. Its basic strategy is to follow the dot-dashed lines, which are supposed to be axis lines, and to identify technical entities along these axes. The interpretation module continues the analysis by identifying surrounding objects. The module studies the disassembling possibilities and the transmission of motion in the setup to provide high-level knowledge rules for the final stage of interpretation.

Transfer to CAD. Once Celesstin has recognized the drawing's high-level technical elements, it must transfer the information to the CAD system. Most CAD systems compose a drawing as the superposition of several layers. At any time a user can manipulate only a subset of these layers. The highest layer contains the elements recognized as instances of library models. The text layer contains the nomenclature and other annotations, and a dimensioning layer gathers all the dimensions found during analysis. Descending the hierarchy, other layers contain identified blocks that are not part of higher level entities; lines that are not part of any recognized block, themselves classified according to their thickness (thin lines and thick lines); and, in the lowest layer, the residues of segmentation (the small connected components not recognized as being part of graphics or text).

User interface. Celesstin's user interface provides interactive tools and easy access to the whole system. As Figure 2 shows, the hierarchical, icon-based interface lets users choose the tools they want for image cleanup, vectorization, decomposition into blocks, interpretation by the expert system, and export to the CAD system. Most tools combine an interactive part, often reduced to the input of parameters, and an automated part for processing information. Some tools are much more interactive. For example, the vector editor allows manual postprocessing of the vectorization result.

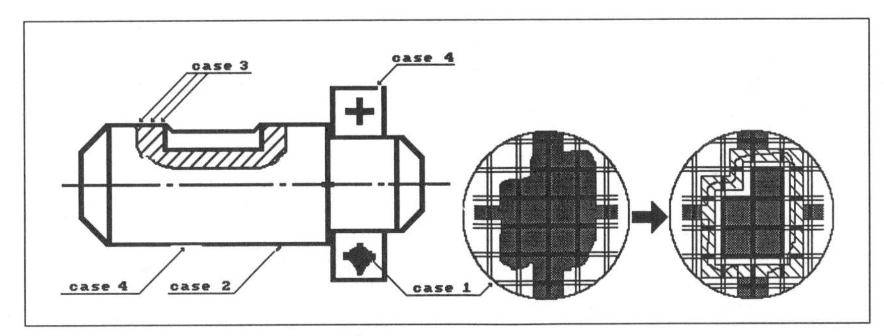

Figure 3. Line structure corrections.

Vectorization and line recognition

Vectorization consists of converting the bitmap into a set of line vectors. For our vectorization, we implemented and improved the method proposed by Lin et al.[3] (see the sidebar on vectorization). After vectorization, Celesstin performs polygonal approximation[4] on the resulting structure to reduce the number of segments. This step also includes a first classification of the line segments into thin and thick lines.

Celesstin corrects the resulting lines by eliminating small irregularities that would disrupt further analysis. Case 1 in Figure 3 shows the correction for contours of black blobs. The black areas must be precisely delineated and represented by the corresponding contour. Case 2 shows the removal of small barbs. If there is no line nearby, Celesstin removes isolated short line segments connected to other segments only at one end. In case 3, all couples of aligned segments having the same thickness and meeting at a junction point are merged into a single line. In a technical drawing, a line doesn't arbitrarily change from thick to thin, and case 4 shows the application of continuity rules to correct the line thickness in different configurations.

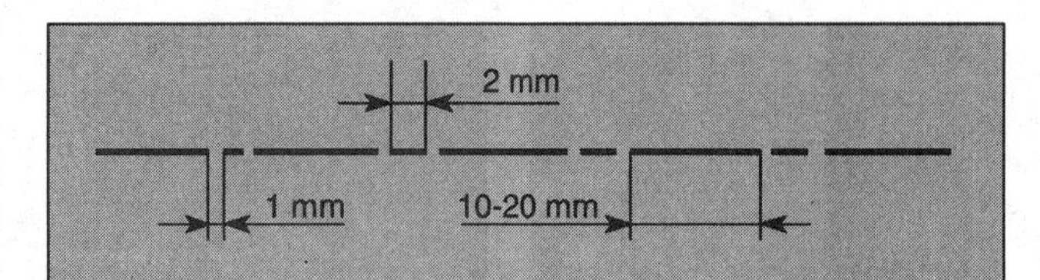

Figure 4. French standard NF E 04-103 for dot-dashed lines.

The next step is to detect dot-dashed lines, representing axis lines, and dashed lines, representing hidden lines. Dot-dashed and dashed lines are very important because they symbolize symmetry lines on which many elements are centered (shafts, screws, and so on). Several methods have been proposed for this purpose; for example, Kasturi et al.[5] use a structural approach based on linking free ends of segments. Our method works in three steps:

(1) location of candidate line segments, that is, segments aligned and at an appropriate distance from one another;
(2) recognition of the line type, which can be dot-dashed (as in Figure 4) or dashed (only short segments); and
(3) replacement of the set of line segments by a single line entity with the attribute dot-dashed or dashed.

We use the standard definition of such lines, as presented in Figure 4. Celesstin classifies candidate line segments into long and short segments and labels the whole line as dashed or dot-dashed, depending on the proportion of short-long-short sequences of segments. As shown in Figure 2b, where a dot-dashed line has been selected (outlined in blue), Celesstin recognizes the line and codes it as a single entity in the final result of line recognition.

In addition, other elements that Celesstin finds during analysis can lead to the hypothesis of a dot-dashed line. Because such lines represent symmetries, Celesstin considers a set of segments to belong to a dot-dashed line if the blocks that this set crosses are symmetrical with respect to it. This allows for better discrimination than the first method in such cases as a dot-dashed line traversing a hatched area.

Of course, the line-recognition phase does not yield perfect results. For instance, the hatching lines in Figure 2 are not all straight, because vectorization is based on local operators that sometimes give a false approximation of the line direction. But Celesstin easily corrects these distortions at the higher levels of interpretation.

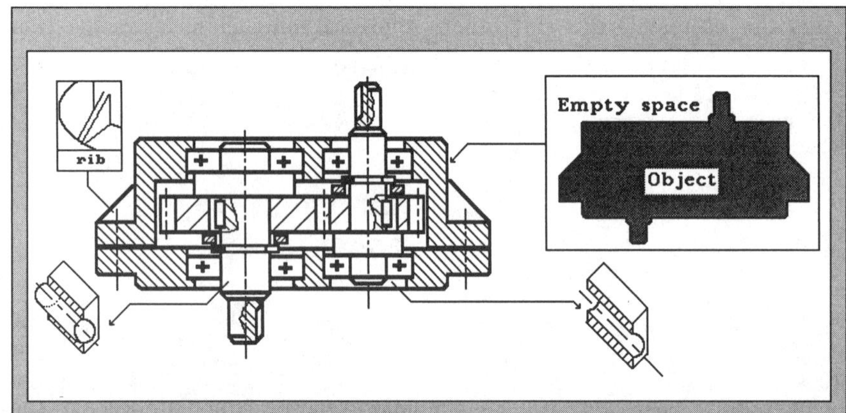

Figure 5. A technical drawing and the search for matter and empty space.

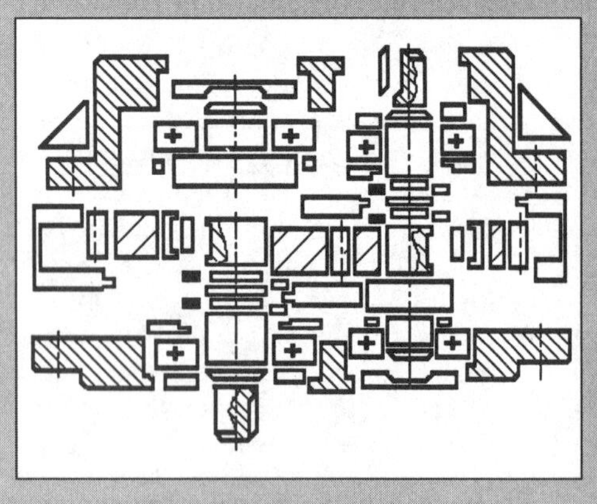

Figure 6. Decomposition of the drawing in Figure 5 into blocks.

Blocks

A principal difficulty in our approach is finding basic elements on which to conduct reasoning. At an earlier stage of this research, we attempted to design techniques for identifying subsets of lines such as hatching and for performing pattern matching between the lines yielded by vectorization and models of CAD objects. We eventually concluded that line-segment features are too low level and do not carry sufficient technical information. We therefore chose to work on a higher level element, the *block*.

In a technical drawing, thick lines represent the orthogonal planar projection of the contours of a mechanical device. Because the thin lines do not outline the device, we decided to ignore them in the first phase. Any minimal closed polygon with thick lines sharing at least one edge with the object's external contour has to enclose matter. If the polygon is hatched, it lies in the section plane; if it is empty, it may represent a piece in front of or behind the section plane (see Figure 5). Any minimal closed polygon with thick lines sharing an edge with one of the polygons previously described will in turn represent a change in geometry (change in diameter of a shaft, for instance), a new piece in contact with the previous one, or empty space. Simple rules allow Celesstin to formulate hypotheses about these blocks using their thin-line content and that of their neighbors. These rules propagate the labeling of all blocks to the whole drawing. Celesstin adds all thick lines not included in this block decomposition to the content of their including block; they represent keyways, ball bearings, or bores. The result is a complete decomposition of the drawing as a set of structures at a level higher than vectors — that is, at the level of minimal closed polygons, or blocks, drawn with thick lines and their thin-line attributes. Figure 6 shows the decomposition of the drawing in Figure 5.

Celesstin's next operation on the blocks determines their technical attributes. Before this analysis, a block's content is only graphical: It is an empty block or filled area (black block), or hatched with parallel thin lines, which may be parallel to a side of the block. The analysis changes this content into an attribute. The assigned attribute may be "black block" or "white" — in other words, unknown. But the analysis usually assigns an attribute such as matter, represented as hatching, threading, or partial section — or empty space that can be tapped.

During this analysis, a block's content can modify its neighbor's attribute. For instance, Figure 7 shows a section in a tapped element. The dot-dashed line indicates that the hole is cylindrical. Both blocks 1 and 3 contain a threading line, but actually it is block 2 that is tapped. The hatchings in blocks 1 and 3 indicate a section in matter; the tapping lines have been put on these blocks because block 2 represents empty space. In such a case, Celesstin's analysis separates the two families of thin lines and associates the tapping lines with block 2. Finally, the system replaces the "unknown" nature of block 2 with its technical attribute: "tapped" (and hence behind the section plane).

The rules used for this analysis are based on typical rules for engineering drawings (French standards NF E 04-104 and NF E 04-012), as illustrated in Figure 8. When the system finds more than two families of parallel lines, it returns to the previous case by labeling two of the families as hatching.

The result of this analysis is a new, complete description of the drawing, no longer as a set of vectors but as a set of blocks characterized by their technical attributes. Even if Celesstin cannot assign a precise attribute to each block, the description contains "confidence islands" from which the interpretation stage can start to identify technical components such as shafts, screws, and bearings.

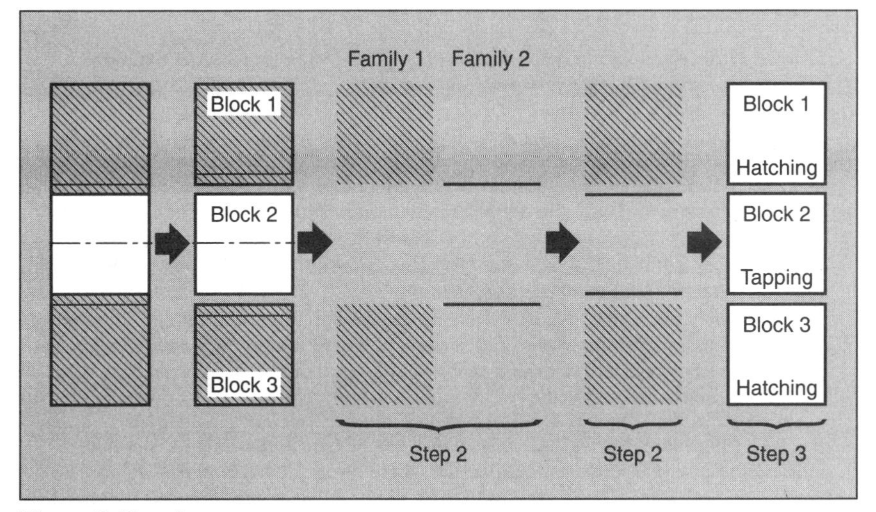

Figure 7. Tapping.

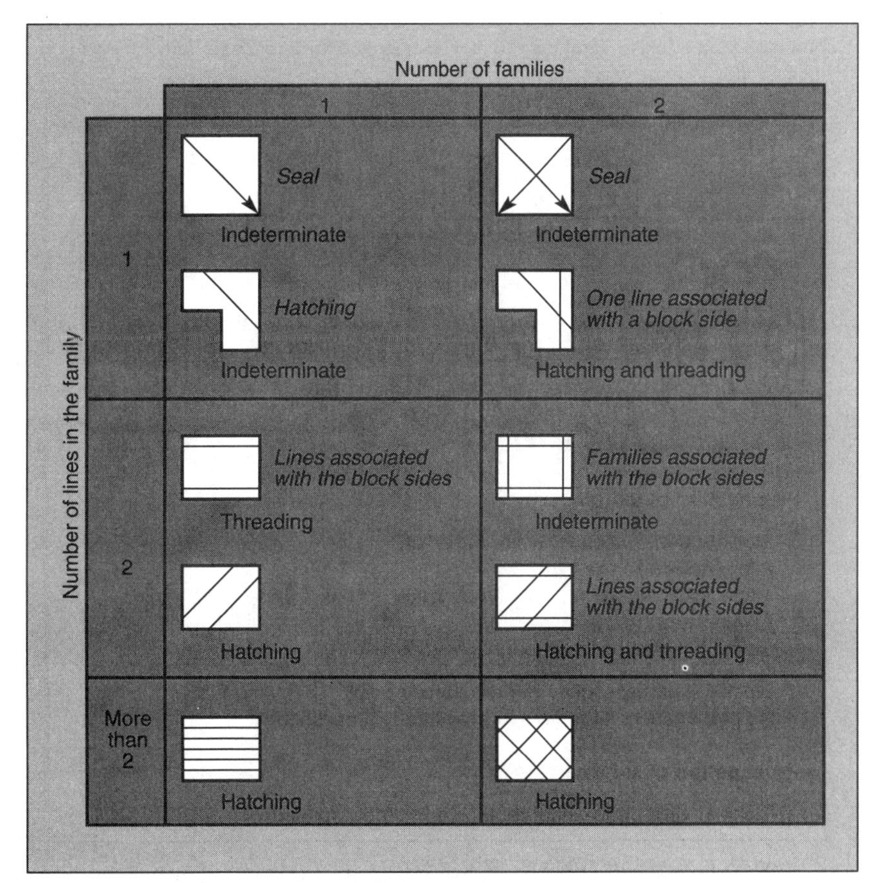

Figure 8. Determining the attributes of blocks.

Interpretation

We started our work on interpretation by analyzing the way engineers look at drawings that have neither dimensioning nor nomenclature. Engineers generally follow the main dot-dashed

Recognition of shafts

The knowledge-based interpretation system in Celesstin uses blocks as its basic elements and combines them into more elaborate components to which it can give semantic labels. Here we illustrate the reasoning process and knowledge rules for recognizing shafts. This reasoning works on all blocks located along a dot-dashed line. The basic idea is to start from islands of confidence found by labeling blocks with their technical attributes and to propagate the matter along the axis line.

Some block attributes are *strong indications* of the presence of matter. A good starting point for the recognition of a shaft is

- a threaded block, or
- a block containing a partial section and located on only one dot-dashed line.

If Celesstin finds no strong indications to initiate the recognition, it can use *weaker indications*, such as the presence of an "enclosing" block representing the background plane of a shaft.

When Celesstin finds a starting point, it applies the following propagation rules:

- Any rectangular or trapezoidal block, symmetrical with respect to the dot-dashed line and neighbor of a block already recognized as matter, is also a matter block.
- Any block containing a partial section (not necessarily rectangular or symmetric) and extending the shaft represents matter.
- Any closing trapezoid (chamfer) on the shaft stops the propagation of matter.
- Any block enclosing a matter block is supposed to be empty, as it usually represents empty space or the casing in the background of the section plane.
- Any tapped block is empty.

The figure below shows the analysis process along the dot-dashed line supporting the gear in Figure 5. Because the whole set of blocks contains at least one partial section, this is a shaft and not a screw. Celesstin thus describes the set of blocks linked in this way with the symbolic notation "shaft."

Such rules are never absolute, but globally such a propagation method remains robust. The strong rules have never failed in our experiments. For recognition failures with weak rules, a last resort is user intervention. In addition, higher level phases of the interpretation may bring into question the proposed labeling and lead Celesstin to propose other hypotheses.

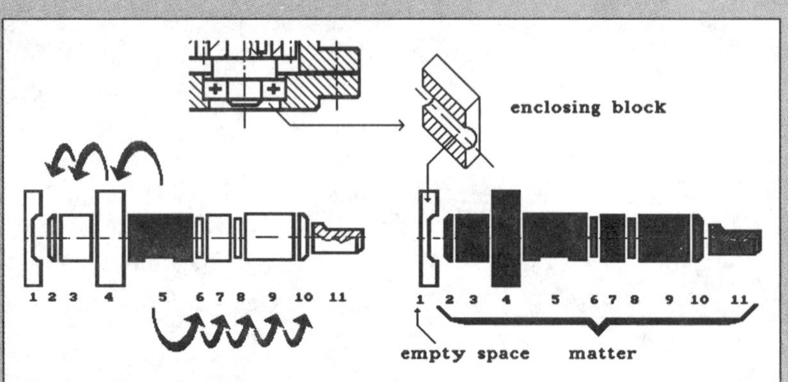

Propagation of matter

Block 5: partial section ⇒ starting point.
Propagation to the left: 4 ⇒ 3 ⇒ 2.
4 and 3 are regular. 2 stops propagation, as it is a closing trapezoid.
Propagation to the right: 6 ⇒ 7 ⇒ 8 ⇒ 9 ⇒ 10.
Same thing happens: propagation stops with 10.

Block 11: partial section ⇒ new starting point.
No propagation; all neighbors are already processed.

Propagation of empty space

Block 1: white "enclosing" rectangle ⇒ empty space.

Propagation of matter along an axis line.

lines, crossing the whole drawing. Then, after they recognize the different components placed on these axis lines, they try to understand the motion-transmission mechanisms. Hence, Celesstin's main interpretation strategy is to focus first on the blocks located on dot-dashed lines, which symbolize a symmetry. The sidebar, "Recognition of shafts," explains how the system applies this strategy.

Celesstin's hypotheses about each technical element must of course be consistent with the remainder of the drawing. Celesstin can analyze a drawing at a higher level by looking at the disassembly and kinematics of the whole setup. But even when Celesstin applies only basic reasoning and simple matching procedures to the blocks it recognizes, it can analyze the drawing in terms of technical elements taken from a CAD library.

However, structuring in terms of blocks is too weak when a system must apply rules such as

- any mechanical device can be disassembled, and
- if a shaft has been recognized, this implies motion and hence a coherent kinematic scheme.

Therefore, we had to devise a new structure, which we call the *entity*. Celesstin creates an entity in two configurations:

- a set of blocks symmetrical with respect to a dot-dashed line and having the same shape, and
- a set of blocks having the same hatching pattern.

After it creates entities, Celesstin can apply higher level knowledge to these new structures. A very strong technological rule is that it *must* be possible to disassemble a mechanical setup. When Celesstin has identified the shafts and screws in the drawing, it creates all possible entities and computes the degree of freedom with respect to one another of all "matter" entities. Celesstin applies a set of knowledge rules using these computational values and tries to progressively disassemble the device. When an entity hinders the disassembling, Celesstin analyzes it more precisely to determine if it represents empty space or a locking device (key, clip).

In addition, a kinematic analysis of the drawing determines the functional role of each entity located around a shaft. The general aim of this analysis is to determine whether an entity touching a

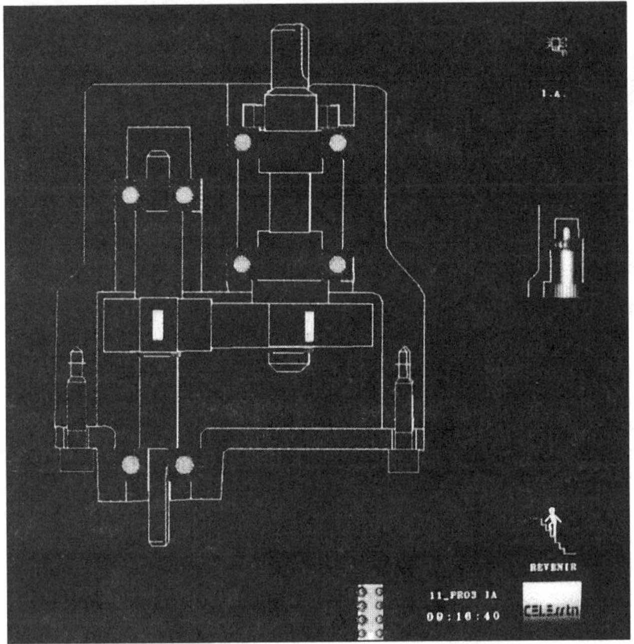

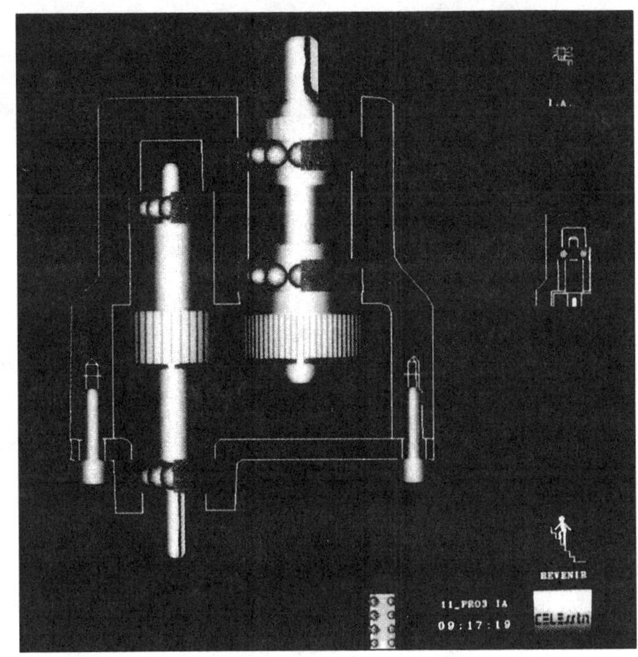

Figure 9. Result of the interpretation process: (a) attributes and recognized entities; (b) two-and-a-half-dimensional display.

shaft rotates with the shaft (for example, the entity could be a gear) or allows the rotation of the shaft with respect to a motionless part such as the casing. This latter configuration identifies bearings and rings.

We implemented the interpretation system in Atome, a blackboard-based multiagent expert system shell.[6] The knowledge rules are relevant only if the drawing represents a usable technical device. Figure 9 shows the result of interpretation of technical entities. In the screen shot in Figure 9a, Celesstin has replaced some entities with the corresponding representation from the CAD library (for instance, the ball bearings) and has associated symbolic attributes such as hatching with some blocks. As this is an interpretation in terms of technical entities, Celesstin can display the recognized drawing in two and a half dimensions (Figure 9b).

Ongoing work and perspectives

The Celesstin system demonstrates the possibilities for formalizing the knowledge mechanical engineers use to read and draw technical documents. We have started studying the different levels of knowledge that can be used in technical document interpretation. As in many other areas, it is possible to distinguish a *structure*, on which a *syntax* can be applied, and *semantics*.[7] For Celesstin, the vectors, blocks, and entities are structural units on which we defined a syntax for combining lower level structures into higher level ones. At the semantics level, we introduced knowledge about mechanical engineering itself — for instance, rules for propagation of matter and for analysis of the drawing in terms of disassembly and motion transmission.

We are adding new modules to Celesstin, including a module that analyzes the structure and syntax of the dimensioning and nomenclature,[8] coupled with a character recognition system for reading the actual numbers of the dimensions and the names in the nomenclature that may be part of the drawing. The character recognition system adds another means of interpretation and transfer to the CAD system: If a drawing annotation refers to an entry in the nomenclature, the system can look up the corresponding entity in the CAD library of models and, by simple pattern matching, check the correspondence between this model and the annotated part of the drawing.

Celesstin is a research product and has some weaknesses. Its representation of knowledge should be better structured, as a "flat" set of rules rapidly becomes hard to manage. Also, its implementation of vectorization is probably less robust and efficient than the vectorization facilities provided by commercial systems. But commercial systems do not provide any high-level analysis of technical entities, which is Celesstin's aim and unique feature. Of course, user interactivity will always be needed to retrieve really usable CAD descriptions of a drawing, but the time and effort necessary for user corrections must be much less than the time and effort needed to create a complete new design in CAD.

Celesstin is in no way a universal, "omniscient" knowledge-based system, knowing everything about any conceivable technical drawing in the world. It is rather a prototype and an integration platform for several research activities (vectorization, knowledge-based interpretation, dimension analysis, and so on). We don't believe that a universal system can be built, as knowledge and aims are very different in different applications. But Celesstin shows a possible path for designing working systems that solve the "transfer from paper to CAD" problem in many specific applications. ■

Acknowledgments

This work was partially funded by a research grant from IBM France to ESSTIN. We thank all the students who worked on the successive versions of Celesstin and all colleagues who have taken part in the work, especially Julian Anigbogu, Yannick Chenevoy, Vincent Chevrier, and Suzanne Collin.

References

1. L.S. Wolfe and J. de Wyze, "An Update on Drawing Conversion," *Computer Aided Design Report*, Vol. 8, No. 6, June 1988, pp. 1-11.

2. J. Serra, "Introduction to Mathematical Morphology," *Computer Vision, Graphics and Image Processing*, Vol. 35, No. 3, Sept. 1986, pp. 283-305.

3. X. Lin et al., "Efficient Diagram Understanding with Characteristic Pattern Detection," *Computer Vision, Graphics and Image Processing*, Vol. 30, No. 1, Apr. 1985, pp. 84-106.

4. T. Pavlidis, *Algorithms for Graphics and Image Processing*, Springer-Verlag, Berlin, 1982.

5. R. Kasturi et al., "A System for Interpretation of Line Drawings," *IEEE Trans. Pattern Analysis and Machine Intelligence*, Vol. 12, No. 10, Oct. 1990, pp. 978-992.

6. H. Lâasri and B. Maître, "Flexibility and Efficiency in Blackboard Systems: Studies and Achievements in ATOME," in *Blackboard Architectures and Applications*, V. Jagannathan, R. Dodhiawala, and L. Baum, eds., Academic Press, Boston, 1989, pp. 309-322.

7. K. Tombre and P. Vaxivière, "Structure, Syntax and Semantics in Technical Document Recognition," *Proc. First Int'l Conf. Document Analysis*, Vol. 1, AFCET, Paris, 1991, pp. 61-69.

8. S. Collin and P. Vaxivière, "Recognition and Use of Dimensioning in Digitized Industrial Drawings," *Proc. First Int'l Conf. Document Analysis*, Vol. 1, AFCET, Paris, 1991, pp. 161-169.

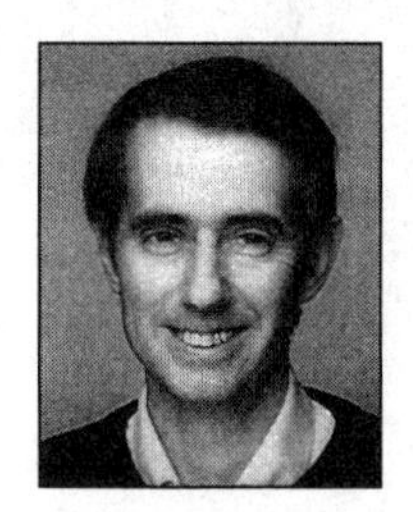

Pascal Vaxivière is an associate professor at the Ecole Supérieure des Sciences et Techniques de l'Ingénieur de Nancy (ESSTIN). His research interests are in technical document image interpretation.

Vaxivière received a Diplôme d'Ingénieur from ESSTIN in 1978 and a DEA degree from the University of Nancy in 1990. He is currently preparing a PhD dissertation at CRIN/CNRS on knowledge-based interpretation of engineering drawings.

Karl Tombre is a research scientist at the Institut National de Recherche en Informatique et Automatique (INRIA) working in a research group common to INRIA Lorraine and CRIN/CNRS in Nancy, France. His research interests are computer vision, image analysis, and document image interpretation.

Tombre received a DEA degree from the University of Nancy in 1983 and a PhD in computer science from the Institut National Polytechnique de Lorraine in 1987.

Address questions concerning this article to Tombre at INRIA Lorraine-CRIN/CNRS, Campus Scientifique, BP 239, 54506 Vandoeuvre lès Nancy Cedex, France.

Knowledge-Directed Interpretation of Mechanical Engineering Drawings

S. H. Joseph and T. P. Pridmore

Abstract— **We present a methodology for the interpretation of images of engineering drawings. Our approach is based on the combination of schemata describing prototypical drawing constructs with a library of low-level image analysis routines and a set of explicit control rules applied by an LR(1) parser. The resulting system (Anon) integrates bottom-up and top-down processing strategies within a single, flexible framework modeled on the human perceptual cycle. Anon's structure and operation are described and discussed, and examples of its interpretation of real mechanical drawings are shown.**

Index Terms—**Artificial intelligence, computer vision systems, document analysis and recognition, document processing, drawing interpretation, engineering drawing analysis, knowledge-based image analysis.**

I. Introduction

COMPUTER-based draughting systems are now widespread in manufacturing industry. Although many new design problems are addressed using CAD tools and environments, most engineering tasks involve modification of existing designs. If, as is frequently the case, the original design was developed and presented on paper, a problem arises: How does one combine CAD model file descriptions with more traditional engineering drawings? A choice must be made between performing a total reworking of the design within the appropriate CAD environment, which is a time-consuming and largely unproductive task, or finding some method of conversion from the old format to the new. Similar problems arise during the input of part geometry to CAD databases and/or numerically controlled production processes and from the use of CAD files as a form of data compression for archiving.

These difficulties have produced a strong commercial demand for automatic interpretation from industries with large stocks of conventional engineering drawings. Moreover, the huge number of paper drawings currently in active use means that the conversion problem will be topical for at least the next decade. In many cases, stores of paper drawings are still growing faster than those of CAD model files. The development of techniques for automatic drawing to CAD conversion would be a major step toward improving productivity in these circumstances.

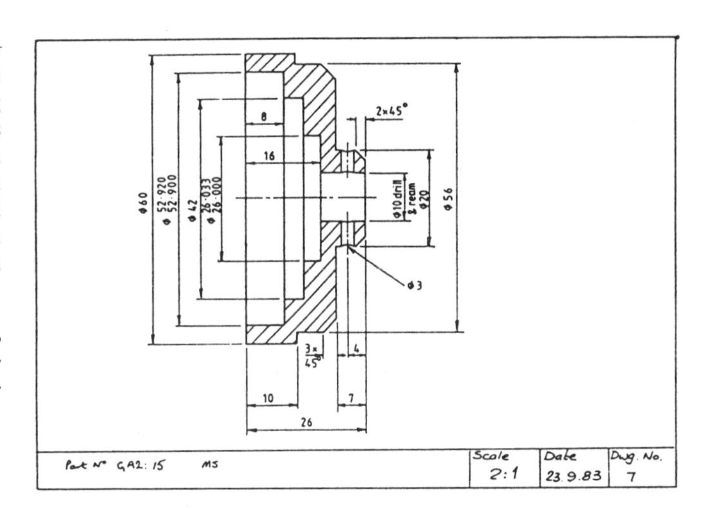

Fig. 1. Small engineering drawing.

Mechanical drawings are complex documents intended to provide a means of communication between trained engineers. The linework conveys a rich mixture of graphical and symbolic information (see Fig. 1). Some lines directly reflect the shape of the desired object, although not necessarily very accurately or to scale; others form large, complex symbols. Dimensioning, for example, is a symbol consisting of text, leader, and witness lines and may be drawn as large as some parts of the physical object. Symbols may overlap, and some control the interpretation of large areas of the drawing. Examining Fig. 1, we can see that dimensions may reference the physical shape directly by their leader (arrowheaded) lines, as in the 8 and 16 dimensions, via witness lines (e.g., the ϕ 60 dimension) or via witness and chain lines (the 4 dimension). Chain lines are one of many special line types; they are most commonly found along axes of symmetry. Crosshatching also combines symbolic and geometric significance; it labels an area (in this case as lying in a sectioning plane), and its extent is consistent with the perimeter of the area.

As the draughtsman is unable to specify objects to the required accuracy graphically, text is critical. This may occur at arbitrary angles and is usually handwritten in a mixture of both lowercase and uppercase characters and numerals. Characters are commonly written over areas of linework. Although much of the text on a mechanical drawing is taken from a restricted lexicon, extended notes are often added when the design is complex or requires special production techniques.

Manuscript received October 10, 1989; revised January 6, 1992. This work was supported by SERC ACME grant GRE 21790. Recommended for acceptance by Editor-in-Chief A. K. Jain.

S. H. Joseph is with the Department of Mechanical and Process Engineering, University of Sheffield, Sheffield, England.

T. P. Pridmore and is with the School of Engineering Information Technology, Sheffield City Polytechnic, Sheffield, England.

IEEE Log Number 9200105.

Reprinted from *IEEE Trans. Pattern Analysis and Machine Intelligence,* Vol. 14, No. 9, Sept. 1992, pp. 928–940.

Drawing conversion comprises the extraction of line structured data from an image of a drawing and its subsequent interpretation and composition into the entities found in CAD systems. The description of this potentially huge collection of lines in terms of expected symbols and acceptable graphics is a task that requires considerable knowledge of the structure and conventions of mechanical engineering drawings. Although drawing standards have been laid down by the appropriate authorities in various countries (BSI, NBS, DIN, etc.) and can aid the designer of an automatic conversion system, they do not completely determine the allowable contents of any given drawing. Both linework and text styles may change from drawing to drawing, depending as they do on both company policy and the style of the individual draughtsman.

Further technical problems arise from the volume of data that must be considered. Even for good-quality drawings, a resolution of more than 10 pixels/mm is required, which implies images over 100 Mbyte for an A0 sheet. In the worst case, 50 pixels/mm may be needed, which suggests 2.5 Gbyte images. Although these can be compressed by binarization and chain or run length, encoding the operation is nontrivial, given the variable quality originals and prints contained in drawing stores.

Current techniques of automatic drawing conversion are discussed in the next section, and approaches adopted by image understanding systems are proposed for their improvement. In Sections III to VI, we give a general description of Anon, the system by which we attempt to achieve such improvements. We first discuss methods of representing knowledge of mechanical drawing constructs. Then, in Section IV, the image analysis techniques are outlined. Section V gives an overview of Anon's control structure, whereas in Section VI, we describe the manner of its implementation. Section VII illustrates the actual operation of the system by a particular case, and Section VIII discusses new insights into the relation between top down control and system size. The overall performance of the system is demonstrated on images of engineering drawings in Section IX, and in Section X, comparisons are made with other image-understanding systems.

II. Approaches: Drawing Conversion or Image Understanding

In reviewing relevant work, we can distinguish attempts to improve the commercial state of the art in drawing conversion from more general research into image understanding systems. It is natural that the former have prioritized the conversion of large complex images to a low-level vector representation, whereas the latter have raised the interpretation to a more adequate level but only on simpler images. If we are ultimately to raise the level of interpretation in commercial systems, the image understanding research should yield methods and ideas. Combining these with a critique of existing conversion techniques can provide the basis for construction of a viable new system. It is less feasible to take image understanding systems and adapt them to succeed on complex drawing images as these systems are not structurally suited to this particular task. Let us therefore look first at drawing conversion and then examine the key methods of image understanding.

Drawing interpretation systems typically view the conversion process as a set of distinct, ordered steps [1], [2]. The grey-level image is first thresholded [3], [4] in an attempt to distinguish black ink from white paper. A vectorization process [5], [7] then generates a description of the binarized drawing in terms of an unstructured set of straight line segments. Their use is limited—CAD descriptions are normally expressed in terms of larger constructs, and consequently, most CAD/CAM applications assume a higher level of representation. More complex graphical elements such as arcs, circles, ellipses, text, arrows, dashed and chained lines, etc., must therefore be obtained by grouping together the appropriate vectors.

There is extensive literature on text and symbol recognition in electrical schematics [8], [11], and considerable progress has been made. The symbols found in mechanical drawings, however, are more complex, These drawings have variable groupings of text and linework (for example, dimensions) that do not fall so easily to the pattern recognition techniques that have been successful elsewhere—the problems encountered are similar to those involved in the extraction of graphical elements. Hence, although research is underway [12], [16], the identification of graphical elements and mechanical symbols are still largely beyond the capabilities of current conversion systems.

When graphical elements and symbols have been made explicit, the 2-D geometry reconstruction process performs any rectification needed to make the graphics consistent with the textual and symbolic information included in the drawing. We are not aware of any current system that attempts this task, although research being carried out on spatial reasoning may make the operation more feasible in the future.

The final step is the combination of views, sections, and details supplied textually and via draughting convention to construct a 3-D model of the drawn object. For complex drawings, this may require knowledge of the type of object under consideration. Several research groups have addressed this problem [17]–[19], although the resulting systems are usually limited to idealized drawings of simple artifacts. This is partly due to a lack of solutions to the intermediate symbol recognition and 2-D geometry reconstruction problems; no competent data exists that can be used as input to a developing 3-D model reconstruction system. In the present work, we seek to obtain that data but do not go on to consider 3-D model reconstruction.

Many of the limitations of the present conversion systems are due to the simple architecture on which they are based. A strictly bottom-up style of processing is neither powerful nor flexible enough for the task at hand [20]. A common approach to vectoriszation is through thinning followed by tracking of the skeletonised image. When this technique is applied, variations in line width often generate short spurs that must be eliminated. Careless removal of short line data, however, can result in the loss of vital information. It is important, for example, not to discard spurs generated by thinning the arrowheads on a dimension line. This sort of mistake can only be avoided by incorporating this data into the processing knowledge of the structure of mechanical drawings.

The entire conversion process should be integrated within a suitable knowledge-based environment. Consider the intersection of broken lines; these may form L or + shapes that conform closely to those found in text strings. A detailed and high-level interpretation of the surrounding drawing is required to resolve this ambiguity. In such cases, specialized image search procedures should be applied under the control of higher processes to extract the significant parts of the structure.

Image-understanding methods provide many valuable insights into these problems. There is a rich source of knowledge about drawing constructs in standards and conventions that we can draw on for a model-based approach [21]. Moreover, the CAD structures we are seeking to generate embody composition and specialization hierarchies that are naturally represented by schema-based systems [22]–[27]. The importance of high-level control of image processing has also been recognized in GOLDIE [28] and SIGMA [27]. There are, however, particular aspects of the present problem that make us selective amongst this wealth of experience. The requirement for handling large complex images militates against methods that maintain all possible interpretations as they proceed (SPAM [25], SIGMA) and favors those that restrict the growth of alternative hypotheses (MAPSEE [22]). For accurate interpretation of technical drawings, the strong and varied geometrical relations between drawing components must be coupled to the image search: this degree of integration remains to be achieved, even in the most advanced systems [29]. The determination of search strategy in general image-understanding systems remains problematic. In the present case, however, our images represent communications that are to be read. This means that the lower levels of interpretation are particularly susceptible to a regular accumulation of cues in a cycle of perception approach [30]. A final and overriding requirement is for a system engineered to be extensible; therefore, the endless variety of drawing constructs can be progressively brought within its scope without the code, or its space and time demands, becoming unmanageable. This is the antithesis of SIGMA, for example, in which the structure of the system is more complex than the image model it handles and where development implies the coordination of seven distinct managers, mechanisms, schedulers, generators, or experts [29].

III. Drawing Entities and Schemata

Anon is a knowledge-based image analysis system intended to extract 2-D graphical elements and symbols from a grey-level image of a mechanical engineering drawing. Although text is identified, no attempt is made to recognize characters. Anon's goal is simply to locate text and provide as much useful information as possible to an external character recognition system [31].

At the heart of Anon is a set of schema[1] classes describing prototypical drawing constructs. In its present form, the system contains classes corresponding to solid, dashed, and chained

[1] We follow Hanson and Riseman [50] in using the term "schema" (with its Greek plural "schemata") when referring to data structures representing stereotypical situations. The alternative "frame" [22] causes some confusion when used in an image analysis context, conflicting as it does with the term commonly used in TV and film. It should be stressed, however, that, like Hanson and Riseman, our "schemata" are directly influenced by Minsky's "frames."

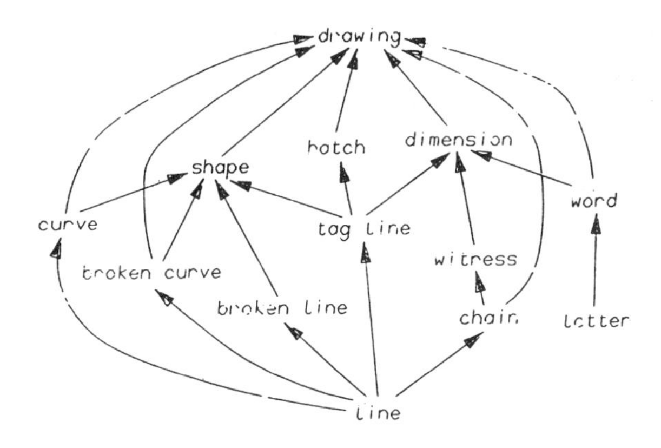

Fig. 2. Possible subpart relations between schemata. Anon's schemata may access functions and data structures associated with their subparts via these connections. Subparts, however, cannot access compound schemata. Note also that the above shows the set of possible part/subpart links. The particular subset of these formed during interpretation of a given drawing depends on local context.

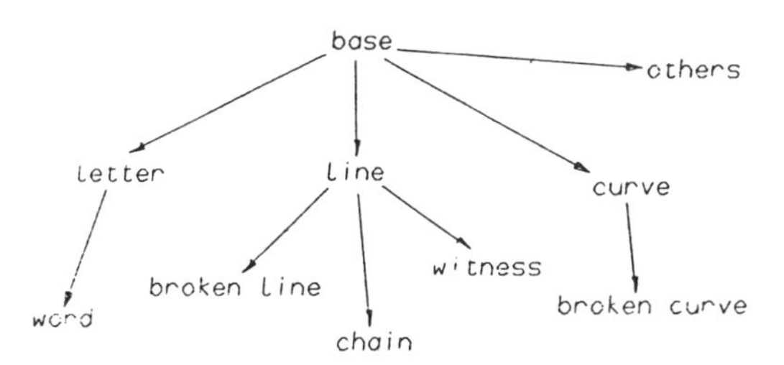

Fig. 3. Subclass relations between schemata. These provide subclasses with a means of access to their superclass.

lines, solid and dashed curves, cross hatching, physical outlines, text (both letters and words), witness and leader lines, and certain forms of dimensioning. Extensions to this set currently under consideration include dashed hatching (which are comparatively rare in mechanical drawings), blocks of text, and angular and radial dimensioning.

Each drawing entity located by Anon is represented by an instance of a particular schema class. Every schema contains a geometrical description of the construct it represents, a set of state variables noting the current condition of that representation, and a number of procedures and functions written in the C language [32]. The latter may usefully be organized into two halves. The first handles administrative tasks, creating an instance of a given class, accessing and adding components, modifying state variables, etc. The second forms the interface to Anon's low-level image analysis operations.

Schemata are not isolated but form the nodes of a network in which arcs correspond to subpart and subclass relations. The inclusion of subpart links means that each schema effectively maintains a structural, as well as geometric, description of the entity it represents. Fig. 2 shows the part/subpart links that may arise in the current implementation of Anon. The subclass links, on the other hand, allow schemata to inherit the properties of related entities in object-oriented fashion. The system's present class/subclass relationships are given in Fig. 3.

Consider an example. For our purposes, a stereotypical chained line is an evenly spaced set of colinear line segments of constant width and contrast. These lines may be divided into two sets (long and short) appearing alternately along the length of the chain. The members of each set are assumed to be of constant length. Corresponding schemata contain a geometrical description of the entity comprising its endpoints, direction, and the mean width and depth (contrast) of its component segments. This data structure is inherited from the schema's superclass: the line (Fig. 3). Additional summary information, particular to chains, includes mean lengths of both long and short lines and the spaces between them. A further structural representation is provided by an ordered set of subpart links to schemata describing the entity's component segments.

Chained line schemata also employ a single state variable, noting whether or not the construct is likely to be ending. This is used to switch between a confident pursuit of the next dash, which tends to ignore distracting linework, and a more cautious exploration of the image that will note and report such distractions. The variable is set, for example, when segments are discovered that, while being positioned and oriented in such a way as to be considered part of the developing line, deviate significantly from the expected length and/or spacing. Such distortions are common when a chain is drawn up to a predefined point.

IV. Knowledge-Directed Image Analysis

It is not our intention here to evaluate the image enhancement available in Anon but rather to describe the techniques employed and refer to reports elsewhere that describe their detailed performance.

Schemata provide a common interface to Anon's library of image analysis routines. This collection of functions and procedures may be subdivided into those that search the image for appropriately placed ink marks and others that use such marks as seed points for the development of low-level descriptive primitives. At present, seeds are located using combinations of circular and linear search patterns. These composite patterns are designed to fall on the part of the image in which the schema predicts its own continuation to lie. Some examples of these patterns are given in Section VII below, and others are reported elsewhere [33]. A schema-specific threshold is then applied to the track of pixels in the pattern, and the dark spots are noted. These are then used to initialize tracking of straight lines [34], circles [35], or area outlines [33]. In all these procedures, the grey-level values are referred to the grey level corresponding to the white paper background and its noise level [36] to give a known statistical basis for decision making at the pixel level.

It would, of course, be possible to apply a knowledge-directed analysis to a preprocessed (e.g., binarized or thinned) image, but this would impede the deployment of a full context sensitive segmentation of the image. In addition, in top-down analysis, predictive knowledge can be used to shorten processing times by avoiding exhaustive operations such as skeletonisation.

Each schema contains functions that interface to the search routines and others that invoke appropriate tracking; both the nature and details of Anon's low-level processing depend on the class, content, and state of the controlling schema. The chained line schema may perform a highly directed search for a predicted line segment involving several linear search patterns and multiple calls to line tracking. The parameters passed to these routines vary with expected line width, length, contrast, etc. (cf. [37]). If, however, the schema's state variable suggests that it might be nearing completion, a more cautious approach is taken. Circular searches (again followed by line tracking) are employed in an attempt to form a more general impression of local context.

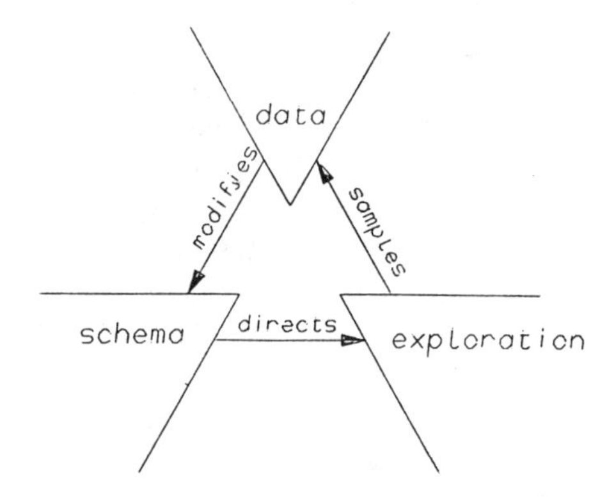

Fig. 4. Neisser's [30] cycle of perception.

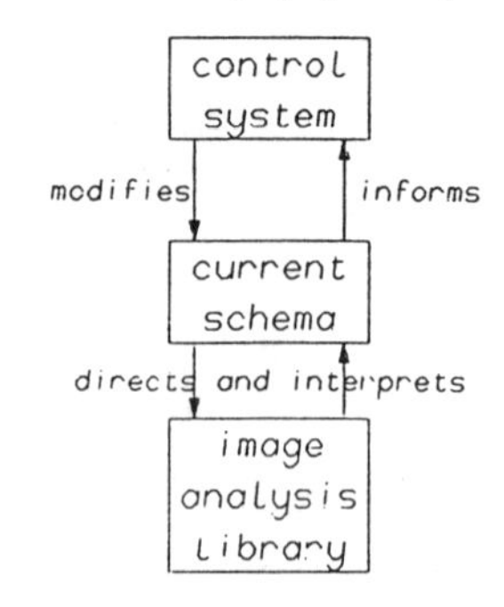

Fig. 5. Anon's cycle of perception.

Note that all of Anon's image analysis is carried out under the control of some given schema—in the context of a particular hypothesis regarding the local content of the drawing. The image analysis library therefore provides an extendible toolbox of procedures whose application varies according to context. Some higher engine is needed, however, to manage the schemata, to decide what class is appropriate, and what its content and state should be.

V. Control Structure

Anon's control structure is modeled on the human "cycle of perception" [38]. The basis of the approach is a continuous loop (Fig. 4) in which a constantly changing world model (or schema) directs perceptual exploration, determines how its findings are to be interpreted, and is modified as a result.

Fig. 5 depicts the perceptual cycle as it is implemented within Anon. Each schema, as well as interfacing to low-level search and tracking routines, contains functions that examine the result of these operations and report on their findings. The controlling, or "current," schema therefore both directs image analysis and interprets the result. A higher level control system is then informed of this interpretation, which takes the form of a symbolic label or "token" and responds by modifying the current schema. Modification may mean updating a state variable, adding new subparts, or replacing the schema with a new one representing a different type of construct. Schemata representing acceptable constructs are stored in the top-level drawing schema (Fig. 2) as they are identified. All such tasks are performed by administrative functions associated with the appropriate schemata.

In Neisser's model of human perception, the cycle continues throughout life. Anon, however, needs some procedure for initiating the system, given a fresh drawing, and terminating it when all the relevant constructs have been found. A bookkeeping module that divides the image into 9×12 subareas and associates with each one an estimate of the background's grey-level and noise characteristics has therefore been designed. From these, a target number of significantly black pixels in each area is estimated. As each new schema is stored, the number of black pixels for which it accounts is calculated and subtracted from the appropriate target figure(s). The drawing schema, which is current on startup, directs attention toward areas with large numbers of unaccounted pixels and terminates Anon when no significantly black areas remain unexamined.

An interesting feature of Anon's three-layer structure is its separation of spatial and symbolic processing. The system's spatial focus of attention, i.e., where it looks in the image and with what expectations of contrast, line width, etc, depends entirely on the current schema. Anon's symbolic focus of attention (the type and state of the current hypothesis and how it is expected to develop) is managed by the control system. Data structures passed around the loop of Fig. 5 comprise the above-mentioned tokens (generated by the current schema), which are attached to 2-D geometrical descriptions of drawing primitives. The control system reads only the tokens, whereas schemata are only affected by the geometric component. This means that the control system need only handle a stream of symbols; the way this is done is described in the next section.

VI. A Strategy Grammar

Anon's control system consists of strategy rules written in the form of an LR(1) grammar and applied by a parser that is generated using the unix utility yet another compiler compiler (yacc) [39]. LR(1) parsing provides a rapid and compact method of syntactic analysis, which may be applied straightforwardly to certain pattern recognition tasks [40]. It is important to stress, however, that the grammatical rules incorporated in Anon are not intended to define a legal engineering drawing in any declarative sense; rather, they specify strategies by which the various components of a drawing might be recognized. We do not suggest that an entire drawing image could be interpreted by a single left to right scan producing a string generated by such a grammar. Our use of an LR(1) grammar as a control rule interpreter restricts the rules that can be applied. In this section, we discuss features of the parser and its combination with schema-based token generation that mitigate these restrictions and aspects of the drawing conversion problem that suit it to our appproach.

Control rules, like string grammars, describe acceptable sequences of events; an LR parser is therefore a natural vehicle for their application. In Anon, this sequence of events is the stream of tokens generated by a successful schema-driven analysis of the input image. Note that the grammatical style reflects the naturally repeating structure of drawing interpretation strategies. When extracting a chained line, for example, the system must locate successive line segments and gaps until some termination condition (e.g., finding a corner or white space) is satisfied. This easy mapping between control and grammar leads to a compact but powerful rule set. The example given in Section VII shows how grammar rules for a chain line form a natural description of a strategy for its extraction.

In reviewing the use of logic as a representation in computer vision, Rao and Jain [41] list among its advantages the provision of a "simple, expressive rule format," "formal precision and interpretability," and "guaranteed consistency." Like logic, LR(1) parsing is a well understood, formal method that displays these properties. The existence of well-established software tools based on the technique is an added bonus.

Yacc is a parser generator that is usually employed to create command interpreters and language compilers. It accepts an LR(1) grammar together with an action code in C language (to build parse trees for example) and generates a table driven, stack-based, finite state machine. This calls a user-defined function to obtain the tokens to be parsed and invokes an appropriate action code whenever a grammar rule is reduced. In Anon, tokens are supplied by the current schema, and the action code is replaced by calls to schema-based administration functions.

Although the machine operates in a single pass left to right, the in-built single token lookahead implies a single step backtracking capability. Examining the states of the finite state machine produced by yacc shows how the stream of tokens produced by the schemata's image analysis progressively defines the structure being extracted from the drawing. A given state corresponds to several rules being active and, thus, several hypotheses being supported. As successive tokens are processed, rules are reduced, a selection of hypotheses are validated, and the results are incorporated into the drawing description. The multiplicity of these hypotheses is exactly that needed to support a single step of backtracking.

The yacc software extends basic parsing capability to cover illegal token sequences, which, in the present context, corresponds to failed strategies. The parser reports failure by effectively generating a special "error" token. Although this is introduced by the parser rather than the current schema, it can, like any other token, be incorporated in grammatical rules. As a result, the user can define situations in which failure becomes acceptable, i.e., set breakpoints beyond which any further processing cannot detract from the entity already found. This ability is essential when designing strategies for the interpretation of engineering drawings. For example, although most dimensioning is associated with nearby text, cases do arise in which numerical values are specified elsewhere in the drawing. Failure to locate text should not result in the abandonment of otherwise-acceptable dimension symbols. The error token allows one to write dimensioning strategies in which text is optional but not necessary. The insertion of such break points also means that the system can be tailored to produce partial interpretations of structures that have not conformed exactly to the model, giving Anon the robustness that is necessary for its operation.

Another aspect of uncertainty in the interpretation of drawings is the presence of points from which there are several alternative search paths, for example, at the intersection of several lines. In this case the current schema not only uses its inbuilt priorities but also seeks an interpretation (a token) that satisfies the strategy grammar. This is readily done by applying the yacc machine's error checking code to each available token before returning one to the control system: erroneous tokens are rejected if one acceptable to the grammar can be generated from an alternative tracking path. Should more than one valid token be available, a schema-specific selection function is invoked.

In addition to providing Anon with a portable, efficient rule interpreter, yacc supplies a flexible user interface that allows fast prototyping of strategies. Given a suitable rule set, a new control system can be compiled within minutes. Furthermore, yacc detects and reports inconsistent rules during compilation, which is an ability that has proven to be very useful during the development of the system. Although the greater modularity offered by more traditional rule languages is attractive, eliminating inconsistencies from such environments can be a long and complex task. McKeown *et al.* [42] have recently reported difficulty in extending their aerial image interpretation system (SPAM) for just this reason.

With the above considerations in mind, it would seem that the reduction in generality brought about by our use of an LR(1) parser is compensated by the efficiency of the yacc machine. The customary compromise between speed and generality has been well made. Note also that the use of a high-speed rule interpreter to direct low-level processes ensures the efficiency of the overall system by keeping the amount of unnecessary image analysis to a minimum.

Mechanical drawings are man-made artifacts intended to communicate information according to a loosely predefined convention. They are read rather than perceived. It is therefore not enough simply to define the constructs that might occur in the drawing; we must determine effective strategies for their extraction and interpretation. In this situation, the facilities provided by yacc make LR(1) parsing a particularly suitable method of applying control knowledge.

VII. Identifying a Chained Line

To complement the rather general overview of Anon presented so far, we will consider its detailed operation in a particular case: the extraction of a chained line. Consider the vertical chain marking the center of the rightmost view in the drawing of Fig. 6. Let us suppose that the first segment has been found and that a line schema is therefore current. This segment could be any one in the line as there is no context available to direct search to an end. On the first perceptual cycle, this directs image processing (Fig. 7(a)) and interprets the result as a BREAK, that is, a gap in the linework is found immediately in front of the current line, and a description of it, together with a BREAK token, returned to the control module. Since no rule is as yet fully satisfied (this will not occur until the second line has been identified), no action code can be employed to modify the current schema. One of the six rules

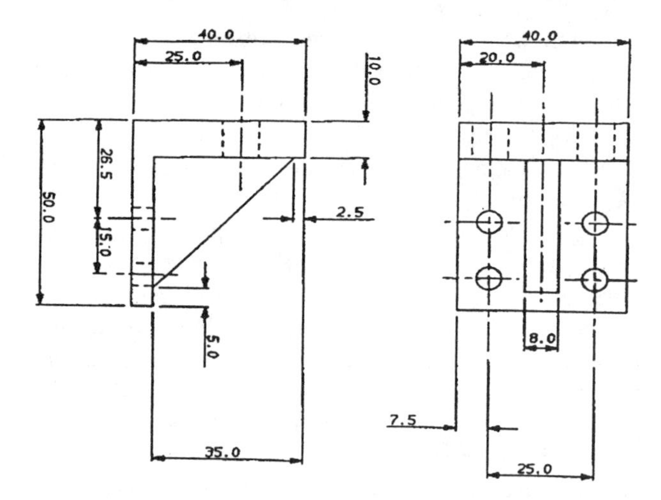

Fig. 6. Sample drawing. Images (128 grey levels) are taken at a resolution of 10 pixels/mm using an EG&G RETICON line-scan camera. The field of view is approximately 20.5 by 14.5 cm, resulting in a 4-Mb image file. Thresholding [36] is applied for ease of display only; it should be stressed that Anon works on the full grey-level image.

that are active is

(1) broken_start: line **BREAK LINE**
{broken_start.instantiate()};

which states that when the second line has been found, a newly instantiated broken_start schema should become current.

In this rule format, terminals (tokens) are shown in bold type, and nonterminals are shown in lower case. Thus, "line" signifies the result of the earlier satisfaction of a rule, causing a simple line schema to be instantiated, whereas LINE signifies a label attached to the results of examination of the image by a schema. The rule becomes active when the current schema corresponds to the nonterminal immediately to the right of the colon separator. The nonterminal to the left of the colon denotes the schema to be current after the rule has fired. The action code is delimited by curly brackets. As all Anon's action code is accessed via an appropriate schema instance, we adopt the convention that the instance type is given first and separated from the function name by a full stop. In the above example, broken_start.instantiate() refers to the instantiate() function of the broken_start schema class.

At this stage, on the second cycle, the original line schema is still current. A break description is, however, now available. The current line schema uses this as a relational constraint to modify its interaction with the image processing library, performing a single circular search at the forward end of the break (Fig. 7(b)). Line tracking from the ink marks returned by this search leads to the extraction of the second line segment (Fig. 7(c)), which the schema acknowledges with a LINE token. When this is received by the control module, rule 1 above is reduced, causing the instantiation of a new broken_start schema, which immediately becomes current.

The use of relational constraints between visual objects in this way is a topic that has received considerable attention in the literature. There now appears to be some consensus among the developers of schema-based systems that relations are equal in importance to the objects they constrain and should therefore be represented as schemata in their own right [43], [22]. This approach has been adopted in Anon. Symbolic

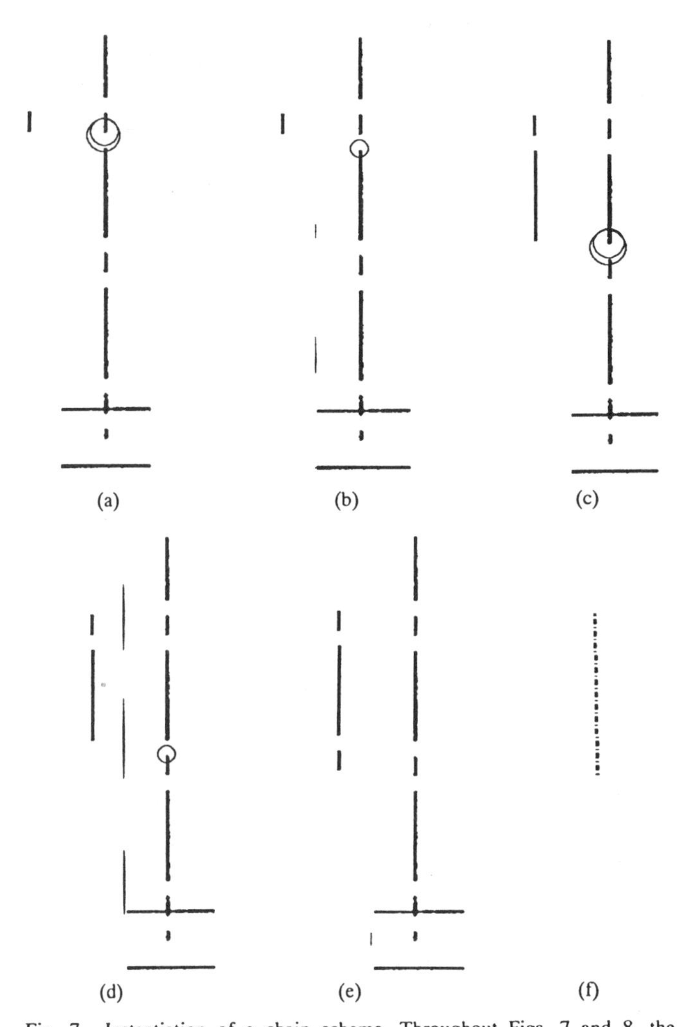

Fig. 7. Instantiation of a chain schema. Throughout Figs. 7 and 8, the contents of the current schema are displayed to the left of each subfigure. Any image searches invoked are shown on the right: (a) Image search under the control of the line schema. Circular patterns spread out from the end of the segment in a torch-like fashion until either black marks are discovered or the distance between the line end and the front of the beam exceeds some predefined maximum value. In this case, the rear end of the next line is located by the second search. The white space between the two lines constitutes a BREAK relation; (b) image analysis by the line schema given additional knowledge of a BREAK. A circular search is performed at the front end of the gap; (c) detection of the long segment by tracking from the seed provided by (b) causes instantiation of a broken_start schema that then instigates another torch-like search; (d) circular search directed by the broken_start schema after detection of the second break. The short segment discovered by line tracking, together with the previous BREAK, allows the broken_start to be extended; (e) extended broken_start. No image analysis is required as the schema decides on the next token by simple examination of its own contents; (f) summarized contents of the newly instantiated chain line.

descriptions of the relations between drawing constructs are stored as schema instances and passed around the perceptual cycle, along with descriptive tokens, as are other graphical entities. The major difference between relational and other schemata is that the former do not contain a procedural component; instead, knowledge regarding how to use relations is maintained within the schema description of the affected drawing constructs.

The developing construct is now recognized as being a broken line of some description; whenever a broken_start schema is current, it examines its contents to determine whether it is straight, curved, or chained. If the broken_start can be classified as chained, the token ISCHAIN is passed to the control system by the broken_start schema and the rule

(2) chain: broken_start ISCHAIN
{chain.instantiate()};

satisfied. This states that when a broken_start returns ISCHAIN, it is specialized as a chain schema, which becomes current. At this stage, the broken_start schema examines its contents but cannot make a reliable decision and, therefore, continues the exploration. First, another BREAK is discovered (Fig. 7(c)), and then, a third line (Fig. 7(d), (e)) is discovered. This reduces rule 3

(3) broken_start: broken_start BREAK LINE
{broken_start.addon()};

and extends the broken_start. Note that the search patterns shown in Fig. 7(c) and (d) are similar to those seen in Fig. 7(a) and (b). As the previously detected line schemata are subparts of the broken_start schema, they are called on to perform the necessary search and tracking operations. The broken_start schema now comprises three line segments joined by two break relations (Fig. 7(e)). No image search is required for the next stage, and the relative lengths of the lines and the regularity of their spacing is sufficient evidence to allow the broken_start schema to return ISCHAIN, reducing rule 1 above and instantiating a chain schema (Fig. 7(f)).

Once a chained line has been recognized, further processing becomes much more directed. The chain schema can predict subsequent breaks and lines and then seek them top-down rather than bottom-up. Fig. 8(a) shows the search pattern used to locate the next (long) line in the example of Fig. 7. A set of three linear search patterns, which are normal to the expected segment, locate ink marks, which seed line tracking. The subsequent LINE, together with the predicted BREAK, satisfies rule 4:

(4) chain: chain BREAK LINE
{chain.addon()};

which extends the schema (Fig. 8(b)). This process continues, where rule 4 is applied several times, until a predicted line cannot be found (Fig. 8(c)). The current schema now returns a different token (END), indicating its failure to locate any further linework. This results in the schema's internal state variable (recall Section VI) being set by the rule

(5) endchain: chain BREAK END
{chain.endset()};.

No other change is made to the content of the chain schema, which remains current. With its only state variable set, the chain tries a more exploratory search on the next cycle (Fig. 8d) but fails to find an acceptable continuation. Instead, it detects the JUNCTION between the chain and the physical outline at the bottom of the figure, which is not legal at this point in the strategy grammar.

Rule 6

(6) bichain: endchain ERROR
{chain.endclear(); chain.invert()};

is therefore reduced, and the direction of the partially developed chain line is inverted to allow the construct to be extended in the reverse direction. Note that the state variable is also reset as the chain is no longer ending. Lines and breaks are predicted in similar fashion during this second

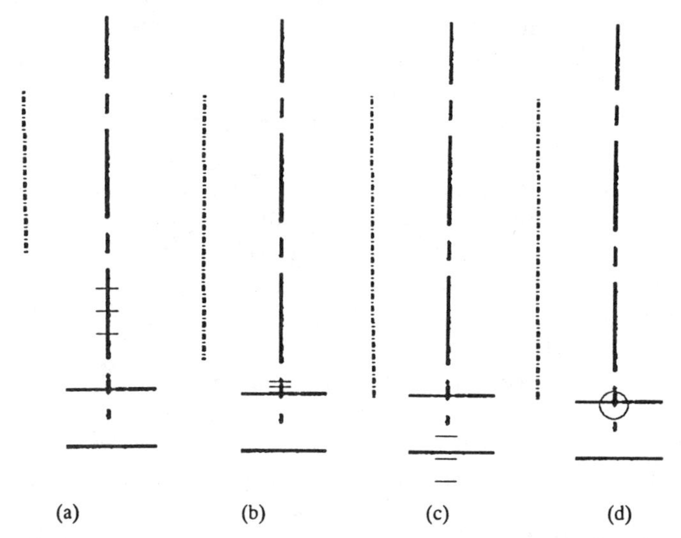

Fig. 8. Developing a chained line. The current schema is shown on the left of each subfigure, and the image searches it invokes are shown to the right: (a) Top-down search for the next (long) line segment in the chain of Fig. 7; (b) top-down searches, which are performed sequentially under the control of the strategy grammar, seek the remainder of the chain's component lines; (c) last search fails to find any linework, causing the chain schema's only state variable to be set; (d) with its state variable set, the chain schema attempts a more tentative search pattern, which also fails to locate an acceptable continuation of the construct. Tracking in this direction is now terminated, the chain schema is inverted, and extensions are sought in the reverse direction.

period of extension (not shown), which once again ends when a predicted line segment cannot be identified. Although a tentative search is tried, after rule 7

(7) endbichain: bichain BREAK END
{chain.endset()};

is reduced, no further segments are discovered, and tracking is terminated by rule 8

(8) drawing: endbichain ERROR
{chain.preserve()};

which makes the completed chain (see Fig. 9) a subpart of the initial drawing schema.

It will be noted that although only three schema instances (one broken_start, one chain, and the drawing) were involved in the above example, six nonterminals have been introduced. This apparent proliferation of nonterminals is not characteristic of the system but arises in this case because we have described only one path of the rule tree. In the full tree, a typical nonterminal represents a collection of schemata in specific states. This many-to-one mapping from hypotheses to nonterminals keeps the size of the grammar to reasonable proportions.

VIII. Issues in Top-Down Control

So far, we have shown how Anon supports the geometrical direction of image search, progressive discrimination between multiple hypotheses, and robust handling of misfit data. These have only been expounded in quite a simple case: a chained line. We now turn to the vital issue of system development and the handling of more complex structures.

Chains are of a low level in that they are formed directly from image primitives and relations: line segments and breaks. Many drawing entities are more complex because they are composed of higher constructs. For example, dimensioning incorporates several components including, sometimes, chains

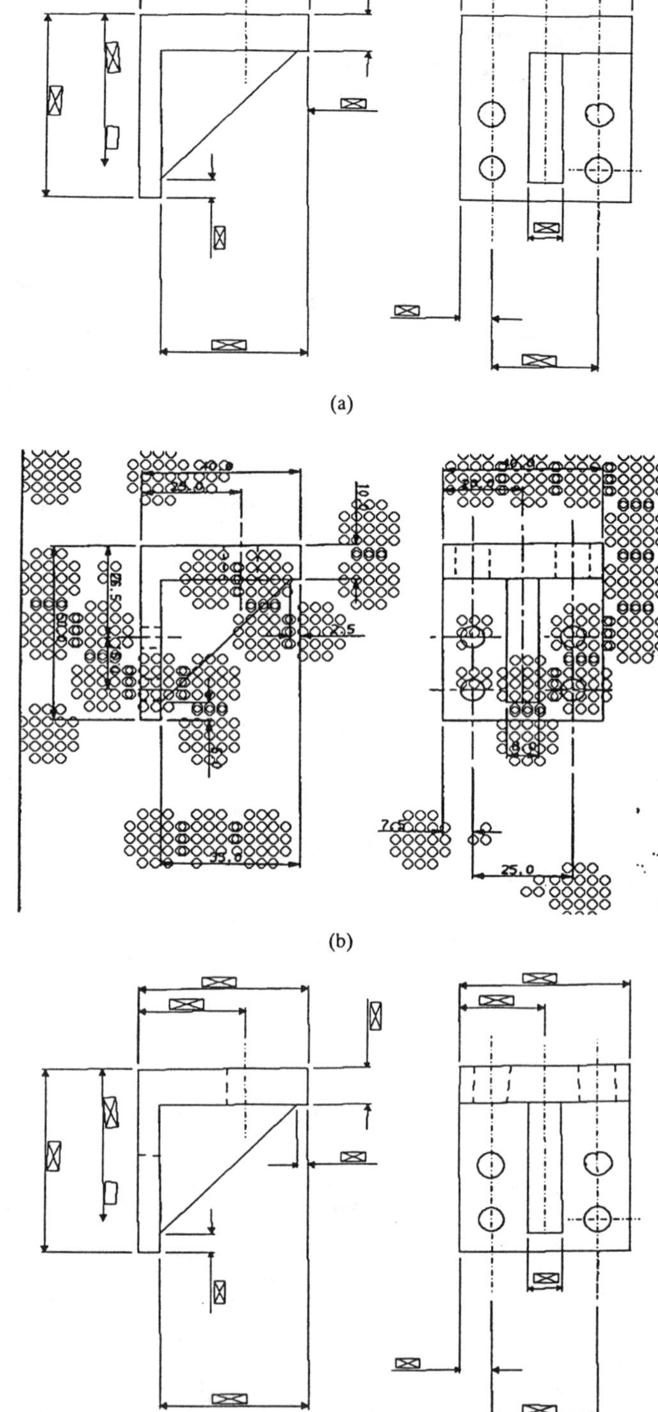

Fig. 9(a). Anon's fully automatic interpretation of the image of Fig. 6. Chains are shown as dot-dashed lines, and text is shown as rectangular boxes. The leader lines of dimensioning are marked by arrowheads. Where text is associated with dimensioning, its box is labeled with a diagonal cross. Physical outlines are simply drawn solid. No other schema classes are involved in this example; (b) initial circular searches generated automatically during the interpretation of Fig. 6; (c) final interpretation, obtained after additional user-supplied starts. The description now includes previously absent text and broken lines denoting hidden detail, which are shown dashed.

(as described in the Introduction). In principle, the combination of high-level schemata to create even higher entities requires only a straightforward reapplication of the techniques discussed above; therefore, it fits naturally into Anon. In practice, the undisciplined creation of such hierarchies could lead to an expansion of system size that would severely restrict its ultimate capabilities.

Consider two rules that contribute to Anon's interpretation of dimensioning:

(9) dimension : leader PARALLEL word
{dimension.instantiate()};

(10) dimension : word PARALLEL leader
{dimension.instantiate()};.

These specify alternative strategies for the identification of annotated leader lines. The first states that a dimension schema can be instantiated and made current by a leader instance returning a PARALLEL relation, i.e., discovering something to the side and similarly oriented and then noting that the particular something is a word. The second requires the word to be found first, followed by a suitably related leader.

At first sight, the need to write out separate strategies in this way might seem to be unwanted labor; it would be preferable to have a single rule describing the parallel relation between word and leader and allow it to be satisfied in whatever order. The means to do this is not available and for good reason. To find a word in the context of a known leader is quite different from a leader in the context of a word. In fact, developing these separate strategies is an essential step in the determination of valid and powerful rules.

It is also evident that we must be able to identify, for example, leader lines in two different circumstances. Rule 9 above provides no contextual information; the leader must be built bottom up in the same way as was the broken_start schema in our earlier example. This type of entity extraction is natural to Anon. Rule 10, however, requires a leader line to be found under the control of a word schema. The word could hypothesize and test directly for a complete leader, arrowheads and all, in the same way that the chain tests for the next line segment. The difference, of course, is that a line segment is a primitive, whereas the leader is a construct of some complexity; to take this route would be to instigate a top-down search for a high-level schema instance, which is an operation that requires a second method for extracting the predicted entity. Nagao *et al.* [23] adopt this approach in their work on aerial imagery. Their program comprises a number of "object-detection subsystems," which are similar to our schemata, which communicate via a blackboard. A given object may be detected by either of two subsystems, where one operates bottom up and the other top down. For domains in which a large number of different objects may appear, this duplication results in a large and potentially unmanageable system. The problem, then, is how to examine higher level hypotheses without constructing separate top down test procedures.

The solution adopted in Anon is to exert top-down control only in modifying the initial search strategy and then to employ bottom-up processing. In the above example, the word schema directs attention toward the region in which the leader is expected to lie, searching for a primitive that may be part of the required structure. This seeds Anon's normal, bottom-up analysis, and if a leader schema is constructed, rule 10 is reduced.

This disciplining of top-down control represents an important new step. It has come to light as a consequence of applying image understanding methods to more complex structures. Without such discipline, excessive duplication of techniques is required. Of course, the solution proposed here is not the only one possible. Future work will be directed to evaluating this solution and proposing alternatives. We expect that there will be a continuing tradeoff between the extent of top-down control and the size of the resulting system. Currently, the present solution appears to provide simplicity without undue restrictions to top-down control.

IX. Performance

The sample interpretation discussed in Section VIII illustrates *how* Anon works rather than *how well*. In the following, we shall attempt to assess the system's performance. This is, unfortunately, a nontrivial task. First, there is no clearly defined target representation for mechanical drawings. The internal structure of a given CAD file usually depends on the manner in which the design was entered by the draughtsman. Second, huge variations can be expected in both the quality and complexity of active engineering drawings. It therefore seems unlikely that any attempt to formally delimit the range of drawings that Anon can be expected to handle would be successful. Direct comparison with previous work is also problematic; we are not aware of any other systems that attempt to produce the entity-level descriptions presented here.

Anon is a large and complex system. The strategy grammar now contains 191 rules (out of a total allowed by yacc of 600), resulting in a 313 state parser (out of an allowed 1000). The parser recognizes 122 terminals and 68 nonterminals, which is yacc's upper limit, where each is 300. In total, Anon comprises some 20 000 lines of C code [32] and, in its present form, supports 15 distinct schema classes. Limitations of space therefore prevent detailed evaluation of each schema in the present report. Future documents will, however, focus on the extraction of particular entities. The line, circle, and boundary tracking algorithms employed have already been assessed [33]–[35].

Given the above considerations, our evaluation of Anon must be qualitative, comparing the system's output to human segmentation of test drawings.

Anon may be run in either interactive or automatic mode, the only difference between the two being the manner in which tracking is initiated. While the system is running interactively, the drawing schema prompts the user each time a new start is required. It should be stressed that the only information provided by the operator is the position and radius of a cursor used to specify an initial circular search. During fully automatic operation, the bookkeeping module described in Section IV is used to generate suitable start positions. A search of fixed radius is then performed at each chosen point. Fig. 9(a) shows Anon's (automatic) interpretation of the drawing of Fig. 6. The initial search patterns used are shown in Fig. 9(b).

The bookkeeping module focuses attention quite well, clustering searches around text, physical outlines, and dimensioning. Two points should be noted. First, no searches are performed around the middle of the chain of Figs. 7 and 8. This is identified early in processing, and the accounts are updated, directing attention elsewhere. Second, although the dark lines running down the sides of the drawing attract a number of searches, these do not generate acceptable schemata. Hence, Anon is capable of discarding any spurious starts that might be induced by image noise.

Comparison of Figs. 6 and 9(a) is encouraging. Anon's description of the original drawing corresponds well to human interpretation and is almost complete. High-level dimensioning, chain, text, and physical outline schemata account for most of the image. Some constructs are, however, missing from the system's output. Perhaps most noticeable is the absence of hidden detail (short dashed lines), which was missed because the pattern of initial searches was too coarse to detect its very short component segments. In some cases, no attempt was made; the lines were so small that the bookkeeping module did not consider them worthy of attention. One piece of text (in the top right of the figure) was also overlooked; this again was due to inappropriate initial search radii.

Although some drawing entities are missed, others are included in more than one schema. The right-most vertical chain, for example, was found twice: first on its own as a center line and then as part of dimensioning. The interaction between these interpretations can be clearly seen in Fig. 9(a). The left-most chain on the right view is also found twice, although this is not obvious from Fig. 9. Two dimension sets at the bottom of the view share a common witness line, which incorporates the chain. The witness, and hence the chain, is therefore described by two separate dimension schemata. Only the (almost) identical representaton of the witness by these two schemata prevents the overlap from showing up in Fig. 9(a). An unfortunate consequence of overlapping schemata is that they disrupt bookkeeping; a section of physical outline was missed because, after accounting for the left chain twice, that part of the drawing did not appear to warrant further attention.

Anon currently makes no attempt to detect overlapping interpretations but deals with the extraction of schemata rather than the resolution of conflicts between them. It is enough to note that each schema represents a valid interpretation of the underlying construct. Similarly, although further work is clearly required to improve Anon's start-up and bookkeeping routines, the artificial generation of start points is at present a peripheral consideration. The main goal of our work is to produce a system capable of using context to direct the interpretation of complex mechanical drawings.

Many of the deficiencies of Fig. 9(a) can be made up by running the system interactively after automatic interpretation. Fig. 9(c) shows the description of Fig. 6 obtained after processing nine additional starts supplied by the user. The missing text has been located, as has most of the hidden detail. Note the inclusion of the previously absent text in a dimension schema that overlaps the one found in automatic mode. Those dashed lines that could not be added are too small to provide the context needed for schema instantiation. Similarly, it will be noted that in Fig. 9(a) and (c), only one of the four horizontal center lines in the right-most view is represented as a chain. The others are absent because interactions with the circles have prevented Anon from obtaining sufficient clean structure to instantiate a chain schema. Omissions of this type, which are typical of schema-based systems, can only be rectified via knowledge of the type of construct expected. Hence, they cannot as yet be overcome in the interactive mode as Anon's user can only specify where tracking should commence. Accurate positioning with an appropriate search radius does, however, give the system the opportunity to apply its knowledge and therefore provides improved output with minimal operator intervention.

Fig. 10 illustrates Anon's performance, given the slightly more complex drawing of Fig. 10(a). The result of automatic interpretation is given in Fig. 10(b), and the final description, after 10 user-supplied starts, is given in Fig. 10(c). Despite the limitations discussed above, the description produced in automatic mode is once again acceptable. Most of the dimensioning and physical outline is represented, along with the larger of the chained lines. The remainder of the text, dimensioning, and physical outline, which was missed due to the coarseness of the automatically generated starts, is successfully added in the interactive mode.

The cross-hatched regions of Fig. 10(a) are quite large and, although they are correctly identified, are represented (cf. Fig. 9) by several overlapping schemata. Conflicts also arise between different drawing entities. Close examination of Fig. 10(b) reveals that four of the parallel lines making up the physical outline also form acceptable hatching (Fig. 11). Several of the shorter cross-hatching lines are similarly (and wrongly) considered to be part of the physical outline. McDermott [44] argues that a knowledge-based system will display erroneous behavior for either of two reasons: 1) inadequacies in its knowledge or 2) inadequacies in its problem-solving behavior. The conflicts depicted here and in Fig. 9 are clearly the result of the latter problem; their resolution requires a global view that Anon's local, schema-based operations cannot be expected to provide. A further example is seen to the far left of Fig. 9(a), where two dimension sets share a common leader line. Anon has created a single schema merging the two together. Only by considering a large part of the image can the system be expected to rectify the error. The development of this capability is the subject of a subsequent report [45].

In contrast, the problems apparent in Fig. 12 are largely due to lack of knowledge. The drawing (Fig. 12(a)) contains a number of new constructs for which Anon has neither schemata nor strategy rules; most of the text touches its underline, and there are several radial dimensions. Performance is further degraded by the use of shared leader lines in all other dimensioning. Even so, automatic interpretation (Fig. 12(b)) correctly identifies most of the physical outline and as much text and dimensioning as can be expected. Interactive help (Fig. 12(c)) adds more text, two of the holes, the dashed arc, the center line, and part of the lower chain.

X. Anon *et al.*

It is useful at this stage to situate our approach with respect to some other knowledge-directed image analysis systems. Two initial points should be made regarding general

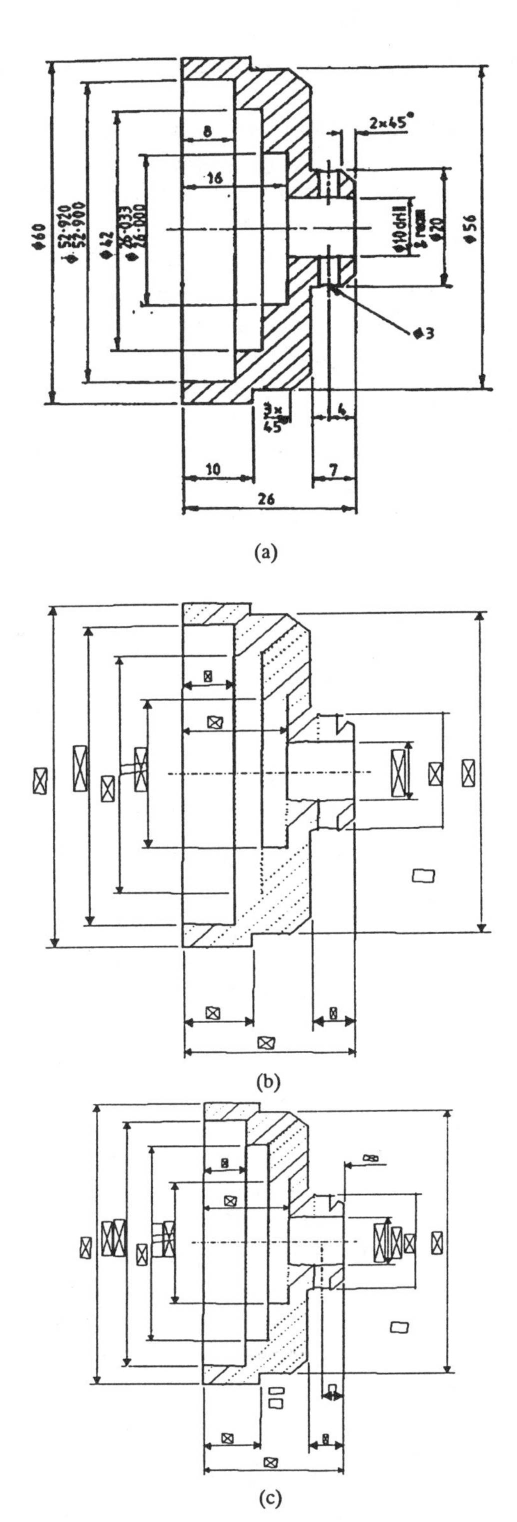

Fig. 10(a). More complex image thresholded, as in Fig. 6, for display purposes only; (b) automatically obtained interpretation of Fig. 10(a). In addition to the schema classes present in Fig. 9(a) and (c) are a number of cross-hatching schemata, which are drawn as sets of parallel finely dotted lines; (c) final description of Fig. 10(a) obtained after processing 10 user-supplied starts. More careful positioning with an appropriate search radius allows Anon to extract the remainder of the text, dimensioning, and physical outline.

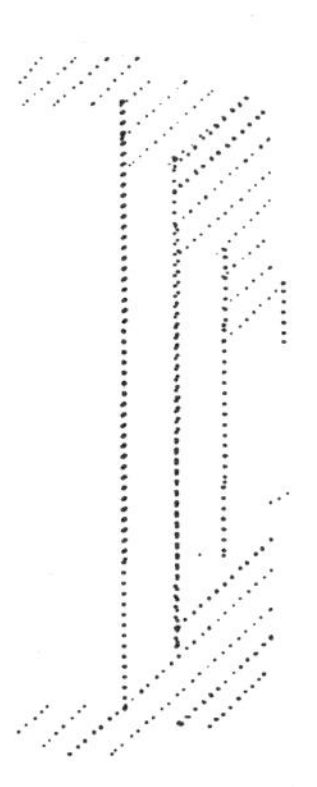

Fig. 11. Areas of cross hatching extracted from the image of Fig. 10(a). All the cross hatching is represented, albeit in several overlapping pieces, but part of the physical outline also forms acceptable hatching.

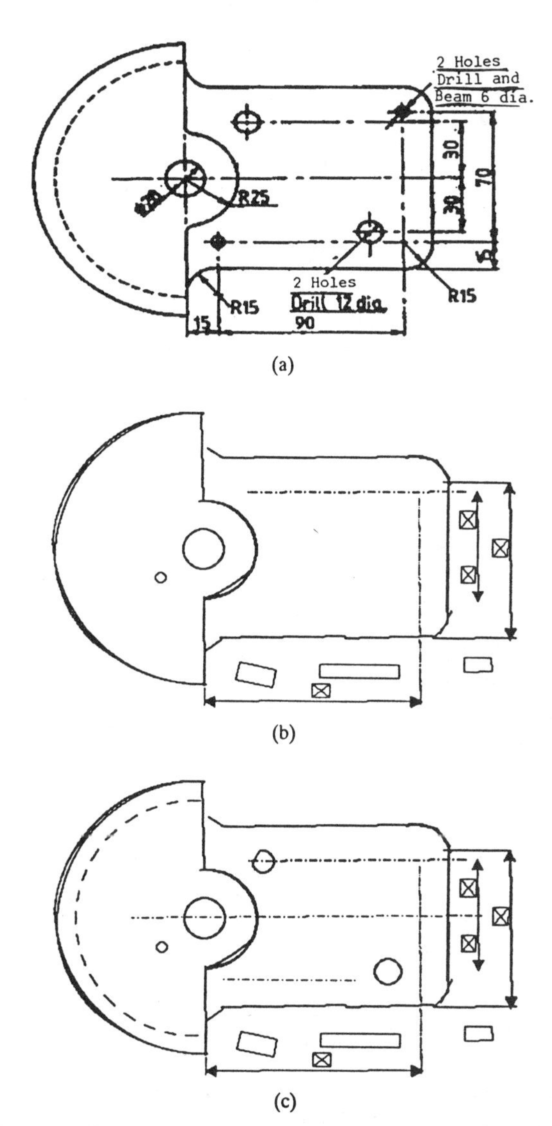

Fig. 12(a). Input drawing image, thresholded for display; (b) result of automatic interpretation; (c) final interpretation. Note the dashed curve to the left of the figure denoting the corresponding broken arc in the drawing.

differences between Anon and other systems. First, ours is an expert task; reading complex drawings requires training. Anon is not, therefore, intended to be a model for general vision. Second, as drawings are naturally read sequentially, strategic knowledge is of particular importance, hence, Anon's reliance on the control rules made explicit by its strategy grammar.

Although parsing is a standard tool in syntactic pattern recognition, its use as a control mechanism is less common. Tropf and Walter [46], however, use an augmented transition network (ATN) to specify strategies for the recognition of modeled industrial parts. Although their system works well, its reliance on procedural knowledge seems excessive, given the task at hand. The ATN specifies the order in which lines and corners extracted from an image may be matched to corresponding features of a geometric model and, therefore, plays a role similar to that of Anon's grammar. However, although they refer to their approach as "analysis by synthesis" [47], [48], the strategies embedded in the ATN bear no relation to the manner in which their images were either created or intended to be perceived. Hence, this type of procedural knowledge seems more suited to the interpretation of communicative documents than images of physical objects.

At an abstract level, the nonterminals of our strategy grammar form the nodes of a (directed) graph in which each arc represents a relation or predicate. This corresponds closely to Mulder's [24] discrimination graphs. The major difference here is that, whereas Mulder's graphs must be acyclic, our LR(1) parser can handle cyclic structures. As interpretation proceeds, the graph is searched, under the control of the parser, for paths from the root to acceptable goal nodes. At each step, some given predicate or relation must be satisfied if attention is to pass from one node to the next.

Despite the advantages afforded by the cycle of perception architecture [38], [30], few computer vision systems have taken this approach explicitly. Glicksman's [49] MISSEE is a notable exception. This is also schema driven, where its goal is to produce a semantic network of schemata describing an aerial image of a small urban scene. MISSEE's schemata communicate directly; therefore, control knowledge and schemata are not distinct but are integrated within the semantic net. This integration raises questions regarding the expandability of the system [49]. A further difference between Anon and MISSEE is the latter's use of a presegmented image; all low-level processing is completed before interpretation commences. This separation of image description from higher level interpretation is commonplace, even in systems employing schemata. VISIONS [50], SPAM [25], MAPSEE [22], [24], and the systems due to Sakai *et al.* [26] and Nagao *et al.* [23] all operate on preprocessed images.

GOLDIE [28] does use schemata to direct image analysis. Moreover, like Anon, its low-level routines are maintained as a separate library of image processing tools. GOLDIE's schemata, however, do not represent expected constructs. Instead, they describe strategies by which different types of image analysis task (region segmentation, line extraction, colinear line grouping, etc.) may be achieved and are invoked in response to goals posted by higher processes.

Matsuyama and Hwang's [27] SIGMA both represents its knowledge of expected objects as schemata and combines low-level processing with interpretation. SIGMA's schemata do not, however, control image analysis directly. This task is delegated to a low-level vision expert (LLVE), which uses knowledge of the available image processing operators (cf. [51]) to satisfy goals posted (cf. GOLDIE) by the higher level components of the system. New goals are posted as schemata are instantiated. At the top level, SIGMA's operation is governed not by explicit strategic knowledge but by a spatial reasoning module that attempts to find consistent sets of schemata.

In its present form, Anon's schema-based operations provide a layer between the grey-level image and the higher level, and more global processes needed to integrate pieces of drawing into a coherent whole. A similar approach is adopted in SPAM [25] and the later versions of MAPSEE [22], [24]. In both systems, schemata (called functional areas in [25]) are created and filled before some more global process (constraint propagation in the case of MAPSEE and higher level rules in SPAM) resolves any conflict between competing interpretations. Although its schemata appear to be more powerful, Anon currently lacks the necessary higher level component. This extension is the subject of continuing research [45].

XI. Conclusion

Anon uses both strategic knowledge and schemata representing prototypical constructs to direct low level image analysis routines in the interpretation of mechanical engineering drawings. The system operates directly on the grey level image, combining the extraction of primitive descriptions with their interpretation. In marked contrast to previous approaches based on a simple pipelined filters architecture, Anon integrates bottom-up and top-down analysis within a framework modeled on the human perceptual cycle.

So far, Anon has been applied to images of piece part drawings. It is hoped, however, that much of the code will prove to be reusable in other engineering domains. The image analysis library, based as it is on sequential tracking of linework, should be applicable to most types of drawing, and many of the constructs represented by Anon's schemata are commonly used elsewhere. The system's modular structure makes it easily extendable, and new image processing techniques, schemata, and/or control rules may be incorporated as required. Software tools provided by yacc ease the addition of the latter considerably, allowing compile-time detection of inconsistent strategies.

Even though our evaluation of Anon has been rather informal, the results obtained to date are encouraging. Figs. 6 to 12 demonstrate the system's ability to provide high-level descriptions that correspond well to human segmentation. Many of the deficiencies of the automatically produced representations are the result of inappropriate initial searches. There are also deficiencies in the handling of multiple interpretations of entities. These parts of Anon are yet to be developed; the existing parts of Anon (forming an integrated extraction/interpretation scheme that is powerful and consistent) provide the requirements to which they can be constructed.

We do not consider that the current Anon system would provide useful fully automatic conversion of drawings to CAD format. It would, however, provide a valuable automatic element to combine with a manual interface and, more importantly, does provide a methodology for the future development of an increasingly automatic system.

References

[1] J. Hofer-Alfeis, "Automated conversion of existing mechanical engineering drawings to CAD data structures: State of the art," *CAPE '86: Conf. Comput. Applications Production Eng.* (Copenhagen), 1986.

[2] M. Karima, K. S. Sadhal, and T. O. McNeil, "From paper drawings to computer-aided design," *IEEE Comput. Graphics Applications*, pp. 27–39, Feb. 1985.
[3] J. S. Weszka and A. Rosenfeld, "Histogram modification for threshold selection," *IEEE Trans. Syst. Man. Cybern.*, vol. 9, pp. 38–52, 1979.
[4] J. Kittler, J. Illingworth, and J. Foglein, "Threshold selection based on a simple image statistic," *Comput. Vision Graphics Image Processing*, vol. 30, pp. 125–147, 1985.
[5] R. W. Smith, "Computer processing of line images: A survey," *Patt. Recogn.*, vol. 20, pp. 7–15, 1987.
[6] W. Black, T. P. Clement, J. F. Harris, B. Llewellyn, and G. Preston, "A general purpose follower for line structured data," *Patt. Recogn.*, vol. 14, pp. 33–42, 1981.
[7] T. Pavlidis, " A vectoriser and feature extractor for document recognition," *Comput. Vision Graphics Image Processing*, vol. 35, pp. 111–127, 1986.
[8] X. Lin, S. Shimotsuji, M. Minoh, and T. Sakai, "Efficient diagram understanding with characteristic pattern detection," *Comput. Vision Graphics Image Processing*, vol. 30, pp. 107–120, 1985.
[9] D. S. Tudhope and J. V. Oldfield, "A high-level recognizer for schematic diagrams," *IEEE Comput. Graphics Applications*, pp. 33–40, 1983.
[10] H. Bley, "Segmentation and preprocessing of electrical schematics using picture graphs," *Comput. Vision Graphics Image Processing*, vol. 28, pp. 271–288, 1984.
[11] H. Bunke, "Experience with several methods for the analysis of schematic diagrams," in *Proc. 6th Int. Conf. Patt. Recogn. IEEE Comput. Soc.*, 1982, pp. 710–712.
[12] K. Iwata, M. Yamamoto, and M. Iwasaki, "Recognition system for three-view mechanical drawings," *Lecture Notes Comput. Sci.*, vol. 31, pp. 240–249, 1988.
[13] D. Dori, and A. Pnueli, "The grammar of dimensions in machine drawings," *Comput. Vision Graphics Image Processing*, vol. 42, pp. 1–18, 1988.
[14] D. Dori, "A syntactic/geometric approach to recognition of dimensions in engineering machine drawings," *Comput. Vision Graphics Image Processing*, vol. 47, pp. 271–292, 1989.
[15] L. A. Fletcher and R. Kasturi, "A robust algorithm for text string separation from mixed text/graphics images," *IEEE Patt. Anal. Machine Intell.*, vol. 10, pp. 910–918, 1988.
[16] R. Kasturi, R. Raman, C. Chennubhotla, and L. O'Gorman, "Document image analysis: An overview of techniques for graphic recognition," in *Proc SSPR90* (Murray Hill, NJ), 1990, pp. 192–230.
[17] M. A. Wesley and G. Markowsky, "Fleshing out projections," *IBM J. Res. Devel.*, vol. 25, pp. 934–954, 1981.
[18] R. M. Haralick and D. Queeney, "Understanding engineering drawings," *Comput. Graphics Image Processing*, vol. 20, pp. 244–280, 1982.
[19] K. Preiss, "Constructing the solid representation from engineering projections," *Comput. Graphics*, vol. 8, no. 4, pp. 381–389, 1984.
[20] S. H. Joseph, T. P. Pridmore, and M. E. Dunn, "Toward the automatic interpretation of mechanical engineering drawings," to be published in *Computer Vision and Image Processing* (A. Bartlett, Ed.). New York: Kogan Page, 1989.
[21] T. O. Binford, "A survey of model-based image analysis systems," *Int. J. Robotics Res.*, vol. 1, pp. 18–63, 1982.
[22] J. A. Mulder, A. K. Mackworth, and W. S. Havens, "Knowledge structuring and constraint satisfaction: The MAPSEE approach," Tech. Rep. 87–21, Dept. Comput. Sci., Univ. British Columbia, Vancouver, Canada, 1987.
[23] M. Nagao, T. Matsuyama, and H. Mori, "Structural analysis of complex aerial photographs," *Proc. 6th IJCAI*, 1979, pp. 610–616.
[24] J. A. Mulder, "Discrimination vision," *Comput. Vision Graphics Image Processing*, vol. 43, pp. 313–336, 1988.
[25] D. M. McKeown, W. A. Harvey, and J. McDermott, "Rule-based interpretation of aerial imagery," *IEEE Patt. Anal. Machine Intell.* vol. PAMI-7, pp. 570–585, 1985.
[26] T. Sakai, T. Kanade, and Y. Ohta, "Model-based interpretation of outdoor scenes," in *Proc. 3rd IJCPR*, 1976, pp. 581–585.
[27] T. Matsuyama and V. Hwang, "SIGMA: A framework for image understanding—Integration of bottom-up and top-down analyses," in *Proc. 9th IJCAI*, 1985, pp. 908–915.
[28] C. A. Kohl, A. R. Hanson, and E. M. Riseman, "A goal-directed intermediate level executive for image interpretation," in *Proc. 10th IJCAI*, 1987, pp. 811–814.
[29] T. Matsuyama and T. S. -S. Hwang, *"SIGMA: A Knowledge-Based Aerial Image Understanding System.* New York: Plenum, 1990.
[30] A. K. Mackworth, "Vision research strategy: Black magic, metaphors, mechanisms, miniworlds, and maps," in *Computer Vision Systems* (A. Hanson and E. M. Riseman, Eds.). New York: Academic, 1978.
[31] D. G. Elliman and I. T. Lancaster, "A review of segmentation and contextual analysis techniques required for automatic text recognition," to be published in *Patt. Recogn.*.
[32] B. W. Kernighan and D. M. Ritchie, *The C Programming Language*. Englewood Cliffs, NJ: Prentice-Hall, 1978.
[33] S. H. Joseph, "Segmentation and aggregation of text from images of mixed text and graphics," in *Research in Informatics, vol. 5* (Reinhard Klette, Ed.). Akademie Verlag, 1991, pp. 265–271.
[34] , "Tracking lines through noise," in *Proc. IEE 3rd Int. Conf. Image Processing* (Univ. of Warwick), 1989.
[35] S. J. Cheetham, "The automatic extraction and classification of curves from conventional line drawings," Ph.D. thesis, Univ. of Sheffield, 1988.
[36] M. E. Dunn and S. H. Joseph, "Processing of poor quality line drawings by local estimation of noise," *Lecture Notes Comput. Sci.* vol. 31, pp. 153–162, 1988.
[37] M. Yachida, M. Ikeda, and S. Tsuji, "A knowledge-directed line finder for analysis of complex scenes," in *Proc. 6th IJCAI*, 1979, pp. 984–991.
[38] U. Neisser, *Cognition and Reality: Principles and Implications of Cognitive Psychology.* San Francisco: W. H. Freeman, 1976.
[39] S. C. Johnson, "Yacc—Yet another compiler compiler," Comp. Sci. Tech. Rep. 32, Bell Lab., Murray Hill, NJ, 1975.
[40] T. C. Henderson and A. Samal, "Shape grammar compilers," *Patt. Recogn.*, vol. 19, pp. 279–288, 1986.
[41] A. R. Rao and A. K. Jain, "Knowledge representation and control in computer vision systems," *IEEE Expert*, pp. 64–79, Spring 1988.
[42] D. M. McKeown, W. A. Harvey, and L. E. Wixson, "Automating knowledge acquisition for aerial image interpretation," *Comput. Vision Graphics Image Processing*, vol. 46, pp. 37–81, 1989.
[43] R. A. Brooks, "Symbolic reasoning among 3D models and 2D images," *Artificial Intell.* vol. 17, pp. 285–348, 1981.
[44] J. McDermott, "Making expert systems explicit," in *Information Processing '86* (H. J. Kugler, Ed.). Amsterdam: Elsevier (North-Holland), 1986.
[45] T. P. Pridmore and S. H. Joseph, "Integrating visual search with visual memory in a knowledge-directed image interpretation system," in *Proc. BMVC90* (Oxford), 1990, pp. 367–373.
[46] H. Tropf and I. Walter, "An ATN model for recognition of solids in single images," in *Proc. 8th IJCAI*, 1983, pp 1094–1097.
[47] H. Tropf, "Analysis-by-synthesis search for semantic segmentation applied to workpiece recognition," in *Proc 5th ICPR*, 1980, pp. 241–244.
[48] W. Hattich, "Recognition of overlapping workpieces by model directed construction of object contours," *Digital Syst. Industrial Automat.*, vol. 1, nos. 2–3, pp. 223–239, 1982.
[49] J. Glicksman, "Using multiple information sources in a computational vision system," in *Proc. 8th IJCAI*, 1983, pp. 1078–1080.
[50] A. R. Hanson, and E. M. Riseman, "VISIONS: A computer system for interpreting scenes," in *Computer Vision Systems* (A. Hanson and E. Riseman, Eds.). New York: Academic, 1978.
[51] A. M. Nazif and M. D. Levine, "Low-level image segmentation and expert system," *IEEE Trans. Patt. Anal. Machine Intell.*, vol. PAMI-6, pp. 555–577, 1984.

S. H. Joseph received the B.A. degree in natural sciences at Cambridge University in 1971 and the Ph.D. degree in applied physics at Leeds University in 1976. Following research in machinery design at Imperial College, London, his interest in CAD and its industrial applications lead to research at Sheffield University on machine reading of engineering drawings. He is a lecturer in engineering design at Sheffield University.

T. P. Pridmore received the B.Sc. degree in computer science at the University of Warwick in 1982 and the Ph.D. degree from Sheffield University in 1987.

At Sheffield University, he worked on robot vision at the Artificial Intelligence Vision Research Unit in the Department of Psychology. During the course of the work reported here, he was a research associate in the Department of Mechanical and Process Engineering. He is now a senior lecturer in information engineering at Sheffield City Polytechnic.

Detection of Dimension Sets in Engineering Drawings

Chan Pyng Lai and Rangachar Kasturi

***Abstract*— This correspondence presents a system for detecting dimension sets in engineering drawings that are drawn to ANSI drafting standards. A new rule-based text/graphics separation algorithm and a model-based procedure for detecting arrowheads in any orientation have been developed. Arrowhead tracking and search methods are used to extract leaders, tails, and witness lines from segmented images containing only graphics. Text blocks and feature control frames extracted from the segmented images are then associated with their corresponding leaders to obtain complete dimension sets. Experimental results are presented.**

***Index Terms*—Document image analysis, engineering drawings, graphics recognition, image processing, pattern recognition, text segmentation**

I. Introduction

Document image analysis systems [1] have many applications. Systems are being designed to process mail pieces to automate their handling and sorting, to read musical scores, to recognize and verify signatures, to process bank checks and other financial documents, to facilitate browsing of library holdings, and to understand newspaper articles.

Increasing use of computer-aided design (CAD) and computer-aided manufacturing (CAM) systems has necessitated conversion of existing archives of paper drawings into a standard CAD/CAM format. Since transferring paper drawings into such a format by means of conventional interactive input systems is labor-intensive, it has become extremely desirable for the computer to process a wide variety of paper drawings and automatically create CAD databases. Development of accurate techniques for such a process will be a major contribution towards improving productivity in engineering design and drafting environments.

In this work, we assume that the drawing has been prepared as per the ANSI [2] drafting standard. A typical engineering drawing drawn as per ANSI standard is shown in Fig. 1. This drawing when digitized at 300 dots per inch resolution resulted in a 2,048 x 2,048 pixels image. The scanner has a built-in binarization step which was found to be adequate to accurately digitize such clean drawings without introducing breaks and other artifacts. However, for poor quality drawings it would be necessary to use a gray scale scanner followed by an adaptive thresholding step. The contents of such a drawing can be grouped into two major classes:

- object lines representing orthographic projections of 3-D objects;
- dimensioning lines and associated text (known as dimension sets) which provide exact definition of object dimensions.

Since dimension sets play an important role in engineering drawings, their recognition is a key component in any machine drawing understanding system [3].

Several researchers [4], [5] have reported 3-D object reconstruction from a given set of 2-D views. They assume that separation of object lines from dimension sets is already completed by a preprocessor stage. Because separation of object lines from dimension sets was an unsolved problem, we began developing techniques to recognize dimension-set components. Results of this effort are reported here.

A dimension-set consists of arrowheads and tails, witness lines, and text. The tail combined with the arrowhead is called a leader. These components are illustrated in Fig. 2. There is a wide variety of ways in which the dimensioning lines are drawn even in a simple drawing as shown in Figs. 3(a)–(h) (all the parts of this figure are taken from Fig. 1). Note that the dimensioning lines frequently intersect object lines (e.g., lines associated with the dimensions "⌀86" and "⌀ $^{57.6}_{56.6}$" in Fig. 1.) making it difficult to separate them. There are also datum markers (e.g., | −A− |) and text within boxes known as feature control frames (e.g., | ⊕ | ⌀ 0.1Ⓜ | BⓂ |). In order to detect such complex dimension-sets, a robust three-phase system has been developed. A block diagram of the system is shown in Fig. 4. The first phase is segmentation, which includes a new text/graphics separation algorithm. This algorithm is described in Section II. In the second phase, vectorization and feature extraction procedures are used to detect text and graphics primitives. Critical-point features are extracted by means of the k-curvature [6], [7] method. These well-known graphics processing methods are not discussed in this correspondence. The final phase includes arrowhead detection, leader detection, leader-pair matching, witness-line/tail extraction, and association of text blocks with leaders for detecting the dimension-sets. Each of these steps are described in detail in Section III. Assembling 3-D enclosure from object lines and processing of object lines are described elsewhere [5], [8].

II. Separation of Text from Graphics

In order to efficiently interpret engineering drawings, it is necessary to separate text from graphics. There are many algorithms described in the literature [9], [10] for text-graphics separation. The method described in [9] is for omni-directional text/graphics segmentation. The algorithm is based on the analysis of connected components and

Manuscript received January 27, 1993; revised June 10, 1993. Recommended for acceptance by A. K. Jain.

C. P. Lai was with the Computer Engineering Program, Department of Electrical and Computer Engineering, Pennsylvania State University, University Park, PA 16802 USA. He is now with CTC Corporation, Johnstown, PA 15903 USA.

R. Kasturi was with the Computer Engineering Program, Department of Electrical and Computer Engineering, Pennsylvania State University, University Park, PA 16802 USA. He is now with the Computer Science and Engineering Department, Pennsylvania State University, University Park, PA 16802 USA; e-mail: kasturi@csc.psu.edu.

IEEE Log Number 9403160.

Reprinted from *IEEE Trans. Pattern Analysis and Machine Intelligence,* Vol. 16, No. 8, Aug. 1994, pp. 848–855.

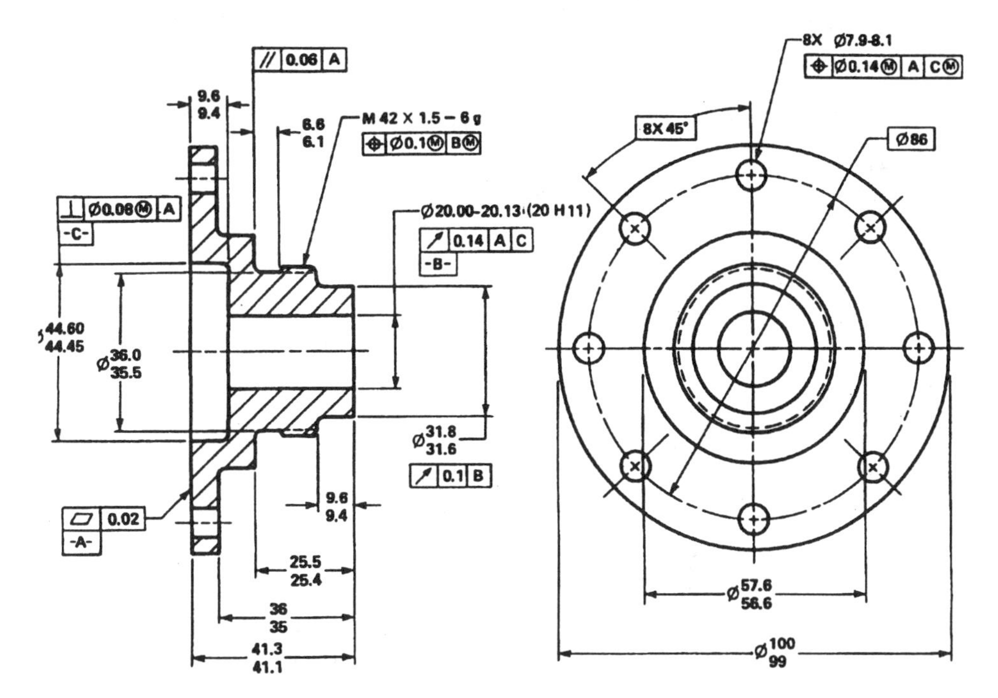

Fig. 1. A typical engineering drawing.

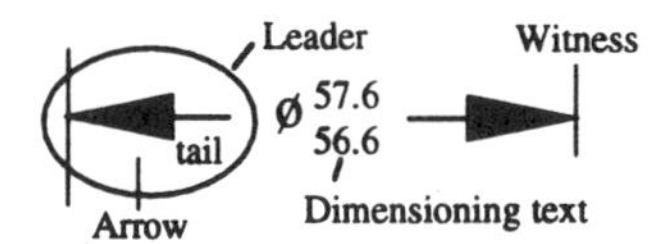

Fig. 2. Components of a dimension-set specifying a diameter.

the application of Hough transform to group components into text strings and hence separate text from graphics. It requires at least three collinear text components. Dashed lines are occasionally mislabeled as text strings by this algorithm. In order to overcome these limitations, a new text/graphics segmentation algorithm specially designed for engineering drawings, described here, was developed. The algorithm is described below.

A. Connected Component (CC) Generation

The CC generation algorithm described in [9] is applied to the image in Fig. 1. The results include upper X,Y coordinates of each CC (X_m,Y_m), lower X,Y coordinates of each CC (X_M,Y_M), width (W) of each CC, and height (H) of each CC. The Y_m is arranged in a nondecreasing order. The centroid of each CC (X_c,Y_c) is also calculated.

B. Letter Height Estimation

In order to estimate letter height on a drawing, a histogram of CC's (the number of occurrences vs. height) is obtained. Components with small heights are considered as noise, and are filtered out during this step. The peak of this histogram is taken as the typical letter height L_h. The estimated letter height is 24 pixels in the drawing in Fig. 1. Recommended letter heights for engineering drawings are specified in the ANSI standard [2]. If the selected peak is significantly different from the recommended letter heights, processing is interrupted for operator interaction.

C. Horizontal Text Extraction

An unclassified element, denoted by $C_{[x]}$, is enclosed by the corresponding CC on the drawing. It may be classified as a text element (denoted by $T_{[x]}$) or a graphics element (denoted by $G_{[x]}$), based on a set of heuristic rules. A search region bounded by two lines at a distance of $0.5L_h$ from (X_m,Y_m) is defined (see Fig. 5). If the top left corner of an element $C_{[n]}$ is found in the search region and if all of the following conditions are satisfied, then $C_{[x]}$ is classified as $T_{[x]}$.

$$0.5L_h < H_{C_{[x]}}, (H_{C_{[n]}}) < 3L_h \tag{1}$$

$$|Yc_{C_{[x]}} - Yc_{C_{[n]}}| < 0.25L_h \tag{2}$$

$$|Xc_{C_{[x]}} - Xc_{C_{[n]}}| < 4L_h \tag{3}$$

$$|H_{C_{[x]}} - H_{C_{[n]}}| < 0.5L_h. \tag{4}$$

The first condition requires each component to be within a specified range of letter height. The lower bound of this inequality helps to avoid labelling dashed lines as text. The upper bound is based on the recommended letter heights for various fields in mechanical drawings [2]. Conditions 2 and 3 ensure that $C_{[x]}$ and $C_{[n]}$ are near each other. The last condition requires components to be of approximately the same height. If the conditions are not satisfied, $C_{[x]}$ remains unclassified. This process continues using the next element in the list until all the elements are considered. After the horizontal text extraction is performed, the numbers 5, 6, and 7 in Fig. 5 are classified as text. Symbol ϕ and periods are still unclassified, which are processed separately.

D. Extraction of Other Horizontal Textual Components

This step deals with text elements which are not collinear with respect to the center line of text orientation (e.g., symbol ϕ and periods in Fig. 5, degree symbols, etc.). Such elements remain unclassified after the above steps. To label such elements, a search box is created with reference to the center of an unclassified element. The width and height of this search box are set to $3L_h$ and L_h, respectively. If the center of any text element is located within the search box, the unclassified element is classified as a text element as long as both width and height of the unclassified element are no more than $2L_h$. After this procedure is performed, symbol ϕ and periods in Fig. 5 are classified as text.

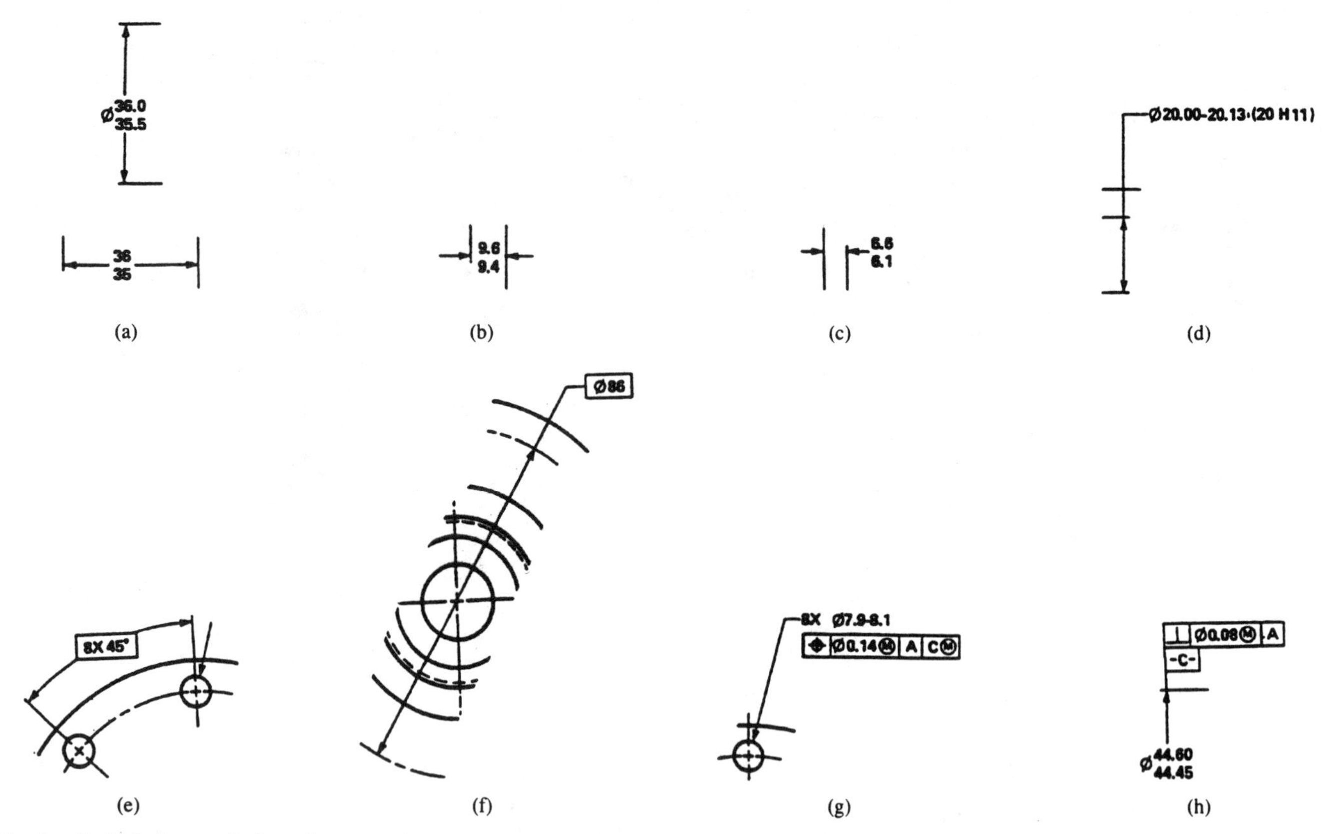

Fig. 3. Typical classes of dimension sets.

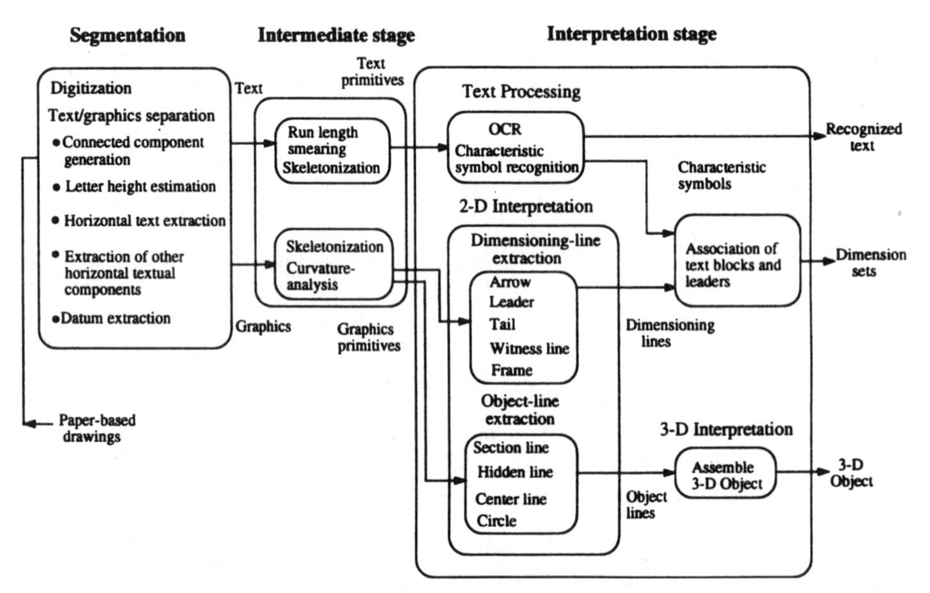

Fig. 4. A three-phase system for interpretation of engineering drawings.

E. Datum Feature Extraction

A datum is a set of two short lines separated by a character as shown in Fig. 6. The datum is used to denote different orthogonal planes in the drawings (e.g., -A-, -B-, -C- in Fig. 1). Two steps to detect a datum are designed. The first step is to search for two unclassified elements, $C_{[x1]}$ and $C_{[x2]}$, which satisfy the following heuristics:

$$H_{C_{[x1]}} < 0.5L_h,\ H_{C_{[x2]}} < 0.5L_h \tag{5}$$

$$|H_{C_{[x1]}} - H_{C_{[x2]}}| < 0.25L_h,$$

$$|W_{C_{[x1]}} - W_{C_{[x2]}}| < 0.25L_h \tag{6}$$

$$|Yc_{C_{[x1]}} - Yc_{C_{[x2]}}| < 0.25L_h,$$

$$|W_{C_{[x1]}}| > 0.25L_h,\ |W_{C_{[x2]}}| > 0.25L_h \tag{7}$$

$$|Xc_{C_{[x1]}} - Xc_{C_{[x2]}}| < 4L_h \tag{8}$$

where H_C, W_C, X_C, and Y_C are height, width, and X and Y coordinates of the centroid of connected component C, respectively, and L_h is the estimated average letter height. Components satisfying these conditions are potential dashes of a datum feature. After two dashed segments are detected, we look for another unclassified element $C_{[x]}$ between the extracted dashed segments. The final step

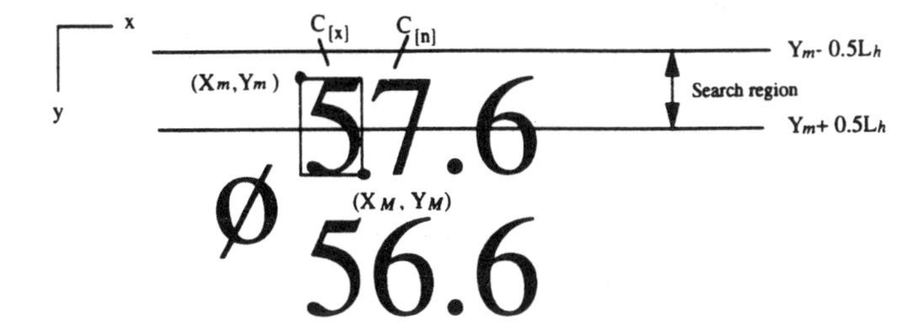

Fig. 5. The search region for component classification.

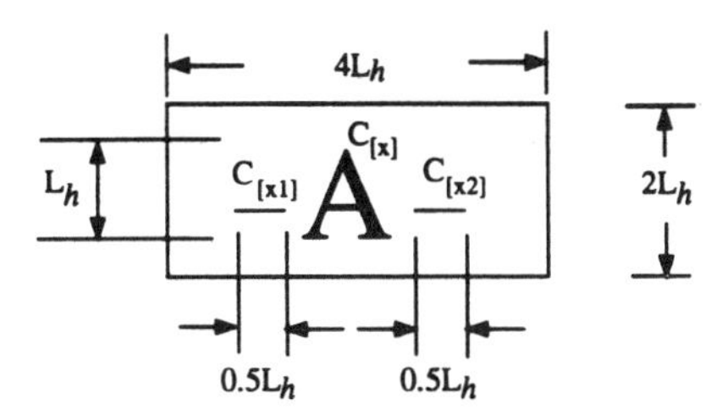

Fig. 6. A datum and its components: Typical dimensions as a function of letter height, L_h, are shown.

is to find a $C_{[x]}$ which satisfies the following conditions:

$$0.5L_h < H_{C_{[x]}}, \quad |H_{C_{[x]}} - 2W_{C_{[x1]}}| < 0.25L_h \tag{9}$$

$$Xc_{C_{[x1]}} < Xc_{C_{[x]}} < Xc_{C_{[x2]}}, \quad |Yc_{C_{[x]}} - Yc_{C_{[x1]}}| < 0.25L_h \tag{10}$$

Vertical text is extracted in a similar manner. After extracting horizontal text and vertical text, any remaining unclassified elements are labeled as graphics elements. Figure 7 shows components labeled as graphics in Fig. 1. Note that the above method is based on the assumption that text seldom touches object lines or other graphics in well-drawn engineering drawings. However, in situations when this assumption is violated, algorithms designed specifically to handle touching characters such as those described in [8] can be applied.

III. Detection of Dimensioning Lines

Antoine *et al.* [11] describe a model arrowhead in engineering drawings. The angle between two lines of the model arrowhead is computed, and the whole image is processed by looking for pairs of connected "small" segments with similar angles. In this way, a set of possible arrows is obtained. By looking at the adjoining neighborhood, the hypothesis for arrows is confirmed using *a priori* knowledge. A polygonal approximation of the contours must be processed before the arrows can be identified by their system. Dori and Dowell [12] describe a two-stage self-supervised approach to identify arrows in engineering drawings. The first stage involves determination of the basic parameters of the arrows. From these parameters, a general search can be carried out on all line endpoints in the second stage to produce a list of potential arrowheads. Further search is made to determine the lines associated with arrowheads. A list of lines, including straight lines and arcs, is used as input to their algorithm.

The algorithm we have designed is capable of handling all of the ANSI standard dimensioning methods illustrated in Fig. 3. Our method first detects arrowheads and then backtracks from the detected arrowheads to find other components of the dimension set. We now describe our dimension set detection algorithm in detail.

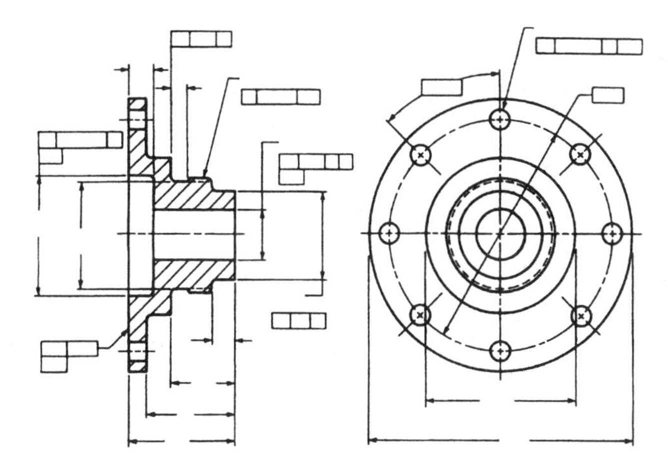

Fig. 7. The components labeled as graphics in Fig. 1.

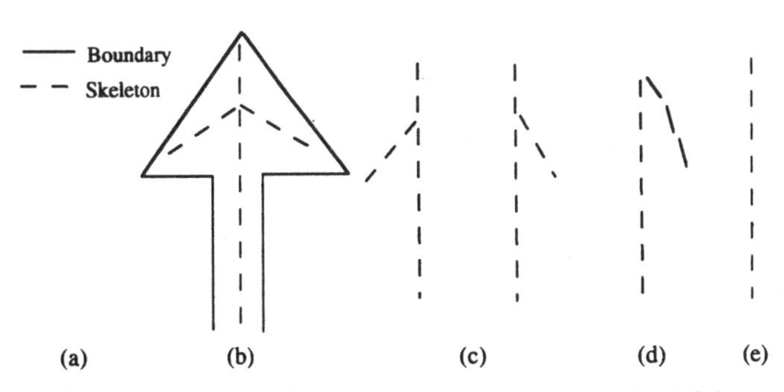

Fig. 8. An ideal arrowhead with its skeleton and actual arrowhead skeletons observed in experiments.

A. Arrowhead Detection

An arrowhead model, including upper corner, lower corner, arrow tip, and arrow back, is created to match potential arrows. In the preprocessing steps, the image is skeletonized using the algorithm described in [13]. Pixels along the skeleton are labeled 1 for skeleton end pixels, 2 for link pixels, and 3 for pixels at junctions of three or more segments. Line types 11, 13, 33, and 22 are obtained. Ideally, an arrowhead after skeletonization should have a straight line and three short segments at the end as shown in Fig. 8(a). However, in practice the short segments are not always present after skeletonization as shown in Figs. 8(b)–(e)). As a consequence we can not rely on recognizing the arrowheads directly from the skeleton. Thus, the following steps must be taken.

B. Pruning Step

Following the skeletonization process, the short line type 13 is trimmed and the junction label is updated to a link pixel if appropriate. The junction label and the chain code of line segments are also updated. After this step, all line segments (types 11, 13, 33) are checked for the presence of an arrowhead at their ends using the model matching described below.

C. Arrowhead Model Generation

To build the arrowhead model, the thickness of line segments is measured in four directions, i.e., horizontal, vertical, and in two diagonal directions at each skeleton pixel. The minimum length measured in the four directions is considered to be the thickness of the line at that point. The arrowhead model is shown in Fig. 9(a). The points along the skeleton where there is a significant change in

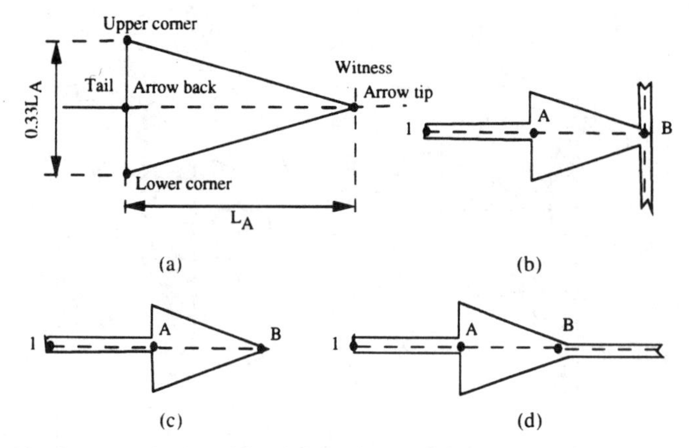

Fig. 9. Arrowhead model and three classes of arrow tip (b)–(d).

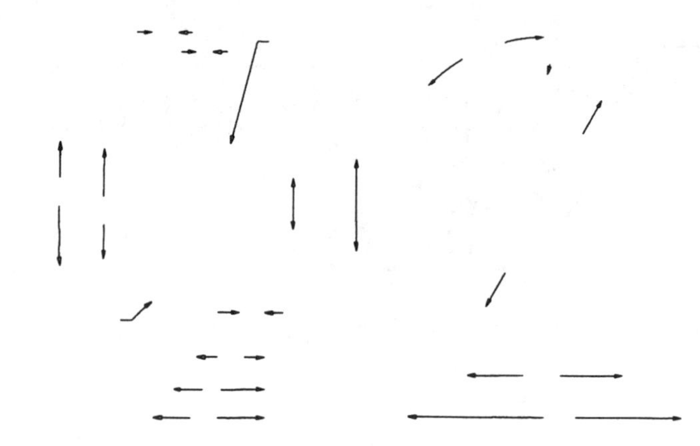

Fig. 10. Arrowheads detected in Fig. 1.

thickness are marked as potential arrow backs. For example, in Fig. 9(b), starting from the left 1 pixel we come to point A where there is a significant change in thickness from low to high. Such a point is marked as the arrow back. The arrow tip is then identified as any one of the following: junction (Fig. 9(b)), end of line (Fig. 9(c)), or start of constant thickness (Fig. 9(d)). Note that arrowheads such as those in Fig. 9(b) would be reduced to that of Fig. 9(c) during the pruning step if the vertical segments are of short length and open-ended. Arrowheads shown in Fig. 9(d) occur if the line continues beyond junction B and the other segments at junction B are short open ends.

The orientation angle of the line connecting the arrow tip and the arrow back with reference to the horizontal line as well as its length L_A are measured. Two points, one on either side of this line, and at a distance of $0.167L_A$ from arrow back are located (this choice is based on the ANSI standard recommendation of 3:1 aspect ratio) to mark the upper and lower corners of the arrowhead model. These two points along with the arrow tip and arrow back complete the model for arrowhead.

D. Model Matching

The model created is used to match a corresponding region of the original binary image. If the number of matching black pixels in this region is at least 75%, the current region is a potential arrowhead. If the potential arrowhead satisfies the triangular-shape requirement, it is recognized as an arrowhead. To satisfy this requirement, the thickness must be in decreasing order from arrow back to arrow tip. To allow for some variation, we check for this decrease at intervals equal to the average thickness of the arrow tail. Arrows detected in Fig. 1 are shown in Fig. 10. Note that this two-stage process of model creation and model matching is necessary to avoid classifying other potential pattern as arrowheads. For example, if a solid rectangle is connected to two lines, one on either side of the rectangle, the model generation process would identify two junctions as potential arrow back and tip. However, such a pattern is rejected during matching stage. The model matching step also helps to distinguish between arrowheads (which are required to have a 3:1 aspect ratio as per ANSI standard) and other triangular-shape patterns which may be present in the drawing.

E. Extraction of Complete Leaders

Detected arrows are tracked towards the arrow back until an open end or a junction is found. If an open end is detected a complete single path leader has been identified (Figs. 3 (a)–(c)). If a junction is detected, further tracking is necessary to extract the complete leader. For example, in Fig. 3(d) backtracking from one of the arrows terminates at another arrowhead. In this case we have connected arrowhead pairs. However, in Fig. 3(f) we need further tracking from the first junction of the arrowhead leader to identify the complete connected arrowhead pair. Additional processing is also necessary to extract complete leaders in Fig. 3(e) because the leaders are connected to the box representing a basic dimension. Similarly, in Fig. 3(g), the leader must be extended beyond its first junction to complete the process.

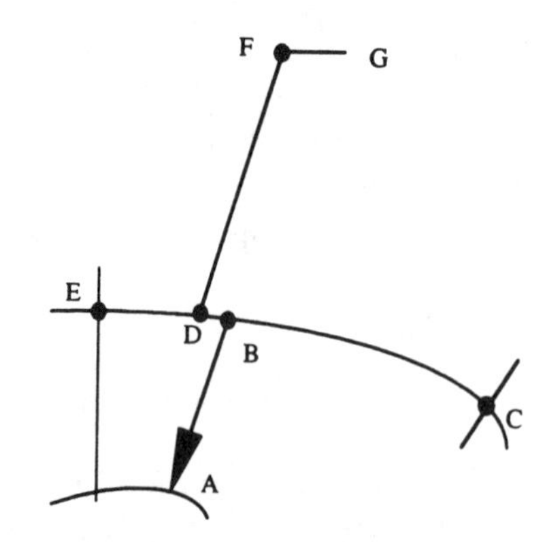

Fig. 11. Illustration of priority assignment for leader tracking.

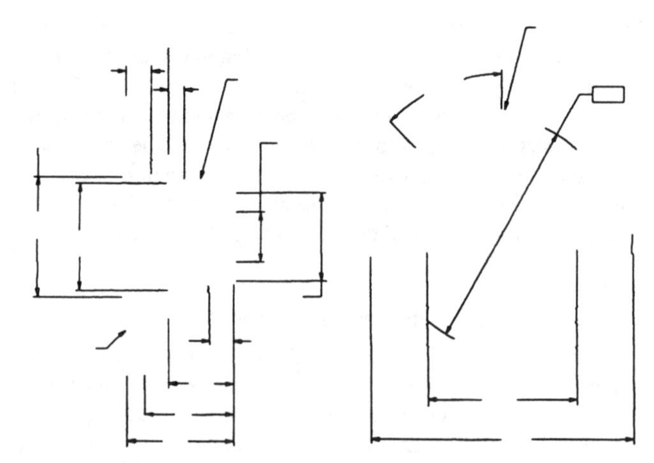

Fig. 12. Complete dimensioning lines detected in Fig. 1.

F. Leader Pair Detection

Complete leader extraction is done by first labeling leaders as either leader-pairs or as single leaders. This is done by pair matching. For each arrowhead, three points are noted. These are arrow tip, T, arrow back, B, and the terminating point (first junction or open end), P. Orientation of every pair of leaders is checked to see if they belong to a pair; if they do, then the pair is labeled as either pointing-in or

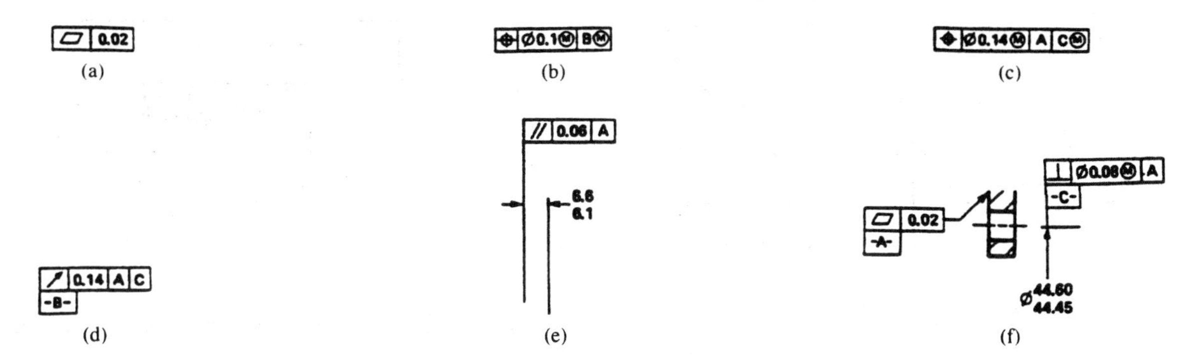

Fig. 13. Typical classes of frames.

pointing out. These rules are summarized below:

$$\text{if } (\Theta_{P_1T_1} \simeq \Theta_{T_1T_2} \simeq \Theta_{T_2P_2}) \text{ and } (D(B_1, B_2) > D(T_1, T_2)) \text{ then Pointing-in} \quad (12)$$

$$\text{if } (\Theta_{T_1P_1} \simeq \Theta_{P_1P_2} \simeq \Theta_{P_2T_2}) \text{ and } (D(B_1, B_2) < D(T_1, T_2)) \text{ then Pointing-out.} \quad (13)$$

where Θ represents the angle of orientation and D is the euclidean distance operator. Note that this rule may or may not identify the connected pair shown in Fig. 3(e) because the dimensioning here represents an angle and hence the leaders are not collinear. However, since the two junctions are connected by a simple loop representing the basic dimension, it is easy to pair these leaders. If the box is not present, then the pair may not be identified until the last step when single leaders are considered for text association (in this case two single leaders would point to the same text block indicating a leader pair).

G. Extraction of Multi-Segment Leaders

For leaders which have several segments (e.g., Figs. 3(f) and (g)), we apply a priority ordered search until complete leaders are extracted. This is done by starting at the first junction of an arrowhead segment and appending other segments based on a simple rule: at each junction we select the segment which causes minimum change in the direction of the leader. For example, consider Fig. 11 which is an enlarged view of the dimensioning lines corresponding to the eight small holes in Fig. 1. The short segment BD is an artifact of skeletonization. In this figure, starting from the arrowhead A we reach junction B. At this junction we have two segments which end at C and D. The segment BD is selected since Θ_{DAB} is less than Θ_{BAC}. A threshold of 20 degrees is also used to ensure that the leader is approximately straight. Then at junction D we have two segments, DE and DF. DF is chosen since Θ_{FAD} is less than Θ_{EAD}. This process is continued until the linearity condition is violated, or an open end or a junction of another leader is reached.

H. Tail Extraction

Leader pairs are occasionally extended with a tail (e.g., Fig. 3(d)). Such extensions may also intersect other lines forming a multi-segment tail. These tails are extracted by applying a priority search rule similar to that described above. The only difference is that we begin at the tip of an arrow and proceed away from the arrow back. Note that the tails may terminate at a horizontal segment since text is always written in a horizontal orientation in ANSI standard drawings. A tail may connect a frame (see Fig. 3(h)). This type of tail is terminated at the place where it loses its linearity. The end of the tail is either an open end (Fig. 3(d)) or a junction (Figs. 3(f) and (h)).

I. Witness Line Extraction

Witness lines are always associated with leader pairs and they touch arrow tips at right angles. Thus, these lines are extracted by starting at an arrow tip and extracting open segments orthogonal to its leader (e.g., witness lines associated with ⌀31.8 in Fig. 1). If the witness line has any junctions other than the one at the arrow tip, the line is continued through the junction until an open end is reached (e.g., witness lines associated with ⌀57.6 in Fig. 1) or loses its linearity condition. Since single leaders do not have witness lines, the object lines in Fig. 3(g) are not mislabeled as witness lines. Complete dimensioning lines extracted from Fig. 1 are shown in Fig. 12. Note that all the witness lines are extracted correctly; however, two segments of the dashed circle connecting the centers of the eight holes are mislabled as witness lines for the dimension ⌀86. Higher-level interpretation will be necessary to handle such special cases.

J. Association of Text with Dimensioning Lines

After detecting all the dimensioning lines, we associate each leader with its corresponding text regions. A run-length smearing algorithm [10] is applied to enclose each text region in its own rectangular block. These blocks are associated with their nearest dimensioning lines as follows.

1) Text blocks that are in between a pair of open ended leaders are associated with their respective leader pair (Figs. 3(a) and (b)).
2) Text blocks which are on one side of an open ended leader pair pointing inward are associated with that leader pair (Fig. 3(c)). Text blocks near an extended tail (Fig. 3(d)) or a single leader (Fig. 3(g)) are also labeled in a similar manner. The search for such text blocks is limited to a small distance from the open end as determined by the average letter height.
3) Text blocks enclosed in a rectangle representing a basic dimension are associated with the corresponding leader pair.

Note that the only text regions which are not included after this step are those corresponding to the so called feature control blocks. These are extracted separately by detecting the rectangular boxes which enclose the text and datum symbols.

K. Frame Extraction

Text within rectangular boxes are known as feature control frames (see Figs. 13(a)–(f)). All of the parts in Fig. 13 are taken from Fig. 1. Fig. 13(a) shows a frame containing two compartments (rectangular boxes) which contain a geometric characteristic symbol and a tolerance. A frame may contain more than two rectangular boxes (Figs. 13(b)–(c)). A frame may also combine a datum feature symbol as shown in Fig. 13(d). A frame may connect to a witness line (Fig. 13(e)) or to a leader (Fig. 13(f)). The information contained in

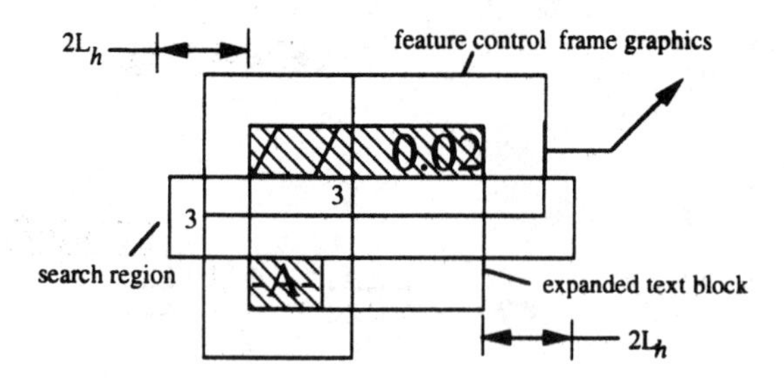

Fig. 14. Search region for merging blocks.

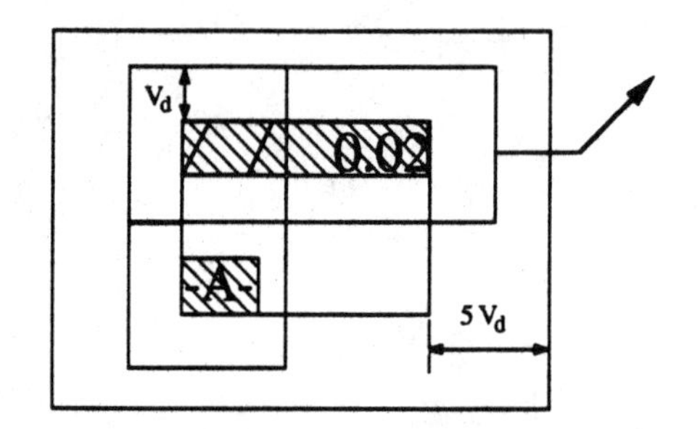

Fig. 15. Extraction step for nonisolated feature control frames.

these frames augment the information contained in the dimensioning text. Thus, it is necessary to separate these frames from the object lines and other graphics for proper interpretation.

The feature control frame contains at least one vertical line which separate the frame into two or more rectangles. This property is userful to distinguish feature control frames from simple rectangles which enclose basic dimensions (e.g., ⌀86 in Fig. 1). Text and symbols enclosed by a feature control frame are merged into a text block during the run-length smearing step described earlier. Clearly, the vertical line(s) in a feature control frame passes through its enclosed text block. Thus, our algorithm looks for one or more lines of types 33 which cross (at approximately right angle) any unclassified text string. Once a feature control frame is detected, the entire rectangular box along with the corresponding text block is extracted.

Occasionally, a feature control block includes text strings in several lines one below the other. A simple example is the feature control text string and the datum marker shown in Figs. 13(d) and (f). Such blocks are detected by first ensuring that the two (or more) text strings are indeed enclosed by a single feature control frame. This is done by defining a search region which separates the two text blocks as shown in Fig. 14. Within this region if a horizontal line segment of type 33 is detected, then the two text strings are labeled as being related and a new expanded text block which includes both text blocks is defined. Then, if the feature control frame graphics (lines forming the boxes of feature control frame) is isolated from other lines (e.g., Fig. 13(d)), then it is easily detected and labeled as such. To make sure that very large boxes are not labeled as feature control frames, we require that the size of the connected component which correspond to the feature control frame should be no larger than twice the size of the corresponding text block. When the feature control frame graphics is not isolated (e.g., Figs. 13(e) and (f)), then we expand the rectangle enclosing the text block in small increments until the entire feature control frame graphics is enclosed. Note that, when this happens there will be only one line which crosses the outer boundary of the expanded box (see Fig. 15). Again, to ensure correct segmentation, the box is not expanded by more than $5V_d$ where V_d is the distance between the text block and its nearest line segment. After this stage, complete dimensioning set has been extracted from Fig. 1 as shown in Fig. 16. Merged text blocks and the feature control frames are clearly shown in this figure.

IV. Summary and Conclusion

We have described a system to extract dimension sets, which includes dimensioning lines and associated annotations, in engineering drawings drawn to ANSI drafting specifications. Separation of dimension sets from object lines is a critical problem for automated conversion of drawings from paper medium to CAD databases. Algorithms for separation of text from graphics, extraction of arrowheads, leaders, tails, and witness lines, association of text blocks to dimensioning lines, and extraction of feature control frames were presented. If drawings contain dimensioning features that do not follow the drafting ANSI standard, such features may not be classified correctly by our system. For example, witness lines which do not touch arrow tips may be mislabeled as object lines. Human intervention or more sophisticated rules would be necessary to handle such drawings. Real engineering drawings digitized to a size of 2,048 x 2,048 pixels were used to demonstrate the performance of these algorithms. These algorithms, when combined with those described in [8] for graphics recognition and in [5] for 3-D interpretation form a complete system for intelligent interpretation of paper-based drawings for integration with CAD/CAM systems.

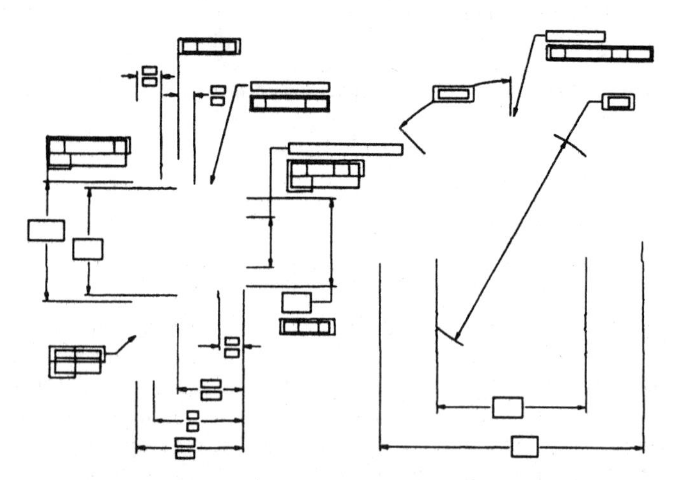
Fig. 16. Complete dimension sets and frames detected in Fig. 1.

References

[1] R. Kasturi and L. O'Gorman, guest editors, "Document image analysis systems and techniques," in special issues of *Machine Vision and Applicat.*, vol. 5, no. 3, 1992, and *Computer*, vol. 25, no. 7, July 1992.

[2] ANSI Y14.5, "Dimensioning and tolerancing," and ANSI Y14.2M, "Line conventions and lettering," *Am. Soc. Mech. Eng.*, New York, 1982.

[3] D. Dori, "A syntactic/geometric approach to recognition of dimensions in engineering drawings," *Comput. Vision, Graphics, and Image Processing*, vol. 47, pp. 271–291, 1989.

[4] M. A. Wesley and G. Markowski, "Fleshing out projections," *IBM J. Res. Develop.* 25, pp. 934–953, 1981.

[5] D. B. Lysak and R. Kasturi, "Interpretation of engineering drawings of polyhedral and nonpolyhedral objects," *Proc. First Int. Conf. Doc. Anal. and Recognit.*, Saint-Malo, France, pp. 79–87, 1991.

[6] L. O'Gorman, "An analysis of feature detectability from curvature estimation," *Proc. IEEE Conf. Comput. Vision and Pattern Recognit.*, 1988, pp. 235–240.

[7] C. H. Teh and R.T. Chin, "On the detection of dominant points on digital curves," *IEEE Trans. Pattern Anal. Machine Intell.*, vol. 11, pp. 859–872, 1989.

[8] R. Kasturi, S.T. Bow, W. El-Masri, J. Shah, J. Gattiker, and U. Mokate, "A system for interpretation of line drawings," *IEEE Trans. Pattern Anal. Machine Intell.*, vol. 12, pp. 978–992, 1990.

[9] L. A. Fletcher and R. Kasturi, "A robust algorithm for text string separation from mixed text/graphics images," *IEEE Trans. Pattern Anal. Machine Intell.*, vol. 10, pp. 910–918, 1988.
[10] F. M. Wahl, M.K.Y. Wong, and R.G. Casey, "Block segmentation and text extraction in mixed text/image documents," *Comput. Vision, Graphics, Image Processing*, vol. 20, pp. 375–390, 1982.
[11] D. Antoine, S. Collin, and K. Tombre, "Analysis of technical document: The REDRAW system," in *Structured Document Image Analysis*. New York: Spring-Verlag, 1992, pp. 385–402.
[12] D. Dori and J. Dowell, "Detection of arrows in engineering drawings: A self-supervised approach to pattern recognition," *Machine Vision and Applicat. on Doc. Image Anal. Syst.* , vol. 6, pp. 69–82, 1993.
[13] J. F. Harris, J. Kittler, B. Llewllyn, and G. Preston, "A modular system for interpreting binary pixel representation of line-structured data," in *Pattern Recognition: Theory and Applications*. Boston, MA: D. Reidel Publishing, 1982, pp. 311–351.

An Interpretation System for Land Register Maps

Luca Boatto, Vincenzo Consorti, Monica Del Buono, Silvano Di Zenzo, Vincenzo Eramo, Alessandra Esposito, Francesco Melcarne, Marco Meucci, Andrea Morelli, Marco Mosciatti, Stefano Scarci, and Marco Tucci

IBM Southern Europe Middle East and Africa Scientific and Technical Solution Center

The semantics of land register maps drive this document conversion system. However, its methods of image representation, vectorization, and symbol recognition :an be generalized to other classes of line drawings.

Creating a cartographic database often involves the acquisition of huge amounts of data from paper drawings. The acquisition is usually performed with hand-operated digitizing tablets, following procedures that are time-consuming, costly, and error-prone. Effective techniques for the automatic input of drawings into a database are difficult to implement, and the many efforts in this direction over the past 20 years have found limited success. Only recently have substantial advances been achieved in this field.[1-5]

We have developed a system for the automatic acquisition of land register maps. The system converts paper-based documents for the Italian Land Register Authority into digital form for integration into an existing database. Compliance with Land Register Authority standards was a basic design issue. These standards are strictly prescribed, and geometric entities (for example, areas relevant from the viewpoint of land taxation) must be computed within narrow tolerances.

Processing a map begins with its digitization by a scanning device. A key step in our system is the conversion from raster format to *graph representation*, a special binary image format suitable for processing line structures. Subsequent steps include vectorization of the line structures and recognition of the symbols interspersed in the drawing. The system requires operator interaction only to resolve ambiguities and correct errors in the automatic processes. The final result is a set of descriptors for all detected map entities, which is then stored in a database.

This system relies upon, and is driven by, the semantics of land register maps. However, the image representation, vectorization techniques, and optical character recognition subsystem are quite general. The methodology implemented in the system can be generalized to the acquisition of other classes of line drawings.

Land register cartography

Figure 1 shows an example of a land register map, also called a cadastral map. Cadastral maps describe the geometry of land properties and buildings in a geographical context. They divide territory into a number of polygons, each one representing a piece of land, a building, a street, or a body of water. The scale can range from 1:500 to 1:5,000. A set of adjacent polygons representing land or buildings belonging to the same owner is called a *parcel*. As shown in Figure 1, each parcel is labeled with a unique number. A separate file records cadastral information related to parcels, such as owner, area, and usage.

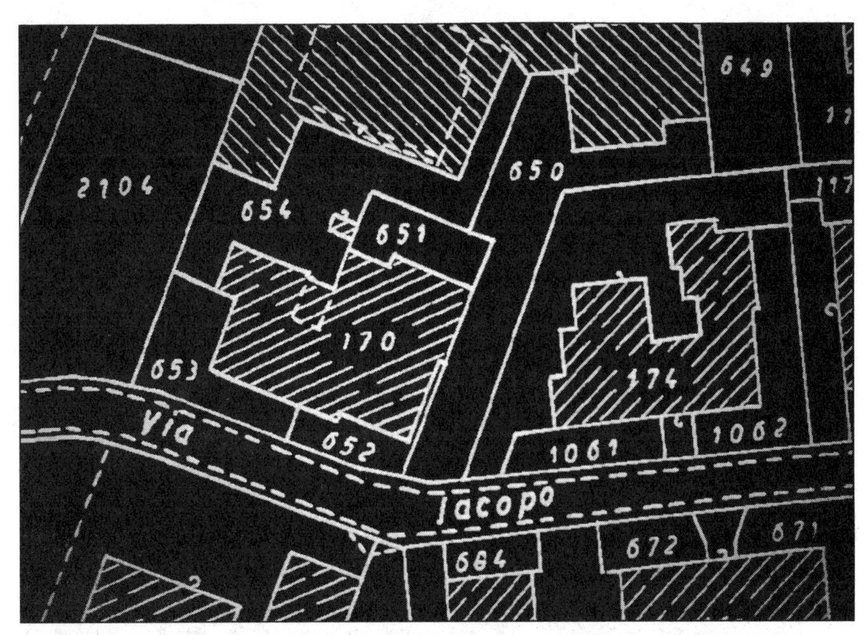

Figure 1. Portion of an Italian Land Register Authority map. Polygons represent pieces of land (unfilled areas), buildings (shaded areas), and streets (labeled by a name). Two adjacent polygons belonging to the same owner are joined by a tilde (~) crossing the common boundary. A set of adjacent pieces of land and buildings belonging to the same owner is called a parcel and labeled by a number placed inside one of the polygons (or alongside, with an arrow pointing to it). For example, parcel 174 consists of a building and two pieces of land. Cadastral symbols (arrow, tilde, bone, etc.) and dashed lines have various meanings, depending on the context.

Italian land register maps are monochrome drawings, in A0 format (100 × 70 centimeters), containing continuous lines, dashed lines, shading patterns, and symbols (alphanumeric characters and cadastral symbols). Strings of characters identify high-level entities such as parcels, streets, and water. These character strings usually do not touch lines. On the other hand, cadastral symbols, which have special meanings, usually do overlap lines.

The Italian Land Register Authority issues a set of guidelines for drawing a cadastral map. These rules form a graphic language and allow the reader to understand the drawing. However, it is not possible to rely entirely on these guidelines for the automatic interpretation of a map, since draftspeople do not always follow them strictly.

About 300,000 paper-based docu-

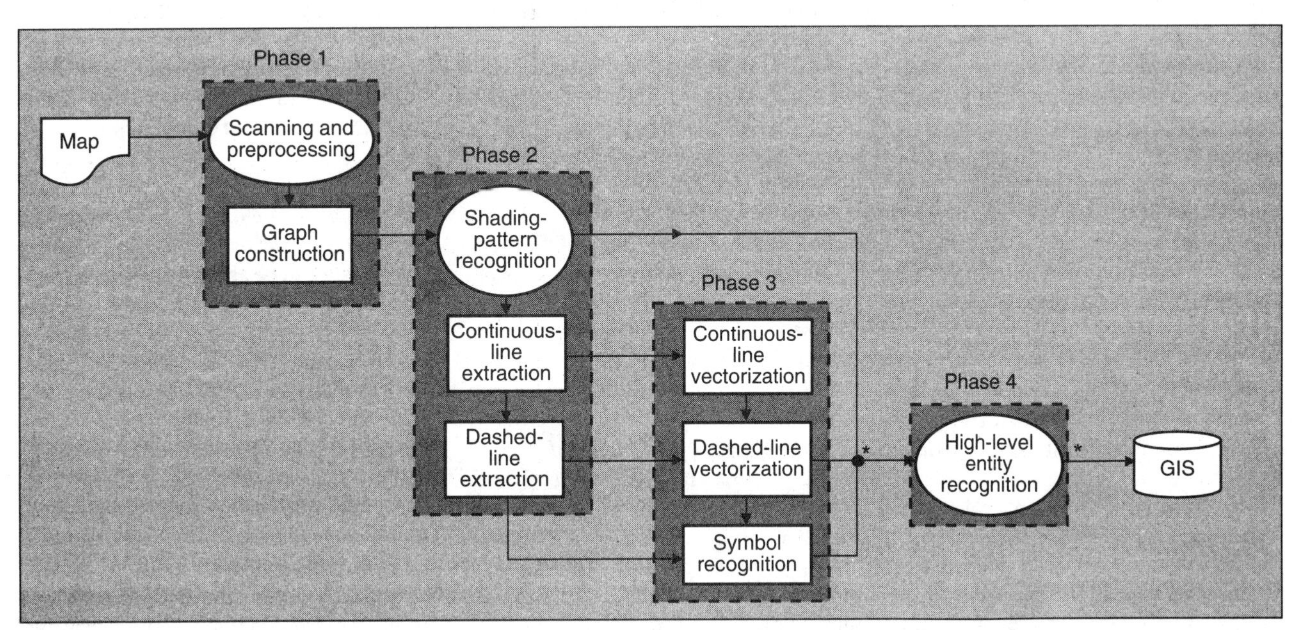

Figure 2. Logical scheme of map interpretation. Boxes and ellipses correspond to the processing steps: Those steps performed automatically by the system are represented by boxes; those requiring operator's interaction are represented by ellipses. Arrows show the dataflow. Asterisks indicate an interactive session, during which the operator checks the output of previous steps.

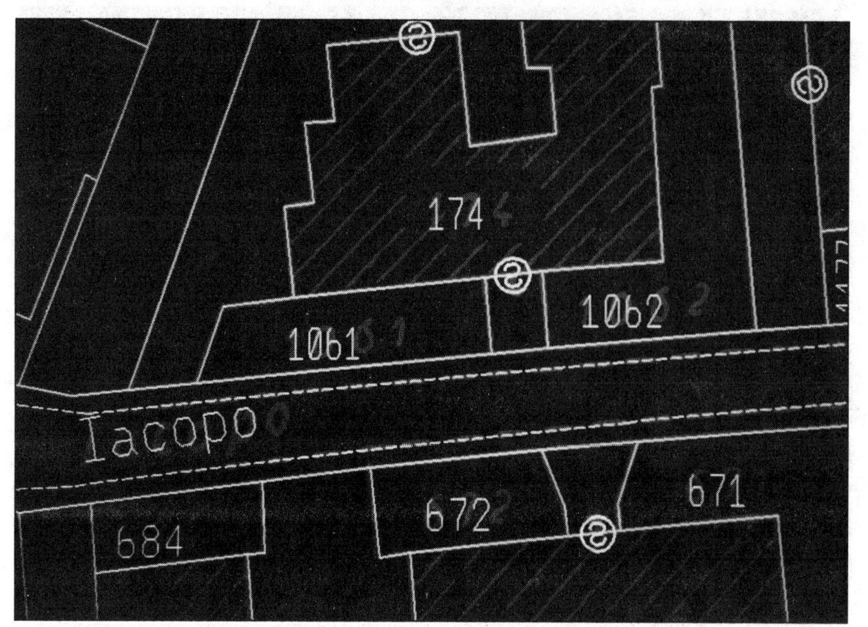

Figure 3. Graphic output of the system. Lines, dashed lines, characters, and cadastral symbols recognized by the system are superimposed on the original image.

ments are being replaced by a computer-based geographic information system. This GIS supports the query, display, and updating of geometric and alphanumeric data describing high-level entities.

System overview

A variety of approaches[6] have been proposed to solve the problems of interpreting line drawings automatically. Text is usually separated from graphics by extracting and classifying connected components in the image.[5,7] This procedure fails, however, when symbols touch or overlap lines — something that occurs often in land register maps. The main problem in recognizing symbols is the variation in their size and orientation. A template-matching approach is inadequate for resolving this problem, but a feature-based approach can work.[2]

In our opinion, the choice of image representation plays a fundamental role in the automatic interpretation of drawings. Accordingly, the first phase in our system generates the graph representation we use in all subsequent processing.

Figure 2 illustrates the phases and steps of the interpretation process. In phase 1, the document is scanned and its graph-encoded binary image is produced. The result is an image decomposed into independent elements, namely, edges and nodes of the graph. In phase 2, each element of the image graph is classified as belonging to one of a set of basic categories such as continuous lines, dashed lines, symbols, and shading patterns.

Phase 3 provides proper descriptions for each labeled image piece, given its category. For example, if an image piece has been recognized as a continuous line, its description will consist of its polygonal approximation by a sequence of vectors. A dashed line would undergo a different vectorization process. Characters and cadastral symbols are also recognized in this phase.

In these first three phases, the system builds a description of the overall geometry of the map or, more specifically, a set of detailed descriptions of the individual geometric entities. System performance in phase 2 is enhanced by relying on the semantics of land register maps, while phases 1 and 3 are largely independent of the class of line drawings being processed.

Finally, in phase 4, the system establishes spatial and semantic relations among geometric entities and also recognizes high-level entities.

Because it is used in legal matters, the cadastral database must have no errors. The operator, therefore, has complete supervision of all decisions made by the system. Operator intervention is required either to resolve ambiguities found by the system during interactive steps or to check the output of fully automatic steps. Additional data validation is performed automatically by checking that the descriptions produced for cadastral entities satisfy logical and geometric constraints.

Figure 3 shows results of interpretation for part of the cadastral map in Figure 1. The remainder of this article describes the main processing steps in Figure 2.

Scanning and preprocessing

This step converts the map into a digitized image suitable for the subsequent processing. The maps are scanned at a sampling rate of 20 points per linear millimeter, with a dynamic range of 256 gray levels. A thresholding operation is then used to obtain a two-level, or binary, image. Each pixel is classified as belonging to the foreground or the background according to whether its gray level is greater or lower than a suitable threshold. The output of this step is a list of *runs* representing the image (see the sidebar titled "Representing the structure of line-like images"). The average cadastral map in this representation occupies about four megabytes of storage.

It is crucial for the thresholded image to reproduce the original as accurately as possible. In particular, the actual width of lines should be preserved everywhere in the image. The threshold is set to the gray level shared by those pixels that differ most in brightness from their neighbors. In fact, these pixels are most likely to be on the boundary between foreground and background.

Thresholding can inject noise into the image. For example, it can create small holes along lines and various irregularities in the figure boundaries. A smoothing process is used to eliminate most of this noise, but some inevitably remains. Consequently, robustness to digitization noise is a basic design requirement for all subsequent processing.

Representing the structure of line-like images

The binary image of a drawing can be represented as a list of vertical or horizontal *runs* (that is, maximal sequences of black pixels in a vertical column or horizontal row, respectively). A vertical run is identified by three integers: its column, the row of the first pixel, and the row of the last pixel. (A horizontal run is identified by its row, the column of the first pixel, and the column of the last pixel.) A subrun is a nonmaximal sequence of black pixels.

The graph construction in our system is based on the run representation.[1] In a *vertical simple graph representation*, the graph is derived using only vertical runs. Figure A shows the image partitioning into edges and nodes for a vertical graph representation. The partitioning is based on the following intuitive observations:

(a) A run adjacent to only one run on each side is likely to be part of a line portion.
(b) A run adjacent to more than one run on a side is likely to be at a crossing point.
(c) A run not adjacent to any run on one side is likely to be at the extreme of a segment.

Such considerations can be formalized and a procedure derived[1] so that each *edge* of the graph is a set of adjacent runs of class (a), while each *node* consists of a single run of class (b) or (c).

This representation can be improved by introducing the notions of crossing area and extreme area as generalizations of the notions of crossing point and extreme point. The notion of areas allows the extension of nodes to include runs previously belonging to edges, so that the lengths of runs in an edge are as uniform as possible (within a certain tolerance due to noise). In Figure B, part (a) highlights the horizontal crossing area and part (b) the vertical crossing area between two lines.

The simple graph representation may fail to associate an edge with a line segment that forms a narrow angle with the direction of runs. The *mixed graph representation* has been introduced to solve this problem. In the mixed graph representation, edges consist of either horizontal or vertical runs, according to the slope of the corresponding line portion, while nodes consist of vertical runs and subruns. In Figure B, part (c) illustrates a mixed graph representation.

The first step in constructing a mixed graph representation is to build both horizontal and vertical simple graph representations, as shown in Figure B, parts (a) and (b). Second, edges are built as sets of adjacent short runs that belong neither to crossing areas nor to extreme areas in the simple representations. (We define a vertical and horizontal run as *conjugate* if they cross, and we define a run as *short* if it is strictly shorter than all of its conjugates.) The remaining pieces of the image, encoded as lists of vertical runs and subruns, are the nodes of the mixed graph (Figure B, part (c)).

The shape of each node is then refined by a heuristic procedure that attempts to minimize the area of the node and to maximize the lengths of the connected edges. (This technique often requires splitting runs into subruns.) Finally, edges undergo a polygonal approximation by generating additional nodes so that each resulting edge corresponds to an approximately straight portion of line.

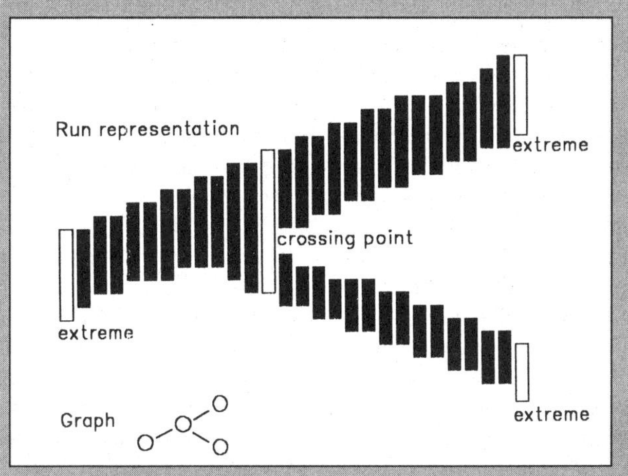

Figure A. First step in the construction of a vertical simple graph representation. Nodes of the graph represent crossing points and extreme points, and correspond to single runs (highlighted in the figure). Edges correspond to remaining sets of adjacent runs.

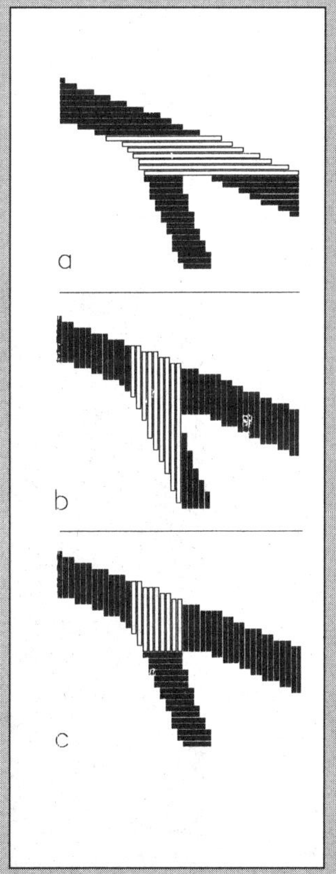

Figure B. Representations of the intersection of two lines. In a horizontal simple graph representation (a) and a vertical simple graph representation (b), the crossing area (corresponding to a node in the graph) is highlighted. In a mixed graph representation (c), each edge corresponds to a set of either vertical or horizontal adjacent runs, according to its slope, and the node is represented by vertical runs and subruns.

Reference

1. S. Di Zenzo and A. Morelli, "A Useful Image Representation," *Proc. Fifth Int'l Conf. Image Analysis and Processing*, World Scientific Publishing, Singapore, 1989.

Graph construction

The drawing interpretation algorithms that we describe here are based on the special image representation that we call *graph representation*. This format is especially convenient for line-like images because it decomposes the line structure into *edges* and *nodes* that formalize the intuitive notions of "lines" and "crossing points between lines." During graph construction, the binary image is partitioned into connected sets of foreground pixels, each corresponding to a graph edge or a graph node. We also use the terms edge and node to mean the image pieces associated with the graph elements.

Our system uses the graph representation for several purposes:

- To detect basic elements of the image (for example, line segments, symbols, and shading patterns) by recognizing corresponding patterns in the graph. (This makes the search for possibly rotated patterns an affordable task, whereas it is almost prohibitively expensive on the raster image.)
- To partition the image into its basic elements (such as segments, symbols, and shaded areas) by recognizing corresponding patterns in the graph.
- To vectorize line segments.
- To help image editing during interactive sessions. (The operator is allowed to pick individual edges and nodes as atomic elements of the image.)

Shading-pattern recognition

In Italian Land Register Authority maps, areas corresponding to buildings are shaded with broken parallel hatching lines spaced 0.5 to 0.7 millimeters apart and forming an angle, preferably 45 degrees, with the perimeter. It is crucial to recognize buildings before the raster-to-vector conversion. Only perimeters have to be vectorized. Hatching lines should be removed from the image graph because they carry no additional information and they introduce noise in the vectorization process. This task is not trivial, because hatching lines touch the perimeter and often overlap symbols or other lines internal to the perimeter.

In detecting a building, the system searches the image graph for subgraphs typically belonging to buildings. Figure 4 gives an example. Nodes 1, 2, and 3 within the chain of collinear edges A, B, C, and D (part of the perimeter) are connected respectively to edges E, F, and G (part of the hatching lines). The edges E, F, and G have no further connections in the graph. When such a comb-like structure is found, the system assumes that a building exists and uses heuristics to search the image graph locally for its perimeter. If the search fails to find a closed polygon, the operator is asked either to reject the candidate building or to interactively help the system identify the perimeter. Finally, all nodes and edges internal to the perimeter, which are recognized as parts of hatching lines, are removed from the image graph.

Continuous-line extraction and vectorization

A component-labeling algorithm is used to derive the connected components of the image, which are then classified as "line" or "sign" based on their radius. (Radius is defined as the maximum distance of the component's pixels from the centroid.) This classification process identifies isolated dashes and symbols.

The system then searches the graph of "line" components to extract symbols overlapping lines. Their subgraphs are usually composed of chains or cycles of short edges, thick edges, or edges for which one of the two endpoints is a terminal node (that is, not connected to other edges). The search locates such small subgraphs and changes the labels of their elements from "line" to "sign." The remaining components of the image graph labeled as "lines" are routed to the vectorization process.

In most drawing analysis systems, the raster-to-vector conversion of a line structure is achieved by thinning the image, thus reducing thick lines to linear sequences of pixels, and then finding an approximating polygonal line. In the latter process, approximation algorithms operate on linearly ordered sets of pixels, which requires each stage of the process to know the next pixel to be approximated.

The graph representation avoids the problems of thinning,[8] since the edges of the image graph are linearly ordered

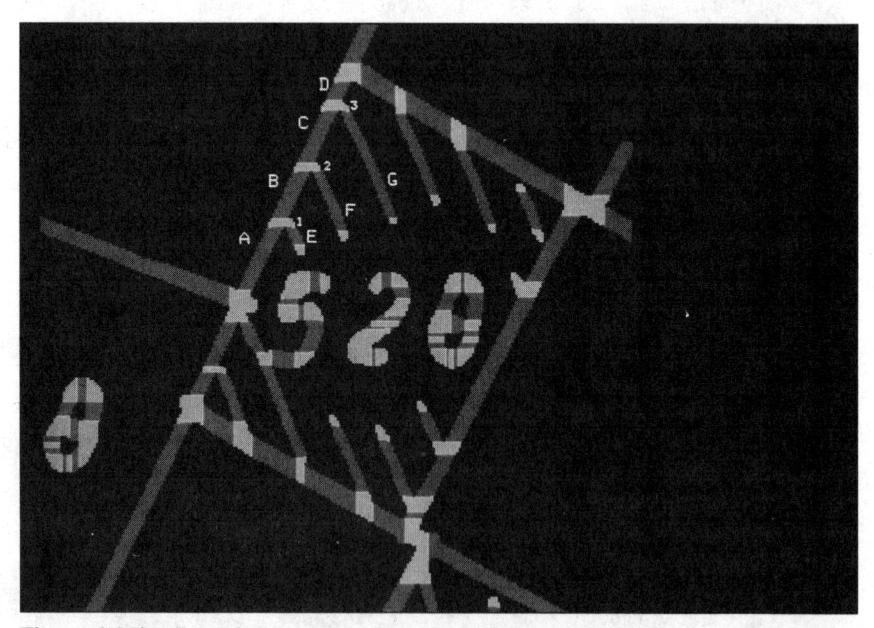

Figure 4. Mixed graph representation of a building in a cadastral map. The subgraph identified by edges A, B, C, D, E, F, and G typically belongs to a building. (Buildings are shaded by broken parallel hatching lines forming preferably 45-degree angles with the perimeter.) Once such a structure is found in the image graph, the chain of collinear edges A, B, C, and D is used as the starting point for a local search to identify the perimeter of the building.

sets of runs. Various polygonal approximation algorithms[9,10] can be modified to work with sequences of runs. In our system, the approximation actually takes place at an early stage, during the construction of the mixed graph representation (see the sidebar on representing line-like images). The approximating elements are rhomboids, rather than ideal segments of zero width. The rhomboids can reflect the actual width of the line being approximated.

The vectorization process relies on the graph representation. First, sequences of collinear edges are searched according to the following "fitting condition": A straight line exists such that (1) the maximum distance between the line and the middle points of the runs of the edges is lower than a certain threshold and (2) the average value of this distance is lower than a threshold. Then vertices are computed as intersections of such lines. Vectors are derived from vertices. When more than two edges converge into a node, the corresponding lines are modified to create a single intersection point. However, if the modified lines violate the fitting condition, new vectors — not associated with any edge — are introduced to connect those intersection points that cannot be merged.

This technique satisfies the Land Register Authority's strict requirements (0.4-mm tolerance). Even higher accuracy has been reached. In the maps converted for the LRA thus far, all vectors were placed inside the raster image, near the medial axis, and no false vectors were generated.

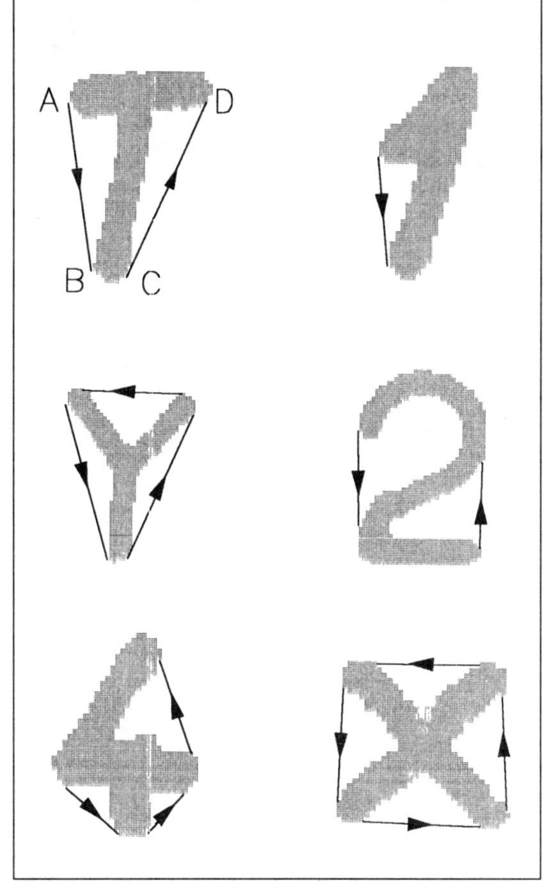

Figure 5. Examples of lid configurations. Lids can be used to derive reliable features for symbol recognition.

Dashed-line extraction and vectorization

Dashes are selected from the set of signs based on the analysis of geometric properties and the recognition of typical patterns in the image graph (for example, node-edge-node). The pool of candidate dashes is then searched for sequences of dashes in which the spacing of adjacent items is lower than a threshold T_D and the angle formed by adjacent items is lower than a threshold T_A. Such a process extracts sequences forming dashed lines that may not be straight. Dashed lines are subsequently approximated by sequences of vectors.

This process, however, is not adequate for the recognition of dashed lines in all possible situations. For example, at the intersection of two dashed lines, different patterns can occur: X, T, L, +, etc. Such patterns are generally labeled as signs, but they are not considered during dashed-line extraction, since they are not recognized as candidate dashes. Hence, they are separately recognized on the basis of their graph structure, and are included in dashed lines if their edges, considered as separate dashes, belong to already identified portions of dashed lines.

After dashed-line extraction, all remaining signs (including unused candidate dashes) are labeled as "symbols."

Symbol recognition

Characters and cadastral symbols are recognized based on geometrical and morphological properties called *features*. Symbols are mostly handprinted, and are arbitrarily rotated and scaled. Features must therefore be invariant under shifts, rotations, and scaling, and robust to distortions and noise. Some reliable features are derived from *lids* (the vectors **AB** and **CD** are the lids of the symbol T in Figure 5) and *sides* (as known from the geometry of plane figures). The patterns formed by lids and sides convey much information about the shape of a figure. Figure 5 shows some examples of possible lid configurations.

With reference to Figurc 5, the following is a very simplified set of requirements that a symbol should satisfy for consideration as a candidate T: (1) There are two lids, (2) neither lid is twice as long as the other, and (3) the angle between **AB** and **CD** is greater than 110 degrees.

More generally, each class of symbols has a set of requirements that a shape should satisfy to be considered as a candidate for that class. Each requirement is so formulated that it is satisfied with probability 1 by every shape actually belonging to the class. This strategy allows adding as many requirements as needed for actual discrimination without performance degradation.

Di Zenzo et al.[11] give a detailed description of the optical character recognition system we use.

Many of the symbols found in a map are part of a string. In such cases, the context provides useful information about orientation, size, syntax, etc., which the system uses to resolve ambiguities (for example, 6 versus 9 or O versus 0). The system uses proximity and collinearity criteria to identify strings.

High-level entity recognition

The output of the previous processing steps is a low-level description of the drawing — except for buildings (see previous discussion under "Shading-pattern recognition"). Segments are described in a *vector graph;* characters and cadastral symbols have been recognized. This representation completely replaces the raster image and is the input for the next processing step, which recognizes the high-level entities of the map such as parcels, streets, and bodies of water.

A cadastral entity is composed of segments and symbols drawn according to specific guidelines. A key point in the design of this part of the interpretation process is the choice of appropriate representations of such guidelines. Our choices were driven by the objective of breaking down the task into a set of well-known problems in graph theory and operational research.

As an example, we describe the technique for identifying cadastral parcels. A parcel is drawn as a set of adjacent polygons, labeled with a number placed inside (when possible) or close to one of the polygons. Two polygons belong to the same parcel if a tilde (~) is drawn on the common edge (see parcel 174 in Figure 1). The first step of parcel recognition locates all the minimal cycles of the vector graph (that is, polygons are identified). Then a new graph is built in which the polygons are the nodes and an edge is created between two nodes if the corresponding polygons are connected by a tilde. The connected components of this graph are the candidate parcels. To find actual cadastral parcels, the procedure looks for an optimal assignment of the set of recognized numbers to the pool of candidate parcels.

The system has been implemented on two IBM platforms. Performance results for Land Register Authority maps are reported in the sidebar titled "System performance."

The graph image representation is a major advantage of our system for interpreting land register maps. It provides a formal description of the image's topological and metrical properties, which has proved useful in solving many problems in the automatic interpretation of line drawings.

The system successfully handles real maps that often contain formal errors and semantic ambiguities. It is fault tolerant and robust to noise. Moreover, its basis in quite general concepts and methodologies makes it suitable for the interpretation of other kinds of line drawings. In fact, this is the focus of our current work. Initial experiments on technological networks and engineering drawings show promising results. ■

System performance

The system consists of a set of programs that communicate through files: The output of one step is the input to the next one. This pipeline structure ensures high software modularity. Furthermore, it allows a batch of maps to be processed simultaneously, thus avoiding operator inactivity during fully automatic steps. Two platforms are currently available: (1) S/370 or S/390 architecture running VM/CMS, equipped with IBM 5080 graphic stations, and (2) IBM RS/6000 workstation running AIX.

The most relevant system-performance measures are elapsed time and operator's intervention time for complete processing of a map sample. However, the accuracy of automatic steps is a key factor, since improving accuracy in these steps can reduce the requirements for interactive operations.

Table A reports the results of shading-pattern recognition (a partially automatic step). Table B reports the results of character recognition (a fully automatic step). Table C shows the average amount of operator's time needed to produce the final database. The results were obtained over a 40-map sample. (On average, a map contains about 600 parcels, 130 buildings, 1,800 characters, and 700 cadastral symbols.) Manual conversion, using digitizing tablets and alphanumeric keyboards, required an average of about 20 hours of operator's time per sample — or 5.7 times more operator's time than the system.

Table A. Results of shading-pattern recognition over a 40-map sample. Both global performance of the step and details of operator intervention are reported. "False buildings" are those parts of the image graph that the system interpreted wrongly as buildings. "Perimeters found" refers to detected buildings.

Description	Number	(percentage)
Total buildings in the maps	7,181	
Buildings detected		
Automatically	6,564	(91.4)
With operator intervention	617	(8.6)
False buildings detected	57	
Perimeters found		
Automatically	6,350	(96.7)
With operator intervention	214	(3.3)
Total	6,564	

Table B. Results of automatic character recognition over a 40-map sample.

Characters in the Maps	Number	(percentage)
Correctly recognized	66,387	(92.47)
Unrecognized	4,927	(6.84)
Misrecognized	497	(0.69)
Total	71,796	

Table C. Average elapsed times needed to process 100 × 70-cm cadastral map on an IBM 4381 processor. Results were obtained over a 40-map sample. Interactive steps account for operator's intervention time.

Description	Time
Phase 1: Preprocessing and graph construction	17 min.
Phases 2 / 3: Construction of low-level geometric description	2 hrs. 39 min.
Phase 4: Recognition of high-level logical entities	32 min.
Interactive steps	2 hrs. 59 min.
Automatic steps	29 min.
Total elapsed time	3 hrs. 28 min.

Acknowledgments

We gratefully acknowledge the encouragement of Herbert Freeman and George Nagy. We also thank Aldo Spirito for his important contribution to our work and Roberto Cremonini for his comments on an earlier draft.

References

1. M. Karima, K.S. Sadhal, and T.O. McNeil, "From Paper Drawings to Computer-Aided Design," *IEEE Computer Graphics and Applications,* Vol. 5, No. 2, 1985, pp. 27-39.

2. M. Ejiri et al., "Automatic Recognition of Engineering Drawings and Maps," *Image Analysis Applications*, R. Kasturi and M.M. Trivedi, eds., Marcel Dekker, New York, 1990.

3. S. Suzuki and T. Yamada, "Maris: Map Recognition Input System," *Pattern Recognition*, Vol. 23, No. 8, 1990, pp. 919-933.

4. M.T. Musavi et al., "A Vision-Based Method to Automate Map Processing," *Pattern Recognition*, Vol. 21, No. 4, 1988, pp. 319-326.

5. S. Bow and R. Kasturi, "A Graphics-Recognition System for Interpretation of Line Drawings," *Image Analysis Applications*, R. Kasturi and M.M. Trivedi, eds., Marcel Dekker, New York, 1990.

6. H. Freeman, "Computer Processing of Line-Drawing Images," *Computing Surveys*, Vol. 6, No. 1, Mar. 1974, pp. 57-97.

7. L.A. Fletcher and R. Kasturi, "A Robust Algorithm for Text String Separation from Mixed Text/Graphics Images," *IEEE Trans. Pattern Analysis and Machine Intelligence*, Vol. 10, No. 6, Nov. 1988, pp. 910-918.

8. B. Jang and R.T. Chin, "Analysis of Thinning Algorithms Using Mathematical Morphology," *IEEE Trans. Pattern Analysis and Machine Intelligence*, Vol. 12, No. 6, June 1990, pp. 541-551.

9. J. Sklansky and V. Gonzalez, "Fast Polygonal Approximation of Digitized Curves," *Pattern Recognition,* Vol. 12, No. 5, Aug. 1980, pp. 327-331.

10. K. Wall and P. Danielsson, "A Fast Sequential Method for Polygonal Approximation of Digitized Curves," *Computer Vision, Graphics, and Image Processing*, Vol. 28, No. 2, Nov. 1984, pp. 220-227.

11. S. Di Zenzo et al., "Recognizing Handprinted Characters of Any Size, Position, and Orientation," to be published in *IBM J. Research and Development*, Vol. 36, No. 2, 1992.

Luca Boatto is a researcher at the IBM Southern Europe Middle East and Africa (SEMEA) Scientific and Technical Solution Center. His research interests include automatic understanding of cadastral maps, engineering drawings, and maps representing technological networks. Boatto was a student fellow at Telettra Digital Signal Processing Laboratory in Rieti and received his degree in electrical engineering from the University of Rome, La Sapienza.

Vincenzo Consorti is a researcher at the IBM SEMEA Scientific and Technical Solution Center. Until 1988, he was with Telettra Digital Signal Processing Laboratory in Chieti, where he worked on electromagnetic compatibility and the theory and design of digital filters. His main research interests involve image recognition and understanding. Consorti received his degree in electrical engineering from the University of Pisa.

Monica Del Buono is a researcher at the IBM SEMEA Scientific and Technical Solution Center. Her research interests include optical character recognition and document analysis. She previously worked in the area of document handling and office automation. Del Buono received her degree in mathematics from the University of Rome, La Sapienza.

Silvano Di Zenzo is director of research at the IBM SEMEA Scientific and Technical Solution Center. He has been assistant professor of mathematics at the University of Genoa, an IBM systems engineer, and manager of an IBM industry project in image processing. His technical interests include machine vision and pattern recognition. Di Zenzo received his degree in electrical engineering from the University of Genoa. He is coeditor of *Image and Vision Computing*.

Vincenzo Eramo is a senior researcher at the IBM SEMEA Scientific and Technical Solution Center, which he joined in 1987. His technical interests include image processing, pattern recognition, and technical drawing understanding. Eramo received his degree in physics from the University of Bologna.

Alessandra Esposito is a researcher at the IBM SEMEA Scientific and Technical Solution Center. She is a member of the Image Processing Group currently involved in the technical drawings processing project—particularly in pattern recognition and image understanding. Esposito received her degree in electrical engineering from the University of Naples, Frederico II.

Francesco Melcarne is a researcher at the IBM SEMEA Scientific and Technical Solution Center. He is a member of the Image Processing Group involved in the technical drawings processing project. Until 1990, he worked at the Institute of Systems Analysis and Computer Science of the Italian National Council of Research on a project concerning the use of artificial intelligence techniques for traffic control. Melcarne received his degree in electrical engineering from the University of Rome, La Sapienza.

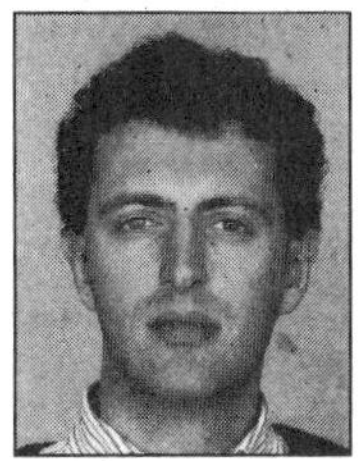

Marco Meucci is a system engineer in the IBM SEMEA Geographical Information Systems Solutions Group. His technical interests are in the research and development of image understanding and user interfaces. Meucci received his degree in mathematics at the University of Rome, La Sapienza.

Andrea Morelli is a researcher at the IBM SEMEA Scientific and Technical Solution Center. His research interests involve image recognition and understanding. Morelli received his degree in physics from the University of Milano.

Marco Mosciatti is a system engineer in the IBM SEMEA GIS Solutions Group. Until 1987 he worked as a consulting engineering in the Mechanical Engineering Department of the University of Rome. His research interests are in image understanding and user interfaces. Mosciatti received his degree in mechanical and industrial engineering from the University of Rome, La Sapienza.

Stefano Scarci is manager of engineering applications at the IBM SEMEA Scientific and Technical Solution Center. Until 1991, he was manager of image and knowledge processing, and responsible for the map interpretation project. He has worked on several research projects in signal processing and pattern recognition, both at the IBM SEMEA Scientific and Technical Solution Center and at the IBM T.J. Watson Research Center in the US. Scarci received his degree in electrical engineering from the University of Naples, Frederico II. He is a member of the IEEE.

Marco Tucci is a member of the Image Processing Group at the IBM SEMEA Scientific and Technical Solution Center. Until 1990, he was a researcher with the Computer Science Department of the University of Rome. His research interests include artificial intelligence, software engineering, pattern recognition, and image understanding. Tucci received his degree in electrical engineering from the University of Rome, La Sapienza.

Readers can contact Vincenzo Consorti at IBM SEMEA Scientific and Technical Solution Center, Viale Oceano Pacifico 171, 00144 Rome, Italy; e-mail consorti@romesc.vnet.ibm.com.

Segmentation and Preliminary Recognition of Madrigals Notated in White Mensural Notation*

Nicholas P. Carter
Departments of Physics and Music, University of Surrey, Guildford, Surrey GU2 5XH United Kingdom

Abstract: An automatic music score-reading system will facilitate applications including computer-based editing of new editions, production of databases for musicological research, and creation of braille or large-format scores for the blind or partially-sighted. The work described here deals specifically with initial processing of images containing early seventeenth century madrigals notated in white mensural notation. The problems of segmentation involved in isolating the musical symbols from the word-underlay and decorative graphics are compounded by the poor quality of the originals which present a significant challenge to any recognition system. The solution described takes advantage of structural decomposition techniques based on a novel transformation of the line adjacency graph which have been developed during work on a score-reading system for conventional music notation.

Key Words: structural pattern recognition, printed music, white mensural notation

1 Introduction

The availability of an automatic music score-reading system will facilitate applications including computer-based editing of new editions, production of databases for musicological research, and creation of braille or large-format scores for the blind or partially sighted. The availability of electronic databases of music representational language encodings will also hasten the development of point-of-sale printing systems for sheet music and open up possibilities of electronic distribution. For background information regarding the field of acquisition, representation and reconstruction of printed music by computer, the reader is referred to Carter et al. (1988) and Hewlett and Selfridge-Field (1991).

Automatic recognition of printed music has been the subject of research since the late 1960s (when hardware limitations restricted progress—see Pruslin 1967 and Prerau 1970). More recent work at Waseda University on the WABOT-2 keyboard-playing robot used mask-matching implemented in hardware to read nursery song sheets (Matsushima 1985). A team at Osaka University aims to produce an overall "music information processing system" which attempts automatic transcription (producing a score from a soundtrack) and sentiment identification, in addition to recognition of piano scores (Katayose 1989). Other work in score recognition is being undertaken at the University of Ottawa, using projection profiles (Fujinaga 1988) and at University College Cardiff, using simple, localised measurements aimed at producing a low-cost solution to the problem of automatic acquisition for publishers and engravers (Clarke 1988).

The work described here moves beyond the scope of conventional music notation and deals with one particular form of early notation, i.e., seventeenth century white mensural notation. An overview of the note and rest symbols used in white mensural notation, together with their modern equivalents, can be found in Figure 1. In this context the advantages of an automatic score-reading system are supplemented by the facility to convert such obsolete notation into conventional notation ready for publication. Related research into recognition of an even earlier form of notation has been described by McGee and Merkley (1991) in discussing their work with medieval music.

Address offprint requests to: Nicholas P. Carter, Departments of Physics and Music, University of Surrey, Guildford, Surrey GU2 5XH, United Kingdom.

*This research was undertaken with support from Oxford University Press.

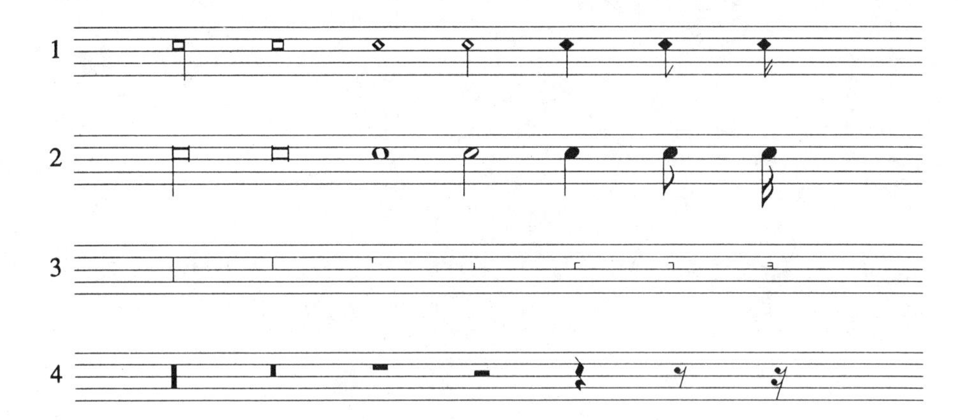

Figure 1. An overview of the note and rest symbols used in white mensural notation (staves 1 and 3), together with their modern equivalents (staves 2 and 4).

Figure 2. An example of seventeenth century white mensural notation.

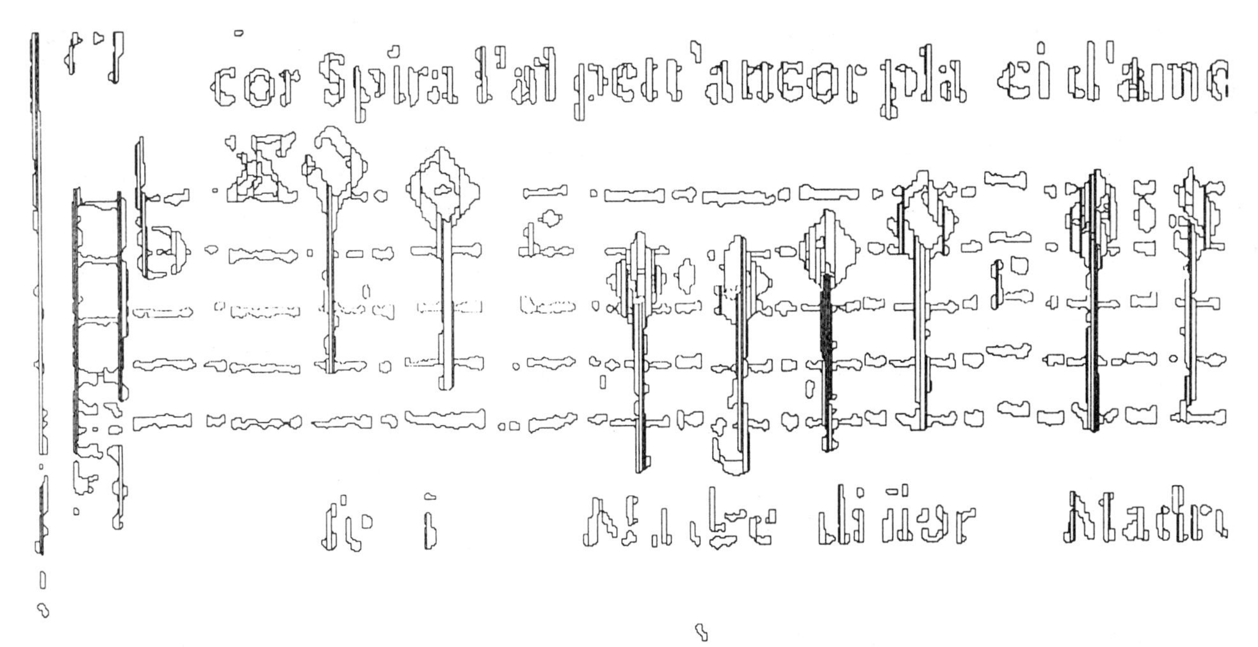

Figure 3. Outlines of the sections which are the nodes of the transformed line adjacency graph for the bottom left-hand corner of Figure 2.

2 Image Acquisition and Structural Decomposition

Binary images are acquired using a conventional flat-bed CCD-based scanner at a resolution of 300 dots per inch. Some of the images had previously been transmitted by fax, so sampling had already taken place at a lower resolution and some additional degradation occurred (Figures 2 and 7). Figure 9 is a more typical example of an image, from a different publication to those of the other figures, with less noise and degradation.

The first operation on each image involves the production of an original transformation of the line adjacency graph (LAG), the details of which have been described elsewhere (Carter 1989; Carter and Bacon 1992), in connection with work on a recognition system for conventional music notation. Subsequent processing is based on manipulation of the nodes of the transformed LAG (termed "sections") which correspond to particular regions of adjacent black pixels (Figure 3). This avoids further operations at the pixel level, thus speeding processing, and also provides a first approximation to the structurally significant portions of the image, such as horizontal line fragments. By making use of attributes of the sections such as area, aspect ratio, and average thickness, significant tolerance of image distortions is built into the system. For example, in order to reduce the noise present in the image, all sections with area ≤ five pixels are removed. In order to achieve recognition of the text when this has been isolated from the original image, some of these so called "noise" sections may have to be restored, as they are in fact structurally important features of the constituent letters.

After removal of noise sections, groups of connected sections are formed into "objects" by a depth-first traversal of the transformed line-adjacency graph. Due to the method of production used in creating the original seventeenth century scores, where each symbol would have been printed using a separate piece of type, a fragmented appearance results and as a consequence the objects formed by the above process normally consist of a single symbol. In terms of score recognition, this is the principal difference (apart from the symbol designs as shown in Figure 1) between these images and conventional music notation. In the latter case, it can normally be assumed that the stavelines will run continuously for the majority of their extent across the page, whereas with the madrigal scores the stavelines consist of a large number of fragments.

3 Isolation of Ornamental Graphic and Text

The pages of notation consist of three main elements: a number of five-line staves with overlaid musical symbols (including notes, rests, and clefs), word underlay (lyrics) and an ornamental graphic situated at the top left of each page. In order to isolate the ornamental graphic, the largest object in the top left hand quarter of the image is found. Then,

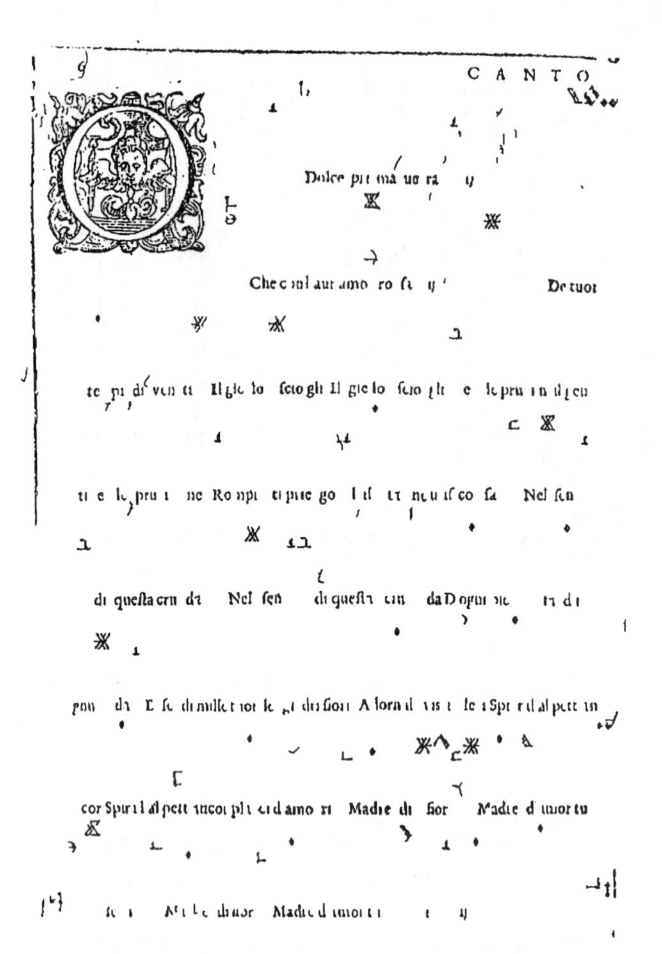

Figure 4. Potential text objects, the ornamental graphic, and some shadow lines extracted from Figure 2.

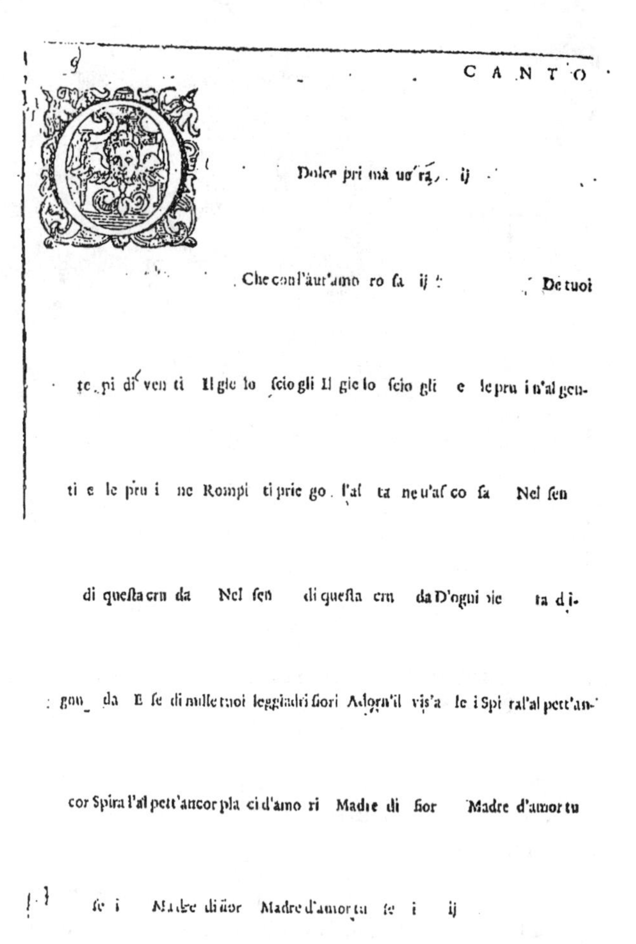

Figure 5. The final selection of text extracted from Figure 2.

both it and all other objects which overlap the region delineated by the graphic object's lowest and rightmost extremes and the top and leftmost edges of the page, are removed. This is necessary in order to include all fragments of the ornamental graphic (which is not guaranteed to be a single-connected component) and some of the shadow lines present at the borders of the image.

The next operation makes use of a maximum line thickness threshold which is set to 2.5 × the most common section thickness. This relies on the fact that most sections are horizontal line fragments which make up the stavelines. The inter-staveline spacing (subsequently referred to as α) is then calculated by finding the minimum spacing between each section with average thickness less than the maximum line thickness and a horizontally-overlapping section with similarly limited thickness. The most common spacing between these lines is taken as the inter-staveline spacing. Potential text objects are categorized as those which satisfy the following criteria:

width $< 1.5 \times \alpha$,
maximum line thickness < height $< 1.5 \times \alpha$
and width/height < 2.

The potential text objects extracted by this method from Figure 2 are shown in Figure 4. The potential text objects are then linked together into strings by testing for vertical overlap between neighbouring objects. Strings of less than four characters are removed as erroneous. Also, text strings which are vertically spaced by less than the maximum line thickness value are combined.

Objects with height of between three and ten times α and width greater than the maximum line thickness are then classfied as "significant" objects. These objects are then examined in pairs and where they overlap vertically, the extent covered by the two objects is recorded to build up a picture of stave location. Horizontal stave dividing lines are then situated where there are gaps in this vertical extent array which are wider than the maximum line thickness. The text strings are then matched to these

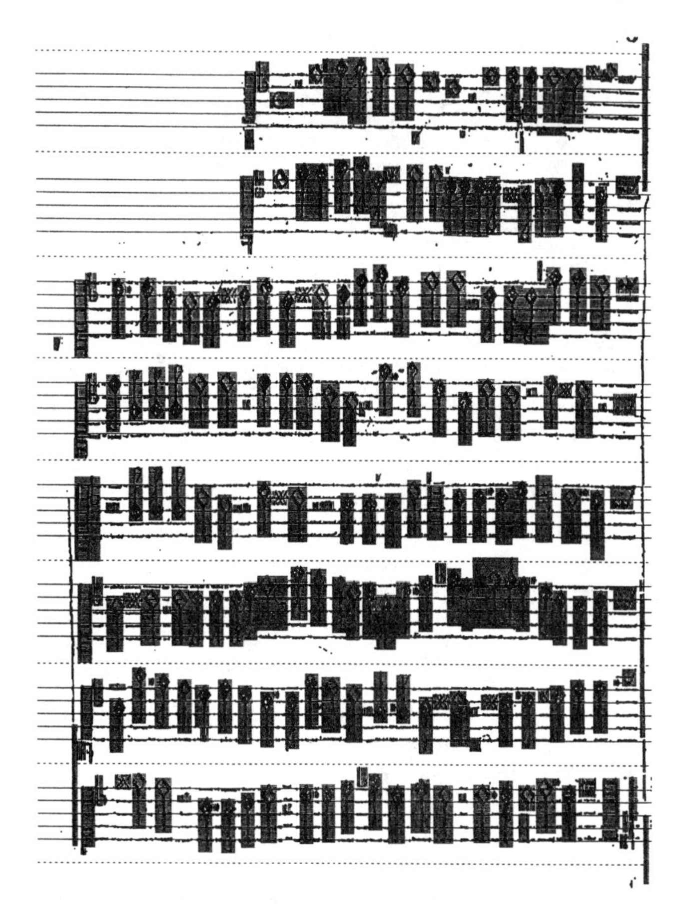

Figure 6. Processed output for Figure 2. The dotted lines are the stave separation lines; solid lines are the staveline approximations; the bounding rectangles of large and small objects are shaded with horizontal and vertical lines respectively.

Figure 7. A further example of seventeenth century white mensural notation.

stave dividing lines and any which are left over are removed. All objects which are finally to be categorized as part of the word underlay are found by locating all objects which are of appropriate size and overlap a horizontal strip of height $1.5 \times \alpha$ centered on a stave dividing line. The results of this process can be seen in Figure 5. All remaining objects are then allocated to a particular stave depending on their vertical position relative to the stave dividing lines.

4 Location of Stavelines

Identifying the location of stavelines is made difficult by the fragmented nature of the image and also by steps in the stavelines caused by the vertical displacement of some of the original engraving blocks. Currently, a global straight line fit is performed based on the location and length of near-horizontal line fragments between each pair of stave dividing lines. The current technique needs to be improved to take account of local distortions such as the "steps" described above, and also curvature in the original (i.e., bowing of the stave). The results of fitting five horizontal lines equally spaced by the value for α can be seen in Figure 6. This information will be used when note pitches are being established, and so needs to be accurate enough to enable distinction between notes superimposed on a line or in a space.

5 Preliminary Musical Symbol Analysis

The first processing operation divides all objects associated with the current stave into 'large' (height $\geq 3 \times \alpha$) and 'small' ($0.5 \times \alpha <$ height $< 3 \times \alpha$). Any object of height less than $0.5 \times \alpha$ is ignored as noise. A small object which has height less than $0.8 \times \alpha$ and width of between 0.1 and $0.6 \times \alpha$ is marked as a possible dot. This will be subjected to syntactic

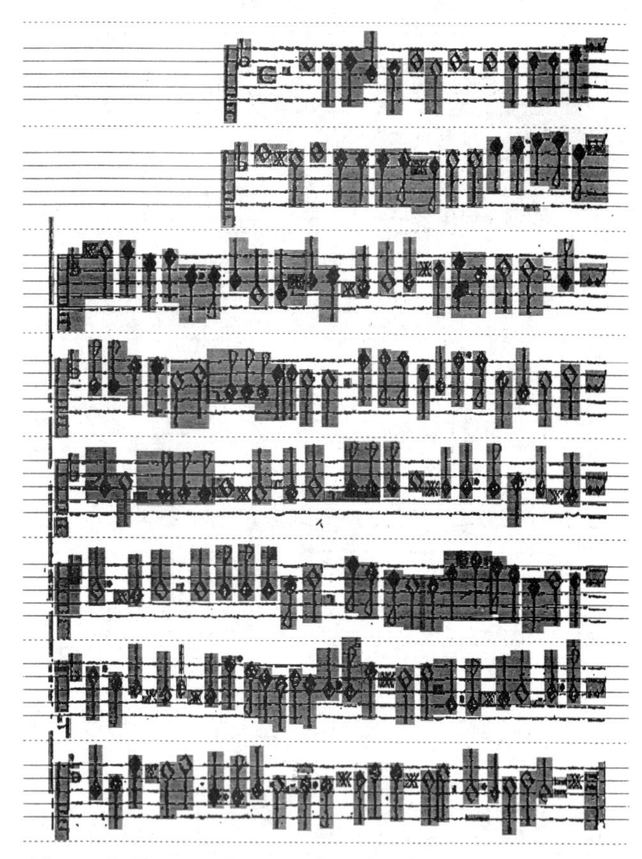

Figure 8. Processed output for Figure 7.

Figure 9. An image of seventeenth century white mensural notation from a different publication—perhaps more typical in terms of image quality.

tests at a later stage to confirm that it is in the appropriate position next to a notehead. All small objects are further tested for horizontal overlap with large objects using an array of horizontal extents compiled during the search for large objects. The use of the horizontal extent of objects is based on the fact that the music on each stave is basically a one-dimensional string of symbols unlike a large proportion of conventional notation which has a far more complex, two-dimensional structure.

Initially, a search is made for a clef at the left hand end of the current stave by testing for an object containing multiple vertical lines and width $> 0.5 \times \alpha$. This is done by producing the transformed line adjacency graph of the object but in this case based on horizontally-oriented run-length encoding. Those sections which have aspect ratio greater than two and length greater than or equal to α are identified as line fragments. All large and small objects which are to the left of the clef are then removed. This will typically include shadow lines at the edge of the page, separated fragments of the clef itself and noise.

The next stage of analysis takes advantage of the guaranteed presence at the end of each line of a subsequent pitch indicator (shaped line a ~) or, in the case of the lowest line on the page, a long (a note with a hollow, square notehead and a stem). The search for the former is undertaken by examining the list of small objects in reverse order, i.e., right to left, for one of width greater than $0.5 \times \alpha$. If this object is less than $1.5 \times \alpha$ in width, a further search is undertaken within the proximity of the object to attempt to locate other fragments which have become detached. Similarly, the long is found on the lowest of the staves by searching backwards through the list of large objects for an object of width greater than or equal to the interstaveline spacing. All large and small objects to the right of the long can then be removed as these can only be barlines or shadow lines. A double barline will be assumed to be present and therefore added by default to the final output data.

All large objects are examined next, with the eventual aim of distinguishing natural signs from stemmed notes and extracting a pitch and rhythm value for each of the latter. At present only some of the necessary tests are implemented, for example, for measuring the "whiteness" of a note head. This particular test will be used to differentiate hollow noteheads. Firstly, vertical line fragments are found (again using horizontally oriented run length encoding) and subjected to a collinearity test in order to locate the potential note stem. Then the white run lengths which cross a vertical line prolonging the note stem are summed to give a figure for "whiteness." This area will be thresholded in order to make the distinction between the two types of note head.

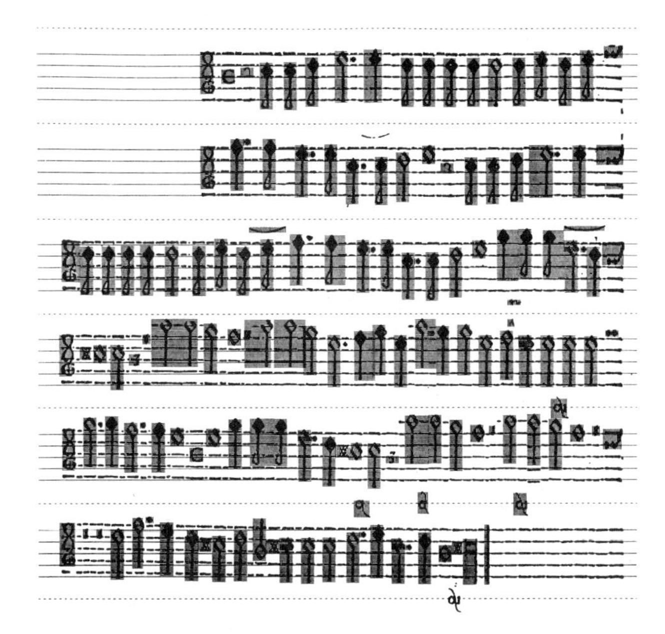

Figure 10. Processed output for Figure 9.

The complete set of small objects on each stave is then examined in adjacent pairs for horizontal overlap. If two small objects are closely spaced vertically and, when combined, would form an object of height greater than $3 \times \alpha$, then they are classified as a fragmented large object. In this case the two small objects would be combined and treated as a single large object. Otherwise, either the two small objects are closely spaced vertically, in which case they are combined into one small object, or one of them must be noise.

A vertical profile is then formed for each small object and where a peak exists which is less than half α in width and greater than half α in height, the object is deemed to be a rest. Additionally, if the height of the peak is between 0.75 and $1.25 \times \alpha$, the rest is categorized as a breve rest (a vertical stroke joining adjacent stavelines).

6 Conclusions

Although the original images for this work were of poor quality, progress has been made in isolating the different categories of symbol present. The word underlay and ornamental graphic have been isolated successfully and some of the necessary tests for categorizing the musical symbols have been implemented (Figures 7–10). Use has been made where possible of syntactic information, such as the guaranteed presence of a clef symbol at the beginning of each stave.

Successful use has been made of the low-level processing operations which have been developed during work on a system for recognition of conventional music notation, including construction of a novel transformation of the line adjacency graph to provide a structural breakdown of the image and object formation for symbol recognition purposes. A variety of thresholds are used which, although their exact values are not crucial, represent domain-specfic knowledge hard-wired into the system. It would be preferable as part of a complete recognition system to separate out this knowledge so that information relating to a different domain, such as conventional music notation or perhaps even circuit diagrams, could be substituted as required.

The main problem revolves around the choice of features for recognition, whether structural or otherwise, in order to produce a reliable system despite the variability of the musical symbols. So far the solution has been found in a mixture of techniques, which also take advantage of the relative simplicity of the musical content of the images.

References

Carter NP, Bacon RA, Messenger T (1988) Acquisition, representation and reconstruction of printed music by computer: A review. Computers and the Humanities 22(2):27–46

Carter NP (1989) Automatic recognition of printed music in the context of electronic publishing. PhD thesis, University of Surrey

Carter NP, Bacon RA (1992) Automatic recognition of printed music. In: Structured document image analysis. Baird HS, Bunke H, Yamamoto K (eds) Springer-Verlag, Heidelberg

Clarke AT (1988) Inexpensive optical character recognition of music notation: A new alternative for publishers. Proceedings of the Computers in Music Conference, Lancaster

Fujinaga I (1988) Optical music recognition using projections. MA dissertation, McGill University, Montreal

Hewlett WB, Selfridge-Field E (1991) Computing in musi cology: A directory of research. Center for Computer Assisted Research in the Humanities, Menlo Park, CA

Katayose H, Inokuchi S (1989) The Kansei music system. Computer Music Journal 13(4):72–77

Matsushima T et al. (1985) Automated recognition system for musical score. Bulletin of Science and Engineering Research Laboratory, Waseda University 112:25–52

McGee W, Merkley P (1991) The optical scanning of medieval music. Computers and the Humanities 25:47–53

Prerau DS (1970) Computer pattern recognition of standard engraved music notation. PhD dissertation, Massachusetts Institute of Technology

Pruslin DH (1967) Automatic recognition of sheet music. ScD dissertation, Massachusetts Institute of Technology

Bibliography

This bibliography includes publications relevant to Document Image Analysis. The papers included here were published in the following journals during the period 1986–1993:

Computer Vision, Graphics, and Image Processing
CVGIP: Graphical Models and Image Processing
CVGIP: Image Understanding
IEEE Transactions on Pattern Analysis and Machine Intelligence
Machine Vision and Applications
Image and Vision Computing
International Journal of Pattern Recognition and Artificial Intelligence
Pattern Recognition
Pattern Recognition Letters
Special issues of *Computer, July 1992* and *Proceedings of IEEE, July 1992.*

A few papers published in other journals are also included to the extent that they are relevant to these topics. It also includes papers presented at the following conferences:

First International Conference on Document Analysis and Recognition, 1991
Second International Conference on Document Analysis and Recognition, 1993

Papers presented in earlier conferences are not included here since significant papers from such conferences have been published in journals. The bibliography is organized as follows:

Books and Edited Volumes
Conferences and Workshops
Special Issues of Journals
Databases and Video
Papers organized into the following sections:

I. Pixel-Level Processing

1. Thresholding and Noise Reduction
2. Thinning and Contour Detection
3. Chain Codes and Other Representations

II. Feature-Level Analysis

1. Linear Approximation and Polygonalization
2. Critical Point Detection
3. Line and Curve Fitting, Hough Transform
4. Shape Representation, Description, and Recognition

III. Text Processing

1. Layout Analysis and Segmentation
2. Character Recognition
   i. Survey Papers

ii. Printed Character Recognition
iii. Handwritten Character Recognition, Handwriting Analysis
iv. Character/Font Generation
3. Signature Analysis and Verification, Document Authentication
4. Document and Form Processing

IV. Graphics Recognition and Interpretation

1. Engineering Drawings and Graphics
2. Electronic Circuit Diagrams
3. Maps and Geographic Information
4. Other Applications: Fingerprints, Music, Chess, Shorthand

Books and Edited Volumes

Adams, G.L., R.L. Serey, and J.P. Thode, *Comparison of Commercially Available Electronic Image Management Systems,* Association for Information and Image Management, Silver Spring, Md.

Avedon, D., *Introduction to Electronic Imaging,* Association for Information and Image Management, Silver Spring, Md., 1992.

Baird, H.S., H. Bunke, and K. Yamamoto, eds., *Structured Document Image Analysis,* Springer-Verlag, Berlin, 1992.

Black, D.B., *Document Capture for Document Imaging Systems,* Association for Information and Image Management, Silver Spring, Md., 1992.

D'Alleyrand, M., *Image Storage and Retrieval Systems,* Intertext Publications, New York, N.Y., 1989.

D'Alleyrand, M., *Handbook of Image Storage and Retrieval Systems,* Van Nostrand Reinhold, New York, N.Y., 1992.

Dougherty, E.R., and C.R. Giardina, *Morphological Methods in Image and Signal Processing,* Prentice Hall, Englewood Cliffs, N.J., 1987.

Gonzalez, R.C., and R.C. Woods, *Digital Image Processing,* Addison-Wesley, Reading, Mass., 1992.

Green, W.B., *Introduction to Electronic Document Management Systems,* Academic Press, Boston, Mass., 1993.

Helgerson, L., *Introduction to Scanning Technology,* Association for Information and Image Management, Silver Spring, Md., 1987.

Jain, A.K., *Fundamentals of Digital Image Processing,* Prentice Hall, Englewood Cliffs, N.J., 1989.

Haralick, R., and L.G. Shapiro, *Computer and Robot Vision,* Addison-Wesley, Reading, Mass., 1992.

Kasturi, R., and M.M. Trivedi, eds., *Image Analysis Applications,* Marcel Dekker, New York, N.Y., 1990.

Kasturi, R., and R.C. Jain, *Computer Vision, Vol. 1: Principles,* IEEE Computer Society Press, Los Alamitos, Calif., 1991.

Kasturi, R., and R.C. Jain, *Computer Vision, Vol. 2: Advances and Applications,* IEEE Computer Society Press, Los Alamitos, Calif., 1991.

Muller, N.J., *Computerized Document Imaging Systems: Technology and Applications,* Artech House, Boston, Mass., 1993.

Pavlidis, T., *Algorithms for Graphics and Image Processing,* Computer Science Press, Rockville, Md., 1982.

Pennebaker, W.B., and J.L. Mitchell, *Still Image Data Compression Standard,* Van Nostrand Reinhold, New York, N.Y., 1993.

Pratt, W.K., *Digital Image Processing,* Second Edition, John Wiley & Sons, New York, N.Y., 1991.

Rabbani, M. and P.W. Jones, *Digital Image Compression Techniques,* SPIE Tutorial Text, Bellingham, Washington, 1991.

Rosenfeld, A. and A. Kak, *Digital Picture Processing, Volumes 1 & 2,* Academic Press, New York, N.Y., 1982.

Russ, J.C., *The Image Processing Handbook,* CRC Press, Boca Raton, Florida, 1992.

Saffady, W., *Electronic Document Imaging Systems: Design, Evaluation, and Implementation,* Meckler, Westport, Connecticut, 1993.

Schalkoff, R.J., *Digital Image Processing and Computer Vision,* John Wiley & Sons, New York, N.Y., 1989.

Schalkoff, R.J., *Pattern Recognition, Statistical, Structural and Neural Approaches,* J. Wiley & Sons, New York, N.Y., 1992.

Suen, C.Y., and R. DeMori, eds., *Computer Analysis and Perception, Vol. 1,* Visual Signals, CRC Press, Boca Raton, Florida, 1982.

Sugihara, K., *Machine Interpretation of Line Drawings,* MIT Press, Cambridge, Mass., 1989.

Weztler, F., *Desktop Image Scanners and Scanning,* Association for Information and Image Management, Silver Spring, Md., 1989.

Conferences and Workshops

IAPR International Workshops on Frontiers in Handwriting Recognition, 1990, 1991, 1993.
Biannual International Conferences on Document Analysis and Recognition, 1991, 1993.
IEEE Conferences on Computer Vision and Pattern Recognition, 1983–1994, published by IEEE Computer Society Press.
Biannual International Conferences on Pattern Recognition, published by IEEE Computer Society Press.
Scandinavian Conferences on Image Analysis, World Scientific, Singapure, 1980–1994.
SPIE Conferences on Character Recognition and Document Recognition, 1992–1994.
Symposiums on Document Analysis and Retrieval, University of Nevada, Las Vegas, 1992–1994.
International Workshop on Advances in Structural and Syntactic Pattern Recognition, 1990–1994.

Special Issues of Journals

Document Image Analysis Techniques, R. Kasturi and L. O'Gorman, eds., *Machine Vision and Applications*, Vol. 5, No. 3, 1992 and Vol. 6, Nos. 2–3, 1993.
Document Image Analysis Systems, L. O'Gorman and R. Kasturi, eds., *Computer*, Vol. 25, No. 7, 1992.
Optical Character Recognition, T. Pavlidis and S. Mori, eds., *Proc. IEEE*, Vol. 80, No. 7, IEEE Service Center, Piscataway, N.J., 1992.
Handwriting Processing and Recognition, R. Plamondon, ed., *Pattern Recognition*, Vol. 26, No. 3, 1993.
Postal Processing and Character Recognition, H. Tominaga, ed., *Pattern Recognition Letters*, Vol. 14, No. 4, 1993.

Databases and Video

1. A database for handwritten word recognition, Center of Excellence for Document Analysis and Recognition, State University of New York, Buffalo, N.Y., 1992. See Hull, J.J., "Database for Handwritten Word Recognition Research," *IEEE Trans Pattern Analysis and Machine Intelligence*, 1994.
2. University of Washington English Document Image Database I, 1993. See Phillips, I.T., J. Ha, R.M. Haralick, and D. Dori, "The Implementation Methodology for a CD-ROM English Document Database," *Proc. Second Int'l Conf. Document Analysis and Recognition*, IEEE CS Press, Los Alamitos, Calif, 1993, pp. 484–487.
3. *Proc. Second Int'l Conf. Document Analysis and Recognition*, IEEE CS Press, Los Alamitos, Calif, 1993 Commemorative CD-ROMS:
   CD-ROM A: ETL-1 and ETL-2; handwritten alphanumerics and printed Kanji.
   CD-ROM B: ETL-3 thru ETL-7; handwritten alphanumerics, symbols, Hiragana and Katakana.
   CD-ROM C: ETL-8 and ETL-9; handwritten Kanji.
   CD-ROM D: Document page images from ten different kinds of documents, mostly Japanese, including newspaper (A2 full-page in 600 dpi), novels, technical papers, patent gazette, etc.

   The above four CDs published by Japanese Technical Committee for Optical Character Recognition, Japan Electronic Industry Development Corporation, Contact: Document Image Database JEIDA '93 Secretariate, Electrotechnical Laboratories, 1-1-4, Umezono, Tsukuba, Ibarki, 305 Japan.

4. Handprinted Character Database, National Institute of Standards and Technology, Advanced Systems Division, Contact: Wilson, C.L. and Garris, M.D., Gaithesburg, Md.
5. Optical Character Recognition and Document Image Analysis, by George Nagy, Rensselaer Video, Clifton Park, N.Y., 1992.
6. DIMUND (Document IMage UNDerstanding) Information Server on-line database, available through DIMUND Gopher or contact Document-Request@cfar.umd.edu.

Technical Papers

I. Pixel-level Processing

1. Thresholding and Noise Reduction

Billawala, N., P.E. Hart, and M. Peairs, "Image Continuation," *Proc. Second Int'l Conf. Document Analysis and Recognition,* IEEE CS Press, Los Alamitos, Calif., 1993, pp. 53–57.

Cheng, J.-C., and H.-S. Don, "Segmentation of Bilevel Images Using Mathematical Morphology," *Int'l J. Pattern Recognition and Artificial Intelligence,* Vol. 6, No. 4, 1992, pp. 595–628.

Cho, S., R. Haralick, and S. Yi, "Improvement of Kittler and Illingworth's Minimum Error Thresholding," *Pattern Recognition,* Vol. 22, 1989, pp. 609–617.

Hertz, L., and R.W. Schafer, "Multilevel Thresholding Using Edge Matching," *Computer Vision, Graphics, and Image Processing,* Vol. 44, 1988, pp. 279–295.

Kittler, J., and J. Illingworth, "Minimum Error Thresholding," *Pattern Recognition,* Vol. 19, 1986, pp. 41–47.

Liu, Y., R. Fenrich, and S.N. Srihari, "An Object Attribute Thresholding Algorithm for Document Image Binarization," *Proc. 2nd Int'l Conf. Document Analysis and Recognition,* IEEE CS Press, Los Alamitos, Calif., 1993, pp. 78–81.

Parker, J.R., C. Jennings, and A.G. Salkauskas, "Thresholding using an Ilumination Model," *Proc. 2nd Int'l Conf. Document Analysis and Recognition,* IEEE CS Press, Los Alamitos, Calif., 1993, pp. 270–273.

Pavlidis, T., "Threshold Selection Using Second Derivatives of the Gray Scale Image," *Proc. 2nd Int'l Conf. Document Analysis and Recognition,* IEEE CS Press, Los Alamitos, Calif., 1993, pp. 274–281.

Sahoo, P.K., S. Soltani, and A.K.C. Wong, "A Survey of Thresholding Techniques," *Computer Vision, Graphics, and Image Processing,* Vol. 41, 1988, pp. 233–260.

Ye, Q.Z., and P.E. Danielsson, "On Minimum Error Thresholding and its Implementations," *Pattern Recognition Letters,* Vol. 7, No. 4, 1988, pp. 201–206.

Zhou, X., and R. Gordon, "Generation of Noise in Binary Images," *CVGIP: Graphical Models and Image Processing,* Vol. 53, No. 5, 1991, pp. 476–478.

2. Thinning and Contour Detection

Arcelli, C., and G. Sanniti di Baja, "Finding Local Maxima in a Pseudo-Euclidean Distance Transform," *Computer Vision, Graphics, and Image Processing*, Vol. 43, 1988, pp. 361–367.

Arcelli, C., and G. Sanniti di Baja, "A One-Pass Two-Operation Process to Detect the Skeletal Pixels on the 4-Distance Transform," *IEEE Trans. Pattern Analysis and Machine Intelligence*, Vol. 11, No. 4, 1989, pp. 411–414.

Arcelli, C., and G. Sanniti di Baja, "Ridge Points in Euclidean Distance Maps," *Pattern Recognition Letters*, Vol. 13, 1992, pp. 237–243.

Baruch, O.,"Line Thinning by Line Following," *Pattern Recognition Letters*, Vol. 8, 1988, pp. 271–276.

Borgefors, G., "Distance Transformations in Digital Images," *Computer Vision, Graphics, and Image Processing*, Vol. 34, 1986, pp. 344–371.

Bourbakis, N.G., "A Parallel-Symmetric Thinning Algorithm," *Pattern Recognition*, Vol. 22, 1989, pp. 387–396.

Brandt, J.W., and V.R. Algazi, "Continuous Skeleton Computation by Voronoi Diagram," *CVGIP: Image Understanding*, Vol. 55, No. 3, 1992, pp. 329–338.

Chen, Y.S., and W.H. Hsu, "A Comparison of Some One-pass Parallel Thinnings," *Pattern Recognition Letters*, Vol. 11, 1990, pp. 35–41.

Chen, C.S., and W.H. Tsai, "A New Fast One-pass Thinning Algorithm and Its Parallel Hardware Implementation," *Pattern Recognition Letters*, Vol. 11, 1990, pp. 471–477.

Chen, Y.S., and W.H. Hsu, "A Modified Fast Parallel Algorithm for Thinning Digital Patterns," *Pattern Recognition Letters*, Vol. 7, No. 2, 1988, pp. 99–106.

Chen, Y.S., and W.H. Hsu, "A Systematic Approach for Designing 2-Subcycle and Pseudo 1-Subcycle Parallel Thinning Algorithms," *Pattern Recognition*, Vol. 22, 1989, pp. 267–282.

Chin, R.T., et al., "A One-Pass Thinning Algorithm and Its Parallel Implementation," *Computer Vision, Graphics, and Image Processing*, Vol. 40, 1987, pp. 30–40.

Cinque, L., C. Guerra, and S. Levialdi, "Computing Shape Description Transforms on a Multiresolution Architecture," *CVGIP: Image Understanding*, Vol. 55, No. 3, 1992, pp. 287–295.

Crabtree, Jr., S.J., L.-P. Yuan, and R. Ehrlich, "A Fast and Accurate Erosion-Dilation Method Suitable for Microcomputers," *CVGIP: Graphical Models and Image Processing*, Vol. 53, No. 3, 1991, pp. 283–290.

Eckhardt, U., and G. Maderlechner, "Thinning for Document Processing," *Proc. 1st Int'l Conf. Document Analysis and Recognition*, AFCET-IRISA/INRIA, France, 1991, pp. 490–498.

Gökmen, M., and R.W. Hall, "Parallel Shrinking Algorithms Using 2-Subfields Approaches," *Computer Vision, Graphics, and Image Processing*, Vol. 52, 1990, pp. 191–209.

Guo, Z., and R. W. Hall, "Fast Fully Parallel Thinning Algorithms," *CVGIP: Image Understanding*, Vol. 55, No. 3, 1992, pp. 317–328.

Hall, R.W., "Comments on 'A Parallel-Symmetric Thinning Algorithm' by Bourbakis," *Pattern Recognition*, Vol. 25, No. 4, 1992, pp. 439–441.

Haralick, R.M., "Performance Characterization in Image Analysis: Thinning, A Case in Point," *Pattern Recognition Letters*, Vol. 13, 1992, pp. 5–12.

Hayat, L., A. Naqvi, and M. B. Sandler, "Comparative Evaluation of Fast Thinning Algorithms on a Multiprocessor Architecture," *Image and Vision Computing*, Vol. 10, No. 4, 1992, pp. 210–218.

Hazout, S., and N.Q. Nguyen, "Image Analysis by Morphological Automata," *Pattern Recognition*, Vol. 24, No. 5, 1991, pp. 401–408.

Jaisimha, M.Y., R.M. Haralick, and D. Dori, "A Methodology for the Characterization of the Performance of Thinning Algorithms," *Proc. 2nd Int'l Conf. Document Analysis and Recognition*, IEEE CS Press, Los Alamitos, Calif., 1993, pp. 282–286.

Jang, B.K., and R.T. Chin, "Analysis of Thinning Algorithms Using Mathematical Morphology," *IEEE Trans. Pattern Analysis and Machine Intelligence*, Vol. 12, No. 6, 1990, pp. 541–551.

Jang, B.K., and R.T. Chin, "One-Pass Parallel Thinning: Analysis, Properties, and Quantitative Evaluation," *IEEE Trans. Pattern Analysis and Machine Intelligence*, Vol. 14, No. 11, 1992, pp. 1129–1140.

Jenq, J.-F., and S. Sahni, "Serial and Parallel Algorithms for the Medial Axis Transform," *IEEE Trans. Pattern Analysis and Machine Intelligence*, Vol. 14, No. 12, 1992, pp. 1218–1224.

Kumar, P., D. Bhatnagar, and P.S.U. Rao, "Pseudo One Pass Thinning Algorithm," *Pattern Recognition Letters*, Vol. 12, 1991, pp. 543–555.

Kundu, M.K., B.B. Chaudhuri, and D.D. Majumder, "A Parallel Graytone Thinning Algorithm (PGTA)," *Pattern Recognition Letters*, Vol. 12, 1991, pp. 491–496.

Lam, L., S.-W. Lee, and C.Y. Suen, "Thinning Methodologies—A Comprehensive Survey," *IEEE Trans. Pattern Analysis and Machine Intelligence*, Vol. 14, No. 9, 1992, pp. 869–885.

Lam, L., and C.Y. Suen, "Evaluation of Thinning Algorithms from a OCR Viewpoint," *Proc. 2nd Int'l Conf. Document Analysis and Recognition*, IEEE CS Press, Los Alamitos, Calif., 1993, 287–290.

Lee, S.-W., L. Lam, C.Y. Suen, "Performance Evaluation of Skeletonization Algorithms for Document Image Processing," *Proc. 1st Int'l Conf. Document Analysis and Recognition*, AFCET-IRISA/INRIA, France, 1991, pp. 260–271.

Leymarie, F., and M.D. Levine, "Simulating the Grassfire Transform Using an Active Contour Model," *IEEE Trans. Pattern Analysis and Machine Intelligence*, Vol. 14, No. 1, 1992, pp. 56–75.

Li, X., and A. Basu, "Variable-Resolution Character Thinning," *Pattern Recognition Letters*, Vol. 12, 1991, pp. 241–248.

Liow, Y.-T., "A Contour Tracing Algorithm That Preserves Common Boundaries between Regions," *CVGIP: Image Understanding*, Vol. 53, No. 3, 1991, pp. 313–321.

Lu, S.W., and H. Xu, "False Stroke Detection and Elimination for Character Recognition," *Pattern Recognition Letters*, Vol. 13, 1991, pp. 745–755.

Lunscher, W.H.H.J., and M.P. Beddoes, "Fast Binary-Image Boundary Extraction," *Computer Vision, Graphics, and Image Processing*, Vol. 38, 1987, pp. 229–257.

Mahmoud, S.A., I. AbuHaiba, and R.J. Green, "Skeletonization of Arabic Characters Using Clustering Based Skeletonization Algorithm (CBSA)," *Pattern Recognition*, Vol. 24, No. 5, 1991, pp. 453–464.

Martínez-Pérez, M.P., J. Jiménez, and J.L. Navalón, "A Thinning Algorithm Based on Contours," *Computer Vision, Graphics, and Image Processing*, Vol. 39, 1987, pp. 186–201.

Milgram, M., and T. De Saint Pierre, "Boundary Detection and Skeletonization with a Massively Parallel Architecture," *IEEE Trans. Pattern Analysis and Machine Intelligence*, Vol. 12, No. 1, 1990, pp. 74–78.

Niblack, C.W., P.B. Gibbons, and D.W. Capson, "Generating Skeletons and Centerlines from the Distance Transform," *CVGIP: Graphical Models and Image Processing*, Vol. 54, No. 5, 1992, pp. 420–437.

O'Gorman, L., "k × k Thinning," *Computer Vision, Graphics, and Image Processing*, Vol. 51, 1990, pp. 195–215.

Paglieroni, D.W., "Distance Transforms: Properties and Machine Vision Applications," *CVGIP: Graphical Models and Image Processing*, Vol. 54, No. 1, 1992, pp. 56–74.

Paglieroni, D.W., "A Unified Distance Transform Algorithm and Architecture," *Machine Vision and Applications J.*, Vol. 5, 1992, pp. 47–55.

Pal, S., and P. Bhattacharyya, "Analysis of Template Matching Thinning Algorithms," *Pattern Recognition*, Vol. 25, No. 5, 1992, pp. 497–505.

Parui, S.K., and A. Datta, "A Parallel Algorithm for Decomposition of Binary Objects Through Skeletonization," *Pattern Recognition Letters*, Vol. 12, 1991, pp. 235–240.
Plamondon, R., et al., "Validation of Preprocessing Algorithms: A Methodology and its Applications to the Design of a Thinning Algorithm for Handwritten Characters," *Proc. 2nd Int'l Conf. Document Analysis and Recognition*, IEEE CS Press, Los Alamitos, Calif., 1993, pp. 262–269.
Ragnemalm, I., "Neighborhoods for Distance Transformations Using Ordered Propagation," *CVGIP: Image Understanding*, Vol. 56, No. 3, 1992, pp. 399–409.
Ragnemalm, I., "Fast Erosion and Dilation by Contour Processing and Thresholding of Distance Maps," *Pattern Recognition Letters*, Vol. 13, 1992, pp. 161–166.
Saha, P.K., B. Chanda, and D.D. Majumder, "A Single-Scan Boundary Removal Thinning Algorithm for 2-D Binary Object," *Pattern Recognition Letters*, Vol. 14, 1993, pp. 173–179.
Sanniti di Baja, G., "O(N) Computation of Projections and Moments from the Labeled Skeleton," *Computer Vision, Graphics, and Image Processing*, Vol. 49, 1990, pp. 369–378.
Shih, C.C., and R. Kasturi, "Extraction of Graphic Primitives from Images of Paper Based Line Drawings," *Machine Vision and Applications J.*, Vol. 2, 1989, pp. 103–113.
Sinha, R.M.K., "A Width-Independent Algorithm for Character Skeleton Estimation," *Computer Vision, Graphics, and Image Processing*, Vol. 40, 1987, pp. 388–397.
Smith, R.W., "Computer Processing of Line Images: A Survey," *Pattern Recognition*, Vol. 20, 1987, pp. 7–15.
Sossa, J.H., "An Improved Parallel Algorithm for Thinning Digital Patterns," *Pattern Recognition Letters*, Vol. 10, 1989, pp. 77–80.
Sur, A., and A.K. Datta, "Single-Pass Pragmatic Algorithm for Thinning (SPAT) for Syllabic Script," *Proc. 1st Int'l Conf. Document Analysis and Recognition*, AFCET-IRISA/INRIA, France, 1991, pp. 559–584.
Tang, Y.Y., T. Li, and S.-W. Lee, "VLSI Implementation for HVRI Algorithm in Pattern Recognition," *Proc. 2nd Int'l Conf. Document Analysis and Recognition*, IEEE CS Press, Los Alamitos, Calif., 1993, pp. 460–463.
Trahanias, P.E., "Binary Shape Recognition using the Morphological Skeleton Transform," *Pattern Recognition*, Vol. 25, No. 11, 1992, pp. 1277–1286.
Williams, D.J., and M. Shah, "A Fast Algorithm for Active Contours and Curvature Estimation," *CVGIP: Image Understanding*, Vol. 55, No. 1, 1992, pp. 14–26.
Wu, A.Y., S.K Bhaskar, and A. Rosenfeld, "Computation of Geometric Properties from the Medial Axis Transform in O(n log n) Time," *Computer Vision, Graphics, and Image Processing*, Vol. 34, 1986, pp. 76–92.
Wu, R.-Y., and W.-H. Tsai, "A New One-pass Parallel Thinning Algorithm for Binary Images," *Pattern Recognition Letters*, Vol. 13, 1992, pp. 715–723.
Xia, Y., "Skeletonization Via the Realization of the Fire Front's Propagation and Extinction in Digital Binary Shapes," *IEEE Trans. Pattern Analysis and Machine Intelligence*, Vol. 11, No. 10, 1989, pp. 1076–1086.
Xu, W, and C. Wang, "CST: A Fast Thinning Algorithm Implemented on a Sequential Computer," *IEEE Trans. Systems, Man, and Cybernetics*, Vol. SMC-17, No. 5, 1987, pp. 847–851.
Yu, S.S., and W.H. Tsai, "A New Thinning Algorithm for Gray-Scale Images by the Relaxation Technique," *Pattern Recognition*, Vol. 23, 1990, pp. 1067–1076.

3. Chain Codes and Other Representations

Ali, S.M., and R.E. Burge, "A New Algorithm for Extracting the Interior of Bounded Regions Based on Chain Coding," *Computer Vision, Graphics, and Image Processing,* Vol. 43, 1988, pp. 256–264.
Cai, Z., "Restoration of Binary Images Using Contour Direction Chain Codes Description," *Computer Vision, Graphics, and Image Processing,* Vol. 41, 1988, pp. 101–106.
Kim, S.D., J.H. Lee, and J.K. Kim, "A New Chain-Coding Algorithm for Binary Images Using Run-Length Codes," *Computer Vision, Graphics, and Image Processing,* Vol. 41, 1988, pp. 114–128.
Lindenbaum, M., and J. Koplowitz, "A New Parameterization of Digital Straight Lines," *IEEE Trans. Pattern Analysis and Machine Intelligence,* Vol. 13, No. 8, 1991, pp. 847–852.
Manohar, M., P.S. Rao, and S.S. Iyengar, "Template Quadtrees for Representing Region and Line Data Present in Binary Images," *Computer Vision, Graphics, and Image Processing,* Vol. 51, 1990, pp. 338–354.

II. Feature-Level Analysis

1. Linear Approximation and Polygonalization

Aoyama, H., and M. Kawagoe, "A Piecewise Linear Approximation Method Preserving Visual Feature Points of Original Figures," *CVGIP: Graphical Models and Image Processing,* Vol. 53, No. 5, 1991, pp. 435–446.

Boxer, L., et al, "Polygonal Approximation by Boundary Reduction," *Pattern Recognition Letters,* Vol. 14, 1993, pp. 111–119.
Boxer, L., "Computing Deviations from Convexity in Polygons," *Pattern Recognition Letters,* Vol. 14, 1993, pp. 163–167.
Bhattacharya, P., and A. Rosenfeld, "Contour Codes of Isothetic Polygons," *Computer Vision, Graphics, and Image Processing,* Vol. 50, 1990, pp. 353–363.
Bimal, K.R., and S.R. Kumar, "A New Approach to Polygonal Approximation," *Pattern Recognition Letters,* Vol. 12, 1991, pp. 229–234.
Bimal, K.R., and S.R. Kumar, "An Algorithm for Polygonal Approximation of Digitized Curves," *Pattern Recognition,* Vol. 13, 1992, pp. 489–496.
Hakimi, S.L., and E.F. Schmeichel, "Fitting Polygonal Functions to a Set of Points in the Plane," *CVGIP: Graphical Models and Image Processing,* Vol. 53, No. 2, 1991, pp. 132–136.
Han, M.H., D. Jang, and J. Foster, "Identification of Cornerpoints of Two-Dimensional Images Using a Line Search Method," *Pattern Recognition,* Vol. 22, 1989, pp. 13–20.
Hemminger, T.L., and C.A. Pomalaza-Ráez, "Polygonal Representation: A Maximum Likelihood Approach," *Computer Vision, Graphics, and Image Processing,* Vol. 52, 1990, pp. 239–247.
Imai, H., and M. Iri, "Computational-Geometric Methods for Polygonal Approximations of a Curve," *Computer Vision, Graphics, and Image Processing,* Vol. 36, 1986, pp. 31–41.
Lindenbaum, M., and J. Koplowitz, "Compression of Chain Codes Using Digital Straight Line Sequences," *Pattern Recognition Letters,* Vol. 7, No. 3, 1988, pp. 167–171.
Parker, J.R., "Extracting Vectors from Raster Images," *Computer Vision, Graphics, and Image Processing,* Vol. 12, 1988, pp. 75–79.
Yu, B., et al., "Isothetic Polygon Representation for Contours," *CVGIP: Image Understanding,* Vol. 56, No. 2, 1992, pp. 264–268.

2. Critical Point Detection

Abe, K., C. Arcelli, and A. Held, "Split and Merge for a Hierarchical Contour Sketch via Dominant Point Detection," *Proc. 1st Int'l Conf. Document Analysis and Recognition,* 1991, pp. 392–400.
Abe, K., et al., "Comparison of Methods for Detecting Corner Points from Digital Curves—A Preliminary Report," *Proc. 2nd Int'l Conf. Document Analysis and Recognition,* IEEE CS Press, Los Alamitos, Calif., 1993, pp. 854–857.
Ansari, N., and E.J. Delp, "On Detecting Dominant Points," *Pattern Recognition,* Vol. 24, No. 5, 1991, pp. 441–451.
Ansari, N., and K.-W. Huang, "Non-Parametric Dominant Point Detection," *Pattern Recognition,* Vol. 24, No. 9, 1991, pp. 849–862.
Bimal, K.R., and K.S. Ray, "Detection of Significant Points and Polygonal Approximation of Digitized Curves," *Pattern Recognition Letters,* Vol. 13, 1992, pp. 443–452.
Bimal, K.R., and K.S. Ray, "An Algorithm for Detection of Dominant Points and Polygonal Approximation of Digitized Curves," *Pattern Recognition Letters,* Vol. 13, 1992, pp. 849–856.
Cheng, F.H., and W.H. Hsu, "Parallel Algorithm for Corner Finding on Digital Curves," *Pattern Recognition Letters,* Vol. 8, No. 1, 1988, pp. 47–53.
Espelid, R., and I. Jonassen, "A Comparison of Splitting Methods for the Identification of Corner-Points," *Pattern Recognition Letters,* Vol. 12, 1991, pp. 79–83.
Illing, D.P., and P.T. Fairney, "Determining Perceptually Significant Points on Noisy Boundary Curves," *Pattern Recognition Letters,* Vol. 12, 1991, pp. 557–564.
Mehrotra, R., S. Nichani, and N. Ranganathan, "Corner Detection," *Pattern Recognition,* Vol. 23, No. 11, 1990, pp. 1223–1233.
Mohan, S.K., "An Adaptive Dominamt Point Detection Algorithm for Digital Curves," *Pattern Recognition Letters,* Vol. 14, 1993, pp. 385–390.
Pan, X., and D.M. Lane, "A Parallel Method for Locating and Representing 2D Contours," *Pattern Recognition Letters,* Vol. 13, 1992, pp. 629–637.
Pavlidis, T., "A Vectorizer and Feature Extractor for Document Recognition," *Computer Vision, Graphics, and Image Processing,* Vol. 35, 1986, pp. 111–127.
Pei, S.-C., and C.-N. Lin, "The Detection of Dominant Points on Digital Curved by Scale-Space Filtering," *Pattern Recognition,* Vol. 25, No. 11, 1992, pp. 1307–1314.
Rangarajan, K., M. Shah, and D. Van Brackle, "Optimal Corner Detector," *Computer Vision, Graphics, and Image Processing,* Vol. 48, 1989, pp. 230–245.
Teh, C.H., and R.T. Chin, "On the Detection of Dominant Points on Digital Curves," *IEEE Trans. Pattern Analysis and Machine Intelligence,* Vol. 11, No. 8, 1989, pp. 859–872.

Ventura, J.A., and J.-M. Chen, "Segmentation of Two-Dimentional Curve Contours," *Pattern Recognition,* Vol. 25, No. 10, 1992, pp. 1129–1140.

Wu, W.-Y., and M.-J.J. Wang, "Detecting the Dominant Points by the Curvature Based Polygonal Approximation," *CVGIP: Graphical Models and Image Processing,* Vol. 55, No. 2, 1993, pp. 79–88.

Wuescher, D.M., and K.L. Boyer, "Robust Contour Decomposition Using a Constant Curvature Criterion," *IEEE Trans. Pattern Analysis and Machine Intelligence,* Vol. 13, No. 1, 1991, pp. 41–51.

3. Line and Curve Fitting, Hough Transform

Barry, P.J., and R.N. Goldman, "Interpolation and Approximation of Curves and Surfaces Using Pslya Polynomials," *CVGIP: Graphical Models and Image Processing,* Vol. 53, No. 2, 1991, pp. 137–148.

Bhandarkar, S.M., and M. Suk, "Qualitative Features and the Generalized Hough Transform," *Pattern Recognition,* Vol. 25, No. 9, 1992, pp. 987–1006.

Chen, L.H. and W.H.Tsai, "Moment-Preserving Curve Detection," *IEEE Trans. Systems, Man, and Cybernetics,* Vol. 18, 1988, pp. 148–158.

Connelly, S., and A. Rosenfeld, "A Pyramid Algorithm for Fast Curve Extraction," *Computer Vision, Graphics, and Image Processing,* Vol. 49, 1990, pp. 332–345.

Günther, O., and E. Wong, "The Arc Tree: An Approximation Scheme to Represent Arbitrary Curved Shapes," *Computer Vision, Graphics, and Image Processing,* Vol. 51, 1990, pp. 313–337.

Illingworth, J. and J. Kittler, "A Survey of the Hough Transform," *Computer Vision, Graphics, and Image Processing,* Vol. 44, 1988, pp. 87–116.

Howell, G.W., D.W. Fausett, and L.V. Fausett, "Quasi-Circular Splines: A Shape-Preserving Approximation," *CVGIP: Graphical Models and Image Processing,* Vol. 55, No. 2, 1993, pp. 89–97.

Kao, T.-W., et al., "A Constant Time Algorithm for Computing Hough Transform," *Pattern Recognition,* Vol. 26, No. 2, 1993, pp. 277–286.

Kass, M., A. Witkin, and D. Terzopoulos, "Snakes: Active Contour Models," *IJCV,* 1988, pp. 321–331.

Kierkegaard, P., "A Method for Detection of Circular Arcs Based on the Hough Transform," *Machine Vision and Applications J.,* Vol. 5, 1992, pp. 249–263.

Kiryati, N., and A.M. Bruckstein, "Antialiasing the Hough Transform," *CVGIP: Graphical Models and Image Processing,* Vol. 53, No. 3, 1991, pp. 213–222.

Kiryati, N., M. Lindenbaum, and A.M. Bruckstein, "Digital or Analog Hough Transform?" *Pattern Recognition Letters,* Vol. 12, 1991, pp. 291–297.

Kiryati, N., Y. Eldar, and A.M. Bruckstein, "A Probabilistic Hough Transform," *Pattern Recognition,* Vol. 24, No. 4, 1991, pp. 303–316.

Kropatsch, W.G., and H. Tockner, "Detecting the Straightness of Digital Curves in O(N) Steps," *Computer Vision, Graphics, and Image Processing,* Vol. 45, 1989, pp. 1–21.

Landraud, A.M., "Image Processing for Primitive Image Features Recognition in the Case of Discontinuous Line Images of Printed Characters," *Image and Vision Computing,* Vol. 7, 1989, pp. 225–232.

Leavers, V.F., "The Dynamic Generalized Hough Transform: Its Relationship to the Probabilistic Hough Transforms and an Application to the Concurrent Detection of Circles and Ellipses," *CVGIP: Image Understanding,* Vol. 56, No. 3, 1992, pp. 381–398.

Leung, D.N.K., L.T.S. Lam and W.C.Y. Lam, "Diagonal Quantization for the Hough Transform," *Pattern Recognition Letters,* Vol. 14, 1993, pp. 81–189.

Leung, M.K., and Y.H. Yang, "Dynamic Strip Algorithm in Curve Fitting," *Computer Vision, Graphics, and Image Processing,* Vol. 51, 1990, pp. 146–165.

Leyton, M., "Symmetry-Curvature Duality," *Computer Vision, Graphics, and Image Processing,* Vol. 38, 1987, pp. 327–341.

Li, Z.N., B. Yao, and F. Tong, "Linear Generalized Hough Transform and its Parallelization," *Image and Vision Computing,* Vol. 11, No. 1, 1993, pp. 11–24.

Lowe, D.G., "Organization of Smooth Image Curves at Multiple Scales," IJCV, Vol. 3, 1989, pp. 119–130.

Medioni, G., and Y. Yasumoto, "Corner Detection and Curve Representation Using Cubic B-Splines," *Computer Vision, Graphics, and Image Processing,* Vol. 39, 1987, pp. 167–178.

Niblack, W., and D. Petkovic, "On Improving the Accuracy of the Hough Transform," *Machine Vision and Applications J.,* Vol. 3, 1990, pp. 87–106.

Paglieroni, D.W., and A.K. Jain, "Control Point Transforms for Shape Representation and Measurement," *Computer Vision, Graphics, and Image Processing,* Vol. 42, 1988, pp. 87–111.

Pao, D.C.W., H.F. Li, and R. Jayakumar, "Shapes Recognition Using the Straight Line Hough Transform: Theory and Generalization," *IEEE Trans. Pattern Analysis and Machine Intelligence,* Vol. 14, No. 11, 1992, pp. 1076–1089.

Parent, P., and S.W. Zucker, "Trace Inference, Curvature Consistency, and Curve Detection," *IEEE Trans. Pattern Analysis and Machine Intelligence,* Vol. 11, No. 8, 1989, pp. 823–839.

Pham, B., "Quadratic B-Splines for Automatic Curve and Surface Fitting," *Computer Vision, Graphics, and Image Processing,* Vol. 13, 1989, pp. 471–475.

Potier, C., and C. Vercken, "Geometric Modelling of Digitized Curves," *Proc. 1st Int'l Conf. Document Analysis and Recognition,* 1991, pp. 152–160.

Rosin, P.L., and G.A.W. West, "Segmentation of Edges into Lines and Arcs," *Image and Vision Computing,* Vol. 7, 1989, pp. 109–114.

Stephens, R.S., "Probabilistic Approach to the Hough Transform," *Image and Vision Computing,* Vol. 9, 1991, pp. 66–71.

Thazhuthaveetil, M.J., and A.V. Shah, "Parallel Hough Transform Algorithm Performance," *Image and Vision Computing,* Vol. 9, 1991, pp. 88–92.

Thomas, A.D.H., "Compressing the Parameter Space of the Generalized Hough Transform," *Pattern Recognition Letters,* Vol. 13, 1992, pp. 107–112.

Tong, F., and Z.-N. Li, "On Improving the Accuracy of Line Extraction in Hough Space," *Int'l J. Pattern Recognition and Artificial Intelligence,* Vol. 6, No. 5, 1992, pp. 831–847.

Vinod, V.V., et al., "A Connectionist Approach for Peak Detection in Hough Space," *Pattern Recognition,* Vol. 25, No. 10, 1992, pp. 1253–1264.

Wang, L., and T. Pavlidis, "Detection of Curved and Straight Segments from Gray Scale Topography," *CVGIP: Image Understanding,* Vol. 58, No. 3, 1993, pp. 352–365.

West, G.A.W., and P.L. Rosin, "Techniques for Segmenting Image Curves into Meaningful Descriptions," *Pattern Recognition,* Vol. 24, No. 7, 1991, pp. 643–652.

Xu, L., and E. Oja, "Randomized Hough Transform (RHT): Basic Mechanisms, Algorithms, and Computational Complexities," *CVGIP: Image Understanding,* Vol. 57, No. 2, 1993, pp. 131–154.

Yip, R.K.K., P.K.S. Tam, and D.N.K. Leung, "Modification of Hough Transform for Circles and Ellipses Detection Using a 2-Dimensional Array," *Pattern Recognition,* Vol. 25, No. 9, 1992, pp. 1007–1022.

4. Shape Representation, Description, and Recognition

Bengtsson, A., and J.-O. Eklundh, "Shape Representation by Multiscale Contour Approximation," *IEEE Trans. Pattern Analysis and Machine Intelligence,* Vol. 13, No. 1, 1991, pp. 85–93.

Chetverikov D., and A. Lerch, "A Multiresolution Algorithm for Rotation-Invariant Matching of Planar Shapes," *Pattern Recognition Letters,* Vol. 13, 1992, pp. 669–676.

Dai, M., P. Baylou, and M. Najim, "An Efficient Algorithm for Computation of Shape Moments from Run-Length Codes or Chain Codes," *Pattern Recognition,* Vol. 25, No. 10, 1992, pp. 1119–1128.

De Stefano, C., F. Tortorella, and M. Vento, "Using Entropy for Drawing Reliable Templates," *Proc. 2nd Int'l Conf. Document Analysis and Recognition,* IEEE CS Press, Los Alamitos, Calif., 1993, pp. 345–348.

Flusser, J., and T. Suk, "Pattern Recognition by Affine Moment Invariants," *Pattern Recognition,* Vol. 26, No. 1, 1993, pp. 167–174.

Fu, C.-W., J.-C. Yen, and S. Chang, "Calculation of Moment Invariants via Hadamard Transform," *Pattern Recognition,* Vol. 26, No. 2, 1993, pp. 287–294.

Goshtasby, A., "Gaussian Decomposition of Two-Dimensional Shapes: A Unified Representation for CAD and Vision Applications," *Pattern Recognition,* Vol. 25, No. 5, 1992, pp. 463–472.

Grimson, W.E.L., "On the Recognition of Parameterized 2D Objects," IJCV, Vol. 3, 1988, pp. 353–372.

He, Y., and A. Kundu, "2-D Shape Classification Using Hidden Markov Model," *IEEE Trans. Pattern Analysis and Machine Intelligence,* Vol. 13, No. 11, 1991, pp. 1172–1184.

Hamada, A.H., "Structural Recognition of Disturbed Symbols Using Discrete Relaxation," *Proc. 1st Int'l Conf. Document Analysis and Recognition,* 1991, pp. 170–178.

Jiang, X.Y., and H. Bunke, "Simple and Fast Computation of Moments," *Pattern Recognition,* Vol. 24, No. 8, 1991, pp. 801–806.

Jagadish, H.V., and A.M. Bruckstein, "On Sequential Shape Descriptions," *Pattern Recognition,* Vol. 25, No. 2, 1992, pp. 165–172.

Kanaoka, T., et al., "A Higher-Order Neural Network for Distortion Invariant Pattern Recognition," *Pattern Recognition Letters,* Vol. 13, 1992, pp. 837–841.

Krzyzak, A., S.Y. Leung, and C.Y. Suen, "Reconstruction of Two-Dimensional Patterns from Fourier Descriptors," *Machine Vision and Applications J.,* Vol. 2, 1989, pp. 123–140.

Leu, J.-G., "Computing a Shape's Moments from its Boundary," *Pattern Recognition,* Vol. 24, No. 10, 1991, pp. 949–957.

Leavers, V.F., and J.F. Boyce, "The Radon Transform and its Application to Shape Parameterization in Machine Vision," *Image and Vision Computing,* Vol. 5, 1987, pp. 161–166.

Leavers, V.F., "Use of the Radon Transform as a Method of Extracting Information about Shape in Two Dimensions," *Image and Vision Computing,* Vol. 10, No. 2, 1992, pp. 99–107.

Li, Y., "Reforming the Theory of Invariant Moments for Pattern Recognition," *Pattern Recognition,* Vol. 25, No. 7, 1992, pp. 723–730.

Li, Z.C., et al., "Shape Transformation Models and Their Applications in Pattern Recognition," *Int'l J. Pattern Recognition and Artificial Intelligence,* Vol. 4, 1990, pp. 65–94.

Li, Z.C., et al., "Harmonic Models of Shape Transformations in Digital Images and Patterns," *CVGIP: Graphical Models and Image Processing,* Vol. 54, No. 3, 1992, pp. 198–209.

Li, B.-C., and J. Shen, "Fast Computation of Moment Invariants," *Pattern Recognition,* Vol. 24, No. 8, 1991, pp. 807–813.

Li, B.-C., "A New Computation of Geometric Moments," *Pattern Recognition,* Vol. 26, No. 1, 1993, pp. 109–113.

Ma, J., C.K. Wu, and X.R. Lu, "A Fast Shape Descriptor," *Computer Vision, Graphics, and Image Processing,* Vol. 34, 1986, pp. 282–291.

Marshall, S., "Review of Shape Coding Techniques," *Image and Vision Computing,* Vol. 7, 1989, pp. 281–294.

Milios, E.E., "Shape Matching Using Curvature Processes," *Computer Vision, Graphics, and Image Processing,* Vol. 47, 1989, pp. 203–226.

O'Gorman, L., and G.A. Story, "Subsampling Text Images," *Proc. 1st Int'l Conf. Document Analysis and Recognition,* 1991, pp. 219–227.

Ohtaki, Y., et al., "A Method of Automatically Compressing Fonts with High Resolution," *Proc. 1st Int'l Conf. Document Analysis and Recognition,* 1991, pp. 251–259.

Picton, P.D., "Hough Transform References," *Int'l J. Pattern Recognition and Artificial Intelligence,* Vol. 1, 1987, pp. 413–425.

Pitas, I., and A.N. Venetsanopoulos, "Morphological Shape Representation," *Pattern Recognition,* Vol. 25, No. 6, 1992, pp. 555–565.

Prokop, R.J., and A.P. Reeves, "A Survey of Moment-Based Techniques for Unoccluded Object Representation and Recognition," *CVGIP: Graphical Models and Image Processing,* Vol. 54, No. 5, 1992, pp. 438–460.

Reiss, T.H., "The Revised Fundamental Theorem of Moment Invariants," *IEEE Trans. Pattern Analysis and Machine Intelligence,* Vol. 13, No. 8, 1991, pp. 830–834.

Safaee-Rad, R., et al., "Application of Moment and Fourier Descriptors to the Accurate Estimation of Elliptical-Shape Parameters," *Pattern Recognition Letters,* Vol. 13, 1992, pp. 497–508.

Shepherd, T.S., et al., "A Method for Shift, Rotation, and Scale Invariant Pattern Recognition Using the Form and Arrangement of Pattern-Specific Features," *Pattern Recognition,* Vol. 25, No. 4, 1992, pp. 343–356.

Simon, J.C., "Invariance in Pattern Recognition: Application to Line Images," *Image and Vision Computing,* Vol. 4, No. 1, 1986, pp. 11–23.

Simon, J.C., and K. Zerhoumi, "Robust Description of a Line Image," *Proc. 1st Int'l Conf. Document Analysis and Recognition,* 1991, pp. 3–14.

Sirjani, A., and G.R. Cross, "On Representation of a Shape's Skeleton" *Pattern Recognition Letters,* Vol. 12, 1991, pp. 149–154.

Stockman, G., "Object Recognition and Localization via Pose Clustering," *Computer Vision, Graphics, and Image Processing,* Vol. 40, 1987, pp. 361–387.

Teh, C.H., and R.T. Chin, "On Digital Approximation of Moment Invariants," *Computer Vision, Graphics, and Image Processing,* Vol. 33, 1986, pp. 318–326.

Trahanias, P.E., "Binary Shape Recognition Using the Morphological Skeleton Transform," *Pattern Recognition,* Vol. 25, No. 11, 1992, pp. 1277–1288.

Tsay, Y.T., and W.H. Tsai, "Model-Guided Attributed String Matching by Split-and-Merge for Shape Recognition," *Int'l J. Pattern Recognition and Artificial Intelligence,* Vol. 3, 1989, pp. 159–179.

Xie, M., and M. Thonnat, "An Algorithm for Finding Closed Curves," *Pattern Recognition Letters,* Vol. 13, 1992, pp. 73–81.

Yuen, H.K., J. Illingworth, and J. Kittler, "Detecting Partially Occluded Ellipses Using the Hough Transform," *Image and Vision Computing,* Vol. 7, 1989, pp. 31–37.

Yuen, H.K., et al., "Comparative Study of Hough Transform Methods for Circle Finding," *Image and Vision Computing,* Vol. 8, 1990, pp. 71–77.

Zetzsche, C., and T. Caelli, "Invariant Pattern Recognition Using Multiple Filter Image Representations," *Computer Vision, Graphics, and Image Processing,* Vol. 45, 1989, pp. 251–262.

III. Text Processing

1. Layout Analysis and Segmentation

Akiyama, T., and N. Hagita, "Automated Entry System for Printed Documents," *Pattern Recognition,* Vol. 23, No. 11, 1990, pp. 1141–1154.

Akindele, O.T., and A. Belaid, "Page Segmentation by Segment Tracing," *Proc. 2nd Int'l Conf. Document Analysis and Recognition,* IEEE CS Press, Los Alamitos, Calif., 1993, pp. 341–344.

Amamoto, N., S. Torigoe, and Y. Hirogaki, "Block Segmentation and Text Area Extraction of Vertically/Horizontally Written Document," *Proc. 2nd Int'l Conf. Document Analysis and Recognition,* IEEE CS Press, Los Alamitos, Calif., 1993, pp. 739–742.

Aubert, M., A. Chehikian, and L. Delaporte, "Mixed Image/Text Office Documents Processing," *Proc. 1st Int'l Conf. Document Analysis and Recognition,* 1991, pp. 192–200.

Bayer, T.A., "Understanding Structured Text Documents by a Model Based Document Analysis System," *Proc. 2nd Int'l Conf. Document Analysis and Recognition,* IEEE CS Press, Los Alamitos, Calif., 1993, pp. 448–453.

Belaod, A., and O.T. Akindele, "A Labeling Approach for Mixed Document Blocks," *Proc. 2nd Int'l Conf. Document Analysis and Recognition,* IEEE CS Press, Los Alamitos, Calif., 1993, pp. 749–752.

Bloomberg, D.S., "Multiresolution Morphological Approach to Document Image Analysis," *Proc. 1st Int'l Conf. Document Analysis and Recognition,* 1991, pp. 963–971.

Campigli, P., et al., "A Reading System for Printed Documents," *Proc. 1st Int'l Conf. Document Analysis and Recognition,* 1991, pp. 585–593.

Casey, R.G., and G. Nagy, "Document Analysis—A Broader View," *Proc. 1st Int'l Conf. Document Analysis and Recognition,* 1991, pp. 839–849.

Chenevoy, Y., and A. Belaod, "Hypothesis Management for Structured Document Recognition," *Proc. 1st Int'l Conf. Document Analysis and Recognition,* 1991, pp. 121–129.

Conway, A., "Page Grammars and Page Parsing—A Syntactic Approach to Document Layout Recognition," *Proc. 2nd Int'l Conf. Document Analysis and Recognition,* IEEE CS Press, Los Alamitos, Calif., 1993, pp. 761–764.

Cullen, J.F., and K. Ejiri, "Weak Model-Dependent Page Segmentation and Skew Correction for Processing Document Images," *Proc. 2nd Int'l Conf. Document Analysis and Recognition,* IEEE CS Press, Los Alamitos, Calif., 1993, pp. 757–760.

Dengel, A., "Initial Learning of Document Structure," *Proc. 2nd Int'l Conf. Document Analysis and Recognition,* IEEE CS Press, Los Alamitos, Calif., 1993, pp. 86–90.

Dengel, A., and G. Barth, "High Level Document Analysis Guided by Geometric Aspects," *Int'l J. Pattern Recognition and Artificial Intelligence,* Vol. 2, 1988, pp. 641–655.

Derrien-Peden, D., "Frame-Based System for Macro-Typographical Structure Analysis in Scientific Papers," *Proc. 1st Int'l Conf. Document Analysis and Recognition,* 1991, pp. 311–319.

Doermann, D.S., and R. Furuta, "Image Based Typographic Analysis of Documents," *Proc. 2nd Int'l Conf. Document Analysis and Recognition,* IEEE CS Press, Los Alamitos, Calif., 1993, pp. 769–773.

Downton, A.C., and C.G. Leedham, "Preprocessing and Presorting of Envelope Images for Automatic Sorting Using OCR," *Pattern Recognition,* Vol. 23, 1990, pp. 347–362.

Enguehard, C., P. Trigano, and P. Malvache, "ANA: Automatic Natural Acquisition," *Proc. 1st Int'l Conf. Document Analysis and Recognition,* 1991, pp. 375–383.

Fletcher, L. A., and R. Kasturi, "A Robust Algorithm for Text String Separation from Mixed Text/Graphics Images," *IEEE Trans. Pattern Analysis and Machine Intelligence,* Vol. 10, No. 6, 1988, pp. 910–918.

Fisher, J.L., "Logical Structure Descriptions of Segmented Document Images," *Proc. 1st Int'l Conf. Document Analysis and Recognition,* 1991, pp. 302–310.

Forchhammer, S., and K.S. Jensen, "Parallel Processing of Scanned Documents Containing Halftones," *Proc. 1st Int'l Conf. Document Analysis and Recognition,* 1991, pp. 201–209.

Hao, X., J.L.T. Wang, and P.A. Ng, "Nested Segmentation: An Approach for Layout Analysis in Document Classification," *Proc. 2nd Int'l Conf. Document Analysis and Recognition,* IEEE CS Press, Los Alamitos, Calif., 1993, pp. 319–322.

Hermann, P., and G. Schlagetar, "Retrieval of Document Images Using Layout Knowledge," *Proc. 2nd Int'l Conf. Document Analysis and Recognition,* IEEE CS Press, Los Alamitos, Calif., 1993, pp. 537–540.

Hiarayama, Y., "A Block Segmentation Method for Document Images With Complicated Column Structures," *Proc. 2nd Int'l Conf. Document Analysis and Recognition,* IEEE CS Press, Los Alamitos, Calif., 1993, pp. 91–94.

Hönes F., and J. Lichter, "Text String Extraction Within Mixed-Mode Manuscript," *Proc. 2nd Int'l Conf. Document Analysis and Recognition,* IEEE CS Press, Los Alamitos, Calif., 1993, pp. 655–659.

Hönes, F., et al., "A Hybrid Approach for Document Image Segementation and Encoding," *Proc. 1st Int'l Conf. Document Analysis and Recognition,* 1991, pp. 444–453.

Imade, S., S. Tatsuta, and T. Wada, "Segmentation and Classification for Mixed Text/Image Documents Using Neural Network," *Proc. 2nd Int'l Conf. Document Analysis and Recognition,* IEEE CS Press, Los Alamitos, Calif., 1993, pp. 930–934.

Ingold, R., and D. Armangil, "A Top-Down Document Analysis Method for Logical Structure Recognition," *Proc. 1st Int'l Conf. Document Analysis and Recognition,* 1991, pp. 41–49.

Itonori, K., "Table Structure Recognition Based on Textblock Arragement and Ruled Line Position," *Proc. 2nd Int'l Conf. Document Analysis and Recognition,* IEEE CS Press, Los Alamitos, Calif., 1993, pp. 765–768.

Ittner, D.J., and H.S. Baird, "Language-Free Layout Analysis," *Proc. 2nd Int'l Conf. Document Analysis and Recognition,* IEEE CS Press, Los Alamitos, Calif., 1993, pp. 336–340.

Iwane, K., M. Yamaoka, and O. Ikawi, "A Functional Classification Approach to Layout Analysis of Document Images," *Proc. 2nd Int'l Conf. Document Analysis and Recognition,* IEEE CS Press, Los Alamitos, Calif., 1993, pp. 778–781.

Iwata, K., and G. Marcu, "A Color Classification Algorithm," *Proc. 2nd Int'l Conf. Document Analysis and Recognition,* IEEE CS Press, Los Alamitos, Calif., 1993, pp. 726–729.

Jain, A.K., and S.K. Bhattacharjee, "Text Segmentation Using Gabor Filters for Automatic Document Processing," *Machine Vision and Applications J.,* Vol. 5, No. 3, 1992, pp. 169–184.

Jain, A.K., and S.K. Bhattacharjee, "Address Block Location on Envelopes Using Gabor Filters," *Pattern Recognition,* Vol. 25, No. 12, 1992, pp. 1459–1477.

Joseph, S.H., "On the Extraction of Text Connected to Linework in Document Images," *Proc. 1st Int'l Conf. Document Analysis and Recognition,* 1991, pp. 993–999.

Kamel M., and A. Zhao, "Extraction of Binary Character/Graphics Images from Grayscale Document Images," *CVGIP: Graphical Models and Image Processing,* Vol. 55, No. 3, 1993, pp. 203–217.

Kelly, D.P.M., and D.M. Abrahamson, "Document Structure Recognition," *Proc. 1st Int'l Conf. Document Analysis and Recognition,* 1991, pp. 525–532.

Kerpedjiev, S.M., "Automatic Extraction of Information Structures from Documents," *Proc. 1st Int'l Conf. Document Analysis and Recognition,* 1991, pp. 32–40.

Kise, K., et al., "Incremental Acquisition of Knowledge About Layout Structures From Example of Documents," *Proc. 2nd Int'l Conf. Document Analysis and Recognition,* IEEE CS Press, Los Alamitos, Calif., 1993, pp. 668–671.

Kreich, J., A. Luhn, and G. Maderlechner, "An Experimental Environment for Model Based Document Analysis," *Proc. 1st Int'l Conf. Document Analysis and Recognition,* 1991, pp. 50–58.

Krishnamoorthy, M., et al., "Syntactic Segmentation and Labeling of Digitized Pages from Technical Journals," *IEEE Trans. Pattern Analysis and Machine Intelligence,* Vol. 15, No. 7, 1993, pp. 737–747.

Lam, S.W., and S.N. Srihari, "Multi-Domain Document Layout Understanding," *Proc. 1st Int'l Conf. Document Analysis and Recognition,* 1991, pp. 112–120.

Lefevre, P., and Y. Pedron, "Document Segmentation Software Implemented on a Transputer Network," *Proc. 1st Int'l Conf. Document Analysis and Recognition,* 1991, pp. 975–983.

Lii, J., P.W. Palumbo, and S.N. Srihari, "Address Block Location Using Character Recognition and Address Syntax," *Proc. 2nd Int'l Conf. Document Analysis and Recognition,* IEEE CS Press, Los Alamitos, Calif., 1993, pp. 330–335.

Lin, Y.-S., and J.L. Wang, "Characters Extraction in Chinese Documents," *Proc. 1st Int'l Conf. Document Analysis and Recognition,* 1991, pp. 934–942.

Machii, K., H. Fukushima, and M. Nakagawa, "On-Line Text/Drawings Segmentation of Handwritten Patterns," *Proc. 2nd Int'l Conf. Document Analysis and Recognition,* IEEE CS Press, Los Alamitos, Calif., 1993, pp. 710–713.

Mojahid, M., and J. Virbel, "Towards a Cognitive Approach of Control Strategies for Document Structures Recognition," *Proc. 1st Int'l Conf. Document Analysis and Recognition,* 1991, pp. 683–691.

Nagy, G. and M. Viswanathan, "Dual Representation of Segmented Technical Documents," *Proc. 1st Int'l Conf. Document Analysis and Recognition,* 1991, pp. 141–151.

O'Gorman, L., "The Document Spectrum for Page Layout Analysis," *IEEE Trans. Pattern Analysis and Machine Intelligence,* Vol. 15, No. 11, 1993, pp. 1162–1173.

Okamoto, M., and M. Takahashi, "A Hybrid Page Segmentation Method," *Proc. 2nd Int'l Conf. Document Analysis and Recognition,* IEEE CS Press, Los Alamitos, Calif., 1993, pp. 743–748.

Pavlidis, T. and J. Zhou, "Page Segmentation by White Streams," *Proc. 1st Int'l Conf. Document Analysis and Recognition,* 1991, pp. 945–953.

Pavlidis, T., and J. Zhou, "Page Segmentation and Classification," *CVGIP: Graphical Models and Image Processing,* Vol. 54, No. 6, 1992, pp. 484–496.

Saitoh, T., M. Tachikawa, and T. Yamaai, "Document Image Segmentation and Text Area Ordering," *Proc. 2nd Int'l Conf. Document Analysis and Recognition,* IEEE CS Press, Los Alamitos, Calif., 1993, pp. 323–329.

Shapiro, V., G. Gluchev, and V. Sgurev, "Handwritten Document Image Segmentation and Analysis," *Pattern Recognition Letters,* Vol. 14, No. 1, 1993, pp. 71–78.
Spitz, A.L., "Style Directed Document Recognition," *Proc. 1st Int'l Conf. Document Analysis and Recognition,* 1991, pp. 611–619.
Srihari, S.N., and V. Govindaraju, "Analysis of Textual Images Using the Hough Transform," *Machine Vision and Applications J.,* Vol. 2, 1989, pp. 141–153.
Takizawa, K., et al., "Extraction of Character Strings from Unformed Document Images," *Proc. 2nd Int'l Conf. Document Analysis and Recognition,* IEEE CS Press, Los Alamitos, Calif., 1993, pp. 660–663.
Tang, Y.Y., and C.Y. Suen, "Document Structures: A Survey," *Proc. 2nd Int'l Conf. Document Analysis and Recognition,* IEEE CS Press, Los Alamitos, Calif., 1993, pp. 99–102.
Tang, Y.Y., et al., "Document Analysis and Understanding: A Brief Survey," *Proc. 1st Int'l Conf. Document Analysis and Recognition,* 1991, pp. 17–31.
Taxt, T., P.J. Flynn, and A.K. Jain, "Segmentation of Document Images," *IEEE Trans. Pattern Analysis and Machine Intelligence,* Vol. 11, 1989, pp. 1322–1329.
Tombre, K., and P. Vaxivière, "Structure, Syntax and Semantics in Technical Document Recognition," *Proc. 1st Int'l Conf. Document Analysis and Recognition,* 1991, pp. 61–69.
Trigano, P., et al., "Acquisition of Words in DOCAL, a Document Understanding System," *Proc. 1st Int'l Conf. Document Analysis and Recognition,* 1991, pp. 384–391.
Wang, C.H., and S.N. Srihari, "A Framework for Object Recognition in a Visually Complex Environment and its Application to Locating Address Blocks on Mail Pieces," *IJCV,* Vol. 2, 1988, pp. 125–151.
Wang, D., and S.N. Srihari, "Classification of Newspaper Image Blocks Using Texture Analysis," *Computer Vision, Graphics, and Image Processing,* Vol. 47, 1989, pp. 327–352.
Watanabe, T., Q. Luo, and N. Sugie, "Structure Recognition Methods for Various Types of Documents," *Machine Vision and Applications J.,* Vol. 6, Nos. 2–3, 1993, pp. 163–176.
Wieser, J., and A. Pinz, "Layout Analysis: Finding Text, Titles, and Photos in Digital Images of Newspaper Pages," *Proc. 2nd Int'l Conf. Document Analysis and Recognition,* IEEE CS Press, Los Alamitos, Calif., 1993, pp. 774–777.
Yamashita, A. et al., "A Model Based Layout Understanding Method for the Document Recognition System," *Proc. 1st Int'l Conf. Document Analysis and Recognition,* 1991, pp. 130–138.
Yeh, P.S., et al., "Address Location on Envelopes," *Pattern Recognition,* Vol. 20, 1987, pp. 213–227.

2. Character Recognition

i. Survey Papers

Baird, H.S., "Recognition Technology Frontiers," *Pattern Recognition Letters,* Vol. 14, 1993, pp. 327–334.
Cheng, F.H., and W.H. Hsu, "Research on Chinese OCR in Taiwan," *Int'l J. Pattern Recognition and Artificial Intelligence,* Vol. 5, 1991, pp. 139–164.
Elliman, D.G., and I.T. Lancaster, "A Review of Segmentation and Contextual Analysis Techniques for Text Recognition," *Pattern Recognition,* Vol. 23, 1990, pp. 337–346.
Davis, R.H., and J. Lyall, "Recognition of Handwritten Characters—A Review," *Image and Vision Computing,* Vol. 4, 1986, pp. 208–218.
Gilloux, M., "Research Into the New Generation of Character and Mailing Address Recognition Systems at the French Post Office Research Center," *Pattern Recognition Letters,* Vol. 14, No. 4, 1993, pp. 276–276.
Govindan, V.K., and A.P. Shivaprasad, "Character Recognition—A Review," *Pattern Recognition,* Vol. 23, 1990, pp. 671–683.
Impedovo, S., L. Ottaviano, and S. Occhinegro, "Optical Character Recognition—A Survey," *Int'l J. Pattern Recognition and Artificial Intelligence,* Vol. 5, Nos. 1&2, 1991, pp. 1–24.
Mantas, J., "An Overview of Character Recognition Methodologies," *Pattern Recognition,* Vol. 19, 1986, pp. 425–430.
Matsui, T., et al., "State of the Art of Handwritten Numeral Recognition in Japan—The Results of the First IPTP Character Recognition Competition," *Proc. 2nd Int'l Conf. Document Analysis and Recognition,* IEEE CS Press, Los Alamitos, Calif., 1993, pp. 391–396.
Mori, S., C.Y. Suen, and K. Yamamoto, "Historical Review of OCR Reasearch and Development," *Proc. IEEE,* Vol. 80, No. 7, IEEE Service Center, Piscataway, N.J., 1992, pp. 1029–1058.
Nagy, G., "At the Frontiers of OCR," *Proc. IEEE,* Vol. 80, No. 7, IEEE Service Center, Piscataway, N.J., 1992, pp. 1093–1100.
Nouboud, F., and R. Plamondon, "On-Line Recognition of Handprinted Characters: Survey and Beta Tests," *Pattern Recognition,* Vol. 23, 1990, pp. 1031–1044.

Tai, J.W., "Some Research Achivements on Chinese Character Recognition in China," *Int'l J. Pattern Recognition and Artificial Intelligence,* Vol. 5, 1991, pp. 199–206.

ii. Printed Character Recognition

Akiyama, T., and N. Hagita, "Automated Entry System for Printed Documents," *Pattern Recognition,* Vol. 23, 1990, pp. 141–1154.

Anigbogu, J.C., and A. Belaïd, "Application of Hidden Markov Models to Multifont Text Recognition," *Proc. 1st Int'l Conf. Document Analysis and Recognition,* 1991, pp. 785–793.

Backmutsky, V., and V. Zmudikov, "Some Ergonomic Improvements of Text Error Detection and Prevention in DTP-Systems," *Proc. 2nd Int'l Conf. Document Analysis and Recognition,* IEEE CS Press, Los Alamitos, Calif., 1993, pp. 947–950.

Bai, G., "Multifont Chinese Character Recognition Using Side-Stroke-End Feature," *Proc. 2nd Int'l Conf. Document Analysis and Recognition,* IEEE CS Press, Los Alamitos, Calif., 1993, pp. 794–797.

Baird, H.S., "Feature Identification for Hybrid Structural/Statistical Pattern Classification," *Computer Vision, Graphics, and Image Processing,* Vol. 42, 1988, pp. 318–33.

Baird, H.S. and R. Fossey, "A 100–Font Classifier," *Proc. 1st Int'l Conf. Document Analysis and Recognition,* 1991, pp. 332–340.

Bayer, T.A. and U.H.G. Kressel, "Cut Classification for Segmentation," *Proc. 2nd Int'l Conf. Document Analysis and Recognition,* IEEE CS Press, Los Alamitos, Calif., 1993, pp. 565–568.

Bokser, M., "Omnidocument Technologies," *Proc. IEEE,* Vol. 80, No. 7, IEEE Service Center, Piscataway, N.J., 1992, pp. 1066–1078.

Bradford, R. and T. Nartker, "Error Correlation in Contemporary OCR Systems," *Proc. 1st Int'l Conf. Document Analysis and Recognition,* 1991, pp. 516–524.

Bunke, H., "A Fast Algorithm for Finding the Nearest Neighbor of a Word in a Dictionary," *Proc. 2nd Int'l Conf. Document Analysis and Recognition,* IEEE CS Press, Los Alamitos, Calif., 1993, pp. 632–637.

Caesar, T., J.M. Gloger, and E. Mandler, "Utilization of Large Disordered Sample Sets for Classifier Adaptation in Complex Domains," *Proc. 2nd Int'l Conf. Document Analysis and Recognition,* IEEE CS Press, Los Alamitos, Calif., 1993, pp. 790–793.

Cash, G.L., and M. Hatamian, "Optical Character Recognition by the Method of Moments," *Computer Vision, Graphics, and Image Processing,* Vol. 39, 1987, pp. 29–310.

Carter N.P., "Segmentation and Preliminary Recognition of Madrigals Notated in White Measural Notation," *Machine Vision and Applications J.,* Vol. 5, 1992, pp. 223–230.

Casey, R., et al, "Optical Recognition of Chemical Graphics," *Proc. 2nd Int'l Conf. Document Analysis and Recognition,* IEEE CS Press, Los Alamitos, Calif., 1993, pp. 627–631.

Chang, F., et al., "Stroke Segmentation as a Basis for Structural Matching of Chinese Characters," *Proc. 2nd Int'l Conf. Document Analysis and Recognition,* IEEE CS Press, Los Alamitos, Calif., 1993, pp. 35–40.

Chen, C.H., and J.L. DeCurtins, "Word Recognition in a Segmentation-Free Approach to OCR," *Proc. 2nd Int'l Conf. Document Analysis and Recognition,* IEEE CS Press, Los Alamitos, Calif., 1993, pp. 573–576.

Chen, F.R., L.D. Wilcox, and D.S. Bloomberg, "Detecting and Locating Partially Specified Keywords in Scanned Images Using Hidden Markov Models," *Proc. 2nd Int'l Conf. Document Analysis and Recognition,* IEEE CS Press, Los Alamitos, Calif., 1993, pp. 133–138.

Chen, P.N., Y.S. Chen, and W.H. Hsu, "Stroke Relation Coding-A New Approach to the Recognition of Multi-Font Printed Chinese Characters," *Int'l J. Pattern Recognition and Artificial Intelligence,* Vol. 2, 1988, pp. 149–160.

Cho, S.-B., and J.H. Kim, "Recognition of Large-Set Printed Hangul (Korean Script) by Two-Stage Backpropagation Neural Classifier," *Pattern Recognition,* Vol. 25, No. 11, 1992, pp. 1353–1360.

Cho, S.,-B. and J.H. Kim, "A Two-Stage Classification Scheme with Backpropagation Neural Network Classifiers," *Pattern Recognition Letters,* Vol. 13, 1992, pp. 309–313.

Crowner, C., and J.J. Hull, "A Hierarchical Pattern Matching Parser and its Application to Word Shape" *Proc. 1st Int'l Conf. Document Analysis and Recognition,* 1991, pp. 323–331.

d'Acierno, A., C. De Stefano, and M. Vento, "A Structural Character Recognition Method Using Neural Networks," *Proc. 1st Int'l Conf. Document Analysis and Recognition,* 1991, pp. 803–810.

De Luca, P.G., and A. Gisotti, "Printed Character Preclassification Based on Word Structure," *Pattern Recognition,* Vol. 24, No. 7, 1991, pp. 609–615.

El-Dabi, S.S., R. Ramsis, and A. Kamel, "Arabic Character Recognition System: A Statistical Approach for Recognizing Cursive Typewritten Text," *Pattern Recognition,* Vol. 2, 1990, pp. 485–495.

El-Khaly, F., and M.A. Sid-Ahmed, "Machine Recognition of Optically Captured Machine Printed Arabic Text," *Pattern Recognition,* Vol. 23, No. 11, 1990, pp. 1207–1214.

Fogelman, F., E. Viennet, and B. Lamy, "Multi-Modular Neural Network Architectures: Applications in Optical Character and Human Face Recognition," *Int'l J. Pattern Recognition and Artificial Intelligence,* Vol. 7, No. 4, 1993, pp. 721–756.

Gan, K.W., and K.T. Lua, "A New Approach to Stroke and Feature Point Extraction in Chinese Character Recognition," *Pattern Recognition Letters,* Vol. 12, 1991, pp. 381–387.

Goshtasby, A., and R.W. Ehrich, "Contextual Word Recognition Using Probabilistic Relaxation Labeling," *Pattern Recognition,* Vol. 21, 1988, pp. 455–462.

Gosselin, B., "Boolean Neural Network Applied to the Recognition of Typographic Characters," *Proc. 1st Int'l Conf. Document Analysis and Recognition,* 1991, pp. 426–434.

Guyon, I., "Applications of Neural Networks to Character Recognition," *Int'l J. Pattern Recognition and Artificial Intelligence,* Vol. 5, 1991, pp. 353–382.

Ho, T.K., and H.S. Baird, "Perfect Metrics," *Proc. 2nd Int'l Conf. Document Analysis and Recognition,* IEEE CS Press, Los Alamitos, Calif., 1993, pp. 593–597.

Ho, T.K., J.J. Hull, and S.N. Srihari, "Word Recognition With Multi-Level Contextual Knowledge," *Proc. 1st Int'l Conf. Document Analysis and Recognition,* 1991, pp. 905–915.

Ho, T.K., J.J. Hull, and S.N. Srihari, "A Computational Model for Recognition of Multifont Word Images," *Machine Vision and Applications J.,* Vol. 5, No. 3, 1992, pp. 157–168.

Ho, T. K., J.J. Hull and S.N. Srihari, "A Word Shape Analysis Approach to Lexicon Based Word Recognition," *Pattern Recognition Letters,* Vol. 13, 1992, pp. 821–826.

Hoch, R., and T. Kieninger, "On Virtual Partitioning of Large Dictionaries for Contextual Postprocessing to Improve Character Recognition," *Proc. 2nd Int'l Conf. Document Analysis and Recognition,* IEEE CS Press, Los Alamitos, Calif., 1993, pp. 226–231.

Hong, T., and J.J. Hull, "Text Recognition Enhancement with a Probabilistic Lattice Chart Parser," *Proc. 2nd Int'l Conf. Document Analysis and Recognition,* IEEE CS Press, Los Alamitos, Calif., 1993, pp. 222–225.

Hong, Y.-S., and S.-S. Chen, "Character Recognition in a Sparse Distributed Memory," *IEEE Trans. Systems, Man, and Cybernetics,* Vol. 21, No. 3, 1991, pp. 674–678.

Horiuchi, T. et al., "Stamped Character Recognition Method Using Range Image," *Proc. 2nd Int'l Conf. Document Analysis and Recognition,* IEEE CS Press, Los Alamitos, Calif., 1993, pp. 782–785.

Huang, J.S., and P.M. Huang, "Machine-Printed Chinese Character Recognition Based on Linear Regression," *Int'l J. Pattern Recognition and Artificial Intelligence,* Vol. 5, 1991, pp. 165–173.

Huang, X., J. Gu, and Y. Wu, "A Constrained Approach to Multifont Chinese Character Recognition," *IEEE Trans. Pattern Analysis and Machine Intelligence,* Vol. 15, No. 8, 1993, pp. 838–end.

Hull, J.J., and Y. Li, "Interpreting Word Recognition Decisions with a Document Database Graph," *Proc. 2nd Int'l Conf. Document Analysis and Recognition,* IEEE CS Press, Los Alamitos, Calif., 1993, pp. 488–492.

Jiang, J., W. Kim, and H. Tominaga, "Recognition and Representation," *Proc. 2nd Int'l Conf. Document Analysis and Recognition,* IEEE CS Press, Los Alamitos, Calif., 1993, pp. 959–962.

Ju, R.H., I.C. Jou, and M.K. Tsay, "Zooming Techniques on Digital Chinese Character Patterns: A Further Study and Improvement," *Image and Vision Computing,* Vol. 9, 1991, pp. 194–200.

Kahan, S., T. Pavlidis, and H.S. Baird, "On the Recognition of Printed Characters of any Font and Size," *IEEE Trans. Pattern Analysis and Machine Intelligence,* Vol. PAMI-9, No. 2, 1987, pp. 274–288.

Kigo, K., "Improving Speed of Japanese OCR Through Linguistic Preprocessing," *Proc. 2nd Int'l Conf. Document Analysis and Recognition,* IEEE CS Press, Los Alamitos, Calif., 1993, pp. 214–217.

Konno A., and Y. Hongo, "Postprocessing Algorithm Based on the Probabilistic and Semantic Method for Japanese OCR," *Proc. 2nd Int'l Conf. Document Analysis and Recognition,* IEEE CS Press, Los Alamitos, Calif., 1993, pp. 646–649.

Kundu, A., and Y. He, "On Optimal Order in Modeling Sequence of Letters in Words of Common Language as a Markov Chain," *Pattern Recognition,* Vol. 24, 1991, pp. 603–608.

Lam, S.W., and W.K. Hui, "Reading Constrained Text Using Hierarchical Hidden Markov Models," *Proc. 2nd Int'l Conf. Document Analysis and Recognition,* IEEE CS Press, Los Alamitos, Calif., 1993, pp. 151–154.

Lambert, G., "A Projection Reducing the Constraints of Direction," *Proc. 2nd Int'l Conf. Document Analysis and Recognition,* IEEE CS Press, Los Alamitos, Calif., 1993, pp. 557–560.

Lebourgeois, F., and J.L. Henry, "A Contextual Processing for an OCR System, Based on Pattern Learning," *Proc. 2nd Int'l Conf. Document Analysis and Recognition,* IEEE CS Press, Los Alamitos, Calif., 1993, pp. 861–865.

Lee, H.J., C.H. Tung, and C.H.C. Chien, "A Markov Language Model in Chinese Text Recognition," *Proc. 2nd Int'l Conf. Document Analysis and Recognition,* IEEE CS Press, Los Alamitos, Calif., 1993, pp. 72–75.

Lee, H.J., and M.C. Lee, "Understanding Mathematical Expressions in a Printed Document," *Proc. 2nd Int'l Conf. Document Analysis and Recognition,* IEEE CS Press, Los Alamitos, Calif., 1993, pp. 502–505.

Lee, K.H., K.-B. Eom, and R.L. Kashyap, "Character Recognition Based on Attribute-Dependent Programmed Grammar," *IEEE Trans. Pattern Analysis and Machine Intelligence,* Vol. 14, No. 11, 19??, pp. 1122–1128.

Li, R.Y., and M. Xu, "Character Recognition using a Fast Neural-net Classifier," *Pattern Recognition Letters,* Vol. 13, 1992, pp. 369–374.

Liang S., M. Ahmadi, and M. Shridhar, "Segmentation of touching Characters in Printed Document Recognition," *Proc. 2nd Int'l Conf. Document Analysis and Recognition,* IEEE CS Press, Los Alamitos, Calif., 1993, pp. 569–572.

Lopresti, D.P., and J.S. Sandberg, "Certifiable Optical Character Recognition," *Proc. 2nd Int'l Conf. Document Analysis and Recognition,* IEEE CS Press, Los Alamitos, Calif., 1993, pp. 432–435.

Lu, Y., "On the Segmentation of Touching Characters," *Proc. 2nd Int'l Conf. Document Analysis and Recognition,* IEEE CS Press, Los Alamitos, Calif., 1993, pp. 440–443.

Lua, K.T. and K.W. Gan, " Recognizing Chinese Characters Through Interactive Activation and Competition," *Pattern Recognition,* Vol. 23, No. 12, 1990, pp. 1311–1321.

Mandler, E., "AdlatiX—A Computer-Trainable OCR-System," *Proc. 1st Int'l Conf. Document Analysis and Recognition,* 1991, pp. 341–349.

Miyahara, K., and F. Yoda, "Printed Japanese Character Recognition Based on Multiple Modified LVQ Neural Network," *Proc. 2nd Int'l Conf. Document Analysis and Recognition,* IEEE CS Press, Los Alamitos, Calif., 1993, pp. 250–253.

Nagy, G., S. Seth, and K. Einspahr, "Decoding Substitution Ciphers by Means of Word Matching With Application to OCR," *IEEE Trans. Pattern Analysis and Machine Intelligence,* Vol. PAMI-9, No. 5, 1987, pp. 710–715.

Nagahashi, H., and M. Nakatsuyama, "A Pattern Description and Generation Method of Structural Characters," *IEEE Trans. Pattern Analysis and Machine Intelligence,* Vol. PAMI-8, No. 1, 1986, pp. 112–118.

Nakayama, T., and A.L. Spitz, "European Language Determination from Image," *Proc. 2nd Int'l Conf. Document Analysis and Recognition,* IEEE CS Press, Los Alamitos, Calif., 1993, pp. 159–162.

Nartker, T.A., S.V. Rice, and J. Kanai, "OCR Accuracy: UNLV's 2nd Annual Test," *Inform magazine,* Association for Information and Image Management, Silver Spring, Md., Jan. 1994, pp. 40–45.

O'Hair, M.A., and M. Kabrisky, "Recognizing Whole Words as Single Symbols," *Proc. 1st Int'l Conf. Document Analysis and Recognition,* 1991, pp. 350–358.

Okamoto, M., and B. Miao, "Recognition of Mathematical Expressions by Using the Layout Structures of Symbols," *Proc. 1st Int'l Conf. Document Analysis and Recognition,* 1991, pp. 242–250.

Oommen, B.J., and J.R. Zgierski, "Breaking Substitution Cyphers Using Stochastic Automata," *IEEE Trans. Pattern Analysis and Machine Intelligence,* Vol. 15, No. 2, 1993, pp. 185–192.

Ozawa, H., and T. Nakagawa, "A Character Image Enhancement Method with Various Background Images," *Proc. 2nd Int'l Conf. Document Analysis and Recognition,* IEEE CS Press, Los Alamitos, Calif., 1993, pp. 58–61.

Pai, H.F., and H.C. Wang, "A Two-Dimensional Cepstrum Approach for the Recognition of Mandarin Syllable Initials," *Pattern Recognition,* Vol. 26, No. 4, 1993, pp. 569–578.

Pavlidis, T., "Recognition of Printed Text under Realistic Conditions," *Pattern Recognition Letters,* Vol. 14, 1993, pp. 317–326.

Ramesh, S.R., "A Generalized Character Recognition Algorithm: A Graphical Approach," *Pattern Recognition,* Vol. 22, 1989, pp. 347–350.

Rocha J., and T. Pavlidis, "A Solution to the Problem of Touching Characters," *Proc. 2nd Int'l Conf. Document Analysis and Recognition,* IEEE CS Press, Los Alamitos, Calif., 1993, pp. 602–605.

Saiga, H. et al., "An OCR System for Business Cards," *Proc. 2nd Int'l Conf. Document Analysis and Recognition,* IEEE CS Press, Los Alamitos, Calif., 1993, pp. 802–805.

Sakoda, B., J. Zhou, and T. Pavlidis, "Refinement and Testing of a Character Recognition System Based on Feature Extraction in Grayscale Space," *Proc. 2nd Int'l Conf. Document Analysis and Recognition,* IEEE CS Press, Los Alamitos, Calif., 1993, pp. 464–469.

Shlien, S., "Multifont Character Recognition for Typeset Documents," *Int'l J. Pattern Recognition and Artificial Intelligence,* Vol. 2, 1988, pp. 603–620.

Sinha, R.M.K., "Rule Based Contextual Post-Processing for Devanagari Text Recognition," *Pattern Recognition,* Vol. 20, 1987, pp. 475–485.

Sinha, R.M.K., "On Using Syntactic Constraints in Text Recognition," *Proc. 2nd Int'l Conf. Document Analysis and Recognition,* IEEE CS Press, Los Alamitos, Calif., 1993, pp. 858–861.

Sinha, R.M.K., and B. Prasada, "Visual Text Recognition Through Contextual Processing," *Pattern Recognition,* Vol. 21, 1988, pp. 463–479.

Sinha, R, et al., "Hybrid Contextual Text Recognition With String Matching," *IEEE Trans. Pattern Analysis and Machine Intelligence,* Vol. 15, No. 9, 1993, pp. 915–925.

Srihari, S.N., "From Pixels to Paragraphs: The Use of Contextual Models in Text Recognition," *Proc. 2nd Int'l Conf. Document Analysis and Recognition,* IEEE CS Press, Los Alamitos, Calif., 1993, pp. 416–422.

Stringa, L., "A New Set of Constraint-Free Character Recognition Grammars," *IEEE Trans. Pattern Analysis and Machine Intelligence,* Vol. 12, No. 12, 1990, pp. 1210–1217.

Takahashi, H., "A Neural Net OCR Using Geometrical and Zonal-Pattern Features," *Proc. 1st Int'l Conf. Document Analysis and Recognition,* 1991, pp. 821–828.

Takahashi, H. et al., "A Spelling Correction Method and Its Application to an OCR System," *Pattern Recognition,* Vol. 23, 1990, pp. 363–377.

Tseng, L.Y., and C.T. Chuang, "An Efficient Knowledge-Based Stroke Extraction Method for Multi-Font Chinese Characters," *Pattern Recognition,* Vol. 25, No. 12, 1992, pp. 1445–1458.

Tsirkolias, K., and B. G. Mertzios, "Statistical Pattern Recognition Using Efficient Two Dimensional Moments With Applications to Character Recognition," *Pattern Recognition,* Vol. 26, No. 6, 1993, pp. 877–882.

Tsujimoto, S., and H. Asada, "Resolving Ambiguity in Segmenting Touching Characters," *Proc. 1st Int'l Conf. Document Analysis and Recognition,* 1991, pp. 701–709.

Tsujimoto, S., and H. Asada, "Major Components of a Complete Text Reading System," *Proc. IEEE,* Vol. 80, No. 7, IEEE Service Center, Piscataway, N.J., 1992, pp. 1133–1149.

Voisin, J., and P.A. Devijver, "An Application of the Multiedit-Condensing Technique to the Reference Selection Problem in a Print Recognition System," *Pattern Recognition,* Vol. 20, 1987, pp. 465–474.

Wang, J., and J. Jean, "Resolving Multifont Character Confusion with Neural Networks," *Pattern Recognition,* Vol. 26, No. 1, 1993, pp. 175–187.

Wang, L., and T. Pavlidis, "Direct Gray-Scale Extraction of Features for Character Recognition," *IEEE Trans. Pattern Analysis and Machine Intelligence,* Vol. 15, No. 10, 1993, pp. 1053–1065.

Wells, C.J., et al., "Fast Dictionary Look-Up for Contextual Word Recognition," *Pattern Recognition,* Vol. 23, 1990, pp. 501–508.

Wolberg, G., "A Syntactic Omni-Font Character Recognition System," *Int'l J. Pattern Recognition and Artificial Intelligence,* Vol. 1, 1987, pp. 303–322.

Wong, P.-K., and C. Chan, "A Robust Real-Timed Recognizer of Printed Chinese Characters," *Pattern Recognition,* Vol. 25, No. 10, 1992, pp. 1211–1215.

Yang, Y., T. Horiuchi, and T. Toraichi, "Automatic Seal Identification using Fluency Function Approximation and Relaxation Matching Method," *Proc. 2nd Int'l Conf. Document Analysis and Recognition,* IEEE CS Press, Los Alamitos, Calif., 1993, pp. 786–789.

iii. Handwritten Character Recognition, Handwriting Analysis

Abbink, G.H., H.L. Teulings, and L.R.B. Schomaker, "Description of On-line Script Using Hollerbach's Generation Model," *Proc. Int'l Workshop Frontiers in Handwriting Recognition, 1993, pp. 217–224.*

Abdulla, W.H., A.O.M. Saleh, and A.H. Morad, "A Preprocessing Algorithm for Hand-Written Character Recognition," *Pattern Recognition Letters,* Vol. 7, No. 1, 1988, pp. 13–18.

Abuhaiba, I.S.I., and P. Ahmed, "Restoration of Temporal Information in Off-Line Arabic Handwriting," *Pattern Recognition,* Vol. 26, No. 7, 1993, pp. 1009–1028.

Abuhaiba, I.S.I., and P. Ahmed, "A Fuzzy Graph Theoretic Approach to Recognize the Totally Unconstrained Handwritten Numerals," *Pattern Recognition,* Vol. 26, No. 9, 1993, pp. 1335–1350.

Adi, L.S., F. Miraim, and B. Jacob, "On-Line Recognition of Hebrew Script Using Multilevel Knowledge Base," Proc 1st Int'l Conf. Document Analysis and Recognition, 1991, pp. 863–875.

Ahmed, P., and C.Y. Suen, "Computer Recognition of Totally Unconstrained Handwritten Zip Codes," *Int'l J. Pattern Recognition and Artificial Intelligence,* Vol. 1, 1987, pp. 1–15.

Al-Emami, S., and M. Usher, "On-Line Recognition of Handwritten Arabic Characters," *IEEE Trans. Pattern Analysis and Machine Intelligence,* Vol. 12, No. 7, 1990, pp. 704–710.

Alimi, A., and R. Plamondon, "Performance Analysis of Handwritten Strokes Generation Models," *Proc. Int'l Workshop Frontiers in Handwriting Recognition, 1993, pp. 272–283.*

Almuallim, H., and S. Yamaguchi, "A Method of Recognition of Arabic Cursive Handwriting," IEEE Trans. Pattern Analysis and Machine Intelligence, Vol. PAMI-9, No. 5, 1987, pp. 715–722.

Al-Yousefi, H. A., and S.S. Udpa, "Recognition of Arabic Characters," *IEEE Trans. Pattern Analysis and Machine Intelligence,* Vol. 14, No. 8, 1992, pp. 853–857.

Amin, A., and W.H. Wilson, "Hand-Printed Character Recognition System Using Artificial Neural Networks," *Proc. 2nd Int'l Conf. Document Analysis and Recognition,* IEEE CS Press, Los Alamitos, Calif., 1993, pp. 943–946.

Baptista, G., and K.M. Kulkarni, "A High Accuracy Algorithm for Recognition of Handwritten Numerals," *Pattern Recognition,* Vol. 21, 1988, pp. 287–291.

Bellegarda, E.J., et al., "A Probabilistic Framework for On-line Handwriting Recognition," *Proc. Int'l Workshop Frontiers in Handwriting Recognition,* 1993, pp. 225–234.

Bercu, S., and G. Lorette, "On-Line Handwritten Word Recognition: An Approach Based on Hidden Markov Models," Proc. Int'l Workshop Frontiers in Handwriting Recognition, 1993, pp. 385–390.

Bernard, G., "Multilayer Perceptron and Uppercase Handwritten Characters Recognition," *Proc. 2nd Int'l Conf. Document Analysis and Recognition,* IEEE CS Press, Los Alamitos, Calif., 1993, pp. 935–938.

Bertille, J.M., "An Elastic Matching Approach Applied to Digit Recognition," *Proc. 2nd Int'l Conf. Document Analysis and Recognition,* IEEE CS Press, Los Alamitos, Calif., 1993, pp. 82–85.

Blackwell, K.T. et al., "A New approach to Hand-written Character Recognition," *Pattern Recognition,* Vol. 25, No. 6, 1992, pp. 655–666.

Boccignone, G. et al., "Recovering Dynamic Information from Static Handwriting," *Pattern Recognition* Vol. 26, No. 3, 1993, pp. 409–418.

Bozinovic, R.M., and S.N. Srihari, "Off-Line Cursive Script Word Recognition," *IEEE Trans. Pattern Analysis and Machine Intelligence,* Vol. 11, No. 1, 1989, pp. 68–83.

Bramala, P.E., and C.A. Higgins, "A Blackboard Approach to On-line Cursive Handwriting Recognition for Pen Based Computing," *Proc. Int'l Workshop Frontiers in Handwriting Recognition,* 1993, pp. 295–304.

Breuel, T.M., "Recognition of Handprinted Digits using Optimal Bounded Error Matching," *Proc. 2nd Int'l Conf. Document Analysis and Recognition,* IEEE CS Press, Los Alamitos, Calif., 1993, pp. 493–496.

Brossman, C., and G.R. Cross, "Model-Based Recognition of Characters in Trademark Artwork," *Pattern Recognition Letters,* Vol. 11, 1990, pp. 363–370.

Brown, R.M., T.H. Fay, and C.L. Walker, "Handprinted Symbol Recognition System," *Pattern Recognition,* Vol. 21, 1988, pp. 91–118.

Burges, C.J.C., et al., "Off Line Recognition of Handwritten Postal Words Using Neural Network," *Int'l J. Pattern Recognition and Artificial Intelligence,* Vol. 7, No. 4, 1993, pp. 689–704.

Caesar, T., J.M. Gloger, and E. Mandler, "Preprocessing and Feature Extraction for a Handwritten Recognition System," *Proc. 2nd Int'l Conf. Document Analysis and Recognition,* IEEE CS Press, Los Alamitos, Calif., 1993, pp. 408–411.

Caesar, T. et al., "Recognition of Handwritten Word Images by Statistical Methods," *Proc. Int'l Workshop Frontiers in Handwriting Recognition,* 1993, pp. 409–416.

Cai, Z., "A Handwritten Numeral Recognition System Using a Multi-Microprocessor" *Pattern Recognition Letters,* Vol. 12, 1991, pp. 503–509.

Chang, H.D., and J.F. Wang, "Preclassification for Handwritten Chinese Character Recognition by a Peripheral Shape Coding Method," *Pattern Recognition,* Vol. 26, No. 5, 1993, pp. 711–720.

Chen, C.H., and G.K. Myers, "Hypothesis Evaluation for Word Recognition," *Proc. Int'l Workshop Frontiers in Handwriting Recognition,* 1993, pp. 379–384.

Chen, K.J., K.C. Li, and Y.L. Chang, "A System for On-Line Recognition of Chinese Characters," *Int'l J. Pattern Recognition and Artificial Intelligence,* Vol. 2, 1988, pp. 139–148.

Chen, L.-H., and J.-R. Lieh, "Handwritten Character Recognition Using a 2-Layer Random Graph Model by Relaxation Matching," *Pattern Recognition,* Vol. 23, No. 11, 1990, pp. 1189–1205.

Chen, M.Y., and A. Kundu, "An Alternative to Variable Duration HMM in Handwritten Word Recognition," *Proc. Int'l Workshop Frontiers in Handwriting Recognition,* 1993, pp. 82–91.

Chen, Y.-S., "Primitives Segmentation and Association for a Line Character," *Proc. 2nd Int'l Conf. Document Analysis and Recognition,* IEEE CS Press, Los Alamitos, Calif., 1993, pp. 23–26.

Cheng, F.H., W.H. Hsu, and M.Y. Chen, "Recognition of Handwritten Chinese Characters by Modified Hough Transform Techniques," *IEEE Trans. Pattern Analysis and Machine Intelligence,* Vol. 11, No. 4, 1989, pp. 429–439.

Cheng, F.H., W.H. Hsu, and C.A. Chen, "Fuzzy Approach to Solve the Recognition Problem of Handwritten Chinese Characters," *Pattern Recognition,* 1989, pp. 133–141.

Cheng, F.H., W.H. Hsu, and M.G. Kuo, "Recognition of Handprinted Chinese Characters Via Stroke Relaxation," *Pattern Recognition,* Vol. 26, No. 4, 1993, pp. 579–594.

Cheriet, M., "Reading Cursive Script by Parts," *Proc. Int'l Workshop Frontiers in Handwriting Recognition,* 1993, pp. 403–408.

Chhabra, A.K., et al., "High-Order Statistically Derived Combinations of Geometric Features for Handprinted Character Recognition," *Proc. 2nd Int'l Conf. Document Analysis and Recognition,* IEEE CS Press, Los Alamitos, Calif., 1993, pp. 397–401.

Chou, S.C., and W.H. Tsai, "Recognizing Handwritten Chinese Characters by Stroke-Segment Matching Using an Iteration Scheme," *Int'l J. Pattern Recognition and Artificial Intelligence,* Vol. 5, 1991, pp. 175–197.

Chou, S.L., and S.S. Yu, "Sorting Qualities of Handwritten Chinese Characters for Setting Up a Research Database," *Proc. 2nd Int'l Conf. Document Analysis and Recognition,* IEEE CS Press, Los Alamitos, Calif., 1993, pp. 474–477.

Chouinard, C, and R. Plamondon, "Thinning and Segmenting Handwritten Characters by Line Following," *Machine Vision and Applications J.,* Vol. 5, No. 3, 1992, pp. 185–198.

Cohen, E., J.J. Hull, and S.N. Srihari, "Understanding Spatially Structured Handwritten Text," *Proc. 1st Int'l Conf. Document Analysis and Recognition,* IEEE CS Press, Los Alamitos, Calif., 1991, pp. 984–992.

Cohen, E., J.J. Hull, and S.N. Srihari, "Understanding Handwritten Text in a Structured Environment: Determining Zip Codes From Addresses," *Int'l J. Pattern Recognition and Artificial Intelligence,* Vol. 5, Nos. 1&2, 1991, pp. 221–264.

Dimauro, G., S. Impedovo, and G. Pirlo, "A Signature Verification System based on Dynamical Segmentation Technique," *Proc. Int'l Workshop Frontiers in Handwriting Recognition,* 1993, pp. 262–271.

Dimauro, G. et al., "A System for Bankchecks Processing," *Proc. 2nd Int'l Conf. Document Analysis and Recognition,* IEEE CS Press, Los Alamitos, Calif., 1993, pp. 454–459.

Dimitriadis, Y.A., J.L. Coronado, and C. de la Maza, "A New Interactive Mathematical Editor, Using Handwritten Symbol Recognition, and Error Detection-Correction with an Attribute Grammar," *Proc. 1st Int'l Conf. Document Analysis and Recognition,* 1991, pp. 885–893.

Doermann, D., and A. Rosenfeld, "The Interpretation and Recognition of Interfering Strokes," *Proc. Int'l Workshop Frontiers in Handwriting Recognition,* 1993, pp. 41–50.

Dooijes, E.H., and E. Hamstra-Bletz, "A Topological Approach to Handwriting Understanding," *Proc. 1st Int'l Conf. Document Analysis and Recognition,* 1991, pp. 594–602.

Downton, A.C., and R.W.S. Tregidgo, "The Use of a Trie Structured Dictionary as a Contextual Aid to Recognition of Handwritten British Postal Addresses," *Proc. 1st Int'l Conf. Document Analysis and Recognition,* 1991, pp. 542–550.

Duneau, L., and B. Dorizzi, "An Improved Classification for Cursive Script Recognition," *Proc. 2nd Int'l Conf. Document Analysis and Recognition,* IEEE CS Press, Los Alamitos, Calif., 1993, pp. 842–845.

Edelman, S., T. Flash, and S. Ullman, "Reading Cursive Handwriting by Alignment of Letter Prototypes," IJCV, Vol. 5, No. 3, 1990, pp. 303–331.

El-Sheikh, T.S., and S.G. El-Taweel, "Real-Time Arabic Handwritten Character Recognition," *Pattern Recognition,* Vol. 23, 1990, pp. 1323–1332.

El-Sheikh, T.S., and R.M. Guindi, "Computer Recognition of Arabic Cursive Scripts," *Pattern Recognition,* Vol. 21, 1988, pp. 293–302.

El-Wakir, M.S., and A.A. Shoukry, "On-Line Recognition of Handwirtten Arabic Characters," *IEEE Trans. Pattern Analysis and Machine Intelligence,* Vol. 12, No. 7, 1990, pp. 704–710.

Fenrich, R., and J. Hull, "Concerns in Creation of Image Databases," *Proc. Int'l Workshop Frontiers in Handwriting Recognition,* 1993, pp. 112–121.

Franke, J., et al., "Experiments with the CENPARMI Data Base Combining Different Classification Approaches," *Proc. Int'l Workshop Frontiers in Handwriting Recognition,* 1993, pp. 305–311.

Freund, R., "Recognition of Handwritten Characters by Quasi-monotonic Programmed Array Grammars with Attribute Vectors," *Proc. Int'l Workshop Frontiers in Handwriting Recognition,* 1993, pp. 355–360.

Fujisaki, T., et al., "On-Line Run-On Character Recognizer: Design and Performance," *Int'l J. Pattern Recognition and Artificial Intelligence,* Vol. 5, Nos. 1&2, 1991, pp. 123 137.

Fujisaki, T., et al., "On-line Unconstrained Handwriting Recognition by Probabilistic Method," *Proc. Int'l Workshop Frontiers in Handwriting Recognition,* 1993, pp. 235–241.

Fukushima, K., "Connected Character Recognition with a Neural Network," *Proc. 2nd Int'l Conf. Document Analysis and Recognition,* IEEE CS Press, Los Alamitos, Calif., 1993, pp. 240–243.

Gader, P.D., and M.A. Khabou, "Automated Feature Generation for Handwritten Digit Recognition by Neural Networks," *Proc. Int'l Workshop Frontiers in Handwriting Recognition,* 1993, pp. 21–30.

Gader, P., et al, "Recognition of Handwritten Digits Using Template and Model Matching," *Pattern Recognition,* Vol. 24, No. 5, 1991, pp. 421–431.

Gan, K.W., and K.T. Lua, "Chinese Character Classification using an Adaptive Resonance Network," *Pattern Recognition,* Vol. 25, No. 8, 1992, pp. 877–882.

Gilloux, M., "Research into the New Generation of Character and Mailing Address Recognition Systems at the French Post Office Research Center," *Pattern Recognition Letters,* Vol. 14, 1993, pp. 267–276.

Gilloux, M., J.M. Bertille, and M. Leroux, "Recognition of Handwritten Words in a Limited Dynamic Vocabulary," *Proc. Int'l Workshop Frontiers in Handwriting Recognition,* 1993, pp. 417–422.

Gilloux, M., M. Lerox, and J.M. Bertille, "Strategies for Handwritten Words Recognition using Hidden Markov Models," *Proc. 2nd Int'l Conf. Document Analysis and Recognition,* IEEE CS Press, Los Alamitos, Calif., 1993, pp. 299–304.

Goraine, H., M. Usher, and S. Al-Emami, "Off-Line Arabic Character Recognition," *Computer,* Vol. 25, No. 7, 1992, pp. 71–74.

Govindaraju, V., A. Shekhawat, and S.N. Srihari, "Interpretation of Handwritten Addresses in US Mail Stream," *Proc. Int'l Workshop Frontiers in Handwriting Recognition,* 1993, pp. 197–206.

Guerfali, W., and R. Plamondon, "Normalizing and Restoring On-line Handwriting," *Pattern Recognition,* Vol. 26, No. 3, 1993, pp. 419–431.

Guillevic, D., and C.Y. Suen, "Cursive Script Recognition: A Fast Reader Scheme," *Proc. 2nd Int'l Conf. Document Analysis and Recognition,* IEEE CS Press, Los Alamitos, Calif., 1993, pp. 311–314.

Gluhchev, G., V. Shapiro, and V. Sgurev, "An Interactive Approach for Handwritten Material Analysis," *Proc. 1st Int'l Conf. Document Analysis and Recognition,* 1991, pp. 401–409.

Gupta, A., et al., "An Integrated Architecture for Recognition of Totally Unconstrained Handwritten Numerals," *Int'l J. Pattern Recognition and Artificial Intelligence,* Vol. 7, No. 4, 1993, pp. 757–774.

Guyon, I., et al., "Design of a Neural Network Character Recognizer for a Touch Terminal," *Pattern Recognition,* Vol. 24, No. 2, 1991, pp. 105–119.

Ha, J.Y., et al, "Unconstrained Handwritten Word Recognition with Interconnected Hidden Markov Models," *Proc. Int'l Workshop Frontiers in Handwriting Recognition,* 1993, pp. 455–460.

Hamanaka, M., K. Yamada, and J. Tsukumo, "On-Line Japanese Character Recognition Experiments by an Off-Line Method Based on Normalization-Cooperated Feature Extraction," *Proc. 2nd Int'l Conf. Document Analysis and Recognition,* IEEE CS Press, Los Alamitos, Calif., 1993, pp. 204–207.

Hamanaka, M., K. Yamada, and J. Tsukumo, "Normalization-cooperated Feature Extraction Method for Handprinted Kanji Character Recognition," *Proc. Int'l Workshop Frontiers in Handwriting Recognition,* 1993, pp. 343–348.

Hendrawan, A., C. Downton, and C.G. Leedham, "A Fuzzy Approach to Handwritten Address Verification," *Proc. Int'l Workshop Frontiers in Handwriting Recognition,* 1993, pp. 207–216.

Heutte, L., et al., "Handwritten Numeral Recognition Based on Multiple Feature Extractors," *Proc. 2nd Int'l Conf. Document Analysis and Recognition,* IEEE CS Press, Los Alamitos, Calif., 1993, pp. 167–170.

Higgins, C.A., and D.M. Ford, "Stylus Driven Interfaces—The Electronic Paper Concept," *Proc. 1st Int'l Conf. Document Analysis and Recognition,* 1991, pp. 853–862.

Hildebrandt, T.H., and W. Liu, "Optical Recognition of Handwritten Chinese Characters: Advances since 1980," *Pattern Recognition,* Vol. 26, No. 2, 1993, pp. 205–225.

Ho, T.K., "Recognition of Handwritten Digits by Combining Independent Learning Vector Quantizations," *Proc. 2nd Int'l Conf. Document Analysis and Recognition,* IEEE CS Press, Los Alamitos, Calif., 1993, pp. 818–821.

Horiuchi, T., et al., "On Method of Training Dictionaries for Handwritten Character Recognition Using Relaxation Matching," *Proc. 2nd Int'l Conf. Document Analysis and Recognition,* IEEE CS Press, Los Alamitos, Calif., 1993, pp. 638–641.

Houle, G., et al., "Handwritten Word Recognition Using Collective Learning Systems Theory," *Proc. Int'l Workshop Frontiers in Handwriting Recognition,* 1993, pp. 92–101.

Hsieh, C.-C., and H.-J. Lee, "Off-Line Recognition of Handwritten Chinese Characters by On-Line Model-Guided Matching," *Pattern Recognition,* Vol. 25, No. 11, 1992, pp. 1337–1352.

Huang, J.S., and M.L. Chung, "Separating Similar Complex Chinese Characters By Walsh Transform," *Pattern Recognition,* Vol. 20, 1987, pp. 425–428.

Huang, H.S., and K. Chuang, "Heuristic Approach to Handwritten Numeral Recognition," *Pattern Recognition,* Vol. 19, 1986, pp. 15–19.

Huang, Y.S., and C.Y. Suen, "Combination of Multiple Classifiers with Measurement Values," *Proc. 2nd Int'l Conf. Document Analysis and Recognition,* IEEE CS Press, Los Alamitos, Calif., 1993, pp. 598–601.

Huang, Y.S., and C.Y. Suen, "An Optimal Method of Combining Multiple Classifiers for Unconstrained Handwritten Numeral Recognition," *Proc. Int'l Workshop Frontiers in Handwriting Recognition,* 1993, pp. 11–20.

Irie, B., "A New Pattern Recognition Method Using Nonlinear Transformation," *Proc. Int'l Workshop Frontiers in Handwriting Recognition,* 1993, pp. 337–342.

Kaltenmeier, A., et al., "Sophisticated Topology of Hidden Markov Models for Cursive Script Recognition," *Proc. 2nd Int'l Conf. Document Analysis and Recognition,* IEEE CS Press, Los Alamitos, Calif., 1993, pp. 139–142.

Karls, I., et al., "Segmentation and Recognition of Cursive Handwriting with Improved Structured Lexica," *Proc. Int'l Workshop Frontiers in Handwriting Recognition,* 1993, pp. 437–442.

Kawatani, T., "Handprinted Numeral Recognition with the Learning Quadratic Discriminant Function," *Proc. 2nd Int'l Conf. Document Analysis and Recognition,* IEEE CS Press, Los Alamitos, Calif., 1993, pp. 14–17.

Kawatani, T., "Handprinted Numeral Recognition with the Learning Distance Function," *Proc. Int'l Workshop Frontiers in Handwriting Recognition,* 1993, pp. 324–330.

Kawatani, T., and N. Miyamoto, "Verification of Personal Handwriting Characteristics for Numerals and its Application to Recognition," *Pattern Recognition Letters,* Vol. 14, 1993, pp. 335–343.

Keenan, F.G., L.J. Evett, and R.J. Whitrow, "A Large Vocabulary Stochastic Syntax Analyses for Handwriting Recognition," *Proc. 1st Int'l Conf. Document Analysis and Recognition,* 1991, pp. 794–802.

Kerrick, D.D., and A.C. Bovik, "Microprocessor-Based Recognition of Handprinted Characters from a Tablet Input," *Pattern Recognition,* Vol. 21, 1988, pp. 525–537.

Kim, D.H., et al., "Handwritten Korean Character Image Database," *Proc. 2nd Int'l Conf. Document Analysis and Recognition,* IEEE CS Press, Los Alamitos, Calif., 1993, pp. 470–473.

Kimura, F., and M. Shridhar, "Handwritten Numerical Recognition Based on Multiple Algorithms," *Pattern Recognition,* Vol. 24, No. 10, 1991, pp. 969–983.

Kimura, F., and M. Shridhar, "Recognition of Connected Numeral Strings," *Proc. 1st Int'l Conf. Document Analysis and Recognition,* 1991, pp. 731–739.

Kimura, F., et al., "Modified Quadratic Discriminant Functions and the Application to Chinese Character Recognition," *IEEE Trans. Pattern Analysis and Machine Intelligence,* Vol. 9, 1987, pp. 149–153.

Kimura, F., and M. Shridhar, "Segmentation-Recognition Algorithm for Zip Code Field Recognition," *Machine Vision and Applications J.,* Vol. 5, No. 3, 1992, pp. 199–210.

Kimura, F., M. Shridhar, and Z. Chen, "Improvements of a Lexicon directed Algorithm for Recognition of Unconstrained Handwritten Words," *Proc. 2nd Int'l Conf. Document Analysis and Recognition,* IEEE CS Press, Los Alamitos, Calif., 1993, pp. 18–22.

Kimura, F., M. Shridhar, and N. Narasimhamurthi, "Lexicon Directed Segmentation - Recognition Procedure for Unconstrained Handwritten Words," *Proc. Int'l Workshop Frontiers in Handwriting Recognition,* 1993, pp. 122–131.

Kita, N., "Making A Personal Recognition Dictionary from Characters Automatically Generated by Using Handwriting Model," *Proc. 2nd Int'l Conf. Document Analysis and Recognition,* IEEE CS Press, Los Alamitos, Calif., 1993, pp. 76–81.

Kleinberg, E.M., and T.K. Ho, "Pattern Recognition by Stochastic Modeling," *Proc. Int'l Workshop Frontiers in Handwriting Recognition,* 1993, pp. 175–183.

Krtolica, R., and S. Malitsky, "Two-Stage Box Connectivity Algorithm for Optical Character Recognition," *Proc. 2nd Int'l Conf. Document Analysis and Recognition,* IEEE CS Press, Los Alamitos, Calif., 1993, pp. 179–182.

Kumamoto, T., et al., "On Speeding Candidate Selection in Handprinted Chinese Character Recognition," *Pattern Recognition,* Vol. 24, No. 8, 1991, pp. 793–799.

Kundu, A., Y. He, and P. Bahl, "Recognition of Handwritten Word: First and Second Order Hidden Markov Model Based Approach," *Pattern Recognition,* Vol. 22, 1989, pp. 283–297.

Kuo, H.-H., and J.-F, Wang, "A New Method for the Segmentation of Mixed Handprinted Chinese/English Characters," *Proc. 2nd Int'l Conf. Document Analysis and Recognition,* IEEE CS Press, Los Alamitos, Calif., 1993, pp. 810–813.

Kwon, O.S., et al., "A Cursive On-Line Recognition System Based on the Combination of Line Segments," *Proc. 2nd Int'l Conf. Document Analysis and Recognition,* IEEE CS Press, Los Alamitos, Calif., 1993, pp. 200–203.

Lam, L., and C.Y. Suen, "Structural Classification and Relaxation Matching of Totally Unconstrained Handwritten Zip-Code Numbers," *Pattern Recognition,* Vol. 21, 1988, pp. 19–31.

Lecolinet, E., and J.P. Crettez, "A Grapheme-Based Segmentation Technique for Cursive Script Recognition," *Proc. 1st Int'l Conf. Document Analysis and Recognition,* IEEE CS Press, Los Alamitos, Calif., 1991, pp. 740–748.

Lee, D.S., and S.N. Srihari, "Handprinted Digit Recognition: A Comparison of Algorithms," *Proc. Int'l Workshop Frontiers in Handwriting Recognition,* 1993, pp. 153–163.

Lee, H.-J., and B. Chen, "Recognition of Handwritten Chinese Characters via Short Line Segments," *Pattern Recognition,* Vol. 25, No. 5, 1992, pp. 543–552.

Lee, S., and J.C. Pan, "Invariant Handwritten Numeral Recognition with Spatio-Temporal Feature Representation," *Proc. Int'l Workshop Frontiers in Handwriting Recognition,* 1993, pp. 331–336.

Lee, S., and Y. Choi, "Recognition of Unconstrained Handwritten Numerals Based on Dual Cooperative Neural Network," *Proc. Int'l Workshop Frontiers in Handwriting Recognition,* 1993, pp. 1–10.

Lee, S.W., E.S. Kim, and B.W. Min, "Efficient Postprocessing Algorithms for Error Correction in Handwritten Hangul Address and Human Name Recognition," *Proc. 2nd Int'l Conf. Document Analysis and Recognition,* IEEE CS Press, Los Alamitos, Calif., 1993, pp. 232–235.

Lee, S.W., and H.H. Song, "Optimal Design of Reference Models Using Simulated Annealing Combined with Improved LVQ3," *Proc. 2nd Int'l Conf. Document Analysis and Recognition,* IEEE CS Press, Los Alamitos, Calif., 1993, pp. 44–249.

Lee, S.W., J.S. Park, and Y.Y. Tang, "Performance Evaluation of Nonlinear Shape Normalization Methods for the Recognition of Large-Set Handwritten Characters," *Proc. 2nd Int'l Conf. Document Analysis and Recognition,* IEEE CS Press, Los Alamitos, Calif., 1993, pp. 402–407.

Legault, R., and C.Y. Suen, "Refining Curvature Feature Extraction to Improve Handwriting Recognition," *Proc. Int'l Workshop Frontiers in Handwriting Recognition,* 1993, pp. 31–40.

Lemairé, B., "Practical Implementation of a Radial Basis Function Network for Handwritten Digit Recognition," *Proc. 2nd Int'l Conf. Document Analysis and Recognition,* IEEE CS Press, Los Alamitos, Calif., 1993, pp. 412–415.

Leroux, M., J.C. Salome, and J. Badard, "Recognition of Cursive Script Words in a Small Lexicon," *Proc. 1st Int'l Conf. Document Analysis and Recognition,* 1991, pp. 774–782.

Leung, C.H., Y.S. Cheung, and Y.L. Wong, "A Knowledge-Based Stroke-Matching Method for Chinese Character Recognition," *IEEE Trans. Systems, Man and Cybernetics,* Vol. 17, 1987, pp. 993–1003.

Li, H.F., R. Jayakumar, and M. Youssef, "Parallel Algorithms for Recognizing Handwritten Characters Using Shape Features," *Pattern Recognition,* Vol. 22, 1989, pp. 641–652.

Li, X., and N.S. Hall, "Corner Detection and Shape Classification of On-Line Handprinted Kanji Strokes," *Pattern Recognition,* Vol. 26, No. 9, 1993, pp. 1315–1334.

Liao, C.-W., and J.S. Huang, "A Transformation Invariant Matching Algorithm for Handwritten Chinese Character Recognition," *Pattern Recognition,* Vol. 23, No. 11, 1990, pp. 1167–1188.

Liao, H.Y., J.S. Huang, and S.T. Huang, "Stroke-Based Handwritten Chinese Character Recognition Using Neural Networks," *Pattern Recognition Letters,* Vol. 14, No. 10, 1993, pp. 833–840.

Lin, C.K., K.C. Fan, and F.T.-P. Lee, "On-Line Recognition by Deviation-Expansion Model and Dynamic Programming Matching," *Pattern Recognition,* Vol. 26, No. 2, 1993, pp. 259–268.

Liu, Y.J., L.Q. Zhang, and J. Tai, "A New Approach to On-Line Handwritten Chinese Character Recognition," *Proc. 2nd Int'l Conf. Document Analysis and Recognition,* IEEE CS Press, Los Alamitos, Calif., 1993, pp. 192–195.

Lopresti, D., and A. Tomkins, "Approximate Matching of Hand-Drawn Pictograms," *Proc. Int'l Workshop Frontiers in Handwriting Recognition,* 1993, pp. 102–111.

Lorette, G., and Y. Lecourtier, "Is Recognition and Interpretation of Handwritten Text A Scene Analysis Problem?," *Proc. Int'l Workshop Frontiers in Handwriting Recognition,* 1993, pp. 184–196.

Lu, S.W., Y. Ren, and C.Y. Suen, "Hierarchical Attributed Graph Representation and Recognition of Handwritten Chinese Characters," *Pattern Recognition,* Vol. 24, No. 7, 1991, pp. 617–632.

Lu, Y., S. Schlosser, and M. Janeczko, "Fourier Descriptors and Handwritten Digit Recognition," *Machine Vision and Applications J.,* Vol. 6, 1993, pp. 25–34.

Madhvanath, S., and V. Govindaraju, "Holistic Lexicon Reduction," *Proc. Int'l Workshop Frontiers in Handwriting Recognition,* 1993, pp. 71–81.

Makwana, R., "Handwritten Word Recognition Using Local Line Orientations and Associative Memories," *Proc. 1st Int'l Conf. Document Analysis and Recognition,* 1991, pp. 576–584.

Marcelli, A., N. Likhareva, and T. Pavlidis, "A Structural Indexing Method for Character Recognition," *Proc. 2nd Int'l Conf. Document Analysis and Recognition,* IEEE CS Press, Los Alamitos, Calif., 1993, pp. 175–178.

Marukawa, K., et al., "An Error Correction Algorithm for Handwritten Chinese Character Address Recognition," *Proc. 1st Int'l Conf. Document Analysis and Recognition,* 1991, pp. 916–924.

Marukawa, K., et al., "A Post-Processing Method for Handwritten Kanji Name Recognition Using Furigana Information," *Proc. 2nd Int'l Conf. Document Analysis and Recognition,* IEEE CS Press, Los Alamitos, Calif., 1993, pp. 218–221.

Matan, O., et al., "Reading Handwritten Digits: A ZIP Code Recognition System," *Computer,* Vol. 25, No. 7, 1992, pp. 59–63.

Matic, N., et al., "Writer-Adaptation for On-Line Handwritten Character Recognition," *Proc. 2nd Int'l Conf. Document Analysis and Recognition,* IEEE CS Press, Los Alamitos, Calif., 1993, pp. 187–191.

McKeeby, J.W., R.S. Heller, and Y. Moses, "The Writing Instruction Script Hebrew (WISH) System," *Proc. Int'l Workshop Frontiers in Handwriting Recognition,* 1993, pp. 367–372.

Mitchell, B.T., and A.M. Gillies, "A Model-Based Computer Vision System for Recognizing Handwritten ZIP Codes," *Machine Vision and Applications J.,* Vol. 2, 1989, pp. 231–243.

Morasso, P., et al., "Recognition Experiments of Cursive Dynamic Handwriting with Self-Organizing Networks," *Pattern Recognition,* Vol. 26, No. 3, 1993, pp. 451–460.

Morasso, P., et al., "SCRIPTOR: An On-Line Recognition Engine of Cursive Handwriting with Incremental Learning Capabilities," *Proc. Int'l Workshop Frontiers in Handwriting Recognition,* 1993, pp. 431–436.

Moreau, J.-V., et al., "A Postal Check Reading System," *Proc. 1st Int'l Conf. Document Analysis and Recognition,* 1991, pp. 758–764.

Morishita, T., M. Ooura, and Y. Ishii, "A Kanji Recognition Method which Detects Writing Errors," *Int'l J. Pattern Recognition and Artificial Intelligence,* Vol. 2, 1988, pp. 181–195.

Murase, H., "On-Line Recognition System for Free-Format Handwritten Japanese Characters," *Int'l J. Pattern Recognition and Artificial Intelligence,* Vol. 5, 1991, pp. 207–220.

Nadal, C., and C.Y. Suen, "Applying Human Knowledge to Improve Machine Recognition of Confusing Handwritten Numerals," *Pattern Recognition,* Vol. 26, No. 3, 1993, pp. 381–389.

Nagaishi, M., "Identifying Ability of a Recognition Method Based on the Field of Induction," *Proc. 2nd Int'l Conf. Document Analysis and Recognition,* IEEE CS Press, Los Alamitos, Calif., 1993, pp. 926–929.

Nakano, Y., "A Pattern Retrieval System for Character Recognition Using a Very Large Pattern Database," *Proc. 1st Int'l Conf. Document Analysis and Recognition,* 1991, pp. 410–416.

Nakagawa, M., et al., "Principles of Pen Interface Design for Creative Work," *Proc. 1st Int'l Conf. Document Analysis and Recognition,* 1993, pp. 718–721.

Nakayima, Y., and S. Mori, "A Model-Based Classifier in a Scheme of Recognition Filter," *Proc. 2nd Int'l Conf. Document Analysis and Recognition,* IEEE CS Press, Los Alamitos, Calif., 1993, pp. 68–71.

Nishida, H., "Structural Feature Extraction on Multiple Bases with Application to Handwritten Character Recognition System," *Proc. 2nd Int'l Conf. Document Analysis and Recognition,* IEEE CS Press, Los Alamitos, Calif., 1993, pp. 27–30.

Nishida, H., and S. Mori, "A Model of Structural Pattern Transformation of Handwritten Characters," *Proc. Int'l Workshop Frontiers in Handwriting Recognition,* 1993, pp. 62–70.

Nouboud, F., and R. Plamondon, "A Structural Approach to On-Line Character Recognition: System Design and Applications," *Int'l J. Pattern Recognition and Artificial Intelligence,* Vol. 5, 1991, pp. 311–335.

Ohkura, M., et al., "On Discrimination of Handwritten Similar Kanji Characters by Subspace Method Using Several Features," *Proc. 2nd Int'l Conf. Document Analysis and Recognition,* IEEE CS Press, Los Alamitos, Calif., 1993, pp. 589–592.

Ohmori, K., "On-line Handwritten Kanji Character Recognition Using Hypothesis Generation in the Space of Hierarchical Knowledge," *Proc. Int'l Workshop Frontiers in Handwriting Recognition,* 1993, pp. 242–251.

Palumbo, P.W., et al., "Postal Address Block Location in Real Time," *Computer,* Vol. 25, No. 7, 1992, pp. 34–42.

Paquet, T., and Y. Lecourtier, "Handwriting Recognition: Application on Bank Cheques," *Proc. 1st Int'l Conf. Document Analysis and Recognition,* 1991, pp. 749–757.

Paquet, T., and Y. Lecourtier, "Recognition of Handwritten sentences using a Restricted Lexicon," *Pattern Recognition,* Vol. 26, No. 3, 1993, pp. 91–407.

Paquet, T., and Y. Lecourtier, "Automatic Reading of the Literal Amount of Blank Checks," *Machine Vision and Applications J.,* Vol. 6, Nos. 2–3, 1993, pp. 151–162.

Parizeau, M., and R. Plamondon, "Allograph Adjacency Constraints for Cursive Scripts Recognition," *Proc. Int'l Workshop Frontiers in Handwriting Recognition,* 1993, pp. 252–261.

Park, H.S., and S.W. Lee, "Large-Set Handwritten Character Recognition with Multiple Stochastic Models," *Proc. 2nd Int'l Conf. Document Analysis and Recognition,* IEEE CS Press, Los Alamitos, Calif., 1993, pp. 143–146.

Park, H.S., and S.W. Lee, "Off-Line Recognition of Large-Set Handwritten Hangul with Hidden Markov Models," *Proc. Int'l Workshop Frontiers in Handwriting Recognition,* 1993, pp. 51–61.

Pereira, N., and N. Bourbakis, "Recognition of Handwritten Characters Using a Character Reduction Methodology," *Proc. Int'l Workshop Frontiers in Handwriting Recognition,* 1993, pp. 361–366.

Pettier, J.C., and J. Camillerapp, "An Optimal Detector to Localize Handwriting Strokes," *Proc. 1st Int'l Conf. Document Analysis and Recognition,* 1991, pp. 710–718.

Pettier, J.C., and J. Camillerapp, "Script Representation by a Generalized Skeleton," *Proc. 2nd Int'l Conf. Document Analysis and Recognition,* IEEE CS Press, Los Alamitos, Calif., 1993, pp. 850–853.

Pflug, V., "Using N-Grams for the Definition of a Training Set for Cursive Handwriting Recognition," *Proc. 2nd Int'l Conf. Document Analysis and Recognition,* IEEE CS Press, Los Alamitos, Calif., 1993, pp. 295–298.

Plamondon, R., "Handwritten Processing and Recognition," *Pattern Recognition,* Vol. 26, No. 3, 1993, pp. 379–380.

Plamondon, R., et al., "On the Automatic Extraction of Biomechanical Information from Handwriting Signals," *IEEE Trans. Systems, Man, and Cybernetics,* Vol. 21, No. 1, 1991, pp. 90–101.

Plessis, B., et al., "A Multiclassifier Combination Strategy for the Recognition of Handwritten Cursive Words," *Proc. 2nd Int'l Conf. Document Analysis and Recognition,* IEEE CS Press, Los Alamitos, Calif., 1993, pp. 642–645.

Powalka, R.K., et al., "Multiple Word Segmentation with Interactive Look-Up for Cursive Script Recognition," *Proc. 2nd Int'l Conf. Document Analysis and Recognition,* IEEE CS Press, Los Alamitos, Calif., 1993, pp. 196–199.

Raafat, H., and M.A.A. Rashwan, "A Tree Structured Neural Network," *Proc. 2nd Int'l Conf. Document Analysis and Recognition,* IEEE CS Press, Los Alamitos, Calif., 1993, pp. 939–942.

Rao, P.V.S., "Shape Vectors for On-Line and Off-Line Recognition of Cursive Script Words," *Proc. 1st Int'l Conf. Document Analysis and Recognition,* 1991, pp. 568–575.

Ray, S., "A Heuristic Noise Reduction Algorithm Applied to Handwritten Numeric Characters," *Pattern Recognition Letters,* Vol. 7, No. 1, 1988, pp. 9–12.

Revow, M., C.K.I. Williams, and G.E. Hinton, "Using Mixtures of Deformable Models to Capture Variations in Hand Printed Digits," *Proc. Int'l Workshop Frontiers in Handwriting Recognition,* 1993, pp. 142–152.

Rhee, P., and J.H. Yoo, "$\Theta(1)$ Feature Extraction for a Handwritten Character Recognition," *Proc. Int'l Workshop Frontiers in Handwriting Recognition,* 1993, pp. 312–317.

Ristad, E., "A Principled Performance Measure for Online Handwriting Recognition," *Proc. Int'l Workshop Frontiers in Handwriting Recognition,* 1993, pp. 397–402.

Rose, T.G., and L.J. Evett, "Semantic Analysis for Large Vocabulary Cursive Script Recognition," *Proc. 2nd Int'l Conf. Document Analysis and Recognition,* IEEE CS Press, Los Alamitos, Calif., 1993, pp. 236–239.

Rose, T.G., L.J. Evett, and R.J. Whitrow, "The Use of Semantic Information as an Aid to Handwriting Recognition," *Proc. 1st Int'l Conf. Document Analysis and Recognition,* 1991, pp. 629–637.

Sabourin, M., et al., "Classifier Combination for Hand-Printed Digit Recognition," *Proc. 2nd Int'l Conf. Document Analysis and Recognition,* IEEE CS Press, Los Alamitos, Calif., 1993, pp. 163–166.

Schomaker, L., "Using Stroke- or Character-Based Self-Organizing Maps in the Recognition of On-line, Connected Cursive Script," *Pattern Recognition,* Vol. 26, No. 3, 1993, pp. 443–450.

Seki, Y., and N. Takasawa, "Relationship Between the Construction of Chinese Character and the Correct Ratio of Writer Identification," *Proc. 2nd Int'l Conf. Document Analysis and Recognition,* IEEE CS Press, Los Alamitos, Calif., 1993, pp. 838–841.

Sekita, I., et al., "Feature Extraction of Handwritten Japanese Characters by Spline Functions for Relaxation Matching," *Pattern Recognition,* Vol. 21, 1988, pp. 9–17.

Seni, G., and S. Ng, "Towards and On-line Cursive Word Recognizer," *Proc. Int'l Workshop Frontiers in Handwriting Recognition,* 1993, pp. 449–454.

Senior, A.W., and F. Fallside, "Using Constrained Snakes for Feature Spotting in Off-Line Cursive Script," *Proc. 2nd Int'l Conf. Document Analysis and Recognition,* IEEE CS Press, Los Alamitos, Calif., 1993, pp. 305–310.

Senior, A.W., and F. Fallside, "An Off-Line Cursive Script Recognition System Using Recurrent Error Propagation Networks," *Proc. Int'l Workshop Frontiers in Handwriting Recognition,* 1993, pp. 132–141.

Seshadri, S., and D. Sivakumar, "A Technique for Segmentating Handwritten Digits," *Proc. Int'l Workshop Frontiers in Handwriting Recognition,* 1993, pp. 443–449.

Shridhar, M., and A. Badreldin, "Recognition of Isolated and Simply Connected Handwritten Numerals," *Pattern Recognition,* Vol. 19, 1986.

Shridhar, M., and A. Badreldin, "Context-Directed Segmentation Algorithm for Handwritten Numeral Strings," *Image and Vision Computing,* Vol. 5, 1987, pp. 3–9.

Simard, B., B. Prasanda, and R.M.K. Sinha, "On-Line Character Recognition Using Handwriting Modelling," *Pattern Recognition,* Vol. 26, No. 7, 1993, pp. 993–1008.

Simon, J.C., "Off-Line Cursive Word Recognition," *Proc. IEEE,* Vol. 80, No. 7, IEEE Service Center, Piscataway, N.J., 1993, pp. 1150–1161.

Sin, B.K., and J.H. Kim, "A Statistical Approach With HMMs for On-Line Cursive Hangul (Korean Script) Recognition," *Proc. 2nd Int'l Conf. Document Analysis and Recognition,* IEEE CS Press, Los Alamitos, Calif., 1993, pp. 147–150.

Singer, Y., and N. Tishby, "Cursive Handwriting Word Spotting Using a Discrete Dynamic Approach," *Proc. Int'l Workshop Frontiers in Handwriting Recognition,* 1993, pp. 373–378.

Sizov, K.A., "Recognition of Symbols and Words Written by Hand," *Proc. 1st Int'l Conf. Document Analysis and Recognition,* 1991, pp. 692–700.

Srihari, R.K., et al., "Use of Language Models in On-line Sentence/Phrase Recognition," *Proc. Int'l Workshop Frontiers in Handwriting Recognition,* 1993, pp. 284–294.

Srihari, S.N., "Recognition of Handwritten and Machine Printed Text for Postal Address Interpretation," *Pattern Recognition Letters,* Vol. 14, 1993, pp. 291–302.

Srihari, S.N., V. Govindaraju, and A. Shekhawat, "Interpretation of Handwritten Addresses in US Mailstream," *Proc. 2nd Int'l Conf. Document Analysis and Recognition,* IEEE CS Press, Los Alamitos, Calif., 1993, pp. 291–294.

Srikantan, G., "Gradient Representation for Handwritten Character Recognition," *Proc. Int'l Workshop Frontiers in Handwriting Recognition,* 1993, pp. 318–323.

Strathy, N.W., C.Y. Suen, and A. Krzyzak, "Segmentation of Handwritten Digits Using Contour Features," *Proc. 2nd Int'l Conf. Document Analysis and Recognition,* IEEE CS Press, Los Alamitos, Calif., 1993, pp. 577–580.

Stringa, L., "Efficient Classification of Totally Unconstrained Handwritten Numerals with a Trainable Multilayer Network," *Pattern Recognition Letters,* Vol. 10, 1989, pp. 273–280.

Suen, C.Y., et al., "Computer Recognition of Unconstrained Handwritten Numerals," *Proc. IEEE,* Vol. 80, No. 7, IEEE Service Center, Piscataway, N.J., 1992, pp. 1162–1180.

Suen, C.Y., et al., "Building a New Generation of Handwriting Recognition Systems," *Pattern Recognition Letters,* Vol. 14, 1993, pp. 303–315.

Tai, J.-W., Y.-J. Liu, and L.-Q. Zhang, "A Model Based Detecting Approach for Feature Extraction of Off-Line Handwritten Chinese Character Recognition," *Proc. 2nd Int'l Conf. Document Analysis and Recognition,* IEEE CS Press, Los Alamitos, Calif., 1993, pp. 826–829.

Takahashi, H., and T.D. Griffin, "Recognition Enhancement by Linear Tournament Verification," *Proc. 2nd Int'l Conf. Document Analysis and Recognition,* IEEE CS Press, Los Alamitos, Calif., 1993, pp. 585–588.

Takeshita, T., S. Nozawa, and F. Kimura, "On the Bias of Mahalanobis Distance Due to Limited Sample Size," *Proc. 2nd Int'l Conf. Document Analysis and Recognition,* IEEE CS Press, Los Alamitos, Calif., 1993, pp. 171–174.

Tappert, C.C., "Speed, Accuracy, and Flexibility Trade-Offs in On-Line Character Recognition," *Int'l J. Pattern Recognition and Artificial Intelligence,* Vol. 5, Nos. 1&2, 1991, pp. 79–95.

Tappert, C.C., C.Y. Suen, and T. Wakahara, "The State of the Art in On-Line Handwriting Recognition," *IEEE Trans. Pattern Analysis and Machine Intelligence,* Vol. 12, No. 8, 1990, pp. 787–808.

Taxt, T., J.B. Ólafsdóttir and M. Dæhlen, "Recognition of Handwritten Symbols," *Pattern Recognition,* Vol. 23, No. 11, 1990, pp. 1155–1166.

Toyokawa, K., et al., "An On-Line Character Recognition System for Effective Japanese Input," *Proc. 2nd Int'l Conf. Document Analysis and Recognition,* IEEE CS Press, Los Alamitos, Calif., 1993, pp. 208–213.

Tsay, Y.-T., and W.-H. Tsai, "Attributed String Matching by Split-and-Merge for On-Line Chinese Character Recognition," *IEEE Trans. Pattern Analysis and Machine Intelligence,* Vol. 15, No. 2, 1993, pp. 180–184.

Tsuruoka, S., et al., "Segmentation and Recognition for Handwritten 2-Letter State Names," *Proc. 2nd Int'l Conf. Document Analysis and Recognition,* IEEE CS Press, Los Alamitos, Calif., 1993, pp. 814–817.

Tung, C.-H., and H.-J. Lee, "2-Stage Character Recognition by Detection and Correction of Erroneously-Identified Characters," *Proc. 2nd Int'l Conf. Document Analysis and Recognition,* IEEE CS Press, Los Alamitos, Calif., 1993, pp. 834–837.

Tung, C.-H., Y.J. Chen, and H.-J. Lee, "Performance Analysis of an OCR System via an Artificial Handwritten Chinese Character Generator," *Proc. 2nd Int'l Conf. Document Analysis and Recognition,* IEEE CS Press, Los Alamitos, Calif., 1993, pp. 315–318.

Verikas, A.A., M.I. Bachauskene, S.J. Vilunas and D.R. Skaisgiris, "Adaptive Character Recognition System," *Pattern Recognition Letters,* Vol. 13, 1992, pp. 207–212.

Wakahara, T., "Handwritten Numeral Recognition Using LAT with Structural Information," *Proc. Int'l Workshop Frontiers in Handwriting Recognition,* 1993, pp. 164–174.

Wang, A.-B., J.S. Huang, and K.-C. Fan, "Optical Recognition of Handwritten Chinese Characters by Partial Matching," *Proc. 2nd Int'l Conf. Document Analysis and Recognition,* IEEE CS Press, Los Alamitos, Calif., 1993, pp. 822–825.

Wang, J.-H., and S. Ozawa, "Automated Generation of Chinese Character Structure Data Based on Extracting the Strokes," *Proc. 2nd Int'l Conf. Document Analysis and Recognition,* IEEE CS Press, Los Alamitos, Calif., 1993, pp. 806–809.

Ward, J.R., and T. Kuklinski, "A Model for Variability Effects in Handprinting with Implications for the Design of Handwriting Character Recognition Systems," *IEEE Trans. Systems, Man and Cybernetics,* Vol. 18, 1988, pp. 438–451.

Wakahara, T., H. Murase, and K. Odaka, "On-Line Handwriting Recognition," *Proc. IEEE,* Vol. 80, No. 7, IEEE Service Center, Piscataway, N.J., 1992, pp. 1181–1194.

Wakahara, T., "Toward Robust Handwritten Character Recognition," *Pattern Recognition Letters,* Vol. 14, 1993, pp. 345–354.

Webster, R.G., and M. Nakagawa, "The Feasibility of a Parallel Processing Oriented Character Recognition Method," *Proc. 2nd Int'l Conf. Document Analysis and Recognition,* IEEE CS Press, Los Alamitos, Calif., 1993, pp. 714–717.

Wells, C.J., L.J. Evett and R.J. Whitrow, "Word Look-up for Script Recognition—Choosing a Candidate," *Proc. 1st Int'l Conf. Document Analysis and Recognition,* IEEE CS Press, Los Alamitos, Calif., 1991, pp. 620–628.

Westall, J.M., and M.S. Narasimha, "Vertex Directed Segmentation of Handwritten Numerals," *Pattern Recognition,* Vol. 26, No. 10, 1993, pp. 1473–1486.

Xie, S.L., and M. Suk, "On Machine Recognition of Hand-Printed Chinese Characters by Feature Relaxation," *Pattern Recognition,* Vol. 21, 1988, pp. 1–7.

Xia, Y., C. Sun, "Recognizing Restricted Handwritten Chinese Characters by Structure Similarity Method," *Pattern Recognition Letters,* Vol. 11, 1990, pp. 67–73.

Yamada, K., "Feedback Pattern Recognition by Inverse Recall Neural Network Model," *Proc. 2nd Int'l Conf. Document Analysis and Recognition,* IEEE CS Press, Los Alamitos, Calif., 1993, pp. 254–257.

Yamada, H., K. Yamamoto, and T. Saito, "A Nonlinear Normalization Method for Handprinted Kanji Character Recognition-Line Density Equalization," *Pattern Recognition,* Vol. 23, 1990, pp. 1023–1029.

Yan, H., "Design and Implementation of Optimized Nearest Neighbor Classifiers for Handwritten Digit Recognition," *Proc. 2nd Int'l Conf. Document Analysis and Recognition,* IEEE CS Press, Los Alamitos, Calif., 1993, pp. 10–13.

Yasuda, M., K. Yamamoto, and H. Yamada, "Effect of the Perturbed Correlation Method for Optical Character Recognition," *Proc. 2nd Int'l Conf. Document Analysis and Recognition,* IEEE CS Press, Los Alamitos, Calif., 1993, pp. 830–833.

Yong, Y., "Handprinted Chinese Character Recognition Via Neural Networks," *Pattern Recognition Letters,* Vol. 7, No. 1, 1988, pp. 19–25.

Yoshimura, I., M. Yoshimura, and H. Uno, "Partition of Documents Based on a Complexity Measure for the Purpose of Text-Independent Writer Identification," *Proc. Int'l Workshop Frontiers in Handwriting Recognition,* 1993, pp. 391–396.

Yoshimura, M., T. Shimizu, and I. Yoshimura, "A Zip Code Recognition System Using the Localized Arc Pattern Method," *Proc. 2nd Int'l Conf. Document Analysis and Recognition,* IEEE CS Press, Los Alamitos, Calif., 1993, pp. 183–186.

Zahour, A., B. Taconet, and A. Faure, "A New Method for Recognition of Arabic Cursive Scripts," *Proc. 1st Int'l Conf. Document Analysis and Recognition,* IEEE CS Press, Los Alamitos, Calif., 1991, pp. 454–462.

Zhao, M., "Two-Dimensional Extended Attribute Grammar Method for the Recognition of Hand-Printed Chinese Characters," *Pattern Recognition,* Vol. 23, 1990, pp. 685–695.

Zhao, P., T. Yasuda, and Y. Sato, "Cursivewriter: On-Line Cursive Writing Recognition System," *Proc. 2nd Int'l Conf. Document Analysis and Recognition,* IEEE CS Press, Los Alamitos, Calif., 1993, pp. 703–706.

Zhou, J., V. Govindaraju, R.S. Acharya, and S.N. Srihari, "Recognition of State Name Abbreviations," *Proc. Int'l Workshop Frontiers in Handwriting Recognition,* 1993, pp. 423–430.

iv. Character/Font Generation

Cabrelli, C.A., and U.M. Molter, "Automatic Representation of Binary Images," *IEEE Trans. Pattern Analysis and Machine Intelligence,* Vol. 12, 1990, pp. 1190–1196.

Govindan, V.K., and A.P. Shivaprasad, "Artificial Database for Character Recognition Research," *Pattern Recognition Letters,* Vol. 12, 1991, pp. 645–648.

Haruki, R., K. Toraichi, and Y Ohtaki, "A Multi-Stage Algorithm of Extracting Joint Points for Generating Function-Fonts," *Proc. 2nd Int'l Conf. Document Analysis and Recognition,* IEEE CS Press, Los Alamitos, Calif., 1993, pp. 31–34.

Ju, R.-H., I.-C. Jou, and M.-K. Tsay, "Zooming Techniques on Digital Chinese Character Patterns: A Further Study and Improvement," *Image and Vision Computing,* Vol. 9, 1991, pp. 194–200.

Moon, Y.S., and W.K. Hui, "High Quality Chinese Fonts Generation for Desktop Publishing—A Computer Vision Approach," *Pattern Recognition Letters,* Vol. 9, 1989.

Nakashima, K., et al., "A Contour Fill Method for Alpha-Numeric Character Image Generation," *Proc. 2nd Int'l Conf. Document Analysis and Recognition,* IEEE CS Press, Los Alamitos, Calif., 1993, pp. 722–725.

Namane, A., and M.A. Sid-Ahmed, "Character Scaling by Contour Method," *IEEE Trans. Pattern Analysis and Machine Intelligence,* Vol. 12, 1990, pp. 600–606.

Sinha, R.M.K., and H.C. Karnick, "Plang Based Specification of Patterns with Variations for Pictorial Data Bases," *Computer Vision, Graphics, and Image Processing,* Vol. 43, 1988, pp. 98–110.

3. Signature Analysis and Verification, Document Authentication

Ammar, M., "Progress in Verification of Skillfully Simulated Handwritten Signatures," *Int'l J. Pattern Recognition and Artificial Intelligence,* Vol. 5, 1993, pp. 337–351.

Ammar, M., Y. Yoshida, and T. Fukumura, "Off-Line Preprocessing and Verification of Signatures," *Int'l J. Pattern Recognition and Artificial Intelligence,* Vol. 2, 1988, pp. 589–602.

Ammar, M., Y. Yoshida, and T. Fukumura, "Structural Description and Classification of Signature Images," *Pattern Recognition,* Vol. 23, 1990, pp. 697–710.

Brault, J.J., and R. Plamondon, "Segmenting Handwritten Signatures at their Perceptually Important Points," *IEEE Trans. Pattern Analysis and Machine Intelligence,* Vol. 15, No. 9, 1993, pp. 953–956.

Bromley, J., et al., "Signature Verification Using a 'Siamese' Time Delay Neural Network," *Int'l J. Pattern Recognition and Artificial Intelligence,* Vol. 7, No. 4, 1993, pp. 669–688.

Brzakovic, D., and N. Vujovic, "Document Recognition/Authentication Based on Medium-Embedded Random Patterns," *Proc. 2nd Int'l Conf. Document Analysis and Recognition,* IEEE CS Press, Los Alamitos, Calif., 1993, pp. 95–98.

Chang, H.D., J.F. Wang, and H.M. Suen, "Dynamic Handwritten Chinese Signature Verification," *Proc. 2nd Int'l Conf. Document Analysis and Recognition,* IEEE CS Press, Los Alamitos, Calif., 1993, pp. 258–261.

Gazzolo, G., and L. Bruzzone, "Real Time Signature Recognition: A Method for Personal Identification," *Proc. 2nd Int'l Conf. Document Analysis and Recognition,* IEEE CS Press, Los Alamitos, Calif., 1993, pp. 707–709.

Goodson, K.J., and P.H. Lewis, "A Knowledge Based Line Recognition System," *Pattern Recognition Letters,* Vol. 11, 1990, pp. 295–304.

Kameda, T., S. Pilarski, and A. Ivanov, "Notes on Multiple Input Signature Analysis," *IEEE Trans. Computers,* Vol. 42, No. 2, 1993, pp. 228–234.

Lam, C.F., and D. Kamins, "Signature Recognition Through Spectral Analysis," *Pattern Recognition,* Vol. 22, 1989, pp. 39–44.

Likforman-Sulem, L., H. Maítre, and C. Sirat, "An Expert Vision System for Analysis of Hebrew Characters and Authentication of Manuscripts," *Pattern Recognition,* Vol. 24, No. 2, 1991, pp. 121–137.

Parizeau, M., and R. Plamondon, "A Comparative Analysis of Regional Correlation, Dynamic Time Warping, and Skeletal Tree Matching for Signature Verification," *IEEE Trans. Pattern Analysis and Machine Intelligence,* Vol. 12, No. 7, 1990, pp. 710–716.

Pham, B., "Conic B-Splines for Curve Fitting: A Unifying Approach," *Computer Vision, Graphics, and Image Processing,* Vol. 45, 1989, pp. 117–125.

Plamondon, R., and G. Lorette, "Automatic Signature Verification and Writer Identification—The State of the Art," *Pattern Recognition,* Vol. 22, 1989, pp. 107–131.

Revillet, M.J., "Signature Verification on Postal Cheques," *Proc. 1st Int'l Conf. Document Analysis and Recognition,* 1991, pp. 767–773.

Risse, T., "Hough Transform for Line Recognition: Complexity of Evidence Accumulation and Cluster Detection," *Computer Vision, Graphics, and Image Processing,* Vol. 46, 1989, pp. 327–345.

Sabourin, R., M. Cheriet, and G. Genest, "An Extended-Shadow-Code Based Approach for Off-Line Signature Verification," *Proc. 2nd Int'l Conf. Document Analysis and Recognition,* IEEE CS Press, Los Alamitos, Calif., 1991, pp. 1–5.

Shapiro, V., G. Gluhchev, and V. Sgurev, "Automatic Document Authentication: Approach and Methods," *Proc. 1st Int'l Conf. Document Analysis and Recognition,* 1991, pp. 894–902.

Yoshimura, I., and M. Yoshimura, "Off-Line Writer Verification Using Ordinary Characters as the Object," *Pattern Recognition,* Vol. 24, No. 9, 1991, pp. 909–915.

4. Form and Document Processing

Agazzi, O.E., and S. Kuo, "Joint Normalization and Recognition of Degraded Document Images Using Pseudo-2D Hidden Markov Models," *Proc. 2nd Int'l Conf. Document Analysis and Recognition,* IEEE CS Press, Los Alamitos, Calif., 1991, pp. 155–158.

Amano, T., et al, "DRS: A Workstation-Based Document Recognition System for Text Entry," *Computer,* Vol. 25, No. 7, 1992, pp. 67–71.

Baird, H.S., "Anatomy of a Versatile Page Reader," *Proc. IEEE,* Vol. 80, No. 7, IEEE Service Center, Piscataway, N.J., 1992, pp. 1059–1065.

Baird, H.S., "Document Image Defect Models and Their Uses," *Proc. 2nd Int'l Conf. Document Analysis and Recognition,* IEEE CS Press, Los Alamitos, Calif., 1993, pp. 62–67.

Boccignone, G., et al., "Building and Object-Oriented Environment for Document Processing," *Proc. 2nd Int'l Conf. Document Analysis and Recognition,* IEEE CS Press, Los Alamitos, Calif., 1993, pp. 436–439.

Burbaud, J.-C., "Applied Development of Advanced Techniques in Post Offices," *Pattern Recognition Letters,* Vol. 14, No. 4, 1993, pp. 259–265.

Campbell-Grant, I.R., and P.J. Robinson, "An Introduction to ISO DIS 8613, "Office Document Architecture," and its Application to Computer Graphics," *Computer & Graphics,* Vol. 11, 1987, pp. 325–341.

Casey, R.G., and D.R. Ferguson, "Intelligent Forms Processing," IBM Systems J., Vol. 29, 1990, pp. 435–450.

Casey R., et al., "Intelligent Forms Processing System," *Machine Vision and Applications J.,* Vol. 5, No. 3, 1992, pp. 143–156.

Chandran, S., and R. Kasturi, "Structural Recognition of Tabulated Data," *Proc. 2nd Int'l Conf. Document Analysis and Recognition,* IEEE CS Press, Los Alamitos, Calif., 1993, pp. 516–519.

Collin, S., K. Tombre, and P. Vaxivihre, "Don't Tell Mom I'm Doing Document Analysis; She Believes I'm in the Computer Vision Field," *Proc. 2nd Int'l Conf. Document Analysis and Recognition,* IEEE CS Press, Los Alamitos, Calif., 1993, pp. 619–622.

Cortelazzo, G.M., et al., "On the Application of Geometric Form Description Techniques to Automate Key-Section Recognition," *Pattern Recognition,* Vol. 26, No. 1, 1993, pp. 89–94.

Dengel, A., et al., "From Paper to Office Document Standard Representation," *Computer,* Vol. 25, No. 7, 1992, pp. 63–67.

Dengel, A., "The Role of Document Analysis and Understanding in Multi-Media Information Systems," *Proc. 2nd Int'l Conf. Document Analysis and Recognition,* IEEE CS Press, Los Alamitos, Calif., 1993, pp. 385–390.

Dinan, R.F., L.D. Painter, and R.R. Rodite, "Image Plus High Performance Transaction System," IBM Systems J., Vol. 29, 1990, pp. 421–434.

Doermann, D.S., and A. Rosenfeld, "The Processing of Form Documents," *Proc. 2nd Int'l Conf. Document Analysis and Recognition,* IEEE CS Press, Los Alamitos, Calif., 1993, pp. 497–501.

Esposito, F., D. Malerba, and G. Semeraro, "Automated Acquisition of Rules for Document Understanding," *Proc. 2nd Int'l Conf. Document Analysis and Recognition,* IEEE CS Press, Los Alamitos, Calif., 1993, pp. 650–654.

Franke, J., and M. Oberländer, "Writing Style Detection by Statistical Combination of Classifiers in Form Reader Applications," *Proc. 2nd Int'l Conf. Document Analysis and Recognition,* IEEE CS Press, Los Alamitos, Calif., 1993, pp. 581–584.

Fujisawa, H., Y. Nakano, and K. Kurino, "Segmentation Methods for Character Recognition: From Segmentation to Document Structure Analysis," *Proc. IEEE,* Vol. 80, No. 7, IEEE Service Center, Piscataway, N.J., 1992, pp. 1079–1092.

Futrelle, R.P., et al., "Document Analysis, Understanding & Knowledge Access," *Proc. 1st Int'l Conf. Document Analysis and Recognition,* 1991, pp. 101–111.

Ingold, R., "A Document Description Language to Drive Document Analysis," *Proc. 1st Int'l Conf. Document Analysis and Recognition,* 1991, pp. 294–301.

Ishitani, Y., "Document Skew Detection Based on Local Region Complexity," *Proc. 2nd Int'l Conf. Document Analysis and Recognition,* IEEE CS Press, Los Alamitos, Calif., 1993, pp. 49–52.

Kanai, J., et al., "Performance Metrics for Document Understanding Systems," *Proc. 2nd Int'l Conf. Document Analysis and Recognition,* IEEE CS Press, Los Alamitos, Calif., 1993, pp. 424–427.

Kanungo, T., R.M. Haralick, and I. Phillps, "Global and Local Document Degradation Models," *Proc. 2nd Int'l Conf. Document Analysis and Recognition,* IEEE CS Press, Los Alamitos, Calif., 1993, pp. 730–734.

Kopec, G.E., and P.A. Chou, "Automatic Generation of Custom Document Image Decoders," *Proc. 2nd Int'l Conf. Document Analysis and Recognition,* IEEE CS Press, Los Alamitos, Calif., 1993, pp. 684–687.

Kreich, J., "Robust Recognition of Document," *Proc. 2nd Int'l Conf. Document Analysis and Recognition,* IEEE CS Press, Los Alamitos, Calif., 1993, pp. 444–447.

Lam, S.W., L. Javanbakht, and S.N. Srihari, "Anatomy of a Form Reader," *Proc. 2nd Int'l Conf. Document Analysis and Recognition,* IEEE CS Press, Los Alamitos, Calif., 1993, pp. 506–509.

Monger, D., G. Leedham, and A. Downton, "An Interactive Document Image Description for OCR of Handwritten Forms," *Proc. 2nd Int'l Conf. Document Analysis and Recognition,* IEEE CS Press, Los Alamitos, Calif., 1993, pp. 524–527.

Myka, A., and U. Güntzer, "Using Electronic Facsimilies of Documents for Automatic Reconstruction of Underlying Hypertext Structures," *Proc. 2nd Int'l Conf. Document Analysis and Recognition,* IEEE CS Press, Los Alamitos, Calif., 1993, pp. 528–532.

Nagy, G., S. Sheth, and M. Vishwanathan, "A Prototype Document Image Analysis System for Technical Journals," *Computer,* Vol. 25, No. 7, 1992, pp. 10–22.

Nardelli, E., M. Fossa, and G. Proietti, "Raster to Object Conversion Aided by Knowledge Based Image Processing," *Proc. 2nd Int'l Conf. Document Analysis and Recognition,* IEEE CS Press, Los Alamitos, Calif., 1993, pp. 951–954.

Paik, J., S. Jung, and Y. Lee, "Multiple Combined Recognition System for Automatic Processing of Credit Card Slip Applications," *Proc. 2nd Int'l Conf. Document Analysis and Recognition,* IEEE CS Press, Los Alamitos, Calif., 1993, pp. 520–523.

Phillips, I.T., S. Chen, and R.M. Haralick, "CD-ROM Document Database Standard," *Proc. 2nd Int'l Conf. Document Analysis and Recognition,* IEEE CS Press, Los Alamitos, Calif., 1993, pp. 478–483.

Phillips, I.T., et al., "The Implementation Methodology for a CD-ROM English Document Database," *Proc. 2nd Int'l Conf. Document Analysis and Recognition,* IEEE CS Press, Los Alamitos, Calif., 1993, pp. 484–487.

Sakai, T., "A History and Evolution of Document Information Processing," *Proc. 2nd Int'l Conf. Document Analysis and Recognition,* IEEE CS Press, Los Alamitos, Calif., 1993, pp. 377–384.

Satoh, S., A. Takasu, and E. Katsura, "A Collaborative Supporting Method Between Document Processing and Hypertext Construction," *Proc. 2nd Int'l Conf. Document Analysis and Recognition,* IEEE CS Press, Los Alamitos, Calif., 1993, pp. 533–536.

Schürmann, J., et al., "Document Analysis—From Pixels to Contents," *Proc. IEEE,* Vol. 80, No. 7, IEEE Service Center, Piscataway, N.J., 1992, pp. 1101–1119.

Senda, S., M. Minoh, and K. Ikeda, "Document Image Retrieval System Using Character Candidates Generated by Character Recognition Process," *Proc. 2nd Int'l Conf. Document Analysis and Recognition,* IEEE CS Press, Los Alamitos, Calif., 1993, pp. 541–546.

Srihari, R.K., "Intelligent Document Understanding: Understanding Photographs with Captions," *Proc. 2nd Int'l Conf. Document Analysis and Recognition,* IEEE CS Press, Los Alamitos, Calif., 1993, pp. 664–667.

Srihari, S.N., "High-Performance Reading Machines," *Proc. IEEE,* Vol. 80, No. 7, IEEE Service Center, Piscataway, N.J., 1992, pp. 1120–1132.

Taylor, S.L., R. Fritzson, and J.A. Pastor, "Extraction of Data from Preprinted Forms," *Machine Vision and Applications J.,* Vol. 5, No. 3, 1992, pp. 211–222.

Taylor, S.L., et al., "An Intelligent Document Understanding System," *Proc. 2nd Int'l Conf. Document Analysis and Recognition,* IEEE CS Press, Los Alamitos, Calif., 1993, pp. 107–110.

Tokunaga, Y., "History and Current State of Postal Mechanization in Japan," *Pattern Recognition Letters,* Vol. 14, No. 4, 1993, pp. 277–280.

Wada, M., "Proposal for Fully Automated Mail Processing System for the 21st Century," *Pattern Recognition Letters,* Vol. 14, No. 4, 1993, pp. 281–290.

Wang, D., and S.N. Srihari, "Analysis of Form Images," *Proc. 2nd Int'l Conf. Document Analysis and Recognition,* 1991, pp. 181–191.

Watanabe, T., et al., "Structure Analysis of Table-Form Documents on the Basis of the Recognition of Vertical and Horizontal Line Segments," *Proc. 1st Int'l Conf. Document Analysis and Recognition,* 1991, pp. 638–646.

Watanabe, T., Q. Luo, and N. Sugie, "Toward a Practical Document Understanding of Table-Form Documents: Its Framework and Knowledge Representation," *Proc. 2nd Int'l Conf. Document Analysis and Recognition,* IEEE CS Press, Los Alamitos, Calif., 1993, pp. 510–515.

Yan, C.D., Y.Y. Tang, and C.Y. Suen, "Form Understanding System Based on Form Description Language," *Proc. 1st Int'l Conf. Document Analysis and Recognition,* 1991, pp. 283–293.

Yan, H., "Skew Correction of Document Images Using Interline Cross-Correlation," *CVGIP: Graphical Models and Image Processing,* Vol. 55, No. 6, 1993, pp. 538–543.

Yu, C.L., Y.Y. Tang, and C.Y. Suen, "Document Architecture Language (DAL) Approach to Document Processing," *Proc. 2nd Int'l Conf. Document Analysis and Recognition,* IEEE CS Press, Los Alamitos, Calif., 1993, pp. 103–106.

Yuan, J., L. Xu, and C.Y. Suen, "Form Items Extraction By Model Matching," *Proc. 1st Int'l Conf. Document Analysis and Recognition,* 1991, pp. 210–218.

Zidouri, A.B.C., S. Chinveeraphan, and M. Sato, "Classification of Compound Document Image Patterns by MCR Stroke Index," *Proc. 2nd Int'l Conf. Document Analysis and Recognition,* IEEE CS Press, Los Alamitos, Calif., 1993, pp. 753–756.

IV. Graphics Recognition and Interpretation

1. Engineering Drawings and Graphics

Arias, J.F., et al., "Interpretation of Telephone System Manhole Drawings," *Proc. 2nd Int'l Conf. Document Analysis and Recognition,* IEEE CS Press, Los Alamitos, Calif., 1993, pp. 365–368.

Bixler, J.P., L.T. Watson, and J.P. Sanford, "Spline-Based Recognition of Straight Lines and Curves in Engineering Line Drawings," *Image and Vision Computing,* Vol. 6, 1988, pp. 262–269.

Collin, S., and P. Vaxivière, "Recognition and Use of Dimensioning in Digitized Industrial Drawings," *Proc. 1st Int'l Conf. Document Analysis and Recognition,* 1991, pp. 161–169.

Cooper, M.C., "Interpretation of Line Drawings of Complex Objects," *Image and Vision Computing,* Vol. 11, No. 2, 1993, pp. 82–90.

De Stefano, C., F. Tortorella, and M. Vento, "Using Entropy for Drawing Reliable Templates," *Proc. 2nd Int'l Conf. Document Analysis and Recognition,* IEEE CS Press, Los Alamitos, Calif., 1993, pp. 345–348.

Dori, D., "A Syntactic/Geometric Approach to Recognition of Dimensions in Engineering Machine Drawings," *Computer Vision, Graphics, and Image Processing,* Vol. 47, No. 3, 1989, pp. 271–291.

Dori, D., "Syntax Enhanced Parameter Learning for Recognition of Dimensions in Engineering Machine Drawings," *Int'l J. Robotics and Automation,* Vol. 5, No. 2, 1990, pp. 59–67.

Dori, D., "Symbolic Representation of Dimensioning in Engineering Drawings," *Proc. 1st Int'l Conf. Document Analysis and Recognition,* 1991, pp. 1000–1010.

Dori, D., and A. Pnueli, "The Grammar of Dimensions in Machine Drawings," *Computer Vision, Graphics, and Image Processing,* Vol. 42, 1988, pp. 1–18.

Dori D., and K. Tombre, "Paper Drawings to 3-D CAD: A Proposed Agenda," *Proc. 2nd Int'l Conf. Document Analysis and Recognition,* IEEE CS Press, Los Alamitos, Calif., 1993, pp. 866–869.

Dori, D., et al., "Sparse-Pixel Recognition of Primitives in Engineering Drawings," *Machine Vision and Applications J.,* Vol. 6, Nos. 2–3, 1993, pp. 69–82.

Filipski, A.J., and R. Flandrena, "Automated Conversion of Engineering Drawings to CAD form." *Proc. IEEE,* Vol. 80, No. 7, IEEE Service Center, Piscataway, N.J., 1992, pp. 1195–1209.

Futrelle, R.P. et al, "Understanding Diagrams in Technical Documents," *Computer,* Vol. 25, No. 7, 1992, pp. 75–78.

Gude, T., and V. Märgner, "Automatic Interpretation of Graphical Drawings," *Proc. 1st Int'l Conf. Document Analysis and Recognition,* 1991, pp. 656–664.

Hamada, A.H., "A New System for the Analysis of Schematic Diagrams," *Proc. 2nd Int'l Conf. Document Analysis and Recognition,* IEEE CS Press, Los Alamitos, Calif., 1993, pp. 369–372.

Hori, O., et al., "Line-Drawing Interpretation Using Probabilistic Relaxation," *Machine Vision and Applications J.,* Vol. 6, No. 2–3, 1993, pp. 100–109.

Hori, O., and S. Tanigawa, "Raster-to-Vector Conversion by Line Fitting Based on Contours and Skeletons," *Proc. 2nd Int'l Conf. Document Analysis and Recognition,* IEEE CS Press, Los Alamitos, Calif., 1993, pp. 353–358.

Joseph. S.H., "Processing of Engineering Line Drawings for Automatic Input to CAD," *Pattern Recognition,* Vol. 22, 1989, pp. 1–11.

Joseph, S.H., and T.P. Pridmore, "Knowledge-Directed Interpretation of Mechanical Engineering Drawings," *IEEE Trans. Pattern Analysis and Machine Intelligence,* Vol. 14, No. 9, 1992, pp. 928–940.

Kaneko, T., "Line Structure Extraction from Line-Drawing Images," *Pattern Recognition,* Vol. 25, No. 9, 1992, pp. 963–973.

Kasturi, R., et al., "A System for Interpretation of Line Drawings," *IEEE Trans. Pattern Analysis and Machine Intelligence,* Vol. 12, No. 10, 1990, pp. 978–992.

Kim, S.H., J.W. Suh, and J.H. Kim, "Recognition of Logic Diagrams by Identifying Loops and Rectilinear Polylines," *Proc. 2nd Int'l Conf. Document Analysis and Recognition,* IEEE CS Press, Los Alamitos, Calif., 1993, pp. 349–352.

Lai, C.P., and R. Kasturi, "Detection of Dashed Lines in Engineering Drawings and Maps," *Proc. 1st Int'l Conf. Document Analysis and Recognition,* 1991, pp. 507–515.

Lai, C.P., and R. Kasturi, "Detection of Dimension Sets in Engineering Drawings," *Proc. 2nd Int'l Conf. Document Analysis and Recognition,* IEEE CS Press, Los Alamitos, Calif., 1993, pp. 606–613.

Lamb, D., and A. Bandopadhay, "Shape From Line Drawings: Beyond Huffman-Clowes Labeling," *Pattern Recognition Letters,* Vol. 14, No. 3, 1993, pp. 213–220.

Lee, I., et al., "A Study on a Method of Dividing Machine-Parts into Functional Groups for Technical Illustrations," *Proc. 2nd Int'l Conf. Document Analysis and Recognition,* IEEE CS Press, Los Alamitos, Calif., 1993, pp. 886–889.

Lysak, D.B., Jr., and R. Kasturi, "Interpretation of Engineering Drawings of Polyhedral and Non-Polyhedral Objects," *Proc. 1st Int'l Conf. Document Analysis and Recognition,* 1991, pp. 79–87.

Marill, T., "Emulating the Human Interpretation of Line-Drawings as Three-Dimensional Objects," IJCV, Vol. 6, No. 2, 1991, pp. 147–161.

Martí, E., et al., "A System for Interpretation of Hand Line Drawings as Three-Dimensional Scene for CAD Input," *Proc. 1st Int'l Conf. Document Analysis and Recognition,* 1991, pp. 472–480.

Min, W., Z. Tang, and L. Tang, "Recognition of Dimensions in Engineering Drawing Based on Arrowhead-Match," *Proc. 2nd Int'l Conf. Document Analysis and Recognition,* IEEE CS Press, Los Alamitos, Calif., 1993, pp. 373–376.

Min, W., Z. Tang, and L. Tang, "Using Web Grammar to Recognize Dimensions in Engineering Drawing," *Pattern Recognition,* Vol. 26, No. 9, 1993, pp. 1407–1416.

Monagan, G., and M. Roosli, "Appropriate Base Representation Using a Run Graph," *Proc. 2nd Int'l Conf. Document Analysis and Recognition,* IEEE CS Press, Los Alamitos, Calif., 1993, pp. 623–626.

Nagasamy, V., and N.A. Langrana, "Engineering Drawing Processing and Vectorization System," *Computer Vision, Graphics, and Image Processing,* Vol. 49, 1990, pp. 379–397.

Nakamura, Y., R. Furukawa, and M. Nagao, "Diagram Understanding Utilizing Natural Language Text," *Proc. 2nd Int'l Conf. Document Analysis and Recognition,* IEEE CS Press, Los Alamitos, Calif., 1993, pp. 614–618.

Ogawa, H., "Symmetry Analysis of Line Drawings Using the Hough Transform," *Pattern Recognition Letters,* Vol. 12, 1991, pp. 9–12.

Pao, D., H.F. Li, and R. Jayakumar, "Graphic Features Extraction for Automatic Conversion of Engineering Line Drawings," *Proc. 1st Int'l Conf. Document Analysis and Recognition,* 1991, pp. 533–541.

Pasternak, B., and B. Neumann, "Adaptable Drawing Interpretation Using Object-Oriented and Constraint-Based Graphic Specification," *Proc. 1st Int'l Conf. Document Analysis and Recognition,* 1991, pp. 359–364.

Satoh, S., H. Mo, and M. Sakauchi, "Drawing Image Understanding System With Capability of Rule Learning," *Proc. 2nd Int'l Conf. Document Analysis and Recognition,* IEEE CS Press, Los Alamitos, Calif., 1993, pp. 119–124.

Starovoitov, V.V., et al., "Binary Texture Border Extraction on Line Drawings Based on Distance Transform," *Pattern Recognition,* Vol. 26, No. 8, 1993, pp. 1165–1176.

Vaxivière, P., and K. Tombre, "Celesstin: CAD Conversion of Mechanical Drawings," *Computer,* Vol. 25, No. 7, 1992, pp. 46–54.

Wang, P.S.P., "Machine Visualization, Understanding and Interpretation of Polyhedral Line-Drawings in Document Analysis," *Proc. 2nd Int'l Conf. Document Analysis and Recognition,* IEEE CS Press, Los Alamitos, Calif., 1993, pp. 882–885.

Wu, W., and M. Sakauchi, "A Multipurpose Drawing Understanding System with Flexible Object-Oriented Framework," *Proc. 2nd Int'l Conf. Document Analysis and Recognition,* IEEE CS Press, Los Alamitos, Calif., 1993, pp. 870–873.

2. Electronic Circuit Diagrams

Fahn, C.S., J.F. Wang, and J.Y. Lee, "A Topology-Based Component Extractor for Understanding Electronic Circuit Diagrams," *Computer Vision, Graphics, and Image Processing,* Vol. 44, 1988, pp. 119–138.

Ito, M., et al., "Pattern Analysis and Evaluation of Printed Circuit Boards," *Proc. 2nd Int'l Conf. Document Analysis and Recognition,* IEEE CS Press, Los Alamitos, Calif., 1993, pp. 798–801.

Okazaki, A., et al., "An Automatic Circuit Diagram Reader with Loop-Structure-Based Symbol Recognition," *IEEE Trans. Pattern Analysis and Machine Intelligence,* Vol. 10, No. 3, 1988, pp. 331–341.

3. Maps and Geographic Information

Ablameyko, S.V., B.S. Beregov, and A.N. Kryuchov, "Computer-Aided Cartographical System for Map Digitizing," *Proc. 2nd Int'l Conf. Document Analysis and Recognition,* IEEE CS Press, Los Alamitos, Calif., 1993, pp. 115–118.

Amin, T.J., and R. Kasturi, "Map Data Processing: Recognition of Lines and Symbols," Optical Eng., Vol. 26, 1987, pp. 354–358.

Antoine, D., "CIPLAN: A Model-Based System with Original Features for Understanding French Plats," *Proc. 1st Int'l Conf. Document Analysis and Recognition,* 1991, pp. 647–655.

Arai, H., S. Abe, and M. Nagura, "Intelligent Interactive Map Recognition using Neural Networks," *Proc. 2nd Int'l Conf. Document Analysis and Recognition,* IEEE CS Press, Los Alamitos, Calif., 1993, pp. 922–925.

Boatto, L., et al, "A Prototype Document Analysis System for Land Register Maps," *Computer,* Vol. 25, No. 7, 1992, pp. 25–33.

Consorti, V., L.P. Cordella, and M. Iaccarino, "Automatic Lettering of Cadastral Maps," *Proc. 2nd Int'l Conf. Document Analysis and Recognition,* IEEE CS Press, Los Alamitos, Calif., 1993, pp. 129–132.

Deseilligny, M.P., H. Le Men, and G. Stamon, "Map Understanding for GIS Data Capture: Algorithms for Road Network Graph Reconstruction," *Proc. 2nd Int'l Conf. Document Analysis and Recognition,* IEEE CS Press, Los Alamitos, Calif., 1993, pp. 676–679.

Freeman, H., and J. Ahn, "On the Problem of Placing Names in a Geographic Map," *Int'l J. Pattern Recognition and Artificial Intelligence,* Vol. 1, 1987, pp. 121–140.

Gao, Q., et al., "A Color Map Processing System PU- CMPS," *Proc. 2nd Int'l Conf. Document Analysis and Recognition,* IEEE CS Press, Los Alamitos, Calif., 1993, pp. 874–877.

Holmes, P D., and E.R.A. Jungert, "Symbolic and Geometric Connectivity Graph Methods for Route Planning in Digitized Maps," *IEEE Trans. Pattern Analysis and Machine Intelligence,* Vol. 14, No. 5, 1992, pp. 549–565.

Janssen, R.D.T., R.P.W. Duin, and A.M. Vossepoel, "Evaluation Method for an Automatic Map Interpretation System for Cadastral Map," *Proc. 2nd Int'l Conf. Document Analysis and Recognition,* IEEE CS Press, Los Alamitos, Calif., 1993, pp. 125–128.

Kasturi, R., et al., "Map Data Processing in Geographical Information Systems," *Computer,* Vol. 22, No. 12, 1989, pp. 10–21.

Kasturi, R., and J. Alemany, "Information Extraction from Images of Paper-based Maps for Query Processing," *IEEE Trans. Software Eng. (special section on Image Databases),* Vol. 14, No. 5, 1988, pp. 671–675.

Laine, A., S. Schuler, and V. Girish, "Orthonormal Wavelet Representations for Recognizing Complex Annotations," *Machine Vision and Applications J.,* Vol. 6, Nos. 2–3, 1993, pp. 110–123.

Lin, Y.M., et al., "An Object-Oriented Kernel for Geographical Information Systems," *Proc. 2nd Int'l Conf. Document Analysis and Recognition,* IEEE CS Press, Los Alamitos, Calif., 1993, pp. 878–881.

Madej, D., "An Intelligent Map-to-CAD Conversion System," *Proc. 1st Int'l Conf. Document Analysis and Recognition,* 1991, pp. 603–610.

Madej, D., and A. Sokolowski, "Towards Automatic Evaluation of Drawing Analysis Performance—A Statistical Model of Cadastral Map," *Proc. 2nd Int'l Conf. Document Analysis and Recognition,* IEEE CS Press, Los Alamitos, Calif., 1993, pp. 890–893.

Maio, D., and S. Rizzi," Clustering by Discovery on Maps," *Pattern Recognition Letters,* Vol. 13, 1992, pp. 89–94.

Musavi, M.T., et al., "A Vision Based Method to Automate Map Processing," *Pattern Recognition,* Vol. 21, 1988, pp. 319–326.

Nakamura, A., et al., "A Method for Recognizing Character Strings from Maps Using Linguistic Knowledge," *Proc. 2nd Int'l Conf. Document Analysis and Recognition,* IEEE CS Press, Los Alamitos, Calif., 1993, pp. 561–564.

Ogier, J.M., et al., "Attributes Extraction for French Map Interpretation," *Proc. 2nd Int'l Conf. Document Analysis and Recognition,* IEEE CS Press, Los Alamitos, Calif., 1993, pp. 672–675.

Seemuller, W.W., "The Extraction of Ordered Vector Drainage Networks from Elevation Data," *Computer Vision, Graphics, and Image Processing,* Vol. 47, 1989, pp. 45–58.

Shimada, S., et al., "Paralleled Automatic Recognition of Maps and Drawings for Constructing Electric Power Distribution Databases," *Proc. 2nd Int'l Conf. Document Analysis and Recognition,* IEEE CS Press, Los Alamitos, Calif., 1993, pp. 688–691.

Shimotsuji, S., et al., "A Robust Recognition System for Drawing Superimposed on a Map," *Computer,* Vol. 25, No. 7, 1992, pp. 56–59.

Suzuki, S., and T. Yamada, "Maris: Map Recognition Input System," *Pattern Recognition,* Vol. 23, 1990, pp. 919–933.

Tanaka, N., T, Kamimura, and J. Tsukumo, "Development of a Map Vectorization Method Involving a Shape Reforming Process," *Proc. 2nd Int'l Conf. Document Analysis and Recognition,* IEEE CS Press, Los Alamitos, Calif., 1993, pp. 680–683.

Xu, T., and X. Lin, "A New Algorithm Separating Text Strings from Map Images," *Proc. 2nd Int'l Conf. Document Analysis and Recognition,* IEEE CS Press, Los Alamitos, Calif., 910–913, 1993.

Yamada, H., K. Yamamoto, T. Saito and S. Matsui, "Map: Multi-Angled Parallelism for Feature Extraction From Topographical Maps," *Pattern Recognition,* Vol. 24, No. 6, 1991, pp. 479–488.

Yamada, H., K. Yamamoto and K. Hosokawa, "Directional Mathematical Morphology and Reformalized Hough Transformation for the Analysis of Topographic Maps," *IEEE Trans. Pattern Analysis and Machine Intelligence,* Vol. 15, No. 4, 1993, pp. 380–387.

Yamamoto, K., H. Yamada, and S. Muraki, "Symbol Recognition and Surface Reconstruction from Topographic Map by Parallel Method," *Proc. 2nd Int'l Conf. Document Analysis and Recognition,* IEEE CS Press, Los Alamitos, Calif., 1993, pp. 914–917.

Yan, H., "Color Map Image Segmentation Using Optimized Nearest Neighbor Classifiers," *Proc. 2nd Int'l Conf. Document Analysis and Recognition,* IEEE CS Press, Los Alamitos, Calif., 1993, pp. 111–114.

4. Other Applications: Fingerprints, Music, Chess, Shorthand

Armand, J.-P., "Musical Score Recognition: A Hierarchical and Recursive Approach," *Proc. 2nd Int'l Conf. Document Analysis and Recognition,* IEEE CS Press, Los Alamitos, Calif., 1993, pp. 906–909.

Baird, H.S., and K. Thompson, "Reading Chess," *IEEE Trans. Pattern Analysis and Machine Intelligence,* Vol. 12, No. 6, 1990, pp. 552–559.

Carter, N.P., "Segmentation and Preliminary Recognition of Madrigals Notated in White Mensural Notation," *Machine Vision and Applications J.,* Vol. 5, No. 3, 1992, pp. 223–230.

Cheng, T., et al., "A Symbol Recognition System," *Proc. 2nd Int'l Conf. Document Analysis and Recognition,* IEEE CS Press, Los Alamitos, Calif., 1993, pp. 918–921.

Chong, M.M.S., et al., "Automatic Representation of Fingerprints for Data Compression by B-Spline Functions," *Pattern Recognition,* Vol. 25, No. 10, 1992, pp. 1199–1210.

Doermann, D.S., E. Rivlin, and I. Weiss, "Logo Recognition Using Geometric Invariants," *Proc. 2nd Int'l Conf. Document Analysis and Recognition,* IEEE CS Press, Los Alamitos, Calif., 1993, pp. 894–897.

Dori, D., et al., "Radiographs as Medical Documents: Automating Standard Measurements," *Proc. 2nd Int'l Conf. Document Analysis and Recognition,* IEEE CS Press, Los Alamitos, Calif., 1993, pp. 553–556.

Fahmy, H., and D. Blostein, "A Graph Grammar Programming Style for Recognition of Music Notation," *Machine Vision and Applications J.,* Vol. 6, Nos. 2–3, 1993, pp. 83–99.

Hrechak, A.K., and J.A. McHugh, "Automated Fingerprint Recognition Using Structural Matching," *Pattern Recognition,* Vol. 23, 1990, pp. 893–904.

Jain, A.K., and Y. Chen, "Bar Code Localization Using Texture Analysis," *Proc. 2nd Int'l Conf. Document Analysis and Recognition,* IEEE CS Press, Los Alamitos, Calif., 1993, pp. 41–44.

Lee, S.W., J.H. Kim, and F.C.A. Groen, "Translation-, Rotation-, and Scale- Invariant Recognition of Hand-Drawn Symbols in Schematic Diagrams," *Int'l J. Pattern Recognition and Artificial Intelligence,* Vol. 4, 1990, pp. 1–25.

Leedham, C.G., and A.C. Downton, "Automatic Recognition and Transcription of Pitman's Handwritten Shorthand—An Approach to Shortforms," *Pattern Recognition,* Vol. 20, 1987, pp. 341–348.

Leplumey, I., and J. Camillerapp, "Comparison of Region Labelling For Musical Scores," *Proc. 1st Int'l Conf. Document Analysis and Recognition,* 1991, pp. 674–682.

Leplumey, I., J. Camillerapp, "Comparison of Region Labelling For Musical Scores," *Proc. 1st Int'l conf. Document Analysis and Recognition,* 1991, pp. 674–682.

Leplumey, I., J. Camillerapp, and G. Lorette, "A Robust Detector for Music Staves,: *Proc. 2nd Int'l Conf. Document Analysis and Recognition,* IEEE CS Press, Los Alamitos, Calif., 1993, pp. 902-905.

Mardia, K.V., etal., "Techniques for On Line Gesture Recognition on Workstations," *Image and Vision Computing,* Vol. 11, No. 4, 1993, pp. 283-294.

Martin, P., and C. Bellissant, "Low-Level Analysis of Music Drawing Images." *Proc. 1st Int'l Conf. Document Analysis and Recognition,* 1991, pp. 417–425.

Martin, P., and C. Bellissant, "Neural Networks for the Recognition of Engraved Musical Scores," *Int'l J. Pattern Recognition and Artificial Intelligence,* Vol. 6, No. 1, 1992, pp. 193–208.

Mehtre, B.M., and B. Chatterjee, "Segmentation of Fingerprint Images—A Composite Method," *Pattern Recognition,*, Vol. 20, 1987, pp. 429–435.

Mennens, J., et al., "Optical Recognition of Braille Writing," *Proc. 2nd Int'l Conf. Document Analysis and Recognition,* IEEE CS Press, Los Alamitos, Calif., 1993, pp. 428–431.

Modayur, B.R., et al., "Muser: A Prototype Musical Score Recognition System Using Mathematical Morphology," *Machine Vision and Applications J.,* Vol. 6, Nos. 2–3, 1993, pp.

Murase, H., and T. Wakahara, "Online hand-Sketched Figure Recognition," *Pattern Recognition,* Vol. 19, 1986, pp. 147–160.

Nair, A., and C.G. Leedham, "Evaluation of Dynamic Programming Algorithms for the Recognition of Shortforms in Pitman's Shorthand," *Pattern Recognition,* Letters, Vol. 13, 1992, pp. 605–612.

Qiao, Y., and G. Leedham, "Segmentation and Recognition of Handwritten Pitman Shorthand Outlines Using and Interactive Heuristic Search," *Pattern Recognition,* Vol. 26, No. 3, 1993, pp. 433–441.

Randriamahefa, R., et al., "Printed Music Recognition," *Proc. 2nd Int'l Conf. Document Analysis and Recognition* IEEE CS Press, Los Alamitos, Calif., 1993, pp. 898–901.

Roach, J.W., and J.E. Tatem, "Using Domain Knowledge in Low-Level Visual Processing to Interpret Handwritten Music: An Experiment," *Pattern Recognition,* Vol. 21, 1988, pp. 33–44.

Srinivasan, V.S., and N.N. Murthy, "Detection of Singular Points in Fingerprint Images," *Pattern Recognition,* Vol.25, No. 2, 1992, pp. 139–153.

Taxt, T., J.B. Ólafsdóttir, and M. Dæhlen, "Recognition of Handwritten Symbols," *Pattern Recognition,* Vol. 23, 1990, pp. 1155–1166.

Viard-Gaudin, C., N. Normand, and D. Barba, "A Bar Code Location Algorithm Using a Two-Dimensional Approach," *Proc. 2nd Int'l Conf. Document Analysis and Recognition*IEEE CS Press, Los Alamitos, Calif., 1993, pp. 45–48.

Xiao, Q., and H. Raafat, "Fingerprint Image Postprocessing: A Combined Statistical and Structural Approach," *Pattern Recognition,* Vol. 23, No. 10, 1991, pp. 985–992.

Key Word Index

About the authors

Rangachar Kasturi

Rangachar Kasturi joined Penn State University in 1982 after completing his graduate studies at Texas Tech University. His primary research focus in recent years has been in the area of Document Image Analysis (DIA). His group's main contribution has been the design of efficient algorithms to generate intelligent interpretations of engineering drawings and maps to facilitate automatic conversion from paper medium to computer databases.

Kasturi has been named as the editor-in-chief of the *IEEE Transactions on Pattern Analysis and Machine Intelligence*. He is the managing editor of *Machine Vision and Applications* journal. He was a guest editor for several special issues on DIA (*Computer*, July 1992, *Machine Vision and Applications*, Vol. 5, No. 3, 1992 and Vol. 6, Nos. 2-3, 1993). He is a coauthor of the two-volume tutorial text, *Computer Vision: Principles and Applications*, published by IEEE Computer Society Press in 1991. He served as the chairman of the Technical Committee on Graphics Recognition (TC-10) of the International Association for Pattern Recognition from 1988 to 1992. He is a program cochair for the third International Conference on Document Analysis and Recognition and a general chair for the First International Workshop on Graphics Recognition.

Kasturi received his BE degree in electrical engineering from Bangalore University in 1968 and his MSEE and PhD from Texas Tech University, Lubbock, in 1980 and 1982, respectively. During 1989-90 he was a visiting scholar at the Artificial Intelligence Laboratory of the University of Michigan. Before entering graduate school he had worked as a research and development engineer with several companies in India for ten years. He is a senior member of the IEEE.

Kasturi can be reached via e-mail at kasturi@cse.psu.edu

Lawrence O'Gorman

Lawrence O'Gorman has been at AT&T Bell Laboratories, Murray Hill, NJ, in the Information Systems Research Laboratory since 1984. He is currently a Distinguished Member of Technical Staff. From 1980 to 1981 he was with Computing Devices Company in Ottawa, where he worked on digital signal processing and filter design. His research interests include document image analysis, image processing, pattern recognition, security aspects of networked documents, and machine vision precision.

He is the author of 18 journal papers, 25 conference proceedings papers, four book chapters, has been co-editor of special issues on document image analysis in *Computer* and in *Machine Vision and Applications*, and has nine patents in application or granted. He has been active on several program committees of the IEEE and IAPR and is a Senior Member of the IEEE.

O'Gorman received the BA.Sc. degree from the University of Ottawa, Ontario, in 1978, the MS degree from the University of Washington, Seattle, in 1980, and the PhD degree from Carnegie Mellon University, Pittsburgh, in 1983, all in electrical engineering.

O'Gorman can be reached via e-mail at log@research.att.com

Offices of the IEEE Computer Society

Headquarters Office
1730 Massachusetts Avenue, N.W.
Washington, DC 20036-1903
Phone: (202) 371-0101 — Fax: (202) 728-9614
E-mail: hq.ofc@computer.org

Publications Office
P.O. Box 3014
10662 Los Vaqueros Circle
Los Alamitos, CA 90720-1264
Membership and General Information: (714) 821-8380
Publication Orders: (800) 272-6657 — Fax: (714) 821-4010
E-mail: cs.books@computer.org

European Office
13, avenue de l'Aquilon
B-1200 Brussels, BELGIUM
Phone: 32-2-770-21-98 — Fax: 32-2-770-85-05
E-mail: euro.ofc@computer.org

Asian Office
Ooshima Building
2-19-1 Minami-Aoyama, Minato-ku
Tokyo 107, JAPAN
Phone: 81-3-408-3118 — Fax: 81-3-408-3553
E-mail: tokyo.ofc@computer.org

IEEE Computer Society Press Publications

CS Press publishes, promotes, and distributes over 20 original and reprint computer science and engineering texts annually. Original books consist of 100 percent original material; reprint books contain a carefully selected group of previously published papers with accompanying original introductory and explanatory text.

Submission of proposals: For guidelines on preparing CS Press books, write to Manager, Press Product Development, IEEE Computer Society Press, P.O. Box 3014, 10662 Los Vaqueros Circle, Los Alamitos, CA 90720-1264, or telephone (714) 821-8380.

Purpose

The IEEE Computer Society advances the theory and practice of computer science and engineering, promotes the exchange of technical information among 100,000 members worldwide, and provides a wide range of services to members and nonmembers.

Membership

All members receive the monthly magazine *Computer*, discounts, and opportunities to serve (all activities are led by volunteer members). Membership is open to all IEEE members, affiliate society members, and others interested in the computer field.

Publications and Activities

Computer Society On-Line: Provides electronic access to abstracts and tables of contents from society periodicals and conference proceedings, plus information on membership and volunteer activities. To access, telnet to the Internet address info.computer.org (user i.d.: guest).

***Computer* magazine:** An authoritative, easy-to-read magazine containing tutorial and in-depth articles on topics across the computer field, plus news, conferences, calendar, interviews, and product reviews.

Periodicals: The society publishes 10 magazines and seven research transactions.

Conference proceedings, tutorial texts, and standards documents: The Computer Society Press publishes more than 100 titles every year.

Standards working groups: Over 200 of these groups produce IEEE standards used throughout the industrial world.

Technical committees: Over 29 TCs publish newsletters, provide interaction with peers in specialty areas, and directly influence standards, conferences, and education.

Conferences/Education: The society holds about 100 conferences each year and sponsors many educational activities, including computing science accreditation.

Chapters: Regular and student chapters worldwide provide the opportunity to interact with colleagues, hear technical experts, and serve the local professional community.